ISBN 978-1-5278-4886-3
PIBN 10917611

FB &c Ltd, Dalton House, 60 Windsor Avenue, London, SW19 2RR.
Company number 08720141. Registered in England and Wales.

For support please visit www.forgottenbooks.com

CIRCULATES IN EVERY PROVINCE OF CANADA AND ABROAD

MARINE ENGINEERING of Canada

A monthly journal dealing with the progress and development of Merchant and Naval Marine Engineering, Shipbuilding, the building of Harbors and Docks, and containing a record of the latest and best practice throughout the Sea-going World. Published by
The MacLean Publishing Co., Limited
MONTREAL, Southam Building TORONTO 143-153 University Ave. WINNIPEG, 1207 Union Trust Bldg. LONDON, ENG., 88 Fleet St.

Vol. VIII. Publication Office, Toronto—January, 1918 No. 1

The Publisher's Page

TORONTO JANUARY 1918

The Economy of Business Paper Advertising

By H. E. CLELAND, NEW YORK

Awarded the Higham prize at the convention of Associated Advertising Clubs, at St. Louis, June 1917. This prize is given annually to the one delivering the most constructive address in the fewest words.

(Continued from last month)

THE time is coming when the principle of intensive advertising will be universally recognized and followed.

Manufacturers of machines and material and equipment which are sold direct to the consumer in technical fields will advertise them in technical papers.

Manufacturers of merchandise sold through retailers will advertise in trade papers first and will no longer be lured into attempting to sidestep the dealer. They will do the first thing first.

Manufacturers of merchandise selling to the general public will establish consumer demand (after distribution has been properly taken care of) by the zone system, using newspapers and outdoor advertising.

Advertisers to the general public with a thoroughly established distributing system will use newspapers and national mediums of many different kinds.

And advertising will reach the high plane that it is destined for only when all advertising men place their business along these lines, and only when the salesmen of different branches of advertising recognize that predatory invasion of the other man's field is the deadliest way to destroy confidence in all advertising.

We cannot force the super-tax of advertising extravagance upon the manufacturers of the world without eventually paying for it ourselves.

No wise salesman knowingly overstocks a dealer. No wise manufacturer of machinery ever recommends the use of his machine in order to make a sale if that machine will not fit the consumer's conditions. No up-standing and honest and far-seeing man will make a contract that does not benefit his customer as well as himself.

Advertising men must follow these methods or kill the goods.

Therefore, advertising men will advocate the use of mediums that present the utmost in efficient economy for the advertiser.

Which means that for certain classes of services, machinery, material, equipment and merchandise, the business papers will be used to their fullest extent. They must be used up to but not beyond the line of their efficiency.

And the business papers must continue the practice that all of the worth-while of them follow now—they must decline advertising which is not in the business line of the paper. To do otherwise is to destroy the principle of the specialized publication.

For those things which "belong" in a business paper—but for no others—the business paper presents the most economical method of efficient advertising.

Thor Iron Works Limited
Toronto, Canada

If what you need is not advertised, consult our Buyers' Directory and write advertisers listed under proper heading.

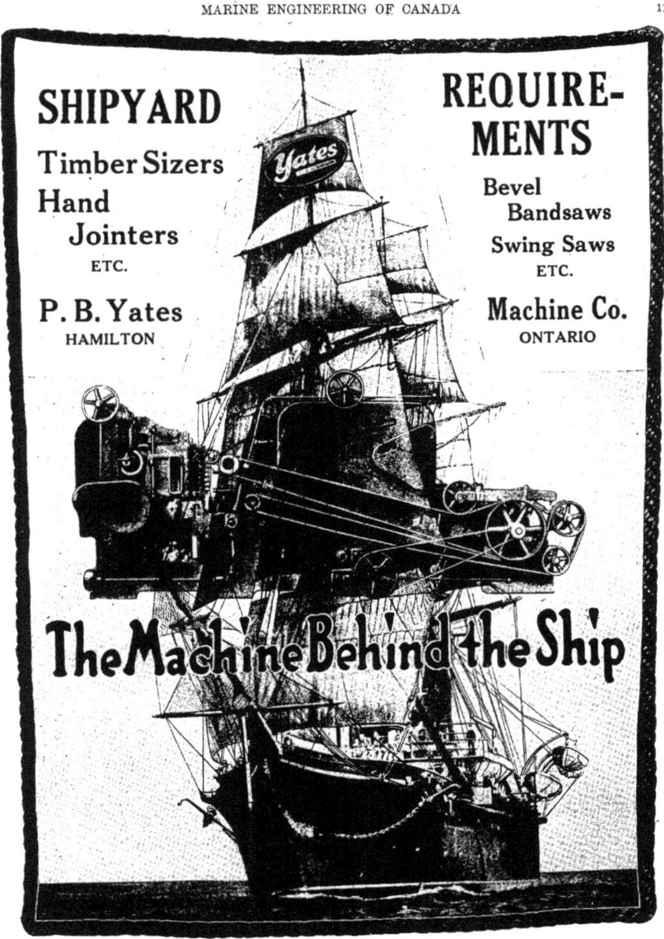
SHIPYARD
Timber Sizers
Hand
Jointers
ETC.
P. B. Yates
HAMILTON
REQUIRE-
MENTS
Bevel
Bandsaws
Swing Saws
ETC.
Machine Co.
ONTARIO
Yates
The Machine Behind the Ship

OUR GUARANTEE

goes with every

"CORBET"

Automatic Double Cylinder Steam Towing Machine

The satisfaction these machines are giving and the large number of testimonials we have received, from those who have installed them on their tugs speaks for itself. Anyone wishing to increase the efficiency and earning power of their tugs or barges should place their order immediately, in order to secure delivery by May 1st, 1918.

WRITE NOW for prices, testimonials and information sheet.

The Corbet Foundry & Machine Company, Limited

OWEN SOUND ONTARIO

Made in four sizes accommodating Steel Hawser from $\frac{5}{8}$" dia. up to $1\frac{1}{2}$" dia.

Midland, Ont., August 16th, 1917.

The Corbet Foundry and Machine Co., Limited,
Owen Sound, Ont.

Dear Sirs,—

We are pleased to be able to report to you that your No. 2 Automatic Steam Towing Machine, which has 1200 ft. of 1-inch dia. Steel Hawser, which you installed on our tug D. S. Pratt is giving us first-class satisfaction. We have been using this machine two years and there is no doubt but that it is far ahead of the old manilla rope, both in cost and trouble of handling. We take pleasure in recommending same.

Yours truly,
Canadian Dredging Co., Limited.
Norman L. Playfair, Sec.-Treas.

Ship Repairs

and

ALTERATIONS OF ALL KINDS

General Machinists and Manufacturers

ENQUIRIES SOLICITED

Hyde Engineering Works

27 William St. MONTREAL

P.O. Box 1185. Telephones Main 1889, Main 2527

Air Compressors
Boilers
Winches
Plate Work
Gray Iron Castings

Address enquiries to nearest Sales Office:

Sherbrooke	St. Catharines
Montreal	Toronto
908 E. T. Bank Bldg.	710 C. P. R. Bldg.
Cobalt	Vancouver
	616 Standard Bank Bldg.

The Jenckes Machine Co. LIMITED

Works Sherbrooke, Que. St. Catharines, Ont.

If what you need is not advertised, consult our Buyers' Directory and write advertisers listed under proper heading.

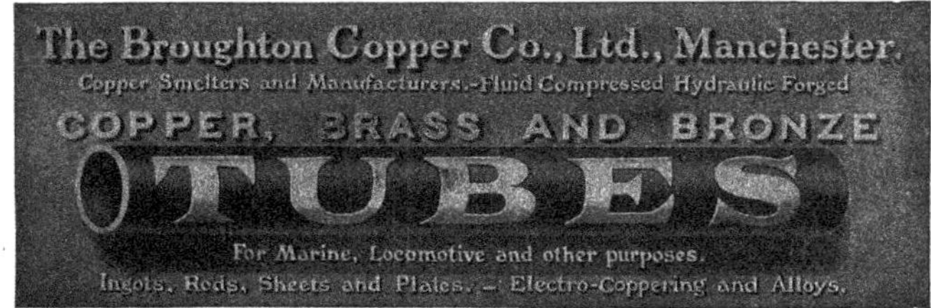
The Broughton Copper Co., Ltd., Manchester.
Copper Smelters and Manufacturers.-Fluid Compressed Hydraulic Forged
COPPER, BRASS AND BRONZE
TUBES
For Marine, Locomotive and other purposes.
Ingots, Rods, Sheets and Plates.—Electro-Coppering and Alloys.

START THE NEW YEAR RIGHT
Make sure of the ACCURACY OF YOUR GAUGES and then arrange to make PERIODICAL TESTS during the year to see that this accuracy is maintained.
THE AMERICAN DEAD WEIGHT GAUGE TESTER
simplifies gauge testing
Agents for Canada: CANADIAN FAIRBANKS-MORSE CO., Limited
Montreal Toronto Quebec Calgary
AMERICAN STEAM GAUGE & VALVE MFG. COMPANY
New York Chicago BOSTON Atlanta Pittsburgh

Constructing Fabricated Ships on Manufacturing Lines*

By R. H. M. Robinson **

A timely definition and explanation of a type of ship construction not previously attempted in any country and a historical resume of the development of the fabricated ship make the author's remarks instructive and interesting. The problems which arose and the manner in which they were overcome promise interesting developments in the future.

RECENTLY the term "fabricated ships" has become one of public interest. A fabricated ship may be defined, briefly, as a ship on which the work of punching and shaping the plates and shapes and, to some extent, assembling the riveting, is done in a fabricating shop, ordinarily employed for bridge or tank work, as distinguished from the usual practice of doing it in a shipyard punch shop. So far as the writer knows, the fabricated ship is a product of American progressiveness. The writer claims no credit for the idea or for its development, and hardly knows why he should prepare a paper on the subject unless it be that he has the honor to be connected with two companies now constructing ships on this system, one of which may reasonably claim to be the first to do so. The principal credit for carrying out the fabricating idea is due to Mr. C. P. M Jack, now consulting engineer, and to Mr. Max Willemstyn, engineering manager of the Chester Shipbuilding Company and of the Merchant Shipbuilding Corporation, and it would be improper to neglect to state that the ideas of Messrs. Jack and Willemstyn could probably not have been carried out except for the the interest and assistance given them by Mr. Jas. A. Farrell, president of the United States Steel Corporation, and by the officials of the American Bridge Company.

FIG. 1—VIEW OF A FABRICATED SHIP IN COURSE OF CONSTRUCTION.

Scope of System

The construction of fabricated ships makes it possible to have the steelwork done by those who are specialists in the line of fabricating steel with all of the special tools and labor-saving devices at their disposal. It relieves the mind of the shipyard operator from the multitude of detail that goes with such work, reducing his problems to those of engineering, erection, riveting and assembling by the workmen, and to installation of other units of outfit and equipment.

The type of ship which the Chester Shipbuilding Company has built, and which it and the Merchant Shipbuilding Corporation are now building, provides for the use in all places of conventional ship shapes and ship plates, subject to the rules, inspection and survey of the registration society, both as to quality and material, dimensions, riveting and other matters. These vessels are cargo carriers of about 8,800 tons to 9,000 tons deadweight capacity with poop, topgallant forecastle and bridge, the latter being about 124 ft. long. The decks, with the exception of the poop and topgallant forecastle, are without sheer at the centre line. Complete double bottom for carrying of fuel oil or water ballast is fitted. The machinery is located amidships and consists of high and low-pressure turbines driving a single screw through reduction gears.

In the emergency which has recently arisen there have been proposed ships constructed from bridge shapes and universal mill plates with regard to which the writer has no personal detailed knowledge. It would appear to him that there might arise certain difficulties in using this character of material in the construction of ships, but he would prefer to leave to those who are personally cognizant of the matter any questions in regard to this.

Tanker Developments

During the late winter of 1912 there was a sudden demand for oil tankers, while freight charters were low and cargo ships plentiful, and almost anything which could be made into an oil tanker was used. Mr. Jack then conceived the idea of using a system of vertical cylindrical tanks which were to be built and tested ashore, then installed bodily in the steamer. Two cantilever-type colliers, the Borgestad and Fritzoe, were acquired under long-term charter. Contracts with Pittsburg tank builders and New York pipe fitters and shipwrights were entered into. The tank builders started to erect the steel (amounting to about 900 tons) long before the arrival of the steamer, and by June the loading of the tested tanks in the Borgestad was begun, and the actual loading was finished in 10 days. It is needless to say that almost unsurmountable difficulties had to be overcome—in the first place, on account of the use of cylindrical tanks for ships; secondly, because of the manufacture of these tanks by tank-builders, the riveting by house-smiths and several other "objectionable" features which were not done according to old-time custom. Mr. Jack, during this time, spent several months in Norway to adjust matters with the classification society and had to withdraw the second ship from their class and change to Lloyds. The first ship sailed in September, 1913, without class, which was only again assigned after a trial of three months.

During 1913 the Escalona, then building at a British yard, was changed into a

---

*A paper presented before the Society of Naval Architects and Marine Engineers, at New York, November, 1917.

**Member of Council.

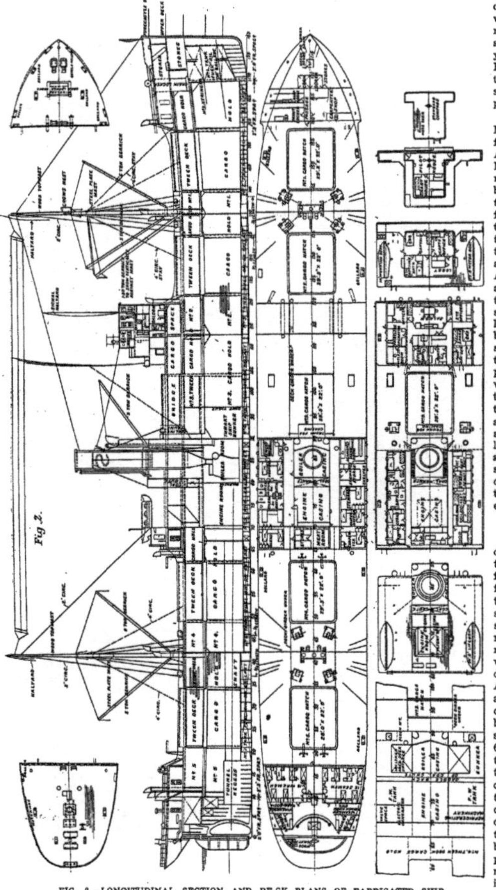

FIG. 2.—LONGITUDINAL SECTION AND DECK PLANS OF FABRICATED SHIP.

cylindrical tanker, but, due again to difficulties of classification, horizontal tanks were installed. These tanks were also manufactured and tested complete outside the shipyard. In the fall of that year three ships were started in Great Britain on the general design of the tankers now building at Chester, but the shipyard would extend the fabricating portion only to the cylindrical tanks.

In 1914 the Chris Knudsen and the Mills were converted on this side. The former was on the type of the Borgestad. A new departure was made with the Mills. The tanks in this ship were connected to tanktop and deck, a trunk built on deck and connections made from tank to tank and to the ship's side, thus forming cylindrical tanks and spaces between tanks and shell, with oiltight centre line bulkhead between tanks. Practically all the material was ready for assembly when shipped from Pittsburg. All holes were punched and countersunk, and stiffeners riveted to trunk sides, centre line, &c., wherever possible before shipping. The only pieces which were left blank were the outstanding leg of bounding angles connecting to the old ship's structure, and one edge of the transverse bulkheads. On account of the peculiar dome roof construction of tanks, which dome formed into rectangular trunk, only about 60 per cent. of the material was multiple-punched and the balance single-punched. The ships converted at New York were all erected with regular tank builders' outfits, consisting of a gantry or stiffleg derrick, air compressor outfit, blacksmith's forge and one or two ships' fitters. All piping, ventilating, &c., was fabricated beforehand, except the closing pieces.

Initial Efforts

The above experience is given in some detail as it was gradually leading toward the fabricated ship, but in taking up the fabricated ship as a whole another difficulty arose on account of the size of the job involved in fabricating complete ships, involving thousands of tons of steel instead of hundreds. The tank manufacturers previously engaged could not entertain the proposition, and others were quoting prices which made the undertaking impossible commercially. The American Bridge Company, however, after several months of search and discussion, decided to fabricate two ships, but would only undertake to do the absolutely parallel midship body. About 60 per cent. of the total steel weight of the first two ships was thus manufactured. This was increased to 70 per cent. on the following ships, and to-day the bridge shops contracted for 85 per cent. of the hull steel and will fabricate also part of the curved portion of the ship. Of course a 100 per cent. fabricated ship is possible, but with the present bridge-shop equipment this is not obtainable. They will first have to increase their furnace facilities to take care of the bending of frames and the beveling of angles, which may be done in the near future if the standard ship is here to stay.

Duplication Necessary

The manufactured ship is a commercial possibility only if duplicate ships are built and, therefore, this policy was adhered to as far as possible from the beginning in contracting for the ships at Chester, there being but four different classes. These classes are as follows:—

1. Tank ship of about 9,000 tons deadweight capacity, fitted with cylindrical tanks on Jack's system. The vessel is built with topgallant forecastle, poop and short bridge. The main deck at centre line and top of expansion trunk are without sheer. Cylindrical tanks extend from tank top to under side of main deck, and a continuous trunk extending between poop and topgallant forecastle takes care of expansion and provides a working deck between the bridge, poop and forecastle decks. Machinery is located aft and consists of a high and low-pressure turbine driving a single screw through reduction gears.

2. Two-decked freighter to about 8,800 tons to 9,000 tons deadweight capacity, having poop, topgallant forecastle and bridge, the latter being about 124 ft. in length. This vessel is built with sheer. Complete double bottom for the carrying of fuel oil or water ballast is fitted. Machinery is located amidships and consists of high and low-pressure turbines driving a single screw through reduction gears.

3. Shelter-deck freighter of about 9,300 tons deadweight capacity, having poop, topgallant forecastle and bridge. Complete double bottom for the carrying of fuel oil and water ballast is fitted. Machinery is located amidships and consists of high and low-pressure turbines driving a single screw through reduction gears.

4. Two-decked freighter of about 8,800 tons to 9,000 tons deadweight capacity, having poop, topgallant forecastle and bridge. The decks, with the exception of the poop and forecastle decks, are without sheer at the centre line. Complete double bottom for the carrying of fuel oil or water ballast is fitted. Machinery is located amidships and consists of high and low-pressure turbines, driving a single screw through reduction gears.

Riveting Details

Except as to details, which may be modified to provide ease of fabrication, the fabricated ship does not differ essentially in structure from any ship of a similar class. One of the prime questions involved is that of multiple punching. The greatest difficulty with the arrangement for multiple punching lies in the fact that the minimum distance from centre to centre of punch dies in a multiple punch is regularly 2 in., which distances can occasionally be reduced to 1⅞ in. The term "multiple punch" implies that the spacing of rivets is to be laid out so that the spacings of rivets in different successive rows are in multiples of each other or in multiples of 2 in. Take, for example, a butt spacing of four diameters of ⅞-in. rivets, equal to 3½ in. The frame riveting in this case could be six diameters, or 5¼ ins., which is no multiple of 3½ in., and neither is 3½ in. a multiple of 2 in. A compromise is made in this case by making an average butt spacing of 3½ in., which can be done by spacing in groups of three, viz., 3¼ in., 4 in., 3¼ in. The punches for the butt and frame spacing would be spaced in the machine as follows: 3¼ in., 2 in., 2 in., 3¼ in. The first, second, fourth and fifth punches would be used for the butt spacing. The first, third and fifth for the frame spacing, allowing for a 5¼ in. spacing. Where the butt spacing is 3½ diameters and the frame spacing 7 diameters, such as on smaller size ships, there is not much trouble, but the multiple punch is hardest to apply on the 9,000-ton, clear-hold, single deck ships, with their 5½ and 6 diameters frame riveting. Again, on the larger ships, with 4-in. butt spacing it is easier to be according to rule, and it should be here noted that Lloyd's Register of Shipping has been very progessive and broadminded in allowing the use of the average spacing instead of the strict and absolute adherence to stated rules.

The first thing to do on a ship designed for multiple punch is to arrange the midship section in such a way that all longitudinal rows of holes fall in line with the principles just explained; second, to arrange the plate width so that the greatest number of equal width plates are obtained; third, to see that if possible each plate is symmetrical about the long centre line of the plate, so as to avoid any possible countersinking errors. No templates are required for setting the multiple punch, which is entirely done from the typical rivet spacing drawing.

Human Element

In staggered riveting, the minimum longitudinal spacing between the rivets of one row and those of the other is 1 in., but 1⅛ in. is preferable. This space is required for the stops and pawls, which stop the table during the downward stroke of the punch-head. It is, of course, possible to reduce the minimum space to small fractions, but then the human element of error is introduced, as a man has to insert a liner between stop and pawl, equal in thickness to the change in spacing wanted. Otherwise the machine is electrically controlled, except the throwing in of the clutch for punching.

A "pole," being a batten about 3 in. wide by ½ in. thick, and of such length as required, is used as template for the longitudinal spacing. Steel stops are inserted in the holes of the batten corresponding with rivet spacing, or, on the newest machines, steel stops are set up in a steel rack to suit the pole. Different color of light is shown where butts, edge-seams, frames &c., occur, and the man on the levers is acting solely according to these colors

The multiple punches have, as a rule, only three or four set-ups, that is to say, one set-up for butts, one for edge-laps, one for frames, one for, say, intercostal angles in conjunction with the edge-laps. In this case the watertight bulkheads could not be punched if the holes would not entirely coincide with one of the four set-ups. It is possible to omit a few holes in some of the set-ups, but this is done only by hand. The man on the punch can release any number of punches temporarily by disconnecting the gag-operating gear for these punch dies.

The Chester Shipbuilding Co., Ltd., has, wherever possible, carried out the idea of duplication and standardisation, but very often has had to abandon it in detail. The hull, engines, boilers, and in fact all large items for a group of ships have been duplicated. The dimensions for all but two ships contracted for by the Chester Shipbuilding Co. are: length, 401 ft.; moulded breadth, 54 ft.; moulded depth, 32 ft. 9 in. Some have had sheer, but in the latest type sheer is dispensed with and a raking forcastle and poop substituted, giving the effect of sheer but simplifying the work of manufacture. The results of the fabrication in the Bridge Co.'s mills as outlined above is beyond criticism. Work is fair, rivet holes require no reaming, and the resulting fit of joints and water tightness is excellent. Constructing a ship becomes to a great extent a problem of routing, handling and erection of material and riveting.

The Chester Shipbuilding Co. and the Merchant Shipbuilding Corporation are now building some 68 ships on the manufacturing principles outlined above. It is expected to embody even more than has been possible in the past the manufacturing idea, thus making it possible, in this day of labor shortage, to do a maximum of skilled shipyard labor of the conventional kind.

BRITISH COLUMBIA SHIPBUILDING

EVERY shipyard in British Columbia engaged in constructing wooden steamers for the Imperial Munitions Board is well ahead of its schedule, according to E. C. Walsh, who is the official responsible for the lumber going into the vessels, and whose department is one of the most important under the Munitions Board. The building of these twenty-seven steamers has meant a great deal for the lumbering industry of the province, for about 1,400,000 feet of timber is needed for each ship, and ten million feet has yet to be delivered. Not only are old-established logging camps engaged on getting out ship timber, but four new camps have been started for this purpose. Merrill & King, on Thurlow Island, get out 100,000 feet a day, and a similar quantity is logged on Thurlow Island at the P. D. Anderson camp. The Hansen camp at Granite Bay is getting out 100,000 feet, and so does the Jackson Bay camp, which is a Munitions Board plant, operated by the Colonial Trust Co. The needs of the shipping has resulted in lumbermen getting out the long lengths necessary for the programme, and the production of ship knees is also a big feature. These knees are very essential in order to give the requisite strength to the vessels, and the fleet now under construction will be a big advertisement for British Columbia timber.

Lumber Inspected

The lumber is very carefully inspected and it must come up to a grading set by Mr. Walsh, who was with the Hastings mill for twenty-five years. The Inspectors' Bureau Association also checks up the quality and issues a certificate if it comes up to grade, and there is a representative at every mill cutting timber for the ships. The shipyards are also held responsible for what they receive, so there is small chance of faulty lumber going into the vessels.

At the Western Canada Shipyards, of which W. W. Clark is superintendent, there are four ways. Number One hull is completed and should be launched at any time, but she was finished so far ahead of the scheduled time, which was February 1, that the shafting and other engine room gear has not arrived. Otherwise she could be sent into the water. When it is considered that the ground for the yard was broken as recently as June 5, it will be seen that rapid work has been done. Work on the second vessel will be about 20 per cent. faster than on the first ship as the workmen get more efficient and more men are trained. The ceiling has been started on the third vessel, and the fourth is in frame, while the saws are whining on the lumber for the fifth.

False Creek is beginning to look like a miniature Clyde, for right across from the Western Canada yard is the big Coughlan plant, where steel steamers are under construction and the hull of one is looming large, for she will be launched this month.

Launching Dates

The launching dates set for the six ships at the Western Canada yard are: Feb. 1, March 15, April 21, May 15, June 7, and June 13. The dates at Lyall's yard, North Vancouver, are: Jan. 20, Feb. 10, March 2, March 23 and April 11. The Cameron-Genoa, Victoria, launching dates are: Feb. 1, March 1, April 1 and May 1. The Foundation yard, Victoria, dates are: Jan. 20, Feb. 20, March 20 and May 20. The dates for the New Westminster Construction and Engineering Co. are: Feb. 1, March 1, May 15 and June 15, and for the Pacific Construction Co., Coquitlam, March 1 and May 1.

All the yards engaged in the work have been fortunate in that no labor troubles have interfered with the progress, and Mr. Walsh states that the greatest harmony has prevailed, all hands apparently realizing the need for turning out the tonnage as rapidly as possible. This feeling is prevalent from the logger who drops the trees to the men who build the ship, and the six yards are working in a spirit of friendly rivalry.

EXTERIOR VIEW. NOTE PROPELLERS AND RUDDER.

The steamers under construction are 250 feet long, 43½ feet beam, and a moulded depth of 25 feet. They will carry 2,800 tons deadweight on a draught of 21 feet, and most of the lumber used in construction is Douglas fir.

Are Strong Vessels

A visit to the Western Canada yard shows that the specifications of the Imperial Munitions Board are high. From the vessel in frame to the completed No. 1 hull every timber seems to be of beautiful lumber of immense strength, and the vessels are very solidly built. There is no evidence of any weakness, and the magnificent timbers are a great testimony to British Columbia's great lumbering asset. The ships are to be built to Lloyd's A1 classification and the British Board of Trade's requirements for cargo steamers.

The frames, of course, vary in size. Some are as big as 21 x 23 inches, and the smallest is 12 x 14 inches. The rudder-post is 37 feet long and 20 x 20 inches, and is in one piece, as is the propeller post, which is 30 x 36 inches and 37 feet long. The beams are 14 x 12 and 14 x 13, and the stem is in one piece, 45 feet long and 18 x 26 inches.

It will be seen from the few instances quoted that strength is insisted on, and there is no reason to fear that this British Columbia fleet built of B. C. lumber will bring discredit on the province.

SUBMARINES TO RAISE SUNKEN SHIPS

THAT the submarine considered as an engineering creation may be put to practical constructive use in salvaging sunken ships seems likely from the details which have been received regarding an invention by W. D. Sisson, an American engineer. Full particulars of the contrivance are not at present available, but from such as are given the general working scheme may be deduced.

The four accompanying engravings

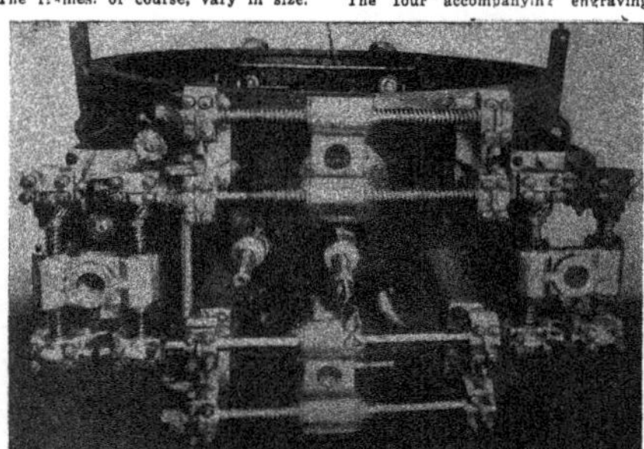

ARRANGEMENT OF DRILL MAGNETS AND SHIFTING DEVICES.

are from photographs taken in the shop where the diving bell was built, and it is uncertain whether it has yet received its initial test in actual working conditions.

The hull is a vanadium steel sphere, 8 ft. in diameter, and consists of two halves with a water-tight joint, by means of which they are bolted together. It is built sufficiently strong to withstand the high pressure which will be encountered when submerged to great depths and is roomy enough for the two operators and the working apparatus.

The object of the invention is to provide a device which will drill holes in the sides of the sunken ships, thus affording a means of attaching a series of sunken pontoons to the vessel to be lifted, so that when the requisite number of pontoons is in position the water in them may be pumped out. and the resulting buoyancy will lift the ship. The drills used are 2 in. in dia. and driven by electric motors inside the shell. The sphere is held tightly against the sides of the vessel on which it is operating by means of a series of electro-magnets attached to the outside of the sphere by spindles passing through holes in the four adjustable saddles, which may be seen in the two engravings. Four 3,000 candle-power incandescent electric lamps enclosed within a 2 in. glass protector, reinforced with steel net, are attached near the magnets, outside the shell, for throwing light on the work and the operators are enabled to see what they are doing by looking through lenses 4 in. thick in the sphere.

It may be explained that the steel pontoons which will be used are 40 ft. long and 15 ft. dia. and each has a lifting power of 300 tons when exhausted. It is proposed to have the submarine guide the pontoons to their places after drilling the holes, and for the purpose there are four propellers and a rudder so arranged as to propel the globe in a horizontal or vertical direction, as desired, at a speed of two miles per hour. Just how the actual attachment of the pontoons to the ship will be accomplished we are unable to state.

The submarine and equipment weigh 6 tons in the air and are lowered into the water by means of a cable strong enough to support a weight of 56 tons. Through this cable also run the wires for carrying current to the electric machinery and lights as well as the telephone wires. The atmosphere within the sphere is replenished by oxygen from a cylinder capable of supplying sufficient oxygen over a working period of 72 hours. Chemicals are doubtless provided to absorb the carbonic acid gas produced by respiration.

EXTERIOR VIEW SHOWING MAGNETS AND DRILLING MECHANISM.

LOOKING INTO THE LOWER HALF, SHOWING LAYOUT OF MACHINERY.

SOME ASPECTS OF SHIP DESIGN

THE present activity in shipbuilding combined with the fact that most of the tonnage is for ocean use has directed considerable attention to the various bodies under whose supervision the vessels are constructed. Attention was drawn recently to an agreement between the leading registration societies whereby greater uniformity of specifications would be obtained. The difficulties of operating under the previously existing variety of requirements was exemplified by a writer in the Times Engineering Supplement who dealt with the possibilities of saving steel in ships.

Draught and Strength

In designing warships or special types of merchant vessels calculations are made to decide whether the vessels will be strong enough for their intended work. These calculations are very complicated and involve a large amount of labor. The ship is treated as if she were a beam supported by her buoyancy, which is equal to her displacement, and loaded with weights made up of her own weight and that of any loads she may be carrying. This gives the bending moment and shearing force at any point in her length.

The value of these factors will vary in accordance with the distribution of the buoyancy and weight, but the maximum bending moment will occur at or near amidships, and the shearing force will have a maximum at about the quarter lengths of the ship. The strength of the section can be estimated at any point so that the stresses acting can at once be determined. Usually if the strength is sufficient amidships it will be so throughout. In special cases, however, the strength is estimated at any doubtful point and the material disposed accordingly. It should, of course, be understood that these calculations are not to be regarded as absolute. They really form a means of comparison between vessels, and it is doubtful if the stresses obtained by this means really represent those actually existing in a vessel at sea.

In ordinary merchant vessels calculations are not required. Registration societies have drawn up tables of scantlings for steel vessels by means of which the necessary strength in any ship can be provided. These tables have been derived to some extent by calculation, but for the most part they are based on experience. It is not surprising, therefore, that similar vessels built under different societies' rules vary in strength.

Work of the Load Line Committee

The greater the displacement of a ship the larger will be the forces acting on her. The question of the relation between draught and strength of ships was very fully dealt with by the Load Line Committee. This committee estimated the strength of various ships at their amidships sections with steel scantlings as fixed by the four principal registration societies. The values were found to vary, and they were plotted as curves having for their base the length to depth ratio of the ship. Fair curves were drawn for each society, representing the minimum values, and a minimum curve was then drawn which was taken as a standard by the Load Line Committee. It was found that the strength of the section varied directly as the draught of the ship, the beam, and as a factor depending on the length of the ship. A standard of transverse strength was also investigated by the committee and determined in a similar way to the longitudinal strength. It follows, therefore, that in most cases the scantlings at present adopted by the registration societies will give a strength greater than the standard. Hence it is to be expected that some modifications will be brought about in the rules of the societies with the object of reducing the amount of steel in merchant ships.

The rules, however, have a further application which is important. The standard of strength set up by the Load Line Committee refers to vessels running at a draught equal to the maximum laid down by the Load Line rules. Many ships are built the carrying capacity of which is determined by the cargo space available rather than by the cargo weight. Such vessels cannot stow a sufficient weight of cargo to load them down to the maximum permissible draught. In consequence, the forces brought to bear on them are less, and their strength could accordingly be reduced in the direct ratio of their actual draught to the maximum. The registration societies have partly recognized this consideration already. Lloyd's Register will allow a reduction in scantlings for vessels carrying a limited amount of cargo with a fixed freeboard. Germanischer Lloyd specifically state the reduction in scantlings for reduced draught, but fix a maximum reduction which can be applied to the longitudinal and transverse structural parts. These reductions apply to the half length of the vessel amidships. At the ends only half the amount is allowed. The principle used is that the strength of the ship must vary as the draught. Bureau Veritas adopt a similar procedure to the German society, and state that all the midship tubular scantlings, that is the scantlings of the strength parts, may be reduced in the direct ratio of the proposed draught of water to the full draught allowable, while the scantlings at the ends are to be those which correspond with the reduced midship scantlings. A limit below which the scantlings are not to be reduced is fixed.

Subdivision

In addition to the reason already given for many ships running at draughts below the maximum permissible there is now the operation of the International Convention Rules to be taken into account. Any vessel carrying 12 or more passengers comes under this convention, which fixes the subdivision in accordance with the freeboard ratio—that is, the ratio between the freeboard and the draught. While it is theoretically possible to put a sufficient number of bulkheads into a ship to enable her to run at her maximum draught and still fulfil the convention rules, the result will often be the introduction of so many bulkheads into the ship that she would not be satisfactory commercially. In consequence, in order not to diminish unduly the length of holds, fewer bulkheads will be fitted and the draught reduced below the maximum. It will be some compensation for this reduction in draught if the registration societies allow the scantlings to be reduced.

In fixing the scantlings for a ship from the rules of a registration society reference is made only to the principal dimensions, and these take no account of the deck erections. On the other hand by the Load Line rules the greater the erections the greater can the maximum draught be. In considering Lloyd's rules, for instance, the midship scantlings for a ship with erections will be the same whatever the extent of these, provided the centre or bridge erection is not less than the minimum length given by Lloyd's Register for long bridges. Taking a particular ship, say, 510 ft. long, the load draught would be about 31 ft. 4 in. if the bridge was of the minimum length of 102 ft. If the length of the bridge were increased to 306 ft. and a poop and forecastle added, each 64 ft. in length, the load draught would be 33 ft. 1 in. Therefore, although the two ships would have the same strength amidships, the calculated forces brought to bear on the latter would be 5½ per cent. greater than on the first. If the three erections were now merged into one, that is, the ship made into an awning decker, the permissible load draught would be 34 ft. 1 in. Here again the midship scantlings would be unaltered, but the forces acting on the ship would be about 9 per cent. greater than in the first case. If the process of increasing the erections is still continued and a centre erection added on the awning deck the load draught must still remain the same, although in this case the strength would have been increased.

Types and Freeboard

To fix the scantlings of a ship so that she may be suitable for her work and still fulfil the rules of the registration societies attention should be given to the relation between her draught and depth. An awning-deck vessel must run at a less draught than a similar vessel having an upper deck in the same position as the awning deck. In the first case the load draught is fixed from considerations of strength, but in the second from considerations of reserve buoyancy. The latter vessel, however, can have a greater draught than the former because she is a stronger ship. If the strength of the awning-deck ship is increased her draught can be increased in proportion. At least, this was the rule, but nowadays it is no uncommon thing to find awning-deck ships running at the same draught as full scantling or upper-deck ships. Shelter-deck vessels are of the same strength as awning-deck vessels, and it will be remembered that some time ago the Shipping Controller gave permission for these vessels to run at a load draught equal to that of a full scantling ship, provided certain rules were complied with. The carrying out of these rules, however, did not increase the strength of the vessels.

Perhaps the best illustration of how steel can be saved by careful attention to the relation between the draught and depth is given by the case of an awning-deck vessel having a complete deck added over her, the draught at which the vessel is required to run being that given by assessing the freeboard from the awning-deck. In the ordinary way the deck over the awning-deck—that is, the shade deck—would be taken as an awning deck. This would considerably increase the scantlings, as compared with the assumption that the second deck down is an awning-deck. The scantlings for the shade deck would then be the same as for an awning-deck. The awning-deck itself would be treated as a first deck, and so on. A further reduction in the scantlings would be brought about by the fact that the length to depth ratio can be taken to the shade deck, so that altogether an appreciable weight of steel would be saved.

It is quite clear from what has been said that there are many merchant vessels running at the present time of greater strength than they need be if existing knowledge on the subject were properly applied, and it may be anticipated that after the war a revision of the rules will be brought about which will effect considerable savings of steel.

Wants More Ships Built at Coast.—H. S. Clements, member-elect for Comox-Alberni, has left for the East. It is his intention to make every endeavor to forward the interests of his constituency in regard to shipbuilding. He claims that both Comox and Alberni contain excellent sites for shipbuilding and he will present their advantages to the Government and to firms who may be interested. Mr. Clements will take a short trip to the South before returning, as his health has not been at its best lately.

G. T. P. May Buy S.S. Kilburn.—The steamer F. A. Kilburn, owned by the Independent Steamship Co., may soon be in the service of the Grand Trunk Pacific, plying in the Alaska trade. Capt. C. H. Nicholson, manager of the G. T. P. steamers, is now in California, and it is reported that he will close the deal for the vessel's transfer before returning. The Kilburn is now at Astoria, Ore., undergoing repairs. A few days previous to that she was towed to port in a leaking condition and her officers were charged with having conspired to sink the vessel. Subsequent investigation showed the allegations to be without basis.

Notes on the Organization of a Shipyard Service Department*

By Samuel W. Wakeman

Demands for help beyond all preconceived ideas are the source of not a little anxiety to those who have assumed responsibility for making good the ship supply of the Allies. The satisfactory results attending the activities outlined by the author are not due to any new or radical ideas but represent the application to shipyard work of what has been done by successful concerns in many other and varied lines of manufacturing activity.

THE object of this paper is to describe a method of handling certain phases of shipyard work which has been successfully tried out in one large eastern plant. Although some of the functions of this department are probably being carried out in the different yards to-day, we do not know of anyone who has treated this problem in just this manner. About two years ago the president of the plant became interested in the enormous number of men that it was necessary to hire each month in order to maintain the working force then needed to carry out the work in hand. He determined that this problem must be solved, and a committee was selected from the plant, composed of workmen, foremen and others. They made a tour of the country, visiting the following manufacturing concerns: Link Belt Company, Sherwin Williams Company, Jeffery Manufacturing Company, Cleveland Hardware Company, Curtis Publishing Company, Chalmers Motor Company, Cleveland Foundry Company, Goodyear Tyre and Rubber Company, Cadillac Motor Company, National Cash Register Company, Dodge Motor Company, Packard Motor Company, Ford Motor Company, Westinghouse Manufacturing Company, Carnegie Steel Company. The most courteous treatment was accorded this committee at each of the above-mentioned plants, and the fullest information was given by them on all of the points which this committee investigated.

Centralized Activities

On their return they prepared a report of their investigations, together with such recommendations as to the betterment of the then existing conditions as they thought should be made, these recommendations including such ideas as the centralizing of employment instead of allowing each one of the different department heads to carry this on themselves. It also recommended that a new building be erected for this purpose; that a different point of view be taken with regard to applicants for positions, namely, their treatment should be as courteous and as fair as was possible to be given; that the housing and boarding situation be investigated; that all men leaving the employ of the company should be interviewed and given a chance to air their grievances. They further recommended a great many lines of activities to be encouraged in the plans and touched on many lines that had to do with the physical, mental and social condition of the employees. The recommendations of this committee were printed in pamphlet form and sent to about 200 people in the plant. They were asked to freely criticise and, if possible, to make further recommendations. All of this information was then collected and formed the basis for the policy which is being pursued at the present time in regard to the service work in this plant.

A brick building was erected near the main gates of the plant to take care of the hospital work, the employment department, and the general activities of that nature. These activities were grouped under the name of Service Department which, as its name signifies, is to render service to the employees of the plant. This department was put under the supervision of the general superintendent of the plant. The scope of its activities are as follows:—Employment, hospital, safety work, educational work, apprentices, employees' club, works paper, suggestion committee work, yard restaurant, shop training. The personnel of this department consists of the following: Supervisor, assistant, clerks, stenographer, supervisor of instructors, employment agent, employment clerk, doorman, doctors, nurses, messenger boys for conducting applicants to shops.

Head of Service Department	Employment Agent		Interviewing and hiring new employees. Recruiting labor outside the plant.
	Correspondence Clerk		Housing new employees. Interviewing and recording departing employees.
	Messengers (3)		Monthly analysis of employees hired and departed.
	Recording Clerk		Recording on cards new employees. Daily report of men hired.
	Door Man		Filing record cards of new employees. Charge of main waiting room.
	Assistant Stenographer 2 Clerks	Employees' Magazine	Associate editors. Collection, compilation, proof-reading, publication and distribution.
		Safety Work	Monthly statement of hours lost and percentage; accidents; recommendations approved; by General Safety Committee. Investigating and reporting special cases, bulletin posting. Shop committees, organization and reporting. General Safety Committee minutes.
		Apprentices	Applications, tools, bonuses and certificates. Shop Credit reports. General welfare, social activities, etc. Book apprentices' time and quarterly reports.
		Employees' Club	Minutes of meetings. Publicity of activities. Membership and dues. Athletic activities.
		General	Circulation of monthly periodicals. Arrangement of bulletin boards. Restaurant com'tee, supervision of yard restaurant Naturalization assistance. Navy League, rifle company. Assistance to employees in promoting outings. Information of any kind to employees.
		Educational	Keep records of attendance. Outline of curriculum. Receive applicants for classes.
		Housing	Listing of lodging and rooming houses available for new men. Information regarding houses for sale and to rent. Advice to employees wishing to buy houses.
		Suggestion Committee	Receipt and referring of suggestions to parties directly concerned. After receipt of comments, action upon same by committee. Recommendation of an award. Notification to sender.

Interdepartmental Service

In order to see clearly the relationship between the Service Department and the superintendents and foremen in the yard in regard to the supply of labor, the following illustration is used: The ship carpenter may want 10 pieces of 12-in.

*Read before the Society of Naval Architects and Marine Engineers, at New York, November, 1917.

by 12 in. yellow pine, 30 ft. long. His requisition comes to the purchasing agent, who buys this material and has the same delivered to the ship carpenter. If on inspection by him three or four pieces out of the 10 prove defective, he condemns this material and the purchasing agent replaces it. The ship carpenter decides what class of work this lumber can be used for and decides also whether the price paid for it is justified.

The same procedure holds for the supply of labor. A foreman finds he will need a certain number of men at a given time. He sends his requisition for this labor to the employment division of the service department. They send the required number of men to the foreman after interviewing these men and weeding out, so far as is possible, such applicants as would be incapable of carrying out the work for which they are required. These men are sent to the foreman of the department, who takes or rejects whatever number he may see fit and rates those he keeps, and those he rejects are sent back to the employment department, where an effort is made to place them in some other department in the yard where they may find work which they can do.

If the men hired are not satisfactory to the foreman, he may discharge them. In no particular are any of the foreman's prerogatives in regard to the handling of labor taken from him. The advantage of this method, however, is that the foreman is not compelled to interview a large number of men who have no qualifications for the work he is supervising, and his time is conserved for the more important duties he has to perform.

Physical Inspection

Each applicant for a position is put through a careful physical examination by the company's doctor, and only such men are employed as are capable of doing the work for which they are hired. This has resulted in getting a very good class of men, and, where slight physical defects are found which can easily be remedied, the man is sent to the proper medical authority and taken care of. At first there was considerable objection on the part of some of the men, but at the present time practically no one ob-
prerogatives in regard to the handling of
jects to the physical examination.

The employment department keeps an up-to-date requisition of all labor requirements of the plant, and takes such steps as are necessary, through its various sources of supply, to keep this requisition filled. It attempts to learn, by keeping in close touch with the various departments, the probable future needs for labor, which is a virtually essential factor in keeping a steady flow of men into the yard. All applicants for work are quartered in a waiting-room in the service building, equipped with chairs, and are interviewed in turn by the employment agent, being given a private interview in a separate room. Each applicant when hired is given a book which contains all the information about the plant that it is considered a new man should have. This book has been carefully prepared and gives the hours of labor, the method of tool delivery, certain safety suggestions, times and methods of payment of wages and, in general, information in a condensed form which the new employee generally gets only after long association in the plant. This book has proven to be a great help to new men coming into the plant.

All employees transferred or leaving the employ of the company are required to be interviewed by the employment agent before they can obtain their wages. This provides an opportunity to place the man in a department where he may fit to better advantage than the one in which he was working, and also gives him opportunity to state any grievances he may have in connection with the company, his foreman or his work. These grievances are carefully tabulated and looked into and have been found to be a great source of valuable information in connection with controlling the feeling the employees have toward the company. All records for present and past employees are filed alphabetically in the employment bureau, and these records contain the name, age, trade, physical qualifications, address, place of past employment, rate, &c. There is a daily report prepared by the employment department, showing the number of men hired and leaving by trades, also daily gain or loss in the working force. The department also makes a monthly report of the turnover of labor, including the number of men hired and leaving by trades, showing the length of service and reasons for the man leaving.

Housing Facilities

One of the most important phases of the employment work is to provide proper housing facilities, and to this end the employment department lists, for the information of all applicants, boarding houses and available houses for sale or for rent. If there is no position open at the time the applicant applies in person or by letter, an application card is filled out and filed by trades. All applications for employment are filed and, as far as possible, the applicant is informed as to what date his services will be required, giving the minimum and maximum rate for the trade desired and the hours of labor.

Under the service department is a competent surgeon and an up-to-date hospital where all of the accident cases in the yard are treated. The families of the employees may obtain free medical attention during certain hours of the day. The safety work in the plant, which has reached a fair state of development, was combined with the work of the service department and the different shop committees, together with the General Safety Committee, use this building as a meeting place, where frequent talks by men who are specialists along this line of work are given to these men. Each department has its own safety committee, which makes periodical inspections and reports directly to the General Safety Committee. A great many ideas for work of this nature have originated in these committees and numerous appliances and safeguards have been adopted, due to suggestions received from them. All accidents are carefully investigated and the causes are immediately remedied; responsibility for accidents as far as possible is placed and steps taken to rectify them.

All of the educational work in connection with the training of apprentices and the training of men in the shops and different departments throughout the yard have been placed under the jurisdiction of the service department. During the winter months, evening schools are conducted which provide training in naval architecture, mould-loft work, plumbing, coppersmith work, machine shop design, mathematics and blue-print reading. The instructors for these evening classes are, for the most part, drawn from the various departments in the yard and the instructions carried out under the direction of the local school board. These classes are open to any male beyond the age of 16 in this locality and are attended by mechanics, apprentices and helpers to the number of about 200.

All of the work in connection with the apprentices in the plant is carried out under the supervision of the foremen in the different departments, but along the lines as laid down by the educational part of the service department, and the head of this department has been very successful in this connection with the attitude of the apprentice toward the company. Each year a dinner, to which prominent officials of the company are invited, is given by the apprentices, and efforts are made to promote the best of feeling among the young men who are being trained for future service in the company.

Social Services

The service department supervises the publication of a works paper, which is issued every three months by the employees. This paper has developed to such an extent that its issue is awaited with a great deal of interest by all of the men in the plant. The company presented to the employees of the plant a club house equipped with bowling alleys, rifle range, assembly and dancing hall, pool table, a band room and an athletic field where the different athletic teams practice and have their games. Quarters are provided for the athletic teams of the company in this building with suitable shower-baths, &c. This club is run entirely by the employees of the company, and only such necessary supervision is given to it by the service department as is absolutely necessary. A restaurant is maintained inside the plant for the benefit of the employees. The supervision of this restaurant, although run by an outside caterer, is left to a committee of the employees who are in close touch with the service department.

One very important part of this service department work is the handling of the suggestions sent in by employees. Boxes are placed throughout the yard, and prizes are offered for suggestions sent in by the men. There is a great deal of work in connection with the proper understanding of what the employee means, the making of sketches, the referring of these suggestions to the proper people so that they may be acted on, and the attitude which the men have toward this phase of the work must be carefully handled in order that the greatest amount of good may come from it. The work of this committee, of which the head of the service department is chairman, is one of the very important functions of this department.

Tracing Results

All of this work has begun to show results which it is felt can directly be traced to the work of this department. Statistics for many of the large manufacturing plants throughout the country that have attempted to increase their working force have shown that it is necessary to hire anywhere from seven to ten men in order to obtain one permanent employee. The latest results here have shown that is has been necessary to hire two men instead of from seven to ten in order to obtain one. When you consider that it is estimated that it costs somewhere between 20 dols. and 35 dols. to hire a new employee the result of cutting down the turnover in any plant means a tremendous saving in dollars and cents, although it may not always be possible to measure this directly. Much of the reason for this decrease in turnover can be credited to the selection on the part of the service department; the hearty co-operation of the foremen, who have allowed a very liberal system of transfer from one department to another; the training of men into trades which they were not familiar with and, in the main, to the general broad policy of the man directing the company.

Before the above plan was in operation, in order to maintain a force of about 4,000 employees it was found necessary to hire, in the course of a year, 8,000 men. During the last six months the present force has been increased from 4,000 to 8 000, and this has been done by hiring 9 000 men. While these results cannot all be attributed to the work of this department, nevertheless it is felt that, in a large measure, the work which has been accomplished by them is responsible for a large percentage of this showing.

It is not claimed that any new or radical ideas have been adopted in connection with the handling of this situation, but what has been done is to find out what successful concerns in other lines of work have done, and to combine them together in such a way that they might be adapted to the work of a shipyard. A man was employed who was thoroughly familiar with the sources of labor supply in the district in which this plant is located, and much credit is due him for the enthusiasm with which he approached this difficult problem. An outline diagram showing the relationship of the different functions of this department to each other is given in the preceding page.

SILVER LININGS TO THE SUBMARINE PIRACY

By Capt. Geo. S. Laing

THE greater part of the world looks down on our enemies' U-boat performance with unmitigated horror and calls it fiendish, diabolical, piratical, and so on. No man on earth has oratorical powers sufficient to justify such foul procedure. No writer's quill can make thinkers believe that German seamanship represents "freedom of the seas."

The subaqueous craft had its limitations set out in Hague rules of warfare, but the most unscrupulous interpreter must quake when a naval command embraces the sinking of peaceful passenger ships, hospital ships, neutral shipping and fishermen of all kinds.

Hundreds of women and children, trusting to the chivalry of the sea were murdered on board the Lusitania and other passenger boats that were similarly dealt with, such as the Falaba, etc. Hundreds of wounded soldiers, some in delirium, some unconscious, all helpless, were tossed into the cold waters of the English Channel to drown, when Hun torpedoes riddled their hospital ship, a craft that is no safter now than a troopship—which is the legitimate prey of the underwater vessel.

Then such cases as the Belgian Prince have been numerous. Laughing and jeering at drowning men is another Teutonic branch of naval tactics, and towing seamen under in their open life-boats is still another, and there are more; in fact it was the absolute abandonment of all that is fair and human in sea warfare that brought President Wilson to line up for world freedom.

While all this has to be lamented, and whilst these awful acts will be burned into the histories of the allied nations for future weighing, the zenith of the professional pirates' work has been reached, and the writer now wishes to show a silver lining to the cloud—the greatest cloud of clear-cut hellish murder that has ever besmirched the sea. Yes! it's the same old sea that we thought had only hallowed associations, such as exploration, colonization, missionary effort, trade expansion, etc., and in other ways, the source of all rain and snow, lake, river, and spring waters.

The Silver Lining

When Von Tirpitz conjured up his "sink everything in sight policy" he forgot that compensating forces often attend ruthless measures. Now let us count up some of the good things that have arisen from the moaning spirits of non-combatants, whose bodies lie strewn in the pierced hulls of thousands of craft which dot the sea-floors of British and European waters. A world wide sentiment is in our favor and it is well known that all neutrals have at some time or other protested against the inhuman and of course illegal destruction of sacred life and valuable property on the high seas by the Hun.

These protests on the surface do no more good than shaking your fist in your pocket, but away deep down in the consciences of the nations that have had to swallow the naval outrages, a coral reef of opinion is being formed that must have an expression sooner or later.

What will this expression be?

A scorching and branding denuniciation of loathsomeness towards the German navies, both the war and the mercantile fleets. This rightful punishment will be reflected in the Atlantic and Pacific for many years and Germany will bear the Cain mark to some extent for murdering non-combatants as long as waves recoil from rock-bound coasts.

There are times when forgiveness in all its beauty enters the sanctum of minds both national and individual, but to actually forget altogether this bloodiest of stains, where helpless women and babes, unarmed seamen of belligerent and neutral states, and male passengers old and young were intentionally sacrificed on the savage altar of piracy, is absolutely impossible.

Through the submarine ghoulishness however will arise the power of democracy which will stand for all time in the way of Kaiser-like fiends using the glorious seas for such satanic purposes.

Thus from the viewpoint of character, nations will learn more and more to respect and love right and scorn brutal force—which is never allied to fair play, that most powerful ingredient so necessary in the make up of great men or nations.

From Another Angle

It was never properly understood and seldom acknowledged in Britain, Canada or the United States, how vital the ship and sailor question is to great nations, until the Hun pirates began to sink everything in sight. In this light the destruction of shipping has caused so many economic difficulties that the nations who have had to tighten their belts have also been forced to scratch their heads and think of sea values hitherto buried in oblivion.

Frank T. Bullen, whose body is under hatches and whose soul has gone aloft, tried both by pen and tongue to arouse interest in consideration of British merchant ships and their officers and men, but he might as well have lectured to Thames bargemen on the Sumner method of finding a ship's position.

Serving the nation under the Red Ensign as a certificated officer was almost equivalent to a wasted life. The pay, instead of being in accordance with the responsibilities and education of a navigator and seaman, was about on a par with a policeman or unskilled worker. Very often the living quarters for officers were mere boxes far removed from the delusive name of "state room." At the present moment a wheelsman on a Yankee lake freighter draws as much pay as the chief officer on an ocean tramp steamer. Canada has not come up to the Yankee scale yet, but she will have to if she wishes to draw intelligent lads from the beach. As regards Canadian certificated men, much better results would beset the profession if they were all under the shelter of one society. Whilst on this subject, it would be well if Canada looked ahead now for boy material to work up in the merchant vessels, thus laying the foundation for a real Canadian mercantile marine—every captain, mate and engineer, every cook and steward, fireman and sailor a Canadian.

Oh yes! The British Empire is falling over itself now to make amends for her past indifference to "those that go down to the sea in ships," but it took the Kaiser with his tin sharks to bring about this righteous adjustment.

Further, every man, woman and child in the British Empire will soon understand that their country is dependent on sea power—which, boiled down, means ships and crews. Premiers and eminent statesmen have had to admit it, and the daily press has run short of type eulogising Jack and his "floating coffin" as grand old Plimsoll called it. We seldom hear this man's name now, although he was the seaman's Wilberforce.

We find then that the British Empire as well as the U.S. has awakened to realize the real value of a merchant fleet and means to act on a higher plane in the matters of food, accommodation and remuneration, three drawing powers that will enhance the whole marine business.

Another Silver Lining

Canada's foreign trade and actually some of her coasting has depended to a large extent on the Old Country or foreign bottoms, and why? Simply because some folks thought that the ocean was only for bathing in.

What are we doing now?

"The scene is changed." We are building ships galore and not before time. With the losses of British, French, and Italian vessels through the Hun policy, the shortage of tonnage was fast approaching a calamity, and had Canada and Uncle Sam not been wooed back to Neptune's chariots, the U-boats' havoc might have ended the war ere now. Every yard on the Great Lakes and on the beautiful seaboards of our freedom loving country is feverishly hammering wood, iron, and cement into floating structures, to wit, ships, for as Lloyd George and others of his ilk say "it's a question of ships."

Never again will our Dominion be almost wholly dependent on the mother country and Norwegian tonnage for her ocean transportation, at least not to the detriment of our national stability. After all, how can a nation be mistress of her export and import values unless she owns and operates her own ships to and from all the ends of the earth? Sea transportation is the greatest commercial asset in times of peace and the most essential thing in war.

A Commercial Silver Lining

What has the depletion of tonnage taught us in some industrial avenues?

To sharpen our wits and create home markets. Generally speaking a vast amount of the manufactured stuff that was purchased in Canada came from Germany and Austria. This condition was very apparent also in all other parts of our far-flung empire. To those who knew Hamburg, Stettin, Bremen and Bremerhaven in their real export days, the thoughful question must be,—"Where has such industrial activity gone to?" In these few ports one could find the largest sailing ships and steamers of the long voyage fleet of various flags, and most of them loading for North and South America, Africa, India and Australia.

Some one says: "Yes, and similar conditions will grow again."

Reader! it is impossible! you might as well talk of trying to give a man another chance after he has been hanged for murder.

The sea as an element has ever stood before man's judgment as a wonder, an indefinable representation of strength, a thing that God gave to all mankind as a source of food, as means of travel, and the scientific base of all life, from the human being to the vegetable kingdom. There is no room here to fully explain sea values, but if I put the reader on the road to analyze for himself, he will immediately find out that everything that grows on the earth gets sustenance from the ocean. Our great luminary "old sol" is constantly loading up aerial ships (clouds) with distilled water from the sea and chartering these celestial carriers to all parts of the sphere. Without this natural source and law how could life be possible?

Privateering Was Honorable

Then if we understand the purpose of the sea to mean all this, how can the rest of the world look upon Germany's infamous submarine deeds as anything but downright piracy—by which I mean plunder and murder on the high seas? Even in privateering there was an element of honor and sport, but not so with this calculated work of the Hun seamen. It is sheer deviltry and surpasses the conjured plans of Herod, who may now be boosted into Madame Tussaud's plaster saint community in Baker St., London.

Will Canada, the United States, and Britain be prepared to resume a full speed ahead trade policy with the Teutons when peace comes to hang these awful horrors of the sea on the walls of our minds and in the histories which our children must read?

No sir, human clay is not plastic enough for that. There will be a time of retribution, and during that period, as at the present moment, the nations who fought to strangle the amphibious octopus will have found better and pleasanter paths of interchange in goods, manufactured goods and all commercial by-products; thus leaving the pirates in sackcloth and ashes as far as their trading on the high seas is concerned.

This fair Dominion and our great ally Uncle Sam, joined in territory and fast becoming joined with bruised blood and mutual mourning, will be inclined to hew most of their own wood and draw most of their own water independent of Germanic trade and merchandise for at least a decade. By that time Teutonic minds may be built up on cleaner pillars of national character.

Sea transporation, sea fisheries, sea institutions of chivalry, our sister colonies, the laws that govern international canals and a thousand indirect bearings that have their solution in the freedom of the seas, must not, cannot, fall under the brute rule of a piratical nation.

A Valued Heritage?

Yet, did we ever value these things and try to safeguard them beyond all possibilities until the Kaiser conceived his sink everything and murder everything afloat policy?

I leave the shore end of Hun hellishness to be analyzed by a soldier, it also contains many barbaric facts that will act as nightmares in the nurseries of our descendants.

The writer has witnessed the German mercantile marine invasion of the Calcutta, Chittagong and Rangoon trade, the Australian and New Zealand trade, the Cape Town trade and so on, and the men of the British merchant fleet never uttered a word of protest. Will it come back to that again?

The answer to that question is connected with the further query:—Will those that go down to the sea under the Red Ensign and Stars and Stripes forget the Falaba and Belgian Prince brand of sea expression?

The Capt. Fryatt incident alone sealed Germany's fate on the sea. I can even imagine the cave fishermen under the cliffs of Rosario on the Rio de la Plata, or the "launcharos" at Valparaiso exclaiming "Aleman vapor no bueno."

Unmasked in Time

While the silver linings are hard lessons, let us use them and thank Providence that we found the sea murderer out as soon as we did, for God knows what would have happened if German naval expansion had gone on for a few more years.

Other nations can see now that they practically lived on the heated approach to the crater of a volcano. The British Empire and the United States have just the proper geographical positions to make them worthy keepers of the freedom of the seas, with Australia and New Zealand keeping guard in the Southern Hemisphere.

American Engine Builders Restore Damaged Ships

Special Correspondence *

Scientific sabotage is indeed a mild term to use in describing the malicious damage done by their crews to German ships in American harbors when it was seen that war was becoming inevitable. All the resources of the devil were utilized on a wholesale scale, and the equally ingenious and efficacious remedies applied by U. S. engineers are a matter for congratulation and mutual satisfaction amongst the Allied nations at this time.

ALL the damage done to 109 German ships by their crews, prior to their seizure by the United States Government when war was declared, has been repaired and these ships are to-day in service, adding more than 500,000 gross tonnage to the transport cargo fleets in war service for the United States.

There is evidence that a German central authority gave an order for destruction on these ships, effective on or about February 1, 1917, simultaneous with the date set for unrestricted submarine warfare and that the purpose was to inflict such vital damage to the machinery of all German ships in our ports that none could be operated for from 18 months to 2 years.

German Purpose Signally Defeated

This purpose has been defeated in signal fashion. In less than eight months all the ships were in service.

The destructive campaign of the German crews cunningly comprehended a system of ruin which they believed would necessitate the shipping of new machinery to substitute for that which was ruthlessly battered down or painstakingly damaged by drilling or dismantlement. There is documentary proof that the enemy believed the damage irreparable.

To obtain new machinery would have entailed a prolonged process of design, manufacture, and installation. Urged by the necessity of conserving time the engineers of the Navy Department succeeded, by unique means, in patching and welding the broken parts and replacing all of the standard parts which the Germans detached from their engines and destroyed or threw overboard.

*Published by permission of the Committee of Public Information, Washington, D.C.

BROKEN VALVE CHEST AND SCORED PISTON ROD OF THE OSTWEGO, FORMERLY THE PRINZ EITED FRIEDRICH.

INTERIOR OF CYLINDER ON U.S.S. AGAMEMNON, FORMERLY THE KAISER WILHELM II. THE LINER WAS CRACKED AND IS BEING PREPARED FOR REPAIRS BY WELDING.

Scheme of Ruin Shrewdly Devised

The mechanical evidence is that the campaign of destruction was operated on these ships for more than two months and that the Germans were convinced that they were making a thorough job of it. Their scheme of ruin was shrewdly devised, deliberately executed, and it ranged from the plugging of steam pipes to the utter demolition of boilers by dry firing.

When the United States Shipping Board experts first surveyed the ruin the belief was expressed that much new machinery would have to be designed, manufactured, and installed, making 18 months a fair minimum estimate of the time required. However, at the Navy Department, where the need of troop and cargo ships was an urgent issue, officers of the Bureau of Steam Engineering, having faith that the major portion of the repairs could be accomplished by patching and welding, declared it was possible to clear the ships for service by Christmas, and the last of the fleet actually took her final sea test and was ordered into service as a Thanksgiving gift to the nation.

Special Workmen on the Job

To accomplish this end the Navy Department secured the services of all available machinery welders and patchers, many of them having been voluntarily offered by the railroads.

Although explosives were not used in the process of destruction, the engineers of the Navy Department were always conscious of the danger of hidden charges of high explosives which might become operative and disastrous when the machinery was put to a test. Instances of artful pipe plugging, of concealing steel nuts and bolts in delicate cylinders, of depositing ground glass in oil pipes and bearings, of cunningly changing indicators, of filling fire extinguishers with gasoline and similar means of spoliation, were common enough to induce the engineers to make a rule calling for thorough overhauling. On each ship there was no boiler that was not threaded through every pipe for evidence of plugging, no mechanism of any sort that was not completely dismantled, inspected, and reassembled before it was finally passed as safe.

Record of Damage Found on One Ship

A memorandum written in German was picked up on one of the ships which gave a complete record of the destruction on that ship. Investigation revealed that the list, which had evidently been left through an oversight, was correct in every detail. The following is a translation of excerpts from this memorandum:

"Starboard and port high pressure cylinder with valve chest: Upper exhaust outlet flange broken off (can not be repaired)."

"Starboard and port second intermediate pressure valve chest: Steam inlet flange broken off (can not be repaired)."

"First intermediate pressure starboard: Exhaust pipe of exhaust line to second intermediate pressure flange broken off (can not be repaired)."

"Starboard and port low pressure exhaust pipe damaged (can not be repaired)."

The parenthetical optimism of the German who was so confident of the thoroughness of his mutilation is now the source of much glee among naval engineers, inasmuch as every one of the supposedly irreparable parts was in fact speedily repaired and those engines are to-day as powerful and serviceable as when they left the hands of their makers.

The method of patching and welding broken marine engines had never before been practised, although the art has been known in the railroad industry for 15 years. Three methods of patching were used; electric welding, oxy-acetylene welding, and ordinary mechanical patching, the latter often later being welded. Following the repairs tests of the machinery were first made at the docks, where the ships were lashed firmly to the piers while the propellers were driven at low speed; and later each ship was taken to sea for vigorous trial tests. The patches and welds were reported as having given complete satisfaction.

BROKEN VALVE, BROKEN VALVE CHEST AND BROKEN FLANGE ON THE OSTWEGO, FORMERLY THE PRINZ EITEL FRIEDRICH OF HAMBURG-AMERICAN LINE.

When the Leviathan, formerly the Vaterland and the largest ship afloat was put into commission by the United States Government and sent to sea for a trial run, her commander, a young American naval officer, was ordered to "exert every pound of pressure that she possesses, for if there is any fault we want to know it now." The Leviathan stood the test. She was one of the ships least mutilated, due to the fact that she was in bad repair and it was believed that she would not be fit to put to sea for many months. The navy engineers found it necessary to overhaul and partially redesign and reconstruct many important parts of the engines.

Ships Now in United States Navy

The larger German ships which have been repaired and are to-day in commission as a part of the United States Navy,

TYPICAL BREAK IN CYLINDER. THE PHOTOGRAPH SHOWS IT AFTER THE PART HAD BEEN PREPARED FOR WELDING.

NEW CAST IRON PATCH INSERTED IN BROKEN CYLINDER BY OXY-ACETYLENE WELDING. PHOTO TAKEN ON U.S.S. ANTIGONE, FORMERLY THE S.S. NECKAR

SHOWING THE DESTRUCTION OF A STEAM PIPE ON THE U.S.S. MADAWASKA FORMERLY THE S.S. KOENIG WILHELM II.

with their former German and their new American names are as follows:

German Name	American Name
Vaterland	U.S.S. Leviathan.
Amerika	U.S.S. America.
Andromeda	U.S.S. Bath.
Barbarossa	U.S.S. Mercury.
Breslau	U.S.S. Bridgeport.
Cincinnati	U.S.S. Covington.
Frieda Leonhart	U.S.S. Astoria.
Frederic der Grosse	U.S.S Huron.
Geier	U.S.S. Schurz.
George Washington	U.S.S. George Washington.
Grosser Kurfurst	U.S.S. Aeolus.
Grunewald	U.S.S. Gen. George W. Goethals.
Hamburg	U.S.S. Powhatan.
Hermes	U.S.S. Hermes.
Hohenfelde	U.S.S. Long Beach.
Keil	U.S.S. Camden.
Kaiser Wilheim II.	U.S.S. Agamemnon.
Koenig Wilheim II.	U.S.S. Madawaska.
Kronprinz Wilhelm	U.S.S. von Steuben.
Kronprinzessin Cecille	U.S.S. Mount Vernon.
Leibenfels	U.S.S. Houston.
Locksun	U.S.S. Gulfport.
Neckar	U.S.S. Antigone.
Nicaria	U.S.S. Pensacola.
Odenwald	U.S.S. Newport News.
Praesident	U.S.S. Kittery.
Praesident Grant	U.S.S. President Grant.
Praesident Lincoln	U.S.S. President Lincoln
Prinzess Irene	U.S.S. Pocahontas.
Prinze Eitel Friedrich	U.S.S. Dekalb
Rhein	U.S.S. Susquehanna.
Rudolph Blumberg	U.S.S. Beaufort.
Saxonia	U.S.S. Savannah.
Staatssekraetar Solf	U.S.S. Samoa.
Vogensen	U.S.S. Quincy.

Repaired by Shipping Board

The ships taken over and repaired by the Shipping Board, with their German and American names, are as follows:

German Name	American Name
Allemannia	Owasco.
O. J. D. Ahlers	Monticello.
Adamsturm	Actaeon.
Arnallas Vinnon	Chillicothe.
Atlas	(No name.)
Armenia	(No name.)
Arcadia	(No name.)
Andalusia	(No name.)
Adelheid	(No name.)
Bulgaria	(No name.)
Borneo	Nipsic.
Bohemia	Artemis.
Bochum	Montpelier.
Bavaria	(No name.)
Calabaria	(No name.)
Carl Diederichsen	Raritan.
Clara Jebsen	Tioga.
Clara Mennig	Yodkin.
Cobelenz	Sachem.
Constantia	(No name.)
Dobek	Mononghela.
Darvel	Wamsutta.
Elmshorn	Casco.
Elsass	Appelas.
Esslingen	Nyanza.
Farn Gerraux	Farn Gerraux.
Gouverneur Jaeschke	Watoga.
Holsatia	Tippecanoe.
Harburg	Pawnee.
Indra	Tonawanda.
Johanne	Iosco.
Koln	Anphion.
Kurt	Mochulu.
Loongmoon	Coosa.
Lyceemoon	Quantico.
Mark	Suwanne.
Mia	Oconee.
Madgeburg	Neuse.
Mattedor	Montauk.
Marudu	Yazoo.
Nassovia	Isonomia.
Neptun	Minnow.
Ottawa	Mus ota
Olivant	(No name.)
Ockenfels	Pequot.
Prinz Eitel Friedrich (Hamburg-American Line)	Ostewgo.
Prinzess Alice	Matoika.
Pennsylvania	Manasemond.
Pisa	Ascutney.
Pongtong	Quinnebaug.
Tortonia	Yucca.
Prinz Joachim	Moccasin.
Prinz Oskar	Orion.
Prinz Sigismund	General Gorgas.
Prinz Waldemar	Wacouta.
Pommern	Rappahannock.
Rajah	Rajah.
Rheatia	Black Hawk.
Sachsen	Chattahoochee.
Sachsenwald	General Ernst.
Sambia	Tunica.
Savoia	General Hodges.
Serrapis	Osage.
Setos	Itasca.
Staatssekretar Kraetke	Tacony.
Steinbek	Arapahoe.
Suevia	Wachusett.
Camilla Rickmers	Ticonderoga.
Tsin Tau	Yuma.
Jubingen	Seneca.
Wasgenwald	Wasgenwald.
Wiegand	Midget.
Willehad	Wyandotte.
Wittekind	Iroquois.

Every one of these ships was found to be either deliberately damaged or rendered useless through ravages of neglect before they fell into the hands of the United States Government.

The most serious typical damage was done by breaking cylinders, valve chests, circulating pumps, steam and exhaust nozzles on main engines, and by dry firing boilers and thus melting the tubes and distorting the furnaces, in at least one instance probably using thermit to make the destruction complete.

There were many instances of minor and easily detectable destruction, such as cutting piston and connecting rods and stays with hack saws, smashing engine-room telegraph systems, and the removal and destruction of parts which the Germans evidently believed could not be replaced. The most insidious sabotage was that which was concealed. In plugging a steam pipe the method was to disjoint the pipe and insert a solid piece of brass which would be sawed off flush with the joint. The pipe would then be reconnected, showing no evidence of having been tampered with.

Fire Extinguishers Filled With Gasoline

Indicators were astutely reversed in

INTERIOR OF BOILER OF THE S.S. POMMERN NOW THE U.S.S. RAPPAHANNOCK. SHOWING HOW THE GERMAN CREW MELTED DOWN THE BOILER BY DRY FIRING PROBABLY USING "THERMIT" TO INTENSIFY THE HEAT. THESE BOILERS WERE REMOVED.

many instances. Numerous fire extinguishers were found to be filled with kerosene and gasoline. Piles of shavings and refuse were strewn about where fires might be started, open cans of kerosene being found in several of these incendiary traps. There had evidently been a plan to burn the ships under certain conditions and it is believed that the German crews were seized and interned somewhat in advance of their expectations.

Case of Vaterland Distinctive

The case of the Vaterland was quite different from that of any other ship.

TYPICAL VANDALISM ON THE GERMAN SHIPS. WITH A BATTERING RAM OR A SLEDGE THE VANDALS SMASHED THIS CIRCULATING PUMP CASING WHICH WAS DULY REPAIRED BY THE ELECTRIC WELDING PROCESS

Engineers of the Navy Department who examined the big liner declared that inferior engineering had been practised in her construction. She has four turbine engines ahead and four astern on four shafts. All of the head engines were found in good condition and all of the astern engines were found damaged.

The major portion of the damage was credited more to faulty operation than to malicious intent. Cracks were found in the casing of the starboard high-pressure backing turbine of such size as to make it certain that the engine had not been used on the vessel's last run. Certain documentary evidence found on the ship corroborated this belief. It also indicated that the Vaterland on her last trip had made less than 20 knots.

There was just enough evidence of mutilation to warrant full investigation and the vast mass of machinery, electric apparatus, and piping in the Vaterland was patiently and doggedly examined before she was sent to sea. Original defects in her engine equipment were corrected, she was overhauled and in many respects refitted, and on the whole she was declared a better ship when she entered the service of the United States than when she took her maiden voyage.

OFFICIAL NARRATIVE OF GERMAN U-BOAT SINKING BY U.S. DESTROYERS

THE event related below took place in November last and resulted in the destruction of a German submarine and the capture of the crew by American destroyers. According to an official statement recently made by Secretary Daniels, the capture was made by the U.S.S. Fanning, assisted by the U.S.S. Nicholson. The Fanning was commanded by Lieut. A. S. Carpender, Lieut. G. H. Fort being the executive officer.

About 4.10 p.m., while escorting a convoy, the lookout of the Fanning, Coxswain David D. Loomis, sighted a small periscope about a foot above water some distance off the port bow. The periscope was visible for only a few seconds. The destroyer immediately headed for the spot and three minutes after the periscope had been sighted dropped a depth charge. The Nicholson also speeded to the position of the submarine, which appeared to be headed toward a merchant vessel in the convoy, and dropped another depth charge. At that moment the conning tower of the U-boat came to the surface between the Nicholson and the convoy. The Nicholson fired three shots from her stern gun. The bow of the submarine came up rapidly. She was down by the stern, but righted herself and seemed to increase her speed. As the Nicholsan cleared, the Fanning headed for the U-boat, firing from the bow gun. After the third shot the crew of the submarine all came on deck and held up their hands, the submarine surrendering at 4.28 p.m.

Submarine Goes Down

The Fanning approached the submarine to pick up the prisoners, both destroyers keeping their batteries trained on the boat. A line was got to the submarine, but in a few minutes she sank, the line was let go and the crew of the U-boat jumped into the water and swam to the Fanning.

Although the crew all wore life preservers, a number of them were exhausted when they reached the side of the destroyer. As the submarine sank, five or six men were caught by the radio aerial and carried below the surface before they disentangled themselves. Ten of the men were so weak that lines had to be passed under their arms to haul them aboard. One man was in such a condition that he could not even hold the line thrown him. Chief Pharmacist's Mate Elzer Harwell and Coxswain Francis G. Connor (N. N. V.), jumped overboard after this man and secured a line under his arms. When he was hauled aboard every effort was made to resuscitate him, but he died in a few minutes. The four officers of the submarine and the 35 members of the crew were all taken prisoners.

After being taken on board the prisoners were given hot coffee and sandwiches. Though kept under strict guard, they seemed contented and after a short time commenced to sing. To make them comfortable the crew of the destroyer gave them their warm coats and heavy clothing.

The German officers said the first depth charge had wrecked the machinery of the submarine and caused her to sink to a considerable depth.

The submarine bore no number nor distinguishing marks. She was identified by life belts and by statements of an officer and men of the crew. One of the life belts had "Kaiser" marked on one side and "Gott" on the other.

Cheered by Germans

The Fanning proceeded to port and transferred her prisoners under guard. As they were leaving in small boats, the Germans gave three cheers. The commanding officer of the Fanning read the burial service over the body of the dead German sailor, and the destroyer proceeded to sea and buried him with full honors.

The commander of the Fanning reports that the conduct of all his officers and crew was excellent. He gives particular credit, to the officer of the deck, Lieut. Walter O. Henry, and to Coxswain Loomis, who sighted the periscope, and highly commends Chief Pharmacist's Mate Harwell and Coxswain Connor, who jumped overboard to save the drowning German.

Praised by British

The British commander in chief, under whom the destroyers were operating, in a report to the British Admiralty, said: "The whole affair reflects credit on the discipline and training of the United States' flotilla," and added that the incident showed that the Fanning "is a man-of-war in the best sense of the term, well disciplined and organized and ready for immediate action." He concludes that the credit for this must be given her commander. He mentions as worthy of particular commendation Lieut. Henry, deck officer of the Fanning, the coxswain who sighted the periscope, and the chief pharmacist's mate and coxswain who jumped overboard to rescue the drowning German.

The British admiral also commends the prompt action of the Nicholson, which completed the success of its sister ship.

The success of shipbuilding and marine engineering enterprise is largely dependent on its "Wheelsmen." This series of articles has for its object the featuring in a racy, interesting and instructive fashion, the personal training, experience and achievement of those who to-day in Canada are energetically and effectively navigating the twin craft to higher degree prominence in their capacity as designers, constructors, outfitters, etc.

WILLIAM JAMES McSHANE

NOT the least prominent of those sections of the Dominion affected by the industrial activity of the past three years is the province of New Brunswick. While affected to a considerable degree by the demand for munitions, the urgency-inspired development of shipbuilding has, to a greater degree perhaps, been responsible for the present state of activity in this particular section of the busy East. When the influence of agricultural and allied interests in a generally prosperous district are considered, the ultimate demands on establishments doing a general engineering business are both large and varied. Considerable interest, therefore, attaches to the personality of the subject of these remarks, whose lifetime association with a leading New Brunswick engineering firm has resulted in his reaching a responsible and trusted position therewith.

Canadian born of Canadian parents, William James McShane first saw the light in St. John, in the year 1865, and to-day, after 35 years in the service of T. McAvity & Sons, discharges the duties of chief draftsman and patternmaking superintendent. While much of Mr. McShane's success is no doubt due to the guiding influence and tuition of his father who held the position before him, it is no disparagement to his parent to remark that the developments of recent years have brought greatly increased responsibilities and demanded a degree of judgment and resource largely exceeding the old days when our Spoke was absorbing the rudiments of his lifework.

The day schools of St. John were his source of knowledge until the start of his apprenticeship with McAvitys in 1883, at the age of eighteen. The regular 5-year period of training was gone through by Mr. McShane, and the fact that this formative period of his life was spent under the supervision of his father has doubtless been a matter of mutual benefit to our Spoke and his employers on many occasions since. Doubtless Mr. McShane reasoned that a part of the country which was good enough for a firm like McAvity (they had been established almost half a century when he joined them) was quite good enough for him, and the determination to do well whatever came to his hand accounts for his following the example of his employers and becoming also, so far as lay in his power, long established.

WILLIAM JAMES McSHANE.

The name of McAvity is a household word in the Maritimes, as will be gathered from the nature of their business. Established in 1837, they developed with the country, operating the Vulcan Iron Works and a separate plant devoted to brass work, producing plumbers', steam fitters' and engineers' supplies while recently a machine shop of considerable size has been rendered necessary to meet their requirements. The wide field of activity open to a studious, resourceful and energetic person in the position filled by our Spoke is apparent, and that Mr. McShane fully availed himself of the opportunities offered is evidenced by the numerous inventions, etc., in the various lines produced by McAvity. Numerous types of valves, fittings, electric railway semaphores and corporation specialties are the fruit of his efforts or bear the impress of his influence in some form or other.

The close attention to work necessary to insure such a record may account for the comparatively small amount of traveling done by Mr. McShane but the requirements of his position did not prevent him from following the advice contained in the old saying about "all work and no play, etc.," and his record in rowing matches and similar sports is evidence of his physical as well as mental ability.

Our Spoke is of the Liberal persuasion in politics, is Roman Catholic in creed, and beyond being a Knight of Columbus has few society affiliations. His belief in the value of temperance is well seconded by his efforts for the cause, and he acknowledges that much of his success has been due to his close ties with the Father Matthews Temperance Society. He has been bereaved through the death of his wife, and resides at 27 Duke St., St. John, N.B.

In speaking of his career, Mr. McShane attributes most of his success to having "acquired practical experience, at the same time developing the ability to back it up by the study of technical books and journals, the former to keep the rudiments of knowledge always fresh, and the latter to maintain close fresh, and the latter to maintain close touch with recent developments in products and equipment, and acquire a knowledge of the methods and devices employed by others."

BLACK SEA—BALTIC CANAL PROJECTED BY CENTRAL POWERS

By Mark Meredith

THE importance of inland waterways is appreciated in all countries where cheap transportation is an absolute necessity, and the latest project in this line that will be carried out is the construction of a chain of canals to connect the Austro-Hungarian rivers with the Vistula and the Elbe to form a great inland waterway system, with the Baltic for an outlet. By this Hamburg will be in touch with Vienna, Bulgarian and Roumanian points, and with the Black Sea, giving the enterprise stategical as well as commercial value.

A glance at the map of Hungary shows how much of the most navigable portion of the Danube flows through that country and no other waterway in Europe is so suitable for large steamer traffic. Obviously, too, there are great advantages in quick conveyance between Silesia, with its mines and industries, and Vienna, and between the Austrian capital, and the coal fields and tin mines of Bohemia.

The project contemplated provides for a canal for large vessels, branching off from the Danube at Ganserndorf, going north-west to Pardubitz, Bohemia, and north-east through the Silesian mining and industrial districts to Oderberg. Another canal will connect the Danube above Vienna with Budweis and Bohemia, and join the Moldau River and a subsidiary waterway will branch off at Manrisch-Ostrau, Moravia, connecting with the Vistula and leading to Cracow. Work is also now in progress on canalizing the Elbe between Meinik, Bohemia, and Jaromir for ships of 600 tons.

It is calculated that freight could be shipped by these inland waterways from Cologne to the centre of the Balkans a distance of 1,560 miles at 8/- per ton. With a Rhine-Danube Canal goods could be conveyed from Germany to Turkey much more cheaply than now, and such a waterway would make the Central Powers much more independent of British sea power, which is perhaps the prime reason why such a costly undertaking is being pushed now.

LACHINE CANAL TRAFFIC

STATISTICS available as to the freight shipped through the Lachine Canal for the season of 1917 show that 607 vessels made 7,936 trips this season of 294,773 tons. 1916 totalled 585 vessels 278,760 tons burthen with the vessels making 7,579 trips through the canal. In the year just closed there were more than 70,000 tons of grain and other merchandise, shipped above the amount in 1916. Of wheat there were 13,201,732 bushels, an increase of over 2,000,000 bushels.

There were 265,411 more tons of coal; 8,000 more cords of pulpwood; 24,000 more tons of pulp; 85,000 more sacks of flour; 38,000 more tons of sand and 165 more cargoes through the canal. The passenger list for the season tötalled 69,910, a slight increase over 1916 figures. The total combined tonnage of all the ships through the canal amounted to 4,145,836 tons an increase of 144,663 tons. The cargo tonnage was 70,000 more than that the previous season's figures.

There was a big decrease in all grains with the exception of wheat. In corn, oats, barley, rye and flaxseed, the season of 1917 showed a falling off of 5,518,100 bushels. There were cargoes 25,168,452 in 1916, and 19,350,352, in 1917. Eggs, butter, cheese and apples, all showed big decrease in 1917. Flour of which, there was none shipped last year, passed through the canal this season in 85,800 sacks. In lumber there has been a decrease of 1,289,400 feet board measure brought down from Ottawa for United States ports by American canal boats.

JAPAN IS RUSHING SHIPBUILDING

E. F. CROWE, Canadian Trade Commissioner at Yokohama, in a report to the Department of Trade and Commerce, states that Japan is launching upon a shipbuilding campaign which will involve the construction of 250 ships per annum. He states that at the end of September there were in Japan 113 shipbuilding slips owned by 42 firms. In each slip a ship of 1,000 tons can be built. This is more than three times the number of ships Japan owned before the war. Many more are also building, and 24 are expected to be completed before the end of the war. When all these berths are put into full operation, subject to a supply of steel and iron materials, Japan will be able to build more than 250 ships, aggregating one million tons, yearly. This is more than three times over Canada's maximum shipbuilding proposals as recently announced.

SHIPBUILDING PROGRAM IN U. S.

PLANS for a $2,000,000,000 Government shipbuilding programme were revealed at Washington on Friday when the Shipping Board asked Congress for authority to place $701,000,000 worth of additional ship contracts. At the same time an immediate appropriation of $82,000,000 was asked for the extension of shipyards and for providing housing facilities for workmen.

Thus far the board has been authorized to spend for shipbuilding $1,234,000,000, contracts for most of which have been awarded. To-day's request for a further authorization and an additional appropriation brings the estimates of funds needed for shipbuilding to $2,018.000,000. If the additional funds are made available they will be put largely into fabricated steel ship contracts, although some contracts for ordinary steel ships will be let, as will a few for wooden ships on the Pacific Coast.

No new shipyards are planned by the board. The fabricating yards have contracts for about 500 ships now, which will keep them constantly employed into 1919, but the board is anxious to place more contracts and begin on the task of providing materials.

The board's housing plans call for the expenditure of about $35 000.000. The remainder of the $82 000 000 asked for will be used to expand shipyards already built.

BUSINESS ON SOO CANALS

BELOW are shown comparative figures for the traffic passing through the American and Canadian Soo canals during 1916 and 1917. In 1917 the United States Canal was opened April 24 and closed December 17; season, 238 days. The Canadian Canal was opened April 25 and closed December 17; season, 237 days. In 1916 the United States Canal was opened April 20 and was closed December 19; season, 244 days. The Canadian Canal was opened April 18 and closed December 18; season, 245 days.

SHIPBUILDING PROGRESS IN B. C.

THE steamer "War Songhee," first of the Imperial Munitions Board ships to be launched in Victoria, will be ready shortly for the installation of boilers and engines. Work on the superstructure and cabins has been rushed ever since the vessel was sent down the ways, and it is probable that the Foundation Co. will have her in shape to start for the assembling plant at Ogden Point before the second I. M. B. ship is launched.

The laying of the fifth keel at the Foundation Co. yard was finished recently, and a start on the square framing is now being made. The planking of hull No. 2 will probably have been carried out by the end of the month, although the Imperial Munition Board's recent order that no hull be launched until all the woodwork is accomplished will delay its introduction to the sea.

The "Beatrice Castle" and "Jean Steedman" will be ready for sea in ten days. All the equipment for the two vessels is now at the Cameron-Genoa Mills Shipbuilding, Ltd., yards and is being assembled and installed as rapidly as possible.

Hull No. 6, at the Cameron-Genoa yards, is fast approaching the launching stage and is considerably further advanced than any other I. M. B. vessel still out of the water.

IN October, 1917, eight vessels were launched in Japan aggregating 38,420 tons, according to "Eastern Commerce," a Japanese paper printed in English.

	United States Canal		Canadian Canal	
	1916	1917	1916	1917
EASTBOUND				
Articles—				
Copper, short tons	122,161	110.188	3 882	8.624
Grain, bushels	47,098 808	40,045.944	46,086.506	18.369.851
Flour, barrels	6.887.129	5.105.511	3,339.331	3,344,438
Iron ore, short tons	51,935 809	50,201.456	11,487,371	11,107,205
Pig iron, short tons	35,032	10 624	....	....
Lumber, M., feet B.M.	327.663	342 051	13.689	8 558
Wheat, bushels	139,237.877	120,751,356	86,825.438	65,148.093
General merchandise, short tons	281.470	196,203	57.443	68.635
Passengers, number	12,064	6,339	16,055	12,541
WESTBOUND				
Coal, hard, short tons	2.122.509	2,389.449	87.710	172.750
Coal, soft, short tons	12,970,073	14,648,567	942.827	1,088,087
Flour, barrels	421	80	12,960	....
Grain, bushels	9.245	8.185	....	....
Manufactured iron, short tons	144,091	78.313	23.071	13.145
Iron ore, short tons	14.479	46 866	14.448	18 853
Salt, barrels	663.058	452.727	66.787	117,229
General merchandise, short tons	959,354	926,087	850.753	300,789
Passengers, number	11,243	5,651	15,560	13,898
SUMMARY				
Vessel passages, number	13.716	17.536	6.691	5,349
Registered tonnage, net	57,038,207	53,413,807	12,786,256	11,893,426
Freight—				
Eastbound, short tons	58.775.064	56.204.519	15.373.354	13.841.966
Westbound, short tons	16,310,155	18,157,331	1,429,646	1,610,082
Total freight, short tons	75,085,219	74,361,850	16,803,000	15,452,048

BRASS AND OTHER COPPER ALLOYS USED IN MARINE ENGINEERING*

By J. T. Milton

SINCE reading the paper on the above subject, my attention has been drawn by one of its authors to a paper upon the Inspection of Brass and Bronze, presented in June last to a meeting in America of the American Society for Testing Materials by Messrs. Alfred D. Flinn and Ernst Jonson. Also to two papers by Mr. Flinn upon experiences with brass in civil engineering works. One of these latter was read at a meeting of the Municipal Engineers of the city of New York in November, 1914, the other at a meeting of the American Institute of Metals in October, 1915.

Mr. Flinn is deputy chief engineer of the board of water supply of the city of New York, and Mr. Jonson is an engineer inspector of the same board.

It is considered that the papers are the most remarkable and most valuable of any which have been published on the subject of "Brass."

Both of these latter papers contain descriptions of the brass-work used in the construction of the new water works of New York City, commonly referred to as the Catskill Aqueduct, and they record some of the experiences with the brass from 1908 onwards.

Mr. Flinn states that in America, as is the case here also, considerable confusion exists to the use of the terms "brass" and "bronze" to describe alloys of copper, and it is worthy of note that in the title of his paper he uses the word brass, and in general refers to the metals of which he recounts the experience as "brass," whereas the great bulk of the metal used was manganese bronze, and was termed "bronze" in the specifications.

In arranging for the enormous works in connection with the water supply of New York City it was necessary to use immense weights of some strong incorrodible material, and very great consideration was given not only to the designs of details, but also to the composition of the metal used. There were castings to be employed ranging from a weight of a few ounces each to 22,000 lbs. (say, ten tons). Altogether there were about 2,000,000 lbs. of copper alloy castings and about 1,000,000 lbs. of bolts and other parts. Some of the pieces were forgings, ranging from small bolts up to valve rods 6 inches diameter, 31 feet long, weighing 3,200 lbs. The castings included valve chests for valves 72 inches diameter, some of them weighing 44,000 lbs. each when completed with covers, etc. The total expenditure for the purchase of brass was approximately 1,000,000 dollars.

Before ordering this vast amount of metal for one undertaking, several alloys, such as manganese bronze, naval brass, phosphor bronze, etc., were all investigated. Information was sought from responsible metallurgists and experienced engineers regarding designs and specifications, and, as a result, "manganese bronze" was selected for general use, but not to the exclusion of all other alloys. All the articles were obtained under specification and were carefully inspected. Most of the work was carried out by very experienced manufacturers by their own methods, some of the firms having had experience for a very large number of years. Whenever possible hydrostatic and other tests were made, in addition to the customary tensile and bend tests. The specifications stipulated the desired physical qualities, but not methods, of production, nor metallurgical composition.

*Being the author's contribution to the discussion on paper read before the Institute of Marine Engineers. Appeared Nov. issue, page 282.

The following extracts from the specifications are taken from Mr. Finn's papers:—

The Bronze Castings.—All bronze castings shall be made of new metal, shall be free from objectionable imperfections, and shall conform accurately to patterns. When the castings are being machined, if the metal shews signs of imperfect mixing, they shall be rejected. Unless otherwise called for in the specifications, or on the drawings, bronze where indicated upon the drawings shall mean manganese bronze.

Manganese Bronze.—All manganese bronze shall be equal to Spare's, Parson's or Hyde's manganese bronze, and shall have a tensile strength of not less than 65,000 lbs. per square inch, an elastic limit of not less than 45 per cent. of the ultimate tensile strength and an elongation of not less than 25 per cent. in two inches.

Brass Rivet Rod.—Tensile strength of brass rivet rods shall be not less than 55,000 lbs. per square inch. The elastic limit shall be not less than 30,000 lbs. per square inch and the elongation not less than 20 per cent.

Stems.—The main gate stem shall be of manganese bronze or other bronze of approved composition, of such dimensions that when placed in tension under a load in pounds, equal to the area of the valve opening in square inches, multiplied by 125, the minimum cross section of the stem shall have a resultant unit stress not exceeding two-thirds the elastic limit of the material used. These stems shall be turned straight and true, and shall have all threads lathe-cut.

Rolled Bronze.—Whenever the term bronze is used in these specifications in a general way or on the drawings without qualification it shall be manganese or vanadium bronze or monel metal. Whenever the characteristics of any material are not particularly specified such material shall be used as is customary in first-class work of the nature for which the material is employed.

The minimum physical properties of bronze shall, except as otherwise specified, be as follows:—

Castings—
Ultimate tensile strength ... 65.000 lbs. sq. in.
Yield point .... 32,000 " " "
Elongation .... 25 per cent.

Rolled material, thickness one inch and below—
Ultimate strength .... 72,000 lbs. sq. in.
Yield point .... 36,000 " " "
Elongation .... 28 per cent.

Rolled material, thickness above one inch—
Ultimate strength .... 70,000 lbs. sq. in.
Yield point .... 35,000 " " "
Elongation .... 28 per cent.

After being forged into a bar, rolled, or forged bronze shall stand: First, hammering hot to a fine point. Second, bending cold through an angle of 120 degrees to a radius equal to the thickness of the bar.

Bronze Pipes.—The pipe shall be seamless drawn, semi-annealed, iron-pipe-size tubing of the sizes shewn on the contract drawings or ordered. It shall be made of the bronze specified in the general sections or of an approved commercial bronze. All pipe furnished under this item shall be free from season cracks, surface cracks or other defects.

When the pipe is finished, ready for shipment, the engineer will subject about 1 per cent. of the lot, taken at random, to the following tests:—

1st.—Each test pipe shall stand threading perfectly, with a die with the usual thread for the size of pipe.

2nd.—After annealing, the end of each test pipe shall stand being flattened by hammering until the sides are brought parallel with a curve on the inside of the ends not greater than twice the thickness of the metal, without shewing cracks or flaws.

3rd.—After annealing, each test piece shall have a piece 3 inches long cut from it, which, when split, shall stand opening out flat without shewing cracks or flaws.

TYPICAL RESULTS OF PHYSICAL TESTS OF BRASSES USED ON THE CATSKILL AQUEDUCT.

Yield Pt. lbs. sq. in.	Ultimate Stren. lbs. sq. in.	Ratio of Yield to Net	Elongation per cent.	Reduc. of area per cent.	Fracture
Forgings					
36,500	73,150	50	41.5	46.8	Irregular
37.500	75,750	50	35.5	43.9	"
38.250	76,900	50	35.5	46.8	"
52,500	77,100	68	31.0	...	Irreg. Silky
49,300	76,150	54	33.5	...	Silky Cup
50,000	75,350	67	31.0	...	Irreg. Silky
43,500	70,000	62	34.0	47.0	Irregular
36,000	67,500	53	40.5	43.5	"
Castings					
39,650	66,000	60	40.0	...	Irreg. Grain
43,100	68,850	53	25.0	...	" "
39,950	68,050	59	29.0	...	" "
33,000	66,250	50	25.0	25.4	"
32,750	67,500	48	30.0	27.7	"
32,250	67,200	48	36.0	34.1	"
39,500	69,100	57	39.0	...	Irreg. Silky
47,500	71,000	67	40.5	...	" Gran.
50,000	70,000	71	34.0	...	Ang. Silky
33,250	73,000	45	30.0	28.5	Irregular
33,000	73,500	45	35.0	34.0	"
34,500	73,250	47	40.0	36.5	"
43;250	69,950	62	40.5	...	Ang. Silky
44,000	70,050	63	34.0	...	Irreg. Cup
42,500	69,500	61	37.5	...	Ragged Cup
46,000	69,850	66	33.0	...	Ang. Cup
44,000	69,600	63	40.5	...	Ag. Fine Gr.
41,500	72,350	57	36.5	...	Irred. ½-Cp
40,000	68,000	59	38.0	...	Forked Silky
42,000	67,000	63	37.0	...	Ang. Silky
43,750	69,050	63	37.0	...	Forked Silky

Mr. Flinn states that in spite of the care exercised, the brass furnished proved to be distinctly unsatisfactory in most of the uses to which it was put. No suspicions of definite troubles developed until late in 1913, when numerous bolts and rods were found to be cracked, and since early in 1915 some of the castings from at least three different foundries have also been found to be cracked. All these castings had, before acceptance, been subjected to hydrostatic pressures of 200 and 300 lbs. per square inch for a half-hour or more, yet some months later, after having been erected in position, they leaked under pressure of only a few pounds. In some castings the cracks have extended since they were first observed, and in some additional cracks have been discovered as time has elapsed. In nearly all of the cases of the cracked castings the defects appeared to be close to or in a repair made by "burning."

In the case of defects in plates, bolts, rods, sides and rungs of ladders, etc., some of the articles had not been taken out of their packing cases. In some cases the cracks were fine and superficial, but in others they were deep. In some cases in bolts from ½ inch to 2¼ inches in diameter the whole cross section was affected, some of the bolts being found to be severed. Two or three years have passed in some cases before the defects developed sufficiently to be detected.

Mr. Flinn states that but for the seriousness of the matter "bronze" or "brass" would be but a laughing-stock among the engineers of the aqueduct.

In the 1914 paper Mr. Flinn mentions that only a few small brass pipes had failed. He says, indeed, that there is but small excuse for supplying other than dependable brass pipe nowadays, as correct methods of manufacture are well known in the trade; but in the 1915 paper he states that serious failures had been experienced with pipes also.

He states that the defects were found to be so general and so distributed through the output of various manufacturers, and were so common in the different kinds of brass that all wrought brass fell under suspicion. After the first discovery of the cracking, replacements were made with brass produced by methods avoiding cold working in the hope that by this further trouble would be avoided, but unfortunately this did not prove to be the case.

It was well known that cold working produced initial strains in the material, and after the discovery of the defects it was decided to use plain, extended or hot rolled rods wherever practicable and to anneal all material which had to be drawn or rolled cold, in the hope that further trouble would not occur, but, as stated, this hope was not realised, as plain extended, hot forged and annealed brass rods, supposedly free from initial strains, have also failed in disturbingly large quantities.

It is noteworthy that at the time of reading the paper in 1915 Mr. Flinn was able to state that **large forgings**, which include chiefly the items for large sluice gates and valves, have shewn no sign whatever of failure. He states: No brand or make of brass or bronze has wholly escaped. Manganese bronze, naval brass (including a well-known bronze and its imitation) and Muntz metal from **all** the manufacturers who have furnished any considerable quantity, **all have failed.**

Mr. Flinn makes some very drastic remarks as to the want of such knowledge of the physical characters and capacities of this group of alloys as is required to be a safe and dependable guide to their manufacture, inspection and use. As his investigations and experience progressed, he found not only unfortunate foundry and shop ignorance of important details of the manufacture of copper alloys, but also seemingly a lack of definite and complete knowledge of some of the fundamental characteristics of these alloys. He says: Before civil and mechanical engineers can venture to use brass extensively for important works where it would be subjected to other than very low stress, some questions will have to be conclusively answered by manufacturers and experts. The following are the questions he submits:—

Can brass or bronze of high tensile strength be reliably produced which can be used safely for important permanent structures in such parts as bolts and other rolled, drawn, extruded or forged shapes?

What should be the specifications for such brasses or bronzes?

What inspection methods and tests should be used?

By what tests can the tendency to subsequent failure be detected at any time after manufacture?

What working stresses can be safely used for these various alloys?

Will these brasses or bronzes deteriorate by reason of constantly applied or frequently repeated stress—i.e., will they fail from fatigue?

Can large hollow manganese bronze castings be made of such forms as water and steam valves?

If manganese bronze is not suitable for such castings, what composition can be used?

Can any repairs of brass castings to be subjected to hydrostatic pressure be safely made by any process of "burning in" or welding now commonly used? If so, what relation has the shape and size of the casting to the methods and means to be employed?

What tests can be applied to prove that such repair has been successfully made, and that later a crack will not develop because of it?

Mr. Jonson made some experiments to show that cold worked brass was always in a state of initial stress. He subsequently found that very similar methods of test had been adopted by Professors Heyn and Martens in Germany. An account of these latter was given by Professor Heyn in his May lecture to the Institute of Metals in 1914. Mr. Jonson in one set of experiments took 6-inch lengths of round rods, faced their ends into parallelism and accurately measured their length. He then reduced the diameter of 4 inches of the length by successive steps of ⅛ inch, measuring the length of the turned specimen. This became longer at each turning, until a small central core only was left. The experiments shewed that initially the outer parts of the piece were in tension and the centre in compression. As the portions in tension were successively turned off, the compression became somewhat relieved, and the piece consequently elongated.

Messrs. Martens and Heyn developed a method for determining the tendency of brass to crack by immersing specimens in a solution of a mercury salt. In much strained samples cracks would then occur in a short time, but in other samples, presumably not so much stressed, the cracks would occasionally not become apparent for a number of days. This method was found to be useful, but not infallible, as some rods which within a reasonable period of observation developed no cracks by the application of the mercury solution subsequently spontaneously cracked.

This method of test was thought to depend upon the corrosion of the surface, and many experiences indicated that corrosion had much to do with the cracking under some circumstances. This was amply proved by some experiments upon samples, which were subjected to considerable strain and then submitted to corrosive influences. Some pieces of old corroded brass rods were also tested. They would stand bending only 45 degrees to 80 degrees before fracture, but other pieces of the same rod, when cleaned with emery cloth could be bent through 160 degress, or even flat on themselves without breaking. On the other hand, to show that corrosion is not a necessary factor in the cracking of brass, it is only necessary to recall that many specimens were found to be cracked when they were removed from the packing cases before being placed in use.

Mr. Flinn, in both papers, states that the trouble experienced may be classified as follows:—

1.—Breaks from stress, (a) initial stress due to methods of manufacture and (b) applied stress due to use.

2.—Damage by wrong heat treatment as in forging, bending, flanging, upsetting and annealing.

Damage of class 2 results entirely from lack of skill, knowledge or care on the part of the manufacturer.

Damage of class 1 (b) is dealt with in the joint paper of Messrs, Flinn and Jonson. They point out that manufacturers claimed for some of their special bronzes greater strength and toughness than steel possesses, besides being so much less liable to deterioration from corrosion. Ultimate strength and ultimate extension, however, are not the qualities which should determine the greatest stress which might be used in structures; this should be determined by the yield point of the material, the ultimate strength having little to do with the safe working stress when comparing different sorts of material.

The authors, however, are careful to refer to the "true" yield point of the metal, and state that a high yield point signifies nothing unless it is known what portion of this yield point is cancelled by initial stress, and, therefore, not available for working stress. They point out that if a test piece is continuously stretched by a gradually increasing load, when the true yield point is reached the metal flows until it is sufficiently hardened to carry the load. This, of course, is well known to be so in the case of all ductile materials. If a piece of ductile metal, whether it is a test piece or a piece of structural material, is subjected to a stress which deforms it, its yield point as regards further deformation due to the same kind of stress is raised, and it will not further deform until it is subjected to a greater stress than that initially put upon it. When it is seen that in the tests of the material used in the aqueduct work quoted in Mr. Flinn's papers that the ratio of yield point to ultimate strength ranges from 50 to 68 per cent. in the case of forgings, and from 45 to 71 per cent. in castings, one is forced to conclude that some of the results were obtained from specimens which were either subjected to high initial stresses or which had been submitted to high stress before test, and that, therefore, the recorded yield points were not "true." It is precisely this difficulty which has led our engineers to omit conditions as to yield point from acceptance specifications for ordinary steel for constructive purposes.

The authors of the paper upon the Inspection of Brass and Bronze consider that acceptance tests should include measurements of initial stress, and they suggest methods for determining this. They state that initial stress may be relieved in two ways: either by stretching the metal near the surface by mechanical work or by temporarily reducing its elastic limit by heating, and thus allowing it to be stretched by the compressive stress of the interior of the metal. The mechanical working used for the purpose is either squeezing between the rolls of a straightening machine or bending the bar successively in four directions at right angles to one another. This operation in America is given the name of "springing." Annealing, however, appears to be the more reliable method. Thev state that it has been proved that if sufficient time is allowed for annealing the temperature need not be so high as is required where initial stress has to be eliminated by quick annealing so that it is possible to eliminate initial stress by slow annealing without appreciably lowering the yield point given by cold drawing.

The question of "burning on" tips of propeller blades was referred to when the previous paper on brass was read at this institute. In this connection the remarks of Messrs. Flinn and Jonson upon foundry repairs will be of considerable interest.

They say that as brass and bronze castings are expensive to make it is not desirable to reject such castings on account of defects which can be remedied. Minor leaks in a casting may be stopped by peening. Which, however, they consider to be a questionable method of treatment. For very small defects they prefer to drill, and then to screw in a plug of the same metal as the casting. If this is not practicable the defective portions should be cut out by chisel or drill till all the defective metal is removed and the cavity then filled by pouring in metal from a crucible in such a way that the entire surface of the cavity is melted, and then consolidated by the added metal. If a gas flame is used for this there is a risk of the filling not being solid, besides the possibility of oxidation. When this method is adopted the cooling of the metal will be accompanied by contraction, which sets up severe tensile stresses in the added metal and in its surroundings. Further, the rapid cooling of the added metal makes its elastic limit higher than that of the neighboring portions. The state of tension so set up may be so severe that after a time cracks will appear.

The stresses may be lessened by keeping the casting heated to a high temperature while the repair is being made, but the best way to prevent cracking is to anneal the casting immediately the repair has been made. The annealing temperature should be maintained for several hours, so as to give the metal time to flow. This, they say, is the only way to insure safety in castings in which "burning in" has been adopted.

On the *Engineering Record* of August 12th, 1914, there appeared a communication from Mr. L. D. Van Aken, the Superintendent of the National Brass and Copper Tube Company, of Hastings-on-Hudson, New York, on the subject of the cracking of brass and bronze. He states that the frequency of these cracks has alarmed metallurgists, engineers, etc., and he attributes the failures mainly to what he terms the greed for higher tensile strength, which has led to a disregard for safety. He mentions that the U. S. Navy has adopted reasonable specifications for the elimination of the trouble. He records some experiments which he had made upon a brass having a composition of copper 61.33, zinc 37.66, lead 0.16, tin 0.83, iron 0.02. (This is very nearly the composition of our well-known Naval brass.) This was first made into an extruded rod 1⅛ inch diameter, allowed to cool slowly in the air and then tested. The result is given as No. 1. The rod was then cold drawn to 1 inch diameter, with the result given as No. 2. After this the cold drawn rod was annealed at 1,300 deg. F. for 30 minutes, and then allowed to air cool; the results of tests in this condition are given as No. 3. A portion of the rod after the same annealing was quenched in cold water instead of being allowed to cool in the air; the results are given in No. 4. Mr. Van Aken states that the annealing as in No. 3 is sufficient to prevent crackage of any description.

No.	Ultimate stren. lbs. per sq. in.	Elas. limit lbs. per sq. in.	Rat. Elas. to Ult. %	Elongat. in 2" %	Reduction of Area %	Hardness Scleroscope
1	58,000	26,550	46.	57.5	59.5	23.
2	73,100	70,100	96	27	45.2	29½, sur. 53
3	58,000	27,200	47	53	81.5	20.
4	63,800	29,900	47	38	38.5	20.

It will be noticed how enormously the cold drawing has affected the tensile strength, and more especially the elastic limit, and that the cold worked material still possesses as much ductility as measured by the extension and by the reduction area as would be shewn by a sample of a mild steel forging. Judged from the tensile test, it might be thought to be an excellent constructive material. Its high tensile strength, however, is dearly bought if it is accompanied by the tendency to spontaneously crack. The tests show also that annealing has restored the qualities of the metal to their original amounts.

It is to be noted that Mr. Van Aken only tested extended rods. In Mr. Flinn's papers extended rods are mentioned in such a way as to lead to the inference that extruded metal may be received with the same confidence as rolled or forged material. However, no comparative tests are given of the results of the two methods of manufacture. It is extremely desirable that some exhaustive experiments should be undertaken with different compositions of metals to show the effect upon the metal of the different processes, viz., casting pure and simple, forging, hot rolling, extrusion and also cold work, say, by drawing. These tests should include the effects of annealing. They should also include the resistance to repeated applications of stress commonly called fatigue tests. In fact, this kind of test is of greater importance than any other. It is evident that at present there is a lack of knowledge of the physical properties of many of the much used special alloys, although we have been using them for many years for very important structural purposes.

In the same volume of Transactions of the American Institute of Metals, which contains Mr. Flinn's paper, there is another communication by Dr. P. D. Merica and Mr. R. W. Woodward upon "The Failure of Structural Brasses." Dr. Merica is a Physicist of the Bureau of Standards, Washington.

In this paper there is a description of some work done in the Bureau of Standards in investigating the causes of spontaneous cracking in brass. The authors consider that failures are due primarily to the presence of internal stresses, and state that manufacturers should endeavour to produce material free from these stresses, and yet possessing the physical properties produced by the mechanical treatment which is responsible for these stresses. They state that it is possible, for instance, to make naval brass practically free from internal stress and yet having an ultimate strength of 60,000 lbs. and an elastic limit of from 30,000 to 35,000 lbs. per square inch.

The paper gives the results of numerous tests made with samples, some of which were new material, and others of brass which had been put into service. The authors tested these samples for internal stresses by Heyn's method. They state that their results may be relied upon to have a maximum error of not more than 1,000 to 2,000 lbs. per square inch, or 5 per cent.

The figures they give showed that all the samples they tested were more or less initially stressed, some of them to an enormous extent. Some of the samples have tension stresses on the surface, and others compression stresses.

They investigated the method of testing for initial stress by immersing the samples in solutions such as mercuric chloride, mercurous nitrate, ammonium hydroxide, etc. Cracks were often produced, sometimes in two minutes, sometimes in 24 hours. They found, however, that while those samples which had a high tension on the surface cracked, those which had large compressive stresses did not. This test, therefore, is not a sufficient one to enable defective material to be detected.

The following are the main conclusions arrived at by the authors:—

1. Failures occur, so far as is known, only in brass which has been worked, i.e., forged, hot or cold rolled, drawn, or extruded.

2. These failures are independent of composition.

3. The fractures occur, in general, transversely to the direction of working.

4. Failures occur in brass materials which are under initial stress provided these stresses average over 5,000 to 6,000 lbs. per square inch.

5. These stresses can be almost entirely removed by annealing at temperatures of about 300 deg. C. to 400 deg. C., which temperatures do not affect sensibly the ordinary physical properties.

6. No certain indication of the magnitude, or even the presence, of such stresses has been found in studies of the structure or of etching or corroding agents.

7. Brass may be manufactured practically free from stresses and still retain excellent mechanical properties, i.e., will not be too soft.

Dr. Merica stated that investigations are being continued at the Bureau of Standards, and invited the co-operation of both manufacturers and users of brass.

It will be noticed that neither in this present communication nor in the former paper has any reference been made to corrosion in general. This is, of course, a very important matter, and so far as condenser tubes are concerned is the subject of research by a committee of the Institute of Metals, assisted by a Government grant. Without wishing to detract from this very useful work, it is thought that the subject of the physical properties of copper alloys, especially as regards their reliability and their adaptability for use for important structural purposes, is quite as worthy of a thorough research as is that of corrosion.

ENGINE ROOM UPKEEP PROBLEMS —II

Speaking of stripping threads brings me to another point. When I find an engine, pump, or other such device provided with thumbscrews or cap screws instead of studbolts, I like to see these removed and studbolts put in. Why? Because in doing and undoing a joint where the cap screws are used, there is wear on the threads in the cast iron, and these will give out more quickly than will the threads on the cap screws. Then new threads must be made in the casting and larger screws used. Now, if we use studbolts, the wear comes on the bolts. The threads in the cast iron remain undamaged. When the threads on the outer end of the bolts become damaged, these bolts may be removed, still leaving in the cast iron good threads fit to hold another bolt. On the same principle we occasionally find that in places where pipes are frequently connected and disconnected there is a bushing screwed into the casting. This is perhaps especially noticeable in some rock drills. When the threads in the bushing become defective, a new bushing can be substituted and the threads used are again good throughout. I have seen this hint used to advantage in steam plants also, but of course, it applies most to places where a union alone does not fully serve.

My first experience in replacing studbolts was with regard to broken ones. New ones were provided by a blacksmith, but were 1/32 in. over size. A plug tap to match was provided and all went O.K. My next was the replacing of thumbscrews by studbolts. As the threads in the cast iron were bad, I wanted a larger size of bolts, and as some of the old screws were 1/32 in. over size, I decided to have the new ones an eighth of an inch larger and having the standard thread for its own size. There was in this case ample cast iron remaining for the required strength, and I am somewhat of a crank on the subject of standard threads. My employer took the order to the nearest machinist, who, by the way, was a very exact man and delighted in work which required much use of the micrometer and precision to the thousandth of an inch. He refused to furnish the studs, claiming that the new thread in the casting would not be a full thread owing to the size the old hole had been tapped. He seemed to think that what was needed was studbolts having same thread (pitch) as the old bolts, but enough larger to give good full new threads. This latter dimension was to be determined by accurate measurements made of the holes themselves, consequently the order was held over. When my employer explained to me how matters stood, I objected on the ground that this would call for one or more special sized taps, which might never be required again, and therefore were an undesirable investment and perhaps delay also unless the machinist had them on hand. No! Mr. Machinist hadn't any, but could supply one or more at a reasonable price. Next I set out and procured studbolts of the size and thread I had desired, then came back and put them in. I do not think anyone but the machinist has had cause for complaint concerning the affair yet. Since then I have replaced half-inch thumb or cap screws by five-eighths studbolts, five-eighths screws by three-quarter bolts, and three-quarter screws by seven-eighths bolts, using standard threads in each case, and I have not yet regretted doing so. I do not remember using a tap drill for these except in the case of the seven-eighths ones. The idea was to remove as little metal as possible and not to remove any except with the tap providing the said tap could work safely without being preceded by a drill.

By comparing each bolt with the correct size tap drill for the next larger size bolt mentioned, it is easy to see that in some cases we will fall one, two or possibly three thirty-seconds of an inch short of having a full thread for the new bolt unless we deviate from the standard. Even with the 1/32 in. oversize, we may fall short. This is less noticeable with the U.S. thread and with the Whitworth thread than with the V thread, and is most likely to occur when we try to put an exact size bolt in a worn hole tapped for an oversize bolt. Assuming that the thread occupies 4/32, and we are 2/32 shy of metal, then our thread is weakened to one-half, if, however, we are shy 2/32 on a thread occupying 5/32, we do two-fifths, and if shy, only 1/32 in five, we lose only the points of the V's, and they hold comparatively little anyway. To make up for this, if the old hole is retapped without drilling the old threads out, some of the new threads will come into the old ones in a way that gives us a full thread part of the distance, and as the pitch line of our new thread describes a larger circle than did the pitch line of the old thread, we have that much more hold in the casting than our calculations in thread depth alone would lead us to believe. For instance, with a new ⅝in. bolt, the pitch line of our new thread is a twenty-five per cent. larger circle than was that of the former half-inch screw, and this makes an appreciable difference in our favor. Then again, the ⅝in. bolt with No. 11 thread calls for a ½in. tap drill. If the old bolt was ½in. with No. 12 thread, a 27/64 in. tap drill would have been correct for it. With 3/64 in. worn from the inner circle of these threads, we still have in places 1/32 in. to spare after having obtained full ⅝in. thread in those places. To diagram these features for the sizes given would require too much space here, and as there is much difference in the condition of old threads in various stages of wear, it is really better for each to use his own judgment in the matter, taking into consideration the actual state of existing conditions and pressure carried in the individual cases. The idea is to discriminate between instances where the special taps are necessary and those where the standard size and thread can be safely used if it means a saving of time and expense in making the repair.

WORK OF SHIPPING BOARD OF GREAT MAGNITUDE

THE first annual report of the United States Shipping Board, brief mention of which was made in last week's issue, says among other things that the Emergency Fleet Corporation, which is the most important of the board's war powers, is now engaged in what is probably the greatest construction task ever attempted by a single institution. It has developed for this purpose within a period of six months an organization of more than 1,000 employees, including a large force of technical experts. The corporation has sixteen offices in various parts of the United States. It is supervising the building of 1,118 vessels in 116 shipyards distributed throughout the United States. It is disbursing for the construction of those ships something in excess of a billion dollars per annum. It is controlling substantially all the shipbuilding of the country other than of naval vessels, and its program calls for the completion in 1918 of eight times the tonnage delivered in 1916.

Meeting the Situation Fairly

When the United States entered the war, says the report, American shipbuilders were principally engaged in the construction of ships for foreign account, principally British and Norwegian. The yards were working substantially at capacity. The problem of the corporation, therefore, was not merely to build ships, but also to build new yards in which more ships might be constructed.

To this end, contracts were placed for a large number of wooden ships, most of them to be built in new yards, it being found that such ships could be built, within certain limits, without interfering with the steel ship program. The corporation has experienced some difficulty in arranging for supplies of proper lumber, but it is believed that this difficulty has been largely overcome and that the wooden ships may be expected to be completed with a fair degree of promptness.

2,500 Ton Steel Ships Requisitioned

As to steel ships the corporation has made every effort to standardize designs for steel construction, but it is deemed inadvisable at this time to publish the details of those designs.

In order to speed up construction and to assure unity of control over shipyards, the corporation, on August 3, 1917, in accordance with President Wilson's order of July 11, 1917, requisitioned all steel ships under construction of over 2,500 tons dead-weight capacity. As a result of this order the United States became possessed of 413 ships in various stages of construction, of a total tonnage when completed of 2,937,808 tons. Of these, 33, of a total tonnage of 257,575, have been completed, and it is estimated that the remainder will be delivered within the next 18 months.

Training the Workers

The corporation is making rapid progress in placing direct contracts for ship construction. In the last analysis it is man-power that builds ships; and the mobilization of a large, competent, trained and willing force of workers for the shipyards of the country has been among the most important of the corporation's activities.

Other excerpts from the report follow:

The board has recently adopted a policy of requisitioning for government account all power-driven steel cargo vessels of 2,500 tons dead-weight or over, and all passenger vessels of 2,500 tons gross register or over, adapted to ocean service. The operation of these vessels is entrusted in large part to the companies by whom they were formerly controlled, but all receipts have been for government account, the owners being credited with the requisition rate fixed by the board. The vessels, moreover, are under full control of the board as to voyages, cargoes, and rates and as to safeguards for their protection against the hazards of war-zone service. It is hoped that this policy will be the means of achieving two results which in the opinion of the board are necessary: First, complete unity of control over the distribution of shipping, so that war needs may be filled in the order of their emergency; and second, just and effective regulation of rates, so that the nations at war with Germany may not be financially exhausted by extortionate transportation charges.

Further experience may develop that, consistent with the purposes above outlined, some of the steamers not adapted to war service may profitably be released from requisition. As to other lines, more effective methods of regulation and control may prove to be necessary. It should be borne in mind that aside from the requisition power the board at present has no jurisdiction over rates in foreign trade and no jurisdiction over interstate rates, except where vessels are common carriers operating on regular routes. If experience shall prove it to be necessary, the board will recommend drastic legislation to meet more effectively the situation arising from the acute dearth of tonnage.

The freight rates to be charged to shippers on requisitioned vessels left in commercial service or assigned to foreign governments have been the subject of careful consideration. The board, of course, has no desire to profiteer at the expense of any government associated with the United States in the war against Germany, or of the general public. On the other hand existing contracts, particularly those on a c.i.f. basis, and special market conditions often present a danger that a reduction in rate will not insure to the benefit of the consumer but will merely transfer from the shipowners to favored shippers the monopoly earnings which it is intended to eliminate. The policy of the board therefore will be to charge freight rates based on the requisition rate whenever it can receive assurance that the benefit of these low rates will accrue to the American or allied governments or to the consuming public; and to charge higher rates whenever this fact shall not be so established. The application of this policy will no doubt present difficulties in individual cases but it is believed in principle to be sound.

The requisition program as yet has been in effect for such a short period of time that no accurate statement of receipts and expenditures can now be made. Accounts, however, are being kept with scrupulous care along sound and conservative lines, and in due season will be presented to the Congress.

Ships From Great Lakes Utilized

The Great Lakes, shut off as they are from sea-level connection with the ocean, have nevertheless contributed materially to the work of procuring tonnage for war purposes. An unprecedented number of newly constructed ships have been brought down from the Lakes, and in addition 21 steel steamers have been withdrawn from Lake commerce and are being brought down for war service. Of these, 16 were too long to pass through the locks, and it was necessary to cut them in two, bring them down in parts, and reassemble them on the St. Lawrence. Twelve of the vessels were put together afloat, an achievement never before accomplished. Ten steel tugs have also been commandeered from the Lake fleet, and are being fitted with surface condensers for ocean use.

A committee has been appointed to advise the board as to the matter of just compensation to be paid for vessels from the Lakes to which title has been taken, and that committee is now at work on its task.

Chartering Committee's Duties Outlined

The program of control of shipping through requisition covers only American steam vessels over 2,500 tons deadweight. To exercise a salutary degree of control over other American and neutral shipping in American foreign trade, a chartering committee was established, with offices in New York; and with the assistance of the War Trade Board all charters of neutral ships, of American steamers under 2,500 tons, and of sailing ships of any size loading from ports of the United States to foreign ports, must be approved by this committee. The committee has exercised this control with two primary objects in mind: First, so far as possible, to induce neutral tonnage to assume its fair share of trans-Atlantic trade; and, second, to effect a material reduction in the high charter rates prevailing, not only in trans-Atlantic but in South American and oriental trades.

Under Schedules 1591 and 1593 the Bureau of Supplies and Accounts of the Navy Department is calling for bids for furnishing various navy yards with large quantities of electrical material. Among the material required is cable, generator, motor sets, high-tensioned wire, magnet wire, etc.

The War Trade Board has passed a resolution authorizing the director of the bureau of exports to license freely cargoes originating in the United Kingdom, France and Italy or their colonies, possessions or protectorates when shipped via United States ports to United Kingdom, France and Italy or their colonies, etc.

As regards Canadian shipments, cargo of the above character is of two kinds—that which leaves Canada by rail, passing through the United States in bond, and then shipped from a United States port, and cargo shipped from Canada by steamer which later touches at a United States port.

With regard to shipments from Canada to Japan made by steamers which call at a port of the United States, the same ruling will apply as to shipments from Canada to England, and such cargo will be licensed by the collectors of customs in exchange for copies of the manifest.

Eighty commercial automobiles, valued at $69,709, and 1,020 passenger motor cars, valued at $676,483, were exported from the United States to Canada during October, according to the latest returns of the Department of Commerce. During the ten months' period ending October, 1917, there were 696 commercial motor cars, valued at $913,088, and 14,261 passenger automobiles, valued at $10,791,081, imported into Canada from the United States. The combined exports of commercial and passenger automobiles from the United States to the Dominion in October a year ago numbered 1,021 cars, valued at $718,962, while during the ten months' period of that year they were 11,118 cars, valued at $8,025,079.

STEAM POWER DEVELOPMENT

A SHORT time ago V. H. Manning, Director of the Bureau of Mines, Department of the Interior, Washington, D.C., made the statement that 25 per cent. of the coal production of the United States was wasted through inefficient use. His point was that a large proportion of the steam engine and boiler plants of the country are not up-to-date and are consuming coal wastefully as compared with results given by plants of the latest and most approved design.

Since engineering methods are always advancing, it is out of the question that all of the power plants of the country should be at any one time of the latest design, but Mr. Manning's statement is very suggestive as to the importance to the country of improvements in its power equipment. The annual production of coal in the United States is now about 600,000,000 tons, and 150,000,000 tons of this would be saved if all consuming plants were as good as those of the best type. Perhaps the most impressive idea of this saving can be obtained if the reader will think of the labor involved in mining, handling, and transporting 150,000,000 tons of coal, and of shoveling it into the fire-boxes; and of the benefits that might be derived by having that labor distributed in the other industries.

Generation Back Comparison

If it is interesting to calculate the savings that may be effected in the future by bringing all equipment up to the standard of what at this date is recognized as the best, it is also interesting to compare not only the best practice but the common practice of to-day with the best practice of a generation ago. Mr. Manning's statement has led us to make an inquiry along this line, says a statement issued by the National City Bank, New York, and looking backward for a starting point we fixed upon the Centennial Exhibition at Philadelphia in 1876. Doubtless many of our mature readers will remember viewing, as youths, with awe and wonderment, the great Corliss engine which stood at the centre of Machinery Hall, and furnished the driving power for all the exhibits in the building. There were seats for the public about the engines; the motion of the great flywheels created a breeze, and the engine was the admiration not only of the non-technical public, but of the members of the engineering profession from all over the world.

With the Corliss engine standing out prominently as a landmark in engine development, we appealed to President Alexander C. Humphreys, of Stevens Institute of Technology, for information as to progress in engine-building since that time. It will add a touch of human interest to the story to state that it then developed that Professor Robert M. Anderson, of the Department of Engineering, Stevens Institute, was in part influenced to his choice of a profession by seeing, as a boy, the big Corliss engine in Machinery Hall. By the courtesy of Stevens Institute we have the interesting sketch which follows:

To summarize in the briefest possible manner the progress which it records, the Corliss engine at the Centennial occupied about 54,000 cubic feet of space and consumed 2.2 pounds of coal to develop one horse-power; while a modern steam turbine of equal capacity will occupy 115 cubic feet of space and consume nine-tenths of a pound of coal to develop a horse-power, giving a saving in coal consumption of 59 per cent. According to Mr. Manning's estimate, that the average engine efficiency at the present time is 25 per cent. below the standard of the best type, the average now is well above the best of 1876; and, of course, the average efficiency in 1876 was much below the Corliss standard.

Electric Power Transmission

Even more notable than the improvement in engine and boiler efficiency is the gain accomplished by the development of power transmission through the electric current. The transmission of power by electricity was unknown when the Centennial Exhibition was held, but in the year 1915 51 per cent. of all the power applied to industry in the United States, outside of transportation, was delivered by electric current, and as a result of all this development the total horse-power in use in the United States, to each million inhabitants, was 482,392 in 1915, as compared with 68,182 in 1879, or an increase of more than seven-fold. The facility with which power can be transmitted by electric current does away with the necessity of generating power where it is to be used. It can be generated at the most economical spot, as at the mouth of a coal mine or the site of a water-fall, and the industries which use it may be located in a spot more suitable to them.

High Speed Reciprocating Engines

The introduction of the high speed reciprocating engine dates from shortly after Mr. Edison's invention of the incandescent lamp and development of the direct-current generator in 1881. The isolated lighting plants, followed by small lighting and power plants, gave an impetus to shaft-governed automatic cut-off engines, allowing of higher rotative speeds than were attainable with drop cut-off mechanisms. Among these were: the Buckeye, the Porter-Allen, the Fitchburg, the Armington and Sims, the McIntosh and Seymour, the Ball, the Westinghouse, the Willans, and many others. These were built simple and compound and were operated either non-densing or condensing. They varied in economy from 3.6 lbs. of coal per hour per B.H.P. for simple non-condensing to 2.15 lbs. of coal per hour per B.H.P. in compound condensing types.

Up to date the highest economy developed in reciprocating engines has been attained in the Stumpf (Straight) or Uniflow engine, obtaining a B.H.P. on 1.08 lbs. of coal per hour, same being in a 300 h.p. engine with an initial steam pressure of 130 lbs. per square inch absolute, and superheat of 261 deg. F. This record has been equalled by the best performances of stationary triple-expansion engines using steam at 170 lbs. per sq. in. absolute and 203 deg. F. superheat. The highest of all steam engine efficiencies so far reached has been obtained by the Wolf tandem-compound "locomobile," which consists of an internally fired through-tube boiler, having the engine combined so that the cylinders are jacketed by the final pass of the hot gases flowing to the stack—in other words, the cylinder projects into the smoke box. The initial steam pressure is from 175 to 225 lbs. per sq. in. absolute, is superheated to 800-850 deg. F. before entering the high pressure cylinder and again superheated to 450-500 deg. F. before entering the low pressure cylinder. The feed water is heated in an economizer in the breeching. The coal consumption of 1 lb. per hour per B.H.P. is common practice and 0.8 lb. per hour per B.H.P. has been attained.

The Steam Turbine

The reaction type of steam turbine, known under his name, was first introduced in England by the Hon. Charles A. Parsons in 1884, and was introduced in the United States by the Westinghouse Co. in 1895, but it was not until about 1900 that it met with general approval. The DeLaval impulse turbine was introduced in its present form in 1889. The compound velocity impulse turbine, known as the Curtis, was introduced into the United States by Charles G. Curtis about 1896.

Since their introduction there have

been many modifications and combinations of these types, better fitting their application to various power developments. For central station work the steam turbine has replaced the reciprocating engine. In fact, the turbine has the advantage in all electric generation, especially in the larger outputs. This is not, however, on account of better fuel economy, since for the same grades of machinery and equal ranges of temperature, pressure, etc., there has been found no appreciable difference in economy betwewen the best turbines and the best reciprocating engines, other characteristics influencing the preference. The real reason for the replacement of the engine by the turbine has been the greater compactness of the latter, thus resulting in a smaller investment in real estate.

Boiler Efficiency

There has been marked improvement in boiler efficiency since 1876. The combined efficiency of furnace and boiler, being the ratio of delivered energy in the steam to that available in the fuel, ranges in coal-fired boilers from about 55 per cent. to 75 per cent. With liquid (oil) fuel and powdered coal, 80 per cent. is attainable. The most remarkable and promising development, the restriction being that the fuel must first be transformed into a gas, is the Bone system of so-called "surface combustion." This consists in burning a perfect mixture of gas and air in boiler tubes filled with porous refractory material, the result being incandescent combination, with no radiation or excess air losses. An efficiency of 94 per cent. has been attained.

The development of the steam engine illustrates very clearly the part which capital plays in community progress. A steam engine is the servant of the community, no matter who owns it, and the improvements upon the engine, the changing over of power plants to bring them up-to-date, and the construction of new power plants, constantly absorb an enormous amount of new capital. Into such enterprises the profits of business largely go, while the resulting benefits go to the entire community. The man without property, who spends every dollar of his income the same week he receives it, is able to buy more with his wages, because steam engine efficiency has been doubled in the last forty years.

The world is suffering frightful losses through the war, and people are wondering what the effects of these gigantic debts will be upon the earnings of industry and the welfare of the peoples. We may learn something about this from the experience of the past. "It is our improved steam engine," wrote Francis Jeffrey in the "Edinburgh Review" in 1819, "that has fought the battles of Europe, and exalted and sustained, through the late tremendous contest, the political greatness of our land. It is the same great power which enables us now to pay the interest on our debt and to maintain the arduous struggle in which we are still engaged with the skill and enterprise of other countries less oppressed with taxation."

The steam engine which helped England bear the burdens of the wars with Napoleon was a poor affair compared with the engine of to-day, but even the latter, as Mr. Manning and Professor Anderson have indicated, may have its average efficiency in practice greatly increased. Beyond these prospective gains are those which lie in other sources of power, and in all the improvements in industrial equipment which are only partially introduced, or as yet only forming in the minds of inventors. More important than all these are the possibilities which lie in more thoroughly organized and harmonious industry, the development of intelligence and efficiency among the people, and the spread of a better understanding of common interests. We must look to these changes for gains in production which will enable society to overcome the losses of the war, and repeat the progress accomplished by the original application of the steam engine to industry.

MOTOR DRIVEN DRY VACUUM PUMP

THE makers of the dry vacuum pump illustrated in the accompanying engraving are the Wheeler, Condenser and Engineering Co. of Carteret, N.J., who have endeavored, in its design and construction, to provide an air pump for use on condensers, which will satisfactorily fulfill the requirements of such an apparatus in both marine and stationary applications.

It is especially adapted for use in that type of installation where it is desirable to maintain a high vacuum in the condenser, in which case the use of separate condensate and air pumps is approved practice.

MOTOR-DRIVEN DRY VACUUM PUMP.

As the cut shows the pump is of the vertical type and is driven by an electric motor direct connected to the crank shaft. The inlet valves of the pump are of the semi-rotative type, which are so manipulated by the valve gear as to draw air from the condenser during its full stroke. Clearance difficulties are eliminated in this design of valve gear by providing ports which register with an equalizing air passage. The discharge valves are of the poppet type and are easily accessible, as are all the wearing parts, and their method of arrangement ensures adequate drainage of condensed vapor.

Advantages claimed for the electric driven rotative dry vacuum pump over the reciprocating steam drive are: ease of installation; less attention required; no steam pipe radiation losses; and improvement of plant load factor.

TORONTO FIRM BUILDS MOTOR LIFE-BOATS

SEVERAL large motor-powered life-boats are now under construction at the Canadian Beaver Company at the foot of York Street. They are to be used on the Norwegian Government ships, which are being built at the Polson Iron Works.

The life-boats are 24 feet over all, with a beam of seven feet. They are being fitted with a 12 horse-power Foreman gasoline motor, and have a seating capacity of 30, and when loaded are expected to attain a speed of seven miles an hour. Fitted with two belts of cork 12 inches in diameter amidships and tapering off to nothing at bow and stern, they are absolutely unsinkable. The boats are also fitted for towing, and have the luxury of a canvas hood to enclose the whole cockpit. The frames are made of the finest white oak and are planked with cypress. Four pairs of oars are provided. The oak water-casks are aft, and the provision locker in the bows. The fuel tank is amidships, and has a capacity for 50 gallons.

The Canadian Beaver Company is also constructing a number of life-rafts for the same ships, the tanks of which are made of regulation government copper, the frames of white oak, and will, in case of emergency, be taken in tow by the life-boats. The rafts are 8 feet by 10 feet, and have a capacity of 25 persons. The rafts are always carried loose on the decks, except in very rough weather, when they are lashed.

Vancouver Sends First Shipment of Lumber.—The first shipment of lumber from Vancouver to the Orient left Vancouver on January 15 on a steamer of the Dollar Line, and consisted of 300,000 feet cut at the new Dollar sawmill at Roche Point. It was lightered down the inlet to the Dollar dock. The motor schooner May, now loading 1,500,000 feet at the Dollar mill for San Pedro, has, it is understood, been sold to New York interests, who will take delivery of her when she arrives at San Pedro. The May is a new ship recently completed on the Columbia River, and this is her first voyage. She was bought on the stocks by the Canadian Robert Dollar Co., who have now turned her over at the usual profit prevailing in ship deals of these days.

Steam Saving Auxiliaries of the Engine and Boiler Rooms

By C. T. R.

In view of the circumstance that steam-saving auxiliaries aboard ship continue to increase in number, and that they are being designed and constructed to meet, in the most effective manner, both ordinary and special service applications, this series of articles describing and illustrating at least the more important types of such apparatus seems to us more or less timely, both from the point of view of familiarizing engine and boiler room staffs with the products of different manufacturers, and that of their acquiring a closer intimacy with specific detail arrangement, relative to operation, maintenance and periodic overhaul.

BOILER FEED REGULATORS—II.

THE flexibility possessed by some types of boiler together with the occasional desirability of reducing the water level in proportion to the load carried suggested the idea of producing a device which would enable such operating conditions to be automatically obtained. The development of the idea was doubtless further prompted by the customary method of locomotive operation, whereby it is usual to raise the water level to the safe high limit before reaching any particular grade. When climbing the grade, the rate of feed is reduced or sometimes discontinued for a short period so that all the heat from the furnace may be spent in making steam and none in heating feed water. The variable type of feed-water heater duplicates in stationary practice that which was found advantageous in locomotive practice, and also provides a uniform rate of feed at any particular level, provided the load is also uniform. In other words the greater the load the lower the level, with correspondingly drier steam and reduced risk of priming.

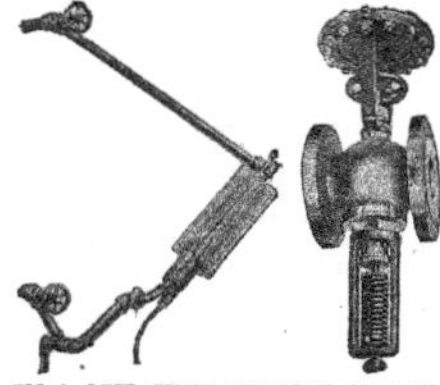

FIG. 3. LEFT, GENERATOR OF "S.-C." REGULATOR. RIGHT, REGULATING VALVE OPERATED BY GENERATOR.

'S'-C" Regulator

This apparatus possesses as its source of motive power, a certain fixed quantity of water in a closed system, the pressure of the liquid varying with the amount of heat imparted to it through variations of the water level. It is designed to give a continuous graduated feed, increasing as the level decreases, and decreasing as the level rises until at high level it shuts off the feed entirely. Figs. 3 and 4 show the appearance and construction of the device which consists of a generator R, a regulating valve in the feed pipe (not shown), and a hydraulic tube M, connecting the generator with the regulator.

The generator is installed on the same level as normal water level in the boiler and is adjustable to various angles so that the vertical height from top to bottom of generator may vary from 2 in. to 8 in. Through the generator passes a ⅝ in. tube D, which is in series with the steam and water connections, but which does not communicate with the internal space of the generator. The only connection into the generator is the tube M from regulator, and the entire system of generator, tubing and regulator is filled up with water and sealed with plug I. The level of the boiler water in the ⅝ in. tube D will now rise or fall with the boiler. The water in the generator, however, is being cooled all the time by radiation through the extenrnal fins, and kept below boiling point, while a portion of tube D is filled with steam at boiler pressure and tends to heat up the upper part of the water in the generator. According as the boiler level falls, so will more of the tube D heat up the generator and increase the pressure so that the feed pipe regulator will be opened correspondingly more.

The sectional illustration, Fig. 4, shows the condition of the generator liquid when the boiler level is half way down. The upper portion is changed into steam and maintains a constant pressure in the controlling system; while, if the boiler level rises, the valume of steam decreases with the decreased amount of steam in tube D until, at the high limit, the feed is closed off completely.

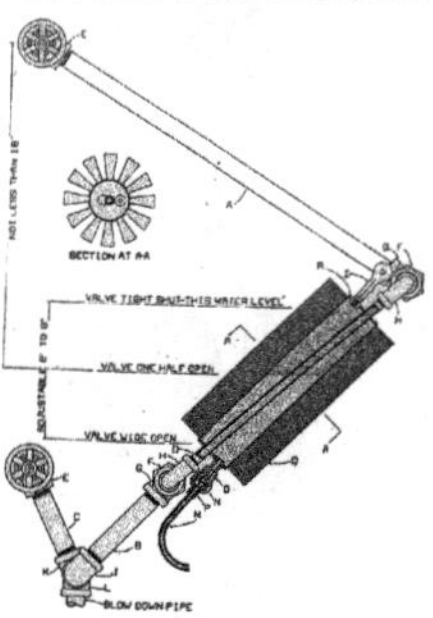

FIG. 4. SECTIONAL VIEW OF "S.-C." GENERATOR SHOWING CONTAINED FLUID UNDER PRESSURE.

FIG. 5.—SHOWING ARRANGEMENT FEED "ELLIOT" FLOAT OPERATED FEED REGULATOR WITH CONTROLLING VALVE.

The flexible fittings used allow of the "S-C" regulator being adjusted when the boiler is in operation, and by varying the vertical height referred to the limits of the boiler level may be set to suit a particular type of boiler. The system so operates that on an increasing load it will always feed at a lower rate than boiler output until peak load level is reached, and on a decreasing load it will feed above the boiler output until light load level is reached. The only time the regulating valve will close is when there is no load.

This apparatus is manufactured by the "S-C" Regulator Co., Fostoria, O.

Elliot Regulator

The objects aimed at by the designers of this regulator are reliability, accessibility, least number of working parts, and ability to make adjustment under full steam pressure. Illustrations Figs. 5 and 6 show one of the designs, which is characterized by the use of a long body water column and a diaphragm type of controlling valve. In actual operation, their relative positions are changed, as the water column is mounted on the boiler, and the controlling valve is located in any convenient place in the feed line.

The regulating device consists of a float operated pilot valve, which operates with the fluctuations in the water level and causes the controlling valve to be closed or opened, depending upon whether the water level is above or below

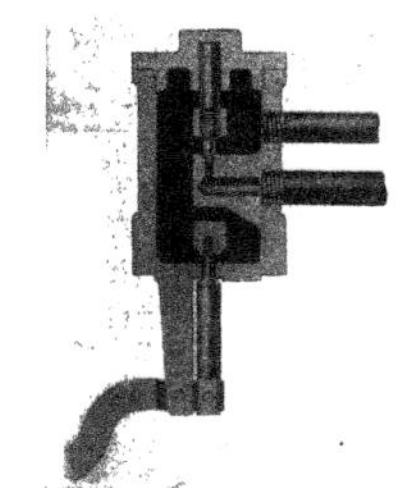

FIG. 6.—ENLARGED VIEW OF PILOT VALVE OF "ELLIOT" REGULATOR

the normal. The pilot valve is shown in detail in Fig. 6, the chamber which contains it being mounted on the top of the water column. An arm extends downward from the chamber and supports the float lever.

The pilot valve is of the double type, the upper valve which controls the exhaust to atmosphere being carried in a yoke, which in turn is mounted on the stem of the lower or steam valve. The exhaust valve is guided on its upper end by a cap which closes the chamber, and which can be removed to enable adjustment of the valve to be made after long service. The upper pipe from the valve chamber, as shown in Fig. 5, is connect- to the diaphragm chamber of the controlling valve.

Assuming the water level in boiler to be above normal, the float rises and lifts up the long arm of the lever, thus opening the steam valve downward and closing the exhaust valve on top. The diaphragm chamber which is filled with water is now under boiler pressure, and the controlling valve spindle is forced down till the valve is closed and the feed stopped. As the water level begins to drop, the weight of the float is gradually transferred to the lever until the weight of the float overcomes the steam pressure which holds the exhaust valve on its seat. When the exhaust valve is unseated the weight of the float seats the admission valve, and the steam that is contained in the chamber and pipe line passes through the exhaust valve to atmosphere, relieving the pressure and allowing the controlling valve to be opened by the spring provided for that purpose.

As the water level again rises in the boiler, the float has a tendency to rise, but the pressure of the steam on the admission valve keeps it closed until the float has acquired sufficient buoyancy to lift the lever when the cycle of operations is repeated. With ordinary boiler pressures, the necessary buoyancy of the float is obtained with not more than ⅝ in. variation in water level, while the construction of the pilot valve chamber enables adjustment to be made under full steam pressure.

These regulators are made by the Elliott Company, Pittsburgh, the various types being suited for use on water tube and fire-tube boilers of all types.

Foster Regulator

This regulator is of the intermittent type and depends for its action on the alternate expansion and contraction of a length of tube which is filled with steam or water, according to the variation of water level. Its action is slightly different from those regulators which depend on the buoyancy of floats or other bodies, as the time necessary for its coming into action depends on the speed with which the expansion tube radiates its heat. The resulting lapse of time between the change of level and action of the regulator allows it to remain longer in each position with less frequent opening and closing.

As illustrated in Fig. 6 the expansion tube 1 is screwed into the main bracket 2, in which are also fixed two guide rods 4. The upper ends of rods 4 are tied with a bracket 6 on which is hinged a bell crank 8, the short arm of which rests on top of adjusting screw 9 in the upper end of expansion tube. At the lower ends of rods 4 is clamped a bracket 11 which carries bell crank 10, one arm of which is flexibly connected to the long arm of bell crank 8, while the other arm extends horizontally and engages with the stem of the regulating valve 14. Positive closing of this valve is insured by an ad-

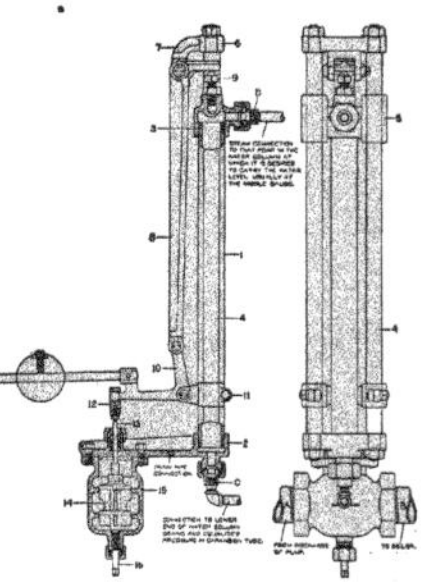

FIG. 7. "FOSTER" REGULATOR WITH EXPANSION TUBE AND MECHANICALLY OPERATED CONTROLLING VALVE.

justable counter weight carried on an extension of the bell crank arm.

The alternate admission of steam and water to the expansion tube is obtained by the method of connecting up the two ends C and D to the boiler. The expansion tube is first of all located at a point completely above the high level of the water in the boiler, after which connection C is carried to a point below the desired level, and connection D to a point just at the desired height of the water level in the column. Whenever the water level falls below connection from D, the expansion tube will be filled with steam, as its lower end is also above water level. The expansion due to the heat of the steam causes adjusting screw 9 to raise the short arm of the upper bell crank, so that the long arm moves over toward the expansion tube and opens up the regulating valve by means of the lower bell crank. The feed gradually attains a maximum as the expansion tube gets hotter, and remains so until the water level rises high enough to close connection from D. Water will now enter both ends of the expansion tube, replacing the steam which condenses as the heat is radiated, and causing tube to contract, so that the feed regulating valve is allowed to close, due to weight of ball on rod, which also causes the short arm of upper bell crank to follow down in top of adjusting screw as the expansion tube contracts.

While a very small variation of water level is sufficient to operate this device, the time necessary for the tube to warm up and cool down, causes less frequency of operation than with certain other intermittent types. There is an entire absence of stuffing boxes, the whole action is mechanical and therefore positive, and the position of the counterweight ball indicates instantly whether the feed valve is open or closed. This regulator is made by the Foster Engineering Co., Newark, N.J.

Economiser Explosions.—The idea that an economiser is safe from explosion because it is only exposed to water pressure is probably due to the fact that vessels which have to be worked under steam or air pressure are invariably tested by means of water. The reason for so doing is that danger is eliminated of the parts flying violently asunder should any part fail, as would happen if steam or air were employed. There is, however, a great difference between an economiser under pressure and a steam boiler or other vessel under cold water pressure. In the latter case, in the event of any part of the structure failing, the pressure is instantly relieved, with very little increase in the volume of the fluid, and no great harm is done; in the former case much of the water in the economiser is usually at a temperature much greater than 212 deg. Fah., so that should the pressure be suddenly relieved due to failure of the containing vessel, a portion of the water would immediately flash into steam, the accompanying violent expansive force of which may cause entire disruption of the economiser.

The MacLean Publishing Company
LIMITED
(ESTABLISHED 1888)

JOHN BAYNE MACLEAN - - - - - - - President
H. T. HUNTER - - - - - - - - - Vice-President
H. V. TYRRELL - - - - - - - - - General Manager

PUBLISHERS

MARINE ENGINEERING
of Canada

A monthly journal dealing with the progress and development of Merchant and Naval Marine Engineering, Shipbuilding, the building of Harbors and Docks, and containing a record of the latest and best practice throughout the Sea-going World.

J. M. WILSON, Editor. B. G. NEWTON, Manager.

OFFICES

CANADA—
Montreal—Southam Building, 128 Bleury St., Telephone Main 1004.
Toronto—143-153 University Ave., Telephone Main 7324.
Winnipeg—1207 Union Trust Building. Telephone Main 3449.
Eastern Representative, E. M. Pattison.
Ontario Representative, S. S. Moore.
Toronto and Hamilton Representative, M. H. Wood.

UNITED STATES—
New York—R. B. Huestis, 111 Broadway, New York. Telephone 8971 Rector.
Chicago—A. H. Byrne, 900 Lytton Bldg., 14 E. Jackson Street. Phone Harrison 1147.
Boston—C. L. Morton, Room 733, Old South Bldg., Telephone Main 1024.

GREAT BRITAIN—
London—The MacLean Company of Great Britain, Limited, 88 Fleet Street, E.C. E. J. Dodd, Director, Telephone Central 12960. Address: Atabek, London, England.

SUBSCRIPTION RATE

Canada, $1.00; United States, $1.50; Great Britain, Australia and other colonies, 4s. 6d. per year; other countries, $1.50. Advertising rates on request.

Subscribers, who are not receiving their paper regularly, will confer a favor by telling us. We should be notified at once of any change in address giving both old and new.

Vol. VIII JANUARY, 1918 No. 1

PRINCIPAL CONTENTS

BUILDING A CANADIAN MERCHANT MARINE

A MOMENTOUS event in the industrial history of Canada was foreshadowed in the recent announcement by the Hon. C. C. Ballantyne of the national shipbuilding policy which is to be carried out by the Union Government. Some such action of this nature could hardly be otherwise than expected in view of parallel developments in allied fields both in Britain and the States.

It is now many weeks since the United States Government took over all the tonnage in course of construction in American yards for both home and foreign owners, and the pending action of the Canadian Government is but a natural sequel to the systematic transfer to British registry of ships which had originally been contracted for with foreign owners.

That such action on the part of the Government will meet with unanimous approval throughout the country is a foregone conclusion. One of the chief handicaps of the future and which till now seemed insuperable would have been the lack of tonnage wherewith to insure the world-wide distribution of Canadian products when commerce resumes its natural course. The lack of American-owned tonnage has been a consistent factor in retarding the export trade of that nation for many decades, and the certainty that the termination of hostilities will find the States in possession of a vast fleet of merchantmen is a source of much pleasurable anticipation to the manufacturing interests of that country besides acting as a desirable stimulant to the production of ships in the meantime. Much harmful lassitude may creep in unawares when it is felt that no permanent benefit of an individual nature would be derived from the shipbuilding activity of the Dominion, and visions of foreign-built and owned ships enjoying advantages in cost of production and choice of freights while Canadian yards were undergoing gradual disintegration are not calculated to insure the same degree of efficiency which would be otherwise obtained.

All of this uncertainty is removed, particularly by the fact that negotiations are already under way with the object of installing plate and shape mills at some suitable place or places best calculated to fit in with the present disposition of steel plants and the location of principal shipyards.

The certainty of an ample supply of material insured by these conditions may be offset to a considerable extent by financial consideration which no amount of well-intentioned government assistance can sidetrack. We refer particularly to the cost and availability of imported material which will be ruling when, if not before, such a plant as essayed by the Government could be obtained and put into operation.

Equipment such as is required can only be obtained from the States, and under present conditions contracts for such machinery would have to receive approval and some degree of priority in delivery by the United States authorities if it is to be had inside of two years. A plate mill capable of producing maximum size plates as used in the construction of 10,000-ton boats would have an annual capacity largely in excess of our requirements, and the cost of maintaining the plant idle for perhaps six months in the year would have to be taken care of by Government subsidy. Certainly the cost of producing ships in this manner could not be expected to equal that in other countries where continuous operation was possible.

The deciding factor would therefore be as to whether the shipowners or the nation in general should pay the extra cost involved through using Canadian plates. If conditions could be foreseen accurately enough to determine the certainty and amount of foreign material which could be obtained at any specified time, the advisability of installing plate mills might be seriously questioned. The possibility of securing equipment may be increased through diplomatic channels to a degree which would be most desirable in insuring, through the availability of material at a suitable cost, the continuance of shipbuilding in this country during the most critical time,—the transition period, when competition with the world would begin to assume an important influence on affairs.

CANADIAN MARINE "HEADLIGHTS"

LEWIS DAHLGREN, president and general manager, Thor Iron Works, Ltd., shipbuilders, Toronto; vice-president and manager, Dominion Shipbuilding Co., Ltd., Toronto, was born at Ashland, O., U.S.A., son of Lewis and Olivia (Hanson) Dahlgren.

After receiving his education in public and high schools of Superior, Wisc., he was apprenticed to the Superior Shipbuilding Co., of that city from 1893-

LEWIS DAHLGREN

1900, following which he worked as ship fitter in Union Iron Works, San Francisco, 1900-1902. In 1902 Mr. Dahlgren returned to the Superior Shipbuilding Co. as plater and assistant foreman, remaining with this firm from 1902-1912.

Mr. Dahlgren's connection with Canada extends from 1912, in which year he came to Port Arthur, Ont., as assistant superintendent with the Western Drydock & Shipbuilding Co., later becoming superintendent and general superintendent. In 1916 he was appointed superintendent, Thor Iron Works, Toronto; manager, 1917; general manager, August, 1917.

Mr. Dahlgren is a member of the Kiwonis Club, and his societies are A.F. & A.M.; F.O.E.; I.O.O.F. In politics he is Conservative and is Baptist in religion.

In 1903 he married Ida Merino, daughter of Joseph Merino, Fort William, Ont., and has a family of two sons and one daughter. Mr. Dahlgren's residence is 1269 King St. W., Toronto, Ont.

—Photograph courtesy International Press.

Suggested Aids to Enable Wounded Ships to Keep Afloat-II*

By Chas. V. A. Eley

The statement has been made on many occasions during the past three years that the great conflict now raging is an engineers' war. No one, I think, will question that it is a war which opens up new fields in which engineers of all classes, whether naval or military, marine or civil, have been, and will be, called upon to give to our Empire the very best their brains are capable of, whether in new invention, construction, or in the use of appliances for the confounding of our enemies. Enabling wounded ships to keep afloat is highly desirable.

IT will be noticed that steam jet pumps show up the best for weight and cost. They would also be the easiest to instal, but they would only be capable of getting rid of one-fortieth of the quantity of water that centrifugal pumps would. The Pulsometer pumps appear to be the most costly, and the heaviest of the three arrangements in proportion to the work they could do, but they have the same advantage as the jet pumps, inasmuch as they can be operated directly by the steam instead of requiring some intermediate power as the centrifugal pumps do, such as driving the separate engines, or off the propeller shafts or countershafts. The Pulsometers are also capable of greater deeds than the steam jet pumps, as they could pump over 11,000,000 gallons per hour instead of 3,000,000 gallons.

The question therefore arises: Would a ship like the Olympic ever be damaged to a greater extent than to require pumping arrangements to deal with more than 3,000,000 gallons per hour beyond the quantity it would be possible to pump with her auxiliary pumps?

Let us suppose that such a vessel has been damaged about 25 ft. below the water line by a submerged rock or an iceberg, and that the captain and engineer have agreed to let her keep a level keel until they see exactly what the damage amounts to. Therefore, they let the water flow level in the ship so that it will reach the suctions of ballast pumps and the centrifugal pumps. For present purposes we will assume that such pumps would be capable of the following duties:—

	Per Hour
5 ballast pumps	1,250 tons
3 bilge pumps	450 tons
4 centrifugal pumps ...	16,456 tons
Tons per hour ..	18,156

And the proposed steam jet salvage pumps 13,392 tons.

We get a total of 31,548 tons per hour, which would deal with all of the incoming water entering the ship through a hole 4 ft. by 2 ft. at 25 ft. below surface of water. If the maximum damage is likely to be greater than this, then the Expulsor form of pump could be considered, in which case we would have:—

Capacity of auxiliary pumps per hour	18,156 tons
Capacity of Pulsometer pumps per hour	50,224 tons
Total	68,380 tons

This total of 68,380 tons per hour would be equal to the incoming water entering the ship through a hole about 6 ft. by 3 ft. at 25 ft. below the surface of the sea, provided the ship was permitted to keep a level keel

If a still greater margin of safety is desired, it could be obtained by either an arrangement of centrifugal pumps driven by separate engines, or by means of clutches or change wheels, putting the propellers out of gear and countershaft in gear, to drive centrifugal pumps by the main engines. Taking the separately driven pumps first, there would be sufficient steam to drive about 234-20-inch pumps capable of delivering about 564,107 tons per hour. Adding the capacity of the auxiliary pumps, we get a total of 582,263 tons per hour, which total would deal with the incoming water through a hole about 15 ft. by 10 ft. at 25 ft. below the surface of the sea; a much larger hole than I believe would be made by a mine or torpedo.

If special light bulkhead centrifugal pumps were used driven off the main engines. weights would be reduced, whilst the efficiency of the plant would be at its best. The only drawback would be if the engines were broken down the pumps could not be worked. It is not likely, however, that all three engines would be placed hors-de-combat at the same time. In this case, if the engines were capable of working at 55.000 H.P., as many as 390-20-inch centrifugal pumps could be operated, giving about 936,000 tons per hour, or with adding the capacities of the auxiliary pumps about 954.156 tons per hour, which would deal with the incoming water through a hole 24 ft. by 10 ft. at 25 ft. below the surface of the sea; or a hole 150 ft. by 5 ft. at the surface of the water, the latter measurement being taken from the water line to 5 ft. below it. It will be seen, therefore, that there is no limit to what could be done in the way of pumping water out of such a ship right up to the huge cavities mentioned, which are very unlikely to occur even though the vessel receives torpedo damage or is riddled with shot-holes.

Assuming that such vessel has a length of 852 ft., a mean breadth or beam of 72 ft. above water, and a freeboard of, say, 35 ft., we would have a maximum capacity of 2,147,040 cubic feet; or 61,344 tons of water would be required to sink the ship on an even keel. As such a state of things would mean that the vessel would sink in about 3 minutes 56 seconds if no pumps were working, we can immediately say it is unnecessary to make such extreme provision. If provision is made against such a vessel sinking at, say, 20 or 30 min. rate, I think it would be all ever needed, no matter what the kind of accident, provided that her double bottom was extended up to the sides to, say, 5 ft. above the water line.

As I estimate that the amount of damage received by the Titanic could not have amounted to more than six square feet of perforation at about 25 ft. below the surface, it would appear to be quite unnecessary to provide for the before-mentioned 24 ft. by 10 ft. = 240 square feet, or 40 times the amount of damage received by the Titanic. I would therefore suggest that in a similar ship the following pumping provision should be made:—

1st. That the bilge suction on the centrifugal circulating pumps be 30 in. diameter.

2nd. Four centrifugal pumps with 50 in. suctions and delivery inlets and outlets arranged to be driven by the four sets of electric light and power engines, each of which, say, are capable of developing 550 B.H.P. These four pumps would be capable of delivering about 650 tons of water out of the ship per minute, making the capacity of the auxiliary pumps up to 1,000 tons a minute. (The emergency electric light engines could be brought into operation whilst the larger ones were employed on pumping duty.)

3rd. On or near the five aftermost bulkheads I would suggest the fixing of sixty 20 in. centrifugal pumps, to be operated directly from the main engine and turbine shafts, and so arranged as to be driven by chain or toother gears, whichever is found most suitable for the positions in which it is best for the pumps to be placed.

These sixty pumps would deliver about 2.500 tons of water per minute, making the total capacity with the auxiliary pumps about 3,500 tons per minute. The same sixty pumps would absorb about

*Institute of Marine Engineers' Paper, Nov., 1917.

about 10,000 B.H.P. of the 55,000 B.H.P. of the main engines and turbine.

4th. If could be arranged, I would also suggest running two very large suction mains right forward and aft to the second bulkhead from either end of the ship. I would also suggest the fittings of large valves or sluice gates on such second bulkheads, and others at each compartment and on the forward and aftermost bulkheads.

5th. The necessary delivery pipes and valves to complete the installation.

I estimate that the foregoing would cost:—

Additions to auxiliary plant, including four 50-in. pumps	£ 2,000
Sixty 20 in. centrifugals	£10,000
Wrought iron pipes, valves and fixing	£ 3,000
	£15,000

This expenditure would ensure that in event of the ship meeting with disaster of the worst kind, the machinery would be capable of getting rid of 3,500 tons of water per minute, and at the same time making for port under her own steam at about three-quarters speed. The extra immersion due to the extra weight would be nearly two inches only. These additions would provide for a 17½ minutes' sinking rate.

As my estimate of the damage to the Titanic represents the water pouring into the ship during the first ten minutes amounted to 318 tons per minute, it is safe to say that the foregoing arrangements would deal with eleven times the amount of damage which happened to the Titanic, without having to close a single water-tight door or destroying the "trim" of the vessel. Further, if the ship is very badly damaged in any two of her water-tight compartments (other than those containing the engines) such compartments could be isolated if deemed necessary, and other compartments flooded, if thought desirable, to keep an even keel. For example, suppose her two forward compartments are badly damaged by a head-on collision with which the pumps could not cope. The water-tight doors, if any, could be closed, and by means of one of the large proposed suction pipes the two aftermost compartments could also be flooded and the vessel kept on a level keel, whilst the pumps could be made to deal with any other incoming water. Again, there is another and a very important factor. When it is found that the pumps are mastering the water as it enters the ship, many a clever engineer or ship's carpenter would quickly devise means of using material on board to reduce the inflow of water to a minimum.

In event of the foregoing scheme appearing too ambitious, the owners of such a vessel would perhaps be prepared to adopt a more modified one by fitting into the vessel, say, 300 3-in. steam jet pumps or 75 6-in. ones. Either would deal with about 220 tons of water per minute, and would cost only £1,500, or, say, with pipes and valves and fixing, £2,000. Add to this the suggested cost of the alterations to auxiliaries, £2,000, it would only cost £4,000 to bring the machinery up to a point to drive out incoming water at the rate of 1,220 tons a minute, a rate four times greater than the water entered the Titanic. But, as the whole of the steam used by the main engines would be required to operate the steam jet pumps, the vessel would have to "heave to" until the holes had been stopped. This would be bad for the vessel if submarines were about, or even in a storm.

Another scheme, not so costly as the first, and not quite so moderate as the second, might therefore find favor. That is to instal two more 50 in. centrifugal pumps with steam engines of 550 B.H.P. each complete. These would cost about £2,000, or, with pipes, valves and fixing, say £3,000. Again, adding the £2,000 for additions to auxiliary pumps, we arrive at a total expenditure of £5,000, and there would be sufficient pumps on board to deal with 1,320 tons of incoming water per minute, which is over four times the rate that the water flowed into the Titanic, but would only be sufficient to cope with a 46 minutes' sinking rate. At the same time the vessel could continue her journey, as there would be sufficient steam left for the vessel to travel at 21 knots per hour. It might even be that the speed of the vessel would be increased, if by an ingenious arrangement of the valves and pipes on the delivery side of the pumps the streams of outgoing water could be directed right astern below the waterline. The extra immersion of the ship due to the added weight of the six 50 in. pumps and two sets of engines and necessary pipes and valves could be ignored in a vessel of such a large size, as the weight would amount to only about 100 tons, giving an extra immersion of the ship of only half an inch.

To sum up, I think that the owners and engineers of any vessel engaged in passenger service should not neglect this engineering side of a ship's life-saving apparatus, for the Titanic might never have sunk had she have had the above proposed £5,000, or even much less, spent upon her; in fact, there would have been no need probably for the officers of the ship to let the passengers know even that any danger existed, as the speed of the ship need not have been reduced materially except as a caution against meeting more ice, and no watertight doors need have been closed.

The saving of a ship from sinking is an engineering matter absolutely, consequently such work should come under the direct control of the chief engineer of the vessel; and the engineer in charge during a "watch" should be fully qualified to act just as his chief would do in cases of emergency. The marine engineer should train himself fully in the subject of "How to save a sinking ship." He should arrange, if possible, a full set of indicators at some central point to show at a glance the height of water in any compartment. This will enable him to act with promptitude in times of great danger. He should see that all the necessary appliances for restricting and stopping the flow of water into a ship are kept in some place or places in the ship in the same manner as is done with apparatus for fire service. He should also make himself thoroughly acquainted with the pump connections in the ship, and see that every possible pump is arranged to pump out of the bilge to the sea. If any pumps are not so arranged, it should be his duty to at once inform the responsible individual in writing, pointing out the necessity of such an arrangement. It may seem to some not necessary to point out such matters as these, but one cannot take too great precautions. The marine engineer should therefore make this question of salvage pumps one of his special hobbies. It may at some time or other be the means of saving many lives besides his own. He should also make himself master of the water-tight bulkhead and the water-tight deck questions, and try to arrange problems as to what would be likely to happen if his vessel was perforated at certain points, either through running head-on to rocks, or scraping rocks or icebergs, or by collision, by projectiles, or torpedoes, or in any way which may suggest itself to him. He would then be more ready to act in cases of great emergency, and not be so likely to do the wrong thing, as he would probably do if he did not make a study of such matters.

Let us now see how he would be expected to act in order to prevent the sea entering the vessel. Then we will see at the same time what is likely to be required in the way of material to be used for stopping leaks. Let us take cases of collision first. Suppose his vessel has just been run into by another, and the other vessel has backed out instead of keeping her bows in the breach made until the crew have got a collision mat ready to place over the hole. The water would be pouring into the ship chiefly through the upper side of the vessel. The salvage pumps would be started, the valves on the circulating pumps could be closed to sea suction and opened to bilge suction, and all boiler-feed pumps could be set to pump from the bilge if necessary, if they had been arranged for such duty.

Whilst this work was going on in the engine and boiler rooms, the ship's carpenter should at once prepare on the weather deck a collision mat of strong materials, say 3 in. planks with sufficient channel iron or plates fastened thereto on the outer side to sink the mat and give it necessary strength, and an old sail or rubber sheeting nailed on the inner side of the mat. Whilst the mat is being made an old sail could be suspended over the hole, a couple of iron bars being pushed through it so as to prevent it being drawn into the ship. Members of the crew could be preparing the necessary ropes to suspend the mat, to guide it into position, and to keep it in its position after it is placed.

There will be plenty of scope for the engineer to use any material he may have at hand, also according to the exigencies of the occasion. For instance, with internal planking he could nail rubber or insertion strips on one edge of each plank, and when all the planks were in position, by wedging down from the top, he could prevent the water entering at the joints. Or, if he considered it necessary, either for strength or greater protection, he could plate the outside of each plank with iron or steel. Again, if he had an iron plate on board large enough, and there was sufficient headroom to enter the plate into slots, he could put such plate into position instead of the planks, afterwards making it tight round the edges by wedging the plate close to the side, driving in packing where necessary. By any of the foregoing methods the engineer could make a good temporary patch, which would last until the vessel reached a port where proper repairs could be carried out; therefore, he should carry materials on board, and keep them handy so that he could make a patch at short notice. If his vessel is arranged with water-tight doors, and water-tight scuttles in the decks, they will require very careful manipulation. If it is found that the pumps can deal with all of the incoming water, no doors should be closed. If it is found that the water is gaining on the pumps, even to a small extent, all doors should be left open, as in a short time the inrush of water would be restricted, and then the pumps would gain on the water. If the indicators show that the water is rising rapidly in any compartment, and that the pumps are not capable of dealing with the situation, then the compartment must be isolated:—

1. If the orlop deck is a water-tight one, with water-tight scuttles, and the damage is below such deck, the water-tight scuttles in same should be closed.

But if the damage is above such deck, the water-tight scuttles in same should not be closed. If they are closed, the water will build up on top of such deck and probably destroy the stability of the vessel, and cause it to "turn turtle." Only the scuttles in the decks above the damage should be closed.

2. All water-tight doors on the bulkheads of the compartment to be isolated should be closed. Then, if the damage is below the orlop deck, and the orlop deck is a water-tight one, a very small quantity of water will be locked up.

If the serious damage extends to several compartments, and it is found necessary to close up more of them, do not close any that the pumps are capable of dealing with. If the vessel is arranged with air-locks at the water-tight doors to enable some of the engineering staff to enter the flooded compartments with the "Fleuss" or similar diving apparatus, such men could take the necessary material for making a temporary patch; and when completed the compartments could be pumped out and all doors opened again.

In the case of the vessel running "head-on" to rocks, and perforating the forward compartments, the speed of the ship had probably not been reduced, and the forward compartments have not only been perforated, but the whole of the forward part of the vessel has been lifted some feet above the normal position. The water rushing into the forward compartments will tend to hold the head of the vessel firmly on the rocks, so much so as to prevent the vessel going astern off them. The engineer should immediately put his pumps to work to get the water out of the ship as quickly as it enters. He should then flood one or two compartments aft, so as to bring up the head of the ship to a floating position. This should enable the vessel to be run astern off the rocks. Any temporary repairs considered desirable could be effected either before or after the vessel's head is lifted, and the vessel could proceed on her journey.

In cases of contract with icebergs or rocks with a glancing blow, causing perforation of several compartments along the side of the vessel, the engineer might act as suggested in the case of the Titanic.

In cases of perforation by projectiles the engineer could act in a similar way as in cases of collision.

In cases of perforation by torpedo or submerged mine or floating mine, the damage may be so great as to cause the loss of the vessel if it is not a very large one or a very high-powered one, notwithstanding the activity of the captain and crew.

But with a vessel arranged with a proper system of salvage pumps, the attempt could be made to save her, and in many cases no doubt the result would be satisfactory. Very much would depend upon the position and the character of the damage; but I believe that many of the vessels already sunk during the present war might have reached port if they had been provided with apparatus which the engineers could have used to prevent their ship from sinking.

Even if an enemy submarine, when they found that the object of their attack was not sinking, was to fire a second torpedo, probably if the damage made by the first one had been made good quickly, the damage caused by the second one might even be satisfactorily dealt with. In any case, it would mean that there would be one torpedo the less for sinking some other vessel.

SEEK TO SECURE MORE CONTRACTS

H. H. STEVENS, S. J. Crowe and Major Cooper, the members-elect in the Federal House for the three Vancouver constituencies, held a conference recently, at which it was decided to take action towards securing more orders for wooden ships for construction in Vancouver. A joint letter on the subject will be sent to the British Columbia representative of the Imperial Munitions Board, and representations will be made to Premier Borden on the subject.

Several more Vancouver firms are now in position to go ahead with the construction of wooden ships, provided they can secure the contracts. Recently the Vancouver Shipyard, Limited, Captain George Watt, manager, approached the Munitions Board with a view of obtaining contracts for four ships, the firm now being in a position to construct ships similar to those for which already 27 contracts have been let in British Columbia, 12 to local firms. The firm was notified that no more ships were required.

This answer has occasioned considerable surprise among those interested in shipbuilding in the province, on account of the great scarcity of ships, and the fact that it has generally been acknowledged that wooden ships such as are being built here are acceptable to the Imperial authorities.

GOVERNMENT TO CONSIDER STEEL PLANT SCHEME FOR VANCOUVER

VANCOUVER, Jan. 22.—"The Province" to-day has a despatch from Victoria which says an intimation was given by Hon. Wm. Sloan, Provincial Minister of Mines, that Vancouver might become the headquarters of a large electrical smelting and steel plant industry, and enter the world's competition for steel shipbuilding "on the same basis as Belfast and Glasgow."

"For several months past," the story continues, "it is said, the Government has been engaged in negotiations with Eastern capitalists and electrical smelter experts. The negotiations now have reached a point where the Government has agreed to join with the Pacific Steel Company of Eburne and the Aetna Iron and Steel Company of Port Moody—in which companies the Tudhope interests of Orillia, Ont., and R. F. Turnbull, a well-known electrical metallurgist of Welland, Ont., are identified, together with other interested parties—inviting an eminent outside metallurgist of international standing to visit Vancouver and make a final report on the proposal to establish a steel plant here. It is understood if the final report proves satisfactory the Government will extend the company its aid and influence."

This information, adds the "The Province," was confirmed by Hon. J. W. deB. Farris, Attorney-General. The Tudhope-Turnbull interests, he said, for some time had been active in sizing up the iron and steel situation on the Pacific Coast.

Victoria, B.C.—The first set of engines for the Imperial Munitions Board ships building in British Columbia have arrived. The machinery is intended for the steamer now approaching the launching stage at the yards of the Cameron-Genoa Mills Shipbuilders, Ltd., it is said. It is being stored at the Ogden Point assembling plant.

ASSOCIATION AND PERSONAL

A Monthly Record of Current Association News and of Individuals Who Have Been More or Less Prominent in Marine Circles

W. B. Fortune, one of the engineers in charge of construction of the Quebec Bridge, has been appointed by the American International orporation to an important position in connection with ship construction at the yard at Philadelphia.

G. T. Milne, British Trade Commissioner for Australia has been appointed senior British Trade Commissioner in Canada, in succession to Hamilton Wickes, who is for the present remaining at headquarters in England. Mr. Milne's headquarters will be in Montreal.

Fred W. Field, editor of the *Monetary Times*, and a leading authority on financial and commercial matters in Canada, has been appointed British Trade Commissioner at Toronto with a territory that will probably embrace the entire Province of Ontario.

Edward E. Roberts, inventor of the Roberts marine water tube boiler, died recently at his home in Brooklyn, aged 76 years. He came of New England stock and was born in Manchester, England, while his parents were temporarily abroad.

War Nootka Launched.—The wooden steamer War Nootka was launched at the Western Canada Shipyards, at the foot of Carroll St., Vancouver, Jan. 14. Mrs. A. R. Mann, wife of the president of the Western Canada Shipyards, acted as sponsor and made no mistake when the time came for her to smash the traditional bottle of wine on the bows of the vessel.

Guy E. Tripp, chairman of the board of directors of the Westinghouse Electric & Mfg. Co., has been appointed head of the production division of the ordnance bureau, in which capacity he will follow up, supervise and stimulate the production of all articles purchased. This is a further step in the reorganization of the ordnance bureau.

McGill Professor Now Cabinet Minister.—Sir Auckland C. Geddes, K.C.B. who has been Minister of National Service since last summer in the British Government, was for three years Professor of Anatomy at McGill University, Montreal. Professor Geddes came to Montreal from Dublin in 1913 but returned to England in 1915 and later received a commission in the Northumberland Fusiliers. He was later appointed to the General Staff and raised to the rank of Grigadier General.

Lieut. George H. Forster, who, before going overseas was manager of the Linde Canadian Refrigeration Co., Montreal, has been badly gased in the recent fighting while serving with the British army. Although his condition is serious, there is hope of his recovery. Lieut. Forster joined the 148th McGill Battalion, under the command of Lieut.-Col. Magee, and left in 1916 for England, but later was transferred to the Imperial Forces.

H. B. Pickering, of Seattle, has been appointed engineer of construction for the Foundation Co., and will have direct charge of the engineering work in the shipyards now established in Victoria, Tacoma and Portland, and will have direction of all construction work in connection with the new ten-way shipyard which the company is establishing on the North Pacific Coast. Mr. Pickering will be assistant to Bayly Hipkins, who has full charge of the Foundation Company's operations on this coast, with headquarters at Seattle. Mr. Pickering was born in Seattle and is a graduate of the University of Michigan. He was employed by the United States Government in the naval yards at Philadelphia and Brooklyn, and also was employed in engineering work before coming out West.

LICENSED PILOTS

ST. LAWRENCE RIVER

Captain Walter Collins, 43 Main Street, Kingston, Ont.; Captain M. McDonald, River Hotel, Kingston, Ont.; Captain Charles J. Martin, 13 Balaclava Street, Kingston, Ont.; Captain T. J. Murphy, 11 William Street, Kingston, Ont.

ST. LAWRENCE RIVER, BAY OF QUINTE, AND MURRAY CANAL

Captain James Murray, 106 Clergy Street, Kingston, Ont.; Capt. James H. Martin, 259 Johnston Street, Kingston, Ont.; John Corkery, 17 Rideau Street, Kingston, Ont.; Captain Daniel H Mills, 272 University Avenue, Kingston, Ont.

ASSOCIATIONS

DOMINION MARINE ASSOCIATION

President—A. A. Wright, Toronto. Secretary—Francis King, Kingston, Ont.

GREAT LAKES AND ST. LAWRENCE RIVER RATE COMMITTEE

Chairman—W. F. Herman, Cleveland, Ohio. Secretary—Jas. Morrison, Montreal.

INTERNATIONAL WATER LINES PASSENGER ASSOCIATION.

President—O. H. Taylor, New York.
Secretary—M. R. Nelson, 1184 Broadway, New York.

SHIPPING FEDERATION OF CANADA.

President—Andrew A. Allan, Montreal; Manager and Secretary—T. Robb, 218 Board of Trade, Montreal; Treasurer, J. R. Binning, Montreal.

SHIPMASTERS' ASSOCIATION OF CANADA.

Secretary—Captain E. Wells, 45 St. John Street, Halifax, N.S.

GRAND COUNCIL N.A.M.E. OFFICERS.

A. R. Milne, Kingston, Ont., Grand President.
J. E. Belanger, Bienville, Levis, Grand Vice-President.
Neil J. Morrison, P.O. Box 238, St. John, N.B., Grand Secretary-Treasurer.
J. W. McLeod, Owen Sound, Ont., Grand Conductor.
Lemuel Winchester, Charlottetown, P.E.I., Grand Doorkeeper.
Alf. Charbonneau, Sorel, Que., and J. Scott, Halifax, N.S., Grand Auditors.

THE MERCHANT NAVY IN PEACE AND WAR

AN illustrated lecture on the above subject will be given by Captain George S. Laing, in Orange Hall, corner of Euclid Avenue and College Street, Toronto, on the evening of February 14, at 9 p.m. Captain Laing is a close student of marine matters from various points of view, and his world-wide experience in all classes of boats insures an enjoyable discourse for his hearers.

1918 Directory of Subordinate Councils, National Association of Marine Engineers.

Name.	No.	President.	Address.	Secretary.	Address.
Toronto,	1	Arch. McLaren,	324 Shaw Street	E. A. Prince,	108 Chester Ave.
St. John,	2	W. L. Hurder,	209 Douglas Avenue	G. T. G. Blewett,	36 Murray St.
Collingwood,	3	John Osburn,	Collingwood, Ont.	Robert McQuade,	Collingwood, Ont.
Kingston,	4	Joseph W. Kennedy,	305 Johnston Street	James Gillie,	101 Clergy St.
Montreal,	5	Eugene Hamelin,	Jeanne Mance Street	O. L. Marchand,	93 Fifth Avenue, Lachine, P.Q.
Victoria,	6	John E. Jeffcott,	Esquimault, B.C.	Peter Gordon,	808 Blanchard St.
Vancouver,	7	Isaac N. Kendall	319 11th St. E., Vanc.	E. Read,	Room 10-12, Jones Bldg.
Levis,	8	Michael Latulippe,	Lauzon, Levis, Que.	J. E. Belanger,	Bienville, Levis, Que.
Sorel,	9	Nap. Beaudoin,	Sorel, Que.	Alf. Charbonneau,	Box 204, Sorel, Que.
Owen Sound,	10	John W. McLeod	570 4th Ave.	J. Nicoll,	714 4th Ave. East
Windsor,	11	Alex. McDonald,	28 Crawford Ave.	Neil Maitland,	[illegible] London St. W.
Midland,	12	Geo. McDonald	Midland, Ont.	Roy N. Smith,	Box 178
Halifax,	13	Robert Blair	29 Parrsboro Street	Chas. E. Pearce,	Portland St., Dartmouth, N.S.
Sault Ste. Marie,	14	Charles H. Innes,	27 Euclid Road	Geo. S. Biggar,	43 Grosvenor Ave.
Charlottetown,	15	J. A. Rowe	176 King Street	Chas. Cumming,	27 Easton St.
Twin City,	16	H. W. Cross.	[illegible] Ambrose St	E. L. Williams	142 Secord St., Port Arthur, Ont.

Marine News From Every Source

St. John-Yarmouth Service. — The new steamer Keith Cann was expected to start on the bay route early in January.

Vancouver, B.C.—The first of the wooden steamers building at the Western Canada Shipyards will be launched some time this week.

Monday, Jan. 7, was "Harbor night" at Board of Trade Chambers, St. John, N.B., when papers on harbor development were read and discussed.

Cuban trade with St. John during November amounted to $52,729, consisting of 8,831 barrels of potatoes valued at $51,379 and fish to the value of $1,350.

Sydney, N.S.—The Sydney Foundry & Machine Co. contemplate moving their plant to a new site on the water front. A shipyard will probably be established at the new site.

Esquimalt, B.C.—The Princess Victoria is still on the slip-way at Yarrows, Ltd., getting a shaft renewed and undergoing a general overhaul. She will soon be ready to resume the triangular run.

Toronto, Ont.—The Toronto Ferry Co's. docks at the foot of Bay street have been destroyed by fire entailing a loss of $35,000. The steamer Chippewa was damaged to the extent of $1,000.

West Vancouver, B.C.—The City Council has endorsed the action of North Vancouver City Council and its mayor in urging that a steel plant be established on the Pacific Coast to deal with ores from the province.

The Canadian Steel Foundries, Ltd., Montreal have recently recieved a large order from the Imperial Munitions Board for stern frame castings for a number of vessels now being constructed in different parts of the Dominion.

St. Lawrence Shipbuilding & Steel Co. has been incorporated at Ottawa to build and operate ships, docks, etc., at Sorel, Que., with a capital stock of $1,000,000. The incorporators are P. Bercovitch, E. Lafontaine and N. Gordon, all of Quebec.

Standard Ship Launched.—The second of the 8,800-ton standardized steel merchantmen under construction for the U. S. Emergency Fleet Corporation was launched recently at a Pacific port thirty days after the first launching and sixty days after the keel was laid.

Yarrows, Ltd., of Esquimalt, B.C., has a contract with the Imperial Munitions Board for producing thirty propellers. A large number of propellor contracts have been given to Yarrows. Each propellor has a diameter of 14 ft. 6in., and weighs roughly six and three-quarter tons.

Big Lake Fleet for Ocean Service.—The United States Shipping Board has decided to remove from the Great Lakes an additional thirty ships for ocean service. The vessels will be cut in half this winter and will be removed through the Welland Canal and reassembled when navigation is resumed in the spring. Already 42 ships have been brought out.

Quebec, Que.—The Canada Steamship Lines are establishing a plant at Cap de la Madeleine, near Three Rivers, for the repair of ships. It is intended that this shall replace the one destroyed some time ago at Sorel, but it will be larger and better equipped. The municipality is being asked for exemption from taxation. Some of the Steamships' fleet is wintering there.

Lloyds Cut Insurance Rate.—Lloyds underwriters last Saturday accepted war insurance on transatlantic cargoes at rates much below the flat rate of the British Government, according to the "Times." This action probably is connected with the proposal to extend the Government's scheme, which in effect, would make such insurance on cargoes in British ships a Government monopoly.

Chatham, Ont.—The Chatham, Wallaceburg and Lake Erie Railway was operated last Friday for the first time with the use of hydro-electric power. For some time the road has been practically tied up, as a result of the low gas pressure, which is used to generate the current for the road. So successful was the test that it is quite likely the road will continue using this source of power.

St. John winter service.—Winter port activities have been [illegible] on an active scale, and the outlook for a big winter business at the port is good. From November 23rd to December 18th twenty-four steamers, aggregating 103,634 tons, have entered the port for cargoes.

No definite announcement has yet been made as to the date of opening the new C. G. R. elevator.

S. S. Angouleme Wrecked.—The steel freighter Angouleme which was built at the Thor Iron Works, Toronto, last summer is reported a wreck off the Scarlet Island on the Nova Scotia coast. The Angouleme was christened "Orleans" and was built for the Playfair interests but was later renamed and sold to the Overland Navigation Co. The vessel was 261 feet long, 43 feet wide, 28 feet deep and had a deadweight capacity of 4,500 tons.

Steel Freighter Launched at Vancouver.—The steel freight steamer Alaska, built at the Coughlan yards, Vancouver, B.C., was launched successfully on Saturday morning in the presence of a large crowd. The bottle of wine was broken over the vessel's bow by Mrs. Koldrup, wife of the Norwegian Consul at Seattle. The Alaska has been under construction since last March. Her length is 427 feet and her gross tonnage 5,730.

Vancouver, B.C.—The wooden steamship Warnootaka was launched at the Western Canada Shipyards here last Friday. Mrs. R. A. Mann, wife of the president of the company, broke the bottle of wine over the vessel's bows. The Warnootaka, named after the first white settlement in British Columbia, is said to be the first vessel of any size ever launched in False Creek. Her keel was laid July 20 last. She is 250 feet long, with a deadweight capacity of 2,800 tons.

German Steel Output.—The annual report of the German Steel League, the central organization of the industry, makes the significant admission that although the entire industry during the past year was "speeded up" to the limit of its capacity the output was seriously curtailed by "the well-known productive difficulties"—that is to say, increasing shortage of labor and raw materials. Export business, owing to the inordinate demands of the Army and Navy, is practically at a standstill.

Would Make St. John a Dominion Port.—A proposal for the Dominion Government to take over the port of St. John, N.B., make it a Dominion property, and place it under a commission similar to the Harbor Commission at Montreal, has been considered by some of the members of the Dominion Government. Hon. F. B. Carvell is taking the matter up with the Dominion Cabinet, and while it may take some time to complete the negotiations, it looks as if something may come of it.

Customs Receipts Drop in December.—Customs receipts for the Dominion for the month of December show a decrease of $945,028, when compared with the same month a year ago. The total receipts for December 1917 were $10,939,056, as against $11,884,084 in 1916. The total receipts for the nine months

of the fiscal year, ending December 31, however, show an increase of $19,650,-882 over the same period in 1916. In 1916 they totalled $106,613,081, while the total to date in 1917 was $126,263,-963.

Halifax, N.S.—L. M. Trask of this city, who recently purchased the Milton Foundry at Yarmouth, has purchased the Bridgetown Foundry, which is located in the centre of the town near the river front. With his foundries and machine shops connected therewith, Mr. Trask expects to manufacture stoves, ships' castings, hoists and winches, and a complete line of fuel oil engines up to 200 h.p. suitable for vessels of all kinds. He will also make a specialty of traction engines.

B. C. Schooners on Australian Trade.—Of the Canada West Coast Navigation Co.'s fleet the following have done a voyage from British Columbia: Mabel Brown to Sydney; Margaret Haney, to Bombay; Geraldine Wolvin, to Sydney; Laura Whalen, to Adelaide; Jessie Norcross, to Adelaide; and the Malahat to Sydney. On passage are the Janet Carruthers, which sailed on August 21 for Adelaide; the Esquimalt, sailing the same day for Melbourne, and the Marie Barnard, which left November 13 for Sydney.

Marine Engineers Meet.—New wage schedules showing increases were considered by the Great Lakes Executive of the National Association of Marine Engineers, who were in conference in Toronto a few days ago. The new schedule has not yet been issued, but it is understood that it will approximately be from $1,440 to $2,100 a year for chief engineers, and from $1,152 to $1,800 a year for second engineers. The present wage scale is as follows: $960 to $1,500 a year for second engineers, and $1,200 to $1,800 a year for chief engineers.

Vancouver, B.C.—First of the fleet of twelve wooden schooners built in British Columbia to return to this Province, the Jessie Norcross reached here on Jan. 10, with a cargo of sugar for the British Columbia Sugar Refinery at Vancouver. She made a quick passage from Fiji, encountering favorable winds most of the way across. The Jessie Norcross was launched at North Vancouver by the Wallace Shipyards on April 29, 1917, and loaded lumber at Genoa Bay for Port Adelaide, Australia. She is to load another lumber cargo in British Columbia for the Antipodes.

British Government Extend Trade Commissioner Service.—The appointment of F. W. Field as British Trade Commissioner for Ontario is part of a scheme of the British Government to extend and develop the trade commissioner service in the Empire. It is now proposed to increase the number of trade commissioners in several of the dominions and to appoint commissioners in India and some of the Crown colonies and protectorates. Canada has hitherto been served by only one commissioner, who has headquarters at Montreal. G. T. Milne has been appointed to the Montreal office.

Peterboro, Ont.—On Jan. 6, fire destroyed the factory of the Peterboro Canoe Co. For some months past the company has been manufacturing shell boxes, and the plant was stocked with machinery and materials used in this line of manufacture, in addition to the regular equipment for manufacturing canoes and power boats. The entire plant, with the exception of a warehouse, was badly gutted. The exact cause of the fire is not known. The loss to the building will be in the neighborhood of $30,000, and the machinery and stock will run the total loss considerably higher.

Toronto Customs Record Revenue.—Customs returns at the port of Toronto for 1917 made a new high record. The total was $35,677,000, an increase of $3,-897,800 over the returns for 1916. That such a substantial increase was recorded is all the more gratifying by reason of the fact that the 1916 returns were an increase of some thirteen million dollars over those of 1915. Compared with pre-war customs revenue, the returns now pouring in represent an increase of 100 per cent. The largest monthly return in 1917 was in May, when it amounted to $3,546,256. The gain in customs revenue is due to a combination of circumstances. Actual importations have increased to a considerable extent and the total has been added to by the general increase in the cost of goods of all kinds.

New Shipbuilding Plant at Toronto.—Work is now proceeding on the contruction of a two-million-dollar shipbuilding plant at Toronto on the west side of the harbor. The plant is being erected for the Dominion Shipbuilding Co., which was recently granted a provincial charter. It is expected that the entire plant will be finished by next summer, when work will commence on the construction of five steel freighters for salt water service. The vice-president and general manager is L. Dahlgren of Toronto. The company has leased for a term of 21 years slightly over 15 acres of reclaimed land at the foot of Spadina avenue. They are erecting a modern shipbuilding plant, the largest building being about 800 feet by 100 feet. Five shipbuilding berths capable of taking ships of canal size, that is, up to 261 feet in length by 43 feet beam, are being erected.

SUCCESSFUL SEA TRIP BY CONCRETE SHIP

THE first concrete ship has made a round trip and sailed into its home port, after a successful voyage. It is a Norwegian vessel, caled the Nansenfjord, and has just returned from the British Isles, after having steamed two thousand miles. This ship was built several months ago, and is made of structural steel and concrete. It is said to have answered every test satisfactorily, unaffected by the vibration of the engines and notwithstanding the jamming which a ship frequently receives when it is docked. Also it shows no ill-effects caused by contraction and expansion. The vessel is of the single-screw type, driven by a heavy oil engine, and has a displacement of about five hundred tons.

RESISTANCE OF AN OIL TO EMULSIFICATION

WHEN oil is used over and over again, as is the usual practice in large power plants, it may emulsify in a few days, if it is not of good quality, and have to be thrown away, as it will not pass through the filters. In order to arrive at a definite knowledge of the emulsifiability of oils, the United States Bureau of Standards has carried out investigations, the results being embodied in Technologic Paper No. 86, entitled "Resistance of an Oil to Emulsification."

To avoid the waste of time and the inconvenience of actually trying the oil in a turbine or other machine, a laboratory test has been devised by means of which it may be predicted whether or not the oil would give satisfaction in service.

The test consists of violently stirring a small sample of the oil with water under standard conditions. Then the oil that will separate out from the water the most rapidly is the best oil for turbine lubrication, other things being equal. It is found that most of the oils on the market are either very bad or very good, and that there is no difficulty in finding turbine oils that will settle out from the water in a minute or less after stirring as described.

For some other purposes it is desirable that oils should emulsify readily; and the test shows that oils of this class, when subjected to the same treatment as the turbine oils, will not separate at all in one hour, nor completely separate in one year. When turbine oils are used for months or years without renewal, they gradually deteriorate until they emulsify readily, even though they are very good oils to start with. The conditions of use in different plants are so different that it would be impossible to predict just how long a given oil would last, but it is believed that any power-house engineer, by the use of this test, could keep track of the deterioration of his oil and thus discover when it is time to clean out his lubricating system.

U. S. COAL EXPORTS TO CANADA

REDUCTION in exportation of coal from the United States suggested by Fuel Administrator Garfield will probably not affect the rank of the United States as a coal exporting country. A compilation by the National City Bank of New York shows that the United States, prior to the war, held third rank as a coal exporting nation, but, with Germany now cut off from the rest of the world, it has taken second place.

Great Britain, Germany and the United States have for many years supplied the bulk of the coal used in international trade. In 1913 the total amount of coal

exported from coal producing countries was in round figures about 200,000,000 tons, of which approximately 40,000,000 tons was bunker coal, while a considerable portion of that classed as exported went to the world's coaling stations to be supplied to vessels. The value of the bunker coal used in normal times aggregates $250,000,000 out of a total of $700,000,000 worth passing out of the coal producing countries. In 1913, the year preceding the war, Great Britain exported coal of all kinds amounting to 93,000,000 tons; Germany, 40,000,000 tons; the United States, 29,000,000 tons; while Austria-Hungary was fourth with about 9,000,000 tons.

Much of the 200,000,000 tons exported before the war went to adjacent territories and a much smaller amount than is generally supposed went overseas. In 1917 the U. S. exported about 32,000,000 tons, of which more than 17,000,000 tons went to Canada, and about 8,000,000 passed into the bunkers of ocean vessels. Exports of coal to Canada for the past year were the largest in the history of this country, and the value is placed at $58,000,000, about one-fourth of the quantity being anthracite. The total value of all the coal exported and passed into bunkers for the year amounted to $113,000,000.

INCREASED TRAFFIC AND FEWER SLIDES ON PANAMA CANAL

TRAFFIC through the Panama Canal last year increased largely over the previous years, according to the annual report of Col. Chester Harding, governor of the canal. A total of 1,876 vessels of all classes passed through the canal from July 1, 1916, to June 30, 1917, inclusive.

Of these, 905 passed from the Atlantic to the Pacific, and 971 from the Pacific to the Atlantic. In the fiscal year 1915, 1,088 vessels passed through the canal, and in 1916, 787. The total number of vessels transiting the canal since it opened for commercial traffic in August, 1914, is 3,751. The total net tonnage, canal measurements, for the several years is as follows: 1915, 3,849,035; 1916, 2,479,762; and 1917, 6,009,358. The cargo tonnage transported was, for 1915, 4,969,792; 1916, 3,140,046; and 1917, 7,229,255.

Loss From Present Traffic Rules

The traffic for the year yielded a revenue of $5,631,781.66 from tolls. The rules for levying tolls have thus far remained unchanged.

It will be remembered, says the report, that under present law and regulation tolls are based on canal tonnage, at the rate of $1.20 per net ton, except when this product exceeds the registered net tonnage, United States rules, at a rate of $1.25 per ton, in which case the lesser amount is collected. The confusion, lack of uniformity, and loss in revenue to the canal resulting from the present arrangement have been fully discussed in previous reports, and remedial legislation is now pending in Congress.

In this connection, attention is invited to the fact that the revenues from tolls for the past fiscal year would have been $6,668,247.32, if canal rules alone had governed, which is $1,036,465.66 more than the amount actually collected.

Dredging Work of the Year

The year's dredging chargeable to construction, which includes all the excavation in the canal prism at locations where the full widths and depths have not been once obtained is as follows: Gaillard Cut, 183,904 cubic yards; between Gamboa and Pedro Miguel Lock, 246,998 cubic yards; and Pacific entrance, 221,138 cubic yards. At the end of the year there remained 1,409,140 cubic yards of original excavation to be done within the limits of the canal prism.

Cucaracha slide has given no further trouble since the large movement that blocked the canal in August, 1916, as described in the report of last year. To reduce the chances of interruptions to traffic due to future similar movements if they occur, the material in the slide was removed for a distance of 100 feet outside the canal prism.

It is believed that in the future the great slides of the canal will be of historic interest only.

CANADA'S PRODUCTION

The value of production in Canada during the year ending to-day from field crops, forests, mines and fisheries, according to the *Monetary Times*, was approximately $1,507,667,000, compared with $1,275,734,812 in 1916. The value of production (field crops, mines, fisheries and forest products) in 1908 was $703,590,000.

For the first time on record, in 1917 the value of Canada's field crops was over $1,000,000,000. This was due to the high prices prevailing.

THE reaction type of steam turbine, known under his name, was first introduced in England by the Hon. Charles A. Parsons in 1884, and was introduced in the United States by the Westinghouse Co. in 1895.

If any advertisement interests you, tear it out now and place with letters to be answered

Steam-Tight Condulets

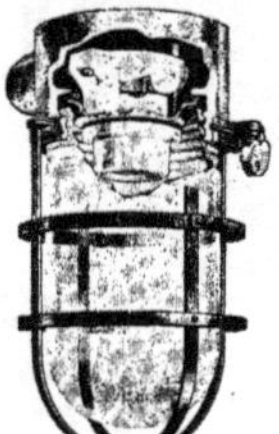

Condulet broken away to show parts.

Electric Light Fittings

whose design, material and workmanship insure long and satisfactory service,

Furnished in either iron or brass for ½, ¾ and 1-inch conduit.

Made in two sizes, to take 40 and 100-watt lamps respectively.

Condulet broken away to show parts.

OIL TANKER "REGINOLITE" WIRED THROUGHOUT IN CONDULETS

Catalogs giving complete information on all Condulets mailed Free upon request.

CROUSE-HINDS COMPANY
OF CANADA, LIMITED
TORONTO, ONT., CAN.

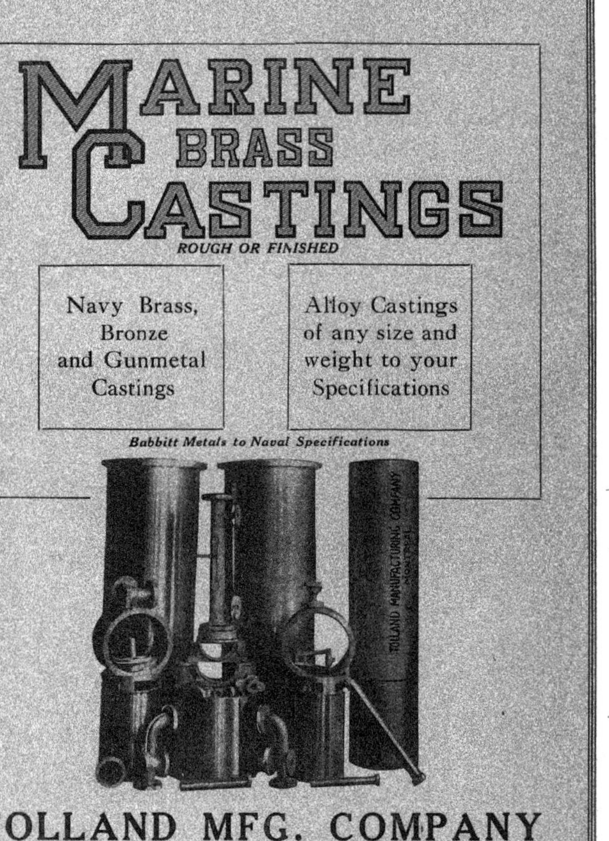
MARINE
BRASS
CASTINGS
ROUGH OR FINISHED
Navy Brass, Bronze and Gunmetal Castings
Alloy Castings of any size and weight to your Specifications
Babbitt Metals to Naval Specifications
TOLLAND MANUFACTURING COMPANY
TOLLAND MFG. COMPANY
1165 Carriere Rd.
Montreal, Que., Can.

If what you need is not advertised, consult our Buyers' Directory and write advertisers listed under proper heading.

BERTRAM
MACHINE TOOLS
42-inch Vertical Boring and Turning Mill
Motor driven through speed box. Built in sizes from 42-inch to 100-inch swing
Structural Bridge Shop and Shipbuilding Plant Machinery
Locomotive and Car Shop Equipment
Repair Shop Machinery
General Machine Shop Equipment
Drop us a line for Photographs and full particulars.
M 111 PHOTO 1057
The John Bertram & Sons Company Limited
DUNDAS, ONTARIO, CANADA
MONTREAL 723 Drummond Bldg.
TORONTO 1002 C.P.R. Bldg.
VANCOUVER 609 Bank of Ottawa Bldg.
WINNIPEG 1205 McArthur Bldg.

WILKINSON & KOMPASS
TORONTO HAMILTON WINNIPEG
IRON AND STEEL
HEAVY HARDWARE
MILL SUPPLIES
AUTOMOBILE ACCESSORIES
WE SHIP PROMPTLY

10" Special Double-Suction Circulating Pump with 8 x 8 Vertical Engine.

MORRIS

is specializing on

Circulating Pumps

FOR

Surface Condensers

Have you our Catalog?

MORRIS MACHINE WORKS

BALDWINSVILLE, N.Y., U.S.A.

Canadian Sales Agents:

STOREY PUMP & EQUIPMENT COMPANY
TORONTO

Marconi Wireless Apparatus

Puts your vessel in touch with over 4,000 other craft fitted with the same reliable system; saves docking delays and a hundred-and-one inconveniences which cost time and money; banishes anxiety by providing an unrivalled means of communication between ship and shore; carries with it the incalculable benefits of world-wide Marconi service,—including the furnishing of trained and certificated operators at short notice anywhere, and free consultation with expert radio engineers in all principal seaports;—and is undoubtedly the best and

Cheapest Form of Marine Insurance

Marconi Apparatus may be purchased outright or rented under a yearly maintenance contract. The reasons that make the rental of a telephone preferable to the purchase of your own instrument apply also to wireless equipment. Connection with an organized system is absolutely essential if you want satisfaction. Shipowners endorse our recommendation of a yearly maintenance arrangement because experience has proved the value of "Service."

The Marconi Wireless Telegraph Co. of Canada, Limited

Contractors to the British, Canadian and the Allied Governments

Telegrams: "ARCON" Montreal

HEAD OFFICE:
Shaughnessy Bldg., 137 McGill St., Montreal

Telelphone Main 8144

BRANCH OFFICES:
Halifax, Toronto, Vancouver, St. John's, Newfoundland

TURBOLITE

Steam Turbine and Electric Generator Direct Connected

Weatherproof and Suitable for Out-of-Door Work

The *Turbolite*, built in one complete unit, free from complicated parts, is composed of a small Steam Turbine Engine built integral with a Direct Current Dynamo. It does not require any special foundation as the entire outfit is assembled on one base and is therefore always in perfect alignment which insures smooth operation.

The workmanship and quality of material used in the construction of the *Turbolite* are the highest grade that can be obtained; every part is interchangeable, which makes it an easy matter to replace a part should it become necessary. Each outfit is thoroughly tested under actual service conditions before it leaves our shops, and is shipped intact ready to run as soon as the steam pipe and wires are connected to it.

Anyone of ordinary mechanical ability can install and operate the *Turbolite*, making it serviceable in Steam Plants of every description.

As an electric lighting plant for Tug Boats and small Steamships the *Turbolite* is unequalled. The costly installations and expensive maintenance have been eliminated by this compact, simple, but efficient lighting system; making it the ideal plant for any small steamship or vessel.

It is also particularly adapted to lighting industries dealing in combustible materials where great risks of fire and explosions by other forms of illumination, make electric lights almost a necessity, such as Oil Drilling, Rigs and Pumping Plants, Oil Refineries, Cotton Gins and Compressors, Threshing Machines, Saw Mills and similar plants.

The portability of the *Turbolite* makes it an ideal lighting system for contracting work and temporary operations of all kinds, as Steam Shovels, Wrecking Cranes, Excavating Machines, Dredges, Traction Engines and work of like nature.

With each outfit we furnish a complete set of instructions which will enable anyone to make the installation complete.

THE **Turbolite** is entirely enclosed, having no exposed parts, and can therefore be placed out-of-doors, as there is no possibility of damage from the weather.

The only moving parts of this machine are the turbine wheel and dynamo armature which are mounted on one shaft revolving on three ball bearings of the most approved design, and which require very little lubrication. On account of the revolving motion and the absence of reciprocating parts, there is absolutely no vibration to the **Turbolite.**

Proper speed regulation is had by a centrifugal governor mounted directly on the wheel as a unit, which operates a simple valve through the medium of one lever. Adjustment of the governor is seldom necessary, but can be easily and quickly accomplished without taking any portion of the outfit apart.

Will operate on 100 Pounds Steam Pressure or Higher. Reliable and Economical in Operation.

The turbine wheel is made from a high-grade crucible steel casting, and the buckets against which the steam impinges are made from a special quality of bronze and are designed for the utmost efficiency and durability.

The Dynamo is of the two-pole type and is easily understood. The armature is built upon a sleeve and can be easily removed from the shaft should occasion demand.

The workmanship and quality of material used in the construction of the **Turbolite** are the highest grade that can be obtained; every part is interchangeable, which makes it an easy matter to replace a part should it become necessary. Each outfit is thoroughly tested under actual service conditions before it leaves our shops, and is shipped intact ready to run as soon as the steam pipe and wires are connected to it.

Model	Size K.W.	Size Steam Pipe Inches		Revolutions per Minute	Number of Lamps	Capacity Watts C.P.	Dimensions Inches			Boxed Cubic Feet	Approximate Weight Pounds	
		Intake	Exhaust				Length	Width	Height		Net	Boxed
H L	½	⅜	1¼	3000	22 14 9	25—20 40—35 60—56	28	12¾	15	4.7	200	225
E	1½	½	1½	2500	60 37 24 15	25—20 40—35 60—56 100—80	34½	17	18	8.7	374	400

The Dominion General Equipment Company

CRAGG BUILDING, HALIFAX, N.S.

BUYERS' DIRECTORY

ACCUMULATORS, HYDRAULIC
Jenckes Machine Co., Ltd., Sherbrooke, Que.
Smart-Turner Mach. Co., Hamilton, Ont.

AERATING RESERVOIRS
Spray Engineering Co., Boston, Mass.

ALLOYS, BRASS AND COPPER
Dominion Copper Products Co., Ltd., Montreal, Que.

ANCHORS
Kearfott Engineering Co., New York, N.Y.
Henry Rogers Sons & Co., Wolverhampton, Eng.
Wm. C. Wilson & Co., Toronto, Ont.

ASBESTOS GOODS
Wm. C. Wilson & Co., Toronto, Ont.

BABBITT METAL
Aikenhead Hardware, Ltd., Toronto, Ont.
Can. B. K. Morton Co., Toronto, Montreal.
Hoyt Metal Company, Toronto, Ontario.
Wilkinson & Kompass, Hamilton, Ont.
Wm. C. Wilson & Co., Toronto, Ont.

BARGES
Can. B. K. Morton Co., Toronto, Montreal.
Polson Iron Works, Toronto, Ont.

BAROMETERS
Wilson & Co., Wm. C., Toronto, Canada.

BARS, GRATE
Babcock & Wilcox, Ltd., Montreal, Que.

BARS, IRON AND STEEL
Topping Brothers, New York, N.Y.

BEARINGS, BRASS
Empire Mfg. Co., London, Ont.
Murray & Fraser, New Glasgow, N.S.
United Brass & Lead Co., Toronto, Ont.

BELLS, SHIPS, ENGINE ROOM, ETC.
Aikenhead Hardware, Ltd., Toronto, Ont.
Cory & Son, Inc., Chas., New York, N.Y.
Empire Mfg. Co., London, Ont.
Morrison Brass Mfg. Co., James, Toronto, Ont.

BELTING, LEATHER
Aikenhead Hardware, Ltd., Toronto, Ont.
Can. B. K. Morton Co., Toronto, Montreal.
Wm. C. Wilson & Co., Toronto, Ont.

BIBBS, COMPRESSION
Empire Mfg. Co., London, Ont.

BINNACLES
Morrison Brass Mfg. Co., James, Toronto, Ont.
Topping Brothers, New York, N.Y.

BLOCKS, CARGO, HEEL AND TACKLE
Aikenhead Hardware, Ltd., Toronto, Ont.
Loveridge, Ltd., Docks, Cardiff, Wales.
Topping Brothers, New York, N.Y.

BLOWERS, TURBO-
Can. Allis-Chalmers, Toronto, Ont.
Mason Regulator & Engin. Co., Montreal, Que.

BOILER COMPOUND
Wm. C. Wilson & Co., Toronto, Ont.

BOILER FEED PUMPS
Can. Allis-Chalmers, Toronto, Ont.
Can. Ingersoll-Rand Co., Ltd., Sherbrooke, Que.
Smart-Turner Mach. Co., Hamilton, Ont.
Williams Machinery Co., A. R., Toronto, Ont.

BOILER FITTINGS
Empire Mfg. Co., London, Ont.
McAvity & Sons Ltd., T., St. John, N.B.

BOILERS, MARINE
Babcock & Wilcox, Ltd., Montreal, Que.
Can. Allis-Chalmers, Toronto, Ont.
Can. Fairbanks-Morse Co., Montreal, Que.
Can. Vickers, Ltd., Montreal, Que.
Collingwood Shipbuilding Co., Collingwood, Ont.
Doxford & Sons, William, Sunderland, England.
Goldie & McCulloch, Ltd., Galt, Ont.
Hall Engineering Works, Montreal, Que.
Jenckes Machine Co., St. Catharines, Ont.
Mason Regulator & Engin. Co., Montreal, Que.
Montreal Dry Docks & Shipbuilding Co., Montreal, Que.
National Shipbuilding Co., Goderich, Ont.
The John McDougall Caledonian Iron Works Co., Montreal, Que.
Marsh & Henthorn, Ltd., Belleville, Ont.
Polson Iron Works, Toronto, Ontario.
Port Arthur Shipbuilding Co., Port Arthur, Ont.
Thor Iron Works, Toronto, Ont.
Williams Machinery Co., A. R., Toronto, Ont.

BOOKS, TECHNICAL, MARINE
MacLean Publishing Co., Toronto, Ont.

BOLTS
Topping Brothers, New York, N.Y.
Wilkinson & Kompass, Hamilton, Ont.

BROKER
Fuller, Louis, N., 163 Hollis St., Halifax, N.S.
Page & Jones, Mobile, Ala., U.S.A.

BUCKETS, CLAMSHELL
Beatty & Sons, Welland, Ont.
Jenckes Machine Co., Ltd., Sherbrooke, Que.

BUCKETS, COALING
Beatty & Sons, Welland, Ont.
Marsh & Henthorn, Ltd., Belleville, Ont.

CABLE
Wm. C. Wilson & Co., Toronto, Ont.

CABLE, LEAD COVERED AND ARMORED
Standard Underground Cable Co., Hamilton, Ont.

CABLE, ACCESSORIES
Standard Underground Cable Co., Hamilton, Ont.

CAPSTANS
Advance Mach. & Welding Co., Montreal, Que.
Dake Engine Co., Grand Haven, Mich.
Jenckes Machine Co., Ltd., Sherbrooke, Que.
Kennedy & Sons, Wm., Owen Sound, Ont.
Wm. C. Wilson & Co., Toronto, Ont.

CALKING TOOLS
Empire Mfg. Co., London, Ont.
Topping Brothers, New York, N.Y.

CALKING TOOLS, ELECTRIC
Aikenhead Hardware, Ltd., Toronto, Ont.

CALKING TOOLS, PNEUMATIC
Aikenhead Hardware, Ltd., Toronto, Ont.
Can. Ingersoll-Rand Co. Sherbrooke, Que.

CALORIFIERS
Low & Sons, Ltd., Archibald, Glasgow, Scotland

CASTINGS
Collingwood Shipbuilding Co., Collingwood, Ont.
Jenckes Machine Co., St. Catharines, Ont.
Kennedy & Sons, Wm., Owen Sound, Ont.
Marsh & Henthorn, Ltd., Belleville, Ont.
The John McDougall Caledonian Iron Works Co., Montreal, Que.

CASTINGS, ALLOY
Tolland Mfg. Co., Montreal, Que.

CASTINGS, ALUMINUM
Empire Mfg. Co., London, Ont.
United Brass & Lead Co., Toronto, Ont.

CASTINGS, BRASS
Booth-Coulter Copper & Brass Co., Toronto, Ont.
Crouse-Hinds Co. of Canada, Ltd., Toronto, Ont.
Empire Mfg. Co., London, Ont.
McAvity & Sons Ltd., T., St. John, N.B.
United Brass & Lead Co., Toronto, Ont.
Tolland Mfg. Co., Montreal, Que.

CASTINGS, GREY IRON, MALLEABLE, ALUMINUM
Crouse-Hinds Co. of Canada, Ltd., Toronto, Ont.

CASTINGS, ELECTRIC
Crouse-Hinds Co. of Canada, Ltd., Toronto, Ont.

CEMENT SOLUTION
Can. Bitumastic Enamels Co., Toronto, Ont.

CHAINS
Aikenhead Hardware, Ltd., Toronto, Ont.
Kearfott Engineering Co., New York, N.Y.
Henry Rogers, Sons & Co., Wolverhampton, Eng.
Topping Brothers, New York, N.Y.
Wm. C. Wilson & Co., Toronto, Ont.

CHANDLERY, SHIP
Leckie, Ltd., John, Toronto, Ont.

CHUTES, COAL
Can. Allis-Chalmers, Toronto, Ont.

CLAMPS
Topping Brothers, New York, N.Y.

CLOCKS
American Steam Gauge & Valve Mfg. Co. Boston, Mass.
Morrison Brass Mfg. Co., James, Toronto, Ont.
Williams Machinery Co., A. R., Toronto, Ont.

COAL
Nova Scotia Steel & Coal Co., New Glasgow, N.S.

COCKS, BILGE, DISCHARGE, INDICATOR, ETC.
McAvity & Sons Ltd., T., St. John, N.B.
Morrison Brass Mfg. Co., James, Toronto, Ont.

COCKS, BASIN
Empire Mfg. Co., London, Ont.

COMPASSES
Wm. C. Wilson & Co., Toronto, Ont.

COMPRESSORS, AIR
Can. Allis-Chalmers, Toronto, Ont.
Can. Fairbanks-Morse Co., Montreal, Que.
Canadian Ingersoll-Rand Co., Sherbrooke, Que.
Jenckes Machine Co., St. Catharines, Ont.
Smart-Turner Mach. Co., Hamilton, Ont.
Williams Machinery Co., A. R., Toronto, Ont.

CONCRETE MIXERS
St. Clair Bros., Galt, Ontario.
The John McDougall Caledonian Iron Works Co., Montreal, Que.

CONDENSERS
Can. Allis-Chalmers, Toronto, Ont.
Jenckes Machine Co., Ltd., Sherbrooke, Que.
Kearfott Engineering Co., New York, N.Y.
Morris Machine Works, Baldwinsville, N.Y.
Smart-Turner Mach. Co., Hamilton, Ont.
Weir Ltd., G. & J., Cathcart, Glasgow, Scotland.
Williams Machinery Co., A. R., Toronto, Ont.

CONDULETS, MARINE
Crouse-Hinds Co. of Canada, Ltd., Toronto, Ont.

CONVEYORS, ASH, COAL
Babcock & Wilcox, Ltd., Montreal, Que.

COPING MACHINES
Bertram & Sons, Ltd., John, Dundas, Ont.

COPPERSMITHS
Booth-Coulter Copper & Brass Co., Toronto, Ont.

COPPER TUBES, SHEETS AND RODS
Dominion Copper Products Co., Ltd., Montreal, Que.

COTTON, CALKING
Wilson & Co., Wm. C, Toronto, Canada.

COVERS, CANVAS, FOR HATCHES, LIFE-BOATS, ETC.
Leckie, Ltd., John, Toronto, Ont.

COUNTERS, REVOLUTION
American Steam Gauge & Valve Mfg. Co., Boston, Mass.

CRANES
Aikenhead Hardware, Ltd., Toronto, Ont.
Can. Fairbanks-Morse Co., Montreal, Que.
Williams Machinery Co., A. R., Toronto, Ont.

CRANES, ELECTRIC
Babcock & Wilcox, Ltd., Montreal, Que.
Smart-Turner Mach. Co., [illegible]

CRANES, GANTRY
Smart-Turner Mach. Co., Hamilton, Ont.

CRANK SHAFTS
Canada Foundries and Forgings, Welland, Ont.

DAVITS, BOAT
Babcock & Wilcox, Ltd., Montreal, Que.

DECK PLUGS, ELECTRIC
Crouse-Hinds Co. of Canada, Ltd., Toronto, Ont.

DERRICKS
Aikenhead Hardware, Ltd., Toronto, Ont.
Dake Engine Co., Grand Haven, Mich.
Jenckes Machine Co., Ltd., Sherbrooke, Que.
Marsh & Henthorn, Ltd., Belleville, Ont.

DISTILLERS
Kearfott Engineering Co., New York, N.Y.

DREDGES
Collingwood Shipbuilding Co., Collingwood, Ont.
Norbom Engineering Co., Philadelphia, Pa.
Polson Iron Works, Toronto.

DRILLS, AIR
Aikenhead Hardware, Ltd., Toronto, Ont.
Can. Ingersoll-Rand Co. Sherbrooke, Que.

DRILLS, TWIST
Aikenhead Hardware, Ltd., Toronto, Ont.
Can. B. K. Morton Co., Toronto, Montreal.
Williams Machinery Co., A. R., Toronto, Ont.

DRILLS, ELECTRIC
Aikenhead Hardware, Ltd., Toronto, Ont.
Wilkinson & Kompass, Hamilton, Ont.

DRY DOCKS
Can. Vickers, Ltd., Montreal, Que.
Collingwood Shipbuilding Co., Collingwood, Ont.
Doxford & Sons, William, Sunderland, England
Georgian Bay Shipbuilding & Wrecking Co., Midland, Ont.
Montreal Dry Docks & Shipbuilding Co., Montreal
Muir Bros. Dry Dock Co., Port Dalhousie, Ont.
National Shipbuilding Co., Goderich, Ont.
Polson Iron Works, Toronto, Ont.
Port Arthur Shipbuilding Co., Port Arthur, Ont.
Thor Iron Works, Toronto, Ont.
Yarrows, Limited, Victoria, B.C.

ECONOMIZERS, FUEL
Babcock & Wilcox, Ltd., Montreal, Que.

EJECTORS
Empire Mfg. Co., London, Ont.
Morrison Brass Mfg. Co., James, Toronto, Ont.
Smart-Turner Mach. Co., Hamilton, Ont.

ELECTRICAL SUPPLIES
Can. General Electric Co., Toronto, Ont.

ELECTRIC LAMPS
Wm. C. Wilson & Co., Toronto, Ont.

ELECTRO-PLATING
United Brass & Lead Co., Toronto, Ont.

ELEVATING MACHINERY
Goldie & McCulloch, Ltd., Galt, Ont.

ENAMEL
Can. Bitumastic Enamels Co., Toronto, Ont.

ENGINES, DIESEL
Can. Allis-Chalmers, Toronto, Ont.

ENGINES, HOISTING
Advance Mach. & Welding Co., Montreal, Que.
Can. Allis-Chalmers, Toronto, Ont.
Jenckes Machine Co., Ltd., Sherbrooke, Que.
Kennedy & Sons, Wm., Owen Sound, Ont.
Marsh & Henthorn, Ltd., Belleville, Ont.
Port Arthur Shipbuilding Co., Port Arthur, Ont.
Williams Machinery Co., A. R., Toronto, Ont.

ENGINE, INTERNAL COMBUSTION
Doxford & Sons, William, Sunderland, England.

ENGINES, MARINE
Bolinders Co., New York, N.Y.
Can. Allis-Chalmers, Toronto, Ont.
Can. Fairbanks-Morse Co., Montreal, Que.
Can. Vickers, Ltd., Montreal, Que.
Doxford & Sons, William, Sunderland, England.
Goldie & McCulloch, Ltd., Galt, Ont.
Iron Works, Ltd., Owen Sound, Ont.
Mason Regulator & Engin. Co., Montreal, Que.
The John McDougall Caledonian Iron Works Co., Montreal, Que.
Montreal Dry Docks & Shipbuilding Co., Montreal, Que.
Murray & Fraser, New Glasgow, N.S.
National Shipbuilding Co., Goderich, Ont.
Norbom Engineering Co., Philadelphia, Pa.
Polson Iron Works, Toronto, Ont.
Port Arthur Shipbuilding Co., Port Arthur, Ont
Trout Co., H. G., Buffalo, N.Y.
Williams Machinery Co., A. R., Toronto, Ont.

ENGINES, STEERING
Dake Engine Co., Grand Haven, Mich.
Jenckes Machine Co., Ltd., Sherbrooke, Que.
Kennedy & Sons, Wm., Owen Sound, Ont.
Wm. C. Wilson & Co., Toronto, Ont.

Nova Scotia Steel & Coal Company Limited

New Glasgow, Nova Scotia, Canada

FINISHED COUPLING SHAFT, 18 IN. DIAMETER BY 21 FT. LONG.

Heavy Marine Engine Forgings in the Rough or Finish Machined

Our Steel Plant at Sydney Mines, N.S., together with our Steam Hydraulic Forge Shop and modernly equipped Machine Shop at New Glasgow, N.S., place us in position to supply promptly Marine Engine Crank and Propeller Shafting, Piston and Connecting Rods; also Marine and Stationary Steam Turbine Shafting of all diameters and lengths, either as forgings or complete ready for installation, and equal to the best on the American Continent.

EVAPORATORS
- Kearfott Engineering Co., New York, N.Y.
- Mason Regulator & Engin. Co., Montreal, Que.
- Weir Ltd., G. & J., Cathcart, Glasgow, Scotland

EXTRACTORS, GREASE
- American Steam Gauge & Valve Mfg. Co. Boston, Mass.

EYE BOLTS AND NUTS
- Canada Foundries and Forgings, Welland, Ont.

FANS
- Aikenhead Hardware, Ltd., Toronto, Ont.
- Empire Mfg. Co., London, Ont.
- Smart-Turner Mach. Co., Hamilton, Ont.
- Williams Machinery Co., A. R., Toronto, Ont.

FENDERS, ROPE
- Leckie, Ltd., John, Toronto, Ont.
- Wilson & Co., Wm. C., Toronto, Canada.

FERRULES, CONDENSER
- Booth-Coulter Copper & Brass Co., Toronto, Ont.

FERRO-MANGANESE
- Mitchells, Ltd., Glasgow, Scotland.

FILES
- Aikenhead Hardware, Ltd., Toronto, Ont.
- Can. B. K. Morton Co., Toronto, Montreal.
- Williams Machinery Co., A. R., Toronto, Ont.

FIRE BRICKS
- Mitchells, Ltd., Glasgow, Scotland.
- Williams Machinery Co., A. R., Toronto, Ont.

FILTERS, FEED WATER
- Mason Regulator & Engin. Co., Montreal, Que.

FITTINGS, MOTOR BOAT
- Empire Mfg. Co., London, Ont.
- Murray & Fraser, New Glasgow, N.S.

FLOW METERS
- Spray Engineering Co., Boston, Mass.

FLUE CLEANERS
- Wm. C. Wilson & Co., Toronto, Ont.

FIXTURES, ELECTRIC
- Can. General Electric Co., Toronto, Ont.
- Cory & Son, Inc., Chas., New York, N.Y.
- Crouse-Hinds Co. of Canada, Ltd., Toronto, Ont.

FORGES
- Aikenhead Hardware, Ltd., Toronto, Ont.

FLOODLIGHTS, ELECTRIC
- Crouse-Hinds Co. of Canada, Ltd., Toronto, Ont.

FLUORSPAR
- Mitchells, Ltd., Glasgow, Scotland.

FORGINGS, ALL KINDS
- Aikenhead Hardware, Ltd., Toronto, Ont.
- Collingwood Shipbuilding Co., Collingwood, Ont.
- The John McDougall Caledonian Iron Works Co., Montreal, Que.
- Nova Scotia Steel & Coal Co., New Glasgow, N.S.
- St. Clair Bros., Galt, Ont.

FORGINGS, STEEL AND IRON
- Canada Foundries and Forgings, Welland, Ont.

GAUGES, RECORDING
- American Steam Gauge & Valve Mfg. Co., Boston, Mass.
- Empire Mfg. Co., London, Ont.

GASKETS
- Wm. C. Wilson & Co., Toronto, Ont.

GAUGE GLASSES
- Wm. C. Wilson & Co., Toronto, Ont.

GAUGES, WATER, PRESSURE, COMPOUND AND VACUUM
- Aikenhead Hardware, Ltd., Toronto, Ont.
- Babcock & Wilcox, Ltd., Montreal, Que.
- Empire Mfg. Co., London, Ont.
- Morrison Brass Mfg. Co., James, Toronto, Ont.

GENERATORS, STEAM TURBO
- Can. General Electric Co., Toronto, Ont.

GENERATORS AND CONVERTORS
- Can. General Electric Co., Toronto, Ont.
- Can. Fairbanks-Morse Co., Montreal, Que.

GRAPHITE
- Wm. C. Wilson & Co., Toronto, Ont.

GRINDERS, ELECTRIC
- Wilkinson & Kompass, Hamilton, Ont.

GRINDERS, PNEUMATIC
- Can. Ingersoll-Rand Co. Sherbrooke, Que.

GUY RODS AND ANCHORS, ELECTRIC
- Crouse-Hinds Co. of Canada, Ltd., Toronto, Ont.

HAMMERS
- Canada Foundries and Forgings, Ltd., Welland, Ont.

HAMMERS, DROP AND BELT DRIVEN
- Beaudry & Co., Boston, Mass.

HAMMERS, MOTOR DRIVEN
- Beaudry & Co., Boston, Mass.

HAMMERS, NAIL MACHINE
- Beaudry & Co., Boston, Mass.
- United Hammer Co., Boston, Mass.

HAMMERS, PNEUMATIC
- Barr, H. Edsil, Palace Bldg., Erie, Pa.

HAMMERS, POWER
- United Hammer Co., Boston, Mass.

HEADLIGHTS, ELECTRIC
- Crouse-Hinds Co. of Canada, Ltd., Toronto, Ont.

HEATING EQUIPMENT
- Empire Mfg. Co., London, Ont.
- Low & Sons, Ltd., Archibald, Glasgow, Scotland

HEATERS, FEED, WATER
- Babcock & Wilcox, Ltd., Montreal, Que.
- Can. Allis-Chalmers, Toronto, Ont.
- Kearfott Engineering Co., New York, N.Y.
- Mason Regulator & Engin. Co., Montreal, Que.
- Weir Ltd., G. & J., Cathcart, Glasgow, Scotland.

HOISTS, ASH
- Beatty & Sons, Welland, Ont.
- St. Clair Bros., Galt, Ont.

HOISTS, CHAIN
- Aikenhead Hardware, Ltd., Toronto, Ont.
- Can. Fairbanks-Morse Co., Montreal, Que.
- Dake Engine Co., Grand Haven, Mich.
- Fopping Brothers, New York, N.Y.
- Williams Machinery Co., A. R., Toronto, Ont.

HOISTS, CARGO, MOVING, ETC.
- Dake Engine Co., Grand Haven, Mich.
- Jenckes Mach. Co., Ltd., Sherbrooke, Que.

HOISTING MACHINERY
- Beatty & Sons, Welland, Ont.
- Can. Ingersoll-Rand Co. Sherbrooke, Que.
- Jenckes Machine Co., Ltd., Sherbrooke, Que.
- Marsh & Henthorn, Ltd., Belleville, Ont.
- Williams Machinery Co., A. R., Toronto, Ont.

HOSE
- Wm. C. Wilson & Co., Toronto, Ont.

HYDRAULIC MACHINERY
- The John McDougall Caledonian Iron Works Co., Montreal, Que.

INDICATORS, ENGINE
- American Steam Gauge & Valve Mfg. Co., Boston, Mass.
- Cory & Son, Inc., Chas., New York, N.Y.

INDICATORS, SPEED
- Aikenhead Hardware, Ltd., Toronto, Ont.
- Cory & Son, Inc., Chas., New York, N.Y.

INJECTORS
- Aikenhead Hardware, Ltd., Toronto, Ont.
- Can. Allis-Chalmers, Toronto, Ont.
- Empire Mfg. Co., London, Ont.
- Morrison Brass Mfg. Co., James, Toronto, Ont.
- Williams Machinery Co., A. R., Toronto, Ont.

INGOTS
- Broughton Copper Co., Ltd., Manchester, Eng.

INSULATORS, ELECTRIC
- Crouse-Hinds Co. of Canada, Ltd., Toronto, Ont.

INSTRUMENTS, NAUTICAL
- Leckie, Ltd., John, Toronto, Ont.

JOINTS, BALL SLIP
- Norbom Engineering Co., Philadelphia, Pa.

INSURANCE, MARINE
- Toronto Insurance & Vessel Agency, Toronto, Ont.

JOURNAL WEDGES
- Canada Foundries and Forgings, Ltd., Welland, Ont.

LAMPS, ARC, INCANDESCENT
- Can. General Electric Co., Toronto, Ont.

LATHES
- Foss & Hill Machinery Co., Montreal, Que.
- Bertram & Sons, Ltd., John, Dundas, Ont.

LEATHER, LACE
- Wm. C. Wilson & Co., Toronto, Ont.

LIFEBUOYS, BELTS AND PRESERVERS
- Fosbery Co., Barking, Essex, Eng.
- Wm. C. Wilson & Co., Toronto, Ont.

LIFEBOATS
- Wm. C. Wilson & Co., Toronto, Ont.

LIFE BOAT EQUIPMENT
- Leckie, Ltd., John, Toronto, Ont.
- Wm. C. Wilson & Co., Toronto, Ont.

LIFE JACKETS
- Fosbery Co., Barking, Essex, Eng.
- Leckie, Ltd., John, Toronto, Ont.

LIGHTS, ALL KINDS
- Can. Fairbanks-Morse Co., Montreal, Que.
- Crouse-Hinds Co. of Canada, Ltd., Toronto, Ont.
- Morrison Brass Mfg. Co., James, Toronto, Ont.

LIGHTS, SIDE, PORT
- Crouse-Hinds Co. of Canada, Ltd., Toronto, Ont.
- McAvity & Sons Ltd., T., St. John, N.B.

LIGHTING SETS
- Can. Fairbanks-Morse Co., Montreal.
- Cory & Son, Inc., Chas., New York, N.Y.
- Crouse-Hinds Co. of Canada, Ltd., Toronto, Ont.
- Kearfott Engineering Co., New York, N.Y.
- Williams Machinery Co., A. R., Toronto, Ont
- Wilson & Co., Wm. C., Toronto, Canada.

LUBRICATORS
- Empire Mfg. Co., London, Ont.

MACHINISTS
- Can. Allis-Chalmers, Toronto, Ont.
- Hyde Engineering Works, Montreal, Que.

MACHINE AND FORGE WORK
- Thor Iron Works, Toronto, Ont.

MANGANESITE, PASTE
- Wm. C. Wilson & Co., Toronto, Ont.

METERS, STEAM, WATER
- Can. Allis-Chalmers, Toronto, Ont.

MOTORS
- Aikenhead Hardware, Ltd., Toronto, Ont.
- Can. General Electric Co., Toronto, Ont.
- Can. Fairbanks-Morse Co., Montreal, Que.
- Williams Machinery Co., A. R., Toronto, Ont.

NAUTICAL INSTRUMENTS
- Wm. C. Wilson & Co., Toronto, Ont.

NET LIFTERS
- Dake Engine Co., Grand Haven, Mich.
- Leckie, Ltd., John, Toronto, Ont.

NOZZLES, ALL KINDS
- Empire Mfg. Co., London, Ont.
- Tolland Mfg. Co., Boston, Mass.

NUTS
- Wilkinson & Kompass, Hamilton, Ont.

OAKUM
- Wm. C. Wilson & Co., Toronto, Ont.

OIL CUPS
- Empire Mfg. Co., London, Ont.
- Murray & Fraser, New Glasgow, N.S.

OILCLOTHING
- Wm. C. Wilson & Co., Toronto, Ont.

OILS AND GREASE
- Wm. C. Wilson & Co., Toronto, Ont.

OIL SEPARATORS
- Can. Allis-Chalmers, Toronto, Ont.

OAKUM
- Leckie, Ltd., John, Toronto, Ont.

OXY-ACETLYENE OUTFITS
- Can. Fairbanks-Morse Co., Montreal, Que.
- McAvity & Sons Ltd., T., St. John, N.B.
- Williams Machinery Co., A. R., Toronto, Ont.

PAINT
- Ault & Wiborg Co. of Can., Ltd., Toronto, Ont.
- Brandram-Henderson, Ltd., Montreal, Que.
- Can. Bitumastic Enamels, Ltd., Toronto, Ont.,
- Wm. C. Wilson & Co., Toronto, Ont.

PAINTING EQUIPMENT, AUTOMATIC
- Spray Engineering Co., Boston, Mass.

PACKING, AMMONIA
- Can. B. K. Morton Co., Montreal.
- France Packing Co., Philadelphia, Pa.
- Wm. C. Wilson & Co., Toronto, Ont.

PACKING, MARINE
- Aikenhead Hardware, Ltd., Toronto, Ont.
- Can. B. K. Morton Co., Montreal.
- France Packing Co., Philadelphia, Pa.
- Wm. C. Wilson & Co., Toronto, Ont.

PACKING, METALLIC
- Aikenhead Hardware, Ltd., Toronto, Ont.
- Can. B. K. Morton Co., Montreal.
- France Packing Co., Philadelphia, Pa.

PACKING, STEAM
- Aikenhead Hardware, Ltd., Toronto, Ont.
- France Packing Co., Philadelphia, Pa.
- Wm. C. Wilson & Co., Toronto, Ont.

PINIONS, CUT
- Hamilton Gear & Machine Co., Toronto, Ont.

PANELBOARD AND CABINETS, ELECTRIC
- Crouse-Hinds Co. of Canada, Ltd., Toronto, Ont

PIPE, LAPWELD, CAST IRON, RIVETED
- Empire Mfg. Co., London, Ont.
- Norbom Engineering Co., Philadelphia, Pa.

PIPE JOINT CEMENT
- Wm. C. Wilson & Co., Toronto, Ont.

PITCH, PINE
- Wilson & Co., Wm. C., Toronto, Canada.

PLANERS
- Bertram & Sons, Ltd., John, Dundas, Ont.

PLANERS, STANDARD AND ROTARY
- Preston Woodworking Machinery Co., Preston, Ont.
- Yates Mach. Co., P. B., Hamilton, Ont.

PLATE PUNCH TABLES
- Norbom Engineering Co., Philadelphia, Pa.

PLATE WORK
- Jenckes Machine Co., St. Catharines, Ont.

PLUMBING EQUIPMENT
- Low & Sons, Ltd., Archibald, Glasgow, Scotland.
- Empire Mfg. Co., London, Ont.
- Wm. C. Wilson & Co., Toronto, Ont.

PROPELLOR BLADES, BRONZE
- Empire Mfg. Co., London, Ont.
- Murray & Fraser, New Glasgow, N.S.
- Yarrows, Limited, Victoria, B.C.

PROPELLER WHEELS
- Kennedy & Sons, Wm., Owen Sound, Ont.
- Trout Co., H. G., Buffalo, N.Y.

PIPING, COPPER
- Booth-Coulter Copper & Brass Co., Toronto, Ont.

PIPING, STEAM
- Babcock & Wilcox, Ltd., Montreal, Que.

PULLEYS
- Smart-Turner Mach. Co., Hamilton, Ont.
- Williams Machinery Co., A. R., Toronto, Ont.

PORT LIGHTS
- Wm. C. Wilson & Co., Toronto, Ont.

PUMPS
- Can. Fairbanks-Morse Co., Montreal, Que.
- Canada Foundries & Forgings, Welland, Ont.
- Goldie & McCulloch, Ltd., Galt, Ont.
- Williams Machinery Co., A. R., Toronto Ont.

PUMP, AIR
- Can. Ingersoll-Rand Co. Sherbrooke, Que.
- Smart-Turner Mach. Co., Hamilton, Ont.
- Weir Ltd., G. & J., Cathcart, Glasgow, Scotland.
- Williams Machinery Co., A. R., Toronto, Ont.

PUMPS, CIRCULATING
- Kingsford Fdy. & Mach. Works, Oswego, N.Y.
- Morris Machine Works, Baldwinsville, N.Y.

PUMPS, CENTRIFUGAL
- Can. Allis-Chalmers, Toronto, Ont.
- Can. Ingersoll-Rand Co. Sherbrooke, Que.
- Kearfott Engineering Co., New York, N.Y.
- The John McDougall Caledonian Iron Works Co., Montreal, Que.
- Norbom Engineering Co., Philadelphia, Pa.
- Smart-Turner Mach. Co., Hamilton, Ont.
- Williams Machinery Co., A. R., Toronto. Ont.

PUMPS, FEED WATER
- Can. Allis-Chalmers, Toronto, Ont.
- Weir Ltd., G. & J., Cathcart, Glasgow, Scotland.
- Williams Machinery Co., A. R., Toronto, Ont.

PUMPS, HAND AND POWER
- Aikenhead Hardware, Ltd., Toronto, Ont.
- Smart-Turner Mach. Co., Hamilton, Ont.
- Williams Machinery Co., A. R., Toronto, Ont.

PUMPS, HIGH PRESSURE
- Canadian Ingersoll-Rand Co., Sherbrooke, Que.
- Smart-Turner Mach. Co., Hamilton, Ont.

PUMPS, STEAM TURBO
- Can. Allis-Chalmers, Toronto, Ont.
- Wm. C. Wilson & Co., Toronto, Ont.

PUMPING MACHINES
- Norbom Engineering Co., Philadelphia, Pa.

PUNCHES
- Bertram & Sons, Ltd., John, Dundas, Ont.

PUNCHES, SINGLE, DOUBLE AND MULTIPLE
- Norbom Engineering Co., Philadelphia, Pa.

PURIFIERS, WATER
- Babcock & Wilcox, Ltd., Montreal, Que.
- Can. Allis-Chalmers, Toronto, Ont.

RADIO ENGINEERS
- Cutting & Washington, Inc., Cambridge, Mass.

RADIATORS, STEAM, ELECTRIC
- Empire Mfg. Co., London, Ont.
- Low & Sons, Ltd., Archibald, Glasgow, Scotland
- Wm. C. Wilson & Co., Toronto, Ont.

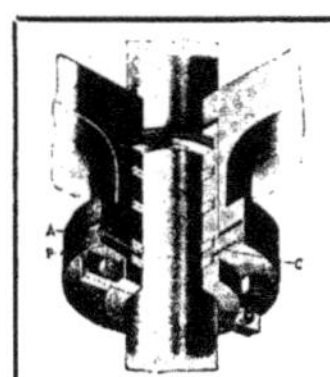

RANGES
Wm. C. Wilson & Co., Toronto, Ont.

RECEIVERS, AIR
Can. Allis-Chalmers, Toronto, Ont.
Can. Ingersoll-Rand Co. Sherbrooke, Que.

REGULATORS, FEED WATER
American Steam Gauge & Valve Mfg. Co. Boston, Mass.

REGULATORS, PRESSURE
Mason Regulator & Engin. Co., Montreal, Que.

REPAIRS, MARINE
Corbet Foundry & Mach. Co., Owen Sound, Ont.
Can. Vickers, Ltd., Montreal, Que.
Collingwood Shipbuilding Co., Collingwood, Ont.
Georgian Bay Shipbuilding & Wrecking Co., Midland, Ont.
Hyde Engineering Works, Montreal, Que.
Iron Works, Ltd., Owen Sound, Ont.
Kennedy & Sons, Wm., Owen Sound, Ont.
Montreal Dry Docks & Shipbuilding Co., Montreal Que.
Muir Bros. Dry Dock Co., Port Dalhousie, Ont.
National Shipbuilding Co., Goderich, Ont.
Port Arthur Shipbuilding Co., Port Arthur, Ont.
Yarrow, Limited, Victoria, B.C.

RIGGING, WIRE ROPE
Leckie, Ltd., John, Toronto, Ont.

RIVETS
Wilkinson & Kompass, Hamilton, Ont.

RIVETERS, PNEUMATIC
Can. Ingersoll-Rand Co., Ltd., Sherbrooke, Que.

RODS, COPPER, BRASS, BRONZE
Standard Underground Cable Co., Hamilton, Ont.

ROLLS, STRAIGHTENING, BENDING
Bertram & Sons, Ltd., John, Dundas, Ont.

ROOFING
Wm. C. Wilson & Co., Toronto, Ont.

ROPE BLOCKS
Aikenhead Hardware, Ltd., Toronto, Ont.
Can. Fairbanks-Morse Co., Montreal, Que.
Thor Iron Works, Toronto, Ont.

ROPE
Leckie, Ltd., John, Toronto, Ont.
Wm. C. Wilson & Co., Toronto, Ont.

RUBBER COATS
Wm. C. Wilson & Co., Toronto, Ont.

SAWS, BAND
Preston Woodworking Machinery Co., Preston, Ont.

SAW MILL MACHINERY
Preston Woodworking Machinery Co., Preston, Ont.
Yates Machine Co., P. B., Hamilton, Ont.

SCALES, BOILERS, ENGINES
Can. Fairbanks-Morse Co., Montreal, Que.

SCOWS
Collingwood Shipbuilding Co., Collingwood, Ont.
Polson Iron Works, Toronto, Ont.

SHAPES
Topping Brothers, New York, N.Y.

SEPARATORS, OIL, STEAM
Mason Regulator & Engin. Co., Montreal, Que.
Smart-Turner Mach. Co., Hamilton, Ont.

SHAFTING
Wilkinson & Kompass, Hamilton, Ont.

SHAFTING, BRONZE
Empire Mfg. Co., London, Ont.
Murray & Fraser, New Glasgow, N.B.

SHEARS
Bertram & Sons, Ltd., John, Dundas, Ont.
Norbom Engineering Co., Philadelphia, Pa.

SHIPBUILDING TOOLS
Aikenhead Hardware, Ltd., Toronto, Ont.
Can. Ingersoll-Rand Co. Sherbrooke, Que.

SHIPS, BUILDERS OF
Can. Vickers, Ltd., Montreal, Que.
Collingwood Shipbuilding Co., Collingwood, Ont.
Doxford & Sons, William, Sunderland, England.
Georgian Bay Shipbuilding & Wrecking Co., Midland, Ont.
Montreal Dry Docks & Shipbuilding Co., Montreal, Que.
National Shipbuilding Co., Goderich, Ont.
Polson Iron Works, Toronto, Ont.
Port Arthur Shipbuilding Co., Port Arthur, Ont.
Thor Iron Works, Toronto, Ont.
Yarrows, Limited, Victoria, B.C.

SHIP BROKERS
Page & Jones, Mobile, Ala.

SHIP PLATES
Nova Scotia Steel & Coal Co., New Glasgow, N.S.
Topping Brothers, New York, N.Y.

SHIPS' TELEGRAPHS
Cory & Son, Inc., Chas., New York, N.Y.
Morrison Brass Mfg. Co., James, Toronto, Ont.
Wm. C. Wilson & Co., Toronto, Ont.

SIDE LIGHTS
Wm. C. Wilson & Co., Toronto, Ont.

SLEDGES
Wilkinson & Kompass, Hamilton, Ont.

SPECIAL MACHINERY
Smart-Turner Mach. Co., Hamilton, Ont.

SMOOTH-ON
Wm. C. Wilson & Co., Toronto, Ont.

SPIKES
Topping Brothers, New York, N.Y.
Wm. C. Wilson & Co., Toronto, Ont.

SPRAY COOLING SYSTEMS
Spray Engineering Co., Boston, Mass.

SPRINKLERS, PARK
Spray Engineering Co., Boston, Mass.

STANDPIPES
Jenckes Machine Co., Ltd., Sherbrooke, Que.
Thor Iron Works, Toronto, Ont.

STAYBOLTS, FLEXIBLE
Can. Allis-Chalmers, Toronto, Ont.

STEAMSHIP AGENTS
Page & Jones, Mobile, Ala.

STEAM SEPARATORS
Can. Allis-Chalmers, Toronto, Ont.

STEAM SPECIALTIES
Can. Fairbanks-Morse Co., Montreal, Que.
Empire Mfg. Co., London, Ont.

STEAM TRAPS
Aikenhead Hardware, Ltd., Toronto, Ont.
American Steam Gauge & Valve Mfg. Co., Boston, Mass.
Empire Mfg. Co., London, Ont.
Mason Regulator & Engin. Co., Montreal, Que.
Smart-Turner Mach. Co., Hamilton, Ont.

STEEL, HIGH SPEED
Can. B. K. Morton Co., Toronto, Montreal.
Nova Scotia Steel & Coal Co., New Glasgow, N.S.
Wilkinson & Kompass, Hamilton, Ont.

STEEL WORK, STRUCTURAL
Babcock & Wilcox, Ltd., Montreal, Que.

STEERING GEARS
Wm. C. Wilson & Co., Toronto, Ont.

STOKERS, MECHANICAL
Babcock & Wilcox, Ltd., Montreal, Que.

SUPERHEATERS, STEAM
Babcock & Wilcox, Ltd., Montreal, Que.

SWITCHBOARDS, ELECTRIC
Crouse-Hinds Co. of Can., Ltd., Toronto, Ont.

TALLOW
Wm. C. Wilson & Co., Toronto, Ont.

TANKS, STEEL
Jenckes Mach. Co., Ltd., Sherbrooke, Que.
The John McDougall Caledonian Iron Works Co., Montreal, Que.
Port Arthur Shipbuilding Co., Port Arthur, Ont.
Thor Iron Works, Toronto, Ont.

TANKS
Corbet Foundry & Mach. Co., Owen Sound, Ont.
Goldie & McCulloch, Ltd., Galt, Ont.
Jenckes Machine Co., Ltd., Sherbrooke, Que.
Marsh & Henthorn, Ltd., Belleville, Ont.

TELEPHONES, MARINE
Cory & Son, Inc., Chas., New York, N.Y.

THUMB SCREWS AND NUTS
Canada Foundries & Forgings, Welland, Ont.

TREENAILS
Topping Brothers, New York, N.Y.

TROLLEYS
Can. Fairbanks-Morse Co., Montreal, Que.

TRUCKS, HAND, ELECTRIC
Aikenhead Hardware, Ltd., Toronto, Ont.
Can. Fairbanks-Morse Co., Montreal, Que.

TUBES, BOILER
Babcock & Wilcox, Ltd., Montreal, Que.
Broughton Copper Co., Ltd., Manchester, Eng.

TUBES, COPPER AND BRASS
Standard Underground Cable Co., Hamilton, Ont.

TUGS
Collingwood Shipbuilding Co., Collingwood, Ont.
Polson Iron Works, Toronto, Ont.

TURBINES, DIRECT-DRIVING AND GEARED
Can. Allis-Chalmers, Toronto, Ont.
Doxford & Sons, William, Sunderland, England.

TURNBUCKLES
Canada Foundries & Forgings, Welland, Ont.

UNIONS, ALL KINDS
Dart Union Company, Toronto, Ont.

VALVE, DISCS
Wm. C. Wilson & Co., Toronto, Ont.

VALVES
American Steam Gauge & Valve Mfg. Co., Boston, Mass.
Babcock & Wilcox, Ltd., Montreal, Que.
Can. Fairbanks-Morse Co., Montreal, Que.
Empire Mfg. Co., London, Ont.
McAvity & Sons, Ltd., T., St. John, N.B.
Mason Regulator & Engin. Co., Montreal, Que.
Norbom Engineering Co., Philadelphia, Pa.
Williams Machinery Co., A. R., Toronto, Ont.
Wm. C. Wilson & Co., Toronto, Ont.

VALVES, FOOT
Aikenhead Hardware, Ltd., Toronto, Ont.
Smart-Turner Mach. Co., Hamilton, Ont.

VALVES, STOP, REDUCING, SAFETY CHECK, ETC.
Aikenhead Hardware, Ltd., Toronto, Ont.
American Steam Gauge & Valve Mfg. Co., Boston, Mass.
Empire Mfg. Co., London, Ont.
McAvity & Sons Ltd., T., St. John, N.B.
Morrison Brass Mfg. Co., James, Toronto, Ont.

VARNISHES
Aikenhead Hardware, Ltd., Toronto, Ont.
Ault & Wiborg Co. of Can., Ltd., Toronto, Ont.
Leckie, Ltd., John, Toronto, Ont.
McAvity & Sons Ltd., T., St. John, N.B.

VENTILATION EQUIPMENT
Empire Mfg. Co., London, Ont.
Low & Sons, Ltd., Archibald, Glasgow, Scotland

WASTE
Wm. C. Wilson & Co., Toronto, Ont.

WATER COLUMNS
Morrison Brass Mfg. Co., James, Toronto, Ont.

WATER HEATERS
Empire Mfg. Co., London, Ont.
Morrison Brass Mfg. Co., James, Toronto, Ont.

WATER SOFTENERS
Babcock & Wilcox, Ltd., Montreal, Que.
Can. Allis-Chalmers, Toronto, Ont.

WEDGES, OAK
Topping Brothers, New York, N.Y.

WELDING, ELECTRIC
Hall Engineering Works, Montreal, Que.
Marine Welding Co., Buffalo, N.Y.
Beatty & Sons, M., Welland, Ont.

WELDING, OXY-ACETYLENE
Advance Mach. & Welding Co., Montreal, Que.

WHISTLES AND SYRENS
Empire Mfg. Co., London, Ont.
Morrison Brass Mfg. Co., Jas., Toronto, Ont.

WINCHES, CARGO
Advance Mach. & Welding Co., Montreal, Que.
Aikenhead Hardware, Ltd., Toronto, Ont.
Jenckes Machine Co., St. Catharines, Ont.

WINCHES, DOCK, SHIP
Advance Mach. & Welding Co., Montreal, Que.
Beatty & Sons, M., Welland, Ont.
Jenckes Machine Co., St. Catharines, Ont.
Marsh & Henthorn, Ltd., Belleville, Ont.
Wilson & Co., Wm. C., Toronto, Canada.

WINCHES, TOWING
Corbet Foundry & Mach. Co., Owen Sound, Ont.
Jenckes Mach. Co., Ltd., Sherbrooke, Que.

WINCHES, TRAWL
Beatty & Sons, M., Welland, Ont.
Jenckes Mach. Co., Ltd., Sherbrooke, Que.
Wm. C. Wilson & Co., Toronto, Ont.

WINDLASSES
Advance Mach. & Welding Co., Montreal, Que.
Dake Engine Co., Grand Haven, Mich.
Jenckes Mach. Co., Ltd., Sherbrooke, Que.
Wilson & Co., Wm. C., Toronto, Canada.

WIPER CAPS, OILER BOXES, ETC.
Morrison Brass Mfg. Co., James, Toronto, Ont.

WIRE, COPPER CLAD STEEL
Standard Underground Cable Co., Hamilton, Ont.

WIRE, COPPER, BRASS, BRONZE
Standard Underground Cable Co., Hamilton, Ont.

WIRE, INSULATED
Standard Underground Cable Co., Hamilton, Ont.

WIRELESS OUTFITS
Cutting & Washington, Inc., Cambridge, Mass.
Marconi Wireless Telegraph Co., Montreal, Que.

WOOD WORKING MACHINERY
Aikenhead Hardware, Ltd., Toronto, Ont.
Can. Fairbanks-Morse Co., Montreal, Que.
Preston Woodworking Machinery Co., Preston.
Williams Machinery Co., A. R., Toronto, Ont.
Yates Mach. Co., P. B., Hamilton, Ont.

WOOD BORING TOOLS
Aikenhead Hardware, Ltd., Toronto, Ont.
Can. Ingersoll-Rand Co. Sherbrooke, Que.

WOODITE GAUGE GLASS WASHERS
Wm. C. Wilson & Co., Toronto, Ont.

WRENCHES
Canada Foundries & Forgings, Welland, Ont.

INDEX TO ADVERTISERS

MARINE ENGINEERING of Canada

THE MACLEAN PUBLISHING COMPANY, LIMITED

Toronto Montreal Winnipeg Boston New York Chicago London, Eng.

Vol. VIII. Publication Office, Toronto, February, 1918 No. 2

WILLIAM DOXFORD AND SONS
LIMITED

SUNDERLAND, ENGLAND

Shipbuilders **Engineers**

13-Knot, 11,000-Ton Shelter Decker for
Messrs. J. & C. Harrison Ltd., London

Builders of all Types of Vessels up to 20,000 Tons, D.W.
Builders of Reciprocating Engines and Boilers of all Sizes.
Builders of Turbines, Direct-Driving and Geared.
Builders of Internal Combustion Engines, Doxford's Opposed Piston Type
Builders of Special Coal and Ore Carriers.
Builders of Special Oil Tank Steamers.
Builders of Special Self-Discharging Colliers.
Builders of Special Bunkering Craft.
Builders of Special Floating Oil Storage Tanks.

The Publisher's Page

TORONTO FEBRUARY, 1918

CONDITIONS AFTER THE WAR*

By LT.-COL. THOMAS CANTLEY

Chairman of the Board, Nova Scotia Steel and Coal Co.

ALTHOUGH the transition from war to peace conditions—whenever it may come—must cause some temporary difficulty and disorganization, there is a confident feeling that there will rapidly develop a huge demand for the products of the steel industry in order to make good the devastation of war, and also to cover the vast arrears of commercial work which have been so long held up.

There is the further fact that the world scarcity of food, the devastated farms, manufacturing plants, towns and homes of Europe will make for an enormous immigration to Canada, to accommodate which will cause the opening up of vast new agricultural areas, with a consequent demand for railways, bridges, buildings, tools and implements. The prosperity enjoyed by the agricultural population should also result in a very great demand for farming implements. Railway equipment and rolling stock in Canada, and in fact the shortage of railway equipment throughout the world, is great, while needed replacements are not being made, and there is a veritable famine for equipment of this character everywhere.

These facts seem to lead to the inevitable conclusion that while there may be a slight hiatus immediately following the war's end, within a very short time thereafter Canada will embark on an era of unexampled and unprecedented development and prosperity, while the iron and steel industry of the Dominion will be one of the first to participate.

May the day soon dawn when Canada's heroic sons, who have so gloriously battled for liberty and world freedom, will be triumphant and free to return to their homes and again take their place in their old-time peaceful pursuits, honored by all for what they dared and what they have done.

*Extract from an article in "Industrial Canada"

Say Good-Bye

To Pipe Coupling Troubles

Shake them for a lifetime by using

Dart Union Pipe Couplings

TEE UNION ON THE RUN
Female

Dart Unions give faithful service day in and day out, year after year—they never loosen up and leak because both faces are of Bronze (a non-deteriorating metal) and the heavy iron parts will not stretch or pull apart.

No upkeep cost — No replacement cost — No loss from leaks

Sold by all Supply Houses. Positively guaranteed.

With Dart Unions you can make connections easily and quickly, whether pipes are in or out of alignment, and you can break and remake connections as often as you desire without affecting the Dart's efficiency.

The Dart Union Company Limited, Toronto, Ont.

FLANGE UNION

UNION ELBOW—Female

TEE UNION ON THE OUTLET
Female

If what you need is not advertised, consult our Buyers' Directory and write advertisers listed under proper heading.

Stern Casting
for Ice Breaker
"JOHN D. HAZEN"
STEEL
CASTINGS
Ship Castings are a Specialty.
CANADIAN STEEL FOUNDRIES
LIMITED
General Offices:
Transportation Bldg.,
Montreal
Works:
Montreal and
Welland

We are manufacturers of single, reversible and double drum hoists, used for elevating coal, sahes, mortar, etc., and also used for hoisting material in shipyards and dockyards, etc.

CONCRETE MIXERS

Our batch concrete mixers are acknowledged by leading contractors to have the finest blade mixing system on the market to-day.

OUR MARINE FORGING DEPT. AND MACHINE SHOP in connection are fully equipped for turning out cranks, connecting rods, drop forgings, etc. Forgings, either rough forged or machined, rough cut or finished up. Our forging department personally superintended by one of the best steel experts in Canada. We also do all kinds of machine work by the time method, or according to agreed prices.

ST. CLAIR BROTHERS

Galt, Ontario, Canada

Rush Repairs

EVEN the best of boats meet misfortune. Any day a broken part may send your ship limping to port to lay up while a new part is ordered and installed.

The wireless equipped ship telegraphs her orders for repair parts while far at sea and finds them waiting on the dock upon arrival. Time is money.

Cutting & Washington Wireless

sets are reliable, powerful and fool-proof. They can be installed on any boat — old or new, operated by a member of the crew and paid for out of the savings they effect.

CUTTING & WASHINGTON, Inc.
20 Portland St., CAMBRIDGE, MASS.

If what you need is not advertised, consult our Buyers' Directory and write advertisers listed under proper heading.

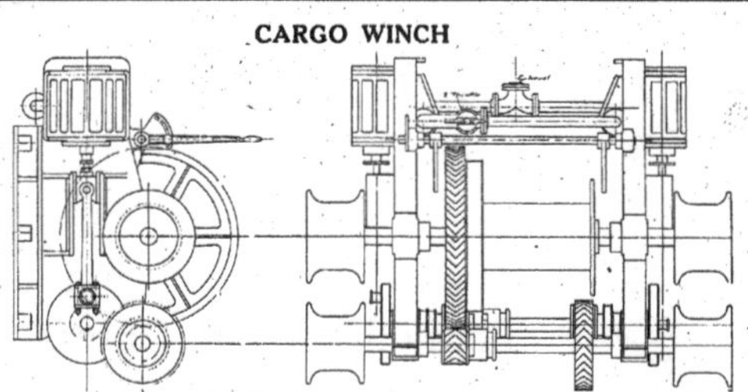

THE quality, strength and design of our lines enable us to meet the up-to-the-minute demand in construction and service.

Cargo Winches, Trawl Winches, Dock Winches, Ash Hoists, Coaling Buckets, Clamshell Buckets, Hoisting Machinery

Our facilities place us in a position to fill orders with despatch. Let us quote you on any line or lines in which you are interested.

M. BEATTY & SONS, LIMITED - Welland, Ontario

BELMONT SUPERHEAT STEAM PACKING

Specially adapted for Marine and all High Speed Engines.

Every strand is thoroughly lubricated before braiding and positively cannot become hard in service.

Get a box and if it doesn't prove to be a better packing than you have ever used you need not pay one cent.

Write for Booklet

WM. C. WILSON & CO.

21 Camden Street - TORONTO, Canada

Engineers and Machinists
Brass and Ironfounders
Boilermakers and Blacksmiths

SPECIALTIES

Electric Welding and Boring Engine Cylinders in Place.

The Hall Engineering Works, Limited
14-16 Jurors Street, Montreal

Ship Building and Ship Repairing in Steel and Wood.

Boilermakers, Blacksmiths and Carpenters

The Montreal Dry Docks & Ship Repairing Co., Limited

DOCK—Mill Street OFFICE—14-16 Jurors Street

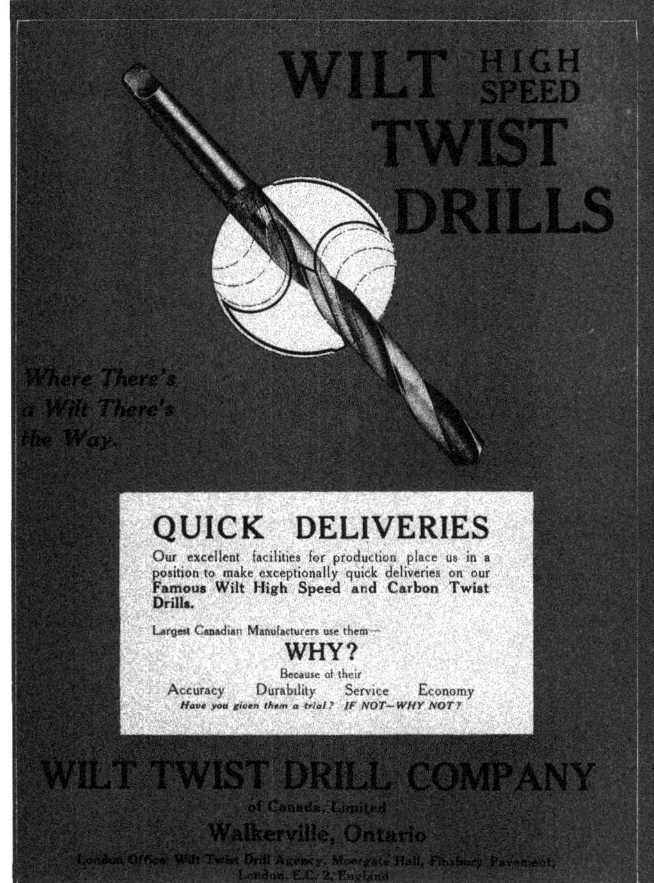
WILT HIGH SPEED
TWIST
DRILLS
Where There's
a Wilt There's
the Way.
QUICK DELIVERIES
Our excellent facilities for production place us in a position to make exceptionally quick deliveries on our Famous Wilt High Speed and Carbon Twist Drills.
Largest Canadian Manufacturers use them—
WHY?
Because of their
Accuracy Durability Service Economy
Have you given them a trial? IF NOT—WHY NOT?
WILT TWIST DRILL COMPANY
of Canada, Limited
Walkerville, Ontario
London Office: Wilt Twist Drill Agency, Moorgate Hall, Finsbury Pavement,
London, E.C. 2, England

Possibilities of Designing Cargo Ship Lines of Simple Form*

By Naval Constructor Wm. McEntee, U.S.N.

The investigation, the results of which the author presents, was made to show how the power required for propulsion by different types of cargo vessels is affected by simplification of lines. He arrives at the conclusion that cargo vessels can be built of simplified lines which will give practically as good results from the resistance standpoint as those built from the will give practically as good results from resistance standpoint as those ordinarily built.

THE recent wide discussion and general interest in the possibilities of standard cargo ships to be built in large numbers from standard plans has naturally brought into prominence the matter of simplifying the form of ships. The tendency in ship construction for a long period has been in the direction of simplification. If, for example, the sailing vessels which formed the Spanish Armada, with their high poops and forecastles and overhanging galleries, are compared with the present-day ship, the effect of this tendency is shown very clearly. A part of the change has, no doubt, been due to the introduction of steam power and the increase in size of ships made possible by their construction of iron and steel, but even before the era of steam propelled iron and steel ships the tendency toward simplification was having a decided effect. In this gradual change there has been a greater variation in the form of that part of the shed above the water-line. This is possibly due to the fact that there was more latitude in the upper works for the expression of individual ideas, or for artistic embellishment than there was with the under-water body.

Development of Body Forms

In order to make any reasonable progress through water the under-water form had to be of some reasonably good shape, and the practical limits of form variation were more restricted. An examination of the lines of one of the large boats, or small ships, used by the Norsemen nearly 1,000 years ago shows that for their purpose the shape of the under-water body was almost equal to the best we have to-day. In general the simplification of form requires the use of plain surfaces at right angles instead of curved surfaces wherever possible. Where curvature is unavoidable, the simplest of curves, the circle, should be used. All curvature should be in one direction only, that is, surfaces of double curvature should be avoided so far as possible.

These considerations lead, in the forming of a ship, to the adoption of a midship section formed by two vertical lines and a horizontal line at the bottom, the corners or bilges being rounded off by circles of suitable radius. The sides and bottom of the ship will be flat surfaces carrying the full midship section well forward and aft, making what is known as the parallel middle body extend as far as possible without excessive increase in resistance. Forward, where it is necessary to fine the lines to form an entrance, it can be done by gradually decreasing the width of the ship, keeping the bottom flat and the sides vertical, the bilge being formed by a quarter circle of constant radius equal to that used for the middle part of the vessel.

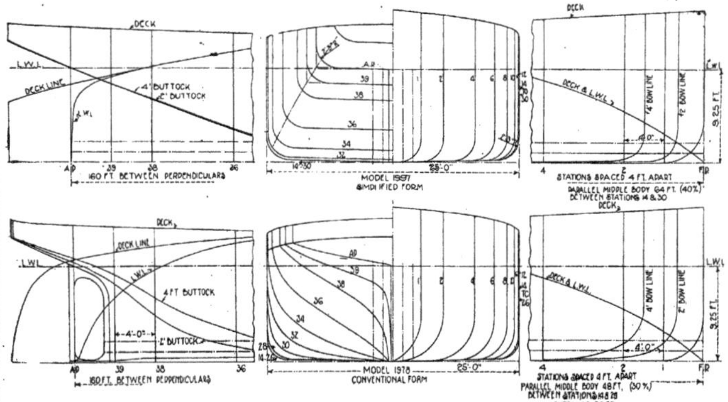

FIG. 1.—LINES OF 160-FT. OIL-FUEL BARGES. MODELS 1978 AND 1997 OF 830 AND 850 TONS AND CONVENTIONAL AND SIMPLIFIED FORM RESPECTIVELY.

Aft, the possibilities of simplification are limited by the necessity for providing space for the propellers. Two different methods of simplifying the after body are shown in model 1,997 and model 2,056, Figs. 1 and 2, annexed. The general result in these forms is to make the water-line somewhat wider and place the displacement higher up. This arrangement seems favorable from both the resistance and the propulsion standpoints, but may have disadvantages as regard sea-going qualities.

Scope of Paper

In the present investigation an attempt has been made to show how the

*A paper presented at the November meeting of the Society of Naval Architects and Marine Engineers in New York.

power required for propulsion is affected by simplification of the lines of three different types of cargo vessels.

Models 1,978 and 1,997, lines of which are shown on Fig. 1, are models of 160-ft. oil-fuel barge of 830 tons and 850 tons respectively. Through a slight error in making model 1,997, the cylindrical or prismatic co-efficient and displacement are somewhat larger than for the model with which it is compared, that is, No. 1,978, of conventional form. The effective horsepower curves for the estimated power are shown in Fig. 4. It will be seen that the curves are very nearly the same, the simplified form requiring about 35 effective horse-power at 6 knots as compared with 34 effective horse-power for the conventional form.

Models 2,023 and 2,056, line of which are shown in Fig. 2, are of a 400-ft. 10.5-knot cargo ship of the usual single-screw type. At 10.5 knots it will be seen from the power curves, shown in Fig. 5, that the simplified form requires something less than 2 per cent. greater effective horse-power than for the conventional form. As model 2,023 gave results which were considered to be good for this type of vessel, it follows that the simplified form, model 2,056, would require appreciably less power than many cargo ships of convenient form in this general class now built and building.

In Fig. 3 are shown the lines of the United States collier Neptune, and of a 500-ft. cargo vessel of simplified lines and somewhat greater displacement. The effective horse-power curves for these two models are given in Fig. 6. There is also shown the estimated effective horse-power curve for a 500-ft. ship of conventional form having the same displacement, 20,000 tons, and the same cylindrical or prismatic coefficient, 0.74. This latter curve is estimated from tests of one of a series of models of conventional form. The simplified form requires a little more power than the conventiona up to 13 knots. From that up to a speed of about 17 knots the simplified form is the better, the advantage at speed of 1[illegible] knots being in the neighborhood of 7 per cent. On the same diagram it will be seen that the Neptune, which is of the

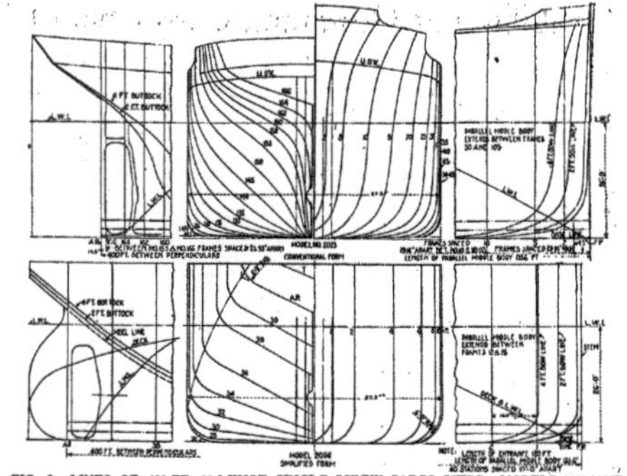

FIG. 2.—LINES OF 400-FT. 10.5-KNOT SINGLE SCREW CARGO SHIPS. MODELS 2023 AND 2056, BEING CONVENTIONAL AND SIMPLIFIED FORMS RESPECTIVELY.

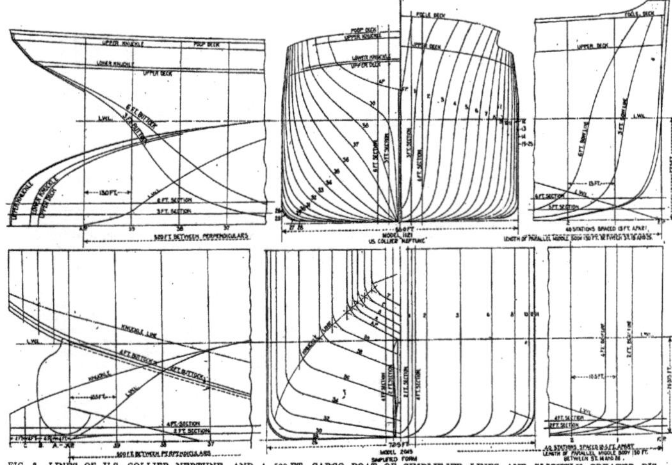

FIG. 3.—LINES OF U.S. COLLIER NEPTUNE AND A 500-FT. CARGO BOAT OF SIMPLIFIED LINES AND SLIGHTLY GREATER DISPLACEMENT.

TABLE I.—DIMENSIONS.

Model No.	Type.	Length, feet	Beam, feet	Draught feet	Displacement, tons	Speed knots	E.H.P.	Long. Coefficient	Midship Section Coefficient	Parallel Middle Body %
1978	Oil fuel barge, conventional form........	160	25.0	9.25	830	6.0	34.4	0.80	0.98	30.0
1997	Oil fuel barge, simplified lines	160	25.0	9.25	850	6.0	35.6	0.82	0.98	40.0
2023	400-ft. freighter, conventional form	400	57.3	26.0	13,137	10.5	1,080	0.788	0.987	33.4
2056	400-ft. freighter, simplified lines	400	57.3	26.0	13,137	10.5	1,120	0.788	0.987	33.4
*Series	500-ft. cargo vessel, conventional form ..	500	73.6	26.77	20,000	14.0	3,440	0.74	0.96	30.0
2045	500-ft. cargo vessel, simplified lines	500	72.5	26.375	20,000	14.0	3,290	0.74	0.989	30.0
1121	U.S. collier Neptune	520	65.0	27.5	19,340	14.0	3,460	0.738	0.983	25.0

*Estimated from Experimental Model Basin's Parallel Middle Body Series (No. 26).

conventional form, but 520 ft. in length and of somewhat less displacement, that is, 19,340 tons, requires in general more power than is found by experiment to be necessary in a ship of corresponding size constructed on the simplified lines.

The main dimensions of the models, or rather of the ships represented by them, are given in table I, which is printed in the adjoining columns.

In general, it appears to be safe to conclude that cargo vessels can be built of simplified lines which will give practically as good results from the resistance standpoint as those built from the present conventional lines. If propulsive efficiency is also taken into account, it is believed that the simplified form will have, at least in certain cases, advantages over the more complicated forms of the present type.

FIG. 4. FIG. 5 FIG. 6.

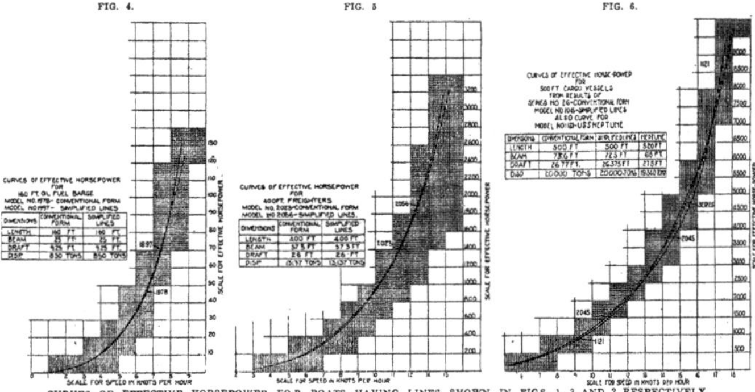

CURVES OF EFFECTIVE HORSEPOWER FOR BOATS HAVING LINES SHOWN IN FIGS. 1, 2 AND 3 RESPECTIVELY.

COOLING OF ENCLOSED SPACES BY VENTILATION

By S. B. Crosby.

ON board some vessels there are magazine spaces or other similarly enclosed spaces which are required to be maintained at a lower temperature than that which prevails in the spaces adjacent to the said enclosed space or magazine.

The decks and bulkheads surrounding the enclosed space are provided with insulation and an adequate ventilation system is installed to supply the air to maintain the specified temperature.

A case will be assumed as follows: Temperature of air in surrounding spaces, 110 degrees F.; temperature on deck, 80 degrees F.; temperature to be maintained inside of enclosed space, 90 degrees F.

It will be necessary first to have a line on the rate of heat transmission through the insulated walls. Tests on cork insulation conducted at the Washington Navy Yard show that with 3-inch cork the transmission of heat per hour in B.t.u. per square foot per degree difference in temperature is 0.19 B.t.u., while with 4-inch cork it is 0.16 B.t.u. Also the transmission of heat through a 1-inch pipe passing through the bounding walls into a given insulated space is 0.282 B.t.u. per hour per degree difference of temperature, while a 2-inch pipe transmits 0.297 B.t.u., and a 3-inch pipe 0.305 B.t.u., thus showing that the losses through a pipe or other small projection is not necessarily proportionate to the size.

Assume that the loss through pipes, hangers, poor joints, etc., is 10 per cent. additional, and if this percentage is added to the figures given by the Washington Navy Yard experiments, the transmission of heat per hour through 3-inch cork will be 0.209 B.t.u., and through 4-inch cork 0.176 B.t.u. per square foot per degree difference of temperature, and the total transmission will be:

For 3-inch cork $0.209 \times 20 = 4.18$ B.t.u. per hour $= 0.07$ B.t.u. per minute per square foot.

For 4-inch cork $0.176 \times 20 = 3.52$ B.t.u. per hour $= 0.059$ B.t.u. per minute per square foot.

One B.t.u. will raise the temperature of 57.4 cubic feet of dry air 1 degree F. at about 80 degrees temperature, but allowance must be made for the humidity in the air, and it is customary to use 55 cubic feet instead of 57.4 cubic feet, hence 1 B.t.u. will be taken up by each 5.5 cubic feet of air, and with the heat transmission previously given, the amount of incoming air at 80 degrees to be supplied per minute to neutralize radiation through 1 square foot of bulkhead or deck surface will be:

For 3-inch cork $5.5 \times 0.07 = 0.385$ cubic feet.

For 4-inch cork $5.5 \times 0.059 = 0.325$ cubic feet.

The quantities given are the minimum required, and for practical purposes about 15 per cent. should be allowed for unequal distribution; accordingly, for 3-inch cork it is recommended that 0.44 cubic feet of air be supplied per square foot of radiation, and 0.37 cubic feet for 4-inch cork.

It will also be noted that the use of the 4-inch cork insulation is superior to the 3-inch cork and requires a lesser quantity of air and a smaller ventilating system.

Alderman W. S. Weldon of Montreal, has been appointed Collector of Customs of the Port of Montreal. He was at one time president of the Montreal Amateur Athletic Association.

LAUNCHING A FREIGHTER IN MID-WINTER

An unusual event took place at the Polson Iron Works, Toronto, on Feb. 11, when the 3,500-ton steel freighter Asp was launched in spite of the severe weather and thick ice in the bay. As a result of the exceptionally severe winter the ice in Toronto Bay was unusually thick and a large sheet had to be removed before the launching could take place. The method adopted for removing the ice was decidedly ingenious and perfectly successful.

Ice Feature

A large sheet of ice, 300 feet by 80 feet, and approximately 28 inches thick, was first cut away and broken up, beginning at a point about 150 feet from the end of the dock. Two narrow channels, 450 feet long, including 300 feet in front of the ways, were then cut in the ice, forming a sheet 450 feet long by 70 feet wide. This sheet was towed into the large space prepared for it, leaving open water in front of the ways. The ice in front of the ways on the far side was then blown up with dynamite, the loose blocks forming a cushion for the ship when launched. The ice in the dock in front of the ways ranged from 48 inches to 54 inches thick. Altogether approximately 6,000 tons of ice had to be removed before the launching could take place.

Constructional Features

The Asp is a sister ship to the Tento launched last October, and was the twelfth ship launched at the Polson's yards since last August. The Asp was built for Norwegian interests, but will be taken over by the British Government for the duration of the war. She was christened by Mrs. Winstrup, wife of Captain Winstrup, representing the owners. The Asp, like the Tento, is of standard construction for ocean service, built to the highest class Bureau Veritas. The principal dimensions are as follows: Length over all, 261 feet; length between perpendiculars, 251 feet; beam moulded, 43 feet 6 inches; depth moulded, 23 feet. The mean draft will be 19 feet 6 inches when carrying a total deadweight of 3,500 tons.

THE "ASP" AFTER LAUNCHING. THE LOOSE ICE OBTAINED BY DYNAMITING RECEIVED THE IMPACT OF THE HULL IN EXCELLENT MANNER.

The engine room, boiler room and navigating bridge deck are amidships, with cargo spaces fore and aft. The ship is of single deck type, and was built on the deep frame principle. The deep frame system with auxiliary side frames constitute the basic constructional features of the hull fabrication. She has a cellular double bottom fore and aft, also peak tanks and power watertight bulkheads. The shell plating is overlapped at the edges with overlapped scarf butts. The Asp is a single screw freighter.

Main Propelling Engines

The propelling machinery has been constructed to Bureau Veritas requirements for a working pressure of 180 pounds per sq. inch. The main engines are of inverted, direct-acting, surface condensing, triple expansion type, the cylinders being 20½ in., 33 in., and 54 in. diameter by 36 in. stroke. The condenser and pumps are independent of the main engine structure. The H.P. and I.P. cylinders are served by piston valves, and the L.P. cylinder by a double ported slide valve. All these valves are on the fore and aft centre line of the engine. The bed plate is of box section cast iron, as are also the columns. All shafting is of best open-hearth forged steel. The crank shaft is of built-up type, with cast steel crank arms. The connecting rods, cross-heads, and piston rods are of open-hearth, forged steel. The valve gear is of Stephenson link motion type, with direct acting steam reversing gear. The propellers are four bladed of cast iron.

Boilers

The boiler installation consists of two 14 ft. diameter by 12 ft. long, single-ended, Scotch marine type units, arranged for natural draft only, and built to pass Bureau Veritas requirements for 180 pounds working steam pressure. Each boiler has three corrugated steel furnaces of 42 in. inside diameter. The tubes are 3½ in. diameter, lap welded and standard gauge. A separate combustion chamber for each furnace and one double funnel are other features.

Auxiliary Machinery Equipment

The pumping equipment includes deck pumps for each hold; vertical duplex, brass-fitted, main boiler feed pump; a duplicate of the latter for use as a general service donkey pump; duplex horizontal piston type bilge pump; a 7-in. duplex, brass-fitted, ballast pump; and a fresh water pump; 4½ in.—4 in.—5 in. of duplex type; air pump and centrifugal circulating pump for main condenser. The feed water heater is of multi-coil type installed in the feed pump discharge line. One 7½ k.w. Enberg electric generating set is installed in the engine room for the ship's lighting system. An evaporator of 15 tons capacity is also installed in the engine room.

Deck Machinery

For handling cargo there are six 7 in. by 10 in. horizontal double cylinder steam winches of Clark-Chapman type, one at each mast and derrick. The

CUTTING AN OPEN SPACE IN THE LAKE TO RECEIVE A 450 FT. BY 70 FT. SHEET OF ICE FROM THE LAUNCHING BASIN.

THE 3,500-TON STEEL FREIGHTER "ASP" TAKING THE WATER AT POLSON IRON WORKS, TORONTO, FEB. 11. ABOVE, CUTTING THE BASIN IN THE ICE. BELOW, DYNAMITING 48 IN. ICE TO ACT AS CUSHION FOR BOAT AFTER ENTERING WATER.

winches have central barrels and drums, and are placed on girders above the deck. Each mast carries one pitch pine, 3-ton derrick, the latter long enough to land cargo 10 ft. away from ship's side. The windlass is of direct-acting type for operation by hand, as well as by steam, and is complete with reversing gear. A hand screw steering gear is installed aft, and a steam steering gear amidships. The total bunker capacity is about 350 tons.

MIDLAND SHIPBUILDING CO.

THE Midland Shipbuilding Co. of Midland, Ont., although devoting its energies principally to repair work, will shortly lay the keel of a steel freighter which will be 261 feet long, 43 ft. 6 inches beam, and 23 feet deep. The company is repairing the C.P.R. steamer Athabasca which was damaged late in the season by ice. Twenty new plates are being replaced in her bow.

The forward end of the steamer Mariska is undergoing extensive alterations, a new forecastle and steel deck-house being built upon her and equipped.

A new steel deck-house is also being built forward on the steamer Glenlyon of the Great Lakes Transportation Co., including captain's and mate's accommodation. A new steel deck-house aft and various minor repairs are being made in the hull.

All the big beams and stanchions are being removed from the Western Star, of the Great Lakes Line, which is being reconstructed to the Herriman type of lake cargo steamers. The wood-work in the after cabin is being rebuilt and the engines and boilers overhauled.

The W. C. Franz, of the Algoma Steamship Co., will appear in the spring as a Herriman type of lake carrying steamer.

The boilers and engines of several other ships in the harbor are being overhauled. Though reconstruction work is taking up a great part of the company's time, the management is looking forward to commencing operations on the new ships as soon as the steel arrives, and it is expected daily.

ATLANTIC CONFERENCE FORMED

ACCORDING to an official announcement from Liverpool, England, dated Jan. 30, British, American and French steamship lines interested in the North Atlantic trade have concluded an alliance for the purpose of governing the passenger business of the lines, according to an official announcement. The alliance will be known as the "Atlantic Conference," replacing a similar organization which existed before the war, but from which the German and Austrian lines are now excluded. The main offices of the organization will be in Paris.

The formation of the new "Conference" may be taken as a definite and far-reaching step towards an after-war shipping policy in which the Entente Allies do not propose to allow Germany to dicate. The old pre-war Atlantic Conference had its offices at Jena, in the heart of Germany, and it held most of its meetings in Berlin and Cologne. The new conference will consist, for the present, only of British, American and French lines, German and Austrian lines are excluded, and no neutral lines have yet been admitted.

The underlying idea of the organizaton is protective. A representative of the conference, in a talk with the newspapermen here, said that "the primary object is to consolidate, in a commercial sense, the interest of the allies, who have stood shoulder to shoulder in the war. We must have everything ready when peace is declared to endeavor to conserve business to the alles and to keep control of the Atlantic trade. We know what we have had to suffer in the past through the insidious German penetration, and how the German lines worked always to the prejudice of other nationalities in spite of their agreements."

The new organization will co-operate with various sectional conferences which are already in existence, such as the Trans-Atlantic Passenger Conference in New York, and the Mediterranean-Atlantic Conference, which includes mainly Italian and British lines.

The Atlantic Conference, as at present constituted, includes the following lines: Cunard, White Star, Allan, Canadian Pacific, Anchor, Anchor-Donaldson and Dominion (British); American Line (American); Compagnie Generale Trans-Atlantique (French).

BUNKER FUEL AND STORES TO BE LICENSED BY BOARD

DIRECTOR L. L. Richards, Bureau of Transportation, War Trade Board, has issued a statement relative to the granting of licenses for bunker fuel, port, sea and ship's stores and supplies for all vessels clearing from ports of the United States or its possessions; the rules and regulations of special interest to the owners, agents and time charterers of vessels, to ship brokers, and the public being embodied in "General Rules No. 1."

On and after Feb. 1, 1918, it will be necessary, in complying with General Rules No. 1, for the owner or time charterer of vessels under the flag of a neutral country to sign an agreement binding himself and those who control his vessels to observe and comply with the General Rules. The rules have been sent to vessel owners and other interested parties and may be secured upon application to the War Trade Board, Bureau of Transportation, Washington, D.C.

The owner or time charterer of a vessel under the flag of a neutral country can on and after Feb. 1, 1918, secure a license for bunker fuel, port, sea, and ship's stores and supplies only after he has given to the War Trade Board, Bureau of Transportation, Washington, D.C., the detailed facts required by the board regarding all vessels owned, managed, chartered, or controlled by the owner or charterer of the vessel for which bunker fuel, port, sea, and ship's stores and supplies are desired.

The foreign owner or charterer of vessels under the flag of a neutral country must vest some one in the United States with authority to enter into an agreement with the War Trade Board, Bureau of Transportation, binding the subscriber to the agreement to comply with and be governed by all the regulations set forth in General Rules No. 1.

The Bureau of Transportation is desirous of doing everything possibe to avoid the detention of vessels when these rules are put into force; for this reason the owners and time charterers of vessels are being given advance notice of the rules in order that they may take such steps as will enable them to secure promptly such licenses for bunker fuel, port, sea, and ship's stores and supplies as it will be the policy of the War Trade Board to grant.

No vesel is allowed to clear from any port of the United States or its possessions without having secured a license or licenses covering all the bunker fuel, port, sea, and ship's stores and supplies aboard the vessel at the time of sailing. The license or licenses must cover not only the fuel, stores, and supplies taken aboard at the port of the United States, but also the fuel, stores, and supplies which the vessel brought into the country when she entered.

It is emphasized that voyages and charters for all neutral vessels and all American vessels not requisitioned by the United States Shipping Board should first be approved by the Chartering Committee of the United States Shipping Board at New York. This Committee will then inform the War Trade Board of the approvals they have given, so that the War Trade Board will have this information in hand when considering each application for license. The above method, if carefully followed, will avoid delays in granting licenses.

It is also to be noted that the rules provide that no application shall be granted for bunker fuel, port, sea, and ship's stores and supplies for a sailing vessel for a voyage into the war zone. This is to prevent the avoidable loss of tonnage.

MATERIAL FROM U.S. FOR SHIPBUILDING

ARRANGEMENTS have been made, it is understood, whereby a supply of ship plates and steel shapes for the construction of the first vessels to be laid down under the Dominion Government's recently announced shipbuilding programme will be procured from the United States. In view of the present lack in Canada of facilities for the rolling of such material, it was necessary to enter into negotiations to obtain it from the United States.

NOTES ON EXPORT PRACTICE

By "Commerce."

IN connection with the f.o.b., f.a.s. and c.i.f. clauses in quotations on goods intended for the export market considerable misunderstanding sometimes arises. We frequently see the term "F.o.b. factory," which means little or nothing to the foreign buyer except that the seller is not particularly anxious for the business, as he does not trouble to give his customer any accurate idea of what the goods will actually cost him when delivered. F.o.b. meaning "free on board," should only be used in connection with export business where the goods are actually delivered stowed in the hold of the vessel. F.a.s. signifies "free alongside," and is somewhat vague in meaning, as it might indicate that demurrage charges on the cars or lighters were up to the buyer. Strictly speaking, this term is the same as f.o.b. minus all dock and loading charges, cranage, primage, etc.

By Way of Illustration

Let us take the case of a manufacturer of woodworking machinery in Toronto quoting against an inquiry from Buenos Aires. A regular line of steamers runs from both New York and Philadelphia, and one of these ports would provide the best route. There are several ways in which the quotation might be made:

F.o.b. Toronto.—This phrase implies, though it does not actually state, that the shipment will be loaded on cars routed for New York or Philadelphia. Every kind of charge, such as rail, freight, insurance, dock charges and all subsequent costs, are to be borne by the buyer. Now it is easy to see that the South American importer will have a most painful time figuring out just what this machinery will cost him by the time it is installed in his factory.

F.a.s. New York.—In this case the shipper undertakes to pay rail freight and lighterage right into the steamship dock. In the case of goods arriving in New York on the Jersey shore, they are lightered across to the steamship dock free of charge by the railway company; nevertheless it is often more convenient and quicker for the shipper to make his own arrangements with a private lighterage company to handle these goods across the North River. However, in either case the charges are all up to the seller of the machines until the cases are placed upon the pier.

F.o.b. Steamer, New York.—This is certainly the most satisfactory way to make quotations, as both parties know exactly how far each is liable. The shipper has to have the goods stowed in the hold, as shown by the bill of lading, which is recognized by the bank as a negotiable receipt for the shipment. In quoting on this basis the seller should add to his offer three important pieces of information:

(i.) Size and number of cases.
(ii.) Prevailing freight rate.
(iii.) Prevailing marine insurance rate.

On this basis the foreign buyer can tell exactly what the machinery is going to cost him laid down by simple addition.

Increased Responsibility

C.i.f. Buenos Aires.—Quoting on this basis carries the responsibility of the shipper a little farther. Cost, insurance and freight to Buenos Aires means exactly what it says. However, there is sometimes a little misunderstanding in regard to the marine insurance. Shippers sometimes imagine that because they contract to insure the shipment up to the port of arrival that they are liable for its safe transit. Nothing, of course, is farther from the truth; the shippers' responsibility ceases as soon as the goods are stowed in the hold, exactly as in the case of an f.o.b. sale. In addition, he merely contracts to pay the ocean freight and insurance to destination. We are considering the case of a shipment to the Argentine and the laws of the republic hold that the insurance in this case must include war risk. Now in shipments to most other countries at the present time it is legitimate for the shipper to add the war risk premium to his invoices, but not in the case of the Argentine. Certain Customs authorities call for a definite certified statement of the freight charges on the invoice; this is necessary to enable them to estimate the "home market value" of the goods and to arrive at the ad valorem duty. It is always wise to include such a statement, whether specifically required or not.

F.o.b. Buenos Aires.—The buyer will sometimes request all quotations to be put in this form. It would not, however, be to the advantage of any shipper to quote f.o.b. destination, because the phrase is too elastic and might include many large and unforeseen charges. It would certainly be held to cover duty and cost of unloading, sundry Customs charges and cartage or teaming. The best alternative would be for the shipper to quote c.i.f. and leave all such local charges to the buyer.

Export Licenses

At the present time all the great exporting countries have embargoes in force covering practically all metals and manufactures of metal. The embargo is designed primarily to keep such material out of enemy countries and further to control the output in the event of there being a shortage of the commodity. Where good reason exists and there is no shortage export licenses or permits are issued to approved persons at the discretion of the authorities. In Canada the Chief Commissioner of Customs at Ottawa issues a license to approved firms for the export to allied countries of embargoed metals and manufactures, provided that a declaration is produced from both the shipper and the buyer that the material will in no case be converted to enemy use. This license attached to the export documents is scrutinized by the Customs officer at the point of export and the inspector has authority to hold any shipments not properly described or covered by a license.

A similar arrangement is in force in the United States. In respect of iron or steel in various forms the embargo is more complete, as no licenses are being issued for the export of these goods unless it is shown that they are primarily intended for the manufacture, directly or indirectly, of war material for the allies. Other materials, such as non ferrous metals, oil, gasoline, coal and coke, fertilizers and explosives, may be exported for ordinary industrial use subject to license, but, of course, in the event of any shortage becoming apparent the same restrictions would probably be applied as already cover iron and steel. Permits for the export of these goods to Canada are to be obtained on application to the Division of Export Licenses, 1435 K Street, Washington, D.C. Any attempt to evade this law renders the parties concerned to a fine of $10,000, or two years' imprisonment, so it will be seen that the Washington Government is taking a very strong stand against the possible shipment of goods for enemy use.

In Great Britain controllers have been appointed to deal with different raw materials, and each of them has authority to issue permits for the export of his particular commodity, provided (i.) that a surplus of such material exists in the country, and (ii.) that shipping space is available for its transport. The chief difficulty which arises is that a machine or other article contains materials coming into the province of two or more different controllers, in which case it is not easy to obtain the endorsation of all these different officials to the same permit.

INCREASED USE OF "ISHERWOOD" SYSTEM OF SHIP CONSTRUCTION

ACCORDING to a recent statement issued by J. W. Isherwood, of London, Eng., an increase in the number of ships contracted for under his system of construction is recorded during the past year, despite the almost total abolition of building for private ownership.

No less than 180 vessels, aggregating in deadweight carrying capacity about 1,334,000 tons, have been placed since the end of 1916. The grand total now stands at 800 vessels, aggregating 6,000,000 tons deadweight carrying capacity, as against 6 vessels, aggregating 31,608 tons in 1908. Of the 800 ships referred to, there are 386 general cargo vessels, colliers and ore steamers, and passenger vessels aggregating 2,891,600 tons deadweight carrying capacity, 24 Great Lakes freighters for 279,600 tons, 310 oil-tank steamers for 2,800,000 tons, 77 barges for 26,800 tons, 2 dredgers for 760 gross register tons, and 1 trawler for 570 gross register tons. The use of the system is greatly on the increase in the United States and in Japan.

160 Brake Horse Power Semi-Diesel Marine Oil Engine

The following article and illustrations pertain to a type of prime mover which has demonstrated its suitability for developing power such as is required for coasting vessels and fishing boats. By operating on the stroke-stroke cycle, the required power is developed at a relatively low speed and the resulting propeller efficiency makes this type of motor of interest to users of commercial craft. Constructional and operating features are fully dealt with.

ONE of the outstanding developments of recent times in marine propulsion is undoubtedly the greatly extending sphere of application of the semi-Diesel or hot-bulb type of oil engine, together with the rapid increase in power developed per unit. Quite recently this type of engine was universally regarded as being suitable for the development of only relatively small powers, say, 25 brake horse-power to 30 brake horse-power per cylinder, but recent modifications in design and the virtues of rigid simplicity have brought it greatly into favor even for powers up to 130 brake horse-power per cylinder.

Field of Application

The great majority of marine semi-Diesel engines have points of similarity in the system upon which they work, and practically all are of the two-stroke cycle, utilising the enclosed crank chamber for the compression of scavenging air. The Beardmore engine has been designed essentially to meet the peculiar requirements of marine propulsion, and is the result of much experience with this type of engine installed in yachts, coasters, lighters and fishing vessels.

It is particularly suited to consume fuels ranging from 0.8 to 0.9 specific gravity, but can be adjusted to use either slightly lighter or heavier oils. Working on the two-stroke cycle, it develops 160-brake horse-power while running at 280 r.p.m. This relatively slow speed of revolution is, of course, conducive to high propeller efficiency. As all bearing surfaces are liberal and materials carefully selected an outstanding feature of the engine is its capacity for long and hard service. The engine is directly reversible by means of compressed air, and requires no disconnecting clutch between the engine and propeller, the latter advantage being one that appeals very strongly to marine engineers in general and oil-engine users in particular. Fig. 1 gives sections and elevations of the engine, Fig. 3 details of the main valves, Fig. 4 diagrams from the cylinder and crank chamber, and Figs. 5 to 8 are general arrangements of an installation on board ship.

Chief Features

In the main engine there are four cylinders, each 11 in. diameter by 15 in. stroke, and the cycle of operations may be briefly described as follows, the reference numbers being those shown on the section of one of the cylinders in Fig. 1:—During the upward stroke of the piston air is drawn through non-return valves into the crank chamber 1, to be compressed on the downward stroke and admitted into the cylinder at the correct moment through scavenging air ports located directly opposite to the exhaust ports 2. These ports are covered and uncovered by the piston, on the top of which a deflector is arranged to give to the air blast the direction most suitable for efficient scavenging. While the under side of the piston is performing the functions of a scavenging air pump the upper side compresses the contents of the previously scavenged

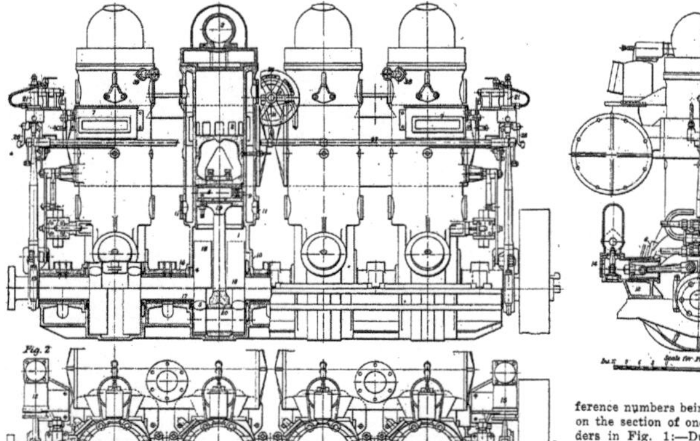

FIG. 1. ELEVATION PLAN AND END VIEW OF "BEARDMORE" MARINE OIL ENGINE.

cylinder on the upward stroke in preparation for the downward or power stroke.

A separate fuel-pump is provided for each cylinder, and once in each revolution of the engine fuel is sprayed directly into the combustion chamber 3, through pulverising nozzles fitted with automatic check valves, and ignition is

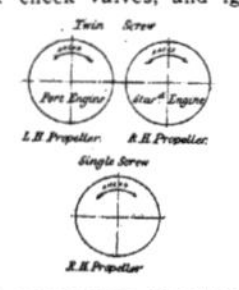

FIG. 2. DIRECTIONS OF ROTATION LOOKING FORWARD.

obtained through the fuel impinging on on the heated surface of the combustion chamber.

The combined heat of compression and combustion sustains the temperature necesary for ignition, and consequently no blow lamps are required except for a few minutes before starting the engine. The various operations of working cycle can be clearly understood from an inspection of the indicator diagrams taken from the cylinder and crank chamber and reproduced in Fig. 4' and the sequence and correlation of the operations will be appreciated. These diagrams show the excellent results obtainable by careful design based on extensive experience.

Scavenging Arrangements

As already stated, scanvenging air is compressed in the enclosed crank chamber, and in order to obtain the necessary quantity of air at a suitable pressure for efficent scavenging the volume of the crank chamber is reduced to a minimum and the whole is made as nearly as possible airtight. The latter condition is difficult to attain where the shaft passes through the walls of the crank chamber, and the method adopted on this engine is interesting. A bronze ring, 4, Fig. 1, is carefully bedded to the body of the shaft close to the crank-web, and is pressed against a machined surface on the inner side of the crank chamber by springs, 5, between the ring and crank-web. The ring revolves with the shaft, and consequently wear is confined to the flat surface and automatically corrected by the springs, which exert sufficient pressure to prevent the ring leaving its face when the air pressure in the crank chamber is below that of the atmosphere.

Since inefficient scavenging is necessarily accompanied by loss of power and waste of fuel, the problem of rendering the crank chamber airtight is of the utmost importance and has received much attention from engineers, but perfect airtightness has not yet been achieved. Channels are therefore provided on this engine between the main-bearing cap and the side of the crank chamber to allow the free escape of any trifling leakage past the ring, and thus to prevent the possibility of interference with the lubrication of the main-bearing bush.

Air is admitted to the crank chamber through grids at the front and back covered by flexible plate valves backed by light springs, and these are mounted on large doors, 6, Fig. 8, the removal of which gives easy access to the crank-pin bushes.

Lubrication Problem

It will be obvious that with this enclosed type of crank chamber, which is utilised as a scavenging pump, forced lubrication cannot be applied to the crank-pin bearings, because a large quantity of the overflow would be taken up by the air and eventually find its way into the working cylinders, with the attendant result of excessive consumption of lubricant. The question of lubrication, therefore, is a difficult one, but has been solved in this engine in a most satisfactory manner. A forced sight-feed system, with separate plunger pump and oil tube to each point where lubrication must be positive and regular, has been adopted, and by this means the quantity of lubricant supplied to any point is definitely controlled by the attendant, while any interruption of the supply due to a choked tube or faulty valve can be instantly detected and the sight glass shows clearly which point is affected.

Two eight-feed lubricator boxes, 7, Fig. 1, are employed on this four-cylinder engine, and the reciprocating drive is obtained from the fuel-pump actuating gear. The 16 feeds are distributed as follow: three to each cylinder and one to each crank-pin. The cylinder feeds are disposed in such a way as to lubricate the entire working surfaces of the piston and cylinder wall, while one feed also serves to lubricate the gudgeon pin.

The gudgeon pin is hollow, as shown at 8, and is secured in place by a screwed pin, which is split at the point to prevent the possibility of falling out. At the other end of the hollow gudgeon pin is a plunger with a scoop end, 9, which is forced against the cylinder wall by means of a spring and collects oil from the lubricating tube already mentioned. The feed for the crank-pin is carried through the side of the crank chamber and led to the interior of a revolving ring, 10, from which the oil is carried by centrifugal force to its destination. The lubrication of the main bearings is effected by means of siphon tubes—an accepted marine practice.

Cylinder Construction

The cylinders are cast separately and are extended below the working barrels to form the upper halves of the crank chambers. The cylinder bodies are completely encircled by ample water jackets, and large access doors, 11, are provided to facilitate the removal of any deposit which may be formed therein. The cylinder covers and silencers are also jacketed, and the circulating water is supplied by a reciprocating pump, 12, placed at one end of the bedplate and actuated by an eccentric, 13, on the crankshaft.

The pump body is of cast iron lined with brass and is provided with brass

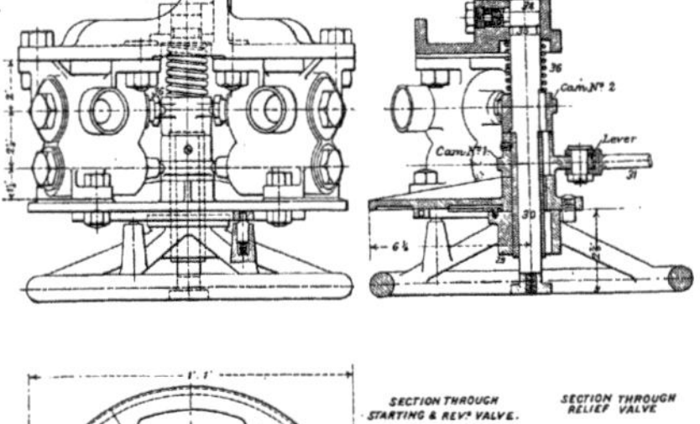

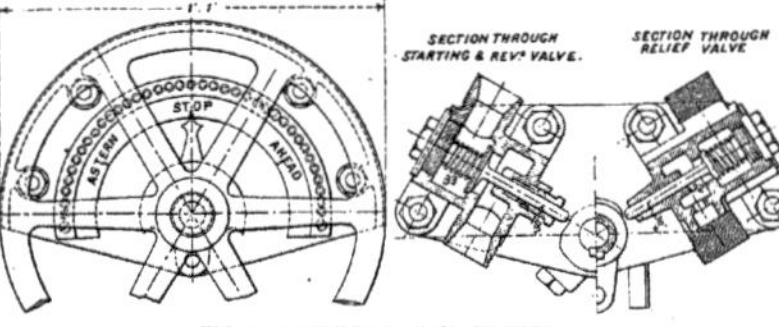

FIG. 3. DETAILS OF MAIN VALVES.

non-return suction and delivery valves, spring-loaded escape valve and a large cast-iron air vessel. Access to the suction valve is obtained by removing door 14, and to the delivery valve by removing the air vessel. A bilge pump, 15, of the same size and design, is placed at the other end of the bedplate and similarly operated.

The bedplate is cast in one piece, with suitable flanges for bolting down to the seating in the ship, and is provided with machined recesses to receive the main bearing bushes. The main bearing caps, 16, Fig. 1, are readily accessible for removal, and when removed the bushes, 17, which are made eccentric to the shaft on their outer diameter, can easily be slipped from under the shaft without disturbing any gear whatever. The bushes are of Admiralty quality gun-metal with white metal lining.

Crank Shaft and Rods

The crankshaft is a solid forging of Siemens-Martin steel and is machined all over. Large balance-weights, 18, Fig. 1, are secured to the crank-webs, and these not only give to the engine an evenness of turning moment and a desirable freedom from vibration, but also serve to reduce the clearance volume of the crank chamber, and thus increase the scavenging efficiency.

The connecting rods are machined steel forgings fitted with adjustable bearings at top and bottom ends. The top end, 19, of each rod is of special design to allow of easy adjustment, and the bearing surface is of phosphor-bronze. Examination of the gudgeon pin and brush, also piston and piston rings, is obtained by removing the cylinder cover and drawing the piston upwards after the lower end of the connecting rod has been disconnected. The marine-type crank-pin bush, 20, is of Admiralty quality gun-metal, with white metal lining, and is accessible for examination or adjustment through the doors, 6, arranged at the front and the back of the crank chamber.

Fuel Pumps

The four fuel pumps, 21, are grouped in pairs at opposite ends of the engine and are operated by the eccentrics, 13, previously mentioned in connection with the water pumps, a horizontal rocking lever, 22, fulcrummed at its centre, being employed to carry the fuel-pump strikers, 23. The stroke of the strikers is, of course, very much in exces of that to be communicated to the fuel-pump plungers, and consequently the discharge of fuel to the combustion chambers is restricted to a very short period. The idle portion of the stroke is utilised for governing purposes, and as the action of one is identical with the others it will be sufficent to describe one. The pump striker is pivoted on the rocking lever and is free to rotate but for the action of a light spring, 24, drawing it towards its guide. On the face of the guide a step is formed which throws the striker off its original course, but as this effect is counteracted by the spring already

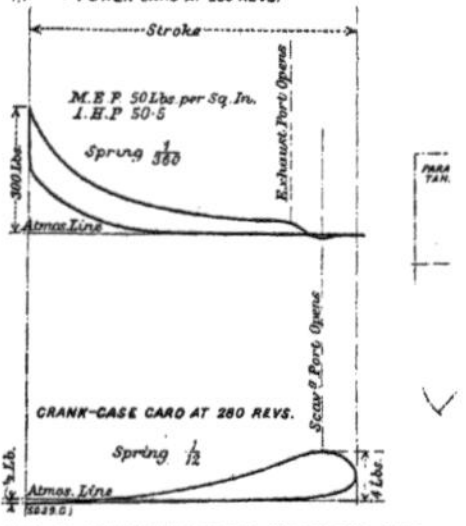

FIG. 4. DIAGRAMS FROM CYLINDER AND CRANK CASE.

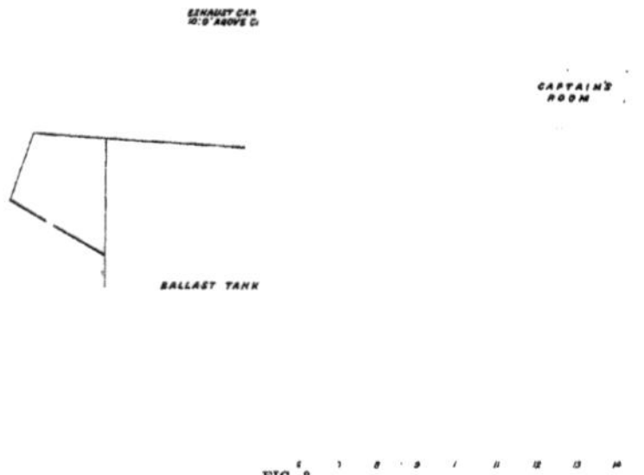

FIG. 8.

mentioned the engine speed determines whether the striker shall hit or miss the pump plunger.

The tension of the spring can be adjusted to cut out at any engine speed, and when that adjustment has been made the following conditions are fulfilled:—At all engine speeds up to and including a predetermined limit the fuel pump delivers a definite quantity of fuel during a definite fraction of each engine stroke, the quantity delivered being entirely under the control of the attendant on the starting platform, and when the engine speed exceeds the predetermined limit the pump automatically ceases to operate until the normal speed is regained. When it is remembered that all four fuel pumps are alike it will be readily understood that the system of governing is very sensitive and quick in action, and it is worthy of note that the tension of the governor springs can be adjusted either individually or collectively while the engine is running.

The maximum quantity of fuel to be delivered at full power is determined by a screwed collar, 25, Fig. 1, on the lower end of the pump plunger. By turning this collar to right or left the plunger is raised or lowered relatively to the striker and consequently the stroke of the plunger is limited, but this adjustment, of course, is only required when tuning the engine to suit a given quality of fuel and should not be used for any other purpose.

Engine Control

The ordinary control of the engine speed is effected through a hand wheel, 26, Fig. 1, placed at the centre of the engine. The fuel-pump strikers are mounted on eccentrics in the rocking lever, and these eccentrics are connected by a system of levers and links to a horizontal shaft, 27, running along the front of the engine. Any rotary movement of this shaft is therefore transmitted to the eccentrics, and the strikers are thus raised or lowered collectively according to the direction of movement imparted to the horizontal shaft by the hand wheel. Very fine graduations of the engine speed are obtainable in this way. A priming handle, 28, is fitted to each fuel pump for charging the pipes before starting, and any pump can be put out of action instantaneously by raising the handle to a horizontal position.

Should it become necessary for a pump to be put out of action permanently the screwed collar already referred to in connection with tuning meets this requirement, and when thus cut out the suction and delivery valves and springs can be removed for examination by simply disconnecting the delivery pipe and screwing out the plug in the top of the pump barrel.

Starting and reversing are effected by means of compressed air and the details of the gear, Fig. 3 show how all the operations of the engine are controlled by a single hand wheel. The hand wheel is mounted on a sleeve, 29, which also carries a lever and cam No. 1, while a central spindle, 30, passing through the sleeve, carries cam No. 2. The lever is connected by a link, 31, Fig. 1 and 3 to the fuel-pump controlling shaft, 27, Figs 1 and 3, the cam No. 1 opens one of the two relief valves, 32, Fig. 3, and the cam No. 2 opens one of the two high-pressure air valves, 33, Fig. 3. The air valves are directly connected with the compressed-air reservoirs, Figs. 5 and 7 on the one side and with mechanically-operated distributing valves on the other side.

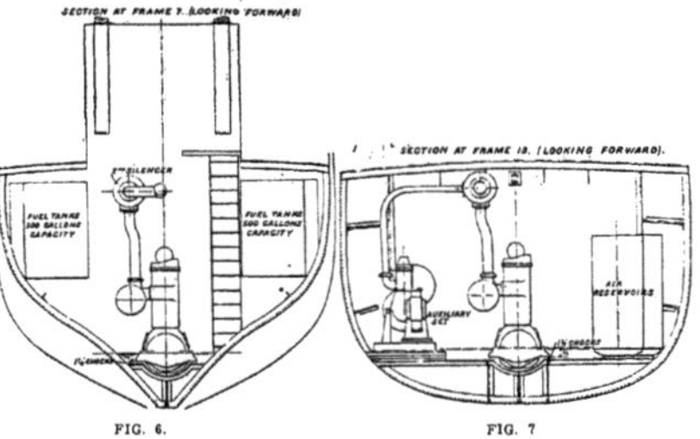

FIG. 6. FIG. 7

Manœuvre Power

It will now be understood that when the hand wheel is turned clockwise the cam No. 2 opens the air valve on the left and admits compressed air to the ahead distributing valves, the cam No. 1 opens the relief valve on the right and places the stern distributing valves in communication with the atmosphere, while the lever brings the fuel pumps into action. The engine then starts on compressed air, but as the fuel pumps operate immediately the engine starts, the air connection can be closed on the completion of one revolution. The latter is the function of the central spindle which carries the cam No. 2. By pushing the central spindle towards the back, the air-valve spindle drops off the cam and a spring plunger, 32 Fig. 3, entering a groove, 35, in the central spindle, locks it in the new position. The hand wheel can now be used to regulate the speed of the engine without having any effect on the closed air valves, but whenever the wheel is placed in the stop position the central spindle is automatically released and is returned by a spring, 36, to its original position in readiness for a fresh start in either direction.

A start in the astern direction is effected by turning the hand wheel anti-clockwise, and thus placing the astern distributing valves in communication with the compressed air reservoirs and opening the ahead distributing valves to the atmosphere. Moving the hand wheel in either direction from the stop position brings the fuel pumps into action, and it is noteworthy that the driver's attention is confined to two simple operations, viz., turning the hand wheel in the proper direction and pushing the central spindle when the engine has started.

The air-distributing valves, 37, Fig. 1, are actuated by the eccentrics provided for operating the water and fuel pumps, but are brought into operation only when compressed air is admitted to them, and consequently there is no wear and tear of the reversing mechanism while the engine is running. It is an admirable feature of this gear that no part of it is in motion except for a few seconds during each start or reverse of the engine, and it is therefore an easy matter to withdraw valves and other parts for examination or overhaul while the engine is at work.

Two of the engine cylinders are fitted with spring-loaded non-return valves, 38, Fig. 1, through which the air reservoirs can be charged while the engine is running. A sufficient supply can be maintained in this way for all ordinary purposes, but an independent power-driven air compressor is supplied to meet the requirements of excessive manoeuvring and to renew the supply should it be lost from any cause.

Ship Installations

Figs. 5 to 8 show a 160 brake horse-power engine installed in a typical British coaster, 75 ft. long between perpendiculars, 18 ft. beam, and 8. ft. 6 in. draught. The auxiliary set consists of an air compressor and general service water pump driven by an independent oil engine, which operates on the same fuel as the main engine. The compressor is capable of charging the air reservoirs to a pressure of 350 lb. per square inch, although the main engine can be started or reversed by air at 100 lb. pressure, and this allows air to be stored in excess of the quantity required for ordinary purposes. In fact the air reservoirs are of such capacity that ordinary manoeuvring requirements can be fully met by one while the other is kept charged for emergencies.

The water pump is capable of performing all the duties of the main bilge and circulating pumps, and is also arranged to deal with water ballast. Fuel is stored in cylindrical tanks having a total capacity of 1,000 gallons, and this quantity gives the vessel an acting radius of 750 nautical miles. The propeller is of cast iron and has three blades of large surface, the propeller shaft is of mild steel encased in a continuous gun-metal liner, and the stern tube is of cast iron fitted with renewable lignum-vitae bearings.

The thrust block is of the most approved marine type, having individually adjustable shoes faced with white metal and a shaft-bearing at each end. The engine in this ship is 10 ft. 5 in. long from the forward face of the flywheel to the aft face of the crankshaft coupling, 5 ft. 4 in. high from centre of shaft to top of combustion chambers, and weighs 145 cwt. with silencers, flywheel, mountings and piping. The weight of the complete installation, including fuel tanks, floor plates, pipes and connections. is 14 tons. and the engine-room bulkheads are 15 ft. apart.

We are indebted to "Engineering" for the data and illustrations contained in this article.

There is an amusing story going round the British Grand Fleet, just the kind of joke which Jack Tar likes to give and take with his friends. It is quite seriously affirmed that when an American squadron, consisting of the U.S.S. Delaware, New York, and Wyoming, with destroyers and other craft came up the Firth, the British flagship signalled to them: "You are to anchor west of the Forth Bridge." But the Americans passed under the bridge and sailed on. Shortly the British admiral made another signal: "We signalled just now that you were to anchor west of the Forth Bridge; why don't you stop?" And the American flagship immediately signalled the reply: "Well, I guess we have only passed one bridge as yet."

NAVIGATION CLASSES AT QUEEN'S UNIVERSITY, KINGSTON, ONT.

THE third three-month session of the School of Navigation at Queen's University, which opened the middle of December last, is open without charge to all seamen who desire to be instructed in navigation and seamanship, and will prepare candidates for the examinations for certificates as coasting master, coasting mate, inland master, inland mate, minor master, tug-boat captain.

All work necessary for the various grades is taught, the school being in charge of Capt. W. Steeve, who has not only had an extensive experience as master, but has also been highly successful in training for the above-mentioned certificates.

Lectures and Examination

During the course of the session a series of lectures on terrestrial magnetism, atmospheric electricity, and storms and weather predictions will be given by A. L. Clark, Ph.D., Professor of Physics, and also a series of lectures on practical astronomy by D. Buchanan, Ph.D., Professor of Astronomy and Mathematics. The examinations for the different certificates are conducted in Kingston by Capt. H. W. King, Dominion supervisor of marine examinations.

Subjects of Instruction

Many seamen have been away from school so many years that they have forgotten much that they knew when they were young. Many men have had to leave school at an early age, so that they have not had a chance to get a good education. Candidates who need to "brush up" their arithmetic and English will be taught these subjects in a straightforward, practical way. The arithmetic will be the kind that is useful to seamen and the English will be the spelling and writing that are necessary to pass the examinations for certificates. If a man who is studying for a certificate is "well up" in arithmetic and English he will not need to take any preparatory work, but most men who go up for their first certificate will be glad to know they can get instruction in these two subjects.

Master's Classes

In navigation the pupils will study the following subjects thoroughly:—How to work a mercator and an amplitude and to find the deviation of the compass by "Time Azimuth," tables, ocean currents, calculations for "dead reckoning," uses and adjustments of the sextant, deviation of the compass, and the international code signals.

In seamanship the subjects taken up are as follows:—How to rig a sea anchor and how to handle disabled vessels, the use of the lead in heavy weather, the mortar and rocket apparatus, how to rig jury rudders, instruction in averages, bottomry bond, charter party, bills of lading, etc.

Mate's Classes

In seamanship the student will be taught thoroughly the rule of the road as it applies to both steamers and sailing vessels, the regulation lights, and signals for fog and distress. He will be instructed practically in the use of the lead, the log, in rigging and stowing cargo. He will also be taught the necessary things about "fore and aft rigged" sailing vessels which he needs to know to get his certificate. If a man is studying to pass for a position on a steamer he will be instructed in the various fittings for fire protection, the working of collision bulk-heads or water-tight compartments, the construction and use of the marine telegraph, the fitting and lowering of lifeboats, and other things which are required in the examinations.

In navigation the student will be shown how to work a day's work, how to find the latitude by the meridian altitude of the sun, how to take a bearing by compass and determine the ship's position on a chart, and how to shape a course and determine the distance run from any given departure.

Inland Waters Classes

For the men who wish to take examinations for certificates on steam or sailing vessels plying on inland waters and minor waters, the different subjects will be taught more simply to prepare students according to the several requirements under the provisions of the Canada Shipping Act.

The following extracts from Canadian regulations relate to the examination of masters and mates of coasting and inland vessels:

Coasting Qualifications

Mate.—A candidate must not be less than nineteen years of age and must have served two years at sea, or,—

(b) He must have served one year as mate of a passenger or freight steamer on the Great Lakes whilst holding a certificate of competency as mate of a passenger steamer on the inland waters.

In Navigation.—A candidate for mate's certificate will be required:—

(a) To write a legible hand and spell correctly; to be tested by a quarter hour's dictation.

(b) To work a day's work.

(c) To find the latitude by the meridian altitude of the sun.

(d) To take a bearing by compass and determine his position by cross bearings on a true or magnetic chart and to shape a course and determine the distance run from any given departure.

In Seamanship.—He must possess a thorough knowledge of the rule of the road as regards both steamers and sailing vessels, their regulation lights and fog, sound and distress signals.

(a) He will be required to repeat from memory any one, or all, of the articles '1' to '32' inclusive, of the rule of the road, and write the answers to not less than ten questions on the rule of the road which the examiner will give him.

(b) He must understand the lead, log, knotting, splicing, rigging and stowing cargo.

duties of master which the examiner may think necessary to ask.

He will also be required to answer in writing on paper which the examiner will give him the following questions, numbering each answer to correspond with the question:—

1. When taking a meridian altitude, how do you know when the sun is on the meridian, or, in other words, when it is noon?

2. How does the sun bear (true or magnetic) when on the meridian of the observer in these latitudes?

3. What do you mean by the deviation of the compass and how is it caused?

4. Having determined the deviation, how do you know when it is easterly and when westerly?

5. Supposing the sun when on the meridian bore . . . by your compass, what would be the deviation of that compass for the direction of the ship's head at the time, the variation given on the chart being . . . .

6. How could you find the deviation of your compass when in port or when sailing along a coast?

7. Name some suitable objects by which you could readily obtain the deviation of your compass when sailing along the coast of Nova Scotia, or the coast of places you have been accustomed to use.

8. The bearing of two objects, when in a line with each other, was found on the chart to be . . . . magnetic but when brought in a line on board they bore . . . . by your compass. Read the deviation of your compass for the direction of the ship's head at the time.

9. What means are there for checking the deviation of your compass by night?

10. Supposing the north star bore . . . by your compass, what would be the deviation of that compass for the direction of the ship's head at the time, supposing the variation given on the chart to be . . . ?

11. Do you expect the deviation to change? If so, state under what circumstances.

12. What is meant by the variation of the compass and what is the cause of it?

Inland Waters Qualifications

Mate.—A candidate must not be less than nineteen years of age and must have served two years at sea or on the inland waters, one year of which he must have been as wheelsman.

In Navigation.—A candidate for a mate's certificate will be required:—

(a) To write a legible hand and spell correctly, to be tested by one quarter hour's dictation.

(b) To determine his position by cross bearings on the chart.

(c) To shape a course and determine the distance run from any given departure.

In Seamanship.—He must possess a thorough knowledge of the rules of the road (International and for the Great Lakes) as regards both steamers and sailing vessels, their regulation lights, for sound and distress signals.

(a) He will be required to repeat from memory any one or all of the articles "1" to "32" inclusive, concerning the rules of the road, and write the answers to not less than ten questions on the rules of the road which the examiner will give him.

(b) He must understand the lead, log, knotting, splicing, rigging and stowing cargo.

(c) He will be examined in seamanship generally, either for "square rig," "fore-and-aft" or "steamer—passenger, freight or ferry," according to the certificate required.

If for steamer:—

(d) He must understand the fittings for fire purposes, bulkhead sluices, if any, and telegraph.

(e) He must understand the securing and lowering of life-boats and rafts.

(f) He will be required to answer any other questions appertaining to the duties of mate which the examiner may think necessary to ask.

Master.—A candidate must be not less than twenty-one years of age and must have been at sea or on the inland waters other than the minor waters three years, one year of which he must have served as mate whilst holding a mate's certicate.

In Navigation.—A candidate for a master's certificate, in addition to the qualifications for a mate, will be required:—

(a) To explain on a magnetic or true chart how to shape a course to counteract the effect of a given current and find the distance made good towards a given point in a certain time.

(b) To answer a few practical questions on the errors of the compass and their causes and the correction of the use of the magnets.

In Seamanship.—In addition to the qualifications for a mate:—

(a) He must know the principal lights on the inland waters, and the currents.

(b) He will be required to explain how he would lay out an anchor in case of stranding.

(c) How to rig a jury rudder.

(d) He will be questioned as to his knowledge of protests, invoices, charter party and bills of lading.

(e) He will be required to answer any other questions appertaining to the duties of master which the examiner may think necessary to ask.

PECULIAR RESULTS OF LOOSE PROPELLER

By A. F. Menzies

A STEEL passenger steamer, employed in the coastwise service on the Pacific coast met with an accident recently which had rather peculiar results The ship is 192 ft. long x 35 ft. beam x 14.9 ft. depth of hold, and is fitted with 18½ x 30" x 50" x 36" stroke triple expansion engines.

The engines had been running ahead for a short time and the vessel had considerable way on her when a full asterr order was given. On the engines being reversed they ran away with such violence that the turning wheel was thrown off, due, of course, to the effects of centrifugal force. The wheel was of the usual type, being made in halves and the two pieces bolted together on the shaft. Fortunately no one was injured by the flying pieces.

After the engines were stopped it was found to be impossible to move them and it was concluded that the shafts had

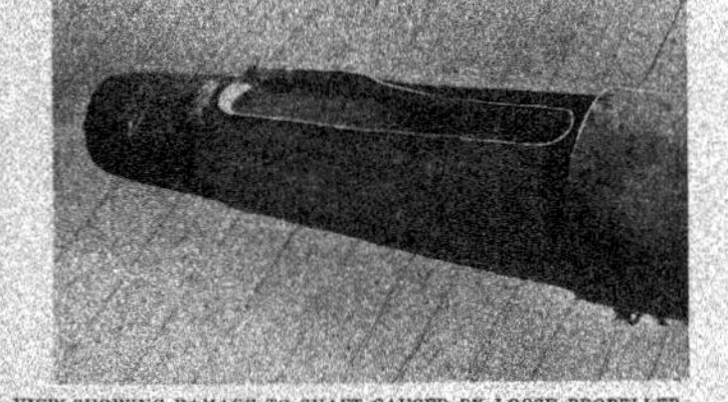

VIEW SHOWING DAMAGE TO SHAFT CAUSED BY LOOSE PROPELLER.

broken in the stern tube and become wedged, or that the wheel had come off and was jammed against the rudder post. When the vessel was docked the latter assumption was found to be correct.

The nut on the end of the tail shaft had evidently worked off and the first revolution astern had forced the propeller off, this phase would be assisted by the headway on the ship. With the propeller locked against the rudder post the shaft still revolving a violent blow was received every revolution when the key in the shaft struck the keyseat in the propeller. The photo of the shaft shows the tapered portion split open by these blows.

Lieut. S. S. Smith, formerly sales and credit manager for the Lake Winnipeg Shipping Co., Winnipeg, Man., who was reported missing early in June, is now officially presumed to have been killed in action. Lieut. Smith went overseas with a Winnipeg battalion in Oct., 1915.

VENTILATOR COWL QUERY

WE HAVE a number of ventilator cowls to make, similar to the one shown in the accompanying sketch. These are to be made of 16 gauge stock, in four sections, and with vertical seams. Owing to the necessity of hammering or swedging the pieces before the same are assembled, the writer would like to obtain some information on the development of the various portions on the flat sheet and the method adopted in the subsequent operations of forming the different sections to get an even and regular contour. Any data in this connection will be greatly appreciated.

• • •

METHOD OF LAYING OFF COWL

FIG. 1 shows a ventilator cowl built up of four pieces with vertical seams. This type of cowl is much preferable from its visual effect to that made up with girth seams, and where pleasure yachts, passenger vessels and Government or naval craft are concerned the plate arrangement here featured is usually insisted upon. Fig. 2 illustrates the construction of the cowl, the layout of which may be drawn to any desired scale, or be made apply to any required size pipe.

Measure off A B equal to 20 inches, bisect at P and raise the vertical line V P equal to 40 inches. Draw the overhang V C at right angles to P V and equal to 18½ inches. With E as centre and radius (equal to 25 inches) draw in the arc C K F J so as to be tangent to the perpendicular A J raised from A B. Raise the perpendicular B L M.

Locate the point G, which is 4 inches above the line A B and 2 inches from B L M. With G as centre and 2 inches radius, strike the quadrant M N D tangent to B L M. Directly above the point G locate D, which is on the quadrant M N D. Connect by a straight line the

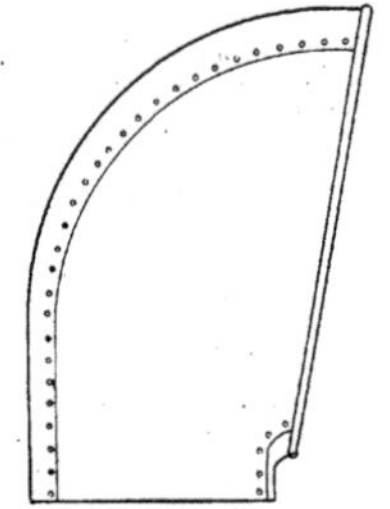

FIG. 1. VENTILATOR COWL WITH VERTICAL SEAMING.

points D and C, thus completing the outline of the cowl. Now divide the back into four equal parts as shown by A J F K C. Similarly divide the throat

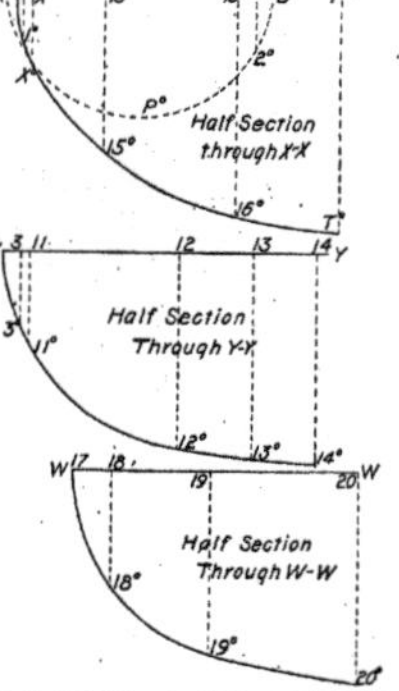

FIG. 2. GENERAL ANALYSIS OF VENTILATOR COWL CONSTRUCTION.

sheet line into the same number of equal parts as shown by the points B L M N, and D. Connect these points by straight lines to their relative points on the back sheet line, as J L, F M, K N. These lines will be used as sectional planes and on them will be constructed the half sectional views.

With P as centre and radius P A, strike the semicircle A P° B, thus defining the half-sectional view through the plane A B. Bisect J L at Q, then with Q as centre and radius Q J strike the semicircle J Q° L. Again, with R as centre, describe the semicircle F R° M. Similarly with S K and T C as radii, strike the half section views K S° N and C T° D respectively.

It is the usual practice in this type of cowl to make the back and throat sheets each equal to one-sixth of the circumference and the side sheets each equal to two-sixths. On this principle our patterns and templets will be developed. Divide the semicircle A P° B into exactly six equal parts and locate the end points 1° and 2°. At right angles to A B project the lines 1° 1 and 2° 2, thus locating the points for the rivet lines. Divide the arc J Q° L into six equal parts, the extreme points 3° and 4° being projected at right angles to the plane J Q L, thus locating the rivet line points 3 and 4. Similarly, the arcs F R° M, K S° N and C T° D are each divided into six equal parts, their extreme division points being projected at right angles to their respective planes, thus locating the rivet points 5 6 7 8 9 and 10.

Erect the perpendicular 1 3 to the line A B, and with a suitable centre as at U and radius U 9, describe an arc through the located points 9 7 5, and tangent to the perpendicular 1 3. The line 1 3 5 7 and 9 defines the rivet line for the back and side sheets. The rivet line for the throat and side sheets is drawn in a similar manner. Erect the perpendicular 2 6, and with a suitable centre draw the arc cutting through the points 6 8 4 and 10, and tangent 2 6.

For the development of the back sheet, take the distances A J, J F, F K and K C, transfer over to Fig. 3 and locate the points A J F K C on the straight line A C. Draw lines at right angles to A C through all these points. Measure off carefully along the curve the distance A 1°, Fig. 2, and transfer this over to Fig. 3 by measuring at each side of 1, locating the points 1° 1°, which will be equal to one-sixth of the circumference, through the plane having A B as diameter. Similarly measure J 3°, Fig. 2, and transfer to J 3°, Fig. 3. Again, transferring all the distances as F 5°, K 7° and C 9°, measured along the curve (or calculated), Fig. 2, to their respective locations in Fig. 3, an even curve is drawn in through these points by the aid of a flexible batten.

It will be observed that in Fig. 2 the line 1 3 5 7 9 is shorter than the line A J F K C; this, then, will have to be shown in Fig. 3. From the located point E, draw a straight line to F, locating the point F^1 where this line intersects the rivet line, Fig. 3. Measure carefully along the curve from this point F^1 to the point 9. This distance is then measured from 5° to 9^1 in Fig. 3. The distance F^1 to 1, Fig. 2, is next transferred to Fig. 3 as from 5° to 1^1. Suitable curves are drawn through the points 9^1 C 9^1 and 1^1 A 1^1 respectively. This completes the templet for the back sheet; laps, of course, to be added as required. If the reader is in doubt about the curves 9^1 C 9^1 and 1^1 A 1^1, perhaps it would be just as well to prolong the rivet line from 9^1 to 9° and from 1^1 to 1°, thus leaving the ends square. It will require in any case to be "dressed" after being "blocked" or hollowed out.

The throat sheet is developed in a manner similar to the back sheet. The distances in Fig. 2, B L and L M, M N and N D measured along the curve, are

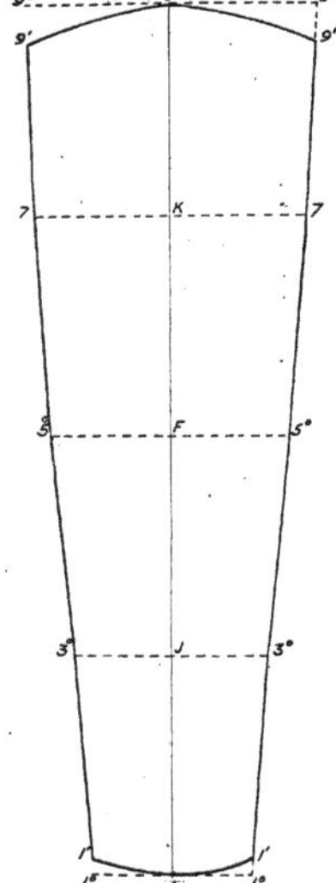

FIG. 3. DEVELOPMENT OF BACK SHEET.

transferred over to Fig. 4 and located on the straight line by the points B L M N D. Parallel lines at right angles to B D are drawn through these points. In Fig. 4, B 2° is made equal to B 2°, Fig. 2, which is also equal to A 1°, Figs. 2 and 3. Again, L 4°, Fig. 4, is made equal to L 4°, Fig. 2. Similarly the distance M 6°, N 8°, D 10°, Fig. 2, are transferred over to their respective positions in Fig. 4. It will be observed that the distances B 2°, L 4°, M 6°, etc., Fig. 4, are equal to the distances respectively A 1°, J 3°, F 5°, etc., in Fig. 3. Through these located points draw an even curve with the flexible wooden batten, as was done in Fig. 3.

It will be noticed that the rivet line 2 4 6 8 10, Fig. 2, is longer than the outline B L M N D. This extra length must be added on to the rivet line in Fig. 4. From the point G, through the centre of the quadrant M N D, draw the line G H. Measure from D to this intersection point on the quadrant, and transfer this distance over to Fig. 4, thus locating the line G H. From G^1, measured along the curve to 10, Fig. 2, transfer this distance to Fig. 4 by measuring from the line G H to 10^1. Similarly, transfer the distance G^1 to 2, Fig. 2, over to Fig. 4 as from the line G H to 2^1. Connect the points 2^1 B 1^1 and 10^1 D 10^1 each by suitable curves.

Fig. 4 shows the templet for the throat sheet, suitable laps to be added. The side sheets are developed in a slightly different manner. Transfer the outline of the side sheet as shown by 1 3 5 7 9 10 8 6 4 2, Fig. 2, to the similarly numbered outline in Fig. 5. Connect the points 1 P 2, 2 Q 4, 5 R 6, etc., Fig. 5 similar to Fig. 2. Measure off P $1''$ and P $2''$, Fig. 5, each equal to 1/6 of the circumference of which 2 B^1, Fig. 2, is the diameter, or measure along the semicircle the distance 1° 2°, Fig. 2, and place half of this at each side of point P, Fig. 5, to locate the points $1''$ and $2''$.

Calculate or measure off half the distance along the arc of 3° to 4°, Fig. 2, and place this distance at each side of the centre Q on line 3 Q 4, thus locating 3^2 and 4^2 in Fig. 5. Again, take the length of the arc 5° 6°, Fig. 2, and place half of this on each side of R on the extended line 5^2 R 6^2 to locate the points 5^2 and 6^2, Fig. 5. Similarly transfer the distances 7° 8° and 9° 8°, Fig. 2, to their relative positions to locate the points 7^2 9^1, and 8^2 $10''$. Connect the points $1''$ 3^2 5^2 7^2 9^1 and $2''$ 4^2 6^2 8^2 $10''$, as shown in Fig. 5. Measure the length of the line 1 3 5 7 and 9 and also the line $1''$ 3^2 5^2 7^2 and 9^1. Take half the distance between these two lengths and locate the points 9^2 and 1^2 by measuring from 9^1 and $1''$ respectively. Similarly measure to find out how much longer the line 2 4 6 8 10 is than the line $2''$ 4^2 6^2 $8^{2''}$. Take this difference and add half from $10''$ to 10^2 and half from $2''$ to 2^2. Connect 1^2 to 2^2 to locate the outline for the bottom of the side sheet.

To obtain the curved line 9^2 to 10^2, it is necessary to find out the points W'' T^1 and Y''. This is done in the following manner: Divide C D, Fig. 2, into four equal parts, and through these points draw the planes W W XX and YY parallel to the line A B. Project the points X^1 and X'' through, and at right angles to A B, thus locating the points X^1 and X'' on X^3 X^3, which is shown on, and is continued on each side of A and B. Where the line F R M intersects the line X X, locate the point 15. Also locate 16 where K S N intersects the line X X. T is already located on X X. Project these points 15 and 16, and T down to and at right angles to X^3 X^3 Continue these projections as shown in the half section view through X X. On the elevation, Fig. 2, raise the perpendicular 15, 15° to its intersection of its arc F R° M. Take this perpendicular length and transfer to the line 15, 15° in the half-

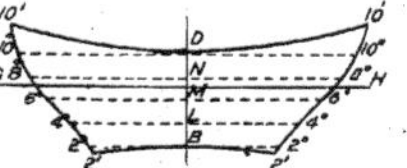

FIG. 4. TEMPLET FOR THROAT SHEET.

sectional view X X, Fig. 2. At right angles to K S N draw the perpendicular 16, 16° intersecting the semicircle K S° N at 16°. Transfer this length 16, 16° to 16 16° in the sectional view for this plane, which is X X. T T° is drawn at right angles to C T D, and its length transferred to T T° in the sectional view.

The distance X'' X° in the sectional view is equal to the distance through X'' on the line X X in the elevation. This distance is not shown in the elevation view, but it is obtained by using a little judgment. The distance through the point 5 is equal to 5 5° and through the point 3 is equal to 3 3°, so if X'' was midway between the points 3 and 5, it would be equal to the mean of the two distances 3 3° and 5 5°. But X'' is nearer to 5 than 3, so it will be seen that it is not difficult to get the distance X'' X° for the cross-section. An even curve drawn through the points T° 16° 15° 5° and X^1, in the section view, will define the half-section for the plane X X.

The sections for YY and W W are constructed in a similar manner.

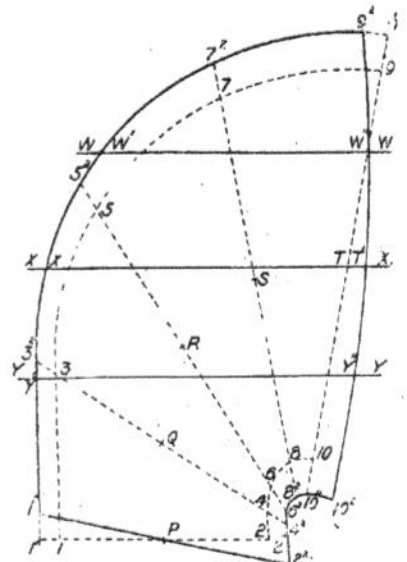
FIG. 5. DEVELOPMENT OF TEMPLETS FOR SIDE SHEETS.

The plane lines WW, X X and Y Y in Fig. 2 are transferred over to their proper positions in Fig. 5. The distance X° 15° 16° and T° in the sectional view, Fig. 2, is equal to the curve of the side sheet; it is, therefore transferred over to Fig. 5, measuring from the point X^1 along the line X X to the point T^1. With the three points 9^2, T^1 and 10^2 located, the curve may be drawn in by a flexible batten. The points W" and Y" on this curve are located by taking the length of the curve 3° 11° 12° 13° and 14°, Fig. 2, and measuring this along the line Y Y, Fig. 5, from the point Y' to locate Y". To locate W", the distance 18° 19° and 20°, Fig. 2, is measured from W' along the line W W, Fig. 5. Five points are shown to locate the curve 9' W" T' Y" and 10' and, although three points would be quite sufficient, the number is left to the option of the reader. If three points are chosen, it will be then only necessary to construct the cross-sectional view through X X in Fig. 2.

The heavy lines in Fig. 5 show outline of the templet, minus the laps, for the side sheet.

Sets or gauges of the sections of the various planes will be used for trying out the templets, during the process of hollowing or blocking out.

The continuation of the cowl below the line A B is a straight cylinder and is developed as such. This cylindrical course is, of course, fitted inside the cowl for water-shed purposes, using the line A B as the girth seam line.

ITALIAN SUBMARINE CHASERS

DURING the early part of the war hostile submarines did not venture very far from their own coast, and consequently for a time the "danger zone" was fairly limited in area. But it soon became evident that the larger class of submarine possessed nautical qualities of no mean order, which, in conjunction with an abundant supply of fuel, enabled it to undertake long cruises, and to remain at sea for weeks instead of days. Thus, from the North Sea the area of operations extended first to the Channel and then to the Atlantic. Five months after the outbreak of war, it will be remembered, H.M.S. Formidable fell a victim to submarine attack more than 700 miles from the nearest German base, and by the spring of 1915 hostile submarines were encountered in the Atlantic at a considerable distance from the Irish coast. This rapid increase in the striking radius of enemy submarines naturally tended to complicate the problem of defence. The motor launches and other small craft which had proved so valuable for patrol work in coastal waters lost much of their effectiveness when the raiders took to blue water, and it has, therefore, become necessary to depend chiefly on destroyers for countering the menace. Nevertheless, the many hundreds of fast motor launches which were built in Britain and in the United States have done good work, and they are still the bane of the smaller submarines which the enemy apparently continues to build wholesale, principally for laying mines in the North Sea. In this issue we are able to illustrate the Italian variation of the "chaser" type. Conditions in the Adriatic are such as to favor the employment of these miniature craft, and we understand that much of the success which has attended the Italian navy in its efforts to keep the waters of the Adriatic open to friendly navigation has been due to the vigilance and skilful handling of the "chaser" flotilla. The external characteristics of this type may be seen from the photograph. The boats appear to be somewhat smaller than the typical British "M.L.," and as the hull is modelled on hydroplane lines they are probably much faster. According to our contemporary, "The Engineer," to whom we are indebted for the photograph and data, some of these tiny auxiliaries are capable of working up to astonishing speeds, accompanied by extraordinary facility of manœuvre.

ITALIAN SUBMARINE CHASERS AT FULL SPEED.

CLERGYMEN INVENTORS

IN referring to the work of the Rev. Robert Stirling, inventor of the heat regenerator, an English contemporary remarks that quite a large number of scientists and a good many inventors, some of marked eminence, have been in holy orders. The very earliest scientists, men like Roger Bacon and Schwartz, who vie for the proud distinction of having invented gunpowder, were naturally churchmen, because knowledge was practically concentrated in ecclesiastical circles. The first really eminent clergyman-inventor that we are able to recall was Lee, of Elizabeth's time, who in an age when mechanical knowledge was very small, laid by his invention of the stocking-frame, the beginning of a great industry. Cartwright, the inventor of another textile machine, the power-loom, was also a parson. Possibly these three, Stirling, Lee and Cartwright, stand out above all other mechanical inventors in holy orders, but in the domain of chemical science Priestly, a Unitarian minister, holds the highest place. Then we have Bell, who invented the reaper; Forsyth, a Scottish divine who produced an exceedingly ingenious percussion cap for fire-arms—it had a magazine of percussive pellets, and unluckily the explosion of one used to fire the others; Desaguliers— familiarly called amongst the old mining engineers of Cornwall, Desagulez, and well known as the philosopher of early steam engineers; Gainsborough, a brother of the famous artist, and an "ingenious mechanick;" C. W. Cecil, an inventor of a gas engine in 1820; the Abbé Arnal who, according to report, first proposed the use of the crank in the propulsion of ships, 1788, and, if our recollection serves us aright, Capel, the inventor of the mining fan. Other names occur to us. The Rev. Edward Berthon, Canon of Romsey, whose collapsible boat is well known, and Dr. Watson, Bishop of Llandaff, who was an ardent experimentalist in electricity and the first man to send charges, from frictional machines, of course, through considerable distances—through 10,000 ft. of wire at Shooter's Hill in 1748 or so was amongst them.

FISHERMEN V.C. AND THE KING

TOM WING, M.P., tells how he accompanied a fisherman from a mine-sweeping trawler, who was awarded the V.C., to Buckingham Palace to receive his decoration from the King.

"He rolled along with me in true seafaring style," says Mr. Wing, "until, as we approached the gates, I remarked, 'We must pull ourselves together, otherwise we shall not be admitted.' My gallant friend who had been chewing a plug of tobacco, took it out of his mouth and carefully put it in his cap.

"It is usual for the King to shake hands with recipients, and when he did it on this occasion my man would not let go. He held on, and to make certain it was all right, he put his other hand on it. Then he looked at his Majesty as much as to say: "Are you the King, or are you only "kidding" me?".

"When we got outside he said it was all like a dream, and he felt for his pipe to make quite sure he was awake."

GRAND COUNCIL OF N. A. M. E. HOLD CONVENTION

THE National Association of the Marine Engineers has just closed the fourteenth session of the Grand Council, conferring at the Hotel Riendeau, Montreal. This convention has been one of the most successful in the history of the organization, and the year has terminated with the balance on the right side of approximately $2,300.

The session has been one of the busiest for many years, largely as a result of the remarkable developments of the past three years and the direct bearing these have had upon this particular branch of marine engineering. Existing conditions have naturally necessitated many changes, apart from the encouraging growth of this association. Many new clauses have been added, and others have been revised, both in the constitution and the ritual. Two new councils have been organized, and addition made to the membership of nearly 250. The exceptional heavy business that required the constant attention of the delegates lengthened the duration of the convention term, and the incidentals generally associated with gatherings of this description were somewhat interfered mith, but the entertainment committee were able to give a very good account of themselves notwithstanding. The feature of the week was undoubtedly the banquet held on Thursday evening—and well into the wee small hours of the following morning. Like good marine engineers, however, very few of them allowed their "holds" to take in too much water. A note of sadness was shown throughout the proceedings by the fact of the recent and sudden decease of the Grand President, Alex. R. Milne, just a few days prior to the opening of the convention. Mr. Milne was a marine engineer of long standing and one of the oldest members of the association. After a brief illness lasting only a few days Mr. Milne died of pneumonia in the Kingston General Hospital. Mr. Milne was born in the county of Frontenac in the year 1841, starting work at the age of eleven in the employ of Wm. Wilkinson, of Kingston, as a saddler and harness maker, serving four years. At this point he started his engineering career with the Kingston Locomotive Works at the age of 15 years. After a four year apprenticeship he went to Montreal, taking a position with E. E. Gilbert in the marine engineering work, returning after two years to Kingston where at the age of nineteen he passed his examinations as a marine engineer, his certificate, however, being withheld until he had attained his majority. He installed the machinery in the steamer Kingston and for some years was chief engineer of the Rockwood Hospital. For many years he was engineer on different steamers operating on the Great Lakes. His loss will be deeply felt by the members of the National Association who have been associated with him for so many years. At the convention just closed representative delegates were present from practically every council in the Dominion from Halifax to Vancouver. The delegates present were: E. A. Prince and W. C. Wood from Toronto, W. J. Morrison (Grand Sec'y-Treas.) and Geo. T. G. Blewett from St. John, N.B., E. A. Reid and H. R. Dixon from Collingwood, Wm. A. McWilliam and A. E. Kennedy from Kingston, E. Hamelin and O. L. Lamoureax from Montreal, Ephraim Read and Jas. C. Adam from Vancouver, B.C., Jeffrey Roe and Albert Theriault from Levis, P.Q., Jas. A. Laundry and Naps Beaudoin from Sorel, P.Q., R. J. McLeod from Owen Sound, Ont., E. Hamelin of Local No. 5 was largely responsible for the success of the convention.

GOVERNMENT PLACE ORDERS FOR SHIPS

FABRICATION and assembling of materials for the first of the fleet of merchantmen to be built by the Dominion Government has been commenced, and a similar work for two other vessels will shortly be undertaken. The first ship, which will be a steel steamer of 4,350 tons burden will be laid down at the Canadian Vickers yard at Montreal. A second of 8,200 tons will follow at the same establishment, and, it is expected, a contract for a third ship of 3,800 tons will soon be made with the Collingwood Shipbuilding Co., which has a vacant berth in its yard. It is hoped that the ships to be built at the Vickers yard will be in commission by September next.

The Government's shipbuilding programme contemplates the laying down of the keels of some forty ships to June, 1919. Of the number, four of 5,000 tons and six of 8,200 tons are to be built on the Pacific Coast. All the vessels, however, are intended to relieve the shortage on the Atlantic and to ply there while the war continues.

Arrangements have been made whereby the Government is assured of a supply of ship plates and steel materials for its shipbuilding programme to June, 1919. Orders for steel for the Canadian Government will, it is understood, be placed by the United States' Purchasing Commission with various steel companies in that country. The Canadian Government, which, by the arrangement will be assured of securing material at as favorable prices as possible, will then make contracts with reference to deliveries and other features of the purchase agreements.

BANQUET OF THE GRAND COUNCIL OF THE NATIONAL ASSOCIATION OF MARINE ENGINEERS AT THEIR 14TH ANNUAL SESSION, HOTEL RIENDEAU, MONTREAL.

The MacLean Publishing Company LIMITED

(ESTABLISHED 1888)

JOHN BAYNE MACLEAN - - - - - - - President
H. T. HUNTER - - - - - - - - - - Vice-President
H. V. TYRRELL - - - - - - - - - - General Manager

PUBLISHERS

MARINE ENGINEERING of Canada

A monthly journal dealing with the progress and development of Merchant and Naval Marine Engineering, Shipbuilding, the building of Harbors and Docks, and containing a record of the latest and best practice throughout the Sea-going World.

J. M. WILSON, Editor. B. G. NEWTON, Manager.

OFFICES

CANADA—

Montreal—Southam Building, 128 Bleury St., Telephone Main 1004.
Toronto—143-153 University Ave., Telephone Main 7324.
Winnipeg—1207 Union Trust Building. Telephone Main 3449.
Eastern Representative, E. M. Pattison.
Ontario Representative, S. S. Moore.
Toronto and Hamilton Representative, J. N. Robinson.

UNITED STATES—

New York—R. B. Huestis, 111 Broadway, New York, Telephone 8971 Rector.
Chicago—A. H. Byrne, 900 Lytton Bldg., 14 E. Jackson Street. Phone Harrison 1147.
Boston—C. L. Morton, Room 733, Old South Bldg., Telephone Main 1024.

GREAT BRITAIN—

London—The MacLean Company of Great Britain, Limited, 88 Fleet Street, E.C. E. J. Dodd, Director, Telephone Central 12960. Address: Atabek, London, England.

SUBSCRIPTION RATE

Canada, $1.00; United States, $1.50; Great Britain, Australia and other colonies, 4s. 6d. per year; other countries, $1.50. Advertising rates on request.

Subscribers, who are not receiving their paper regularly, will confer a favor by telling us. We should be notified at once of any change in address giving both old and new.

Vol. VIII FEBRUARY, 1918 No. 2

PRINCIPAL CONTENTS

PUTTING ESSENTIALS FIRST

WASHINGTON intends to establish a plan by which coal distribution will be made to consumers in the order of their importance, having regard to their bearing on problems of immediate contact with the war.

This is simply another step in the process of grading industries into essentials and non-essentials.

This process must of necessity be done by pressure from without. It must be compulsory, for the very good and simple reason that if it waited on voluntary effort the wait would be long and barren.

A man who has built up his organization, both manufacturing and sales, is not going to step up of his own free will, even in a time of stress, and admit that his enterprise correctly comes in the non-essential class.

The war is turning itself into a mighty big national sieve, and when it comes down to brass tacks a lot of things are apt to drop into the ash heap as non-essentials.

WAR AS A TEACHER

WILLIAM C. REDFIELD, United States Secretary of Commerce, is making an effort to cause United States manufacturers to realize that they have got to get their house in shape for the aftermath of war. Apart from the question of maintaining credits, Mr. Redfield sees something bigger than the financial element in the case.

In brief he advocates: (1) overcoming past failures in applying science to industry, the United States official holding that this was the secret of Germany's tremendous development; (2) a greater application of the principle of technical training for industrial workers, to bring about the happy co-operation of trained minds leading trained workers; (3) greater knowledge of costs, and (4) the elimination of industrial and commercial waste.

The war is teaching individuals, firms and nations more lessons in the matter of stopping leaks than they ever learned before.

When war teaches a lesson it does not stop to ask if the pupil is willing or otherwise.

The willing one learns and comes out on top of the heap.

The man or firm who will not learn or bend goes on and goes down. The bailiff or the assignee conducts the rest of the obsequies.

BRINGING THE WAR HOME

SOME of the apologists for heatless days in United States claimed that it was high time the people of the republic were having a few object lessons that would make them realize that the republic was at war.

They reckoned that heatless and workless days would have this effect. So the first order was for five heatless days, and then an intermittent series, making in all thirteen of them.

From official sources it was suggested that employers should make an effort to pay wages as usual.

It would be quite an object lesson of the horrors of war to have a string of holidays with the pay days coming and going as usual.

If the proportion of employers and employees is 95 to 5, it means that the object lesson is for the five per cent.

Taking a few days off with wages as usual is rather a vague way of feeling the horrors of war.

Fo'c'stle Days
or
Reminiscences of a Wind Jammer
By Capt. Geo. S. Laing

The steamer and the motorship have deprived the seafarer of much of the old-time romance attaching to a sailor's life. The advent of wireless has also increased the safety of it. Deprived of these features, a sailor is immediately dependent on his own resources, and Captain Laing gives us a true and seamanlike reflection of sea life under canvas. The present feverish activity in constructing self-propelled craft affords a suitable background for incidents in a career in which gaffs, booms, tar and canvas held sway before being displaced by mechanical unloaders, derrick ma'ts, fuel oil, coal and wire rope.

OUTWARD BOUND

THE finest poetry of the seas, the greatest of ocean romances, the very foundations of navigation and seamanship—yea, naval warfare too, are inseparably bound up with the old time wind-jammer. If, to a belief in these sentiments, is added the influence of heredity, the claim of the sea as my chosen calling will be readily understood.

Born in Dundee, Scotland, in the zenith of that port's activities in whaling, flax, lumber and jute, taking to the sea was almost contagious. Moreover, my father had been an officer in the merchant marine till the circumstances of shipwreck had drifted him into the customs water service. Then to clinch everything, my mother was a shipowner's daughter and the oldest boy in the family was already afloat.

Imbibing a Love for the Sea

On Saturdays or school holidays I would be found with other boys on a raft or punt in the middle of a dock, with slats of wood for oars and in the background a fat policeman would invariably be in attendance, ever ready to give us a warm welcome to *terra firma*.

At other times you might look aloft at some whaler's mast-head and sure enough our gang were distributed like crows on the yard-arms or sitting straddle-legged over an out-rigger or crosstree, hailing passers-by with "Whale oh," "Where away," "There she blows" and so on.

The very atmosphere of foreign lands and probable adventures could be drunk in by crowds of boys who visited the jetties on the silvery Tay. One could board an old brig, flax laden from Russia, or a steam tramp with timber from Finland or an iron clipper ship from India with jute. Over and above such environment I had met an old salt called Sandy Gilbert, when living in Arbroath, and it was partly due to this old sailor's yarns that I longed to sail abroad.

Now here are a few of the encouraging references to sea life that I had to listen to from my father:

"If I had the least attachment for a cat or dog, my only fear would be that these pets might be smuggled on board a ship."

Again, "The first sea that smashes over the weather rail will knock him senseless," and "He is choosing the toughest life on this planet." Then with a final groan the deprecatory advice would finish up with: "Fancy that boy on a top-sail yard taking in a reef!"

My mother did not say much but she wet most of the articles in the sea kit with tears as she gathered the out-fit for my initial voyage abroad. The first thing to upset her was to see me dressed up in oil-skin suit and sou' wester, with heavy knee boots of leather—the regalia of storms. Over this suit a leather belt carrying a sheathed knife was worn, and the picture of a young sea-dog was complete.

Leaving Home

At the age of 14 I signed indentures and as a real live apprentice seaman has to be indentified from ordinary shore mortals, a brass bound suit of blue serge, was donned and a poop-down-haul cap with the company's house flag on it was used as head gear. With this sea livery on and a rolling gait like an old seaman, I thought I was fit company for such men as laid the Atlantic cable.

On a memorable day orders came for me to take passenger steamer to Hamburg, Germany, there to join the barque Maggie Dorrit, which vessel was loading a general cargo for Australia. To dwell on the parting from home would be depressing to the reader and not too favorable to myself, as the jerk was not so easily neutralized as was at first anticipated. I had my first taste of sea humor while bound to Hamburg on the old "Coblenz." The second day out was a Sunday and the mate, who evidently had literary tastes all his own, asked me if I would like a Sunday book to read. Certainly I would, but what was my surprise when he handed me "The Life of Charles Peace," the notorious thief and murderer. If this was the mate's Sunday reading his week-day literature would have made a tombstone dance.

Unlike the majority of human creatures I was never sea-sick, but I'll admit that I was frequently home-sick. The run to Hamburg was without adventure. I had, however, enjoyed the sight of the brown sailed trawlers on Dogger Bank, and the raising of Heligoland. Inside of this famous island the Cuxhaven pilot boarded us and we were soon inside the Elbe river.

On arrival at Hamburg docks I was met by the Captain of the Maggie Dorrit. The old barque was moored to the dolphins in the sailing ship harbor, so with the aid of a small screw ferryboat nicknamed the "Jolly Farmer" we were soon on board of the vessel that was to be my home for five years.

Description of the "Maggie Dorrit"

After being shown into the apprentices' "half-deck," the name given to the small house abaft the main-mast, time was allowed for unpacking my dunnage, for be it remembered a young sailor in those days had to supply his own bedding and eating utensils. In fact the owner of the ship found nothing for the boy except hard work and suspicious food.

Let us now examine the barque. The Maggie Dorrit was an old vessel, having been built in 1860. In her palmy days she had been a full rigged ship and had some very smart passages to her credit between British ports and India, China and Australia. At that time she was one of the "Cities" and belonged to the Smiths of Glasgow.

The Maggie Dorrit, like most vessels

of her time, was a poor carrier but a good sea boat. Her plates were of ⅜ in. iron and the decks, covering boards, bulwark stauchions, fyfe rails; in fact all such work from monkey-poop to knight-heads was solid teak, with here and there a little greenheart. Although only 900 tons registered, the lower-masts and shrouds, and the anchors and chains would be considered heavy enough nowadays for a vessel double that measurement. The crew numbered 20: three officers, four petty officers, four apprentices and eight A. B's., the captain or "old man" making up the total. Legally the "old man" is not considered as a member of the crew, and in this light he is often nick-named "King of the Island," especially when he surrounds himself with unnecessary pomp and dignity.

Leaving Hamburg for Brisbane, Queensland

At last the cargo, some 1,200 tons of miscellaneous goods, embracing liquors, musical instruments, cement and machinery was on board. With a forward crowd of mixed Europeans, every one of whom tumbled on board "three sheets in the wind," we were soon towing down the river. Heavy ice in the fair-way caused us to anchor for a time and this opportunity was made good use of, for we sent aloft two royal yards and rigged out the jibboom. Brace-bumpkins, whisker-booms and lower topping-lifts were also adjusted and everything made shipshape for the North Sea in the winter. After further towing, the old barque started to rise and fall in answer to a heavy ground swell. The tug saluted a farewell and our 4 in. diam. towrope was hauled in to the chorus of a chanty. Our voyage had really started and as luck would have it, the wind was unfavorable so the vessel stood off the land with the yards braced sharp up on the backstay and the leach bowliness hauled out. As each sail was set the vessel heeled over more and more till the channel plates were awash, and the lee scuppers and ports allowed dead water to enter the main deck.

One of the striking features for the first voyager is to hear chanty singing on the different ropes. Halyards, sheets and tack tackles all have words and tunes to suit the occasion. There were three first voyage boys and none of us looked so neat and trim now, as the old barque climbed over the watery humps and threw the spray around like grape shot.

Picking Watches

All hands were mustered and the watches picked and set. The time worn system of four hours on duty and four hours off, with a dog-watch breaking up the period from 4 to 8 p.m. never appealed to growing boys but there it was—a hard and fast sea institution. Attached to this service rule there is only one certainty, and that is, that you can always depend on your watch on deck, but never be disappointed should you lose the whole or part of your watch below.

I was put in the starboard watch—2nd mate's—and had as watch-mate a parson's son from Tilbury, who also was a first voyager. This boy became so violently sea-sick, that we feared for his life. His condition made the writer more scared than if a hurricane had overtaken us. One boy in the port watch who was a second voyager and whom we may call Fid consoled poor Tilbury with an offer of some greasy hash as an antidote, but this same "ancient mariner" will entertain us at his own expense later on.

Tack Ship Every Four Hours

Amongst the many manoeuvres that a sailing ship can be put through, tacking ship is perhaps the noisiest. Every watch saw the Maggie Dorrit throw her head into the wind's eye and thereby work round on the other tack. When a vessel is in the wind she simply shakes every block and spar on board, and one has to keep clear of running braces and sheets. One morning the wind veered in our favor and we bore down on the East Goodwin light-ship. With the Foreland heights to star-board and the chalk cliffs of Dover well aboard, we laid a course for Beachy Head. Tilbury was just getting over his sea-sickness but it is almost safe to say that he could have been prevailed upon to try and swim ashore. But he was otherwise engaged, for our first lesson in flag signalling happened off Dover. With the "old man" at the international book and the 3rd mate picking out and bending on the flags to the signal halyards, we as novices tried to obey the yells and grunts of our superiors. To keep the loose flags from blowing over the lee rail was quite enough for us, and as the answering pennant is hauled down for the last time, eight bells are struck by your humble servant which means that the starbord watch goes below till noon.

A Place of Heavy Shipping Traffic

Beachy Head is one of the world's busiest corners. Every conceivable kind of craft passes it when bound west or south from the Baltic or North Sea ports. What a panorama is now in view! Three pilot cutters ready to supply pilots for London, Antwerp and Dunkirk, two coasting schooners, two passenger liners, one bound for Hull, and the other for New York, a steam collier, a craft in ballast under tow, and a cattle-boat.

OUTWARD BOUND

The "half deck" must now be described as it is our home. Just abaft the main mast our little house was situated and its approximate inside measurements would be ten feet long, eight feet broad and six feet high. In this wooden box four boys had their belongings, and it was our mess-room, sleeping berth, and everything else combined. Condensation was never better exemplified than it was in the allotment of sailors' apartments. The rules that governed the cubic capacity for each man or boy must have been worked out originally by a society of politicians whose standard measure was, "one living sailor must have as much room as two dead shore people." Two thwartship bunks and two fore and aft bunks took up a side and inner end of the house. No table was provided, the tops of our sea trunks acted as seats and tables too. A small locker for our culinary outfit and a water breaker completed the furniture. Then dangling from the ceiling was a colza oil lamp that made a slight yellowish glimmer in the darkness. Thank God that boys don't go to sea under such poor conditions now.

The Food Question

No son of Neptune ever suffered from the gout or gastric nightmares so common on shore from over-indulgence in rich foods. Until we arrived in Brisbane our stomachs were treated to dishes of conglomerate foods that I will try to enumerate. Salt horse (beef) and pickled squeal (pork) with pantiles (biscuits) fifty per cent. bone-dust and fifty per cent. shorts, loomed up on the menu horizon. Weekly allowances comprised eight ounces of vile butterine, and fourteen ounces of wet brown sugar. Fresh water for all purposes was doled out at three quarts a day per man. Some of the concoctions we had prescribed from the salt meats and biscuits when mixed with grease were called "slumgullion," "crackerhash," "dandy funk," "lobskouse de foremast," and "dog's body." Soup of the pea and bean variety was termed "rib cement." Soft bread was called "ballast" and rice had the awful name of "strike me blind," but if a few raisins or currants were noticeable in the rice, it became "railway stations." Concerning the tea and coffee these liquids could not be distinguished by taste or smell but by the time of day that one got them. Thus in the morning it was coffee and in the evening it was tea. A visual analysis of these muddy waters would disclose both residues—grounds and leaves. Occasionally a pair of socks or mittens that had been drying over the galley fire would be found in the pots when the Dr. (cook) washed up.

Our pantiles soon became full of maggots and weevils. These nuisances were called "fresh meats." The cure for maggoty biscuits was perhaps novel. One had either to eat the biscuits in the dark and avoid seeing the wrigglers or souse the bread in salt water and then place in the oven till the live stock emerged from the heat. Molasses was known as "long tailed sugar" and tinned meats were termed "more Chicago." If the Lord sent the food the devil certainly sent the cook as far as sailing ships were concerned. Soups were served in separate "kids" but other foods were just thrown together like pig swill. As regards messing conditions, the writer can vow that he never saw fo'c'sle hands treated properly till he arrived on Great Lake craft.

Deep-water stewards were often superior beings in every way compared to the cooks, and the former saw that the captain and mates were saved from these awful messes. It is only right to state however, that they must have come through the swill ordeal in the initial stage of their career, for, if there is one redeeming feature about the merchant marine it is the absence of push or pull as regards climbing through the grades. All must come alike through the hard avenues of the adventurous profession. "My uncle is a shareholder" kind of business has no place at sea, it's purely a land institution.

Tacking Down Channel

The Maggie Dorrit had signalled to St. Catharines, Isle of Wight, but the wind again headed us off towards the French coast. Whilst the homeward bounders were staggering along at ten and twelve miles on hour with their crojacks in the gaskets and the weather clews of their mainsails hauled up, our barque was diving furiously enough to wet the jib-tacks and only making about five miles an hour. I didn't look so bright and brave in my oilskin suit now. My watch mate, Tilbury, also cut the figure of an abandoned tramp on the Thames Embankment. With a stiff wind and a clean bottom the Maggie Dorrit kept both watches hustling when tacking. No sooner had the mainyard swung than the boxed head yards required swinging, and I know of no heavier physical exertion than rope hauling. To note the sails fill on the new tack and feel the old girl heel over with groans of satisfaction from the gear aloft was a peculiar sensation to a green boy. We ploughed thus at a six point (67°) angle from the wind for two days.

A Half-Deck Fight

Fid, the senior apprentice, had been rubbing it in too thick, and Hank, a first voyager, called Tilbury and myself as a council of war. It had been found that Fid had very advanced somnolent habits. He could sleep anywhere, in any posture, and under any circumstances. Poor Hank found that the responsibility of getting Fid to relieve the wheel or lookout at eight bells was too much like bringing back to life some drowned person. During the relief of night watches it was the custom to muster on the quarter-deck, where each man and boy answered "sir" to his name as the officer shouted out the list.

When Fid had no immediate duty at the change of a watch Hank had instructions to answer "sir" to his name, thereby allowing the sleepy one to lie back and escape the ordeal of mustering on time. Hank not only feared detection in this matter, but naturally chafed at being the servile unit to one no better than himself.

It was agreed that Fid should answer for himself hereafter, and that was the cloud that carried the squall. It was the gravy-eye watch—4 to 8 in the morning —and when the mate shouted Fid's name no answer came forth. He was the only absentee. Repeated shouts from the poop brought no response. Then a howl of "Did you call that fellow"? Two or three "yes, sirs" made this plain, meantime an atmosphere of mutiny and its consequences came over the three first voyagers.

"There will be no relief of watches till that name is answeded," shouted the mate further. This was like a match to gunpowder, for two A. B's plunged into our house and simply tore Fid from his slumbers. He must have thought it was a collision, for he fell twice to the deck in a stupor as he was literally pushed along to the after capstan. There he stood in the cold Channel weather in his underwear, but he soon answered the roll, and the concert closed for the moment.

To make sure that Hank wouldn't be swallowed up, we had arranged that if Fid struck him the other two boys would join in the fight. In cold latitudes it is possible to get a hot drink at midnight and at four in the morning, so our trunk-tops were strewn with pannikins, sugar cans and the bread "barge." The half-deck door, which looked forward, also opened in two parts, the lower part to keep seas out in moderate weather, whilst both halves could be shut in a gale.

Fid couldn't wait till his pants were on but let fly at Hank in his white anger. In a jiffy the four of us were rolling around, sometimes in a lower bunk and sometimes on the water breaker. The lamp was torn from its socket and went out, as colza oil is a slow vegetable grease with its flash point buried in the clouds. "Lift him over the half-door," someone spluttered, and a thud on the teak wood deck was the answer. Fid was out, and what rags he had on his back were in shreds. My bunk, which was flush with the trunk-tops was full of coffee grounds, broken pantiles and molasses. Photographs of loved ones were on the floor in a mixture of broken glass and delf, while Fid shouted for a truce, which was agreed to.

On the morrow the captain moved in the matter and further hostilities were called off. One stipulation was made, however, that should there be any more fighting amongst the boys it had to be done in his presence on the quarter-deck. The fires in Fid's breast were hard to smother, and my blankets and counterpane were also hard to clean. Hank's photos and Tilbury's butter and sugar mixed with glass had to be thrown into Davy Jones' locker—the receptacle that never says "enough."

Meanwhile the old barque had crawled inshore on a N. W. leg, and raised the land near Start Point. When darkness set in again we were under the flare of the Eddystone lighthouse, that nautical milestone known the world over. With a fair wind we were able to square in the yards a bit. What a difference in comfort, the decks became comparatively dry, the vessel lost her heavy list or heel, and the lights of the Devonshire coast reminded us that another day would put us into the Atlantic with the Old Country a fading memory, but strange and foreign lands ahead.

(To be continued)

SEAMANSHIP AND NAVIGATION

Conducted by "The Skipper" *

Articles of direct interest to mariners, discussions of seamanship, navigation rules, and allied topics. Inquiries from readers are invited and will be replied to promptly through these columns, unless otherwise requested.

RULE OF THE ROAD STUDIES

ONE of the most important branches of seamanship is the safe manœuvring of vessels as regards the rules laid down by the Government for avoiding collisions. Courts of inquiry hand out rulings on the matter month after month, and young men desiring certificates of competency as mates learn the articles from the book. In this writing, however, an effort will be made to systematically study the rule of the road afloat.

To begin with, we must grasp the fact that the green, red, and stern light (white) together represent a circle. The colored lamps cover 20 points of the compass and the stern light 12, thus making the 360°. Any movement of the ship's head then must tend to "open up" or "shut in" these lights in turn. To follow this month's lesson it is not necessary to mention mast-head lights, especially as they cover the same arc as the combined sidelights.

One of the Ottawa booklets setting forth probable questions on rule of the road that may be given by examiners to candidates has at least one question that apepars to be an incentive to study the collision regulations. The question runs along this line: "You see a green light bearing S.S.E. Between what points of the compass must the vessel carrying the green light be heading?". The answer is: The yessel showing the green light must be heading on the arc between S.W. and N.N.W. The way to solve any question of this nature is to reverse the bearing, which gives you one limit, and for green lights allow 10 points to the left for the other limit.

Say, that a red light bears North, then South will be one of the limits and W.N.W. the other. Red lights take their swing to the right. W.N.W. is 10 points to the right of South. There is a six-point method for this kind of reckoning, and if you have learnt it, don't move, as it gives the same results; but the harder method as given here helps those who are weak in boxing the compass.

From right ahead to two points before the beam could be called the real danger zone in matters of collision. Of course, the colored lights cover four more points, i.e., as far as two points abaft the beam.

All vessels meeting opposition colored lights within the zone just referred to must either "stand on" or "keep clear." Where opposition lights are seen abeam the possible lines of collision are almost negligible, although responsibility never ceases even then. The safety with beam lights is the fact that very little helm adjusts matters, or, again, a momentary ease down will put both ships well apart. In this respect it is important to remember that a steamer's telegraph is a great factor in collisions. Rather than telegraph to the engineer for "half speed" or "slow," to get out of a tight place, such as parallel vessels, some men in charge of a bridge prefer plugging right on till she hits.

When confronted with an awkward predicament, such as another steamer showing red on your starboard bow close to, as oftens happens in misty weather, why not go round the long way? Altering your ship's head to port through a swing of perhaps 28 points and avoiding collision is better than swinging to starboard and leaving one of your anchors in the other fellow's fore rigging.

Steamers should not always take for granted that they can reach across all windjammers' bows. Sailing craft have long bowsprits and jibbooms, and canvas is not always second to bunker coal.

The diagram shown herewith is the start of some tall thinking which we must enter into in order to thrash the subject out in text and picture. The questions already spoken of in the Ottawa booklet are here proven by the geometrical method of study.

The student must imagine that he is on board the model, sizing things up from the centre of his compass, and I might add that all seamanship questions relative to steering, bearings, deviation, etc., are most easily digested from this point of observation—the centre of your compass. This diagram shows the ten points of swing that opposition colored sidelights have from one point on the bow to two points abaft the beam, thereby illustrating the possible lines of collision for each bearing.

Get a green crayon or pencil and color in the segments on the left or port hand. Color the right or starboard hand segments in with red. Thus your model will be surrounded by opposition lights, i.e., vessels showing red on your starboard side and vessels showing green on your port side.

Each bearing shows you that the stranger can head on the reverse point to the bearing. Is that clear? The other limit is the short leg of the segment. The arc of horizon between the two lines must contain the ten possible points that the stranger can head on, mark you—as long as she shows an opposition light.

There is nothing to explain and little to learn where colored lights are not opposed to each other, so it's purely the collision lines that we are after and not the red to red and green to green sailing.

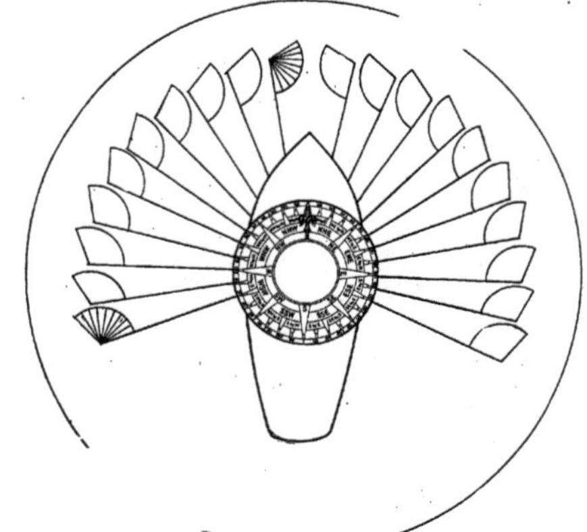

CHART FOR USE IN RULE OF THE ROAD STUDIES.

*Editor's Note.—This department is conducted by a qualified ship's master, and will deal with questions involving seamanship and navigation. The attention of young men desiring to obtain certificates as mates and masters is particulary directed to this department.

Now come back to the mental calculations and we will find that the diagram helps us to more clearly see what is done.

Say a red light is seen three points on the starboard bow of your model, i.e., N.E. by N. Red on that bearing will give the limits as S.W. by S. and N.W. by N. Now put a pair of parallel rulers over the short leg of the segment in question and work down to the compass rose and it will give you N.W. by N. The other limit or the reverse bearing can never be in dispute; it is clear at first sight.

Work around the drawing in other ways and the practice will help you in many directions. For instance, take a segment abaft the beam and one on the bow and fill them in with lines or courses. What do you grasp from this?

In the segment on the bow every collision line aims right at your track, or, in other words, crosses you.

Is it so with the other one? No; a third of the lines are all you would have to worry about, showing that what I said about the real danger zone being from right ahead to two points before the beam is an approximation of value.

As our drawing concerns bearings of lights, it will be well to note that where vessels are crossing each other as to involve risk of collision there is a method which proves what will happen, and its daily practice should never be forgot.

Say that you are on a steamer's bridge and see a sailing ship's green light on your port bow. Take a compass bearing of it, either with your azimuth mirror or your hand. In about a minute take another bearing. If the sailing ship light has "closed in," i.e., come nearer your head, you will either have to alter your course to port or ease down for a minute. But if the green light has altered its bearing towards your stern you are safe in keeping as you were. Vessels cannot collide where an opposition light alters its bearing towards your stern. The greatest danger is where the bearing does not alter at all. In this case the ships' courses are converging to a point of collision. Again, where the opposition light (on either side) closes in on your course the condition is safe, but needs watching, and the reason is this: The fact that the sailing ship with green on your port bow is closing in means that she will cross your bow if her speed is maintained, but as this is an uncertainty, it is best in darkness to help her across by acting as I said before—either ease your steamer for a minute or put your opponent right ahead, following her back till you are again on your course.

Next month's issue will contain further diagrams with explanations. Meantime get interested in this most vital subject, and think of it from different angles.

NOTICES TO MARINERS

THE following notices have been issued by Department of Marine, Ottawa under dates given:

No. 3 of 1918, Jan. 21
(Atlantic No. 2)

NOVA SCOTIA

(6) South Cost—Halifax harbor entrance—Sambro Outer bank—Lightship replaced on her station.

Former notice.—No. 78 (190) of 1917.

Position.—6 1/5 miles 156° (S. 3° E. mag.) from Sambro island light. Lat N. 44° 20′ 25″, Long. W. 63° 30′ 20″

Lightship on station.—Lightship No. 15, maintained in the above position, which has been off her station temporarily while undergoing repairs, has been replaced on her station.

Variation in 1918: 21° W.
Authority: Departmental records.
Admiralty charts: Nos. 2410, 729, 1651, 2666 and 2670.
Publication: Nova Scotia Pilot, 1911, page 141.
Canadian List of Lights and Fog Signals, 1917: No. 322.
Departmental File: No. 21040M.

NEW BRUNSWICK

(7) Northumberland strait—Shemogue—Range lights established.

Range lights have been established at Shemogue as follows:

(1) FRONT RANGE LIGHT

Position.—On the shore of Dugay point inside, the harbour. Lat. N. 46° 10′ 20″, Long. W. 64° 8′ 10″

Character.—Fixed white light, shown from a locomotive headlight reflector lantern.

Elevation.—17 feet.

Visibility.—9 miles in the line of range.

Power.—2000 candles.

Structure.—Pole, with triangular daymark attached, shed at base.

Material.—Wood.

Colour.—White.

(2) BACK RANGE LIGHT

Position.—690 feet 213° 45′ (S. 56° 45′ W. mag.) from the front range light.

Character.—Fixed white light, shown from a locomotive headlight reflector lantern.

Elevation.—25 feet.

Visibility.—10 miles in the line of range.

Power.—2000 candles.

Structure.—Pole, with triangular daymark attached, shed at base.

Material.—Wood.

Colour.—White.

Sailing directions—The lights in one, bearing 213° 45′ (S. 56° 45′ W. mag.), lead from the open strait of Northumberland over the bar at the entrance to the harbour, but as soon as Amos point, that on the east side of the entrance, is passed the range must be left and the buoys followed.

Variation in 1918: 23° W.
Authority: Report from Mr. G. S. Macdonald, Resident Engineer, St. John.
Admiralty charts: Nos. 2034, 1651, 2516 and 2666.
Publication: St. Lawrence Pilot, Vol. 2, 1916, page 228.
Canadian List of Lights and Fog Signals, 1917: To be inserted as Nos. 819 and 820.
Departmental File: No. 20819C.

No. 4 of 1918, Jan. 22
(Atlantic No. 3)

NOVA SCOTIA

(8) South Coast—Egg island—Change in character of light.

Position.—On Egg island, Lat. N. 44° 39′ 51″, Long. W. 62° 51′ 58″.

Date of alteration.—On or about 1st March, 1918, without further notice.

Alteration.—The revolving white light will be replaced by a flashing white light, showing two bright flashes every fifteen seconds, thus:

Flash; eclipse 3 seconds; flash; eclipse 12 seconds.

Order.—Third dioptric.

Power.—100000 candles.

Illuminant.—Petroleum vapour, burned under an incandescent mantle.

Authority: Departmental records.
Admiralty charts: Nos. 2439, 2624, 729, 1651, 2666 and 2670.
Publication: Nova Scotia Pilot, 1911, page 111.
Canadian List of Lights and Fog Signals, 1917: No. 349.
Departmental File: No. 20349A.

NOVA SCOTIA

(9) South Coast—Whitehead island—Change in character of light.

Position.—On Whitehead island. Lat. N. 45° 12′ 2″, Long. W. 61° 8′ 35″.

Date of alteration.—On or about 11th March, 1918, without further notice.

Alteration.—The revolving white light will be replaced by a flashing white catoptric light, showing four flashes, at 3 seconds intervals, every twenty-four seconds, thus: Flash; 3 seconds interval; flash; 3 seconds interval; flash; 3 seconds interval; flash; 15 seconds interval.

For half the time of revolution, or 12 seconds, the light will be totally eclipsed; for the other half a light of 500 candlepower will be visible, through which the stronger flashes will show.

Power.—Naked light 500 candles; flashes 30000 candles.

Illuminant—Petroleum vapour, burned under an incandescent mantle.

Authority: Departmental records.
Admiralty charts: Nos. 2517, 2560, 729, 1651 and 2666.
Publication: Nova Scotia Pilot, 1911, page 54.
Canadian List of Lights and Fog Signals, 1917: No. 392.
Departmental File: No. 20392A.

NOVA SCOTIA

(10) Cape Breton island—South coast St. Esprit island—Change in character of light.

Position.—On St. Esprit island. Lat. N. 45° 37′ 30″, Long. W. 60° 29′ 10″.

Date of alteration.—On or about 15th March, 1918, without further notice.

Alteration.—The revolving white light will be replaced by a flashing white catoptric light, showing three flashes, at 6 seconds intervals, every thirty-six seconds, thus: Flash; 6 seconds intervals; flash; 6 seconds interval; flash; 24 seconds interval.

For half the time of revolution, or 18 seconds, the light will be totally eclipsed; for the other half a light of 500 candlepower will be visible, through which the stronger flashes will show.

Power.—Naked light 500 candles; flashes 40000 candles.

Illuminant.—Petroleum vapour, burned under an incandescent mantle.

Authority: Departmental records.
Admiralty charts: Nos. 2727, 1651, 2516 and 2666.
Publication: St. Lawrence Pilot, Vol. 2, 1916, page 65.
Canadian List of Lights and Fog Signals, 1917: No. 443.
Departmental File: No. 20443A.

* * *

No. 5 of 1918, Jan. 28.
(Pacific No. 1)

BRITISH COLUMBIA

(11) Vancouver island—Victoria harbour — Selkirk Water — Canadian Northern Railway bridge—Lights.

Description of bridge.—A bridge, consisting of two approaches, with a bascule bridge between them, has been constructed by the C. N. R. Co. across Selkirk Water from under the Gore road bridge to the Songhies reserve. The northern approach is a pile trestle partly filled in and runs 210° (S. 4° 30' W. mag.) for a distance of about 1,500 feet from the Gore road bridge. The bascule bridge is of the roller type having a steel girder leaf with a concrete rest pier on each side of the opening, and having a clearance between piers of 77 feet, the bottom of girder being 7 feet above high water. The opening lies about 250 feet west of Halkett island. The southern approach is a pile trestle about 500 feet long and strikes the shore at a point on the Songhies reserve about 500 feet west of Sister rock.

Position of bascule bridge.—The north end of the span of the bascule bridge is in Lat. N. 48° 26' 20", Long. W. 123° 22' 56".

Lights.—A fixed white light on each side of the opening will be maintained by the Railway Company, to mark the channel.

Elevation of lights.—6 feet above high water.

Remarks.—The bridge is not being used at present, and until such time as trains are operated over it the bascule will be kept raised.

Variation in 1918: 25° 30' E.
Authority: Report from Agent of Department of Marine, Victoria.
Admiralty charts: Nos. 1897b, 576, 2840, 2689 and 1911.
Publication: British Columbia Pilot, Vol. 1, 1913, page 64.
Departmental File: No. 29925.

BRITISH COLUMBIA

(12) Stuart channel—False reef—Beacon replaces buoy.

Position.—On the rock that dries 3 feet at southeast end of False reef. Lat. N. 48° 58' 24", Long. W. 123° 42' 10".

Sextant angles fixing the position of the beacon:

W. tangent of Scott island ......0°
Bare point lighthouse...........49° 40'
Sharpe point.... ....94 10
Fraser point.... .. ..79 30

Description.—Concrete base, surmounted by a staff carrying a wooden slatwork ball.

Colour.—White.

Elevation.—Top of beacon is 21 feet above high water mark.

Buoy discontinued.—The red and black can buoy, heretofore moored south of False reef, has been withdrown.

Former notice.—No. 98 of 1900.

Sailing directions.—The beacon should be given a berth of 1/3 cable to the southeastward and of two cables from southwestward to north.

Authority: Report from Agent of Dept. of Marine, Victoria.
Admiralty charts: Nos. 3618, 579 and 1917.
Publication: British Columbia Pilot, Vol. 1, 1913, page 227.
List of Buoys and Beacons in British Columbia, 1916: Page 232.
Departmental File: No. 38867.

UNITED STATES OF AMERICA

(13) Admiralty Inlet — Marrowstone Point—Change in Characteristic and Position of Light—Fog Signal Changed

Position.—On Marrowstone point.
Lat. N. 48° 6' 5", Long. W. 122° 41' 12".

(1) LIGHT

Alteration.—About 1st February, 1918, the characteristic of the light will be changed to group occulting white every 10 seconds, thus: Light 2½ seconds; eclipse 1 second; light 2½ seconds; eclipse 4 seconds.

Elevation.—28 feet.

Power.—1,500 candles.

Remarks.—The light will be exhibited from the fog signal building, located 342° 30' N. 42£ W. mag.) from the old site of light.

(2) FOG SIGNAL

Alteration.—The fog signal has been changed from an acetylene gun to a third-class horn, with three mouths, to sound a group of blasts every 30 seconds.

Remarks.—The horns, 10 feet above ground, project from a small square cement-colored structure.

Variation in 1918: 24° 30' E.
Authority: U.S. Dept. of Commerce N to M No. 52 of 1917 and No. 1 of 1918.
Admiralty charts: Nos. 1792, 1947, 2689, 1917 and 2531.
Publication: British Columbia Pilot, Vol. 1, 1913, pages 79 and 80.

NEW BRUNSWICK

(14) North Coast—Chaleur Bay—Bathurst Harbor—Change in Color of Front Range Lighthouse

Former notice.—No. 47 (159) of 1915.

Position.—On the northwest extremity of Carron point.
Lat. N. 47° 39' 23", Long. W. 65° 36' 50".

Change in color of lighthouse.—Bathurst front range lighthouse has been painted white throughout.

Authority: Departmental records.
Admiralty charts: Nos. 1715 and 2516.
Publication: St. Lawrence Pilot, Vol. 2, 1916, page 307.
Canadian List of Lights and Fog Signals, 1917: No. 930.
Departmental File: No. 20930R.
Departmental File: No. 39971.

* * *

All bearings, unless otherwise noted, are true and are given from seaward in degrees from 0° (North) to 360°, measured clockwise, followed by the magnetic bearing in degrees in brackets, miles are nautical miles, heights are above high water of ordinary spring tides, and all depths are at low water of ordinary spring tides.

Pilots, masters or others interested are earnestly requested to send information of dangers, changes in aids to navigation, notice of new shoals or channels, errors in publications, or any other facts affecting the navigation of Canadian waters to the Chief Engineer, Department of Marine, Ottawa, Canada. Such communications can be mailed free of Canadian postage.

INCREASED TRAFFIC AND FEWER SLIDES ON PANAMA CANAL

TRAFFIC through the Panama Canal last year increased largely over the previous years, according to the annual report of Col. Chester Harding, governor of the canal. A total of 1,876 vessels of all classes passed through the canal from July 1, 1916, to June 30, 1917, inclusive.

Of these, 905 passed from the Atlantic to the Pacific, and 971 from the Pacific to the Atlantic. In the fiscal year 1915, 1,088 vessels passed through the canal, and in 1916, 787. The total number of vessels transiting the canal since it opened for commercial traffic in August, 1914, is 3,751. The total net tonnage, canal measurements, for the several years is as follows: 1915, 3,849,035; 1916, 2,479,762; and 1917, 6,009,358. The cargo tonnage transported was, for 1915, 4,969,792; 1916, 3,140,046; and 1917, 7,229,255.

Loss From Present Traffic Rules

The traffic for the year yielded a revenue of $5,631,781.66 from tolls. The rules for levying tolls have thus far remained unchanged.

It will be remembered, says the report, that under present law and regulation tolls are based on canal tonnage, at the rate of $1.20 per net ton, except when this product exceeds the registered net tonnage, United States rules, at a rate of $1.25 per ton, in which case the lesser amount is collected. The confusion, lack of uniformity, and loss in revenue to the canal resulting from the present arrangement have been fully discussed in previous reports, and remedial legislation is now pending in Congress.

In this connection, attention is invited to the fact that the revenues from tolls for the past fiscal year would have been $6,668,247.32 if canal rules alone had governed, which is $1,036,465.66 more than the amount actually collected.

Dredging Work of the Year

The year's dredging chargeable to construction, which includes all the excavation in the canal prism at locations where the full widths and depths have not been once obtained is as follows: Gaillard Cut, 183,904 cubic yards; between Gamboa and Pedro Miguel Lock, 246,998 cubic yards; and Pacific entrance, 221,138 cubic yards. At the end of the year there remained 1,409,140 cubic yards of original excavation to be done within the limits of the canal prism.

Cucaracha slide has given no further trouble since the large movement that blocked the canal in August, 1916, as described in the report of last year. To reduce the chances of interruptions to traffic due to future similar movements if they occur, the material in the slide was removed for a distance of 100 feet outside the canal prism.

It is believed that in the future the great slides of the canal will be of historic interest only.

PROGRESS IN NEW EQUIPMENT

There is Here Provided in Compact Form a Monthly Compendium of Shipbuilding and Marine Engineering Auxiliary Product Achievement

COMPENSATING LIFE BOAT DAVIT

A SIMPLE, reliable and rapid working life boat davit is a necessity in normal times but, under conditions as they now exist at sea, the possible saving of the lives of a ship's passengers and crew even more often depends upon those features in the life boat davit.

A simple davit is one that is foolproof, contains a miminum of operating parts and has no automatic features. The method of operating the davit should be simple and obvious, thus preventing confusion and delay when the services of the davits are needed. A reliable davit is one that can be depended upon to positively fulfill its requirements under the most adverse conditions of weather, sea and list. A reliable davit must of necessity be efficient and upon its mechanical efficiency depends its ease of operation.

Inasmuch as an interval of two minutes or less may decide whether the life boats can be launched or not, the rapidity of operation is a vital factor. Rapidity of operation depends upon the amount of power required to work the davit—its ease of operation.

The Marten-Freeman davit, an installation of which is shown in the accompanying engraving has proved itself, under service conditions, to be simple, reliable and fast operating. It consists of a cast steel frame which forms both the track and supports, a steel truck of the tandem roller type, which rolls along the track and carries a cast steel boom at its fulcrum, the compensating link, the actuating screw and crank. The screw engages a bronze nut which floats in the truck in such a manner that an obstruction on the track cannot cause the screw to bind. The actuating screw is amply protected from injury by a tee bar.

The feature of the mechanical principle involved is the compensating link which connects the foot of the boom to a fixed centre at the base of the frame. When the boom is vertical, as in the inboard position, the link is nearly horizontal and there is little or no stress in the link member. As the boom is moved outboard there is immediately set up by the weight of the boat an overturning tendency which considerably increases with the degree of angularity in the boom. As the boom moves outboard however the compensating link moves up into a position where it assumes more and more tension thus acting in effect like a counterbalancing load placed upon the foot of the boom and thereby tending to keep the boom much more nearly in equilibrium during all stages of its movement. This action obviously reduces the pull on the actuating screw very materially.

A further reduction in the amount of tension in the screw is secured through the effect of the inclined track on which the truck rolls. The resultant force of the weight of the boat and the stress through the link is brought nearer to a line perpendicular to the track by this means. The inclined track acts as a brake to the outboard movement and as an aid to the inboard movement. The compensating action of the link together with the effect of the inclined track greatly reduces the amount of tension in the operating screw. The turning of the crank therefore requires comparatively small effort and the davit can be operated much faster than would otherwise be the case.

With this type of davit the life boats can be put overside and ready to lower away even against a bad list of the ship, but the great advantage lies in the rapid and positive outboard movement of the boom, which enables the boats to be put overside before the sinking ship has acquired a bad list, thus ensuring a safe lowering away of all of the life boats.

Where the boats are provided with hand falls, a stationery gypsy head is placed on the side of the frame at a convenient point and one man can control the falls gear at each end of the boat with two or three turns of the line around this head. After the boat is paced in her chocks, two or three turns of the falls are taken around the gypsy head and the line then belayed to the cleat. The gear is thus always in readiness to put the boat overside and to lower away immediately, the life boat in the foreground of the illustration having been put overside, ready to

VIEW OF BOAT DECK EQUIPPED WITH COMPENSATING LIFEBOAT DAVITS.

lower away by the two men shown, in less than one minute.

This davit is the product of the Marten-Freeman Company, 301 Manning Chambers, Toronto, and has been passed and approved by the Board of Steamboat Inspection, Marine Department, Ottawa, Lloyd's Register of Shipping for Canada, U. S. Steamboat Inspection Service, Lloyd's Register of Shipping for U. S., American Bureau of Shipping and the Bureau of Construction and Repair, U. S. N.

HEAVY DUTY MARINE MOTOR

THE accompanying illustration shows a heavy duty marine motor which is built in a variety of types and sizes by the Foreman Motor and Machine Co. Ltd. of Toronto. Similar type motors are being supplied by this firm to local ship builders for installation in motor driven life-boats. These engines are built in two, four and six cylinders, ranging from 15 H. P. to 75 H. P.

The cylinders are cast singly, and without valve pockets on either side. This design, allowing for equal expansion, insures a perfectly true cylinder under working conditions. Ample water space is provided, with a plug for cleaning out the jacket. The valves being in the cylinder head, side pockets are done away with, and the whole charge to be ignited is directly over the piston, insuring perfect combustion and giving good economy in operation.

The base is made in two pieces, the upper one being well ribbed throughout to support the cylinders and crank shaft bearings. The supporting of the crank shaft bearings in the upper base casting provides a unique adjustment for wear in that the shaft is raised to its original position and is not being gradually lowered as the bearings wear. Large plates are provided on the side of the case for the easy removal or adjustment of the connecting rod and crank shaft bearings. The lower base casting is extended to support the reverse gear and thrust bearings and is provided with troughs to keep the oil at a proper working level.

The forged steel crank shaft is 2½ ins. in diameter and is provided with bearings of liberal length, the end bearings being 5½ ins. long and the intermediate bearings 3 ins. The forged steel connecting rods are fitted with babbitt lined bronze bearings of liberal size for the piston pins. The piston pins are hardened, ground, and are held in place by two set screws and a retaining device. The pistons are extra long, and have four rings of semi-steel fitted above the piston pin, and the necessary oil groves below.

The flywheel is bolted to a flange forged solid with the crank shaft and is easily removable if required when installing.

Plunger pumps are used on the slow and medium speed motors for water circulation and gear pumps on motors built for high-speed. All water fittings being of brass or bronze every motor is suitable for salt water use without any alteration.

The chrome nickel steel valves are provided with ample cooling arrangements and are so designed that they cannot drop into the cylinder should the retaining pin become broken. The cam shaft is of liberal size 1¼ inches in diameter and is driven by hardened steel gears. Ball bearings are provided on the gear end.

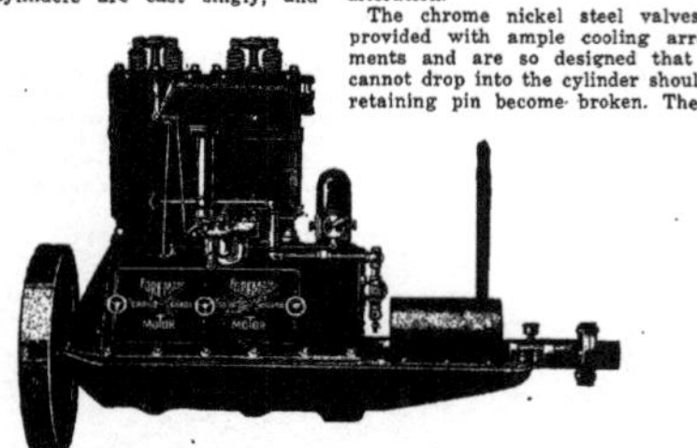

HEAVY DUTY MARINE MOTOR WITH OVERHEAD VALVES, SELF-CONTAINED REVERSE GEAR AND MAGNETO IGNITION.

The exhaust manifold is completely water jacketed except on the under side which contains the intake manifold, kept warm by the exhaust. A standard float feed carburetor is provided of a size most suitable for each motor.

Regular ignition on all sizes is provided by a high tension magneto system with impulse starter. The motor can be started by the magneto from the slowest speeds and unfailing ignition is provided when running. Make and break ignition can be supplied if desired.

On the four cylinder and larger sizes a governor is built into the motor; it is fitted to the cam shaft, is completely enclosed and is attached to the gas throttle which is especially desirable for one man controlled boats.

An oil pump operated from the cam shaft pumps oil from the lowest part of the base through a strainer and floods the main bearings and crank pins with oil; oil troughs keep the proper oil level under each connecting rod for splash to pistons and other working parts. A separate oil supply to cam shaft gears insures adequate lubrication to these working parts.

The two cylinder type which is rated at 15 H. P. at 500 R. P. M. is suitable for speeds considerably higher with a corresponding increase in power. The four cylinder, rated at 30 to 50 H. P. has a good margin over its rated horse-power. The six cylinder is rated at 45 H. P. at 500 R. P. M. and at 75 H. P at 800 R. P. M.

These motors are well adapted for running on kerosene and give good satisfaction. The kerosene equipment consists of a jacketed part of the exhaust pipe extending below the manifold on the cylinder. The carburetor is attached to the bottom of this chamber, and as the kerosene vapor comes in contact with the hot exhaust pipe it is thoroughly gasified and remains in this state while it is drawn into the cylinder. Arrangement is made to generate a small quantity of steam, which is drawn in with the kerosene gas; this practically eliminates carbon and softens the action of the motor, as well as increase the efficiency.ciency.

AUTOMATIC STEAM TOWING WINCH

BY permitting the use of flexible steel cable instead of Manila rope, the Corbet Automatic Steam Towing winch effects a considerable saving in renewal expense for towing lines, steel cables having a life of approximately fifteen years as against the frequent renewal cost of ropes.

Tug operation costs are reduced as one or two men less are required for operation. An important feature is the saving of time in the operation of the tug, as the raft or scow can be brought close to the tug while the tug is under way. Still another feature is the addition of a spool which can be operated independently from the drum by means of a friction between the gear and the drum, thereby enabling the machine to serve two purposes at one time.

The engines in No. 1 and the two larger sizes are placed outside of the frames, enabling the machine to be built more compact and lower, thereby preventing the tug from being tipped when pulling at right angles with the tow, and engines much easier to get to for adjustments.

The machine is operated by one lever, which starts, stops and reverses, and only requires one man to operate it, and is built by the Corbet Foundry & Machine Co., Owen Sound.

The bed of the winch occupies a deck space of 7 ft. by 7 ft. The engine is composed of two 12 by 12 in. cylinders, which operate the drum and spool by a pinion 11 in. diameter on crank shaft and a gear 55 in. diameter by 5 in. face on the drum shaft, thereby giving the machine abundant power.

These machines are equipped with the automatic hawser leader, which travels to and fro across the front of the drum while the cable is going out or in, thereby preventing the hawser climbing on top of itself and allows the tug to pull at right angles with the tow. It is also equipped with the automatic release and take-up valve, which automatically releases the hawser when an extra strain comes on it and takes it in again as soon as the hawser slackens, thereby preventing parting of the hawser and losing the tow in rough weather.

AUTOMATIC STEAM TOWING WINCH.

The machines have a friction between the drum and the gear which enables the engines and spools to be operated independently from the drum while the tug is under way towing a raft or barge.

The machines are operated by one lever, which starts, stops, and reverses, and these whinches eliminate the old-fashioned link motion, and the tow can be brought up close to the tug while the tug is under way, without having to back the tug up to the tow, thereby giving complete control of the tow while changing the course in rivers or narrow channels, etc.

MANITOBA STEEL & IRON CO.

THE organization meeting of the Manitoba Steel & Iron Co., Ltd., was held in Winnipeg recently, and the following Board of Directors was elected for the ensuing year:—T. R. Deacon, H. B. Lyall, Sir Augustus Nanton, Geo. F. Galt, G. W. Allan, Sir Douglas Cameron, Chas. Pope, Capt. Wm. Robinson, W. H. Cross. At a subsequent meeting of the directors the officers were elected as follows: T. R. Deacon, president; H. B Lyall, vice-president; Walter Stuart secretary.

The company has been incorporated with a Dominion charter, with an authorized capital of $500,000, to take over the merchant end of the business of the Manitoba Bridge & Iron Works, which has grown to considerable dimensions. The new company is strongly financed and will carry on a general merchant business in heavy steel goods, such as structural steel, plates and sheets, bar iron and steel, boiler tubes, rivets, bolts, railway supplies, mining equipment, heavy forging billets and stock for shipbuilding. An entire block of land with suitable warehouse has been secured on Logan Avenue, with railway siding facilities, and business will be commenced on March 1. The business will be carried on along lines similar to the large steel and iron merchant businesses of the United States, contracts for large tonnages being made at one time with the rolling mills. Authority is also given under the charter to build and operate rolling mills and blast furnaces.

The Manitoba Bridge & Iron Works purpose confining their business more to the purely manufacturing side of the business, for which this change will afford them more needed room on their present site. The latter company is also applying for a Dominion charter, with an authorized capital of $1,000,000.

SEA WATER DISTILLING APPARATUS

THOSE who have to deal with the distilling of sea or other water or with evaporation problems of almost any kind will be interested in this design of Lillie Evaporator now being built by the Wheeler Condenser & Engineering Co. of Carteret, N.J. It is, as will be noted, a modification of a regular Lillie sextuple effect sea water distilling apparatus. Two of these now under construction are to be operated by steam up to 60 lbs. per square inch gauge pressure, or at any lower pressure.

The point that will catch the veteran's eye is the employment of four condensers, side by side, as distinctly shown in the illustration.

This unusual arrangement of condensers permits seven different combinations of operation, as follows:

(1) It may be operated as one single effect or more single effects.

(2) It may be operated as one or more double effects, with vapors reversible in each.

(3) It is possible to operate it as a triple effect, or as two triple effects, with vapors reversible in each.

(4) It is impossible, of course, to operate it as two quadruple effects, but every effect may be utilized by grouping as one quadruple effect and one double effect, in both of which the vapors are reversible.

(5) It may be operated as one vapor reversible quadruple effect with both end effects, or either end pair of the section cut out.

(6) With one effect at either end cut out, it may be operated as a vapor reversible quintuple effect.

(7) Lastly, it may be operated as a vapor reversible sextuple effect.

It is evident that should a mishap occur at either end, in the middle, or anywhere else, there is little danger that this evaporator will be put out of commission entirely.

Kingston, Ont.—The City Council has decided to submit a by-law to the people for the purpose of raising by debentures $30,000 for the building of a public wharf.

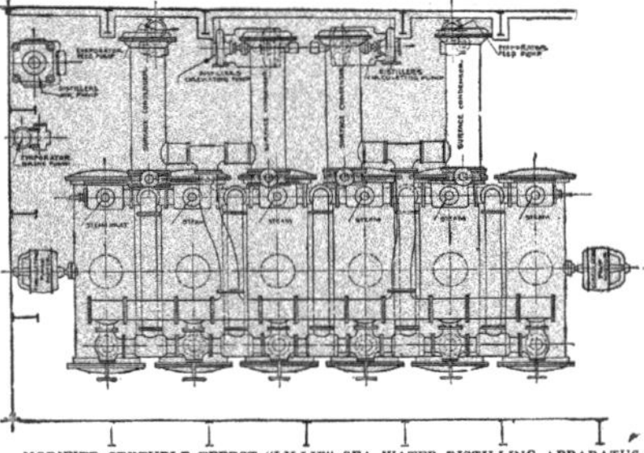

MODIFIED SEXTUPLE EFFECT "LILLIE" SEA WATER DISTILLING APPARATUS.

ASSOCIATION AND PERSONAL

A Monthly Record of Current Association News and of Individuals Who Have Been More or Less Prominent in Marine Circles

Alexander R. Milne, marine engineer, known all over the Great Lakes, died at Kingston, Ont., on Feb. 6, after two days' illness. He was about 75 years of age.

Capt. James Lanaway, a former resident of Woodstock, Ont., died in Seattle on Jan. 27, aged 58. For many years prior to going west seven years ago he was a captain on the Great Lakes.

Lieut. S. S. Smith, formerly sales and credit manager for the Lake Winipeg Shipping Co., Winnipeg, Man., who was reported missing early in June, is now officially presumed to have been killed in action. Lieut. Smith went overseas with a Winnipeg battalion in Oct., 1915.

Arthur B Sweezey, who has been attached to the New York office of the Cunard Line for more than twenty-six years, has left for Vancouver to establish an office in that city for the company This is the first of a series of branch offices which the Cunard Co. will establish in various parts of the world.

H. A. Bayfield, engineer in charge of the building of the Government's big assembling plant at Ogden Point, Victoria, B.C., where all the Imperial Munition Board's ships are to be outfitted, died on February 13. He was formerly Superintendent of Dredges for British Columbia under the Federal Government, and before that was in charge of extensive wharf construction at St. John, N.B., and other Eastern seaports.

F. C. Winterburn, chief engineer of the steamer Quadra, died at St. Paul's Hospital, Vancouver, B.C., on Jan. 12. He was 57 years of age and had done duty in England as inspector of munitions for over two years, having heard the call of his country in 1915 while working at the Yarrow yard at Esquimalt. He returned to British Columbia six months ago and shipped as chief of the Quadra which is in the ore trade between Britannia Beach and Tacoma.

J. H. Bertram, collector of Customs for the Port of Toronto, who has been acting collector for the Port of Montreal during the past year, since the retirement of R. S. White, was given a cabinet of cutlery last Friday by the officials and employees of the Montreal Customs House. Mr. Bertram is leaving to resume his duties in Toronto, having been relieved by the appointment of Ald. W. S. Weldon as collector here

Lieut. Harold H. Vroom has invented an appliance called the Vroom hydrophone which has been accepted by the British Admiralty. Lieut. Harold H. Vroom is a son of Ald. E. G. Vroom, of St Lambert, Montreal, and a native of St. Stephen, N.B. He was educated at McGill University, being graduated in 1910 with the degree of Bachelor of Science. In May 1916, he enlisted as sub-lieutenant in the motor patrol service, and was promoted to the rank of lieutenant in the spring of 1917.

Lieut.-Col. Charles N. Monsarrat, who has just been elected a director of Canada Foundries and Forgings Co., is one of the best known bridge engineers in Canada. Entering the service of the Canadian Pacific Railway as structural draftsman in the office of the chief engineer in 1889, he was inspector of steel bridges, 1896-97; engineer in charge of designing and erection of important structures in British Columbia and elsewhere, 1897-1901; assistant engineer in Montreal, 1901-03, and engineer of bridges, 1903-11. In 1911 he was appointed chairman and chief engineer of the Quebec Bridge Commission, and to that important work he devoted himself until the completion of the great structure.

Albert George Hill, late manager of Babcock & Wilcox, Ltd., Toronto office, has been appointed production engineer in the shipbuilding department of the Imperial Munitions Board, with headquarters in Toronto. Mr. Hill will have charge of production of marine engines and boilers in the Toronto district. Mr. Hill was with Babcock & Wilcox, Ltd., in charge of the Toronto office for five

LICENSED PILOTS

ST. LAWRENCE RIVER

Captain Walter Collins, 43 Main Street, Kingston, Ont.; Captain M. McDonald, River Hotel, Kingston, Ont.; Captain Charles J. Martin, 13 Balaclava Street, Kingston, Ont.; Captain T. J. Murphy, 11 William Street, Kingston, Ont.

ST. LAWRENCE RIVER, BAY OF QUINTE, AND MURRAY CANAL

Captain James Murray, 106 Clergy Street, Kingston, Ont.; Capt. James H. Martin, 259 Johnston Street, Kingston, Ont.; John Corkery, 17 Rideau Street, Kingston, Ont.; Captain Daniel H Mills, 272 University Avenue, Kingston, Ont.

MONTREAL PILOTS' ASSOCIATION

President—Alberic Angers, Montreal.
Secretary—C. B. Hamelin, Champlain, Que.

ASSOCIATIONS

DOMINION MARINE ASSOCIATION

President—A. A. Wright, Toronto. Secretary—Francis King, Kingston, Ont.

GREAT LAKES AND ST. LAWRENCE RIVER RATE COMMITTEE

Chairman—W. F. Herman, Cleveland, Ohio.
Secretary—Jas. Morrison, Montreal.

INTERNATIONAL WATER LINES PASSENGER ASSOCIATION.

President—O. H. Taylor, New York.
Secretary—M. R. Nelson, 1184 Broadway, New York.

SHIPPING FEDERATION OF CANADA.

President—Andrew A. Allan, Montreal; Manager and Secretary—T. Robb, 218 Board of Trade, Montreal; Treasurer, J. R. Binning, Montreal.

SHIPMASTERS' ASSOCIATION OF CANADA.

Secretary—Captain E. Wells, 45 St. John Street, Halifax, N.S.

GRAND COUNCIL N.A.M.E. OFFICERS.

A. R. Milne, Kingston, Ont., Grand President.
J. E. Belanger, Bienville, Levis, Grand Vice-President.
Neil J. Morrison, P.O. Box 238, St. John, N.B., Grand Secretary-Treasurer.
J. W. McLeod, Owen Sound, Ont., Grand Conductor.
Lemuel Winchester, Charlottetown, P.E.I., Grand Doorkeeper.
Alf. Charbonneau, Sorel, Que., and J. Scott, Halifax, N.S., Grand Auditors.

1918 Directory of Subordinate Councils, National Association of Marine Engineers.

Name.	No.	President.	Address.	Secretary.	Address.
Toronto,	1	Arch. McLaren,	324 Shaw Street	E. A. Prince,	49 Eaton Ave.
St. John,	2	W. L. Hurder,	209 Douglas Avenue	G. T. G. Blewett,	36 Murray St.
Collingwood,	3	John Osburn,	Collingwood, Ont.	Robert McQuade,	Collingwood, Ont.
Kingston,	4	Joseph W. Kennedy,	395 Johnston Street	James Gillie,	101 Clergy St.
Montreal,	5	Eugene Hamelin,	Jeanne Mance Street	M. Lasure	120 Ribard St.
Victoria,	6	John E. Jeffcott,	Esquimault, B.C.	Peter Gordon,	808 Blanchard St.
Vancouver,	7	Isaac N. Kendall	319 11th St. E., Vanc.	E. Read,	232 13th St. W.
Levis,	8	Michael Latulippe,	Lauzon, Levis, Que.	Arthur Jolin,	Lauzon, W.
Sorel,	9	Nap. Beaudoin,	Sorel, Que.	Alf. Charbonneau,	Box 204, Sorel, Que.
Owen Sound,	10	John W. McLeod	570 4th Ave.	R. J. McLeod,	862 8th St.
Windsor,	11	Alex. McDonald,	28 Crawford Ave.	Neil Maitland,	[illegible] London St. W.
Midland,	12	Geo. McDonald	Midland, Ont.	A. E. House,	Box 333
Halifax,	13	Robert Blair	29 Parrsboro Street	Chas. E. Pearce,	Portland St., Dartmouth, N.S.
Sault Ste. Marie,	14	Charles H. Innes,	27 Euclid Road	Wm. Hindmarch	34 Euclid Rd.
Charlottetown,	15	J. A. Rowe	176 King Street	Chas. Cumming,	27 Euston St.
Twin City,	16	H. W. Cross,	436 Ambrose St	J. W. Farquharson,	169 College St.
				A. H. Archand,	Champlain, Que.

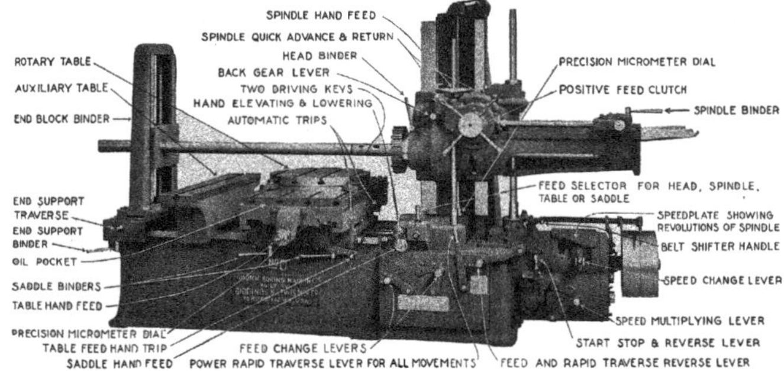

Horizontal Boring, Drilling and Milling Machine

FOSDICK No. 0

Specifications: This Fosdick No. 0 Horizontal Boring, Drilling and Milling Machine has a 3⅛ in. spindle bored to fit a No. 5 Morse Taper. The spindle traverse is 26 in. and the maximum distance from its center to the table is 26 in. The distance from the face of the table to the boring bar support is 60½ in. The cross travel to table is 30 in. and longitudinal traverse to table is 32 in.

There are 16 spindle speeds in each direction, ranging from 12 to 225. The number of feeds in all directions is 16, ranging from .004 in. to .260 in.

The moving parts are incased and a safety friction device, adjustable from the outside, prevents accidents. The control is centralized and all levers are provided with latches to prevent chattering on heavy work.

Attachments: When required we can furnish a plain revolving table, also a revolving table with worm movement—either of these graduated to half degrees, an auxiliary table for work too large for the regular table, boring bars up to 3⅛ in. diameter, and a facing attachment which will face from zero to 18 in. diameter.

For further information and quotations on this or any other manufacturing equipment address our nearest house.

The Canadian Fairbanks-Morse Company, Limited

"Canada's Departmental House for Mechanical Goods"

St. John, N.B. Quebec Montreal Ottawa Toronto Hamilton Windsor Winnipeg Saskatoon Calgary Vancouver Victoria

years, having previously been with the Drummond, McCall & Co., Toronto office, in the engineering department. He was two years with this concern as sales engineer in Toronto district and Western Canada. Prior to coming to Canada Mr. Hill was for some years with Alley & MacLellan, Ltd., of Glasgow, in the marine and high-speed engine department, and previously with Deacon & Foster, consulting engineers, of Birmingham, England.

EXPANSION OF MINERAL INDUSTRY ESSENTIAL

CANADA pays out more money for imported mineral products than she receives from her mines. The value of the mineral production for the calendar years 1913, 1914 and 1915 was $145,600,000, $128,865,000 and $137,100,000 respectively. The imports of products of the mine and manufactures of mine products for the same years were valued at $259,300,000, $181,676,000 and $146,324,000. As the imports also include manufactured, or partly manufactured products, they are much more valuable than the minerals we produce. If, however, Canadian minerals were turned into manufactured products in Canada, the present trade balance in minerals would be reversed.

It is only fair, though, to point out that Canada is under serious disadvantages in the matter of manufacturing. The relatively small and scattered population makes distribution from points of production to points of consumption both difficult and costly. Similarly, where, for example, coal is essential for reducing ore and for manufacturing, the cost of transportation necessary to bring the two raw products together bears heavily on manufacture. Copper, zinc and lead are produced principally in Western Canada, while the manufacturers and chief markets are in Eastern Canada.

In spite of these handicaps, a comparison of the figures for imports and those for production shows the opportunity that exists for developing a home market that will increase as the war goes on. Premier Lloyd George in his recent address stated that "Economic conditions at the end of the war will be in the highest degree difficult. There must follow a world shortage of raw materials, which will increase the longer the war lasts, and it is inevitable that those countries which have control of raw materials will desire to help themselves and their friends first."

The mineral resources of Canada, if developed, could supply not only our own needs but also permit the exportation of a surplus to other parts of the British Empire. There is, in Canada, an urgent need for production to pay for our war debt and borrowings before the war, and if we are to get the greatest value out of our mineral industry it is necessary that our metals and minerals be refined and made into manufactured or partly manufactured products in Canada. The production of certain mineral products in Canada has been stimulated by the war and new industries created. In the period of reconstruction, after the war, it will be necessary to safeguard and provide for the further extension of these industries.—*Conservation.*

CANADA'S MINERAL SUPPLY

PLANS for the development of the mineral resources of Canada are likely to be considered at Ottawa at an early date by the Reconstruction Committee of the Cabinet of which Hon. A. K. MacLean is the chairman. It is stated that the Government has already received expert advice to the effect that Canada, if her mineral resources were properly developed, could supply not only her own needs, but also permit the exportation of a surplus to Europe. At the present time Canada pays out more money for imported mineral products than she receives from her mines. The desirability of having as large a proportion of minerals as possible refined and made into manufactured products in Canada will also be considered by the committee. The production of certain mineral products in Canada has been stimulated by the war and new industries created. Steps will be taken to safeguard and provide for the further extension of these industries.

THE FRENCH STEEL INDUSTRY

BEFORE the war, France was already considered by experts as the richest country of the world in iron ore, after the United States. Besides the basins of Lorraine, which produce 3,000,000,000 tons of ore, she possesses the basin of Normandy, which, in an area of 40.000 square kilometers, contains, according to a recent estimate, nearly a billion tons of a superior quality of ore yielding 50 pèr cent. of iron. In 1914 21 grants in mining districts producing 1,152,000 tons a year, already existed in the departments of the Orne. Manche and Calvados, which yielded 648,000 tons with six grants only under exploitation. About the middle of 1916, a large society, entirely French, the Societe Normande de Metallurgie, was founded, with a capital of 40,000,000 francs, by Messrs, Schneider & Co., the great Creusot manufacturers, the Societe des Acieries de la Marine et Homecourt, and several shareholders belonging to the metallurgic world. This new society leased the foundries and fittings previously organized, or provided for, by the Societe des Hauts-Fourneaux de Caen, formed in 1892, and in July, 1916, started the works afresh now situated at Mondeville-Colombelle near Caen. About 30 kilometers distant is the Soumont mine, capable of yielding even now 4,000 tons of iron ore a month, and which, in the near future will have a larger output. The establishments of the Societe Normande de Metallurgie comprise gas and coke furnaces, large steel works, a flattening mill, smelting works and blast furnaces. Three stations and a railway line 33 kilometers long facilitate the transport service. The line was made by a branch society and runs between the above foundries and the Soumont mines. At present there are four batteries of 42 furnaces each, producing 1,000 tons of coke a day. In January, 1918, two other batteries, now under construction, will be in working order, and will bring the output up to 1,500 tons a day or 500,000 tons a year. The coke, collected by an ingenious system of inclined planes and chains, after being sorted out, is sent in trucks to government metal works. Gas undergoes a series of processes for the purpose of extracting the secondary products obtained by the distillation of coal such as tar, phenol, benzine, tuluol, naphthalin, sulphate of ammonia, used in the manufacture of explosives and for dyeing purposes, and which, prior to the war, were almost exclusively produced by interests in Germany.

The blast furnaces, the first of which has been recently inaugurated, are in themselves 29 meters high, but with their basements tower 45 meters above the foundry ground plot. They can produce 400 tons a day, which is said to be the largest output obtained in Europe up to the present day. To each furnace is affixed a battery of five metal cylinders, or "Cowpers," seven meters in diameter, for heating the air for the combustion of coke, and the reaction of carbon on the oxides of the ore, which is the source of production for cast iron. Three of these gigantic blast furnaces will be ready in the autumn of 1918 and the whole installation, when completed, will allow of an output of 450,000 tons of cast iron a year. The steel works, properly so-called, contain four converters of 30 tons, and five Martin furnaces of 30 tons, allowing of an output of 275,00 tons of Thomas steel and 125,000 tons of Martin steel every year.

The flattening mill can pass 500,000 tons of ingots a year, and is supplied with several reversible trains for the production of all small samples. The Societe already possesses stocks of iron ore sufficient to insure the supplies for its factories without any difficulty. It has had a large water tower, 66 meters in height, built in the neighborhood, with two reservoirs and a canal to the Orne, so that 10,000 cubic meters of water could be supplied hourly. A vast hall 200 meters long contains the electric machinery for distributing motive force, two turbo alternating generators of 3.000 kilowatts, one of 5,000, six gas alternating generators of 6,000 horsepower, and three of 3.000 horsepower. A private harbor has been built on the Orne canal, and it is now ready for vessels of 2,000 tons, and will soon be open to boats of 8,000 tons capacity.

The preceding facts show how the Societe Normande has been able to create in the midst of war one of the most important metallurgic centers in the whole of France.

MARINE NEWS FROM EVERY SOURCE

The Sorel Shipbuilding & Coal Co., of Sorel, Que., has changed its name to that of Tidewater Shipbuilders, Ltd.

Vancouver, B.C.—The first of the vessels being constructed at the Lyall shipyard will be named "War Puget" and the second the "War Cariboo."

Equimalt, B.C.—The British Columbia Salvage Co's. steamer Salvor has been docked at Yarrow's yard for a general overhaul.

Vancouver, B.C.—The Union Co.'s steamer Coquitlam has been hauled out at the B.C. Marine Co's. plant for annual inspection and overhaul.

Victoria, B.C.—The War Songhee, has left her moorings at the Foundation Co's. wharves at Point Hope and has gone to Ogden Point to have the boilers and engines, etc., installed.

St. John's, Nfld.—The keel of the largest vessel yet to be built in Newfoundland has just been laid at a shipyard at Harbor Grace. The ship will be built of wood, and will be 600 tons deadweight.

Vancouver, B.C.—The executors of the John Hendry estate have sold a site on False Creek to John Coughlan & Sons, which will give the shipbuilders an additional 400 feet of frontage which they urgently need for the expansion of their business.

Bridgewater, N.S.—J. N. Rafuse & Sons, of Bridgewater, have launched a schooner of 113 feet keel, named Industrial, at the Foley shipyard, Salmon river and will immediately start the construction of another vessel off the same model.

Collingwood, Ont.—At the annual meeting of the Colingwood Shipbuilding Co., the following officers were elected: President, A. H. Johnson; vice-president, W. G. Smart; managing director, Capt. G. C. Coles and assistant manager G. T. Foulis.

Contract Set for Concrete Ships.—A contract for ten 3,500-ton concrete ships has been let by the U. S. Shipbuilding Board to the Ferrar Concrete Shipbuilding Corporation, of Redondo Beach, Cal., the first to be delivered within six months and the balance in a year.

Vancouver, B.C.—In addition to getting her hull cleaned and painted, the auxiliary schooner Jessie Norcross, Captain McIntyre, is getting two new tail shafts at the Wallace shipyards. She will leave shortly for Chemainus to load a cargo of lumber for Australia.

Victoria, B.C.—The auxiliary schooner Beatrice Castle, which was launched at the Cameron-Genoa Shipbuilders' yards about ten weeks ago, being the last of the schooners built for the H. W. Brown interests, has been rechristened the Stasia. The vessel has been granted provisional French registry.

Steamer Bermudian Sunk.—Word has come to the Marine Department at Ottawa that the steamer Bermudian, taken over by the Admiralty from the Canada Steamship lines, has sunk in Alexandria Harbor in the Mediterranean. It was due to an accident. The ship is to be salved.

Vancouver, B.C.—It is understood that several of the dredgers employed by the Dominion Government will be laid up on March 1, until then end of the fiscal year. The vessels affected by the order are as follows: Ajax, King Edward, Mastodon, Mudlark, Victoria and drill plants Nos. 1 and 2, and rockbreakers Nos. 1 and 2.

Emergency Fleet Co. Laying 25 Keels a Month.— George J. Baldwin, chairman of the Board of the American International Shipbuilding Corporation, which has contracts for the construction of ships for the Emergency Fleet Corporation, told the Senate Commerce Committee the company expected to lay fifty keels during February and March.

Allan Line Discontinues Service From Boston.—The Allan Line, which for more than 25 years has operated steamships between Boston and Glasgow, is to discontinue its Boston service on May 1. The C.P.R. obtained control of the Allan Line about two years ago, and many of its ships have been requisitioned by the British Admiralty for war service.

Vancouver, B.C.—The Lyall Shipyards at North Vancouver are beginning to look forward to another launching, the present indications being that the furthest advanced of the Munitions Board wooden steamers will be ready for the water within the next three weeks. The yard is getting ready to fit tail shafts, which are expected to arrive from the East shortly.

Victoria, B.C.—The auxiliary schooner Jean Steadman, eleventh of the fleet built for the Canada West Coast Navigation Co., is fitting out at Victoria and will run her trials some time this week. She is to load lumber for Australia at the Fraser River Mills and will do her trials while bound up from Victoria to the river. Captain Manning will be the master.

Probe Pilotage System.—The Minister of Marine and Fisheries will, in the course of a few days, create a commission to investigate the pilotage systems in operation at the port of Halifax, and other important ports. In view of the recent disaster at the port of Halifax, the commission will first direct the inquiry into the pilotage system and its operation at that port.

Victoria, B.C.—The name of the wooden steamer, War Tyee, was changed to War Yukon a few days before being launched. The War Yukon will be the first vessel of the Imperial Munitions Board to get her machinery, and will, no doubt, be the first of the twenty-seven under construction to get away to sea. She has reached the stage when she is ready for installation of her engines.

Pictou, N.S.—The Gaspesien has been towed into Pictou harbor and will remain here for the rest of the winter. The Gaspesien when bound from Montreal to New York was caught in the ice in Northumberland Strait and helpless for two weeks. Later she drifted down between Belle River and Point Prim being eventually released by the ice breaker Prince Edward Island.

Cannot Build Ships for Neutrals.—It has been decided, with a view to meeting special British requirements, that no further orders must be taken in Canada for new vessels other than of Canadian or British register; at present a number of new ships are being completed in Canada on Norwegian order. As soon as the ships under construction are completed other vessels will be laid down under the Government programme.

Vancouver, B.C.—The steel steamer Alaska, launched from the Coughlan lards on Jan. 19, has been moved to the fitting-out berth and work will be commenced installing the engines and boilers. The engines are coming from the East, but the boilers were made at the Coughlan yards. The next vessel to be launched from the yard will be the War Camp, which will be sent down the ways some time this month.

Ottawa, Ont.—Tenders will be received until Friday, Feb. 22, for the purchase of the car ferry steamer Leonard, formerly engaged in transferring cars between Levis and Quebec. The Leonard was built in 1914, and designed to fulfill the requirements of the highest class in Lloyd's for a vessel of this type. Her principal dimensions are: Length over all, 313 feet; breadth, 65 feet; depth, 23 feet; draught, 16 feet. The Leonard is now lying at Levis, Que., below the drydock, where she can be examined. Information as to the steamer's engines, boilers, steam pressure, speed, bulkheads, tracks, gangways, ice breaking, etc., etc., etc., and further information can be obtained at the Department of Railways and Canals, Ottawa.

Must Load Cars to Full Capacity.—An order prohibiting acceptance by any Canadian railway of freight for overseas export unless the cars are loaded to full cubic or weight-carrying capacity was issued last Friday by the Canadian War Board. U. E. Gillen, vice-president of the Grand Trunk and chairman of the board, said that appeals to the public spirit of shippers had brought good results in combatting the car shortage.

Halifax, N.S.—Another shipbuilding plant has been established at Meteghan River by the R. H. Rowes Construction Co., of Boston and New York. The president of the company is George D. Morecroft. A lease for term of years has been taken of the shipyards of the late James Cosman at the mouth of the Meteghan River. Already, mills and pattern shops, equipped with the latest improved shipbuilding machinery, have been erected and are now working.

Victoria, B.C.—Capt. T. E. Gilmour, of this city has invented a caulking machine. The machine consists of two V-shaped guides which force the oakum under two revolving steel disks, the latter being hammered by compressed air into the seams, a handle to draw and steer the machine along the seams, and eight tension springs to hold the guides in position make up the device. It is claimed that the machine can caulk at the rate of 600 feet per hour.

North Vancouver, B.C.—The Wallace Shipyard Ltd. will launch the War Power early in March. The War Power is the same size as the War Dog, 4,700 tons deadweight, which the firm put in the water recently. The engines, a triple set of 1,650 i.h.p. are now completed in the yard's machine shop. The boilers, two 15 ft. 6 ins. diameter by 11 ft. 3 ins. long, are being built by the Vulcan Iron Works at their Granville Island plant in Vancouver.

Montreal, Que.—The steamship Percesian," commanded by Captain Joseph Bernier, the Arctic explorer, has been sunk, according to a cable received by the Robert Reford Company here. The "Percesian" left Halifax Jan. 1. She was a vessel of 782 tons, the property of the Gaspe & Baie des Chaleurs Steamship Co. She was well known in the local shipping trade, as she was for years engaged in the service between Montreal and Gulf ports. Latterly she had been transferred to overseas service.

Victoria, B.C. — Construction of two big tugboats for North Coast service and several scows. all for the Whalen pulp interests, is likely to be the next shipbuilding order awarded to the Foundation Company's yard at Point Hope in the inner harbor. The tugboats will have a length of 128 feet, a breadth of about 26 feet and a depth of 16 feet, so that they will be practically the largest vessels of their kind in these waters. and by a wide margin the biggest and most powerful tugboats ever built in the North-west.

Cunard, Anchor and Donaldson Lines Unite.—Announcement of a steamship merger that may mean a great deal to the development of British and Canadian after-the-war trade was made last week at the Montreal office of the Cunard Line. The Cunard, Anchor and the Anchor-Donaldson lines were joined under one management on Jan. 1. They will be operated, so far as their Canadian business is concerned, under the supervision of the Robert Reford Co. instead of through American offices, as heretofore. Joint offices have been established in Winnipeg and Vancouver.

France Would Buy B.C.-built Ships.—The agitation begun in British Columbia for the revocation of the Dominion Government's decision to disallow shipbuilding in Canada to the order of foreign interests makes the fact that ninety-six auxiliary powered schooners have recently been bought in the United States by France of special interest. It is said that if the Imperial Munitions Board would allow ships to be constructed in Canada for registry other than British, the shipyards of this Province would have no difficulty in getting plenty of contracts from French interests.

Vancouver, B.C.—The wooden steamer War Nootka, launched from the Western Canada Shipyards on January 14, will leave False Creek some time this week. She will be towed to Victoria and berthed at the Ogden Point fitting-out plant of the Imperial Munitions Board to receive her engines and boilers. The spans at Granville street and Kitsilano bridges will therefore be open one morning while the ship passes out. She is the Munitions Board fleet to be launched here. The yard is making good progress on the other three hulls and the number five hull, building on the ways vacated by the War Nootka, is now in frame.

U. S. Concerns Would Start Steel Plants in Canada.—It is reported in Ottawa that the Dominion Government has received tentative offers from large American companies desiring to engage in the manufacture of steel for Canadian ships. These companies, however, required encouragement in the way of bonuses and concessions before taking up the manufacture of steel products on a large scale in Canada. The Government, it is understood, has not yet decided to bonus corporations desiring to engage in the work. Companies may, it is stated, submit a proposition and prices to the Government which will consider them and accept or reject them as it sees fit.

Vancouver, B.C.—The wooden steamer *War Nootka*, one of 46 similar standard type vessels contracted for in Canada by the Imperial Munitions Board, has been launched by the Western Canada Shipyards, False Creek, Vancouver. The dimensions of the vessel are as follows: Length between perpendiculars, 250 ft.; length over all, 259 ft.; breadth, extreme, 43½ ft.; breadth, molded, 42½ ft.; depth, molded, 25 feet; depth, over keel, 27 ft.; draft for displacement, 22 ft.; draft over keel, 21 ft.; deadweight on 20 ft. maximum, to Lloyd's summer freeboard, ap-

Champion Steel Rivet Forges and Blowers

Shipbuilders, bridge builders, boiler makers and railroads in the United States use more Champion Forges and Blowers than all other makes put together. Almost 99 out of every 100 erectors of structural steel buildings use the Champion Steel Rivet Forge shown by illustration at the left.

In Canada obtain at

Over 600,000 Champion No. 400 Steel Blacksmith Blowers and Forges are in use to-day. Blowers purchased 12 years ago are still running without a particle of lost motion to the crank and as smooth and noiseless as the day they were placed to the fire.

We would like to tell you why Champion Steel Rivet Forges and Blowers are universally favored.

We would like to explain the Champion patent-protected construction that gives Forges and Blowers ability to serve perfectly for many years.

The Famous 400 Champion Steel Blacksmith Blower

Made with adjustable ball bearings only. The patented high-speed spiral gearing is the only high-speed gearing that can possibly be built to run a blower at the extreme speed necessary to produce a roaring, white heat pressure blast without destruction to the gearing as well as the bearings.

Say you are interested

No. 1½ Champion One-Fire Variable Speed Electric Blower

Has six variable speeds and produces a very strong blast whenever extra heavy fire is needed, or a blast just right for smallest work. Used by many shipbuilders and in many railroad shops.

AIKENHEAD HARDWARE LIMITED

17, 19, 21 TEMPERANCE STREET, TORONTO

War Tyee Launched at Victoria.—The wooden steamer War Tyee was successfully launched at the Cameron-Genoa Mills Shipbuilders, Ltd., Victoria, B.C., yards on Jan. 24. This vessel is the third of the boats building on the coast for the Imperial Munitions Board. She was christened by Mrs. Butchart, wife of R. P. Butchart, representative of the Imperial Munitions Board. The War Tyee is the same size as the War Songhee and War Nootka, being a standard type chosen by the board. The War Tyee is 250 feet long, 43 feet 6 inches beam, and 25 feet deep, with a deadweight capacity of 2,800 tons on a 21 feet draught. The machinery will be installed at the Ogden Point assembling plant. She will be equipped with triple expansion engines of 950 horse-power. The official trials of the War Tyee will probably be run about the middle of March.

Canadian General Electric Co.—The fiscal year of the Canadian General Electric Co. was brought to a close with the end of December. Col. Frederic Nicholls, the president, states that the volume of business transacted during the 12 months in question compares favorably with the previous year. "While under present unsettled conditions it is difficult to look very far ahead," he adds, "the outlook is encouraging. The principal improvement effected by the company in 1917 was the fitting up of its Bridgeburg works at the Niagara River as a shipyard, which is now in operation for the purpose of constructing four steel cargo steamers of 3,500 tons each for the Imperial Munitions Board. At the Davenport works in Toronto the company has contracts for brass cartridge cases and steel shell forgings for the United States Government, and at its Rockfield plant in Montreal it is engaged in the production of 6-inch forgings for the Imperial Munitions Board.

If what you need

Can be obtained from all principal Ship Chandlers, and Merchants

"FOSCO" and "FOSBAR"

Life Jackets or Preservers, in Cork or Kapok. To latest Government requirements.

Also Life Buoys, Ships Rope Fenders

MANUFACTURERS:

FOSBERY CO., BARKING, ESSEX, ENG.

FRANCE

Marine Type

Metallic Packing

For All

Conditions of Service

FRANCE PACKING COMPANY
TACONY—PHILA., PENNA.

"What do they know of England
Who only England know?"
—*Kipling*

SPRACO

Pneumatic Painting Equipment

is helping "Her Lady of the Snows" to more speedily meet the world's great need for ships and more ships.

Spraco—the modern way of applying paint—enables one man to do the work of several skilled painters.

Spraco Equipment applies the paint in coatings thick or thin, free from streaks and brush marks.

Write today, giving us full information regarding your painting problem. The recommendations of our engineers will be made without your incurring charge or obligation.

Spray Engineering Company

93 Federal Street
Boston, Mass.

Cable Address: "Spraco Boston"
Western Union Code

Representatives in Canada:

Rudel-Belnap Machinery Co., 95 McGill St., Montreal

Hoyt's Marine Metals

The British, French and United States fighting fleets use Hoyt Metals in immense quantities.

The Hoyt Quality is famous the world over. You'll invariably find it everywhere where utmost speed and dependability are in demand:

MARINE BABBITT

This alloy is made according to Government formula of the very best selected materials, and carefully alloyed. We can guarantee uniformity throughout. In using this Hoyt alloy you will never have any trouble with your bearings on your engines after once installed.

HOYT'S EAGLE "A" BABBITT

This metal is used extensively by some of the largest shipbuilding concerns in the United States for main bearings on the propeller shaft. It is thoroughly alloyed and made of selected stock and highly anti-friction.

HOYT LEAD PIPE

We manufacture this both in the soft and antimonial hardened pipe. The greatest care is exercised to give uniform thickness of wall, and in the selection of the raw material used.

HOYT SHEET

We not only manufacture what we believe to be the highest grade sheet lead on the market, but a sheet antimonial alloy which has proven, by a great number of years of severe tests, to be far superior to sheet lead. This is due to the fact of its greater tensile strength and greater rigidity and ability to resist corrosive action.

PLEASED TO CALL AND GO INTO OUR REPRESENTATIVE WOULD BETHE MATTER IN DETAIL

Hoyt Metal Company Eastern Avenue and Lewis St. **Toronto**

NEW YORK ST. LOUIS LONDON, ENGLAND

If any advertisement interests you, tear it out now and place with letters to be answered.

10" Special Double-Suction Circulating Pump with 8 x 8 Vertical Engine.

THERE IT IS!

And we absolutely guarantee that it is impossible to manufacture

A BETTER BABBITT
than
IMPERIAL GENUINE
for
MARINE ENGINE BEARINGS

In actual test we have found that under a load of 300 to 1200 lbs. to the square inch the rise in temperature was scarcely perceptible.

The tenacity of the tin and copper mixture combined with ductility renders it best for high speed work and where special service is demanded.

1910

NIAGARA NAVIGATION COY.
April 28th, 1910.

The Canada Metal Co., Ltd.,
Toronto.

Dear Sirs:—In replying to your enquiry of recent date as to how we found our stern bearings on our Twin Screw Steamer Cayuga which were babbitted with your **Imperial Genuine Bearing Metal.** Beg to say the bearing is under water and after two seasons hard service there being no perceptible wear as you know we substituted your **Imperial Genuine Babbitt** to replace the lignum vitæ bearings which were a source of trouble and expense and since the change these particular bearings gave us no concern.

Yours truly,
GEO. M. ARNOLD,
Chief Engineer N.N. Coy.

1918

THE CANADA STEAMSHIP LINES, LIMITED
Toronto, Jan. 14, 1918.

The Canada Metal Company,
Toronto, Ont.

Dear Sirs:—In reply to your inquiry as to the quality of your **Imperial Genuine Babbitt,** I have no hesitation in saying it is the best quality of babbitt for high speed and heavy pressure work that I have handled to date.

This has also been used in making rings for United States Metallic Packing, and find it lasts longer than any other metal I have used.

Yours truly,
(Signed) W. NOONAN,
District Superintendent.

FOR GENERAL MACHINERY BEARINGS
HARRIS HEAVY PRESSURE
WILL GIVE EXCELLENT SERVICE

We manufacture
SHEET LEAD
and have large stocks of Ingot Tin, Zinc Spelter, Pig Lead, Antimony, Ingot Copper.

THE CANADA METAL CO., Limited

HAMILTON MONTREAL **TORONTO** WINNIPEG VANCOUVER

When Writing to Advertisers Kindly Mention this Paper

JB&SCo
BERTRAM
MACHINE
TOOLS
For Structural, Bridge and Shipbuilding Plants
Modern in design and built for heavy service, our line embraces a varied equipment of Punches, Shears, Bending and Straightening Rolls, Coping Machines, Rotary and Plate Planers.
The assistance and advice of our engineers are yours for the asking.
Double Punch and Shear.
Capacity—
Shears 8-in. by 1½-in. plate.
Punches 2½-in. hole in 1½-in. plate.
M 278 - 12653
The John Bertram & Sons Company
Limited
DUNDAS, ONTARIO, CANADA
MONTREAL
723 Drummond Bldg.
TORONTO
1002 C.P.R. Bldg.
VANCOUVER
609 Bank of Ottawa Bldg.
WINNIPEG
1205 McArthur Bldg.
JB&SCo

BUYERS' DIRECTORY

ACCUMULATORS, HYDRAULIC
Smart-Turner Mach. Co., Hamilton, Ont.

AERATING RESERVOIRS
Spray Engineering Co., Boston, Mass.

ALLOYS, BRASS AND COPPER
Dominion Copper Products Co., Ltd., Montreal, Que.

ANCHORS
Kearfott Engineering Co., New York, N.Y.
Henry Rogers Sons & Co., Wolverhampton, Eng.
Wm. C. Wilson & Co., Toronto, Ont.

ARCHITECT, NAVAL
Watts, J. Murray, Philadelphia, Pa.

ASBESTOS GOODS
Wm. C. Wilson & Co., Toronto, Ont.

BABBITT METAL
Aikenhead Hardware, Ltd., Toronto, Ont.
Can. B. K. Morton Co., Toronto, Montreal.
Hoyt Metal Company, Toronto, Ontario.
Wilkinson & Kompass, Hamilton, Ont.
Wm. C. Wilson & Co., Toronto, Ont.

BARGES
Can. B. K. Morton Co., Toronto, Montreal.
Polson Iron Works, Toronto, Ont.

BAROMETERS
Wilson & Co., Wm. C., Toronto, Canada.

BARS, GRATE
Babcock & Wilcox, Ltd., Montreal, Que.

BEARINGS, BRASS
Empire Mfg. Co., London, Ont.
Murray & Fraser, New Glasgow, N.S.
United Brass & Lead Co., Toronto, Ont.

BELLS, SHIPS, ENGINE ROOM, ETC.
Aikenhead Hardware, Ltd., Toronto, Ont.
Cory & Son, Inc., Chas., New York, N.Y.
Empire Mfg. Co., London, Ont.
Morrison Brass Mfg. Co., James, Toronto, Ont.

BELTING, LEATHER
Aikenhead Hardware, Ltd., Toronto, Ont.
Can. B. K. Morton Co., Toronto, Montreal.
Wm. C. Wilson & Co., Toronto, Ont.

BIBBS, COMPRESSION
Empire Mfg. Co., London, Ont.

BINNACLES
Morrison Brass Mfg. Co., James, Toronto, Ont.

BLOCKS, CARGO, HEEL AND TACKLE
Aikenhead Hardware, Ltd., Toronto, Ont.
Loveridge, Ltd., Docks, Cardiff, Wales.

BLOWERS, TURBO-
Mason Regulator & Engin. Co., Montreal, Que.

BOAT CHOCKS
Marten-Freeman Co., Toronto, Ont.

BOILER COMPOUND
Wm. C. Wilson & Co., Toronto, Ont.

BOILER FEED PUMPS
Can. Ingersoll-Rand Co., Ltd., Sherbrooke, Que.
Smart-Turner Mach. Co., Hamilton, Ont.
Williams Machinery Co., A. R., Toronto, Ont.

BOILER FITTINGS
Empire Mfg. Co., London, Ont.
McAvity & Sons Ltd., T., St. John, N.B.

BOILERS, MARINE
Babcock & Wilcox, Ltd., Montreal, Que.
Can. Fairbanks-Morse Co., Montreal, Que.
Can. Vickers, Ltd., Montreal, Que.
Collingwood Shipbuilding Co., Collingwood, Ont.
Doxford & Sons, William, Sunderland, England.
Goldie & McCulloch, Ltd., Galt, Ont.
Hall Engineering Works, Montreal, Que.
Mason Regulator & Engin. Co., Montreal, Que.
Montreal Dry Docks & Shipbuilding Co., Montreal, Que.
National Shipbuilding Co., Goderich, Ont.
The John McDougall Caledonian Iron Works Co., Montreal, Que.
Marsh & Henthorn, Ltd., Belleville, Ont.
Polson Iron Works, Toronto, Ontario.
Port Arthur Shipbuilding Co., Port Arthur, Ont.
Thor Iron Works, Toronto, Ont.
Williams Machinery Co., A. R., Toronto, Ont.

BOOKS, TECHNICAL, MARINE
MacLean Publishing Co., Toronto, Ont.

BOLTS
Wilkinson & Kompass, Hamilton, Ont.

BROKER
Page & Jones, Mobile, Ala., U.S.A.

BRASS GOODS
McAvity & Sons, T., St. John, N.B.
Mueller Mfg. Co., H., Sarnia, Ont.

BUCKETS, CLAMSHELL
Beatty & Sons, Welland, Ont.

BUCKETS, COALING
Beatty & Sons, Welland, Ont.
Marsh & Henthorn, Ltd., Belleville, Ont.

CABLE
Wm. C. Wilson & Co., Toronto, Ont.

CABLE, LEAD COVERED AND ARMORED
Standard Underground Cable Co., Hamilton, Ont.

CABLE, ACCESSORIES
Standard Underground Cable Co., Hamilton, Ont.

CAPSTANS
Advance Mach. & Welding Co., Montreal, Que.
Dake Engine Co., Grand Haven, Mich.
Kennedy & Sons, Wm., Owen Sound, Ont.
Wm. C. Wilson & Co., Toronto, Ont.

CALKING TOOLS
Empire Mfg. Co., London, Ont.
Topping Brothers, New York, N.Y.

CALKING TOOLS, ELECTRIC
Aikenhead Hardware, Ltd., Toronto, Ont.

CALKING TOOLS, PNEUMATIC
Aikenhead Hardware, Ltd., Toronto, Ont.
Can. Ingersoll-Rand Co. Sherbrooke, Que.

CALORIFIERS
Low & Sons, Ltd., Archibald, Glasgow, Scotland

CASTINGS
Can. Steel Foundries, Ltd., Montreal, Que.
Collingwood Shipbuilding Co., Collingwood, Ont.
Kennedy & Sons, Wm., Owen Sound, Ont.
Marsh & Henthorn, Ltd., Belleville, Ont.
The John McDougall Caledonian Iron Works Co., Montreal, Que.
Mueller Mfg. Co., H., Sarnia, Ont.

CASTINGS, ALLOY
Tolland Mfg. Co., Montreal, Que.

CASTINGS, ALUMINUM
Empire Mfg. Co., London, Ont.
United Brass & Lead Co., Toronto, Ont.

CASTINGS, BRASS
Crouse-Hinds Co. of Canada, Ltd., Toronto, Ont.
Empire Mfg. Co., London, Ont.
McAvity & Sons Ltd., T., St. John, N.B.
United Brass & Lead Co., Toronto, Ont.
Tolland Mfg. Co., Montreal, Que.

CASTINGS, GREY IRON, MALLEABLE, ALUMINUM
Crouse-Hinds Co. of Canada, Ltd., Toronto, Ont.
McAvity & Sons, T., St. John, N.B.

CASTINGS, ELECTRIC
Crouse-Hinds Co. of Canada, Ltd., Toronto, Ont.

CELLAR DRAWERS
Mueller Mfg. Co., H., Sarnia, Ont.

CEMENT SOLUTION
Can. Bitumastic Enamels Co., Toronto, Ont.

CHAINS
Aikenhead Hardware, Ltd., Toronto, Ont.
Kearfott Engineering Co., New York, N.Y.
Henry Rogers, Sons & Co., Wolverhampton, Eng.
Wm. C. Wilson & Co., Toronto, Ont.

CHAIN BLOCKS
Morris Crane & Hoist Co., Herbert, Niagara Falls, Ont.

CHANDLERY, SHIP
Leckie, Ltd., John, Toronto, Ont.

CLAMPS
Topping Brothers, New York, N.Y.

CLOCKS
American Steam Gauge & Valve Mfg. Co. Boston, Mass.
Morrison Brass Mfg. Co., James, Toronto, Ont.
Williams Machinery Co., A. R., Toronto, Ont.

CLOSETS
Mueller Mfg. Co., H., Sarnia, Ont.

COAL
Nova Scotia Steel & Coal Co., New Glasgow, N.S.

COCKS, BILGE, DISCHARGE, INDICATOR, ETC.
McAvity & Sons Ltd., T., St. John, N.B.
Morrison Brass Mfg. Co., James, Toronto, Ont.

COCKS, BASIN
Empire Mfg. Co., London, Ont.

COMPASSES
Scythes & Co., Toronto, Ont.
Wm. C. Wilson & Co., Toronto, Ont.

COMPRESSORS, AIR
Can. Fairbanks-Morse Co., Montreal, Que.
Canadian Ingersoll-Rand Co., Sherbrooke, Que.
Smart-Turner Mach. Co., Hamilton, Ont.
Storey Pump & Equipment Co., Toronto, Ont.
Williams Machinery Co., A. R., Toronto, Ont.

COMPRESSED AIR TANKS
Scaife & Sons Co., Wm. B., Oakmont, Pa.

CONCRETE MIXERS
St. Clair Bros., Galt, Ontario.
The John McDougall Caledonian Iron Works Co., Montreal, Que.

CONDENSERS
Kearfott Engineering Co., New York, N.Y.
Morris Machine Works, Baldwinsville, N.Y.
Smart-Turner Mach. Co., Hamilton, Ont.
Storey Pump & Equipment Co., Toronto, Ont.
Weir Ltd., G. & J., Cathcart, Glasgow, Scotland.
Williams Machinery Co., A. R., Toronto, Ont.

CONDULETS, MARINE
Crouse-Hinds Co. of Canada, Ltd., Toronto, Ont.

CONTRACTORS' SUPPLIES
McAvity & Sons, T., St. John, N.B.

CONSULTING ENGINEER
Watt, J. Murray, Philadelphia, Pa.

CONVEYORS, ASH, COAL
Babcock & Wilcox, Ltd., Montreal, Que.

COPING MACHINES
Bertram & Sons, Ltd., John, Dundas, Ont.

COPPER TUBES, SHEETS AND RODS
Dominion Copper Products Co., Ltd., Montreal, Que.

CORDAGE
Scythes & Co., Toronto, Ont.

COTTON, CALKING
Wilson & Co., Wm. C., Toronto, Canada.

COTTON DUCK
Scythes & Co., Toronto, Ont.

COVERS, CANVAS, FOR HATCHES, LIFE-BOATS, ETC.
Leckie, Ltd., John, Toronto, Ont.

COUNTERS, REVOLUTION
American Steam Gauge & Valve Mfg. Co., Boston, Mass.

CRANES
Aikenhead Hardware, Ltd., Toronto, Ont.
Can. Fairbanks-Morse Co., Montreal, Que.
Williams Machinery Co., A. R., Toronto, Ont.

CRANES, ELECTRIC
Babcock & Wilcox, Ltd., Montreal, Que.
Morris Crane & Hoist Co., Herbert, Niagara Falls, Ont.
Smart-Turner Mach. Co., [illegible]

CRANES, GANTRY
Smart-Turner Mach. Co., Hamilton, Ont.

CRANES, JIB
Morris Crane & Hoist Co., Herbert, Niagara Falls, Ont.

CRANES, OVERHEAD TRAVELLING
Morris Crane & Hoist Co., Herbert, Niagara Falls, Ont.

CRANK SHAFTS
Canada Foundries and Forgings, Welland, Ont.

DAVITS, BOAT
Babcock & Wilcox, Ltd., Montreal, Que.

DECK PLUGS, ELECTRIC
Crouse-Hinds Co. of Canada, Ltd., Toronto, Ont.

DERRICKS
Aikenhead Hardware, Ltd., Toronto, Ont.
Dake Engine Co., Grand Haven, Mich.
Marsh & Henthorn, Ltd., Belleville, Ont.

DISTILLERS
Kearfott Engineering Co., New York, N.Y.

DISTRESS OUTFITS
Scythes & Co., Toronto, Ont.

DREDGES
Collingwood Shipbuilding Co., Collingwood, Ont.
Norbom Engineering Co., Philadelphia, Pa.
Polson Iron Works, Toronto.

DRILLS, AIR
Aikenhead Hardware, Ltd., Toronto, Ont.
Can. Ingersoll-Rand Co. Sherbrooke, Que.

DRILLS, CENTRE
Wilt Twist Drill Co., Walkerville, Ont.

DRILLS, BLACKSMITH AND BIT STOCK
Wilt Twist Drill Co., Walkerville, Ont.

DRILLS, ELECTRIC
Aikenhead Hardware, Ltd., Toronto, Ont.
Wilkinson & Kompass, Hamilton, Ont.

DRILLS, HIGH SPEED
Wilt Twist Drill Co., Walkerville, Ont.

DRILLS, TRACK
Wilt Twist Drill Co., Walkerville, Ont.

DRILLS, RACHET AND HAND
Wilt Twist Drill Co., Walkerville, Ont.

DRILLS, TWIST
Aikenhead Hardware, Ltd., Toronto, Ont.
Can. B. K. Morton Co., Toronto, Montreal.
Williams Machinery Co., A. R., Toronto, Ont.
Wilt Twist Drill Co., Walkerville, Ont.

DRY DOCKS
Can. Vickers, Ltd., Montreal, Que.
Collingwood Shipbuilding Co., Collingwood, Ont.
Doxford & Sons, William, Sunderland, England.
Georgian Bay Shipbuilding & Wrecking Co., Midland, Ont.
Montreal Dry Docks & Shipbuilding Co., Montreal.
National Shipbuilding Co., Goderich, Ont.
Polson Iron Works, Toronto, Ont.
Port Arthur Shipbuilding Co., Port Arthur, Ont.
Thor Iron Works, Toronto, Ont.
Yarrows, Limited, Victoria, B.C.

ECONOMIZERS, FUEL
Babcock & Wilcox, Ltd., Montreal, Que.

EJECTORS
Empire Mfg. Co., London, Ont.
Morrison Brass Mfg. Co., James, Toronto, Ont.
Smart-Turner Mach. Co., Hamilton, Ont.

ELECTRIC LAMPS
Wm. C. Wilson & Co., Toronto, Ont.

ELECTRO-PLATING
United Brass & Lead Co., Toronto, Ont.

ELEVATING MACHINERY
Goldie & McCulloch, Ltd., Galt, Ont.

ENAMEL
Can. Bitumastic Enamels Co., Hamilton, Ont.

ENGINES, HOISTING
Advance Mach. & Welding Co., Montreal, Que.
Kennedy & Sons, Wm., Owen Sound, Ont.
Marsh & Henthorn, Ltd., Belleville, Ont.
Port Arthur Shipbuilding Co., Port Arthur, Ont.
Williams Machinery Co., A. R., Toronto, Ont.

ENGINE, INTERNAL COMBUSTION
Doxford & Sons, William, Sunderland, England.

ENGINES, MARINE
Bolinders Co., New York, N.Y.
Can. Fairbanks-Morse Co., Montreal, Que.
Can. Vickers, Ltd., Montreal, Que.
Doxford & Sons, William, Sunderland, England.
Goldie & McCulloch, Ltd., Galt, Ont.
Iron Works, Ltd., Owen Sound, Ont.
Mason Regulator & Engin. Co., Montreal, Que.
The John McDougall Caledonian Iron Works Co., Montreal, Que.
Montreal Dry Docks & Shipbuilding Co., Montreal, Que.
Murray & Fraser, New Glasgow, N.S.
National Shipbuilding Co., Goderich, Ont.
Norbom Engineering Co., Philadelphia, Pa.
Polson Iron Works, Toronto, Ont.
Port Arthur Shipbuilding Co., Port Arthur, Ont.
Trout Co., H. G., Buffalo, N.Y.
Williams Machinery Co., A. R., Toronto, Ont.

ENGINE STARTERS (AIR)
Scaife & Sons Co., Wm. B., Oakmont, Pa.

ENGINES, STEERING
Dake Engine Co., Grand Haven, Mich.
Kennedy & Sons, Wm., Owen Sound, Ont.
Wm. C. Wilson & Co., Toronto, Ont.

ENAMELWARE
Mueller Mfg. Co., H., Sarnia, Ont.

EVAPORATORS
Kearfott Engineering Co., New York, N.Y.
Mason Regulator & Engin. Co., Montreal, Que.
Weir Ltd., G. & J., Cathcart, Glasgow, Scotland.

EXTRACTORS, GREASE
American Steam Gauge & Valve Mfg. Co., Boston, Mass.

EYE BOLTS AND NUTS
Canada Foundries and Forgings, Welland, Ont.

FANS
Aikenhead Hardware, Ltd., Toronto, Ont.
Empire Mfg. Co., London, Ont.
Smart-Turner Mach. Co., Hamilton, Ont.
Williams Machinery Co., A. R., Toronto, Ont.

FENDERS, ROPE
Leckie, Ltd., John, Toronto, Ont.
Wilson & Co., Wm. C., Toronto, Canada.

FERRO-MANGANESE
Mitchells, Ltd., Glasgow, Scotland.

FILES
Aikenhead Hardware, Ltd., Toronto, Ont.
Can. B. K. Morton Co., Toronto, Montreal.
Williams Machinery Co., A. R., Toronto, Ont.

FIRE BRICKS
Mitchells, Ltd., Glasgow, Scotland.
Williams Machinery Co., A. R., Toronto, Ont.

FILTERS, FEED WATER
Mason Regulator & Engin. Co., Montreal, Que.

FITTINGS, MARINE
McAvity & Sons, T., St. John, N.B.

FITTINGS, MOTOR BOAT
Empire Mfg. Co., London, Ont.
Mueller Mfg. Co., H., Sarnia, Ont.
Murray & Fraser, New Glasgow, N.S.

FLAGS
Scythes & Co., Toronto, Ont.

FLOW METERS
Spray Engineering Co., Boston, Mass.

FLUE CLEANERS
Wm. C. Wilson & Co., Toronto, Ont.

FIXTURES, ELECTRIC
Cory & Son, Inc., Chas., New York, N.Y.
Crouse-Hinds Co. of Canada, Ltd., Toronto, Ont.

FORGES
Aikenhead Hardware, Ltd., Toronto, Ont.

FLOODLIGHTS, ELECTRIC
Crouse-Hinds Co. of Canada, Ltd., Toronto, Ont.

FLUORSPAR
Mitchells, Ltd., Glasgow, Scotland.
Scythes & Co., Toronto, Ont.

FORGINGS, ALL KINDS
Aikenhead Hardware, Ltd., Toronto, Ont.
Collingwood Shipbuilding Co., Collingwood, Ont.
The John McDougall Caledonian Iron Works Co., Montreal, Que.
Nova Scotia Steel & Coal Co., New Glasgow, N.S.
St. Clair Bros., Galt, Ont.

FORGINGS, STEEL AND IRON
Canada Foundries and Forgings, Welland, Ont.

GAUGES, RECORDING
American Steam Gauge & Valve Mfg. Co., Boston, Mass.
Empire Mfg. Co., London, Ont.

GASKETS
Wm. C. Wilson & Co., Toronto, Ont.

GAUGES COCKS
McAvity & Sons, T., St. John, N.B.

GAUGE GLASSES
Wm. C. Wilson & Co., Toronto, Ont.

GAUGES, WATER
McAvity & Sons, T., St. John, N.B.

GAUGES, WATER, PRESSURE, COMPOUND AND VACUUM
Aikenhead Hardware, Ltd., Toronto, Ont.
Babcock & Wilcox, Ltd., Montreal, Que.
Empire Mfg. Co., London, Ont.
Morrison Brass Mfg. Co., James, Toronto, Ont.

GENERATORS AND CONVERTORS
Can. Fairbanks-Morse Co., Montreal, Que.

GRAPHITE
Wm. C. Wilson & Co., Toronto, Ont.

GRATINGS
Canada Wire & Iron Goods Co., Hamilton, Ont.

GRINDERS, ELECTRIC
Wilkinson & Kompass, Hamilton, Ont.

GRINDERS, PNEUMATIC
Can. Ingersoll-Rand Co., Sherbrooke, Que.

GUY RODS AND ANCHORS, ELECTRIC
Crouse-Hinds Co. of Canada, Ltd., Toronto, Ont.

HAMMERS
Canada Foundries and Forgings, Ltd., Welland, Ont.

HARDWARE, MARINE
Scythes & Co., Toronto, Ont.

HEADLIGHTS, ELECTRIC
Crouse-Hinds Co. of Canada, Ltd., Toronto, Ont.

HEATING EQUIPMENT
Empire Mfg. Co., London, Ont.
Low & Sons, Ltd., Archibald, Glasgow, Scotland

HEATERS, FEED, WATER
Babcock & Wilcox, Ltd., Montreal, Que.
Kearfott Engineering Co., New York, N.Y.
Mason Regulator & Engin. Co., Montreal, Que.
Weir Ltd., G. & J., Cathcart, Glasgow, Scotland.

HOOKS
Morris Crane & Hoist Co., Herbert, Niagara Falls, Ont.

HOISTS, ASH
Beatty & Sons, Welland, Ont.
St. Clair Bros., Galt, Ont.

HOIST BLOCKS
Morris Crane & Hoist Co., Herbert, Niagara Falls, Ont.

HOISTS, CHAIN
Aikenhead Hardware, Ltd., Toronto, Ont.
Can. Fairbanks-Morse Co., Montreal, Que.
Dake Engine Co., Grand Haven, Mich.
Morris Crane & Hoist Co., Herbert, Niagara Falls, Ont.
Williams Machinery Co., A. R., Toronto, Ont.

HOISTS, CARGO, MOVING, ETC.
Dake Engine Co., Grand Haven, Mich.

HOISTING MACHINERY
Beatty & Sons, Welland, Ont.
Can. Ingersoll-Rand Co. Sherbrooke, Que.
Marsh & Henthorn, Ltd., Belleville, Ont.
Williams Machinery Co., A. R., Toronto, Ont.

HOSE
Wm. C. Wilson & Co., Toronto, Ont.

HYDRAULIC MACHINERY
The John McDougall Caledonian Iron Works Co., Montreal, Que.

INDICATORS, ENGINE
American Steam Gauge & Valve Mfg. Co., Boston, Mass.
Cory & Son, Inc., Chas., New York, N.Y.

INDICATORS, SPEED
Aikenhead Hardware, Ltd., Toronto, Ont.
Cory & Son, Inc., Chas., New York, N.Y.

INJECTORS
Aikenhead Hardware, Ltd., Toronto, Ont.
Empire Mfg. Co., London, Ont.
Morrison Brass Mfg. Co., James, Toronto, Ont.
Williams Machinery Co., A. R., Toronto, Ont.

INGOTS
Broughton Copper Co., Ltd., Manchester, Eng.

INSULATORS, ELECTRIC
Crouse-Hinds Co. of Canada, Ltd., Toronto, Ont.

INSTRUMENTS, NAUTICAL
Leckie, Ltd., John, Toronto, Ont.

JACKS
Morris Crane & Hoist Co., Herbert, Niagara Falls, Ont.

JOINTS, BALL SLIP
Norbom Engineering Co., Philadelphia, Pa.

INSURANCE, MARINE
Toronto Insurance & Vessel Agency, Toronto, Ont.

JOURNAL WEDGES
Canada Foundries and Forgings, Ltd., Welland, Ont.

LATHES
Foss & Hill Machinery Co., Montreal, Que.
Bertram & Sons, Ltd., John, Dundas, Ont.

LEATHER, LACE
Wm. C. Wilson & Co., Toronto, Ont.

LIFEBUOYS, BELTS AND PRESERVERS
Fosbery Co., Barking, Essex, Eng.
Wm. C. Wilson & Co., Toronto, Ont.

LIFEBOATS
Wm. C. Wilson & Co., Toronto, Ont.

LIFE BOAT EQUIPMENT
Leckie, Ltd., John, Toronto, Ont.
Wm. C. Wilson & Co., Toronto, Ont.

LIFE JACKETS
Fosbery Co., Barking, Essex, Eng.
Leckie, Ltd., John, Toronto, Ont.

LIGHTS, ALL KINDS
Can. Fairbanks-Morse Co., Montreal, Que.
Crouse-Hinds Co. of Canada, Ltd., Toronto, Ont.
Morrison Brass Mfg. Co., James, Toronto, Ont.

LIGHTS, SIDE, PORT
Crouse-Hinds Co. of Canada, Ltd., Toronto, Ont.
McAvity & Sons Ltd., T., St. John, N.B.

LIGHTING SETS
Can. Fairbanks-Morse Co., Montreal.
Cory & Son, Inc., Chas., New York, N.Y.
Crouse-Hinds Co. of Canada, Ltd., Toronto, Ont.
Kearfott Engineering Co., New York, N.Y.
Williams Machinery Co., A. R., Toronto, Ont
Wilson & Co., Wm. C., Toronto, Canada.

LOCKERS, METAL
Canada Wire & Iron Goods Co., Hamilton, Ont.

LUBRICATORS
Empire Mfg. Co., London, Ont.

MACHINISTS
Hyde Engineering Works, Montreal, Que.

MACHINE AND FORGE WORK
Thor Iron Works, Toronto, Ont.

MANGANESITE, PASTE
Wm. C. Wilson & Co., Toronto, Ont.

MARINE CASTINGS
McAvity & Sons, T., St. John, N.B.

MARINE SPECIALTIES
McAvity & Sons, T., St. John, N.B.

MOTORS
Aikenhead Hardware, Ltd., Toronto, Ont.
Can. Fairbanks-Morse Co., Montreal, Que.
Williams Machinery Co., A. R., Toronto, Ont.

NAUTICAL INSTRUMENTS
Scythes & Co., Toronto, Ont.
Wm. C. Wilson & Co., Toronto, Ont.

NET LIFTERS
Dake Engine Co., Grand Haven, Mich.
Leckie, Ltd., John, Toronto, Ont.

NOZZLES, ALL KINDS
Empire Mfg. Co., London, Ont.
Tolland Mfg. Co., Boston, Mass.

NUTS
Wilkinson & Kompass, Hamilton, Ont.

OAKUM
Scythes & Co., Toronto, Ont.
Wm. C. Wilson & Co., Toronto, Ont.

OILS
Moore & Co., Benjamin, Toronto, Ont.

OIL CUPS
Empire Mfg. Co., London, Ont.
Murray & Fraser, New Glasgow, N.S.

OILCLOTHING
Wm. C. Wilson & Co., Toronto, Ont.

OILS AND GREASE
Wm. C. Wilson & Co., Toronto, Ont.

OAKUM
Leckie, Ltd., John, Toronto, Ont.

ORNAMENTAL IRONWORK
Canada Wire & Iron Goods Co., Hamilton, Ont.

OXY-ACETLYENE OUTFITS
Can. Fairbanks-Morse Co., Montreal, Que.
McAvity & Sons Ltd., T., St. John, N.B.
Williams Machinery Co., A. R., Toronto, Ont.

PAINT
Ault & Wiborg Co. of Can., Ltd., Toronto, Ont.
Can. Bitumastic Enamels, Ltd., Hamilton, Ont.
Moore & Co., Benjamin, Toronto, Ont.
Wm. C. Wilson & Co., Toronto, Ont.

PAINTING EQUIPMENT, AUTOMATIC
Spray Engineering Co., Boston, Mass.

PACKING, AMMONIA
Can. B. K. Morton Co., Montreal.
France Packing Co., Philadelphia, Pa.
Wm. C. Wilson & Co., Toronto, Ont.

PACKING, MARINE
Aikenhead Hardware, Ltd., Toronto, Ont.
Can. B. K. Morton Co., Montreal.
France Packing Co., Philadelphia, Pa.
Wm. C. Wilson & Co., Toronto, Ont.

PACKING, METALLIC
Aikenhead Hardware, Ltd., Toronto, Ont.
Can. B. K. Morton Co., Montreal.
France Packing Co., Philadelphia, Pa.

PACKING, STEAM
Aikenhead Hardware, Ltd., Toronto, Ont.
France Packing Co., Philadelphia, Pa.
Wm. C. Wilson & Co., Toronto, Ont.

PANELBOARD AND CABINETS, ELECTRIC
Crouse-Hinds Co. of Canada, Ltd., Toronto, Ont

PIPE, LAPWELD, CAST IRON, RIVETED
Empire Mfg. Co., London, Ont.
Norbom Engineering Co., Philadelphia, Pa.

PIPE JOINT CEMENT
Wm. C. Wilson & Co., Toronto, Ont.

PITCH
Scythes & Co., Toronto, Ont.

PITCH, PINE
Wilson & Co., Wm. C., Toronto, Canada.

PLANERS
Bertram & Sons, Ltd., John, Dundas, Ont.

PLANERS, STANDARD AND ROTARY
Yates Mach. Co., P. B., Hamilton, Ont.

PLATE CLAMPS
Morris Crane & Hoist Co., Herbert, Niagara Falls, Ont.

PLATE PUNCH TABLES
Norbom Engineering Co., Philadelphia, Pa.

PLUMBING EQUIPMENT
Low & Sons, Ltd., Archibald, Glasgow, Scotland.
Empire Mfg. Co., London, Ont.
Mueller Mfg. Co., H., Sarnia, Ont.
Wm. C. Wilson & Co., Toronto, Ont.

PROPELLOR BLADES, BRONZE
Empire Mfg. Co., London, Ont.
Murray & Fraser, New Glasgow, N.S.
Yarrows, Limited, Victoria, B.C.

PROPELLER WHEELS
Kennedy & Sons, Wm., Owen Sound, Ont.
Trout Co., H. G., Buffalo, N.Y.

PIPING, STEAM
Babcock & Wilcox, Ltd., Montreal, Que.

PULLEYS
Smart-Turner Mach. Co., Hamilton, Ont.
Williams Machinery Co., A. R., Toronto, Ont.

PORT LIGHTS
Wm. C. Wilson & Co., Toronto, Ont.

PUMPS
Can. Fairbanks-Morse Co., Montreal, Que.
Canada Foundries & Forgings, Welland, Ont.
Goldie & McCulloch, Ltd., Galt, Ont.
McAvity & Sons, T., St. John, N.B.
Williams Machinery Co., A. R., Toronto Ont.

PUMP, AIR
Can. Ingersoll-Rand Co. Sherbrooke, Que.
Smart-Turner Mach. Co., Hamilton, Ont.
Weir Ltd., G. & J., Cathcart, Glasgow, Scotland.
Williams Machinery Co., A. R., Toronto, Ont.

PUMPS, CENTRIFUGAL
Can. Ingersoll-Rand Co. Sherbrooke, Que.
Kearfott Engineering Co., New York, N.Y.
The John McDougall Caledonian Iron Works Co., Montreal, Que.
Norbom Engineering Co., Philadelphia, Pa.
Smart-Turner Mach. Co., Hamilton, Ont.
Williams Machinery Co., A. R., Toronto, Ont.

PUMPS, CIRCULATING
Morris Machine Works, Baldwinsville, N.Y.

PUMPS, FEED WATER
Weir Ltd., G. & J., Cathcart, Glasgow, Scotland.
Williams Machinery Co., A. R., Toronto, Ont.

PUMPS, HAND AND POWER
Aikenhead Hardware, Ltd., Toronto, Ont.
Smart-Turner Mach. Co., Hamilton, Ont.
Williams Machinery Co., A. R., Toronto, Ont.

PUMPS, HIGH PRESSURE
Canadian Ingersoll-Rand Co., Sherbrooke, Que.
Smart-Turner Mach. Co., Hamilton, Ont.

PUMPS, STEAM TURBO
Wm. C. Wilson & Co., Toronto, Ont.

PUMPING MACHINES
Norbom Engineering Co., Philadelphia, Pa.

PUNCHES
Bertram & Sons, Ltd., John, Dundas, Ont.

PUNCHES, SINGLE, DOUBLE AND MULTIPLE
Norbom Engineering Co., Philadelphia, Pa.

PURIFIERS, WATER
Babcock & Wilcox, Ltd., Montreal, Que.
Can. Allis-Chalmers, Toronto, Ont.

RADIO ENGINEERS
Cutting & Washington, Inc., Cambridge, Mass.

RADIATORS, STEAM, ELECTRIC
Empire Mfg. Co., London, Ont.
Low & Sons, Ltd., Archibald, Glasgow, Scotland
Wm. C. Wilson & Co., Toronto, Ont.

RANGES
Wm. C. Wilson & Co., Toronto, Ont.

RECEIVERS, AIR
Can. Ingersoll-Rand Co. Sherbrooke, Que.

REGULATORS, FEED WATER
American Steam Gauge & Valve Mfg. Co. Boston, Mass.

REGULATORS, PRESSURE
Mason Regulator & Engin. Co., Montreal, Que.

REPAIRS, MARINE
Corbet Foundry & Mach. Co., Owen Sound, Ont.
Can. Vickers, Ltd., Montreal, Que.
Collingwood Shipbuilding Co., Collingwood, Ont.
Georgian Bay Shipbuilding & Wrecking Co., Midland, Ont.
Hyde Engineering Works, Montreal, Que.
Iron Works, Ltd., Owen Sound, Ont.
Kennedy & Sons, Wm., Owen Sound, Ont.
Montreal Dry Docks & Shipbuilding Co., Montreal Que.
Muir Bros. Dry Dock Co., Port Dalhousie, Ont.
National Shipbuilding Co., Goderich, Ont.
Port Arthur Shipbuilding Co., Port Arthur, Ont.
Yarrow, Limited, Victoria, B.C.

RIGGING, WIRE ROPE
Leckie, Ltd., John, Toronto, Ont.

RIVETS
Severance Mfg. Co., S., Glassport, Pa.
Wilkinson & Kompass, Hamilton, Ont.

RIVETERS, PNEUMATIC
Can. Ingersoll-Rand Co., Ltd., Sherbrooke, Que.

RODS, COPPER, BRASS, BRONZE
Standard Underground Cable Co., Hamilton, Ont.

ROLLS, STRAIGHTENING, BENDING
Bertram & Sons, Ltd., John, Dundas, Ont.

ROOFING
Wm. C. Wilson & Co., Toronto, Ont.

ROPE BLOCKS
Aikenhead Hardware, Ltd., Toronto, Ont.
Can. Fairbanks-Morse Co., Montreal, Que.
Morris Crane & Hoist Co., Herbert, Niagara Falls, Ont.
Thor Iron Works, Toronto, Ont.

ROPE
Leckie, Ltd., John, Toronto, Ont.
Wm. C. Wilson & Co., Toronto, Ont.

RUBBER COATS
Wm. C. Wilson & Co., Toronto, Ont.

SAW MILL MACHINERY
Yates Machine Co., P. B., Hamilton, Ont.

SCALES, BOILERS, ENGINES
Can. Fairbanks-Morse Co., Montreal, Que.

SCOWS
Collingwood Shipbuilding Co., Collingwood, Ont.
Polson Iron Works, Toronto, Ont.

SEPARATORS, OIL, STEAM
Mason Regulator & Engin. Co., Montreal, Que.
Smart-Turner Mach. Co., Hamilton, Ont.

SHAFTING
Wilkinson & Kompass, Hamilton, Ont.

SHAFTING, BRONZE
Empire Mfg. Co., London, Ont.
Murray & Fraser, New Glasgow, N.B.

SHEARS
Bertram & Sons, Ltd., John, Dundas, Ont.
Norbom Engineering Co., Philadelphia, Pa.

SHIPBUILDING TOOLS
Aikenhead Hardware, Ltd., Toronto, Ont.
Can. Ingersoll-Rand Co. Sherbrooke, Que.

SHIPS, BUILDERS OF
Can. Vickers, Ltd., Montreal, Que.
Collingwood Shipbuilding Co., Collingwood, Ont.
Doxford & Sons, William, Sunderland, England.
Georgian Bay Shipbuilding & Wrecking Co., Midland, Ont.
Montreal Dry Docks & Shipbuilding Co., Montreal, Que.
National Shipbuilding Co., Goderich, Ont.
Polson Iron Works, Toronto, Ont.
Port Arthur Shipbuilding Co., Port Arthur, Ont.
Thor Iron Works, Toronto, Ont.
Yarrows, Limited, Victoria, B.C.

SHIP BROKERS
Page & Jones, Mobile, Ala.

SHIP PLATES
Nova Scotia Steel & Coal Co., New Glasgow, N.S.
Topping Brothers, New York, N.Y.

SHIPS' TELEGRAPHS
Cory & Son, Inc., Chas., New York, N.Y.
Morrison Brass Mfg. Co., James, Toronto, Ont.
Wm. C. Wilson & Co., Toronto, Ont.

SIDE LIGHTS
Wm. C. Wilson & Co., Toronto, Ont.

SLEDGES
Wilkinson & Kompass, Hamilton, Ont.

SLINGS
Morris Crane & Hoist Co., Herbert, Niagara Falls, Ont.

SPECIAL MACHINERY
Smart-Turner Mach. Co., Hamilton, Ont.

SMOOTH-ON
Wm. C. Wilson & Co., Toronto, Ont.

SPIKES
Wm. C. Wilson & Co., Toronto, Ont.

SPRAY COOLING SYSTEMS
Spray Engineering Co., Boston, Mass.

SPRINKLERS, PARK
Spray Engineering Co., Boston, Mass.

STANDPIPES
Jenckes Machine Co., Ltd., Sherbrooke, Que.
Thor Iron Works, Toronto, Ont.

STEAMSHIP AGENTS
Page & Jones, Mobile, Ala.

STEAM SPECIALTIES
Can. Fairbanks-Morse Co., Montreal, Que.
Empire Mfg. Co., London, Ont.

STEAM TRAPS
Aikenhead Hardware, Ltd., Toronto, Ont.
American Steam Gauge & Valve Mfg. Co., Boston, Mass.
Empire Mfg. Co., London, Ont.
Mason Regulator & Engin. Co., Montreal, Que.
Smart-Turner Mach. Co., Hamilton, Ont.

STEEL, HIGH SPEED
Can. B. K. Morton Co., Toronto, Montreal.
Nova Scotia Steel & Coal Co., New Glasgow, N.S.
Wilkinson & Kompass, Hamilton, Ont.

STEEL WORK, STRUCTURAL
Babcock & Wilcox, Ltd., Montreal, Que.

STEEL STAIRWAYS
Canada Wire & Iron Goods Co., Hamilton, Ont.

STEERING GEARS
Wm. C. Wilson & Co., Toronto, Ont.

STOKERS, MECHANICAL
Babcock & Wilcox, Ltd., Montreal, Que.

SUPERHEATERS, STEAM
Babcock & Wilcox, Ltd., Montreal, Que.

SWITCHBOARDS, ELECTRIC
Crouse-Hinds Co. of Can., Ltd., Toronto, Ont.

TALLOW
Wm. C. Wilson & Co., Toronto, Ont.

TANKS, STEEL
The John McDougall Caledonian Iron Works Co., Montreal, Que.
Port Arthur Shipbuilding Co., Port Arthur, Ont.
Scaife & Sons Co., Wm. B., Oakmont, Pa.
Thor Iron Works, Toronto, Ont.

TANKS
Corbet Foundry & Mach. Co., Owen Sound, Ont.
Goldie & McCulloch, Ltd., Galt, Ont.
Marsh & Henthorn, Ltd., Belleville, Ont.
Scaife & Sons Co., Wm. B., Oakmont, Pa.

TANKS (AIR, GAS AND LIQUID)
Scaife & Sons Co., Wm. B., Oakmont, Pa.

TAPPING MACHINES
Mueller Mfg. Co., H., Sarnia, Ont.

TELEPHONES, MARINE
Cory & Son, Inc., Chas., New York, N.Y.

TESTERS, METER
Mueller Mfg. Co., H., Sarnia, Ont.

THUMB SCREWS AND NUTS
Canada Foundries & Forgings, Welland, Ont.

TRAVELLING BLOCKS
Morris Crane & Hoist Co., Herbert, Niagara Falls, Ont.

TROLLEYS
Can. Fairbanks-Morse Co., Montreal, Que.
Morris Crane & Hoist Co., Herbert, Niagara Falls, Ont.

TROLLEY HOISTS
Morris Crane & Hoist Co., Herbert, Niagara Falls, Ont.

TRUCKS, HAND, ELECTRIC
Aikenhead Hardware, Ltd., Toronto, Ont.
Can. Fairbanks-Morse Co., Montreal, Que.

TUBES, BOILER
Babcock & Wilcox, Ltd., Montreal, Que.
Broughton Copper Co., Ltd., Manchester, Eng.

TUBES, COPPER AND BRASS
Mueller Mfg. Co., H., Sarnia, Ont.
Standard Underground Cable Co., Hamilton, Ont.

TUGS
Collingwood Shipbuilding Co., Collingwood, Ont.
Polson Iron Works, Toronto, Ont.

TURBINES, DIRECT-DRIVING AND GEARED
Doxford & Sons, William, Sunderland, England.

TURNBUCKLES
Canada Foundries & Forgings, Welland, Ont.

SPIKES, SMALL RAILROAD
Severance Mfg. Co., S., Glassport, Pa.

UNIONS, ALL KINDS
Dart Union Company, Toronto, Ont.

VALVES, AIR
Mueller Mfg. Co., H., Sarnia, Ont.

VALVE, DISCS
Wm. C. Wilson & Co., Toronto, Ont.

VALVES
American Steam Gauge & Valve Mfg. Co., Boston, Mass.
Babcock & Wilcox, Ltd, Montreal, Que.
Can. Fairbanks-Morse Co., Montreal, Que.
Empire Mfg. Co., London, Ont.
McAvity & Sons, Ltd., T., St. John, N.B.
Mason Regulator & Engin. Co., Montreal, Que.
Norbom Engineering Co., Philadelphia, Pa.
Williams Machinery Co., A. R. Toronto, Ont.
Wm. C. Wilson & Co., Toronto, Ont.

VALVES, FOOT
Aikenhead Hardware, Ltd., Toronto, Ont.
Smart-Turner Mach. Co., Hamilton, Ont.

VALVES, MARINE
McAvity & Sons, T., St. John, N.B.

VALVES, STOP, REDUCING, SAFETY CHECK, ETC.
Aikenhead Hardware, Ltd., Toronto, Ont.
American Steam Gauge & Valve Mfg. Co., Boston, Mass.
Empire Mfg. Co., London, Ont.
McAvity & Sons Ltd., T., St. John, N.B.
Morrison Brass Mfg. Co., James, Toronto, Ont.

VALVES, MIXING
Mueller Mfg. Co., H., Sarnia, Ont

VALVES, REDUCING, PRESSURE
Mueller Mfg. Co., H., Sarnia, Ont.

VARNISHES
Aikenhead Hardware, Ltd., Toronto, Ont.
Ault & Wiborg Co. of Can., Ltd., Toronto, Ont.
Leckie, Ltd., John, Toronto, Ont.
McAvity & Sons Ltd., T., St. John, N.B.
Moore & Co., Benjamin, Toronto, Ont.

VENTILATION EQUIPMENT
Empire Mfg. Co., London, Ont.
Low & Sons, Ltd., Archibald, Glasgow, Scotland
Scythes & Co., Toronto, Ont.

WASTE
Scythes & Co., Toronto, Ont.
Wm. C. Wilson & Co., Toronto, Ont.

WATER COLUMNS
Morrison Brass Mfg. Co., James, Toronto, Ont.

WATER HEATERS
Empire Mfg. Co., London, Ont.
Morrison Brass Mfg. Co., James, Toronto, Ont.

WATER SOFTENERS
Babcock & Wilcox, Ltd., Montreal, Que.
Can. Allis-Chalmers, Toronto, Ont.

WATER SUPPLY SYSTEMS
Mueller Mfg. Co., H., Sarnia, Ont.

WELDING, ELECTRIC
Hall Engineering Works, Montreal, Que.
Marine Welding Co., Buffalo, N.Y.
Beatty & Sons, M., Welland, Ont.

WELDING, OXY-ACETYLENE
Advance Mach. & Welding Co., Montreal, Que.

WHISTLES AND SYRENS
Empire Mfg. Co., London, Ont.
McAvity & Sons, T., St. John, N.B.
Morrison Brass Mfg. Co., Jas., Toronto, Ont.

WINCHES, CARGO
Advance Mach. & Welding Co., Montreal, Que.
Aikenhead Hardware, Ltd., Toronto, Ont.

WINCHES, DOCK, SHIP
Advance Mach. & Welding Co., Montreal, Que.
Beatty & Sons, M., Welland, Ont.
Marsh & Henthorn, Ltd., Belleville, Ont.
Morris Crane & Hoist Co., Herbert, Niagara Falls, Ont.
Wilson & Co., Wm. C., Toronto, Canada.

WINCHES, TOWING
Corbet Foundry & Mach. Co., Owen Sound, Ont.

WINCHES, TRAWL
Beatty & Sons, M., Welland, Ont.
Wm. C. Wilson & Co., Toronto, Ont.

WINDLASSES
Advance Mach. & Welding Co., Montreal, Que.
Dake Engine Co., Grand Haven, Mich.
Wilson & Co., Wm. C., Toronto, Canada.

WIPER CAPS, OILER BOXES, ETC.
Morrison Brass Mfg. Co., James, Toronto, Ont.

WIRECLOTH
Canada Wire & Iron Goods Co., Hamilton, Ont.

WIRE, COPPER CLAD STEEL
Standard Underground Cable Co., Hamilton, Ont.

WIRE, COPPER, BRASS, BRONZE
Standard Underground Cable Co., Hamilton, Ont.

WIRE, INSULATED
Standard Underground Cable Co., Hamilton, Ont.

WIRELESS OUTFITS
Cutting & Washington, Inc., Cambridge, Mass.
Marconi Wireless Telegraph Co., Montreal, Que.

WOOD WORKING MACHINERY
Aikenhead Hardware, Ltd., Toronto, Ont.
Can. Fairbanks-Morse Co., Montreal, Que.
Williams Machinery Co., A. R. Toronto, Ont.
Yates Mach. Co., P. B., Hamilton, Ont.

WOOD BORING TOOLS
Aikenhead Hardware, Ltd., Toronto, Ont.
Can. Ingersoll-Rand Co. Sherbrooke, Que.

WOODITE GAUGE GLASS WASHERS
Wm. C. Wilson & Co., Toronto, Ont.

WRENCHES
Canada Foundries & Forgings, Welland, Ont.

YACHT BROKER
Watt, J. Murray, Philadelphia, Pa.

Nova Scotia Steel & Coal Company

Limited

New Glasgow, Nova Scotia, Canada

RUDDER FOR SS. "LUX." WEIGHT 9 TONS, LENGTH 42 FEET

Ship Forgings of all Shapes, Sizes and Weights up to 75 Tons

We manufacture, of Fluid Compressed Steel, forgings entering into the construction and equipment of steel vessels up to and including the largest building or afloat, and embracing Rudder Frames—sectional or one piece, also Rudders complete; Stern Posts and Stern Brackets for single, Twin and Triple Screw Ships, Rudder Heads, Boat Davits, Derricks, etc.

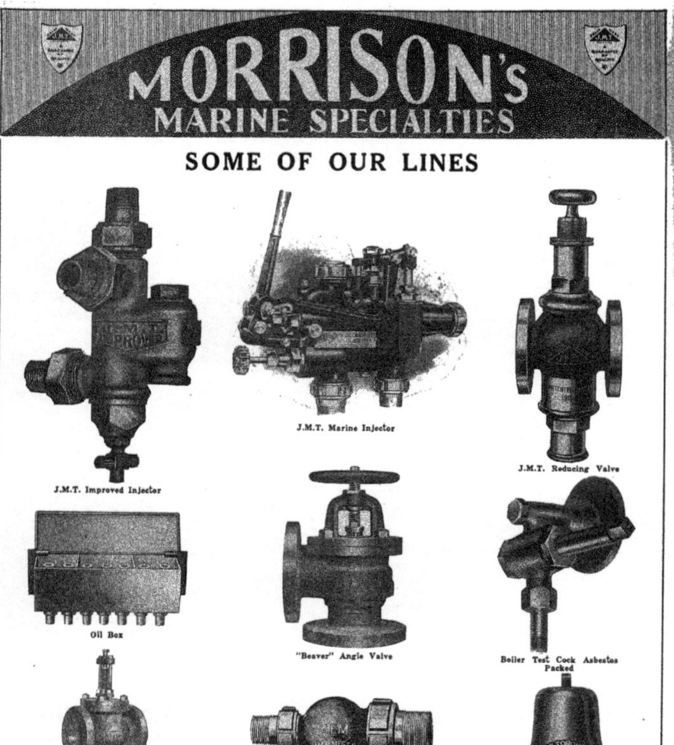
MORRISON'S
MARINE SPECIALTIES
SOME OF OUR LINES
J.M.T. Improved Injector
J.M.T. Marine Injector
J.M.T. Reducing Valve
Oil Box
"Beaver" Angle Valve
Boiler Test Cock Asbestos Packed
"Beaver" Spring Loaded Overboard Discharge Valve
"Gem" Ejector
Bell
SEND FOR OUR NEW CATALOG—and see complete line.
The James Morrison Brass Mfg. Co., Limited
93-97 Adelaide St. West, Toronto, Canada

MORRISON'S
MARINE SPECIALTIES
SOME OF OUR LINES
Cylinder Oil Pump
Ships' Light
"Marine" Globe Valve
MORRISON'S MARINE SPECIALTIES REPRESENT 35 YEARS' SUCCESSFUL EXPERIENCE IN MEETING THE MOST EXACTING DEMANDS.
Approved by the Government Marine Inspection Department.
Every Article is tested before leaving factory and positively guaranteed.
J.M.T. Compound Automatic Whistle Valve
Grease Cup
Bell
Send for New Catalog
One Quality—the Best
Hose End Gate Valve
Asbestos Packed Cock
Ship's Side Cock with Locking Guard
Engine Room Telegraph
J.M.T. Globe Valve
The James Morrison Brass Mfg. Co., Limited
93-97 Adelaide St. West, Toronto, Canada

ADVERTISING INDEX

CIRCULATES IN EVERY PROVINCE OF CANADA AND ABROAD

MARINE ENGINEERING of Canada

A monthly journal dealing with the progress and development of Merchant and Naval Marine Engineering, Shipbuilding, the building of Harbors and Docks, and containing a record of the latest and best practice throughout the Sea-going World. Published by
The MacLean Publishing Co., Limited

MONTREAL, Southam Building TORONTO 143-153 University Ave. WINNIPEG, 1207 Union Trust Bldg. LONDON, ENG., 88 Fleet St

Vol. VIII. Publication Office, Toronto—March, 1918 No. 3

The Publisher's Page

TORONTO MARCH, 1918

Seek an Introduction through Brown

THERE are three men — Brown, Smith and Simpson.

Smith does not know Simpson; Simpson does not know Smith. Brown is a big and influential man. He knows both Smith and Simpson.

The three meet some day and Smith and Simpson are introduced.

Because of the standing of Brown, Smith is satisfied that Simpson is alright and Simpson doesn't have to wonder if Smith is O.K. It is enough that Brown introduced them.

So it goes. Many a fortune has been made through the right introduction to the right man. Many a success could be traced to the occasion when Mr. Brown said: "Mr. Simpson, I would like to introduce a very close friend of mine, Mr. Smith."

And so it is with advertising in a reliable journal. It brings buyer and seller together. Progressive, up-to-date, successful men read papers like this and reliable firms with modern products advertise.

So there you are. Each respects the other. The prospective buyer knows that the advertised product must be good or it wouldn't (and couldn't) be advertised. The advertiser knows that an inquiry from a reader of this journal, for instance, must be worth cultivating and good service is accordingly rendered.

Our Pleasure—to bring you together

If any advertisement interests you, tear it out now and place with letters to be answered.

Marine Applications of Floating Frame Type Reduction Gears*

By John H. Macalpine

The author points out that the attainment of a higher power constant by reduction gears of the floating frame type result in a marked reduction in weight over gears of the rigid type. This allows increased efficiency to be obtained for the same weight of machinery, as a greater ratio allows use of slower propeller rotation and a smaller, more efficient turbine.

THE present paper may be considered to be a sequel to a long article by the author "On Reduction Gears," which appeared in "Engineering" from May 5 to June 16, 1916; and also to a leader in "Engineering" of September 17, 1909. These articles taken together deal very fully with the design, theory, and performance of reduction gears of the floating frame type as originally produced by the late Admiral Geo. W. Melville, U.S.N., and the author, known as the "I" beam type. In "Engineering" of November 17, 1911, there is shown a variety introduced by the late Mr. Geo. Westinghouse, known as the hydraulic floating frame type, which will also be referred to. For a full discussion of the subject the reader is referred to these articles.

The bearings of a pinion and gear are, in most machines, supported very rigidly in a heavy framework or bedplate. Such gears will here be referred to as "rigid gears." Reduction gears designed thus had been considerably experimented with to transmit the power of the turbine, but before Admiral Melville and the author took the question up in 1904 the solution of the problem of gearing for the turbine seems to have been despaired of except for very small powers.

Assuming that the teeth could be cut almost perfectly in a modern gear-cutting machine, it appeared evident to us that one important condition was left which, for the best result, it was essential to remove—the slightest error of alignment of the pinion, which is long in proportion to its diameter, would produce great inequality of distribution of pressure along the teeth. The method we adopted for freeing the design from delicacy of alignment was very simple. It consisted in mounting the pinion bearings in a very stiff frame, supported near the middle of its length by the bedplate, but (where the teeth are not in mesh) very free to move about an axis transverse to the axes of the gear and pinion. Thus the exact position of the pinion axis at any time is in part determined by the interaction of the teeth. Great equality of distribution of tooth pressure results. The figures to be given will illustrate this construction, which will now be referred to very briefly as it is well known.

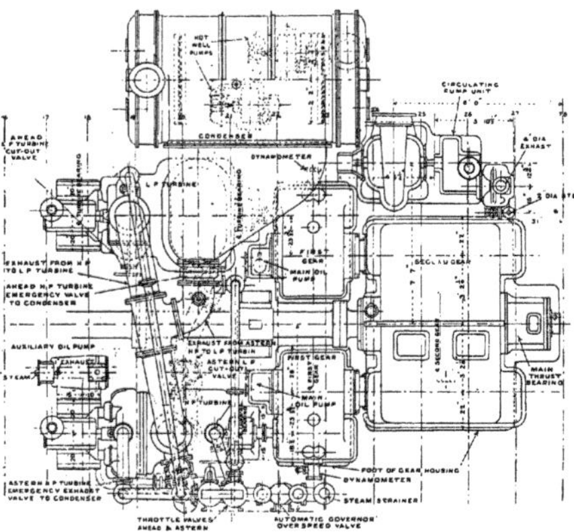

FIG. 1. ASSEMBLY PLAN OF CROSS-COMPOUND TURBINE WITH HIGH PRESSURE AND LOW PRESSURE PARTS, EACH ACTUATING A SEPARATE SINGLE-PINION FIRST REDUCTION GEAR.

Existing Conclusions

The discussion given in the articles referred to above need not be repeated here, but the most important conclusions reached, particularly in the long article of last year, may be briefly given:

1. A floating frame gear will operate perfectly with the pinion sensibly out of line, and, therefore, the great difficulty of rigid gears—bad distribution of tooth pressure from minute errors of alignment—vanishes.

2. That the floating frame corrects from one-half to two-thirds of the effect of torsional yield of the pinion.

3. That the performance of gears of any design should be judged by the value of the "power constant" ‡, it will safely bear.

4. From published data for rigid gears, and from those given here for floating frame gears, it is clear that, as anticipated, the power constant at which the gears are usually run is much higher in the case of floating frame gears than of rigid gears.

The power constant for rigid gears appears usually to be not greater than 2, while 3.5 and 4 are perfectly safe values for floating frame gears. Indeed, in the 12-hour full-power trial of the United States collier Neptune the average power constant was 5.22. It has been in service for over one and a half years, and is frequently used at full power.

5. Comparing gears of different types, with a given ratio of reduction, a rise in the safe power constant quickly reduces the weight per horse-power. This is illustrated in last year's article by the examination of a number of marine installations, and it is there shown that the higher power constant of the floating frame gear effects a marked reduction of weight over the rigid gear. Or, for the same weight of machinery, a greater ratio of reduction may be effected, allowing of a slower rotation and higher efficiency of propeller, or of a faster, smaller, and more efficient turbine.

These are the principal conclusions, although there are others of some importance and many details to which I do not refer here.

I will now illustrate and describe two marine installations: the first of a slow oil-tank, single-screw steamship, with

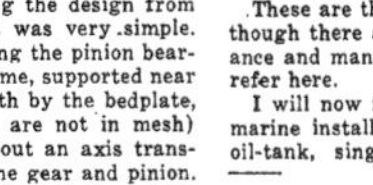

*Read before the Institution of Naval Architects, 1917.

‡ Power constant $= \frac{1000 P}{D^3 R}$

where P = horse-power transmitted
D = diameter of pinion in inches
R = revolutions per minute of pinion

double reduction; the second, a faster passenger and freight twin-screw steamer with single reduction.

Oil-Tank Steamship

This ship is being built by the Chester Shipbuilding Company, Limited. It is one of 45 freight ships of about the same power for which the Westinghouse Ma-

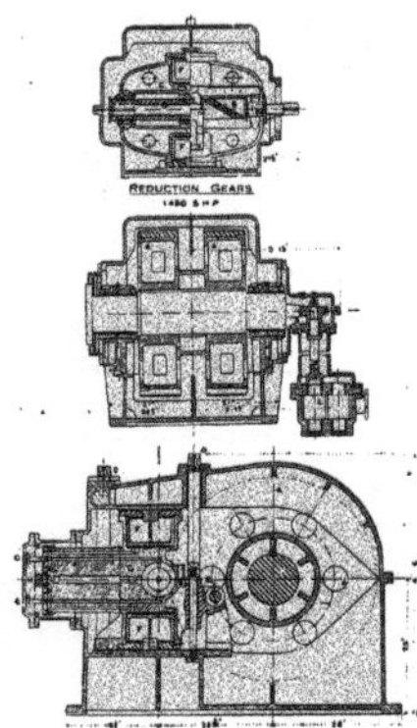

FIGS. 2, 3, 4. DETAILS OF FIRST REDUCTION GEAR.

chine Company are building duplicate or nearly duplicate machinery.

Length	401 ft.
Beam	54 ft.
Draught	25 ft. 9 in.
Displacement	12,650 tons
Cargo capacity	9,000 tons
Full speed, loaded	11.5 knots
Full power	2,900 S.H.P.
Turbines	3,860 R.P.M.
Propeller	75 R.P.M.

The arrangement is sufficiently shown by the plan, Fig. 1. There is a cross-compound turbine, the high-pressure and low-pressure parts each actuating a separate single-pinion first reduction gear which reduce the revolutions per minute from 3,860 to 483. There is one second reduction gear with two pinions, each first reduction gear actuating one of these, and the large gear driving the propeller shaft at 75 r.p.m. The turbines are aft of the gears. This arrangement allows the propeller shaft to be drawn inboard without disturbing the turbines or gears.
composed of a two-row, high-pressure, impulse wheel followed by reaction blad-
The ahead high-pressure turbine is
ing carried on a drum. The ahead low-pressure turbine has only reaction blading. Each of these turbine casings contains an astern turbine consisting of a two-row impulse wheel. The high-pressure and low-pressure turbines are usually operated as a cross-compound turbine; or, by suitable pipe and valve connections, either ahead or astern turbine elements can be operated singly and exhaust direct to the condenser in case of emergency. The astern turbines can develop over 60 per cent. of the full ahead power with the same flow of steam.

With dry saturated steam of 180 lb. per square inch at the high-pressure inlet and 28.5 in. vacuum (30 in. barometer) the guaranteed consumption of the turbine is 11.50 lb. of water per shaft horse-power hour. Various valves, etc., for regulating the turbines are indicated in Fig. 1, but these need not be described.

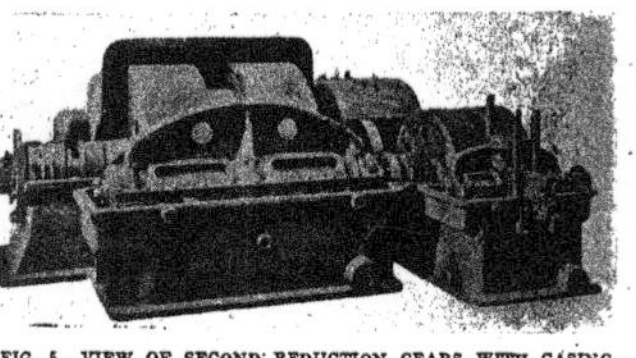

FIG. 5. VIEW OF SECOND REDUCTION GEARS WITH CASING PARTIALLY REMOVED.

The first reduction gear is shown in Figs. 2 to 5. The pinion is driven through a flexible coupling allowing end play, carried on the right-hand end, Figs. 2 and 5, of a flexible shaft B passing through the pinion and fastened to it at the left-hand end. The teeth are cut in right and left-hand helices, as is seen clearly in the second reduction gear at the left-hand side of Fig. 5. The pinion has three bearings held in a very stiff floating frame C.

When it is not desired to measure the power being transmitted the long bolts D, Figs. 4 and 5, are screwed down and fasten the floating frame to the bedplate or housing, but still leave it sufficiently free, so that any inequality of moment of the tooth pressures on the right and left-hand helices about the centre of length of the floating frame would cause the frame to tip about a transverse axis and restore this equality. For a full analysis of this see the articles of 1909 and 1916 in "Engineering," referred to above.

Measuring Horse-power

When it is desired to measure the horse-power when running ahead the lower valve G, Figs. 4 and 5, is opened, which allows oil to flow from the lower side of the centre pinion bearing at considerable pressure to the hydraulic cylinder F through the passages shown. The bolts D having been screwed back slightly, the pressure in F will cause the floating frame to rise about the fulcrum E. This slight rise is indicated through a finger H, actuating a multiplying mechanism which is too small to show in the figure. A pressure gauge indicates the oil pressure in F. As the weight of the floating frame, pinion, etc., are known, these can be deducted from the total downward pressure of the pinion on its bearings. From this and the revolutions of the pinion the horse-power transmitted by the gear can, obviously, be at once calculated. Similarly, when going astern, the power may be measured by means of the upper hydraulic cylinder F.

Thus this floating frame embodies the features of the original I-beam frame and the dynamometer of the hydraulic floating frame explained in the 1911 article in "Engineering," referred to in the first paragraph of the present paper. Although many of these hydraulic floating frames have been built and have acted most satisfactorily for all land work and much of the marine work, the Westinghouse Machine Company have returned to the original and simpler I-beam type. The only changes are:

First, originally the flexibility was provided by the web of separate I-beams. This construction is still retained except in the smaller gears. In these the web is cast on to the lower half of the floating frame, as shown in the second reduction, Fig. 8.

Second, the dynamometer is introduced for marine work. Of course, in double reduction installations this is only re-

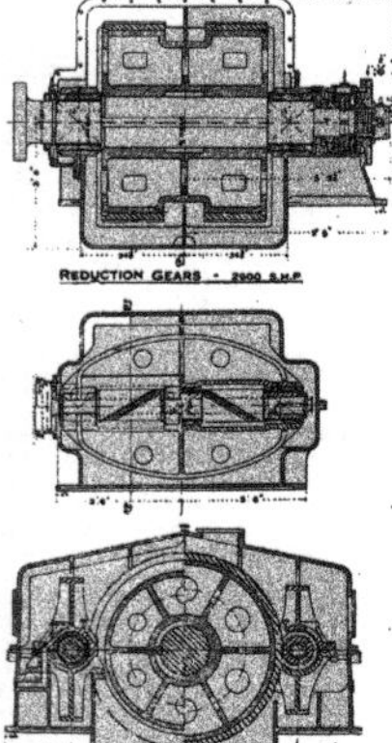

FIGS. 6, 7, 8. DETAILS OF SECOND REDUCTION GEARS WITH MICHEL THRUST BEARING.

quired in the first reduction gear, as shown for this ship.

The floating frame is deprived of freedom of rotation about a vertical axis by two horizontal struts near the ends, the position of one being shown at I, Fig. 2. They are also seen in Fig. 5. Without these the floating frame would have slight instability about this axis. The rims of the large gear A, Fig. 5, are rolled, though cast-steel rims have also been used with perfect success. These are mounted on an iron casting and the centre shaft is of forged steel.

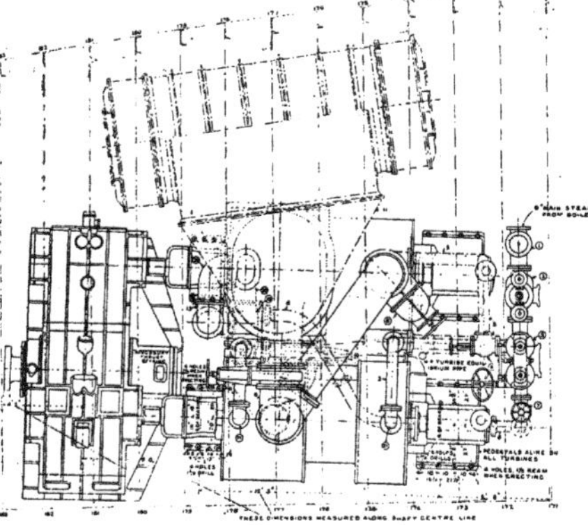

FIG. 9. PLAN OF MACHINERY FOR PORT SIDE.

The second reduction gears, Figs. 5 to 8, call for no further remark. A Kingsbury thrust (known in Britain as the Michel) is mounted at T, the forward end of the shaft.

	First Reduction Gear	Second Reduction Gear
Power per pinion	1,450 h.p.	1,450 h.p.
Pitch diameter of pinion	5.114 in.	10.228 in.
R.P.M. of pinion	3,860	483
Power constant	2.81	2.81

These gears are standard, designed for a maximum rating of 3,400 h.p., with a power constant of 3.3.

The lubricating oil is pumped from a drain tank in the bottom of the ship by a rotary pump, L, Fig. 3, which discharges it into an overhead gravity tank (fitted with strainers) at a height which will give a pressure of about 5 lb. per square inch to the oil at the discharge to the bearings and teeth. After leaving the gravity tank it passes through a surface water cooler and thence discharges to the sides of the bearings, where the bearing pressure is zero, and to the teeth as they are coming into mesh, so that it cannot be thrown out by the high centrifugal forces. If the level of the oil in the gravity tank should fall a short distance, still leaving a full supply for 10 or 15 minutes, an electric bell rings, warning the engineer to start an auxiliary oil pump. When the turbines are running slowly or stopped the oil is circulated by this pump.

The condenser circulating pump is driven by a geared turbine. The vacuum in the condenser is maintained by a Westinghouse Leblanc air ejector, and the condensate removed by a turbine-driven centrifugal pump. The former weighs 108 lb. and the latter 740 lb. They replace an air pump of many times their weight.

The weight of the foregoing installation is 75 tons. This covers from the main turbine throttle and governor valve to the condenser inclusive; also the thrust block and reduction gears (but no shafting), circulating pump, with turbine and its gear, air ejector, condensate pump, auxiliary oil pump, gravity oil tank, oil piping, and cooler.

S.S. Maui

This twin-screw ship is being built by the Union Iron Works, San Francisco, for the Matson Navigation Company. It will carry passengers and freight between San Francisco and the Hawaiian Islands.

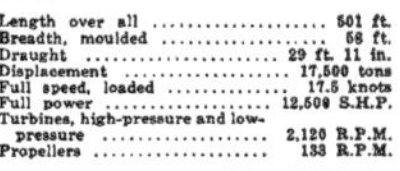

Length over all	501 ft.
Breadth, moulded	58 ft.
Draught	29 ft. 11 in.
Displacement	17,500 tons
Full speed, loaded	17.5 knots
Full power	12,500 S.H.P.
Turbines, high-pressure and low-pressure	2,120 R.P.M.
Propellers	133 R.P.M.

Figs. 9 to 12, pages 322 and 323, show the arrangement of machinery for the port side. Fig. 11 is a sectional elevation showing the turbines, and Fig. 12 one showing the gear. The turbines are cross-compound, reversible, and similar to those in the oil-tank steamer. To conform to the limitations of the engine room the high-pressure turbine actuates a pinion almost on top of the gear. Running either ahead or astern the high-pressure turbines may be used alone, exhausting direct to the condenser, and the astern turbines can transmit 60 per cent. of full ahead power.

The steam pressure at the high-pres-

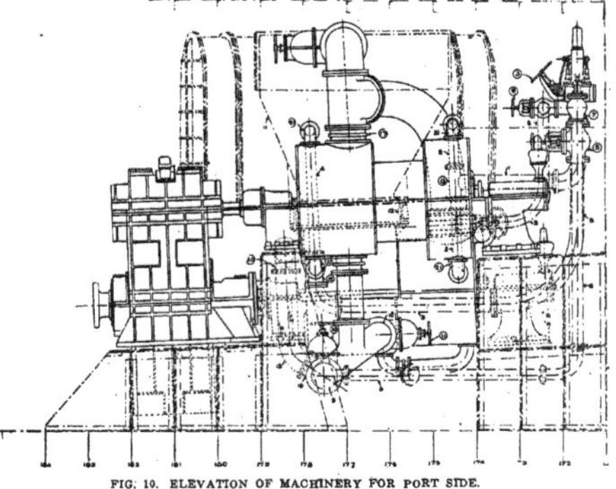

FIG. 10. ELEVATION OF MACHINERY FOR PORT SIDE.

sure turbine inlet is 225 lb. per square inch with 50 deg. F. superheat. The contracted vacuum is 28.5 in. (30 in. barometer). Under these conditions the following performance is guaranteed.:

Total S.H.P.	R.P.M. Propellers	Steam per S.H.P. Hour for Turbines
10,000	125	11
8,500	120	10.7

There is only a single reduction from each turbine to the propeller, and the

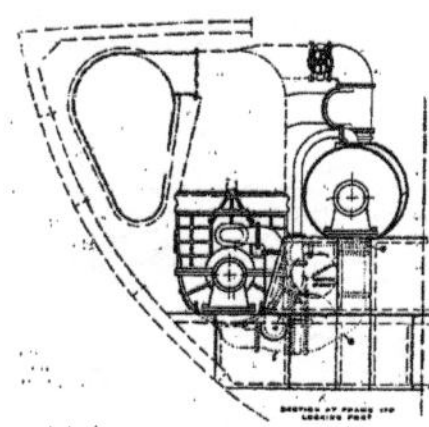

FIG. 11. SECTIONAL ELEVATION SHOWING TURBINES.

floating frame is of the same type as that shown in Fig. 2.

Maximum power per pinion......	3,125 H.P.
Pitch diameter of pinion	7.794 in.
R.P.M. of pinion	2,120
Power constant	3.113

The lubricating system and auxiliaries are of the same type as those for the oil-tank steamer.

The weight of this installation, including the same parts as for the oil-tank steamer, except the condensers, oil pipes, and gravity oil tank, is 204 tons.

The introduction of the floating frame has been much accelerated during the last five or six months. There are in operation 102 gears, with a total of 126,442 h.p.; and on order 155 gears, with a total of 611,727 h.p. Besides the turbines and gears for land work and for the 45 freight ships and the Maui, the Westinghouse Machine Company are building the gears and, in all but two cases, the turbines for six warships, including a United States scout cruiser of 90,000 h.p. and a battleship for a foreign Government of 88,000 h.p. Indeed, principally to cope with this work the Westinghouse Electric Company, who have acquired the Westinghouse Machine Company, have begun to build, near Philadelphia, Pa., works which will be several times the size of their present extensive plant.

The simplicity of adjustment of the gear and pinion is remarkable. The pinion is placed in the floating frame and, by means of two gauges supplied with each gear, the distances of the centres of gear and pinion at the forward and aft ends of the floating frame are made equal; the holes for the bolts holding down the floating frame are then reamed. There is no scraping of bearings to bring the axes accurately into one plane and secure bearing all along the tooth face, as must be done with a rigid gear. The action of the teeth will do that, and, as the theory shows, however the bearings wear or alter under stress, good distribution of pressure along the teeth will be maintained. The gear is then run, usually under quite a light load, and the hard-bearing spots, if any, scraped. The work is then complete.

Practical Results

If the floating frame functions as stated, the wear on the teeth should be extremely small. A remarkable proof of this was given in the article of last year (referred to above) from the operation of the first floating frame gear installed, that at the Commonwealth Steel Company, Granite City, Illinois. This had not nearly so high a power constant as is now used, but it was reported that the maximum load reached 40 per cent. of its rating, under which condition the power constant is 1.68. It was started on March 31, 1911, and, after running for 10 months night and day, five days per week, the load averaging 85 per cent. of its rating, the tool marks were still visible on the teeth. Mr. K. S. Howard, the chief engineer, has informed the author that this gear was last inspected on April 30, 1916, when they found the scraper marks still visible on the gear teeth. Originally these were of imperceptible depth. The driving face of the pinion teeth had taken on a very high polish and the gear was as quiet as at first. This is after more than five years of service. He further reports that since installation it has operated 24 hours per day, six days per week, with loads varying from 50 to 125 per cent. of rating, the average being close to full load, and he sees no reason why it should not operate satisfactorily for an indefinite period. There have been no repairs. The only discrepancy in the report is that, apparently, in the busier times of the last few years the gear has had to work

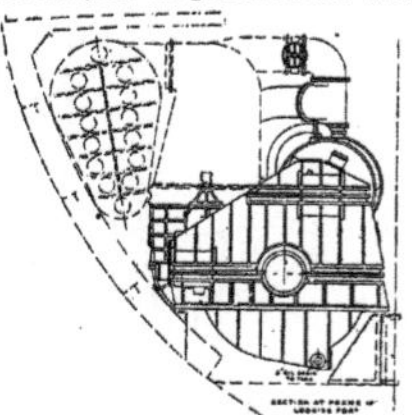

FIG. 12. SECTIONAL ELEVATION SHOWING REDUCTION GEAR.

under a greater average load and for one extra day in the week.

Again, there is a 1,000-kw. gear, guaranteed for 25 per cent. overload, at the San Diego Electric Railway Company's plant in California, which has been running 19 hours per day for five years. The superintendent of motive power states that it shows no sign of wear, and that there has been no maintenance cost whatever. Similarly, two 5,500 h.p. single-pinion gears of the Cleveland Electric Illuminating Company have been in operation for four years, giving no trouble and without wear. These results are not exceptions, they are almost invariable, showing the great safety of the device in spite of frequent, high-power constants. A further proof of the entire absence of point contact of the teeth is that chrome nickel steel has never been used for the pinions, as has frequently been done for rigid gears. Nickel steel has been used twice, to meet the physical characteristics required by the United States Government; but now these requirements have been changed and the pinions are always forged of a good quality of mild carbon steel.

LAUNCH OF S.S. ALASKA

THE steel steamer Alaska which was launched at John Coughlan and Sons shipyard, Vancouver, B.C., recently was built for Norwegian owners but will operate under the direction of the British Admiralty. Capt. Peter Omundsen will be the master of the vessel.

The Alaska is 427 feet long over-all, 410 feet 5½ inches between perpendiculars, moulded beam 54 feet and moulded depth 29 feet 9 inches.

The deadweight capacity of the vessel is 8,800 tons on a draft of 24 feet, 2 inches, and she has a capacity of 490,000 cubic feet. The vessel is built under the survey of Lloyd's Register of Shipping to the highest class for ocean-going vessels.

The S.S. Alaska is a steel, single screw, two-decked cargo steamer of the poop, bridge and forecastle type, with machinery amidships.

There are five main watertight bulkheads and one non-watertight screen bulkhead between the engine and fire-rooms. A double bottom is fitted complete from forward to after peak bulkhead and is constructed on the cellular system. The double bottom is divided by watertight floors and centre line keel into fourteen compartments and two peak tanks.

The decks are supported by two rows of widely spaced strong pillars in lieu of the usual numerous small pillars of light sections, thus facilitating the stowage of cargo in the holds.

Gear For Cargo Work

There are four large and one small cargo hatches through the upper and second decks. Cargo will be handled through each of the four main hatches by a pair of five-ton cargo booms, stepped on derrick tables on the two steel pole masts. For handling cargo through the small hatch, a derrick post with one three-ton boom will be fitted on each side.

In addition to these booms, a thirty-ton boom will be supplied for the handling of extra heavy cargo. The vessel is fitted with ten cargo winches, one warping winch on poop and one quick warping windlass for handling anchors and warping the ship.

The topmasts are made to telescope so that, if required, the vessel can pass under bridges over the Manchester ship canal in England, and for the same purpose the topmost section of the funnel is arranged so that it may be easily removed. A flying bridge is provided between the poop and bridge decks so that the crew may pass in safety in rough weather, even if the decks are flooded.

The steam steering engine is located in the poop deck and operated from the wheel house by telemotor control. Auxiliary hand steering gear is also provided. Water ballast or oil fuel will be carried in the peak tanks or double bottoms and reserved feed water in the double bottom under the machinery space. Drinking water will be carried in separate fresh water tanks located in the bridge space.

The ship is electrically lighted throughout, the current being generated by means of two 15 K. W. direct current generator sets working at 110 volts. All navigation lights are electrically fitted and in addition to those required by law an eighteen-inch searchlight is fitted on the bridge. All accommodations are amply heated by steam.

The boilers are fitted for burning either oil or coal, the permanent coal bunkers are located around the machinery casing, while reserved capacity is located in the after end of the bridge-house, which space can be used alternatively for cargo. The total bunker capacity is 1,255 tons and as the coal consumption is about thirty-eight tons a day the vessel will have a steaming radius of thirty-four days, during which time she should cover a distance of 9,000 nautical miles.

The fuel oil capacity will be about 1,100 tons, which will permit a voyage

SS. ALASKA IMMEDIATELY AFTER TAKING THE WATER AT THE YARD OF J. COUGHLAN AND SONS, VANCOUVER, B.C. SHE IS 427 FEET LONG OVERALL, HAS A DEADWEIGHT CAPACITY OF 8,800 TONS, AND WILL HAVE GEAR-DRIVEN TURBINES OF 2,500 HORSE-POWER.

Accommodation for the officers is located in the after midship house, while the forward house contains the captain's room and bath, large dining saloon, hospital, pantry and steward's rooms. All accommodations for the officers will be comfortably furnished with joiner work and quartered oak.

Men Have Good Quarters

The crew of thirty-six men will be quartered aft in the poop, their quarters will be well furnished and instead of the usual manner of one large cabin, they are arranged in smaller cabins for two men each. These quarters comply with all the regulations of the British Board of Trade and also the requirements of the Seaman's Act of the United States. Separate messrooms are provided for sailors and firemen.

The wheelhouse on the bridge is substantially constructed of teak and is fitted with the usual navigation instruments.

of forty days' duration. During such a voyage the crew can at all times be fed with fresh provisions, made possible by a refrigerating and cold storage plant located on the upper deck at the after end of the engine casing.

Description of Engines

The propelling machinery consists of one high pressure and one low pressure turbine of the Parsons type, both connected by double helical gears to the main shaft, the gears reducing the speed of the turbines from 3,600 revolutions to 100 revolutions per minute for the propellor.

The reversing turbines are built in the same casing as the main turbines and will develop two-thirds of the full speed forward, making manoeuvring easy.

The turbines will develop 2,500 shaft horsepower and will drive the vessel at a speed of eleven knots in a loaded condition. The turbines will drive a four-blade built-up propellor about seventeen feet in diameter.

Steam will be supplied by three single-end Scotch marine boilers, each containing 2,670 square feet of heating surface and having an inside diameter of fourteen feet nine inches and a length of eleven feet, built to conform to Lloyd's requirements, with a working pressure of 190 lbs. per square inch.

In addition there will be provided a main and an auxiliary condenser, circulating and air pumps, main and auxiliary feed pumps, ballast pump, fire and bilge pumps, fresh and salt water sanitary pumps, feed water heaters, evaporator and distilling plant, ash ejector and wireless outfit.

RE STANDARIZATION OF MARINE ENGINES

A CORRESPONDENT of the Institute of Marine Engineers, in criticizing a recent paper on the above subject, remarks that some of the leading engineering firms in Britain still continue to turn out anything but a satisfactory or ideal job. The following abstract from his communication contains many items of timely interest:

Re Government Control

It may be ignorance of what is the best practice; it may be the manufacturer's idea of economy to use old drawings and patterns and antiquated methods of manufacture. Or it may be, as several members suggest, that the shipowner is not always as flush of money as he would like to be, so that that greatest of all sinners, "Initial Expense," does not permit of his having the best. But whatever the reason for it, the fact still remains that we have a big lot of engineering science and experience that we are not making use of.

It is for this reason many of us hope that the Government will retain its control over shipping and shipbuilding for many years to come after the war. Standardization, as arrived at by a competent body of all the different interests concerned, enables you to put to sea with the best practice that the profession is acquainted with, and by turning out the work in large quantities it enables us to produce the best almost as cheap, if not quite as cheap, as a lot of individual inferior stuff.

Although the shipowner gets what he pays for, yet it is rather degrading to our characters as engineers to have to turn out what we know to be inferior work. The manufacturer takes no pleasure in turning it out; the superintendent engineer does not like having to keep it in repair; and as for marine engineers—well, how many poor thirds have lost all hopes of ever getting aloft just because someone thought how nice it would be to have a nice cheap lead-base alloy for the bearing metals.

So whether we can compel the owner to buy the best or not, it should, nevertheless, be the business of our institution to keep before him what, in its collective judgment, is absolutely the best practice obtainable. To me it appears to be the acme of stupidity to use one's own brains as against the considered and reasoned judgment of the whole engineering profession.

Auxiliaries

As regards auxiliaries, the leading firms who specialize in these things will not lead us far wrong. But I think even these and the minor details of other parts should receive consideration. It is frequently the small details that make all the difference between a first-class job and a mediocre one. How often does one see the very best of work ruined for the sake of fitting an oil cup that has a sporting chance of catching at least 50 per cent. of the oil fed to it? Whereas a well-designed cup made to touch a worsted drip every time the engine turned its top centre would give the bearings every drop of the oil supplied. Also, we still see oil boxes bolted on to the cylinders. I have myself found the temperature of the oil as high as 148 deg. F., which, compared with a temperature of about 75 deg. F., would have a reduced viscosity of over 50 per cent. Yet it would be just as easy to bolt the oil-box to an insulated bracket, keeping it a couple of inches away from the cylinder.

Then we still see on modern jobs turning engines with the worm shaft at the bottom, close to the tank top, which takes anything up to a couple of hours to get in and out, instead of a few minutes. And this runs away with a lot of valuable time for overhauling, besides further reducing the third's chances of ever playing on a golden harp. And we still go away to sea in brand new jobs where you have to disconnect two or three pipes before you can get at a leaky bilge pipe, just because some inexperienced draughtsman, who had never heard of a place called the sea, and consequently he did not know that his bilge pipe would ever have to come out again on account of their having a very bad habit of "going in the neck." He simply went to the drawer and fished out an old engine arrangement, and congratulated himself because he had been able to save 6 ft. of copper pipe, which, nevertheless, would have to be paid for over and over again in "time lost."

Design Details

Again, a great deal of time is lost over badly designed top and bottom ends. The simple device of having a tapped hole in the top end bolt for an eye bolt saves much time in rigging up gear for lowering the bottom end brasses; yet you do not often find the tapped holes there.

I have seen a whole day wasted trying to get a top end adrift because the pinching pin had broken off short. Had the pin been screwed into the side where it could have been drilled out, instead of into the end of the keep, a lot of valuable time would have been saved.

I have been on a first-class passenger ship, not yet eight years old, where the top end crown brass was part and parcel of the crosshead, and could not be taken down without lifting the cylinder cover and slinging the piston. This ought never to be, and yet the job was designed by one of our leading firms who turn out first-class battleships. A standardized engine would save us from a lot of this stuff.

Then we still go to sea in brand new ships with main check valves and other boiler mountings and valves, where you have to tighten up the gland nuts with a hammer and caulking tool. This is cheese-paring with a vengeance and wants standardizing out of existence.

One member, speaking of piston rods and glands, says different jobs require different treatment. But surely that is the result either of faulty workmanship or want of specification as to the composition of the material.

There is a lot of sloppy engineering about those sort of things, and a lot of thinking and guessing, when it is possible, by taking a little more trouble, to prove. Take, for example, this case of piston-rod gland packing. Why cannot a competent body of engineers, such as this institute can furnish, pass the leading "makes" through the mill of their criticism and pronounce a verdict that would be worth having? What are the things we require to know? Roughly, they are:

(1) Amount of friction on rods.
(2) Steam tightness.
(3) Amount of abuse they will stand.
(4) Time taken in fitting and overhauling.
(5) Life of packing.
(6) Facilities for renewals.
(7) Initial cost.
(8) Cost of repairs.

Most of these points can be settled with almost mathematical accuracy. Any engineering professor at one of our laboratories could work out experimentally the amount of friction on the rod for any desired steam pressure and the coal consumption thereby represented. The loss of water through leaky glands most of our superintendents would be able to tell us who have had years of experience with many kinds of packing. The initial cost can be found exactly; the repairs or renewals almost exactly. The remaining points can be found with a good degree of accuracy.

The importance of deciding which is the best article from the above points of view is very important. It means that instead of packing, say, 1,000 glands with one type and another, 1,000 with about 50 different types, we pack 2,000 with the best, and which would incidentally have the beneficial effect of reducing the initial expense. Similarly with other things; each part or system should be examined and receive the verdict of the profession. If the owner wants a round condenser let him have one, but let him have no doubt about what marine engineers think of pear-shaped ones, or the contra-flow system of condensation.

Air Pump Possibilities

It would be with many tender regrets if we had to lay aside our Edwards air pump for the Mirrless Watson multi-ejector air pumps, but it might have to be done in the interests of standardization and scientific progress. Most of us would be very sorry to part with our Weir groan, but we might conceivably have to get rid of our old love to take on a motor or turbine driven centrifugal feed pump. A 1 per cent. saving of steam in a feed pump that is always at work is worth a considerable addition to the initial expense. I am conscious of having gone beyond the author's own idea of standardization, but after all a marine engine is far from complete without its auxiliaries just as without the author's pipe arrangement. At least, I think the auxiliary machinery should have to conform to the best known practice as regards steam consumption, simplicity of design, space taken up, weight and reliability.

In conclusion, I would suggest that a standard engine should not be considered complete without being able to feel its pulse. Yet it is seldom you see a ship sent away to sea with anything better than a Richards' indicator. There has been a lot in the press lately about the ignorance of marine engineers (which obviously accounts for our not being able to obtain any advancement in rank as in the deck department); yet it is not altogether our own fault. We can scarcely imagine our American cousins putting to sea in their new merchant fleet without CO_2 recorders, differential draught gauges, steam and water meters, pyrometers, recording gauges, etc.

It is not enough to know that we can get 8,000 tons along at 10 knots for 33 tons of coal per day. We want to know how it is that we are not doing it on 32. Guessing is no use. We must have the proper instruments and make sure that our own ideas and assumptions are the correct ones.

FROM LAKES TO SEA

DURING the last few months a strange procession has been moving through the great lakes to the sea. It was if the victims of some ruthless U-boat campaign had risen to the surface. Bows of ships, sterns of ships, sections from the middle minus bow or stern floated weirdly on their way. The lakes were making their contributions to the new Atlantic fleet of the United States Shipping Board.

The twenty-five locks of the Welland Canal can admit boats of 3,500 tons with 250 feet length and 44 feet beam. Thirty ships not exceeding these dimensions have already passed out to sea, while sixteen more have been cut in sections and so transported through the canal, to be assembled at Montreal. The larger passenger steamers must remain with the fresh water fleet, since their great width precludes the passage of locks, even in sections. This process of cutting boats in two costs about as much as new construction, but effects a great saving in time and in materials. The process may be completed in about 100 days, one-third the time required to build a new boat. During the winter months, when lake navigation is closed the use of this tonnage on the Atlantic is so much clear gain. Before spring new construction on the great lakes will have replaced the loss.

Some Minor Details of Ship Construction and Equipment-I*

By C. Waldie Cairns, M.Sc.

Although the items dealt with by the author do not in the aggregate account for much in the total cost of a ship, they have a very important influence on the subsequent satisfaction given by the vessel in actual service. Practice in regard to these details varies greatly and unexpectedly in different yards, and he suggests the advisability of adopting a "Standard Fittings Specification," which would insure adequate service and life irrespective of origin.

IT will be generally admitted that shipbuilding art and science have reached a fairly satisfactory position, though there must be possibilities, not yet fully realised in the more efficient application both of labor and present materials, towards the production of efficient vessels, and we hear more than rumors of new materials. No doubt, much useful information has yet to be obtained and disseminated as to the form best suited to any desired speed and deadweight, but in the main we may assert that the modern steamer generally "keeps right side up," and gets through the worst weather with fewer serious instances of strain, either to upper works or to bottom, than in the days when the registration societies were struggling with the problems arising from increase of size and of fullness of the ordinary cargo ship. The writer has no intention, in this contribution, of invading the domains of the naval architect in any of the important provinces mentioned, but wishes to direct attention to a number of items which, from his experiences with shipbuilders and ships, he judges receive somewhat scant attention in the cargo steamer equipment.

It is not suggested that all shipbuilders neglect all the items mentioned, attention to these details varies very much. The practice regarding some of these is excellent in certain yards, but the writer has seen important items overlooked in yards that, in most respects, are, at least, up-to-date. None of the items referred to account for much in total cost of a ship, but they may mean a great deal in relation to subsequent satisfaction, safe working or freedom from many irritating small repairs. Certain of them might well be the subjects of standardising agreements, of a "standard fittings' specification," which would eliminate a slight advantage in the market that a "mean" yard may hold over a yard where quality is considered. Practice with regard to some fittings is not even sufficiently standardised in individual yards, choice of details being occasionally left to foremen, apparently without any standard arrangement.

The necessary minimum scantlings and ground and mooring tackle are already imposed on builders in the rules of classification societies. It would be a convenience to owners of ships, and no doubt to builders also, if a well-considered minimum standard in other details of equipment could be formulated, or even a series of grades of equipment which could be referred to by distinguishing titles. There might, for instance, be a minimum "coasting collier equipment" or "Class 1," and "Class 2," "coasting collier equipment"—similarly, one or more grades of "Baltic trades equipment" of "collier and ore trade equipment," of "River Plate trade equipment," of "Eastern trade equipment." Some of these would differ comparatively little in main features.

Possibly, the commencement made by this institution on the North-East Coast Institution engine specification, which is already leading to national developments, may, by its example, induce this or another institution to make an effort to standarise good practice in some of the details that will be referred to. It may be urged:—Why not specify what you think good in the ships in which you are concerned, and let others put up with what they get? Well, laying aside many strong reasons and giving the selfish one, good details are much more cheaply obtained when they are general practice than when they have to be specially bargained for, and perhaps specially designed for one's own ships.

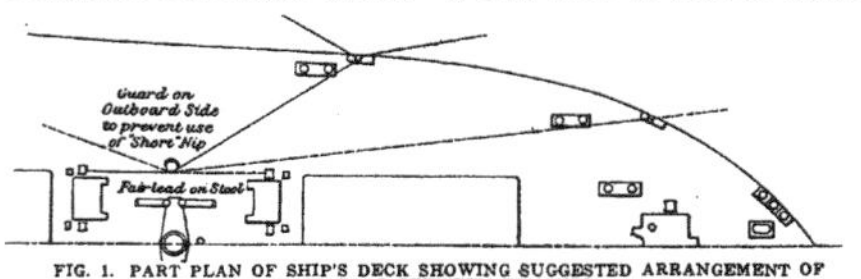

FIG. 1. PART PLAN OF SHIP'S DECK SHOWING SUGGESTED ARRANGEMENT OF FAIRLEADS BETWEEN WINCHES.

Chain Cables

One point regarding these concerns the shipbuilder. Possibly the shipbuilder will insist that the point rather concerns the owner and his advisers, but as it becomes apparent on design of the hawse pipe, the shipbuilder has first cognisance of it. This point is, that with most stockless anchors, in any but the smallest ships, the joining shackle between anchor shackle and end link of the cable is, with the anchors in the "stowed" position, so far down in the pipe that it is impossible to knock out the shackle pin without lowering the anchor out of the hawse pipe. If this were only required in dry dock, little harm would result, but as it is necessary, under certain conditions, to moor ships by the main cables, easy access to an end-shackle is of importance. This is easily arranged for by the provision of a short length of cable, say 4 ft. to 8 ft. long, according to need, on each anchor, thus bringing the shackle into an accessible position above deck, where, after the anchor has been "hung off," the shackle pin can be easily driven out.

The next point, regarding cables, is to some extent a reflection on the makers. Cable makers are supposed to work to some standard as to dimensions of cable links, end links and joining shackles: presumably, again, this is the "Admiralty Cable Scale," of which the windlass makers issue useful tabular particulars. The form of certificates issued from the proving houses records the "length and breadth of links, but no report is made of variation among lengths of links, nor is any permissible "degree of tolerance" stated on the certificate.

Correct forms of links, with consequent correct pitch of links, is of the greatest importance for reliable operation of the windlass. Cable-lifters can, of course, be designed for any pitch of chain, but if the pitches are not reasonably equal throughout a ship's equipment, "jumping" of the cable over the "lifters" may ensue, and undue wear and tear of the "whelps" will certainly occur.

Reference to this matter is not to be considered "finicking" or pedantic. A case came under the notice of the writer during the last three years, where a large part of the cable equipment of a new steamer had to be replaced for this reason, and within the last few months he took an opportunity of checking over five lengths, all just delivered from one maker, and found the following variation in "length overall of ten links" when ranged by hand in dry dock bottom:—

	Ft.	In.
No. 1 Length	6	9¼
No. 2 "	6	6¾
No. 3 "	6	8
No. 4 "	6	6
No. 5 "	6	9

Windlass makers probably do not expect perfection of pitching, and cable makers certainly cannot attain it, but some "limits" might well be agreed, and test house officials might then be directed to check lengths of cable when under light tension and record on the certifi-

*Paper read before the North-East Coast Institution of Engineers and Shipbuilders at Newcastle-on-Tyne.

cate the correctness of end links and the maximum and minimum "lengths over ten links," with power to reject cables not coming within agreed limits of variation.

Mooring and Warping Arrangements

The needs and requirements of our sailor friends deserve the fullest attention in this matter, and if their ideas appear to have expanded recently in connection with the fittings deemed requisite on even "common tramps," we should remember that the dimensions of such ships have continued to grow in most trades, and that at the same time the lengths of jetties and berths have not been growing proportionately. Hence methods of mooring and of handling warps that were possible under bygone conditions may no longer be applicable —for instance, hauling lines often now have to be run out from positions in the middle half-length of a steamer, instead of over bow or stern, as on shorter ships or at berths of great length. Further, at coal tips, or other crowded berths, ships are often "scarphed," or laid with their lengths partly overlapping, making safe and handy means of working "breast ropes" necessary. It will be judged from this that all requirements are not fulfilled by "quick-warping drums" on windlass and a warping winch on the poop, useful and necessary as these helps are.

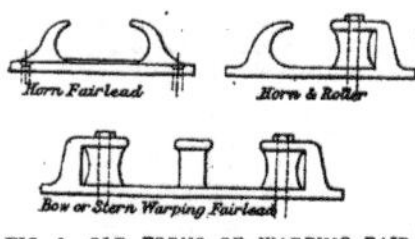

FIG. 2. OLD FORMS OF WARPING FAIRLEADS WHICH ARE STILL COMMON.

Fairlead sheaves on comparatively light cast iron stools are often fitted in the vicinity of windlass and of after winches. These are usually strong enough to act as guides or supports for warping wires between warping chocks and the winch or windlass, provided the angle between the two directions of the wire is nearly 180 deg., and these fairleads are useful in their way, but what is really required in addition is the placing of strong warping guides on deck or on stools, between the winches, and in other positions to deal with breast ropes. In some positions these ropes may have to be led with parts nearly, or quite, parallel, so applying to the fairlead a load equal to double the tension of the rope.

These warping guides would not only save time in working the ship, but would minimise the number of occasions when snatch blocks—easily strained, and always destructive to the ship's wires—have to be used. The arrangement suggested is shown in Fig. 1.

It will be noticed that fairleads are indicated at the ship's side, as well as the extra fairleads between the winches. In too many cases horn fairleads only are fitted at the side; these fairleads should be of ample strength, equal to the old horn fairleads.

It is also important that bitts or bollads should be so placed that the hauling line passes close to them between the fairlead at the ship's side and the fairlead at the winches, as this enables "stoppers" to be used as to avoid loss of tautness of line whilst transferring from winch drum to bitts. For all fairleads for warps, etc., there is a healthy tendency towards the use of sheaves of large diameter, and in most cases these sheaves are now provided with brass bushes.

The warping fairlead of old form, Fig. 2, is still common. In this type, bushing of the roller or sheave is not sufficient, a brass washer should be provided at top and bottom also, otherwise the roller will almost invariably be found to be "rusted up solid" before the ship is many months old.

There are, however, several special forms of fairleads on the market, and already widely used, their cost differs but little from the common form, and as they are all self-lubricating and fitted with sheaves of large diameter, and of good form, they are a distinct advantage to a ship in saving wear and tear of hauling lines. A point to be borne in mind, however, is that they give somewhat poor guidance to a wire led upwards steeply from ship to quay; some special arrangement may be necessary if much work of that kind is expected. These conditions rule in Panama Canal locks and have made necessary a form of mooring chock, which will prove useful both under such conditions as to quays and where, as sometimes happens, a line has to be led from a loaded ship up to a light ship lying alongside.

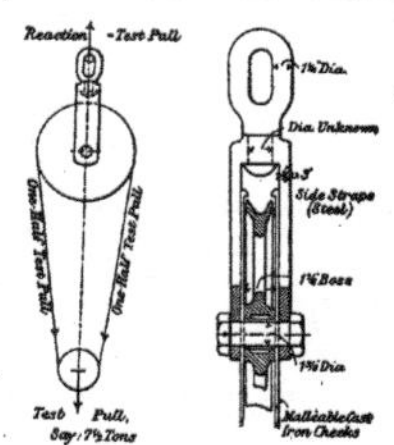

FIG. 3. METHOD OF APPLYING TEST LOAD TO 5-TON TRUCK.

FIG. 5. DETAIL OF CARGO BLOCK EXAMINED BY THE AUTHOR AS DESCRIBED.

Cargo Gear

One always feels that in a well-organized world ships would not require to be fitted with a heavy and expensive outfit of cargo gear, to spend probably four-fifths of its time as ballast, or as to much idle weight to be carried round at sea. However, while there remain ports where vessels may be asked to lie—as frequently at Hull, Rotterdam and elsewhere—with lighters at each side to receive cargo worked out by the ship's own winches, or ports in which, although quay berths are provided, enterprise does not "run to" cranes—ships other than those serving special trades must continue to carry about with them anything from four to thirty winches, with, in the latter case, a forest of masts, derricks, and derrick posts to enable them to be utilised.

The elementary theory, as to loads on derricks, spans, runners, blocks, etc., has been written about, calculated about, and reduced to handy form in numerous text books—some even specially written for easy comprehension by the least mathematical of ship's officers. In spite of this it is rare to find a sound view of cargo gear taken, even in the drawing office. In many yards cargo gear appears to have "just grown" by some system of "trial and error," presided over by the foreman blacksmith. This man is, in most cases, a good man at his trade, but it does not follow that he is consequently an authority on determining dimensions or stresses.

The writer remembers the case of one yard, where, repenting of these ways, the drawing office was set to design derrick fittings, and a quite neat design was evolved. Unfortunately, the designer forgot that derricks are ever swung out over the side, and the span-head connections, while fairly satisfactory for the direct tension produced by a fore-and-aft position of derrick, were stressed to over 16 tons per sq. in. by the bending strain produced by the thwartship direction of spans when the derricks were swung out. Probably the foreman blacksmith's "eye," without calculation, would have been safer than this incomplete consideration.

Examples From Practice

Another example of the insufficient consideration given to cargo gear is evident in the choice usually made of cargo blocks for derrick ends.

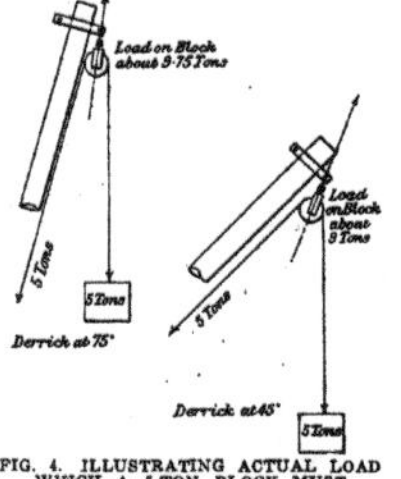

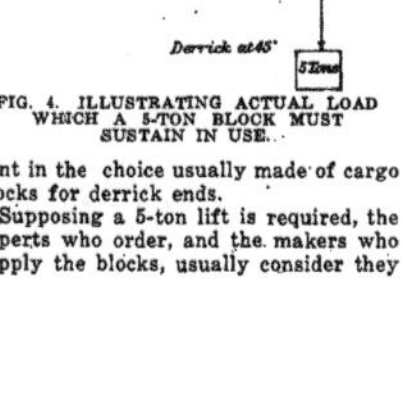

FIG. 4. ILLUSTRATING ACTUAL LOAD WHICH A 5-TON BLOCK MUST SUSTAIN IN USE.

Supposing a 5-ton lift is required, the experts who order, and the makers who supply the blocks, usually consider they have behaved liberally by arranging for a block tested to 7½ or perhaps 10 tons, and the unsuspecting sailorman and stevedore are expected to go happily and trustfully about their business of lifting a 5-ton load. But what is the real margin here? First of all, considering dead load effects, the block under a 7½-

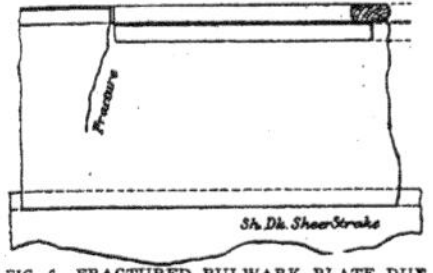

FIG. 6. FRACTURED BULWARK PLATE DUE TO CARELESS NOTCHING.

ton test is loaded as in Fig. 3. When supporting a 5-ton load, in working positions, the method of application is quite different, and results in pressures on the pin and tension on the suspending eye amounting to from about 1.8 times the load with derrick at 45 deg. to 1.95 times the load with derrick at 75 deg. (see Fig. 4). So the margin of test load over working load, instead of being 50 per cent. or 100 per cent., has vanished, even if the working load is a dead load.

The test load on the block, having been carefully applied by a testing machine, was certainly a "dead load," producing stresses directly calculable by elementary methods. Working loads on cargo gear are, however, "live" loads—very much so. No matter how carefully they may be "picked up," there are invariably surgings and shocks produced in lowering and checking; by these the stresses are at least doubled, as compared with those produced by the "dead" loading arising from supporting the same weight at rest. This further discounts the utility of the test load, which, in reality stresses the block much lower than working conditions will.

There is no doubt that we are often saved from breakages of cargo gear by the fact that the system yields when a load is checked or surges, or drops after a check. This yielding is mainly from the elastic stretching of the material of the span, runner, and shrouds, along with the deflection of mast or derrick-post from the unstrained position. The importance of this yielding, as a means of modifying the severe effects of "impact," may be judged by the following comparison:—

Assume a load of 1 ton drops from rest, with a free fall of 1 inch before the cargo gear begins to check its descent. At the moment when the fall has been completely arrested the reaction of the cargo gear is as follows varying with the amount of "give" of the gear:—

Yield of Point of Attachment of Load from First Check to Final Stoppage of Fall In.	Reaction Stressing the Cargo Gear (1 Ton Load), Tons
1	4.00
2	3.00
3	2.66
4	2.50
6	2.33
12	2.16

It will thus be seen how difficult it is to get the stresses down even to double the stress arising from a dead load of same weight.

Tests

Calculation of the expected yielding of the point of attachment would be somewhat tedious, especially as the bending of mast and stretch of shrouds is involved, but as a guide to the order of the dimensions involved it may be pointed out that the extension of 120 ft. of steel wire rope, when stressed up to 6 tons per square inch, is probably between ⅝ and 1 in. This 120 ft. is mentioned as being the approximate total length of runner plus span, taking span as being carried down to cleats or bitts near the deck. If a manilla downhaul tackle is fitted, the yield will be much greater and the multiplication of nominal stress somewhat decreased.

The old Admiralty "tests and loads for chain" quoted in "Mackrow" gives figures on the following basis:—

1 in. chain—"breaking strain," 27 tons,
"proof strain," 12 tons,

and states that the "working strain" for crane chains, etc., should not exceed 2/9th the "breaking strain" or half the "proof strain."

It is to be noted that the stress on material, taken on the sides of chain links would thus be:—

	Tons per square inch.	
At breaking load...	17.2	38,500 lbs./sq. in.
At proof load......	7.63	17,100 lbs./sq. in.
At working load...	3.81	8,550 lbs./sq. in.

The working load stress may appear high, but it must be remembered that a chain load is never reversed.

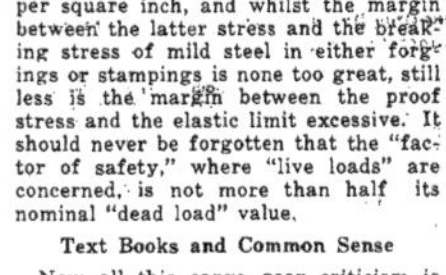

per square inch, and whilst the margin between the latter stress and the breaking stress of mild steel in either forgings or stampings is none too great, still less is the margin between the proof stress and the elastic limit excessive. It should never be forgotten that the "factor of safety," where "live loads" are concerned, is not more than half its nominal "dead load" value.

Text Books and Common Sense

Now all this cargo gear criticism is merely a mixture of elementary textbook and common-sense—or what the writer hopes is common-sense—but unfortunately he finds that the text-book and common-sense points of view are not always applied to cargo blocks. The shipbuilder usually orders from manufacturers an outfit of ready-made blocks, tested to 5 tons, 7½ tons, or whatever occurs to him as a reasonable margin over the owner's desired "lifts." Manufacturers are not infallible either. In the case of one much patronised maker of blocks a request for drawings for inspection brought the reply, "Sorry, we have no drawings. Our work is all done to templates." In such a case there is always a fear that there never have been any drawings, or that ancient error has been copied from another manufacturer. Some alleged specialties are admittedly "designed" by copying those of other makers, mistakes included, without any questioning criticism on the part of the copyist.

The writer has examined stresses of a cargo block dimensioned as in Fig. 5 on the previous page and found stresses

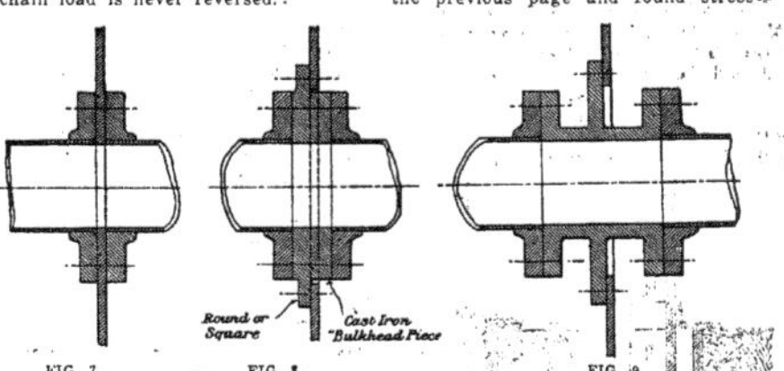

FIG. 7. FIG. 8. FIG. 9.

Presumably this refers to the best class of wrought iron, of which the breaking stress of the unworked material would be sensibly higher than the breaking stress of the chain on test. Even making allowance for such steel as is now used in blocks and other cargo gear forgings or stampings in place of iron as in chain, it would not appear advisable to exceed these "calculated" stresses (that is calculated as if dead loads), particularly in view of the somewhat indeterminate "live load" effects already referred to. This much we can definitely say, that where we design for a nominal stress under "dead load" of about 3.8 tons per square inch, the "live load" effect will raise the actual stress at least up to the proof stress of 7.6 tons

per ton of load at eye (or what is the same, per ton load on pin) as follow:—

	Part	Nature of Stress	Per Square Inch Stress per Ton Load at Eye
1	Pin........	Shearing	0.338 ton or 724 lb.
2	Pin........	Bending taken $\frac{WL}{8}$	0.918 ton or 2,060 lb.
3	Pin........	Pressure on bearing	869 lb.
4	Binding...	Tension at pin-hole.	0.491 ton or 1,100 lb.
5	Oval eye..	Tension on sides	0.4075 ton or 915 lb.
6	Neck of eye	Tension	No data.

The makers' test load for this block is not known; the working load suggested is 5 tons; and after making allowance for method of application of working load, as per Fig. 4, the apparent stresses (leaving live load and shock out of account) are as follows:—

Stress No.	Stress or Load per Square Inch under Proposed Working Load
1	3.29 tons or 7,050 lbs.
2	8.94 tons or 20,000 lbs.
3	8,450 lbs.
4	4.78 tons or 10,700 lbs.
5	3.97 tons or 8,910 lbs.
6	No data.

A double-sheave block proposed by makers for a test load of 20 tons showed stresses under test load as in the next column. Stresses under working load are also tabulated, assuming working load half the test load and that, as is possible, the block is used combined either with a single or a double-sheaved

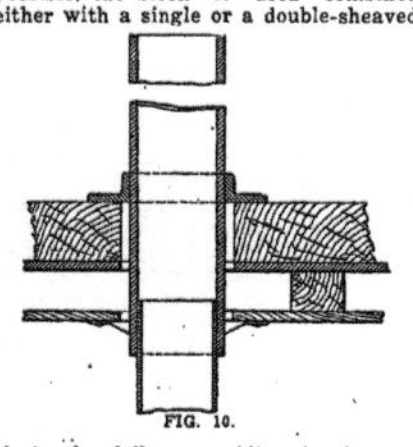

FIG. 10.

block, the fall rope adding in the one case one-third, and in the other one-fourth to the load on pin and on eye.

It will be noticed that in a double block, whether coupled with another double block, or with a single block, the effect of the fall rope in reducing the margin between real effect of working load and the test load is much less than in the case of a single-sheave block already commented on. The writer submits that even allowing for the fact that the loading of blocks, like that of chains, is

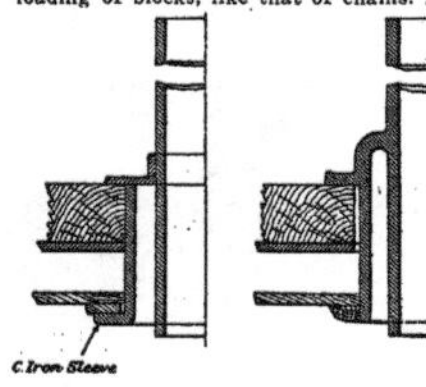

FIG. 11. FIG. 12.

never reversed, some of the stresses tabulated are very much too high.

Part.	Nature of Stress.	Under Test (20 ton)	—Stresses— Tons per Sq. In. under Working Loads Single Block	Double Block
Pin	Shearing	4.81	2.77	2.6
Pin	Bending	2.57	1.70	1.6
Pin	Pressure on bearing	2.86	1.90	1.79
Binding	Tension at pinhole	8.90	5.94	5.55
Oval eye	Tension in sides	10.00	6.70	6.25
Neck of eye	Tension	13.4	9.00	8.43

Torsion Meter for Power Transmitted by Propeller Shafts

The advent of the turbine introduced new problems in the measurement of power developed by marine engines. The conditions under which such apparatus is used on board ship demand simplicity of construction and reliability of operation. The apparatus described in the following article has proven satisfactory in actual service and is the result of efforts to combine extreme accuracy and lightness with the aforementioned features.

EXTREME accuracy, simplicity of construction, and lightness for unit length of shaft combine to insure the success of the Gary-Cummings torsion meter. This meter is the result of long experience with various power measuring instruments on board ship and the makers have been able, through their own efforts as well as through the experience of users with whom they have kept in close touch, to bring this instrument to a high state of prefection.

The personal equation is entirely eliminated from the readings on this type of torsion meter. These are recorded mechanically and form a permanent record for future reference if desired. All parts of the meter which are subjected to centrifugal action are made as light as possible and balanced perfectly. The instrument is largely housed within the shaft entirely out of the way. When space is limited, as on torpedo-boat destroyers, this feature is of great importance. All parts which might possibly get out of order are attached to the outside of the shaft flange where they can be readily inspected and removed or replaced in a very few minutes. The instrument contains no gears whatever. Another feature that contributes largely to its accuracy is the fact that it measures the torsion in one complete section of a shaft, making necessary but one multiplying device. There is but one part to wear and this is easily taken up, giving considerable advantage over those meters which attempt to measure the angular twist in a short section and require numerous multiplying devices all of which are a possible source of error in the reading.

Construction

The instrument is primarily a rigid steel tube extending for one complete shaft section and fastened rigidly at one end but free to move at the other. The fixed end of this tube is firmly attached to an arm which rests in a radial slot in the flange of the shaft where it is securely fastened. Along the steel tube at intervals are spider collars which fit loosely but are held from moving along the tube by long cotter pins placed through the tube on each side of the collar. These spiders eliminate all possibility of any whipping action. The free end of the steel tube is supported and centered in a ball bearing so that it is free to rotate with respect to the shaft. It is therefore free to transmit any angular displacement between the tube and the shaft to an arm which is secured across the end and extends radially through another slot in the shaft flange in which it is free to move angularly. In other words, any angular displacement between the two flanges on the shaft can be noted by the movement of this arm with respect to the flange. An umbrella, equipped with two recording points, is fastened securely to the outside of the shaft flange at the free end of the tube in such a way that it is capable of rotating in a plane at right angles to the shaft. Any movement between the arm and the flange is transmitted through a sliding ball joint to the umbrella, causing it to turn an amount which is directly proportional to the twist in the shaft. The two points on the periphery of the umbrella record on a specially prepared card placed in a holder at the side of the shaft as it revolves and are adjusted so as to cause a small negative reading when no power is being transmitted. When the shaft is transmitting power the distance between the two marks recorded on the card minus the zero reading is directly proportional to the twist in the shaft. From this measurement the actual torque transmitted by the shaft can be readily obtained from the calibration data. Using this and the revolutions per minute of the shaft the horsepower transmitted is easily calculated. If the record of the shaft revolutions is noted on the same card with the torsion meter reading, a permanent record is obtained which can be filed for future reference.

Torsion Member

In constructing this meter unusual care is exercised to secure an absolutely rigid fastening at the fixed end of the meter tube. The supporting spider is turned to fit the inside of the shaft with a tight fit. On the spider is a collar into which the steel tube fits and to which it is riveted solidly. The radial arm is secured to the spider head by machine screws prick-punched to prevent unscrewing. The arm itself fits tightly into the slot cut in the shaft flange and is secured by bolts through the flange. This construction eliminates the possibility of any play at this end of the tube.

The free end of the tube is riveted to a suitable collar which rests in a ball-bearing within the spider which supports and centers this end of the tube. The spider is carefully made to fit within the shaft and is then secured there by tapered plugs in each spider leg which expand the four ears hard against the walls of the shaft. Further stability is added to the spider by two webs connecting opposite adjacent ears. The tube is thus supported and centered at this end but is free to rotate, or to move along the shaft as it expands or contracts. The position of the meter tube is therefore controlled entirely from its fixed end. The relative position of the two flanges is what the meter must measure and what is transmitted to the recording umbrella.

To understand the instrument it must be borne in mind that the tube is secured to one flange of the shaft only. It is free to rotate in the other flange; therefore the meter tube itself cannot possibly be subjected to any twisting action, but it is the shaft that twists about the meter tube, and it is this twist or difference in displacement between the shaft and the tube that is measured.

Recording Device

Probably the most interesting part of this torsion meter is the method of recording the twist of the shaft. A V-shaped or butterfly arm extends from the radial piece to the sliding ball joint which moves the umbrella. This butterfly piece is pivoted, as shown in the sketch, to the radial arm in order to provide for the very slight but positive movement of this arm in a plane vertical to the shaft axis. The pivoting joint can be easily adjusted for wear or for an approximate zero reading by the conical screws at each end. At the sliding ball joint provision also is made for taking up any possible wear. The ball is fastened to the supporting axis of the umbrella so that any angular movement of the radial arm attached to the free end of the meter tube causes a proportionate rotation of the umbrella. The recording points are set on opposite sides at the outer edge of the umbrella. When the shaft rotates these recording points, move in a plane perpendicular to the shaft. As the shaft is twisted, the plane of rotation of one recording point moves aft and the other forward to correspond with the amount of twist in the shaft. The needles record on a card the exact distance between these planes. The prepared card is held in a holder supported from the hull or bearing foundation. The plane of the card must form practically a tangent with the circle described by the points of the needles when recording. When not taking readings a spring keeps the card away from the recording needles. All that is necessary to take a reading is to push the card-holder forward against an adjustable stop and the needles will record as the shaft rotates.

The card-holder itself has to be carefully constructed to obtain the best results. The opening for the needle record is only 3¼ inches wide by 2½ inches high, and in order that the needle readings may be properly centered on the card the holder contains a screw adjustment for moving the opening in the holder forward or aft parallel to the shaft axis a short distance. Provision is also made for adjusting the recording card so that both needles will mark properly. Ootherwise, only one needle would record. With these two adjustments the recording holder can take care of all variations due to temperature and the like.

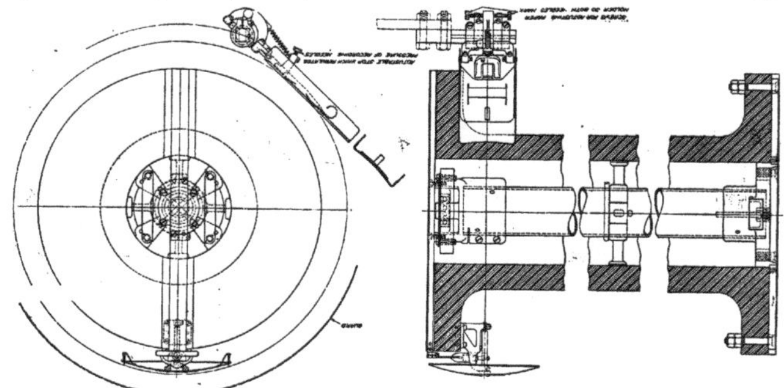

DETAILS OF TORSION METER AND CARD HOLDER.

Calibration

In order to obtain the best results, the shaft should be calibrated after the meter is installed in the section in which it is to be used aboard ship. The torsion meter is installed in the shaft exactly as aboard ship except for the card-holder. The shaft is placed on two suitable supports and the end containing the fixed end of the meter securely fastened to a heavy fixed girder. The opposite end of the shaft is supported by a bearing which allows it to turn in an angular direction with as little friction as possible. To this free end of the shaft another girder it attached for applying the test loads. Whatever apparatus is used it is necessary that the turning forces shall be balanced so that there will be no eccentric twist in the shaft. The usual method consists of applying known weights at one end of the beam and a turn-buckle or jack with a weighing scale at the other end. Suitable corrections are made for the weight pans and the turn-buckle with its attachments. The weights in the pan at one end act downward while the turnbuckle must act with equal force in an upward direction at the other end of the girder. The points at which the turn-buckle and the scale pan act upon the girder should be knife edges carefully measured from and on the same line with the axis of the shaft. For convenience these distances are usually made equal.

Of course the recording umbrella does not rotate around the shaft axis during this calibration test as it does when installed in the ship. Extra provision must, therefore, be made for properly recording the angular twist obtained during this test. The record card is placed in a small holder carried by an arm that rests on a special plate on the shaft itself. This supporting plate is fitted to the circumference of the shaft and when moved around the shaft causes the record card to describe a circle whose center is at the center line of the shaft.

As the card passes the recording needle two parallel lines are traced. The distance between these lines is a measure of the twist or torque applied to the shaft.

Applying Load

In conducting a calibration test the needles should be adjusted so that the zero reading will be from ten to fifteen-hundredths of an inch, and a card taken.

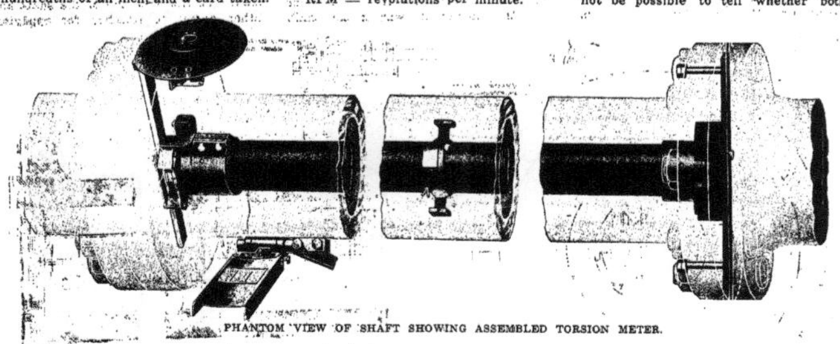

PHANTOM VIEW OF SHAFT SHOWING ASSEMBLED TORSION METER.

A known moment is applied to the shaft which is then rapped sharply with a lead hammer to remove any "set" that may have been caused by binding in its bearing, and a record taken from the recording points on the umbrella. Weights are then added in increments and the turn-buckle adjusted each time for the same upward thrust. The shaft should be thoroughly rapped each time with lead hammers before readings are taken. These operations are continued until the desired load has been reached. To check this, similar records are taken as the load is removed step by step. From these results, with proper corrections for zero readings of the instrument, the calibration curve for the meter is easily plotted. Torsion meter readings are plotted as abscissae and the torque, which is the equivalent load at one-foot radius, as ordinates. If the test has been properly conducted the resulting curve will be a straight line, indicating that the angular twist in the shaft is directly proportional to the load applied; this will, of course, be the case up to the elastic limit of the shaft. If the curve is a straight line it proves that the calibration is correct unless there was an error, such as friction, common to all points.

From the calibration readings the torque per one inch of meter reading is calculated. This will be found practically the same for all the loads applied to the shaft. Theroretically it should be exactly the same. The average of these gives a figure on which to base the horsepower constant for that meter in that particular shaft. To facilitate the ready calculation of the power transmitted, however, the constant used for each shaft contains the additional constants which enter into the calculation of the horsepower.

Calculations

Algebraically, the calculations are as follows:

$$S. H. P. = K \times T \times RPM$$

R = radius on which load is applied in feet.

W = weight of load in pounds.

RPM = revolutions per minute.

M = torque or moment in foot pounds.

Q = foot pounds for 1 inch torsion meter reading.

T = torsion meter reading in inches.

K = torsion meter constant.

From this equation the horsepower transmitted through each shaft in the ship is calculated, and when noted on the sheet containing the torsion meter reading becomes a part of a permanent record of the test.

In the use of this meter aboard ship it is essential that the needles record a small negative reading when the shaft is transmitting no power—that is, when the propellers are dragging, but still turning over. This zero reading is obtained by properly adjusting the set screws joining the butterfly piece with the radial arm when the meter is is installed. If the meter was set for no zero reading, that is, both points marking the same line on the card, it would not be possible to tell whether both needles were recording or not. With the small negative reading, however, it is easy to note that both needles are recording properly.

Setting Recording Needles

A gauge in the spare-part box fits the flange with a tongue projecting out to set the height of the recording needles.

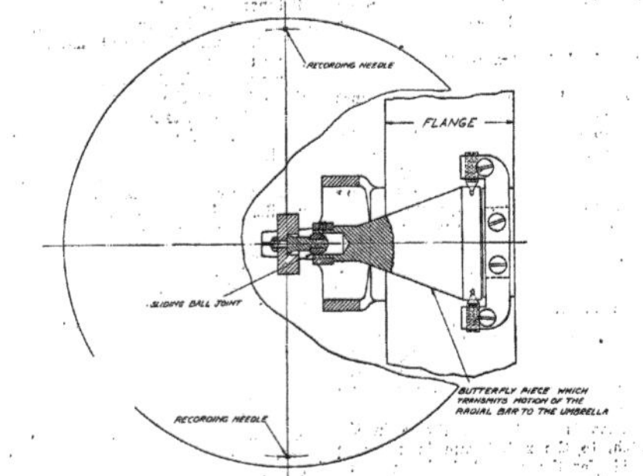

SECTIONAL VIEW OF BUTTERFLY PIECE AND SLIDING BALL JOINT

These should be set with a watch key (No. 4 key in spare-part box) so that they just touch the gauge as it is moved around the flange. Then both recording points are at the same distance from the centre of the shaft and will record when the card is brought in contact. By means of the centring screws on the block attached to the top of the bar the needles are set so that one is about ⅛ inch farther from the flange than the other. The mark on the gauge should be equidistant between the planes of the

THE MOST CONVENIENT WAY OF TAKING CARDS.

needles. This is the approximate zero reading.

The needles are set so tht their planes of rotation do not cross when the shaft twists in ahead direction, otherwise the zero reading would have to be added to the reading taken when under way instead of being subtracted as it should be.

Operation

In running a test the recording apparatus is attached to the shaft, care being taken to scrape the rust and dirt from the flange. The butterfly piece should be adjusted by the centring screws, so that it is freely movable without lost motion.

The Card-Holder

The card-holder is attached by two bolts to its support; which should be built out from the frame of the ship. By loosening the binding screws on the collars over the round bar the holder may be moved forward or aft, so that the centre of the opening is in the plane of the axis of the umbrella as it revolves with the shaft. Otherwise, when the shaft twists one needle would mark on the card and the other would be out of range.

Getting Zero Reading

After tightening all screws the torsion meter is ready for dragging the shaft. This should be done by getting some way on the ship, about one-third speed, then shutting off the steam to the turbine, including all auxiliary exhaust, and permitting the shafts to revolve partly by way of the ship and partly by its own inertia and that of the rotor. Cards should be taken at frequent intervals while the shaft gradually slows down and the minimum of these readings taken as the zero reading. Before dragging it is usual to try taking cards in order to be familiar with the operation so that no cards will be lost during the drag, as the shaft soon stops revolving after the steam is shut off. One man, experienced at taking cards, should be at each torsion meter while the shafts are dragging.

Taking Readings

Since the stop was set so that needles would not touch the card when the ship geats under way, the first thing to do is to adjust this stop so that the needles just make a mark without tearing the card. About one dozen cards should be placed in the holder. Some prefer to adjust the stop so that the needles are about 1/16 inch away from the card. Then, with a slight pressure with the forefinger placed horizontally back of the card, bulge it outward slightly, just enough to make needles work, as shown in the illustration. The holder is sprung forward with the other hand at the same time. This method is used by all experienced operators, because it lessens the danger of the needles striking metal and does not require such a fine adjustment of the stop screw. It is attended, however, with some danger of getting the fingers cut slightly by the revolving needles, so should not be attempted by careless operators. Should one needle mark before the other, the card should not be pushed farther in, but the side thumb screws should be adjusted, backing off one and tightening up the other. (Be sure holder is held tightly between these two adjusting screws before taking another record.) Keep trying this until both needles mark.

To calculate the horsepower a card is taken at the same time the revolutions per minute are being determined. Very often the torsion meter operator will take counter readings of each shaft on the minute, then go aft and take torsion meter cards, then counter readings again, say five minutes later. Each card should be marked with the shaft from which it was taken and also with the date and time. The revolutions per minute of each shaft are worked out for five minutes and noted on the cards also. The linear distance in inches between the two parallel marks on the cards minus the zero reading (in rare cases this is plus because planes of needles cross as shaft twiststs) is the corrected tersion meter reading. A convenient scale is furnished each ship for measuring cards. Multiply the corrected reading (in inches and fractions thereof) by the torsion meter constant and this result by the revolutions per minute of the shaft final result will be the horsepower being transmitted through the shaft. The torsion meter constant is worked out from the shop calibration of the shaft, as already described.

A NEW WOOD, THE LIGHTEST KNOWN

By D. Street.

A NEW wood, apparently little known and called balsa wood, is exceedingly light and promises to have an extended field of usefulness in connection with cold storage structures when heat insulation is important. It is a tropical wood, growing principally in the States of South and Central America.

The wood is remarkable—first, as to its lightness; second, as to its microscopical structure; third, for its absence of woody fibre; fourth, for its elasticity; and fifth, for its heat-insulating qualities. So far as investigation has disclosed, it is the lightest commercial wood known. It has also considerable structural strength, which makes it suitable for many cases. In general appearance balsa wood resembles bass wood.

Until recently Missouri cork wood, weighing 18 lb. per cubic foot, was believed to be the lightest, but recent investigations indicate that balsa wood is much lighter, having a net weight of 7.3 lbs. per cubic foot. The ordinary commercial balsa wood is seldom perfectly dry, and because of the moisture content its weight has been found to be between 8 lbs. and 13 lbs. per cubic foot.

The extreme lightness of this wood suggests its application as a buoyancy material in life-preservers and life-boats. When, however, it was attempted to apply the wood practically it was found to be of little value, because it absorbed water in great quantities, and also because it soon rotted, and also warped and checked when worked. After testing nearly every method that had been suggested, Colonel Marr's method of treating woods, which had been recently patented, was finally successful. In this method the wood is treated in a bath, of which the principal ingredient is paraffin, by a process which coats the interior cells without clogging up the porous system. The paraffin remains as a coating or varnish over the interior cell walls, preventing the absorption of moisture and the ill-effects as to change of volume and decay which would otherwise take place; it also prevents the bad effects of dry rot which follow the application of any surface treatment for preserving wood of the same type. The Marr process tends to drive out all water and makes the wood waterproof.

THE import and export figures for the chief ocean ports of Canada for the six months ending September, 1917, show the following totals:

Montreal	$395,169,388
St. John	113,390,304
Halifax	57,528,777
Vancouver	31,178,558
Quebec	13,675,155

CANADIAN BUILT MOTOR LIFE BOATS

THROUGH the courtesy of the Canadian Beaver Co., and the Foreman Marine & Machine Co., of Toronto, we are enabled to illustrate herewith the lifeboats and power plant which are now being completed by them.

As mentioned in a recent issue, these boats are 24 feet over all, by 7 feet beam, and have a seating capacity of 30. They have buoyancy tanks under the seats and will have cork belts to increase their safety.

The Foreman motors with which these boats are fitted are of the two-cycle type with two cylinders 4¼-inch bore, by 4¼-inch stroke. These motors develop 12 horse-power at 700 revolutions per minute, and are saltwater equipped, as shown in the illustration. Atwater-Kent Unisparker ignition system is employed, and Schebler model D caburetor is fitted.

12 HORSEPOWER 2-CYCLE MOTOR FOR 24 FT. LIFE BOAT.

The water pump and ignition system are driven by steel spiral gears running on enclosed ball bearings submerged in oil. Multiple disc reverse gear is fitted to the motor shaft and supported on the base extension. Heavy ball thrust bearings to take the ahead and astern load at end of base extension, thus preventing any strain on the gear or engine parts.

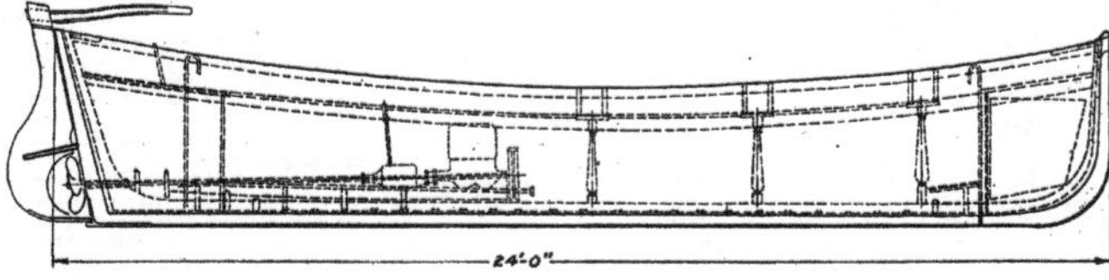

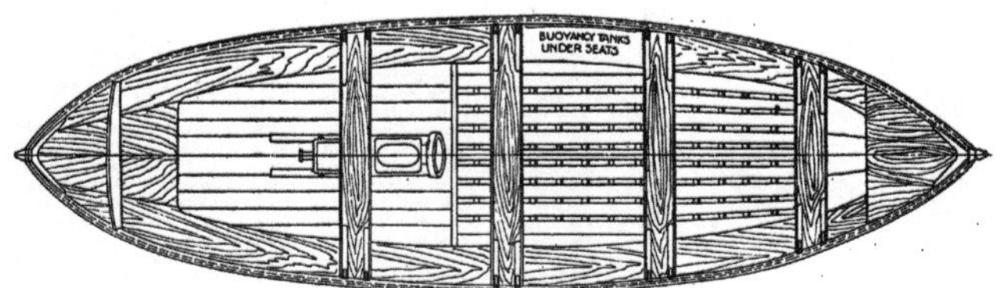

24 FT. SHIP'S LIFE BOAT FOR THIRTY PERSONS. MOTOR DRIVEN BY TWO-CYLINDER TWO-CYCLE GASOLINE ENGINE.

REPORT OF INQUIRY ON "MONT BLANC" EXPLOSION

THE conclusions arrived at as a result of the formal inquiry directed to be held by the Minister of Marine are embodied in a report recently submitted by L. J. A. Drysdale who had associated with him as nautical assessors Captain Demers of Ottawa, Dominion Wreck Commissioner, and Captain Walter Ross, R. C. N., of the city of Halifax. The inquiry was begun on the 13th day of December, 1917, and the report was drawn up after having heard, all the witnesses who could throw any light on the situation.

The conclusions reached by the signatory and concurred in by the nautical assessors are as follows:

1. The explosion on the S. S. "Mont Blanc" on December 6th was undoubtedly the result of a collision in the harbor of Halifax between the S. S. "Mount Blanc and the S. S. "Imo."

2. Such collision was caused by violation of the rules of navigation.

3. That the pilot and master of the S. S. "Mont Blanc" were wholly responsible for violating the rules of the road.

4. That Pilot Mackey by reason of his gross negligence should be forthwith dismissed by the Pilotage Authorities, and his license cancelled.

5. In view of the gross neglect of the rules of navigation by Pilot Mackey and the attention of the Law Officers of the Crown should be called to the evidence taken on this investigation with a view to a criminal prosecution of such pilot.

6. We recommend to the French authorities such evidence with a view to having Captain Lemedec's license cancelled and such captain dealt with according to the law of his country.

7. That it appears that the Pilotage Authorities in Halifax have been permitting Pilot Mackey to pilot ships since the investigation commenced, and since the collisions above referred to, we think the authorities, i.e., Pilotage Authorities, deserving of censure. In our opinion the authorities should have promptly suspended such pilot.

8. The master and pilot of the "Mont Blanc" are guilty of neglect of the public safety in not taking proper steps to warn the inhabitants of the city of a probable explosion.

9. Commander Wyatt is guilty of neglect in performing his duty as C. X. O. in not taking proper steps to ensure the regulations being carried out and especially in not keeping himself fully acquainted with the movements and intended movements of vessels in the harbor.

10. In dealing with the C. X. O.'s negligence in not ensuring the efficient carrying out of traffic regulations by the pilots, we have to report that the evidence is far from satisfactory that he ever took any efficient steps to bring to the notice of the Captain Superinten-

dent neglect on the part of the pilots.

11. In view of the allegations of disobedience of the C. X. O.'s orders by pilots, we do not consider such disobedience was the proximate cause of the collision.

12. It would seem that the pilots of Halifax attempt to vary the well known rules of the road, and in this connection we think Pilot Renner in charge of an American tramp steamer on the morning of the collision, deserving of censure.

13. That the regulations governing the traffic in Halifax Harbor in force since the war were prepared by competent naval authorities; that such traffic regulations do not specifically deal with the handling of ships laden with explosives, and we have to recommend that such competent authority forthwith take up and make specific regulations dealing with such subject; we realize that whilst the war goes on under present conditions explosives must move, but in view of what has happened we strongly recommend that the subject be dealt with specifically by the proper authorities.

VIEW SHOWING THE REINFORCING STEEL AT BOW.

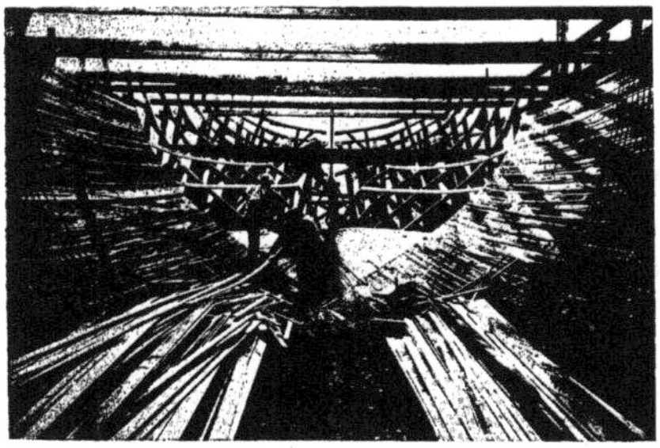

CONCRETE BOAT BUILDING IN SPAIN

SPAIN, the country in which Columbus contrived to get enough money to buy and equip three wooden boats which were used in a little sea jaunt of more or less historic interest along about 1492, has turned to concrete as a shipbuilding material in 1918.

The corporation known as Works & Pavements, of Barcelona, Spain, has accomplished in a relatively short period of time some of the most important public works of reinforced concrete and in view of the developments that it believes will take place in maritime construction has at the present time in course of construction the first cargo boat in this field, some details of which will be seen in accompanying photographs. This ship, while the initial attempt of this kind of this firm, has nevertheless led it to make plans for the construction in 1918 of a gross tonnage of 40,000 corresponding to a displacement of 70,000, and consisting of standard type ships of 300, 500 and 1,000 tons each. Next year they propose to extend their operations and to provide for enlarging their plant have acquired an additional site on the shores of the Mediterranean, having an area of 250,000 square metres with 2,000 metres water frontage. This will permit them to provide ways for the carrying on of the construction of 30 boats at the same time; some of which are planned to be of 6,000 tons capacity.

The company has acquired various patents that will permit it to build rapidly and economically reinforced concrete ships.

It has also been proved that algae and the various forms of sea growths that are developed in sea water may be removed from the surface of concrete very easily.

VIEW SHOWING THE PLACING OF THE FORMING AT STERN.

CANADIAN SHIPS TO HAVE WIRELESS

THE Minister of Naval Service has given notice that on and after the first day of January, 1918, every British steamer registered in Canada, of sixteen hundred tons gross tonnage or upwards, sailing to or from any port in Europe or in the Mediterranean Sea, shall be provided with an efficient radio-telegraph apparatus in good working order, properly installed and maintained, capable of transmitting and receiving messages over a distance of at least one hundred nautical miles by day and by night, with two certificated operators in charge and with suitable accommodation for the apparatus and for the operators. The owner and the master or other person in charge of any British steamer of sixteen hundred tons and upwards registered in Canada which sails to or from any port in Europe or the Mediterranean Sea without being provided with the apparatus, installation, accommodation and operators required by this regulation shall each be guilty of violating the provisions of this regulation."

USE OF OIL TO CALM SEAS

DECEMBER 27, in lat. 38° 40' N., long. 148° 49' W., encountered a gale with winds of hurricane force from SS.W. to WNW. through west. High seas were running and breaking over the vessel forward and amidships. Used old waste engine oil successfully in keeping seas from breaking aboard by pouring it into the water-closet bowls forward and amidships on the weather side after stuffing oakum in the drainpipes. Used about 15 gallons of oil.—Hydrographic Bulletin.

Capt. Andrew Dunlop, aged eighty-seven, who was a marine captain from 1854 till 1908, died at Kingston, Ont., on Feb. 20. He was born in Kingston. For many years he commanded steamers of the Richelieu & Ontario Navigation Co.

The MacLean Publishing Company
LIMITED
(ESTABLISHED 1888)

JOHN BAYNE MACLEAN - - - - - - - President
H. T. HUNTER - - - - - - - - - - Vice-President
H. V. TYRRELL - - - - - - - - - General Manager

PUBLISHERS

MARINE ENGINEERING of Canada

A monthly journal dealing with the progress and development of Merchant and Naval Marine Engineering, Shipbuilding, the building of Harbors and Docks, and containing a record of the latest and best practice throughout the Sea-going World.

J. M. WILSON, Editor. B. G. NEWTON, Manager.

OFFICES

CANADA—
Montreal—Southam Building, 128 Bleury St., Telephone Main 1004.
Toronto—143-153 University Ave., Telephone Main 7324.
Winnipeg—1207 Union Trust Building. Telephone Main 3449.
Eastern Representative, E. M. Pattison.
Ontario Representative, S. S. Moore.
Toronto and Hamilton Representative, J. N. Robinson.

UNITED STATES—
New York—R. B. Huestis, 111 Broadway, New York, Telephone 8971 Rector.
Chicago—A. H. Byrne, 900 Lytton Bldg., 14 E. Jackson Street. Phone Harrison 1147.
Boston—C. L. Morton, Room 733, Old South Bldg., Telephone Main 1024.

GREAT BRITAIN—
London—The MacLean Company of Great Britain, Limited, 88 Fleet Street, E.C. E. J. Dodd, Director, Telephone Central 12960. Address: Atabek, London, England.

SUBSCRIPTION RATE

Canada, $1.00; United States, $1.50; Great Britain, Australia and other colonies, 4s. 6d. per year; other countries, $1.50. Advertising rates on request.

Subscribers, who are not receiving their paper regularly, will confer a favor by telling us. We should be notified at once of any change in address giving both old and new.

Vol. VIII MARCH, 1918 No. 3

PRINCIPAL CONTENTS

RECONSTRUCTION MUST PRECEDE TRADE EXPANSION

WHATEVER the ultimate distribution of the world's markets as a result of the war and despite the indefinite prolongation of involved commercial conditions such as must result from the later German conquests, the need of export business as a means of maintaining Canada's industrial status at the requisite height becomes increasingly apparent with the passage of time.

The whole question of export trade, its possibilities, desirabilities, difficulties, risks, is in a great measure an unknown quantity to many of those who seem to see in it a solution of post-war tribulation.

In considering ways and means to insure a minimum dislocation of industry during the transition period, attention might be directed to certain governmental activities in Britain. Steps have already been taken in advance of conditions, and the Old Country is meeting trouble more than half way with a view to overcoming it more promptly and more effectively.

A committee composed of manufacturers and business men has been appointed to consider the provision of new industries for the engineering trades. Its duties will be to compile a list of the articles suitable for manufacture by British engineers, which were either not made in the United Kingdom or were made in insufficient quantities, and for which there is likely to be a future demand.

Owing to the war industry in this country having been developed along different lines from that of Britain —the absence of controlled factories, etc.,—there is not the same probability of close governmental co-operation, yet such would seem to be desirable and its occurrence would need to be prior as well as being essential to any organized effort to develop export trade as a feature of Canadian industry.

The action of the Dominion Government in arranging with Canadian shipbuilders to lay keels for Government owned ships exclusively, as berths become vacant, is therefore of timely importance. Unrestricted shipping competition immediately after the war would be a complete reversal of all precedents. The shippers would be the competitors and shipowners would go after the highest freights combined with routes selected for safety and economy of operation. Such circumstances combined with existing features of Canadian navigation would tend to keep the supply of bottoms available for our export trade at an undesirable minimum, consisting of regular lines with reduced tonnage.

The number of boats contemplated by the Government is doubtless subject to such future increase as prospects and circumstances justify, but the fact that such action has been taken is of invaluable aid to the nation in removing what would have been a fatal handicap at a critical period.

This development is still more welcome for its effect on the general attitude of the industrial world toward Government co-operation. Having provided shipping facilities which will not be subject to private exploitation it is not a very far removed step to extend co-operation of an allied nature to such industrial interests as are expected to supply the wherewithal to form cargoes.

The necessity for national recognition of such pending conditions is urgent. Public opinion is notoriously slow in its adaptation to rapid changes in commercial conditions. Future export trade cannot reasonably be expected to absorb the capacity of all the war-created metal working plants nor must metal working industries be given undue preference over other activities which are more truly indigenous.

Export trade should be sought after as a source of permanent revenue to the country and such manufacturers as essay activities in this work must be fully aware of this feature otherwise the reaction against such temporary exploitation of other countries' needs will later react against those firms who have been actuated by genuine motives.

It is in guiding the country through such paths as these and effectively lending its prestige to trade co-operation that the Government can be of great assistance from now on. The shipping program is a good augury, and the same spirit developed along obvious, though not identical, lines, will do much to inspire confidence and encourage preparation to meet all future conditions.

Fo'c'stle Days
or
Reminiscences of a Wind Jammer
By Capt. Geo. S. Laing

The steamer and the motorship have deprived the seafarer of much of the old-time romance attaching to a sailor's life. The advent of wireless has also increased the safety of it. Deprived of these features, a sailor is immediately dependent on his own resources, and Captain Laing gives us a true and seaman-like reflection of sea life under canvas. The present feverish activity in constructing self-propelled craft affords a suitable background for incidents in a career in which gaffs, booms, tar and canvas held sway before being displaced by mechanical unloaders, derrick masts, fuel oil, coal and wire rope.

BEFORE daybreak next morning the scene was superb. There on the high Cornish cliffs the Lizard lights shone out, whilst the Wolf Rock blinked his red and white alternate flashes to those who dared pass him in his vigil. Coasting traffic bound to the Bristol Channel hung on to the shore and rounded the faithful Longships Island at Land's End, whilst the "Maggie Dorrit" put a little southing in her western wings. The sky was heavily clouded but now and again the moon would find a thin spot to peep through as our barque rolled along under a main topgallant sail at the rate of twelve miles an hour. Down to leeward a man-o'-war was steaming up for Plymouth, whilst far astern a fleet of fishermen, resembling the tail of a comet, stood off the land.

The "old man" spun one of his best yarns to the mate under the mizen weather-cloth and by the tone of his voice he was glad to see the open ocean, as sailing ship skippers get very little sleep when on the home coasts. Their nervousness when near the land must to some extent be excused, for a windjammer is a much bigger responsibility than a steamship, i.e., as far as manoeuvring is concerned.

To describe one's feelings when, for the first time the sea has swallowed up your native country and nothing remains for the eyes to feast on but the waters beneath you and the firmament overhead, is not an easy matter. There is something uncanny about the disappearance of the terra-firma, and to sea apprentices the physical feeling seems to lay between seasickness and home-sickness—a void. As boys we reflected on our homes and dear ones and suddenly realized that distance from those associations only tends to strengthen the cords of filial duty, and every boy who keeps this acknowledgement in his daily mirror of thought will be safe.

Sea Changes its Color

As we sail to the SW the water changes from its murky color to a clearer shade of blue or green, according to the sun and sky, for the ocean is painted from its overhead surroundings. Coastal water is affected by river pollution and the disturbance of ooze, hence it seldom attains the beautiful transparent hues that we meet with in deep water.

The cables were unbent from the anchors and stowed ship-shape in the chain lockers, as our craft will not be in anchorage waters for at least three months. Plugs were hammered into the towing and hawse pipes to ensure a dry fo'c'sle when the old girl dives into a seaway.

Going Aloft in a Gale

The three first voyagers were just beginning to go aloft with ease and confidence except Tilbury, and it was plain from the first that the makings of a sailor were not in that boy's constitution. To fully initiate us into Atlantic furies a northerly gale sprang up and laid its shivery paws on the "Maggie Dorrit." In one watch we reduced her canvas from royals to upper-top-sails. In the second dog-watch, 6-8 p.m., the mainsail was furled and the leg'o-mutton spanker brailed in, which of course left our mizen-mast bare. A dangerous quarterly sea threw huge "dollops" of crest water into the barque's waist. I had my first experience of being knocked into the lee scuppers when trying to get into the half-deck door at eight bells.

About midnight the sea and wind had increased and the water which had leaked through doors and portholes was washing well up under the lower bunks as the vessel staggered on in her seemingly maddened gait. As green hands we thought that the vessel must be foundering as the swishing and pounding of the seas against our house certainly denoted. Fid dispelled this horrible thought by announcing at the change of watches that "this is fine weather compared to what awaits us at the Cape of Good Hope."

At the relief of watches the second mate whistled for Tilbury and me and we had orders to remain on the poop, as we hadn't sense enough to know how to dodge a sea with sailor-like agility. It may be mentioned that after one is accustomed to the ship's behaviour in stormy weather it becomes possible to watch a chance and get along the deck dozens of times without being caught in an avalanche of water. Moreover, when a vessel is running with the wind and sea on the quarter, the poop and fo'c'sle head are comparatively safe.

Our first gale had lots of discomfort, but its grandeur was magnificent. Daybreak found us reefing the fore-sail. As it was an all hands job the second mate took both watches aloft, but the first voyagers were told to "stay in around the bunt out of the damn'd road." In reefing a squaresail the strenuous and dangerous work is on the yardarms.

The next evening found us setting topgallant sails over reefed topsails, not an uncommon trick when the mercurial and aneroid barometers act in an erratic way and fail to be trusted. The wind was less fanatical in the squalls but the sea had abated little. Paddy's lantern (the moon) was at its full and with its brilliance added to the phosphorescent charge in the waves a good light was cast all around the dark outline of the vessel. This was further augmented when the crest of some extra big sea would literally rush on board like a wild animal and appear as white as snow in its foaming fury.

We are now told in sea-dog style that hereafter we must learn to rough it and take our chances in the well deck "amongst the crowd," but for fear a boy is washed overboard without any initiation to hard weather, any sensible mate allows the youngsters a little shelter at the start.

Heavy Weather Scenes

Looking astern the aspect is fearsome. Each wave can be likened to a smoking mountain of fast moving water, the crests actually boiling and hissing with foam and the aquatic leviathan overtaking the frail ship, for every craft is tiny and insignificant in comparison with the watery elements. Just as the mountain appears to be a few feet from hurling its wrapped-up tonnage on to the vessel's poop, the good "Maggie Dorrit" raises her stern triumphantly whilst her bowsprit and flying-jib-boom point down in the valley. As the wave passes under the keel the positions are reversed. Just for a moment the vessel shudders as she is balanced on the water pinnacle, then down again till her rudder trunk squirts the foam into the helmsman's long boats, whilst the jib-boom points into the clouds ahead.

This switchback motion is also accompanied by heavy rolling as the sails have little side pressure now and the groaning of blocks and rigging adds to the fray. Scuppers and bulwark ports cannot relieve the vessel of the seas that attempt to smother her, so she just rolls and plunges and rises and falls in the seething turmoil. If her hatches keep out the water she is likely to weather the blast and so the night drags on.

Tilbury was sent below to fill the captain's pipe and brought it back so tightly plugged that I had the pleasure of undoing it. This job, by the way, was a pleasant one and our "old man" could smoke like a volcano, especially in dirty weather. His motto seemed to be "more wind, more tobacco," for in between smokes he would eat it. A boy was not allowed to use the weed until his second voyage. During the spell of heavy weather the cook had been generous in his use of salt water both in the tea and coffee and no one seemed to take it amiss. On one occasion the soup looked like a gallon of sea water with an onion dropped in it.

Chanteying

At last the weather became fine and we were greatly entertained with all sorts of chanteys, as every stitch of canvas was piled on to our craft. Sailing-ship men could certainly make some of the heavy-pulling jobs almost enjoyable by this boisterous sort of singing.

Here are a few of Jack Tar's songs or "chanteys" as they are termed at sea. When furling a large square-sail, the boatswain, who is generally in the bunt, will drawl out to a well-known tune:

"We'll rob all the banks in England
To pay Paddy Doyle for his boots."

When the word boots comes everyone fairly yells, and at the same moment puts his brute strength into getting the sail on top of the yard. Thus a concentrated jerk is given to the work until the order floats along to "tie her up with the gaskets." No one has breath to sing chanteys when knocking the wind out of a sail, it is after that ordeal has been survived and the canvas lies dead that the singing is brought down.

In sweating up topsail halyards or the double sheets of courses (lower sails) we might use:

"O do my Johnny Baker, come rock and roll me over
Oh do, my Johnny Baker, do."

With a big long drag on the last word something has got to come.

Anotner such chantey is:

"Haul away Joe—"
"O once I had an Irish gal, and she was fat and lazy O'"
"Away, haul away, oh haul away, Joe,"
"But now I've got a nigger one, she drives me nearly crazy O'"
"Away, haul away, oh haul away, Joe."
"King Louis was the king o' France before the revolution"
"The people cut his head off, which spoilt his constitution,"
"Away, haul away, oh haul away Joe."

There are many other sea melodies to suit capstan heaving, hand over fist, and stamp and go methods of rope-hauling. A few readers will be familiar with:

"Whisky for my Johnny"
"Blow the man down"
"Banks of Sacramento," etc., etc.

Without a doubt this kind of hilarity helped us to forget or laugh at some of the inseparable hardships that belong to sailing ship life. Every ship's fo'c'sle contained perhaps two or three men who could, through voice modulation, make rope-hauling into a concert.

After the Gale is Over

The first job after a gale is to dry out all bedding and wearing apparel in your watch below as the salt spray penetrates everywhere. Hank and I kept a dairy but Tilbury said that his records would only depress, and he was determined to desert the vessel in Australia anyway.

Porpoise Catching

We were looking forward to meeting the NE trade winds when the man at the wheel was the first to spy a school of porpoises. These mammals made right for the ship, where they seemed to enjoy swimming under the jib-boom. The displacement water as it curls up on the stem holds a fascination for the sea pigs. They appear to describe half circles when jumping in and out of the water, and it is a pretty sight to see them bowling along and then make a detour, to come up again blowing and puffing like so many breathless boys after a race. The first mate, whom we disrepectfully called "Ropeyarns" on account of his unruly whiskers, was bent on catching a porpoise, and Fid, the ancient mariner, got out the grains—a kind of four-headed harpoon.

THE FAITHFUL LONG-SHIP ISLAND LIGHTHOUSE AT LANDS' END

In a case like this a fo'c'sle hand does the harpooning from under the jib-boom. With his feet steadied on the back ropes of the martingale, also known as the dolphin striker, the position is ideal, being near the water and ahead of the vessel. The second party in the ordeal stands at the lee cat-head with the hauling end of the fish rope. As the water is literally black with porpoises it would be more difficult to miss than to hit, so with a yell of "watch there, watch," the harpoonist thrusts his spear into the swimming mass. In a minute or so we had a fine specimen hauled on board. Before the fish is dead and black skin of its back runs into different livid shades of blue, while its under part approaches a cream color. The flesh is red and more like ox meat than fish, and it makes a very good meal.

As we get to the southward the temperature of sea and air gradually rise and the evenings and night watches are especially attractive for the boys. It becomes easy now to draw out the fo'c'sle hands in conversation, either when coiling up the braces or "standing by" at the lee of the longboat which sits in chocks forward of the main hatch. Each boy in his turn must of course keep a two hour spell on the lee side of the poop, where he acts as the officer's aide-de-camp.

The first calm day was used to advantage by changing the heavy weather canvas for the "flying-fish suit" that does duty in the tropics. Poor Tilbury showed more signs of timidity in going aloft and was the target of derision. Our third mate made himself famous by promising to call at the half-deck door every other evening and expound on navigation, seamanship and astronomy. A squall brought the NE trades along and the "Maggie Dorrit" sped a course to cross the equator in about 28° west longitude.

Flying Fish on Board

At first the trade wind was well on the beam and flying fish came over the weather rail at odd times. These winged herrings looked exceptionally pretty as they rose from the sea and showed their silvery armor in the sunlight. They frequently rise to a height of twenty to thirty feet and then make a horizontal sweep of perhaps a thousand feet. Flying fish are chased by bonito, albacor and dolphins, so nature has given them the means of aerial locomotion to protect themselves. As a breakfast dish they are sweet and wholesome food.

One of the old shell-backs showed us how to preserve flying-fish as a curio. Here is the procedure. Put the fish in the harness or pickle cask for two days, then gut and partly dry, afterwards stuffing with oakum and tobacco dust. Remove the eyes and put beads in, stretch the wings on skewers to prevent contraction and varnish the whole thing over. A few more sun baths and the preserving is complete. The writer has kept flying-fish for several years that were treated in this simple fashion.

Fid was ordered to give us instruction in knots and splicing, and his utter disgust at our slow learning was amusing. Just because Hank required a little more demonstration in making a running-bowline an altercation took place and the ancient mariner pulled into the house one of the fore royal braces, which was coiled up at our door on the fyfe-rail of the mainmast. With this rope he intended to punish us, but on remembering how things went in the last battle his Kaiser courage forsook him.

The Tropic of Cancer

We were soon within the tropical latitudes and for days at a time the trade wind on the quarter would not vary a degree in force or direction. In the night watches with no yards and sails to trim our spare time was taken up in nature studies. What a lovely sky with its dual system of clouds. The cirrus or feeder clouds of the "mare's tail" type sail majestically into the teeth of the wind away high up in their celestial galleries, while the clumps of cirro-cumulus roll down to the doldrum area at the equator propelled by their aerial master, the NE trades. In this respect it is soon noticed by watchful seamen that all pronounced motions of air or water have counter agencies at work. The trade winds that play such an important part in sailing ship routes are caused mainly through the rising of superheated air at the earth's centre—the equator. This causes the inrush of northerly wind north of the "line" and southerly wind south of the line. The deflection to the eastward in both the trade winds, the NE and SE, is caused by the earth's rotary motion. That means if the earth was stationary the draught of air or wind would simply hold a true north and south direction. These two interesting winds are divided by a belt of calms and torrential rains called "the doldrums," of which we will speak a little later on. The parallelogram of forces is ever at work with wind and water, and the whole system of surface winds and ocean currents is controlled by temperature changes causing energy through elasticity and contraction. Compensation of balance is thus carried on in nature's laboratory.

THE VERNAL EQUINOX

From a Ship's Deck on the Equator
(A Marine Picture)
By Capt. Geo. S. Laing

Rising from his warm bath
On the globe's bisecting path,
Watch old Sol in fiery grandeur
Celebrate his spring meander.

Refraction aids him in his climb,
True dictator of man's time.
See him wink through clouds unrolled,
Painting nimbus into gold.

Up to the zenith point he strives,
Till sextant reads two forty-fives,
Then with his P.M. brakes applied
He glides down western skies with pride.

Cancer shouts for sunnier times
On his broadly scattered climes,
Speaks of Canada's fruit and wheat
Depending on the solar heat.

As ship and sun are on the "line,"
Old Capricorn knows the time,
And whines in goaty style—"Remember
To call this way in late September."

Till June the 21st brings round its mirth,
And birds roll out their tunes to mother earth,
Old Sol will gently scale his northern ground,
And shower sweet blessings all around.

To impressive minds that are drinking in for the first time the wonders of the tropics the predominant pictures are generally the flat calm, the immediate inky-darkness after sunset with the accompanying brilliancy of the star-lit canopy and the contrasts that the old ocean lays before his admirers with his mirror-bosom, cat-paw ruffles and tiny waves that even the frail Portuguese man-o'war (Physalia) can negotiate.

The apprentices of the "Maggie Dorrit" were charmed by their magic-like surroundings and a "hole" in the trades put the following visual feast before us.

Our barque was under full sail but becalmed, not a breath of wind, and absolutely no suspicion of a wave or swell. The ocean was one huge mirror and the canvas hung ghost-like from the yards and stays as motionless as a painting on a wall. With no twilight and the moon absent, the blackest of nights stole over us and the stillness was that of the grave.

Under these conditions one is tempted to lie flat on his back on one of the hatches and gaze up at the faithful watchers: Ontares, Fomalhaut, Centaurii, Spica, Crucis, etc., all dancing and sparkling like diamonds in a sea of pitch.

To those who have never lost sight of the ground under their feet these sights might go unheeded, but remember that the ocean seamen lose contact with the land for weeks and months and are perforce obliged to note the elements around them, and at the same time realize that no human hands or minds can govern the servants of nature.

Another pretty sight was given us when, after such conditions as above, a light breeze came and caused the ship to sail along at about three miles an hour. The seas here are frequently charged with phosphorescent salts and the result is most beautiful. It is pitch dark on the waters and the track of the vessel resembles a silver lane as she glides along in phantom quietness. The presence also of very tiny marine life in mollusca form, each shining like a planet and being somersaulted here and there with the displacement eddy of the vessel adds to the seemingly fairy conditions. Should one draw a bucket of this seawater on deck and spill it in the darkness it would seem as if a hive of fireflies had been hid in the pail and were suddenly resurrected.

(To be continued.)

An Interesting Transmission Cable

For the transmission of electric energy from the Trollhattan power station of Lysekil, on the west coast of Sweden, a cable has just been laid across the Gullmar Firth. The cable is intended for transmission of energy at 20,000 volts, and it is stated that it is the first time a cable of such a weight (17 kg. per metre = 37 lb. per yard) and of such a dimension (diameter 74 mm. = 2.9 in.) has been manufactured in one length; it is 1,104 m. (3,622 ft.) long and has been lowered to a depth of 120 m. (394 ft.), which, it is claimed, is the greatest depth which a high-voltage transmission cable so far has been lowered.

SEAMANSHIP AND NAVIGATION

Conducted by "The Skipper" *

Articles of direct interest to mariners, discussions of seamanship, navigation rules, and allied topics. Inquiries from readers are invited and will be replied to promptly through these columns, unless otherwise requested.

RULE OF THE ROAD STUDIES—II.

ONE of the most critical situations in steamship manœuvring is where a faster steamer with a point or two of crossing in her course sags down on your starboard beam. Remember that this craft is not crossing you at an angle of 45° to 90°, like the one mentioned in the previous article "Showing Red on the Starboard Bow Close To." The present case is where the vessels close on each other in a broadside manner. The steamer that has the outer one on her own starboard side should head in a parallel line with the stranger and reduce her speed until the other craft has shot ahead. Don't confound this predicament with the overtaking rules. Difference in speed and course will bring quite a number of vessels on your beam that never came from the stern light area.

In ship-handling the deck officer on each craft should be able to analyse the other's position and possibilities, and for that reason the thorough understanding of the collision rules is a larger problem than many people are aware of.

While sailing ships are not in the majority now-a-days, the rules governing their movements must be mastered. Government examiners demand it, and after all a prospective mate or master who can show intelligence with windjammer questions on the rule of the road, need have no fear on a steamer's bridge.

Here are a few general aids that will help to solve the sailing ship rules and also train the mind to work out all manner of positions, whether it be mechanical craft, auxiliary craft or windjammers.

Before using the chart shown herewith try and remember that:

1. When you are "closehauled," i.e., "on a wind," the opposite tack vessel, if within your collision zone, will bear down on you from your lee bow.

2. Where danger of collision is most likely to be present in regard to closehauled ships meeting each other, the opposition light will appear approximately from one to four points on the bow. If broader on the bow your opponent will "open up" her green light to your green —which of course removes all danger.

3. When closehauled, opposition lights to windward must be "free" ships.

4. When carrying the wind on your quarter an opposition light to leeward is generally closehauled on the same tack as you have the wind, i.e., if you have a starboard quarter wind, a green light on your lee bow or port bow will be a starboard tack ship.

5. Again, when you have a quarterly wind an opposition light to windward is either dead before the wind or has the wind on opposite quarter.

6. Both for steam and sail opposition lights abaft six points from the head are safe unless carried by faster ships. This predicament is spoken of at the beginning of the article.

7. For sail the starboard tack ship is the safest position as regards the rule of the road, and where beating up a coast line or bay will admit of long tacks an old but good dodge is to sail port tack in daylight and starboard tack at night.

8. Cast iron rules are an impossibility in writing on this vital branch of seamanship as each vessel has characteristics of her own and speed, weather and sea room are to a great extent the framers of sailor practice.

9. In twin-screw craft if anxious about your space to swing round in put your starboard engine astern when swinging on port helm, i.e., coming to starboard. When altering your head to port which is starboard helm put your port propeller astern while swinging. This kind of engine drill is also most valuable when rounding bends in a river, especially where steering is bad through reduced speeds and shallow water.

As the sailing ship rules alter with the directin of the wind it must be apparent that for beginners there is no better field of learning than to master the "keep clear" and "stand on" movements of the non-mechanical craft. Then steamship rule of the road will come along like second teeth.

Always assume that you are in the large model marked "your ship." Put the model lights in by filling the enclosures on the bow with green and red. Do the same to the surrounding craft according to the enclosures only having both sidelights in view for craft right ahead and the X's within the overtaking angle. The two small rings right ahead represent a sailing ship, while the three small rings represent a steamer. They are both bound south or end-on to the model.

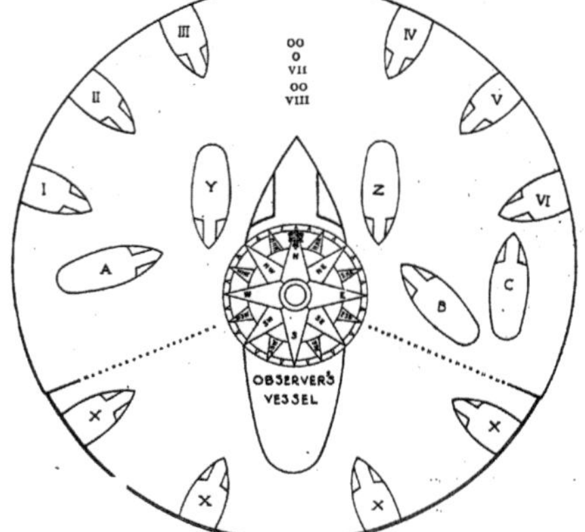

CHART FOR DETERMINING OTHER SHIP'S COURSE FROM OBSERVATION OF ITS LIGHT.

How to use the chart marked "your ship" for sail only.

Lay a pen or pencil on the chart to denote the direction of the wind—which is the key to all questions and answers in sail. Of course the chart is adapted for steamboat practice too.

Put the model on the port tack, carrying the wind at WNW. What do we find then?

I is dead before it or a dead aft ship. II, III and VIII are all free to starboard. IV has the wind abeam, V is on the starboard tack, VI is in stays and likely paying off on the starboard tack, Y and

*Editor's Note.—This department is conducted by a qualified ship's master, and will deal with questions involving seamanship and navigation. The attention of young men desiring to obtain certificates as mates and masters is particularly directed to this department.

Z are "passing" ships. Never mix passing ships with crossing ships. A is free to port; B is going about; C is the lee ship.

Put the model free to starboard, call the wind ESE, what do we find then? Y Z and VIII are closehauled to port, VIII is still in the collision zone, A, I, II and III are in stays, IV, V and VI are free to port, B and C are weather ships.

Make yourself acquainted with the possible movements in this way.

Say that the model has the wind abaft the beam at WSW. What is VIII doing showing both lights right ahead?

Answer.—She is bound south and is closehauled to starboard, i.e., starboard tack.

Give another illustration of how two sailing ships could meet each other head on?

Answer.—When carrying the wind abeam, sometimes called "a soldier's wind." Say that the model has the wind at east, then VIII will have it on the opposite beam and be heading south against the model heading north.

Nothing short of model demonstration will put the practical use of the rule of the road into a man's head. Get the government booklet and you will see the meaning of the cold print in these illustrated articles.

The wind chart shown herewith makes one acquainted with the sailing positions, for rules cannot be understood fully unless the alphabet of the sea is explained. In this respect it will be well to note the following meanings, all of which are connected with ship handling:

Dead before it and dead aft are synonymous.

Passing ships—Green to green or red to red.

Crossing ships—Showing opposition lights.

Lee ship—Going the same way but to leeward of you.

Weather ship—Going the same way but to windward of you.

Tacking and going about are synonomous.

Wearing is going round before the wind and sea.

Luff and helm down bring a ship to the wind.

Let her off, or up helm, or away, is to put a ship from the wind.

Closehauled and on a wind means the ship is on a tack.

Free means that she has the wind abaft the beam, this vessel is thus distinguishable from one that has the wind aft or closehauled, all three names being used in the regulations.

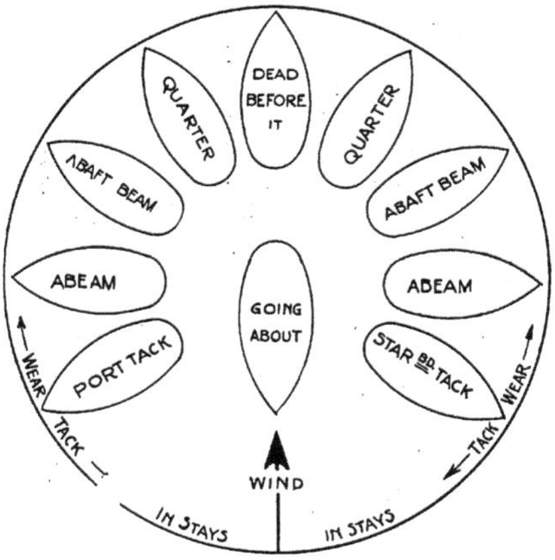

Next month's issue will contain examples of sailing ships and steamers meeting in all directions and will fully illustrate in another form of drawing the various clauses laid down in the collision rules.

WIND CHART FOR STUDYING SAILING POSITIONS.

NOTICES TO MARINERS

THE following notices to mariners have been issued by Department of Marine, Ottawa, under dates given:

No. 7 of 1918, Feb. 9.

(Pacific No. 2.)

BRITISH COLUMBIA

(17) Vancouver Island—East Coast—Nanaimo Harbor—Regulations Relating to Mooring Buoys

Mooring buoys.—The Western Fuel Company maintains three mooring buoys north of its coaling wharves in Nanaimo harbor, as follows:

Position of eastern buoy.—375 feet 37° (N. 11° 30' E. mag.) from the southeast corner of the main (easterly) coaling wharf.

Position of middle buoy. 40° feet 86° (N. 60° 30' E. mag.) from the northwest corner of the main coaling wharf.

Position of westerly buoy.—450 feet 21° (N. 30' W. mag.) from the northwest corner of the main coaling wharf.

Mooring lines.—The company is authorized to moor vessels to the buoys, or to stretch mooring lines in the intervals between the buoys and the wharf or vessels.

Lights.—So long as such moorings exist as obstructions to navigation each end of every line will be marked at night by a fixed white light of sufficient power, and shall be supervised by a watchman charged with the duty of slacking away the lines if necessary.

Sailing directions.—When these lights are exhibited vessels using the south channel should pass on the north side of the buoys.

Harbor Master's permission.—Before any line is placed permission must be obtained from the Harbor Master, as provided in section 27 of the general regulations for the government of harbors.

Penalty.—Any violation of this regulation is punishable by a penalty of twenty dollars.

Variation in 1918: 25° 30' E.
Authority: Order-in-Council No. 39, dated 8th January, 1918.
Admiralty charts: Nos. 573, 2512 and 579.
Publication: British Columbia Pilot, Vol. 1, 1913, pages 311, 312, and 313.
Department File: No. 29467.

ALASKA

(18) Summer Strait—Helm Rock—Gas and Bell Buoy Established

Position.—On the northern side of Helm rock.

Gas and bell buoy established.—Helm rock gas and bell buoy has been established in the position heretofore occupied by the can buoy.

Description.—Cylindrical buoy, with skeleton superstructure.

Color.—Red and black horizontal bands.

Character of light.—Flashing white light every 2 seconds, flash 0.2 seconds duration.

Elevation.—12 feet.

Power.—130 candles.

Depth.—19 fathoms.

Authority: U.S. Dept. of Commerce, N. to M. No. 4 of 1918.
Admiralty charts: Nos. 1432, 2463, and 2431.
Publication: Alaska Pilot, 1908, page 98.

No. 8 of 1918, Feb. 15
(Inland No. 2)

Quebec

(19) River St. Lawrence—Lake St. Louis —Pointe Claire Light-station— Change in Color of Light

Position.—About a mile eastward of Pointe Claire, on a pier located a little northward of the steamboat channel.
Lat. N. 45° 25′ 37″, Long. W. 73° 47′ 43″

Date of alteraton.—Opening of navigation in 1918 without futher notice.

Alteration.—The characteristic of the light will be changed from fixed white to fixed red.

Order.—Fifth dioptric.

Authority: Department records.
Admiralty chart: No. 259a.
Canadian Naval chart: No. 50.
Publication: St. Lawrence Pilot above Quebec. 1912, page 113.
Canadian List of Lights and Fog Signals, 1917: No. 1530.
Departmental File: No. 21530A.

ONTARIO

(20) Lake Erie—Pelee Passage—South-east Shoal—Lightship to Be Temporarily Replaced by Gas Buoy

Position.—½ mile south-easterwardly from the southern extremity of south-east shoal.
Lat. N. 41° 49′ 15″, Long. W. 82° 27′ 32″

Lightship temporarily discontinued.—The south-east shoal lightship will not be replaced on her station immediately upon the opening of navigation in 1918.

Temporary buoy.—Her station will be temporarily marked by a gas buoy, painted red, showing a fixed red light.

Remarks.—Further notice will be given when the lightship is replaced on her station.

Authority: Memo. from Commissioner of Lights.
Admiralty charts: Nos. 332 and 678.
Publication: U.S. H.O. Publication No. 108D. 1907, page 89.
Canadian List of Lights and Fog Signals, 1917: No. 1862.
Departmental File: No. 45095.

No. 9 of 1918, Feb. 18
(Atlantic No. 5)

NEW BRUNSWICK

(21) St. Croix River—Spruce Point Light—Corrections to List of Lights

Position of Lighthouse.—Spruce point lighthouse, St. Croix river, is located on the point one mile west of Oak point, as shown on admiralty chart No. 464, and not on the point ½ mile above Oak point, which is Bluff head.

The georgraphical position as fixed by the International Boundary Surveys, Department of the Interior, is
Lat. N. 45° 10′ 3″, Long. W. 67° 11′ 9″

Sailing directions.—The bracket and remarks in the last column of the Canadian list of lights referring to all St. Croix river lights are to be cancelled, and the following remarks entered for Spruce point lighthouse:—

"A vessel coming up the river should bring Spruce point light to bear 285° (N. 56° W. mag.) and then head on it till within a cable of Bluff head."

Variation in 1918: 19° W.
Admiralty charts: Nos. 464, 2013, 352, 2492 and 2670.
Publication: Nova Scotia and Bay of Fundy Pilot, 1911, page 293.
Canadian List of Lights and Fog Signals, 1917: No. 32.
Departmental File: No. 28502.

NOVA SCOTIA

(22) South Coast—Halifax—Gate Vessel "Diana" Replaced Temporarily By a Buoy During Repairs

The Gate vessel "Diana" will, without further notice, be removed from her position in Halifax harbor for repairs. A buoy, on which the gates can hinge, will replace this vessel temporarily while she is undergoing repairs.

Authority: Dept. of the Naval Service.
Admiralty charts: Nos. 311, 2320, 2410, 729 and 1651.
Publication: Nova Scotia Pilot, 1911, page 133.
Departmental File: No. 45129.

NOVA SCOTIA

(23) Cape Breton Island—South Coast—Louisburg Harbor — Lighthouse Point—Change in Characteristic of Fog Alarm

Former notice.—No. 83 (231) of 1911.

Position.—At lighthouse point light-station on the north side of entrance to Louisburg harbor.

Change in characteristic of fog alarm.—The fog alarm has been changed so as to sound two blasts of 3 seconds duration each every 30 seconds, thus:

Blast	Silent	Blast	Silent interval
3 secs.	2 secs.	3 secs.	22 secs.

Authority: Report from Mr. J. A. Leger, District Engineer, Halifax.
Admiralty charts: Nos. 2692, 2727, 1651, 2516 and 2666.
Publication: St. Lawrence Pilot, Vol. 2, 1916, page 57.
Canadian List of Lights and Fog Signals, 1917: No. 453.
Departmental Files: Nos. 20453C and F.

No. 10 of 1918. Feb. 27.

Nova Scotia

(24) South Coast—Port of Halifax—Public Traffic Regulations

These regulations are issued for the guidance of masters of vessels entering the port, and must be obeyed or their ships will be treated as hostile. They will remain in force until amended or cancelled by a notice to mariners.

This notice is divided into five sections dealing with entrance to port, pilotage regulations, movements within the port, vessels leaving port, and harbor traffic, small yachts and pleasure craft. It contains thirty-three clauses and cancels all previous traffic regulations. Departmental file: No. 32504.

No. 11 of 1918. Feb. 28.
(Atlantic No. 7.)

NOVA SCOTIA

(225) South Coast—Lockeport—Gull Rock—Fog Alarm Established

Date of establishment.—1st May, 1918, without further notice.

Position.—On Gull rock, adjoining south-east side of lighthouse. Lat. N. 43° 29′ 12″, Long. W. 65° 5′ 50″.

Description.—Diaphone, operated with air, compressed by an oil ngine. It will give one blast of four seconds' duration every 30 seconds, thus:

Blast	Silent interval
4 secs.	26 secs.

Structure.—Square wooden building, with a gable roof; on concrete foundation 8 feet high.

Color.—White.

Remarks.—The horn, elevated about 25 feet above high water mark, points 130° 30′ (S. 30° E. mag.)

Variation in 1918: 19° 30′ W.
Authority: Report from Mr. J. A. Leger, District Engineer, Halifax.
Admiralty charts: Nos. 340, 730, 1651 and 2670
Publication: Nova Scotia Pilot, 1911, page 19[illegible]
Canadian List of Lights and Fog Signals, 1917: No. 260.
Departmental File: No. 20260F.

NOVA SCOTIA

(26) South Coast—Halifax—Port open day and night

Until further orders vessels will be allowed to enter the Port of Halifax by day and night.

Previous notice.—No. 108 (362) of 1916.

Authority: Dept. of the Naval Service.
Departmental File: No. 32504.

QUEBEC

(27) River St. Lawrence—Little Metis—Fog alarm established.

Day of establishment.—On or about 1st April, 1918, without further notice.

Position.—On Metis point, 40 feet north of lighthouse. Lat. N. 48° 41′ 0″, Long. W. 68° 2′ 20″.

Description.—Diaphone, operated with air, compressed by an oil engine. It will give three blasts every minute, thus:

Blast	Silent	Blast	Silent
2 secs.	5 secs.	2 secs.	5 secs.

Blast	Silent interval.
4 secs.	42 secs.

Structure.—Square building, with a gable roof.

Material.—Wood.

Color.—White.

Remarks.—Horn, elevated 28 feet above high water mark, points 332° (N. 4° W. mag.)

Variation in 1918: 24° W.
Authority: Report from Mr. J. A. Smith, District Engineer, Quebec.
Admiralty charts: Nos. 307 and 2516.
Publication: St. Lawrence Pilot, Vol. 1, 1916, page 111.
Canadian List of Lights and Fog Signals, 1917: No. 1086.
Departmental File: No. 21086F.

TRADE WITH THE NETHERLANDS

BELOW is published a list of articles the names of the importers of which in the Netherlands have been forwarded by Mr. Ph. Geleerd, Acting Trade Commissioner, Rotterdam. The names of these importers may be obtained by Canadian manufacturers and exporters who may be interested in after the war trade with Holland upon application to the Commercial Intelligence Branch of the Department of Trade and Commerce. (Refer File No. 20411):—Asbestos, automobiles, automobile tires, building material, chemicals, chemical dry colors, electrical specialties, iron and steel, machinery and hardware, magnesite, mica.

Note.—All correspondence with the Netherlands should show that after war business is intended.

PROGRESS IN NEW EQUIPMENT

There is Here Provided in Compact Form a Monthly Compendium of Shipbuilding and Marine Engineering Auxiliary Product Achievement

SHIP LIGHTING UNIT

THE above illustration shows a unit lighting set particularly adapted to the lighting of ships. The Watson seven and one-half kilowatt generator is direct connected to a Canadian Sirocco high-speed vertical engine all mounted on one base.

The engine operates without noise or vibration, is oiled automatically, and is of the enclosed type. Lubrication is ample at any speed, and is not affected by the wearing of bearings, etc. Oil is separated from the water, cooled and filtered at every circuit. Careful analysis of hundreds of repairs ordered for engines lubricated with sight feed oil cups showed conclusively that over 80 per cent. of the repairs were due in one way or another to insufficient lubrication. The forced lubrication system with which these engines are provided enables them to run steadily for from three months to a year at high speed and full load without the adjustment of a bolt or screw or any additional oil after the first filling.

This is made possible by the positive feed of oil to every bearing and wearing surface from a pump in the base of the engine, driven through gears from the crank shaft.

The Watson generator which forms part of this unit is carefully tested during construction, and after completion with regards to mechanical features and electrical characteristics. The field ring is cast from a high grade of dynamo steel and the laminated pole pieces are firmly clamped to it by means of two bolts, insuring a perfect magnetic circuit.

The armature core is built of laminations punched from No. 29 gauge sheet steel, separately pointed to minimize core losses and when assembled are keyed to the armature shaft. The coils are form wound with double cotton-covered wire, and are insulated with fibre, linen tape and high-grade insulating varnishes. The complete armature is given a running balance to eliminate vibration.

The sleeve bearings which form part of this generator are cast from bearing bronze, and are so designed as to align against a shoulder in the head casting. They are held in place by a fixed key instead of by the usual set screw. Oil cups aret provided which serve the double purpose of a filling device and an overflow.

The brush holders are of the box type with radial feed, which prevents the shifting of the point of contact due to brush wear.

The A. R. Williams Machinery Co., Ltd., Toronto, Ont., are the Canadian agents for this equipment.

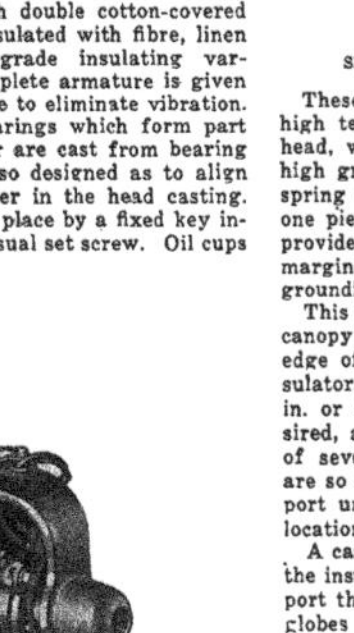

SEVEN AND ONE-HALF KILOWATT, DIRECT CONNECTED SHIP LIGHTING UNIT.

HIGH TENSION STREET LIGHTING FIXTURES

WHEELER high tension series street lighting fixtures are designed to provide a high factor of safety with very substantial mechanical construction and convenient reflectors to properly centre the filament of the lamp so as to give the widest possible light distribution.

SHOWING COMPLETE FIXTURE

These are all built around the Wheeler high tension series porcelain receptacle head, which is a self-contained unit of high grade electrical porcelain having a spring clip series receptacle moulded in one piece with it, and so designed as to provide ample insulation with a wide margin of safety against accidental grounding on high tension circuits.

This head is supported by a cast iron canopy with three set screws in the lower edge of the skirt to firmly grip the insulator. It is threaded at the top for ¾-in. or 1¼-in. iron pipe, as may be desired, and plain or ornamental brackets of several designs are supplied, which are so constructed as to give ample support under wind stress even in exposed locations.

A cast iron skirt (J) securely bolted to the insulator has three set screws to support the different types of reflectors and globes used therewith. The reflectors

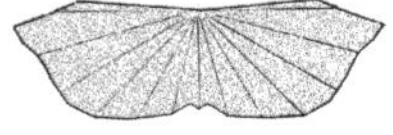

are flat and fluted to give the desired wide light distribution, and are drawn down with a deep neck to properly position the reflecting surface above the filament centre. These reflectors are attached by means of the set screws in the skirt, which pass through holes in the collar of the reflector and bear against a galvanized seating ring (M) with an inverted wedge shape cross section. The tightening of the set screws against this seating ring prevents the reflector from rattling in high winds, and thus prevents chipping of the enamel.

The insulator has ears on opposite sides moulded in one piece with the porcelain. These ears are pierced with holes, through which the lead wires are tied before entering the ports underneath the skirt of the insulator that lead to the terminals inside the head, or the wires may be brought down through the

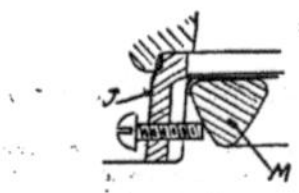

brackets directly to the terminals when the fixtures are used for concealed wiring.

SAFETY GAUGE GLASS

AN improved gauge glass is being manufactured by the Nason Manufacturing Co., 71 Fulton Street, New York, and several important advantages are apparent in its construction.

As will be seen from the illustration the glass is so enclosed and centered in its housing that it is immune from breakage caused by expansion and contraction, and should the glass become broken maliciously or through some unusual accident, the metal netting encasing it renders injury from flying glass fragments impossible. Each gauge-cock fitting is equipped with a ball valve which closes automatically should the gauge glass become broken. The above fitting is made in steel for ammonia refrigeration work and in bronze for steam conditions.

SAFETY GAUGE GLASS AND HOUSING

The safety gauge glass is sold with gauge fittings complete and is also sold separately for insertion in any standard gauge fitting at present manufactured.

PORTABLE ELECTRIC HOIST

THE accompanying engraving shows an electrically-driven hoist which is readily adaptable to the handling of material around yards and stores. It is operated by a standard A. C. motor which is supplied to suit various voltage and cycle requirements.

The motor drives the 12-in. diameter drum through worm gearing, the capacity of the output being 1,500 pounds. The drum is capable of overhauling 50 feet of ½-inch cable, and, as originally designed, the hoist was arranged to elevate material, reverse automatically and come back to charging position.

The motor is provided with standard reverse control, or where special requirements make it necessary specially designed controls can be designed to suit.

A magnetic brake of the horseshoe type is fitted. It is of simple and rugged construction, is not liable to jam, and insures a dead stop by the motor immediately the current is cut off. This hoist is the product of the Volta Manufacturing Co., Welland, Ont.

PORTABLE ELECTRIC HOIST.

CAPTAINS AND ENGINEERS MUST REGISTER

THE Dominion Department of Marine has issued the following regulations affecting the registration of all masters, or mates or engineers in Canada of sea-going or other ships.

1. Every person residing in Canada, not more than sixty-five years of age, who holds a certificate of competency, other than a temporary certificate, as a master or a mate or an engineer whether for sea-going or other ships, shall, on or before the thirtieth day of April, one thousand nine hundred and eighteen, send a statement to the Minister of Marine and Fisheries at Ottawa, giving his full name and address, his nationality, the date of his birth and the date and number of every certificate of competency held by him as a master, a mate or an engineer whether for sea-going or other ships. Such statement shall be made, in the case of masters and mates, in Form A, and by engineers in Form B, in the schedule hereto. Every such person thereafter changing the nature of his employment or his address shall forthwith notify the Minister of Marine and Fisheries of such change. Temporary employment during the winter months or when the ship upon which any such person is employed is undergoing repairs need not, however, be reported.

2. Every person knowing or having reason to believe that any person employed by him or by anybody corporate of which he is the manager or superintendent is under sixty-five years of age and is the holder of a certificate of competency as a master, a mate or an engineer whether for sea-going or other ships shall forthwith ascertain if such person is a holder of any such certificate and, if he is, and does not produce a certificate that he has reported as required by regulation 1 hereof, shall thereupon send a statement to the Minister of Marine and Fisheries containing the full name and address of such person, which statement shall be in Form C in the schedule hereto.

3. Every person when thereto required by the Minister of Marine and Fisheries, or by any person thereto authorized by the said Minister, shall post up and keep so posted up in a prominent place in his premises, or in the premises of which he is the manager or superintendent, where it can easily be read by the employees employed by him or subject to his orders, a notice in form D in the schedule hereto. And every person shall at all reasonable times permit any person thereto authorized by the Minister of Marine and Fisheries to enter such premises and make such enquiries as he may desire for the purpose of ascertaining what, if any, certificates any of such employes may hold and any other information that he may require in connection therewith.

4. Any person refusing or neglecting to make any statement, enquiry or answer required or authorized under the provisions of these regulations, or refusing to post up or keep posted up any notice in accordance with the provisions of these regulations, or refusing to permit any duly authorized person to enter any premises or to make any enquiry authorized by these regulations, and any persons knowingly giving any particulars that are untrue or misleading in any statement or answer authorized by these regulations, shall be guilty of an offence and shall be liable on summary conviction to a fine not exceeding One Hundred Dollars, or to imprisonment for any term not exceeding two months, or to both fine and imprisonment.

RODOLPHE BOUDREAU,
Clerk of the Privy Council.

Automatic Storage Battery Industrial Trucks

Profitable—

If material is to be moved fifty feet one of these trucks will do the job profitably; while on a long haul one man with one of these machines can do the work of fifteen men with hand trucks.

Relieve Congestion
Speed up Production
Protect your Goods

The Elevating Platform Truck here shown has a capacity of 4,000 pounds and a speed up to 5 miles per hour. The truck is run under a small platform, previously loaded, automatically raises it clear of the floor, and after carrying it to its destination, returns for the next platform.

This does away with congestion and the damage to goods due to hurried handling on and off a truck.

We Have a Truck for Any Purpose.

For complete information please write our nearest house.

The Canadian Fairbanks-Morse Company, Limited

"Canada's Departmental House for Mechanical Goods"

St. John, N.B., Quebec, Montreal, Ottawa, Toronto, Hamilton, Windsor, Winnipeg, Saskatoon, Calgary, Vancouver, Victoria

ASSOCIATION AND PERSONAL

A Monthly Record of Current Association News and of Individuals Who Have Been More or Less Prominent in Marine Circles

G. G. Bushby, of Vancouver, has been elected president of the British Columbia Manufacturers' Association.

Hon. W. J. Hanna has been appointed president of the Imperial Oil Co., with headquarters in Toronto. Mr. Hanna succeeds W. C. Teagle.

Captain Robert Fraser, who for over twelve years was marine superintendent of the Montreal Transportation o., at Kingston, ont. died at Long Beach, Cal., on Sunday.

Edward C. Fry, for more than half a century Lloyd's agent at Quebec, died recently after a brief illness. He was born at Bristol, England, and was recognized as one of the best informed shipping men in that city.

Reginald Beaumont has been appointed assistant manager of the G. T. P. Steamship Lines, with headquarters at Prince Rupert, B.C. Mr. Beaumont has been assistant to the manager at Prince Rupert for the past eight years.

S. H. Lunt, who has been assistant to R. J. Young, general manager at Montreal of the Export Association of Canada for the past eighteen months, is now acting as manager in the temporary absence of Mr. Young, who has been appointed as a member of the Canadian War Board, with headquarters in Washington.

Major David Seath, for nineteen years secretary-treasurer of the Montreal Harbor Commission, died at his home, 380 Lansdowne Avenue, Westmount, on Saturday morning. Major Seath was in his seventy-first year and had been a lifelong resident of Montreal. He resigned from the Harbor Board last autumn.

J. J. Scullan, formerly general superintendent of the Davenport Works of Canadian Allis-Chalmers, Ltd., Toronto, and latterly manager of that firm's shipyard at Bridgeburg, Ont., has been appointed manager of hull construction for the Marine Boat Corporation of New York.

James Lynn Rodgers, recently appointed Consul-General in Canada for the United States, has taken up his duties at Montreal, replacing Vice-Consul Bradley, who retired some weeks ago. Mr. Rodgers was consul-general at Havana. He is a native of Columbus, Ohio, and entered the U. S. service in 1915, as consul-general at Shanghai.

Victoria, B.C.—It is understood that the Dominion Government will give $50,000 towards the construction of the shed required on pier No. 2 at the new outer wharves.

W. J. Sergent, who, for about eighteen years has been associated with the Allan Line and the Canadian Pacific Ocean services, is retiring from active duty as chief superintending engineer. He remains associated with the company in a consultative capacity. Kenneth MacKenzie, who has been his chief assistant, has now been appointed chief superintendent engineer of all the company's fleets, with headquarters in Liverpool.

At the annual meeting of the Dominion Marine Association held in Toronto recently the following officers were elected:—President, J. T. Mathews, Toronto; 1st vice-president, W. J. McCormack, Ste. Marie; 2nd vice-president, G. E. Fair, Toronto; general counsel, Francis King, Kingston. Committee:—A. A. Wright, W. E. Burke, A. E. Mathews, C. E. Harris, J. F. M. Stewart (Toronto), L. Henderson, J. Waller (Montreal), J. Sowards (Kingston).

ONE of the biggest difficulties connected with the quantity production of ships is the question of rivet driving. An idea of the magnitude of the task may be obtained when it is said that if a concern were to build one ship a week, a weekly number of 650,000 rivets must be driven. When we consider that the best rivet drive is by a company which drives 250,000 per week, and the next best is by the three largest shipyards on the Atlantic coast, the magnitude of the task involved in driving 650,000 rivets in one week may well be imagined.

LICENSED PILOTS

ST. LAWRENCE RIVER

Captain Walter Collins, 43 Main Street, Kingston, Ont.; Captain M. McDonald, River Hotel, Kingston, Ont.; Captain Charles J. Martin, 13 Balaclava Street, Kingston, Ont.; Captain T. J. Murphy, 11 William Street, Kingston, Ont.

ST. LAWRENCE RIVER, BAY OF QUINTE, AND MURRAY CANAL

Captain James Murray, 106 Clergy Street, Kingston, Ont.; Capt. James H. Martin, 259 Johnston Street, Kingston, Ont.; John Corkery, 17 Rideau Street, Kingston, Ont.; Captain Daniel H Mills, 272 University Avenue, Kingston, Ont.

MONTREAL PILOTS' ASSOCIATION

President—Alberic Angers, Montreal.
Secretary—C. B. Hamelin, Champlain, Que.

ASSOCIATIONS

DOMINION MARINE ASSOCIATION

President—A. A. Wright, Toronto. Secretary—Francis King, Kingston, Ont.

GREAT LAKES AND ST. LAWRENCE RIVER RATE COMMITTEE

Chairman—W. F. Herman, Cleveland, Ohio. Secretary—Jas. Morrison, Montreal.

INTERNATIONAL WATER LINES PASSENGER ASSOCIATION.

President—O. H. Taylor, New York.
Secretary—M. R. Nelson, 1184 Broadway, New York.

SHIPPING FEDERATION OF CANADA.

President—Andrew A. Allan, Montreal; Manager and Secretary—T. Robb, 218 Board of Trade, Montreal; Treasurer, J. R. Binning, Montreal.

SHIPMASTERS' ASSOCIATION OF CANADA.

Secretary—Captain E. Wells, 45 St. John Street, Halifax, N.S.

GRAND COUNCIL N.A.M.E. OFFICERS.

A. R. Milne, Kingston, Ont., Grand President.
J. E. Belanger, Bienville, Levis, Grand Vice-President.
Neil J. Morrison, P.O. Box 238, St. John, N.B., Grand Secretary-Treasurer.
J. W. McLeod, Owen Sound, Ont., Grand Conductor.
Lemuel Winchester, Charlottetown, P.E.I., Grand Doorkeeper.
Alf. Charbonneau, Sorel, Que., and J. Scott, Halifax, N.S., Grand Auditors.

1918 Directory of Subordinate Councils, National Association of Marine Engineers.

Name.	No.	President.	Address.	Secretary.	Address.
Toronto,	1	Arch. McLaren,	324 Shaw Street	E. A. Prince,	49 Eaton Ave.
St. John,	2	W. L. Hurder,	209 Douglas Avenue	G. T. G. Blewett,	86 Murray St.
Collingwood,	3	John Osburn,	Collingwood, Ont.	Robert McQuade,	Collingwood, Ont.
Kingston,	4	Joseph W. Kennedy,	395 Johnston Street	James Gillie,	101 Clergy St.
Montreal,	5	Eugene Hamelin,	Jeanne Mance Street	M. Lasure	120 Ribard St.
Victoria,	6	John E. Jeffcott,	Esquimault, B.C.	Peter Gordon,	808 Blanchard St.
Vancouver,	7	Isaac N. Kendall	319 11th St. E., Vanc.	E. Read,	232 13th St. W.
Levis,	8	Michael Latulippe,	Lauzon, Levis, Que.	Arthur Jolin,	Lauzon, W.
Sorel,	9	Nap. Beaudoin,	Sorel, Que.	Alf. Charbonneau,	Box 204, Sorel, Que.
Owen Sound,	10	John W. McLeod	570 4th Ave.	R. J. McLeod,	662 8th St.
Windsor,	11	Alex. McDonald,	28 Crawford Ave.	Neil Maitland,	221 London St. W.
Midland,	12	Geo. McDonald	Midland, Ont.	A. E. House,	Box 333
Halifax,	13	Robert Blair	29 Parrsboro Street	Chas. E. Pearce,	Portland St., Dartmouth, N.S.
Sault Ste. Marie,	14	Charles H. Innes,	27 Euclid Road	Wm. Hindmarch	34 Euclid Rd.
Charlottetown,	15	J. A. Rowe	176 King Street	Chas. Cumming,	27 Euston St.
Twin City,	16	H. W. Cross,	436 Ambrose St	J. W. Farquharson,	169 College St.
				A. H. Archand,	Champlain, Que.

MARINE NEWS FROM EVERY SOURCE

Toronto, Ont.—The city architect has issued to the Toronto Harbor Commission a permit for the erection of ferry sheds at the foot of Bay street to replace those destroyed by fire. The new buildings will cost $10,000.

Erratum.—An item appeared in the February issue of MARINE ENGINEERING giving a list of officials of the Collingwood Shipbuilding Co. This should have read Collingwood Shipping Co., which, is of course, an entirely different concern.

Port Moody, B.C.—The Aetná Iron and Steel Co. are installing a six-ton electric furnace which will be in operation about the middle of this month. The material used will be scrap iron, and the product will consist of merchant bars, light angles and channels. The company is operating the mill on scrap faggots to make rods for shipbuilding purposes.

Bringing Cardiff Coal into Boston.—Five thousand tons of Cardiff coal was brought to Boston, Mass., last week by a British steamer. The coal, consigned to the British Embassy at Washington, for bunker purposes, was diverted by the British Government to relieve the acute shortage here after it was learned that the British ships held up in New York had been supplied with fuel.

Toronto, Ont.—Permit was granted last Thursday to the Dominion Shipbuilding Co. to construct a one or two-storey main building at the foot of Bathurst street on the Harbor Commission property at a cost of $175,000. The building will be of steel throughout and faced with hollow tile. The site leased by the company covers about 16 acres, and five shipbuilding berths will be laid down.

Fort William, Ont.—The Canadian Car & Foundry Co. has awarded the contract for the buildings at the new shipbuilding plant at Fort William to the Dominion Bridge Co. The Bridge Co. must, under the contract, finish the building in six weeks. This is an indication that the actual shipbuilding operations will be commenced as soon as possible. The contract was for twelve mine sweepers 143 feet long by 22 feet beam, the price being given as $2.500.000. The company is now turning out 25 box cars a day at the Fort William car works.

The R. H. Howes Construction Co., of Boston and New York, has established a shipyard at Meteghan River, N.S. A lease for a term of years has been taken of the late James Cosman's shipyards at the mouth of the Meteghan River. Mills, together with pattern shops, equipped with the latest improved machinery, have been erected and are now working. The company recently closed a deal for a tract of 500 acres of very valuable timber land on the Meteghan River and intends to enter into shipbuilding on a very extensive scale. It already has one schooner entirely in frame and other keels will be laid immediately.

Vancouver, B.C.—The War Puget was launched at the Lyall shipyard, North Vancouver, on Feb. 16, the christening ceremony being performed by Mrs. E. H. Bridgeman. The vessel is being built for the Imperial Munitions Board. The War Puget is a wooden steamer, and has the following dimensions:—Length between perpendiculars, 250 feet; beam moulded, 42 ft. 6 inches; depth moulded, 25 feet; draught, 22 feet, and deadweight capacity on 21 feet maximum draught, 2,800 tons. She will be equipped with a triple expansion engine developing 1,000 i.h.p.

Shortage of Pilots At Port of Halifax.—It is announced in the Canada *Gazette* that owing to the shortage of pilots at the Port of Halifax, due partly to the recent disaster, and the fact that there are at present no apprentice pilots eligible for appointment, power has been granted to the Halifax Pilotage Commissioner to license as pilots any person holding a certificate authorizing him to act as master of a ship, or any apprentice who, in the opinion of the commissioners is fit and competent. Any license issued under the regulation granting this power to the commissioners shall continue in force until otherwise ordered.

Rushing Work At Ft. William.—The Canada Car and Foundry Co. of Fort William have their plant in shape now, and the capacity is up to the point where it is possible to turn out a completed freight car every 20 minutes. This huge car factory, which is now working on an order for 5,000 cars for the Canadian Government railway, is located in west Fort William. Twelve hundred workmen are now employed. It is one of the most modern factories in Canada, being built exclusively of brick, steel and glass. Power is derived by the use of huge producer-gas engines. Contracts have been let for a shipbuilding plant to be erected on a site adjacent to the factories shown above, the company having received large orders to build mine-sweepers for one of the Entente powers.

To Fix Prices on Raw Materials for Steel.—A despatch from Washington states that the steel manufacturers have asked the Government to fix prices on all products entering into the manufacture of steel, in order that the fixed price for steel may be stabilized.

Vancouver, B.C.—This city's first bulk shipment of grain has arrived safely at a British port, according to word received here. A steamer which was loaded at the Government elevator here in November with 100,000 bushels of wheat from the prairies, made the journey safely to the United Kingdom via the Panama Canal, and now word is awaited anxiously as to the condition in which the grain in this test shipment reached its destination. The shipment was accompanied by A. W. Alcock, of Winnipeg, grain expert, and assistant chemist of the Canadian Board of Grain Commissioners, who is to observe the condition of the grain at all stages of the voyage.

To Consider Plans for Export Traffic.—To consider the outlook for export traffic and to lay plans for the handling of the same to Atlantic seaports, representatives of railway and inland navigation companies met in Ottawa early this week. The conference has been called by the Prime Minister at the suggestion of Hon. C. C. Ballantyne, Minister of Marine. Information will be obtained from the railway and steamship men as to the facilities for moving to Montreal, Quebec, Halifax and St. John, N.B., freight intended for shipment overseas. An effort will be made to bring about close co-operation between the transportation companies in the use of those facilities.

Ice Conditions on the Lakes.—The first ice report showing conditions on the Great Lakes, as compiled by the Weather Bureau, Detroit, Mich., and the regular display stations of the Meteorological Service of Canada, indicates that the ice field over Western Lake Superior extends, solid, East from Duluth for more than thirty miles, with some open water East of Grand Marais. From Keweenaw Point East to Whitefish Bay, fields are extensive and heavily windrowed over the East portion. The ice averages from 22 to 26 inches. In comparison with the ten-year average there is about eight inches more ice at Duluth, seven inches at Sault Ste. Marie, twelve inches at Mackinaw, and seven inches at Escanaba.

EMPIRE
MARINE
BRASS GOODS
E
Quality Mark
A 203
A-861
Gate Valve:
Solid Wedge:
Non-rising
Stem
A-2512
Steam
Whistle
with
Valve
PIRE
Prompt Service
We have the capacity and facilities for taking care of orders, big or small, and making quick deliveries.
QUALITY PRODUCTS—Every article bearing the "E" trade-mark is factory tested and guaranteed in quality—heavy in weight and made to formula in our own plant. Empire Valves are fitted with packing gland and can be repacked under pressure.
Empire Brass Goods meet all marine demands. A few of our lines:
Steam Traps
Brass Castings
Steam Whistles
Pressure Gauges
Steam Stops
Lubricators
Brass Fittings
Basin Cocks
Brass Valves
Compression Bibbs
Ship Lavatories and
Water Closets
Our finely illustrated catalog shows the full Empire line. Write for one. You'll find it helpful.
Largest Brass Manufacturers in the British Empire
Empire Manufacturing Company
LIMITED
LONDON and TORONTO

Will Start Building Government Ships.—The Canadian Vickers Co., which has commenced work on the first steel steamer of 4,350 tons burden, is to lay down a second of 8,200 tons and it is expected a contract for a third ship of 9,500 tons will soon be made with the Collingwood Shipbuilding Co. The Government's shipbuilding programme contemplates the laying down of the keels of some forty ships to June, 1919. Of this number, four of 5,000 tons and 6 of 8,200 tons are to be built on the Pacific coast. Arrangements have been made whereby the Government is assured of a supply of ship plates and steel materials for its programme. Orders for steel for the Canadian Government would, it is understood, be placed by the United States Purchasing Commission with various American steel companies.

Steel Contracts for U. S. War Work.—Upwards of 8,000 tons of steel plates have been distributed by the Emergency Fleet Corporation among the steel mills, and about 6,000 more tons are required for constructing 18 lake boats which were recently taken over by the Government for service on the Atlantic coast. Specifications are being revised on 50,000,000 bolts and nuts required by the Government, and these will be in the hands of manufacturers in a few days. The American Bridge Co. will furnish 3,000 tons of steel for a number of buildings to be constructed at Atlanta and Baltimore for the Quartermaster Department of the army. Contracts for 100,000 tons of structural steel required for hangars for the British and French as well as the United States Governments are still held in abeyance until it is determined just what time the structural mills can make deliveries.

Collingwood.—One of the early wooden shipbuilders of the Canadian lakes, William Andrews, died here on March 12, aged seventy-seven years. He was born at Clayton, N.Y., and when a boy came to Port Dalhousie, at which point his father established the Andrews Shipbuilding Yards, which were conducted by the family until the construction of the new Welland Canal, which took up their property. Later Mr. Andrews moved to Welland and then to Port Robinson, coming to Collingwood to continue wooden shipbuilding with his brother, Stephen D. Andrews, who was one of the originators of the shipyards now owned by the Collingwood Shipbuilding Company. During the operations of the Andrews' yard on the Welland Canal, William Andrews took part in the construction of many vessels, among others the steamers Zimmerman, City of Toronto (afterwards the Algoma), Alma, Munro, Isaac May, Lake Ontario and Silver Spray. While at Collingwood he was engaged on the Majestic, Germanic and others. He is survived by his wife and his brother, who is Dominion Inspector of Hulls, with headquarters at Collingwood.

U.S. TO CONTROL FUEL OIL

THE United States Government will probably take over supervision of the nation's fuel oil supply within a short time, through the agency of the Federal Fuel Administration. Plans are now being prepared for the organization of an oil division of the Fuel Administration, and it will work out the scheme of control. No decision has been made as yet concerning the advisability of fixing prices, a step which was taken in connection with the Government control of coal contracts and coal distribution.

The situation as regards fuel oil has become acute within the past few weeks due to the heavy demands of the navies of the United States and Great Britain

MANY VESSELS CHANGE HANDS

THE following are the latest reported shipping sales:

Br. s.s. Competitor, 3,526 tons gross, 2,216 net, carries about 6,020 tons deadweight, built at West Hartlepool in 1907; S.S. No. 2 in 1916; s.s. Monkshaven, 3,357 tons gross, 2,097 net, carries about 5,880 tons deadweight, built at Sunderland in 1911, S.S. No. 5 in 1915; and the s.s. Moorlands, 3,602 tons gross, 2,282 net, about 6,440 tons deadweight, built at Sunderland in 1910, S.S. No. 1 in 1917, and owned by the Eskside Steam Shipping Co., Limited (Messrs. C. Smales & Son), Whitby, have been sold to Messrs. Houlder, Middleton & Co., London, for under £300,000.

Two shelter-deck steamers with a deadweight carrying capacity, with the tonnage openings closed in; dimensions: 420 ft. by 54 ft. by 37 ft. moulded, at present building on the north-east coast, and about ready for delivery, with engines 27 in., 44½ in., and 75 in., by 54-in. stroke, estimated to steam twelve knots, have been sold by Mr. A. Munro Sutherland, Newcastle, to the Clan Line for about £250,000 each. One is now ready and the second is to be delivered in about three months. On the 10,800 tons the price works out at £23 2s. 11d. per ton, and on the 9,300-ton basis £26 17s. 7d. per ton.

Arg. s.s. Americano (ex "Flora"), 1,715 tons gross, 1,143 net, built at Newcastle in 1908, S.S. No. 2 in 1917, and owned by the Soc. Importadora y Exportadora de le Pategonia, Buenos Aires, is reported sold to Messrs. Dodero Bros., Buenos Aires, for £150,000.

Br. s.s. Hova, 4.264 tons gross, 2,753 net, carries about 7,525 tons deadweight on 23 ft. 7 in., built at Newcastle in 1910, S.S. No. 1 in 1914, and owned by Mr. F. S. Holland, London, has been sold to the Union-Castle Mail Steamship Co., Ltd., London, for about £135,000.

Br. s.s. Inveran, 4.380 tons gross, 2.853 net, built at Port Glasgow in 1906, S.S. No. 2 in 1914, and owned by the Inveran Steamship Co., Limited (Messrs. R. J. Rowatt & Co.), Glasgow, is reported sold to Houlder, Middleton & Co., Limited, London, for about £115.000.

Br. s.s. Kilsyth, 2,340 tons gross, 1,498 net, built at Stockton in 1903, and owned by Renfrewshire Steamship Co., Limited, Glasgow, is reported to have been sold to English owners for £60,000.

Br. s.s. Crown of Leon (ex "Yanariva"), 3.391 tons gross, 2,155 net, built at Glasgow in 1894 and owned by the Crown Steamship Co., Limited (Messrs. Prentice, Service & Henderson), Glasgow, has been sold to Italian owners in a damaged condition. She carries about 5,600 tons on a draught of 23 ft. 6 in., and steams 9 knots on 22 tons of Scotch coal.

SHIPBUILDING AT COAST

THE month of January has seen a start made on the biggest shipbuilding programme in the history of Canada, for during this year British Columbia shipyards will launch vessels having a total carrying capacity of 188,700 deadweight tons. Vancouver will contribute most of this, for Burrard Inlet and False Creek will launch 146,700 deadweight tons at least, and it is more than likely that additional orders will come along and swell the programme.

The vessels to be built here are ten 8,800 tonners, one 4,700 tonner, four 5,100 tonners, and twelve 2,800 tonners, while yards at Victoria, New Westminster and Coquitlam will contribute fifteen 2,800 tonners.

The 8,800 tonners are to be built at the Coughlan Shipyard on False Creek. The 5,100 and 4,700 tonners are to come from the Wallace shipyard at North Vancouver, and the 2,800-ton vessels will be the produce of the Lyall yard and the Western Canada Shipyards.

The programme received a decided boost yesterday when the Wallace yard announced the order for four government steamers of 5,100 tons each. These ships are to be 337 feet in length between perpendiculars, with a moulded beam of 46 feet 6 inches, moulded depth 25 feet 6 inches, draft, 21 feet, and speed 12 knots.

Launchings started on December 27, when the War Songhees took the water at Victoria. On January 14 the War Nootka was launched at Vancouver, and on January 19 the 8,800-tonner Alaska went down the ways on False creek. On January 24 the War Yukon was sent into her native element at Victoria.

There are other hulls now ready for launching, and it will be a regular procession of tonnage during the year.

NAPHTHALENE AS MOTOR FUEL

By L. E.

THE effect of the compulsory search for a substitute for petrol has been highly beneficial, as it has stimulated experiment and test in directions which would otherwise have been neglected under normal conditions, and research has revealed that coal provides an excellent hydrocarbon which can be utilized as fuel for the internal combustion engine: benzol. But this product is indispensable for war purposes, so it was promptly ruled out of court. Yet benzol is but one of the many hydro-carbons derived from coal, although it is admittedly the most acceptable substitute for petrol, as a fuel for the explosion motor. But coal possesses other constituents of a similar character, although somewhat heavier, and amongst these must be mentioned naphthalene, a substance which, while

not popularly known under its technical name, is familiar enough.

Naphthalene is a crystal. In its crude condition it might also be compared to dirty, finely-powdered snow, being relatively light and fine in its consistency. In this condition it contains a certain proportion of creosote oil, which imparts an oleaginous character to the substance, but which can be driven out by subjecting the powder to a violent centrifugal or whizzing action, and can be further cleaned or rectified by washing with an acid. But naphthalene is the bugbear of the gas engineer, and of the householder, who depends upon coal gas for illumination. With a slight drop in temperature it condenses in the form of snow and becomes an obstruction in the pipe. A very small quantity suffices to impede the natural passage of the gas, and recourse to the force pump or a liquid solvent are the sole effective palliatives. Apart from its conversion into dyes and carbon blocks and moth balls, naphthalene is of little use except possibly for the preparation of firelighters, to which latter purpose about 50% of the total yield is devoted.

In pre-war days crude naphthalene commanded about $20 per ton, and although the current price is from $25 to $35, its reclamation cannot be described as profitable. The annual yield in the aggregate from the gas and tar works of the British Isles is equivalent to about 10,000,000 gallons. Yet naphthalene is closely allied to the more appreciated hydro-carbon, benzol, and it belongs to the same series. The chemical symbol for benzol is C_6H_6, that for napthalene is $C_{10}H_8$, and this close relationship was responsible for the doping practice which had a certain vogue during the early days of benzol as a motor spirit. This fact was also responsible for another line of action amongst certain French and German investigators. They advocated the possibilities of the compound as a fuel for internal-combustion engines, and carried out experiments to support their assertions, but they never achieved distinct success. The general practice was the dissolving of the naphthalene in a lighter hydrocarbon such as petrol or benzol. These two liquid fuels will readily absorb a certain portion of napthalene, the quantity varying from 20 to 40 per cent. But the experiment did not prove successful in practice.

Success in Sight

Other inventors, confident of their ability to use napthalene as a straight fuel, confined their efforts to achieving success in this direction, but other difficulties arose, and success was not forthcoming. But to-day it seems as though success is within a measurable distance of turning this elusive by-product of coal into a commercial motor fuel, and that without recourse to doping or other expedients.

Some five years ago two English experimenters, Mr. J. H. Willis, and Lieut. G. G. Wilson, attacked the problem. Their investigations elicited one important fact and they resolved to continue their researches from point where text books left off. These had settled that the melting point of the hydro-carbon was 79.8 degrees C., and the laboratory experiments were conducted on methodical lines from this point, and these proved conclusively that it was necessary to raise the temperature to about 133 deg. C., in order so highly to change a small volume of air into naphthalene vapour that sufficient would be present in a form suitable for an explosive mixture with a large proportion of air subsequently introduced. It was also found that the explosive mixture was formed when composed of 12 parts of air to 1 of naphthalene vapour.

A special form of vaporizer had to be devised and subsequently the vaporizer, carburetter and fuel tank were combined. With the experimental installation, starting up is conducted on petrol. After running for about ten minutes by which time the naphthalene has been melted and raised to 275 deg. F., the petrol is shut off and the napthalene vapour switched on.

The engine can be started on the naphthalene, which will retain its liquid form for an appreciable period. The results achieved are so striking that the fact is established that it is possible for an internal-combustion engine to be run exclusively on this new fuel.

MUST DEVELOP B. C. SHIPBUILDING

IN nearly every city of Canada there are men from the Old Country who have had experience in shipbuilding in some of the great yards of the old land. Coming to the industrial centres of this country, they have drifted into other lines of work. They found little difficulty in coming in contact with employment to which they could apply their early training, and so it is that very many of these mechanics—and good mechanics they are, too—are in the factory centres of Canada.

The call of the shipbuilding companies, following the war orders and the big construction programmes of the Canadian yards, is calling a large number of these men back to their first love. In not a few Ontario papers, for instance, are paragraphs to be seen stating that "Mr. Blank and family have left for the coast, where Mr. Blank will follow the shipbuilding trade, in which he was engaged prior to coming to Canada." Ontario shipbuilding centres are also drawing on this same source of labor supply.

A Great Industry

As a matter of fact, shipbuilding is a bigger thing on the Pacific Coast than it has ever been before, and it looks as though the foundations were being laid for a great future. At the annual meeting of the B. C. Manufacturers' Association, Mr. J. A. Cunningham, the retiring president, put the case in a new way. He based his appeal on the inroads that Japan was making on the carrying trade of the Pacific. "Have a British Columbia merchant service," he urged. "Let our

HOISTING and HAULAGE MACHINERY FOR SHIP YARDS
All sizes and styles of DERRICKS and DERRICK ENGINES.
For any power available, either steam, electric, gasoline, belt, or hand power.
Send for Catalog
Marsh Engineering Works
Limited
Belleville, Ont.
(Formerly Marsh & Henthorn) Note New Name.
Standard machine for operating a two-line Derrick. Made also for Electric or Belt Power.
Standard machine for High Speed Lifting or Hauling. Made also for Electric or Belt Power.

LONDON GLASGOW PARIS CHRISTIANIA MILAN
UNDER ALL FLAGS
The McNab
Being Sole Representatives in the United States for prominent British manufacturers of Marine Specialties, and also having extensive manufacturing facilities at Bridgeport, Conn., we are in a position to supply:
SHIP'S TELEGRAPHS (Chadburn's Type)
NAVIGATIONAL OUTFITS
SHIP'S PORTS
STEERING GEARS (Napier Type)
STEERING GEARS (Brown Bros. Type)
MANILA ROPE
COMPASSES
MARINE CLOCKS, 1-day and 8-day
REVOLUTION COUNTERS
DIRECTION INDICATORS
BOILER CIRCULATORS
WHISTLE CONTROLS
BRASS CASTINGS OF ALL DESCRIPTIONS
BRASS CASTINGS MACHINED
ETC., ETC.
Write us your requirements, we will be very pleased to tender quotations with deliveries on anything you may need in the line of marine equipment.
OUR MATERIAL LIST MAILED ON REQUEST
THE McNAB COMPANY, - Bridgeport, Conn.

LOVERIDGE'S IMPROVED STEERING GEAR SPRING BUFFERS

(Pattern B). Art. 655 (Pattern A). Art. 657 (Pattern C). Art. 658 (Pattern D). Art. 659 (Pattern E). Art. 660 (Pattern F).
Also makers of Self-Oiling, Cargo, Heel, and Tackle Blocks.
Particulars and Prices on Application.
LOVERIDGE LIMITED, - - DOCKS, CARDIFF, WALES

and are quite smooth, as they are sheltered from the sea. Printed matter should be sent to the Chairman of the Madras, Port Trust, Madras, India, and also to the following: Secretary Coconada Chamber of Commerce, Coconada, South India; Secretary Pondicherry Chamber of Commerce, Pondicherry, French India, and to G. Zellweger, Secretary Tuticorin Chamber of Commerce, Tuticorin, South India. Tuticorin has considerable shipping business. Publications for South India should all be in English.

WANT MORE SHIPS AND FACILITIES

OTTAWA.—Statements by the Prime Minister, the Minister of Marine and Fisheries, Hon. C. C. Ballantyne, and the Minister of Militia, Hon. S. C. Mewburn, as to the probable transportation requirements during the coming season were made at the conference between members of the Government and representatives of transportation companies and harbor commissions recently.

It was suggested during the discussion that the three interests more particularly involved in the transportation of goods overseas should co-ordinate their efforts to increase the transportation. The three necessary factors to accomplish this, it was agreed, are:

1.—Ships.

2.—Improved terminal facilities at the various Canadian ports.

3.—Better railway facilities for the delivery of goods at the ports.

From statements made by the representatives of the various interests it would appear that a larger volume of traffic will pass through Canadian ports than was the case last year. It was decided that the Atlantic ports of Halifax and St. John would be used to a greater extent this summer than they have been in the past.

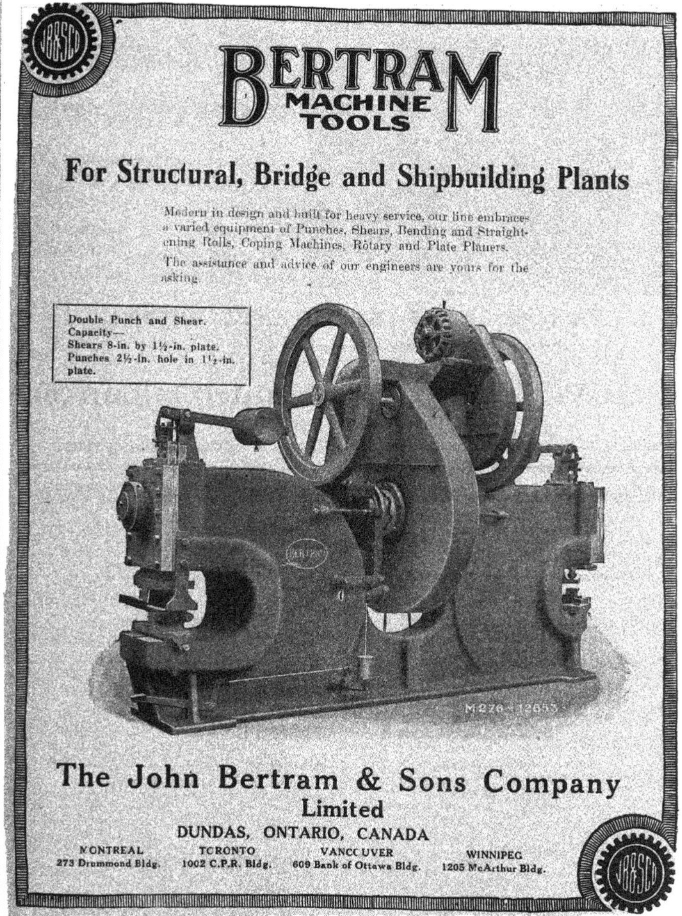
BERTRAM
MACHINE TOOLS
For Structural, Bridge and Shipbuilding Plants
Modern in design and built for heavy service, our line embraces a varied equipment of Punches, Shears, Bending and Straightening Rolls, Coping Machines, Rotary and Plate Planers.
The assistance and advice of our engineers are yours for the asking.
Double Punch and Shear. Capacity— Shears 8-in. by 1½-in. plate. Punches 2½-in. hole in 1½-in. plate.
The John Bertram & Sons Company Limited
DUNDAS, ONTARIO, CANADA
MONTREAL 273 Drummond Bldg.
TORONTO 1002 C.P.R. Bldg.
VANCOUVER 609 Bank of Ottawa Bldg.
WINNIPEG 1205 McArthur Bldg.

Stern Casting
for Ice Breaker
"JOHN D. HAZEN"
STEEL
CASTINGS
Ship Castings are a Specialty.
CANADIAN STEEL FOUNDRIES
LIMITED
General Offices:
Transportation Bldg.,
Montreal
Works:
Montreal and
Welland

Success and Your Associates

ALWAYS associate with men who are earning more money than you are. Such is the advice given by a man whose name is known the world over for his own success and the business which he created. It is good advice of a certainty, but we are not urging that it is the best advice. At the same time it is wise to associate with better or bigger men than you are yourself, because you are likely to be lifted up to their levels.

Most of us would probably like to be the intimates of J. P. Morgan, or of John D, or of Charlie Schwab, or Canny Andrew. If we had a speaking acquaintance with John Wanamaker, or Lord Beaverbrook, or Lord Shaughnessy, and with other distinguished and successful men, we would take comfort from the fact.

And most of us would feel rather small and uncomfortable if we were placed beside President Woodrow Wilson at a dinner table, or beside Sir Robt. Borden, Sir Herbert Holt or Thomas Findley. And the reason would probably be that we know so little of the knowledge that really counts. If these great men engaged us in conversation we would probably find ourselves knowing very little about the things that matter.

This will illustrate the point:

Here's a true story. A Toronto manufacturer found himself on a train going to Albany in company with a number of distinguished Americans about to attend the inaugural ceremonies of a State Governor. A washout led to a delay, and the Canadian was thrown into close association with senators and congressmen and prominent lawyers. They bantered the little Canadian about Canada. He had the pluck of a bigger man. He had been a constant reader of THE FINANCIAL POST, and had absorbed many facts about Canada. He surprised his American traveling companions with his positive, well-informed and wide knowledge of Canada. When the company reached Albany, the Canadian was persuaded to attend the ceremonies, and was introduced as the man who knew all about Canada. The Governor was interested in Canada, and said so, and the little Canadian found himself telling the story of Canada very ardently to a very attentive listener, much to the amusement of his friends who introduced him.

This man made a hit on that journey, and many friends. One of the company gave him a stock market tip worth a fortune.

The point of this story is: You, as a Canadian, can make yourself informed very fully and intelligently if you become a regular reader of THE FINANCIAL POST. You can make yourself a worthy companion and intimate of presidents and magnates. The world likes to listen to men who know something well, and who can talk interestingly and informingly on the subject of their study.

You can become worth listening to if you know your Canada well. And you can get the kind and amount of knowledge concerning Canada which will make you interesting to others if you will read THE FINANCIAL POST regularly each week.

IF you read THE FINANCIAL POST REGULARLY you will find yourself keeping company with the highest paid staff of editors engaged on any publication in Canada—trained men who know how to make others know what they know and learn.

You will find yourself living in a most interesting world—the great, throbbing world of business. You will have your thoughts tremendously stimulated and helpfully directed. You will find yourself becoming a fit table companion for big men — this because you will have knowledge of a quality that will keep you from shame.

What is it that keeps you and THE POST separated? It cannot be its subscription price of $3. Probably it is because you are not very well acquainted with this paper.

We are going to put it to you this way: If you have the desire to be worthy of association with big men, then prepare yourself for such association by reading THE POST, and to make acquaintance easy, we provide the coupon below. It offers you THE POST for four months for a dollar bill.

The MacLean Publishing Company, Limited,
143-153 University Avenue, Toronto.

Send me/us THE FINANCIAL POST for four months for One Dollar. Money enclosed / to be remitted

(Signed)

....

M.E. ....

If what you need is not advertised, consult our Buyers' Directory and write advertisers listed under proper heading.

LOW'S SPECIALITIES FOR SHIPS

We are Specialists in Appliances for

HEATING and VENTILATION

FITTINGS for PLUMBING WORK

Also all kinds of

BRASS and SHEET METAL WORK

REQUIRED IN THE CONSTRUCTION OF SHIPS.

We have supplied many well-kown vessels with all their requirements in the departments referred to.

Low's Gun-Metal Lift and Force Pump

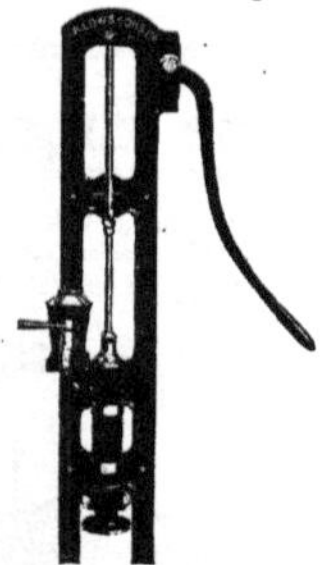

This Pump is made with latest improvements and is very substantially constructed.

Low's Patent Storm Valves

Fitted with Indicating Deck Plates. Approved by the Board of Trade.

New design to meet recent Board of Trade requirements. No sluice or other valve required. Minimum of space occupied.

ARCHIBALD LOW & SONS, LTD.

MERKLAND WORKS, PARTICK, GLASGOW

LIVERPOOL AGENTS: A. J. Nevill & Co., 9 Cook St.

N.E. COAST AGENTS: Ryder, Mumme & Co., Milburn House, Newcastle-on-Tyne.

LONDON OFFICE: 58 Fenchurch Street, E.C. 3.

BUYERS' DIRECTORY

ACCUMULATORS, HYDRAULIC
Smart-Turner Mach. Co., Hamilton, Ont.

AERATING RESERVOIRS
Spray Engineering Co., Boston, Mass.

ALLOYS, BRASS AND COPPER
Dominion Copper Products Co., Ltd., Montreal, Que.
Mueller Mfg. Co., H., Sarnia, Ont.

ANCHORS
Kearfott Engineering Co., New York, N.Y.
McNab Co., Bridgeport, Conn.
Henry Rogers Sons & Co., Wolverhampton, Eng.
Wm. C. Wilson & Co., Toronto, Ont.

ARCHITECT, NAVAL
Watts, J. Murray, Philadelphia, Pa.

ASBESTOS GOODS
Wm. C. Wilson & Co., Toronto, Ont.

BABBITT METAL
Aikenhead Hardware, Ltd., Toronto, Ont.
Can. B. K. Morton Co., Toronto, Montreal.
Hoyt Metal Company, Toronto, Ontario.
Wilkinson & Kompass, Hamilton, Ont.
Wm. C. Wilson & Co., Toronto, Ont.

BARGES
Can. B. K. Morton Co., Toronto, Montreal.
Polson Iron Works, Toronto, Ont.

BAROMETERS
Wilson & Co., Wm. C., Toronto, Canada.

BARS, GRATE
Babcock & Wilcox, Ltd., Montreal, Que.

BEARINGS, BRASS
Empire Mfg. Co., London, Ont.
Murray & Fraser, New Glasgow, N.S.
Mueller Mfg. Co., H., Sarnia, Ont.
United Brass & Lead Co., Toronto, Ont.

BELLS, SHIPS, ENGINE ROOM, ETC.
Aikenhead Hardware, Ltd., Toronto, Ont.
Cory & Son, Inc., Chas., New York, N.Y.
Empire Mfg. Co., London, Ont.
Morrison Brass Mfg. Co., James, Toronto, Ont.
Mueller Mfg. Co., H., Sarnia, Ont.

BELTING, LEATHER
Aikenhead Hardware, Ltd., Toronto, Ont.
Can. B. K. Morton Co., Toronto, Montreal.
Wm. C. Wilson & Co., Toronto, Ont.

BIBBS, COMPRESSION
Empire Mfg. Co., London, Ont.
Mueller Mfg. Co., H., Sarnia, Ont.

BINNACLES
Morrison Brass Mfg. Co., James, Toronto, Ont.

BLOCKS, CARGO, HEEL AND TACKLE
Aikenhead Hardware, Ltd., Toronto, Ont.
Loveridge, Ltd., Docks, Cardiff, Wales.

BLOWERS, TURBO-
Mason Regulator & Engin. Co., Montreal, Que.

BOAT CHOCKS
Corbet Fdry. & Machine Co., Owen Sound, Ont.
Marten-Freeman Co., Toronto, Ont.

BOILER COMPOUND
Wm. C. Wilson & Co., Toronto, Ont.

BOILER FEED PUMPS
Can. Ingersoll-Rand Co., Ltd., Sherbrooke, Que.
Smart-Turner Mach. Co., Hamilton, Ont.
Williams Machinery Co., A. R., Toronto, Ont.

BOILER FITTINGS
Empire Mfg. Co., London, Ont.
McAvity & Sons Ltd., T., St. John, N.B.

BOILERS, MARINE
Babcock & Wilcox, Ltd., Montreal, Que.
Can. Fairbanks-Morse Co., Montreal, Que.
Can. Vickers, Ltd., Montreal, Que.
Collingwood Shipbuilding Co., Collingwood, Ont.
Doxford & Sons, William, Sunderland, England.
Engineering & Machine Works of Canada, St. Catharines, Ont.
Goldie & McCulloch, Ltd., Galt, Ont.
Hall Engineering Works, Montreal, Que.
Mason Regulator & Engin. Co., Montreal, Que.
Montreal Dry Docks & Shipbuilding Co., Montreal, Que.
National Shipbuilding Co., Goderich, Ont.
The John McDougall Caledonian Iron Works Co., Montreal, Que.
Marsh & Henthorn, Ltd., Belleville, Ont.
Polson Iron Works, Toronto, Ontario.
Port Arthur Shipbuilding Co., Port Arthur, Ont.
Thor Iron Works, Toronto, Ont.
Williams Machinery Co., A. R., Toronto, Ont.

BOOKS, MILITARY AND NAVAL
Van Nostrand Co., D., 25 Park Place, New York.

BOOKS, TECHNICAL, MARINE
MacLean Publishing Co., Toronto, Ont.
Van Nostrand Co., D., 25 Park Place, New York.

BOLTS
Wilkinson & Kompass, Hamilton, Ont.

BROKER
Page & Jones, Mobile, Ala., U.S.A.

BRASS GOODS
Corbet Fdry. & Machine Co., Owen Sound, Ont.
McAvity & Sons, T., St. John, N.B.
Mueller Mfg. Co., H., Sarnia, Ont.

BUCKETS, CLAMSHELL
Beatty & Sons, Welland, Ont.

BUCKETS, COALING
Beatty & Sons, Welland, Ont.
Engineering & Machine Works of Canada, St. Catharines, Ont.
Marsh & Henthorn, Ltd., Belleville, Ont.
Reed & Co., Geo., Montreal, Que.

CABLE
McNab Co., Bridgeport, Conn.
Wm. C. Wilson & Co., Toronto, Ont.

CABLE, LEAD COVERED AND ARMORED
Standard Underground Cable Co., Hamilton, Ont.

CABLE, ACCESSORIES
Standard Underground Cable Co., Hamilton, Ont.

CAPSTANS
Advance Mach. & Welding Co., Montreal, Que.
Dake Engine Co., Grand Haven, Mich.
Kennedy & Sons, Wm., Owen Sound, Ont.
Wm. C. Wilson & Co., Toronto, Ont.

CALKING TOOLS, ELECTRIC
Aikenhead Hardware, Ltd., Toronto, Ont.

CALKING TOOLS, PNEUMATIC
Aikenhead Hardware, Ltd., Toronto, Ont.
Can. Ingersoll-Rand Co. Sherbrooke, Que.

CALORIFIERS
Low & Sons, Ltd., Archibald, Glasgow, Scotland

CASTINGS
Can. Steel Foundries, Ltd., Montreal, Que.
Collingwood Shipbuilding Co., Collingwood, Ont.
Kennedy & Sons, Wm., Owen Sound, Ont.
Marsh & Henthorn, Ltd., Belleville, Ont.
The John McDougall Caledonian Iron Works Co., Montreal, Que.
Mueller Mfg. Co., H., Sarnia, Ont.

CASTINGS, ALLOY
Mueller Mfg. Co., H., Sarnia, Ont.
Tolland Mfg. Co., Montreal, Que.

CASTINGS, ALUMINUM
Empire Mfg. Co., London, Ont.
United Brass & Lead Co., Toronto, Ont.

CASTINGS, BRASS
Crouse-Hinds Co. of Canada, Ltd., Toronto, Ont.
Empire Mfg. Co., London, Ont.
McAvity & Sons Ltd., T., St. John, N.B.
McNab Co., Bridgeport, Conn.
Mueller Mfg. Co., H., Sarnia, Ont.
United Brass & Lead Co., Toronto, Ont.
Tolland Mfg. Co., Montreal, Que.

CASTINGS, GREY IRON, MALLEABLE, ALUMINUM
Crouse-Hinds Co. of Canada, Ltd., Toronto, Ont.
Engineering & Machine Works of Canada, St. Catharines, Ont.
McAvity & Sons, T., St. John, N.B.
McNab Co., Bridgeport, Conn.
Mueller Mfg. Co., H., Sarnia, Ont.

CASTINGS, ELECTRIC
Crouse-Hinds Co. of Canada, Ltd., Toronto, Ont.

CELLAR DRAINERS
Mueller Mfg. Co., H., Sarnia, Ont.

CHAINS
Aikenhead Hardware, Ltd., Toronto, Ont.
Kearfott Engineering Co., New York, N.Y.
Henry Rogers, Sons & Co., Wolverhampton, Eng.
Wm. C. Wilson & Co., Toronto, Ont.

CHAIN BLOCKS
Morris Crane & Hoist Co., Herbert, Niagara Falls, Ont.

CHANDLERY, SHIP
Leckie, Ltd., John, Toronto, Ont.

CLAMPS
Topping Brothers, New York, N.Y.

CLAMPS, STEAM AND WATER
Mueller Mfg. Co., H., Sarnia, Ont.

CLOCKS
American Steam Gauge & Valve Mfg. Co. Boston, Mass.
Morrison Brass Mfg. Co., James, Toronto, Ont.
Williams Machinery Co., A. R., Toronto, Ont.

CLOSETS
Mueller Mfg. Co., H., Sarnia, Ont.

COAL
Nova Scotia Steel & Coal Co., New Glasgow, N.S.

COCKS, BILGE, DISCHARGE, INDICATOR, ETC.
McAvity & Sons Ltd., T., St. John, N.B.
Morrison Brass Mfg. Co., James, Toronto, Ont.

COCKS, BASIN
Empire Mfg. Co., London, Ont.

COMPASSES
Scythes & Co., Toronto, Ont.
Wm. C. Wilson & Co., Toronto, Ont.

COMPRESSORS, AIR
Can. Fairbanks-Morse Co., Montreal, Que.
Canadian Ingersoll-Rand Co., Sherbrooke, Que.
Smart-Turner Mach. Co., Hamilton, Ont.
Storey Pump & Equipment Co., Toronto, Ont.
Williams Machinery Co., A. R., Toronto, Ont.

COMPRESSED AIR TANKS
Scaife & Sons Co., Wm. B., Oakmont, Pa.

CONCRETE MIXERS
St. Clair Bros., Galt, Ontario.
The John McDougall Caledonian Iron Works Co., Montreal, Que.

CONDENSERS
Kearfott Engineering Co., New York, N.Y.
Morris Machine Works, Baldwinsville, N.Y.
Smart-Turner Mach. Co., Hamilton, Ont.
Storey Pump & Equipment Co., Toronto, Ont.
Weir Ltd., G. & J., Cathcart, Glasgow, Scotland.
Williams Machinery Co., A. R., Toronto, Ont.

CONDULETS, MARINE
Crouse-Hinds Co. of Canada, Ltd., Toronto, Ont.

CONTRACTORS' SUPPLIES
McAvity & Sons, T., St. John, N.B.

CONSULTING ENGINEER
Watt, J. Murray, Philadelphia, Pa.

CONVEYORS, ASH, COAL
Babcock & Wilcox, Ltd., Montreal, Que.

COPING MACHINES
Bertram & Sons, Ltd., John, Dundas, Ont.

COPPER TUBES, SHEETS AND RODS
Dominion Copper Products Co., Ltd., Montreal, Que.

CORDAGE
Scythes & Co., Toronto, Ont.

COTTON, CALKING
Wilson & Co., Wm. C., Toronto, Canada.

COTTON DUCK
Scythes & Co., Toronto, Ont.

COVERS, CANVAS, FOR HATCHES, LIFE-BOATS, ETC.
Leckie, Ltd., John, Toronto, Ont.

COUNTERS, REVOLUTION
American Steam Gauge & Valve Mfg. Co., Boston, Mass.

CRANES
Aikenhead Hardware, Ltd., Toronto, Ont.
Can. Fairbanks-Morse Co., Montreal, Que.
Williams Machinery Co., A. R., Toronto, Ont.

CRANES, ELECTRIC
Babcock & Wilcox, Ltd., Montreal, Que.
Morris Crane & Hoist Co., Herbert, Niagara Falls, Ont.
Smart-Turner Mach. Co., Ha[illegible]

CRANES, GANTRY
Smart-Turner Mach. Co., Hamilton, Ont.

CRANES, JIB
Morris Crane & Hoist Co., Herbert, Niagara Falls, Ont.

CRANES, OVERHEAD TRAVELLING
Morris Crane & Hoist Co., Herbert, Niagara Falls, Ont.

CRANK SHAFTS
Canada Foundries and Forgings, Welland, Ont.

DAVITS, BOAT
Babcock & Wilcox, Ltd., Montreal, Que.
Corbet Fdry. & Machine Co., Owen Sound, Ont.
Marten, Freeman Co., Toronto, Ont.

DECK PLUGS, ELECTRIC
Crouse-Hinds Co. of Canada, Ltd., Toronto, Ont.

DERRICKS
Aikenhead Hardware, Ltd., Toronto, Ont.
Dake Engine Co., Grand Haven, Mich.
Marsh & Henthorn, Ltd., Belleville, Ont.

DISTILLERS
Kearfott Engineering Co., New York, N.Y.

DISTRESS OUTFITS
Scythes & Co., Toronto, Ont.

DREDGES
Collingwood Shipbuilding Co., Collingwood, Ont.
Norbom Engineering Co., Philadelphia, Pa.
Polson Iron Works, Toronto.

DRILLS, AIR
Aikenhead Hardware, Ltd., Toronto, Ont.
Can. Ingersoll-Rand Co. Sherbrooke, Que.

DRILLS, CENTRE
Wilt Twist Drill Co., Walkerville, Ont.

DRILLS, BLACKSMITH AND BIT STOCK
Wilt Twist Drill Co., Walkerville, Ont.

DRILLS, ELECTRIC
Aikenhead Hardware, Ltd., Toronto, Ont.
Wilkinson & Kompass, Hamilton, Ont.

DRILLS, HIGH SPEED
Wilt Twist Drill Co., Walkerville, Ont.

DRILLS, TRACK
Wilt Twist Drill Co., Walkerville, Ont.

DRILLS, RACHET AND HAND
Wilt Twist Drill Co., Walkerville, Ont.

DRILLS, TWIST
Aikenhead Hardware, Ltd., Toronto, Ont.
Can. B. K. Morton Co., Toronto, Montreal.
Williams Machinery Co., A. R., Toronto, Ont.
Wilt Twist Drill Co., Walkerville, Ont.

DRY DOCKS
Can. Vickers, Ltd., Montreal, Que.
Collingwood Shipbuilding Co., Collingwood, Ont.
Doxford & Sons, William, Sunderland, England.
Georgian Bay Shipbuilding & Wrecking Co., Midland, Ont.
Montreal Dry Docks & Shipbuilding Co., Montreal.
National Shipbuilding Co., Goderich, Ont.
Polson Iron Works, Toronto, Ont.
Port Arthur Shipbuilding Co., Port Arthur, Ont.
Thor Iron Works, Toronto, Ont.
Yarrows, Limited, Victoria, B.C.

ECONOMIZERS, FUEL
Babcock & Wilcox, Ltd., Montreal, Que.

EJECTORS
Empire Mfg. Co., London, Ont.
Morrison Brass Mfg. Co., James, Toronto, Ont.
Smart-Turner Mach. Co., Hamilton, Ont.

ELECTRIC LAMPS
- Wm. C. Wilson & Co., Toronto, Ont.

ELECTRO-PLATING
- United Brass & Lead Co., Toronto, Ont.

ELEVATING MACHINERY
- Goldie & McCulloch, Ltd., Galt, Ont.

ENGINES, HOISTING
- Advance Mach. & Welding Co., Montreal, Que.
- Corbet Fdry. & Machine Co., Owen Sound, Ont.
- Kennedy & Sons, Wm., Owen Sound, Ont.
- Marsh & Henthorn, Ltd., Belleville, Ont.
- Port Arthur Shipbuilding Co., Port Arthur, Ont.
- Williams Machinery Co., A. R., Toronto, Ont.

ENGINE, INTERNAL COMBUSTION
- Doxford & Sons, William, Sunderland, England.

ENGINES, MARINE
- Bolinders Co., New York, N.Y.
- Can. Fairbanks-Morse Co., Montreal, Que.
- Can. Vickers, Ltd., Montreal, Que.
- Doxford & Sons, William, Sunderland, England.
- Goldie & McCulloch, Ltd., Galt, Ont.
- Iron Works, Ltd., Owen Sound, Ont.
- Mason Regulator & Engin. Co., Montreal, Que.
- The John McDougall Caledonian Iron Works Co., Montreal, Que.
- Montreal Dry Docks & Shipbuilding Co., Montreal, Que.
- Murray & Fraser, New Glasgow, N.S.
- National Shipbuilding Co., Goderich, Ont.
- Norbom Engineering Co., Philadelphia, Pa.
- Polson Iron Works, Toronto, Ont.
- Port Arthur Shipbuilding Co., Port Arthur, Ont
- Trout Co., H. G., Buffalo, N.Y.
- Williams Machinery Co., A. R., Toronto, Ont.

ENGINE STARTERS (AIR)
- Scaife & Sons Co., Wm. B., Oakmont, Pa.

ENGINES, STEERING
- Corbet Fdry. & Machine Co., Owen Sound, Ont.
- Dake Engine Co., Grand Haven, Mich.
- Kennedy & Sons, Wm., Owen Sound, Ont.
- Wm. C. Wilson & Co., Toronto, Ont.

ENAMELWARE
- Mueller Mfg. Co., H., Sarnia, Ont.

EVAPORATORS
- Kearfott Engineering Co., New York, N.Y.
- Mason Regulator & Engin. Co., Montreal, Que.
- McNab Co., Bridgeport, Conn.
- Weir Ltd., G. & J., Cathcart, Glasgow, Scotland.

EXTRACTORS, GREASE
- American Steam Gauge & Valve Mfg. Co. Boston, Mass.

EYE BOLTS AND NUTS
- Canada Foundries and Forgings, Welland, Ont.

FANS
- Aikenhead Hardware Ltd., Toronto, Ont.
- Empire Mfg. Co., London, Ont.
- Reed & Co., Geo. W., Montreal, Que.
- Smart-Turner Mach. Co., Hamilton, Ont.
- Williams Machinery Co., A. R., Toronto, Ont.

FENDERS, ROPE
- Leckie, Ltd., John, Toronto, Ont.
- Wilson & Co., Wm. C., Toronto, Canada.

FERRO-MANGANESE
- Mitchells, Ltd., Glasgow, Scotland.

FILES
- Aikenhead Hardware, Ltd., Toronto, Ont.
- Can. B. K. Morton Co., Toronto, Montreal.
- Williams Machinery Co., A. R., Toronto, Ont.

FIRE BRICKS
- Mitchells, Ltd., Glasgow, Scotland.
- Williams Machinery Co., A. R., Toronto, Ont.

FILTERS, FEED WATER
- Mason Regulator & Engin. Co., Montreal, Que.

FITTINGS, MARINE
- McAvity & Sons, T., St. John, N.B.

FITTINGS, MOTOR BOAT
- Empire Mfg. Co., London, Ont.
- Mueller Mfg. Co., H., Sarnia, Ont.
- Murray & Fraser, New Glasgow, N.S.

FLAGS
- Scythes & Co., Toronto, Ont.

FLOW METERS
- Spray Engineering Co., Boston, Mass.

FLUE CLEANERS
- Wm. C. Wilson & Co., Toronto, Ont.

FIXTURES, ELECTRIC
- Cory & Son, Inc., Chas., New York, N.Y.
- Crouse-Hinds Co. of Canada, Ltd., Toronto, Ont.

FORGES
- Aikenhead Hardware, Ltd., Toronto, Ont.

FLOODLIGHTS, ELECTRIC
- Crouse-Hinds Co. of Canada, Ltd., Toronto, Ont.

FLUORSPAR
- Mitchells, Ltd., Glasgow, Scotland.
- Scythes & Co., Toronto, Ont.

FORGINGS, ALL KINDS
- Aikenhead Hardware, Ltd., Toronto, Ont.
- Collingwood Shipbuilding Co., Collingwood, Ont.
- The John McDougall Caledonian Iron Works Co., Montreal, Que.
- Nova Scotia Steel & Coal Co., New Glasgow, N.S.
- St. Clair Bros., Galt, Ont.

FORGINGS, STEEL AND IRON
- Canada Foundries and Forgings, Welland, Ont.

GAUGES, RECORDING
- American Steam Gauge & Valve Mfg. Co., Boston, Mass.
- Empire Mfg. Co., London, Ont.

GASKETS
- Wm. C. Wilson & Co., Toronto, Ont.

GAUGES COCKS
- McAvity & Sons, T., St. John, N.B.

GAUGE GLASSES
- Wm. C. Wilson & Co., Toronto, Ont.

GAUGES, WATER
- McAvity & Sons, T., St. John, N.B.
- McNab Co., Bridgeport, Conn.

GAUGES, WATER, PRESSURE, COMPOUND AND VACUUM
- Aikenhead Hardware, Ltd., Toronto, Ont.
- Babcock & Wilcox, Ltd., Montreal, Que.
- Empire Mfg. Co., London, Ont.
- Morrison Brass Mfg. Co., James, Toronto, Ont.

GENERATORS AND CONVERTORS
- Can. Fairbanks-Morse Co., Montreal, Que.

GRAPHITE
- Wm. C. Wilson & Co., Toronto, Ont.

GRATINGS
- Corbet Fdry. & Machine Co., Owen Sound, Ont.
- Canada Wire & Iron Goods Co., Hamilton, Ont.

GRINDERS, ELECTRIC
- Wilkinson & Kompass, Hamilton, Ont.

GRINDERS, PNEUMATIC
- Can. Ingersoll-Rand Co. Sherbrooke, Que.

GUY RODS AND ANCHORS, ELECTRIC
- Crouse-Hinds Co. of Canada, Ltd., Toronto, Ont.

HAMMERS
- Canada Foundries and Forgings, Ltd., Welland, Ont.

HARDWARE, MARINE
- Scythes & Co., Toronto, Ont.

HEADLIGHTS, ELECTRIC
- Crouse-Hinds Co. of Canada, Ltd., Toronto, Ont.

HEATING EQUIPMENT
- Empire Mfg. Co., London, Ont.
- Low & Sons, Ltd., Archibald, Glasgow, Scotland

HEATERS, FEED, WATER
- Babcock & Wilcox, Ltd., Montreal, Que.
- Kearfott Engineering Co., New York, N.Y.
- Mason Regulator & Engin. Co., Montreal, Que.
- McNab Co., Bridgeport, Conn.
- Weir Ltd., G. & J., Cathcart, Glasgow, Scotland.

HOOKS
- Morris Crane & Hoist Co., Herbert, Niagara Falls, Ont.

HOISTS, ASH
- Beatty & Sons, Welland, Ont.
- St. Clair Bros., Galt, Ont.

HOIST BLOCKS
- Morris Crane & Hoist Co., Herbert, Niagara Falls, Ont.

HOISTS, CHAIN
- Aikenhead Hardware, Ltd., Toronto, Ont.
- Can. Fairbanks-Morse Co., Montreal, Que.
- Dake Engine Co., Grand Haven, Mich.
- Morris Crane & Hoist Co., Herbert, Niagara Falls, Ont.
- Williams Machinery Co., A. R., Toronto, Ont.

HOISTS, CARGO, MOVING, ETC.
- Dake Engine Co., Grand Haven, Mich.

HOISTING MACHINERY
- Beatty & Sons, Welland, Ont.
- Can. Ingersoll-Rand Co. Sherbrooke, Que.
- Corbet Fdry. & Machine Co., Owen Sound, Ont.
- Marsh & Henthorn, Ltd., Belleville, Ont.
- Williams Machinery Co., A. R., Toronto, Ont.

HOSE
- Wm. C. Wilson & Co., Toronto, Ont.

HYDRAULIC MACHINERY
- The John McDougall Caledonian Iron Works Co., Montreal, Que.

INDICATORS, ENGINE
- American Steam Gauge & Valve Mfg. Co., Boston, Mass.
- Cory & Son, Inc., Chas., New York, N.Y.
- McNab Co., Bridgeport, Conn.

INDICATORS, SPEED
- Aikenhead Hardware, Ltd., Toronto, Ont.
- Cory & Son, Inc., Chas., New York, N.Y.

INJECTORS
- Aikenhead Hardware, Ltd., Toronto, Ont.
- Empire Mfg. Co., London, Ont.
- Morrison Brass Mfg. Co., James, Toronto, Ont.
- Williams Machinery Co., A. R., Toronto, Ont.

INGOTS
- Broughton Copper Co., Ltd., Manchester, Eng.

INSULATORS, ELECTRIC
- Crouse-Hinds Co. of Canada, Ltd., Toronto, Ont.

INSTRUMENTS, NAUTICAL
- Leckie, Ltd., John, Toronto, Ont.

JACKS
- Morris Crane & Hoist Co., Herbert, Niagara Falls, Ont.

JOINTS, BALL SLIP
- Norbom Engineering Co., Philadelphia, Pa.

INSURANCE, MARINE
- Toronto Insurance & Vessel Agency, Toronto, Ont.

JOURNAL WEDGES
- Canada Foundries and Forgings, Ltd., Welland, Ont.

LATHES
- Foss & Hill Machinery Co., Montreal, Que.
- Bertram & Sons, Ltd., John, Dundas, Ont.

LEATHER, LACE
- Wm. C. Wilson & Co., Toronto, Ont.

LIFEBUOYS, BELTS AND PRESERVERS
- Fosbery Co., Barking, Essex, Eng.
- Wm. C. Wilson & Co., Toronto, Ont.

LIFEBOATS
- Wm. C. Wilson & Co., Toronto, Ont.

LIFE BOAT EQUIPMENT
- Leckie, Ltd., John, Toronto, Ont.
- Wm. C. Wilson & Co., Toronto, Ont.

LIFE JACKETS
- Fosbery Co., Barking, Essex, Eng.
- Leckie, Ltd., John, Toronto, Ont.

LIGHTS, ALL KINDS
- Can. Fairbanks-Morse Co., Montreal, Que.
- Crouse-Hinds Co. of Canada, Ltd., Toronto, Ont.
- Morrison Brass Mfg. Co., James, Toronto, Ont.

LIGHTS, SIDE, PORT
- Crouse-Hinds Co. of Canada, Ltd., Toronto, Ont.
- McAvity & Sons Ltd., T., St. John, N.B.

LIGHTING SETS
- Can. Fairbanks-Morse Co., Montreal.
- Cory & Son, Inc., Chas., New York, N.Y.
- Crouse-Hinds Co. of Canada, Ltd., Toronto, Ont.
- Kearfott Engineering Co., New York, N.Y.
- Williams Machinery Co., A. R., Toronto, Ont.
- Wilson & Co., Wm. C., Toronto, Canada.

LOCKERS, METAL
- Canada Wire & Iron Goods Co., Hamilton, Ont.
- Reed & Co., Geo. W., Montreal, Que.

LUBRICATORS
- Empire Mfg. Co., London, Ont.

MACHINISTS
- Corbet Fdry. & Machine Co., Owen Sound, Ont.
- Hyde Engineering Works, Montreal, Que.

MACHINE AND FORGE WORK
- Thor Iron Works, Toronto, Ont.

MANGANESITE, PASTE
- Wm. C. Wilson & Co., Toronto, Ont.

MARINE CASTINGS
- McAvity & Sons, T., St. John, N.B.
- Mueller Mfg. Co., H., Sarnia, Ont.

MARINE SPECIALTIES
- McAvity & Sons, T., St. John, N.B.

MOTORS
- Aikenhead Hardware, Ltd., Toronto, Ont.
- Can. Fairbanks-Morse Co., Montreal, Que.
- Williams Machinery Co., A. R., Toronto, Ont.

NAUTICAL INSTRUMENTS
- Scythes & Co., Toronto, Ont.
- Wm. C. Wilson & Co., Toronto, Ont.

NET LIFTERS
- Dake Engine Co., Grand Haven, Mich.
- Leckie, Ltd., John, Toronto, Ont.

NOZZLES, ALL KINDS
- Empire Mfg. Co., London, Ont.
- Tolland Mfg. Co., Boston, Mass.

NUTS
- Wilkinson & Kompass, Hamilton, Ont.

OAKUM
- Scythes & Co., Toronto, Ont.
- Wm. C. Wilson & Co., Toronto, Ont.

OIL CUPS
- Empire Mfg. Co., London, Ont.
- Murray & Fraser, New Glasgow, N.S.

OILCLOTHING
- Wm. C. Wilson & Co., Toronto, Ont.

OILS AND GREASE
- Wm. C. Wilson & Co., Toronto, Ont.

OAKUM
- Leckie, Ltd., John, Toronto, Ont.

ORNAMENTAL IRONWORK
- Canada Wire & Iron Goods Co., Hamilton, Ont.

OXY-ACETLYENE OUTFITS
- Can. Fairbanks-Morse Co., Montreal, Que.
- McAvity & Sons Ltd., T., St. John, N.B.
- Williams Machinery Co., A. R., Toronto, Ont.

PAINT
- Ault & Wiborg Co. of Can., Ltd., Toronto, Ont.
- Wm. C. Wilson & Co., Toronto, Ont.

PAINTING EQUIPMENT, AUTOMATIC
- Spray Engineering Co., Boston, Mass.

PACKING, AMMONIA
- Can. B. K. Morton Co., Montreal.
- France Packing Co., Philadelphia, Pa.
- Wm. C. Wilson & Co., Toronto, Ont.

PACKING, MARINE
- Aikenhead Hardware, Ltd., Toronto, Ont.
- Can. B. K. Morton Co., Montreal.
- France Packing Co., Philadelphia, Pa.
- Wm. C. Wilson & Co., Toronto, Ont.

PACKING, METALLIC
- Aikenhead Hardware, Ltd., Toronto, Ont.
- Can. B. K. Morton Co., Montreal.
- France Packing Co., Philadelphia, Pa.

PACKING, STEAM
- Aikenhead Hardware, Ltd., Toronto, Ont.
- France Packing Co., Philadelphia, Pa.
- Wm. C. Wilson & Co., Toronto, Ont.

PANELBOARD AND CABINETS, ELECTRIC
- Crouse-Hinds Co. of Canada, Ltd., Toronto, Ont
- Preston Woodworking Machy. Co., Preston, Ont.

PIPE, LAPWELD, CAST IRON, RIVETED
- Empire Mfg. Co., London, Ont.
- Norbom Engineering Co., Philadelphia, Pa.

PIPE JOINT CEMENT
- Wm. C. Wilson & Co., Toronto, Ont.

PITCH
- Reed & Co., Geo, Montreal, Que.
- Scythes & Co., Toronto, Ont.

PITCH, PINE
- Wilson & Co., Wm. C., Toronto, Canada.

PLANERS
- Bertram & Sons, Ltd. John, Dundas, Ont.

PLANERS, STANDARD AND ROTARY
- Yates Mach. Co., P. B., Hamilton, Ont.

PLATE CLAMPS
- Morris Crane & Hoist Co., Herbert, Niagara Falls, Ont.

PLATE PUNCH TABLES
- Norbom Engineering Co., Philadelphia, Pa.

PLUMBING EQUIPMENT
- Low & Sons, Ltd., Archibald, Glasgow, Scotland.
- Empire Mfg. Co., London, Ont.
- Mueller Mfg. Co., H., Sarnia, Ont.
- Wm. C. Wilson & Co., Toronto, Ont.

PROPELLOR BLADES, BRONZE
- Corbet Fdry. & Machine Co., Owen Sound, Ont.
- Empire Mfg. Co., London, Ont.
- Murray & Fraser, New Glasgow, N.S.
- Yarrows, Limited, Victoria, B.C.

PROPELLER WHEELS
- Kennedy & Sons, Wm., Owen Sound, Ont.
- Trout Co., H. G., Buffalo, N.Y.

PIPING, STEAM
- Babcock & Wilcox, Ltd., Montreal, Que.

PORT LIGHTS
- Mueller Mfg. Co., H., Sarnia, Ont.
- Wm. C. Wilson & Co., Toronto, Ont.

PROPELLER WHEELS
- McNab Co., Bridgeport, Conn.

PULLEYS
- Smart-Turner Mach. Co., Hamilton, Ont.
- Williams Machinery Co., A. R., Toronto, Ont.

PUMP, AIR
- Can. Ingersoll-Rand Co. Sherbrooke, Que.
- Smart-Turner Mach. Co., Hamilton, Ont.
- Weir Ltd., G. & J., Cathcart, Glasgow, Scotland.
- Williams Machinery Co., A. R., Toronto, Ont.

PUMPS
Can. Fairbanks-Morse Co., Montreal, Que.
Canada Foundries & Forgings, Welland, Ont.
Goldie & McCulloch, Ltd., Galt, Ont.
McAvity & Sons, T., St. John, N.B.
Williams Machinery Co., A. R., Toronto Ont.

PUMPS, CENTRIFUGAL
Can. Ingersoll-Rand Co. Sherbrooke, Que.
Kearfott Engineering Co., New York, N.Y.
The John McDougall Caledonian Iron Works Co, Montreal, Que.
Norbom Engineering Co., Philadelphia, Pa.
Smart-Turner Mach. Co., Hamilton, Ont.
Williams Machinery Co., A. R., Toronto, Ont.

PUMPS, BILGE
Mueller Mfg. Co., H., Sarnia, Ont.

PUMPS, CIRCULATING
McNab Co., Bridgeport, Conn.
Morris Machine Works, Baldwinsville, N.Y.

PUMPS, FEED WATER
McNab Co., Bridgeport, Conn.
Weir Ltd., G. & J., Cathcart, Glasgow, Scotland.
Williams Machinery Co., A. R., Toronto, Ont.

PUMPS, HAND AND POWER
Aikenhead Hardware, Ltd., Toronto, Ont.
Smart-Turner Mach. Co., Hamilton, Ont.
Williams Machinery Co., A. R., Toronto, Ont.

PUMPS, HIGH PRESSURE
Canadian Ingersoll-Rand Co., Sherbrooke, Que.
Smart-Turner Mach. Co., Hamilton, Ont.

PUMPS, STEAM TURBO
Wm. C. Wilson & Co., Toronto, Ont.

PUMPING MACHINES
Norbom Engineering Co., Philadelphia, Pa.

PUNCHES
Bertram & Sons, Ltd., John, Dundas, Ont.

PUNCHES, SINGLE, DOUBLE AND MULTIPLE
Norbom Engineering Co., Philadelphia, Pa.

PURIFIERS, WATER
Babcock & Wilcox, Ltd., Montreal, Que.

RADIO ENGINEERS
Cutting & Washington, Inc., Cambridge, Mass.

RADIATORS, STEAM, ELECTRIC
Empire Mfg. Co., London, Ont.
Low & Sons, Ltd., Archibald, Glasgow, Scotland.
Wm. C. Wilson & Co., Toronto, Ont.

RANGES
Wm. C. Wilson & Co., Toronto, Ont.

RECEIVERS, AIR
Can. Ingersoll-Rand Co. Sherbrooke, Que.

REGULATORS, FEED WATER
American Steam Gauge & Valve Mfg. Co., Boston, Mass.

REGULATORS, PRESSURE
Mason Regulator & Engin. Co., Montreal, Que.

REPAIRS, MARINE
Corbet Foundry & Mach. Co., Owen Sound, Ont.
Can. Vickers, Ltd., Montreal, Que.
Collingwood Shipbuilding Co., Collingwood, Ont.
Engineering & Machine Works of Canada, St. Catharines, Ont.
Georgian Bay Shipbuilding & Wrecking Co., Midland, Ont.
Hyde Engineering Works, Montreal, Que.
Iron Works, Ltd., Owen Sound, Ont.
Kennedy & Sons, Wm., Owen Sound, Ont.
Montreal Dry Docks & Shipbuilding Co., Montreal Que.
Muir Bros. Dry Dock Co., Port Dalhousie, Ont.
National Shipbuilding Co., Goderich, Ont.
Port Arthur Shipbuilding Co., Port Arthur, Ont.
Yarrow, Limited, Victoria, B.C.

RIGGING, WIRE ROPE
Leckie, Ltd., John, Toronto, Ont.

RIVETS
Severance Mfg. Co., S., Glassport, Pa.
Wilkinson & Kompass, Hamilton, Ont.

RIVETERS, PNEUMATIC
Can. Ingersoll-Rand Co., Ltd., Sherbrooke, Que.

RODS, COPPER, BRASS, BRONZE
Standard Underground Cable Co., Hamilton, Ont.

ROLLS, STRAIGHTENING, BENDING
Bertram & Sons, Ltd., John, Dundas, Ont.

ROOFING
Reed & Co., Geo., Montreal, Que.
Wm. C. Wilson & Co., Toronto, Ont.

ROPE BLOCKS
Aikenhead Hardware, Ltd., Toronto, Ont.
Can. Fairbanks-Morse Co., Montreal, Que.
Morris Crane & Hoist Co., Herbert, Niagara Falls, Ont.
Thor Iron Works, Toronto, Ont.

ROPE
Leckie, Ltd., John, Toronto, Ont.
McNab Co., Bridgeport, Conn.
Wm. C. Wilson & Co., Toronto, Ont.

RUBBER COATS
Wm. C. Wilson & Co., Toronto, Ont.

SAW MILL MACHINERY
Preston Woodworking Machy. Co., Preston, Ont.
Yates Machine Co., P. B., Hamilton, Ont.

SAWS, BAND
Preston Woodworking Machy. Co., Preston, Ont.

SCALES, BOILERS, ENGINES
Can. Fairbanks-Morse Co., Montreal, Que.

SCOWS
Collingwood Shipbuilding Co., Collingwood, Ont.
Polson Iron Works, Toronto, Ont.

SEPARATORS, OIL, STEAM
Mason Regulator & Engin. Co., Montreal, Que.
Smart-Turner Mach. Co., Hamilton, Ont.

SHAFTING
Wilkinson & Kompass, Hamilton, Ont.

SHAFTING, BRONZE
Empire Mfg. Co., London, Ont.
Murray & Fraser, New Glasgow, N.B.

SHEARS
Bertram & Sons, Ltd., John, Dundas, Ont.
Norbom Engineering Co., Philadelphia, Pa.

SHIPBUILDING TOOLS
Aikenhead Hardware, Ltd., Toronto, Ont.
Can. Ingersoll-Rand Co. Sherbrooke, Que.

SHIPS, BUILDERS OF
Can. Vickers, Ltd., Montreal, Que.
Collingwood Shipbuilding Co., Collingwood, Ont.
Doxford & Sons, William, Sunderland, England.
Georgian Bay Shipbuilding & Wrecking Co., Midland, Ont.
Montreal Dry Docks & Shipbuilding Co., Montreal, Que.
National Shipbuilding Co., Goderich, Ont.
Polson Iron Works, Toronto, Ont.
Port Arthur Shipbuilding Co., Port Arthur, Ont.
Thor Iron Works, Toronto, Ont.
Yarrows, Limited, Victoria, B.C.

SHIP BROKERS
Page & Jones, Mobile, Ala.

SHIP PLATES
Nova Scotia Steel & Coal Co., New Glasgow, N.S.
Topping Brothers, New York, N.Y.

SIDE LIGHTS
Wm. C. Wilson & Co., Toronto, Ont.

SLEDGES
Wilkinson & Kompass, Hamilton, Ont.

SLINGS
Morris Crane & Hoist Co., Herbert, Niagara Falls, Ont.

SPECIAL MACHINERY
Corbet Fdry. & Machine Co., Owen Sound, Ont.
Smart-Turner Mach. Co., Hamilton, Ont.

SMOOTH-ON
Wm. C. Wilson & Co., Toronto, Ont.

SPIKES
Wm. C. Wilson & Co., Toronto, Ont.

SPRAY COOLING SYSTEMS
Spray Engineering Co., Boston, Mass.

STANDPIPES
Jenckes Machine Co., Ltd., Sherbrooke, Que.

STEAMSHIP AGENTS
Page & Jones, Mobile, Ala.

STEAM SPECIALTIES
Can. Fairbanks-Morse Co., Montreal, Que.
Corbet Fdry. & Machine Co., Owen Sound, Ont.
Empire Mfg. Co., London, Ont.

STEAM TRAPS
Aikenhead Hardware, Ltd., Toronto, Ont.
American Steam Gauge & Valve Mfg. Co., Boston, Mass.
Empire Mfg. Co., London, Ont.
Mason Regulator & Engin. Co., Montreal, Que.
Smart-Turner Mach. Co., Hamilton, Ont.

STEEL, HIGH SPEED
Can. B. K. Morton Co., Toronto, Montreal.
Nova Scotia Steel & Coal Co., New Glasgow, N.S.
Wilkinson & Kompass, Hamilton, Ont.

STEEL WORK, STRUCTURAL
Babcock & Wilcox, Ltd., Montreal, Que.
Corbet Fdry. & Machine Co., Owen Sound, Ont.

STEERING GEARS
Corbet Fdry. & Machine Co., Owen Sound, Ont.
Engineering & Machine Works of Canada, St. Catharines, Ont.
Wm. C. Wilson & Co., Toronto, Ont.

STOKERS, MECHANICAL
Babcock & Wilcox, Ltd., Montreal, Que.

SUPERHEATERS, STEAM
Babcock & Wilcox, Ltd., Montreal, Que.

SWITCHBOARDS, ELECTRIC
Crouse-Hinds Co. of Can., Ltd., Toronto, Ont.

TALLOW
Wm. C. Wilson & Co., Toronto, Ont.

TANKS, STEEL
Corbet Foundry & Mach. Co., Owen Sound, Ont.
Goldie & McCulloch, Ltd., Galt, Ont.
The John McDougall Caledonian Iron Works Co., Montreal, Que.
Marsh Engineering Works, Belleville, Ont.
Port Arthur Shipbuilding Co., Port Arthur, Ont.
Reed & Co., Geo., Montreal, Que.
Scaife & Sons Co., Wm. B., Oakmont, Pa.
Thor Iron Works, Toronto, Ont.
Scaife & Sons Co., Wm. B., Oakmont, Pa.

TANKS (AIR, GAS AND LIQUID)
Scaife & Sons Co., Wm. B., Oakmont, Pa.

TAPPING MACHINES
Mueller Mfg. Co., H., Sarnia, Ont.

TELEGRAPHS, SHIPS
McNab Co., Bridgeport, Conn.
Cory & Sons, Inc., Chas., New York, N.Y.
Morrison Brass Mfg. Co., James, Toronto, Ont.
Wilson & Co., Wm., Toronto, Ont.

TELEPHONES, MARINE
Cory & Son, Inc., Chas., New York, N.Y.
McNab Co., Bridgeport, Conn.

TESTERS, METER
Mueller Mfg. Co., H., Sarnia, Ont.

THUMB SCREWS AND NUTS
Canada Foundries & Forgings, Welland, Ont.

TRAVELLING BLOCKS
Morris Crane & Hoist Co., Herbert, Niagara Falls, Ont.

TROLLEYS
Can. Fairbanks-Morse Co., Montreal, Que.
Morris Crane & Hoist Co., Herbert, Niagara Falls, Ont.

TROLLEY HOISTS
Morris Crane & Hoist Co., Herbert, Niagara Falls, Ont.

TRUCKS, HAND, ELECTRIC
Aikenhead Hardware, Ltd., Toronto, Ont.
Can. Fairbanks-Morse Co., Montreal, Que.

TUBES, BOILER
Babcock & Wilcox, Ltd., Montreal, Que.
Broughton Copper Co., Ltd., Manchester, Eng.

TUBES, COPPER AND BRASS
Mueller Mfg. Co., H., Sarnia, Ont.
Standard Underground Cable Co., Hamilton, Ont.

TUGS
Collingwood Shipbuilding Co., Collingwood, Ont.
Polson Iron Works, Toronto, Ont.

TURBINES, DIRECT-DRIVING AND GEARED
Doxford & Sons, William, Sunderland, England.

TURNBUCKLES
Canada Foundries & Forgings, Welland, Ont.

SPIKES, SMALL RAILROAD
Severance Mfg. Co., S., Glassport, Pa.

UNIONS, ALL KINDS
Dart Union Company, Toronto, Ont.

VALVES, AIR
Mueller Mfg. Co., H., Sarnia, Ont.

VALVE, DISCS
Wm. C. Wilson & Co., Toronto, Ont.

VALVES
American Steam Gauge & Valve Mfg. Co., Boston, Mass.
Babcock & Wilcox, Ltd., Montreal, Que.
Can. Fairbanks-Morse Co., Montreal, Que.
Empire Mfg. Co., London, Ont.
McAvity & Sons, Ltd., T., St. John, N.B.
Mason Regulator & Engin. Co., Montreal, Que
McNab Co., Bridgeport, Conn.
Norbom Engineering Co., Philadelphia, Pa.
Williams Machinery Co., A. R., Toronto, Ont.
Wm. C. Wilson & Co., Toronto, Ont.

VALVES, FOOT
Aikenhead Hardware, Ltd., Toronto, Ont.
Smart-Turner Mach. Co., Hamilton, Ont.

VALVES, STOP, REDUCING, SAFETY CHECK, ETC.
Aikenhead Hardware, Ltd., Toronto, Ont.
American Steam Gauge & Valve Mfg. Co., Boston, Mass.
Empire Mfg. Co., London, Ont.
McAvity & Sons Ltd., T., St. John, N.B.
Morrison Brass Mfg. Co., James, Toronto, Ont.

VALVES, MIXING
Mueller Mfg. Co., H., Sarnia, Ont

VALVES, REDUCING, PRESSURE
Mueller Mfg. Co., H., Sarnia, Ont.

VARNISHES
Aikenhead Hardware, Ltd., Toronto, Ont.
Ault & Wiborg Co. of Can., Ltd., Toronto, Ont.
Leckie, Ltd., John, Toronto, Ont.
McAvity & Sons Ltd., T., St. John, N.B.
Moore & Co., Benjamin, Toronto, Ont.
Reed & Co., Geo., Montreal, Que.

VENTILATORS, COWL
McNab Co., Bridgeport, Conn.

VENTILATION EQUIPMENT
Empire Mfg. Co., London, Ont.
Low & Sons, Ltd., Archibald, Glasgow, Scotland
Scythes & Co., Toronto, Ont.

WASTE
Scythes & Co., Toronto, Ont.
Wm. C. Wilson & Co., Toronto, Ont.

WATER COLUMNS
Morrison Brass Mfg. Co., James, Toronto, Ont.

WATER HEATERS
Empire Mfg. Co., London, Ont.
Morrison Brass Mfg. Co., James, Toronto, Ont.

WATER SOFTENERS
Babcock & Wilcox, Ltd., Montreal, Que.

WATER SUPPLY SYSTEMS
Mueller Mfg. Co., H., Sarnia, Ont.

WELDING, ELECTRIC
Hall Engineering Works, Montreal, Que.
Marine Welding Co., Buffalo, N.Y.
Beatty & Sons, M., Welland, Ont.

WELDING, OXY-ACETYLENE
Advance Mach. & Welding Co., Montreal, Que.

WHISTLES AND SYRENS
Empire Mfg. Co., London, Ont.
McAvity & Sons, T., St. John, N.B.
McNab Co., Bridgeport, Conn.
Morrison Brass Mfg. Co., Jas., Toronto, Ont.

WINCHES, CARGO
Advance Mach. & Welding Co., Montreal, Que.
Aikenhead Hardware, Ltd., Toronto, Ont.
Corbet Fdry. & Machine Co., Owen Sound, Ont.

WINCHES, DOCK, SHIP
Advance Mach. & Welding Co., Montreal, Que.
Beatty & Sons, M., Welland, Ont.
Marsh & Henthorn, Ltd., Belleville, Ont.
Morris Crane & Hoist Co., Herbert, Niagara Falls, Ont.
Wilson & Co., Wm. C., Toronto, Canada.

WINCHES, TOWING
Corbet Foundry & Mach. Co., Owen Sound, Ont.

WINCHES, TRAWL
Beatty & Sons, M., Welland, Ont.
Wm. C. Wilson & Co., Toronto, Ont.

WINDLASSES
Advance Mach. & Welding Co., Montreal, Que.
Corbet Fdry. & Machine Co., Owen Sound, Ont.
Dake Engine Co., Grand Haven, Mich.
Wilson & Co., Wm. C., Toronto, Canada.

WIPER CAPS, OILER BOXES, ETC.
Morrison Brass Mfg. Co., James, Toronto, Ont.

WIRE, COPPER CLAD STEEL
Standard Underground Cable Co., Hamilton, Ont.

WIRE, COPPER, BRASS, BRONZE
Standard Underground Cable Co., Hamilton, Ont.

WIRE, INSULATED
Standard Underground Cable Co., Hamilton, Ont.

WIRELESS OUTFITS
Cutting & Washington, Inc., Cambridge, Mass.
Marconi Wireless Telegraph Co., Montreal, Que.

WOOD WORKING MACHINERY
Aikenhead Hardware, Ltd., Toronto, Ont.
Can. Fairbanks-Morse Co., Montreal, Que.
Preston Woodworking Machy. Co., Preston, Ont.
Williams Machinery Co., A. R., Toronto, Ont.
Yates Mach. Co., P. B., Hamilton, Ont.

WOOD BORING TOOLS
Aikenhead Hardware, Ltd., Toronto, Ont.
Can. Ingersoll-Rand Co. Sherbrooke, Que.

WOODITE GAUGE GLASS WASHERS
Wm. C. Wilson & Co., Toronto, Ont.

WRENCHES
Canada Foundries & Forgings, Welland, Ont.

YACHT BROKER
Watt, J. Murray, Philadelphia, Pa.

INDEX TO ADVERTISERS

Asbestos Packed Boiler Drain Cock, Bronze

J. M. T. Line Covers All Requirements

Morrison's line of Marine Specialties meet with every requirement of the ship. They are of high quality and modern in design. Every article is guaranteed reliable by thorough factory test. Approved by the Dominion Marine Inspection Department, Lloyds' Survey and Munition Board.

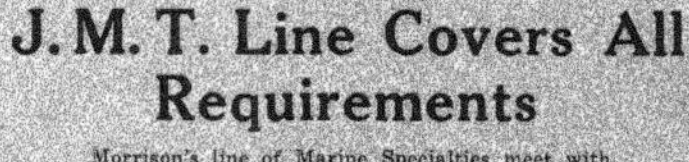
J.M.T. Reducing Valve

SOME OF OUR LINES:

- Stop Valves
- Adjustable Check Valves
- Non-Return and Equalizing Valves
- Check Valves
- Gate Valves
- Safety Valves
- Relief Valves
- Pressure Reducing Valves
- Injectors
- Ejectors
- Pressure Gauges
- Vacuum Gauges
- Ammonia Gauges
- Recording Gauges
- Engineers' Clocks
- Lubricators
- Steam Whistles
- Sirens
- Asbestos Packed Cocks
- Gland Cocks
- Suction and Discharge Valves
- Ships' Telegraphs
- Ships' Side Lights
- Gong Bells
- Gong Pulls
- Chain Guide Brackets
- Telegraph Chain
- Binnacles
- Water Service Cocks
- Hose Valves and Fittings
- Pipe and Fittings, etc., etc.

"Marine" Angle Valve, Bronze

Overboard Discharge Valve, Spring Loaded

OUR MARINE CATALOG consists of 60 pages with 150 cuts. It would be of great assistance to you. Pleased to send it on request.

The James Morrison Brass Mfg. Co.
Limited
Toronto Canada

Ship's Side Cock with Locking Shield, Bronze

Cylinder Oil Pump

Hose-end Valve

Oil Box, Cast Bronze

Grease Cup

Asbestos Packed Salinometer Cock, Bronze

CIRCULATES IN EVERY PROVINCE OF CANADA AND ABROAD

MARINE ENGINEERING of Canada

A monthly journal dealing with the progress and development of Merchant and Naval Marine Engineering, Shipbuilding, the building of Harbors and Docks, and containing a record of the latest and best practice throughout the Sea-going World. Published by
The MacLean Publishing Co., Limited

ONTREAL, Southam Building | TORONTO 143-153 University Ave. | WINNIPEG, 1207 Union Trust Bldg. | LONDON, ENG., 88 Fleet St.

Vol. VIII. | Publication Office, Toronto—April, 1918 | No. 4

Yates

MUELLER

Port Lights
Ships' Bells
Stern Tube Bushings
Propeller Shaft Liners
Air Pump Liners
Hand Bilge Pumps, etc.

CASTINGS

in Brass or Bronze
For Shipbuilders

We are equipped with special facilities for the manufacture of all brass and bronze castings, necessary to shipbuilders. The capacity of our foundries is one million pounds of metal per week. Think of what this means to you in the matter of deliveries. We maintain a chemical laboratory for the analysis of all metals and testing equipment for determining the physical qualities of castings, with the consequence that our product is guaranteed to be of the very highest quality. Send us your specifications to figure on.

We can furnish Stern Tube Bushings and Propeller Shaft liners rough turned and bored if desired. We have a practical expert at your convenience. We will be glad to have him visit you and give you the benefit of his experience in order to facilitate the solution of your problems. Blue prints on application.

Reducing and Regulating Valves

We have the right kind of valve for your particular service. Different conditions in the same kind of service necessitate different types of valves. We make valves for water, steam, gas, air, oil and ammonia. If you have a specially intricate problem to solve, our experts are at your command. Send for the MUELLER REGULATOR catalog.

H. MUELLER MFG. CO., LTD.

SARNIA, ONTARIO, CANADA

WM. B. SCAIFE & SONS CO.
PITTSBURGH, PA.

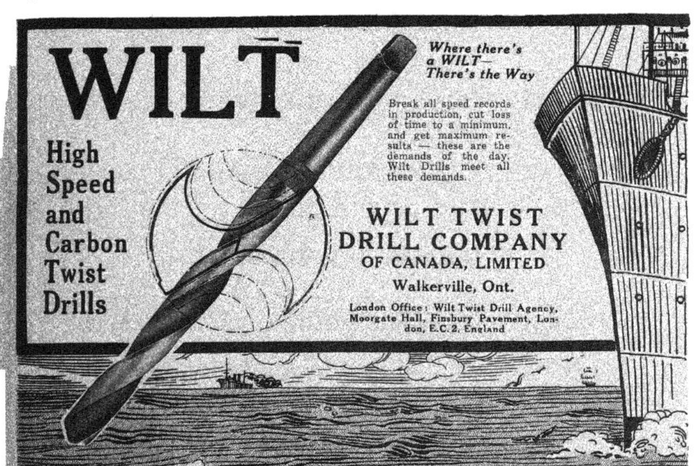
WILT
High Speed and Carbon Twist Drills
Where there's a WILT—There's the Way
Break all speed records in production, cut loss of time to a minimum, and get maximum results — these are the demands of the day. Wilt Drills meet all these demands.
WILT TWIST DRILL COMPANY
OF CANADA, LIMITED
Walkerville, Ont.
London Office: Wilt Twist Drill Agency, Moorgate Hall, Finsbury Pavement, London, E.C.2, England

U. S. Shipping Board: Its Handicaps and Accomplishments

By Edward N. Hurley *

The need for ships—ocean transportation—is the supreme question of the hour. Rumors, wise and otherwise, have created some degree of uncertainty regarding the progress of our southern Ally's marine effort. On March 26 the following address was delivered before the National Marine League at New York. The facts and figures are silently eloquent.

IF by the exercise of magic a bridge could be thrown across the Atlantic over which our armies, their artillery and supply trains could move rapidly and unhampered to the battle lines in France, would any military man in Berlin, Vienna, Rome, Paris, London, or Washington have any doubt but that the world would be made safe for democracy before the year goes out? We have the men, we have the guns, we have the supplies, but without means of getting them to the front we might as well be without them. And unless we get our men to the battle line we will not wih this war.

Start Under a Handicap

So it all comes back to ocean transportation—to the vital need of ships. Fail there and we fail utterly. Upon the Shipping Board has devolved the responsibility of supplying this need, and supplying it under the most extraordinary conditions that ever existed-supplying it at the most crucial period of the war's history—at a time when every other industry is being taxed to its utmost capacity in the matter of materials and labor to provide war necessities.

The handicaps have been many. We were not a maritime nation. Our flag had almost vanished from the seas, and with the exception of a few widely scattered shipyards, merchant marine construction had almost become a lost art with us. Then came this sudden call to outdo the rest of the world in the upbuilding of a merchant marine; a call coming at a moment when the navy was undergoing the greatest expansion in its history, when most, if not all, of the established yards were feverishly engaged in rush construction on dreadnoughts, destroyers, submarines, fuel ships, tenders, and other auxiliary craft, and when munition makers were absorbing that part of skilled labor which had not been called to government navy yards or private shipbuilding plants.

Facts as They Exist

So it was a case of not only working from the ground up, but of first securing the ground upon which to make a start, some of it marsh land which had to be filled in before launching ways could be laid. Therefore we who are engaged in the work appreciate the magnitude of the task. I doubt if its magnitude is generally realized. I am not here to emphasize that magnitude—I am not here as an optimist glorifying an outcome—I am not here as a pessimist saying the task can not be accomplished. I am here to tell you of the situation as it is—to deal with facts as they exist—to lay all of our cards upon the table, and face up.

When we took hold of this job of shipbuilding, we found there was not shipyard in existence with which we could place an order. The old yards, with their trained force of shipbuilders, were filled to capacity. We had to establish the yards first, get the shipbuilders to take charge of them and train the men to build the ships.

VIEW OF BERTH IN SHIPYARD SHOWING TYPE OF TRAVELLING CRANE AND SCAFFOLDING INSTALLED.
—Copyright by Committee on Public Information.

There were 37 steel shipyards in America at the time of our entrance into the war. We have located 81 additional steel and wood yards, while 18 other yards have been expanded. Does America realize what this job means? Does it realize what a tribute is paid to its own initiative in this achievement?

Big Job in a Big Way

We are building in the new and expanded steel yards 235 new steel shipways, or 26 more than at present exist in all the shipyards of England. If we had been content with doing the job in a small way, we might have built a few new yards, and added a little to our capacity. A few ships might have been finished more quickly; but it was the spirit and will of America to do the job in a big way, and the judgment of the country will be vindicated by the results when all these new ways are completed and are turning out ships. Many of these ways have actually been finished. The new industry we have created will make America the greatest maritime nation in the history of the world.

Battle With the Elements

Struggling against something that can not be avoided is more baffling than struggling against something that can be. You can appeal to striking men to go back to work, but you can make no appeal against zero weather. We did what we could. We told the new shipyards to go ahead and use dynamite in locating their pilings. The men in those new yards fought the bitter winter. They had the same spirit and demonstrated the same pluck and unselfishness as the men in the trenches. They have virtually completed the job of building America's new shipyards—the new yards that will make us the greatest shipbuilding nation.

It has been an uphill struggle. I am willing to confess there have been times when we have been discouraged, not at the magnitude of the task, but through a doubt of human ability to accomplish the stupendous work in the short time

But we have had our moments of elation when we have felt that we are making progress. The record made by the Skinner & Eddy Co., of Seattle, is a case in point. That company laid the keel for an 8,800-ton vessel, which was launched in 64 days. She was delivered to the Fleet Corporation on January 5, and started on the first voyage on January 14. This record accomplishment shows what can be done in live, wide-awake efficient American shipyards.

Steadily Taking the Water

Then a few days ago we received a telegram from the Moore Shipbuilding Co., of Oakland, Cal., announcing the

*Chairman of the United States Shipping Board.

successful launching of one of their large vessels. Twenty minutes later we received another telegram from the same company announcing the launching of a second ship of the same type, and 40 minutes afterwards a third telegram saying that a third vessel of similar character had gone overboard.

There are two methods for computing the construction of tonnage to show what is accomplished. One is by showing the tonnage in the water; the other is by showing the tonnage under construction. But when a great many ships are put under construction at the same time the question that should be asked is, How are they all progressing; how near to completion is the vast program? Here is the answer:

Steel Construction Tonnage

The total amount of our steel construction program on March 1 was 8,205,708 dead-weight tons. This is made up of 5,160,300 dead-weight tons under contract with the Emergency Fleet Corporation, and 3,045,408 dead-weight tons of requisition vessels.

Of this total steel construction, 2,121,568 dead-weight tons, or approximately 28 per cent., has been completed. That means that in addition to the building of our big new yards we have also been building ships. That is, the program for steel ships has advanced 28 per cent. toward completion. Of the amount of steel ships under contract and under requisition, 655,456 dead-weight tons, or approximately 8 per cent., were actually completed and in service on March 1 of this year, nearly a month ago. This amount of floating tonnage exceeds our total output in 1916, including steel, wooden, and sailing vessels, by approximately 50 per cent.

In the yards which we have already completed and those which are nearing completion the progress will be cumulative from this time on.

Building Yards and Ships

Thus, while we have been building the yards and training the new forces necessary to construction, we have also been building the ships.

Notwithstanding the difficulties of organization, of the handicaps of bad weather conditions, of transportation embargoes and railroad congestion, nearly as much tonnage has been constructed in American waters in the past three months as by all the other maritime nations of the world combined.

I have referred to the necessity of providing additional facilities for the building of ships. That is, for the creation of new shipyards, for enlarging old ones, for the education of new shipbuilders, and, I may now add, the necessity of providing increased means for obtaining engines, boilers, turbines, and other equipment. At the outset the 37 old steel yards began increasing their capacity until they now have 195 ways as against 162 eight months ago. Other parts of their plants have increased in proportion. We then made provision for additional new steel yards, some of which have been given financial assistance by the Emergency Fleet Corporation. Thirty additional new steel shipyards are thus being created, with a total of 203 shipbuilding ways. Thus we now have in the aggregate 67 steel shipyards either wholly or partly engaged in Fleet Corporation work. These yards will have a total of 398 steel building ways. Of these 35 yards, with 258 ways, are on the Atlantic and Gulf coast; 19 yards, with 66 ways, are on the Pacific; while 13 yards, with 74 ways, are on the Great Lakes.

VIEW OF STEEL HULL IN COURSE OF CONSTRUCTION. THE MEN ARE PLACING FRAMES IN POSITION FOR RIVETTING TO THE KEEL
—Copyright by Committee on Public Information.

Wooden Ship Construction

Our program for building wooden ships has been beset with many difficulties and handicaps which could not well be foreseen. A year ago wooden shipbuilding in the United States was almost a lost art. We found 24 old wooden shipyards with 73 shipways. The capacity for wooden shipbuilding has been increased until we now have 81 wooden shipbuilding yards, with 332 ways completed or nearing completion.

ASSEMBLING THE LOWER PORTIONS OF THE BOW FRAMES OF A STEEL SHIP
—Copyright by Committee on Public Information.

Assuming that these ways will each produce two standard ships per year, we should turn out about 2,300,000 dead-weight tons of wooden ships annually. These 332 wooden shipbuilding ways, now nearing completion, added to our 398 steel building ways, will give us a total of 730 berths upon which to build steel and wooden vessels. When you consider that we had only 162 steel building ways a few months ago and 73 wooden shipbuilding ways—a total of 235—an increase is shown of 495 wooden and steel berths on which we can build ships.

Delayed by Rail Embargo

Our program on wooden ships was delayed by the fact that we were unable to provide the necessary big timber in sufficient quantities from the forests east of the Mississippi River. We found ourselves obliged to go to the west coast and arrange for the transportation of hundreds of carloads of Douglas fir, which it was necessary to substitute for long-leaf yellow pine for keels and other heavy timber. Here again we encountered a serious phase of the transportation problem. Our shipments were held up for weeks by the railroad embargo.

Turbines and Engines

The situation giving us the most concern is the completion of turbines and engines. The very rapid expansion of the shipbuilding program caught the turbine and engine manufacturers totally unprepared. In the past the engines for ships built in this country had been manufactured at the shipbuilding plants. As contracts for new shipyards were given it became necessary to increase the turbine and engine building capacity at the same time. Special tools of all kinds were required for the engine builders' shops, and these tools had to be secured from manufacturing shops already overcrowded with war orders. In addition to this, the severe weather and the transportation tie-up seriously delayed the construction of some of our largest turbine-building plants. We anticipated delay during the earlier months for lack of the turbines and engines, but expect to make up for the early shortage.

RIDING THE LOAD IN A SHIPYARD. THE MEN ARE CARRIED ON A SHIP'S RUDDER.

—Copyright by Committee on Public Information.

First Concrete Vessel

The proposal to build ships of concrete was first regarded as a fascinating absurdity. On March 14 there was launched from the yards of the San Francisco Co. the first concrete steamship, a vessel which the builders christened Faith. We hope she will exemplify her name. The builders believe she will, for in the telegram announcing the successful launching of the vessel were added these words: "Appearance after launching warrants us in saying to you that we believe this form of construction may be safely depended upon."

Now as to labor—our strong right arm. There has been much talk of conscripting labor—of forcing it into shipyards as our soldiers have been brought into the camps. I am fully aware that I am flying in the face of a growing popular sentiment that men should be drafted into the industry which supports the battle lines, but I wish to put myself on record as being opposed to the conscription of labor. I do not believe conscription necessary, for I believe labor itself will produce conditions which will render idle all thought of conscripting workmen. The vast majority of our workmen are men of intelligence, and when they come to a full realization of the fact that any defection on their part now will injure their fellow-workers in almost every field of industrial activity, I feel sure they will respond to all demands made upon them. Unless they fully do their part, their brothers will suffer.

Labor Loyally Expected

It would be useless to manufacture material and supplies and pile up the products in the wharves if there are no ships to transport them. So unless our ship workers do their best, other industries must slow down or halt completely, with the result that thousands of workers throughout the country will suffer for lack of employment.

I believe that labor has begun to realize that fact, but I want to drive it home to them, for there are some, I regret to say, who do not yet sense their responsibility. There are many who are not working to their full capacity. There are many who, because of the high wages they are earning, are prone to take too many holidays. Labor generally throughout our shipyards is to-day receiving the highest rate of wages ever paid for similar work in the history of the world. The additional cost of our ships, due to increased wages in shipyards covering the program we have mapped out, will be in excess of $300,000,000. We expect, and we have a right to expect—the country has a right to expect—that labor will render for this increase of wages a corresponding increase in production—that is, the output of ships.

Confidence in Leaders

All has not gone smoothly in the labor situation, and there have been times when this phase of the problem was enough to cause discouragement. The vast majority of laboring men are patriotic; the leaders whom I have known through close contact in Washington, especially Mr. Gompers and his immediate associates, have my confidence, and the country recognizes their patriotism. With only one exception the leaders of the shipyard crafts generally have shown a spirit of co-operation, ready to sink their personal differences in the common pool of patriotism.

We have established a labor adjustment board whose complete fairness can not be questioned.

We have felt that it was our duty to see to it that the problem of housing the workmen in these vast new plants we have been creating was solved with care. We have not rushed into this work with closed eyes. Our duty is to guard the public expenditures; to see to it that there is no abuse of the liberality of Congress in the matter of appropriations. Every dollar expended must bring a dollar's worth of return to the Government. The cost-plus system has been banned by Congress in the housing op-

ASSEMBLED COMPONENT OF SHIP'S FRAMING AWAITING INCORPORATION WITH THE HULL.

—Copyright by Committee on Public Information.

erations, because Congress itself, as well as the rest of us, have felt that there should be a greater check, not merely upon profits but upon the actual cost of all work done for the Government.

Army of Reserve Workers

The new yards have been established, wherever possible, away from the congested districts, and while this was necessary, it brought with it the problem of transportation, as well as of housing. We are arranging now for proper transportation, as well as for proper housing.

We have recruited a volunteer force of 250,000 highly skilled mechanics who have, with a patriotism that has made us all proud, agreed to hold themselves in readiness for our call. These men are being held in reserve, remaining in their present employment until such time as in the development of our yards the demand arises for their services.

The Fabricating System

This brings me to the point where I desire to make a brief reference to what have been popularly termed our three fabricating shipyards. The term is more or less a misnomer, for these yards, located at Hog Island, Newark Bay, and Bristol, Pa., are in reality assembling yards. The shipbuilding materials which will go into the making of vessels launched at these yards are being fabricated in scores of steel plants, scattered throughout the country as far west as Omaha, Neb. In some instances 95 per cent. of the work on these materials is being done at points far remote from the shipyards. The so-called fabricated ship is almost a new method of ship construction—almost as new to England as it is to us. But from the progress of the work as it has thus far developed we are confident that it will be the means of adding millions of tons to our merchant marine.

When the high point in the curve of production finally is reached and the magnitude of America's shipbuilding program is realized it will be a continuous performance of production and launching.

There is no doubt that we are destined to be one of the leading shipbuilding nations of the world.

STEEL MASTS IN COURSE OF CONSTRUCTION. PNEUMATIC DRILLING, RIVETTING, AND CAULKING MACHINES IN OPERATION.
—Copyright by Committee on Public Information.

BRITISH WEST INDIES TRADE

CLOSER commercial relations when the war is over between Canada, the British West Indies and contiguous British territory is a subject that is receiving some attention at the present time among upper Canadian manufacturers and shippers.

It is one that has a deep interest for the Maritime Provinces, as any development in the direction referred to must inevitably lead to increased shipments through the maritime ports of St. John and Halifax, and must also tend to stimulate the manufacturing industries of these lower provinces.

Tariff Increases Trade

As a result of the operation of the preferential tariff, says the St. John "Board of Trade Journal," there has been a slight increase in the trade between Canada and the British West Indies. In the last fiscal year the imports from the West Indies reached a total of $14,239,595, as against $6,354,991 in 1916. Exports from Canada did not show the same ratio of increase, the figures for 1917 being $5,179,083 and for 1916, $4,134,901. It is quite apparent from the above that the manufacturers and shippers of Canada are not getting all out of the trade that they should get, or that they might obtain under improved transportation conditions.

The "Barbados Advocate," in a recent article on the subject, remarked:

"It cannot be too strongly pointed out that this one market, that of the British West Indies, is comparatively close at hand, is waiting for further development, and possesses these direct advantages that the merchants there are strongly desirous of having more extensive and closer trade relations with us, and that, already, throughout these colonies there are active branches of two leading Canadian banks, ready to assist with information and banking facilities in developing our trade."

Fast Steamship Service

The same paper declares that these advantages would be futile if they were not linked with such transportation conditions as more frequent and quick steamship services between near Canadian ports and the British West Indies, fast through freight services with Canadian business centres timed to meet the arrival and departure of the steamships, and through freight rates and charges between the West Indies and Montreal and Toronto via a Canadian port and not exceeding those via New York and Boston. There must also be cold storage facilities; cool, ventilated accommodation, especially for fruits, in both steamships and cars, rapid sailings and prompt handling of cargoes.

These results can be obtained by the utilization of the advantages possessed by the Canadian ports of Halifax and St. John. Both possess peculiar advantages. St. John has western connection by two transcontinental lines, the Canadian Pacific Railway and the Canadian Government Railway; Halifax has the latter line of railway. There is practically little difference in the ocean distance between the two maritime ports and the West India Islands. The advantage in this respect is with Halifax, but there is only a difference of twenty-one miles. Against this short ocean advantage, St. John has the advantage of a very much shorter land haul by both transcontinental lines. By the Canadian Pacific Railway the combined rail and ocean mileage from Montreal to Bermuda via St. John is 1,254 miles, while via Halifax and the Canadian Government Railway it is 1,588 miles. In the shipments north of the molasses crop of 1917 a much larger proportion came via St. John, illustrating the advantage that the shorter rail haul to the interior gives to St. John.

Clearly, the two ports must work together in this as well as in other national enterprises involving the use of lower province ports.

Concentration of the business thought of Canada upon the whole question is very much needed at the present time, so that when the war is over no time may be lost in putting plans into effect.

SHIP'S AIR PURIFYING APPARATUS

DIFFICULTIES with the attuning apparatus of radio receiving and sending instruments have been the subject of experimentation by William J. Baldwin, of New York, who recently designed an apparatus for a radio room that would admit air not only separated from rain, salt spray and spume, but that would also condition or regulate the humidity within the room by keeping it at a common standard of humidity, regardless of the outside changes.

Cold water in the spray form will do this, provided the spray can be gotten rid of after it has combined with the excess of humidity (steam in the air) and then separated from the air by some practicable form of apparatus that is simple and clean, and that occupies small space.

While a shower of rain will clear the atmosphere, the elements of nature have all outdoors in which to set up the apparatus for such a result. The cabin of a ship or a radio cabinet is infinitely small in comparison with all outdoors, yet a similar result can be achieved in a space as small as a cabin or radio cabinet.

To accomplish this a cold-water spray

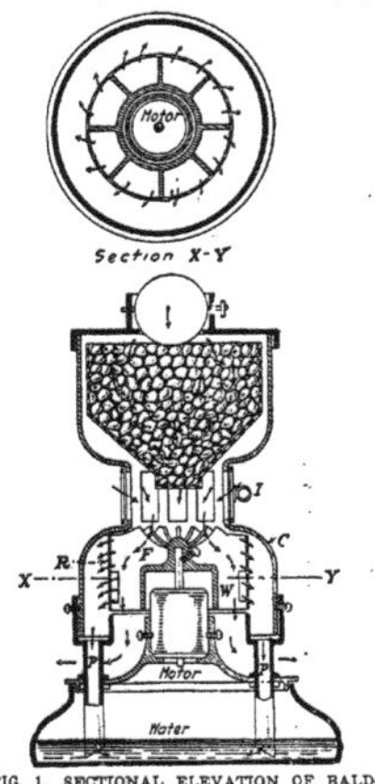

FIG. 1. SECTIONAL ELEVATION OF BALDWIN AIR DRYER AND COOLER.

apparatus was used similar to one designed two years earlier for the purpose of precipitating the CO_2 when found in great excess in a confined space, as in the hold of a submarine, when forced to stay a long time under water.

In this machine a spray of potash water was used in connection with a mechanical dust precipitator for the purpose of seizing on the carbonic acid in the air and thrown it down, so to eliminate the CO_2 at the dust and water discharge of the apparatus.

It was proposed to put the apparatus in a bulkhead; the discharge side of the apparatus coming into the air of the living quarters, the air freed of its CO_2 being forced back again into the chamber of greatest vitiation, thus forming a cycle.

In the general experiments it was found that a prepared spray of chemical liquid or even cold water not only seized on the dust but on the other gases in the air, with which the chemical in the spray would combine, and that a pure-water spray turned into the apparatus would keep the humidity of the air constant by suitable regulation of the temperature of the water spray. In a room of 90° F. and a humidity of almost 90%, it was possible to drop the humidity to 45% by reducing the temperature of the spray to 70° F.

A simple form of the apparatus arranged for cooling a room is shown in Figs. 1 and 2. The apparatus from the inlet to the line X-Y (Fig. 1) is an ordinary fan or blower F, and the downward extensions of the fan blades or wings W are necessary to accelerate the rotary motion of the air.

The rotor R is a rotating hoop of permeable metal, against the inner side of which the air is thrown with all its impurities. If the heavy particles in the air, such as dust, mud or particles of water, strike into the perforations of the hoop, they pass through into a quiet space formed by the outer case C. Or, if they strike on the solid part of the hoop, they are rubbed through the nearest holes by the forward movement of the air. They then pass into the quiet space C, drop down within it, and escape by the pipe or pipes P into the tank.

The air does not escape with the heavy particles, as might appear at first, for the lower ends of the pipe legs are sealed by the water in the tank. The tank may be of any shape, or there may be no tank, the separated particles going to a waste pipe.

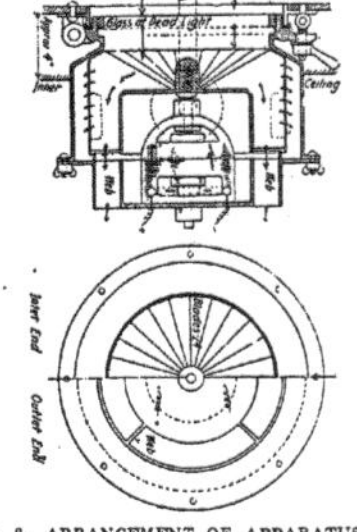

FIG. 2. ARRANGEMENT OF APPARATUS ADAPTED TO SHIP'S PORT-HOLE.

The greater the velocity of the rotor R the more efficient is the apparatus. A speed of 5,000 ft. per min. is very practicable, but 10,000 is not excessive, either from the point of bursting or for any other reason.

The fan gives the same static pressure as any other centrifugal blower of equal diameter and speed and requires only equal power for equal work, and the power required for separation and friction is considerably under 25% of the blower power required to move the air.

The device as described above illustrates the principle of the various types of apparatus, whether used for taking the dust from the air, the CO_2 from a chimney or an enclosed space, the excess of humidity from the air, throwing down fog, cooling and moving air, etc. It will be noted that the ice in the upper tank serves both as a cooling medium and for the supply of cold water, broken into spray by the rapid motion of the fan, for the purification of the air.

STEAMBOAT INSPECTION REPORT FOR 1916-17 ISSUED

THE annual report of the Dominion Government steamboat inspection department for the fiscal year 1916-1917 has just been issued. There are no casualties reported for the Collingwood district. In the Kingston division there are the following:

On June 20, 1916, the steamer Stormount, of Montreal, while on a coasting voyage, stranded on Gull ledge, N.S. No lives lost.

On June 28, 1916, the steamer Meteor, of Ottawa, while on her regular trip from Temiskaming to Haileybury struck a rock and sank, causing some damage to the hull, which was afterwards repaired on dock at Haileybury. No lives lost.

On October 3, 1916, the steamer Simla, of Montreal, while on a trip to Montreal, laden with coal, ran into a fog bank in the narrows above Brockville, at Royal island, where she struck a rock causing serious damage to the hull and sank immediately. She was raised and placed in dry dock, where permanent repairs were effected. No lives lost.

In the Toronto division the list is as follows:

On August 15, 1916, the steamer Otonabee was totally destroyed by fire at Big Bay Point, Lake Simcoe.

On August 19, 1916, the steamer Saronic was partially destroyed by fire at Cockburn Island, Lake Huron, the wreck was sold to parties at Milwaukee, Wis., the bodies and engines removed and the hull converted into a tow barge.

On September 4, 1916, the tug Fred Davidson foundered on Georgian Bay near Byng Inlet.

On September 24, 1916, the pleasure yacht Sky Pilot was totally destroyed by fire near Gravenhurst, Ont.

On October 19, 1916, the fishing tug Gambler was partially destroyed by fire at Port Dover, Ont.

On November 21, 1916, the tug T. F. Battle was partially destroyed by fire at Thorold, Ont.

On November 23, 1916, the fishing tug Ariadne sprang a leak on Lake Erie; she was beached and abandoned on Point Pelee. The engine and boiler were removed.

FIG. 3. ARRANGEMENT OF APPARATUS FOR REMOVING FLUE DUST FROM CHIMNEY GAGSES.

PACIFIC COAST DEVELOPMENTS

Featuring the Record of Progress and Dealing With the Steps Being Taken to Stimulate and Enlarge the Already Established Shipping and Shipbuilding Enterprises

"WAR CAMP" LAUNCHED AT VANCOUVER

THE steamship War Camp, an 8,800-ton vessel, was launched from the shipyard of J. Coughlan and Sons, Vancouver, on March 16th. This is the second vessel of this size which has gone to the sea from this yard.

A little over a year ago Messrs. Coughlan were engaged in the structural steel fabricating business and in this short interval of time they have succeeded in building and equipping a first-class shipyard, and have undertaken the construction of ten vessels of 8,800-tons dead-weight capacity for early delivery. These vessels are fitted with turbines which have been secured from the United States, but the boilers for developing power are being built in the shops of Messrs. Coughlan at Vancouver.

THE LAUNCH OF THE SS. "WAR CAMP" AT VANCOUVER ON MARCH 16.

LAUNCH OF THE "WAR POWER"

ON March 23 there was launched from the Wallace Shipyards' plant at N. Vancouver, B.C., the steel steamer War Power. This is the second deep sea steamer turned out by this firm.

The War Power is 300 feet long b.p. by 45 feet moulded beam by 27 feet moulded depth. She will carry 4,700 tons d.w., and will be driven by a set of triple expansion engines of 1,650 i.h.p.

As she took the water the War Power was christened by Mrs. C. Wallace. In spite of the early hour, 4.30 a.m., of the ceremony, a large number of visitors turned out to witness the event, a special ferry being run to accommodate visitors from Vancouver. Mr. J. Cant, the general superintendent of the yard, took charge of the work of launching.

The War Power is much more a product of the province than any of the previous steel boats which have been launched in British Columbia. All the work on the hull was, of course, done in the yard. In addition, the Wallace Shipyards have built the machinery, which, at the time of launching, was partially installed. The boilers, 15 ft. 6 in. in diameter by 11 ft. 3 in. long, have been built by the Vulcan Iron Works at their Granville Island plant, Vancouver. The winches, etc., were constructed by the North Shore Iron Works, of N. Vancouver. With the exception of a few engine room auxiliaries, the vessel when completed will be entirely a British Columbia product.

Besides the War Power, the Wallace Shipyards have contracts for several more steel vessels, and the yard is now being greatly extended to provide for their construction. The machine shop is exceedingly busy, for besides engines for their own boats they have on hand three sets of machinery for the Imperial Munitions Board wooden ships. And, of course, take care of their coastwise and deep sea repair work.

The Wallace Foundry, on Granville Island, Vancouver, is now in excellent running order, and is turning out all the castings required for the shipyard construction plans as well as a quantity of outside work.

After the launching the visitors gathered in the company's offices and were suitably entertained by Mr. Wallace, Sr., and the staff. Mr. Wallace was in receipt of many congratulations over the successful event and well wishes for the future.

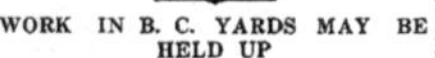

WORK IN B. C. YARDS MAY BE HELD UP

CONSIDERABLE anxiety has been occasioned in British Columbia shipbuilding circles by the refusal by the government of permission to the Vancouver steel shipbuilding yards to proceed with new contracts for vessels of the standard type of 8,800 tons for allied nations, says a Canadian Press despatch, and the issuance of orders by the Imperial Munitions Board at Ottawa for the mak-

S. S. WAR POWER, LAUNCHED FROM THE WALLACE SHIPYARDS AT N. VANCOUVER, B.C., MAR. 23, 1918.

ing of structural alterations in the vessels already near completion, may result in a halt in operations at the Coughlin shipyards.

The government was reported to have planned to adopt a standard size for steel ships, and intended to take control of the plants, contracting for the entire output of a Nova Scotia steel plant.

Mayor Gale of Vancouver is en route to Ottawa and will use every endeavor to have the government give the necessary permission to construct the contract ships.

TORPEDO DAMAGE TO MERCHANT SHIPS

THE effects which explosions of mines and torpedoes have on the structure of merchant ships are of such supreme importance at this time that great interest has been taken in the report (recently published by consent of the Admiralty) which was issued last year as a confidential paper by the council of the Institution of Naval Architects. The committee consisted of S. W. Barnaby (chairman), Admiral Sir Henry Jackson, Prof. Sir John Biles, W. H. Whiting, A. E. Seaton, Prof. T. B. Abell, Prof. J. J. Welch, and R. W. Dand. The report is as follows:

I.—Cargo Ships

Your committee have thought it desirable to issue a preliminary report containing suggestions for certain temporary war expedients which might be adopted at very little expense in the ordinary type of cargo vessel, and which they are of opinion would have the effect of greatly increasing their chances of safety after being mined or torpedoed, provided that only one main compartment is opened up to the sea.

The committee, having examined the information placed at their disposal by the Admiralty and the Board of Trade, are of opinion that the loss of many of these vessels has been due to three causes:

(a) The existence of watertight doors low down in the bulkheads, which could not be closed after the explosion.

(b) Fractures of suction pipes in the attacked compartment, permitting water to flow into adjacent compartments.

(c) The penetration of bulkheads adjacent to the attacked compartments by fragments of plating, frames, rivets, etc.

A large number of cargo vessels have four main holds and six bulkheads arranged as follows: No. 1, collision bulkhead; No. 2 between Nos. 1 and 2 holds; No. 3 at fore-end of boiler room; No. 4 at aft end of engine room; No. 5, between No. 3 and 4 holds; No. 6 at after peak.

With fewer bulkheads than this the chances of keeping a vessel afloat are much reduced.

Taking this six-bulkhead arrangement as a typical case, the recommendations of the committee, as applied to cargo ships with six or more bulkheads, are as follows:

1. All existing watertight doors low down in main bulkheads should be closed up and so screwed that they cannot be opened. If watertight doors are necessary they should be fitted high up in the bulkhead.

This is particularly important when bunker coal is carried forward of the boiler-room bulkhead, as a watertight door through which coal is being trimmed cannot be depended upon.

2. The water-tight door to the shaft tunnel in the engine-room bulkhead should be closed up and so secured that it cannot be opened, access to the tunnel being provided by means of a water-tight trunk carried up to the bulkhead deck.

Cases have occurred when the engine-room has been flooded too quickly to permit of the tunnel door being closed, with the result that considerable quantities of water have entered the after holds through the tunnel, which was not sufficiently water-tight.

An explosion in one of the after holds would be likely to injure the tunnel, resulting in the flooding of the engine-room and boiler-room, unless the tunnel door is closed as proposed.

3. The tunnel, or any other longitudinal passage below the water-line, should be thoroughly water-tight.

4. Each suction pipe, where it enters the compartment from which it drains, should be provided with a screw-down non-return or other suitable valve, which can be worked from the bulkhead deck. These valves should be kept closed at all times, except when the pumps are in use.

5. The amount of injury which the bulkheads might receive from flying fragments will depend greatly on the nature and amount of cargo in the hold.

If the hold is well filled the risk of injury is much less than when partly filled or empty. If the cargo is of a heavy nature the upper parts of the bulkheads will probably be exposed.

It is recommended that, wherever possible, the bulkheads should be protected temporarily by means of timber or other suitable material, forming a splinter screen. In some cases the cargo itself could be utilized for this purpose. The necessity for some protection of this sort is particularly great in the case of ships in ballast.

Very little time and money would be expended in giving these measures of protection to a large number of cargo vessels.

II.—Existing Large Mail and Passenger Vessels

These vessels having more numerous bulkheads in proportion to their length than the ordinary cargo steamer already reported on should be better able to withstand attack from mine or torpedo.

nO the other hand, many mail and passenger steamers have water-tight doors in the 'tween decks giving communication through the water-tight bulkheads between passenger and other spaces. They have also side scuttles for ventilating cabins, and numerous sanitary discharges very near the water-line. These are liable to be submerged and to admit water, should the scuttles be open or the controlling valves to the baths, water-closets, etc., be inefficient. This liability will be greater in the case of vessels having small initial stability or possessing longitudinal bulkheads, if damage occurs in way of such bulkheads. In some vessels also there are passages leading through bulkheads in the ship's hold to allow the firemen and greasers to gain their quarters without traversing any of the passenger decks. All these features have proved to be sources of danger.

Further, the hatchways in such vessels are relatively small, and do not afford such instantaneous relief from air and gas pressure resulting from an explosion as in the case of the larger hatchways of cargo vessels. For this reason it is considered that bulkheads in the neighborhood of the explosion are more liable to distortion in mail and passenger vessels than in cargo ships, and that water-tight doors on such bulkheads cannot be relied upon to close after an explosion. Devices for closing doors from the bridge are likely to fail, both by reason of the above-mentioned distortion and because of the probable destruction of the hydraulic or electric mains provided for the purpose in the region of the explosion.

In view of the foregoing it is suggested that the following measures be adopted in existing vessels during the continuance of the war:

1. The disuse of all firemen's passages and the closing up of all doorways in bulkheads traversed by them when below the bulkhead deck, the firemen passing to and from their quarters by way of an upper deck, where a temporary screened passage could be provided if necessary.

2. All openings from the engine-room into shaft tunnels to be abolished and access to tunnels provided through trunks abaft the engine-room bulkhead; these trunks to be water-tight to the bulkhead deck.

3. All water-tight doors in transverse bulkheads throughout the machinery spaces to be closed and so secured that they cannot be opened, provision being made for efficient supervision in each separate compartment. If this is deemed impracticable in certain cases, it is recommended that in such cases the doors on the level of the stokehold and engine-room floors should be closed and others fitted as high up as possible on these bulkheads.

4. All water-tight doors in transverse bulkheads on decks below the bulkhead deck to be closed and so secured that they cannot be opened, additional exits to the decks above being provided where necessary.

5. All side scuttles situated below the first deck above the bulkhead deck should be closed up and sealed. Special fan ventilation is to be provided if necessary for the cabins affected.

(Continued on page 103.)

TEST PIECES FOR MATERIAL COMING UNDER SUPERVISION OF LLOYD'S REGISTER

By A. F. Menzies.

METHODS and procedures for the accurate testing of materials have assumed great importance in the last three years, consequent upon the expansion in all metal working industries necessitated by the insatiable demands of modern warfare. The impetus given to shipbuilding by the inroads of the submarine and the increased volume of traffic handled has resulted in a widespread demand for knowledge of matters relating to design and construction and the testing of materials entering into ship construction is of considerable interest at the present moment.

Material entering into the construction of ships coming under the supervision of Lloyd's Register of Shipping, is subject to test and the test-pieces illustrated in this article have been standardized by a Canadian plant for their own convenience. Since considerable time can be wasted in the looking up of specifications and requirements for the test-pieces used, tabulated drawings were prepared in such a manner that all the dimensions required by Lloyd's and those governed by the design of the machine were readily accessible together with the requirements which had to be met by the test-pieces under load. By this means the information was always available and the making of rough sketches was obviated.

Fig. 4 shows the test-pieces required for the tensile proving of plates and shapes. The elongation required in the table applies to material ⅜ inch thick and over. For ship-material less than S/16 inch thick the requirements as regards elongation are slightly smaller, the minimum elongation necessary being 16 per cent. For boiler plate the elongation of material less than ⅜ inch thick may be not more than 3 per cent. below the tabulated values.

The cold and temper bend test piece for plates and shapes is shown in the upper view of Fig. 2 and for cold and temper bends the test piece must withstand bending double on itself, the maximum allowable internal radius being 1½ times the thickness of the test piece.

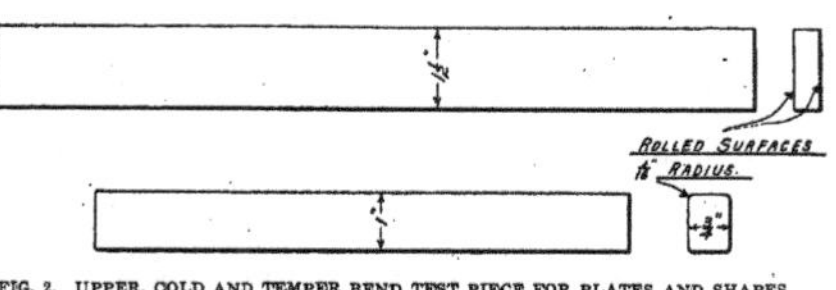

FIG. 2. UPPER, COLD AND TEMPER BEND TEST PIECE FOR PLATES AND SHAPES. LOWER, BEND TEST PIECE FOR STEEL CASTINGS AND FORGINGS.

Cold and Temper Bend Tests For Plates and Shapes

In all cold bend tests and in temper bend tests for plates and shapes on samples ½ in. to 1 in. in thickness the rough edge may be removed by filing or grinding. In samples 1 in. in thickness or over the edges may be machined. For both cold and temper bends the test piece shall withstand, without fracture, being doubled over until the internal piece and the sides are parallel.

Bend Test Steel Castings and Forgings.

For steel castings the test piece must withstand without fracture, being bent cold through an angle of 120° the internal radius being not greater than 1 in.

For steel forgings the test piece must withstand, without fraction, being bent cold through an agle of 180é the internal radius being not greater than ¼ in.

FIG. 1. TEST PIECES A FOR PLATES AND SHAPES

TABLE I.—	Tensile Strength	Elongation
Plates, not flanging	28-32 tons	20%
Plates, flanging	26-30 tons	23%
Shapes (ship)	28-33 tons	20%
Shapes (boiler)	28-32 tons	20%

The elongation applies to material ⅜ in. and over; for ships' material under ⅜ in. thick, for boiler material of the same thickness not more than 3 per cent. belowe tabulated value.

The tensile test pieces for bars are shown in Fig. 3. Test piece B is used for bars one inch or less in diameter and C for bars over one inch. Test piece B may be used for bars over one inch in diameter provided that the diameter of the gauge length is turned down to one inch or less diameter.

For steel castings and forgings the test pieces, shown in Fig. 4, are threaded. This enables a shorter over all piece to be used and economizes in material. The third piece would have required an 1½ inch thread which would have been too large for the jaws of the testing machine. This piece is therefore shown with a square end, which also enables a shorter grip to be taken than if the piece were cylindrical.

The bend test pieces for steel forgings and castings are shown in the lower part of Fig. 2. The test piece for steel castings must withstand without fracture being bent cold through an angle of 120°, the internal radius being not greater than one inch.

The bend test piece for steel forgings must withstand being bent cold through an angle of 180° without fracture, the internal radius being not greater than ¼ inch.

PICKLING SHIP STEEL

By D. D. Street

THE preparing or "pickling" of ship steel is becoming more common. Whatever compounds may be used for coating steel, it is very necessary that the mill scale should be thoroughly cleaned from the plates and the bars beforehand. The British Admiralty specify the removal of scale by immersing the steel in a bath of dilute hydrochloric acid, which contains one part of acid to nineteen of water. After remaining in some time the steel is taken out, hosed with fresh water and scrubbed with steel brooms. The material is then coated with linseed oil and allowed to stand in the weather for some days. In merchant work this has usually been considered an unnecessary expense, the usual practice being to put the material together without removing the mill scale and allow it to rust up. This rusting process is going on whilst the ship is on the stocks and until she is ready for painting. The rust

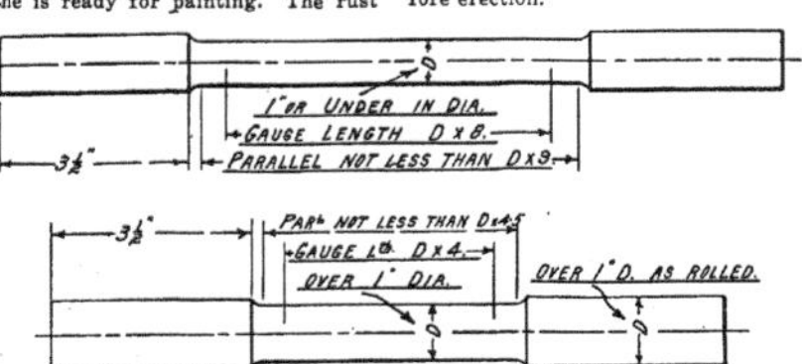

FIG 3. UPPER TENSILE TEST PIECE B FOR BARS. LOWER TENSILE TEST PIECE C FOR BARS.

NOTE.—Test piece B may be used for any dia. bar provided gauge length is turned to less than 1 in. in diameter

TABLE 2—	Tensile Strength	Elongation B	Elongation C
Combination Chamber Stays	28-32 tons	20%	24%
Stay Bars	26-30 tons	23%	28%
Rivet Bar, Boiler	26-30 tons	25%	30%
Rivet Bar, Ship	25-30 tons	25%	30%

and scale is then scraped off and the paint applied. The underwater portion of the ship is not painted before the launch, and the sea-water accelerates the rusting up with the result that when the vessel is placed in dry dock a very good surface can be obtained for the application of the anti-corrosive paint.

There are three principal methods of cleaning forgings, namely, pickling, tumbling and sand-blasting. The pickling is much better than sand-blasting. In the former method all surfaces are cleaned thoroughly from oxides or foreign substances which are detrimental to the machining operation, whilst in the tumbling and sand-blasting fine sand is hammered into the surface. This has a big effect in the machining operations, the cost of machining being three or four times as great with a forging cleaned by sand-blasting as in a pickled forging. In addition, drop-forgings that have small holes or indentations are not properly cleaned except by pickling. It has been found out that the additional cost of pickling is more than effective and is offset by the saving in the cost of machining.

It is not uncommon to find nowadays merchant shipowners specifying that the steel used in their ships should be pickled. It is well-known that the rusting process is not absolutely effective, and the extra expense in the pickling is probably easily saved in the less amount of corrosion occurring and consequent smaller upkeep cost for renewal of steel. At the present time there is another strong argument in favor of pickling. Ships are rushed through very quickly, and if possible, dry docking near or on completion is dispensed with. This means that the bottom of the vessel has to be properly painted before leavin[illegible] the stocks, and it is very doubtful if the rusting-up process has gone far enough such cases, therefore, it is better to be to ensure a good application of paint. In such cases, therefore, it is better to be on the safe side and pickle the steel before erection.

BRITISH SHIPBUILDING ACTIVITY

NEVER has there been greater activity in the shipbuilding yards of GGreat Britain than exists at present. On the four principal shipbuilding rivers of the British Isles every available foot of adjoining land has a ship more or less in construction on it. Nearby engine works are turning out motive power for the vessels at a tremendous rate. Shipbuilders of various sorts—men, women, boys, girls—are all working at top speed. Great Britain has two great objects in all this activity. The first is to help to beat the German U-boat destructiveness and the second is to recoup her greatly theatened prestige as the first maritime nation of the world.

Nowhere is the growing importance of the United States and Japan as maritime powers after the war so seriously taken as in the British Isles, and when the time comes for readjustment on the seas the latter is evidently intending to reassume their maritime precedence, and the British navy, which has been kept intact during the war, will enable authority to be given to these demands. It is well known in America that British builders are rushing out countless destroyers, cruisers and other naval ships, which must for the present remain as "mysteries." Standardization of construction has been adopted in the shipbuilding yards to the greatest possible extent. All parts of hulls, engines—in fact everything—are standard and may be used where first needed. Hitherto thirty-five sizes of steel sections were used in the construction of ordinary vessels. an[illegible] standardization process has reduced the number to eight. Before the war Germany was the only real competitor on the sea and now there are two—friendly may be, but nevertheless competitors—Japan and the United States.

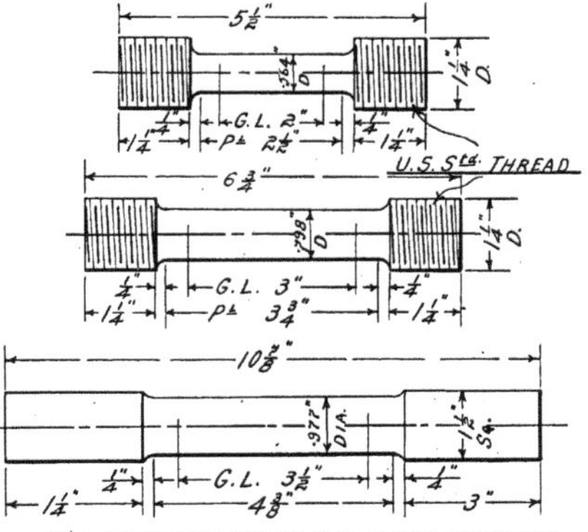

FIG. 4. TENSILE TEST PIECES FOR STEEL CASTINGS AND FORGINGS.

TABLE 3—	Tensile Strength	Elongation
Steel Castings	26-35 tons	20%
Forgings	28-32 tons	29% for 28 tons T. 25% for 32 tons T.

In no case must the sum of the tensile strength and elongation be less than 57 for steel forgings

SHIPYARD FOR BUILDING CONCRETE SHIPS AT MOSS, CHRISTIANIA, NORWAY.

NORWEGIAN PROGRESS IN CONCRETE SHIP DESIGN

As an interesting example of the use of reinforced concrete in shipbuilding, the construction in a Norwegian shipyard of motor-driven concrete ships of 200 tons dead weight capacity may be of interest. The advisability of the use of this material in shipbuilding has been widely discussed of late and during recent years many experimental ships and barges have been constructed by various firms.

The Fougner Concrete Shipbuilding Co., of Christiania, have for some few years been experimenting along these lines and have recently launched the first concrete motor-driven ship, the Nansenfjord, here described.

Numerous obstacles had to be overcome by Mr. Fougner, the founder of the company, and much adverse criticism had to be met, as it was thought that concrete, even when reinforced, was not a suitable material when the high stresses and constant vibration encountered were considered.

The first ship built by this company, the Namsenfjord, is of 200 tons deadweight capacity, and is driven by a crude oil engine, developing 80 brake horsepower and driving the ship at 7½ miles per hour. Three months ago the ship was thoroughly tested out in a heavy gale and has since been employed in the coasting service, carrying cargoes of sand, flour and lumber.

As shown by one of the illustrations the ship was built close to the water's edge and launched sideways.

It is claimed that the concrete ship has the advantages of lower cost, greater rapidity of construction and reduced upkeep expenses, as no painting is required for the hull except that required for the prevention of fouling.

As a result of its initial success the company has now on hand a number of contracts for sea-going ships of various sizes. It is now building standard types of ships of 600 and 1000 tons deadweight, motor-driven, or oversea and coasting trade. In a couple of months the two first vessels of these types will be completed, and we gather that when the yard has been finished as designed it will have a yearly output of about 30,000 tons deadweight.

One of the illustrations shown herewith gives a good view of the present yard and it will be noticed that besides the ship already described, shown on the stocks, several concrete lighters and a reinforced concrete floating dock have also been constructed. The building of these docks is contemplated in sizes large enough to lift vessels up to 15,000 tons deadweight.

The Fougner Co. has recently obtained Lloyd's class A1 for Scandinavian coasting service for its motor ships.

REGARDING THE DESIGNING OF SCREW PROPELLERS

By Raymond E. Lovekin.*

The fact that many marine engineers and naval architects use a great deal of guess work in connection with designing propellers is well known, and is admitted to be so by many of the aforementioned individuals. Realizing that such a condition exists, it indeed seems most discouraging that such an important part of a vessel as the screw propeller does not receive the careful study and attention which it deserves. Of course there have been a great many fairly efficient propellers designed by comparative and guess methods, but where the deviations of hull and power conditions are great, these methods resolve themselves into a conglomeration of useless and unreliable data.

With the above facts confronting all those interested in making a vessel a great success, there are still those who believe that a "good enough" propeller, which will merely produce the desired speed with a given horse-power, can be designed by the comparative or guess methods, in many instances such a propeller has been designed, but it is not realized that a propeller designed by the proper authorities would attain the desired speed with less horse-power, thus cutting down fuel consumption, initial

*Managing Director, American Screw Propeller Company.

THE REINFORCED CONCRETE SHIP NAMSENFJORD ON HER TRIAL TRIP.

weight of propulsive machinery and space required for the same; all of which résolve into a saving of money. For the information of all those interested in efficiency, may it be stated that "the propeller should be the starting point for the designing of marine propulsive machinery," for it is only after this element is correct that the designed engines and boilers will follow along in logically efficient lines.

The many failures which have existed in connection with propulsive problems have been due to the ignorance of the propeller designer in not realizing the importance of his work, and may it be stated that it is a known fact that approximately sixty per cent. of the vessels now in use are capable of being made more propulsively efficient. This fact at first may seem rather alarming, but it has been found to be true by modern authorities who have analyzed many of the screw propellers designed up until a short time ago. Such failures are not generally known, due to the fact that the vessel attains her required speed, and furthermore, because an efficient trial trip (from which reliable data can be collected) was never run. This latter fact is in most cases true in connection with wooden motorships, auxiliary schooners and other less important vessels.

The most important and useful item in connection with screw propeller designing is "the effective horse-power curve," which gives the necessary horse-power to pull a hull through the water at different speeds. Of course, it requires time and a slight expenditure of money to secure this curve, but the value of it can be readily attested to by modern authorities. If such a curve was more generally used, and in the hands of proper authorities, the many failures which have existed could be greatly reduced.

In summing up the above as a whole, the question as to "who can design propellers of the highest efficiency" will no doubt arise. The answer to this is quite simple, as there are several individuals and one company which have for their sole occupation the estimation of power for propulsion of vessels and the designing of screw propellers. The services of the aforementioned can be secured very readily and at a very moderate cost, and it would be advisable for all those who are interested in eliminating waste of power, time and money to get in touch with said parties and profit by their experience, knowledge and equipment.

The designing of a proper screw propeller, as every engineer knows, is a difficult task and requires the above mentioned assets; for without these, it is almost impossible to obtain the high propulsive efficiency sought for by most engineers and attained by so few.

This article is by no means meant as a reflection upon the ability or knowledge of the average marine engineer, as the designing of a proper screw propeller is to-day recognized as a very scientific problem and requires much data, knowledge and experience. Furthermore, the average engineer or naval architect does not have time to specialize upon one subject, due to the many other branches of his profession which he is required to handle.

May this article serve to make more engineers and naval architects take a greater interest in screw propellers and propulsive problems.

THE NORWEGIAN SHIP NAMSENFJORD READY TO LAUNCH. CONCRETE FLOATING DOCK IN BACKGROUND

THE LIFE OF MARINE MOTORS

EVEN until quite recently the average life of a marine motor seemed to be greatly in doubt, and it was generally thought to be much less than that of a steam engine. One is beginning to question whether this is so, for the performances of internal-combustion engines, both large and small, have in point of durability surprised even their manufacturers.

So far it is somewhat difficult to give any reasoned opinion on the actual durability of any type of engine for the simple reason that it usually lasts much longer than the time during which it is economical to run it. In other words, advances in the design and construction of marine motors of all classes have been made so rapidly that before an engine is worn out and fit only to be scrapped, new developments have arisen and new motors have been constructed that are so much more efficient in operation and convenient in use that it pays to do away with the old machine before it is actually worn out.

In spite of this, however, we hear frequently of individual cases of long service; of motors built 15 years ago, when construction was more imperfect, running to-day just as well as when they were first purchased. There is in fact little reason why an internal-combustion engine should not last as long as a steam engine. It is true that the higher speed tends to a greater wear of parts, but only to a minor degree, since lubrication is correspondingly improved and, so far as experience goes, the cost of renewals in motor vessels is not higher than in steamers.

With motor ships, experience extends over seven years only, but in the very earliest vessels that were constructed it has been found that the wear is negligible, and according to present indications a life may be anticipated at least as long as that of the steam engine. Moreover, the motor has a great advantage over the steam engine, since the fuel consumption scarcely varies 2 or 3 per cent., whereas in the latter the steam consumption increases tremendously with age.—*Motor Ship and Motor Boat.*

STEEL SHIPBUILDING IN AUSTRALIA

P. H. ROSS, Canadian trade commissioner at Melbourne, Australia, reports that considerable progress is being made in the shipbuilding industry in Australia. Recently the expert engaged in England to supervise the Commonwealth Government's shipbuilding scheme arrived in Australia and is now completing his examination of the various federal, state and privately-owned plants and yards.

The Commonwealth Government has extensive yards at Sydney; the State Government of New South Wales has a plant at Newcastle and the State Government of Victoria has extensive yards at Melbourne, all of which (independent of the federal shipping advisor) are managed by competent steel shipbuilders with practical experience in British yards.

It is announced that in February there would be sufficient steel in Australia to build seven steamers. At the outset some imported materials will be used but the bulk of the steel will be supplied by the Broken Hill Company's steel works at Newcastle, N.S.W., which, with subsidiary or associated companies, will ultimately be in a position to furnish all the structural steel for the framing and plates required for the construction of steamers in Australia.

The work will be apportioned to the yards in the several states and, while no authoritative statement has yet been made as to the size and class of the steamers, it is considered by experts having knowledge of the facilities available that the initial vessels will be about 330 feet long by 48 feet beam, with a carrying capacity of about 3,500 tons.

Some General Applications of Electric Welding Process*

Special Correspondence

It is now some years since the first use of the various processes of welding iron and steel electrically, although their extended application has been comparatively recent. Much work impossible of accomplishment by other means can be done efficiently and economically by this method, the instances illustrated being of wide variety and considerable utility.

THE electric arc welding process is attaining an ever-widening field of usefulness in practically all branches of industry in which iron or steel are involved in any considerable quantity, and which thus afford opportunities for its application to many purposes incident to manufacture, repair and maintenance.

Although the several processes of welding iron and steel by means of the electric arc have been known for years, their application to useful service in the various industries has been comparatively recent. But since the necessary apparatus has been greatly simplified compared to what it was a few years ago and standard machines and switchboard equipment have been developed and introduced, the process is growing in favor and being generally adopted. In every case its application is favored by reasons of economy and efficiency and often because by means of it work may be accomplished which would be impossible for any blacksmith to do.

Perhaps its chief field of application is in plants such as iron and steel foundries, steel mills, railways, both steam and electric, general repair shops,

FIG. 2. VIEW OF PATCHED TUBE SHEET IN LOCOMOTIVE BOILER.

shipyards, and marine repair shops as well as in general manufacturing work. A few typical instances of the nature of the work to which electric arc welding is put in several of the above classes of service may prove of interest.

Steel Mills and Foundries

One of the most serious mishaps to blast furnaces is to have the tap holes clogged with frozen metal. This necessitates shutting down the furnace and immense loss finally if the trouble is not immediately remedied. A heavy carbon arc provides a prompt and efficient means for overcoming this trouble. Such an arc, using 800 amperes or more, will clear out tap holes or tuyeres at the rate of about 3 feet per hour.

In steel foundries where heavy castings are made the question of cutting off the large risers becomes of considerable importance. This work is quickly and cheaply performed by the electric arc using the carbon electrodes with current as heavy as can be economically obtained. Many large castings are found to be slightly defective, due to misplaced cores, blow holes and cracks. The electric arc provides a means of reclaiming these castings speedily and with low cost by building on metal where needed, such as pads and lugs, also burning out and filling any cracks and holes in the castings. Scrap metal is economically cut by the electric arc into sizes suitable for the cupola.

Rolling mills have found that the electric arc provides a means for saving worn machinery and parts such as shafts, frames, cracked spindles, ladles, rolls with worn wobblers and other parts of rolling mill machinery. These types of equipment involve considerable expenditure if replaced and in many cases the expensive part is not damaged. For example a roll may have its wobblers as good as new. These repairs are easily made at a cost of less than 15 per cent. of the cost of a new roll. In Fig. 1 is shown a repair job of this nature which was accomplished by means of electric arc welding.

FIG. 1. WORN ROLLING MILL WOBBLERS REPAIRED BY ELECTRIC WELDING.

Railway Shops

In equipments used by railroads, both steam and electric, the construction is such that many of the repair operations are well suited for welding by the electric arc. On steam locomotives, for instance, there are a great many wearing parts which can be easily and cheaply made as good as new by the building up process. Worn spots on the wheels are built up in this manner, saving considerable time and money in the actual repair as well as additional saving due to the fact that the diameter of the wheel is not reduced. Worn mud rings can also be built up by welding on new material.

FIG. 3. REPAIRING HULL OF STEEL SHIP, SHOWING SCOPE OF APPLICATION.

Undoubtedly the welding of flues in

*By permission Canadian General Electric Co.

locomotive boilers offers the best field for electric welding in steam railway shops. The larger railroads have adopted as a standard construction the welding of the fire ends of locomotive tubes and since the electric arc affords the only successful means for working on vertical or overhead surfaces, the electric process is depended upon to weld practically all the locomotive tubes. Locomotives when so built, have been known to run the full life of the tube without leakage at the tube sheet.

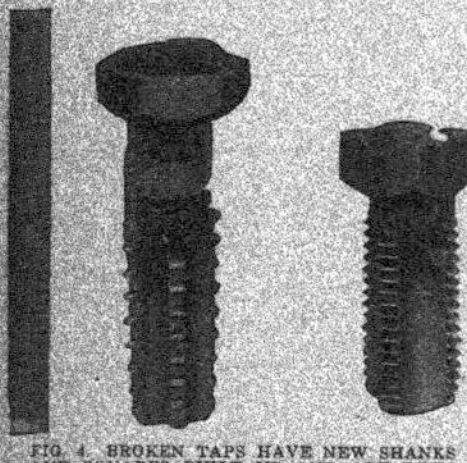

FIG. 4. BROKEN TAPS HAVE NEW SHANKS AND SQUARES BUILT UP AND ARE THEN REMOVED WITH LITTLE DIFFICULTY.

Other repairs to the locomotive which may be made are the insertion of patches on tube sheets or even entire tube sheets and the caulking of various seams and stay bolts to make them steam tight. A view of a patched tube sheet is shown in Fig. 2. Worn piston rods cause leakage of steam and consequent loss of power. This condition is easily remedied by arc welding, additional metal being built on the low spots and the rod then put in a lathe and turned down to proper size. Also fractures in the framework, either of the locomotive or of the cars themselves, can be welded by this process and a successful weld practically ensured.

FIG. 3. CAST IRON GEAR ON PUNCHING MACHINE, REPAIRED WITHOUT BEING REMOVED FROM THE SHAFT.

Marine Service

Inasmuch as ships at the present time are constructed almost entirely of steel, repairs may be made to practically any part of the ship structure by means of an arc welding set.

In a number of large ports, barges, which have been provided with arc welding equipments, are towed to a ship and by means of cable led through port holes repairs may be made to the boilers or in any part of the ship.

The saving by this method consists partly in the fact that the ship need not be docked and partly in that the repairs can be effected without removing the defective part. Broken stern frames, propellers and damaged plates in the hull can be repaired quickly, although it is generally necessary to dry dock the vessel for repairs of this nature. The view of the damaged ship shown in Fig. 3 will afford some idea of the scope of the work it is possible to do with the electric arc.

Miscellaneous Repair Work

One cause of considerable loss in manufacturing establishments is caused by taps or tap bolts breaking off in castings. Ordinarily this would mean either scrapping the casting or cutting out the tap with the gas flame and later building up again, re-drilling and re-tapping. By means of the electric arc the metal is built on to the tap and brought up to the level of the casting and there welded fast to a nut. After cooling, a wrench is used and the tap or bolt removed without damaging the threads already cut. See Fig. 4.

Smoke stacks, ventilating flues, and similar structures are built and repaired economically by the electric process. The arc can be used for cutting the metal to the required size and later to weld it to the desired shape. Complicated joints are made without particular difficulty.

Cutting points of high speed steel are welded on to less expensive shanks of soft steel by this method.

Large parts of machine tools of various kinds are often broken and considerable expense and loss of time is involved if the ordinary methods of repair are followed or if replacement is depended upon. The electric arc, however, provides means for repairing such parts at a minimum expense of time and cost. Fig. 5 shows a cast iron punch gear which has been successfully repaired by the arc weld without being removed from the machine.

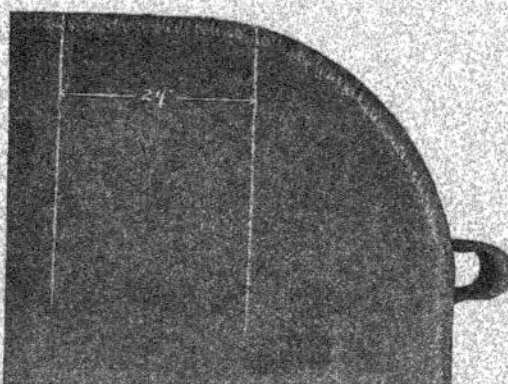

FIG. 6. TANK BOTTOM WELDED WITH METALLIC ELECTRODE ARC.

General Manufacturing

The electric arc welding process is just as desirable in a manufacturing plant as it is in repair work. These equipments are used in automobile plants, manufacturing plants, locomotive shops, tank shops and plants of practically any description where metal is to be joined. A good illustration of this class of work is Fig. 6., showing the bottom of a tank welded by means of

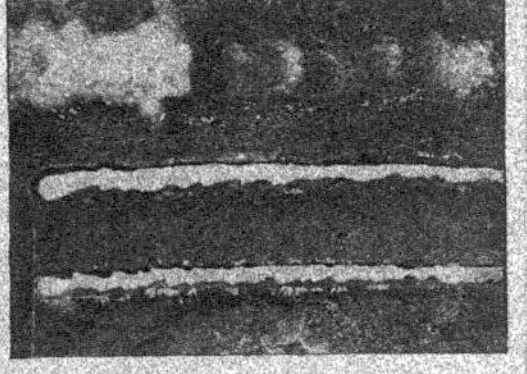

FIG. 7. ILLUSTRATING THE JOINING OF PLATES AND SHOWING THE APPEARANCE OF CUTS MADE BY ARC.

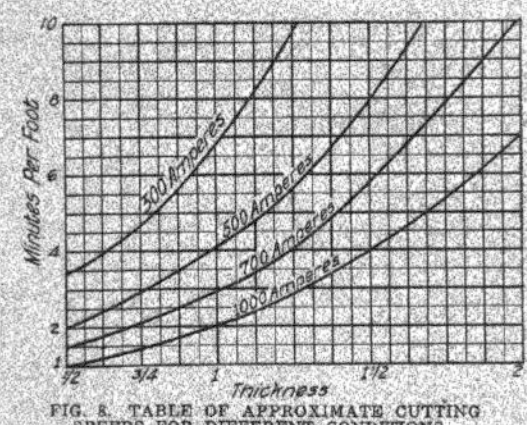

FIG. 8. TABLE OF APPROXIMATE CUTTING SPEEDS FOR DIFFERENT CONDITIONS.

the metallic electrode arc. Fig. 7 shows the weld made in joining two plates of steel by the carbon electrode process, also the appearance of cuts made by the carbon electrode.

Where cutting is to be done by the carbon arc the current used will depend on the cutting speed required. For light metal and where speed is not important 300 amperes is sufficient, but where the metal is 2 in. thick or more it is desirable to use heavier currents and up to 1,000 amperes can be used. Curves are shown in Fig. 8 giving the approximate cutting speed for different sections of metal and using different values of current.

CYLINDER BORING MACHINE

By A. E. Menzies.

THE lessons which have been taught by the war are as varied in character as they are legion in number. Our machine shops are, perhaps, the largest gainers from the lecture halls of necessity.

The shop where the accompanying illustration was taken was for some time engaged in the manufacture of high explosive shells. As the work handled before that was steamboat repairs the staff had to commence at the grass roots. When the shell contracts were completed and the shipping shortage was realised the management decided to get back to their own line and accepted a contract to turn out tramp steamers. As they decided to build their own engines fresh problems were presented to the machine shop for solution.

A New Lease of Life

The first set of cylinders were bored out on a bull lathe; but like every other tool in the shop, this machine had more work than it could handle. The subject of the illustration, a car wheel lathe, built long, long ago by Wm. Collier & Co., Ltd., of Salford, Manchester, had been lying by resting and was shortly to be sent to the foundry when the machine shop superintendent had a vision. The photo is what he saw.

To get the required height above the bed cast iron packing blocks about 3 ft. high were made and the head and tail stocks packed up on them. The shaft, gear and pinion on top of the front head were added, the pinion meshing with the gear built into the face plate.

An 8-in. boring bar secured to the face plate by a flanged coupling and running in bearing placed where the tail stock spindle used to be carries the travelling boring head. The head is driven by a key and travelled by a screw driven by a star wheel. The head used for boring the 62-in. dia. cylinder shown in the illustration has three arms, each arm carrying a tool. Setting up the work is not a difficult job as the bed affords ample surface to block and wedge from. The bar, of course, is used to set up by.

Besides being used for boring cylinders this machine has successfully faced up the ends of a large cylindrical condenser shell. It also came in extremely handy for boring out the recess for the main bearings in the bedplates of engines which the firm is building.

CYLINDER BORING MACHINE MADE FROM CAR WHEEL LATHE.

USE OF OIL TO CALM SEAS

ON ——, the U.S.S. Vulcan, while doing special duty in assisting members of the crew of the ——, had occasion to devise some method of getting of men off the said steamer in a very rough sea when lifeboats would not have succeeded. Owing to the complete success of the method used, it is thought worthy of attention, and the following report

Some Minor Details of Ship Construction and Equipment II

By C. Waldie Cairns, M.Sc.

Although the items dealt with by the author do not in the aggregate account for much in the total cost of a ship, they have a very important influence on the subsequent satisfaction given by the vessel in actual service. Practice in regard to these details varies greatly and unexpectedly in different yards, and he suggests the advisability of adopting a "Standard Fittings Specification," which would insure adequate service and life irrespective of origin.

ONE might examine the whole range of details of cargo gear and find much to criticize; probably many stresses even above those tabulated could be discovered. It may now be asked, "How are there so few accidents with cargo gear?" One help is that even where, say 5-ton gear is specified, a ship may make many successive voyages with only 2-½ ton or 3-ton lifts to make, and when a 5-ton lift is made, a distrustful chief officer will often take special precautions, such as using purchase blocks, preventers, etc. It will be admitted that if gear is installed as 5-ton gear, these precautions should not be necessary; a little care would ensure this at comparatively small expense, if good practice were, by a standard specification or otherwise made common. Most of us know of accidents under present conditions of design.

Much Cause for Criticism

Meanwhile, may the writer suggest that a reasonable basis for stresses in cargo-gear forgings of steel would be that where loads in a single direction only are to be applied, stresses should not exceed a nominal 8,000 lbs. per square inch under working loads, that where reversal of stresses occurs, 5,000 lbs. per square inch should not be exceeded, and that both these figures should be reduced where sectional area of any part is less than 1 sq. in., or a part is less than ⅝ in. thick. These stresses would allow some margin for live load effects, but parts subject to wear should, of course, be diminished in excess.

From this short consideration of stresses on cargo gear, the writer would like to pass to consider for a few minutes some points in cargo-gear arrangements. Many ship general arrangements, when examined, show a fore stay and a back stay to each lower mast, and it will generally be found that these stays have been taken into consideration in the staying of the most against cargo loads.

In view of the fact that derricks, even where fitted in pairs, must swing across the centre, these stays have to be "let go" during working of cargo, unless work is never to be carried on simultaneously at two adjacent hatchways. These stays are useful to prevent undue vibration of the mast at sea, but they should never be counted on for cargo discharging purposes; the necessary support to the mast should be got by spreading out the lower ends of the shrouds sufficiently, and where 10-ton to 20-ton lifts have to be provided for by fitting additional shrouds for temporary use on the opposite side of the mast to the heavy lifting gear with their lower ends still further spread out. These will not seriously hamper discharging from adjacent hatchways, especially as during extra heavy lifts work in adjacent hatches will generally be suspended.

A Neglected Point

A point about heavy lifting-gear sometimes neglected is that the step for the heavy derrick should be got as close in to the mast as possible, so that it may be, as directly as possible, under the point of suspension of the span at lower-mast head, otherwise the guying of the heavy derrick requires heavy tackle. Details of all this gear require careful consideration. Some of the commonly used arrangements of long links, shackles, etc., for span heads are not satisfactory, stresses being excessive in certain positions of the derrick.

Another important point with regard to cargo gear is the means for handling derrick spans. In some cases a purchase block is fitted above the derrick with a guide block on the cross-tree of the mast, forming a purchase tackle, the fall end of the same wire being led to belaying cleats or bitts about the base of the mast. To raise or lower the derrick in this way is rather a dangerous operation as the wire has to be "stoppered" and the end taken to the winch barrel or warping drum, and when the derrick has been lifted to somewhere above the desired position it has to be again stoppered, whilst the wire end is again made fast to cleats or bitts.

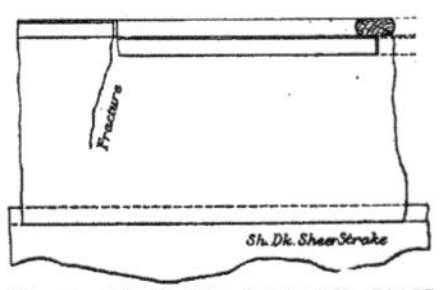

FIG. 6.—FRACTURED BULWARK PLATE DUE TO CARELESS NOTCHING.

To avoid the danger and inconvenience of these operations a topping lift should be fitted to each derrick, the simplest arrangement consisting of a single part span passing over a guide block at the cross threes of the mast with a downhauling tackle, either of light wire or of manilla connected to an eye in the span end, which is at the cross-trees when the derrick is down. With this arrangement the derrick can be lifted or lowered to any desired position and the working span fall then made fast to bitts or bollards.

The writer feels that in touching on cargo gear he has taken a bigger subject than can be reasonably dealt with as part of a general paper. There is undoubtedly room for an examination of the whole subject, and plenty of material for a long and important contribution to the proceedings of the institution.

Erections, etc.

It will be remembered by many of the members that there was a time when so-called erections on a ship were practically neglected in strength calculations. It seemed to be taken for granted that because they were lightly constructed they did not take part in the straining of the ship.

Whilst shipbuilders and others neglected this, natural laws did not, with the result that these erections were subjected to stresses roughly proportioned to their distance from the neutral axis of the vessel, although possibly they did not add very much to the strength, in view of their lightness. The results showed themselves in various ways—upper decks tearing, houses working, and in some cases weakness showing on a strong deck at the abrupt end of a house which stiffened it locally.

In the main parts of the structure of a ship these bad effects are now prevented by suitable means, but there are some minor ways in which the operation of the laws of elasticity show themselves. The writer has in mind a case which he saw recently, where a bulwark plate was fitted abreast deck houses on a shelter-deck ship. The ends of the bulwark plates were shaped for sightliness, and there was no great indication of strain there. The upper edge of the plate, however, at about its mid-length had been stepped, so that a wood rail abreast the house might run with its top at the same level as the bulb-angle stiffener fitted on the remainder of the length of the top edge of the plate. There was thus an abrupt stoppage of the light bulb-angle in addition to the square cut in the plate edge; the consequent concentration of stresses in this corner has fractured the bulwark plate, as in Fig. 6.

It is also found that detached houses well above a shelter deck may acutely feel the straining of a ship in bad weather to the extent that the seams of their wooden top decks will cease to be watertight. These often take up again when bad weather and consequent straining and working of the ship cease. It seems evident that better arrangements on

*Paper read before the North-East Coast Institution of Engineers and Shipbuilders at Newcastle-on-Tyne.

these house tops are necessary—possibly the addition of tieplates would be beneficial.

House Top Considerations

Another position in which house tops do not receive sufficient care in design occurs alongside engine casings. It is common here to attach beam ends to the engine casing side by simple lugs, or even by a strap piece bent at right angles, with a couple of rivets each way, the rivets through the casing side sometimes being so far from the beam that the connection lacks stiffness. A wood deck on such supports certainly does not get fair play, and would give much more satisfaction if a tieplate with angle to casing side were fitted fore and aft on top of the beams before laying any wood. The beams too would probably be better if somewhat stiffer than usual in such positions, and if fitted with small knee brackets. The writer's views on this matter are strengthened by his knowledge of the case of a ship in which such a deck is quite satisfactory on one side of the ship, and is very unsatisfactory on the other side; on the satisfactory side a steel bulkhead parallel to the casing forms an intermediate support for the beams, whilst the beams on the other side of the ship have no such support, there being on that side only a wood bulkhead. This plainly points to lack of stiffness of the beams being one of the principal causes of this deck leaking. There is, of course, not much weight on the deck above, but the beams and their connection evidently take part in the general working of the ship.

A reasonable camber on beams such as those supporting house tops is also desirable. The writer thinks that in many cases the camber of these is too flat. Boiler casing tops too should have very pronounced camber. It is common to see such tops built flat and ultimately sag somewhat, with consequent rapid corrosion.

Deck Machinery—Windlasses

Better provision might well be made for means of closing the chain pipes through the windlass bed-plate when a vessel gets away into deep water. If the windlass makers could give us a little higher lip on the chain pipes on the top of the bed-plate to enable a canvas coat to be lashed to it and also to the cable, there would be more security against getting unnecessary water down into the chain lockers when the steamer "digs her nose into a head sea." At the present time makeshift methods of covering up the closing plate usually supplied for the top of these chain pipes with Portland cement are sometimes adopted. These methods would be more effective if the plate was so cut as to make a more complete closing and to drop a little way into the chain pipe and rest there on an internal lip.

Winches

Shipbuilders are generally shy about lifting the winches high enough off decks to enable chipping and painting to be carried out underneath. Nine inches is a very moderate height for this purpose.

Hand Steering Gear

Most cargo ships with steering engine amidships are still fitted with a hand steering gear aft. This is in practically all cases so slow and clumsy as to be almost useless. In some cases reported to the writer where, through breaking of steering chains, the hand gear aft has had to be put into operation, a smash-up of the gear has been the result. Possibly straining during coupling up with the rudder lashing about may have had something to do with the trouble. At the same time the writer was never very strongly impressed by the amount or disposition of material in these gears; the nuts always appear so very weak in proportion to the large screws employed.

Some builders have, in recent examples, abolished these hand gears in favor of an arrangement for steering by means of the after winches, and with a good arrangement this would appear to the writer likely to produce better control of the ship than could be maintained by even a large number of men at the hand wheel, which must really be too slow for effective control, whether the ship is under her own steam, or broken down and in tow.

Probably an efficient arrangement of this kind would obtain sanction as an alternative method of steering from the registration societies. This system seems worth working out.

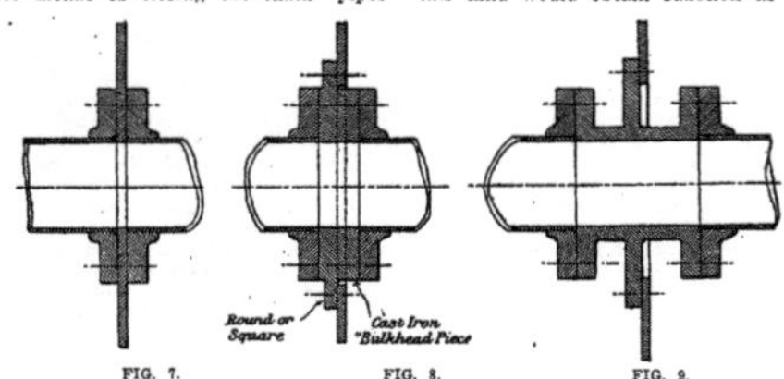

FIG. 7. FIG. 8. FIG. 9.

Steam and Exhaust Pipes

Very objectionable methods of carry-

Small Fittings

tails of ship's equipment causes more trouble during the life time of a ship than port lights or scuttles. As a mechanical contrivance the ordinary side scuttle leaves much to be desired. Tightening is effected by a single screw, the balancing reaction at the opposite side of the glass frame being the inadjustable support of a hinge bolt. Hence; if the rubber packing is hard and rather thick, the means taken by "Jack" to cure the leak that is soaking his surroundings with sea-water—putting the poker-end or a spanner into the loop of the nut, and heaving up tight—is likely to result in a serious straining of the glass frame. Another frequent cause of strain is the effect of a gradual filling up by successive paintings of the slight clearance between spigot part of glass frame and inner diameter of the cast-iron or main frame. Here "Jack" never dreams of the cause of the trouble, and "heaves her up tight"—tight enough, but not watertight.

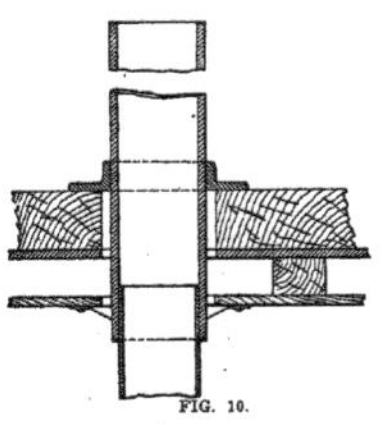

FIG. 10.

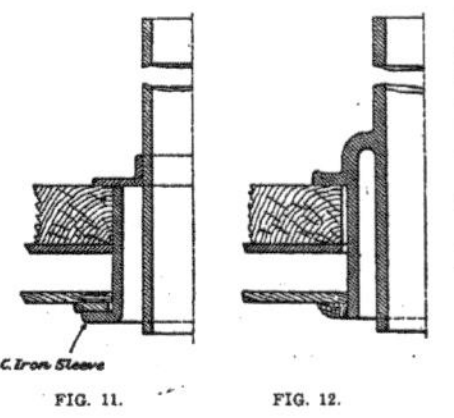

FIG. 11. FIG. 12.

No care in design will ever make a sidelight absolutely "fool-proof." They would certainly be more satisfactory if a hinged screw were fitted on the hinge pin, with sufficient play in the hinge eyes to permit the screw to apply some pressure. The writer thinks this system was at one time on the market, but did not "catch on." A double-jointed hinge would be rather better, with a similar screw, as it would not depend on any play in hinge eyes.

To withstand blows of seas and pressure from fenders, etc., sidelight frames applied by some builders to forecastle and poop sides should be much stronger, particularly when, as usual in cargo-boat practice, cast-iron flanges are used. Any guidance specification in referring to this matter might well stipulate a minimum weight of brass for the glass frame and of cast-iron for the main flange, as well as giving sections of these parts and thickness of glass.

A paper of this length cannot by any means exhaust the list of matters requiring greater attention in ship design and construction. Possibly much that is taken for granted and carried out without question under present practice would not stand criticism if looked into. Take, for instance, a hatch coaming with heavy flange on the lower edge. Considered as a girder, its section is appallingly inefficient, its neutral axis is most unsuitably placed, and the edge which has to stand compressional stress is of the poorest form for the purpose, the hollow half-round moulding being useless except to prevent chafing of the tarpaulin; its rivets are not even in shear to enable it to help the plate efficiently.

Recently builders and the registration societies after a Royal Commission woke up, and have applied bulb angle stiffeners to coamings in certain stress, both by raising the neutral axis and by adding lateral stiffness to the plate. Possibly there are other details where equally radical treatment would be justified, and this paper will not have been useless if it leads other members, in the discussion which the writer hopes to evoke, to draw attention to and, if possible, indicate the necessary corrections for such errors in practice.

(Concluded.)

U. S. AUTHORITIES URGE CO-OPERATION BETWEEN WOOD SHIPBUILDERS

THE possibilities of increasing output from shipyards engaged in construction for the Emergency Fleet Corporation have been subject to investigation recently, with the result that efforts are being made to secure closer co-operation amongst builders along the lines of methods of construction, plant and storage equipment, etc.

According to a recent statement by Mr. James O. Heyworth, manager of the division of wood ship construction, some yards can work 300 to 400 men per ship at the time when the peak of construction of the ship is ready for them; other yards can work rarely more than 160 to 200 men per ship; even some of the old shipbuilders believe 100 to 125 men per ship is the limit for good construction.

Can Rely on Drawings

It must be remembered that in the Ferris standard type the plans are so accurate that much of the woodwork can be done with safety by sawing to the exact dimensions. For your information there are yards where the old shipbuilder was doubtful whether he dared saw to the exact size, leaving from 1 to 3 inches that could be hand reduced after the timber was in place, taking this precaution to be sure that the alignment of the ship would come out right. Experience shows that the plans and drawings can be relied upon.

We have records of yards where they have put as high as 500 men on a ship to advantage. In one yard this has recently caught up previous delay by over two months. We have numerous yards where 300 to 400 men are worked to advantage on one ship. Other shipbuilders that I have personally talked to were fearful of either economic or from the production standpoint that over 150 men per ship could be worked.

This wide variation in conception of the possibilities in construction can be very quickly remedied, as is being done in many yards to-day, and all future production materially increased if the following suggestion is carried out.

The shipbuilders in each district should form an association and meet at least once in every two weeks; remember that a week from now is equal to one month in ordinary times. The district supervisor should meet with this association; the meeting should be formal and carried on in a business-like manner, straight to the point, viz.: To illustrate by photographs and drawings, or papers, or brief talks, the advantages certain yards are obtaining by their methods and equipment and general handling of construction. The minutes of these meetings should be immediately put in proper form and distributed among the shipbuilders of the district, as well as a copy to the home office.

Must Speed Up Production

This is no time to work on the old basis for wood construction. The country needs ships more than anything else. We must have them. Production must be speeded up.

It is logical that a general comparison can be made where so many yards are at work, and it is also logical that many of the yards are being operated along better lines of production than other yards. It is your duty and obligation to the soldiers abroad to take the utmost advantage of this fact and act accordingly.

This division is developing engineers, travelling principally for the purpose of arriving at how the yards are doing at best. Co-operation with these engineers will help, but real co-operation between the shipbuilders of each district, worked out in a proper manner, can double our production in the next six months.

Detroit.—"In comparison with the same period last year, there is about the same quantity of ice in Lake Superior, less in Lake Michigan, about the same quantity in Lake Huron, less in Lake Erie, and more in Lake Ontario," says the second ice bulletin of the season, issued by the United States Weather Bureau. On comparison with the ten-year average there is nine inches more of ice at Duluth, seven more at Sault St. Marie, thirteen more at Mackinaw, and eight at Escanaba.

The MacLean Publishing Company
LIMITED
(ESTABLISHED 1888)

JOHN BAYNE MACLEAN - - - - - - - President
H. T. HUNTER - - - - - - - - - - Vice-President
H. V. TYRRELL - - - - - - - - - - General Manager

PUBLISHERS

MARINE ENGINEERING of Canada

A monthly journal dealing with the progress and development of Merchant and Naval Marine Engineering, Shipbuilding, the building of Harbors and Docks, and containing a record of the latest and best practice throughout the Sea-going World.

B. G. NEWTON, Manager.
J. M. WILSON, Editor. A. R. KENNEDY, Asst. Editor.
Associate Editors:
A. G. WEBSTER J. H. RODGERS (Montreal) W. F. SUTHERLAND

OFFICES

CANADA—
Montreal—Southam Building, 128 Bleury St., Telephone Main 1004.
Toronto—143-153 University Ave., Telephone Main 7324.
Winnipeg—1207 Union Trust Building. Telephone Main 3449.
Eastern Representative, E. M. Pattison.
Ontario Representative, S. S. Moore.
Toronto and Hamilton Representative, J. N. Robinson.

UNITED STATES—
New York—R. B. Huestis, 111 Broadway, New York, Telephone 8971 Rector.
Chicago—A. H. Byrne, 900 Lytton Bldg., 14 E. Jackson Street. Phone Harrison 1147.
Boston—C. L. Morton, Room 733, Old South Bldg., Telephone Main 1024.

GREAT BRITAIN—
London—The MacLean Company of Great Britain, Limited, 88 Fleet Street, E.C. E. J. Dodd, Director, Telephone Central 12960. Address: Atabek, London, England.

SUBSCRIPTION RATE

Canada, $1.00; United States, $1.50; Great Britain, Australia and other colonies, 4s. 6d. per year; other countries, $1.50. Advertising rates on request.

Subscribers, who are not receiving their paper regularly, will confer a favor by telling us. We should be notified at once of any change in address giving both old and new.

Office of Publication, 143-153 University Avenue, Toronto, Ontario.

Vol. VIII APRIL, 1918 No. 4

PRINCIPAL CONTENTS

STEEL DEMAND WILL CONTINUE BRISK

THE question of steel supply is of such ever-increasing gravity that too much importance cannot be attached to the recent agreement arrived at between the Canadian Government and th Dominion Iron & Steel Co. for the erection by the latter of a plate mill.

When the proposal was first put forward some months ago we expressed the opinion that present benefits might be offset later by having to meet renewed competition from the States.

Events continue to show, however, that present benefits are so absolutely essential that future conditions can very well be left to themselves. The immense war program of the United States is gradually assuming definite shape and one of its features is the constantly increasing requirements in tonnage, not only for shipbuilding and munitions, but for military construction work.

Some few weeks ago it was thought that the completion of certain projects would relieve tonnage requirements, but the opposite has been the case.

Prospects of a sufficiency of steel plate in Canada are thereby weakened, while the wisdom of the Government's action is accordingly strengthened. Long before the time plates are flooding the market, the new mill will have justified itself, not only as a war measure but as an economic feature of Canadian industry.

SHOULD CANADA HAVE HER OWN CLASSIFICATION SOCIETY?

THE certainty of a great merchant marine being possessed by the United States in the near future and, presumably permanently afterward, has already given rise to an agitation in favor of an American classification society so that the insurance business resulting from marine activity may be handled by native companies.

A special committee on merchant marine of the Boston Chamber of Commerce recently addressed a report to the Senators and Representatives in Congress, in which they stated that one essential of complete success in American shipbuilding and navigation was a thoroughly American inspection, survey and classification service capable of performing for the United States a work which Lloyd's has long rendered for the British Empire.

In voicing a demand for attention from the officials of the United States Shipping Board, Winthrop L. Marvin states that "under the law creating it, the Shipping Board will hold its huge merchant fleet only temporarily, as a trustee for the American people ,with the avowed purpose of transferring it to citizen owners to be operated by private enterprise as soon as possible after this present war has ended. And the Shipping Board is under obligation to arrange for the inspection and insurance of these hundreds of American ships in such a way that after the war they can remain under the American flag as carriers of peaceful commerce.

"Such a prospect is not very bright if their classification remains under the control of that nation which will be their chief competitor."

In view of the instances of discrimination cited by Mr. Marvin, the parallel of Canada's shipping situation is immediately suggested. Discussion is howover limited by the ultimate intention of the Canadian Government as to whether it will operate the new fleet on a permanent basis.

Such a circumstance might possibly result in it carrying part of its own insurance so as to offset any effort of Lloyd's in favor of British owned vessels, but the discouraging effect on private ownership of such a policy will doubtless be given full consideration at the proper time.

The advent of a Canadian merchant marine under government auspices should carry with it the germ of a future Canadian Bureau of Shipping, which, by placing some of the burdens and handicaps of the past on the shoulders of the country as a whole, would do much to insure the continued benefit of our new fleet in the hands of private owners, should be Government's present action be limited to a certain after war period.

Fo'c'stle Days
or
Reminiscences of a Wind Jammer
By Capt. Geo. S. Laing

The steamer and the motorship have deprived the seafarer of much of the old-time romance attaching to a sailor's life. The advent of wireless has also increased the safety of it. Deprived of these features, a sailor is immediately dependent on his own resources, and Captain Laing gives us a true and seaman-like reflection of sea life under canvas. The present feverish activity in constructing self-propelled craft affords a suitable background for incidents in a career in which gaffs, booms, tar and canvas held sway before being displaced by mechanical unloaders, derrick masts, fuel oil, coal and wire rope.

PART III.

DAYLIGHT PICTURES

IN daylight the flat calm gives two other phases of its character. Provided that the sky is still cloudless, the sea will appear beautifully transparent, and should a piece of broken delf or bright tin be thrown into the water it may be traced to a very great depth as it sinks. Again, if you go out on the end of the jib-boom you can see quite plainly the outline of the ship under water from her forefoot to a point beneath her mainmast. In this way the wily shark is often discovered, for he seems to think that the shade of a vessel's keel makes a good dug-out, while fish of a humbler type get food from the slimy sea-grass that soon grows on a ship's bottom.

Suddenly a series of dark spots on the sunlit waters will denote the coming of wind. These ruffles on the ocean mirror are called cat's-paws owing to their resemblance to a reflection. One puff comes tripping along as if on ice and only has enough wind in it to blow the sails out for a few seconds. Then another puff from an opposite direction has its innings and so on. All points of the compass seem to have a wind each within an hour, yet our barque has never answered her helm to any of the would-be zephyrs.

One day as lime-juice was being served out at noon the word was passed along that a shark was hovering alongside. In a jiffy the shark-hook was baited with a piece of fat pork from the harness cask and the one inch diameter line dangled from the taff-rail. A shoe-block was lashed to the spanker-boom to clear the ship's overhange should we be lucky enough to haul the pirate on board. As sharks have a reputation for taking big bites, a foot of steel chain is attached to the long barbed hook for fear he swallows the outfit entirely and closes his jaws on the Manila rope, which would be easy chewing to him.

Hauling a Shark on Board

Naturally the apprentices were high strung over the fate of this sea monster, and it was not long before the shark obliged everyone by hooking on. Being very sluggish in his previous movements, he now tugged and jerked frantically, but his line was no more than thirty fathoms in length, so his radius was confined. The deck officer called to the watch to "lay aft and get this sardine on board." One of the seamen, who was sick of the calm weather, suggested taking the shark forward, where he could act as a tow-boat. This was only one of the many jocular comments that escaped from the usually stolid fo'c'sle hands.

A 500-pound shark says goodbye to his native lair in a very forceful manner, but we soon had him close up under the boom. The rope was then belayed to a pin and every man took a minute's rest after the heavy pull. The carpenter soon appeared with an axe and a block of wood, having in mind that all sharks must part with their sledge-hammer tails. A shark's "rudder" is equal to a bullock's horns. After slipping a running bowline over the pirate's tail, the real fish line was eased away, and in that manner we pulled our trophy off the poop on to the quarter-deck.

Every deep-water ship carries a shark's tail on the jib-boom end, which is renewed from time to time, so our old seasoned appendage was thrown to the sea and the latest arrival "nailed to the mast." With a stab or two at the heart, our enemy got less vicious, although his heavy thrusts fully demonstrated his abnormal power. Some choice steaks were cut from the fish, and, as fresh food was entirely absent from our menu, the new dish was welcome. Sailors prefer baby sharks about three or four long for cooking purposes, but "any port in a storm."

Superstition generally causes some member of the crew to view the contents of the shark's stomach, on the pretence that a good suit of oilskins and a handy pair of leather sea boots might be found therein. Of course, the inference is that our fish might have taken in some drowning sailor into his inner chambers. Then someone cuts out the spine or backbone, and after drying, the pieces are threaded on a wire string through the marrow holes.

The shark carries very high rank when in his home elements, for he is attended to by a small army of servants. This retinue consists of perhaps a dozen sea slugs, who look after his outer adornments and keep his skin clean and clear of irritants. His chief secretary, however, is a beautifully striped fish weighing perhaps three or four pounds. Sailors call this fish "the pilot," as it would seem to observers that he actually goes on as an advance agent in search of food for his never-satisfied master. The slugs or cleaners came on board on their majesty's back, but the more intelligent "pilot" stayed behind dodging hither and thither about the ship's rudder post in a disconsolate manner. Guess he must look for another job.

It must now be admitted that calms, star-gazing and shark excitement have less captivating attractions for skippers than they have for wanderlust boys. The old man was getting so savage at the thoughts of a protracted passage to Australia that everyone avoided him if possible. Even the cat that used to stay most religiously on the poop was noticed on the knight-heads looking for flying fish. And the rope drill that we had to go through when an imaginary puff of wind hit the old barque almost brought smoke from the sheave holes of the brace blocks.

Baffling winds took the place of the regular trades, and the "Maggie Dorrit" headed when possible for the St. Paul rocks, which are situated about 1° north of the equator and in longitude 27° west. This small jagged reef only raises its volcanic teeth a few feet above the ocean, but is in reality the summit of a giant subaqueous mountain that spreads its broad base miles below on the dark and secret ocean bed that man still knows little about.

"The Doldrums"

In about 5° N. the doldrums set in. This doldrum area has a variable breadth of a hundred to three hundred miles, according to the sun's declination. It is a belt of meteorological samples that floats around the centre of the globe and acts as broken stowage between the two steady trade winds. Incidentally it also has a barometric effect on skippers' and mates' tempers, but outside of that it is a real good drill shed for apprentices who are learning to handle a ship with her labyrinth of ropes, sails and rigging, and her divers manoeuvres.

Moreover, the doldrum area is mostly wet from above, the most delicious and torrential rains in the world. With a sea temperature of 82° F. and the humid atmosphere hovering around 95° F., these equatorial rains are a blessing to the sailing ship crews. Remember our allowance of fresh water was three quarts per man per day for all purposes. Real doldrum rain comes down in blobs the size of a 50-cent piece, and if you look aloft it would appear that you were being thrashed over the face with a fly-swatter, such is the force of the downpour.

The old man came to our half-deck door and laid the law down to the effect that all dirty clothes must be washed, and all bodies scrubbed as long as the opportunity was with us. The scuppers were plugged for a watch while the blankets and other heavy goods were tramped. When the scuppers were freed the ocean looked defiled with the pollution of dirty soap suds, the result of long soiled linen. Sailors' linen is comprised mostly of a dungaree cotton of over-all quality. Oilskins are discarded in the doldrums and the boys trot around bootless and capless, and in some cases without a shirt in this natural Turkish bath. Fid, the ancient mariner, carried a piece of soap in his pocket, and between bracehauling he lathered his clothes that were on him, while a stream of dirt and dye ran out on his bare feet from the leg of his pants. Such is ocean procedure at the equator and one can never forget it.

At last we have worked the old "Maggie Dorrit" through the wet patches and expect to cross the "line" or equator tomorrow in fine, dry weather. Puffs of all magnitudes have helped to blow us here, and we are looking forward with more or less fear to the ancient institution known as Neptune Sports, for the brunt of this aquatic battle must be borne by Hank, Tilbury and your humble servant.

Neptune Sports

This rough-house branch of seamanship at the crossing of the equator was a nonsensical drama, but it was part of windjammer life and cannot be separated from it. It was especially welcome in the passenger carrying sailing ships that took out the pioneer colonists to Australia. In fact, any diversion that lifted the awful monotony and often misery of the men, women and children who undertook the fifteen thousand-mile passage acted like a clear sky after a cyclone. In passages that ran from four to five months, and in many instances the land was never sighted till perhaps Tasmania loomed up, games of the tamest kind appeared to be concerts of distinction to those who for the first time realized the vastness of the ocean and its unhomelike surroundings.

The old man had sent the sailmaker aloft to the fore topsail yard at eight bells in the second dog-watch, and in a muffled voice he hailed the deck thus:

Sailmaker—"What ship is that?"

The Old Man—"The Maggie Dorrit, of Dundee."

The Sailmaker—"Where are you bound for?"

The Old Man—"Brisbane, Australia."

Sailmaker—"Have you any on board who have not been this way before, and are not recognized sons of the sea?"

The Old Man—"Yes, sir, I have three first voyagers who wish to join your venerable society."

Sailmaker—"Then I'll come over your bows to-morrow and give you a clearance document into the open Southern Hemisphere."

The Old Man—"Thank you, Father Neptune; good-night."

The initial palaver was over and next day brought the real game with its crude staging. Under the after skids a topgallant sail was stretched with earings and clues apart to make an improvised font. This was loaded with sea water; the irony of it, for had we not been baptized in the North Atlantic; yes, we had, in fact, tasted the North Sea and English Channel, too. A little advice from the third mate amounted to "take it bravely and you'll make a quick passage." We had orders to retire to the half-deck, where we were blindfolded and told to reduce our clothing to lower topsails, which meant pants only.

Not one of us had the least knowledge of what trend the game was to take, but I was hailed first and led out of the house. The air was simply charged with sympathy; one could hear such remarks as: "This boy will make a very creditable member," and "He has the build and rake of a Columbus"; but the truest of all was: "His mother wouldn't know him."

Over the bows came old Neptune, whose insignia consisted of an oilskin coat and sou'wester cap, long oakum whiskers taken from tarred rigging lanyards, and a harpoon on his long staff. Three pork barrels acted as the officiating chairs, and Neptune (the sailmaker) took the middle seat, being supported by "Chips" (the carpenter) on his left and the third mate on the right. Chips was the doctor and the third mate was the judge. An officer was usually put in the trio for fear the horse play went too far, in which case his commands would be repected and the crisis smoothed over.

I was soon at the foot of the gangway plank, which rested on the quarter-hatch and overlooked the font, and also was within reach of the Neptune council board. A few questions of an impossible nature were asked, such as "Can you tie a sheep-shank in a brace bumpkin?" Answer—"No, sir." "Did you ever see a yardarm sheave get jammed with grease?" Answer—"No, sir."

"I doubt you lack qualifications for this equatorial club. Just give him a smell of your retentive fluid, doctor."

At this juncture I had an evil-smelling mixture shoved under my nose, whilst the rope yarn cork of the bottle was used on my chops after the manner of a shaving brush. The questions that followed were even more dangerous to answer.

"Have you a great longing to remain at sea all your life?" Answer—"Yes, sir."

"Do you ever wish for fresh eggs and milk since leaving the abominable dry land?" Answer—"No, sir."

"Did you ever throw a stone at a sparrow?" Answer—"Yes, sir."

"That saves you from disgrace; now climb up and pay homage to your chief, Father Neptune."

I was partly pushed up the plank, and when Neptune was greeting me with profuse hand-shaking and blessing me with the most extravagant honors, his fist suddenly lost its fratenal grip and with an ungraceful plunge I was in the font.

The thing is done so quickly that one is apt to think that he is overboard, but after a rough handling in the sail tank your eyes are unbandaged and you attempt to grin through the grease and brine that is on you.

First aid comes in the shape of a razor made from hoop iron. With this instrument an attendant scrapes some of the sticky stuff from your face, announcing at the same time that you are now an accepted "son of the sea.". The eye bandages generally come off in the water fight, and you are allowed to view the ceremony at some other boy's expense.

Everyone of us kept our tempers and came off in good standing. The whole thing takes many different forms, mainly depending on the humor and disposition of the captain, who authorizes its execution, while steamboat men pooh-pooh the whole affair and often cross the equator without even passing a salutation to the geographical circle that bisects our globe. Of course, the men always looked for grog at such times and generally got it. These tropical sights and games have at least put a little cheer into the apprentices' lives; and now the Maggie Dorrit is in south latitude showing all her canvas to the south-east trade winds. Our yards are braced sharp up on the port tack, and we head over towards the islands of Fernando Norohno, off the Brazilian coast at Pernambuco. These islands are used as a penal colony for Brazilian criminals, and are under military rule. Nevertheless it is on record that more than one party has got clear of the island and landed on the nearest beach, some 300 miles off.

The "dead horse" was now worked off, which meant that our barque was over a month out from Hamburg. This "dead horse" period is equivalent to the month's "advance note" which poor Jack has to hand over to the shipping and boarding-house keepers, who fleece him during his "half-seas" entanglement on shore. The practice is greatly out of the hands of the "land sharks" now, and seamen have every incentive to save their hard-earned money and behave in a rational manner.

Although a month seems a long time to cover four thousand miles, the reader must remember that in such passages as we are on now, viz., from Europe to Australia, the real sailing takes place on the lower hoops of the South Indian Ocean, from Tristan da Cunha Island in latitude 37° S. and longitude 12° W. to Tasmania. The old barque had travelled this wild track, which is known as the "Roaring Forties," many times, and had daily records of 240, 250 and 300 miles to her credit. She will do it on this trip, too, as we will find out later on.

A few migratory birds rested on our spars as we pushed to the southward. They were almost tame with hunger and soon found out that our galley was infested with cockroaches. It is apropos that I mention here that ocean birds do not avail themselves of this resting opportunity on a ship's spars, but take to the water instead when in need of rest.

The S.E. trades soon freshened and the Maggie Dorrit heeled over to starboard, while the old man was heard to say with glee: "Take your medicine, old girl, you need it after being asleep on the equator for nearly a week."

Forecastle a Misnomer

As the Maggie Dorrit cut a S.S.W. furrow into the South Atlantic and kept a constant deluge of salt spray flying over her weather bow, the shell-backs took each day's task as a matter of course. And looking back at the conditions in which these men found themselves in respect to food, pay and quarters, is it any wonder that the majority of them were too fond of liquor when on shore? No, sir, it is not. Although the apprentices were forbidden to enter the fo'c'sle, there came a time when curiosity overcame obedience—and what a magnificent sight a topgallant fo'c'sle is.

Someone has said: "Pigs is pigs," and shipowners in the middle of last century must have had such a saying put into practice when it came to the builders making the men's quarters. The full and dignified name of "forecastle" is never understood till one enters such a hovel on an old-time windjammer. After that the emphasis on "castle" becomes pronounced, and a qualifying adjective, which is often used when outside the "three-mile limit," is quite appropriate if the conversation has reference to "Jack's palace."

The Maggie Dorrit's fo'c'sle could hardly be described, but it was almost dark, it was always damp, and the aroma could sometimes be defined as a fog.

Every dive that the barque made caused the sea to swish through broken port holes, leaky hawse plugs and scuttle combings. Little ocean rivers wound their courses all over the floor. The overhead deck wept copiously where the Sampson-posts, knight-heads and bowsprit broke through. The pantile barge or biscuit box, with its point-line lanyard and its Matthew Walker knots, swung merrily from a beam above the pall-bits of the windlass. A bully-beef can, half filled with salt slush and having a wick made from unwoven canvas, did duty as a lamp. Among the ocean relics in the British Museum such a lamp should find a place, lest the horrors of the sea go unrecorded. The vile smoke from this outfit would have asphyxiated a Seven Dials cat.

The sides of the fo'c'sle were, of course, the bows of the ship, with no attempt to line or cover in the beams, bolts, nuts, etc.

You could find evidence of paint in every part of the vessel, even to the lower fore-peak, but not in our fo'c'sle. The reader may have seen lumber shanties, railroad camps, and other rough abodes of men, but the sea-dogs' kennel of the old windjammer had everything in the way of human apartments eclipsed. I have seen as much attempt at home surroundings in a downstairs morgue. And on the same old packet we had a beautiful solid teak and highly varnished quarter-hatch leading to the cargo hold aft, which hold was as trim and nicely painted as a new grocery store. The one place was for cargo and the other place for human beings. These same A.B.'s, each one as valuable to a maritime nation as a skilled workman on shore, and maybe more so, were paid $13 per month. Something like 20c a day would pay for the rotten grub that was thrown at the same creatures.

And everyone skins a sailor, not only ashore, but afloat, for at this hard life—and the truth must be told—the clothes (slops) and tobacco that Jack purchases from or through the captain is sold at 200 and 300 per cent. profit and sometimes more. A sailor's tobacco bought duty free from the bonded warehouse at 30c a pound takes on an ocean value of 75c, whilst Blucher boots, bucko caps and gaudy-colored shirts picked up in the slum Jewish houses of London and Liverpool at "knock-me-down" figures were unblushingly given away at "West-end" prices. This is, of course, looking back to a class of vessel and conditions almost obsolete, but these unfair conditions were a real part of the sea life being portrayed.

And yet, with such quarters to live in and food not worthy of the name, these jack tars found occasion to sing songs and play mouth organs as we staggered through the fresh S.E. trades. Each second dog-watch would find the forward crew at some amusement, and a few of them could spin yarns that made the apprentices green with envy. None of them hid the fact that Australia was their final destination, although by law and agreement they had signed articles to stick to the Maggie Dorrit for three years, or until she returned to a home port in the British Isles.

(To be continued.)

SAILORS OF NEW U. S. SAILING SHIPS LEARNING OLD CHANTEY SONGS

THE United States Shipping Board Recruiting Service at Boston issues the following:

Stanton H. King, of Boston, has the only war job of its kind. He is official chantey man for the American merchant marine. His work will be to revive chantey singing among merchant sailors who will join the country's new cargo ships through the United States Shipping Board Recruiting Service, national headquarters of which are at Boston.

While chantey singing has declined on all seas, owing to the change in recent years from sailing vessels to steamers—there not being much opportunity to "heave and haul" on board a steamer—its revival is considered important for two reasons.

Chanteys insure team work when a crew are pulling on ropes, even aboard steamers; while the building of large numbers of American schooners means an increased demand for men who can "reef, hand and steer" on sailing vessels, where chantey singing used to flourish.

The Shipping Board trains men to serve on steamers, but if a certain percentage ship on sailing vessels and carry with them the almost lost knack of chantey singing, they will be the better equipped for their work, according to sharps on the seafaring game.

Stanton H. King probably is the country's best known chantey singer. Chantey singing is part of a weekly entertainment he gives Jack ashore at a mission of which he is head. The programme is usually varied, and to hear Mr. King lead his sailor friends in "Shenandoah," "Bound for the Rio Grande," or "Blow the Man Down" is to understand the psychologic punch of the well-sung chantey.

Mr. King is an old salt, and learned chantey singing in its home, on deep-water vessels. He began going to sea 38 years ago.

TORPEDO DAMAGE TO MERCHANT SHIPS

(Continued from page 89.)

6. Each suction pipe where it enters the compartment from which it drains should be provided with a screw-down non-return or other suitable valve which can be worked from the bulkhead deck. These valves are to be kept closed except when the pumpse are in use.

7. Where ventilation trunks, etc., pass through watertight bulkheads below the bulkhead deck, valves operated from above that deck should, in all cases, be fitted at the bulkheads, and if any trunk pierces the bulkhead at a low level it is strongly recommended that a new lead should be devised to carry it through the bulkhead at as high a level as possible.

8. Valves on all sanitary discharges at the ship's side should be such as will prevent water passing inboard through them when the vessel has considerable trim or list. These valves should be frequently examined and kept in good order.

9. Any ash or rubbish shoots, etc., having upper ends opening on decks below the bulkhead deck should be provided with watertight covers, and instructions should be issued to ensure that these covers are always in place when the shoots are not in use.

Prince Rupert, B.C.—All work at the big dry dock here has been ordered shut down. Although no official announcement has been made, it is stated this move foreshadows the leasing of the plant to American interests.

SEAMANSHIP AND NAVIGATION

Conducted by "The Skipper" *

Articles of direct interest to mariners, discussions of seamanship, navigation rules, and allied topics. Inquiries from readers are invited and will be replied to promptly, through these columns, unless otherwise requested.

RULE OF THE ROAD STUDIES.—III.

IT is well to remember that the rule of the road will always remain a debatable subject. Even lawyers themselves have thought the rules a good enough ground on which to rear voluminous expressions of opinion, when collisions between vessels are sifted under the cold calculations of the Admiralty courts.

This fact should warn young mariners not to split hairs in interpreting the rules but rather assume responsibility where another vessel seems to act in an immovable position. Railroad traffic is tied to the lines, but ships are not, and for this reason alone elasticity of action and interpretation must ever be present in handling vessels under a few peculiar circumstances that apparently no printed rules can set forth.

Both the Great Lakes rule book and the international one as well make it very plain that if you find it an impossibility to abide by the rules in a tight corner, you are at liberty to do something else as long as your motive is to avoid collision—that horrible word that every sailor fears at some time or other in his hazardous calling.

A study of the accompanying chart will help to impress on the mind the meaning of the rules which every prospective candidate for a license must be able to commit to memory. After you get the diagram idea of mastering the rules, a set of home-made cardboard models will be as welcome to your sight as the much-worn checker-board for evening sport.

Sail—Arrows Denote Wind

Clause (a) A vessel which is running free shall keep out of the way of a vessel which is closehauled.

Look to example 1 on the chart.

The lower ship has the wind WSW which is two points abaft her port beam, while the upper vessel must be closehauled to starboard.

Ruling: The lower craft keeps clear, while the upper craft stands on.

Example 2 is another case covering the same clause.

The lower craft has the wind SSE or two points on her starboard quarter, while the upper vessel is closehauled to starboard and steering east.

Ruling: The lower craft keeps clear and the upper one stands on.

Example 3 shows how a vessel in stays or going about can appear to a ship that is running dead before the wind. The lower craft must fall off on one tack or the other, but the free ship in any case must keep clear. Wind is from the NE.

Clause (b) A vessel which is closehauled on the port tack shall keep out of the way of a vessel which is closehauled on the starboard tack.

Example 4. The lower craft has wind at WNW, i.e., she is on the port tack. The upper ship is heading SW and must be on the starboard tack.

Ruling: The lower craft keeps clear and the upper one stands on.

Example 5 gives the lower ship a starboard tack position and on her lee bow you find the port tack vessel.

Ruling: The lower vessel stands on, while the other one keeps clear.

Clause (c) When both are running free with wind on different sides, the vessel which has the wind on the port side shall keep out of the way of the other.

Example 6. Wind at SE. The lower craft has wind free to starboard and the upper vessel is free to port.

Ruling: The upper craft keeps clear, the lower one stands on.

Example 7. Wind at W. The lower craft is free to port and the upper one free to starboard.

Ruling. The lower vessel keeps clear and the upper one stands on.

Clause (d) When they are running free with the wind on the same side, the vessel which is to windward shall keep out of the way of the vessel which is to leeward.

Example 8.—Both vessels have the wind free on the port side. The wind being SSW the westerly ship will be to windward.

Ruling: The last-mentioned craft keeps clear while the lee or easterly ship stands on.

What about two vessels closehauled on the same tack? The book says nothing of such a case. True, but you must use your sailor initiative and remember that a weather ship always keeps clear of a lee one when they are heading on the same course, as vessels closehauled on the same tack must be.

Example 9.—Wind is ESE. All three are free to starboard.

Ruling: The most easterly craft keeps clear of the other two, while middle ship keeps clear of the most westerly one, but stands on for the weather or easterly vessel.

Clause (e) A vessel which has the wind aft shall keep out of the way of the other vessel.

Example 10.—Wind is west, so the westerly ship is running dead before it, i.e., she has the wind aft. How are the other craft going? They are closehauled on opposite tacks.

Ruling: The westerly ship keeps clear while the other two stand on. In fact the vessel with the wind aft keeps clear of all sailing ships with of course the

CHART ILLUSTRATING PROCEDURE OF VESSELS APPROACHING EACH OTHER.

*Editor's Note.—This department is conducted by a qualified ship's master, and will deal with questions involving seamanship and navigation. The attention of young men desiring to obtain certificates as mates and masters is particularly directed to this department.

exception of where she may be overhauled.

This (e) clause is not used on the Great Lakes, but it must be understood by all seamen on our coasts and in all salt water practice. Now that a number of Canadian mates and masters have to get a coastwise certificate too, as well as a lake paper, the division between ocean and inland sailors is becoming thinner and thinner each year. Perhaps in another decade it will have disappeared.

Steamboats

These craft are distinguishable by the mast head light forward. The dots show the general and probable movements.

When two steam vessels are meeting end on or nearly end on, so as to involve risk of collision, each shall alter her course to starboard so that each will pass on the port side of the other.

Example 11.—Ruling: Each steamer swings to starboard till their positions become red to red.

Example 12.—This shows what the end-on rule does not mean. These steamers cannot collide. They are showing green to green.

To be end-on you must see the other steamer's funnel and masts in a line ahead of you, or his masthead lights and sidelights if dark.

When two steam vessels are crossing so as to involve risk of collision the vessel which has the other on her starboard side shall keep out of the way of the other.

Example 13 shows the predicament from one point of view, and example 14 gives you the other impression.

Rulings: In 13 the lower steamer generally alters her course to starboard and passes red to red. In 14 the upper steamer acts in a similar fashion.

When a steam vessel and a sailing vessel are proceeding in such directions as to involve risk of collision, the steam vessel shall keep out of the way of the sailing vessel.

Example 15.—The lower craft being a steamer must keep clear of the other two as they are both sail.

Notwithstanding anything contained in these rules, every vessel overtaking another shall keep out of the way of the overtaken vessel.

Example 16.—The overtaking craft, irrespective of everything else, must keep clear of the overtaken ship. Here we find that a sailing ship must keep clear of a steamer—the upper figure.

The writer has only one experience of this condition really happening to him, but the uncommon predicament is always useful in "sheeting home' 'the meaning of a rule that is apt to out-wit beginners.

This is the third consecutive article on the rule of the road, and the writer feels sure that if the young men who aspire to fill positions as mates and masters will study the Government rules and use either diagrams or cardboard models, the interpretation of safe manoeuvring of ships will come their way and stay there. To commit the rules to memory is only one part of the lesson.

Next month we must take up the compass and gradually work into the dead reckoning methods of navigational procedure.

When you get puzzled or want an explanation write to this office.

NOTICES TO MARINERS

THE following notices to mariners have been issued by the Department of Marine, Ottawa, under dates given:

No. 12 of 1918, March 12
(Inland No. 3)

ONTARIO

(28) Lake Ontario—Beamsville machine gun ranges—Caution.

Aviation range.—Target practice on land and from the air over Lake Ontario by the Royal Flying Corps will begin in the spring of 1918 at the Beamsville machine gun ranges, located about eight miles west of Port Dalhousie lighthouse.

Prohibited area.—The surface of Lake Ontario in front of part of the First Concession of the Township of Clinton, in the Province of Ontario, containing an area of six square miles, bounded on the east and west by lines running due north astronomically into the lake one mile on each side of the point where the median line of Lot 11 cuts the shore, on the south by the lake shore line, and on the north by an east and west line distant three miles north from the said median point, forms the danger area for target practice for machine guns and other firearms for the Royal Flying Corps in Canada and for any other of His Majesty's forces.

Danger flags.—Flagstaffs have been erected on the shore of Lake Ontario, one at or near the extreme east boundary and one at or near the extreme west boundary of the danger zone, and red flags will be hoisted and kept flying on both these flagstaffs whenever firing is taking place.

Buoys.—Fifteen spar buoys spaced half a mile apart have been placed to mark the boundaries of the said danger area, extending northward from the shore to a distance of three miles from the stop butts along the east side of the said danger zone; along the end for a width of two miles; and along the west side for three miles; the said buoys mark the extreme limits of the danger zone on the north, east and west. The said buoys are painted white and are surmounted by a red sign having the words "Danger, Machine Gun Ranges" printed thereon.

Caution.—Notice is hereby given that any vessel or boat passing inside these buoys during the hours of practice incurs serious risk, and no attempt should, under any circumstances, be made to cross the aforesaid area as long as the red flags hoisted on the flagstaffs at the lake shore are left flying.

Penalties.—Any person who, without lawful authority, destroys, injures, removes or alters the position of any flagstaff, buoy or flag above referred to, or moors or fastens any vessel, boat, raft or other thing to any of the said buoys, is guilty of an offence, and liable on summary conviction to a fine not exceeding one hundred dollars, or to imprisonment for any term not exceeding thirty days, or to both fine and imprisonment.

Authority: Order in Council No. 506, dated 7th March, 1918.
Admiralty Charts: Nos. 1152, 678, and 797.
Canadian Naval Chart: No. 63.
Publication: U.S. H. O. Publication No. 108D, 1907, page 133.
Departmental File: No. 39508.

* * *

No. 13 of 1918, March 19
(Inland No. 4)

ONTARIO

(29) Lake Ontario, east end—Simcoe island—Lighthouse moved from Snake island bank to Fourmile point.

The lighthouse heretofore located on the south-east side of Snake island bank has been moved to Fourmile point, Simcoe island.

Former notices.—No. 13 of 1900, and No. 50 (128) of 1909.

New position.—On top of limestone bluff on the west extremity of Fourmile point. Lat. N. 44° 10′ 53″, Long. W. 76° 31′ 37″.

The following sextant angles fix the position of the lighthouse: Ninemile point lighthouse, 0°; former site of lighthouse on Snake island bank, 89° 30′; cupola of Rockwood asylum, 57° 40′; dome of St. George's Cathedral, 29° 49′.

Character.—Fixed red light.

Elevation.—47 feet.

Visibility.—8 miles from all points of approach by water.

Order.—Fourth dioptric.

Structure.—Octagonal tower, with sloping sides; octagonal lantern.

Material.—Tower, wood; lantern, iron.

Color.—Tower, white; lantern, red.

Height.—37 feet, from base to top of ventilator on the lantern.

Hand foghorn.—The hand foghorn operated from a shed on the point has been removed to the lighthouse, and will be continued in use to answer signals from steamers in the vicinity of the station in thick weather.

Authority: Report from Mr. W. H. Carson, District Engineer.
Admiralty Charts: Nos. 1152 and 797.
Canadian Naval Chart: No. 68.
Publication: St. Lawrence Pilot above Quebec, 1912, page 201.
Canadian List of Lights and Fog Signals, 1917: Nos. 1747 and 1747.2.
Departmental File: No. 21747.2C.

* * *

No. 14 of 1918, March 22
(Inland No. 5)

ONTARIO

(30) Lake Erie—Port Burwell—Dredging

Dredging has been done by the Department of Public Works of Canada at Port Burwell during 1917, as follows:

(1) Through the bar which forms at the south end of the breakwater an area 600 feet long and 180 feet wide was dredged to a minimum depth of 15.6 feet below the zero of the gauge, which is 571.8 feet above mean tide, New York.

(2) From the south end of the breakwater to the south end of the harbor entrance piers the area was dredged to a

minimum depth of 18 feet below the zero of the gauge, the outer 350 feet to a width of 140 feet, and the remainder up to the entrance piers to a width of 180 feet.

(3) From the south end of the entrance piers to the car ferry slip the areas dredged were of an irregular shape, on shoals that had formed. Depths provided 16.7 to 23 feet below the zero of gauge.

Authority: Report from Mr. A. J. Stevens, Acting District Engineer, P. W. Dept.
Admiralty Charts: Nos. 332 and 678.
Canadian Naval Chart: No. 80.
Publication: U.S. H. O. Publication No. 108D, 1907, page 92.
Departmental File: No. 37256.

ONTARIO

(31) Lake Erie — Port Stanley — Dredging.

Dredging has been done by the Department of Public Works of Canada at Port Stanley during 1917, as follows:

(1) Through a shoal which forms at the south end of the west breakwater an area 480 feet long and 80 feet wide was dredged to a depth of 23 feet below the zero of the gauge, which is 571.8 feet above mean tide, New York.

(2) From the south end of the breakwater to the south end of the harbor piers the area was dredged to a minimum depth of 22 feet below the zero of the gauge, the outer portion to a width of 80 feet and the inner portion to a width of 160 feet.

(3) From the south end of the harbor entrance piers northward to a point at George street (about 400 feet southward of highway bridge) the area was dredged to a minimum depth of 22 feet below the zero of the gauge, the outer portion for a length of 1,200 feet to a width of 50 feet, and the inner portion to a width of 40 feet.

Authority: Report from Mr. A. J. Stevens, Acting District Engineer, P. W. Dept.
Admiralty Charts: Nos. 332 and 678.
Canadian Naval Chart: No. 80.
Publication: U.S. H. O. Publication No. 108D, 1907, page 91.
Departmental File: No. 37257.

ONTARIO

(32) Lake Erie—Rondeau—Dredging

Dredging has been done by the Department of Public Works of Canada at Rondeau during 1917, as follows:

(1) An area from 40 to 80 feet wide in the east half of the channel between the harbor entrance piers from a point 480 feet north of the outer end of the east pier for a length northward of 300 feet, to a depth of 19 feet below the zero of the gauge, which is 571.8 feet above mean tide, New York.

(2) The dock along the wharf belonging to the Lake Erie Coal Company was dredged to a depth of 19 feet below the zero of the gauge for a length of 975 feet, the width of this dredged area being 50 feet, with the exception of the easterly 150 feet and the westerly 100 feet where the width is 40 feet.

Authority: Report from Mr. A. J. Stevens, Acting District Engineer, P. W. Dept.
Admiralty Charts: No. 332 and 678.
Canadian Naval Chart: No. 80.
Publication: U.S. H. O. Publication, No. 108D, 1907, page 89.
Departmental File: No. 37,287.

No. 15 of 1918. March 27.
(Atlantic No. 8)

NOVA SCOTIA

(33) Halifax Harbor—Bedford basin—Chart issued

New chart.—A chart of Bedford basin, numbered 410 of the Canadian Hydrographic Survey, has just been published by the Hydrographic Survey, Department of the Naval Service of Canada.

Copies may be obtained from the Hydrographic Survey, Department of the Naval Service, Ottawa, for fifteen cents per copy, payable in advance.

Departmental File: No. 28499.

QUEBEC

(34) River St. Lawrence—Off Longue Pointe—Position of buoy—Correction

Previous notice.—No. 118 (308) of 1917.

Buoy moved.—Spar buoy No. 172M, to be placed on the opening of navigation in 1918 in a position 760 feet 75° (E. mag.) from Longue Pointe church, is merely the red and black banded buoy moored heretofore 550 feet 96° (S. 69° E. mag.) from the church, moved 250 feet 26° (N. 41° E. mag.) from its old position.

Lat. N. 45° 35′ 10″, Long. W. 73° 30′ 11″.

Variation in 1918: 16° W.
Admiralty Charts: No. 1127, 2788 and 2830b.
Canadian Naval Charts: Nos. 1, 2 and 22.
Publication: St. Lawrence Pilot above Quebec, 1912, page 97.
Departmental File: No. 25577.

NEWFOUNDLAND

(35) West coast—Port au Port approach—Existence of a rock

Position.—10 4/10 cables 290° (N. 42° W. mag.) from Fox island triangulation station.

Lat. N. 48° 44′, Long. W. 58° 43½′.

Depth.—2¼ fathoms.

Variation in 1918: 28° W.
Authority: British Admiralty N. to M. No. 246 of 1918.
Admiralty Charts: Nos. 422, 2876, 283 and 232a.
Publication: Newfoundland Pilot, Vol. 1, 1917, page 397.

ALL WATERS

(36) Warning to vessels to keep clear of convoys

Masters of vessels are hereby warned that all steam vessels are to keep clear of Convoys that they may meet or overtake.

"War Instructions for British Merchant Vessels" are to be carefully observed. The practice of cutting through a convoy is not permissible.

Authority: British Admiralty N. to M. No. 302 of 1918.

UNITED KINGDOM

(37) Names of Vessels to be displayed on entering Ports

Notice is hereby given that on and after 1st March, 1918, each and every merchant ship entering a port within the United Kingdom shall display her name painted in white letters on a black board on the side on which she is approaching the examination steamer and at such other times, and in such manner as may be directed by the Port Authorities.

The name shall be painted in block letters of such size as to enable the name to be read by the naked eye at a distance of twice the vessels' own length, in ordinary clear weather (visibility=0 in the scale) whether by day or under searchlight beam at night.

Vessels of under 500 tons gross shall display one such board placed over the side in the vicinity of the vessel's bridge.

Vessels of 500 tons gross and over shall display two such boards, one in the vicinity of the vessel's bridge, and the other in the next most conspicuous position over the side.

Note.—The above order does not relieve a vessel of the necessity of complying with the Board of Trade requirements as to the proper equipment of signal flags.

Authority: British Admiralty N. to M. No. 309 of 1918.

* * *

Pilots, masters or others interested are earnestly requested to send information of dangers, changes in aids to navigation, notice of new shoals or channels, error in publications, or any other facts affecting the navigation of Canadian waters to the Chief Engineer, Department of Marine, Ottawa, Canada. Such communications can be mailed free of Canadian postage.

ABERRATION OF SOUND SIGNAL

THE following instance of a silent zone near a fow signal is published for the benefit of those navigators who have not encountered this phenomenon or who have failed to take note of the caution published in the Light Lists issued by the Bureau of Lighthouses and by the Hydrographic Office under the heading "Fog Signals" in the introduction to those books:

"Proceeding to sea from Tompkinsville, Staten Island, Jan. 17, we cleared the Ambrose Channel about 2.45 p.m., and about 500 yards to the westward of the whistling buoy Ambrose Channel Light Vessel was picked up. The wind was light westerly, that is, blowing from the ship toward the light vessel, and the distance from the whistling buoy to the light vessel is about 2½ miles. The weather was somewhat hazy. Soon after sighting the light vessel it was noted that she was operating her steam fog signal and that the sound of the fog signal was inaudible. We passed the whistling buoy within 100 yards and proceeded toward the light vessel, half a dozen observers on the bridge being cautioned to listen for the fog signal and report when it was first heard. Four of these observers were officers. The fog signal was not heard until the light vessel bore 15°, distant about 700 yards, when four observers distinguished the sound. When the light vessel bore 0°, we turned to starboard and the course was set 180°. The signal was not heard after the completion of the turn."

END OF AN INTERESTING CRAFT

THE English convict ship Success, built in 1790, which had been on exhibition in the United States for several years, while making an exhibition tour of inland waters, got into an ice gorge at Carrollton, Kentucky, recently, which tore a hole in her stout teak timbers, causing her to sink. During her eventful history the Success lay for several years at the bottom of Sydney Harbor-Australia, after her career as a convict ship had ended. She was later raised and then started out in her new guise as a museum of the horrors of the time when she was a prison ship.

SPEED REDUCTION GEAR

RECENT developments in the use of prime movers having high rotational speeds have necessitated the employment, in many cases, of some form of speed reduction gear. The difficulties surrounding this application of gearing are well known. High peripheral speeds, proper lubrication, floor space occupied and lack of convenience and general adaptability combine to form a problem trying the designers' ingenuity.

FIG. 1. LOWER HALF OF CASING SHOWING INTERNAL GEAR IN POSITION.

A new form of speed reduction gear has recently been developed by the Poole Engineering Co., of Baltimore, the design being such that it may be used for either step up or step down service.

The high efficiency of the gear has been shown in a recent test where the gear gave an efficiency of 98.5 per cent. as an average of five readings.

The torque of a direct current motor was measured by means of a Prony brake at a certain number of armature amperes, volts and field amperes, the speed also being measured. The motor was then coupled to the gear and the Prony brake attached to the gear. The load was then adjusted to give the same electrical readings and the torque then read. The ratio of this second torque to the first multiplied by the gear ratio gave the efficiency.

$$\text{Efficiency} = T_1 \div T_2 \times 6.666$$

The totally enclosed feature makes it impossible for the workman to get his fingers or clothing caught in any rotating part or gear. This feature also makes for low maintenance cost and permits its use in wet, dirty, dusty or gritty places.

Low Maintenance Cost

After installation there is little or no maintenance expense attached to the operation of this gear. The only attention needed is to see that an adequate supply of oil is in the reservoir. This may be contrasted with a belt or chain drive where constant attention is needed due to the stretching of belts or the wear of chains and sprockets.

Owing to the self-contained nature of this gear the cost of installation is very slight. It is only necessary to provide a firm foundation and to insure that the alignment of the gear and the driving and driven unit is perfect. When mounted in accordance with sound engineering practice this reduction gear is practically noiseless and works without vibration.

The casing is horizontally split, so that by the removal of the casing bolts the tops may be removed for inspection without disturbing the gears or shafting in any way.

As the high speed and low speed shafts are in axial alignment and all tooth pressures are balanced there is no tendency to deflect the shafts and produce unequal stresses on the bearings. The bearings merely have to support the weight of the rotating members. In the case of motors connected to chains or belts there is always considerable side pull on the bearings, this causing wear and often heating.

The high and low speed shafts are each independently supported by the casing. The low speed shaft is supported at each end by heavy duty S.K.F. ball bearings. One end of the high speed shaft is supported in a babbitt split bearing and the other end in a bronze bushing inside the cage, which forms a part of the low speed member. The pinion is cut integral with the high speed shaft and is allowed to float and seek its equilibrium between the intermediate gears.

Three Line Tooth Contact

Another advantage of the construction adopted is that there are three lines of tooth contact as compared to the one line of contact of the ordinary herringbone and pinion. This results in a lower tooth pressure for equal horse-power transmitted and consequently in a longer life.

FIG. 2. GEAR MEMBERS.

The gear will drive in either direction without any changes and both the high and low speed shafts revolve in the same direction. Speeds as high as 6,000 r.p.m. and over are successfully handled.

Turbine Drive

The high economy of the modern steam turbine makes it a desirable prime mover for the operation of various machines. Steam turbines to be economical must necessarily run at high speeds, yet many machines, such as centrifugal pumps, fans, generators, etc., require a speed much slower than that of the steam turbine. To reduce the speed of one or increase the speed of the other is a compromise which materially lowers the efficiency of both units. The mechanical speed reduction gear offers an admirable solution of this problem.

The extensive application of electricity throughout the industrial world has demonstrated the efficiency of motor

FIG. 3. GENERAL ASSEMBLY SHOWING PINION AND INTERMEDIATE GEARS (ONE INTERMEDIATE GEAR BEING REMOVED FOR CLEARNESS).

drive. It is impossible, however, to secure a really slow speed motor and it will often be found advisable to purchase a high speed one and to drive through a reduction gear.

Construction

The gear members are cut on a gear generator which produces a true involute stub tooth of great accuracy and finish. These gears are not subjected to any process such as filing, scraping or polishing after being cut, as this is unnecessary.

This speed reduction gear consists of an internal double helical gear, see Fig. 1, made of a special analysis open hearth steel, forging heat treated to increase its ductility and to insure uniform hardness; a double helical pinion, Fig. 2, cut integral with the high speed shaft; intermediate double-helical gears shown in the same illustration made of high carbon steel forgings with large bronze bearings, mounted on hardened and ground steel shafts secured to the cast steel slow speed member.

The high speed shaft and pinion as well as the slow speed members carrying the intermediate gears are accurately ground all over and are carefully tested for static and dynamic balance before being assembled.

The casing is made of tough, grey, cast iron, horizontally split. It is very substantial and well ribbed to provide a rigid support for the gear members, this insuring efficient and quiet operation.

The high-speed shaft has a central passage through which the oil is pumped, and continuous radial passages in the pinion. The oil after lubricating the gears and bearings is drained to the main oil reservoir in the base where it is filtered, cooled, returned to the pump and used over again.

Operation

The internal gear is held rigidly in place in the casing and is kept from turning by means of two keys which slip into two keyways in the gear and casing. The pinion meshes with the intermediate gears and the intermediate gears mesh with the internal gears.

FIG. 5. STANDARD TURBINE INSTALLATION.

Let us assume that we wish to drive a low speed machine with a high speed mover. The prime mover should be connected with the pinion shaft; this causes the pinion shaft to rotate and turn the intermediate or planetary gears. As the internal gear is held stationary, the intermediate gears must not only rotate but must of necessity revolve about the pinion shaft. This causes the low speed shaft, connected to the spider housing the intermediate gears, to rotate at the desired speed.

FIG. 4. TURBO-GEAR MOTOR LINE SHAFT DRIVE. 100 H.P. GEAR. SPEED REDUCTION 730-122 R.P.M.

Motor Line Shaft Drive

A very interesting application is found in Fig. 4, which shows two 100 h.p. motors direct connected to the line shafts through this device. It will be noticed that the gears and motors are mounted on base plates set upon concrete pillars.

An interesting application of a direct connected reduction gear and steam turbine is shown in Fig. 5. This illustrates a 100 horsepower turbo-gear direct connected to a Terry steam turbine reducing the turbine speed from 3,600 to 440 rev. per min., which is the speed of the driven unit—in this case a blower.

Two of the most noticeable features of this outfit are its compact and symmetrical appearance, together with the absence of offset shafts, the driving and driven shafts being in the same straight line. These features result in a considerable saving of floor space, as well as permitting a better grouping of the machines.

It is frequently found desirable to drive centrifugal pumps, fans, etc., with a low speed prime mover such as a gas engine. Fig. 6 shows an installation of this character where the engine speed is stepped up to the desired pump speed.

FIG. 6. 100 H.P. INSTALLATION STEPPING UP ENGINE SPEED TO THAT REQUIRED FOR PUMP.

LOSS OF HEAT BY RADIATION

A UNIQUE test for determining loss of heat by radiation from boiler settings and steam drums has been run by Mr. C. A. Eastwood, superintendent of Station A, Pacific Gas and Electric Company, San Francisco, Cal. For his experiment Mr. Eastwood selected a 560 h.p. B. & W. boiler, the steam drums of which were covered with one course of common brick. A rectangular can containing a measured quantity of water was placed on top of one drum and the boiler was run at its rated capacity for a period of three days. During this time the rise in temperature of the water was carefully noted, and with these data it was a comparatively simple matter to determine: (a) the amount of heat radiated per square foot of surface per hour; (b) the amount of heat radiated from the total exposed area of the drums; and (c) the quantity of fuel which was being burned to make up for this loss. Mr. Eastwood's figures showed that the loss totalled 390 barrels of oil per year.

To determine how much of this fuel-loss could be saved by the use of an effective heat insulating material, one 2½-inch course of Nonpareil Insulating Brick, manufactured by the Armstrong Cork & Insulation Company, Pittsburg, Pa., was then placed on top of the common brick covering and readings were taken as in preceding tests. The result showed that with the Nonpareil insulating brick in place the loss of heat amounted to 144 barrels of oil per year —a saving of 246 barrels, or 63 per cent. With oil costing 70 cents per barrel, the saving amounted to $172.20 per year.

In considering these figures there are two points in connection with them that might be mentioned. First, the total amount of heat lost represents that radiated from the steam drums alone; second, only 2½ inches of Nonpareil brick were used. Had the radiation from the exposed boiler walls been included as well, the total heat-loss would have been considerably increased. Furthermore, if the common brick had been removed entirely and replaced with a course of Nonpareil brick set on edge, giving 4½ inches of insulation instead of 2½ inches, a still greater saving would have been effected, since, according to tests, Nonpareil brick are said to be ten times better heat insulators than either common brick or fire brick.

COTS OPEN AND SUPPORTED BY CHAINS.

FOLDING COTS FOR MARINE USE

ONE of the most interesting details regarding the naval preparations being made by the United States has to do with the accommodation provided for the crew of the submarines.

Life on a submarine is arduous in the extreme. The cramped quarters, impossibility of taking proper exercise or recreation and the damp and at times impure atmosphere make the life somewhat unenviable from the standpoint of comfort.

It is a commonly known fact that floor space or space of any kind is many times more valuable in the submarine than anywhere else, due to the unusually large amount of intricate mechanism required to operate the vessel and its offensive armament.

This lack of space seriously affects the amount of room available for quarters for the men and results in the placing of the berths or hammocks in any space available. This has resulted in the devising of many ingenious arrangements in order to conserve the precious floor space.

COTS COLLAPSED AND FOLDED AGAINST SHIP WALL

Of these, the swinging cot provided for each member of the crew, and recently adopted by the U. S. Government, is worthy of mention.

A late type of these cots, for which the Englander Spring Bed Co., Bush Terminals, Brooklyn, N.Y., recently secured the contract, are to be suspended by chains, thus avoiding the possibilities of broken ropes.

As will be seen from the illustrations these cots fold up against the side of the ship when not in use and in this position take up very little room.

The first illustration shows how the cots appear when slung in position for sleeping. The backs are provided with special hinge fastenings which are bolted to the bulkhead, while the fronts of the cots are supported by means of the chains. The hinge arrangement allows the cots to be folded flat against the bulkheads whenever they are not in use by simply unhooking the chains and folding the upper cot down and the lower one up. This is very clearly shown in the second illustration. These cots should be of considerable use wherever sleeping quarters were restricted.

POWER HAMMER WITH ADJUSTABLE TAPER GIB

THE accompanying illustration shows a recent design of Fairbanks power hammer as built by the United Hammer Co., Boston, Mass., in which provision has been made to take up wear on the ram guides, so that a much finer ram adjustment can be obtained than hitherto.

The hammer is of the overhead crank type with adjustable stroke; the capacity under the ram being altered to suit requirements by adjustment provided on the connecting rod. The detachable anvil block is securely strapped to the column, which has an opening to accommodate work of unusual length or shape.

The ram guides are fitted with an adjustable bronze taper gib taken up accurately and quickly, and the gib and face plates have been so designed that they may be fitted to hammers of this type, which are already in use.

LIFE-SAVING SHIP DAVIT

THE accompanying illustration shows a ship's davit which, while primarily designed to save life during accidents and fires at sea, is also adaptable for all general purposes that the ordinary ship davit is used for.

In accidents at sea, whether the ship is in a listing condition or on fire, it is of the utmost importance that the life-boats, when full of people, be launched as far out at sea away from the sinking ship as possible, and with this davit, this is quickly and easily done. The construction of this davit is such that the occupants in the lifeboat may lower themselves safely into the water, where

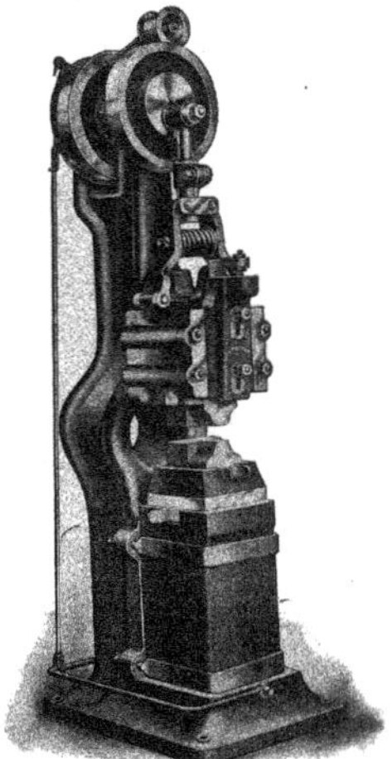

POWER HAMMER WITH ADJUSTABLE TAPER GIB

the boat, when the controllers are ready, automatically releases itself from the supporting cables.

Inasmuch as the occupants of a life-boat lower themselves to the water, this feature enables the last one on a sinking ship to be saved, since no one is required to remain on deck to assist in lowering the last boat and then take their chances by jumping into the sea and swimming to the lifeboat (which they may and they may not succeed in doing). During accidents at sea more or less excitement will prevail—personal accidents are liable to occur and deprive a man of his usefulness—he may lose his hold on the lowering cables and let the lifeboat fall into the sea; with this davit, such an accident is practically impossible, since the lowering mechanism is at all times under the control of quick-acting brakes, and if necessary the lifeboat may be halted in mid-air on its downward trip.

The inventor, Edward S. Jones, of Mobile, Ala., further states that parties leaving the ship may lower themselves, and on their return connect the lifeboat with the cables and raise themselves without assistance from anyone on the ship, although assistance from the deck can be given if required.

NOVEL DESIGN OF SHIPS' DAVITS FOR LAUNCHING LIFE-BOATS

THE IMPROVEMENT OF EUROPEAN SMALL CRAFT

By D. J.

NOT the least important of the many problems in industrial efficiency is that dealing with local transport. Vessels from overseas bearing the raw materials for manufacturing industries are discharged often at great distances from the exact locality where the raw material is required; commodities for foreign markets are manufactured inland and must be transferred to the seaports for shipment abroad. That the railways are capable of dealing rapidly with a large volume of this traffic goes without saying, but for non-perishable goods there is no question but that water transport is always cheaper. Hence there has grown up in recent years a steadily increasing volume of trade around coasts and upon inland waterways.

The coastal trade is of a totally different nature from the canal trade, although perhaps both methods are available for transport of goods between two specified places; but whereas, if the material is transported by sea, vessels capable of withstanding the effect of storms and rough seas are required, for inland traffic the frailest craft, provided it is watertight, can be employed. Nevertheless, whilst the problems of supplying suitable craft for these diverse systems of transport are so vastly different from one another, the problem of the improvement of existing craft of both types is, strangely enough, the same. The great drawback to the employment of local waterborne transport has been that the time taken is generally much greater than when railway transport is utilized, and any attempt to improve the former must of necessity be directed towards the reduction of the time taken.

Considering first of all the coastal trade. It will be found that whilst in one or two specified trades are steamships employed in this class of work, such as the coal traffic and the fish traffic, quite a large proportion of it is still carried on in small sailing craft, and even where steamships are employed these are nearly always of small dimensions and fitted with engines of such power that only low speeds are possible. Since the value of a ship can be measured by the number of complete voyages it can make per annum, it will be apparent that if by employment of machinery of higher power, rendering higher speeds possible, the time of each round trip can be materially reduced, the earning capacity of the vessel will be greater, other things remaining the same. The difficulty has been, however, that more powerful machinery, i.e., larger boilers and engines, have required more space and in consequence the hold space and therefore the earning capacity of the vessel is reduced, the economical compromise being reached with vessels of such power that the speed is somewhere in the region of eight knots.

But the advent of the motor engine would appear to have changed all that. In the internal-combustion engine, working on the Diesel cycle—or on the semi-Diesel cycle for the smallest coasting vessels—we have a source of power occupying but a small space, capable of utilizing almost any kind of crude oil, simple to manage, and particularly suited for the development of small power; in short, exactly the type of engine required for small coastal vessels. Hence it may be asserted that in the future there should be a wide and profitable field for the employment of moderately sized coasting vessels fitted with oil engines of sufficient power to give them a speed of at least 12 knots. One other factor will be essential for the efficient employment of this means of transport, and that is, the means for the rapid handling of cargoes at the terminal ports.

Inland Craft

Turning next to the type improvements possible in craft used upon inland waterways, here development in size is impossible. The size of craft is fixed to a large extent by the size of the canals, the locks and the tunnels through which the canals pass, and save in the case of new canals, which could, of course, be built upon more generous lines, any improvement in speed must be accomplished without any increase in dimensions of the vessel. Existing types of barges can, however, be fitted with small engines, capable of rendering them more speedy than the horse-drawn barge, and so that barges not fitted with engines may be utilized, but at a far greater speed than hitherto, tugs of extremely small dimensions can be fitted with powerful engines. It is the internal-combustion engine that has rendered this development possible, and though of smaller type than for coastal vesels, it is the same type of oil engine that is utilized.

But there is another source of power available for use on inland waterways which could not be employed on coastal vessels. Electricity has already been utilized and it possesses amongst many others, the advantage of not giving off poisonous gases, such as are given off in the exhaust of oil engines, and which, upon canal systems where the tunnels are long, may render the employment of oil engines impossible. Two systems of electric propulsion have been employed, one in which a storage battery carried on a barge gives electric energy to a motor, or motors driving propellers, or operating a winding gear, and the other in which energy is obtained from an overhead wire by a method similar to that adopted on many tramway systems. Neither of these systems has been developed to any great extent, but both are capable of being economically employed in any great scheme of rapid canal transport.

THE highest steam pressure now in ordinary use for steam turbines is 300 lb. per square inch, but boilers are under construction in America for a pressure of 350 lb. per square inch, and one has been in successful operation in the British Thomson-Houston Company's works for more than a year. The superheat is chosen so as to make the initial steam temperature fall between 600 deg. and 700 deg. Fah.

ASSOCIATION AND PERSONAL

A Monthly Record of Current Association News and of Individuals Who Have Been More or Less Prominent in Marine Circles

E. P. Lamport, of Lamport & Holt, Ltd., whose death took place March 9, was a well-known member of British shipping circles.

C. N. Schrag, of the Bawden Machine Co., is leaving there the 1st of May, to go into partnership with O. W. Meissner, manufacturers' agent, Montreal, handling a line of railroad, pulp mill, marine and engineering supplies. Mr. Schrag will open a Toronto office in the Royal Bank Building.

James M. Smith has arrived at Fort William to take charge of the shipbuilding and of the plant of the Canadian Car and Foundry Company, which is under contract to turn out twelve trawlers for the French Government before the close of navigation. Most of the steel for their construction has been assembled at the works, and the building in which they will be built is almost completed.

Owen Sound.—One of the oldest residents of Owen Sound section passed away recently in the person of Captain Alexander McKenzie, East Linton. For many years he has been known on the Great Lakes, where he has commanded many boats during his 54 years as a sailor in Canada. He was born in Ross-shire, Scotland, 74 years ago, and came to Canada when 20 years of age, living first at Toronto. The late Capt. McKenzie was for many years in command of various schooners and was a sailor of the old school, one who felt for years that sailing on anything except a schooner was not sailing at all. He was a strict disciplinarian with his crews, but always commanded their respect.

DIRECTOR-GENERAL OF FLEET CORPORATION

Sincere gratification is expressed in steel circles concerning the appointment of Charles M. Schwab, chairman of the Board of Directors of the Bethlehem Steel Corporation, to the post of Director General of the Emergency Fleet Corporation. Mr. Schwab has the confidence of the entire country and is known as a competent organizer of labor. He has thorough knowledge of men and his wide experience in the manufacturing and business affairs, is assurance that he will conduct the affairs of the Emergency Fleet Corporation to the satisfaction of the Government and of the country at large. His knowledge of steel manufacturing and shipbuilding is second to none in the nation. President Wilson is to be congratulated upon his selection of Mr. Schwab to head the Emergency Fleet Corporation.

Captain McGough, one of the veterans of the old sailing vessels, has been appointed acting harbor master in place of Captain Murray, who was killed in the Halifax disaster last December.

John Hole, Sr., who for the past two years has been assistant superintendent of construction for the Toronto Harbor Commission, is leaving the commission to enter private practice as a general engineer and contractor.

H. B. Smith was re-elected president of the Owen Sound Shipbuilding Co. on Friday, March 8. This was the time of the annual meeting, when a satisfactory year's work was reported. The officers: President, H. B. Smith, Owen Sound; vice-president, J. W. Norcross; second vice-president, R. M. Wolvin; secretary-treasurer, S. H. Lindsay; directors, Messrs. H. B. Smith, J. W. Norcross, R. M. Wolvin, H. W. Cowan, F. S. Isard, Capt. McDougall and H. Dyment.

The British-American Shipbuilding Co., who leased part of Beatty & Sons' plant at Welland, Ont., will shortly launch the first 3,500-ton steel freighter, which they are building for the Imperial Munitions Board. The vessel is 261 feet long, 43 feet broad, 23 feet deep. On the second hull the keel and floor frames are down and about 50 per cent. of the steel has been fabricated. Arrangements are being made for a third berth. The officials of the company are as follows: H. F. S. Estrup, president; H. A. Sifton, vice-president; H. M. Balfour, general manager; H. D. Davison, treasurer, and G. A. Leonard, secretary.

HUGE SUM FOR U. S. NAVY

The Naval Appropriation Bill, carrying approximately $1,312,000 immediately available to meet the navy's war requirements, was passed unanimously by the House late Saturday, April 20, without a record vote.

LICENSED PILOTS

ST. LAWRENCE RIVER

Captain Walter Collins, 43 Main Street, Kingston, Ont.; Captain M. McDonald, River Hotel, Kingston, Ont.; Captain Charles J. Martin, 13 Balaclava Street, Kingston, Ont.; Captain T. J. Murphy, 11 William Street, Kingston, Ont.

ST. LAWRENCE RIVER, BAY OF QUINTE, AND MURRAY CANAL

Captain James Murray, 106 Clergy Street, Kingston, Ont.; Capt. James H. Martin, 259 Johnston Street, Kingston, Ont.; John Corkery, 17 Rideau Street, Kingston, Ont.; Captain Daniel H Mills, 272 University Avenue, Kingston, Ont.

MONTREAL PILOTS' ASSOCIATION

President—Alberic Angers, Montreal.
Secretary—C. B. Hamelin, Champlain, Que.

ASSOCIATIONS

DOMINION MARINE ASSOCIATION

President—A. A. Wright, Toronto. Secretary—Francis King, Kingston, Ont.

GREAT LAKES AND ST. LAWRENCE RIVER RATE COMMITTEE

Chairman—W. F. Herman, Cleveland, Ohio. Secretary—Jas. Morrison, Montreal.

INTERNATIONAL WATER LINES PASSENGER ASSOCIATION.

President—O. H. Taylor, New York.
Secretary—M. R. Nelson, 1184 Broadway, New York.

SHIPPING FEDERATION OF CANADA.

President—Andrew A. Allan, Montreal; Manager and Secretary—T. Robb, 218 Board of Trade, Montreal; Treasurer, J. R. Binning, Montreal.

SHIPMASTERS' ASSOCIATION OF CANADA.

Secretary—Captain E. Wells, 45 St. John Street, Halifax, N.S.

GRAND COUNCIL N.A.M.E. OFFICERS.

A. R. Milne, Kingston, Ont., Grand President.
J. E. Belanger, Bienville, Levis, Grand Vice-President.
Neil J. Morrison, P.O. Box 238, St. John, N.B., Grand Secretary-Treasurer.
J. W. McLeod, Owen Sound, Ont., Grand Conductor.
Lemuel Winchester, Charlottetown, P.E.I., Grand Doorkeeper.
Alf. Charbonneau, Sorel, Que., and J. Scott, Halifax, N.S., Grand Auditors.

1918 Directory of Subordinate Councils, National Association of Marine Engineers.

Name.	No.	President.	Address.	Secretary.	Address.
Toronto,	1	Arch. McLaren,	324 Shaw Street	E. A. Prince,	49 Eaton Ave.
St. John,	2	W. L. Hurder,	209 Douglas Avenue	G. T. G. Blewett,	36 Murray St.
Collingwood,	3	John Osburn,	Collingwood, Ont.	Robert McQuade,	Collingwood, Ont.
Kingston,	4	Joseph W. Kennedy,	395 Johnston Street	James Gillie,	101 Clergy St.
Montreal,	5	Eugene Hamelin,	Jeanne Mance Street	M. Lasure	120 Ribard St.
Victoria,	6	John E. Jeffcott,	Esquimault, B.C.	Peter Gordon,	808 Blanchard St.
Vancouver,	7	Isaac N. Kendall	819 11th St. E., Vanc.	E. Read,	232 13th St. W.
Levis,	8	Michael Latulippe,	Lauzon, Levis, Que.	Arthur Jolin,	Lauzon, W.
Sorel,	9	Nap. Beaudoin,	Sorel, Que.	Alf. Charbonneau,	Box 204, Sorel, Que.
Owen Sound,	10	John W. McLeod	570 4th Ave.	R. J. McLeod,	662 8th St.
Windsor,	11	Alex. McDonald,	28 Crawford Ave.	Neil Maitland,	1 London St. W.
Midland,	12	Geo. McDonald	Midland, Ont.	A. E. House,	Box 333
Halifax,	13	Robert Blair	29 Parrsboro Street	Chas. E. Pearce,	Portland St., Dartmouth, N.S.
Sault Ste. Marie,	14	Charles H. Innes,	27 Euclid Road	Wm. Hindmarch	34 Euclid Rd.
Charlottetown,	15	J. A. Rowe	176 King Street	Chas. Cumming,	27 Euston St.
Twin City,	16	H. W. Cross,	436 Ambrose St	J. W. Farquharson,	189 College St.
				A. H. Archand,	Champlain, Que.

MARINE NEWS FROM EVERY SOURCE

PORT COLBORNE.—The steamer Wyoming cleared here on the 17th for Buffalo dry-dock. There is no ice in sight.

Toronto.—It is expected that the steamer Cayuga, of the Canada Steamship Line, will go into commission on May 18th between Toronto, Niagara-on-the-Lake, Queenston, and Lewiston.

Victoria.—The C. P. R. steamers Empress of Asia and Empress of Russia have been commandeered by the Dominion Government. The order has already gone forth that the sailings of the ships have been cancelled.

Newcastle, N.B.—At the shipyard at Rosebank, the work on the vessel under construction is proceeding satisfactorily. This vessel is a four-masted schooner of 540 tons. It is expected that she will be launched about August 1st. The keel for another vessel is also partly laid.

Halifax.—The steamer City of Wilmington was burned south of Sable Island on Saturday. The fire broke out suddenly early in the morning, and, despite the efforts of the crew, it spread rapidly throughout the ship. "S.O.S." signals were flashed out and preparations were made to abandon the steamer. The weather was fairly favorable and the boats were launched, provisioned and the crew entered them.

Cleveland.—The steamer Harvester, ter, which left South Chicago for Port McNicoll with a cargo of oats, was the first boat of the grain fleet to sail, and she may have trouble making port, as it was reported early in the week that the ice in Georgian Bay is pretty heavy. The Harvester has 640,000 bushels of oats, which is the largest grain cargo ever loaded at a Lake Michigan port.

Toronto, Ont.—Reclamation work involving an expenditure of $1,732,000 according to the gross estimates which have been approved for the year will be undertaken by the Board of Harbor Commissioners of Toronto this summer. So far as is known at present no work will be done over at the Island, and the entire activities this summer will be in reclamation work in the industrial area east of Cherry street and along the water front from Bathurst street east, according to the officials of the commission. The commission is at present engaged in rebuilding the passenger sheds of the Toronto Ferry Co., which were burned this winter.

WM. NEWMAN JOINS THE U.S. FLEET CORPORATION

Well Known Toronto Architect Takes Responsible Position on War Work

With the reorganization of the plant of the United States Fleet Corporation at Hog Island, Philadelphia, the largest shipbuilding company in the United

WILLIAM NEWMAN

Naval architect and works manager of the Polson Iron Works, Toronto, who has left that company to join the United States Fleet Corporation.

States, Toronto loses Wm. Newman, naval architect and works manager of the Polson Iron Works & Steel Shipbuilding Company, who will in the course of a few days become connected with the American company.

"I am leaving for the coast in a few days," said Mr. Newman, "but the chances are that I will remain in the East when I make my final decision as to where I will locate myself."

Given a Purse of Gold

Upon severing his connection with the Polson Company, Mr. Newman was made the recipient of a handsome purse of gold presented by Col. J. B. Miller on behalf of the company.

Detroit.—Built in 1879 and for many years considered one of the finest vessels on the Great Lakes, the sidewheeler passenger boat Idlewild was sold to-day to Texas parties, who will operate her in freight trade out of Texas ports. The boat has been out of commission ten years. She formerly ran between Detroit, Toledo and Port Huron, and later was placed on a St. Lawrence River route. The Idlewild is an iron hull vessel, 150 feet in length, 26 feet wide and 47.4 feet deep. She was constructed at the Wyandotte yards.

Victoria.—That 4,001 men are being employed in the shipyards of British Columbia controlled by the Imperial Munitions Board was a fact brought out at the afternoon session of the shipyard wage inquiry recently.

Kingston.—The Pyke Towing and Salvage Company shipped pumps and outfit to Trenton to raise the steamer "Rideau Queen," sunk in fifteen feet of water. As the boat lies partly on her side and in an eight-mile current, considerable difficulty is expected.

Ottawa, Ont.—Appropriations for public works chargeable to capital include $550,000 for Toronto harbor improvements. The sum to be voted is $450,000 less than was voted last year, and moreover includes a revote of $250,000, unexpended, of last year's appropriation.

Port Wade.—Capt. Howard Anderson, H. B. Short, of Digby; Joseph Gaskill, of Grand Manan, and Capt. John W. Snow, of Port Wade, are establishing a shipbuilding plant at Port Wade, just across the harbor from Digby, and will commence at once a three-mast schooner.

Sault Ste. Marie.—Ferry service between the two Soos was resumed on the 22nd by the ferry Algoma breaking through the icebound channel after an interruption of two months. Very little trouble was encountered, the trip being made in ten minutes, about the usual running time. Regular schedule trips will begin shortly.

Vancouver, B.C.—The British-American Shipbuilding Co. is negotiating for a site for a plant on the Kitsilano Indian Reserve. Plans provide for the laying of eight keels at once and a building programme of about $12,000,000. Dredging to cost $60,000 may be required. S. Mathieson, of Vancouver, is interested.

Kingston, Ont.—Just as soon as the harbor is cleared of ice the Kingston Shipbuilding Co., owned by the Canada Steamship Lines, will launch another trawler. Within a year the company has turned out four trawlers. They will be put into commission on the sea, the estimated cost of each being $75,000. The company has a big staff of men at work.

Victoria, B.C.—That the Victoria Machinery Depot has been offered contracts by the Dominion Government for the building of three steel ships of 5,100 tons each and that it has declined the offer is stated by C. V. G. Spratt, head of the company.

Toronto.—The city has been given permission by the Private Bill Committee to make a temporary $1,750,000 loan to the Harbor Board. The City Solicitor said the board planned to spend between $12,000,000 and $14,000,000, but was held by its contract with the Canadian Stewart Co. to provide this amount.

Sandwich.—Backed by the Border Chamber of Commerce, the Town Council will petition the Dominion Government to establish shipbuilding yards here. Local financiers are willing to help build a plant, providing the Government will guarantee orders for merchant ships for service between Canada and Great Britain. River front property between Hill and Brock Streets has already been leased.

Annapolis Royal, N.S.—The launching of the new three-mast schooner Hilda M. Stark, 172 feet long and 648 tons gross, at the local shipyard, was successfully accomplished at 11.35 March 4, having been unavoidably postponed from Tuesday on account of the storm. The laying of the keel for another three-mast schooner of about the same size will begin at once at this yard, and a larger four-mast vessel will be put in frame at the same time.

Yarmouth, N.S.—The old time shipbuilding property here, which in the years when Yarmouth became known the world over by her shipping was owned by the late Mr. James Jenkins, is once more to become a shipyard. A lease of the property immediately north of the old Jenkins homestead has been taken by a company now being promoted, and work preparatory to building will commence next week. Their first vessel will be a schooner of 300 tons, and Mr. Deveau expects to have the keel stretched by the first of April.

Halifax, N.S.—Capt. Lemedec, master of the Mont Blanc, with which the Imo collided in Halifax harbor on December 6 last, causing the explosion which wrecked a portion of the city, has left the province, so the Supreme Court of Nova Scotia was informed recently. L. A. Lovett, who appeared on behalf of Captain Lemedec, stated that following his discharge under the writ of habeas corpus, and not having been notified of any possible review of the case, he had left the city and the jurisdiction of the court.

Windsor.—Capt. Fred. J. Trotter, of Amherstburg, has been awarded the contract of raising the lightship Falken, which was sunk by the ice during the winter and is now lying off Scudder's dock at the north end of Pelee Island. The lightship was formerly a Norwegian whaling steamer. It remained at its station last fall so late that the ice prevented it making port at Amherstburg. Capt. Trotter, who is now assisting the crew of the lighthouse tender Aspen to recover gas buoys lost in the ice last winter, will leave for Pelee Island next week and make arrangements for raising the Falken.

Ottawa.—The Dominion Government will resume control of the harbor at Port Dover, will repair damage done to it by storms, and will place it in condition to accommodate cargo and other vessels, Hon. F. B. Carvell, Minister of Public Works told a large delegation which waited upon him. The cost of repairs immediately required would, Hon. William Charlton estimated, amount to from $60,000 to $75,000, and a larger scheme of improvements would cost from $234,000 to $575,000, according to the plan adopted.

Vancouver.—The Canadian Explosives Limited has bought the yacht, White Cloud, from Captain Pybus, formerly commander in the C. P. R. transpacific service, and will use the craft as a ferry between James Island, Sydney and Saanichton. She has left Vancouver for Victoria, where she is undergoing some alterations to fit her for the service. The White Cloud was built at Hongkong and attracted much attention when she arrived here because she was rigged after the style of a Chinese junk and carried junk sails. She has a powerful auxiliary engine.

North Vancouver, B.C.—This city had a most encouraging programme in shipbuilding last year and launched one steel steamer and six auxiliary schooners. This year will see six wooden steamers and several steel steamers sent into the water and the industry has meant a great deal to the north shore city. Last year's launchings were the steel steamer War Dog and the five-master auxiliary schooners, Mabel Brown, Janet Geraldine. Wolvin, Jessie Norcross, Janet Carruthers. Mabel Stewart and Marie Barnard. The War Puget heads the 1918 programme with five other wooden steamers to follow from the Lyall yard and several steel steamers from the Wallace yard; the ambitious city will have good cause to congratulate itself on doing its bit in solving the tonnage problem.

Owen Sound.— Robert McMenemy, lighthouse keeper at Otter Head, has disappeared. McMenemy had been lightkeeper there for several years, being appointed from the Port Arthur district when the light was first placed there. He had carried on his work regularly and without accident until the close of navigation last fall; but after the lights had been ordered out he failed to report in as usual. Little was thought of it for a few days, as delays might have occurred; but when time passed with no news from him a party was sent out by the Sault Paper Co. from their camp at Pukosoi, a short distance from Otter Head, in search of the missing man. When the searchers reached Otter Head they found the lighthouse deserted; but a note had been left by McMenemy, which read:—December 11, 1917. "Am leaving to-day. Am a very sick man, and do not expect to reach home alive." That was all. Since then no trace has been found of the light-keeper.

Toronto.—H. W. Cowan, operating manager of the Canada Steamship Lines, announces the following appointments to the different freight steamers on the lines of the company for the season of 1918:—(J. W. Grainger is the mechanical superintendent). Steamer, W. Grant Morden, captain, Neil Campbell; engineer, Robert Chalmers, J. H. G. Hagarty, G. W. Pearson, Charles Robertson; Emperor, D. W. Burke, George N.

Toronto.—Until April 25 the western entrance of the harbor will be closed to navigation while dredging operations are conducted, according to an announcement made by the harbor master. The closing of the Western Gap means that about thirteen minutes must be added to the time which boats coming from Hamilton require for the trip, as they must sail round outside the island and enter by the eastern channel. It is announced that should the weather be stormy, ships may use the western entrance, and on stormy days the order does not apply. Two coal boats from Oswego are expected to arrive next week.

Detroit.—In comparison with the similar date last year, there is less ice reported on Lake Superior and Michigan, somewhat more in Lake Huron over the southern portion, in Lake Erie over the eastern portion, and also in Lake Ontario, says the sixth of the season's bulletins on lake ice conditions issued by the United States Weather Bureau in connection with reports from regular and display stations of the Weather Bureau and the Meteorological Service of Canada. During the week the ice has broken up and disappeared from most of the lakes. In Lake Superior the field has moved back to the extreme west end, and extends from Duluth beyond vision. Over the central portion the fields have moved beyond vision. At Whitefish Point the fields in the lake are moving in and out with the winds, and the ice in the bay is breaking up and becoming soft.

Collingwood.—The boilermakers and iron shipbuilders in the Collingwood shipyards have reached an agreement with the Collingwood Shipbuilding Company, by which the men obtain an average increase in wages of 25 to 30 per cent., a nine-hour day, instead of a ten-hour, recognition of the grievance committee of the men, and a signed agreement with the company for a year, commencing April 9. This is the outcome of the Board of Conciliation and Investigation, which recently convened in Collingwood. The chairman of the board was Hammett P. Hill, of Ottawa; Fred Bancroft, of Toronto, represented the men, and Capt. J. B. Foote the commany. The agreement provides the following rates of wages for new work, which are increased 2½ cents per hour for old work; new work rates, boilermakers, ship fitters and hand rivetters, 50 cents an hour; holders-on, 54c; heaters, 35c; chippers and caulkers, 50c; reamers, 35c; drillers, 40c; bolting up, 32½c; punch and shear men, 40c; punch

You know Aikenhead's
Marine Hardware and
Ship Supplies
Are Good
because—
you see on this page "named" products that you well know to be the best obtainable.
You want our Catalog—and we want you to have it. Please send your name.
GENUINE JENKINS VALVES
Globe, Angle, Check and Gate Valve, also Iron Body Gate Valves.
GENUINE "KEYSTONE" RATCHET
We carry a complete line of Ratchets. Full assortment of Kekstone and Armstrong.
GENUINE SIMPLEX CHAIN HOIST
Manufactured by J. G. Speidel. Full assortment and other makes as well.
IMPROVED STEEL TACKLE BLOCK
All sizes for manilla rope. Also Diamond Shell Tackle Blocks for Wire Rope.
GENUINE VANDORN ELECTRIC DRILL
Big stock carried in Toronto, also Electric Grinders. Superior to all other electric drills. Guaranteed to stand up to HARD SERVICE.
GENUINE OSTER BULL-DOG DIE STOCKS
We carry a full line of Pipe Stocks and Dies in Oster, Borden, Jardine, Toledo, and Armstrong. Also full line of Bolt Stocks and dies.
GENUINE STILLSON WRENCH
Full stock up from 6" to 48" carried. Also Trimo Pipe Wrenches and Chain Tongs.
AIKENHEAD HARDWARE LIMITED
17, 19, 21 Temperance Street
Toronto, Canada

shed helpers, 33½c; furnace-men, 50c; slab helpers, 40c; plate hangers, 40c; general helpers, 32½c; acetylene burners, 47½c; acetylene welders, 50c. Twenty occupations are included in the schedule.

Montreal.—Notices were yesterday served on the Board of Trade and other business interests by the Montreal Harbor Commission that their charges on wharfage on goods handled over the Montreal wharves and on cars, switched by the Harbor Commissioners' Railway, would be increased by at least 25 per cent. on the wharfage charges, and 50 cents a car on the switching movement of cars handled by the harbor and front railway. This, it was stated, will mean as to wharfage charges that on all general merchandise there will be an increase of at least 5 cents a ton, making the charge 25 instead of 20 cents. On special commodities, such as grain, the wharfage charge is increased from 3 to 5 cents per ton; on flour, from 6 to 8 cents per ton on wood pulp, from 6 to 8 cents per ton; on asbestos and asphalt and whiting, from 8 to 10 cents per ton; on hay, news print, sulphur and tar, from 12 to 15 cents per ton; on pig and scrap iron, from 16 to 20 cents per ton; on lumber and timber, from 8 to 10 cents per 1,000 feet, board measure; and on bricks, from 8 to 10 cents per 1,000.

SHIPPING BOARD WILL TEST MAXIM "NON-SINKABLE" SHIP

THE Shipping Board issues the following:

The Shipping Board has decided to build a vessel along the line designed by Hudson Maxim. Mr. Maxim's device can be placed in a ship and later, if it is found not to be as effective as he claims, it can be removed and the ship used in regular service. The Maxim device possesses many points which seem practicable. The Shipping Board feels that some start should be made toward producing non-sinkable ships.

Thousands of designs have been submitted to the ship protection committee, which has worked faithfully in trying to arrive at conclusions and make recommendations thereon. Now that an inventor of Mr. Maxim's reputation has submitted a plan it is thought wise to build the first non-sinkable ship of the type proposed by him.

Mr. John A. Donald, chairman of the ship protection committee, said that several methods for protecting vessels from submarine attack had been considered, and that the board is now engaged in making special tests of these.

Two Methods of Protection

"There are," he said, "two systems of protecting vessels against underwater attack. One method is to prevent the torpedo from reaching the hull. Another, the inside method, is designed to make a vessel unsinkable if a torpedo should reach the hull. Mr. Maxim's principle is along the line of inside protection—that is, intended to keep a vessel afloat after it has been damaged by a torpedo."

ONE OF MANY INSTALLATIONS OF MARTEN-FREEMAN COMPENSATING DAVITS

FOOL PROOF—RELIABLE—FAST OPERATING

TIME IS LIFE

WHAT GUARANTEE HAVE YOU THAT IT WILL BE POSSIBLE TO PUT ALL YOUR LIFEBOATS OVERSIDE IN THE FEW MINUTES AVAILABLE?

THE MARTEN-FREEMAN COMPANY, LIMITED
301 MANNING CHAMBERS, TORONTO, ONTARIO.

Hoisting and Haulage Machinery

OF ALL KINDS — FOR ALL PURPOSES

Standard 2 Drum Hoist for Derrick Use

Winches, or Hoisting and Haulage Drums. For either Steam, Electric, Gasoline, Belt, Horse or Hand Power.
Derrick Irons for all sizes of Derricks.
Dump Cars. Steel or Wood Body, Side or End Dump.
Other Small Cars, any design, made to your liking, with any type of trucks, and with any style of body.
Car Wheels, Axles and Boxes. Either Chilled Tread Wheels or Light Spoked Wheels. Any size or style of axle or box.
Steel Buckets, for handling concrete or any kind of excavated material. Great variety of designs and sizes to choose from.
Ship Winches, steam steering gear, Ash hoists, speed lifting gears, etc.

LET US ESTIMATE ON YOUR NEEDS.

MARSH ENGINEERING WORKS, LIMITED - Belleville, Ontario
(FORMERLY MARSH & HENTHORN, LIMITED.—NOTE NEW NAME)

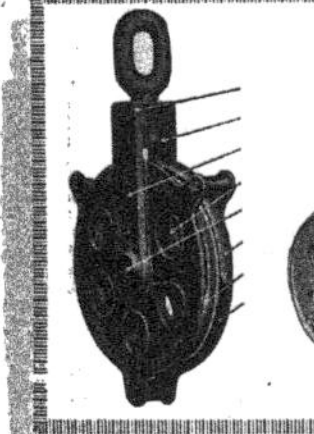

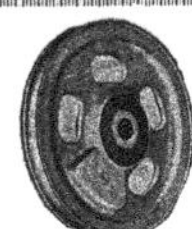

Loveridge's

Self-oiling, Cargo, Heel, and Tackle Blocks

Also makers of Steering Gear Spring Buffers

Particulars and Prices on Application

LOVERIDGE, LIMITED
DOCKS, CARDIFF, WALES

PORT HOPE MAY HAVE SHIP PLANT

Board of Trade There Has Already Taken This Matter Up

Special to CANADIAN MACHINERY.

PORT HOPE, April 16.—There is some likelihood that Port Hope will have a shipbuilding plant. The Port Hope Sanitary Mfg. Co. have decided to abandon the western section of their factory there, confining their enterprise to the large central building, which is bounded on the east, west and north by the harbor. The building thus made available has a fine location, having water depth, sufficient for launching basins, on three sides. It is understood that the Port Hope Board of Trade is already negotiating with the object of securing such an industry. Port Hope at one time had the finest harbor on the north shore of Lake Ontario and it would now appear that its location and facilities were to be utilized in a purpose to which they seem well adapted.

ST. LAWRENCE OPEN

The Signal Service reported navigation open in the St. Lawrence on April 21. Buoys are being placed in position by the Government steamer above and below Quebec. The ice is disappearing rapidly from the river. Several steam barges will leave here shortly for Montreal, and a number are expected here soon.

"Science, when well digested, is nothing but good sense and reason."

The Science of Chemistry
AND
Harris Heavy Pressure
Babbitt Metal

Chemistry is the keystone to modern industrial efficiency. It reveals the constituent parts of substances and can transform a weakness into great strength.

In no industry does the science of chemistry play a more important part than it does in the manufacture of high grade Babbitt Metals.

In the making of the famous HARRIS HEAVY PRESSURE BABBITT an expert metallurgical chemist guarantees every particle of the various ingredients that are included in its composition.

This never-ending maintenance of high quality has its reward in producing a babbitt "WITHOUT A FAULT."

We manufacture a Babbitt Metal for every purpose. Every one reliable and used the world over.

WRITE FOR COMPLETE LIST.

OUR BOOK ON BABBITT METALS MAILED FREE

The Canada Metal Co., Limited

MONTREAL HAMILTON **TORONTO** WINNIPEG VANCOUVER

Four Edition. Pp. i-xvi+332. With 18 Plates and 237 other Illustrations, including 59 Folding Diagrams. 21s. net.

STEEL SHIPS

Their Construction and Maintenance

A Manual for Shipbuilders, Ship Superintendents, Students and Marine Engineers.

By THOMAS WALTON, Naval Architect

Contents—I. Manufacture of Cast Iron, Wrought Iron, and Steel—Composition of Iron and Steel, Quality, Strength, Tests, etc. II. Classification of Steel Ships. III. Consideration in making choice of Type of Vessel—Framing of Ships. IV. Strains experienced by Ships—Methods of Computing and Comparing Strengths of Ships. V. Construction of Ships—Alternative Modes of Construction—Types of Vessels—Turret, Self Trimming, and Trunk Steamers, etc.—Rivets and Rivetting, Workmanship. VI. Pumping Arrangements. VII. Maintenance—Prevention of Deterioration in the Hulls of Ships—Cement, Paint, etc.—Index.

LONDON, ENGLAND

CHARLES GRIFFIN & CO., LTD., EXETER STREET, STRAND, W.C. 2.

Be a Marine Officer!

Navigation

By George L. Hosmer,
Associate Professor of Topographical Engineering,
Massachusetts Institute of Technology

An aid to those studying for officers' licenses. Contains only the simple working rules required for daily routine, yet nothing essential is omitted.

214 pages. 4¼ by 6⅞. 52 cuts, including signal code in color. Cloth, $1.25 net, postpaid.

Renouf Publishing Company
Dept. A., 25 McGill College Avenue
MONTREAL, CANADA

Somewhere in Canada where Shop Foremen see this list of High Grade Machinery, these will be selected quickly as most of this Machinery cannot be duplicated anywhere else for immediate shipment.

Largest Stock of Mining Machinery in Canada.

Coal, Coke, High Speed Steel, High Speed Drills.

Send Us Your Inquiries.

Air Compressors, Alley & McLellan 600 ft. at 100 lb. pressure.
Air Compressor, Rand 800 ft.
Motors Heavy Duty 10-25-40-75-125.
Tube Mills, Fraser Chalmers, Power & Mining 22 ft. Ball Mill Hardinge.
Tanks, Rails, Pumps, Aiken Classifiers, Crushers, Holman Drills, Oliver Filterers, Hoists, Boilers, Pumps, Transformers, Lighting Generators, Motors. 2 Ton Alco Truck, Wire Rope, 5 Ton Travelling Crane, 5 Ton Crane, Hoist Engine, Concrete Mixer.
Complete List of Mining Machinery and Lathes.

Send Inquiries for Prices:

Zenith Coal and Steel Products, Limited
1410 Royal Bank Building, Toronto, Ontario
402 McGill Building, Montreal, Quebec

out now and place with letters to be answered.

To Inventors, Engineers, etc.

Messrs. T. G. BEATLEY & SON

Steamship Owners and Brokers

(Sole Partner—Thomas Ernest Brooke)

57 & 58 Leadenhall Street, London, E.C. 3

offer **1,000** guineas

to any person of British, Allied or friendly Nationality and Origin who in the opinion and to the satisfaction of that firm invents and produces the best appliance or device for increasing the speed of steamers at a moderate cost.

(Proposals to be marked "Speed")

also

offer **1,000** guineas

to any person of British, Allied or friendly Nationality and Origin who in the opinion and to the satisfaction of that firm invents and produces the best appliances or devices for the rapid and efficient loading and discharging of Timber, Coal, Iron-Ore, and General Cargoes, capable of being applied at moderate cost to the ordinary type of steamer now in use, or a proportion of that sum in the case of an invention applicable to one species of Cargo only.

(Proposals to be marked "Cargo Appliances")

The Principal of T. G. Beatley & Son to be the sole judge of the merits and suitability of the inventions, and his decision shall be final, binding, and unassailable in any Court.

The above-mentioned firm to have the first option of acquiring the world rights of the inventions if accepted at a price to be agreed on. The above sums of 1,000 guineas respectively, or proportion thereof, as above, shall not in any case be payable unless the rights are acquired by T. G. Beatley & Son.

Proposals accompanied by complete particulars of suggested appliances or devices, if of prima facie utility, will be considered by T. G. Beatley & Son, but they do not bind themselves to accept any of them.

This offer to hold good until the end of the year 1918.

Send Us Your Inquiries

SHIPS BELLS

Made from Pure Bell Metal Complete with Attachments

C. O. CLARK & BROS.

1510 ST. PATRICK STREET MONTREAL, QUE.

MARINE WELDING CO.

Electric Welding, Boiler Marine Work a Specialty,

Reinforcing Wasted Places, Caulking Seams and Welding Fractures.

Plants: BUFFALO, CLEVELAND, MONTREAL

HEAD OFFICE:

36 and 40 Illinois St., BUFFALO

FRANCE

Marine Type

Metallic Packing

For All

Conditions of Service

FRANCE PACKING COMPANY

TACONY—PHILA., PENNA.

CRAFT OF THE HAMMERSMITH

SHIP AND ENGINE PARTS

PERFECTLY FORGED

QUAINT AND UNUSUAL SHAPES ALWAYS INTEREST US.

FORGING CRITICS ARE ESPECIALLY INVITED TO VISIT AND INSPECT OUR PLANTS.

FRITTERING MONEY AWAY AT THIS CRITICAL PERIOD IS INJURIOUS TO THE PUBLIC INTERESTS. THE GOVERNMENT ASKS YOU TO BUY WISELY.

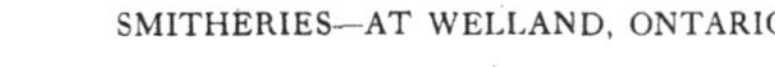

SMITHERIES—AT WELLAND, ONTARIO

Cuts Cost of Painting Ships

One man can do the work of several men, and do thorough, efficient painting with the

SPRACO

Pneumatic Painting Equipment

It applies the paint in coatings thick or thin to the hull of the ship. No streaks, no brush marks where "Spraco" is used.

Some users have saved as much as 50% on labor and 15% on material by the use of "Spraco" equipment. Get in touch with us regarding your ship painting problems—our engineers will be pleased to advise you free of charge.

Spray Engineering Company

New York **93 Federal St., Boston, Mass.** Chicago

Cable Address—SPRACO BOSTON Code—WESTERN UNION

Representatives in Canada:

Rudel-Belnap Machinery Co., 95 McGill St., Montreal

TUBES
COPPER and BRASS
Superior quality, reasonable prices, prompt deliveries. Let us quote on your requirements.
Standard Underground Cable Company of Canada, Limited
Hamilton, Ont.
MONTREAL TORONTO SEATTLE

MILLER BROS. & SONS LIMITED

120 Dalhousie St. Montreal

GREY IRON CASTINGS

SHIPS WINCHES

H. B. FRED KUHLS

MANUFACTURER, 6411-23 Third Avenue, Brooklyn, N. Y.

AND

ELASTIC SEAM PAINT

Guarantees Tight Decks.

Recently adopted by the Government for decks of Submarine Chasers.

ELASTIC GLAZING COMPOSITION

For Side and Bottom Seams and General Glazing.

ELASTIC

ANTI-CORROSIVE PAINT

ANTI-FOULING PAINT

Give Perfect Satisfaction

COPPER PAINT

BRIGHT RED AND GREEN

Last Entire Season

TROWEL CEMENT, WHITE AND GRAY

For Smoothing Rivets and Hulls

BUYERS' DIRECTORY

ACCUMULATORS, HYDRAULIC
Smart-Turner Mach. Co., Hamilton, Ont.

AERATING RESERVOIRS
Spray Engineering Co., Boston, Mass.

AIR PORTS
Turnbull Elevator Mfg. Co., Toronto, Ont.

ALLOYS, BRASS AND COPPER
Dominion Copper Products Co., Ltd., Montreal, Que.
Mueller Mfg. Co., H., Sarnia, Ont.
United Brass & Lead, Ltd., Toronto, Ont.

ANCHORS
Hopkins & Co., F. H., Montreal, Que.
Kearfott Engineering Co., New York, N.Y.
McNab Co., Bridgeport, Conn.
Henry Rogers Sons & Co., Wolverhampton, Eng.
Wm. C. Wilson & Co., Toronto, Ont.

ARCHITECT, NAVAL
Watts, J. Murray, Philadelphia, Pa.

ASBESTOS GOODS
Wm. C. Wilson & Co., Toronto, Ont.

BABBITT METAL
Aikenhead Hardware, Ltd., Toronto, Ont.
Hoyt Metal Company, Toronto, Ontario.
Tallman Brass & Metal Co., Hamilton, Ont.
Wilkinson & Kompass, Hamilton, Ont.
Wm. C. Wilson & Co., Toronto, Ont.

BARGES
Polson Iron Works, Toronto, Ont.

BAROMETERS
Wilson & Co., Wm. C., Toronto, Canada.

BARS, GRATE
Babcock & Wilcox, Ltd., Montreal, Que.

BEARINGS, BRASS
Empire Mfg. Co., London, Ont.
Mueller Mfg. Co., H., Sarnia, Ont.
Tallman Brass & Metal Co., Hamilton, Ont.
United Brass & Lead Co., Toronto, Ont.

BELLS, SHIPS, ENGINE ROOM, ETC.
Aikenhead Hardware, Ltd., Toronto, Ont.
Clarke & Bros. Co., Montreal, Que.
Cory & Son, Inc., Chas., New York, N.Y.
Empire Mfg. Co., London, Ont.
Morrison Brass Mfg. Co., James, Toronto, Ont.
Mueller Mfg. Co., H., Sarnia, Ont.
United Brass & Lead, Ltd., Toronto, Ont.

BELTING, LEATHER
Aikenhead Hardware, Ltd., Toronto, Ont.
Wm. C. Wilson & Co., Toronto, Ont.

BIBBS, COMPRESSION
Empire Mfg. Co., London, Ont.
Mueller Mfg. Co., H., Sarnia, Ont.
United Brass & Lead, Ltd., Toronto, Ont.

BINNACLES
Hopkins & Co., F. H., Montreal, Que.
Morrison Brass Mfg. Co., James, Toronto, Ont.

BLOCKS, CARGO, HEEL AND TACKLE
Aikenhead Hardware, Ltd., Toronto, Ont.
Hopkins & Co., F. H., Montreal, Que.
Loveridge, Ltd., Docks, Cardiff, Wales.

BLOWERS, TURBO.
Mason Regulator & Engin. Co., Montreal, Que.

BOAT CHOCKS
Corbet Fdry. & Machine Co., Owen Sound, Ont.
Marten-Freeman Co., Toronto, Ont.

BOILER COMPOUND
Wm. C. Wilson & Co., Toronto, Ont.

BOILER FEED PUMPS
Can. Ingersoll-Rand Co., Ltd., Sherbrooke, Que.
Smart-Turner Mach. Co., Hamilton, Ont.
Williams Machinery Co., A. R., Toronto, Ont.

BOILER FITTINGS
Empire Mfg. Co., London, Ont.
McAvity & Sons Ltd., T., St. John, N.B.
Wager Furnace Bridge Wall Co., Inc., 149 Broadway, New York, N.Y.

BOILERS, MARINE
Babcock & Wilcox, Ltd., Montreal, Que.
Can. Fairbanks-Morse Co., Montreal, Que.
Can. Vickers, Ltd., Montreal, Que.
Collingwood Shipbuilding Co., Collingwood, Ont.
Doxford & Sons, William, Sunderland, England.
Engineering & Machine Works of Canada, St. Catharines, Ont.
Goldie & McCulloch, Ltd., Galt, Ont.
Hall Engineering Works, Montreal, Que.
Mason Regulator & Engin. Co., Montreal, Que.
Montreal Dry Docks & Shipbuilding Co., Montreal, Que.
National Shipbuilding Co., Goderich, Ont.
Marsh Engineering Works, Belleville, Ont.
Polson Iron Works, Toronto, Ontario.
Port Arthur Shipbuilding Co., Port Arthur, Ont.
Williams Machinery Co., A. R., Toronto, Ont.

BOOKS, MILITARY AND NAVAL
Van Nostrand Co., D., 25 Park Place, New York.

BOOKS, TECHNICAL, MARINE
MacLean Publishing Co., Toronto, Ont.
Renouf Publishing Co., Montreal, Que.
Van Nostrand Co., D., 25 Park Place, New York.

BOLTS
London Bolt & Hinge Works, London, Ont.
United Brass & Lead, Ltd., Toronto, Ont.
Wilkinson & Kompass, Hamilton, Ont.

BROKER
Page & Jones, Mobile, Ala., U.S.A.

BRASS GOODS
Corbet Fdry. & Machine Co., Owen Sound, Ont.
McAvity & Sons, T., St. John, N.B.
Mueller Mfg. Co., H., Sarnia, Ont.
Tallman Brass & Metal Co., Hamilton, Ont.
United Brass & Lead, Ltd., Toronto, Ont.

BUCKETS, CLAMSHELL
Beatty & Sons, Welland, Ont.
Morris Crane & Hoist Co., Herbert, Niagara Falls, Ont.

BUCKETS, DUMP
Morris Crane & Hoist Co., Herbert, Niagara Falls, Ont.

BUCKETS, COALING
Beatty & Sons, Welland, Ont.
Engineering & Machine Works of Canada, St. Catharines, Ont.
Hopkins & Co., F. H., Montreal, Que.
Marsh Engineering Works, Belleville, Ont.
Reed & Co., Geo., Montreal, Que.

CABLE
Hopkins & Co., F. H., Montreal, Que.
McNab Co., Bridgeport, Conn.
Wm. C. Wilson & Co., Toronto, Ont.

CABLE, LEAD COVERED AND ARMORED
Standard Underground Cable Co., Hamilton, Ont.

CABLE, ACCESSORIES
Standard Underground Cable Co., Hamilton, Ont.

CAPSTANS
Advance Mach. & Welding Co., Montreal, Que.
Dake Engine Co., Grand Haven, Mich.
Hopkins & Co., F. H., Montreal, Que.
Kennedy & Sons, Wm., Owen Sound, Ont.
Wm. C. Wilson & Co., Toronto, Ont.

CALKING TOOLS, ELECTRIC
Aikenhead Hardware, Ltd., Toronto, Ont.

CALKING TOOLS, PNEUMATIC
Aikenhead Hardware, Ltd., Toronto, Ont.
Can. Ingersoll-Rand Co. Sherbrooke, Que.

CALORIFIERS
Low & Sons, Ltd., Archibald, Glasgow, Scotland

CASTINGS
Can. Steel Foundries, Ltd., Montreal, Que.
Collingwood Shipbuilding Co., Collingwood, Ont.
Kennedy & Sons, Wm., Owen Sound, Ont.
Marsh Engineering Works, Belleville, Ont.
Mitchell Co., Ltd., Robt., Montreal, Que.
Mueller Mfg. Co., H., Sarnia, Ont.
Tallman Brass & Metal Co., Hamilton, Ont.
United Brass & Lead, Ltd., Toronto, Ont.

CASTINGS, ALLOY
Mueller Mfg. Co., H., Sarnia, Ont.
Tolland Mfg. Co., Montreal, Que.
Tallman Brass & Metal Co., Hamilton, Ont.
United Brass & Lead, Ltd., Toronto, Ont.

CASTINGS, ALUMINUM
Empire Mfg. Co., London, Ont.
Tallman Brass & Metal Co., Hamilton, Ont.
United Brass & Lead Co., Toronto, Ont.

CASTINGS, BRASS
Crouse-Hinds Co. of Canada, Ltd., Toronto, Ont.
Empire Mfg. Co., London, Ont.
McAvity & Sons Ltd., T., St. John, N.B.
McNab Co., Bridgeport, Conn.
Mitchell Co., Ltd., Robt., Montreal, Que.
Mueller Mfg. Co., H., Sarnia, Ont.
Tolland Mfg. Co., Montreal, Que.
Tallman Brass & Metal Co., Hamilton, Ont.
United Brass & Lead Co., Toronto, Ont.

CASTINGS, GREY IRON, MALLEABLE, ALUMINUM
Crouse-Hinds Co. of Canada, Ltd., Toronto, Ont.
Engineering & Machine Works of Canada, St. Catharines, Ont.
McAvity & Sons, T., St. John, N.B.
McNab Co., Bridgeport, Conn.
Mueller Mfg. Co., H., Sarnia, Ont.

CASTINGS, MANGANESE BRONZE
Tallman Brass & Metal Co., Hamilton, Ont.

CASTINGS, ELECTRIC
Crouse-Hinds Co. of Canada, Ltd., Toronto, Ont.

CELLAR DRAINERS
Mueller Mfg. Co., H., Sarnia, Ont.

CHAINS
Aikenhead Hardware, Ltd., Toronto, Ont.
Kearfott Engineering Co., New York, N.Y.
Hopkins & Co., F. H., Montreal, Que.
Morris Crane & Hoist Co., Herbert, Niagara Falls, Ont.
Henry Rogers, Sons & Co., Wolverhampton, Eng.
Wm. C. Wilson & Co., Toronto, Ont.

CHAIN BLOCKS
Morris Crane & Hoist Co., Herbert, Niagara Falls, Ont.

CHAIN SLINGS
Morris Crane & Hoist Co., Herbert, Niagara Falls, Ont.

CHANDLERY, SHIP
Hopkins & Co., F. H., Montreal, Que.
Leckie, Ltd., John, Toronto, Ont.

CLAMPS, STEAM AND WATER
Mueller Mfg. Co., H., Sarnia, Ont.

CLOCKS
American Steam Gauge & Valve Mfg. Co. Boston, Mass.
Morrison Brass Mfg. Co., James, Toronto, Ont.
Williams Machinery Co., A. R., Toronto, Ont.

CLOSETS
Mueller Mfg. Co., H., Sarnia, Ont.
United Brass & Lead, Ltd., Toronto, Ont.

COAL
Nova Scotia Steel & Coal Co., New Glasgow, N.S.

COAL HANDLING MACHINERY
Morris Crane & Hoist Co., Herbert, Niagara Falls, Ont.

COCKS, BILGE, DISCHARGE, INDICATOR, ETC.
McAvity & Sons Ltd., T., St. John, N.B.
Morrison Brass Mfg. Co., James, Toronto, Ont.

COCKS, BASIN
Empire Mfg. Co., London, Ont.
United Brass & Lead, Ltd., Toronto, Ont.

COMPASSES
Scythes & Co., Toronto, Ont.
Wm. C. Wilson & Co., Toronto, Ont.

COMPRESSORS, AIR
Can. Fairbanks-Morse Co., Montreal, Que.
Canadian Ingersoll-Rand Co., Sherbrooke, Que.
Hopkins & Co., F. H., Montreal, Que.
Smart-Turner Mach. Co., Hamilton, Ont.
Williams Machinery Co., A. R., Toronto, Ont.

COMPRESSED AIR TANKS
Scaife & Sons Co., Wm. B., Oakmont, Pa.

CONDENSERS
Kearfott Engineering Co., New York, N.Y.
Morris Machine Works, Baldwinsville, N.Y.
Smart-Turner Mach. Co., Hamilton, Ont.
Weir Ltd., G. & J., Cathcart, Glasgow, Scotland.
Williams Machinery Co., A. R., Toronto, Ont.

CONDULETS, MARINE
Crouse-Hinds Co. of Canada, Ltd., Toronto, Ont.

CONTRACTORS' SUPPLIES
McAvity & Sons, T., St. John, N.B.

CONSULTING ENGINEER
Watt, J. Murray, Philadelphia, Pa.

CONVEYORS, ASH, COAL
Babcock & Wilcox, Ltd., Montreal, Que.
Hopkins & Co., F. H., Montreal, Que.

COPING MACHINES
Bertram & Sons, Ltd., John, Dundas, Ont.

COPPER TUBES, SHEETS AND RODS
Dominion Copper Products Co., Ltd., Montreal, Que.
Tallman Brass & Metal Co., Hamilton, Ont.

CORDAGE
Scythes & Co., Toronto, Ont.

COTTON, CALKING
Wilson & Co., Wm. C., Toronto, Canada.

COTTON DUCK
Scythes & Co., Toronto, Ont.

COVERS, CANVAS, FOR HATCHES, LIFE-BOATS, ETC.
Leckie, Ltd., John, Toronto, Ont.

COUNTERS, REVOLUTION
American Steam Gauge & Valve Mfg. Co., Boston, Mass.

CRANES
Aikenhead Hardware, Ltd., Toronto, Ont.
Can. Fairbanks-Morse Co., Montreal, Que.
Hopkins & Co., F. H., Montreal, Que.
Williams Machinery Co., A. R., Toronto, Ont.

CRANES, ELECTRIC
Babcock & Wilcox, Ltd., Montreal, Que.
Morris Crane & Hoist Co., Herbert, Niagara Falls, Ont.
Smart-Turner Mach. Co., Hamilton, Ont.

CRANES, GOLIATH AND PNEUMATIC
Morris Crane & Hoist Co., Herbert, Niagara Falls, Ont.

CRANES, GANTRY
Morris Crane & Hoist Co., Herbert, Niagara Falls, Ont.
Smart-Turner Mach. Co., Hamilton, Ont.

CRANES, JIB
Morris Crane & Hoist Co., Herbert, Niagara Falls, Ont.

CRANES, PORTABLE
Morris Crane & Hoist Co., Herbert, Niagara Falls, Ont.

CRANES, OVERHEAD TRAVELLING
Morris Crane & Hoist Co., Herbert, Niagara Falls, Ont.

CRANK SHAFTS
Canada Foundries and Forgings, Welland, Ont.

DAVITS, BOAT
Babcock & Wilcox, Ltd., Montreal, Que.
Corbet Fdry. & Machine Co., Owen Sound, Ont.
Hopkins & Co., F. H., Montreal, Que.
Marten, Freeman Co., Toronto, Ont.

DEAD LIGHTS, BRASS
Morris Crane & Hoist Co., Herbert, Niagara Falls, Ont.
Mitchell Co., Ltd., Robt., Montreal, Que.

DECK LIGHTS
Turnbull Elevator Mfg. Co., Toronto, Ont.

DECK PLUGS, ELECTRIC
Crouse-Hinds Co. of Canada, Ltd., Toronto, Ont.

DERRICKS
Aikenhead Hardware, Ltd., Toronto, Ont.
Dake Engine Co., Grand Haven, Mich.
Hopkins & Co., F. H., Montreal, Que.
Marsh Engineering Works, Belleville, Ont.
Morris Crane & Hoist Co., Herbert, Niagara Falls, Ont.

DISTILLERS
Kearfott Engineering Co., New York, N.Y.

DISTRESS OUTFITS
Scythes & Co., Toronto, Ont.

DREDGES
Collingwood Shipbuilding Co., Collingwood, Ont.
Norbom Engineering Co., Philadelphia, Pa.
Polson Iron Works, Toronto.

DRILLS, AIR
Aikenhead Hardware, Ltd., Toronto, Ont.
Can. Ingersoll-Rand Co. Sherbrooke, Que.
Hopkins & Co., F. H., Montreal, Que.

DRILLS, CENTRE
Wilt Twist Drill Co., Walkerville, Ont.

DRILLS, BLACKSMITH AND BIT STOCK
Wilt Twist Drill Co., Walkerville, Ont.

DRILLS, ELECTRIC
Aikenhead Hardware, Ltd., Toronto, Ont.
Wilkinson & Kompass, Hamilton, Ont.

DRILLS, HIGH SPEED
Wilt Twist Drill Co., Walkerville, Ont.
Zenith Coal & Steel Products, Ltd., Montreal, Que.

DRILLS, TRACK
Wilt Twist Drill Co., Walkerville, Ont.

DRILLS, RACHET AND HAND
Wilt Twist Drill Co., Walkerville, Ont.

DRILLS, TWIST
Aikenhead Hardware, Ltd., Toronto, Ont.
Williams Machinery Co., A. R., Toronto, Ont.
Wilt Twist Drill Co., Walkerville, Ont.

DRY DOCKS
Can. Vickers, Ltd., Montreal, Que.
Collingwood Shipbuilding Co., Collingwood, Ont.
Doxford & Sons, William, Sunderland, England.
Georgian Bay Shipbuilding & Wrecking Co., Midland, Ont.
National Shipbuilding Co., Goderich, Ont.
Polson Iron Works, Toronto, Ont.
Port Arthur Shipbuilding Co., Port Arthur, Ont.
Yarrows, Limited, Victoria, B.C.

ECONOMIZERS, FUEL
Babcock & Wilcox, Ltd., Montreal, Que.

EJECTORS
Empire Mfg. Co., London, Ont.
Morrison Brass Mfg. Co., James, Toronto, Ont.
Smart-Turner Mach. Co., Hamilton, Ont.

ELECTRIC LAMPS
Wm. C. Wilson & Co., Toronto, Ont.

ELECTRO-PLATING
Tallman Brass & Metal Co., Hamilton, Ont.
United Brass & Lead Co., Toronto, Ont.

ELEVATING MACHINERY
Goldie & McCulloch, Ltd., Galt, Ont.
Morris Crane & Hoist Co., Herbert, Niagara Falls, Ont.

ELEVATORS
Turnbull Elevator Mfg. Co., Toronto, Ont.

ENGINES, HOISTING
Advance Mach. & Welding Co., Montreal, Que.
Corbet Fdry. & Machine Co., Owen Sound, Ont.
Hopkins & Co., F. H., Montreal, Que.
Kennedy & Sons, Wm., Owen Sound, Ont.
Marsh Engineering Works, Belleville, Ont.
Port Arthur Shipbuilding Co., Port Arthur, Ont.
Williams Machinery Co., A. R., Toronto, Ont.

ENGINE, INTERNAL COMBUSTION
Doxford & Sons, William, Sunderland, England.

ENGINES, MARINE
Bolinders Co., New York, N.Y.
Can. Fairbanks-Morse Co., Montreal, Que.
Can. Vickers, Ltd., Montreal, Que.
Doxford & Sons, William, Sunderland, England.
Goldie & McCulloch, Ltd., Galt, Ont.
Hopkins & Co., F. H., Montreal, Que.
Iron Works, Ltd., Owen Sound, Ont.
Mason Regulator & Engin. Co., Montreal, Que.
National Shipbuilding Co., Goderich, Ont.
Norbom Engineering Co., Philadelphia, Pa.
Polson Iron Works, Toronto, Ont.
Port Arthur Shipbuilding Co., Port Arthur, Ont
Trout Co., H. G., Buffalo, N.Y.
Williams Machinery Co., A. R., Toronto, Ont.

ENGINE STARTERS (AIR)
Scaife & Sons Co., Wm. B., Oakmont, Pa.

ENGINES, STEERING
Corbet Fdry. & Machine Co., Owen Sound, Ont.
Dake Engine Co., Grand Haven, Mich.
Kennedy & Sons, Wm., Owen Sound, Ont.
Wm. C. Wilson & Co., Toronto, Ont.

ENAMELWARE
Mueller Mfg. Co., H., Sarnia, Ont.

EVAPORATORS
Kearfott Engineering Co., New York, N.Y.
Mason Regulator & Engin. Co., Montreal, Que.
McNab Co., Bridgeport, Conn.
Weir Ltd., G. & J., Cathcart, Glasgow, Scotland.

EXTRACTORS, GREASE
American Steam Gauge & Valve Mfg. Co. Boston, Mass.

EYE BOLTS AND NUTS
Canada Foundries and Forgings, Welland, Ont.
United Brass & Lead, Ltd., Toronto, Ont.

FANS
Aikenhead Hardware Ltd., Toronto, Ont.
Empire Mfg. Co., London, Ont.
Reed & Co., Geo. W., Montreal, Que.
Smart-Turner Mach. Co., Hamilton, Ont.
Williams Machinery Co., A. R., Toronto, Ont.

FENDERS, ROPE
Hopkins & Co., F. H., Montreal, Que.
Leckie, Ltd., John, Toronto, Ont.
Wilson & Co., Wm. C., Toronto, Canada.

FERRO-MANGANESE
Mitchells, Ltd., Glasgow, Scotland.

FILES
Aikenhead Hardware, Ltd., Toronto, Ont.
Williams Machinery Co., A. R., Toronto, Ont.

FIRE BRICKS
Mitchells, Ltd., Glasgow, Scotland.
Williams Machinery Co., A. R., Toronto, Ont.

FILTERS, FEED WATER
Mason Regulator & Engin. Co., Montreal, Que.

FITTINGS, MARINE
Hopkins & Co., F. H., Montreal, Que.
McAvity & Sons, T., St. John, N.B.
United Brass & Lead, Ltd., Toronto, Ont.

FITTINGS, MOTOR BOAT
Empire Mfg. Co., London, Ont.
Mueller Mfg. Co., H., Sarnia, Ont.
United Brass & Lead, Ltd., Toronto, Ont.

FLAG POLES, STEEL
Dennis Wire & Iron Works Co., London, Ont.

FLAGS
Scythes & Co., Toronto, Ont.

FLOW METERS
Spray Engineering Co., Boston, Mass.

FLUE CLEANERS
Wm. C. Wilson & Co., Toronto, Ont.

FIXTURES, ELECTRIC
Cory & Son, Inc., Chas., New York, N.Y.
Crouse-Hinds Co. of Canada, Ltd., Toronto, Ont.
Tallman Brass & Metal Co., Hamilton, Ont.

FORGES
Aikenhead Hardware, Ltd., Toronto, Ont.
Hopkins & Co., F. H., Montreal, Que.

FLOODLIGHTS, ELECTRIC
Crouse-Hinds Co. of Canada, Ltd., Toronto, Ont.

FLUORSPAR
Mitchells, Ltd., Glasgow, Scotland.
Scythes & Co., Toronto, Ont.

FORGINGS, ALL KINDS
Aikenhead Hardware, Ltd., Toronto, Ont.
Collingwood Shipbuilding Co., Collingwood, Ont.
Nova Scotia Steel & Coal Co., New Glasgow, N.S.

FORGINGS, STEEL AND IRON
Canada Foundries and Forgings, Welland, Ont.

FURNACE BRIDGE WALLS
Wager Furnace Bridge Wall Co., Inc., 149 Broadway, New York, N.Y.

GAUGES, RECORDING
American Steam Gauge & Valve Mfg. Co., Boston, Mass.
Empire Mfg. Co., London, Ont.

GASKETS
Wm. C. Wilson & Co., Toronto, Ont.

GAUGES COCKS
McAvity & Sons, T., St. John, N.B.

GAUGE GLASSES
Wm. C. Wilson & Co., Toronto, Ont.

GAUGES, WATER
McAvity & Sons, T., St. John, N.B.
McNab Co., Bridgeport, Conn.
Morris Crane & Hoist Co., Herbert, Niagara Falls, Ont.

GAUGES, WATER, PRESSURE, COMPOUND AND VACUUM
Aikenhead Hardware, Ltd., Toronto, Ont.
Babcock & Wilcox, Ltd., Montreal, Que.
Empire Mfg. Co., London, Ont.
Morrison Brass Mfg. Co., James, Toronto, Ont.

GENERATORS AND CONVERTORS
Can. Fairbanks-Morse Co., Montreal, Que.

GLAZING COMPOSITION
Kuhls, H. B. Fred., 6411 3rd Ave., Brooklyn, N.Y.

GONGS
Mitchell Co., Ltd., Robt., Montreal, Que.

GRAPHITE
Wm. C. Wilson & Co., Toronto, Ont.

GRATINGS
Corbet Fdry. & Machine Co., Owen Sound, Ont.
Canada Wire & Iron Goods Co., Hamilton, Ont.

GRINDERS, ELECTRIC
Wilkinson & Kompass, Hamilton, Ont.

GRINDERS, PNEUMATIC
Can. Ingersoll-Rand Co. Sherbrooke, Que.

GUARDS, MACHINERY
Dennis Wire & Iron Works Co., London, Ont.

GUY RODS AND ANCHORS, ELECTRIC
Crouse-Hinds Co. of Canada, Ltd., Toronto, Ont.

HAMMERS
Canada Foundries and Forgings, Ltd., Welland, Ont.

HARDWARE, MARINE
Hopkins & Co., F. H., Montreal, Que.
Scythes & Co., Toronto, Ont.

HEADLIGHTS, ELECTRIC
Crouse-Hinds Co. of Canada, Ltd., Toronto, Ont.
Hopkins & Co., F. H., Montreal, Que.
Tallman Brass & Metal Co., Hamilton, Ont.

HEATING EQUIPMENT
Empire Mfg. Co., London, Ont.
Low & Sons, Ltd., Archibald, Glasgow, Scotland

HEATERS, FEED, WATER
Babcock & Wilcox, Ltd., Montreal, Que.
Kearfott Engineering Co., New York, N.Y.
Mason Regulator & Engin. Co., Montreal, Que.
McNab Co., Bridgeport, Conn.
Weir Ltd., G. & J., Cathcart, Glasgow, Scotland.

HINGES
London Bolt & Hinge Works, London, Ont.
Mitchell Co., Ltd., Robt., Montreal, Que.

HOOKS
Hopkins & Co., F. H., Montreal, Que.
Morris Crane & Hoist Co., Herbert, Niagara Falls, Ont.

HOISTS, ASH
Beatty & Sons, Welland, Ont.
Marsh Engineering Works, Belleville, Ont.
St. Clair Bros., Galt, Ont.

HOIST BLOCKS
Morris Crane & Hoist Co., Herbert, Niagara Falls, Ont.

HOISTS, CHAIN
Aikenhead Hardware, Ltd., Toronto, Ont.
Can. Fairbanks-Morse Co., Montreal, Que.
Dake Engine Co., Grand Haven, Mich.
Hopkins & Co., F. H., Montreal, Que.
Morris Crane & Hoist Co., Herbert, Niagara Falls, Ont.
Williams Machinery Co., A. R., Toronto, Ont.

HOISTS, CARGO, MOVING, ETC.
Dake Engine Co., Grand Haven, Mich.
Hopkins & Co., F. H., Montreal, Que.
Marsh Engineering Works, Belleville, Ont.

HOISTING MACHINERY
Beatty & Sons, Welland, Ont.
Can. Ingersoll-Rand Co. Sherbrooke, Que.
Corbet Fdry. & Machine Co., Owen Sound, Ont.
Marsh Engineering Works, Belleville, Ont.
Morris Crane & Hoist Co., Herbert, Niagara Falls, Ont.
Williams Machinery Co., A. R., Toronto, Ont.

HOSE
Wm. C. Wilson & Co., Toronto, Ont.

HYDRAULIC MACHINERY
The John McDougall Caledonian Iron Works Co., Montreal, Que.

INDICATORS, ENGINE
American Steam Gauge & Valve Mfg. Co., Boston, Mass.
Cory & Son, Inc., Chas., New York, N.Y.
McNab Co., Bridgeport, Conn.

INDICATORS, SPEED
Aikenhead Hardware, Ltd., Toronto, Ont.
Cory & Son, Inc., Chas., New York, N.Y.

INJECTORS
Aikenhead Hardware, Ltd., Toronto, Ont.
Empire Mfg. Co., London, Ont.
Morrison Brass Mfg. Co., James, Toronto, Ont.
Williams Machinery Co., A. R., Toronto, Ont.

INGOTS
Broughton Copper Co., Ltd., Manchester, Eng.

INSULATORS, ELECTRIC
Crouse-Hinds Co. of Canada, Ltd., Toronto, Ont.

INSTRUMENTS, NAUTICAL
Leckie, Ltd., John, Toronto, Ont.

JACKS
Hopkins & Co., F. H., Montreal, Que.
Morris Crane & Hoist Co., Herbert, Niagara Falls, Ont.

JOINTS, BALL, SLIP
Norbom Engineering Co., Philadelphia, Pa.

INSURANCE, MARINE
Toronto Insurance & Vessel Agency, Toronto, Ont.

JOURNAL WEDGES
Canada Foundries and Forgings, Ltd., Welland, Ont.

LATHES
Foss Machine & Supply Co., Geo. F., Montreal, Que.
Bertram & Sons, Ltd., John, Dundas, Ont.

LEATHER, LACE
Wm. C. Wilson & Co., Toronto, Ont.

LIFEBUOYS, BELTS AND PRESERVERS
Fosbery Co., Barking, Essex, Eng.
Hopkins & Co., F. H., Montreal, Que.
Wm. C. Wilson & Co., Toronto, Ont.

LIFEBOATS
Hopkins & Co., F. H., Montreal, Que.
Wm. C. Wilson & Co., Toronto, Ont.

LIFE BOAT EQUIPMENT
Hopkins & Co., F, H., Montreal, Que.
Leckie, Ltd., John, Toronto, Ont.
Wm. C. Wilson & Co., Toronto, Ont.

LIFE JACKETS
Fosbery Co., Barking, Essex, Eng.
Hopkins & Co., F. H., Montreal, Que.
Leckie, Ltd., John, Toronto, Ont.

LIGHTS, ALL KINDS
Can. Fairbanks-Morse Co., Montreal, Que.
Crouse-Hinds Co. of Canada, Ltd., Toronto, Ont.
Hopkins & Co., F. H., Montreal, Que.
Morrison Brass Mfg. Co., James, Toronto, Ont.
Tallman Brass & Metal Co., Hamilton, Ont.

LIGHTS, SIDE, PORT
Crouse-Hinds Co. of Canada, Ltd., Toronto, Ont.
McAvity & Sons Ltd., T., St. John, N.B.
Tallman Brass & Metal Co., Hamilton, Ont.
Turnbull Elevator Mfg. Co., Toronto, Ont.

LIGHTING SETS
Can. Fairbanks-Morse Co., Montreal.
Cory & Son, Inc., Chas., New York, N.Y.
Crouse-Hinds Co. of Canada, Ltd., Toronto, Ont.
Kearfott Engineering Co., New York, N.Y.
Williams Machinery Co., A. R., Toronto, Ont
Wilson & Co., Wm. C., Toronto, Canada.

LOADERS, WAGON AND TRUCK
Morris Crane & Hoist Co., Herbert, Niagara Falls, Ont.

LOCKERS, METAL
Dennis Wire & Iron Works Co., London, Ont.
Reed & Co., Geo. W., Montreal, Que.

LUBRICATORS
Empire Mfg. Co., London, Ont.

MACHINISTS
Corbet Fdry. & Machine Co., Owen Sound, Ont.
Hyde Engineering Works, Montreal, Que.

MANGANESITE, PASTE
Wm. C. Wilson & Co., Toronto, Ont.

MARINE CASTINGS
Mitchell Co., Ltd., Robt., Montreal, Que.
McAvity & Sons, T., St. John, N.B.
Mueller Mfg. Co., H., Sarnia, Ont.
Tallman Brass & Metal Co., Hamilton, Ont.
United Brass & Lead, Ltd., Toronto, Ont.

MARINE SPECIALTIES
Hopkins & Co., F. H., Montreal, Que.
McAvity & Sons, T., St. John, N.B.
Mitchell Co., Ltd., Robt., Montreal, Que.
United Brass & Lead, Ltd., Toronto, Ont.

MOTORS
Aikenhead Hardware, Ltd., Toronto, Ont.
Can. Fairbanks-Morse Co., Montreal, Que.
Williams Machinery Co., A. R., Toronto, Ont.
Zenith Coal & Steel Products, Ltd., Montreal, Que.

NAUTICAL INSTRUMENTS
- Scythes & Co., Toronto, Ont.
- Wm. C. Wilson & Co., Toronto, Ont.

NET LIFTERS
- Dake Engine Co., Grand Haven, Mich.
- Leckie, Ltd., John, Toronto, Ont.

NOZZLES, ALL KINDS
- Empire Mfg. Co., London, Ont.
- Tolland Mfg. Co., Boston, Mass.

NUTS
- Wilkinson & Kompass, Hamilton, Ont.
- United Brass & Lead, Ltd., Toronto, Ont.

OAKUM
- Scythes & Co., Toronto, Ont.
- Wm. C. Wilson & Co., Toronto, Ont.

OIL CUPS
- Empire Mfg. Co., London, Ont.
- Mitchell Co., Ltd., Robt., Montreal, Que.
- Murray & Fraser, New Glasgow, N.S.

OILCLOTHING
- Wm. C. Wilson & Co., Toronto, Ont.

OILS AND GREASE
- Wm. C. Wilson & Co., Toronto, Ont.

OAKUM
- Leckie, Ltd., John, Toronto, Ont.

ORNAMENTAL IRONWORK
- Canada Wire & Iron Goods Co., Hamilton, Ont.

OXY-ACETLYENE OUTFITS
- Can. Fairbanks-Morse Co., Montreal, Que.
- Hopkins & Co., F. H., Montreal, Que.
- McAvity & Sons Ltd., T., St. John, N.B.
- Williams Machinery Co., A. R., Toronto, Ont.

PAINT
- Ault & Wiborg Co. of Can., Ltd., Toronto, Ont.
- Wm. C. Wilson & Co., Toronto, Ont.

PAINTS, ANTI-FOULING
- Kuhls, H. B. Fred., 6411 3rd Ave., Brooklyn, N.Y.

PAINTS, ANTI-CORROSIVE
- Kuhls, H. B. Fred, 6411 3rd Ave., Brooklyn, N.Y.

PAINTING EQUIPMENT, AUTOMATIC
- Spray Engineering Co., Boston, Mass.

PACKING, AMMONIA
- France Packing Co., Philadelphia, Pa.
- Wm. C. Wilson & Co., Toronto, Ont.

PACKING, MARINE
- Aikenhead Hardware, Ltd., Toronto, Ont.
- France Packing Co., Philadelphia, Pa.
- Wm. C. Wilson & Co., Toronto, Ont.

PACKING, METALLIC
- Aikenhead Hardware, Ltd., Toronto, Ont.
- France Packing Co., Philadelphia, Pa.

PACKING, STEAM
- Aikenhead Hardware, Ltd., Toronto, Ont.
- France Packing Co., Philadelphia, Pa.
- Wm. C. Wilson & Co., Toronto, Ont.

PANELBOARD AND CABINETS, ELECTRIC
- Crouse-Hinds Co. of Canada, Ltd., Toronto, Ont
- Preston Woodworking Machy. Co., Preston, Ont.

PIPE, LAPWELD, CAST IRON, RIVETED
- Empire Mfg. Co., London, Ont.
- Norbom Engineering Co., Philadelphia, Pa.
- United Brass & Lead, Ltd., Toronto, Ont.

PIPE RAILING, IRON AND BRASS
- Dennis Wire & Iron Works Co., London, Ont.

PITCH
- Reed & Co., Geo., Montreal, Que.
- Scythes & Co., Toronto, Ont.

PITCH, PINE
- Wilson & Co., Wm. C., Toronto, Canada.

PLANERS
- Bertram & Sons, Ltd., John, Dundas, Ont.

PLANERS, STANDARD AND ROTARY
- Yates Mach. Co., P. B., Hamilton, Ont.

PLATE CLAMPS
- Morris Crane & Hoist Co., Herbert, Niagara Falls, Ont.

PLATE PUNCH TABLES
- Norbom Engineering Co., Philadelphia, Pa.

PLUMBING EQUIPMENT
- Low & Sons, Ltd., Archibald, Glasgow, Scotland.
- Empire Mfg. Co., London, Ont.
- Mueller Mfg. Co., H., Sarnia, Ont.
- Tallman Brass & Metal Co., Hamilton, Ont.
- Wm. C. Wilson & Co., Toronto, Ont.
- United Brass & Lead, Ltd., Toronto, Ont.

PROPELLOR BLADES, BRONZE
- Corbet Fdry. & Machine Co., Owen Sound, Ont.
- Empire Mfg. Co., London, Ont.
- Tallman Brass & Metal Co., Hamilton, Ont.
- United Brass & Lead, Ltd., Toronto, Ont.
- Yarrows, Limited, Victoria, B.C.

PROPELLER WHEELS
- Kennedy & Sons, Wm., Owen Sound, Ont.
- Trout Co., H. G., Buffalo, N.Y.

PIPING, STEAM
- Babcock & Wilcox, Ltd., Montreal, Que.

PORT LIGHTS
- Mitchell Co., Ltd., Robt., Montreal, Que.
- Mueller Mfg. Co., H., Sarnia, Ont.
- Turnbull Elevator Mfg. Co., Toronto, Ont.
- Tallman Brass & Metal Co., Hamilton, Ont.
- Wm. C. Wilson & Co., Toronto, Ont.
- United Brass & Lead, Ltd., Toronto, Ont.

PROPELLER WHEELS
- McNab Co., Bridgeport, Conn.

PULLEYS
- Smart-Turner Mach. Co., Hamilton, Ont.
- Williams Machinery Co., A. R., Toronto, Ont.

PUMP, AIR
- Can. Ingersoll-Rand Co. Sherbrooke, Que.
- Smart-Turner Mach. Co., Hamilton, Ont.
- Weir Ltd., G. & J., Cathcart, Glasgow, Scotland.
- Williams Machinery Co., A. R., Toronto, Ont.

PUMPS
- Can. Fairbanks-Morse Co., Montreal, Que.
- Canada Foundries & Forgings, Welland, Ont.
- Goldie & McCulloch, Ltd., Galt, Ont.
- Hopkins & Co., F. H., Montreal, Que.
- McAvity & Sons, T., St. John, N.B.
- Williams Machinery Co., A. R., Toronto Ont.

PUMPS, CENTRIFUGAL
- Can. Ingersoll-Rand Co. Sherbrooke, Que.
- Kearfott Engineering Co., New York, N.Y.
- Norbom Engineering Co., Philadelphia, Pa.
- Smart-Turner Mach. Co., Hamilton, Ont.
- Williams Machinery Co., A. R., Toronto, Ont.

PUMPS, BILGE
- Mitchell Co., Ltd., Robt., Montreal, Que.
- Mueller Mfg. Co., H., Sarnia, Ont.

PUMPS, CIRCULATING
- McNab Co., Bridgeport, Conn.
- Morris Machine Works, Baldwinsville, N.Y.

PUMPS, FEED WATER
- McNab Co., Bridgeport, Conn.
- Weir Ltd., G. & J., Cathcart, Glasgow, Scotland.
- Williams Machinery Co., A. R., Toronto, Ont.

PUMPS, HAND AND POWER
- Aikenhead Hardware, Ltd., Toronto, Ont.
- Smart-Turner Mach. Co., Hamilton, Ont.
- Williams Machinery Co., A. R., Toronto, Ont.

PUMPS, HIGH PRESSURE
- Canadian Ingersoll-Rand Co., Sherbrooke, Que.
- Smart-Turner Mach. Co., Hamilton, Ont.

PUMPS, STEAM TURBO
- Wm. C. Wilson & Co., Toronto, Ont.

PUMPING MACHINES
- Norbom Engineering Co., Philadelphia, Pa.

PUNCHES
- Bertram & Sons, Ltd., John, Dundas, Ont.

PUNCHES, SINGLE, DOUBLE AND MULTIPLE
- Norbom Engineering Co., Philadelphia, Pa.

PURIFIERS, WATER
- Babcock & Wilcox, Ltd., Montreal, Que.

RADIO ENGINEERS
- Cutting & Washington, Inc., Cambridge, Mass.

RADIATORS, STEAM, ELECTRIC
- Empire Mfg. Co., London, Ont.
- Low & Sons, Ltd., Archibald, Glasgow, Scotland.
- Wm. C. Wilson & Co., Toronto, Ont.

RAILS, OVERHEAD
- Morris Crane & Hoist Co., Herbert, Niagara Falls, Ont.

RANGES
- Hopkins & Co., F. H., Montreal, Que.
- Wm. C. Wilson & Co., Toronto, Ont.

RECEIVERS, AIR
- Can. Ingersoll-Rand Co. Sherbrooke, Que.

REGULATORS, FEED WATER
- American Steam Gauge & Valve Mfg. Co., Boston, Mass.

REGULATORS, PRESSURE
- Mason Regulator & Engin. Co., Montreal, Que.

REPAIRS, MARINE
- Corbet Foundry & Mach. Co., Owen Sound, Ont.
- Can. Vickers, Ltd., Montreal, Que.
- Collingwood Shipbuilding Co., Collingwood, Ont.
- Engineering & Machine Works of Canada, St. Catharines, Ont.
- Georgian Bay Shipbuilding & Wrecking Co., Midland, Ont.
- Hyde Engineering Works, Montreal, Que.
- Iron Works, Ltd., Owen Sound, Ont.
- Kennedy & Sons, Wm., Owen Sound, Ont.
- Muir Bros. Dry Dock Co., Port Dalhousie, Ont.
- National Shipbuilding Co., Goderich, Ont.
- Port Arthur Shipbuilding Co., Port Arthur, Ont.
- Yarrow, Limited, Victoria, B.C.

RIGGING SCREWS
- Hopkins & Co., F. H., Montreal, Que.

RIGGING, WIRE ROPE
- Leckie, Ltd., John, Toronto, Ont.

RIVETS
- London Bolt & Hinge Works, London, Ont.
- Severance Mfg. Co., S., Glassport, Pa.
- Wilkinson & Kompass, Hamilton, Ont.

RIVETERS, PNEUMATIC
- Can. Ingersoll-Rand Co., Ltd., Sherbrooke, Que.

RODS, COPPER, BRASS, BRONZE
- Standard Underground Cable Co., Hamilton, Ont.
- Tallman Brass & Metal Co., Hamilton, Ont.

ROLLS, STRAIGHTENING, BENDING
- Bertram & Sons, Ltd., John, Dundas, Ont.

ROOFING
- Reed & Co., Geo., Montreal, Que.
- Wm. C. Wilson & Co., Toronto, Ont.

ROPE BLOCKS
- Aikenhead Hardware, Ltd., Toronto, Ont.
- Can. Fairbanks-Morse Co., Montreal, Que.
- Morris Crane & Hoist Co., Herbert, Niagara Falls, Ont.

ROPE
- Hopkins & Co., F. H., Montreal, Que.
- Leckie, Ltd., John, Toronto, Ont.
- McNab Co., Bridgeport, Conn.
- Wm. C. Wilson & Co., Toronto, Ont.

RUBBER COATS
- Wm. C. Wilson & Co., Toronto, Ont.

SAW MILL MACHINERY
- Preston Woodworking Machy. Co., Preston, Ont.
- Yates Machine Co., P. B., Hamilton, Ont.

SAWS, BAND
- Preston Woodworking Machy. Co., Preston, Ont.

SCALES, BOILERS, ENGINES
- Can. Fairbanks-Morse Co., Montreal, Que.

SCOWS
- Collingwood Shipbuilding Co., Collingwood, Ont.
- Polson Iron Works, Toronto, Ont.

SCREENS, WIRE
- Dennis Wire & Iron Works Co., London, Ont.

SCREWS, COACH
- London Bolt & Hinge Works, London, Ont.

SEAM PAINT
- Kuhls, H. B. Fred., 6411 3rd Ave., Brooklyn, N.Y.

SEPARATORS, OIL, STEAM
- Mason Regulator & Engin. Co., Montreal, Que.
- Smart-Turner Mach. Co., Hamilton, Ont.

SHAFTING
- Wilkinson & Kompass, Hamilton, Ont.

SHAFTING, BRONZE
- Empire Mfg. Co., London, Ont.

SHEARS
- Bertram & Sons, Ltd., John, Dundas, Ont.
- Norbom Engineering Co., Philadelphia, Pa.

SHIPBUILDING TOOLS
- Aikenhead Hardware, Ltd., Toronto, Ont.
- Can. Ingersoll-Rand Co. Sherbrooke, Que.

SHIPS, BUILDERS OF
- Can. Vickers, Ltd., Montreal, Que.
- Collingwood Shipbuilding Co., Collingwood, Ont.
- Doxford & Sons, William, Sunderland, England.
- Georgian Bay Shipbuilding & Wrecking Co., Midland, Ont.
- National Shipbuilding Co., Goderich, Ont.
- Polson Iron Works, Toronto, Ont.
- Port Arthur Shipbuilding Co., Port Arthur, Ont.
- Yarrows, Limited, Victoria, B.C.

SHIP BROKERS
- Page & Jones, Mobile, Ala.

SHIP PLATES
- Nova Scotia Steel & Coal Co., New Glasgow, N.S.

SIDE LIGHTS
- Hopkins & Co., F. H., Montreal, Que.
- Wm. C. Wilson & Co., Toronto, Ont.

SLEDGES
- Wilkinson & Kompass, Hamilton, Ont.

SLINGS
- Hopkins & Co., F. H., Montreal, Que.
- Morris Crane & Hoist Co., Herbert, Niagara Falls, Ont.

SPECIAL MACHINERY
- Corbet Fdry. & Machine Co., Owen Sound, Ont.
- Miller Bros. & Sons, Ltd., Montreal, Que.
- Smart-Turner Mach. Co., Hamilton, Ont.

SMOOTH-ON
- Wm. C. Wilson & Co., Toronto, Ont.

SPIKES
- Wm. C. Wilson & Co., Toronto, Ont.

SPRAY COOLING SYSTEMS
- Spray Engineering Co., Boston, Mass.

STEAMSHIP AGENTS
- Page & Jones, Mobile, Ala.

STEAM SPECIALTIES
- Can. Fairbanks-Morse Co., Montreal, Que.
- Corbet Fdry. & Machine Co., Owen Sound, Ont.
- Empire Mfg. Co., London, Ont.

STEAM TRAPS
- Aikenhead Hardware, Ltd., Toronto, Ont.
- American Steam Gauge & Valve Mfg. Co., Boston, Mass.
- Empire Mfg. Co., London, Ont.
- Mason Regulator & Engin. Co., Montreal, Que.
- Smart-Turner Mach. Co., Hamilton, Ont.

STEEL, HIGH SPEED
- Hopkins & Co., F. H., Montreal, Que.
- Nova Scotia Steel & Coal Co., New Glasgow, N.S.
- Wilkinson & Kompass, Hamilton, Ont.
- Zenith Coal & Steel Products, Ltd., Montreal, Que.

STEEL SHELVING
- Dennis Wire & Iron Works, London, Ont.

STEEL WORK, STRUCTURAL
- Babcock & Wilcox, Ltd., Montreal, Que.
- Corbet Fdry. & Machine Co., Owen Sound, Ont.

STEERING GEARS
- Corbet Fdry. & Machine Co., Owen Sound, Ont.
- Hopkins & Co., F. H., Montreal, Que.
- Engineering & Machine Works of Canada, St. Catharines, Ont.
- Wm. C. Wilson & Co., Toronto, Ont.

STOCK RACKS FOR BARS, PIPING, ETC.
- Morris Crane & Hoist Co., Herbert, Niagara Falls, Ont.

STOKERS, MECHANICAL
- Babcock & Wilcox, Ltd., Montreal, Que.

SUPERHEATERS, STEAM
- Babcock & Wilcox, Ltd., Montreal, Que.

SWITCHBOARDS, ELECTRIC
- Crouse-Hinds Co. of Can., Ltd., Toronto, Ont.

TALLOW
- Wm. C. Wilson & Co., Toronto, Ont.

TANKS, STEEL
- Corbet Foundry & Mach. Co., Owen Sound, Ont.
- Goldie & McCulloch, Ltd., Galt, Ont.
- Hopkins & Co., F. H., Montreal, Que.
- Marsh Engineering Works, Belleville, Ont.
- Port Arthur Shipbuilding Co., Port Arthur, Ont.
- Reed & Co., Geo., Montreal, Que.
- Scaife & Sons Co., Wm. B., Oakmont, Pa.

TANKS (AIR, GAS AND LIQUID)
- Marsh Engineering Works, Belleville, Ont.
- Scaife & Sons Co., Wm. B., Oakmont, Pa.

TAPPING MACHINES
- Mueller Mfg. Co., H., Sarnia, Ont.

TELEGRAPHS, SHIPS
- McNab Co., Bridgeport, Conn.
- Cory & Sons, Inc., Chas., New York, N.Y.
- Morrison Brass Mfg. Co., James, Toronto, Ont.
- Wilson & Co., Wm., Toronto, Ont.

TELEPHONES, MARINE
- Cory & Son, Inc., Chas., New York, N.Y.
- McNab Co., Bridgeport, Conn.

TESTERS, METER
- Mueller Mfg. Co., H., Sarnia, Ont.

THUMB SCREWS AND NUTS
- Canada Foundries & Forgings, Welland, Ont.
- United Brass & Lead, Ltd., Toronto, Ont.

TRACK SYSTEMS
- Morris Crane & Hoist Co., Herbert, Niagara Falls, Ont.

TRAVELLING BLOCKS
Morris Crane & Hoist Co., Herbert, Niagara Falls, Ont.

TROLLEYS
Can. Fairbanks-Morse Co., Montreal, Que.
Morris Crane & Hoist Co., Herbert, Niagara Falls, Ont.

TROLLEY HOISTS
Morris Crane & Hoist Co., Herbert, Niagara Falls, Ont.

TRUCKS, HAND, ELECTRIC
Aikenhead Hardware, Ltd., Toronto, Ont.
Can. Fairbanks-Morse Co., Montreal, Que.

TUBES, BOILER
Babcock & Wilcox, Ltd., Montreal, Que.
Broughton Copper Co., Ltd., Manchester, Eng.

TUBES, COPPER AND BRASS
Mueller Mfg. Co., H., Sarnia, Ont.
Tallman Brass & Metal Co., Hamilton, Ont.
Standard Underground Cable Co., Hamilton, Ont.

TUGS
Collingwood Shipbuilding Co., Collingwood, Ont.
Polson Iron Works, Toronto, Ont.

TURBINES, DIRECT-DRIVING AND GEARED
Doxford & Sons, William, Sunderland, England.

TURNBUCKLES
Canada Foundries & Forgings, Welland, Ont.
Hopkins & Co., F. H., Montreal, Que.

TURNTABLES
Morris Crane & Hoist Co., Herbert, Niagara Falls, Ont.

SPIKES, SMALL RAILROAD
Severance Mfg. Co., S., Glassport, Pa.

UNIONS, ALL KINDS
Dart Union Company, Toronto, Ont.

VALVES, AIR
Mueller Mfg. Co., H., Sarnia, Ont.

VALVE, DISCS
Wm. C. Wilson & Co., Toronto, Ont.

VALVES
American Steam Gauge & Valve Mfg. Co., Boston, Mass.
Babcock & Wilcox, Ltd., Montreal, Que.
Can. Fairbanks-Morse Co., Montreal, Que.
Empire Mfg. Co., London, Ont.
McAvity & Sons, Ltd., T., St. John, N.B.
Mason Regulator & Engin. Co., Montreal, Que
McNab Co., Bridgeport, Conn.
Norbom Engineering Co., Philadelphia, Pa.
Williams Machinery Co., A. R., Toronto, Ont.
Wm. C. Wilson & Co., Toronto, Ont.
United Brass & Lead, Ltd., Toronto, Ont.

VALVES, FOOT
Aikenhead Hardware, Ltd., Toronto, Ont.
Smart-Turner Mach. Co., Hamilton, Ont.

VALVES, STOP, REDUCING, SAFETY CHECK, ETC.
Aikenhead Hardware, Ltd., Toronto, Ont.
American Steam Gauge & Valve Mfg. Co. Boston, Mass.
Empire Mfg. Co., London, Ont.
McAvity & Sons Ltd., T., St. John, N.B.
Morrison Brass Mfg. Co., James, Toronto, Ont.

VALVES, MIXING
Mueller Mfg. Co., H., Sarnia, Ont.

VALVES, REDUCING, PRESSURE
Mueller Mfg. Co., H., Sarnia, Ont.

VARNISHES
Aikenhead Hardware, Ltd., Toronto, Ont.
Ault & Wiborg Co. of Can., Ltd., Toronto, Ont.
Leckie, Ltd., John, Toronto, Ont.
Moore & Co., Benjamin, Toronto, Ont.
Reed & Co., Geo., Montreal, Que.

VENTILATORS, COWL
McNab Co., Bridgeport, Conn.

VENTILATION EQUIPMENT
Empire Mfg. Co., London, Ont.
Hopkins & Co., F. H., Montreal, Que.
Low & Sons, Ltd., Archibald, Glasgow, Scotland
Scythes & Co., Toronto, Ont.

WASHERS
London Bolt & Hinge Works, London, Ont.

WASTE
Scythes & Co., Toronto, Ont.
Wm. C. Wilson & Co., Toronto, Ont.

WATER COLUMNS
Morrison Brass Mfg. Co., James, Toronto, Ont.

WATER HEATERS
Empire Mfg. Co., London, Ont.
Morrison Brass Mfg. Co., James, Toronto, Ont.

WATER SOFTENERS
Babcock & Wilcox, Ltd., Montreal, Que.

WATER SUPPLY SYSTEMS
Mueller Mfg. Co., H., Sarnia, Ont.

WELDING, ELECTRIC
Hall Engineering Works, Montreal, Que.
Marine Welding Co., Buffalo, N.Y.
Beatty & Sons, M., Welland, Ont.

WELDING, OXY-ACETYLENE
Advance Mach. & Welding Co., Montreal, Que.

WHISTLES AND SYRENS
Empire Mfg. Co., London, Ont.
McAvity & Sons, T., St. John, N.B.
McNab Co., Bridgeport, Conn.
Mitchell Co., Ltd., Robt., Montreal, Que.
Morrison Brass Mfg. Co., Jas., Toronto, Ont.

WINCHES, CARGO
Advance Mach. & Welding Co., Montreal, Que.
Aikenhead Hardware, Ltd., Toronto, Ont.
Corbet Fdry. & Machine Co., Owen Sound, Ont.
Hopkins & Co., F. H., Montreal, Que.
Marsh Engineering Works, Belleville, Ont.

WINCHES, DOCK, SHIP
Advance Mach. & Welding Co., Montreal, Que.
Beatty & Sons, M., Welland, Ont.
Marsh Engineering Works, Belleville, Ont.
Miller Bros. & Sons, Ltd., Montreal, Que.
Morris Crane & Hoist Co., Herbert, Niagara Falls, Ont.
Wilson & Co., Wm. C., Toronto, Canada.

WINCHES, TOWING
Corbet Foundry & Mach. Co., Owen Sound, Ont.

WINCHES, TRAWL
Beatty & Sons, M., Welland, Ont.
Wm. C. Wilson & Co., Toronto, Ont.

WINDLASSES
Advance Mach. & Welding Co., Montreal, Que.
Corbet Fdry. & Machine Co., Owen Sound, Ont.
Dake Engine Co., Grand Haven, Mich.
Hopkins & Co., F. H., Montreal, Que.
Wilson & Co., Wm. C., Toronto, Canada.

WIPER CAPS, OILER BOXES, ETC.
Morrison Brass Mfg. Co., James, Toronto, Ont.

WIRE, COPPER CLAD STEEL
Standard Underground Cable Co., Hamilton, Ont.

WIRE, COPPER, BRASS, BRONZE
Standard Underground Cable Co., Hamilton, Ont.
Tallman Brass & Metal Co., Hamilton, Ont.

WIRE, INSULATED
Standard Underground Cable Co., Hamilton, Ont.

WIRELESS OUTFITS
Cutting & Washington, Inc., Cambridge, Mass.
Marconi Wireless Telegraph Co., Montreal, Que.

WIRE ROPE
Zenith Coal & Steel Products, Ltd., Montreal.

WOOD WORKING MACHINERY
Aikenhead Hardware, Ltd., Toronto, Ont.
Can. Fairbanks-Morse Co., Montreal, Que.
Preston Woodworking Machy. Co., Preston, Ont.
Williams Machinery Co., A. R., Toronto, Ont.
Yates Mach. Co., P. B., Hamilton, Ont.

WOOD BORING TOOLS
Aikenhead Hardware, Ltd., Toronto, Ont.
Can. Ingersoll-Rand Co. Sherbrooke, Que.

WOODITE GAUGE GLASS WASHERS
Wm. C. Wilson & Co., Toronto, Ont.

WRENCHES
Canada Foundries & Forgings, Welland, Ont.

YACHT BROKER
Watt, J. Murray, Philadelphia, Pa.

INDEX TO ADVERTISERS

CIRCULATES IN EVERY PROVINCE OF CANADA AND ABROAD

MARINE ENGINEERING of Canada

A monthly journal dealing with the progress and development of Merchant and Naval Marine Engineering, Shipbuilding, the building of Harbors and Docks, and containing a record of the latest and best practice throughout the Sea-going World. Published by The MacLean Publishing Co., Limited

MONTREAL, Southam Building · TORONTO 143-153 University Ave. · WINNIPEG, 1207 Union Trust Bldg. · LONDON, ENG., 88 Fleet St.

Vol. VIII. Publication Office, Toronto— May, 1918 No. 5

BOLINDER'S

The Engine that is *NOT* a Diesel—The Engine that is *NOT* a Semi-Diesel—The Engine that is the Standard for Hot Bulb Engines

Present Sales and Yearly Output 70,000 B. H. P.

Present U. S. A. Bolinder Installations 43,000 B. H. P.

A. S. "Mabel Brown," first of twelve Auxiliary Schooners fitted with twin 160 B. H. P. Bolinder, built for Messrs. H. W. Brown & Company, Ltd., Vancouver, B. C

BOLINDERS COMPANY, 30 Church St., New York

ESTABLISHED 1860

Sole Canadian Rights to Manufacture the

"HYDE"

Anchor-Windlasses

Steering-Engines

Cargo-Winches

Which have stood the test of 50 YEARS

310,

The "HYDE" Spur-Geared Steam Windlass

Propeller Wheels

Largest Stock in Canada.

Heavy Gears, Mill Repairs, Water Power Plant Machinery

Steel Castings

Manufactured by

The Wm. Kennedy & Sons, Limited

Owen Sound, Ontario

Shipbuilders, Attention!

Ship Chandlery

Our stock consists of:

Brass and Galvanized Hardware
Nautical Instruments
Heavy Deck Hardware
Rope, Oakum, Marline
Paints and Varnishes
Lamps of all types to meet inspectors' requirements, for electric or oil.
Ring Buoys, Life Jackets
Rope Fenders
Life-boat Equipment to Board of Trade specifications

Wire Rope rigging fitted to plan and specification a specialty

Let us estimate on your Block requirements, canvas work, including sails, awnings, hatch covers, nautical instrument and boat covers.

Our Catalogue needed to complete your files. Mailed promptly on request.

JOHN LECKIE, LIMITED
LECKIE BUILDING, TORONTO, ONT.

If what you need is not advertised, consult our Buyers' Directory and write advertisers listed under proper heading.

BERTRAM MACHINE TOOLS

For Structural, Bridge and Shipbuilding Plants

Modern in design and built for heavy service, our line embraces a varied equipment of Punches, Shears, Bending and Straightening Rolls, Coping Machines, Rotary and Plate Planers.

The assistance and advice of our engineers are yours for the asking.

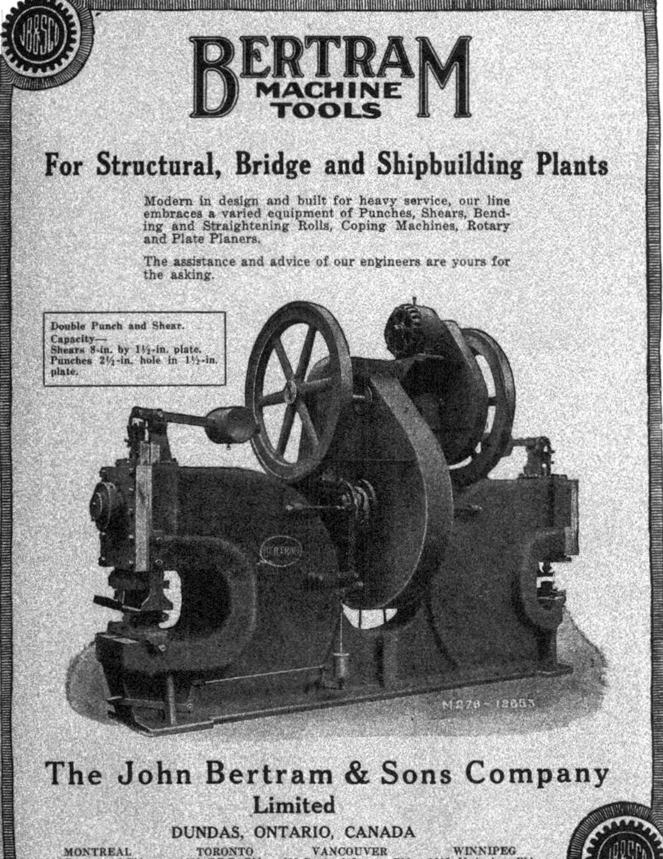

Double Punch and Shear.
Capacity—
Shears 8-in. by 1½-in. plate.
Punches 2½-in. hole in 1½-in. plate.

The John Bertram & Sons Company Limited

DUNDAS, ONTARIO, CANADA

MONTREAL 273 Drummond Bldg. TORONTO 1002 C.P.R. Bldg. VANCOUVER 609 Bank of Ottawa Bldg. WINNIPEG 1205 McArthur Bldg.

MASON'S

MADE-IN-CANADA PRODUCTS

We are prepared to meet your demand for quick deliveries on any of the well-known lines of auxiliary material shown here. All high quality and dependable lines, made in Canada by Canadian workmen. Observe the simplicity and correctness of design in the illustrations.

Sole Licensees and Distributors for:

The Mason Regulator Co.
Griscom-Russell Co.
Nashua Machine Co.
Coppus Engineering and Equipment Co.

Mason I.B. Reducing Valve

Our salesmen are practical engineers and can give sound advice on the efficient and economical application of Mason products. Bulletins and full information on request.

Reilly Multi-screen Feed Water Filter

The Mason Regulator and Engineering Co.
Limited

153 DAGENAIS STREET
Montreal

506 KENT BUILDING
Toronto

WORKS: 135 to 153 Dagenais Street, MONTREAL

Reilly Multi-coil Marine Feed Water Heater

Mason No. 55 Style Pump Pressure Regulator

Mason No. 126 Style Marine Reducing Valve

Reilly Marine Evaporator, Submerged Type

The CORBET Line of Marine Machinery

IS BUILT to Government specifications and is being used by the Naval Service, which is the best guarantee that our machines are up-to-date in every respect. When you place your order with us you **get delivery when promised.** When you instal our machines you **get satisfaction.** That is what you want. Is it not?

Our Line Includes	Cargo Winches Anchor Windlasses Steering Engines Hydraulic Freight Hoists	Automatic Steam Towing Machines for tugs and barges Special machinery built to specifications. Get our prices and delivery before placing your order.

The Corbet Foundry & Machine Co., Ltd.

OWEN SOUND, CANADA

HYDRAULIC EQUIPMENT
OF A
MODERN SHIPYARD
BY J. H. RODGERS†

Through the courtesy of a prominent Canadian shipbuilding company we are enabled to present the accompanying article which illustrates modern practice in a fully equipped plant.

MODERN steel ship construction would be practically impossible without the aid of hydraulic equipment. The nature of the enterprise involves the forming and bending of heavy steel plates for the hull and the framework that would be difficult to accomplish by any other method. The developments that have taken place in this country during the past two years, and more particularly in the last twelve months, have resulted in the establishment of many new shipyards, but owing to the great difficulty in obtaining ship plates, a great number of these yards are constructing wooden vessels. In view of later expansion when the necessary plates are available, it is of interest to study the class of hydraulic machinery installed by one of the large Canadian shipyards.

Weather conditions in Canada are not particularly favorable for the successful maintenance of such equipment, as it is generally located in large sheds or buildings which are difficult to heat during the winter months. Destruction of hydraulic appliances through the agency of frost must be guarded against, especially in northern climates, where the possibilities of freezing are much greater, owing to the frequent and often sudden changes of temperature and atmospheric conditions. This plant, although situated on the shores of the St. Lawrence River, has experienced little difficulty in this connection, due largely to the efficient piping installations and its method of delivery from the pumps in the engine room to the various departments.

From the accumulator, located in the engine room, the piping is carried through a spacious tunnel well below the ground level, the top of the tunnel being about six feet below the surface. This tunnel also contains all electrical conduits, water pipes, steam pipes, etc., which assist in maintaining a uniform temperature that prevents the possibility of frost. The total length of the tunnel is nearly 3,000 feet, and it extends from the power plant to the iron workers' shed, and thence alongside one of the quay walls. From the main tunnel the piping is distributed to the various buildings and to the different machines. This system of delivering the water from the pumps to the presses has been the chief help in avoiding trouble from frost. During the past winter, however, a slight accident was experienced on the horizontal bulldozer, shown in Fig. 1, when the flange of the return cylinder was broken off, through the freezing of the water and the subsequent expansion. This indicates the necessity of always draining the water from hydraulic equipment whenever it is out of service, and also the precaution that must be exercised to guard against low temperatures where such machinery is used.

†Associate Editor Canadian Machinery.

Portable Hydraulic Forming Press

One of the disadvantages of hydraulic equipment is that it is not very suitable for portable use owing to the special facilities that would be required for such purposes, so that in few instances will it be found that appliances are adapted for service other than in a permanent position in the shop, where the work can be conveniently brought to the machine for the necessary operations. However, occasions sometimes arise where it is almost impossible to accomplish certain operations unless the machine can be transferred to different locations on the work, where the same is held in a fixed position while one or more bends are being formed. This condition applies to the shaping of many pieces in connection with the angle irons that constitute certain portions of a ship's framework. By the use of a small portable press for this purpose it is much easier to work to the template, which can be more conveniently handled when it is not required to move the piece being formed.

For efficient work of this character it is necessary to have suitable equipment; on the discharge end of the furnace, where the work is heated, it is the general practice to have a considerable space fitted with heavy cast iron floor plates, the top surface of which is made as level as possible. These plates are provided with a large number of cored holes into which are placed the steel pins that act as fulcrums for the bending operations, and also a heel for taking the thrust of the hydraulic ram.

In Fig. 2 is shown one of the few hydraulic units adapted for portable service. In the operation of this press it is necessary to utilize two lengths of hose, one for the supply and the other for the discharge. For moving from one location to another the appliance is provided with a pair of handles, the lower ends of which are connected to a pair of wheels in such a manner that the cylinder is raised from the floor when the handles are pressed downwards, thus facilitating the movement of the machine to another position. The ram of this particular press is 4 inches in diameter and the working pressure is 1,500 lbs. per sq. inch.

The chief difficulty experienced in the operation of portable hydraulic equipment is the inability to get flexible connections that will stand up to the continual strain of the high pressure required. The pressure itself is not the most serious factor, but if particular care is not exercised in moving the machine the liability of kinking the hose, first in one direction and then in the other, soon destroys its usefulness, as cracks frequently develop and prevent further use for effective work. The reinforced hose used in this plant is principally of English manufacture, as others have not

FIG. 1.—HORIZONTAL BULLDOZER.

given very good results. Even the best, however, will have a short life unless great precaution is taken to protect the hose from injury.

Hydraulic Press With Plate Frame

A special design of press is shown in Fig. 3, where the frame is composed of steel plates, I beams and angle irons. The cast iron cylinders are supported between the overhanging arm with their axis in a vertical position, the main cylinder being 9 ins. in diameter and the return cylinder 4 ins. This press can be used for a large variety of light work, owing to the double arrangement of the working of cylinders. It is generally used for the making and trimming of small forgings. Presses built after this pattern are not as satisfactory as those constructed of cast iron, owing to the greater elasticity of the metal.

Vertical Plate Rolls

Forming the plates for the cylindrical portions of marine boilers is one of the chief details in connection with the work of a shipbuilding plant. In many establishments, particularly those of the smaller type where the plates are of the lighter sizes, the three-roll horizontal system is extensively used. In this plant the heavy plates for the high pressure boilers are formed in the large vertical plate roll shown in Fig. 4, manufactured by Hugh Smith and Company of Glasgow. With the three-roll type the bend is made by the action of the rolls in motion, forcing the plate against the third or shaping roll. In the machine here illustrated the bend is accomplished by the application of pressure while the plate is held stationary, this action being repeated at regular intervals as the plate is revolved intermittently, by the movement of the large main roll.

The machine can be operated at will or automatically, as desired, and the movement can be obtained in either direction. When the plate is first placed in position in the machine, the small rolls A A A that are carried in the forked end of the piston connected to the ram of the small hydraulic cylinders contained in the main cross beam, are moved back to permit the insertion of the plate between the small rolls and the main roll B. The three small rolls A are for the purpose of retaining the plate in constant contact with the main roll, and maintain it in a parallel position while it is being revolved for the following impression. The diameter of these small cylinders is about 3 ins., so that with an initial pressure of 1,500 lbs. per sq. inch the power exerted against each of the three rolls will be about 10,000 lbs. When in the desired position, the movement of the large roll—which is obtained through a system of gearing below the floor—is stopped, so that the bending pressure is applied when the plate is at rest.

The front face of the cross beam C is formed of two long narrow machined strips on either side of the central recess, into which the roll B enters when the cross-head C is forced outwards. This movement is obtained by the action of the toggle F at the top and a similar one (not shown) located below the level of the floor. One end of these toggles is secured to the movable cross-beam C, while the other end is pivoted to the fixed frame member G. The knuckles are connected to the links extending from either end of the operating ram, the cross-head of which is seen at E; the hydraulic cylinders D being located about midway of the height and supported on the central tie rods. The automatic ratchet mechanism is shown at H, and the main control handle at I. When the auxiliary handle is locked in the position shown, the machine will operate automatically during the period desired. When the end of the plate has been reached the reverse is set by hand and

FIG. 3.—60-TON HYDRAULIC PRESS WITH PLATE FRAME.

FIG. 2.—SMALL PORTABLE HYDRAULIC UNIT.

the machine operated automatically in the opposite direction. On the heavy plates several reversals are necessary before the required operation is completed. This machine has a capacity for plates about 10 ft. wide and 1½ in. thick.

Four-Cylinder Flanging Press

A very interesting design of hydraulic press is illustrated in Fig. 5. This is the single frame type and is provided with four rams; two in the upper portion of the overhanging arm, parallel and adjacent to one another; one in a horizontal position at the junction of the arm and the base; and the fourth in a vertical position in the base, in line with the outer of the top two. The arrangement of these four cylinders is such that almost any operation of a flanging, forming, or bending nature can be readily performed. When flanging, the forward of the upper two rams is used as a clamp, while the inner one is bending the flange. The horizontal ram is then used for finishing the piece to the desired shape of the forming block, the latter being bolted to the bed of the press. Each ram can be operated independently, the control levers being shown in the right foreground. For such operations as fire hole or man hole forming, special dies are constructed so that the impression is made at one setting of the dies. Large furnaces are located nearby for the heating of the materials.

FIG. 5.—FOUR-CYLINDER FLANGING PRESS.

Large Hydraulic Gap Riveter

In the manufacture of boilers and the construction of steel vessels the uniting of the various parts is mostly accomplished by means of riveted joints, with subsequent caulking of the seams to make them perfectly steam or water tight. Hand riveting is fast becoming a lost art, as power appliances are now extensively used for this purpose. Pneumatic tools are probably recognized as being of greater service than those operated by hydraulic pressure; this, however, is due to the portable character of the air operated tools and the adaptability to conditions, where it would be impossible to use the hydraulic equipment, rather than the inefficiency of the latter, as in many respects the steady pressure of the hydraulic riveter will accomplish much better results than the rapid fire of the pneumatic tool.

For fabrication work the hydraulic riveter is generally given an important place in boiler shop and shipyard equipment, although necessity demands that the air riveter be utilized for erection purposes. Another feature that restricts the hydraulic machine to certain limits is the fact that the resistance of the operating pressure must be taken by the same frame that supports the hydraulic cylinder. It will therefore be readily seen that the effective use of the hydraulic riveter is confined within certain limits, whereas the air-operated tool can even be used to good advantage where it would be impossible to use the hand hammer.

FIG. 4.—VERTICAL BOILER PLATE ROLLS.

In Fig. 6 is illustrated a large Southwark 12-foot gap riveter, used for the construction of marine boiler shells. The base of the machine is placed in a pit which allows of work to be conveniently handled up to 20 ft. diameter. It will be noticed that the forward portion of the ram that carries the rivet set is supported in slides to maintain alignment with the set in the anvil. The cylinder has a diameter of 9 ins., which gives a compression pressure on the rivet of about 45 tons. A platform is provided for the convenience of the operator. An advantage that the hydraulic method of riveting has over the pneumatic system is that the applied pressure has a greater tendency to spread the rivet in the hole, a factor that adds to the stability of the work.

Lambie's Hydraulic Joggler

Joggling of bars and angle irons is a frequent operation in connection with steel ship construction. For this purpose it is necessary to provide powerful machines, owing to the sharp offset required. Fig. 7 shows a Lambie's hydraulic joggler adapted for this operation. This press is of special heavy design, the operating cylinder having a diameter of 24 ins., providing a pressure, when fully loaded, of 300 tons. The return cylinders, 6 ins., dia., are located on either side of the main cylinder, and are used for discharging the water from the central cylinder and elevating the ram after the operation has been performed. The press is provided with special dies for the rapid handling of the work; these dies are so constructed that they can be adjusted laterally for the different thicknesses of metal.

Keel Plate Bending Machine

An interesting machine, both as regards its dimensions and its operation, is that of the keel plate bending machine shown in Fig. 8. The overall length of this machine is nearly 38 ft. and its total weight is approximately 90 tons. The main braking beam, over which the stock is bent, is in two sections and is supported—or suspended—from four massive arms extending up from the back of the machine. These are of the box type to reduce the weight and yet provide ample strength and rigidity for the heavy duty required. The forming roll, made of forged steel, has a diameter of 17 ins. and a length of 30 ft., the weight of the roll being about 12 tons. The roll is freely supported at a distance of about ¼ of its total length from each end in the bearings of the two swinging arms, the lower end of each arm being attached to the ram of a hydraulic cylinder, which is located below the floor level. The diameter of each cylinder is 18 ins., so that a total pressure of 180 tons can be exerted in the bending of the work, with the exception of that required to raise the roll and its supports. The roll is not fixed in its bearings, so that in the process of forming the work the shaft has a tendency to roll upon the material, thus eliminating the friction that would otherwise be present.

To guide the swinging arms in their proper path and also maintain them in a parallel position, two links are provided on either arm, the inner end being secured to lugs on the machine base. The clamping of the stock is accomplished by hydraulic pressure from a 6 in. dia. cylinder located at the extreme right of the machine. The axis of this cylinder is set at an angle of 6 deg. with the horizontal to coincide with the taper of the wedge on the clamping beam. The upper edge of this beam is parallel with the under face of the brake beam, and the lower surface is divided into four equal steps or wedges, having a taper of about 1 in. per ft. The two main cylinders can be operated jointly or independently as desired, so that tapered work may be easily accomplished. This machine, while primarily designed and intended for the bending of keel plates, can be adapted to a variety of purposes in connection with steel ship construction.

FIG. 6.—12-FT. GAP HORIZONTAL RIVETER.

FIG. 7.—300-TON HYDRAULIC JOGGLER.

250-Ton Press for Large Work

Of the many designs of hydraulic presses, none is more adaptable to such a wide range of operations than that shown in Fig. 9. This machine is especially suited to the forming of large plates into shapes that would be very difficult to obtain in any other manner. By the use of various blocks, either of wood or iron, any desired shape can be readily attained, the workmen in time becoming very expert in its manipulation. The base of the machine, with the operating cylinders, is located below the floor level, and extending upwards are the four main corner shafts for supporting the upper head. The four corner screws are threaded for over half their length for adjusting the position of the top head. The screws are 8 ins. in diameter, and 2½ threads per in., the nuts being 6 ins. thick. These nuts take the entire thrust of the hydraulic pressure, minus the weight of the part supported by the ram together with that of the top head. The lower table is on a level with the shop floor so that it is comparatively easy to move the work to and from the machine. The operating cylinders are four in number, the diameter of the main cylinder being 21¾ ins.; the internal cylinder 14 ins., the push up cylinders 7½ ins., and the clamping or vising cylinders, 6 ins. The control levers are shown to the extreme left.

EXTINGUISHING PETROL FIRES

By M. E. L.

It is generally known, or ought to be, that a petrol fire cannot be put out by means of water, owing to the petrol floating on the water, but a small fire can be suppressed with any heavy fabric which will exclude the air and not take fire, such as a sack or rug, preferably damped. Sand is an excellent medium, especially if damped, although if a petrol fire starts near the engine, sand may get into the parts. Some experiments have recently been carried out which prove that dry sawdust will suppress a petrol fire, and it is even more effective if it is mixed with bicarbonate of soda. Contrary to theory, it is found that the sawdust does not burn.

POSSIBILITIES OF ELECTRIC WELDING IN SHIP CONSTRUCTION

ELECTRIC welding for ship construction has recently been gone into very thoroughly by Arthur J. Mason at the plant of the Federal Shipbuilding Co., Newark, for the Emergency Fleet Corporation, and the report which follows, in part, has been made to Charles Piez, vice-president of the corporation.

Scope of Contemplated Work

Electric welding in its various phases has for years been employed in shipyards and in the arts generally, but for a number of reasons the work has been confined to odd jobs and repairs. The proposal to extend its use to the major part of ship construction has met with gratifying approval from the shipbuilder. It remains for us through this large test to demonstrate its economy in time and money and its adequacy to build a stanch ship.

The purpose of this test is to demonstrate these advantages—to do it in such a way that all may see and contribute, and finally to test the structure itself so completely that there will follow a heart-whole and unanimous belief in the method.

The test itself will take the form of building part of a hull at the Federal Shipbuilding Company's plant, Newark, N.J.

It has been necessary to design a ship to suit the material available, without encroaching on that needed for the regular ship construction at the plant. This has been done. The hull will have the outline, dimensions and strength conforming to the ships the Federal company is building.

It has been thought best to conduct the work at a site apart from the shipways, so as not to interfere with that programme.

A 10,000-ton ship, costing $2,000,000, now costs but $70,000 to rivet. It must be plain that if electric welding only promises to modify this amount no very substantial gain offers.

Splendid benefits we all feel do offer themselves in the possible change in the whole regime of shipbuilding. Our test has in view abolishing or greatly diminishing:

1. The railroad journey from rolling mill to fabricating plant, when the latter is not at the shipyard.
2. The templet makers' work.
3. The markers' work.
4. The punching.
5. Much of the work of the fitters and bolters who flog and pull the pieces to fit on the ways. There lies in the above items an excellent likelihood to save a month's time in construction and a saving of no less than $40 a ton in the cost of steel structure, at least $100,000 a hull on a 10,000-ton vessel.

Briefly the programme is to assemble a hull rapidly by spot welding, tacking the ship together much as a tailer bastes his work in assembling a suit of clothes. The structure then becomes a house favorable for work in all weather and at night in which the completion of the ship may go on.

After the material is thus assembled and fastened with spot welds, so that it is sufficiently strong to hold its shape, the work is completed by arc welding all seams to insure strength and render the work watertight. Roughly we expect the spot welds to be about 10 in. (25.4 cm.) apart.

The preparation of the site is well under way; the pile driving will be completed within ten days. The severe tests of strength contemplated needed about 300 piles.

One quarter of the structure will be riveted, the other three-fourths welded, so that the tests of strength will afford a basis of comparison.

Electric welding offers a great field for lightening a ship. In this design various views of this opportunity will be tried out. The field here is very great—ultimately 10 per cent. of the steel may be eliminated.

One derrick will bring material, the other derrick support the spot-welding yoke, whose function is to tack the material together, fastening the plates either to the frames or to the adjoining plates.

If one visits the ways at any shipyard, it becomes obvious that at any instant only a modicum of the men are for the moment at work. This is unavoidable under the present system. We hope to establish a plan of assembly with more continuity and less waiting on one another.

Only a fifth of the men on a hull are riveters. The spot-weld yoke will forthwith pull the parts to place with a much more vigorous agency than flogging and pulling to place by numerous bolts, now done by the other four-fifths.

The problems of fitting in place the parts of a hull are almost wholly problems arising out of the necessity to make a number of little holes in a plate made by one man at one time and place fit a number of little holes made by another man at another time and another place.

Once all holes are left out of the material, all parts fit. The creeping and kindred problems so perpexing to the shipbuilder disappear. Every plate becomes closer. Every plate justifies itself.

The manufacture of the spot-welding yoke and appliances is placed in the hands of the Universal Electric Welding Company of Long Island City. The design of the yoke is completed, the patterns are made and steel castings will be forthcoming in the next ten days. The early stages of the arc welding are to be accomplished by the Wilson Electric Company, which was so successful in the work on the German ships' repairs, but it is the intention to call in all men with ideas and apparatus and to give them a field to test out in actual work. To this end Professor Adams' committee is searching out all available talent.

An adequate system of testing the work when done is under consideration. The primary test will consist of filling the hull with water and shifting the points of support under continual and close scrutiny, as one-quarter of the whole will be riveted in the normal manner. There will be always a gauge of comparison between this portion and the portion which is welded.

Likewise there will be a chance for comparison of the two forms when subjected to abuse by bumping with rams and in various other ways.

FIG. 9.—250-TON HYDRAULIC PRESS.

Albert Deschamps, of the Atlas Construction Co., Montreal, and one of the engineers in charge of construction in the building of the concrete boat "Concretia," is about to leave on a vacation of several weeks into Northern Ontario.

High Pressure Air Compressor Design and Application--I

By Joseph M. Ford

Compressed air at high pressure is becoming an increasingly important medium in modern engineering practice and in naval warfare. The accompanying paper deals with the machines which produce high pressure air, leaving on one side the question of power transmission by compressed air. In passing, the principal advantages of this system are referred to, and will be seen to be of considerable value. They consist of the facility with which energy can be stored; the unlimited rate at which accumulated energy may be converted into useful work; and, apart from loss of efficiency, leakage cannot cause any awkward consequences.

ONE of the best instances of the adoption of high-pressure air as a medium for power storage is to be found in torpedo work. The torpedo is launched from the tube by means of compressed air, usually stored at a high pressure, and when in the water is propelled by an engine deriving most of its power from compressed air.

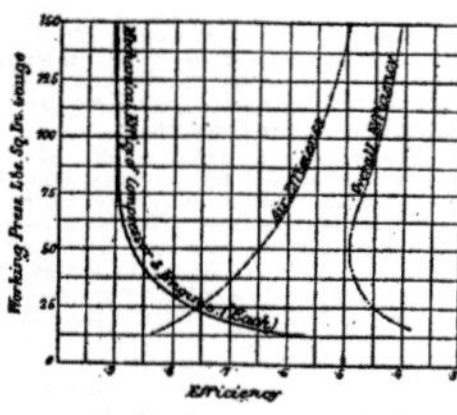

FIG. 1. SINGLE STAGE COMPRESSED AIR POWER. TRANSMISSION. NO PRE-HEATING. MECHANICAL EFFICIENCY OF ENGINE ASSUMED TO BE SAME AS THAT OF COMPRESSOR. COMPRESSION IN LATTER AND EXPANSION IN ENGINE ASSUMED TO BE ADIABATIC.

The starting of prime movers of the internal combustion type of considerable size is almost invariably carried out by compressed air, except perhaps in very special cases, such as some generating sets, where the dynamo can be "motored."

In marine work, with engines of the Diesel and other types, reversing and manoeuvring involve the displacement of a considerable amount of the valve-driving mechanism—most conveniently accomplished by compressed air.

One of the principal applications, however, is in the Diesel engine itself, where compressed air is used not as a medium for energy storage in the usually accepted sense of the term, but to inject the liquid fuel into the working cylinders against the compression pressure of about 30 atmospheres (430 lbs. per sq. inch.)

In some oil-engined ships, high-pressure air is used for driving some auxiliaries, such as steering gear, main engine turning gear, small pumps, etc., and in submarines the air is used, among other things, for blowing the ballast tanks.

Pressures Adopted in Practice

Within limits, the lower the working pressure of the air, the higher the efficiency, but exigencies of space and weight may demand that the storage reservoirs shall occupy minimum space, in which case, for a given amount of energy stored, the pressure should be as high as possible—particularly, for instance, in a torpedo, where the dimensions and weight of power plant must be cut down to a minimum, almost irrespective of cost, and the pressures employed range accordingly from 2,000 to 4,000 lbs. per square inch.

With internal combustion engines the design and type determine the pressure of the starting air. For example, gas, paraffin and petrol engines are frequently started with air at 100 to 300 lbs. per square inch. With Diesel engines of the four-cycle type about 300 lbs. per square inch usually suffices, whereas in engines of the two-stroke type the pressure may be as high as 1,000 lbs. per square inch. One of the disadvantages of using high pressures in this connection is that when starting a refractory engine—which does not readily pick up on fuel—the continued expansion of high pressure air in the working cylinders lowers the temperature to such an extent that the heat of compression is insufficient to ignite the injected oil. With reversing mechanism the obvious advantage accruing from the use of air at high pressures is that the dimensions of the servo-motors may be kept within small limits, which is of moment in some installations, although the higher the pressure the more difficult it is to maintain tightness and the better must be the fits, etc.

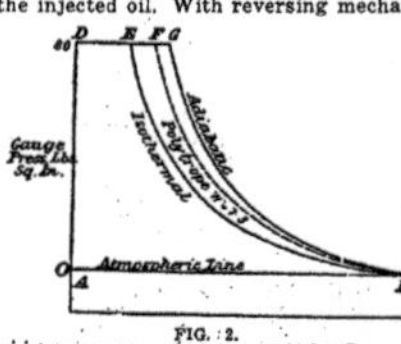

FIG. 2.

Initial temperature of air.. 60 deg. F.
Initial pressure .... 14.7 lb. per sq. in. abs.
Final pressure .... 94.7 " " "

WORK REQUIRED TO COMPRESS AND DELIVER 1 CUB. FT. OF FREE AIR.

(i) Isothermal compression .... 3950 ft.-lb.
(ii) Polytropic compression ("N"=1.3) 4950 "
(iii) Adiabatic compression .... 5275 "

In this case 30 per cent. more work is required for adiabatic compression than for isothermal compression. Hence note importance of cooling during compression.

TEMPERATURES ATTAINED AT END OF COMPRESSION

(i) Isothermal compression .... 60° F.
(ii) Polytropic compression ("N"=1.3).. 335° F.
(iii) Adiabatic compression .... 435° F.

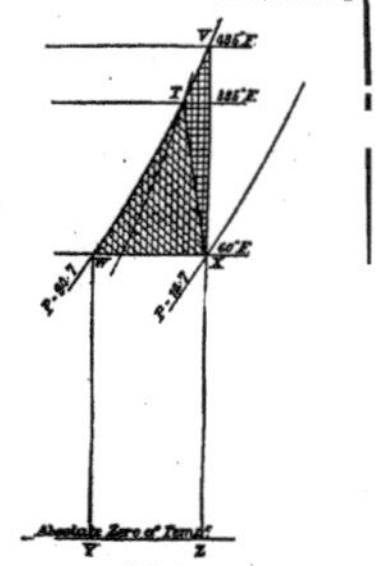

FIG. 3.

Temperature entropy diagram for compression of 1 lb. of air from 14.7 to 94.7 per sq. in. abs.
Heat equivalent of work required for isothermal compression = area WYXZ.
Heat equivalent of extra work required if compression is adiabatic = area VWX (vertical shading).
Heat equivalent of extra work required if compression is polytropic = area TWX (diagonal shading).
Heat equivalent of work saved by cooling shaded horizontally.

For fuel injection, the air pressure is determined by the theoretical cycle of operations in the working cylinder, and varies from 600 to 900, and in extreme cases even to 1,000 lbs. per square inch, according to the particular design of engine, kind of fuel and load.

The question of driving auxiliaries by compressed air is a vexed one. Unless the air can be heated before expansion in the engine cylinder, low initial pressures should be used. With high pressures the re-expansion lowers the air temperatures so much, due to work being done by the air, that snow forms and interferes with the working of the valves, throttle passages, etc. Fig. 1 shows the theoretical efficiency of a transmission system when there is no pre-heating. The mechanical efficiency of the engine

*Part I. of a paper read before the Greenock (Scotland) Association of Shipbuilders and Engineers.

at the different pressures is assumed to be the same as that of the compressor at the corresponding pressures. The "air efficiency" is the ratio of work obtained from the air in the engine to the work done on the air in the compressor. A considerable gain in efficiency may be effected if the air can be heated before use, especially when this can be inexpensively done—e.g., by the exhaust gases of an internal combustion engine. The advantages of this system are nullified, more or less according to particular circumstances, by the low efficiency. Practical considerations, such as the question of leaks, etc., favor low pressures, whereas such considerations as the size of pipes, cylinders and storage generally demand a compromise, although it is cheaper to make an efficient large low-pressure reservoir than a satisfactory small high-pressure one.

Simple Air Compression

By way of introduction a few points on the elementary thermodynamical aspect of air compression may assist towards the understanding of the special features of high-pressure compressors. If air be compressed isothermally, a certain amount of work has to be done (see Fig. 2), giving a theoretical indicator diagram for a machine delivering air at 80 lbs. per square inch. With isothermal compression the work required is represented by the area ABED. If the compression is adiabatic, then the additional work required is represented by the area BGE, from which it will be seen that the aim should be to secure isothermal compression—that is, to prevent increase of the temperature of the air during compression. Not only from the point of view of power absorbed should the temperature rise be small, but also for mechanical reasons, as later explained.

The index "n" in the standard equation $PV^n = C$ is in some ways a measure of the amount of heat abstracted during compression. With adiabatic compression its value for diatomic gases such as air is 1.41, and with isothermal compression its value is unity. In actual practice the amount of heat removed by a water-jacket on an air compressor cylinder is such that its value lies between about 1.28 and 1.35, the latter value being more applicable to large high-speed machines.

The line BF on Fig. 2 represents compression according to the law.

$$PV^{1.3} = \text{constant.}$$

This means that the saving in work effected by cooling the air during compression is represented by the area BGF. Moreover, the reduction in the final delivery temperature is to be noted.

The diagram in Fig. 3 is the "theta" "phi" diagram corresponding to Fig. 2. This shows rather more clearly the difference in the work required and temperature attained by compressing under the various conditions. The figure is self-explanatory.

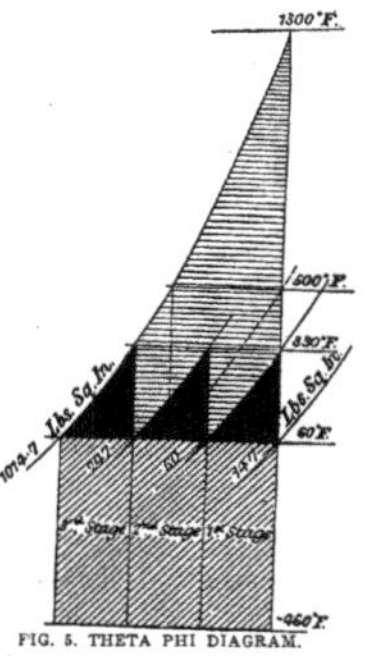

FIG. 4. Pv. DIAGRAM.

FIG. 5. THETA PHI DIAGRAM.

THEORETICAL DIAGRAMS SHOWING SAVING EFFECTED BY STAGE COMPRESSION. COMPRESSION TO 1,000 LBS. PER SQUARE INCH GAUGE.

In an ideal compressor, i.e., one in which compression is isothermal, the work required per unit of air is represented by the diagonally shaded area of each diagram. If compression were carried out in one stage the work required is represented by the total area of the diagram. The horizontally shaded area represents the work saved by compressing in three stages and intercooling. The darkly shaded area represents the amount by which the work required for three-stage compression is still in excess of the ideal. Note the huge saving of three-stage over one-stage compression. In this case the saving is 29½ per cent.

Compression in Stages

When pressures of 1,000 lbs. per square inch and over are required, the temperatures attained by single-stage compression would be extremely high. If lubricating oil could be eliminated from the problem, careful design might succeed in producing a compressor that would stand the stresses set up by the high temperatures.

Lubricating oil burns in air at such temperatures, with the result that the pistons would become carbonized, the valves clogged and the pipes sooted up. In addition, high temperatures have deleterious effects on tempered steel valves and springs. With large high-speed machines the temperature, coupled with the peculiar fatigue phenomenon, makes it well-nigh impossible to get satisfactory working with temperatures higher than those consequent upon about six compressions in one stage.

If the excessive work of compression and high temperatures are to be avoided, the heat must be abstracted from the air in the period between the start of compression and the commencement of delivery. Jacketing alone is insufficient. The method that suggests itself is to compress the air a small amount, then withdraw it from the cylinder, cool it to atmospheric temperature, and return it to the cylinder. Next move the piston further to compress the air still more, and repeat the cooling process, and so on, until the required pressure is attained. This is virtually the method adopted in practice, except that in lieu of carrying out the whole process in a single cylinder, one cylinder is devoted to each stage of compression, and between each stage the air is passed through an intercooler to reduce its temperature to the initial atmospheric temperature. The saving in work effected by stage compression with intercooling is shown in Figs. 4 and 5. The PV diagram, Fig. 4, has been drawn for a three-stage machine compressing to 1,000 lbs. per sq. inch.

If the heat of compression could be removed as fast as it is generated, as with isothermal compression, the work required to compress unit volume would be represented by the diagonally-shaded area. This is the ideal condition. If the compression were carried out in one stage and no heat were removed—that is, if the compression were adiabatic — then the work required to compress unit volume of air would be represented by the total area of the diagram, that is, the diagonally, plus the black, plus the horizontally shaded part. The ratio of this total work to that required in isothermal compression will be noted. If, again, the air were compressed adiabatically to 60 lbs. per square inch only and then cooled to its initial temperature, its condition will then be the same as it would have been had it been compressed isothermally. From this point the second stage of compression is caried out and the cooling process repeated, meaning that the area of the second stage diagram is only the diagonally shaded portion plus the black portion, the horizontally shaded part being completely saved, due to the reduction in the volume of air dealt with, as a result of the intercooling.

Similar conditions obtain in the third stage, the horizontally shaded part again representing work saved. The work required for adiabatic compression in three stages is thus represented by the diagonally shaded portion plus the black portion, which latter represents the amount by which the work required is still in excess of the ideal.

Fig. 5 is the "theta" "phi" diagram corresponding to the PV diagram in Fig. 4. The corresponding areas are similarly shaded. In addition to the work required, this diagram shows the temperatures attained in compression under the various conditions. The reduction of the temperature due to staging is clearly shown. The 1,300 deg. F. is a theoretical value, 500 deg. F. is the temperature which would be attained were the compression carried out in two stages, and 330 deg. F. is the result of three-stage compression. In the case under consideration the total work is equally divided between the stages, whereas in actual practice this condition does not generally hold exactly.

The temperatures attained in the cylinder when air is compressed to various pressures in one, two or three stages, with intercooling, are given in Fig. 6. Piston friction and the effect of the cylinder walls have been neglected. As previously stated, the high temperatures could not be dealt with in practice. The temperatures given for two and three-stage compression are based on the assumption of perfect intercooling, which condition holds for well-designed machines. In some plants where there is a copious supply of cold water, and where intercoolers of liberal dimensions are fitted, the air can be cooled between the stages to less indeed than the initial temperature.

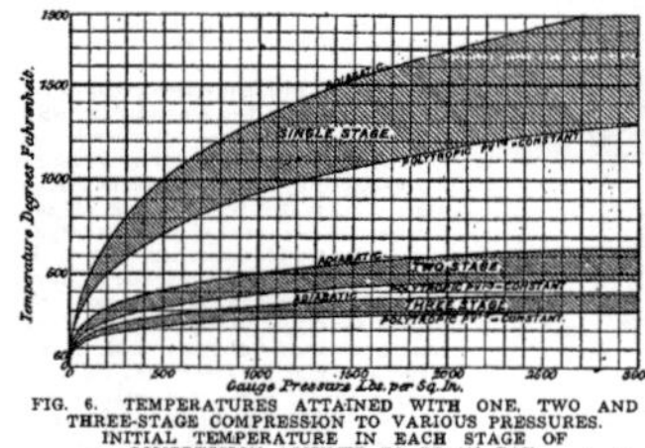

FIG. 6. TEMPERATURES ATTAINED WITH ONE, TWO AND THREE-STAGE COMPRESSION TO VARIOUS PRESSURES. INITIAL TEMPERATURE IN EACH STAGE OF COMPRESSION ASSUMED TO BE 60° FAH.

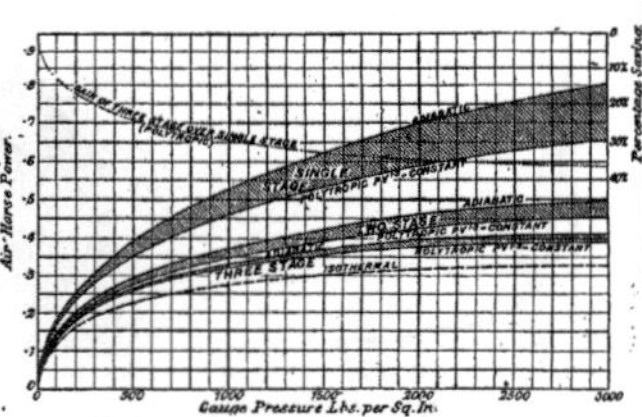

FIG. 7. AIR HORSE-POWER REQUIRED PER CUBIC FOOT FREE AIR PER MINUTE WITH ONE, TWO AND THREE-STAGE COMPRESSION TO VARIOUS PRESSURES.

Fig. 7 gives the air horse-power required to compress 1 cub. ft. of free air per minute in one, two and three stages to various pressures, and shows how the power required approaches the ideal (isothermal compression), and the ill-effects of inefficient jacketing diminish as the number of stages is increased. For example, with air compressed to 1,000 lb. per square inch in one stage, according to the law $PV^{1.3}$=constant, the power required would be 0.46 a.h.p. per cubic foot, but if the cooling were very inefficient, so that the compression were adiabatic, the power would rise to 0.53 a.h.p. per cubic foot—an increase of 15 per cent. With three-stage compression, according to the same law, the power required would be 0.32 a.h.p. per cubic foot, and in this latter case, assuming that the cooling were inefficient, the power would only rise at the most to 0.335 a.h.p. per cubic foot—an increase of 5 per cent., against the 15 per cent. obtained when compression is carried out in a single stage.

Apart from the questions of efficiency and temperatures, an advantage of staging is the reductionof crankpin loads and stresses generally—the reasoning being exactly the same in this connection as for a compound or triple-expansion steam engine. Theoretically an infinite number of stages is required, but practical considerations, such as the multiplicity of parts, the low mechanical efficiency, weight, cost, etc., limits the number of stages.

Pressures and Number of Stages

The best practice is as follows: The maximum terminal pressures are, for one stage, 80 to 100 lb. per square inch; two stages, 600 lb. per square inch; three stages, 1,500 lb. per square inch; four stages, 2,500 to 3,000 lb. per square inch; five stages, 4,500 lb. per square inch; and above 4,500 lb. per square inch, six or seven stages to 6,000 lb. per square inch, which is about the highest pressure for which an air compressor has been constructed.

In the case of very small machines these pressures, for a given number of stages are often exceeded in order to reduce complication. For example, machines have been constructed and work quite satisfactorily when compressing to 2,500 lb. per square inch in two stages. The temperatures attained, however, make it necessary to carry out the cylinder lubrication by soapy water.

Effects of Cylinder Clearance Volumes

An important consideration in the design of a multi-stage compressor is the computation of the clearance volumes of the various cylinders. By clearance volume is understood not merely the volume of the space which separates the piston from the end of the cylinder when the crank is on dead centre, but, in addition to this, the volume of all valve ports and pockets, space behind piston rings if these are air-packed, and, in short, any space into which the air can be compressed, without passing through the delivery valve. With an ordinary single-stage compressor, such as would be used to drive pneumatic tools in a shipyard, the designer aims to reduce the cylinder clearance to the minimum, for by so doing a greater quantity of air is delivered per stroke than would be the case with a larger clearance. Fig. 8, which is a theoretical card for a cylinder compressing to 80 lb. per square inch, is given i[n] explanation. With zero cylinder clea[r]ance the volume of air delivered is repre[-] sented by CD, whereas with 10 per cen[t] clearance and the same cylinder dimer[n]sions the volume delivered becomes FI[D]. Less air is drawn into the cylinder pe[r] stroke, since the air left in the clearanc[e] space re-expands along DE, for a portio[n] of the suction stroke, as shown, and th[e] suction valves do not open to admit [a] fresh supply of air until the point E i[s] the stroke is reached. In this way th[e] volume of air aspired is only EB, as com[-] pared with AB—the volume dealt with i[n] the same cylinder without clearance. Th[e] ratio $\frac{EB}{AB}$ is known as the Indicated Vol[-] umetric Efficiency, and in this particula[r] case is about 69 per cent. The smalle[r] the clearance space, the nearer will E b[e] to A, i.e., $\frac{EB}{AB}$ will be greater, and a large[r] volume of air will consequently be dealt with per stroke.

With a large clearance a larger cylinder is required to deliver a given amount of air than with a small clearance, yet the indicated work done per unit volume of air dealt with remains the same, although the larger cylinder makes for a heavier and more costly machine, with greater piston loads, so that in general practice the clearances are made as small as possible. Referring again to Fig. 8, the higher the compression is carried, the further will E be removed from A, that is, the higher the delivery pressure at which a given cylinder is working, the less will be the indicated volumetric efficiency. Further, if the clearance air receives heat from the cylinder and piston whilst expanding, the law of expansion will approach the isothermal and the pressure of the clearance air will take the longer to drop to the suction pressure, or E will be still further removed from A to G.

It is to be emphasized that the air should not receive heat whilst re-expanding, or, in other words, the clearance space should be as efficiently cooled as possible. The cooler the walls of the clearance space, the more rapidly will the pressure of the re-expanding air fall to the suction pressure and the greater will the ratio $\frac{EB}{AB}$ become. Moreover, if the clearance air receives heat whilst re-expanding, its temperature at point E

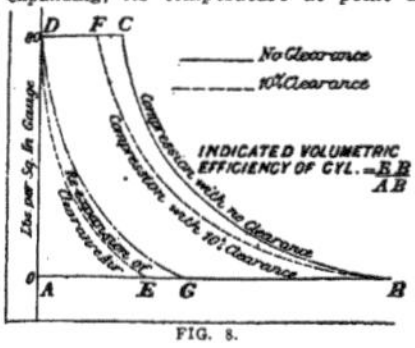

FIG. 8.

will be considerably higher than the temperature of the incoming air. As a result of the mixing of the new supply with the hot clearance air, the temperature at the end of the suction stroke, i.e., at B, will be higher than it would have been had there been no hot air left in the cylinder at point E, thus reducing the weight of air dealt with per stroke and increasing the temperature of compression. In any case the temperature of the air at B must be higher than that of the suction air, owing to contact with the hot walls and piston.

In modern practice, the cylinder clearance volume for a single-stage compressor or low-pressure cylinder of a multi-stage machine is seldom less than 3 to 4 per cent. of the volume swept out by the piston. It is usually found that, within limits for practical reasons, the larger the machine, the smaller the percentage clearance.

If a single-stage machine were constructed to compress to a moderately high pressure, say, 165 lb. per square inch, the indicated volumetric efficiency at the best would not be more than about 80 per cent., whilst the ratio of volume of free air delivered to volume swept by piston would be very much lower still, due to the addition of heat before compression commenced. Further, the cross-head loads would be very great, due to the final delivery pressure being exerted on the whole area of the large piston. The twisting moment diagram for the machine would be uneven, and, perhaps of most consequence, the final delivery temperature would be excessively high.

If, on the other hand, the same size of cylinder were used with a high-pressure cylinder added thereto, so that the first only compressed to 37 lb. per square inch, the indicated volumetric efficiency would be increased to about 94 per cent., due to the lower delivery pressure bringing the point of the low-pressure diagram corresponding to E in Fig. 8 nearer to A. With the high-pressure cylinder mounted in tandem with the low-pressure, it is possible that the weight of the machine per unit of air delivered would not be more than when the compression was carried out in one cylinder. Staging nullifies to a large extent the effects of clearance volume on the volumetric efficiency of a compressor, besides obviating the difficulties with loads and temperatures.

In both the French and German languages the expression for clearance volume, literally translated, means "injurious space," yet the effects of clearance in some cases are beneficial. In a cylinder with no clearance the whole load at the commencement of the stroke due to the acceleration of the moving part comes upon the crank-pin. With clearance, the re-expansion of the clearance air helps largely to accelerate the masses, and an interesting case in this connection may be cited.

With steam-driven compressors it is common practice to mount the steam cylinders in tandem with the air cylinders, in which case the setting of the steam valves must be so arranged that admission is late instead of early, as is usual. If the full steam pressure were applied to the piston before the clearance air pressure in the air cylinder had become reduced by re-expansion, unnecessarily high loads due to these simultaneous full pressures, less the inertia effects, would be imposed upon the connecting rods, etc. High speeds with increased inertia forces tend generally to reduce crank-pin loads. With steam-driven compressors of the foregoing type the greater the clearance volume of the air cylinder the later must the steam admission be timed.

Effects of Clearance in the Higher Stages of a Multi-Stage Compressor

In a multi-stage compressor the work done in the various stages may be varied within fairly wide limits merely by adjusting the clearance volumes of the various cylinders, excepting the low pressure. Suppose, for example, it is required to reduce the proportion of work done in the high-pressure stage of a three-stage compressor in which the high-pressure clearance volume is 10 per cent., and with stage pressures—low pressure 45 lb. per square inch, intermediate pressure 300 lb. per square inch—working against a final delivery pressure of 900 lb. per square inch.

The work done in compression varies as the ratio of final to initial pressure, so that any desired reduction in any given stage can only be effected by reducing this ratio, or in this case by increasing the high-pressure suction pressure, since the final delivery pressure is fixed. The compressor capacity being unaltered, an increase in the high-pressure suction pressure necessarily involves a corresponding reduction in actual volume of high-pressure suction. Since the volume swept by the high-pressure piston remains the same, it is clear that the only way to ensure a reduced suction volume is to decrease the volumetric efficiency of the high-pressure cylinder by increasing the clearance volume, with the result as shown in Fig. 9, where the high-pressure diagram with the original 10 per cent. clearance, together with the diagram resulting from an addition of 20 per cent. extra clearance, are given. The area of the diagram is considerably reduced, the high-pressure suction pressure (the intermediate-pressure delivery pressure) rises from 300 lb. per square inch to 375 lb. per square inch (gauge-pressure), this increase being necessary in order to get the whole of the air into the high-pressure cylinder. The effect on the low-pressure delivery pressure is slightly to increase it, due to the reduced volumetric efficiency of the intermediate pressure cylinder, resultant upon the expansion of the intermediate pressure clearance air from a higher pressure. The actual increase in low-pressure delivery pressure depends upon the intermediate pressure clearance volume. If the intermediate pressure cylinder had no clearance volume, which is impracticable, the low-pressure pressure would remain unaltered. The slight increase in low-pressure delivery pressure somewhat reduces the volumetric efficiency of the compressor.

The simple method of altering the distribution of work would be convenient where the high-pressure delivery temperature was found to be excessive (see later); or in cases where the turning moment was unsatisfactory. The clearance volume can be increased by the addition of a hollow pocket, but cannot so easily be reduced.

Some multi-stage compressors are required to work for long periods at varying delivery pressures, or, in the case of a charging compressor, at a constantly increasing pressure from atmospheric to the full bottle pressure. With Diesel engines it is usual to regulate the blast pressure by throttling the fuel-injection air compressor suction, so that instead of

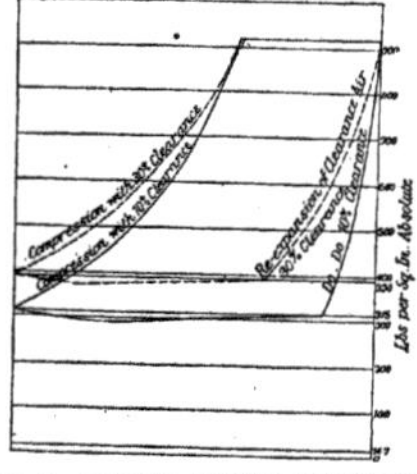

FIG. 9. DIAGRAM SHOWING RESULT OF INCREASING CLEARANCE VOLUME IN HIGH PRESSURE CYLINDER.

compressing and blowing the excess in waste, only the requisite amount is dealt with. The effect on the distribution of work, temperatures, etc., occasioned by this throttling should be considered—especially with marine engines, where the compressor may run for long periods in this condition.

In all machines there are certain stage pressures which give the best all-round running, as regards temperatures, loads, balance, etc., for each delivery pressure. If these stage pressures are accurately determined for the normal final delivery pressure, then it is safe generally to assume that the particular stage division of work so obtained will also be the best proportion for any other delivery pressure at which the machine is likely to work. With Diesel engine fuel-injection compressors this state of affairs rarely obtains. In a three-stage compressor designed for 800 lb. per square inch working pressure, and run at, say, 900 lb. per square inch, the low-pressure and intermediate-pressure pressures will not increase in proportion to the final delivery pressure—actually in the majority of cases they alter little. By suitably proportioning the cylinder clearance volumes of the various stages the machine can be made automatically to divide the total work between the stages in a given proportion, no matter what the delivery pressure may be.

Figs. 10 and 11 show approximately the stage pressures which would be obtained in three machines of the same capacity when working at different pressures. Each machine is designed for a delivery pressure of 900 lb. per square inch, and at this pressure the stage pressures are: low-pressure, 45 lb. per square inch (gauge pressure). The heavy lines on the diagram show what the pressures should be to secure an approximately constant proportional division of work between the stages. The lines marked (a) show the pressures which would obtain in a compressor having cylinder clearances of: low-pressure, 4 per cent.; intermediate-pressure, 5 per cent.; high-pressure, 10 per cent.

The rise in the stage pressures as the final delivery pressure increases is seen to be comparatively small. The lines marked (b) show the pressures with a compressor of the same capacity when the cylinder clearances are: low-pressure, 4 per cent.; intermediate-pressure, 10 per cent.; high-pressure, 20 per cent. Here it is seen that the rise in stage pressure is much more marked than in the first case, that is to say, with these increased clearance volumes the proportional division of work between the stages is more nearly constant than with the smaller clearances. The lines marked (c) give the pressures which would obtain with a machine having clearances: low-pressure, 4 per cent.; intermediate-pressure, 15 per cent.; high-pressure, 30 per cent.

In this case the rise in stage pressures as the delivery pressure increases approximates much more nearly to the "ideal;" that is, the proportional division of work between the stages is very nearly constant, so that the balance of the machine is more likely to be maintained than in the first case, and, further, the final delivery temperature will be lower, for in the first case, when the machine is working at a pressure higher than its designed pressure of 900 lb. per sq. inch, the high-pressure stage is taking more than its designed share of the work. For example, if the first machine (case "a") were run at 1,200 lb. per square inch, the compression ratio in the high-pressure cylinder would be:—

$$\frac{1215}{335} = 3.63$$

If the last machine (case "c") were run at 1,200 lb. per square inch, the high-pressure compression ratio would be:—

$$\frac{1215}{370} = 3.28$$

This means that the temperature attained in the machine with the large clearance would be less than in the machine with the small clearance on account of the smaller compression ratio. Furthermore, if the large clearance is composed of a connecting pipe within a water-jacket, as is sometimes the case, the cooling during compression is improved, resulting in a decrease of the value of "n" in $P V^n =$ constant, still further reducing the temperature. Again, the clearance space being so efficiently cooled results in an increase in the value of "n" for the re-expansion of the clearance air. The incoming air mixes with this comparatively cool air, and consequently the initial temperature is not unduly raised.

The main point, however, is that by correctly arranging the clearance volumes the machine can be made to divide the total work into the best proportions for any delivery pressure with which the compressor may have to deal. A minor point is that with increasing delivery pressures piston leakage becomes greater, and the volumetric efficiency at the higher pressures is less than at the lower. If the low-pressure pressure increases with the delivery pressure, the indicated volumetric efficiency of that cylinder is decreased in addition—or, in short, the overall volumetric efficiency of the machine with the big clearance is slightly less than that of the machine with the small clearances when the delivery pressure is higher than that for which the compressor is designed.

The difference, however, is not worth considering. (It may be mentioned that the lines in Fig. 10 should not actually be quite straight. They are sufficiently accurate, however, as in practice it is impossible to estimate the pressures to within a few pounds.)

(To be continued)

MARINE SCHOOL CHIEFS TO MEET

The United States Shipping Board Recruiting Service at Boston has issued the following:

A conference of all the section chiefs of the free Government navigation and marine engineering schools operated by the United States Shipping Board, will be held Friday and Saturday, May 24 and 25, at the headquarters of Henry Howard, director of recruiting service at the Boston custom house.

A similar meeting was held last fall.

"I feel that this second conference will be very materially beneficial to our ser-

What the Marine Engineer of To-day Must Know*

Requirements of American, English and German Governments Pertaining to Seagoiny Engineers are Outlined

By Robert Haig

THE seagoing engineer of to-day should possess a sound educated, mechanical knowledge of the construction and principle of the machinery placed in his charge. He should have a knowledge of fuel values in relation to coal or oil and a comprehensive grasp of what combustion actually means. He should know what corrosion means to metals and the preventable measures to adopt. He should understand the value of feed heating and the economies to be gained. He should understand what expansion of metals means and the risks that are run by turning live steam into cylinders or cold pipe lines not properly drained, or in raising steam in boilers in insufficient time. He should know what specific gravity, density of liquids and flash point of oils are in relation to his daily tasks, and to qualify him for the important duties he has to discharge.

For the purpose of comparison and interest, I have taken the requirements of the maritime bureaus of the American, English and German governments pertaining to seagoing engineers, which are as follows:

American (Department of Commerce, Steamboat-Inspection Service)

Requirements of Board of Supervising Inspectors:

34. Chief Engineers of Ocean Steamers—Any person holding chief engineer's license shall be permitted to act as first assistant on any steamer of double the tonnage of same class named in said chief's license.

Engineers of all classifications may be allowed to pursue their profession upon all waters of the United States in the class for which they are licensed.

First Assistant — Engineers of lake, bay and sound steamers who have actually performed the duties of engineer for a period of three years shall be entitled to examination for engineer of ocean steamers, applicant to be examined in the use of salt water, method employed in regulating the density of the water in boilers, the application of the hydrometer in determining the density of sea water, and the principle of constructing the instrument; and shall be granted such grade as the inspectors having jurisdiction on the Great Lakes and seaboard may find him competent to fill.

And first assistant engineer of steamers of 1,500 gross tons or over, having had actual service in that position for one year, may, if the local inspectors in their judgment deem it advisable, be licensed as chief engineer of lake, bay, sound or river steamers of 750 gross tons or under, in which case license shall be issued on shall be endorsed with authority to act chief engineer's form of license, which as first assistant engineer of steamers of any tonnage for which he is qualified.

In addition to a sound education and constructional and theoretical knowledge of his machinery, the engineer of to-day is called upon to acquire a wide general knowledge of fuels, chemistry and physics. The necessary workshop practice is extensive.

Any person holding a license as first assistant engineer, and having had experience as first assistant engineer for a portion of the year required for raise of grade, may substitute experience as second assistant engineer, while holding first assistant engineer's license, which experience as second assistant engineer shall only count as one-half; provided that any person having had a first assistant engineer's license for two years, and having had two years' experience as second assistant engineer, shall be eligible for examination for chief engineer's license.

Second Assistant—Any person holding a license as second assistant engineer, and having had experience as second assistant engineer for a portion of the year required for raise of grade, may substitute experience as third assistant engineer while holding second assistant engineer's license, which experience as third assistant engineer shall only count as one-half; provided, that any person having had a second assistant engineer's license for two years, and having had two year's experience as third assistant engineer, shall be eligible for examination for first assistant engineer's license.

Third Assistant — First, second and third assistant engineers may act as such on any steamer of the grade of which they hold license, or as such assistant engineer on any steamer of a lower grade than those to which they hold a license.

Any person holding a license as third assistant engineer and having had twelve months' experience as junior engineer, or twelve months' combined service as third assistant engineer and junior engineer, or two years' experience as oiler or water tender, or two years' combined service as oiler and water tender, since receiving said license, shall be eligible for examination for license as second assistant engineer.

Inspectors may designate upon the certificate of any chief or assistant engineer the tonnage of the vessel on which he may act. (Sec. 4441, R.S.)

British Board of Trade Requirements

Qualifications required by British Board of Trade for the various grades of marine engineers (ordinary certificates):

25. Value of Ordinary Certificates—Ordinary certificates will entitle the holders to go to sea in the grade certified as engineers of any vessel in the British mercantile marine.

26. Second-Class Engineers—A candidate for a second-class engineer's certificate must not be less than 21 years of age.

(a) He must have served as an apprentice engineer for four years at least, and prove that during the period of his apprenticeship he has been employed in the manner set forth.

Journeyman's time will be considered as equivalent to apprenticeship.

Every applicant must produce testimonials of ability as an engineer workman to the satisfaction of the Board of Trade.

If the candidate has served as an apprentice engineer or as journeyman, under the conditions above prescribed, for less than four years, he will be required to make up the deficiency or to complete this period of four years by service as engineer at sea on regular watch on the main engines or boilers of a foreign-going steamer.

If the candidate has not served at all as apprentice engineer or as journeyman, he will be required to have served at sea, in lieu thereof, as engineer on regular watch on the main engines or boilers, six years in a foreign-going steamer.

(b) In addition to the apprenticeship as above described, or the alternative sea service, the applicant must have served one year at sea as engineer on regular watch on the main engines or boilers of a foreign-going steamer, or 18 months in a home-trade steamer.

On and after Jan. 1, 1915, the applicant will be required, in addition to the apprenticeship above described or the alternative sea service, to have served 18 months at sea as engineer on regular watch on the main engines or boilers of a foreign-going steamer, or 27 months in a home-trade steamer.

(c) He must be able to give a satisfactory description of boilers and the methods of staying them, together with the use and management of the different valves, cocks, pipes and connections.

(d) He must understand how to correct defects from accident, decay, etc., and the means of repairing such defects.

(e) He must understand the use of the water guage, pressure guage, barometer, thermometer and salinometer, and the principles on which they are constructed.

(f) He must be able to state the causes, effects and usual remedies for incrustation and corrosion.

(g) He must be able to explain the methods of testing and altering the setting of the slide valves, and method of

*Read before the Society of Naval Architects and Marine Engineers in New York.

testing the fairness of shafts and adjusting them.

(h) He must be able to calculate the suitable working pressure for a steam boiler of given dimensions and the stress per square inch on crank and tunnel shafts when the necessary data are furnished.

(i) He must understand the construction of steering engines, evaporators, feed filters and feed heaters.

(j) He must understand the construction of centrifugal, bucket and plunger pumps, and the principle on which they act.

(k) He must be able to state how a temporary or permanent repair could be effected in case of derangement of a part of the machinery or total breakdown.

(l) He must write a legible hand and have a good knowledge of arithmetic up to and including vulgar and decimal fractions and square root. He must also understand the application of these rules to questions about safety valves, coal consumption, consumption of stores, capacities of tanks, bunkers, etc.

(m) He must be able to pass a creditable examination as to the various construction of paddle and screw engines in general use; as to the details of the different working parts, external and internal, and the use of each part.

(n) He must possess a creditable knowledge of the prominent facts relating to combustion, heat and steam.

27. First-Class Engineer—A candidate for a first-class engineer's certificate must be not less than 22 years of age.

Additional Qualifications for First Engineer

In addition to the qualifications required for a second-class engineer, (a) He must:

(1) Have served at sea for 12 months with a second-class certificate of competency or service, on regular watch on the main engines or boilers of a foreign-going steamer.

(2) Have served at sea for 18 months with a second-class certificate of competency or service, as first engineer of a home-trade steamer, or two years with a second-class certificate of competency or service as second engineer of a home-trade steamer.

(3) Have served 2½ years with a second-class certificate of competency or service as third engineer of a home-trade steamer, if, during the whole of that period, he has been the senior engineer in charge of the whole of a watch on the main engines and boilers; or

(4) Possess, or be entitled to, a first-class certificate of service.

On and after Jan. 1, 1915, the candidate will be required, in addition to the qualifications required for a second-class engineer:

(1) To have served at sea for 18 months with a second-class certificate of competency or service, on regular watch on the main engines or boilers of a foreign-going steamship, as senior engineer in charge of the whole watch; or

(2) To have served at sea for 27 months with a second-class certificate of competency or service as first engineer of a home-trade steamer, or three years with a second-class certificate of competency or service as second engineer of a home-trade steamer.

(3) To have served three years nine months with a second-class certificate of competency or service as third engineer of a home-trade steamer, if, during the entire period, he has been the senior engineer in charge of the whole of a watch on the main engines or boilers; or

(4) To possess, or be entitled to, a first-class certificate of service.

(b) He will be required to make an intelligible hand sketch or a working drawing of some one or more of the principal parts of a steam engine, and to mark in without copy all the necessary dimensions in figures, so that the sketch or drawing could be worked from.

(c) He must be able to take off and calculate indicator diagrams.

(d) He must be able to calculate safety-valve pressures and the strength of the boiler shell, stays and riveting.

(e) He must be able to state the general proportions borne by the principal parts of the machinery to each other, and to calculate the direct stress, the torsional stress and the bending stress in round bars, and the direct stress and the bending stress in rectangular bars with given loads.

(f) He must be able to explain the method of testing and altering the setting of the slide valves, and to sketch about what difference any alteration in the slide valve will make in the indicator diagram and also the method of testing the fairness of shafts, and of adjusting them.

(g) He must be conversant with surface condensation, superheating and the working of steam expansively.

(h) His knowledge of arithmetic must include the mensuration of superficies and solids and the extraction of the square and cube roots, and the application of these rules to questions relating to the power, duty and economy of engines and boilers, and to the stresses in rods, shafts and levers of the engine. He should also be able to calculate the effect of the application of the lever, pulley, inclined plane and other mechanical powers.

(i) He must understand the construction of, and be able to maintain in working condition, the auxiliary machinery which is placed under his charge; namely, refrigerating machinery, electric light engines and dynamos, electric motors fitted to ships' boats, hydraulic machinery and the various descriptions of steering engines, etc.

28. First-Class Certificates Without Second—The Board of Trade may see fit to allow an applicant who, in consequence of service abroad, has had no opportunity to obtain a second-class certificate, to be examined for a first-class certificate although he does not possess a certificate of the lower grade, provided he is able to satisfy them as to the satisfactory character of his services, but these should be

(d) He must be acquainted with the principles of expansion and the modern theory of heat, and be able to solve, with the assistance of his own books or without books, according as the examination papers may be set, questions in economy and duty in connection with engines and boilers.

(e) He must understand how to apply the indicator and to draw the proper conclusions from the diagrams, and to construct the approximate diagrams for any given data.

(f) He must be able to produce, without a copy, a fair working drawing of any part of the machinery with figured dimensions fit to work from.

(g) He must understand the principles of the action of the screw propeller and the paddle-wheel, and must be able to estimate numerically the effect in speed of ship and consumption of fuel due to any alteration in pitch, diameter, revolutions, etc.

(h) He must be able to give a description of boilers and the methods of staying them, and must show that he possesses a knowledge of the theoretical principles which regulate their construction, and that he is able to calculate the strength of the boiler shell, stays and riveting.

(i) He must understand the general nature of the strains and stresses produced by the steam pressure, and by the expansions due to unequal temperatures in boiler shells.

(k) He must have a knowledge of safety-valve construction, and the principles involved in determining the size of a safety valve, and the construction of spring-loaded and dead-weight valves.

Theory and Science

(l) He must possess a thorough knowledge of the theory of combustion, the chemical composition of fuels, the evaporative duty of fuels of given composition; the production of draft; the effect in regard to economy, safety and wear and tear, of increasing or diminishing the proportion of heating surface of grate-bar surface, of area of section of air passages, of area of water surface, of steam-space capacity and water capacity.

(m) He must be able to explain the formation of scale and the precipitation of salt, and the precautionary means adopted in respect thereto, with jet condensers and with surface condensers.

(n) He must understand the general principles involved in the construction of the barometer, thermometer, salinometer and steam and vacuum gages.

(o) He must be familiar with the general results obtained from past experience in relation to corrosion, pitting and galvanic action in boilers, and the use of zinc and of soda in boilers.

(p) He must be able to give a variety of illustrations of how defects have arisen from accident, imperfect construction or deterioration, and how these defects may have been prevented, and the best way of repairing such defects.

(q) He must be familiar with the properties and processes of manufacturing and testing the ordinary materials used in the construction of machinery and must possess an intelligent knowledge of the properties of the lubricants, boiler cements and india rubber in general use in steamers.

(r) He must understand the causes of spontaneous combustion and the formation of explosive gases in coal holds, and the precautionary measures proper to prevent accidents from these causes.

(s) He must be acquainted with the principles and practice of the generation and application of electricity to various purposes on board ship.

(t) He must be able to explain the construction and working of the refrigerating machinery in use on board ship, the electric-lighting plant, the steering engines, hydraulic and pneumatic engines, the pumps, and all other auxiliary machinery placed under the chief engineer's control.

(u) In order to deal intelligently with ballast tanks, the cocks, valves and pumps of which are under the chief engineer's control, and to co-operate the more readily with the master in keeping the vessel in a safe condition, especially when she is light, and when coaling operations are proceeding, candidates are expected to possess an elementary knowledge of the stability of floating bodies.

(v) He must possess a practical knowledge of ship construction, and understand the elementary principles involved, so as to be able to deal with engine and boiler seatings and to supervise and direct any repairs that may be required to an iron or steel ship.

If the candidate does not obtain 67 per cent. of the total number of marks allotted for the papers he will be declared to have failed. The papers will be founded chiefly on the foregoing subparagraphs.

Equivalent Experience

33. On and after January 1, 1915, a candidate for either a second-class or a first-class certificate, who within two years from the date of application to be examined has attended an approved course comprising general mathematical and scientific instruction at a technical school recognized by the Board of Trade as suitable for the training of marine engineers, will be allowed to count time so spent as equivalent to sea service in the ratio of three months at the technical school to two months at sea. Time so spent cannot be accepted as equivalent to more than one-sixth of the total sea service required for either certificate, but a candidate who has been allowed to count such time on examination for a second-class certificate will not be debarred from counting similar subsequent time on examination for a first-class certificate.

Time spent in an approved marine technical school subsequent to obtaining a first-class certificate and within two years from the date of application to be examined may also be accepted as forming part of the qualifying service required under paragraph 32, in the case of candidates for extra first-class certificates, but if such time is substituted for sea service, it will only count as equivalent thereto in the ratio of three months at the school to two months at sea.

In every case in which an allowance is made for time spent at a marine technical school, the candidate will be required to produce the principal's certificate for continuous and regular attendance at all the approved classes and for satisfactory progress.

Home-Trade Oil Engine Ships

Certificates for engineers of home-trade pessenger ships propelled by oil engines:

34. Oil-Engine Certificates.—Candidates may be examined for second-class certificates of competency as engineer of vessels propelled by oil engines.

These certificates will entitle the holders to go to sea as second-class engineers of home-trade passenger ships propelled by oil engines, but will not entitle them to go to sea as second-class engineers of foreign-going ships or of home-trade passenger steamships.

35. Second-Class Engineer (Oil Engines)—A candidate for a second-class oil-engine certificate must be at least 21 years of age. (a) He must prove:

(1) Four years' experience at the making or repairing of machinery, of which at least six months must be at the making or repairing of oil engines; or

(2) Three and a half years in charge of engines and boilers at sea or an equivalent suitable experience on shore.

And in addition to either (1) or (2):

Six months' experience with oil engines at sea.

Note: Alternative service to the above may be considered, but it is essential that the candidate should have experience with oil engines, and have spent at least six months at sea in the engine room of a seagoing vessel.

(b) He must write a legible hand, and have a good knowledge of arithmetic up to and including vulgar and decimal fractions and square root.

He must also understand the application of these and other rules to problems relating to spring-and lever-loaded relief valves, and be able to solve questions about the relative speeds of a vessel at different revolutions, the capacity of oil tanks, etc.

(c) He must be able to give a clear explanation of the principle on which an oil engine works, and to show by means of illustrative sketches and otherwise that he understands the construction of those in general use.

(d) He must be able to describe the chief causes which may make the engine difficult to start, and to explain how he would proceed to remedy any defects.

(e) He must be able to show that he understands the mechanism of the starting and reversing arrangement, and is competent to deal with defects which may lead to failure in the prompt handling or reversing of the engine.

(f) He must have sufficient mechanical ability to be able to overhaul the engine, to adjust the working parts, and to put the engine together again in good working condition.

(g) He must be able to give satisfactory answers to the elementary questions numbered 305 to 310 inclusive in Appendix C.

(h) He must be able to prove (by actual trial if practicable) that he is competent to manipulate an oil engine when under way by starting, running the engine, stopping, reversing or slowing down.

(i). He must understand what is meant by the flash point and have a knowledge of the explosive properties of the oil generally used in engines when exposed in the open air and the danger of exposing any vapor from the oil to a light or of allowing any leak from the oil tanks, particularly into the vessel's bilges.

(j) He must understand the action of wire gauze diaphragms when placed in pipes and connections to oil tanks, etc., for the purpose of preventing the explosion or ignition of oil vapor therein.

(k) He must be able to take the necessary precautions to guard against the escape of inflammable vapor from the vaporizers when the engines are stopped.

(1) He must be able to explain the principle and construction of a dynamo and the construction and arrangement of primary and secondary batteries and induction coils, so far as is necessary for the efficient management of an oil engine.

Shop Experience Requirements

36. Workshop Service—Workshop service must have been performed in works where steam engines, boilers, etc., are made or repaired, but no time served before the age of 15 will be counted.

Not less than two years of the apprentice time must have been spent at fitting, erecting or repairing engines and machinery either in the works or outside. The remaining two years may be made up of time spent in engine works at fitting, erecting or repairing engines and machinery or at one of the other branches of the trade given below, or at an approved technical school (day), the time so spent to count as follows:

Fitting, erecting, repairing or turning	Full time.
Working in drawing office	Full time up to one year, and beyond one year one-half time.
Patternmaking	One-half time with a maximum allowance of one year.
Planing, sloting, shaping and milling	One-third time.
Boilermaking or repairing	One-half time.
Smith work	One-half time with a maximum allowance of one year.
Coppersmith work	One-third time with a maximum allowance of six months.

In the event of the apprentice time being extended to five years or more, four years at machining followed by one year aterecting may be accepted as qualifying.

37. Workshop Service Other Than the Above—When the workshop service has been performed in a place where engines are made, but not in the manner specified in paragraph 36, the case must be referred to the Board of Trade with a report upon the service performed. If the service be such as is useful training for an engineer, the board may accept the service; but in every case the applicant must prove additional engine-room or marine-engine workshop service as required in the succeeding paragraph.

38. Workshop Service Where Engines are not Made—When the workshop service has been performed in a place where steam engines are not made or repaired, and the class of work done is similar to that required in engine making, the service may be acecpted with an additional year of qualifying service; that is, four years' workshop service and either two years at sea on regular watch, or one year at engine fitting in a suitable marine-engine workshop and one year at sea in the engine room. The approval of the Board of Trade must be obtained in every such case before the candidate is examined.

39. Technical Schools—Time spent after the age of 15 at a technical school (recogniezd by the Board of Trade as suitable), where there is an engineering laboratory, may be taken into account and accepted as equivalent to artisan service, usually at the ratio of three years in the technical school to two in artisan service, provided the applicant has taken the full engineering course and can produce the principal's certificate for regular attendance at all the approved classes and for satisfactory progress; and provided also that the remaining portion of the time has been spent in works where steam engines, boilers, etc., are made or repaired, in accordance with the scale of values indicated above.

40. Sea Service—The sea service required by these regulations is, unless otherwise stated, service performed in foreign-going ships of at least the nominal horsepower specified for the respective grades of certificates. The nominal horsepower, as given on the vessel's certificate of registry, may in all cases be accepted by the examiners.

41. Qualifying Service Defined—In the case of candidates for first-class certificates qualifying service means, as a rule, service on regular watch on the main engines or boilers as senior engineer in charge of the whole watch. When, however, three or four engineers are on watch at the same time the service of the two seniors may be allowed to count, and when not less than six engineers are on watch together that of three seniors, provided the certificate from the superintending engineer is produced stating the number of engineers on watch, the rank held by the applicant and that he was in full charge of a watch on a definite section of the main engines or boilers. During the whole of the period claimed, candidates must have been in possession of second-class sertificates.

In the case of candidates for second-class certificates qualifying, service means service as engineer on regular watch on the main engines or boilers.

In no case wil time spent in clerical work be allowed to count.

42. Further as to Qualifying Service—Only such service as gives the experience required to make a man thoroughly competent as a seagoing engineer is accepted as qualifying service. Even for a second-class certificate the candidate must prove to the satisfaction of the examiner that he is qualified by experience and knowledge to act as chief engigneer in an under-powered steamer of 99 nominal horsepower on a voyage say from England to Egypt, taking full responsibility for engines and boilers.

German (Marine Engineering Bureau Requirements)

For admission to the examination for engineers of the second class it is required as follows:

During apprenticeship, must attend evening classes to study the theory of engineering and learn English.

After the fifteenth year of age: Either (a) One must serve 72 months in steam-engine or motor-construction works, in steam-engine or motor-repair works and on the engineering staff of ocean steamers; at least 36 months must be spent in steam-engine or motor works, and at least 24 months on the engineering staffs in voyages of suitable steamers as assistant engineers; or

(b) After passing the examination for third-class engineer one must spend at least 24 months as engineer, and either before or after the examination for third-class engineer must have spent at least 36 months as machinist and blacksmith in steam-engine or motor works.

For admission to examinations for first-class engineer one must serve for at least 24 months after passing the examination for second-class engineer, in voyages of suitable steamers as engineer in short, medium or great voyages.

For admission to the preliminary examination for marine engineer (chief engineer) it is required:

After passing the fifteenth year, one must have spent at least 66 months in machinery construction work in one of the larger construction firms for marine engines, and at least 36 months on the engineering staff of steamers. At least 36 months must be spent in one of the larger steam-engine works, of which six months must be spent in the blacksmith and brazier shops. At least 30 months must be spent on the engineering staff of suitable ocean steamers as assistant or in a higher position in short, medium or long voyages.

The time spent in short voyages is limited to 12 months.

Besides, a 12 months' course at a technical marine engineering school recognized by the state is required.

For final examination for chief engineer, it is required:

After passing the examinotion for engineer of the first class, or after passing the preliminary examination for chief, one must spend 24 months in voyages of suitable steamers as engineer in medium and long voyages.

Besides, a 12 months' course of the upper class of a marine-engineering school recognized by the state is required.

The approval of the technical schools is in the hands of the Central State Board in conjunction with the Imperial Chancellor.

For candidates for second-class engineer, only that time is counted which is spent in steam-engine construction or repair works in its occupation of smith, brazier, machine builder and fitter. The time of service in the forging and brazing shops is limited to six months.

If one has passed the preliminary examination for chief engineer, he receives the certificate of engineer of the second class, and after ocean service of 24 months as engineer in short, medium or long voyages without further examination he receives the certificate of engineer of the first class.

Authority for the bestowal of the certificate of machinist of the first class belongs to a qualified State Board for the examination of qualifications; which considers in addition to the qualifications shown by passing the preliminary examination for marine engineers, the proficiency as a second-class engineer, as well as the time served at sea.

Note: The writer regrets that for obvious reasons it has been quite impossible to get reliable detailed particulars of the German engineering examinations.

Proper Training Facilities Necessary

It is not intended that we should produce an engineering theorist, but it is claimed that facilities should be provided for men, offering both educational and mechanical training of such a nature as to properly equip them for the duties of a marine engineer.

To properly equip men to become lawyers or doctors, splendid schools and institutions are provided in every large city which insist upon years of the highest educational training ,and it is right that it should be so, otherwise a continuous supply of men of the proper caliber could not be assured.

Although education in general in this country is easy to obtain and is of the finest kind, yet extremely little is required of the engineer when he submits himself for examination as a test of his qualifications for the position he aspires to hold. The examinations and tests he has to pass are elementary and perfunctory to the last degzee, entirely out of date in their requirements and out of touch with modern engineering. The tests and examinations should more adequately represent the educational and engineering necessities of the present day. Facilities should be provided by the Federal Government in those cities that are important shipping centers, so that the proper training be placed within the reach of those who desire to study marine engineering. It would then be proper to require that the Government examinations should be uniform and to the point aimed at, and further, should be such a standard and of such a searching nature that only men of at least a fair education and good mechanical training could hope to pass the examinations.

This would not entail any hardship if certain specified educational institutions included in their day and evening classes (with a view to training marine engineers) such subjects as mechanical drawing, algebra, physics, taking up heat values, combustion, refrigeration, strength of materials, electric light, powers, etc., all with direct relation to and including an exhaustive study of the various types of marine engines, boilers, etc.

Facilities should be provided at a nominal rate to make it possible for this technical training to be acquired during the apprenticeship period, and thereby produce a better-informed mechanic, and also for the seagoing engineer when he comes ashore to prepare and qualify himself for further examinations.

It should be required that men desiring to follow the profession of seagoing engineers meet the following requirements before going up for examination:

1. The candidate for examination should be a citizen of the United States.

2. He should not be less than 21 years of age.

3. He should be able to read and write English fluently and be educated equal to the senior classes in the public schools.

4. He should have served not less than 3½ years learning to be a machinist or as a machinist, inclusive (if possible) of one year in the drawing office, with a firm building marine engines and boilers; or

He should have served not less than four years learning to be a machinist or as a machinist with a firm engaged in repairing marine engines and boilers; or

He should have served not less than three years learning to be a machinist or as a machinist with a firm engaged in general engineering work with 1½ years additional in a shop engaged in marine engineering.

This would go a long way to insure a man whose hands had been trained and who had learned the functions of the essential parts of the marine engine.

The complex and exacting duties of the engineer on watch amply demonstrate that we can no longer afford to recruit our engineers from the men who come up through the stokehold. Such a condition is neither desirable nor fair to the men, the shipowner, or the costly property placed in charge of the engineer.

To-day American engineers are paid the highest wages, and the shipowner has a right to expect in return that an engineer holding a government certificate should be a highly capable and efficient officer worthy of the responsibilities of his position.

It would be well for the engineering industry to establish on a reasonable basis, conditions both as to actual training and remuneration during a period of apprenticeship that would attract good types of young men. It is unfortunately all too true that apprenticeship in the past has largely meant only a period of work, with very meager remuneration and little or no attempt made to train him, with the result that he became a good or bad mechanic according to his natural ability.

Lately, conditions have somewhat improved, but much can be done; the apprentice ought to get actual instruction in his work, and encouragement to study should be given so as to develop his taste for the work. The employer will quickly get a return for this effort, as he will get the product of a man who has been taught to use his head. Had this line been more wddely followed in the past, there would not be such a dearth of skillel mechanics to-day.

It is respectfully submitted that this society and other kindred societies should by discussion, representation and every other means within their power, impress upon the Government the extreme importance that the examinations, tests, equalifications and training of the seagoing engineer of the mercantile marine should be brought up to date and placed on a plane with that of other professions, whose responsibilities may be different in kind, but do not exceed in degree that of the men in whose hands are placed the care and maangement of the machinery of our merchant ships, so vital to the country's welfare at the present time.

SCHOONER LAUNCHED FOR CHINESE FIRM IN AUSTRALIA

THERE was launched at Liverpool, N.S., on March 21, from the shipyard of the Nova Scotia Shipbuilding & Transportation Company a four-hundred-ton tern schooner named Abemama, built under contract for Peter Yee Wing and Co., of Sydney, Australia. This vessel is the second of this size launched from this yard within eight months. The name is given in honor of the island where this Australian firm is trading.

The vessel will trade under the British flag, and under British control, owned chiefly by Chinese merchants who are British subjects of Australia. The Chinese republican flag was flown with that of Britain as an honor to the cause of the Allies, and was presented to the owners by the builders.

The vessel was christened by Miss Lillian Bennet, daughter of Dr. Bennet, pastor of the Christian Church, at Milton. Mr. Samuel Wong, director of that far-away business house, who has been here for several months, is well pleased with the launching and the construction of the ship. This Liverpool firm of shipbuilders have two other boats of the same size under construction, and all being erected by contract. J. S. Gardner is the master builder.

The MacLean Publishing Company
LIMITED
(ESTABLISHED 1888)

JOHN BAYNE MACLEAN - - - - - - - President
H. T. HUNTER - - - - - - - - - Vice-President
H. V. TYRRELL - - - - - - - - - General Manager

PUBLISHERS

MARINE ENGINEERING of Canada

A monthly journal dealing with the progress and development of Merchant and Naval Marine Engineering, Shipbuilding, the building of Harbors and Docks, and containing a record of the latest and best practice throughout the Sea-going World.

B. G. NEWTON, Manager.
J. M. WILSON, Editor. A. R. KENNEDY, Asst. Editor.
Associate Editors:
A. G. WEBSTER J. H. RODGERS (Montreal) W. F. SUTHERLAND

OFFICES

CANADA—
Montreal—Southam Building, 128 Bleury St., Telephone Main 1004.
Toronto—143-153 University Ave., Telephone Main 7324.
Winnipeg—1207 Union Trust Building. Telephone Main 3449.
Eastern Representative, E. M. Pattison.
Ontario Representative, S. S. Moore.
Toronto and Hamilton Representative, J. N. Robinson.

UNITED STATES—
New York—R. B. Huestis, 111 Broadway, New York, Telephone 8971 Rector.
Chicago—A. H. Byrne, 900 Lytton Bldg., 14 E. Jackson Street. Phone Harrison 1147.
Boston—C. L. Morton, Room 733, Old South Bldg., Telephone Main 1024.

GREAT BRITAIN—
London—The MacLean Company of Great Britain, Limited, 88 Fleet Street, E.C. E. J. Dodd, Director, Telephone Central 12960. Address: Atabek, London, England.

SUBSCRIPTION RATE

Canada, $1.00; United States, $1.50; Great Britain, Australia and other colonies, 4s. 6d. per year; other countries, $1.50. Advertising rates on request.

Subscribers, who are not receiving their paper regularly, will confer a favor by telling us. We should be notified at once of any change in address giving both old and new.

Office of Publication, 143-153 University Avenue, Toronto, Ontario.

Vol. VIII MAY, 1918 No. 5

PRINCIPAL CONTENTS

SHIPPING NEEDS OF THE FUTURE

CANADIAN exports to the amount of many thousands of tons are, according to Commissioner Ross of the Trade and Commerce Department at Melbourne, held up in this country owing to lack of shipping facilities to Australia. Ships are required elsewhere owing to the exigencies of war. For the present Canadian manufacturers must be prepared to face such conditions—the military campaign is the first consideration—but what of the future?

A nation which hopes to supply other nations with her products, particularly manufactured products, should be prepared to make delivery—and delivery means ships. Canada as an agricultural country may for the present, and in the future while there is an insistent demand for food stuffs, be in the position of being able to make buyer nations take delivery at the seaboard with their own ships. This will not be the case with manufactured goods, particularly as those nations which take our food exports will not only refuse to carry other products, but they will expect to bring manufactured products to us. This on the economic basis of settling for purchases and for the business reason that successful transportation demands cargo in both directions.

All indications are that if Canada wishes to play a part in the Pacific trade after the war, and to extensively export manufactured products, especially where there is national competition, she will require her own vessels to deliver the goods. This should be considered as a factor in relation to any shipbuilding program which on the face of it may not promise to be a financial success. It is under such circumstances that national industry looks to the Government for assistance.

CANADIAN PROJECT MEETS WITH WIDE APPROVAL

ADVICES from Washington justify the commendatory opinions which have been expressed throughout the Dominion regarding the projected plate mill plant in Nova Scotia. Such a step will afford considerable relief to the mills in the States with ultimate beneficial effect on congestion generally.

Announcement has also been made that the first order for steel has been placed in the States for the Canadian shipping program, calling for 80,000 tons of plate and sections. With the gradual increase of government ship work, as the Munitions Board vessels are completed, the volume of material required will assume considerable proportions, and in this respect much interest attaches to a recent statement by the management of the immense Hog Island yard that an average of two months elapses from the time the steel leaves the mill until it reaches the shipyard in fabricated form.

The additional four months required from the time the keel is laid until fitting out is completed mean that six months is required from the time the steel leaves the mill till it is afloat ready for duty.

An idea of the prospective load on transportation facilities in the States may be gathered from the fact that thirty-five shops, widely distributed, are engaged on the work of fabrication: this for one yard's output alone.

The Canadian program we understand calls for the adoption of the fabricate steel system of ship construction and if the plans adopted correspond to those now in use in the States, a high degree of output may be rapidly attained in the Dominion yards.

While the completion of the plate mill is quite some time distant, possibly eighteen months, some advantage may be gained by the fact that two-thirds of this time will be normal weather conditions allowing maximum rate of preparation. It is also not unlikely that in view of its desirable effect on traffic conditions south, the work of building the new equipment will be afforded every facility by United States authorities for rapid completion by the builders.

The project marks an epoch in Canadian industry, and its advent at the present juncture justifies the importance with which it is regarded in the eyes of the industrial leaders of the United States.

REGULATIONS have now been approved by order-in-council whereby the registration of the man and woman-power of Canada will be effected. Every person over sixteen years of age will be affected and provision is made to add those who afterward attain the age of sixteen or are discharged from active service.

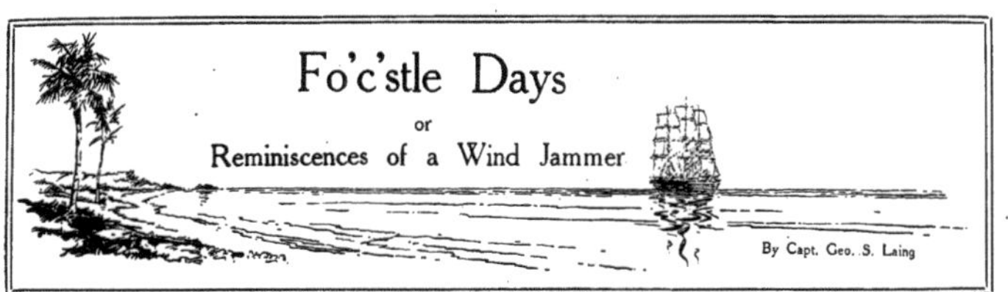
Fo'c'stle Days
or
Reminiscences of a Wind Jammer
By Capt. Geo. S. Laing

The steamer and the motorship have deprived the seafarer of much of the old-time romance attaching to a sailor's life. The advent of wireless has also increased the safety of it. Deprived of these features, a sailor is immediately dependent on his own resources, and Captain Laing gives us a true and seaman-like reflection of sea life under canvas. The present feverish activity in constructing self-propelled craft affords a suitable background for incidents in a career in which gaffs, booms, tar and canvas held sway before being displaced by mechanical unloaders, derrick masts, fuel oil, coal and wire rope.

Part IV.

THE ROARING FORTIES

TILBURY was longing for the end of the trip as he hated sea life worse every day. He said that going aloft would never become natural to him, and we were all in sympathy with the disappointed lad. A secret pact was made amongst the other apprentices to help Tilbury desert in Brisbane. Even Fid was with us in the deal.

Working aloft in sailing-ships had many dangers attached to it, and one can look back to wonder that more accidents did not occur at such monkey manoeuvres. Our royal yards were about a hundred and sixty feet above the deck, and even at half that height, when a vessel is pitching and rolling it becomes a feat to hang on for your life, let alone work for the ship-owner.

Of course the yards were provided with foot-ropes, back-ropes and grummets for the arms, but if sailors paid too much attention to these preventors the sails would never be furled or reefed. The first law of nature—self-preservation—makes one realize that a fall from a weather yard means killed on deck, while a drop from a lee yard spells "man overboard," or "drowned at sea." The "Maggie Dorrit" like all other sailing craft of her day had exceptionally heavy fore and main "tops," and in getting over the spider rigging one was literally climbing in a similar position to a fly on a roof.

Fid being the senior apprentice, was taken on the poop occasionally and instructed in the use of the sextant, the navigator's faithful instrument for measuring the angular heights of the celestial bodies—sun, moon and stars. The other boys being juniors were denied this semi-dignified role until we could successfully pass the tar and grease-pot stage of our noble but tough profession. It was really a quadrant that was used by the boys as they became old enough for such practice, and it was made of ebony with a bone arc and vernier. A brass disc the size of a cent, with a pin hole through it, did duty as a telescope. Even with this crude affair it was possible to get a reading of degrees and minutes of arc that compared favorably with the captain's and mates' observations found of course with their more modern sextants. Sailors often refer to these measuring instruments as "horse-heads" and "ham-bones."

How Vessels are Navigated

As the science of taking a vessel over the trackless ocean is most interesting to young and old alike, it may not be amiss to explain how it is done in our story as all sea apprentices have to grapple with "day's works" for dead reckoning, morning, noon and p.m. sights for latitude and longitude and compass amplitudes and azimuths for course corrections. Coasting work, such as the inland waters of Canada, belongs to the "dead reckoning" branch of navigation and its problems are solved by plane trigonometry. The higher work, as one finds in crossing oceans, touches on spherical trigonometry, nautical astronomy and meteorology. The modus operandi as worked on windjammers was boiled down to something like the following: First of all the dead reckoning will be explained.

Dead Reckoning

As its name implies this is the finding of a ship's position without reference to the heavenly bodies or chronometer. The chart, compass, sounding lead and speed log are the essentials, and each of these articles deserve a little definition. A chart is a marine map giving the exact lay of the coast line with its shoals, spits, reefs, bars, buoys, beacons, light-houses, light-ships, wireless and flag stations, etc. Then the depths of the waters are laid down with snake lines of shallows and deeps, and the maximum and minimum rise and fall of tides, velocity of currents and prevalent drifts. By this map we can tell how much water should be under us and in what direction a reef may bear from our frail craft. The compass is of course our pointer and if kept clear of complaints known as deviation and other magnetic disturbances, this finger of direction never fail us.

The sounding lead is a long manilla line or wire with lead attached for finding the depth of water as we approach the land. A hollow part in the bottom of the sounder is "armed" with tallow and in this way the nature of the sea bottom is also determined which aids in locating the exact position of the ship, the nature of the bottom being given on the chart.

A ship's distance over the ground or through the water is found by means of the log line which is either a clock-like mechanical device or a "hand" log which has a marked line based on proportional quantities related to foot mileage and seconds of time. Neither of these logs, however, can gauge the effect of tides and currents unless the ship is aground or at anchor.

When running along a coast or approaching one from the lake or sea the bearings of conspicuous promontories, mountains, or if near a town, church spires, are all used as lines of bearing with their intersections denoting the vessel's position on the chart. This is called the cross bearing method and is only one of many similar ways of getting the ship's position by compass and chart.

Ocean Navigation

This is much more scholarly work and has its main dependance on sextants, chronometers and azimuth mirrors, with their application to astronomical data. Logarithms lashed hard and fast to sailor geometry also enter the field with tangents, sines and co-sines in the offing. Team work is also carried on with angular heights of the sun, moon and stars above the ocean rim as related to celestial declinations and Greenwich time.

All this paraphernalia is shaken up in a bucket as it were and when the lid is opened out jumps the latitude and longitude. Latitude is the distance a place or ship is north or south of the equator. Longitude is the distance a place or ship is east or west of an index meridian. In our Imperial charts the index spot is a line drawn from pole to pole through Greenwich, cutting the equator at right angles. Thus we have parallels of latitude and meridians of longitude, which could be likened to imaginary hoops encircling our sphere at cross purposes. The Mercator chart is the one most easily adapted for navigators.

Now the sextant is one of the simplest but most important instruments of reflection in the world. In fact the world has been explored and the oceans charted by the sailor's sextant. For measuring vertical and horizontal angles from a ship's deck the sextant has neither rival nor competitor. The instruments used on shore by the surveyors for similar purposes would be utterly useless at sea, as they depend on plumb lines and plane positions of rigidity, whereas the ship can be rolling her top bridge into the foam and diving bows under when the deck officers are using their wonderful sextants. To find the real horizon in a heavy sea way becomes part of a navigator's accomplishments.

In conjunction with the chronometer, which is the finest timepiece made, the ship's position is thus formed. Having dragged the chronometer out, it might be said of that wonderful clock that its delicacy is always being mentioned, yet it can only be wound up every morning by turning it upside down. It should also be said that longitude and time are synonymous. To further explain this, if a ship or a person sails or walks fifteen miles east, a minute is gained, or if it be sixty miles west, then four minutes are lost. Then a chronometer keeps the exact time for Greenwich, and the difference between a ship's local time and her chronometer time is an approximate longitude. The chronometer however measures in a uniform way the path of an assumed sun, producing what is known as M.T.G., or mean time at Greenwich. Hence an equation which is a known quantity for each day of the year has to be applied one way or another to all navigational problems. When it is known that an error of four seconds in our chronometer time can put the ship's position out a mile the fineness of the problems and the instruments can be imagined.

Azimuth mirrors and pelorus are instruments that can be placed on the compass or in its proximity. Their chief uses are in connection with course correction or finding the deviation—an evasive part of the compass error—for all points of the circle. All chart work for instance is worked by two classes of bearings—magnetic and true. Now a bearing by compass must be put in either of these two forms before it can be plotted on the chart. In connection with azimuth, amplitude, and pelorus work we figure in conjunction with ship's apparent time, sun's declination, and an approximate latitude.

Just as a school is full of slide-rules, hand-books, formulas and data curves, so is a sailor surrounded by condensed tables, astronomical almanacks and touch-the-button methods of finding his latitude and longitude. On that account I have not deemed it necessary to describe things in anything but a general way.

THE CAPE PIGEON, AN OCEAN BIRD, FOUND IN THE OPEN WATERS OF THE SOUTHERN HEMISPHERE. IT IS OFTEN SEEN ACCOMPANYING THE LARGER BIRDS.

The "Maggie Dorrit's" latitude was often worked up on the end of the sailmaker's bench, whilst her azimuths and amplitudes were either taken by hand bearing or shadow pin which was simply a piece of brass wire fixed in the centre of the compass bowl. Granted, the longitude was a little more of a study, so it was figured up at the saloon table. Now we must get back to the old barque as the S.E. trades are falling light and the grandest part of our trip to Australia has yet to be told and sailed through.

Trinidad Off South Brazils

One afternoon at one bell or four-thirty p.m. when we were getting our allowance of fresh water from the poop, the ocean rock of Trinidad was sighted about twenty miles to leeward and on our beam. This mountain top is about 800 miles E.N.E. from Rio de Janeiro. As it was the first land we had seen for six weeks it was looked at with curiosity, and the old man immediately took chronometer sights to get as near as possible a reliable longitude. Latitude is never in such important demand, as it can be found accurately any day or night as the sun, moon or stars pass over the observer's meridian sextant work.

Boy Strategy in Getting Food

About this stage of the trip the apprentices got in bad breed with the steward, who accused us of helping ourselves to eatables when passing through the cabins at night for the purpose of calling the mates or getting the log glass. Now, sea apprentices have an abnormal appetite and we were all more or less guilty of this pardonable offence.

A special marmalade pie was made up and left to trap us, and sure enough we were all affected, for the steward had put more than baking powder in his bait. The old man came to see us at the half-deck next noon and seemed very interested in our health. Nothing was said to the flunkey, but we vowed that a chicken or two would likely go amissing before the trip was over, and concerning this we will hear more later on.

Four hungry boys can generally beat one steward when it comes to stomach strategy and after the medicine trick we got away with half a dozen tins of assorted food. The thing was done in this way. The boys in each watch had been told off to do some cleaning in the lazaret, which was a sort of an after-peak for holding stores. Although under the eyes of the captain and steward, the various tins of food were brought on deck submerged in buckets of dirty water, which we carried up to empty over the lee rail. Naturally we had to go along abreast of our half-deck to smuggle the sealed cans into safety, but that was easily done at odd intervals and no suspicion aroused.

It is well to remember that the ocean life with its dustless atmosphere and its strong ozone would almost make a plaster saint rave for food. Then imagine youngsters in their growing teens having for the bulk of their food salt meat and maggoty pantiles. If sharks have an everlasting hunger sea apprentices come next as the following fact will prove.

For some time the old man had persistently maintained that an unwholesome odor haunted the cabins, both dining room and state rooms. The usual talk of dead rats was considered. At last the steward admitted that a box of red herrings had gone bad, so the "Glasgow magistrates" were tried at the saloon table, but neither mate nor master would eat them. Then a most brilliant idea seemed to tell everyone that "the boys might like them," and lo and behold such deep-sea charity was really appreciated. Yes, we ate the whole box of evil-smelling fish and glowed with satisfaction, praying that other boxes of the same food would take a similar trend.

Making Ready for the "Roaring Forties"

After the S.E. trades petered out we headed in the direction of Tristan da Cunha, a lonely island far to the west of Cape of Good Hope. In this vicinity a vessel bound to Australia, India, China or New Zealand, begins to "run the easting down," a sailor term applied to this open stretch of water. "Roaring Forties" is another appellation given to the route in question, and it can easily be seen by a look at the illustration that the Southern Hemisphere has three times the water area that our hemisphere has. For instance one could circumnavigate the globe on the 53rd parallel of south latitude and only come in contact with the land at Magellan Straits which separate Terra-del-Fuego from Chili. In the open waters of the south we frequently steam or sail for 8,000 miles and more without sighting a vessel or an island, our only touch with anything outside of the elements, air and sea, comes through the companionship of our valued friends the albatross, Cape hens and Cape pigeons. Those birds follow us for months and have apparently been put there by nature to cheer up the brine-soaked sailors as they furl, reef, and set the wings of their vessels in accordance with the howling westerly gales.

Ships are Well Tested

Speaking generally, anything that is likely to happen by way of accident in the line of broaching to, foundering, being dismasted, washing men overboard, etc., may be experienced on this long and stormy ocean track. Its heavy swells, its fast moving hills of green granite and its hurricane squalls have no duplication in any other part of the globe. Even the firmament is unique with its jewels and Magellan clouds—peculiar starless patches which appear black in a strongly lighted sky.

On the "Maggie Dorrit," the preparations for "running the easting down," consisted in moving the jury spars from the scupper-ways in amidships, thus protecting the main hatch from the poundage of extra heavy seas. This breakwater idea is a splendid preventor as the spars just mentioned break the seas into foam before they can damage such vital spots as a hatch-way. A fore-hatch or quarter-hatch being near the ends of a ship are in a much safer position than the hatches in the waist of the vessel.

Then the life-boats, gig and longboat were provided with extra lashings for fear a sea would float them over the rail or a squall blow them out of the gripes. All heavy weather sails of number one canvas were bent and the flying-fish suit stowed away. Reef tackles, brace runners and double sheets were examined and renewed if found weak with nips or chafes. The fo'c'sle head scuttle was battered down and a weather board was loose caulked into the lower half of the boys' half-deck door to prevent the roaring cataracts of sea water having free access to that humble home. One must remember that a house on a ship's quarter-deck resembles an island in stormy weather, as it is literally surrounded by water and the motion of the ship throwing the seas from one side to another makes conditions similar to what one finds on a half-tide rock with its seething recoil, surge and spray.

Broaching Cargo

Each stage of our voyage seemed to have a new development and one that very much surprised the boys was soon in evidence. The forward crew had managed to crawl aft under the main deck and secure a case of schnapps from the cargo. A good night's debauch failed to hide its effects in strong sunlight and the wheel had no sooner been relieved at eight bells in the morning than it was noticed that the helmsman was describing snake-like curves on the waters with the old barque.

Then it was found that two inebriated tars had picked out the fore and main tops as sleeping quarters. Those two beauties were lashed where they lay rather than attempt to get them through the lubber holes—which was a doubtful feat at any time. Fid was put to the wheel and four of the men placed in irons, whilst the old man and the mates searched high and low in fo'c'sle, fore-hatch and fore-peak for any stray liquor.

A deposition taken from two of the seamen who were least effected told of how the stuff had been broached by entering the hold through a ventilator at the main fyfe rail on an exceptionally dark night. As this ventilator was only 12 feet from our half-deck door it was concluded that the boys might know something about the misdemeanor, but fortunately we proved our innocence that time.

As the sailors had not become mutinous over the affair the old man deemed it sufficient punishment to "log" them, that is write up the incident in his official log-book and attach a fine to each man's name. This amount would be deducted from their pay, but if the offence happened again, the jail in Brisbane would feature in the ocean spree. As desertion was their aim in the colonies, the fine never worried them a bit.

Signs of Learning

From time to time we had to exhibit some proof of our learning as the third mate had shown us how to make the more advanced knots such as the man-rope, Matthew Walker, stopper and other designs. These knots are made by a different manipulation of the crown and wall formations of strands. I made an awful hash of a long-splice and old "Ropeyarns" kept me up in my watch below till the job looked a little more seamanlike. Fid had his little joke anent this, but "pride goeth before a fall," as in working up the longitude next day by the captain's orders, he got all adrift in his figures. The addition in his altitude,

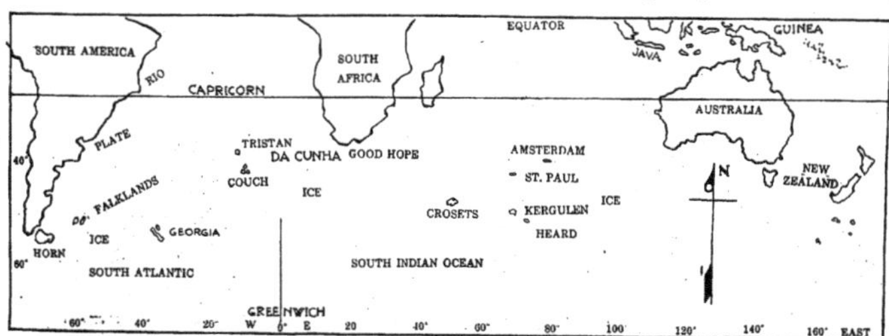

THE ROARING FORTIES: IT IS POSSIBLE HERE TO SAIL FOR 10,000 MILES WITHOUT SIGHTING LAND.

latitude, and polar distance column was wrong, while his equation of time was applied the wrong way. This resulted in a very erroneous longitude which quite humiliated the ancient mariner. Even table 24 in Norie's Epitome was subjected to a wordy squall of derision in his efforts to get the right answer.

Hank and I had several spells at steering, especially when the "Maggie Dorrit" was closehauled or on a bowline, as this is easier than steering by compass. Tilbury put the vessel aback purposely so that he would be put off the poop, for, as he said, "If I learn to steer I might be tempted to stick to the ship." He was more than ever determined to desert the vessel on arrival at Brisbane.

Cleanliness is Next to Godliness

Owing to dry weather we have only been able to wash our faces twice a week of late and this style of toilet by no means pleases Tilbury. What a contrast the sea-apprentice as he kisses his mother and sisters good-bye, and the Hanks and Fids as they are seen a thousand miles from land—oh ye gods of the mysterious ocean! Every one has heard of how sailors holy-stone the decks, but sailing-ship crews frequently required a similar treatment on their own hides through no fault of their own. A few vessels possessed condensers for making extra fresh water, but not so the old rattletrap of the "Maggie Dorrit" persuasion.

Gradually we find ourselves nearing Tristan da Cunha and our real fine weather is finished for at least a month. But there is something fascinating about gales and high seas from quarterly or dead aft directions that buffet and coax you on the lone trail.

(To be continued.)

SEAMANSHIP AND NAVIGATION

Conducted by "The Skipper" *

Articles of direct interest to mariners, discussions of seamanship, navigation rules, and allied topics. Inquiries from readers are invited and will be replied to promptly through these columns, unless otherwise requested.

THE MARINER'S COMPASS

THE compass is certainly the most important instrument on board a ship. Its uses are mainly in connection with chart work and steering. Whether it is the bearing of a lighthouse or a star or another ship's lights, we must trust to the compass for the direction of these objects from our vessel. Then the quartermaster never lifts his eye from the compass as he steers the vessel on a set course given by the officer in charge.

Prospective mates and masters should realize that the compass differs from all other nautical instruments in one vital respect—that it is governed by laws of magnetism, the cause and effect of which are not thoroughly understood by the greatest scientists. This being the case, it easily follows that the position of a compass, its adjustment and protection, should constantly be under competent supervision.

The directive power of the compass, then, is present through the medium of magnets. Certain substances, natural or artificial, have the power of attracting iron. The cause of this attraction is the magnetism spoken of. Lodestone is the name of the natural magnet, but any bar of steel or iron can be magnetized either through contact with the lodestone or by electrical treatment. A freely suspended magnet thus obtained will take up a position coinciding with the magnetic meridian. This North and South direction being approximately found, then all other points of the horizon are also known.

*Part of a paper, "Progress of Marine Engineering and the Education of the Marine Engineer," presented at the New York meeting of the Society of Naval Architects and Marine Engineers, Nov., 1917.

The standard or navigating compass as shown in the illustration has a pedestal and base of teak wood and four brass deck screws. The binnacle part is of brass and surmounts the wooden part. On two thwartship brackets you find the soft iron spheres which play a part in adjusting the compass to a condition of accuracy that only requires the further tabulation of minor corrections such as 1°, 2° or at most 3° deviation on any course to be steered. These final corrections are found daily as the vessel proceeds on her voyage and are computed by working out problems called azimuths and amplitudes. I will come to such work in due time.

In the centre of the pedestal you have the door opening into the lighting cabinet as the compass card is illuminated from below. The other two doors in evidence secure the cabinets containing the magnets. These magnets are about the diameter of an ordinary lead pencil and have their poles painted red and blue, which denotes whether they attract or repel. The north pole of a magnet is painted red and the south pole blue. Magnet poles containing the same nature of polarity repel each other while opposite colored poles attract each other. For the purposes of this article the student must just remember that our earth being itself a magnet its influence on other magnets works out as the above description explains. A little thinking must then unfold the fact that a magnet or compass needle will take up a position parallel with the magnetic meridian.

The oval brass covers on the binnacle top allow the compass card to be viewed from ahead, astern or abeam, or again, the glare of the lighting device can be screened on dark nights where its presence is undesirable when one is peering ahead for shore or ship lights.

Compass cards are generally made of mica paper held in true circular form with silk strings and the suspended needles—of which there may be six, three on each side of the card. A jeweled cap takes up the exact centre of the card and this cap fits over a nicely pointed pivot which rises from the bottom of the bowl.

This bowl must have a centre and thwartship line coinciding with the vessel's keel (fore and aft) and beam (right angles). This ensures the lubber point being in its true place. The lubber point is shown to advantage in the cut of the steering compass, the lubber point and the north point being in a straight line, in other words the ship is heading north.

Steering compasses are generally of the liquid type or "wet." In this case the compass card and needles are hermetically sealed in spirits of wine. In a heavy sea the card is thus much quieter and has not the same chance of deceiving the helmsman that a "dry" card has. During a gale a vessel's head will "yaw" as much as two points each way, this apparent movement of the card naturally causes some steersmen to use far too much helm. It should also be mentioned that the compass bowl is put into a double set of gimbals or brass rings. These rings have a fore and aft and thwartship axis respectively, thus taking care of the rolling and diving motions of the vessel in respect to the compass keeping in a true horizontal plane.

How to Get Practical Knowledge on the Compass

Remember, if you are a "green hand" that you cannot possibly understand every word in the compass definition

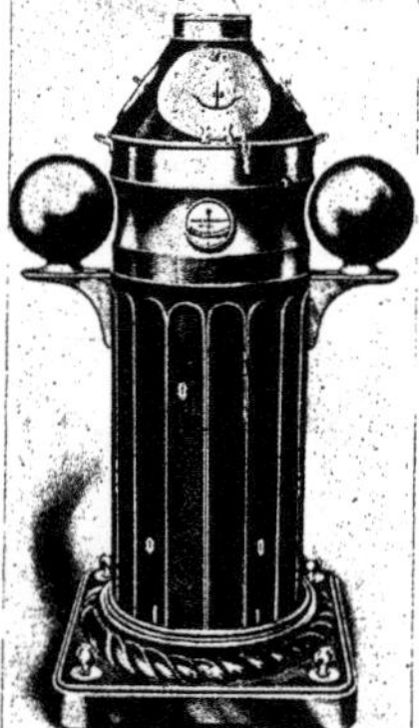

SHIP'S BINNACLE.

just read, but solutions will unfold themselves as we go ahead. First of all I want you to learn to "box" the compass, that is, be able to repeat every point by name from north round through east, south and west. There are the expressions:

North, north by east, north north-east, north-east by north, north-east, north-east by east, east north-east, east by north, east. = 90° or quadrant.

East by south, east south-east, south-east by east, south-east, south-east by south, south-south-east, south by east, south. = 180° or semi-circle.

South by west, south-south-west, south-west by south, south-west, south-west by west, west south-west, west by south, west = 270° or ¾ of a circle.

West by north, west north-west, north-west by west, north-west, north-west by north, north north-west, north by west—north. = 32 points or 360°. Each point equals 11¼°. Thus four points = 45° and so on.

The cardinal points are north, south, east and west. The quadrantal points are north-east, south-east, south-west and north-west. It is almost a general rule to steer by degrees, that is, the wheelsman is given a course in this way: "Keep her" N. 60° E. and S. 42° W." In a few coasting schooners or fishing craft these expressions might be given out in the old way, N. E. x E. ½ E. and S. W. ¼ S. Plainer still, north-east by east a half east, and south-west a quarter south. You note that the compass card on board your ship shows ¼, ½ and ¾ points as well as the full point. Modern compass cards carry a continuous circle of degrees from 0 to 360 and the rose impressions on the charts representing the true and magnetic directions are printed in the same way. As you may come across both ways there is nothing to fear, the naming is different but it has no other effect. You may also hear seamen or see in books the expression north-east by north a half north, which is reading backwards. In such cases it is better to say north-north-east a half east = N. N. E. ½ E. In a few cases the backing up expression is right, but not when it jumps more than a point. When we come to chart work this chance of going wrong through having two names for the same point is eliminated, as in using degrees only there can be no confusion.

There are three kinds of courses or directions that the navigator has to use in compass work, and a short definition of each will help to introduce the practice which will appear in next month's issue.

The compass course is simply the direction that the ship heads on without corrections such as leeway, deviation and variation.

MARINER'S COMPASS IN GIMBAL RINGS.

The magnetic course is a compass course corrected for leeway and deviation, or a course taken from a magnetic chart, or a true course corrected for variation.

A true course is a compass course corrected for leeway, deviation and variation, or a course from the parallel rulers on a true chart.

These courses are all related to each other then, according to circumstances. Another set of definitions may help further.

A compass course is the angle between the direction of the ship's head and the compass needle.

The magnetic course is the angle a ship's track makes with the magnetic meridian.

True course is the angle a ship's track makes with the true meridian.

Diagrams and course corrections will follow in next issue.

(To be continued.)

NOTICE TO MARINERS

The following notices to mariners have been issued by the Department of Marine, Ottawa, under dates given:

No. 16 of 1918, April 3
(Inland No. 6)

ONTARIO

(38) Trent Canal System—List of Lights.

The lights of the Trent Canal system and its branches are maintained by the Department of Railways and Canals by arrangement with the Department of Marine.

The canal is designed when completed to leave the Bay of Quinte at Trenton and follow the chain of Kawartha lakes and connecting waters to Lake Simcoe, crossing Lake Simcoe and Lake Couchiching, and thence follow the course of the Severn River into Matchedash Bay, Georgian Bay, through locks at Port Severn.

The outlets in both directions to the Great Lakes are not yet navigable, but the system in the interior is in operation from the upper reaches of the Trent River through the Kawartha lakes to Lake Couchiching, and the completed sections have been buoyed and lighted.

A complete list of the lights in operation is given in detail in the official copy of this notice.

* * *

No. 17 of 1918, April 3
(Pacific No. 3)

BRITISH COLUMBIA

(39) Canadian list of lights and fog signals—New Edition.

A list of all the lights and fog signals on the Pacific Coast of the Dominion of Canada corrected to the 1st April, 1918, has just been published. Copies will be supplied to mariners free on application

Departmental File: No. 28502.

ALASKA

(40) Revillagigeda channel—Black rock—Light established.

Position.—On highest part of Black rock.

Character of light.—Flashing white every 3 seconds, flash 0.3 second duration.

Elevation.—55 feet.

Power.—130 candles.

Structure.—Small cylindrical house on pyramidal skeleton tower.

Color.—Grey.

Authority: U.S. Dept. of Commerce N. to M. No. 12 of 1918.
Admiralty charts: Nos. 2458 and 2431.
Publication: Alaska Pilot. 1908, page 131.

* * *

No. 18 of 1918, April 6
(Pacific No. 4)

BRITISH COLUMBIA

(41) Burrard inlet—First Narrows—Vancouver harbor—Prospect point—Change in traffic signals—Brockton point—Traffic signals discontinued.

Former notice.—No. 48 (127) of 1910.

System changed.—The system of signals heretofore operated at Prospect point and Brockton point, First Narrows, Vancouver harbor, to inform mariners of the presence in the Narrows of other vessels will be replaced by a system of signals exhibited from Prospect point only.

Location.—The signal pole on the bluff will be moved about 100 feet northward from its present position and will be surmounted by a cross-arm, the whole painted white. Lat. N. 49° 18′ 49″, Long. W. 123° 8′ 33″.

Signals.—Inbound vessels: Signals exhibited on south arm. One black ball denotes single vessel. Two black balls in horizontal position denotes vessel with a tow. Three black balls forming a triangle, apex upwards, denotes both single vessel and vessel with a tow. At night white lights will be substituted for the balls.

Outbound vessels: Signals exhibited on north arm. One black cone denotes single vessel. Two black cones in horizontal position denotes vessel with a tow. Three black cones forming a triangle, apex upwards, denotes both single vessel and vessel with a tow. At night red lights will be substituted for the cones.

Visibility.—These signals can be seen immediately after rounding Brockton point outbound and from all points seaward inbound.

Dates.—The new signals will be put in operation on or about June 1st, 1918.

The existing signals will be discontinued May 15th, 1918.

Authority: Report from Agent of Dept. of Marine, Victoria.
Admiralty charts: Nos. 922, 1922 and 2689.
Publication: British Columbia Pilot, Vol. 1, 1913, page 294.
Departmental File: No. 30296.

* * *

No. 19 of 1918, April 9
Atlantic No. 9)

MARITIME PROVINCES AND QUEBEC

(42) Canadian list of lights and fog signals—New edition.

A list of all the lights and fog signals on the Atlantic Coast of the Dominion of Canada, including the Gulf of St. Lawrence and the River St. Lawrence to Montreal, corrected to the 1st April, 1918, has just been published. Copies will be supplied to mariners free on application.

Departmental File: No. 28502.

NOVA SCOTIA

(43) West coast—Cape St. Mary breakwater—Light improved.

Former notice.—No. 8 (18) of 1916.

Position.—On Cape St. Mary breakwater, 13 feet from its outer end. Lat. N. 44° 5′ 15″, Long. W. 66° 12′ 28″.

Character.—Fixed white light, shown from an anchor lens lantern.

Elevation.—18 feet.

Visibility.—5 miles.

New structure.—Pole, with shed at base.

Material.—Wood.

Color.—White.

Height of pole.—18 feet.

Authority: Report from Mr. J. A. Leger, District Engineer, Halifax.
Admiralty charts: Nos. 352, 2538, 1651, 2670.
Publication: Nova Scotia and Bay of Fundy Pilot, 1911, page 240.
Canadian List of Lights and Fog Signals, 1918: No. 1015.
Departmental File: No. 20194.5C.

QUEBEC

(44) Gulf of St. Lawrence—Gaspe coast—L'Anse au Beaufils—Outer end of wharf carried away by ice—Pole light moved.

Former notice.—No. 83 (210) of 1911.

Position.—Lat. N. 48° 28′ 44″, Long. W. 64° 17′ 44″.

Outer end of wharf carried away.—The outer end of l'Anse au Beaufils wharf has been carried away by ice.

Light.—The fixed red pole has been moved about 20 feet in on the wharf from the position it formerly occupied.

Authority: Report from Supt. of Lights, Quebec.
Admiralty charts: Nos. 1168, 1621 and 2516.
Publication: St. Lawrence Pilot, Vol. 1, 1916, page 94.
Canadian List of Lights and Fog Signals, 1918: No. 978.
Departmental File: No. 20978R.

QUEBEC

(45) River St. Lawrence below Quebec—Lower Traverse lightship—Change in rig.

Former notice.—No. 21 (70) of 1913, and No. 87 (224) of 1917.

Position of lightship.—On the south side of the south traverse at its lower end. Lat. N. 47° 22′ 5″, Long. W. 70° 14′ 36″.

Masts.—The mizzen mast has been removed from the vessel.

Ball.—The red ball shown when the ship is on her station is now hoisted on the stay between the fore and main masts.

Authority: Departmental records.
Admiralty charts: Nos. 3734, 314 and 2516.
Canadian Naval charts: Nos. 20 5and 207.
Publication: St. Lawrence Pilot below Quebec, 1916, page 111.
Canadian List of Lights and Fog Signals, 1918: No. 1174.
Departmental File: No. 21175M.

* * *

No. 20 of 1918, April 10
(Inland No. 7)

ONTARIO

(46) Canadian list of lights and fog signals—New edition.

A list of all the lights and fog signals on the inland waters of the Dominion of Canada, corrected to the 1st April, 1918, has just been published. Copies will be supplied to mariners free on application.

Departmental File: No. 28502.

ONTARIO

(47) Lake Superior—Otter Island—Light temporarily discontinued.

Position.—On the north-west extremity of Otter Island.

Light temporarily discontinued.—Otter Island and hand foghorn are temporarily discontinued.

It is possible that an unwatched light may be established about midsummer without previous notice.

Authority: Memo, from Commissioner of Lights.
Admiralty chart: No. 320.
Publication: U.S. H.O. Publication No. 108A, 1906, page 91.
Canadian List of Lights and Fog Signals, 1918: No. 2177.
Departmental File: No. 22177K.

ONTARIO

(48) Lake Superior—Thunder Bay—Port Arthur harbor—Changes in buoyage.

The following changes will, on the opening of navigation in 1918, be made in the buoyage of Port Arthur harbor:

(1 Change in position of buoy.—Red spar buoy No. 8A moved to a new position 325 feet 346° (N. 17° W. mag.) from the lighthouse on the north breakwater.

Former notice.—No. 23 (64) of 1917.

(2) Buoy discontinued.—Black spar buoy No. 29A, heretofore moored 2550 feet 249° (S. 66° W. mag.) from the south entrance light, has been withdrawn.

Former notice.—No. 17 (57) of 1916.

Variation in 1918: 3° E.
Authority: Departmental records.
Canadian Naval Chart: No. 10.
Publication: U. S. H. O. Publication No. 108A, 1906, page 83.
Departmental File: No. 29214.

THE OPEN BOAT

"When this here war is done," says Dan, "and all the fightin's through,
There's some'll pal with Fritz again as they was used to do;
But not me," says Dan the sailor-man, "not me," says he—
"Lord knows its nippy in an open boat on winter nights at sea."

* * * *

"When the last battle's lost an' won, an' won or lost the game,
There's some'll think no 'arm to drink with squareheads just the same;
But not me," says Dan the sailor-man, "an' if you ask me why—
Lord knows its thirsty in an open boat when the water-breaker's dry."

* * * *

"When all the bloomin' mines is swep' an' ships are sunk no more,
There's some'll set them down to eat with Germans as before;
But not me," says Dan the sailor-man, "not me, for one—
Lord knows it's hungry in an open boat when the last biscuit's done."

* * * *

"When peace is signed and treaties made an' trade begins again,
There's some'll shake a German's hand an' never see the stain;
But not me," says Dan the sailor-man, "not me, as God's on high—
Lord knows it's bitter in an open boat to see your shipmates die."

—C. F. S. in "Punch"

No. 21 of 1918. April 11.
(Atlantic No. 10.)

QUEBEC

(49) River St. Lawrence—Ship channel between Quebec and Montreal—Becancour Traverse—Cap Madeleine village range lights moved.

Change in channel.—The axis of Becancour traverse dredged ship channnel has been changed through an angle of 1° 11′ to the right from a point between buoys 33C and 34C. To mark the new axis the range lights below Cap Madeleine village have been moved northwardly as follows:

(1) Front Range Light.

New position.—217 feet 5° (N. 21° 30′ E. mag.) from the old site of the light.
Lat. N. 46° 22′ 26″, Long. W. 27° 29′ 35″
Character.—Fixed white light.
Elevation.—50 feet.
Visibility.—4 miles in the line of range.
Order.—Catoptric.
Power.—7,000 candles.
Structure.—Square building; square lantern.
Material.—Wood.
Color.—White.
Height.—26 feet, from base to top of ventilator on the lantern.

(2) Back Range Light

New position.—2420 feet 258° (N. 85° 30′ W. mag.) from the front range light.
Character.—Fixed white light.
Elevation.—109 feet.
Visibility.—4 miles in the line of range.
Order.—Catoptric.
Power.—E,000 candles.
Structure.—Skeleton tower, square in plan, with sloping sides, surmounted by an enclosed watchroom and square lantern; the upper portion of the side of the skeleton framework facing the alignment is covered with wooden slats.
Material.—Skeleton frame, steel; watchroom and lantern, wood.
Color.—Skeleton frame, red; slats, watchroom and sides of lantern, white; roof of lantern, red.
Height.—86 feet, from base to top of ventilator on the lantern.

Sailing directions.—The lights, in their new position, mark the axis of Becancour Traverse dredged cut, and should be brought on after passing Batture Francoeur gas buoy No. 30C, and kept in one bearing 258° (N. 85° 30′ W. mag.) until gas buoy No. 39C is abeam. The new alignment intersects the old one between buoys Nos. 33C and 34C.

Variation in 1918: 16° 30′ W.
Authority: Records, Chief Engineer's office, Dept. of Marine.
Admiralty charts: Nos. 2780, 2781, 2830a and 797.
Canadian Naval Charts: Nos. 1 and 23.
Publication: St Lawrence Pilot above Quebec, 1912, page 52.
Canadian List of Lights and Fog Signals, 1918: Nos. 1314 and 1315.
Departmental Files: Nos. 21314C and 21315R.

No. 22 of 1918. April 15.
(Atlantic No. 11.)

QUEBEC.

(50) Gulf of St. Lawrence—Egg island—Change in character of light.

Position.—On Egg island.
Lat. N. 49° 38′, Long. W. 67° 10′.
Date of alteration.—On or about 1st May, 1918, without further notice.

Alteration.—The revolving white light will be replaced by a flashing white catoptric light, showing two flashes, with an interval of 6 seconds between them, every twenty-four seconds, thus: Flash; 6 seconds interval. flash; 18 seconds interval.

For half the time of revolution, or 12 seconds, the light will be totally eclipsed; for the other half a light of 500 candle power will be visible, through which the stronger flashes will show.

Power.—Naked light 500 candles; flashes 50,000 candles.
Illuminant.—Petroleum vapor, burned under an incandescent mantle.

Authority: Report from Mr. J. A. Smith, District Engineer, Quebec.
Admiralty charts: Nos. 307 and 2516.
Publication: St. Lawrence Pilot, Vol. 1, 1916, page 221.
Canadian List of Lights and Fog Signals, 1918: No. 1074.
Departmental File: No. 21074A.

QUEBEC

(51) River St. Lawrence—Godbout—Hand fog horn at light-station.

Position.—At Godbout light-station, about ½ mile eastward of mouth of Godbout river.
Lat. N. 49° 19′ 0″, Long. W. 67° 35′ 43″.
Description.—Hand fog horn.
Remarks.—It is used to answer signals from steamers in the vicinity of the station in thick weather.

Authority: Departmental records.
Admiralty charts: Nos. 307 and 2516.
Publication: St. Lawrence Pilot, Vol. 1, 1916, page 225.
Canadian List of Lights and Fog Signals, 1918: No. 1078.
Departmental File: No. 21078F.

No. 23 of 1918. April 22.
(Pacific No. 5.)

BRITISH COLUMBIA

(52) Vancouver island—East coast—Nanaimo harbor—Change in position of buoys.

Former notice.—No. 135 (443) of 1914.
Buoys changed in position.—The following black platform buoys in Nanaimo harbor have been moved to new positions in 25 feet water:

(a) Entrance buoy No. 1 is now moored 260 feet 299° (N. 86° W. mag.) from Harbor entrance gas-lighted beacon.

(b) South channel buoy No. 3 is now moored 1,100 feet 267° 30′ (S. 62° 30′ W. mag.) from Harbor entrance gas-lighted beacon.

Variation in 1918: 25° E.
Authority: Report from Agent of Dept. of Marine, Victoria.
Admiralty charts: Nos. 573 and 2512.
Publication: British Columbia Pilot, Vol. 1, 1913, page 311.
List of Buoys and Beacons in British Columbia, 1916: Nos. 473 and 478.
Departmental File: No. 29467.

BRITISH COLUMBIA

(53) Portland canal—Lion point; Eagle point; and Stewart—Buoys withdrawn.

Former notice.—No. 51 (125) of 1911.
Buoys discontinued.—The following buoys, heretofore maintained in the approach to the town of Stewart, at the head of Portland canal, have been withdrawn:

(1) The red conical buoy off the westerly extreme of the shoal ground off Lion point.

(2) The black can buoy on the outer edge of shoal off Eagle point.

(3) The black platform buoy on the inner edge of shoal off Eagle point.

(4) The red platform buoy off the shoal ground at the head of Portland canal.

Authority: Report from Agent of Dept. of Marine, Victoria.
Admiralty charts: Nos. 2458, 2431.
Publication: British Columbia Pilot, Vol. 2, 1913, pages 134 and 135.
List of Buoys and Beacons in British Columbia, 1916: Nos. 795, 796, 798 and 800.
Departmental File: No. 31115.

JAPAN

(54) Bungo channel, east side—Province of Iyo—Sada Misaki—Lighthouse established.

Position.—On Sada Misaki.
Lat. N. 33° 20′ 30″, Long. E. 132° 9′ 36″.
Character.—Group flashing white light showing a triple flash every 20 seconds.
Elevation.—149 feet.
Visibility.—18 miles, over an arc of 309° from 264° 40′ (S. 89° W. mag.) through West, North, E. and S. to 213° 40′ (S. 38° W. mag.)
Power.—70,000 candles.
Order.—Third dioptric.
Structure.—Octagonal tower.
Material.—Concrete.
Color.—White.
Height.—50 feet from base to light.

Variation in 1918: 4° 20′ W.
Authority: Notification No. 354 of Dept. of Communications, Japan.

No. 24 of 1918.—April 24.
(Inland No. 8.)

QUEBEC

(55) Ottawa river—Caron point—Light improved.

Position.—On Caron point.
Lat. N. 45° 24′ 16″, Long. W. 73° 55′ 18″.
Light improved.—The occulting white light has been improved by the substitution of a fifth order dioptric illuminating apparatus for the seventh order lens heretofore used.

Authority: Departmental records.
Admiralty charts: Nos. 259a and 797.
Canadian Naval Charts: Nos. 50 and 54.
Publication: St. Lawrence Pilot above Quebec, 1912, page 114.
Canadian List of Lights and Fog Signals, 1918: No. 1533.
Departmental File: No. 21533A.

ONTARIO

(56) Lake Erie—Pelee passage—South-east shoal—Lightship replaced on her station.

Former notice.—No. 8 (20) of 1918.
Position.—⅛ mile southeastwardly from the southern extremity of South-east shoal.
Lat. N. 41° 49′ 15″, Long. W. 82° 27′ 32″.
Lightship on station.—Southeast shoal lightship has been replaced on her station.

Authority: Departmental records.
Admiralty charts: Nos. 332 and 678.
Publication: U. S. H. O. Publication No. 108D, 1907, page 89.
Canadian List of Lights and Fog Signals, 1918: No. 1862.
Departmental File: No. 45095.

No. 25 of 1918. April 25.
(Atlantic No. 12.)

NEW BRUNSWICK

(57) South coast—Bay of Fundy—St. Martins—Light moved to west breakwater.

Former notice.—No. 116 (300) of 1917.
New position.—On the outer end of the west breakwater at St. Martins.
Lat. N. 45° 21′ 18″, Long. W. 65° 32′ 0″.

Character.—Fixed red light, shown from an anchor lens lantern.

Elevation.—31 feet.

Visibility.—7 miles.

New structure.—Pole, with shed at base.

Material.—Wood.

Color.—White.

Height of pole.—29 feet.

Note.—The maintenance of a light on the east breakwater has been discontinued.

Authority: Report from Mr. G. S. Macdonald, Resident Engineer, St. John.
Admiralty charts: Nos. 353, 1651, 2516 and 2670.
Publication: Nova Scotia and Bay of Fundy Pilot, 1911, page 317.
Canadian List of Lights and Fog Signals, 1918: No. 120
Departmental File: No 20120R

ENGLAND

(58) Southwest coast—Scilly isles—St. Mary's Sound—Peninnis head—Fog signal established.

Position.—At Peninnis head lightstation.

Lat. N. 49° 54¼, Long. W. 6° 18¼'.

Description.—A siren giving two blasts every forty-five seconds, thus:

Blast	Silent	Blast	Silent interval
3 secs.	3 secs.	3 secs.	36 secs.

Authority: British Admiralty N. to M. No. 412 of 1918.
Admiralty charts: Nos. 883, 34, 2565, 1123, 2675a and 1598.
Publications: Channel Pilot, Part 1, 1908, page 36; and W. C. England Pilot, 1910, page 42.

No. 26 of 1918. April 30.

(Atlantic No. 13)

NEW BRUNSWICK

(59) South coast—Bay of Fundy—Mispek bell buoy discontinued.

Position.—Off Mispek.

Lat. N. 45° 12' 9", Long. W. 65°, 58' 25".

Buoy discontinued.—The maintenance of Mispek bell buoy has been discontinued and the buoy removed.

Authority: Memo. from Commissioner of Lights.
Admiralty charts: Nos. 352, 353, 1651 and 2670.
Publication: Nova Scotia and Bay of Fundy Pilot, 1911, page 308.
Canadian List of Lights and Fog Signals, 1918: No. 114.
Departmental File: No. 27242.

NEW BRUNSWICK

(60) South coast—Bay of Fundy—Chignecto channel—Off Matthews head—Whistle buoy discontinued.

Former notice.—No. 17 (50) of 1914.

Position.—1 1/3 miles 109 (S. 50° E. mag.) from Matthews head.

Lat. N. 45° 33' 35", Long. W. 64° 56' 30".

Buoy discontinued.—The maintenance of Matthews head whistle buoy has been discontinued and the buoy removed.

Variation in 1918: 21° W.
Authority: Memo. from Commissioner of Lights.
Admiralty charts: Nos. 353, 1651, 2516 and 2670.
Publication: Nova Scotia and Bay of Fundy Pilot, 1911, page 319.
Canadian List of Lights and Fog Signals, 1918: No. 121.5.
Departmental File: No. 19544.

QUEBEC

(61) River St. Lawrence—Ship channel between Quebec and Montreal—Cap a la Roche channel—Changes in buoyage.

The following changes will be immediately made in the buoyage of the ship channel in the vicinity of Deschaillons without further notice:

86Q.—Cap Charles course red conical buoy No. 86Q. will be moved 450 feet upstream, to a new position opposite black can buoy No. 85 Q.

Lat. N. 46° 34' 0", Long. W. 72° 6' 7".

87 Q.—Cap a la Roche curve black can buoy No. 87 Q. will be moved 350 feet upstream.

Lat. N. 46° 33' 55", Long. W. 72° 6' 28".

88 Q.—A new red iron conical buoy, numbered 88 Q., will be moored opposite buoy No. 87 Q. in its new position.

Lat. N. 46° 33' 59", Long. W. 72° 6' 28".

89 Q.—A new black gas buoy, numbered 89 Q., showing an occulting white light, will be moored opposite Cap a la Roche curve red gas buoy No. 90 Q.

Lat. N. 46° 33" 54", Long. W. 72° 6' 50".

90½ Q.—A new red iron conical buoy, numbered 90½ Q., will be temporarily moored 800 feet upstream from Cap a la Roche curve red gas buoy No. 90 Q.

Lat. N. 46° 33' 56'. Long. W. 72° 7' 2".

91 Q.—Cap a la Roche course black can buoy No. 91 Q., will be replaced by a gas buoy, showing an occulting white light.

Lat. N. 56° 33' 49", Long. W. 72° 7' 13".

92 Q.—Cap la Roche course red gas buoy No. 92 Q. will be moved 150 feet upstream to a new position opposite gas buoy No. 91 Q.

Lat. N. 46° 33' 53", Long. W. 72° 7' 15".

Authority: Departmental records.
Admiralty charts: Nos. 2779 and 2830a.
Canadian Naval charts: Nos. 15 and 24.
Publication: St. Lawrence Pilot above Quebec, 1912, page 45.
Canadian List of Lights and Fog Signals, 1918: Nos. 1286.9, 1287.4, and 1287.5.
Departmental File: No. 25577.

No. 27 of 1918. May 1.

(Inland No. 9.)

ONTARIO

(62) Lake Erie—Mohawk island—Change in character of light.

Position.—On Mohawk island.

Lat. N. 42° 50' 2", Long. W. 79° 31' 23".

Alteration.—The revolving white light has been replaced by a flashing white catoptric light, showing two flashes, with an interval of 6 seconds between them, every twenty-four seconds, thus: Flash; 6 seconds interval; flash; 18 seconds interval.

For half the time of revolution, or 12 seconds, the light will be totally eclipsed; for the other half a light of 500 candle power will be visible, through which the stronget flashes will show.

Power.—Naked light 500 candles; flashes 50,000 candles.

Illuminant.—Petroleum vapor, burned under an incandescent mantle.

Authority: Departmental records.
Admiralty charts: Nos. 1605, 332 and 678.
Publication: U. S. H. O. Publication No. 108D, 1907, page 99.
Canadian List of Lights and Fog Signals, 1918: No. 1842.
Departmental File: No. 21842A.

ONTARIO

(63) Rainy river—Long Sault rapids to Singleton island—Range day beacons erected.

Range day beacons.—Five pairs of range day beacons have been erected on the Canadian shore of Rainy river to mark portions of the channel between Long Sault rapids and Singleton island, as follows:

(1) A pair at Long Sault rapids, near the east boundary of Indian Reserve No. 13.

(2) A pair at Long Sault rapids near the middle of the south boundary of Indian Reserve No. 12.

(3) A pair at Manitou rapids.

(4) A pair on the shore north of Singleton island.

(5) A pair on the north end of Singleton island.

Authority: Report from Agent of Dept. of Marine, Kenora.
Departmental File: No. 11981.

[Pilots, masters or others interested are earnestly requested to send information of dangers, changes in aids to navigation, notice of new shoals or channels, errors in publications, or any other facts affecting the navigation of Canadian waters to the Chief Engineer, Department of Marine, Ottawa, Canada. Such communications can be mailed free of Canadian postage.]

AN ELECTRICALLY STEERED SCHOONER

By M. Mark

No innovation in the deck machinery of ships has caused more favorable comment than the simple and efficient electrical steering gear which has been installed on the schooner "Allard." of the Charles R. McCormick fleet. There has been a steady demand for these steering systems, as they fill a long-felt want, especially in those vessels which are just large enough to make the handling of the ship's wheel by hand extremely trying when the ship is negotiating a narrow, crooked channel, or a fast rudder is required in crossing bars. The great attraction about this gear are its extreme simplicity and foolproofness and the very small amount of power it absorbs. Current from the ordinary electric lighting set in use on the auxiliary steamer is found to be more than sufficient to meet all the needs of the operating gear motor. The wheelman simply moves a vertical lever to port or starboard and the gear is engaged and turns the wheel as required, stopping the moment the lever is moved back to the neutral position. In this way the rudder can be thrown hard over in twenty seconds, and the man or men at the wheel are relieved of the arduous pulling or hauling when the vessel is operating in a choppy sea or making a difficult channel or bar. The power absorbed varies, of course, with the speed of the ship and the condition of the sea, but on the "Allard," which is a 2,000,000 feet lumber carrier, the maximum power absorbed at any time was less than three horsepower, whilst in still water this was negligible. Another simple device for the remote control of the steering engine is the Herzog electric steering gear, in which three wires running to the bridge take the place of the pipes in the usual telemotor control. Here again electric power takes the place of the old arduous pulling and hauling and enables the steersman to keep the vessel's course more accurate.

Eastern Canada Marine Activities

Doings of the Month at Montreal and on St. Lawrence River

May Correspondence

OPENING of navigation on the St. Lawrence this year was very much delayed owing to the severe cold of the past winter and the heavy ice upon the river. Lake and river traffic began about the first of the month, but an ocean boat did not put in an appearance until the 7th of May, the latest first arrival from overseas in thirty-three years. The fortunate captain to bring the first vessel to the port was Benjamin Dowse, and, according to custom, he was duly presented with a gold-headed cane and a silk hat by the Montreal Harbor Commissioners, the ceremony taking place in the company's offices at noon on May 9th.

The signs of the times are very noticeable about the harbor this year in the additional number of khaki-clad guards; the force has been increased from 64 to 128, and is under the command of Capt. Baxter Clegg. Another war picture is the camouflage appearance of some of the Transatlantic vessels now in the harbor.

The first mishap of the season on the Lachine Canal occurred on the last day of April, when the failure of the machinery in the collier Henry B. Hall caused the vessel to collide with the upper gate of the lower south lock, and the rush of the water carried the boat down upon the lower gates, which were also torn away. Little interference was caused to navigation, as it was possible to transfer the boats through the north passage of the locks. When the lower gates were replaced the south lock could be operated by operating the basin in conjunction with the lock itself. The four new gates are now in position.

The inaugural meeting of the newly-chartered Province of Quebec division of the Navy League of Canada was held on May the 7th at the offices of the Montreal Harbor Commissioners. In addition to the election of officers, a ladies' committee was appointed from the members of the Montreal Women's Branch of the Navy League of Canada. The question of establishing a naval training ship at the port was considered, also that of an official organ to give greater publicity to the activities of the League. The officers of the new division are: Jas. Carruthers, president; E. W. Beaty, J. McKinnon and Hon. D. O. Lesperance, vice-presidents; D. F. Glass, secretary-treasurer.

Ship repairs and alterations will likely be a large factor in this year's operations, as indicated by the activities at the Montreal Drydock. Among the vessels already repaired or undergoing repairs are the "Cataract," of the Montreal Transportation Company; this vessel has been used for a barge and the refitting now under way will convert her into a lake steamer. The hull is being reinforced with steel frames and many new plates will be added. A locomotive type of boiler is being installed that will develop about 500 h.p. The boat will be practically be a new boat from the deck upward. The "Canobie," belonging to Willson & Patterson, has been fitted with a new stern post and stern tube, also several plate replacements to the hull and caulked all over. The "Calgarian," a steel vessel of 1,300 tons, with a record of long service on the lakes, arrived in Montreal on May the 1st, and will be placed in the drydock to have some of the frames and plates replaced, together with a thorough overhauling, to fit her for ocean service. She has been sold to the French Government. The "Hamiltonion" will undergo similar treatment for the same service. One of the vessels awaiting repairs is the "Lake Crescent," one of the fleet of the United States Shipping Board. She will be docked to have the plates in her bottom straightened, which were damaged after going ashore. It is expected that several others of the same description will shortly be docked for similar repairs. Two boats of the Montreal Light, Heat and Power Company, the "Keyvive," with damaged rudder, and the "Keyport," with damaged bow; also the "Colin W." and the "Stewart," of the Canadian Imports Company, are booked for new stern posts. The T.R.9 is also undergoing her finishing touches.

Fraser, Brace Company, of Montreal, anticipate the launching of two of the 3,000-ton I.M.B. wooden steamers on or about the 25th of May. The hulls of these two boats are rapidly nearing completion, and the mouth of the basin is now being fitted with lock gates that will subsequently make the basin a serviceable dry dock. This "launching" will in reality be a floating operation, as these two vessels have been constructed on the floor of the basin so that the incoming water will float them. Two similar vessels are under construction on the upper level of the basin, and the launching of these are expected in June and July respectively.

Work on the concrete vessel, the "Concretia," is nearing completion, and the builders are awaiting the finishing of the superstructure before giving the boat her initial trial under her own steam. Mr. Jarvis, of the Atlas Construction Co., anticipates this will take place before the end of the month.

PROGRESS ON THE FOUR 3,000-TON WOODEN VESSELS BEING BUILT BY THE FRASER BRACE CO. FOR THE IMPERIAL MUNITIONS BOARD. THE TWO ON THE LOWER LEVEL ARE EXPECTED TO BE LAUNCHED ABOUT THE 25th OF THIS MONTH. THE OPERATION IS AWAITING THE PLACING OF THE GATES AT THE ENTRANCE TO THE BASIN. WHEN THE PHOTO WAS TAKEN THE LACHINE CANAL WAS EMPTY FOR THE ANNUAL SPRING CLEANING.

The Royal Commission appointed to investigate into the pilotage conditions of the principal Canadian harbors have completed their work along the St. Lawrence and in the Maritime Provinces, and Thos. Robb and James N. Bales, two of the officiating members, have left for the Pacific Coast to study the conditions existing at the ports of British Columbia.

Under the examination of Captain Demers, Wrecking Commissioner, and Capt. P. L. Lachance, of Ottawa, the following new men have been granted St. Lawrence River and coast certificates as pilots:—J. R. Gandreau, Montmagny;

Eudore Langlois, Quebec; Willie Pouliot, C. Lachance, Rodrique Lachance, Hermegilde Lachance, St. Jean, Island of Orleans; Leo. Labricque, St. Laurent, Island of Orleans; and Edmond Paquet, Levis.

A representative delegation of the shipbuilding trades of Montreal and district interviewed a committee of the Ottawa Cabinet on May 1st with the purpose of urging the Government to take measures of more appropriate adjustment of the alien labor now employed in Canadian shipyards. It was pointed out that in many cases women could be satisfactorily employed on work now being performed by these neutral aliens.

One of the standard 3,000-ton wooden vessels being constructed for the Imperial Munitions Board was launched from the ways at the shipyard of Quinlan & Robertson, at Quebec, on Saturday, the 11th of May. Two others are being constructed by this firm, and they are both expected to take the water before August.

At a recent meeting of the Quebec Board of Trade Mr. Leon Fiset, of Eastern Harbor, Cape Breton, explained in detail the operation of a proposed shipping company, with a fleet of four steel vessels, plying between the Gulf ports and the harbor of Montreal, engaging in cold storage and the transportation of the same, the plants to be located on Prince Edward Island.

A contract for improvements to the harbor of St. John, N.B., including the completion of the dry dock, the extension of the breakwater, and the establishment of a steel shipbuilding plant, has been signed by the authorities at Ottawa. Ships of 10,000 tons will be constructed, and it is estimated that the total expenditure will be approximately $7,000,000.

Early on the morning of May 5th the Dominion Coal Co.'s steamer "Louisburg," with a cargo of coal, ran on the rocks of English Cape, St. Mary's Bay, Newfoundland, and in a short time became a total wreck. The vessel, originally named the "Thormholme," built in 1881, was 206 feet long, and had a gross tonnage of 1,816 tons. Capt. Jas. Hemp was in charge of the boat at the time of the accident; all on board were saved.

THE "WAR WIZARD" ON THE WAYS AT COLLINGWOOD.

COLLINGWOOD SHIPBUILDING CO. LAUNCH THE "WAR WIZARD"

The Collingwood Shipbuilding Company launched from their yard, May 8th, at 11.30 a.m., the ocean-going steamer "War Wizard," the first of two vessels ordered by the Imperial Munitions Board. The vessel is of the poop, bridge and forecastle type, with engines amidships, and her dimensions are: 261 ft. over all, 251 ft. b.p., 43 ft. 6 in. beam, 20 ft. depth moulded; to carry approximately 2,900 tons deadweight.

The vessel and her equipment have been constructed to the highest classification of the British Corporation Registry. She has two large holds and four large hatchways. The cargo gear is of the most modern type and arranged for quick handling.

Accommodation for the officers and engineers is provided for in a large steel deckhouse on the bridge deck. The petty officers' accommodation is situated under the forecastle, while the crew has commodious accommodation aft. Like all other vessels of this type, provision is made for mounting a gun on the poop.

The engine is of the triple expansion type, the cylinders being 18, 30 and 50 in. by 36 in. stroke, taking steam from two Scotch boilers, 14 ft. diameter by 10 ft. 9 in. long, working at 180 lbs. pres sure, with Howden's forced draught, an the auxiliary machinery and equipmen is of the latest and most complete kin for ocean service. The vessel will b operated under the management of M E. C. Downing, Cardiff. The vessel wa launched with the machinery and boiler on board, and will be ready for sea in very short time. A second ship, th "War Witch," is now on the stocks.

Immediately after the launch of th "War Wizard" the keel of a 3,800-to deadweight steamer was laid. This ves sel is being built to the order of the De partment of Marine, Ottawa, and is also intended for ocean service.

Vancouver.—Fire in the J. Coughlin & Sons' shipyard on False Creek did damage on May 15 to the extent of between $1,500,000 and $2,000,000. A fireman was killed. The steel steamer War Chariot, about two-thirds completed, was engulfed in the flames. The ways, which are built on piles, fell through and the hull twisted by the heat, is now half-submerged in the water beneath. The hull of the steamer War Charger on No. 3 ways still stands, but the plates are badly buckled by the heat.

THE LAUNCH OF THE "WAR WIZARD" AT COLLINGWOOD.

THE "WAR WIZARD" IN THE SLIP AFTER LAUNCHING.

Work in Progress in Canadian Shipyards

A record of Work in Process of Completion — Principal Features of Specification — Approximate Launching Dates—Vessels on Order.

YARROWS, LIMITED, ESQUIMALT, B.C.

One stern-wheel steamer, 185 ft. x 37 ft., for undisclosed interests, 400 tons.

Four shallow-draught vessels for river navigation in India.

H. VOLLMERS, NANAIMO, B.C.

One gasoline tug, 9 tons.

NEW WESTMINSTER CONSTRUCTION & ENGINEERING CO., NEW WESTMINSTER, B.C.

Four wooden steamers, Nos. 10-11-12-13, for undisclosed interests 250 ft. x 43 ft. 6 in. x 25 ft., moulded depth, 2,800 tons d. w., 1,000 h. p., single screw engines, 12 knots, naval architect, I. Alexander. Launchings March 14, April 14, May 15 and June 15, 1918.

B. C. CONSTRUCTION & ENG. CO., POPLAR ISLAND, NEW WESTMINSTER, B. C.

Four wooden vessels, 1,800 gr. tons, 2,800 tons d. w.

NEW WESTMINSTER MARINE CO., FT. OF FURNESS ST., NEW WESTMINSTER, B. C.

STAR SHIPYARD CO., NEW WESTMINSTER, B. C.

J. A. CROLL, PORT ALBERNI, B. C.

PACIFIC CONSTRUCTION CO., PORT COQUITLAM, B. C.

Three wooden freight steamers, Nos. 20-21-22, 250 ft. b. p. x 42 ft. 6 in. x 25 ft., 1,800 gr. tons, 2,800 tons d. w., trip. exp. engines, water tube boilers, 12 knots, 1,000 h. p. Two for undisclosed interests, one for builder's account. Launchings May 20, Jun. 15, Sept., 1918.

Four new berths under construction for steel ships up to 10,000 tons.

Taken over the plant of the Coquitlam Shipbuilding Co.

Gasoline yacht, 50 ft., June, 1918.

C. E. BAINTER, PRINCE RUPERT, B. C.

Several wooden fishing steamers, for undisclosed interests.

GRAND TRUNK DRYDOCK SHIP REPAIR CO., LTD., PRINCE RUPERT, B. C.

S. A. MOULTON, PRINCE RUPERT, B.C.

STANDARD SHIPBUILDING CO., RUSKIN, B.C.

Closed order March, 1918, 10 composite vessels for undisclosed interests.

BRITISH-AMERICAN SHIPBUILDING & ENG. CO., VANCOUVER, B.C.

Establishing plant on Kitsilano Reserve.

Twenty wooden vessels being built for undisclosed interests, 3,500 tons each.

JOHN COUGHLAN & SONS, N. VANCOUVER, B.C.

"War Camp," sister ship to "Alaska," launched in March, 1918.

One steel freight steamer, 427 ft. 9. x 54 ft. x 29 ft. 9 in., draft 24 ft. 2 in., speed 11½ knots, 5,730 gross tons. Launched Jan. 10, 1918, for B. Stolt Nielsen, Christiania, Norway. Both sold to the Cunard Line.

Three steel steamers, "War Chariot," "War Chief" and "War Noble," Nos. 4-5-6, 425 ft. x 54 ft., 8,800 tons cap., for the Cunard Line.

Six steel freight steamers, 425 ft. x 54 ft. x 24 ft. 2 in., 8,800 tons cap., 11 knots, for undisclosed interests. Delivery Jan., Feb., March, May and July, 1918.

One steel freight steamer, "Alaska," 427 ft. 9 in. x 54 ft. x 29 ft. 9 in., draft 24 ft. 2 in., speed 11½ knots, 5,730 gr. tons, 8,800 d. w. Launched Jan. 10, 1918, for B. Stolt Nielsen, Christiania, Norway. Sold to the Cunard Line.

THE FOUNDATION CO., VANCOUVER, B.C.

FRASER VALLEY SHIPBUILDING CO., VANCOUVER, B.C.

GRANT, SMITH & CO., VANCOUVER, B.C.

Six wooden cargo vessels, 250 ft. x 43 ft. 6 in. x 25 ft., 3,000 tons cap., 9½ knots, for the Canadian Government.

HARRISON & LAMOND, SHIPBUILDERS, LTD., VANCOUVER, B.C.

One wooden auxiliary schooner, 225 ft. x 44 ft. x 21 ft. 4 in., 1,600 gr. tons, 2,550 tons cap., for undisclosed interests.

LYALL SHIPBUILDING CO., NORTH VANCOUVER, B.C.

"War Puget," Feb. 15, 1918, 250 ft., wood, 2,080 tons.

Ten wooden cargo ships, for undisclosed interests. Launchings Jan. 20, Mar. 2, Mar. 23 and April 11, 1918.

Building six wooden vessels for own account.

"War Caribou," April 10, 1918, 3,080 tons, wood.

W. R. MANCHION, VANCOUVER, B.C.

1 tug, 100 tons, wood.

NORTHERN CONSTRUCTION CO., VANCOUVER, B.C.

STANDARD SHIPBUILDING CO., DOMINION BANK BUILDING, VANCOUVER, B.C.

Plant to be built at Ruskin, B.C.

Ten composite freight steamers for undisclosed interests.

Eight wooden ships, two for Brazilian Govt., and six for French Govt.

Donohoe reinforced type, wood composite construction, 3,500 and 4,500 tons.

TAYLOR ENGINEERING CO., VANCOUVER, B.C.

Number of small vessels being constructed at total value of $300,000.

4,500-ton floating drydock 352 ft. long.

VANCOUVER SHIPYARD, LTD., VANCOUVER, B.C.

One motor freighter, 125 ft. x 24 ft. x 12 ft.; wood, 160 h. p. Bolinder's crude oil engine. To be completed Aug., for Taylor Engr. Co.

Also repairing several steamers and schooners.

THE WALLACE SHIPYARDS, LTD., N. VANCOUVER, B.C.

"War Dog," 4,500 tons, steel, May 18, 1917.

"War Power," 4,600 tons, steel, March 23, 1918.

Two steel steamers, 315 ft. x 45 ft. x 27 ft., 4,600 tons d. w., trip. exp. vertical engines, two S. E. Scotch boilers, 10 knots, single screw, for the Canadian Government. Deliveries Dec., 1917, and Aug., 1918.

Two wooden auxl. schooners, 255 ft. o. a., 225 ft. keel x 44 ft x 21 ft. 4 in., five masts, 1,500 gr. tons, cap. 2,500 tons or 1,500,000 ft. lumber, two hatches, twin 160 h. p. Bolinders hot-bulb engines, twin screws, for the Canadian Government.

Four wooden cargo vessels, aggregate tonnage of 17,500 tons for undisclosed interests.

WESTERN CANADA SHIPYARDS, LTD., VANCOUVER, B.C.

"War Nootka," Jan. 4, 1918, 3,080 tons (machinery installed at Ogden Point Assembly Sheds).

Six wooden freight steamers, 250 ft. h. p., 259 ft. o. a., 43½ ft. extreme breadth, molded 42½ ft., 25 ft. molded depth, draft 22 ft., 2,800 tons d. w., for undisclosed interests. "War Nootka," launched January 4, 1918. "War Selkirk" launched Mar. 6, 1918.

"War Tatla," "War Casco," "War Chilkat," "War Tanoo, 4 standard vessels now under construction.

CAMERON-GENOA MILLS CO., VANCOUVER, B.C.

"War Yukon," Jan. 24, 1918, 3,080 tons, wood (machinery installed at Ogden Point Assembling Sheds.

THE FOUNDATION CO, OF B. C., LTD., VICTORIA, B. C.

Building ships for undisclosed interests.

Five wooden steamers, 250 b. p. x 42 ft. 6 in. x 25 ft., 1,800 gr. tons, 2,800 tons d. w., one trip. exp. engine, 1,000 h. p., "Howden" water tube boilers, furnished and installed by owners, 10 knots, for undisclosed interests. Esplen & Son & McNaught, N.Y.C., designers.

No. 1—"War Sonchee," launched Dec. 27, 1917.

No. 2—"War Masset," launched April 11, 1918.

No. 3—"War Babine," ready for launching, awaiting propeller and rudder.

No. 4—"War Camchin," ready for launching, awaiting propeller and rudder.

No. 5—"War Nanoose," ready for launching in about five weeks.

CLARENCE HOARD, VICTORIA, B.C.

Wooden car barge for C.P.R., Jan., 1918.

VICTORIA SHIPBUILDING CO., VICTORIA, B.C.

Constructing wooden ships for British Govt., 3,500 tons each.

* * *

C. T. WHITE & SON, LTD., ALMA, N.B. (Office at Sussex, N.B., also).

Two schooner aux. power, 143 ft., 900 tons d. w., first to be launched in April; second in June.

JAMES X. LENTEIGNE, LOWER CARARQUET, N.B.

One schooner, 28 tons, wood.

INTERNATIONAL SHIPBUILDING CO., NORDIN, N.B., (Head Office at Newcastle, N.B.).

One four-masted aux. schooner, 168 ft. o. a., 155 ft. x 37 ft. x 13 ft., 570 gross tons, for builder's account. Naval architect, G. M. Cochrane. Launching May, 1918. Oil engines.

One four-masted schooner 155 ft. x 37 ft. x 13 ft., 570 gross tons, for builder's account. Naval architect, G. M. Cochrane. Launching May, 1918.

EUREKA SHIPBUILDING CO., REXTON, N.B. (KENT CO.).

PORT COLBORNE BUILDING & REALTY CO., LTD., REXTON, N.B. (Head Office, Welland, Ont.).

Four-mast wooden schooner. Has cap. of 3 schooners per year.

GRANT & HORNE, ST. JOHN, N.B.

Two wooden cargo steamers, 250 ft. x 42 ft. 5 in. x 25 ft. 5 in., 1,000 h. p., speed 9½ knots, 2,800 tons d. w., for undisclosed interests. One keel laid Aug., 1917.

MARINE CONSTRUCTION CO., CANADA LTD., ST. JOHN, N.B.

Four wooden auxiliary, four-masted schooners, 185 ft. o. a., 165 ft. keel x 40 ft., 1,100 d. w. tons, for builder's account. J. Murray Watts, Philadelphia, naval architect. Launching April, 1918.

One wooden schr. 900 tons. Launching June, 1918.

PETER McINTYRE, ST. JOHN, N.B.

One schooner, 450 tons, wood.

ST. JOHN SHIPBUILDING CO., ST. JOHN, N.B. (Plant at Courtenay Bay.)

Ten five-masted auxl. schooners, oil burning engines, for undisclosed interests.

Establishing plant for steel and wood construction.

ST. MARTIN'S SHIPBUILDING CO., ST. MARTIN'S, N.B.

One wooden schooner, 450 tons, to be launched in the spring. This vessel to be followed by others.

* * *

ANGLO-NEWFOUNDLAND DEVELOPMENT CO., BOTWOOD, NFLD.

Two three-mast aux. schrs., wood, 450 tons, 150 H.P. for the builders' account.

NEWFOUNDLAND SHIPBUILDING CO., HARBOR GRACE, NFLD.

To build ships of 1,200 tons d.w. Steel ships later.

M. E. MARTIN, MORRIS ARM, NR. ST. JOHN'S, NFLD.

Two wooden schooners, 300 tons each, for builder's account.

One wooden schooner, 500 tons, for builder's account.

UNION SHIPBUILDING CO., ST. JOHN, NFLD.

ANNAPOLIS SHIPPING CO., ANNAPOLIS ROYAL, N.S.

"Hilda M. Clark," March 4, 1918, 187 ft., 640 tons reg.

Two schrs. three-masted, wood 500 tons, 170 ft. long. To be completed August and December, 1918.

B. L. TUCKER, BASS RIVER, N.S.

One schooner, 350 tons, wood.

JAS. S. CREELMAN, BASS RIVER, N.S.

One three-masted schooner, wood, 170 ft. long, 500 tons. To be completed October, 1918, for builders' account.

HANKINSON SHIPBUILDING CO., BELLIVEAU'S COVE, N.S.
One schooner, 360 tons, wood.

A. A. THERIAULT, BELLIVEAU'S COVE, N.S.
Two wooden three-masted tern schooners, 150 ft. x 33 ft. x 11 ft, 11 in., 400 gr. tons, gasoline engines, for builder's account. Launchings April and June, 1918.

One wooden tern auxl. schooner, No. 2, 143 ft. x 33 ft. x 12 ft., 400 gr. tons. Launching Sept., 1918. For the Hankinson Shipping Co., Belliveau's Cove.

One wooden one-masted schr.

ELI PUBLICOVER, BLANDFORD, N.S.

BRIDGEWATER SHIPPING CO., BRIDGEWATER, N.S.
One tern schooner, 450 tons gr., wood. Adapted for aux. power. To be completed Sept. 1918.

One tern schooner, 275 tons gr., wood. Adapted for aux. power. To be completed No., 1918.

Both built on owner's account and for sale.

L. S. CANNING, BROOKVILLE, N.S.
One schooner, three-masted, 300 tons.

W. A. NAUGLER, BRIDGEWATER, N.S.
One wooden three-masted schooner, "Wm. A. Nauglen," 146 ft. o. a., 116 ft. keel x 32 ft. x 11 ft. 6 in., 360 gr. tons, 600 tons cap., for builder's account.

One schooner, 300 tons, wood.

H. MAC ALONEY, CANNING, N.S.
Two wooden schooners, 400 tons each. For builder's account. Launchings July and October, 1918.

S. M. FIELDS, CAPE D'OR, N.S.
One three-masted schooner, 350 tons.

Keel laid for another three-master.

CHESTER BASIN SHIPBUILDERS, LTD., CHESTER BASIN, N.S.
Two schooners, 100 tons, wood. Launchings Aug., 1917, Dec., 1917.

One four-masted schooner, 600 tons. Launching Sept., 1918.

One two-masted schooner, 100 tons. Launching June, 1918.

MORTIMER PARSONS, CHEVERIE, N.S.
One wooden three-masted schooner, 170 ft. x 38 ft. x 14 ft., 575 gr. tons, 900 tons cap., for builder's account. Launching spring, 1918.

Two wooden three-masted schooners for builder's account.

One tern schooner, "Donald Parsons," 140 ft. keel x 34 ft. 4 in. x 12 ft. 4 in., 500 gross tons, 825 tons d.w. for Mortimer Parsons. Launched May 11, 1918.

Another vessel, same size, to be launched Nov., 1918, for builder's account.

MOISE BELLIVEAU, CHURCH POINT, N.S.
One schooner, 450 tons, wood.

FIDELE BOUDREAU, CHURCH POINT, N.S.
One schooner, 350 tons, wood.

CONEAU SHIUBUILDING CO., COMEAUVILLE, N.S.
One three-masted wooden schooner, for builder's account, 450 tons.

J. W. COMEAU, COMEAUVILLE, N.S.
One schooner, 329 tons, Jan., 1918.

J. N. RAFUSE & SONS, CONQUERALL BANKS, DIGBY CO., N.S. (Other plants at Salmon River and Shelburne, N.S.)
Two wooden three-masted schooners, 120 ft. x 32 ft. x 12 ft., 700 tons cap., for builder's account. For the market. Launchings Oct. 25, 1917, and Jan. 10, 1918.

One three-masted wooden schooner, "Integra," 112 ft. x 30 ft. x 11 ft. 6 in., 600 tons. cap., for J. O. Williams & Co., St. John's, Newfoundland. Launching Nov. 25, 1917.

E. F. WILLIAMS, DARTMOUTH, N.S.
One schooner, 350 tons, wood.

ROBAR BROTHERS, DAYSPRING, N.S.
One schooner, 130 tons, Jan., 1918.

MAURICE LEARY, DAYSPRING, N.S.
Schooner for Capt. Ivan Creaser, Dayspring.

J. NEWTON PUGSLEY & CHAS. ROBERTSON, DILIGENT RIVER, N.S.
One schooner, three-masted, 475 tons.

McLEAN & McKAY, ECONOMY, N.S.
One schooner, 150 tons, wood.

S. J. SOLEY, FOX RIVER, N.S.
One schooner, three-masted, 121 ft. keel, 300 tons. Built of native wood. To be launched Sept., 1918. For builder's account. For sale.

ALLAN & FRASER, FRASERVILLE, N.S.
One schooner, 350 tons, wood.

BERNARD W. MELANSON, GILBERT'S COVE, N.S.
270-ton schooner, three-masted, for builder's account. To be completed Oct., 1918.

AMOS BLINN, GROSSES COQUES, N.S.
One schooner, 375 tons. Jan., 1918.

One schooner, 360 tons, wood.

BINN BROS., GROSSES COQUES, N.S.

F. K. WARREN & CO., GROSSES COQUES, N.S. (Offices in Union Bank Chambers, Halifax, N.S. also.)
350-ton tern schooner, under construction Jan., 1018.

THE HALIFAX SHIPBUILDING CO., HALIFAX, N.S.

JOHN McLEAN & SON, HALIFAX, N.S.
One tug, 370 tons, wood.

FAUQUIER & PORTER, HANTSPORT, N.S.
Two wooden four-masted schooners, 178 ft. x 40 ft. x 18 ft., 1,000 gross tons, 2,000 tons d. w. twin 100 h.p. oil engines. Launchings Aug. 1, 1918, and Sept. 1, 1918. Both for sale.

FOLEY BROS., WELCH, STEWART & FAUQUIER, HANTSPOROT, N.S.
Two wooden schooners, for builder's account.

Improving North's shipyard, recently purchased.

Three four-masted schooners, 900 tons each.

SHIPBUILDING & TRANSPORTATION CO., HANTSPORT, N.S.

J. WILLARD SMITH, HILLSBURN, N.S. (Head Office, St. John, N.B.)
One tern schooner, wood, 140 ft. keel, 35 ft. beam, 450 tons net reg. tons.

HENRY COVEY, INDIAN HARBOR, N.S.

CHARLES GRIFFIN, ISAACS HARBOR, N.S.
One schooner, 40 tons, wood.

LEWIS SHIPBUILDING CO., LEWISTON, N.S.
One schooner, 670 tons. Jan., 1918.

INNOCENT COMEAU, LITTLE BROOK, N.S.
One three-masted schooner of 350 tons cap., for J. E. Gaskill, of Grand Manan, N.S., Can.

One wooden schooner, 140 ft. x 35 ft. 6 in. x 17 ft. 6 in., 650 gr. tons, for the Weymouth Shipping Co., Weymouth, N.S., Can. Delivered Jan. 18, 1918.

S. ST. C. JONES, LITTLE BROOK, N.S.

J. W. RAYMOND, LITTLE BROOK, N.S.
One wooden schooner, No. 4, 140 ft. x 36 ft. x 17 ft., 575 gr. tons, for Jones Bros., Weymouth, Digby Co., N.S.

H. A. FRANK, LIVERPOOL, N.S.
Two wooden schooners, 500 tons, for builder's account.

McKEAN SHIPBUILDING CO., LIVERPOOL, N.S.
One wooden three-masted schooner, 126 ft. keel x 33 ft. x 12 ft. 9 in., 700 tons d. w., for builder's account. Launching May, 1918.

W. F. McKEAN & CO., LIVERPOOL, N.S.
One schooner 400 tons, Jan., 1918.

D. C. MULHALL, LIVERPOOL, N.S.
Two wooden schooners, one 300 gross tons, other 400 gross tons, for undisclosed interests.

N.S. SHIPBUILDING & TRANS. CO., LIVERPOOL, N.S.
One wooden tern schooner No. 2; 122 ft. x 3 3ft. 12 ft. 6 in., 450 gr. tons, 70 d. w. tons, 7 knots, for Peter Yee Wing & Co., Ltd., Sydney, Australia. R. McLeod and J. S. Gardner. Launching Feb. 1, 1918.

One two-masted fishing schooner, "Sadie A. Nickle," 130 ft. o. a. x 26 ft. 10 ft. 6 in. hold, 250 d. w., two 50 h.p. oil engines, for Rafuse & Sons. Launching May 15, 1918.

One 3 mast schooner, 420 tons, F-M Co. oil burning engines, for Job Bros., St. Johns, Nfld. Launching July 15.

One 3-masted schooner, 270 tons, for builders' account. Launching Sept. 1.

One two-masted fishing schooner 120 tons for builders account. Launching Oct. 1.

ROBIN, JONES & WHITMAN, LIVERPOOL, N.S.
One schooner 340 tons, wood.

SOUTHERN SALVAGE CO., LIVERPOOL, N.S.
One wooden schooner, three masts, 300 gross tons, 600 tons capacity, for a West Indian firm. Launching Jan., 1918.

One wooden steamer, No. 5, 250 ft. x 43 ft. 6 in. x 25 ft., 2,900 gr. tons, 10 knots, 1,000 h.p., for undisclosed interests. Launching Fall, 1918.

One two-masted schooner, "Win-the-War," 137 ft. o. a. x 26 ft. 2 in. x 11 ft. 6 in., 187 gr. tons, for the builder's account. Launched Nov., 1917.

CONRAD & REINHARDT, LUNENBURG, N.S.

LUNENBURG MARINE RAILWAYS LUNENBURG, N.S.

McLEAN CONSTRUCTION CO., LUNENBURG, N.S.

FRED A. ROBAR, LUNENBURG, N.S.

SMITH & RHULAND, LUNENBURG, N.S.
Two schrs. 225 tons each. Jan., 1918.

Two schooners 100 tons each.

"Donald Cook," 112 ft. beam, depth of hold, 11 ft., for Capt. William Cook. Launched April, 1918.

Two schooners, 168 gr. tons, for W. C. Smith.

One schooner, 150 gr. tons, for Lun. Outfitting Co.

J. B. YOUNG, LUNENBURG, N.S.

ERNST SHIPBUILDING CO., MAHONE BAY, N.S.
One two-masted schooner, 120 ft. o. a. x 25 ft. 6 in. x 10 ft. 6 in., 200 tons d. w., for Lunenburg Outfitting Co. April, 1918 Christened "Madeline Adams."

One three-masted schooner 125 ft. keel x 32 ft. x 12 ft., 600 tons, d. w. To be delivered July, 1918.

Two-master schooner, "Agnes D. McGlaston," 130 ft. x 26 ft. 8 in. x 10 ft. 8 in. Launched Nov. 13, 1917.

Keel laid for 3-masted wooden schr., 200 tons.

J. ERNEST & SON, MAHONE BAY, N.S.
One schooner, 520 tons. Jan., 1918.

O. A. HAM, MAHONE BAY, N.S.
One schooner, "Dozle," 128 tons. Launching May 30. One motor boat about 20 tons.

McLEAN SHIPBUILDING CO., MAHONE BAY, N.S. (Leased to Montague Mahaffy, Toronto.)
Two wooden schooners.

Plant enlarged to build 3 ships at one time.

JOHN McLEAN & SONS, MAHONE BAY, N.S.
One schooner, 95 tons, wood.

J. A. BALSOM CO., LTD., MARGARETSVILLE, N.S.
One schooner, 409 tons. Jan., 1918.

CLARE SHIPBUILDING CO., METEGHAN RIVER, N.S.
One wooden schr. for builder's account. 400 tons.

A. H. COMEAU & Co., METEGHAN, N.S.
One schooner, 400 tons, wood.

AGAPIT COMEAU, METEGHAN, N.S.

JOHN F. DEVEAU, METEGHAN, N.S.
A 300 ton schooner was launched recently for Ritcey & Co., Lunenburg, N.S., and named Charles A. Ritcey.

1 schooner, 425 tons. Jan., 1918.

1 schooner, 400 tons, wood.

J. E. GASKELE, METEGHAN, N.S.

THOMAS GERMAN, METEGHAN, N.S.

One schooner, 350 tons, wood.

R. H. HOWES CONSTRUCTION CO., METEGHAN, N.S. (Leased James Cosman's shipyards.)

Building several wooden schooners for own account.

DR. F. H. MACDONALD, METEGHAN, N.S.

One wooden four-masted schooner, "Rebecca L. Macdonald," 201 ft. c. a. x 36 ft. x 16 ft., 800 gr. tons, 1,500 tons d. w., for builder's account. Launching January 1, 1918.

One wooden fishing schooner, No. 3, 84 ft. x 1 7ft. x 7 ft., 50 gr. tons, 35 h.p. Launching August, 1918.

One schooner, 544 tons, wood.

METEGHAN RAILWAY & SHIPBUILDING CO., METEGHAN, N.S.

One wooden three-masted schooner, 182 ft. x 35 ft., 450 gr. tons.

One schooner, 470 tons, wood.

CHAS. McNEIL, NEW GLASGOW, N.S.

One wooden 3-masted schooner, 103 ft. x 30 ft. x 10 ft. 9 in., 300 tons, for builder's account. Launching July, 1918.

One wooden 3-masted schooner, 142 ft. x 35 ft. 4 in. x 13 ft., 500 gr. tons, for builder's account. Launching Oct., 1918.

THE NOVA SCOTIA STEEL & COAL CO., NEW GLASGOW, N.S.

Two steel freight steamers, 257 ft. 9 in. x 35 ft. x 20 ft., 1,700 gr. tons, 2,850 tons cap., trip. exp. engines, boilers 10 ft. 6 in. x 11 ft. 6 in., 800 h.p., 8.5 knots, for undisclosed interests Launching of one Jan., 1918.

"War Wasp," July 9, 1917, 1,800 tons, steel.

Two steel cargo strs. Raised quarter deck type, 248 ft. 9 in. long, 1,649.68 gross tons. Triple expansion engines, 17, 28, 46 x 33 in. stroke. One, the "War Bee," for undisclosed interests to be completed June, 1918. Other for Steel Co., to be completed Oct., 1918.

O'BRIEN BROS., NOEL, N.S.

One schooner, 325 tons, wood.

W. R. HUNTLEY & SON, PARRSBORO, N.S.

One wooden schooner, 175 ft. x 39 ft. x 17 ft., 900 gross tons, for C. T. White & Son, Sussex, N.S., Can.

Two schooners, 325 tons each. Jan., 1918.

One schooner, 490 tons, wood.

One schooner, 850 tons, four masted.

WAGSTAFF & HATFIELD, PARRSBORO, N.S.

One schooner, 400 tons. Jan., 1918.

SIDNEY ST. C. JONES, PLYMPTON, N.S., (Head Office, Weymouth North, N.S.).

One tern schooner, 200 tons, net reg., 107 ft. x 28 ft. 8 in. x 10 ft. 8 in., to be launched Nov., 1918.

S. SALTER, PARRSBORO, N.S.

One 200-ton schooner, wood.

DOWLING & STODDART, PORT CLYDE, N.S.

One gas boat, 27 tons, wood.

One schooner, 300 tons, wood.

SWIME BROS., PORT CLYDE, N.S.

Several small motor trawlers, for builder's account.

G. M. COCHRANE, PORT GREVILLE, N.S.

One tern schooner, "Alfredock Hedley," 152 ft. 6 in. x 36 ft. x 12 ft. 6 in., 461 tons reg., for Adam B. Mackay, of Hamilton, Ont. Launched Jan., 1918.

One 4-masted schr., 850 tons, 155 ft. x 37 ft. x 18 ft., 2 decks, wood. To be completed Oct., 1918, for builder's account.

H. ELDERKIN & CO., PORT GREVILLE, N.S.

One wooden schooner, four masts, 550 tons.

ELLIOTT GRAHAM, PORT GREVILLE, N.S.

One wooden three-masted schooner, for undisclosed interests.

L. E. GRAHAM, PORT GREVILLE, N.S.

Is building a schooner, "Khaki Lad," of about 325 tons, at Port Greville, N.S., for J. W. Kirkpatrick and others. She will be launched early in October, and is reported to have been sold to Newfoundland parties.

One schooner, 360 tons. Jan., 1918.

SMITH CANNING CO., PORT GREVILLE, N.S.

One schooner, 350 tons, wood.

WAGSTAFF & HATFIELD, PORT GREVILLE, N.S.

One schooner, 400 tons, wood.

WILLIAM CROWELL, PORT LATOUR, N.S.

J. W. RAYMOND, PORT MAITLAND, N.S.

One schooner, 375 tons. Jan., 1918.

PORT WADE SHIPBUILDING CO., PORT WADE, N.S. (Head Office, Digby.)

One 850-ton schooner, wood.

JOHN BROWN, PUBLIC LANDING, N.S.

One tow barge, 50 tons, wood.

THE CUMBERI AND SHIPBUILDING CO., PUGWASH, N.S.

Establishing plant for the construction of wooden ships.

MACKENZIE SHIPPING CO., RIVER JOHN, N.S.

One 4-masted schooner 600 tons, wood. About half built. Fitted for aux. engines.

CHARLES McLELLAN, RIVER JOHN, N.S.

One schooner, 100 tons, wood.

W. J. FOLEY, SALMON RIVER, N.S.

J. N. RAFUSE & SONS, SALMON RIVER, N.S. (Other plants at Shelburne and Conquerall Banks.)

Launched recently a three-masted schooner named "Industrial," at W. J. Foley's ship yard at Salmon River, N.S., of the following dimensions: length 113 ft., breadth 30 ft., depth 11½ ft., 325 tons.

ACADIAN SHIPPING CO., SAULNIERVILLE, N.S.

One schooner, 400 ton, wood.

SAULNIERVILLE SHIPBUILDING CO., LTD., SAULNIERVILLE, N.S.

J. LEWIS & SONS, SHEET HARBOR, N.S.

One schoner, wood, four-masted, 725 gr. tons. To be completed Sept., 1918. For private account.

GEO. A. COX, SHELBURNE, N.S.

One schooner, 200 tons, wood.

One schooner, 322 tons, wood.

JOSEPH McGILL SHIPBUILDING & TRANSPORTATION CO., SHELBURNE, N.S.

"Sparkling Glance," 246 tons register, for Harvey & Co., St. John's, Nfld.

One schooner, 160 tons, wood.

W. C. McKAY & SON, SHELBURNE, N.S.

One wooden schooner, 130 ft. o. a. x 28½ ft. x 11 ft. depth of hold, 140 tons register. Launched Dec. 1, 1917.

One three-masted schooner, for Messrs. Hallet, of Newfoundland.

One schooner, 620 tons, wood.

J. N. RAFUSE & SONS, SHELBURNE, N.S. (Other plants at Salmon River and Conquerall Banks.)

One wooden three-masted schooner, 120 ft. x 32 ft. x 12 ft., 700 tons cap., for builder's account. For the market. Launching Jan. 10, 1918.

One wooden three-masted schooner, 118 ft. x 33 ft. x 12 ft., 700 tons cap., for builder's account. For the market. Launching Jan. 15, 1918.

Two three-masted wooden schooners, 122 ft. 6 in. keel x 32 ft. x 12 ft., 275 net tons, for J. O. Williams & Co., St. John's, Newfoundland. One to be built at Salmon River yard and one at Ship Harbor.

THE SHELBURNE SHIPBUILDERS, LTD., SHELBURNE, N.S.

One tern schooner, "Misty Star," 145 ft. x 31 ft. x 11 ft. 6 in., 600 tons d. w., 330 tons gross, for Harvey & Co., St. John's, Newfoundland. Launching Mar. 15, 1918.

One tern schooner, 154 ft. x 32 ft. x 12 ft. 4 in., 400 gr. tons, 700 tons d. w. cap. Launching June, 1918. Not sold yet.

Two schooners, one 349 tons, other 400 tons.

EASTERN SHIPBUILDING CO., SHIP HARBOR, N.S.

One wooden schooner, 600 tons d. w., for J. N. Rafuse & Sons, Halifax, N.S., Can.

One wooden schooner, 300 tons.

STEPHEN MORASH & CO., SHIP HARBOR, N.S.

One wooden schooner, 160 ft. x 37 ft. x 13 ft., four masts, for Canadian interests. Keel laid Feb. 5, 1917.

JAMES E. PETTIS, SPENCER'S ISLAND, N.S.

One schooner, 425 tons, wood, three-masted.

THE TUSKET SHIPBUILDING CO., TUSKET, N.S.

AMOS H. STEVENS, TANCOOK, N.S.

ALVIN STEVENS, TANCOOK, N.S.

STANLEY MASON, TANCOOK, N.S.

ALBERT PARSONS, WALTON, N.S.

One schooner, 400 tons. Jan., 1918.

H. T. LaBLANC, WEDGEPORT, N.S.

One wooden schooner for undisclosed interests.

600 ton schooner, 165 feet, steam.

WEDGEPORT NAVIGATION & TRANSPORTATION CO., WEDGEPORT, N.S.

One 200-ton steamer.

T. K. BENTLEY, WEST ADVOCATE, N.S.

One 4-masted schooner, wood, 511 tons. To be launched September or October. For owner's account.

J W. KIRKPATRICK, WEST ADVOCATE, N.S.

One 350-ton schooner, wood.

HOEHNER BROS., WEST LA HAVE, N.S.

BEAZLEY BROS., WEYMOUTH, N.S. (Head Office Roy Bldg., Halifax.)

One schooner, 394 tons, three masts.

E. R. GAUDET, WEYMOUTH, N.S.

One 350-ton schooner, three-masted.

E. P. RICE, WEYMOUTH, N.S.

One three-masted schooner, 350 tons.

RICE, WARREN & CO., WEYMOUTH, N.S.

One three masted schooner, 334 tons.

WESTPORT SHIPBUILDING CO., WHITE'S COVE, N.S., DIGBY CO.

Three-masted schooner, building.

FALMOUTH SHIPBUILDING & TRANSPORTATION CO., WINDSOR, N.S.

One wooden schooner, 350 gr. tons, 700 tons cap., for undisclosed interests.

NOEL SHIPBUILDING & TRANSPORTATION CO., WINDSOR, N.S.

One wooden schooner, three masts, 138 ft. x 35 ft. x 13 ft., 450 net tons. Oil auxiliary engines. Launching August, 1918, for builder's account.

C. A. NICKERSON, WOODS HARBOR, N.S.

MILTON SHIPBUILDING CO., YARMOUTH, N.S.

Taken over old plant of James Jenkins. Enlarging plant.

W. O. SWEENY, YARMOUTH, N.S.

100-ton fishing schooner.

YARMOUTH SHIPBUILDING CO., YARMOUTH, N.S.

Two wooden schooners, 150 ft. o. a. x 33 ft., for the builder's account. One to be launched October 1, 1918. 800 tons d. w., 3 masts.

* * *

CAN. ALLIS-CHALMERS (Head Office, Toronto), BRIDGEBURG, ONT.

Six standard freight steamers, 3,500 tons, 261 feet long, for undisclosed interests.

COLLINGWOOD SHIPBUILDING CO., COLLINGWOOD, ONT.

Building ships for undisclosed interests.

Two steel cargo steamers, Nos. 51-52, 50 ft. x 43 ft. x 25 ft., 2,500 gr. tons 2,900 tons d. w., tripple expansion engines 18-30-50 x 36, 2 boilers 14 ft. x 11 ft., 10 knots, for undisclosed interests. Delivery May and Aug., 1918.

Four deep-sea trawlers, Nos. 53-58, inclusive, 125 ft. x 23 ft. 6 in. x 13 ft. 6 in., 288 gross tons, 12¾-21½-35/24 engines, 1 cylinder 13 ft. 6 in. x 10 ft. 6 in., 10 knots, 500 h.p. for undisclosed interests.

One steel freight steamer of 3,800 tons d. w. for undisclosed interests. Keel laid May 8, 1918.

"War Wizard," May 8, 1918, 3,000 tons d. w., 261 ft. x 43 ft. 6 in. x 20 ft. depth. For undisclosed interests.

"War Witch" same size, now building. To be launched ——

R. MORRILL, COLLINGWOOD, ONT.

"Windsor," Aug. 10, steam tug, 105 ft., for Ontario Gravel & Freighting Co.

* * *

PORT ARTHUR SHIPBUILDING CO., PORT ARTHUR, ONT.

Seven steel freight steamers, 261 ft. x 251 ft. o. a. keel x 43 ft. x 23 ft., 2,030 gr. tons, 3,400 tons cap., single deck with poop, bridge and forecastle, two Scotch boilers 14 ft. 8 in. x 11 ft., trip. exp. engines 20-33-54 x 40, 1,200 h.p., six reversible steam winches 7 in. x 12 in., six derrick booms of 4 tons cap., two for the Great Lakes Transportation Co., five are for undisclosed interests.

Six steel trawlers, 125 ft. x 23 ft. 6 in. x 13 ft. 6 in., trip. exp. engines, Scotch boiler, for the Canadian Government.

*Five steel freighters, 261 ft. x 43 ft. x 23 ft.

"Ugelstad."

"War Isis," April 3, 1918, 3,400 tons d. w. Trawler launched same day. April 3, 1918, keel laid for "War Hatha," sister ship to "War Osiris," of same construction, nearly completed.

"War Fish," Aug. 19, 1917, 4,300 tons, steel.

"War Dance," Nov. 3, 1917, 3.400 tons, steel.

Their report:

Four steel screw steamers, ocean freight service—"War Osiris," "Wa Hathor," "War Karma," "War Horus," 3,400 tons d. w. each, 261 ft. o.a. 251 ft. b.p. in length. Triple expansion engines, 2 Scotch boilers each, approx. 1,250 h.p. For completion, two Aug. 31st, and two Nov. '20 1918. For undisclosed interests.

CAN. CAR & FOUNDRY CO., FORT WILLIAM, ONT. (Head Office Transportation Bldg., Montreal.)

Have orders for:

Twelve steel mine sweepers for French Government.

145 ft. o.a. steel construction, value $2,500,000.

Plant under construction.

GREAT LAKES DREDGING CO., FORT WILLIAM, ONT.

"War Sioux," May 12, 1918, 1,700 tons, wood.

THUNDER BAY CONTRACTING CO., FORT WILLIAM, ONT.

One wooden freight steamer, 261 ft. long, for undisclosed interests.

NATIONAL SHIPBUILDING CO., GODERICH, ONT.

GODERICH SHIPBUILDING CO., GODERICH, ONT.

DAVIS DRYDOCK CO., FT. OF BAY ST., KINGSTON, ONT.

One motor pass. boat, 64 ft. x 18 ft. x 5 ft., 52 gr. tons, wood. For ferry service, Gananoque to Clayton. Delivery June 1, 1918.

Twenty-eight lifeboats, regulation size, wood and metal for Government and private interests.

KINGSTON SHIPBUILDING CO., KINGSTON, ONT.

Several steel trawlers for undisclosed interests. First one launched Dec. 22, 1917.

SELBY & YOULDSON, KINGSTON, ONT.

GEORGIAN BAY SHIPBUILDING & WRENCKING CO., MIDLAND, ONT.

Tug, 50 tons, wood.

One tug 40 tons, wood.

MIDLAND DRY DOCK CO., MIDLAND, ONT.

Three steel freight steamers, 261 ft. x 43 ft. 6 in. x 23 ft., 3,400 tons d. w., 10 knots, for undisclosed interests. Deliveries, one in July, 1918, and two others before close of navigation, 1918.

Alterations on steamers "Mariska" and "Glenlyon."

"Western Star" being reconstructed.

MIDLAND SHIPBUILDING CO., LTD., MIDLAND, ONT.

Building ships for undisclosed interests.

Three steel freight steamers, 261 ft. x 43 ft., 6 ft x 23 ft., 3,400 tons d. w.. 10 knots for undisclosed interests. Deliveries, one in July, 1918, and two others before opening of navigation, 1919.

MUIR BROS. DRYDOCK CO., LTD., PORT DALHOUSIE, ONT.

J. W. GEROW, ROSSPORT, ONT.

REID WRECKING CO., SARNIA, ONT.

*One steel tug, 157 ft. x 32 ft. x 19 ft., trip. exp. engines, for lake or ocean service. Keel blocks laid March 5. Launching July, 1918. For Reid Wrecking Co.

WEST, PEACHEY & SONS, SIMCOOE, ONT.

Number of tugs for lumbering interests.

DOMINION SHIPBUILDING CO., DINNICK BLDG., TORONTO.

One steel cargo vessel, 261 ft. x 43 ft., 3,500 tons d. w. cap. for Department of Marine, Ottawa.

Have orders for five other vessels same dimensions.

THE POLSON IRON WORKS, LTD., TORONTO, ONT.

Eight steel cargo steamers, Nos. 133-4-5-6-7-9-40, 216 ft. o. a., 251 ft. b. p. x 43 ft. 6 in. x 22 ft. 11 in., draft loaded 19 ft. 6 in., single decked, 2,350 approx. gross tons, 3,500 tons cap., single screw trip, exp. engine 20½-33-54 x 36, 1,250 h.p., 10 knots, two 14 ft. x 12 ft. Scotch R. T. boilers, 180 h.p., to cost $600,000 each, for undisclosed interests. Launchings throughout 1918.

Four building and four on order.

"Asp," Feb. 11, 1918, 261 ft., 3,500 tons d.w., steel.

"Tento," Oct. 22, 1917; 261 ft., 3,500 tons d.w. steel. For Norwegian interests. Transferred to British registry.

THE THOR IRON WORKS, TORONTO, ONT.

Two steel cargo vessels, 261 ft. long, 2,437 tons, for American interests. John Inglis Co., Toronto, supplying machinery.

Two steel trawlers, 270 tons.

TORONTO SHIPBUILDING CO., LTD., TORONTO, ONT. (Toronto Dry Dock Co., Ltd., under same management.)

Two wooden cargo steamers, 250 ft. B.P. 42 ft. 6 in., moulded breadth 25 ft., moulded depth 22 ft., draft 2,500 tons d. w. Triple exp. engines $\frac{20\times33\times54}{40}$. Howden boiler. Launchings July and Sept., 1918. For undisclosed interests.

THE BRITISH AMERICAN SHIPBUILDING CO., WELLAND, ONT.

Two steel frs., 3,500 tons d. w., 261 ft. O.A., 43 ft. beam, 23 ft. moulded depth, Westinghouse steam turbine engines. Delivery 1918.

WELLAND SHIPBUILDING CO., WELLAND, ONT.

One steel freight steamer, "War Weasel," 261 ft. x 43 ft. 6 in. x 23 ft., 3,300 tons d. w., trip. exp. engines, 14 ft. x 12 in. Scotch boilers, 10 knots, 1,250 h. p. Launching April, 1918.

Two deep frame cargo vessels, "War Badger" and "No. 3," 261 ft. x 43 ft. 6 in. x 23 ft., 3,300 d. w. tons, geared turbines, Howden's forced draught boilers, 10 knots, 1,250 h.p. Launchings June and August, 1918.

* * *

J. A. McDONALD, CARDIGAN, P.E.I. (G. A. Thompson, Montague, and Chas. Lyons, Charlottetown, also interested. Purchased Annandale Lumber Co. plant and removed to Cardigan.)

One three-masted schooner, 325 tons, to be completed November, 1918.

Two more to be built.

TIDEWATER SHIPBUILDERS, LTD., THREE RIVERS, QUE.

(Formerly Sorel Shipbuilding & Coal Co.)

Three steel trawlers for undisclosed interests

One steel steam barge for the Canada Steamships Lines.

R. N. LE BLANC, BONAVENTURE, QUE.

J. Z. DEGAGNE, EBOULEMENTS, QUE.

DAVIE SHIPBUILDING & REPAIRING CO., LEVIS, QUE.

Six military barges, 130 feet long.

One steel car ferry, 5,000 tons d. w.

Eight steel trawlers, 2,035 tons.

One floating crane, 350 tons.

*One steel cargo vessel of 5,000 tons cap.

Building several steel lighters and several wooden drifters.

ATLAS CEMENT CONSTRUCTION CO., LTD., MONTREAL, QUE.

One concrete cargo steamer, "Concretia," 125 ft. x 22 ft. x 13 ft., steel ribbed, hull to be from 3 in. to 5 in., thick, for the builder's account.

CANADIAN VICKERS, LTD., MONTREAL, QUE.

Two cargo steamers, 9.400 tons, steel; 1 dredge, 2,364 tons, steel; 12 trawlers, 3,050 tons, steel; 23 drifters, 3,350 tons, wood.

Six freight steamers, 24 ft. draught, 7,000 tons cap., 11 knots, for undisclosed interests. Delivery, 1917, of two for Norwegian interests. "Porsanger," 394 ft. 6 in. o. a. x 40 ft. 4 in. x 30 ft., triple exp. engines, launched Nov. 29, 1917. Two are for the Imperial Munitions Board.

One steel freight steamer, 2,360 tons, cap., for undisclosed interests.

Three steel cargo vessels, 8,200 tons, to be laid down in May, Aug. and Sept., 1918, for undisclosed interests.

FRASER BRACE & CO., MONTREAL, QUE.

Four wooden cargo steamers, 3,000 tons, for undisclosed interests. Two keels laid during Oct., 1917.

HALL ENGINEERING CO., MONTREAL, QUE.

MONTREAL DRY DOCK & SHIP REPAIRING CO., MONTREAL, QUE.

Operate dock 428 ft. long 30 ft. deep.

MONTREAL SHIPBUILDERS LTD., 37 Belmont St., MONTREAL, QUE. (Associated with Atlas Construction Co.

THE QUEBEC SHIPBUILDING AND REPAIR CO., Board of Trade Bldg., MONTREAL, QUE.

2 vessels for undisclosed interests 3,000 tons each.

QUINLAN & ROBERTSON, MONTREAL, QUE. (Yards at Quebec.)

Four wooden steamers, for undisclosed interests.

LOUIS GAUNDRY, 12 St. Peter St., QUEBEC, QUE.

QUEBEC SHIPBUILDING & REPAIR CO., ST. LAURENT, QUE.

Building ships for undisclosed interests

2 schrs. 1,400 tons and 1,200 tons.

One wooden four-masted auxl. schooner "Martin Connolly," 223 ft. x 42 ft. x 20 ft., 2,100 tons d. w., for undisclosed interests. Launched Oct. 28, 1917.

QUINLAN & ROBERTSON, QUEBEC, QUE.

Four wooden steamers, totalling 6,400 tons, for undisclosed interests.

CANADIAN GOVERNMENT SHIPYARDS, SOREL, QUE.

One steel vessel for undisclosed interests.

LECLAIRE SHIPBUILDING CO., SOREL, QUE.

Building 6 steel ships at value of $1,500,000.

Six trawlers, 125 ft. single screw type expansion engines, 500 I.P.H. steel. To be completed this year for private acct.

H. H. SHEPHERD, SOREL, QUE.

Rebuilding seven drifters. Particulars withheld owing to govt. restrictions.

SINCENNES-McNAUGHTON LINES, SOREL, QUE.

One tug 410 tons, wood.

CAN. GOVERNMENT SHIPYARD, SOREL, QUE.

One steel vessel for undisclosed interests.

Building steam trawlers and wooden drifters for undisclosed interests.

ST. LAWRENCE SHIPBUILDING & STEEL CO., SOREL, QUE.

THE THREE RIVERS SHIPYARDS LTD., THREE RIVERS, QUE.

Two cargo steamers, 3,100 d. w., wood, 260 ft. x 43 ft. 6 in., 1,000 H.P. engines. Names, "War Mingan" and "War Nicolet." To be completed end of navigation season for undisclosed interests.

The NEW "EMCO" GLOBE VALVE

AND THE PLACE WHERE IT'S MADE

Just take a look at our **NEW EMCO** Globe Valve and note carefully a few of its good features. The long, full threads on spindle, the uniform thickness of metal.

It can be packed when open, is fitted with metal packing gland and is complete in every detail.

Every valve is packed with high grade packing before it leaves the factory.

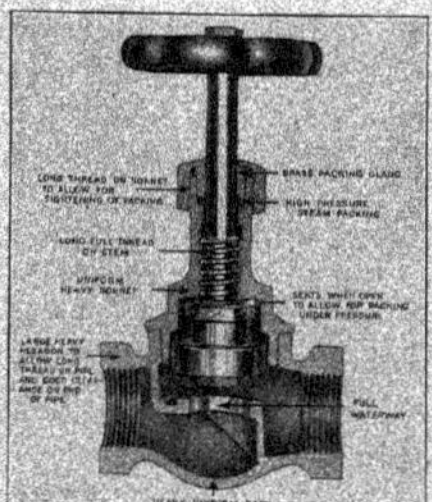

This is only one of the many new lines we have made recently to fill some long-felt wants for high-grade valves.

We have one of the most modernly equipped brass manufacturing plants under the British flag.

We can make prompt shipments from stock. Let us quote you our prices on your next brass goods requirements.

When writing us or ordering from your jobber, ask for the EMCO GLOBE VALVE. It's a winner and we're proud of it.

EMPIRE MANUFACTURING COMPANY, LIMITED

LONDON TORONTO

ASSOCIATION AND PERSONAL

A Monthly Record of Current Association News and of Individuals Who Have Been More or Less Prominent in Marine Circles

N. W. VanWyck has been appointed purchasing agent of the Canada Steamship Lines at Montreal, vice P. Paton, resigned to engage in other business.

J. G. Reid, assistant superintendent of dredging of the Montreal Harbor has just completed his 24th year in the service of the Montreal Harbor Commissioners.

James M. Smith now has charge of the shipbuilding plant of the Canadian Car & Foundry Co., Fort William, Ont., which expects to turn out 12 trawlers for the French Government before the end of the 1918 season.

Capt. T. S. Scott, No. 5 F.C.C.E., has been ordered to report at Halifax to aid in the reconstruction work in that city. Capt. Scott has been professor of civil engineering at Queen's University, Kingston, for several years.

Capt. John B. Forrest, sixty-nine years old, master mariner for more than thirty years, died recently at his home in Walkerville, Ont. Death was due to heart disease. A son of the late Capt. Robert Forrest, he was born at Sandwich, Ont.

J. T. Edmond, formerly with the West Coast Navigation Co., has been appointed ferry superintendent at North Vancouver, B.C. Mr. Edmond is 37 years of age, and served his apprenticeship as marine engineer with the firm of Hepple & Co., South Shields, England. For over a year he was marine engineer with the West Coast Navigation Company, and also served with Moore & Scott, San Francisco, and other prominent firms.

U. Valiquet, M. Can. Soc. C. E., who is Superintendent of Public Works at Ottawa, was to have read a paper on the "Champlain Dry Dock for Quebec Harbor," but owing to illness and urgent business at the capital was prevented from attending the last meeting of the Society of Engineers. In his absence the paper was ably presented by F. H. McGuigan, jr.

Fort William.—Captain Jordan, master of the steamer Franz, the first of the lake carriers to make Fort William harbor since the opening of navigation, was invited to the Kam. Club, where he was presented on behalf of the Fort William Board of Trade with the customary high silk hat, by W. F. Hogarth, chairman of the harbor committee of the board.

LICENSED PILOTS

ST. LAWRENCE RIVER

Captain Walter Collins, 43 Main Street, Kingston, Ont.; Captain M. McDonald, River Hotel, Kingston, Ont.; Captain Charles J. Martin, 13 Balaclava Street, Kingston, Ont.; Captain T. J. Murphy, 11 William Street, Kingston, Ont.

ST. LAWRENCE RIVER, BAY OF QUINTE, AND MURRAY CANAL

Captain James Murray, 106 Clergy Street, Kingston, Ont.; Capt. James H. Martin, 259 Johnston Street, Kingston, Ont.; John Corkery, 17 Rideau Street, Kingston, Ont.; Captain Daniel H Mills, 272 University Avenue, Kingston, Ont.

MONTREAL PILOTS' ASSOCIATION

President—Alberic Angers, Montreal.
Secretary—C. B. Hamelin, Champlain, Que.

ASSOCIATIONS

DOMINION MARINE ASSOCIATION

President—A. A. Wright, Toronto. Secretary—Francis King, Kingston, Ont.

GREAT LAKES AND ST. LAWRENCE RIVER RATE COMMITTEE

Chairman—W. F. Herman, Cleveland, Ohio.
Secretary—Jas. Morrison, Montreal.

INTERNATIONAL WATER LINES PASSENGER ASSOCIATION.

President—O. H. Taylor, New York.
Secretary—M. R. Nelson, 1184 Broadway, New York.

SHIPPING FEDERATION OF CANADA.

President—Andrew A. Allan, Montreal; Manager and Secretary—T. Robb, 218 Board of Trade, Montreal; Treasurer, J. R. Binning, Montreal.

SHIPMASTERS' ASSOCIATION OF CANADA.

Secretary—Captain E. Wells, 45 St. John Street, Halifax, N.S.

GRAND COUNCIL N.A.M.E. OFFICERS.

A. R. Milne, Kingston, Ont., Grand President.
J. E. Belanger, Bienville, Levis, Grand Vice-President.
Neil J. Morrison, P.O. Box 238, St. John, N.B., Grand Secretary-Treasurer.
J. W. McLeod, Owen Sound, Ont., Grand Conductor.
Lemuel Winchester, Charlottetown, P.E.I., Grand Doorkeeper.
Alf. Charbonneau, Sorel, Que., and J. Scott, Halifax, N.S., Grand Auditors.

CANADA STEAMSHIP 1918 APPOINTMENTS

H. W. COWAN, operating manager of the Canada Steamship lines, announces the following appointments to the different freight steamers on the lines of the company for the season of 1918. J. W. Grainger is the mechanical superintendent:

Steamer W. Grant Morden, captain, Neil Campbell; engineer, Robert Chalmers. J. H. G. Hagarty, G. W. Pearson, Charles Robertson. Emperor, D. W. Burke, George N. Smith. E. B. Osler, C. E. Robinson, Wallace Robertson. Midland Prince, A. B. McIntyre, J. A. Pickard. Sir Trevor Dawson, H. Hinslea, W. W. Norcross, Stadacona, George H. Page, W. L. Shay. Martain, R. McIntyre, R. R. Foote. Midland King, P. McKay, James McGregor. W. D. Matthews, N. McGlennon, William Reid. Sarnian, R. Pyette, I. J. Boynton. Cadillac, W. Beatty, Hugh Myler. Calgarian, W. H. Montgomery, A. L. Black. Haddington, R. J. Wilson, Chas. Lericher. Hamiltonian, N. McKay, A. E. Kennedy. Ionic, O. Wing, A. E. Crosthwaite. Wyoming, T. B. Greenway, George Schroeder. Seguin, W. Brian, John B. McLaren. Belleville, ——, John Kennedy. Bickerdike, T. H. Johnston, D. S. LaRue. City of Hamilton, O. Patenaude, Wm. Dungan. City of Ottawa, J. L. Baxter, Joseph Aston; Fairfax, M. Heffernan, F. Patterson. Wiley, M. Egan, N. Hudgins, Charles LaVallee. Water Lily, John Hudgin, G. Rand.

Vancouver, B.C. — The Taylor Engineering Company, of Vancouver, has been awarded the contract, at a figure of over $750,000, for the construction of a floating drydock which will be built in the vicinity of Vancouver. The dock will be identical in design with one which was built a few years ago by the Interisland Steam Navigation Company of Honolulu. The structure will be capable of completely docking vessels of 4,500 tons, but repairs will be carried out on vessels of a tonnage of 12,000.

1918 Directory of Subordinate Councils, National Association of Marine Engineers.

Name.	No.	President.	Address.	Secretary.	Address.
Toronto,	1	Arch. McLaren,	324 Shaw Street	E. A. Prince,	49 Eaton Ave.
St. John,	2	W. L. Hurder,	209 Douglas Avenue	G. T. G. Blewett,	36 Murray St.
Collingwood,	3	John Osburn,	Collingwood, Ont.	Robert McQuade,	Collingwood, Ont.
Kingston,	4	Joseph W. Kennedy,	306 Johnston Street	James Gillie,	101 Clergy St.
Montreal,	5	Eugene Hamelin,	Jeanne Mance Street	M. Lazure	120 Ribard St.
Victoria,	6	John E. Jeffcott,	Esquimault, B.C.	Peter Gordon,	808 Blanchard St.
Vancouver,	7	Isaac N. Kendall	319 11th St. E., Vanc.	E. Read,	232 13th St. W.
Levis,	8	Michael Latulippe,	Lauzon, Levis, Que.	Arthur Jolin,	Lauzon, W.
Sorel,	9	Nap. Beaudoin,	Sorel, Que.	Alf. Charbonneau,	Box 204, Sorel, Que.
Owen Sound,	10	John W. McLeod	570 4th Ave.	R. J. McLeod,	662 8th St.
Windsor,	11	Alex. McDonald,	28 Crawford Ave.	Neil Maitland,	[illegible] London St. W.
Midland,	12	Geo. McDonald	Midland, Ont.	A. E. House,	Box 333
Halifax,	13	Robert Blair	29 Parrsboro Street	Chas. E. Pearce.	Portland St., Dartmouth, N.S.
Sault Ste. Marie,	14	Charles H. Innes,	27 Euclid Road	Wm. Hindmarch	34 Euclid Rd.
Charlottetown,	15	J. A. Rowe	176 King Street	Chas. Cumming,	27 Euston St.
Twin City,	16	H. W. Cross,	436 Ambrose St	J. W. Farquharson,	169 College St.
				A. H. Archand,	Champlain, Que.

Vancouver.—The Way Cayuse, a 2.500 ton wooden ship, was launched successfully at North Vancouver, the third of its kind to take the water from the Lyall shipyards.

More Ships Launched in U. S. Yards.—Ten steel ships of 57,695 tons and six wooden ships of 21,500 tons were launched by American yards in the week ending May 5, the Shipping Board announced recently. Twelve steel ships of 80,180 tons were delivered to the board complete in the same period.

Montreal, Que.—The Canada Steamship Lines' vessel, Calgarian, 1,300 tons, has arrived light from Toronto and will proceed to the Montreal drydock for repairs prior to being placed into commission for other work. When ready for sea the Calgarian will in all probability leave for an Atlantic port to start upon her new duties.

Halifax, May 9.—After a two weeks' struggle against the drift ice in the Gulf of St. Lawrence the Dominion Government steamer Stanley, which broke her rudder en route from the Magdalen Islands to Louisburg. C.B.. arrived at North Sydney, C.B., recently. The Stanley made port under her own steam. Her captain reported his arrival to the Marine and Fisheries Department here this morning, and stated that after affecting temporary repairs he would proceed to Halifax for permanent repairs.

Launchings from U. S. Shipyards This Year.—Exact figures on the progress of the merchant shipbuilding program in the United States this year were disclosed last Friday for the first time. They show that under direction of the Shipping Board there have been launched 236 steel and wooden vessels, with an aggregate tonnage of 1,440,622. There now are operating 157 shipyards with 753 ways in use. There are 398 steel ways, 332 wooden ways, either completed or under construction, four concrete ways, and 19 ways devoted to naval work. The launchings of steel ships include 18 vessels with a total tonnage of 136,250, constructed under Government contracts, and 183 with an aggregate tonnage of 1,195,887, obtained by requisition. The grand total of contract and requisitioned ships already launched is 1,332,127 tons. Already delivered are 138 steel vessels, with a total tonnage of 977,371. Wooden ships launched number 35, with an approximate tonnage of 108.500.

Washington, May 6.—Ship production during the week ending May 4th, which added more than 80,000 tons of new steel ships to the country's rapidly growing merchant marine, averaged approximately 350,000 tons a month, or more than 4,000,000 tons a year. Unofficial reports received by the Shipping Board from shipyards along both the Atlantic and Pacific coasts placed the output of tonnage at 92,000 tons. All of the officially reported new ships launched during this period were requisitioned by the Emergency Fleet Corporation in early stages of construction.

In addition to the building of steel merchant vessels, two tankers of 10,475 tons each also were launched during the past week. These vessels were built by the Union Iron Works, San Francisco. On the list of reported vessels completed were two steel vessels of 8,800 tons each, constructed on the Pacific Coast. Of the total output announced to-day, Great Lakes shipyards completed and delivered twelve ships.

Washington.—Twelve steel ships, totaling 80,180 tons were completed and 13 steel and wooden ships, aggregating 89,195 tons, were launched during the week ending March 5th, the Shipping Board announced to-day on receipt of reports from various shipyards.

This production is inclusive, it was said, of two tankers of 10,455 tons each.

Camden, N.J.—Breaking all war records in the building of ships, the New York Shipbuilding Company recently launched the single screw collier Tuckahoe, 27 days after her keel was laid on the ways. Ten days from now the company promises to turn the new vessel over to the Government fully completed for service.

Admiral von Tirpitz, at a meeting in Hamburg, gloomily contemplated the vision of economic retaliation, says the Hamburger "Nachrichten," and declared: "Without existence of that vigorous industry, which, after Germany was isolated, we converted mainly into a war industry, we should long ago have lost this war. This kind of war industry must shrink, however, when peace comes, while millions of our fellow countrymen will stream back from the trenches without finding sufficient work here, or, in any case, wages corresponding to enormously increased cost of living. Imagine, if we had to simultaneously bear the taxation which must fall on every German, even the poor—for the greatest exaction from property would not be sufficient even remotely to meet it; and further if, in spite of the fallen value of the German mark, we must still buy the most necessary raw materials and food supplies from abroad, notwithstanding all political and other hindrances the situation would produce for all. Can any one believe that under these circumstances, without an increase of power, without an indemnity, without security, we could avoid Germany's ruin?"

Shipyards for India.—The India Munitions Board has established a shipbuilding section at Calcutta, and will establish shipyards at suitable positions in British India.

Vancouver.—The Faith, the world's largest concrete steamer, and the latest innovation in the ocean-carrying trade, which completed its trial trip at San Francisco a few days ago, will make its first ocean voyage from California to Vancouver with a load of salt.

Quebec.—Three vessels launched within three days, such is the record of old Quebec. Two vessels, both built of steel, at the Davis shipbuilding plant, at Lauzon, have been launched and one wooden vessel. A large wooden vessel built at the Quinlan and Robertson yards on the St. Charles River will be launched. More vessels are on the ways at the various shipyards about Quebec.

Hamilton.—At the last meeting of the Board of Trade, G. E. Main submitted a list of 120 shipbuilding concerns doing business in Canada, and stated that Hamilton was particularly well situated for this class of industry, both on account of its harbor and dockage facilities, and its nearness to the Welland Canal. He suggested that an effort be made to secure industries of this nature for Hamilton.

Collingwood.—Another new steamer has been launched at the yards of the Collingwood Shipbuilding Company here. The vessel is of the poop, bridge and forecastle type, with engines amidships, and her dimensions are: 261 feet overall, 251 feet b.p., 43 feet 6 in. beam, 20 feet depth moulded, and she will carry approximately 2,900 tons deadweight.

Quebec.—It is learned on good authority that Chief Justice Duff in Ottawa has granted exemption from military service to all the shipyard employees working on shipbuilding at the Davis yards in Lauzon, Que. It was agreed that should the draftees working at the yards be enlisted their leaving would deprive various departments of experienced men and the entire plant would have been hurt.

BAY OF FUNDY SAFE

THERE was a time in the history of the Bay of Fundy when some fears respecting its safe navigation were justifiable, but these days have long since gone by. With the protection that is thrown around the mariner in the way of lighthouses, lightships, fog alarms, automatic buoys, bell-buoys and other forms of precautionary signals, the perils of Bay of Fundy navigation have now been almost wholly eliminated. This is borne out by the fact that for the past four years, though 22,614 ships, of a tonnage of 13,660,866 tons, have passed in and out of St. John, coming from or going overseas, there has not been a single casualty of any consequence to any of them in the Bay of Fundy.

Some years ago, when the question came before the St. John Board of Trade, a careful study was made of the wreck record for a period of eighteen years subsequent to 1896, with the result that it was shown that with a total tonnage of 42,029,262 tons entering the port, the casualty average was only .033 of 1 per cent. This statement was based on the record for the whole Bay of Fundy from Cape Sable up.

The statements above made as to the safety of the Bay of Fundy were fully established in the evidence submitted to the Special Pilotage Commission (which sat in St. John recently) by several masters of ocean steamers using this port. The pilotage service in the Bay of Fundy was also favorably commented upon. To make the pilotage service more effective it was recommended to the Commission that the present three districts be consolidated into one and a pilotage station established about eight miles below Partridge Island.

ONE OF MANY INSTALLATIONS OF MARTEN-FREEMAN COMPENSATING DAVITS

FOOL PROOF—RELIABLE—FAST OPERATING

TIME IS LIFE

WHAT GUARANTEE HAVE YOU THAT IT WILL BE POSSIBLE TO PUT ALL YOUR LIFEBOATS OVERSIDE IN THE FEW MINUTES AVAILABLE?

THE MARTEN-FREEMAN COMPANY, LIMITED
301 MANNING CHAMBERS, TORONTO, ONTARIO.

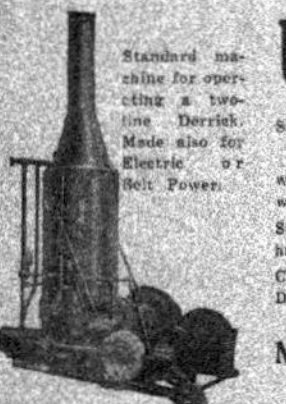

Standard machine for operating a two-line Derrick. Made also for Electric or Belt Power.

Use Machinery in Place of Men

Ship builders will find more difficulty than ever in getting men owing to the new conscription regulations.

HOISTING AND HAULAGE MACHINERY

will handle your heavy materials more rapidly than man power, and at a lower cost. We can supply you with the right machinery for your work. Our experience is at your service.

Steam Hoists, Electric Hoists, Belt Hoists, made in 7 sizes, from 10 to 60 horsepower; will lift your heavy loads, or haul loaded cars quickly and at low cost.

Cars, steel or wood body, any size, shape, or capacity, to dump end, side, bottom or rotary. Also platform cars. **Derricks,** any size, style or capacity.

Large catalog on request.

MARSH ENGINEERING WORKS, LIMITED - Belleville, Ontario
ESTABLISHED 1846.

LOVERIDGE'S IMPROVED STEERING GEAR SPRING BUFFERS

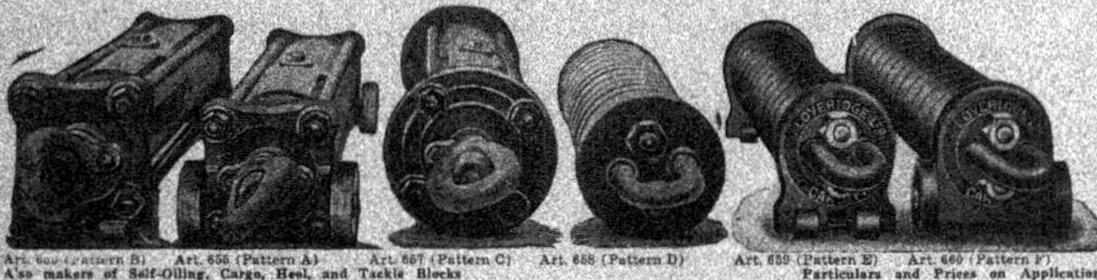

Art. [illegible] (Pattern B) Art. 656 (Pattern A) Art. 657 (Pattern C) Art. 658 (Pattern D) Art. 659 (Pattern E) Art. 660 (Pattern F)

Also makers of Self-Oiling, Cargo, Heel, and Tackle Blocks

Particulars and Prices on Application

LOVERIDGE LIMITED - - - - DOCKS, CARDIFF, WALES

SPEED RECORD MADE IN BUILDING STEEL COLLIER

The United States Shipping Board authorizes the following:

The steel collier Tuckahoe was launched from the Camden ways of the New York Shipbuilding Co. at 10.30 Sunday forenoon, thereby establishing a world's record in rapid ship construction.

The record was 27 days, 2 hours, and 50 minutes. This means that within that period a 5,550-ton steel steamship was built from keel to truck and launched practically complete in every detail—boilers in place, engines installed, masts stepped, funnel in place, propeller fitted, rudder hung, and only finishing touches to be put on.

The Tuckahoe was to have been delivered on June 15. The company therefore got her out 41 days ahead of time.

The vessel has a length of 330 feet, a beam of 50 feet, and will be endowed with a speed of 10½ knots.

This accomplishment of building a steel steamship of this bulk in a fraction more than 27 days is one of the marvels of the day. It surpasses any record hitherto made in any shipyard of the world. The best record of any British shipyard is said to be about two and a half months. Previous to our entrance into the war American shipyards had not outstripped British competitors in the rapid construction of merchant tonnage.

The Tuckahoe was christened by Miss Helen Hurley, the young daughter of the chairman of the Shipping Board.

NEW WOOD SHIP RECORD IN PLACING FRAMES

A new American wood ship construction record is reported to the Shipping Board by the Supple & Ballin Shipbuilding Corporation of Portland, Ore. Hereafter the mark for other wood shipyards to aim for in getting their frames into position on the ways will be 44 hours. This is the Supple & Ballin achievement.

Manager Heyworth, of the Wood Ship Construction Division of the Emergency Fleet Corporation, gave out to-day the following telegram from the Portland shipbuilders:

"Our crew on hull 232 broke a record by assembling and placing all full frames, 79 in all, in 44 hours."

Steam-Tight Condulets

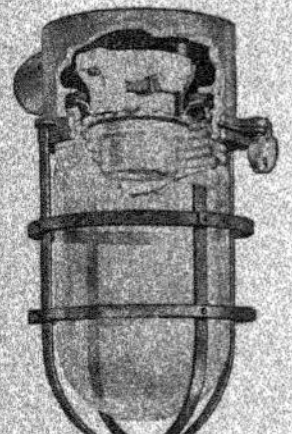

Condulet broken away to show parts.

Electric Light Fittings

whose design, material and workmanship insure long and satisfactory service.

Furnished in either iron or brass for ½, ¾ and 1-inch conduit.

Made in two sizes, to take 40 and 100-watt lamps respectively.

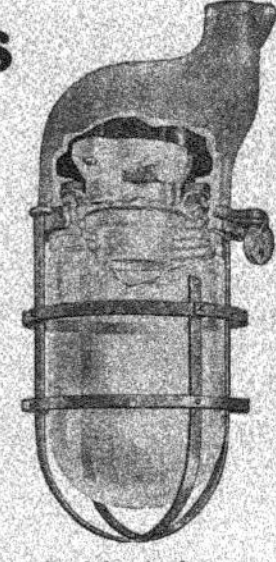

Condulet broken away to show parts.

Canada's Largest Ocean Freighter, the 7000-Ton Steamer Porsanger. Wired Throughout in Condulets.

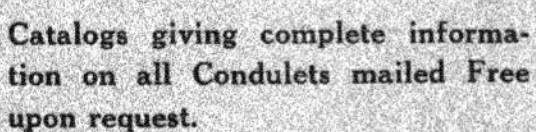

Catalogs giving complete information on all Condulets mailed Free upon request.

CROUSE-HINDS COMPANY
OF CANADA, LIMITED
TORONTO, ONT., CAN.

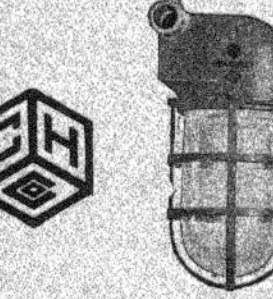

ADVERTISING to be successful does not necessarily have to produce a basketful of inquiries every day.

The best advertising is the kind that leaves an indelible, ineffaceable impression of the goods advertised on the minds of the greatest possible number of probable buyers, present and future.

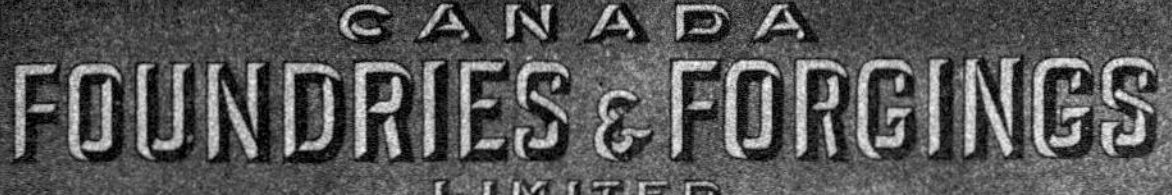

DROP FORGED STEEL

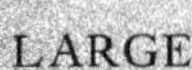

LARGE

SIZES

MARINE BOILER STAY NUTS

FORGED FROM THE SOLID BAR TO LLOYD'S SPECIFICATIONS.
EVERY NUT NEEDS A WRENCH.

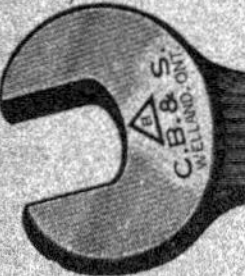

FORGED STEEL WRENCHES

EVERY NUT NEEDS A WRENCH.
PRICE ON APPLICATION.

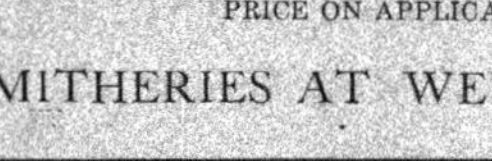

SMITHERIES AT WELLAND, ONT.

Established 1834 Incorporated 1907

MARINE SPECIALTIES

BRASS and IRON

Port Lights, with and without storm doors.
Marine Valves and Cocks, all sizes and kinds.
Water Columns, Asbestos Packed.
Water Gauges.
Gauge Cocks.

Throttle Valve

Main Feed and Shut-off Valve Combined

Rudder Braces,
Dumb Braces,
Dove Tails,
Ships' Bells
Steering Wheel,
Caps, Diamonds, etc.
Ships' Pumps,
Sheaves and Bushings
Ships' Hardware.

T. McAVITY & SONS, LIMITED

Brass and Iron Founders

Wholesale and Retail Hardware, Marine Specialties, etc.

ST. JOHN, N.B., CANADA

Branches at MONTREAL, TORONTO, WINNIPEG, VANCOUVER, LONDON, England

DAKE ENGINE CO.

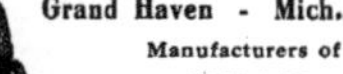

Grand Haven - Mich., U.S.A.

Manufacturers of

STEAM

Steering Engines	Cargo Hoists
Anchor Windlasses	Drill Hoists
Capstans	Spud Hoists
Mooring Hoists	Net Lifters

Write for New Catalog Just Out.

Toronto Agents: Wm. C. Wilson & Co.
Montreal Agents: Mussens Limited

Now Ready.—Fourth Edition. Pp. i-xvi+332. With 18 plates and 237 other illustrations, including 59 Folding Diagrams. 21s. net.

STEEL SHIPS:

THEIR CONSTRUCTION AND MAINTENANCE.

A Manual for Shipbuilders, Ship Superintendents, Students, and Marine Engineers

By THOMAS WALTON, Naval Architect
Author of "Know Your Own Ship"

Contents—I. Manufacture of Cast Iron, Wrought Iron, and Steel.—Composition of Iron and Steel, Quality, Strength, Tests, etc. II. Classification of Steel Ships. III. Considerations in making choice of Type of Vessel.—Framing of Ships. IV. Strains experienced by Ships—Methods of Computing and Comparing Strengths of Ships. V. Construction of Ships.—Alternative Modes of Construction.—Types of Vessels.—Turret, Self Trimming, and Trunk Steamers, etc.—Rivets and Rivetting, Workmanship. VI. Pumping Arrangements. VII. Maintenance.—Prevention of Deterioration in the Hulls of Ships.—Cement, Paint, etc.—Index.

"So thorough and well written is every chapter in the book that it is difficult to select any of them as being worthy of exceptional praise. Altogether, the work is excellent, and will prove of great value to those for whom it is intended."—The Engineer.

London, England: Charles Griffin & Co., Ltd., Exeter Street, Strand, W.C. 2

OAKUM

W. O. DAVEY & SONS, MANUFACTURERS

Jersey City, N.J., U.S.A.

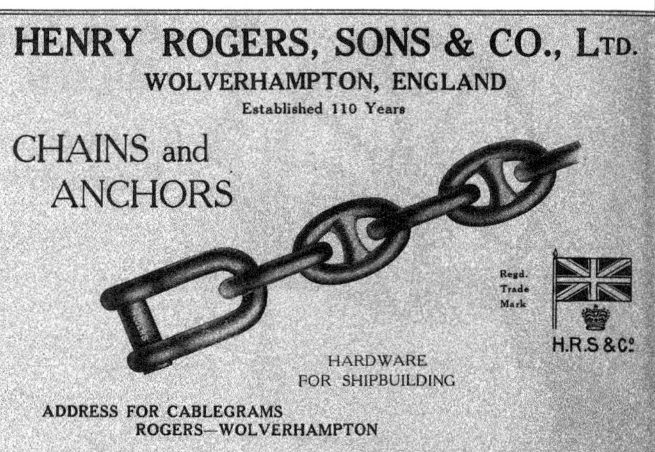

HENRY ROGERS, SONS & CO., LTD.
WOLVERHAMPTON, ENGLAND
Established 110 Years
CHAINS and
ANCHORS
Regd.
Trade
Mark
H.R.S & Co.
HARDWARE
FOR SHIPBUILDING
ADDRESS FOR CABLEGRAMS
ROGERS—WOLVERHAMPTON

LONGMANS' BOOKS ON SHIPBUILDING

PRACTICAL SHIPBUILDING. A Treatise on the Structural Design and Building of Modern Steam Vessels

By A. CAMPBELL HOLMS, *Surveyor to Lloyd's Register of Shipping. Third Edition, Revised and Enlarged. 2 Vols. $20.00 net.*

The revision now made for the third edition has been extensive and thorough. Large portions have been entirely re-written and the whole brought up to date by numerous additions, alterations and amendments. Three new chapters have been added on longitudinal framing, damage repairs, and lifeboats and davits. A large amount of new matter has also been added in connection with oil vessels, oil fuel, fire extinguishing, freeboard regulations, and bulkhead subdivision.

"The book, as a whole, represents the most complete work in existence on the subject of practical shipbuilding."—*Marine Engineering.*

SHIPYARD PRACTICE. As Applied to Warship Construction

By NEIL J. McDERMAID, *Member of the Royal Corps of Naval Constructors, late Instructor on Practical Shipbuilding at the Royal Naval College, Devonport. With Diagrams. Second Edition. Medium 8vo. $5.00 net.*

This book deals with the various operations at the building-slip, from the laying of the keel-blocks, including details of the structure, the making of the patterns for large castings, e.g., stem and stern posts, shaft brackets, etc., working of armor, construction of barbettes, engine and boiler seatings, water-testing, etc., etc.

"The information given bears internal evidence of accuracy and brings together probably more information than has ever before been made public as to the details and fittings in man-of-war practice. * * *"—*Engineering News.*

A COMPLETE CLASS-BOOK OF NAVAL ARCHITECTURE. Practical, Laying-off, Theoretical

By W. J. LOVETT, *Lecturer on Naval Architecture at the Belfast Municipal Technical Institute. With 173 Illustrations and almost 200 fully worked-out Answers and Questions. 8vo. $3.50 net.*

Intended to supply shipwrights, platers, draughtsmen, and others in the shipbuilding world with a sufficiency of naval architecture for the ordinary and everyday needs and to enable them afterwards to study higher works on the subject with intelligence and profit.

Second Edition Just Ready

A TEXT-BOOK ON LAYING-OFF; or, The Geometry of Shipbuilding

By EDWARD L. ATTWOOD *and* I. C. G. COOPER, *Senior Loftsman, H.M. Dockyard, Chatham, Lecturer in Naval Architecture at Chatham, Lecturer in Ship Carpentry at Whitstable. With Diagrams. 8vo. $2.00 net.*

TEXT BOOK OF THEORETICAL NAVAL A[illegible]

By EDWARD L. ATTWOOD, *M. Inst. N. A., Member of Royal Corps of Naval Constructors, formerly Lecturer of Naval Architecture at the Royal Naval College, Greenwich. New Edition, Revised and Enlarged. With 159 Diagrams and 5 Folding Tables. Crown 8vo. $3.00 net.*

Clearness and conciseness of expression characterize this book. A notable feature is the direct practical application of the methods taught. Two new chapters have been added to this edition, one on launching calculations, and one on the turnings of ships.

WARSHIPS. A text-book on the Construction, Protection, Stability, Turning, etc., of War Vessels

By the same author. With numerous Diagrams. Sixth Edition, Medium 8vo. $4.00 net.

Though intended primarily to provide naval officers with authoritative data on the subject, the work will also prove a useful introduction to naval architecture for apprentices and students at dockyards and elsewhere.

The various changes of practice made in recent years will be found embodied in this edition.

STRENGTH OF SHIPS

By ATHOLE J. MURRAY, *Grad. R. N. C., Assist. M. I. N. A. With Diagrams. 3 Folding Plates, and 1 Folding Table.* 8vo. *$5.00 net.*

This book is devoted exclusively to the systematic treatment of the strength of the structures and detail fittings of ships. The treatment is not academical, but essentially the outcome of practical experience in ship design.

"* * * a work which fills a distinct gap in the literature of the subject, and is sure to take a leading place amongst kindred text-books."—*Belfast Northern Whig.*

Write for our Catalogue of Pure and Applied Science.

Longmans, Green & Co., Publishers, Fourth Avenue and Thirtieth Street NEW YORK.

BUYERS' DIRECTORY

ACCUMULATORS, HYDRAULIC
Smart-Turner Mach. Co., Hamilton, Ont.

AERATING RESERVOIRS
Spray Engineering Co., Boston, Mass.

AIR PORTS
Turnbull Elevator Mfg. Co., Toronto, Ont.

ALLOYS, BRASS AND COPPER
Dominion Copper Products Co., Ltd., Montreal, Que.
Mueller Mfg. Co., H., Sarnia, Ont.
United Brass & Lead, Ltd., Toronto, Ont.

ANCHORS
Hopkins & Co., F. H., Montreal, Que.
Kearfott Engineering Co., New York, N.Y.
McNab Co., Bridgeport, Conn.
Henry Rogers Sons & Co., Wolverhampton, Eng.
Wm. C. Wilson & Co., Toronto, Ont.

ARCHITECT, NAVAL
Watts, J. Murray, Philadelphia, Pa.

ASBESTOS GOODS
Wm. C. Wilson & Co., Toronto, Ont.

BABBITT METAL
Aikenhead Hardware, Ltd., Toronto, Ont.
Hoyt Metal Company, Toronto, Ontario.
Tallman Brass & Metal Co., Hamilton, Ont.
Wilkinson & Kompass, Hamilton, Ont.
Wm. C. Wilson & Co., Toronto, Ont.

BARGES
Polson Iron Works, Toronto, Ont.

BAROMETERS
Wilson & Co., Wm. C., Toronto, Canada.

BAR IRON
Mitchells Ltd., Glasgow, Scotland.

BARS, GRATE
Babcock & Wilcox, Ltd., Montreal, Que.

BEARINGS, BRASS
Empire Mfg. Co., London, Ont.
Mueller Mfg. Co., H., Sarnia, Ont.
Tallman Brass & Metal Co., Hamilton, Ont.
United Brass & Lead Co., Toronto, Ont.

BELLS, SHIPS, ENGINE ROOM, ETC.
Aikenhead Hardware, Ltd., Toronto, Ont.
Clarke & Bros. Co., Montreal, Que.
Cory & Son, Inc., Chas., New York, N.Y.
Empire Mfg. Co., London, Ont.
Morrison Brass Mfg. Co., James, Toronto, Ont.
Mueller Mfg. Co., H., Sarnia, Ont.
United Brass & Lead, Ltd., Toronto, Ont.

BELTING, LEATHER
Aikenhead Hardware, Ltd., Toronto, Ont.
Wm. C. Wilson & Co., Toronto, Ont.

BIBBS, COMPRESSION
Empire Mfg. Co., London, Ont.
Mueller Mfg. Co., H., Sarnia, Ont.
United Brass & Lead, Ltd., Toronto, Ont.

BINNACLES
Hopkins & Co., F. H., Montreal, Que.
Morrison Brass Mfg. Co., James, Toronto, Ont.

BLOCKS, CARGO, HEEL AND TACKLE
Aikenhead Hardware, Ltd., Toronto, Ont.
Hopkins & Co., F. H., Montreal, Que.
Loveridge, Ltd., Docks, Cardiff, Wales.

BLOWERS, TURBO-
Mason Regulator & Engin. Co., Montreal, Que.

BOAT CHOCKS
Corbet Fdry. & Machine Co., Owen Sound, Ont.
Marten-Freeman Co., Toronto, Ont.

BOILER COMPOUND
Beveridge Paper Co., Montreal, Que.
Wm. C. Wilson & Co., Toronto, Ont.

BOILER COVERING
Beveridge Paper Co., Montreal, Que.

BOILER CIRCULATORS
Eckliff Circulator Co., Detroit, Mich.

BOILER FEED PUMPS
Can. Ingersoll-Rand Co., Ltd., Sherbrooke, Que.
Smart-Turner Mach. Co., Hamilton, Ont.
Williams Machinery Co., A. R., Toronto, Ont.

BOILER FITTINGS
Empire Mfg. Co., London, Ont.
McAvity & Sons Ltd., T., St. John, N.B.
Wager Furnace Bridge Wall Co., Inc., 149 Broadway, New York, N.Y.

BOILERS, MARINE
Babcock & Wilcox, Ltd., Montreal, Que.
Can. Fairbanks-Morse Co., Montreal, Que.
Can. Vickers, Ltd., Montreal, Que.
Collingwood Shipbuilding Co., Collingwood, Ont.
Doxford & Sons, William, Sunderland, England.
Engineering & Machine Works of Canada, St. Catharines, Ont.
Goldie & McCulloch, Ltd., Galt, Ont.
Hall Engineering Works, Montreal, Que.
Mason Regulator & Engin. Co., Montreal, Que.
Montreal Dry Docks & Shipbuilding Co., Montreal, Que.
National Shipbuilding Co., Goderich, Ont.
Marsh Engineering Works, Belleville, Ont.
Polson Iron Works, Toronto, Ontario.
Port Arthur Shipbuilding Co., Port Arthur, Ont.
Waterous Engine Works Co., Brantford, Ont.
Williams Machinery Co., A. R., Toronto, Ont.

BOOKS, MILITARY AND NAVAL
Van Nostrand Co., D., 25 Park Place, New York.

BOOKS, TECHNICAL, MARINE
MacLean Publishing Co., Toronto, Ont.
Renouf Publishing Co., Montreal, Que.
Van Nostrand Co., D., 25 Park Place, New York.

BOLTS
London Bolt & Hinge Works, London, Ont.
United Brass & Lead, Ltd., Toronto, Ont.
Wilkinson & Kompass, Hamilton, Ont.

BROKER
Page & Jones, Mobile, Ala., U.S.A.

BRASS GOODS
Corbet Fdry. & Machine Co., Owen Sound, Ont.
McAvity & Sons, T., St. John, N.B.
Mueller Mfg. Co., H., Sarnia, Ont.
Tallman Brass & Metal Co., Hamilton, Ont.
United Brass & Lead, Ltd., Toronto, Ont.

BUCKETS, CLAMSHELL
Beatty & Sons, Welland, Ont.
Morris Crane & Hoist Co., Herbert, Niagara Falls, Ont.

BUCKETS, DUMP
Morris Crane & Hoist Co., Herbert, Niagara Falls, Ont.

BUCKETS, COALING
Beatty & Sons, Welland, Ont.
Engineering & Machine Works of Canada, St. Catharines, Ont.
Hopkins & Co., F. H., Montreal, Que.
Marsh Engineering Works, Belleville, Ont.
Reed & Co., Geo., Montreal, Que.

CABLE
Hopkins & Co., F. H., Montreal, Que.
McNab Co., Bridgeport, Conn.
Wm. C. Wilson & Co., Toronto, Ont.

CABLE, LEAD COVERED AND ARMORED
Standard Underground Cable Co., Hamilton, Ont.

CABLE, ACCESSORIES
Standard Underground Cable Co., Hamilton, Ont.

CAPSTANS
Advance Mach. & Welding Co., Montreal, Que.
Dake Engine Co., Grand Haven, Mich.
Hopkins & Co., F. H., Montreal, Que.
Kennedy & Sons, Wm., Owen Sound, Ont.
Wm. C. Wilson & Co., Toronto, Ont.

CALKING TOOLS, ELECTRIC
Aikenhead Hardware, Ltd., Toronto, Ont.

CALKING TOOLS, PNEUMATIC
Aikenhead Hardware, Ltd., Toronto, Ont.
Can. Ingersoll-Rand Co. Sherbrooke, Que.

CALORIFIERS
Low & Sons, Ltd., Archibald, Glasgow, Scotland

CASTINGS
Can. Steel Foundries, Ltd., Montreal, Que.
Collingwood Shipbuilding Co., Collingwood, Ont.
Kennedy & Sons, Wm., Owen Sound, Ont.
Marsh Engineering Works, Belleville, Ont.
Mitchell Co., Ltd., Robt., Montreal, Que.
Mueller Mfg. Co., H., Sarnia, Ont.
Tallman Brass & Metal Co., Hamilton, Ont.
United Brass & Lead, Ltd., Toronto, Ont.
Waterous Engine Works Co., Brantford, Ont.

CASTINGS, ALLOY
Mueller Mfg. Co., H., Sarnia, Ont.
Tolland Mfg. Co., Montreal, Que.
Tallman Brass & Metal Co., Hamilton, Ont.
United Brass & Lead, Ltd., Toronto, Ont.

CASTINGS, ALUMINUM
Empire Mfg. Co., London, Ont.
Tallman Brass & Metal Co., Hamilton, Ont.
United Brass & Lead Co., Toronto, Ont.

CASTINGS, BRASS
Crouse-Hinds Co. of Canada, Ltd., Toronto, Ont.
Empire Mfg. Co., London, Ont.
McAvity & Sons Ltd., T., St. John, N.B.
McNab Co., Bridgeport, Conn.
Mitchell Co., Ltd., Robt., Montreal, Que.
Mueller Mfg. Co., H., Sarnia, Ont.
Tolland Mfg. Co., Montreal, Que.
Tallman Brass & Metal Co., Hamilton, Ont.
United Brass & Lead Co., Toronto, Ont.
Waterous Engine Works Co., Brantford, Ont.

CASTINGS, GREY IRON, MALLEABLE, ALUMINUM
Crouse-Hinds Co. of Canada, Ltd., Toronto, Ont.
Engineering & Machine Works of Canada, St. Catharines, Ont.
McAvity & Sons, T., St. John, N.B.
McNab Co., Bridgeport, Conn.
Mueller Mfg. Co., H., Sarnia, Ont.
Waterous Engine Works Co., Brantford, Ont.

CASTINGS, MANGANESE BRONZE
Tallman Brass & Metal Co., Hamilton, Ont.
United Brass & Lead Co., Toronto, Ont.

CASTINGS, ELECTRIC
Crouse-Hinds Co. of Canada, Ltd., Toronto, Ont.

CELLAR DRAINERS
Mueller Mfg. Co., H., Sarnia, Ont.

CEMENT, HIGH TEMPERATURE
Beveridge Paper Co., Montreal, Que.

CHAINS
Aikenhead Hardware, Ltd., Toronto, Ont.
Kearfott Engineering Co., New York, N.Y.
Hopkins & Co., F. H., Montreal, Que.
Morris Crane & Hoist Co., Herbert, Niagara Falls, Ont.
Henry Rogers, Sons & Co., Wolverhampton, Eng.
Wm. C. Wilson & Co., Toronto, Ont.

CHAIN BLOCKS
Morris Crane & Hoist Co., Herbert, Niagara Falls, Ont.

CHAIN SLINGS
Morris Crane & Hoist Co., Herbert, Niagara Falls, Ont.

CHANDLERY, SHIP
Hopkins & Co., F. H., Montreal, Que.
Leckie, Ltd., John, Toronto, Ont.

CLAMPS, STEAM AND WATER
Mueller Mfg. Co., H., Sarnia, Ont.

CLOCKS
American Steam Gauge & Valve Mfg. Co., Boston, Mass.
Morrison Brass Mfg. Co., James, Toronto, Ont.
Williams Machinery Co., A. R., Toronto, Ont.

CLOSETS
Mueller Mfg. Co., H., Sarnia, Ont.
United Brass & Lead, Ltd., Toronto, Ont.

COAL
Nova Scotia Steel & Coal Co., New Glasgow, N.S.

COAL HANDLING MACHINERY
Morris Crane & Hoist Co., Herbert, Niagara Falls, Ont.
Waterous Engine Works Co., Brantford, Ont.

COCKS, BILGE, DISCHARGE, INDICATOR, ETC.
McAvity & Sons Ltd., T., St. John, N.B.
Morrison Brass Mfg. Co., James, Toronto, Ont.

COCKS, BASIN
Empire Mfg. Co., London, Ont.
United Brass & Lead, Ltd., Toronto, Ont.

COMPASSES
Scythes & Co., Toronto, Ont.
Wm. C. Wilson & Co., Toronto, Ont.

COMPRESSORS, AIR
Can. Fairbanks-Morse Co., Montreal, Que.
Canadian Ingersoll-Rand Co., Sherbrooke, Que.
Hopkins & Co., F. H., Montreal, Que.
Smart-Turner Mach. Co., Hamilton, Ont.
Williams Machinery Co., A. R., Toronto, Ont.

COMPRESSED AIR TANKS
Scaife & Sons Co., Wm. B., Oakmont, Pa.

CONCRETE HARDENER AND WATERPROOFER
Beveridge Paper Co., Montreal, Que.

CONDENSERS
Kearfott Engineering Co., New York, N.Y.
Morris Machine Works, Baldwinsville, N.Y.
Smart-Turner Mach. Co., Hamilton, Ont.
Weir, Ltd., G. & J., Cathcart, Glasgow, Scotland.
Williams Machinery Co., A. R., Toronto, Ont.

CONDULETS, MARINE
Crouse-Hinds Co. of Canada, Ltd., Toronto, Ont.

CONTRACTORS' SUPPLIES
McAvity & Sons, T., St. John, N.B.

CONSULTING ENGINEER
Watt, J. Murray, Philadelphia, Pa.

CONVEYORS, ASH, COAL
Babcock & Wilcox, Ltd., Montreal, Que.
Hopkins & Co., F. H., Montreal, Que.

COPING MACHINES
Bertram & Sons, Ltd., John, Dundas, Ont.

COPPER TUBES, SHEETS AND RODS
Dominion Copper Products Co., Ltd., Montreal, Que.
Tallman Brass & Metal Co., Hamilton, Ont.

CORDAGE
Scythes & Co., Toronto, Ont.

COTTON, CALKING
Wilson & Co., Wm. C., Toronto, Canada.

COTTON DUCK
Scythes & Co., Toronto, Ont.

COVERS, CANVAS, FOR HATCHES, LIFEBOATS, ETC.
Leckie, Ltd., John, Toronto, Ont.
Waterous Engine Works Co., Brantford, Ont.

COUNTERS, REVOLUTION
American Steam Gauge & Valve Mfg. Co., Boston, Mass.

CRANES
Aikenhead Hardware, Ltd., Toronto, Ont.
Can. Fairbanks-Morse Co., Montreal, Que.
Hopkins & Co., F. H., Montreal, Que.
Williams Machinery Co., A. R., Toronto, Ont.

CRANES, ELECTRIC
Babcock & Wilcox, Ltd., Montreal, Que.
Morris Crane & Hoist Co., Herbert, Niagara Falls, Ont.
Smart-Turner Mach. Co., [illegible]

CRANES, GOLIATH AND PNEUMATIC
Morris Crane & Hoist Co., Herbert, Niagara Falls, Ont.

CRANES, GANTRY
Morris Crane & Hoist Co., Herbert, Niagara Falls, Ont.
Smart-Turner Mach. Co., Hamilton, Ont.

CRANES, JIB
Morris Crane & Hoist Co., Herbert, Niagara Falls, Ont.

CRANES, PORTABLE
Morris Crane & Hoist Co., Herbert, Niagara Falls, Ont.

CRANES, OVERHEAD TRAVELLING
Morris Crane & Hoist Co., Herbert, Niagara Falls, Ont.

CRANK SHAFTS
Canada Foundries and Forgings, Welland, Ont.

DAVITS, BOAT
- Babcock & Wilcox, Ltd., Montreal, Que.
- Corbet Fdry. & Machine Co., Owen Sound, Ont.
- Hopkins & Co., F. H., Montreal, Que.
- Marten, Freeman Co., Toronto, Ont.
- Waterous Engine Works Co., Brantford, Ont.

DEAD LIGHTS, BRASS
- Morris Crane & Hoist Co., Herbert, Niagara Falls, Ont.
- Mitchell Co., Ltd., Robt., Montreal, Que.

DECK LIGHTS
- Turnbull Elevator Mfg. Co., Toronto, Ont.

DECK PLUGS, ELECTRIC
- Crouse-Hinds Co. of Canada, Ltd., Toronto, Ont.

DECKING FOR SHIPS
- Marine Decking & Supply Co., Philadelphia, Pa.

DERRICKS
- Aikenhead Hardware, Ltd., Toronto, Ont.
- Dake Engine Co., Grand Haven, Mich.
- Hopkins & Co., F. H., Montreal, Que.
- Marsh Engineering Works, Belleville, Ont.
- Morris Crane & Hoist Co., Herbert, Niagara Falls, Ont.

DISTILLERS
- Kearfott Engineering Co., New York, N.Y.

DISTRESS OUTFITS
- Scythes & Co., Toronto, Ont.

DREDGES
- Collingwood Shipbuilding Co., Collingwood, Ont.
- Norbom Engineering Co., Philadelphia, Pa.
- Polson Iron Works, Toronto.

DRILLS, AIR
- Aikenhead Hardware, Ltd., Toronto, Ont.
- Can. Ingersoll-Rand Co. Sherbrooke, Que.
- Hopkins & Co., F. H., Montreal, Que.

DRILLS, CENTRE
- Wilt Twist Drill Co., Walkerville, Ont.

DRILLS, BLACKSMITH AND BIT STOCK
- Wilt Twist Drill Co., Walkerville, Ont.

DRILLS, ELECTRIC
- Aikenhead Hardware, Ltd., Toronto, Ont.
- Wilkinson & Kompass, Hamilton, Ont.

DRILLS, HIGH SPEED
- Wilt Twist Drill Co., Walkerville, Ont.
- Zenith Coal & Steel Products, Ltd., Montreal, Que.

DRILLS, TRACK
- Wilt Twist Drill Co., Walkerville, Ont.

DRILLS, RACHET AND HAND
- Wilt Twist Drill Co., Walkerville, Ont.

DRILLS, TWIST
- Aikenhead Hardware, Ltd., Toronto, Ont.
- Williams Machinery Co., A. R., Toronto, Ont.
- Wilt Twist Drill Co., Walkerville, Ont.

DRY DOCKS
- Can. Vickers, Ltd., Montreal, Que.
- Collingwood Shipbuilding Co., Collingwood, Ont.
- Doxford & Sons, William, Sunderland, England.
- Georgian Bay Shipbuilding & Wrecking Co., Midland, Ont.
- National Shipbuilding Co., Goderich, Ont.
- Polson Iron Works, Toronto, Ont.
- Port Arthur Shipbuilding Co., Port Arthur, Ont.
- Yarrows, Limited, Victoria, B.C.

ECONOMIZERS, FUEL
- Babcock & Wilcox, Ltd., Montreal, Que.

EJECTORS
- Empire Mfg. Co., London, Ont.
- Morrison Brass Mfg. Co., James, Toronto, Ont.
- Smart-Turner Mach. Co., Hamilton, Ont.

ELECTRIC LAMPS
- Wm. C. Wilson & Co., Toronto, Ont.

ELECTRO-PLATING
- Tallman Brass & Metal Co., Hamilton, Ont.
- United Brass & Lead Co., Toronto, Ont.

ELEVATING MACHINERY
- Goldie & McCulloch, Ltd., Galt, Ont.
- Morris Crane & Hoist Co., Herbert, Niagara Falls, Ont.
- Waterous Engine Works Co., Brantford, Ont.

ELEVATORS
- Turnbull Elevator Mfg. Co., Toronto, Ont.

ENGINES, HOISTING
- Advance Mach. & Welding Co., Montreal, Que.
- Corbet Fdry. & Machine Co., Owen Sound, Ont.
- Hopkins & Co., F. H., Montreal, Que.
- Kennedy & Sons, Wm., Owen Sound, Ont.
- Marsh Engineering Works, Belleville, Ont.
- Port Arthur Shipbuilding Co., Port Arthur, Ont.
- Williams Machinery Co., A. R., Toronto, Ont.

ENGINE, INTERNAL COMBUSTION
- Doxford & Sons, William, Sunderland, England.

ENGINES, MARINE
- Bolinders Co., New York, N.Y.
- Can. Fairbanks-Morse Co., Montreal, Que.
- Can. Vickers, Ltd., Montreal, Que.
- Doxford & Sons, William, Sunderland, England.
- Goldie & McCulloch, Ltd., Galt, Ont.
- Hopkins & Co., F. H., Montreal, Que.
- Iron Works, Ltd., Owen Sound, Ont.
- Mason Regulator & Engin. Co., Montreal, Que.
- National Shipbuilding Co., Goderich, Ont.
- Norbom Engineering Co., Philadelphia, Pa.
- Polson Iron Works, Toronto, Ont.
- Port Arthur Shipbuilding Co., Port Arthur, Ont.
- Trout Co., H. G., Buffalo, N.Y.
- Waterous Engine Works Co., Brantford, Ont.
- Williams Machinery Co., A. R., Toronto, Ont.

ENGINE STARTERS (AIR)
- Scaife & Sons Co., Wm. B., Oakmont, Pa.

ENGINES, STEERING
- Corbet Fdry. & Machine Co., Owen Sound, Ont.
- Dake Engine Co., Grand Haven, Mich.
- Kennedy & Sons, Wm., Owen Sound, Ont.
- Wm. C. Wilson & Co., Toronto, Ont.

ENAMELWARE
- Mueller Mfg. Co., H., Sarnia, Ont.
- United Brass & Lead Co., Toronto, Ont.

EVAPORATORS
- Kearfott Engineering Co., New York, N.Y.
- Mason Regulator & Engin. Co., Montreal, Que.
- McNab Co., Bridgeport, Conn.
- Weir Ltd., G. & J., Cathcart, Glasgow, Scotland.

EXTRACTORS, GREASE
- American Steam Gauge & Valve Mfg. Co. Boston, Mass.

EYE BOLTS AND NUTS
- Canada Foundries and Forgings, Welland, Ont.
- United Brass & Lead, Ltd., Toronto, Ont.

FANS
- Aikenhead Hardware Ltd., Toronto, Ont.
- Empire Mfg. Co., London, Ont.
- Reed & Co., Geo. W., Montreal, Que.
- Smart-Turner Mach. Co., Hamilton, Ont.
- Williams Machinery Co., A. R., Toronto, Ont.

FENDERS, ROPE
- Hopkins & Co., F. H., Montreal, Que.
- Leckie, Ltd., John, Toronto, Ont.
- Wilson & Co., Wm. C., Toronto, Canada.

FERRO-MANGANESE
- Mitchells, Ltd., Glasgow, Scotland.

FIBRE, VULCANIZED
- Beveridge Paper Co., Montreal, Que.

FILES
- Aikenhead Hardware, Ltd., Toronto, Ont.
- Williams Machinery Co., A. R., Toronto, Ont.

FIRE BRICKS
- Beveridge Paper Co., Montreal, Que.
- Mitchells, Ltd., Glasgow, Scotland.
- Williams Machinery Co., A. R., Toronto, Ont.

FILTERS, FEED WATER
- Mason Regulator & Engin. Co., Montreal, Que.

FITTINGS, MARINE
- Hopkins & Co., F. H., Montreal, Que.
- McAvity & Sons, T., St. John, N.B.
- United Brass & Lead, Ltd., Toronto, Ont.

FITTINGS, MOTOR BOAT
- Empire Mfg. Co., London, Ont.
- Mueller Mfg. Co., H., Sarnia, Ont.
- United Brass & Lead, Ltd., Toronto, Ont.

FLAG POLES, STEEL
- Dennis Wire & Iron Works Co., London, Ont.

FLAGS
- Scythes & Co., Toronto, Ont.

FLOW METERS
- Spray Engineering Co., Boston, Mass.

FLUE CLEANERS
- Wm. C. Wilson & Co., Toronto, Ont.

FIXTURES, ELECTRIC
- Cory & Son, Inc., Chas., New York, N.Y.
- Crouse-Hinds Co. of Canada, Ltd., Toronto, Ont.
- Tallman Brass & Metal Co., Hamilton, Ont.

FORGES
- Aikenhead Hardware, Ltd., Toronto, Ont.
- Hopkins & Co., F. H., Montreal, Que.

FLOODLIGHTS, ELECTRIC
- Crouse-Hinds Co. of Canada, Ltd., Toronto, Ont.

FLUORSPAR
- Mitchells, Ltd., Glasgow, Scotland.
- Scythes & Co., Toronto, Ont.

FORGINGS, ALL KINDS
- Aikenhead Hardware, Ltd., Toronto, Ont.
- Collingwood Shipbuilding Co., Collingwood, Ont.
- Nova Scotia Steel & Coal Co., New Glasgow, N.S.

FORGINGS, STEEL AND IRON
- Canada Foundries and Forgings, Welland, Ont.

FURNACE BRIDGE WALLS
- Wager Furnace Bridge Wall Co., Inc., 149 Broadway, New York, N.Y.

FURNACE GRATE BARS
- Beveridge Paper Co., Montreal, Que.

GAUGES, RECORDING
- American Steam Gauge & Valve Mfg. Co., Boston, Mass.
- Empire Mfg. Co., London, Ont.

GASKETS
- Wm. C. Wilson & Co., Toronto, Ont.

GAUGES COCKS
- McAvity & Sons, T., St. John, N.B.

GAUGE GLASSES
- Wm. C. Wilson & Co., Toronto, Ont.

GAUGES, WATER
- McAvity & Sons, T., St. John, N.B.
- McNab Co., Bridgeport, Conn.
- Morris Crane & Hoist Co., Herbert, Niagara Falls, Ont.

GAUGES, WATER, PRESSURE, COMPOUND AND VACUUM
- Aikenhead Hardware, Ltd., Toronto, Ont.
- Babcock & Wilcox, Ltd., Montreal, Que.
- Empire Mfg. Co., London, Ont.
- Morrison Brass Mfg. Co., James, Toronto, Ont.

GENERATORS AND CONVERTORS
- Can. Fairbanks-Morse Co., Montreal, Que.

GLAZING COMPOSITION
- Kuhls, H. B. Fred., 6411 3rd Ave., Brooklyn, N.Y.

GONGS
- Clark & Bro., C. O., Montreal, Que.
- Mitchell Co., Ltd., Robt., Montreal, Que.

GRAPHITE
- Wm. C. Wilson & Co., Toronto, Ont.

GRATINGS
- Corbet Fdry. & Machine Co., Owen Sound, Ont.
- Canada Wire & Iron Goods Co., Hamilton, Ont.

GRINDERS, ELECTRIC
- Wilkinson & Kompass, Hamilton, Ont.

GRINDERS, PNEUMATIC
- Can. Ingersoll-Rand Co. Sherbrooke, Que.

GUARDS, MACHINERY
- Dennis Wire & Iron Works Co., London, Ont.

GUY RODS AND ANCHORS, ELECTRIC
- Crouse-Hinds Co. of Canada, Ltd., Toronto, Ont.

HAMMERS
- Canada Foundries and Forgings, Ltd., Welland, Ont.

HARDWARE, MARINE
- Hopkins & Co., F. H., Montreal, Que.
- Scythes & Co., Toronto, Ont.

HEADLIGHTS, ELECTRIC
- Crouse-Hinds Co. of Canada, Ltd., Toronto, Ont.
- Hopkins & Co., F. H., Montreal, Que.
- Tallman Brass & Metal Co., Hamilton, Ont.

HEATING EQUIPMENT
- Empire Mfg. Co., London, Ont.
- Low & Sons, Ltd., Archibald, Glasgow, Scotland

HEATERS, FEED, WATER
- Babcock & Wilcox, Ltd., Montreal, Que.
- Kearfott Engineering Co., New York, N.Y.
- Mason Regulator & Engin. Co., Montreal, Que.
- McNab Co., Bridgeport, Conn.
- Weir Ltd., G. & J., Cathcart, Glasgow, Scotland.

HINGES
- London Bolt & Hinge Works, London, Ont.
- Mitchell Co., Ltd., Robt., Montreal, Que.

HOOKS
- Hopkins & Co., F. H., Montreal, Que.
- Morris Crane & Hoist Co., Herbert, Niagara Falls, Ont.

HOISTS, ASH
- Beatty & Sons, Welland, Ont.
- Marsh Engineering Works, Belleville, Ont.
- St. Clair Bros., Galt, Ont.
- Waterous Engine Works Co., Brantford, Ont.

HOIST BLOCKS
- Morris Crane & Hoist Co., Herbert, Niagara Falls, Ont.

HOISTS, CHAIN
- Aikenhead Hardware, Ltd., Toronto, Ont.
- Can. Fairbanks-Morse Co., Montreal, Que.
- Dake Engine Co., Grand Haven, Mich.
- Hopkins & Co., F. H., Montreal, Que.
- Morris Crane & Hoist Co., Herbert, Niagara Falls, Ont.
- Williams Machinery Co., A. R., Toronto, Ont.

HOISTS, CARGO, MOVING, ETC.
- Dake Engine Co., Grand Haven, Mich.
- Hopkins & Co., F. H., Montreal, Que.
- Marsh Engineering Works, Belleville, Ont.
- Waterous Engine Works Co., Brantford, Ont.

HOISTING MACHINERY
- Beatty & Sons, Welland, Ont.
- Can. Ingersoll-Rand Co. Sherbrooke, Que.
- Corbet Fdry. & Machine Co., Owen Sound, Ont.
- Marsh Engineering Works, Belleville, Ont.
- Morris Crane & Hoist Co., Herbert, Niagara Falls, Ont.
- Waterous Engine Works Co., Brantford, Ont.
- Williams Machinery Co., A. R., Toronto, Ont.

HOSE
- Wm. C. Wilson & Co., Toronto, Ont.

HYDRAULIC MACHINERY
- The John McDougall Caledonian Iron Works Co., Montreal, Que.

INDICATORS, ENGINE
- American Steam Gauge & Valve Mfg. Co., Boston, Mass.
- Cory & Son, Inc., Chas., New York, N.Y.
- McNab Co., Bridgeport, Conn.

INDICATORS, SPEED
- Aikenhead Hardware, Ltd., Toronto, Ont.
- Cory & Son, Inc., Chas., New York, N.Y.

INJECTORS
- Aikenhead Hardware, Ltd., Toronto, Ont.
- Empire Mfg. Co., London, Ont.
- Morrison Brass Mfg. Co., James, Toronto, Ont.
- Williams Machinery Co., A. R., Toronto, Ont.

INGOTS
- Broughton Copper Co., Ltd., Manchester, Eng.

INSULATORS, ELECTRIC
- Crouse-Hinds Co. of Canada, Ltd., Toronto, Ont.

INSTRUMENTS, NAUTICAL
- Leckie, Ltd., John, Toronto, Ont.

IRON AND STEEL
- Mitchells Ltd., Glasgow, Scotland.

JACKS
- Hopkins & Co., F. H., Montreal, Que.
- Morris Crane & Hoist Co., Herbert, Niagara Falls, Ont.

JOINTS, BALL SLIP
- Norbom Engineering Co., Philadelphia, Pa.

INSURANCE, MARINE
- Toronto Insurance & Vessel Agency, Toronto, Ont.

JOURNAL WEDGES
- Canada Foundries and Forgings, Ltd., Welland, Ont.

LATHES
- Foss Machine & Supply Co., Geo. F., Montreal, Que.
- Bertram & Sons, Ltd., John, Dundas, Ont.

LEATHER, LACE
- Wm. C. Wilson & Co., Toronto, Ont.

LIFEBUOYS, BELTS AND PRESERVERS
- Fosbery Co., Barking, Essex, Eng.
- Hopkins & Co., F. H., Montreal, Que.
- Wm. C. Wilson & Co., Toronto, Ont.

LIFEBOATS
- Hopkins & Co., F. H., Montreal, Que.
- Wm. C. Wilson & Co., Toronto, Ont.

LIFE BOAT EQUIPMENT
- Eckliff Circulator Co., Detroit, Mich.
- Hopkins & Co., F. H., Montreal, Que.
- Leckie, Ltd., John, Toronto, Ont.
- Wm. C. Wilson & Co., Toronto, Ont.

LIFE JACKETS
- Fosbery Co., Barking, Essex, Eng.
- Hopkins & Co., F. H., Montreal, Que.
- Leckie, Ltd., John, Toronto, Ont.

LIGHTS, ALL KINDS
- Can. Fairbanks-Morse Co., Montreal, Que.
- Crouse-Hinds Co. of Canada, Ltd., Toronto, Ont.
- Hopkins & Co., F. H., Montreal, Que.
- Morrison Brass Mfg. Co., James, Toronto, Ont.
- Tallman Brass & Metal Co., Hamilton, Ont.

LIGHTS, SIDE, PORT
- Crouse-Hinds Co. of Canada, Ltd., Toronto, Ont.
- McAvity & Sons Ltd., T., St. John, N.B.
- Tallman Brass & Metal Co., Hamilton, Ont.
- Turnbull Elevator Mfg. Co., Toronto, Ont.

LIGHTING SETS
- Can. Fairbanks-Morse Co., Montreal.
- Cory & Son, Inc., Chas., New York, N.Y.
- Crouse-Hinds Co. of Canada, Ltd., Toronto, Ont.
- Kearfott Engineering Co., New York, N.Y.
- Williams Machinery Co., A. R., Toronto, Ont.
- Wilson & Co., Wm. C., Toronto, Canada.

RUDDER FOR SS. "LUX." WEIGHT 8 TONS, LENGTH 42 FEET.
Ship Forgings of all Shapes, Sizes and Weights up to 75 Tons
Rails, Plates Cold Drawn Shafting and Machinery Steel
WE manufacture, of Fluid Compressed Steel, forgings entering into the construction and equipment of steel vessels up to and including the largest building or afloat, and embracing Rudder Frames—sectional or one-piece, also Rudders complete; Stern Posts and Stern Brackets for single, Twin and Triple Screw Ships, Rudder Heads, Boat Davits, Derricks, etc.
NOVA SCOTIA STEEL & COAL COMPANY, Limited.
NEW GLASGOW, N. S., CANADA

LOADERS, WAGON AND TRUCK
Morris Crane & Hoist Co., Herbert, Niagara Falls, Ont.
Waterous Engine Works Co., Brantford, Ont.

LOCKERS, METAL
Dennis Wire & Iron Works Co., London, Ont.
Reed & Co., Geo. W., Montreal, Que.

LUBRICATORS
Empire Mfg. Co., London, Ont.

MACHINISTS
Corbet Fdry. & Machine Co., Owen Sound, Ont.
Hyde Engineering Works, Montreal, Que.
United Brass & Lead Co., Toronto, Ont.

MANGANESITE, PASTE
Wm. C. Wilson & Co., Toronto, Ont.

MARINE CASTINGS
Mitchell Co., Ltd., Robt., Montreal, Que.
McAvity & Sons, T., St. John, N.B.
Mueller Mfg. Co., H., Sarnia, Ont.
Tallman Brass & Metal Co., Hamilton, Ont.
United Brass & Lead, Ltd., Toronto, Ont.
Waterous Engine Works Co., Brantford, Ont.

MARINE SPECIALTIES
Hopkins & Co., F. H., Montreal, Que.
McAvity & Sons, T., St. John, N.B.
Mitchell Co., Ltd., Robt., Montreal, Que.
United Brass & Lead, Ltd., Toronto, Ont.

MOTORS
Aikenhead Hardware, Ltd., Toronto, Ont.
Can. Fairbanks-Morse Co., Montreal, Que.
Williams Machinery Co., A. R., Toronto, Ont.
Zenith Coal & Steel Products, Ltd., Montreal, Que.

NAUTICAL INSTRUMENTS
Scythes & Co., Toronto, Ont.
Wm. C. Wilson & Co., Toronto, Ont.

NET LIFTERS
Dake Engine Co., Grand Haven, Mich.
Leckie, Ltd., John, Toronto, Ont.

NOZZLES, ALL KINDS
Empire Mfg. Co., London, Ont.
Tolland Mfg. Co., Boston, Mass.

NUTS
Wilkinson & Kompass, Hamilton, Ont.
United Brass & Lead, Ltd., Toronto, Ont.

OAKUM
Scythes & Co., Toronto, Ont.
Wm. C. Wilson & Co., Toronto, Ont.

OIL CUPS
Empire Mfg. Co., London, Ont.
Mitchell Co., Ltd., Robt., Montreal, Que.
Murray & Fraser, New Glasgow, N.S.

OILCLOTHING
Wm. C. Wilson & Co., Toronto, Ont.

OIL, LINSEED
Canada Linseed Oil Mills, Montreal, Que.

OILS AND GREASE
Wm. C. Wilson & Co., Toronto, Ont.

OAKUM
Leckie, Ltd., John, Toronto, Ont.

ORNAMENTAL IRONWORK
Canada Wire & Iron Goods Co., Hamilton, Ont.

OXY-ACETLYENE OUTFITS
Can. Fairbanks-Morse Co., Montreal, Que.
Hopkins & Co., F. H., Montreal, Que.
McAvity & Sons Ltd., T., St. John, N.B.
Williams Machinery Co., A. R., Toronto, Ont.

PAINT
Ault & Wiborg Co. of Can., Ltd., Toronto, Ont.
Wm. C. Wilson & Co., Toronto, Ont.

PAINTS, ANTI-FOULING
Kuhls, H. B. Fred., 6411 3rd Ave., Brooklyn, N.Y.

PAINTS, ANTI-CORROSIVE
Kuhls, H. B. Fred. 6411 3rd Ave., Brooklyn, N.Y.

PAINTING EQUIPMENT, AUTOMATIC
Spray Engineering Co., Boston, Mass.

PACKING, AMMONIA
France Packing Co., Philadelphia, Pa.
Wm. C. Wilson & Co., Toronto, Ont.

PACKING, MARINE
Aikenhead Hardware, Ltd., Toronto, Ont.
France Packing Co., Philadelphia, Pa.
Wm. C. Wilson & Co., Toronto, Ont.

PACKING, METALLIC
Aikenhead Hardware, Ltd., Toronto, Ont.
France Packing Co., Philadelphia, Pa.

PACKING, STEAM
Aikenhead Hardware, Ltd., Toronto, Ont.
France Packing Co., Philadelphia, Pa.
Wm. C. Wilson & Co., Toronto, Ont.

PAINTS, CONDENSER AND SMOKESTACK
Beveridge Paper Co., Montreal, Que.

PANELBOARD AND CABINETS, ELECTRIC
Crouse-Hinds Co. of Canada, Ltd., Toronto, Ont
Preston Woodworking Machy. Co., Preston, Ont.

PIPE, LAPWELD, CAST IRON, RIVETED
Empire Mfg. Co., London, Ont.
Norbom Engineering Co., Philadelphia, Pa.
United Brass & Lead, Ltd., Toronto, Ont.

PIPE RAILING, IRON AND BRASS
Dennis Wire & Iron Works Co., London, Ont.

PITCH
Reed & Co., Geo., Montreal, Que.
Scythes & Co., Toronto, Ont.

PITCH, PINE
Wilson & Co., Wm. C., Toronto, Canada.

PLANERS
Bertram & Sons, Ltd., John, Dundas, Ont.

PLANERS, STANDARD AND ROTARY
Yates Mach. Co., P. B., Hamilton, Ont.

PLATE CLAMPS
Morris Crane & Hoist Co., Herbert, Niagara Falls, Ont.

PLATE PUNCH TABLES
Norbom Engineering Co., Philadelphia, Pa.

PLUMBING EQUIPMENT
Low & Sons, Ltd., Archibald, Glasgow, Scotland.
Empire Mfg. Co., London, Ont.
Mueller Mfg. Co., H., Sarnia, Ont.
Tallman Brass & Metal Co., Hamilton, Ont.
Wm. C. Wilson & Co., Toronto, Ont.
United Brass & Lead, Ltd., Toronto, Ont.

PROPELLOR BLADES, BRONZE
Corbet Fdry. & Machine Co., Owen Sound, Ont.
Empire Mfg. Co., London, Ont.
Tallman Brass & Metal Co., Hamilton, Ont.
United Brass & Lead, Ltd., Toronto, Ont.
Yarrows, Limited, Victoria, B.C.

PROPELLER WHEELS
Kennedy & Sons, Wm., Owen Sound, Ont.
Trout Co., H. G., Buffalo, N.Y.

PIPING, STEAM
Babcock & Wilcox, Ltd., Montreal, Que.

PORT LIGHTS
Mitchell Co., Ltd., Robt., Montreal, Que.
Mueller Mfg. Co., H., Sarnia, Ont.
Turnbull Elevator Mfg. Co., Toronto, Ont.
Tallman Brass & Metal Co., Hamilton, Ont.
Wm. C. Wilson & Co., Toronto, Ont.
United Brass & Lead, Ltd., Toronto, Ont.

PROPELLER WHEELS
McNab Co., Bridgeport, Conn.

PULLEYS
Smart-Turner Mach. Co., Hamilton, Ont.
Williams Machinery Co., A. R., Toronto, Ont.

PUMP, AIR
Can. Ingersoll-Rand Co. Sherbrooke, Que.
Smart-Turner Mach. Co., Hamilton, Ont.
Weir Ltd., G. & J., Cathcart, Glasgow, Scotland.
Williams Machinery Co., A. R., Toronto, Ont.

PUMPS
Can. Fairbanks-Morse Co., Montreal, Que.
Canada Foundries & Forgings, Welland, Ont.
Goldie & McCulloch, Ltd., Galt, Ont.
Hopkins & Co., F. H., Montreal, Que.
McAvity & Sons, T., St. John, N.B.
Williams Machinery Co., A. R., Toronto Ont.

PUMPS, CENTRIFUGAL
Can. Ingersoll-Rand Co. Sherbrooke, Que.
Kearfott Engineering Co., New York, N.Y.
Norbom Engineering Co., Philadelphia, Pa.
Smart-Turner Mach. Co., Hamilton, Ont.
Waterous Engine Works Co., Brantford, Ont.
Williams Machinery Co., A. R., Toronto, Ont.

PUMPS, BILGE
Mitchell Co., Ltd., Robt., Montreal, Que.
Mueller Mfg. Co., H., Sarnia, Ont.

PUMPS, CIRCULATING
McNab Co., Bridgeport, Conn.
Morris Machine Works, Baldwinsville, N.Y.

PUMPS, FEED WATER
McNab Co., Bridgeport, Conn.
Weir Ltd., G. & J., Cathcart, Glasgow, Scotland.
Williams Machinery Co., A. R., Toronto, Ont.

PUMPS, HAND AND POWER
Aikenhead Hardware, Ltd., Toronto, Ont.
Smart-Turner Mach. Co., Hamilton, Ont.
Williams Machinery Co., A. R., Toronto, Ont.

PUMPS, HIGH PRESSURE
Canadian Ingersoll-Rand Co., Sherbrooke, Que.
Smart-Turner Mach. Co., Hamilton, Ont.

PUMPS, STEAM TURBO
Wm. C. Wilson & Co., Toronto, Ont.

PUMPING MACHINES
Norbom Engineering Co., Philadelphia, Pa.

PUNCHES
Bertram & Sons, Ltd., John, Dundas, Ont.

PUNCHES, SINGLE, DOUBLE AND MULTIPLE
Norbom Engineering Co., Philadelphia, Pa.

PURIFIERS, WATER
Babcock & Wilcox, Ltd., Montreal, Que.

RADIO ENGINEERS
Cutting & Washington, Inc., Cambridge, Mass.

RADIATORS, STEAM, ELECTRIC
Empire Mfg. Co., London, Ont.
Low & Sons, Ltd, Archibald, Glasgow, Scotland.
Wm. C. Wilson & Co, Toronto, Ont.

RAILS, OVERHEAD
Morris Crane & Hoist Co. Herbert, Niagara Falls, Ont.

RANGES
Hopkins & Co., F. H., Montreal, Que.
Wm. C. Wilson & Co, Toronto, Ont.

RECEIVERS, AIR
Can. Ingersoll-Rand Co. Sherbrooke, Que.

REGULATORS, FEED WATER
American Steam Gauge & Valve Mfg. Co. Boston, Mass.

REGULATORS, PRESSURE
Mason Regulator & Engin. Co., Montreal, Que.

RELEASING GEARS
Eckliff Circulator Co., Detroit, Mich.

REPAIRS, MARINE
Corbet Foundry & Mach. Co., Owen Sound, Ont.
Can. Vickers, Ltd., Montreal, Que.
Collingwood Shipbuilding Co., Collingwood, Ont.
Engineering & Machine Works of Canada, St. Catharines, Ont.
Georgian Bay Shipbuilding & Wrecking Co., Midland, Ont.
Hyde Engineering Works, Montreal, Que.
Iron Works, Ltd., Owen Sound, Ont.
Kennedy & Sons, Wm., Owen Sound, Ont.
Muir Bros. Dry Dock Co., Port Dalhousie, Ont.
National Shipbuilding Co., Goderich, Ont.
Port Arthur Shipbuilding Co., Port Arthur, Ont.
Yarrow, Limited, Victoria, B.C.

RIGGING SCREWS
Hopkins & Co., F. H., Montreal, Que.

RIGGING, WIRE ROPE
Leckie, Ltd., John, Toronto, Ont.

RIVETS
London Bolt & Hinge Works, London, Ont.
Severance Mfg. Co., S., Glassport, Pa.
Wilkinson & Kompass, Hamilton, Ont.

RIVETERS, PNEUMATIC
Can. Ingersoll-Rand Co., Ltd., Sherbrooke, Que.

RODS, COPPER, BRASS, BRONZE
Standard Underground Cable Co., Hamilton, Ont.
Tallman Brass & Metal Co., Hamilton, Ont.

ROLLS, STRAIGHTENING, BENDING
Bertram & Sons, Ltd., John, Dundas, Ont.

ROOFING
Reed & Co., Geo., Montreal, Que.
Wm. C. Wilson & Co., Toronto, Ont.

ROPE BLOCKS
Aikenhead Hardware, Ltd., Toronto, Ont.
Can. Fairbanks-Morse Co., Montreal, Que.
Morris Crane & Hoist Co., Herbert, Niagara Falls, Ont.

ROPE
Hopkins & Co., F. H., Montreal, Que.
Leckie, Ltd., John, Toronto, Ont.
McNab Co., Bridgeport, Conn.
Wm. C. Wilson & Co., Toronto, Ont.

RUBBER COATS
Wm. C. Wilson & Co., Toronto, Ont.

SAW MILL MACHINERY
Preston Woodworking Machy. Co., Preston, Ont.
Waterous Engine Works Co., Brantford, Ont.
Yates Machine Co., P. B., Hamilton, Ont.

SAWS, BAND
Preston Woodworking Machy. Co., Preston, Ont.

SCALES, BOILERS, ENGINES
Can. Fairbanks-Morse Co., Montreal, Que.

SCOWS
Collingwood Shipbuilding Co., Collingwood, Ont.
Polson Iron Works, Toronto, Ont.

SCREENS, WIRE
Dennis Wire & Iron Works Co., London, Ont.

SCREWS, COACH
London Bolt & Hinge Works, London, Ont.

SEAM PAINT
Kuhls, H. B. Fred., 6411 3rd Ave., Brooklyn, N.Y.

SEPARATORS, OIL, STEAM
Mason Regulator & Engin. Co., Montreal, Que.
Smart-Turner Mach. Co., Hamilton, Ont.

SHAFTING
Wilkinson & Kompass, Hamilton, Ont.

SHAFTING, BRONZE
Empire Mfg. Co., London, Ont.

SHEARS
Bertram & Sons, Ltd., John, Dundas, Ont.
Norbom Engineering Co., Philadelphia, Pa.

SHIPBUILDING TOOLS
Aikenhead Hardware, Ltd., Toronto, Ont.
Can. Ingersoll-Rand Co. Sherbrooke, Que.

SHIPS, BUILDERS OF
Can. Vickers, Ltd., Montreal, Que.
Collingwood Shipbuilding Co., Collingwood, Ont.
Doxford & Sons, William, Sunderland, England.
Georgian Bay Shipbuilding & Wrecking Co., Midland, Ont.
National Shipbuilding Co., Goderich, Ont.
Polson Iron Works, Toronto, Ont.
Port Arthur Shipbuilding Co., Port Arthur, Ont.
Yarrows, Limited, Victoria, B.C.

SHIP BROKERS
Page & Jones, Mobile, Ala.

SHIP PLATES
Nova Scotia Steel & Coal Co., New Glasgow, N.S.

SIDE LIGHTS
Hopkins & Co., F. H., Montreal, Que.
Wm. C. Wilson & Co., Toronto, Ont.

SLEDGES
Wilkinson & Kompass, Hamilton, Ont.

SLINGS
Hopkins & Co., F. H., Montreal, Que.
Morris Crane & Hoist Co., Herbert, Niagara Falls, Ont.

SMELTER LININGS
Beveridge Paper Co., Montreal, Que.

SPECIAL MACHINERY
Corbet Fdry. & Machine Co., Owen Sound, Ont.
Miller Bros. & Sons, Ltd., Montreal, Que.
Smart-Turner Mach. Co., Hamilton, Ont.

SMOOTH-ON.
Wm. C. Wilson & Co., Toronto, Ont.

SPIKES
Wm. C. Wilson & Co., Toronto, Ont.

SPRAY COOLING SYSTEMS
Spray Engineering Co., Boston, Mass.

STEAMSHIP AGENTS
Page & Jones, Mobile, Ala.

STEAM SPECIALTIES
Can. Fairbanks-Morse Co., Montreal, Que.
Corbet Fdry. & Machine Co., Owen Sound, Ont.
Empire Mfg. Co., London, Ont.

STEAM TRAPS
Aikenhead Hardware, Ltd., Toronto, Ont.
American Steam Gauge & Valve Mfg. Co., Boston, Mass.
Empire Mfg. Co., London, Ont.
Mason Regulator & Engin. Co., Montreal, Que.
Smart-Turner Mach. Co., Hamilton, Ont.

STEEL, HIGH SPEED
Hopkins & Co., F. H., Montreal, Que.
Nova Scotia Steel & Coal Co., New Glasgow, N.S.
Wilkinson & Kompass, Hamilton, Ont.
Zenith Coal & Steel Products, Ltd., Montreal, Que.

STEEL SHELVING
Dennis Wire & Iron Works, London, Ont.

STEEL WORK, STRUCTURAL
Babcock & Wilcox, Ltd., Montreal, Que.
Corbet Fdry. & Machine Co., Owen Sound, Ont.

STEERING GEARS
Corbet Fdry. & Machine Co., Owen Sound, Ont.
Hopkins & Co., F. H., Montreal, Que.
Engineering & Machine Works of Canada, St. Catharines, Ont.
Wm. C. Wilson & Co., Toronto, Ont.

STOCK RACKS FOR BARS, PIPING, ETC.
Morris Crane & Hoist Co., Herbert, Niagara Falls, Ont.

STOKERS, MECHANICAL
Babcock & Wilcox, Ltd., Montreal, Que.

SUPERHEATERS, STEAM
Babcock & Wilcox, Ltd., Montreal, Que.

SWITCHBOARDS, ELECTRIC
Crouse-Hinds Co. of Can., Ltd., Toronto, Ont.

TALLOW
Wm. C. Wilson & Co., Toronto, Ont.

TANKS, STEEL
Corbet Foundry & Mach. Co., Owen Sound, Ont.
Goldie & McCulloch, Ltd., Galt, Ont.
Hopkins & Co., F. H., Montreal, Que.
Marsh Engineering Works, Belleville, Ont.
Port Arthur Shipbuilding Co., Port Arthur, Ont.
Reed & Co., Geo., Montreal, Que.
Scaife & Sons Co., Wm. B., Oakmont, Pa.

TANKS (AIR, GAS AND LIQUID)
Marsh Engineering Works, Belleville, Ont.
Scaife & Sons Co., Wm. B., Oakmont, Pa.

TAPPING MACHINES
Mueller Mfg. Co., H., Sarnia, Ont.

TELEGRAPHS, SHIPS
McNab Co., Bridgeport, Conn.
Cory & Sons, Inc., Chas., New York, N.Y.
Morrison Brass Mfg. Co., James, Toronto, Ont.
Wilson & Co., Wm., Toronto, Ont.

TELEPHONES, MARINE
Cory & Son, Inc., Chas., New York, N.Y.
McNab Co., Bridgeport, Conn.

TESTERS, METER
Mueller Mfg. Co., H., Sarnia, Ont.

THUMB SCREWS AND NUTS
Canada Foundries & Forgings, Welland, Ont.
United Brass & Lead, Ltd., Toronto, Ont.

TRACK SYSTEMS
Morris Crane & Hoist Co., Herbert, Niagara Falls, Ont.

TRAVELLING BLOCKS
Morris Crane & Hoist Co., Herbert, Niagara Falls, Ont.

TROLLEYS
Can. Fairbanks-Morse Co., Montreal, Que.
Morris Crane & Hoist Co., Herbert, Niagara Falls, Ont.

TROLLEY HOISTS
Morris Crane & Hoist Co., Herbert, Niagara Falls, Ont.

TRUCKS, HAND, ELECTRIC
Aikenhead Hardware, Ltd., Toronto, Ont.
Can. Fairbanks-Morse Co., Montreal, Que.

TUBES, BOILER
Babcock & Wilcox, Ltd., Montreal, Que.
Broughton Copper Co., Ltd., Manchester, Eng.

TUBES, COPPER AND BRASS
Mueller Mfg. Co., H., Sarnia, Ont.
Tallman Brass & Metal Co., Hamilton, Ont.
Standard Underground Cable Co., Hamilton, Ont.

TUGS
Collingwood Shipbuilding Co., Collingwood, Ont.
Polson Iron Works, Toronto, Ont.

TURBINES, DIRECT-DRIVING AND GEARED
Doxford & Sons, William, Sunderland, England.

TURNBUCKLES
Canada Foundries & Forgings, Welland, Ont.
Hopkins & Co., F. H., Montreal, Que.

TURNTABLES
Morris Crane & Hoist Co., Herbert, Niagara Falls, Ont.

SPIKES, SMALL RAILROAD
Severance Mfg. Co., S., Glassport, Pa.

UNIONS, ALL KINDS
Dart Union Company, Toronto, Ont.

VALVES, AIR
Mueller Mfg. Co., H., Sarnia, Ont.

VALVE, DISCS
Wm. C. Wilson & Co., Toronto, Ont.

VALVES
American Steam Gauge & Valve Mfg. Co., Boston, Mass.
Babcock & Wilcox, Ltd., Montreal, Que.
Can. Fairbanks-Morse Co., Montreal, Que.
Empire Mfg. Co., London, Ont.
McAvity & Sons, Ltd., T., St. John, N.B.
Mason Regulator & Engin. Co., Montreal, Que.
McNab Co., Bridgeport, Conn.
Norbom Engineering Co., Philadelphia, Pa.
Williams Machinery Co., A. R., Toronto, Ont.
Wm. C. Wilson & Co., Toronto, Ont.
United Brass & Lead, Ltd., Toronto, Ont.

VALVES, FOOT
Aikenhead Hardware, Ltd., Toronto, Ont.
Smart-Turner Mach. Co., Hamilton, Ont.

VALVES, STOP, REDUCING, SAFETY CHECK, ETC.
Aikenhead Hardware, Ltd., Toronto, Ont.
American Steam Gauge & Valve Mfg. Co., Boston, Mass.
McAvity & Sons Ltd., T., St. John, N.B.
Morrison Brass Mfg. Co., James, Toronto, Ont.

VALVES, MIXING
Mueller Mfg. Co., H., Sarnia, Ont.

VALVES, REDUCING, PRESSURE
Mueller Mfg. Co., H., Sarnia, Ont.

VARNISHES
Aikenhead Hardware, Ltd., Toronto, Ont.
Ault & Wiborg Co. of Can., Ltd., Toronto, Ont.
Leckie, Ltd., John, Toronto, Ont.
Moore & Co., Benjamin, Toronto, Ont.
Reed & Co., Geo., Montreal, Que.

VENTILATORS, COWL
McNab Co., Bridgeport, Conn.

VENTILATION EQUIPMENT
Empire Mfg. Co., London, Ont.
Hopkins & Co., F. H., Montreal, Que.
Low & Sons, Ltd., Archibald, Glasgow, Scotland
Smythes & Co., Toronto, Ont.

WASHERS
London Bolt & Hinge Works, London, Ont.

WASTE
Smythes & Co., Toronto, Ont.
Wm. C. Wilson & Co., Toronto, Ont.

WATER COLUMNS
Morrison Brass Mfg. Co., James, Toronto, Ont.

WELDING, ELECTRIC
Hall Engineering Works, Montreal, Que.
Marine Welding Co., Buffalo, N.Y.
Beatty & Sons, M., Welland, Ont.

WATER SOFTENERS
Babcock & Wilcox, Ltd., Montreal, Que.

WATER SUPPLY SYSTEMS
Mueller Mfg. Co., H., Sarnia, Ont.

WATER HEATERS
Empire Mfg. Co., London, Ont.
Morrison Brass Mfg. Co., James, Toronto, Ont.

WELDING, OXY-ACETYLENE
Advance Mach. & Welding Co., Montreal, Que.

WHISTLES AND SYRENS
Empire Mfg. Co., London, Ont.
McAvity & Sons, T., St. John, N.B.
McNab Co., Bridgeport, Conn.
Mitchell Co., Ltd., Robt., Montreal, Que.
Morrison Brass Mfg. Co., Jas., Toronto, Ont.

WINCHES, CARGO
Advance Mach. & Welding Co., Montreal, Que.
Aikenhead Hardware, Ltd., Toronto, Ont.
Corbet Fdry. & Machine Co., Owen Sound, Ont.
Hopkins & Co., F. H., Montreal, Que.
Marsh Engineering Works, Belleville, Ont.

WINCHES, DOCK, SHIP
Beatty & Sons, M., Welland, Ont.
Marsh Engineering Works, Belleville, Ont.
Miller Bros. & Sons, Ltd., Montreal, Que.
Morris Crane & Hoist Co., Herbert, Niagara Falls, Ont.
Wilson & Co., Wm. C., Toronto, Canada.

WINCHES, TOWING
Corbet Foundry & Mach. Co., Owen Sound, Ont.

WINCHES, TRAWL
Beatty & Sons, M., Welland, Ont.
Wm. C. Wilson & Co., Toronto, Ont.

WINDLASSES
Advance Mach. & Welding Co., Montreal, Que.
Corbet Fdry. & Machine Co., Owen Sound, Ont.
Duke Engine Co., Grand Haven, Mich.
Hopkins & Co., F. H., Montreal, Que.
Wilson & Co., Wm. C., Toronto, Canada.

WIPER CAPS, OILER BOXES, ETC.
Morrison Brass Mfg. Co., James, Toronto, Ont.

WIRE, COPPER CLAD STEEL
Standard Underground Cable Co., Hamilton, Ont.

WIRE, COPPER, BRASS, BRONZE
Standard Underground Cable Co., Hamilton, Ont.
Tallman Brass & Metal Co., Hamilton, Ont.

WIRE, INSULATED
Standard Underground Cable Co., Hamilton, Ont.

WIRELESS OUTFITS
Cutting & Washington, Inc., Cambridge, Mass.
Marconi Wireless Telegraph Co., Montreal, Que.

WIRE ROPE
Zenith Coal & Steel Products, Ltd., Montreal.

WOOD WORKING MACHINERY
Aikenhead Hardware, Ltd., Toronto, Ont.
Can. Fairbanks-Morse Co., Montreal, Que.
Preston Woodworking Machy. Co., Preston, Ont.
Yates Mach. Co., P. B., Hamilton, Ont.

WOOD BORING TOOLS
Aikenhead Hardware, Ltd., Toronto, Ont.
Can. Ingersoll-Rand Co., Sherbrooke, Que.

WOODITE GAUGE GLASS WASHERS
Wm. C. Wilson & Co., Toronto, Ont.

WRENCHES
Canada Foundries & Forgings, Welland, Ont.

YACHT BROKER
Watt, J. Murray, Philadelphia, Pa.

INDEX TO ADVERTISERS

CIRCULATES IN EVERY PROVINCE OF CANADA AND ABROAD

MARINE ENGINEERING of Canada

A monthly journal dealing with the progress and development of Merchant and Naval Marine Engineering, Shipbuilding, the building of Harbors and Docks, and containing a record of the latest and best practice throughout the Sea-going World. Published by
The MacLean Publishing Co., Limited

MONTREAL, Southam Building TORONTO 143-153 University Ave. WINNIPEG, 1207 Union Trust Bldg. LONDON, ENG., 88 Fleet St.

Vol. VIII. Publication Office, Toronto—June, 1918 No. 6

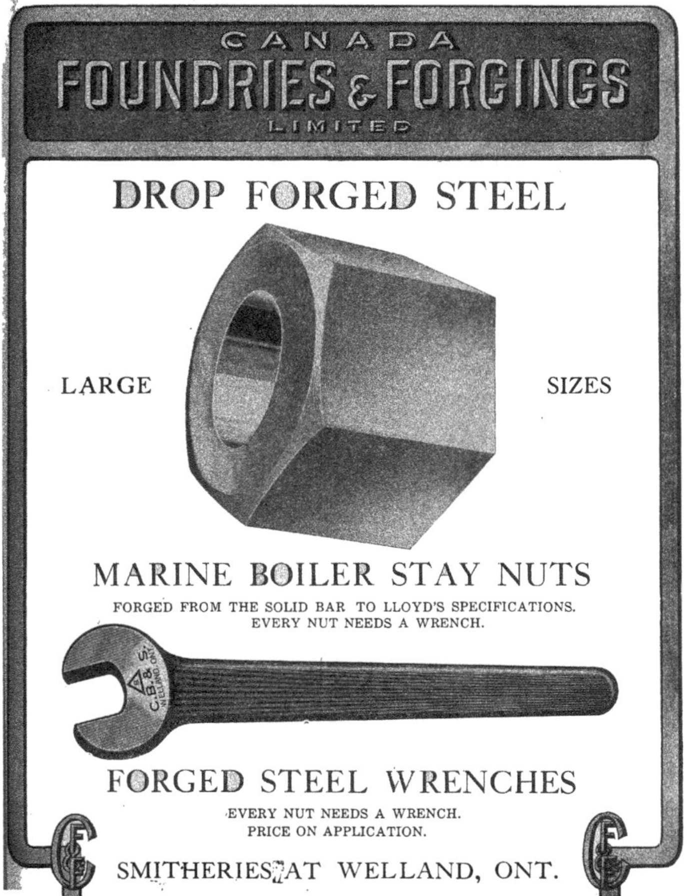
CANADA
FOUNDRIES & FORGINGS
LIMITED
DROP FORGED STEEL
LARGE
SIZES
MARINE BOILER STAY NUTS
FORGED FROM THE SOLID BAR TO LLOYD'S SPECIFICATIONS.
EVERY NUT NEEDS A WRENCH.
C.B.& S.
WELLAND, ONT.
FORGED STEEL WRENCHES
EVERY NUT NEEDS A WRENCH.
PRICE ON APPLICATION.
SMITHERIES AT WELLAND, ONT.

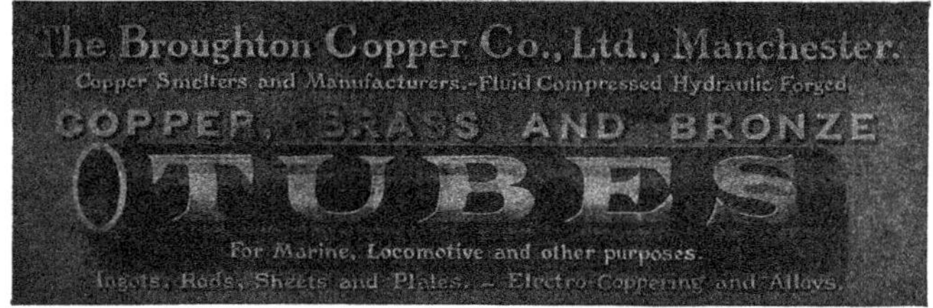
The Broughton Copper Co., Ltd., Manchester.
Copper Smelters and Manufacturers.-Fluid Compressed Hydraulic Forged,
COPPER, BRASS AND BRONZE
TUBES
For Marine, Locomotive and other purposes.
Ingots, Rods, Sheets and Plates, - Electro-Coppering and Alloys.

STEEL TANKS for every requirement
Compressed Air Tanks, Gasoline Tanks, Mufflers, Engine Starter Tanks, Oil and Water Tanks, Gas Receivers, Range Boilers, Etc.
Send for Catalogue
(116 years old—Founded 1802)
Wm. B. Scaife & Sons Co.
NEW YORK OFFICE
26 Cortlandt Street
PITTSBURGH, PA.

CASTINGS Grey Iron and Brass For Shipbuilding

Fast, Efficient Service

We are prepared to supply the shipbuilding trade promptly with good quality Grey Iron and Brass Castings. Any size--any quantity.

MARINE BOILERS AND ENGINES PARTS AND FITTINGS

Get in touch with us. Enquiries and orders given prompt attention.

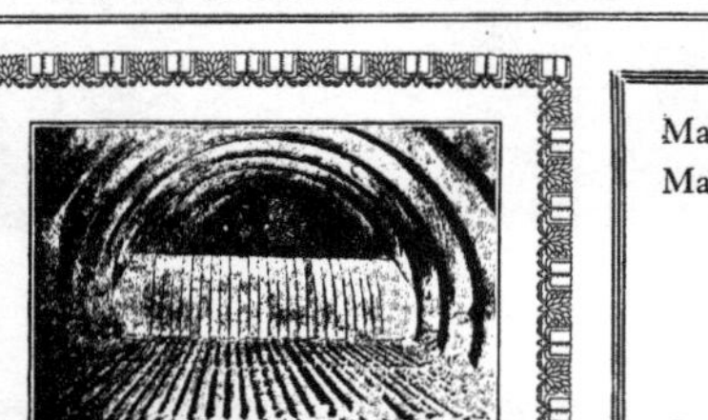

THE WAGER FURNACE BRIDGE WALL

¶ A preferred and valuable feature in marine and stationary boilers—endorsed by governments; steamship, freight and passenger steamer companies; railroads; stationary plants; others.

WAGER FURNACE BRIDGE WALL CO., Inc.
OF NEW YORK SINGER BUILDING
Philadelphia, Detroit, Seattle, Portland
San Francisco -:- Vancouver, B. C.

Mackinnon, Holmes & Co., Limited
Mackinnon Steel Co., Limited
Sherbrooke, Que., Canada

■ ■ ■ ■

Fabricators and Erectors
Specialists in

Structural Steel and Steel Plate Work

of every description
for wooden and steel ships.
A large stock of shapes and plates always on hand.

■ ■ ■ ■

Your enquiries are desired.

MARINE ENGINEERING
of Canada

Volume VIII. TORONTO June 1918

THE PRODUCTION OF LARGE MARINE FORGINGS

BY A. G. WEBSTER*

Suitable Equipment Enabled Urgent Work to be Carried Out in Good Shape —Heavy Tonnage and Quick Delivery Requirements Are Features of Marine Products Activity

ONE of the most interesting industrial developments in Canada since the outbreak of the Great War in 1914 has been the manufacture of marine forgings. With the building of ships on an extensive scale came the demand for material for their construction. It was fortunate therefore that there was a plant in Canada able to undertake this class of work, not only because of the benefits that accrue from having forgings made in the country but also because of the great saving in time owing partly to the fact that the plants in the United States have more work of this character than they can conveniently take care of. To have had to rely on the States for forgings under prevailing conditions would have handicapped the shipbuilders and considerably retarded the rapid construction of merchant vessels.

Equipped for Heavy Work

A well equipped plant had been operated at Welland, Ont., by the Canada Foundries & Forgings Co., who for some years had produced marine and locomotive forgings, but the former on a comparatively small scale, the demand not having been of the proportions calling for a heavy tonnage output. Despite this it was a matter calling for considerable expense and executive ability to organize the plant and labor, skilled and otherwise, to produce forgings of the kind required, to meet the rigid specifications. It requires skill of a high order to take an ingot of steel weighing 30 tons and reduce it down so that it will machine up clean and true without any undue expenditure of time in machining for removing too much surplus metal. A good and true forging means time saved in machining and consequently lower cost of production.

Complete Line of Forgings

The Canada Forge Co. manufacture a complete line of marine forgings for both engine and hull, including principally crank shafts and crank ends, line shafts, tail shafts, thrust shafts, rudder frames, stern posts, etc. The steel used in the manufacture of these forgings comes to the plant in the form of ingots or slabs according to the requirements of the work, and is all open hearth basic, of an analysis to meet Lloyd's requirements. It is obtained from firms making the best material, particular attention being paid to this feature in order to ensure uniform results. The ingots from which the larger forgings are made vary from 14 in. to 30 in. diameter, the latter size ingot weighing about 15 tons.

The steel must be free from piping, blow holes, or segregation and carefully selected in regard to chemical analysis. This is the most important factor in the success or failure of forgings to meet the physical specifications prepared by the inspection companies in relation to marine forgings.

Cutting Off Crop End.

The ingots are shipped to this plant with the crop end. The metal in the bottom end of the ingot is clean and homogenous, so does not need cutting off. The upper or crop end, on the other hand, has to be removed. This is done by first cutting a slot or nick about four inches deep all round the ingot with an oxy-acetylene torch and at a sufficient distance from the end to obtain metal free from piping or

*Associate Editor Marine Engineering.

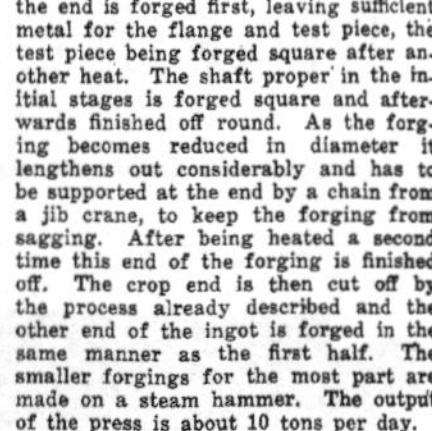

CAST STEEL INGOT BEING FORGED ON 500- TON HYDRAULIC PRESS. INITIAL STAGE IN FORGING A PROPELLOR SHAFT. FURNACES IN BACKGROUND.

other flaws. The ingot is lifted up by an overhead travelling crane and allowed to fall, to break the crop end off after nick has been cut. The broken end is then carefully examined for piping and if any defects are found; which occasionally happens, two or three more pieces are cut off on a power saw to remove the metal containing the piping. The removal of the crop end is done either before the first forging operation or afterwards, according to the requirements of the case. In large forgings the crop end is usually left on until the other end has been forged because it serves as a very useful balance weight during the forging operation; apart from this it is immaterial when it is cut off.

After being inspected the selected ingots are then carefully heated to a uniform temperature ranging between 1900 degrees and 2150 degrees Fah. and are forged down to the desired size and shape under a William Todd 500 ton hydraulic press, care being taken to do the finishing work upon the forgings when the temperature of the steel is within the "critical range" this ensues fine grain structure which in turn ensures greater ductility than would otherwise result from finishing at higher temperatures. Another important feature is to select ingots large enough so that the sectional area of the body of the forging shall not exceed one-fifth the sectional area of the original ingot.

Forging on Large Press

The large forgings and also some of the smaller ones are forged on a 500-ton hydraulic press. This press, it might be mentioned, was installed some years ago and was therefore in operation before this more recent development in the company's activities. The press is served by an electrically operated overhead travelling crane and also by a hand operated jib crane, while alongside the press are a number of large oil fuel furnaces.

In the case of an ingot that is to be forged with the crop end on, a bar is attached to that end for the use of the men during the forging operation. The bottom or sound end is forged first and after being heated for a sufficient length of time is carried over to the press. The ingot is carried by a steel chain sling suspended from an electrically operated overhead crane.

In making a line or tail shaft forging the end is forged first, leaving sufficient metal for the flange and test piece, the test piece being forged square after another heat. The shaft proper in the initial stages is forged square and afterwards finished off round. As the forging becomes reduced in diameter it lengthens out considerably and has to be supported at the end by a chain from a jib crane, to keep the forging from sagging. After being heated a second time this end of the forging is finished off. The crop end is then cut off by the process already described and the other end of the ingot is forged in the same manner as the first half. The smaller forgings for the most part are made on a steam hammer. The output of the press is about 10 tons per day.

On each forging a prolongation of the same sectional area as that of the body of the forging is left for testing purposes.

Heat Treating

All marine forgings are heat treated, or annealed, the company having facilities for handling the largest forgings. As a rule the forgings are heated in a furnace and allowed to cool gradually but in some cases they are oil annealed. For this process large oil tanks are installed, each tank having a compressed air system for agitating the oil to equalize the temperature. The oil is cooled by means of cold water which circulates round the outside of the oil tanks.

The forgings having been allowed to cool are now ready for the heat treating furnace, which is of the car type, over fired furnace, using oil fuel and equipped with Tycos base metal, thermo electric, recording pyrometers, which are of inestimable value in affording a means to check temperatures. The manner in which the forgings are placed upon the furnace is a matter of great importance. The charge is raised above the furnace floor with suitable supports and each forging is placed a few inches

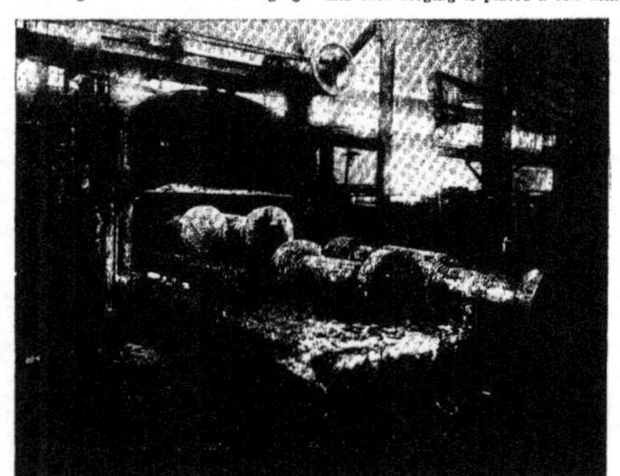

ONE OF THE ANNEALING FURNACES SHOWING SHAFT FORGINGS ON FURNACE FLOOR.

apart, so as to enable the hot gases to freely circulate through the entire charge, so as to insure uniform heating.

The chief reasons for heat treating forgings are to relieve internal stresses or strains which may have occured in the forging operations and to develop the dynamic and static strength of the forgings so as to meet physical tests that are required of them. The results in the heat-treating furnace are governed by the following: Rate of heating, temperature, length of soaking, and rate of cooling, as these features vary with the chemical content of each charge. The desired results can only be obtained with an operator of long experience.

Inspection and Testing

After heat treatment the forgings are ready for inspection by the surveyor or inspector who has been assigned to this duty. All the marine forgings made at this plant have to pass Lloyd's or equally high inspection. The surveyor places the impresion of his private stamp twice on the prolongation of the forgings, the one piece to be used for the bend, the other for the tensil test. The bend test pieces are usually machined to a rectangular section 1 in. wide by ¾-in. thick, with the edges rounded to a radius of 1-16 of an inch. These are required to be bent cold over their thinner section through an angle of 189 degrees without fracture, the internal radius of the bend being not greater than ¼ of an inch. The tensile tests are machined to a standard size, with a diameter of .564 inches, and a gauge length of 2 in.

The test pieces are then gripped between the upper and lower jaws of a testing machine and the total resistance to rupture is measured. Knowing the cross-sectional area of the test pieces and the load required to break them, the strength per square unit may then be determined. The tensile strength is usually given in pounds per spare inch. The elongation is measured in per cent. of the original test section and is determined from the amount of stretch which occurs when the test has been pulled apart by tension.

The unsurpased quality of heat treated forgings manufactured at the Canada Forge Plant is conclusively proven by the ¾ cup and not uncommon full cup silky fractures combined with high elastic limits and large percentages of reduction of area which occur at this static test. The test results having been approved of by the surveyor the forgings are then ready for machining. The company has a large machine shop equipped with modern tools for machining forgings and other work in connection with its business.

ANOTHER VIEW OF ANNEALING FURNACE SHOWING PROPELLOR SHAFT FORGINGS ON FURNACE FLOOR.

INTERIOR OF NEW MACHINE SHOP. GROUP OF MISCELLANEOUS MARINE FORGINGS IN THE FOREGROUND.

A FAMOUS YACHT

IT may be remembered that the famous yacht Germania was at Southampton in August, 1914, having apparently been sent thither with a view to making the British public fancy that there could be no intention in Germany of going to war at that time when her leaders were preparing to compete at so innocent an international gathering as the regatta at Cowes. Be that as it may, the yacht was in due course seized and condemned in prize as enemy property. Being ordered to be sold, she was offered for sale under the direction of the Admiralty Marshal. Being held to be not of commercial value, whilst she was assuredly not fitted for purposes of war, neutral bidders were invited to compete for her, and she passed into the hands of Captain Hans Hannevig at the price of £10,000. This gentleman resides in London, but he is, in fact, a Norwegian subject. So he was called upon to give security to the court that the vessel should not pass into enemy hands. Such security was accordingly given. It took the form of a deposit of £10,000 five per cent. war bonds, which were lodged with the Marshal, Mr. H. W. Lovell. That official has now handed over £5,000 to the British Red Cross Society, and a similar sum to its French sister association. In sending the money, Mr. Lovell explains the circumstances under which he was enabled to hand over so splendid a gift. It appears that in buying the yacht, the actual purchaser was acting for his brother, Mr. Christopher Hannevig, who resides in

INTERIOR OF ONE OF THE MACHINE SHOPS. DRILLING CONNECTING ROD FOR MARINE ENGINE.

United States. It was this latter gentleman who found the money for the guarantee and for the purchase of the war bonds. He subsequently expressed the wish to deal with these war bonds in the way we have indicated, and he was allowed to have them released for his noble purpose on his putting up a bank guarantee for their value, the condition of that guarantee being to insure that the yacht should not get into enemy possession. So the Germania will in future fly the Stars and Stripes, and have her name associated with a princely and humane gift.

PORT OF LAS PALMAS

For years the Port of Las Palmas, in the Canary Islands, has held the first position among Atlantic coaling stations owing to its unique position in the track of shipping between European and African ports, South and Central America, and to its sheltered port, where coaling and cargo operations are possible at all times. Thus the shipping movement which, in 1883 comprised a little over 200 ocean steamers, rose to close on 5,000 in 1913, the year before the war, the quantities of bunker coal supplied being respectively less than 25,000 tons in 1883 and close on 900,000 tons in 1913. An enormous development of trade has been the natural outcome of the shipping facilities afforded by the numerous vessels making this place their port of call and sundry new lines have also been established, bringing the Canary Islands in direct communication with most of the ports of Europe. The population of Las Palmas with its port of La Luz, which was about 20,000 souls in 1883, has grown to near 70,000 at the present time.

This outstanding development of the place has made it clear that the harbor works, considered far too extensive when first created some thirty years ago after the plans of Engineer Juan León y Castillo, have proved inadequate to cope with the present-day traffic. Various plans have been elaborated to enlarge the existing port and to create further facilities for the rapid handling of cargo and improve the bunkering operations, but only to be shelved for lack of support, high cost and opposition and interference from other quarters.

In 1913, however, the commission appointed to examine the various plans resolved to call for a new project, and as a result the present director-engineer of the harbor works has presented plans for a work of the first magnitude entirely on up-to-date lines. These plans are now being actively pushed forward through the usual mill of commissions and reports. Fortunately Spain is at present fully alive to the necessity of improving its harbors and bringing them up to present day requirements, and a general project of granting credits to certain ports for such works is to be brought before the Cortes when they reassemble, and the plans for Las Palmas will be included.

The government grant under this measure will only allow for a minor part of the whole project to be executed, but even this will allow a notable improvement in the existing conditions.

The more immediate works contemplated include:

(1) Dredging the shallower parts of the inner and outer harbors to a minimum depth of 25 feet n.t.

(2) Widening the existing landing pier (Santa Catalina Mole) over the greater part of its length, to double its actual width.

(3) The erection of electrical plant for cargo handling on said pier.

(4) Widening part of the existing breakwater, making it suitable as a cargo-handling quay.

(5) The building of a new breakwater to the south of the Catalina Mole, enclosing the actual outer port.

(6) The erection of new quays with warehouses inside this breakwater, offering a quay length of about 750 metres.

HOW VESSELS ACT WHEN SINKING

By M. L.

In view of the very large number of vessels that have gone to the bottom of the sea in the last three years, more or less, it is of interest that each particular design of craft has its own peculiar way of sinking. The old single-bottom steamers for example, with a minimum number of bulkheads, almost invariably go down on a more or less even keel, which means that they do not stand on end, as did the Titanic and others since, before they disappear beneath the waters. Modern vessels, however, with numerous subdivisions, perform the feat of diving to Davy Jones' locker either with their bows or sterns high out of the water or by first showing a very heavy list to port or starboard. This is caused by the fact that while a sufficient number of compartments become filled to sink these vessels there are other compartments which remain watertight and prevent the water entering from seeking its level.

SUBMARINE AMENITIES

In the course of a controversy on the submarine question before the outbreak of war, Admiral Bacon gave a vivid description of the conditions attending submarine navigation.

"Operations are so simple on paper without sea conditions to contend with. If any one of your readers wishes to appreciate some of the difficulties of submarine work, let him sit down under a chart of the Channel suspended from the ceiling, let him punch a hole through it, and above the hole place a piece of looking glass inclined at 45 deg. Let him further imagine his chair and glass moving sideways as the effect of tide. Let him occasionally fill the room with steam to represent mist. Let him finally crumple the chart in ridges to represent the waves, and then try to carry out some of the manœuvres which look so simple when the chart is spread out upon the table, and looked down upon in the quiet solitude of a well-lit study."

Standard Cargo Ships: Construction and Operating Features

By S r George Carter, K.B.E.

The subject of standardized ships possesscses peculiar interest for marine engineers, apart altogether from the war conditions which have brought about the present development. Previous conditions obtaining in European and American shipyards varied sufficiently to cause different method of approach to be expected from the industries of the various countries involved. The views of Sir George Carter, delivered before the Institute of Naval Architects at its recent meeting, are of value to our shipbuilding concerns in view of ultimate competition which must be faced by Canadian shipyards.

THE object of this paper is to place on record the advantages of standardization in the construction of cargo vessels as a war measure, and to give some account of the work of the Advisory Committee on Merchant Shipbuilding in the design and construction of what are now known as "Standard Ships."

Towards the end of 1915 it was apparent that very great efforts would have to be made to provide sufficient cargo tonnage to supply the needs of the country. During this year and the following one (1916) the workers in shipyards had become very much reduced. Many men had left for the Army, and many others had been transferred to work in connection with the production of munitions of war. Apart from this, the demand for war vessels of all kinds was very great, and, above all, the demands for steel for war purposes increased to such an extent that the supplies available were inadequate. The many demands on the steel supply were all urgent, and it became exceedingly difficult to strike a fair balance between them. As far as shipbuilding was concerned steel for warships obtained priority, with the result that the supply for merchant ships was small and out of all proportion to the necessities of the case. The Board of Trade realized the position and strongly advocated the need for providing for the building of merchant ships, and, in conjunction with the Admiralty, some success was achieved, but far from sufficient to meet the increasing activities of enemy submarines. It soon became apparent that more drastic action was necessary. In December, 1916, the New Ministries and Secretaries Act was passed, which provided for the appointment of a Shipping Controller. The Act gave this Minister wide powers and provided that he should "take such steps as he thinks best for providing and maintaining an efficient supply of shipping." Sir Joseph Maclay became the Shipping Controller, and the Government decided at once that merchant shipbuilding should be under his control.

Merchant Shipbuilding Advisory Committee

In December, 1916, after a conference with shipbuilders and marine engineers, Sir Joseph Maclay appointed the Merchant Shipbuilding Advisory Committee of which I had the honor to be chairman.

Table I.—One-deck Single-screw Cargo Steamers.

Deadweight all told in tons	1,000		2,000		4,000		6,000	
	Steel	Ferro-Concrete.	Steel	Ferro-Concrete.	Steel	Ferro-Concrete.	Steel	Ferro-Concrete.
	ft. in.	ft. in.	ft. in.	ft. in.	ft. in.	ft. in.	ft. in.	ft. in.
Length between perpendiculars	180 0	210 0	220 0	245 0	285 0	305 0	340 0	350 0
Breadth moulded	29 0	33 6	35 0	38 9	41 6	44 3	45 9	48 0
Depth moulded to upper deck	16 6	16 9	20 0	20 3	25 0	25 3	28 3	28 6
Mean draught	14 9	14 9	17 9	17 9	21 3	21 3	23 3	23 3
Displacement in tons	1600	2225	2910	3660	5550	6465	8240	9210

Table II.—Factors for Conversion of Deadweight Carrying Capacity into Total Displacement.

Deadweight all told in tons	1,000	2,000	4,000	6,000
Steel vessels	1.600	1.455	1.387	1.373
Ferro-concrete vessels	2.225	1.830	1.616	1.535

Table III.—Excess Displacement of Ferro-concrete Vessels over Steel Vessels of same Deadweight Carrying Capacity.

Deadweight all told in tons	1,000	2,000	4,000	6,000
Excess per cent.	39	25.8	16.5	11.8

Table IV.—Ratio of Increase of Stress Tons per Square Inch.

Deadweight in tons.	1,000	2,000	3,000	4,000	5,000	6,000
On upper deck	1	1.18	1.33	1.44	1.54	1.64
Bottom plating	1	1.22	1.40	1.58	1.75	1.94

350,700 hexagon nuts, ½", British thread.

The first meeting of the committee was held on December 20, 1916, and after fully considering the whole question it was decided, in collaboration with the Shipping Controller and his other advisers, to proceed at once with an extensive building program of cargo ships of simple design and, as far as practicable, of standard types, both with respect to engines and hull. Although the Government's shipbuilding program gave standardization in shipbuilding concrete form on a large scale, standardization has been widely discussed and urged as for practicable policy both in regard to hull and machinery, and had in fact taken definite shape on a smaller scale by individual firms adopting standard types of their own.

It will be well at this stage to call attention to the great variety of merchant vessels that were under construction at this time. They totalled nearly 500, with an aggregate gross tonnage of over 1,800,000 tons. Amongst them were found fast and slow liners, intermediate passenger and cargo types, cargo boats built for special trades, fast cross-channel steamers, vessels built to carry special cargoes, "tramp" steamers, coasting vessels, and, in fact, all the varieties usually to be found in pre-war days. Even where the types were the same individual vessels differed in size, speed, and arrangements. Many of them were built for special purposes, and practically all of them embodied ideas that were peculiar to particular owners and builders. Whilst many of these specialties undoubtedly served their purpose it is perhaps questionable whether a number of them were absolutely necessary. At any rate, since the standardization of cargo ships has been introduced, it is generally conceded that much of it will continue to be practised even after shipbuilding and shipping conditions have again become normal, the reason being that in many ways economy of time, material, and labor are obtained by its adoption.

It can at once be stated that the greatest variations, both from the point of view of degree and number, were to be found in vessels carrying passengers. The need, however, for refinement of type, comfortable passenger accommodation, record passages, etc., had ceased to exist at the time of the appointment of the Advisory Committee. On the other hand, the need for plain cargo carriers, and plenty of them, was greater than ever it had been before. It was therefore decided that the vessels already under construction which fulfilled to some extent these latter conditions should proceed, and that standard ships should be constructed with all possible speed.

Standardization

The principle of standardization having been accepted, it became necessary to design standard ships to meet the most urgent needs of the Shipping Controller. It was clearly desirable to limit the types as much as possible, but apart from the requirements of the Shipping Controller, the employment of the available facilities in the shipbuilding yards to the greatest advantage had to be considered. For this reason several different lengths were used for the ships, in order to use the greatest number of slips to their best advantage. Needless to say, the practical difficulties were great. No two yards had built identical vessels; each builder had his own special knowledge, his own particular type or types of vessels. Again, practice differed in different districts. The practice in one district was looked upon with disfavor in another. Shipowners, moreover, were anxious, quite naturally, to have their fleets kept up, and preferred vessels of the types similar to those built for them in pre-war days.

It has been said, that if builders had been allowed to proceed in their own way and to their own designs, that the output of ships would have been greater than it has been under standardization. Such a contention will not bear examination when all the facts are considered impartially. There has undoubtedly been delay in the production of standard ships in spite of all the arrangements made to expedite their construction, the chief element being shortage of steel. If each builder had been allowed to proceed with his own type or types of ships, this delay would have been greater, as in addition to not getting the amount of steel required, the multitudinous sections necessary would have caused great delay in rolling at the steel mills, whose output was much increased by the simplification of sections in the standard ships, referred to later. It is not sufficient to compare the rate of production of standard ships with the pre-war rate of production. The limited supply of steel and other adverse conditions must be taken into account.

The argument that delay was caused by asking builders to construct a standard ship differing only slightly from their own type may have some justification in regard to one ship, but not to repeat ships, and it must not be forgotten that never before had a proposition been put forward to build an unlimited number of ships of similar types throughout all the yards. It is interesting also to note that in one yard a standard ship was completed in a little over six months, although the yard had never built a similar ship before. A second similar ship was also completed in practically the same time. In another yard a ship was completed, coaled, and sailed 18 days after launching. In spite, however, of all the difficulties and the diverse opinions held by responsible shipbuilders and engineers the policy of building standard ships has been loyally accepted by them, and I think it may be claimed that as a war emergency measure the policy is being justified, and will in the future be justified to a much greater extent.

Advantages of Standardization

As the policy of building standard ships in bulk is a new one, it will not be without interest to put on record some of the advantages which are claimed for standardization of this kind. They may be summarized as follows:—

1. The detail design for the type can be entrusted to a builder who is conversant with that class of vessel and whose duty it becomes to make the necessary calculations, prepare the specification and all the constructional drawings, for the information of other builders instructed to build vessels of the same type. Small details can, however, still be arranged in accordance with the ordinary practice of the individual builder.

2. The design being in the hands of one firm, all the steel scantlings after being settled in the ordinary manner can be carefully examined to reduce the number of steel sections in the ship as much as possible. This results in producing the fewest changes of rolls at the steel works and in the rolling of much larger parcels of material to the one pattern on account of the large number of ships that are to be built to this design. As an example of this, in the standard ships now built the number of sections are reduced to 8 or 10 as against 30 to 40 used in pre-war days for the same type of ships.

3. All the scrieve board work would be done by one yard giving full-size frame lines, beams, side keelsons, etc., copies of

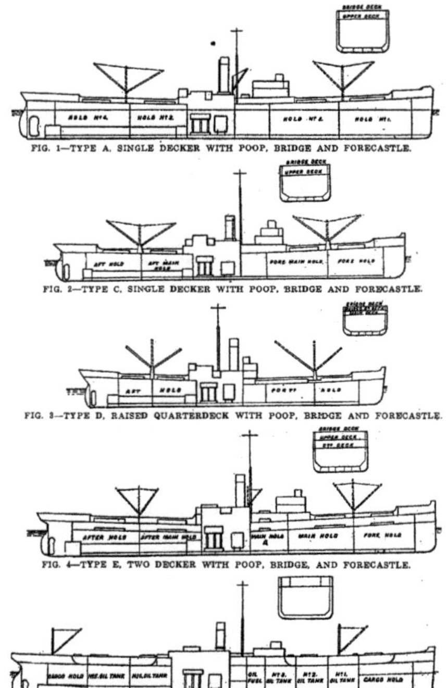

FIG. 1—TYPE A. SINGLE DECKER WITH POOP, BRIDGE AND FORECASTLE.

FIG. 2—TYPE C. SINGLE DECKER WITH POOP, BRIDGE AND FORECASTLE.

FIG. 3—TYPE D. RAISED QUARTERDECK WITH POOP, BRIDGE AND FORECASTLE.

FIG. 4—TYPE E. TWO DECKER WITH POOP, BRIDGE, AND FORECASTLE.

FIG. 5—TYPE Z. SINGLE DECKER WITH POOP, BRIDGE AND FORECASTLE.

which can be supplied to the other firms concerned.

4. Steel-makers and builders can agree to divide the orders for material in accordance with the facilities of the steel works, and having regard to proper times of delivery, so that the material first required is the first in order of delivery.

5. Whilst there is not the same difficulty with steel plates as with sections, arrangements for the proper order of delivery can be made.

6. With machinery the advantages of standardization are probably more generally agreed upon, and more detailed attention is given to this below. It may, however, be remarked here that the elaboration and production of design for auxiliary machinery and detailed fittings can be divided among the various makers, and each can arrange to undertake the whole manufacture of the details of a particular requirement. This does not by any means exhaust all the advantages of standardization, for after these preliminaries have been settled many benefits arise during the construction of the ships and engines. The standard ships, although not all of the same type, have many features in common. For instance, similar sets of engines may be fitted in ships of different types, so that apart from the large number built of each type there is a still larger number into each of which a given set of engines can be placed. In the case of any one vessel and set of machinery built in the same yard, it may so happen that either hull or machinery will be completed some time ahead of the other. In such cases the machinery originally intended for one ship can be transferred to another, no matter where it is being built. These rearrangements are of great assistance in preventing, on the one hand, congestion of the engine works, and on the other, delay to vessels through machinery being behindhand. The auxiliaries and fittings, including forgings and castings, being alike in vesels of each type, can be ordered in large numbers from the same maker and used in any ship or ships which may be ready to receive them.

Standard Ships

Outline profiles and midship sections are given of some of the standard ships in Figs. 1 to 5, annexed, together with their principal dimensions. The general arrangement of the machinery as fitted is shown on page 146.

The first two designs made were the "A" and "B" types. Both these vessels are of the same dimensions, but the first is of the single-deck and second of the two-deck type; their length is 400 ft., and deadweights 8,200 tons and 8,100 tons, respectively.

These types were followed by the "C" type, length 331 ft., and deadweight 5,050 tons, and the "D" type, length 285 ft., 2,980 tons deadweight.

The "C" and "D" types were laid down on the shorter berths as they became available, and later the "E" type was added so that the best use might be made of berths exceeding the "C" vessel's length, but under the "A" and "B" length; the "E" type is 376 ft. in length, and 7,020 tons deadweight.

For these five types of vessels two standard types of engines only were reber of the "A" and "B" type vessels, are all fitted with the standard "A" engines. The "C" and "D" vessels again have a type of engine common to both, although in the case of the "C" type additional boiler-power is provided.

To meet the urgent need for more oil carriers an important alteration was made later in the construction of a number of the "A' and "B" type vessels, through which, if the necessity arose, they could be completed as oilers.

In addition heavy oil carriers of 8,000 tons capacity known as the "Z" type are building. The overall dimensions of the "Z" vesels are the same as the "A" and "B"; the same standard type of engines are used as in the "A," "B," and "F" vessels.

All the types mentioned, except the "Z" oiler, were decided upon, the designs prepared, and details arranged while the Merchant Shipbuilding Advisory Committee was still meeting at the Ministry of Shipping, and other types were projected.

In designing the standard ships a point kept in view was the necessity for producing as quickly as possible plain cargo boats of the maximum carrying capacities with the least expenditure of material and labor, and of sizes most suitable to the majority of building berths and shipping conditions, etc. I feel that it is necessary to emphasize this statement because obviously the Government standard ships were designed to meet urgent war conditions. Whilst they might have a bearing on future standardization, the question has not been dealt with from a peace standpoint.

The detailed design of each type in the first place was entrusted to a builder conversant with the class of vessel required and whose duty it was to carry out the work already mentioned. The engines were dealt with in a somewhat similar manner except that one Scotch and one English engineer used to the class of work collaborated on the one design and its details—these and the design of the ship afterwards meeting with the full assent of the Committee. It was arranged that the Classification Societies should concur in the various scantlings, and undertake, on behalf of the Ministry of Shipping, the entire supervision of the vessels during construction. Two surveyors of Lloyd's Register who had been associated with the Ministry of Munitions in the production of steel were assigned the duty of allocating the orders for steel material to the steel-works and of arranging for its distribution to the builders in conjunction with the Government officials as required. To facilitate the leading and discharging of cargoes large hatchways were arranged with an ample provision of derricks and winches. In the same connection the pillars were specially arranged as mentioned in Appendix I. Further points of detail are discussed in Appendix I. for the hull, and in Appendix II. for the machinery.

General Conditions

As experience was gained with vessels already at sea which had received damage by mine or torpedo, it was only to be expected that modifications would be suggested. For this reason certain alterations were effected in the early standard ships after the plans were prepared. These were principally made on the advice of the anti-submarine authorities; but one other particular alteration is worthy of mention, in that greater safety and comfort is given to the crew by quartering them in the poop rather than in the forecastle, as was the usual practice before the war. It will be remembered that the Council of the Institution formed a Committee last year to consider the question of the effect of mine and torpedo explosions on the structure of merchant ships. The majority of the findings of this Committee, it will interest members to know, were embodied in the standard ships.

Certain precautions were taken to render the visibility of the standard ships as small as possible. Some of these can be mentioned, while others had better be left unnoticed. The derricks and derrick posts were made to hinge down, and there is only one mast for carrying the wireless aerials. This latter is partly telescopic and can be let down to the same level as the top of the funnel. The funnel is also made lower than usual.

Ordinary deck-houses were fitted, and these in conjunction with other outstanding features of the ship can be masked by those on board when desired by the use of arrangements provided.

The speed and horse-power of the standard vessels have purposely been omitted from this paper. There must be many to whom these particulars are familiar. Perhaps there are still some who think that the vessels should have been given greater speeds, although the speeds are somewhat higher than is usual for this type of ship. The direct answer to this is that the actual speeds were concurred in by the Admiralty, who were in a position to know the requirements in that direction better than any one else. This was the guiding factor of the Committee, but it cannot be overlooked that increase of speed in any ship can only be obtained at great cost, and this is particularly the case in cargo ships of full form. Apart from this, at the time the standard ships were designed, consideration was given to the successful commercial use of these vessels after the war. Nevertheless, the great loss in cargo-carrying that would be entailed by fining up the lines, increasing the power of the engines, and the fuel that the vessels would have to carry, etc., must be considered when the question of speed is being viewed from all standpoints.

It may not be out of place here to refer to what are termed "unsinkable" ships. The word "unsinkable" must, of course, be used in a relative sense, because any ship can be sunk provided she receives a sufficient amount of damage. Advocates of unsinkable ships themselves usually specify that the word only applies to some definite amount of damage. Any responsible naval architect can design such an unsinkable ship, and actually many such designs were submitted to the Advisory Committee. There is one objection, however, which is common to all such types of vessels; in every case they require the expenditure of more steel per ton of deadweight carried. At the time the Advisory Committee was formed, as has already been indicated, steel was extremely scarce. Apart from this, all provisions for safety, if of an appreciable value, must complicate the cargo arrangements. Some curtail the amount of cargo carried, and all render it more difficult to handle the cargo. This, apart from increased steel required in construction and lengthened time of building, further limits the carrying capacity of the ship by increasing her time for loading and unloading.

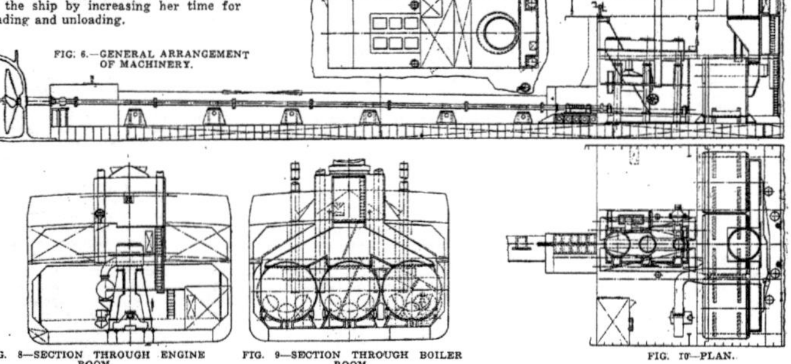

PLAN OF CASING TOP.

FIG. 6.—GENERAL ARRANGEMENT OF MACHINERY.

FIG. 8—SECTION THROUGH ENGINE ROOM.

FIG. 9—SECTION THROUGH BOILER ROOM.

FIG. 10—PLAN.

In May, 1917, the Department of the Controller of the Navy was formed and the work hitherto done by the Ministry of Shipping in connection with Merchant Shipbuilding was transferred to the new Department. The Merchant Shipbuilding Advisory Committee was also transferred, but shortly after this, for reasons which need not be stated here, the Committee resigned, and the Controller of the Navy immediately formed a Shipbuilding Council to assist him in regard to all matters respecting shipbuilding, both merchant and warship building and repairing. The members of the old Advisory Committee became members of this Council. Further types of standard ships have been evolved recently, but I do not propose to include them in this contribution. Enough has been said to illustrate the general principles involved. To introduce any general system of standardization of ships would have been difficult in normal times, it is only to be expected that it has proved to be so even in the midst of the greatest war in history. On the other hand, the need having arisen so acutely called for the attempt to be made. That criticism has arisen as to the Shipping Controller's policy, and the Advisory Committee's action, was only to be expected when one considers the eminent ship designers and shipowners who have special ideas and interests in the design of some ships already built, and the many ideas that have at one time or the other been proposed. That good results have, however, been attained will, I have no doubt, be generally conceded, and the appointment of a Committee to consider the general standardization of marine engines is a sign that we may expect standardization to play a definite part in the future. In post-war work we shall have to consider everything making for cheapness and rapidity of production, and standardization will not be the least among such considerations.

Appendix I.—Hull Details

In all the vessels the longitudinal and transverse scantlings have been designed generally to comply with the standard of strength recommended by the Load Line Committee.

With a view to obtaining simplicity of construction, special consideration was given to the arrangement of the details of the structure, and it was decided to omit the ceiling on the tank top and the stringer on the ship's side, to fit only one keelson on each side in the double bottom, and to have large single angles at the upper and lower edges of the centre girder, suitable structural compensation or arrangements of equivalent strength being fitted in each case. The open bottom method of construction with solid floors on every third frame was adopted in the holds and boiler space, except in types "C" and "D," where on account of the probability of these vessels being engaged in carrying ore and coal cargoes, solid floors are fitted on every frame throughout.

To reduce labor and avoid the necessity of bending the main frame, brackets of extra depth are fitted on the tank margin, and no tumble home was given to the frames at the top sides. Only a centre row of pillars, reeled for shifting boards, was fitted (except for hatch side pillars in types "B" and "E"), and advantage was taken in several instances of the relative position of the hatchways and watertight bulkheads to support the deck by large brackets attached to the bulkheads, the holds being thus as free as possible from obstructions of any kind. To compensate for the absence of pillars at the sides of the hatchways, an arrangement of reinforced hatch-end beams and side coamings was adopted, the latter being also incorporated in the longitudinal strength of the ship in the way of the hatchways nearest amidships. In types "B" and "E," on account of the excessive depth required in the reinforced hatchways, side coamings and end beams at the second deck, when centre pillars only were fitted, it was decided to fit a pillar under each hatchway side coaming at the middle of the length of the hatch.

The officers and engineers, in accordance with the most recent practice, are berthed together in a large house at the fore end of the bridge deck.

Approximate flooding calculations were made for all types of the standard ships.

In vessels "A," "B" and "E," with the main bulkheads in the positions arranged by the builders, it was found that, by making the cross-bunker bulkhead watertight, the vessels would approximately comply with the "two-compartment" standard in way of the machinery space and cross bunker, and the "one-compartment" standard elsewhere. The position of the cross-bunker bulkhead was arranged in order to fulfill the above-mentioned conditions.

In vessel "C," in addition to the bunker bulkhead, an additional bulkhead was fitted in the after-hold, in order to comply with the same condition.

Among the detailed requirements, carried out in these ships which meet the findings of the Committee appointed by this Institution to inquire into the effects of mine and torpedo explosions, might be mentioned the following: No watertight door is fitted between the engine-room and the shaft tunnel, an access trunk being provided from the upper deck to the forward end of the tunnel. In addition, where watertight doors are fitted, they can be worked from an upper-deck position. Each suction-pipe, where it enters the compartment from which it drains, is provided with a screw-down, non-return valve worked from the bulkhead deck.

Appendix II.—Engine Details

The details of the engine design are arranged so that the machinery is rather over-driven for war emergency measures; this policy was followed so that after the war the ships become fairly reasonable commercial propositions. For example, the "A" engine would in peace times probably run at about 80 per cent. of the power obtained from it at present.

As with the hull, all the designs, when settled as previously mentioned, with the exception of the auxiliary machines, were prepared by one firm and issued to the various machinery contractors complete, together with full detail specifications of all the raw material orders and the finished items which it was arranged to obtain from sub-contractors. The drawings and informations issued contained quite an unusual amount of detail, so that firms not familiar with such work would have no difficulty in carrying it out.

The main engine design does not present any new features. There are, however, some outstanding details. The front columns which carry the astern piston-rod guides have been utilized as engine oil-tanks. The moving parts of the engine are of the simplest type, and every attention has been paid to facility for overhaul. The line shafting is all in interchangeable lengths. The reversing gear is of the all-round type, and a turning engine and gear is fitted at the after coupling. Ease of machining and assembling has been provided for in the arrangement and in the details, the piston type of valve for the medium pressure cylinder, for example, being adopted chiefly on that account. The centrifugal circulation pump is mounted with its engine on the main engine bed-plate, so that it can be fitted complete with its pipes and connections in the erecting shop.

Reference has been made in the paper to the advantages derivable from interchangeability, and examples of it have actually occurred. One firm, on discovering a defective sole-plate casting, had it replaced at once by a similar casting from another firm who did not require it at the time; this avoided a delay of several weeks in the erection of the engines. In another case a replaced main steam-pipe branch-piece was immediately dispatched to a vessel ready for her trial trip, the original one having shown leakage on the first steaming of the engine. Sub-contractors in particular were benefited by the multiple work, as they were enabled to proceed with a large number of each article instead of finding differences in small points of designs as invariably occurs when the designs are prepared by independent firms or individuals. The principle of repetition in design was embodied everywhere. Small castings were ordered in multiple.

Apart from the advantage to the sub-contractors in carrying out repeat-work, small items in the engine-room which could be manufactured to better advantage by specialists rather than by large firms were ordered in large numbers for all the firms involved. This would not have been possible had they been different for each vessel. The orders for items such as auxiliary machine valves, valve-boxes, branch-pieces, cast and wrought iron pipes, and so on, were issued in multiple from a single source by firms specializing in this work. For example, the main steam pipes were finished complete by tube-makers, the drawings having been so designed that the unavoidable inaccuracies in fitting boilers and engines in place could be taken up at the joints, thus overcoming the delay which otherwise invariably occurs when pipes have to be made to fit after everything else is in position.

The auxiliary machinery was arranged so as to give the least work in completing the vessel after it was launched, the various connecting-pipes being of the simplest form while providing all the usual conveniences for working. The feed filter and feed heater were embodied in a compact form in the engine hot-well, the feed-filter being easy of access. The winches exhaust to a winch-condenser in the engine-room, the size of which is ample to deal adequately with the winches when they are being used for a rapid discharge of cargo. The general service donkey and harbor-feed donkey are duplicates, and are capable of feeding the boilers in port or at sea. The use of copper piping was avoided as much as possible, and wrought iron was largely used in consequence. except for very small pipes. The result is that the weight of copper piping is less than one-sixth the usual amount in this class of vessel.

The boilers have been designed in accordance with well-established practice, and there is ample room allowed for cleaning purposes and for overhauling. Howden's forced draught is fitted in all cases.

ALLIED SHIPS' BOARD FORMED

A DESPATCH from Washington states that centralized control of transatlantic shipping has been established with the creation of a Ship Control Committee to have supreme charge of the operation of all ships—American, allied and neutral —entering and leaving American ports.

The committee was named by representatives of the Shipping Board, the War and Navy Departments, the Food and Fuel Administrations, the Director-General of Railraods, the British Government and shipowners, who met to devise some plan for speeding up the movement of supplies to Europe. It comprises P. A. S. Franklin, of the International Mercantile Marine, Chairman; H. H. Raymond, head of the Clyde and Mallory Lines, and recently made Shipping Controller at New York, and Sir Cunnop Guthrie, Director of British shipping in this country.

The arrangement, as explained by Shipping Board officials, in effect creates a pool of ships moving supplies to Europe. Goods destined for overseas will be loaded in available ships whether operated by the United States or the allies. With the aid of the Railroad Administration the committee will divert to southern ports much of the supplies that heretofore have clogged the port of New York, and incoming vessels will be directed by wireless to proceed to the ports in which materials of the most importance await shipment.

AN UNFORTUNATE APPOINTMENT

Professor Goodwin, speaking at the annual meeting of the Canadian Manufacturers' Association, quoted from a Canadian technical paper to prove some interesting information he was presenting to the Association, and said: "The fact that we now have a technical paper devoted to the chemical industry in Canada is a most encouraging factor in the development of this most important feature of Canadian life."

Yet, M. E. Nichols, for whom Hon. Wesley Rowell says he hand-picked Canada to fill the $6,000 job of Director of Public Information, after careful investigation says that technical papers are of no value, and that they should not be admitted to the Canadian newspaper mails. A man who made such a gigantic failure of his own newspaper property is hardly an authority on Public Information, much less on Technical Information. Any man who can supply the public with general, special or technical news, such as they want, can make a success. Mr. Nichols failed to give the people of Montreal what they wanted in the way of news. He squandered over $500,000 of friends' money in trying to establish *The Mail* in opposition to *The Gazette*, but the latter came out of the fight stronger and more reputable than ever.

Ottawa.—"There is under construction in Canada," said Sir Joseph Flavelle, chairman of the Imperial Munitions Board, to-day, "one-quarter of the total merchant tonnage produced in the United Kingdom last year. Eighty per cent. of it will be completed this year. There is one British Columbia shipyard where ten ships of 8,800 tons each would have been built but for the recent serious fire. As it is seven of them will be built. If it had not been for the fire this one yard would have turned out an eighteenth of the total British production."

High Pressure Air Compressor Design and Application--II*

By Joseph M. Ford

Compressed air at high pressure is becoming an increasingly important medium in modern engineering practice and in naval warfare. The accompanying paper deals with the machines which produce high pressure air, leaving on one side the question of power transmission by compressed air. In passing, the principal advantages of this system are referred to, and will be seen to be of considerable value. They consist of the facility with which energy can be stored; the unlimited rate at which accumulated energy may be converted into useful work; and, apart from loss of efficiency, leakage cannot cause any awkward consequences.

THROTTLING

IN a compressor which has its suction throttled, the volumes dealt with by the stages are of course reduced, which naturally involves a decrease in stage pressures. In a compressor with no cylinder clearances there would be no limit to the possible compression ratio in any cylinder; that is, the initial pressure in, say, the high-pressure cylinder could be very low and yet the machine would deliver air. In a cylinder with a large clearance, however, there is obviously a definite limit to the ratio of compression if air is to be delivered. If this ratio were exceeded, the air would simply be compressed into the clearance space and re-expanded. In other words, in order that the compressor shall deliver air, the intermediate-pressure delivery pressure must attain a definite figure, so that the total work is still divided between the stages, although not necessarily in correct proportion. The theoretical minimum initial pressure from which a cylinder can compress and just deliver (assuming no leakage) can be obtained from the equation:—

$$c\left[\left(\frac{P_1}{P_2}\right)^{\frac{1}{n}} - 1\right] = 1$$

where

c is the clearance ratio,

P_1 is the delivery pressure (lb. per sq. in. absolute),

P_2 is the minimum initial pressure (lb. per sq. in. absolute),

and

n is the exponent in $P V^n$=constant.

The above expression shows that P_2 increases as c is increased, and that with large clearances the compression ratio in any stage when the compressor is running throttled can be kept within limits, whereas in a machine having very small clearances the high-pressure compression ratio, with excessive throttling, might reach an abnormal figure, resulting in excessive temperatures in that cylinder. The foregoing would tend to show that a compressor with good clearance volumes is preferable to one with small clearances, and for some Diesel engines — especially those of the marine type, where the conditions under which the compressor works may vary between fairly wide limits—this is the case.

As previously mentioned, clearance results in a reduction in the effective capacity of a cylinder, so that larger cylinders must be fitted than would be necessary with no clearance volume. The direct result is an increase in the piston loads. For example, a compressor of 300 cub. ft. per minute displacement at 350 r.p.m., having stage pressures of 45 and 300 lb. per square inch, when working at 900 lb. per square inch delivery pressure, and cylinder clearances for low pressure, intermediate pressure and high pressure of 4, 5 and 10 per cent. respectively, has a maximum air load on the high-pressure piston of about 7,900 lb., whilst the maximum intermediate pressure piston load is 14,300 lb. In a similar machine having, however, low pressure, intermediate pressure and high pressure clearances of 4, 15 and 30 per cent. respectively, the high-pressure piston load is increased to 12,100 lb. and the intermediate pressure load to 22,800 lb. Figs. 12 and 13 give for various clearance ratios the maximum piston loads for compressors of the above capacity and stage pressures.

Fig. 14 shows the combined indicator diagrams for three machines, whose pressure characteristics are shown in Figs. 12 and 13. Inspection of the diagrams super-imposed shows that as the clearance is increased the commencement of delivery and the commencement of suction, in both the intermediate-pressure and high-pressure cylinders, occur later in the stroke. With a large clearance, therefore, the maximum cylinder load is not attained until the crank is at a somewhat smaller angle from the dead centre than would be the case in a machine with no clearances, due to the fact that the actual compression takes place during a longer portion of the stroke, and further, the re-expansion of the greater volume of clearance air occupies a larger portion of the stroke, so that the pressure acting on the piston does work on the crank through a greater angle than it would were the cylinder clearances small. The crosshead load (neglecting inertia and weights of parts) at any point in the stroke, for the particular type of machine under consideration, is the difference between the intermediate-pressure piston load acting upwards and the low-pressure and high-pressure loads acting downwards, so that with the increased clearances the greater loads in the high-

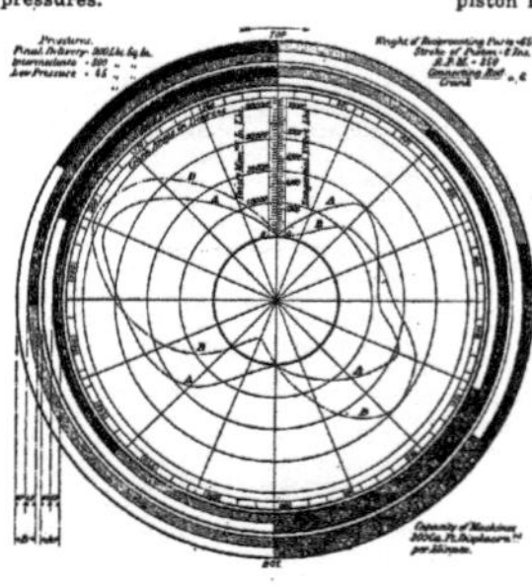

FIG. 15.

Diagram of twisting movements and crank angles for various events of the cycle. A—For a machine with clearance volumes—L.P. 4 per cent., I.P. 5 per cent., H.P. 10 per cent. of volume swept. B—For a machine with clearance volumes—L.P. 4 per cent., I.P. 15 per cent., H.P. 30 per cent. fo volume swept. Final delivery period, black; re-expansion of clearance air, single hatching; suction from intercoolers, circular hatching; transfer periods, cross hatching; compression periods, blank. The diagram from the L.P. points will be practically identical for the two machines, and is therefore not given.

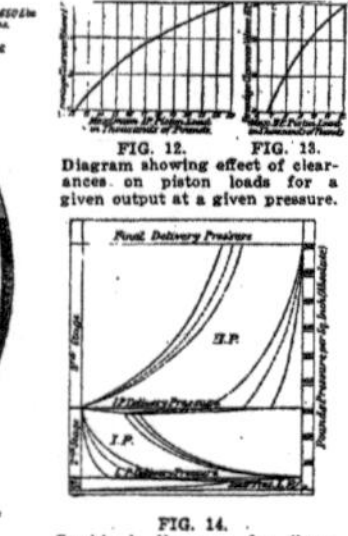

FIG. 12. FIG. 13.

Diagram showing effect of clearances on piston loads for a given output at a given pressure.

FIG. 14.

Combined diagrams for three machines designed for same output and stage pressures.

For a machine with clearance volumes—L.P. 4 per cent., I.P. 5 per cent., H.P. 10 per cent. of volume swept by piston.

For a machine with clearance volumes—L.P. 4 per cent., I.P. 10 per cent., H.P. 20 per cent. of volume swept by piston.

For a machine with clearance volumes—L.P. 4 per cent., I.P. 15 per cent., H.P. 30 per cent. of volume swept by piston.

*Part II. of a paper read before the Greenock (Scotland) Association of Shipbuilders and Engineers.

sor delivering air at 900 lb. per square inch gauge pressure. With equal division of work between the stages and intercooling down to the initial atmospheric temperature, the heat equivalent of the work required per lb. of air is represented by the area under ABHMPGR down to the line of absolute zero temperature (diagonal shading). The cooling effect is assumed to be the same for all stages, and as a result it will be seen that the maximum temperatures attained are the same for each stage of compression, 260 deg. F. The distances a, b and c are, of course, equal, since the compression ratios of the stages are identical.

Supposing that, owing to restricted water supply, climatic conditions, or circumstances connected with the design of the machine, it is found that in the first stage intercooler the air can only be reduced in temperature to 90 deg., and in the second stage intercooler to 120 deg., then the diagram is modified as shown by the dotted lines. The heat equivalent of the work required per lb. of air will then be represented by the area under ADJLOGR down to the line of absolute zero temperature, the excess over the previous case being represented by the horizontally shaded portions. The temperatures attained in the stages are no longer equal, but increase towards the final delivery. In this case they are: low pressure = 260 deg., intermediate pressure = 300 deg., high pressure = 340 deg. The compression ratios of the stages are, however, still equal.

With the poor intercooling, even, the maximum temperature of the cycle can be reduced by changing the distribution of work between the stages.

If the compression ratios in the various stages are made in the proportion x : y : z (see Fig. 16); then the stage temperatures will be the same, and consequently a minimum, although the intercooling is bad and the work distribution unequal.

The modification to the temperature-entropy diagram for these latter conditions is indicated by the chain-dotted lines, and the work in this case is the area under ACFKNER down to the absolute zero of temperature line. This is found to be slightly less than that required for the previous case, in which the compression ratios were equal. In some cases, therefore, better results are obtained from an unequal distribution of work than would be the case were each stage given an equal share, and also it may be pointed out that probably more valve trouble results from excessive temperature than from any other cause.

The design of certain machines is such that the cooling in some stages is likely to be very much better than that in others, in which case the work in the well cooled stages can be increased, and thereby relieve to some extent the stages with inefficient cooling.

Fig. 17 shows approximately the entrophy diagram which would be obtained from a test on one of the machines previously referred to, namely, that with the low-pressure clearance 4 per cent., intermediate-pressure 15 per cent., and high-pressure 30 per cent. of the piston displacement.

It is not proposed to enter into the question of the application of the theta-phi diagram to air compression. The peculiar shape of the lines representing the change from the initial condition to the final condition at end of compression is due to the transfer of heat from the cylinder walls after compression has commenced and to the heat added to the air as a result of piston friction. The deviation to the right of the vertical indicates that the air temperature is rising even faster than with adiabatic compression as a result of the addition of the said heat. Later on in the stroke cooling sets in and the compression becomes further removed from the adiabatic, shown by the line bending over to the left.

The atmospheric temperature is assumed to be 60 deg. F. The point "X" represents the initial condition of the air in the low-pressure cylinder just before compression begins, and it will be observed that the temperature has been increased by the amount delta t, (in this case 20 deg.), the increase being due to the fact that the air on its way to the cylinder has become heated through contact with hot valves, etc. In this particular case, with a mechanically controlled suction valve, the initial pressure is seen to be very slightly above 14.7 lb. per square inch, due to the inertia of the moving column of air in the suction pipe.

The first stage compression is carried to 59.7 lb. per square inch absolute and the temperature rises to 320 deg. F., the large rise being due to the fact that cooling is rather bad in this stage, partly as a result of a large portion of the "cover" being occupied by the high-pressure cylinder and the rest by large delivery valves, so that cooling cannot well be applied to this part. From the diagram it will be seen that the resultant compression is almost adiabatic.

After the first stage of compression the air is cooled in the intercooler to 80 deg. F., but on entering the second stage cylinder it again increases in temperature delta t, due to the hot piston, passages, etc., and also as a result of mixing with the re-expanded clearance air. The restriction in the intercooler tubes and in the valves results in a pressure drop, so that the initial second stage pressure is somewhat lower than the low pressure delivery pressure.

The total effect, then, of the above is that the initial condition in the second stage cylinder is represented by point "Y" instead of point "J," as would be expected.

It is assumed that the large clearance space in the intermediate-pressure cylinder is well cooled, and this, together with the good packeting effect, results in the compression being further removed from the adiabatic than it is in the low-pressure cylinder, so that in spite of the considerably greater share of the work done by the intermediate-pressure, the temperature of compression is not very much in excess of that in the previous stage. A similar sequence of events takes place in the high-pressure cylinder, where, again, owing to the large well-cooled clearance, compression is within the adiabatic. This, together with the fact that the high-pressure stage has the smallest share of the work to do, results in the comparatively low final temperature of 240 deg. F.

Even with a poor distribution of work, as in this case, the temperatures need not necessarily be excessive. There are, however, other aspects of the problem of determining stage proportions. Consider the case of a high-speed Diesel engine compressor, where it is desirable that the balance and twisting moment diagram of the machine should be as good as practicable, in order as far as possible to eliminate vibration.

With a machine of a certain type, it will be obvious that the crosshead load at the top centre will be due to both the low-pressure and high-pressure delivery pressures on their respective pistons, whilst at the bottom centre the load is only that due to the intermediate-pressure pressure. The maximum upwards load is then about twice the maximum downwards load, which would not conduce to the best running.

There are two ways of bettering conditions, one of which is to retain the stage pressures and adjust the intermediate pressure clearance until the load approximates to that desired (see Figs. 12 and 13), and the other is to choose the stage pressures so that the upload is reduced and the down load increased. Increased down loads necessitate larger connecting rod bolts, which in some cases, on account of the restricted space available, might not be easily accommodated.

Provision for Variable Delivery Pressures and Volumes

Another consideration when fixing stage proportions is the possibility of the machine being run at delivery pressures higher than normal.

In the remarks on clearance this question was mentioned, and the advantages of large clearances in this connection were enumerated.

The loads to which large clearances give rise (Figs. 12 and 13) might be inadmissible for various reasons, whilst the increased cost due to the larger cylinders might be a disadvantage, so that the constant proportional division of work between the stages, almost always approximately possible of attainment by choosing suitable clearances, would not in this case be secured.

If the machine runs at a pressure above its normal the division of work will be upset, and the high-pressure stage will take a considerably increased share of the total load, which may result, among other things, in an undue rise of final delivery temperature. If this latter is reasonably low when the machine is working at normal pressure, the rise due to the increased pressure will not be of such great consequence. To meet these conditions it is usual to arrange that the high-pressure stage shall, under normal conditions, take rather less than its proper share of the total work. The temperature in the other stages will be somewhat increased, but the rise in these temperatures due to the small increase in pressures as a result of the greater delivery pressure is inconsidearble.

If the machine has to run throttled in order to reduce the volume delivered, then it is again desirable, if possible, to arrange that the high-pressure stage shall under normal conditions take the smallest share of the work, since throttling reduces the stage pressures without altering the delivery pressure, thereby resulting in an increased high-pressure delivery temperature.

The importance of keeping the high-pressure temperatures low may again be emphasized. Burning and explosion of the lubricating oil here is of great violence when it does occur, which happily is very rare, due, among other things, to the attention given to the selection of the oil itself.

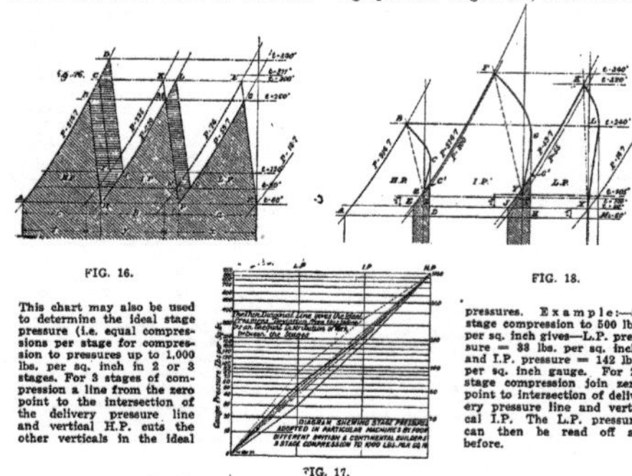

FIG. 16.

FIG. 17.

This chart may also be used to determine the ideal stage pressure (i.e. equal compressions per stage for compression to pressures up to 1,000 lbs. per sq. inch in 2 or 3 stages. For 3 stages of compression a line from the zero point to the intersection of the delivery pressure line and vertical H.P. cuts the other verticals in the ideal pressures. Example:—3-stage compression to 500 lbs. per sq. inch gives—L.P. pressure = 33 lbs. per sq. inch, and I.P. pressure = 142 lbs. per sq. inch gauge. For 2-stage compression join zero point to intersection of delivery pressure line and vertical I.P. The L.P. pressure can then be read off as before.

FIG. 18.

Volumetric Efficiency

One other point which arises in connection with the determination of stage proportions is the question of the volumetric efficiency of the low-pressure cylinder. If the low pressure stage is given a large share of the work, the effective capacity of the machine will be less than it would be with a lower first stage delivery pressure for reasons explained in the notes on the effect of clearance.

From the above it will be seen that multistage compressor design is to a large extent a matter of compromise. Each case should be treated by itself according to the duty of the machine and conditions under which it is to be run, and the best combination under the circumstances of loads, balance, and temperatures should be found, whilst the behaviour when throttler or running at various pressures must also be considered.

The question of cost should not be forgotten, for in some cases the advantages gained by, say, adding clearance, and thereby increasing cylinder sizes, may not warrant the additional expenditure.

Cylinder Proportions

Fig. 18 shows the stage pressures adopted in some of the very latest three-stage compressors for Diesel engines, by four different British and Continental builders, and bears out the remarks regarding the compromise in design of compressor.

Having decided on the stage pressures and cylinder clearance ratios, the actual determination of main dimensions is relatively simple.

When fixing the cylinder sizes, the low-pressure cylinder is settled first, as in a steam engine. A low-pressure delivery pressure is assumed—basing the assumption on previous practice if available—and from this pressure and the clearance volume in the cylinder (depending largely upon the type of valves used) the indicated volumetric efficiency is calculated. The dimensions of the cylinder are chosen such that, after the clearance air has re-expanded, the cylinder can draw in the required amount of air, plus an allowance for leakage through the compressor. Regard should also be paid to the fact that the air on its way to the cylinder, through contact with hot valves, passages, etc., becomes heated—which reduces the actual weight of air dealt with.

Where the workmanship is of the first order, about 85 to 90 per cent. of the air actually taken into the compressor will be delivered when compressing to 1,000 lb. per square inch in three stages.

The second stage cylinder is proportioned so that after its clearance air has expanded it can contain the whole of the air delivered by the low pressure at the low-pressure delivery pressure, and at the temperature resultant upon the mixing of the incoming and the clearance air. Attention should again be given to the effect of the hot cylinder cover, piston, etc. The other stages are similarly treated.

Experience is an important factor in the settling of cylinder proportions, as in the higher stages, at any rate, it is essential that allowance be made for leakage. For example, in a four-stage machine compressing to 3,500 lb. per square inch, possibly not more than 80 per cent. of the air taken in ever reaches the high-pressure cylinder. If this cylinder is made big enough to deal with the whole of the air aspired by the machine, it would actually take considerably more than its intended share of the load.

One method of finding the volumes of the various cylinders is the "constant weight" method, which depends on the fact that (neglecting leakage) the weight of air in all the cylinders at the end of the suction stroke is the same, correction for leakage being made afterwards. Another method, which can be used graphically, is the "equivalent cylinder" method. Here the volumes of the ideal cylinders with no clearance are found, assuming the compression isothermal, and an adjustment is then made for the effect of clearance air and temperature rises.

When the cylinder sizes have been determined, the size, location and number of valves may be fixed, determining to a great extent the way in which the necessary clearance volume is made up.

Small deviations are often made from the theoretical figures for reasons of manufacture. Supposing that the calculated high-pressure cylinder diameter was 3⅜ in., for convenience of manufacture this dimension might possibly be made the more even figure of 3½ in., entailing, however, an increase of 7½ per cent. in the volume. Some builders, however, work to sixteenths of an inch in cylinder diameters.

Modern Multi-Stage Compressor Practice

Comparison with Steam Engines.—The care and consideration which must be given to the question of proportioning a compressor is probably of more importance then that required for, say, steam engine design, in addition to which the engineering practice for the successful construction of compressors must be of the highest order. In the case of a triple marine engine, the cylinder clearance volumes are of comparatively small importance, and the arrangement of the engine need not be altered to suit the clearances. Should the distribution of work be less even than as designed, it is only necessary to alter the cut-off in the different cylinders. The work done by the high-pressure cylinder can readily be reduced by cutting off earlier in the medium-pressure cylinder, but in an air compressor reduction of the high-pressure work entails the addition of clearance volume to the high-pressure cylinder, which addition it is usually almost impossible to effect.

Small errors in the calculation of steam engine design have sometimes been corrected by the introduction of packing pieces, or stepped keys in the valve gear, but with a compressor it is rarely possible to correct mistakes. The workmanship must be of the highest class, especially with such details as valves and high-pressure plungers. Piston rings and cylinder bores also call for particular attention.

A noteworthy difference between steam engine and compressor design is that, whereas with the great majority of steam engines the cylinders are double-acting, in high-pressure compressor work they are invariably single-acting in the generally accepted sense of the term. Another difference is the very frequent use of differential pistons, which are rarely met with in steam engines. These points give rise to a large number of possible arrangements of compressor cylinders, a few of which are shown diagrammatically in Figs. 19 to 29.

Cylinder Arrangements — Fig. 19 shows the most common arrangement of a three-stage compressor. On the down stroke air is drawn into the space marked low pressure through the suction valve, having previously passed through a silencer. On the up stroke this air is compressed, and delivered through the first stage intercooler to the annular space marked intermediate pressure, where, on the next down stroke, the second stage of compression is effected. The air is again cooled in an intercooler before passing to the high-pressure cylinder, where the final compression is carried out. Before the air is delivered to the storage bottles it is customary to pass it through an aftercooler (consisting usually of a submerged pipe), in order that the air in bottles shall not become unduly heated. This compressor arrangement is largely adopted on Diesel engines, as well as for independent plants on account of its comparative simplicity and compactness, but it has several disadvantages.

To examine or renew new-pressure or intermediate-pressure piston rings, it is generally necessary to lift the whole piston and connecting rod up through the cylinders before the gudgeon pin can be removed, which in some cases, such as in submarine engines, is an undesirable feature on account of the limited head-room usually available. It may also be noted that the low-pressure and intermediate-pressure work is all done in the same cylinder, as it were, so that more heat has to pass through the cylinder walls to the jacket than would be the case were an

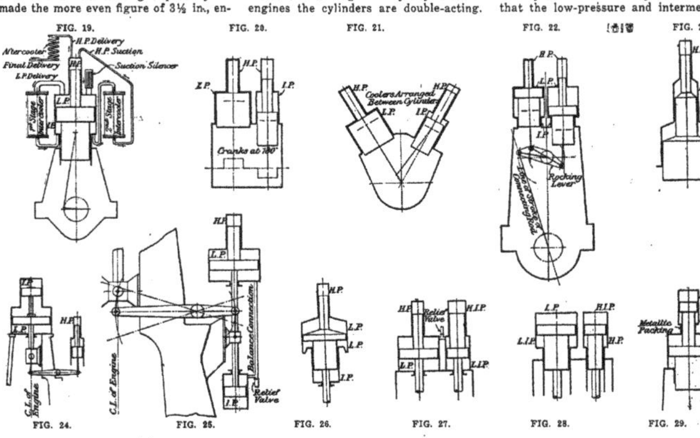

FIG. 19. FIG. 20. FIG. 21. FIG. 22. FIG. 23. FIG. 24. FIG. 25. FIG. 26. FIG. 27. FIG. 28. FIG. 29.

entirely separate cylinder devoted to each stage of compression. This, however, in practice is not of much moment.

Fig. 20 shows a small two-crank machine suitable for a motor-driven plant. With this type of machine the working parts are in constant thrust. A modification of this last arrangement is shown in Fig. 21, where both lines are driven by the same crank. This "V" compressor has been used on several marine Diesel engines, where the effect of the uneven turning moment resulting from this design is of but slight importance. It may be added that the angle of the "V" is usually somewhat less than that shown diagrammatically in the figure.

A curious adaptation of the type shown in Fig. 19 is given in Fig. 22. This arrangement, although expensive, has been used on some of the largest high-speed Diesel engines yet constructed, and is very satisfactory in operation. Instead of employing a single compressor, two entirely separate small machines are used, actuated from the ends of a rocking lever driven from the crank shaft by means of a single connecting rod.

A great advantage gained is a reduction in the size of cylinders necessary, which results in a marked improvement in the cooling and reliability. In addition, the turning moment of the combination is good, whilst the side thrust on the pistons resulting from the obliquity of the connecting rod in the usual design is, in this case, almost entirely absent. Another minor point to which attention may be directed is, that if a compressor valve breaks on, say, the starboard engine in a ship with this arrangement, only 25 per cent. of the main engine air supply is stopped, instead of half, as would be the case were each main engine fitted with a single compressor. As a result, it is quite likely that it would be found possible to carry on without running a big auxiliary compressor, which in many cases, being motor driven, entails the operation of a large generator set. Fig. 23 shows the arrangement occasionally adopted for Diesel engine work. Its obvious feature is its simplicity.

An arrangement which has been employed in one or two instances is shown diagrammatically in Fig. 24. Here the high-pressure stage is entirely separate from the other cylinders, and is driven by means of rocking levers. Accessibility is a feature of this design. To remove the intermediate-pressure piston it is only necessary to lift the intermediate-pressure cylinder cover and take off the nut securing the piston to the rod. By lifting the intermediate-pressure cylinder a little, it will now be a simple matter to withdraw the low-pressure piston. The high-pressure piston is easily withdrawn by putting the crank on top centre and uncoupling the link between the levers and piston. Note that the first stage of compression is effected on the under side of the low-pressure piston. This means that the heat from the intermediate pressure is well removed from the heat of the low pressure, so that the cooling is more efficient. The air in the space between the low-pressure and intermediate-pressure pistons is simply compressed on the up stroke, and re-expands on the down stroke—a relief valve being provided to guard against excessive pressure due to piston ring leakage.

Fig. 25 gives another arrangement of compressor suitable for a Diesel engine, where the air displaced by the idle side of the low-pressure piston is passed to the idle side of the intermediate-pressure piston in a similar manner to the previous case.

In the arrangement shown in Fig. 26 each side of the large piston does a part of the first stage of compression, whilst the second stage is effected in a cylinder at the bottom, as shown. This necessitates the use of a gland subjected to a pressure of perhaps 300 lbs. per square inch as well as a high temperature, but this is of minor importance. The up and down loads are more nearly equal than is the case with the usual arrangement, and, in addition, the maximum crosshead load for a given capacity is reduced as a result of the decreased low-pressure piston diameter. It may be mentioned that this arrangement is adopted in one of the largest compressors yet built for Diesel engine work.

Figs. 27 and 28 show arrangements of four-stage compressors, suitable for torpedo air compressors. The former design necessitates glands, but permits of rather better cooling than the latter, which, however, has an advantage in that crossheads may be dispensed with if required.

A single-line four-stage arrangement is shown in Fig. 29, but this suffers somewhat from lack of accessibility, although the loads up and down can be equalized, as two stages of compression are carried out on each stroke. There are other arrangements to suit various conditions as to accessibility, compactness, balance, etc.

Cylinders and Piston Packing

The conditions under which compressor pistons run are different from those in a steam cylinder, where there is generally a certain amount of condensation and the water so formed acts as packing in an effective manner. In a compressor a little moisture from the atmosphere is present, but this is generally insufficient to permit of using, for instance, solid piston valves, as sometimes met with in steam practice. Special attention should be given to the question of piston packing, which, however perfect, is useless unless the cylinder bore is machined to within extremely fine limits, the required accuracy being greater the higher the pressure which has to be withstood. In the case of high-pressure cylinders opinion is divided as to the best means of producing the nearest approximation to the perfect bore. Some builders finish the bore by an elaborate system of reamers, and the accuracy and uniformity of the bore so produced leave little to be desired. Other manufacturers pin their faith to grinding methods, and in many cases the bore is lapped out to size. The packing used in high-pressure pistons is varied. For machines such as those used for torpedo charging, which only run intermittently, leather or some patent packing of a similar nature is used. Any slight unevenness in the bore or

Automatic valves are essentially of a more or less delicate construction, and the conditions under which they operate are usually severe. The delivery valves have to pass air at very high temperatures, and the heat, unlike that in a steam engine, is a dry heat; whilst in a high-speed machine running at, say, 450 r.p.m., the valves have to lift and seat themselves in a very short space of time.

For example, in a low-pressure cylinder in which compression is carried to 80 lb. per square inch gauge at 450 r.p.m. the delivery valves have to open, pass the air, and completely seat themselves again in a period which is actually less than one-sixtieth part of a second. In order that this operation shall be successfully accomplished, either a powerful spring must be used to reseat the valve, or else the lift must be exceedingly small. The result of using a strong spring is that the pressure required to open the valve is excessive, and the hammering of the valve on its seat under the influence of the spring results in very rapid wear. The alternative, namely, the adoption of a small lift, means that the diameter of the seat must be large in order to obtain the requisite area through the valve for the flow of air. It may be stated that velocities in practice through the valve may be 130 ft. per second for low-pressure valves, and may be 250 ft. per second for high-pressure valves. If the diameter of the seat is increased over that sufficient with a larger lift, the leakage is increased, owing to the larger circumference of the contact circle; and, moreover, since a large valve is more liable to heat distortion than a small one, the possibilities of leakage are still greater.

Whatever the diameter of the valve, the weight should in every case be brought to the minimum. This reduces the inertia forces, and thereby the spring strength. Consequently the wear due to hammering and excess pressure required to open the valve is decreased. The type of valve usually employed for high-pressure work is the thimble valve, which can be used either for suction or delivery, two or more being used per cylinder to obtain the necessary area of flow. The valves proper are usually of nickel steel, and the seats of best quality bronze or gunmetal, although one Diesel engine builder used to fit valves in small compressors with seats of ebony or lignum vitae. The actual seat should be as narrow as possible, in order that the unbalanced load may be reduced to a minimum.

For moderately high pressures small valves of the well-known plate type are frequently used. The plate is usually of special steel and the seat of bronze. It has been found in practice that this type of valve, at any rate in the larger sizes, is liable to fracture if the lift is too great, so that, in order to obtain sufficient area for the passage of the air and at the same time keep down the diameter, it is necessary to adopt multiple seats and to cut ports in the valve for the exit of a part of the air. Given suitable conditions, this type of valve when properly designed and made is superior to the thimble or poppet type, especially for the low-pressure stage, but the plates are somewhat unreliable when the temperature is excessive. One advantage over the other type is that carbonisation has very little effect on the working.

A type of valve to which one prominent Continental firm have pinned their faith is the "Gutermuth" valve, which consists essentially of a strip of steel wrapped round a spindle in a similar manner to a clock spring. This has been used with great success in all the stages of Diesel engine compressors up to quite large sizes.

In high-pressure compressor work the valves must be kept in thoroughly good condition, as otherwise a leakage may take place from any stage to a lower one, thus increasing the pressure and temperature attained in the latter. This would upset the balance, and in extreme cases cause a serious breakdown by adding, perhaps, 250 per cent. to the load on the connecting rod bolts. To guard against such contingencies it is of the utmost importance that ample relief valves be fitted to all stages, and however reliable a machine may appear to be, it is dangerous to run with the relief valves out of action.

It is not proposed to deal with mechanically moved valves, as their application to high-pressure compressor work is limited, usually to the low-pressure suction, with perhaps the addition of the intermediate-pressure suction. The great advantage of a mechanical suction valve for the low pressure is that one can ensure that the cylinder is filled with air at atmospheric pressure, instead of at a lower pressure due to the attenuation consequent upon a spring loaded suction valve. If this mechanical valve is arranged in the piston itself, as is done in some designs, the cylinder clearance volume can be appreciably reduced. The delivery valves are invariably automatic. A valve gear which would adapt itself to the varying point of delivery with different pressures would obviously be somewhat complex. In high pressure work the necessity for this refinement is rather doubtful.

Regarding the disposition of valves, these are usually aranged in the cover of low-pressure cylinders in order to reduce clearance as far as possible, thereby seriously detracting from the cooling surface. Suction valves in the piston are not without disadvantages, since at the end of the suction stroke the inertia of the valve tends to seat it too soon and the cylinder is not fully filled with air. In some cases this arrangement necessitates the suction air being drawn through the crank case. It is argued that this tends to keep the running parts cool, but this a doubtful advantage when it is remembered that the low-pressure compression starts with air which has been already heated, and is probably charged with oil vapor.

THERE is now being developed under the name of the Hunnewell hull a new type of ship's hull, consisting of a steel or wood middle body and concrete ends, especially designed for the rapid and economical building of sea-going cargo vessels, says "Marine Engineering." This new style of hull, for which patents have been applied, is the invention of Constructor F. A. Hunnewell, U.S. Coast Guard, and is receiving attention from both naval constructors and private ship owners and designers.

ECONOMIC PRESSURE

BY FAR the greater part of the civilized world is now in league against Germany. Without the help of France, England, Italy, Belgium, and the countries of the New World, and British possessions and dependencies, Germany cannot carry on an active industrial life. If, then, the Governments of these powers informed the investors, manufacturers and merchants of Germany that in the event of the war being pushed to the inevitable conclusion, they would set up barriers to her trade, it is conceivable that all save the military party in Germany would come to the conclusion that it was wiser to make such better terms as could now be made than continue a conflict which would in the long run leave her in a worse position. Already we see from speeches of her statesmen there are some who are alive to the fact that the submarine campaign bears with it its own vengeance. The more ships that are sunk the fewer there will be after the war to carry on the trade of the world and to convey to Germany the food and raw products she will need. By far the greater amount of tonnage of the world is in the hands of Great Britain and America. If these two countries alone made it clear to the German people, as they well might do, that not a single one of the vessels now in their possession would be allowed to convey food or material to Germany, or to countries which desired to supply her needs, we can hardly doubt that the gloomy economic prospect would have its effect. Germany may hope to compel us to accept a peace of another kind, but the manifesto of the Versailles Conference must have demonstrated to her that there is no sign of weakening in the great powers. She must know also that even though she won a notable military victory in France, Flanders or Italy, she would be no nearer winning the war unless she could also destroy the sea-power of Britain and America. That power is a menace to her economic revival. If it be used as it may be used, then the industrial resuscitation of Germany may be indefinitely deferred. If the peace with Russia becomes indeed a *fait accompli*, it may, after a time, help towards the relief of the starving people of Austria, but it cannot open the way to the resurrection of Germany's economic life. That renaissance depends in no small measure upon the sea, and the control of sea trade we can hold indefinitely. As long as the Allied navies are in existence the resurrection of German trade is controlled by the Allies. It is not for nothing that the navy is called "the first arm." All history shows that in the affairs of Europe he who rules the sea rules the whole.—"Random Reflections" in "The Engineer."

The MacLean Publishing Company
LIMITED
(ESTABLISHED 1888)
JOHN BAYNE MACLEAN - - - - - - - President
H. T. HUNTER - - - - - - - - - Vice-President
H. V. TYRRELL - - - - - - - - - General Manager
PUBLISHERS

MARINE ENGINEERING
of Canada

A monthly journal dealing with the progress and development of Merchant and Naval Marine Engineering, Shipbuilding, the building of Harbors and Docks, and containing a record of the latest and best practice throughout the Sea-going World.

B. G. NEWTON, Manager.
J. M. WILSON, Editor. A. R. KENNEDY, Asst. Editor.
Associate Editors:
A. G. WEBSTER J. H. RODGERS (Montreal) W. F. SUTHERLAND

OFFICES
CANADA—
Montreal—Southam Building, 128 Bleury St., Telephone Main 1004.
Toronto—143-153 University Ave., Telephone Main 7324.
Winnipeg—1207 Union Trust Building. Telephone Main 3449.
Eastern Representative, E. M. Pattison.
Ontario Representative, S. S. Moore.
Toronto and Hamilton Representative, J. N. Robinson.
UNITED STATES—
New York—R. B. Huestis, 111 Broadway, New York, Telephone 8971 Rector.
Chicago—A. H. Byrne, 900 Lytton Bldg., 14 E. Jackson Street. Phone Harrison 1147.
Boston—C. L. Morton, Room 733, Old South Bldg., Telephone Main 1024.
GREAT BRITAIN—
London—The MacLean Company of Great Britain, Limited, 88 Fleet Street, E.C. E. J. Dodd, Director, Telephone Central 12960. Address: Atabek, London, England.

SUBSCRIPTION RATE
Canada, $1.00; United States, $1.50; Great Britain, Australia and other colonies, 4s. 6d. per year; other countries, $1.50. Advertising rates on request.

Subscribers, who are not receiving their paper regularly, will confer a favor by telling us. We should be notified at once of any change in address giving both old and new.

Office of Publication, 143-153 University Avenue, Toronto, Ontario.

Vol. VIII JUNE, 1918 No. 6

PRINCIPAL CONTENTS

PRESENT EMERGENCY INSURES FUTURE ECONOMY

IT is a foregone conclusion that past practice in the matter of shipbuilding will not be allowed to return and dominate output as it has in the past.

The recent statement of Secretary Daniels regarding the expedited building of destroyers in U. S. yards reveals such a tremendous discrepancy between possibilities and accomplishments that it is incredible that any sane man would sanction or approve in the mildest way a return to former conditions.

It would seem that organization is the keynote to the whole situation. Because destroyers were wanted in a hurry, the various officials and yards concerned co-operated to an extent not hitherto equalled, with the result that the actual time from the laying of the keel to the launching was 17½ days. The conditions were such, however, that the time referred to is nearly all assembling time as much of the structural work was prepared in advance ready for erection and assembling before the keel was laid; likewise bulkheads, deckhouses, bridge structure and a section of the keel were riveted up ready for assembling in place on the ways.

With suitable preparation almost any desired record might be put up. What really counts is the actual total time on all the material in a ship, not just in putting them together.

As a matter of production management the instance is illuminating, especially when compared with results of twenty years ago when the average time from the laying of the keel to launching was almost two years. It is against all principles of economy to tie up a building slip with its equipment for twenty-four months when less than three weeks suffices. Probably much of the work now done in shops ashore was formerly carried aboard and fabricated in detail instead of being manufactured in quantities and assembled en masse.

Shipbuilders who look forward to remaining on the active list will find much profitable study in such performances. Quick delivery will always be a factor of importance in securing orders, and if this feature is accompanied by largely reduced overhead charges per hull, the present emergency will alone be memorable for having effected such changes in recognized practice.

NATIONAL CONCENTRATION

ALMOST four years of warfare have removed most of the national hesitation, nervousness, indecision and ignorance which came to light occasionally as the tide of fortune ebbed and flowed in the great struggle. Periods of elation and depression have alternated. When the United States entered the war spirits ran high, to gradually subside awaiting definite accomplishments by our big Ally.

It was presumed by many, that, having the benefit of British and French experience, plus American hustle, things would be doing soon. Despite the disappointment experienced by those expectant enthusiasts, the work of the States during the last year has been on a colossal scale, characterized by a deliberateness of action and actuality of accomplishment which in days to come will be more than ever appreciated. Treachery and treason are best dealt with silently.

The increasing importance of the United States forces, due to the rapid growth of numbers and the development of home sources of war supplies, is of particular interest at this time. Not the least interesting phase is the publicity given to actual accomplishment, and the calculable effects of real facts on the morale of the enemy is probably one reason why such work is given special attention by the authorities at Washington.

An announcement by the Shipping Board shows that one wooden ship per day has been turned out for the past few weeks and steel vessels are being delivered almost as rapidly. In noting this fact it should be observed that one factor has been the overcoming of the lack of transportation. A great speeding up in the delivery of cars for transporting shipbuilding materials is reported by authorities and what has been done in respect of this industry will doubtless be accomplished for the rest of the country's activities in proper rotation.

Fo'c'stle Days
or
Reminiscences of a Wind Jammer

By Capt. Geo. S. Laing

The steamer and the motorship have deprived the seafarer of much of the old-time romance attaching to a sailor's life. The advent of wireless has also increased the safety of it. Deprived of these features, a sailor is immediately dependent on his own resources, and Captain Laing gives us a true and seaman-like reflection of sea life under canvas. The present feverish activity in constructing self-propelled craft affords a suitable background for incidents in a career in which gaffs, booms, tar and canvas held sway before being displaced by mechanical unloaders, derrick masts, fuel oil, coal and wire rope.

Part V.

ALBATROSS REGIONS

THE southern ocean birds soon made their appearance while the barque flew eastwards, with her sheets and braces as tight as fiddle-strings. Light sails, such as royals, gaff-topsail, flying jib, and top-gallant stay-sails, were in the gaskets, while the heavier canvas bellied out from the clues and jack-stays with boiler-plate rigidity. The first voyagers were naturally taken up with the pinioned sailors, and after every meal bits of fat were thrown over-board to feed the birds. Petrel are found in both hemispheres, but the albatross, with its relatives, the mollyhawk and cape hen, are peculiar to the southern ocean only, and so is the cape pigeon.

With the exception of perhaps three months a year the majority of these wonderful birds spend their time on the ocean deserts. The few scattered islands in the South Indian Ocean act as their breeding places, and, apart from that sojourn, it would appear that they hate the sight of land. Cormorants, boobies and pelicans are the very opposite in their traits, as they hug the shore.

Whilst westerly gales are prevalent here, every week or so both wind and sea take a rest for a day, and on such occasions we get to know our bird companions at very close range. During these breaks in the weather the wind, when light, plays a shifting game around the entire compass, and finally starts snoring out of the west with a renewed vigor. The sea meantime loses its crest poundage, but a very treacherous swell buckets the vessel, causing erratic diving and rolling.

One day between gales the "Maggie Dorrit" had almost lost steerage way and the cape pigeons were rising and falling as they rode on the swell close alongside. They are of a black and white checkered plumage, their tiny web feet appearing as shot silk and their dark eyes resemble polished beads. About the size of our domesticated pigeon, the winged sailors are almost tame when there is any fat pork about. The noise they make when fighting for food could almost be called a high-pitched chatter, and their abilities for gorging would be hard to beat. Frequently a bird can be seen with a bulge in the throat, as it actually tries to jump an over-sized piece of fat through its gullet.

Bird Appetites

Sailing ship crews never kill the small birds, but a few are easily caught with an improvised hook and allowed to wobble about on deck. They cannot rise, as the flat deck under their web feet pulls and prevents the upward or paddling motion that such birds use in rising to flight. Concerning the abnormal appetites of these southern birds, it must be remembered that all birds have to keep up a high body temperature, and the pinioned ones of which I speak are practically encased in an inch layer of fat. In such an inclement climate the combustion of food demands a constant stoking of the stomach. With the birds that we caught on the "Maggie Dorrit" it was generally noticed that a sickness overcame each one that was landed on deck. This vomiting must be due to excitement or repulsion to captivity.

The flat calm of the tropics is not in evidence now, but as each capless mound of sea throws its strength against our iron sides the sails flap and bang against the masts with the quick lurch of the ship. Clouds began to bank up in the west, and soon tufts of nimbus sailed along at a low altitude, proclaiming to us that the next breeze would be along in a few hours. Now and again a sharp puff would also remind us that Neptune was getting up steam. These southern waters, winds and clouds have a language all their own.

Catching an Albatross

A pretty sight was taken in as the first of the breeze blew obliquely against the heavy mounds of the ocean swell. This caused each water hill to throw up a curl of spray as the elements fought each other. Later on, as the wind gathered force and its direction fell into its appointed westerly notch, the rising sea became truer to its aerial ally.

We soon had a grand chance to catch an albatross, the king of all sea birds, and perhaps the largest bird that actually soars in the air. The old barque was going about four miles an hour before the first impulse of the coming gale, and our bird tackle was ready. The hook or rather contrivance with which seamen catch this bird is made of a hollow cut piece of diamond-shaped sheet brass, around the continuous edge of which strips of fat pork are lashed with sail twine. A cork float is next attached and a line of signal halyard stuff is paid out astern from the taff-rail.

An albatross has a five or six-inch bill, with a hook at the end almost as pronounced as that of a parrot. When this hook bill makes the acquaintance of our hollow cut diamond the bargain is sealed, for instead of the bird flying ahead or paddling to release itself, its antagonism is shown by spreading its large web feet against the sea. Thus its own resistance to the men on deck, who are pulling on the line, causes a closed link connection, the bird's beak getting jammed in the corner.

The more one pulls, the more persistent this great bird becomes in his back water manœuvres. Nearer and nearer he gets to the ship, and eventually we were dangling him between the sea and the poop deck. He was blindly lashing his wings, but those otherwise mighty pinions were useless in the captive predicament. Just as his head came within reach a slack turn or two of rope-yarn was wound about his bill to prevent any savage outburst when he was released on deck, for remember a bird like this may weigh forty or fifty pounds and have 12 feet of a wing spread.

Look at him as he struts around the quarter-deck in the shelter of the bulwarks, for he cannot rise. Although a good sailor in his own element, he soon becomes sick and wobbles in his gait as the motion of our old barque cheats him in his short perambulations.

Physical Features

The hinged wings of the albatross and his beautiful snowy breast are to be greatly admired. His legs are short, but very stout, whilst his membraneous feet are noteworthy samples of aquatic machinery. The neck of our sea champion is not far behind the swan's in gracefulness. His keen orbed head has long vision and an agile movement, whilst the brute strength that this bird can put into his bill or business end cannot be despised.

The useless slaughter of those birds was never countenanced on the "Maggie Dorrit," so we released a few after wrapping a little sail twine around their leg as a mark of identity should they be hauled on board at some future date. In some cases a wing tip may be touched with a paint brush for the same reason. Then for useful purposes three or four were killed. The flesh is not choice, but the apprentices chanced an allowance of flour on their bird, and when the pie came to be eaten the curiosity of such a feast must have made it palatable, for it disappeared.

Ocean sailors have many uses for the albatross after its death. Besides making an odd-time meal, we claim its head for mounting as a trophy. The web feet can be made into an ornamental tobacco pouch, the breast into a muff, and the hollow bones for pipe stems. A valuable dubbin is made from the fat of the bird by mixing it with a little Stockholm tar. Like all sea birds, the albatross has its special parasite, and this tick infests the inner plumage of the winged giant, although we never came across a bird whose physique seemed to suffer from this source. In this connection it may be remembered that all ocean fish, such as porpoise, bonita, dolphin, baracuda, sharks, etc., are more or less infested with a maggot in their flesh, so it would appear that the great of the air and sea have their troubles in small things, perhaps to offset any majestic importance that may be assumed in their different spheres.

Sea Bird Traits

Some peculiarities that a sailor notices about the large sea birds under discussion may now be cited. As far as the writer has observed, both in boyhood and manhood, the albatross, mollyhawk and cape pigeon never come to rest on a ship's spars or rigging. These birds never cross the ship's bows in a breeze, although they may do so in the temporary spells of broken weather that occur between gales. Their most prominent place is behind the vessel and along her sides searching for food that is thrown overboard. It is common for the same birds to follow a ship for five thousand miles or more, but they leave like the shot out of a gun when the vessel sails north into a lower altitude than 35° south. Again, if a vessel sails too far south into higher latitudes than Cape Horn (56°) the birds become scarce. It would be reasonable to say that our feathered sailors can be found in all the open waters of the Southern Hemisphere between the parallels of 30° and 60°. According to such reliable men as Sir E. Shackleton, the Antarctic waters, to the south of the boundaries just mentioned, have birds and mammals all their own.

At night some of the birds seem to take spells of rest on the billows, whilst a few keep flying behind the ship, but next day generally finds them all at their post overhead. When rising to flight the albatross uses its feet like a steamer's side wheels, and its hinged wings appear in a half-extended posture. If it is calm it takes quite a long horizontal sail in this fashion, but once in the air it is master of its great pinions. It is hard to describe how these birds can float in the air, poised as if they were suspended statues, and looking into the teeth of a gale with never a quiver of a wing.

They will hang thus for minutes at a time; then with a slight cant of the body swoop off over the waste of waters. Their propulsion is generated by this angling of the wings and not by any flapping. They only flap a little when climbing from the sea into the air.

A Lonely Highway

Our old ship is now in 43° S. and 43° E., which puts our nearest continent, uncertainty and scarcity of food whilst South Africa, a thousand miles over the port quarter and Cape Leewin, in Australia, four thousand miles ahead. We think how lonely this "running the easting down" is. The few scattered reefs that the birds claim, such as Amsterdam, St. Paul, Kergulen, Crosets, etc., are places to avoid if possible, and yet shipwrecked crews are glad to steer for them in their small lifeboats after losing the parent vessel. In connection with the abandonment of vessels on such lonely routes the writer has advocated through the marine press how imperative it is that all vessels should carry under compulsion one motor lifeboat to shelter all hands. This would do away with many of the ocean tragedies, where men's pluck and physique give out in long spells of heading for the nearest land under oars or sails.

THE ALBATROSS WEIGHS 40 TO 50 LBS. AND SOMETIMES HAS A WING SPREAD OF 12 FEET.

My own father had a terrible experience in an open boat on this southern route many years ago while serving as an officer on a wooden full-rigged ship, the "Robert Gilroy," bound from Shields to Calcutta with coal. Spontaneous combustion broke out and the vessel was burnt to the water's edge, finally blowing to pieces when 1,400 miles S.E. of Mauritius. Through good seamanship and favorable slants this ship's crew found themselves crawling up the beach of Roderique Island, after covering the above-mentioned distance with oars and sails in nineteen days. With a limited supply of kerosene the uninviting trip could have been made in much more comfort in half the time.

(To be continued.)

SEAMANSHIP AND NAVIGATION

Conducted by "The Skipper" *

Articles of direct interest to mariners, discussions of seamanship, navigation rules, and allied topics. Inquiries, from readers are invited and will be replied to promptly through these columns, unless otherwise requested.

THE MARINER'S COMPASS.—II.

It was mentioned in last month's lesson that the navigator has three kinds of courses to deal with—the compass course, the magnetic course, and the true course. Now in using these terms we come up against three relative items whose application to the three kinds of courses is vital to safe sailing.

Leeway

The items in question are leeway, deviation, and variation. Some beginners may think that leeway is a sailing ship quantity only, but that is not so; it has to be reckoned with and gauged in warship, tramp steamer and tow-barge alike. A definition of leeway is most simple when put in this light: The vessel when pounded with side pressures such as wind and sea does not go where she is looking or heading. She must answer to both forces, forces which are out of parallel, hence the result must be that the ship is put in a position that lays between the line of propulsion and the line of thwartship motion. Is that clear? Remember that leeward is the direction down the wind, while windward is the direction up or towards the wind.

In practice we find the leeway by gauging the angle between the ship's keel and her wake or track as seen by the naked eye. It is done in this fashion. Lay a protractor in position in the fore and aft line of the craft and you can read off at once the point or degrees which lie in the well-defined wake. Of course where leeway is small it can be easily guessed mentally by just glancing astern or peering into your compass with your back to the ship's head. In the old windjammers we often reckoned leeway by reduction of canvas, increasing the leeway as we decreased the sail area. When our ship was "on a bowline" or "on a wind" and we could only show double topsails and reefed foresail and mainsail, the leeway would amount to about three points unless under a weather shore. Remember then that leeway is not a negligible quantity as many officers in steamers have found out, especially when in ballast.

Correcting Compass for Leeway

Say then that you have to correct a compass course for leeway, how is it done? Simply thus: Let the course by compass be S 50° W and a given leeway of 10°. The wind and sea are from NW.

All compass work is done by assuming that the observer stands in the centre of the compass and looks down the working course. Imagine then that you are steaming or sailing in the SW quadrant with the wind and sea where—starboard beam? Yes. With buffeting from that direction the ship will sag towards south won't she? All right, then, simply take the 10° leeway off and the corrected course will become S 40° W, or in the latest way of reckoning it would be 220°. While I have mentioned this new fad of reading the compass degrees continuously from 0 to 360, the old quadrant system must be adhered to in these articles as it will likely be 100 years from now till all seamen, navigators, and teachers get into the habit of calling two, one and one.

Say that you were steering E x N with a strong SE wind, what would your corrected course be after allowing for 2 points of leeway? Heading E x N a SE wind would be on my right or starboard hand. Then the buffeting would be to the left or port hand. Is the answer NE x E? Yes.

Deviation of the Compass

Deviation of the compass is the angle between the magnetic meridian and the compass needle. Or again, you could say that deviation is the disturbing element which is present in all compasses on board iron or steel craft, but is absent in vessels of wood and copper fastenings. Obviously it is caused by the magnetism of the iron in the ship or her cargo. The presence of iron or steel is so damaging to a mariners' compass that the disturbance (deviation) has to be cured, or rather brought within manageable limits by acting on the principle that "like cures like," in the sense that magnetism is its own antidote. Thus we find that compasses are adjusted and corrected with the proper placing of artificial magnets in the vicinity of the binnacle. The objectionable thing to the navigator concerning the elusive disturber called deviation is the fact that the distribution of its effects is unequal and erratic on different parts of the compass, and at different parts of the world. For this reason a constant checking through bearings of the sun,

*Editor's Note.—This department is conducted by a qualified ship's master, and will deal with questions involving seamanship and navigation. The attention of young men desiring to obtain certificates as mates and masters is particularly directed to this department.

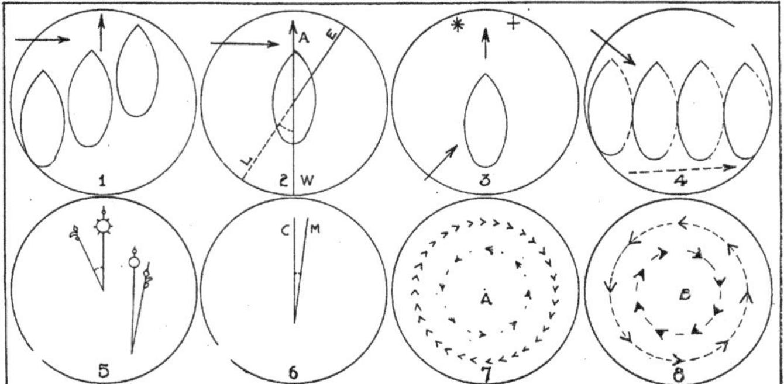

CHART ILLUSTRATING STUDY ON MARINER'S COMPASS.

moon and stars at sea is required to keep a true record of the deviation on the ship's course, and allow the correction when working problems on the chart. At present we are only concerned in the application of deviation to a compass course if it is on its way to the chart from the bridge, or vice versa.

Variation of the Compass

Variation is the angle between the true and magnetic meridians. You know that the earth as a magnet has two magnetic poles, but they do not agree in locality with the geographical poles that our explorers go to. Hence the two directions, the magnetic north and the true north. A compass needle, when not disturbed will point to the magnetic pole, and if there is no deviation on your course your ship is heading so and so magnetic. The only good point about variation of the compass is that it is found for us by the valuable scientists, the meteorologists and hydrographers. These investigators of the elements plot the waving curves of variation on our charts of the world and even give us an approximate idea as to the increase or decrease of the invisible disturber for each locality over a given period of time, for be it known that the variation changes, just as our geologists can prove the slow but sure changes going on above and below the earth's surface. For the present we are only taken up with the application of variation in its relation to correction of compass or true courses.

Rules for Corrections

To turn a true course into a compass course. This is done by applying the variation and deviation, allowing:

Westerly variation and westerly deviation to the right.

And easterly variation and easterly deviation to the left.

If you were correcting the other way, i.e., from compass course to true course, the movement would of course be the reverse.

In the accompanying illustrations and with the worked-out columns you are bound to see my meaning. This class of work is done generally with the aid of a compass card and a table of angles for degrees or vice versa. As a table of angles for one quadrant will suffice for the whole circle it is given below.

The variation and deviation combined make up the compass error. To combine them add them together if names are alike but subtract if names are unlike, thus:

Var. 4° E	Var. 2° E	Var. 6° W
Dev. 8° W	Dev. 2° E	Dev. 6° E
Error 4° W	Error 4° E	Error 0

When names are unlike and you subtract, the error is named after the greater, as in the first example. Then in the last example the fact that the disturbances are equal in quantity, but opposite in point of direction, causes a balance which makes the compass for the time being true.

To keep the student on a sure path the best way to proceed in this work is to recognize that errors or deviations and variations by themselves must have a name apart from easterly and westerly, so the letter R indicating "right," and the letter L indicating "left" come into play as per examples below. The course to be dealt with also gets treated to an L or R.

Examples

True course NE, in degrees
=N 45° 00′ E. R.
Assume 2 points W. error=22° 30′ W. R.

Compass course N67° 30′ E.

Same names add, opposite names subtract. Why did we name the true course R? Because the course being to the east of north is also to the "right" of north. In a south-east course an L would be used to denote that east is to the "left" of south. Is that clear?

Why was a westerly error named R? Because in converting a true course into a compass course westerly errors or westerly deviations or westerly variations all go to the "right."

Don't forget that in converting a compass course into a true course an L would go with a westerly error, because westerly in that case goes the other way. A glance at the arrow circles will explain further. Of course east is always to the right of north and west to the right of south, as west is to the left of north and east to the left of south, that part gives no bother to learn.

Examples Both Ways

From Compass to True:

Comp. Co. SxW=	S11°15′W.R.
Deviation=	6°45′ E.R.
Magnetic Co.	S18°00′W.R.
Variation	5°00′W.L.
True course	S13°00′W.

From True to Compass, bringing same course back:

True Co.	S°13′00W.R.
Variation	5°00′W.R.
Mag. Co.	S18°00′W.R.
Deviation	6°45′ E.L.
Comp. Co.	S11°15′W.

When you get near a cardinal point it may be a little difficult at first, an example follows.

Assume a true course to be W. x N., that is, North seven points or 78°45′W. with a variation of 1½ points or 16° 53′ easterly and a deviation of ½ a point or 5°38′ westerly, to find compass course.

True Co.	N78°45′W.L.
Variation	16°53′ E.L.
(—from 180)	95°38′
Mag. Co.	S84°22′W.R.
Deviation	5°38′W.R.
Comp. Co.	90°00′ or due West.

In such cases if you go over the east or west points the north becomes a south or vice versa.

Again, if you cross at the north or south points the east or west naming must be altered.

How to work a current course will be taken up when we come to chart work proper.

An explanation of the diagrams will now follow:

1. The arrow piercing the circle represents the wind, which is from the West. The arrow pointing north represents the direction of the ship's head. You will note then that the vessel moves like a crab. While she heads north she really "makes" N.N.E. That is leeway. What course would that vessel require to steer if she really wanted to "make" due north? Answer—N.N.W.

2. Another way of studying the leeway. W.A. is the course headed. L.E. is the course "made." The angle between these lines is the leeway. I have dotted the ship's wake.

3. Is a case that often happens with pressures of wind and sea on the quarter. Naturally one would expect a little leeway here too, but it is not always forthcoming; indeed, it may be that she makes a weather course and arrives at the star on the weather bow. The cross to leeward is where she ought to arrive. This is a case of every man know his own craft, but in the majority of cases when wind and sea are on the quarter the ship "gripes," i.e., comes running up to the windward side of her course in spite of the helmsman, unless he is a cracker-jack. This continual "griping" has, of course, the opposite effect of leeway, and hence a special watch on your whereabouts has to be kept, especially when thick weather comes or you are in dangerous waters.

4. There is no mistake likely to happen in this case, as it represents a ves-

ANGLES OF THE POINTS OF THE COMPASS

Name	Points		°	′	″	Name	Points		°	′	″
N	0	=	0	0	0	N E ½ E	4½	=	50	37	30
" ¼ E	¼	"	2	48	45	" ¾ E	4¾	"	53	26	15
" ½ E	½	"	5	37	30	N E by E	5	"	56	15	0
" ¾ E	¾	"	8	26	15	" ¼ E	5¼	"	59	3	45
N by E	1	"	11	15	0	" ½ E	5½	"	61	52	30
" ¼ E	1¼	"	14	3	45	" ¾ E	5¾	"	64	41	15
" ½ E	1½	"	16	52	30	E N E	6	"	67	30	0
" ¾ E	1¾	"	19	41	15	" ¼ E	6¼	"	70	18	45
N N E	2	"	22	30	0	" ½ E	6½	"	73	7	30
" ¼ E	2¼	"	25	18	45	" ¾ E	6¾	"	75	56	15
" ½ E	2½	"	28	7	30	E by N	7	"	78	45	0
" ¾ E	2¾	"	30	56	15	" ¼ E	7¼	"	81	33	45
N E by N	3	"	33	45	0	" ½ E	7½	"	84	22	30
" ¼ E	3¼	"	36	33	45	" ¾ E	7¾	"	87	11	15
" ½ E	3½	"	39	22	30	E	8	"	90	0	0
" ¾ E	3¾	"	42	11	15						
N E	4	"	45	0	0						
" ¼ E	4¼	"	47	48	45						

EACH POINT
being equal to 11 degrees 15 minutes.

sel "hove to" and going "dead" to leeward. She is heading north and making east. I have dotted the lee rails. All classes of ships have to "heave to" when weather is very bad, and this would give one an idea of how a sailing craft would go when braced up on the port tack with wind about N.W. x W. Steamers try if possible to "nose it" more than this, and would lay about N.W. x N. or two points from the wind and sea. If low-powered, however, they just fall off and make more fuss than an old windjammer.

5. Shows the angle of variation which lies between the true or geographical direction and the magnetic direction. The upper figure shows about two points of westerly variation, while the lower illustration shows about one point of easterly variation.

6. Shows the angle of deviation, i.e., the difference between the compass direction and the magnetic direction. In vessels built of wood this part error is absent. Hence you could take a magnetic course from the chart and give it to the man at the wheel. Not so, however, in our modern craft of steel and iron construction. To show how much attention we must put into the deviation question a careful ship's husband or superintendent of construction will insist on a newly launched steel vessel laying with her head in an opposite direction while at the assembling dock to what she headed while on the stocks. This helps to neutralize and tone down the magnetism acquired through the intense pounding and riveting in building. In more advanced lessons the subject will be treated by itself.

7a. The outer circle shows how easterly deviation and variation go. The inner circle shows how westerly deviation and variation go when reckoning from compass to magnetic and true courses.

8b. The outer circle shows how easterly variation and deviation go. The inner circle shows how westerly variation and deviation go when reckoning from true to magnetic and compass courses.

Don't worry about the old captains and mates who never gave this kind of work a moment's notice. They are fast disappearing, and the man of to-morrow cannot study too much. The examiners themselves have to study to keep abreast of the powers behind them that raise the tests for examination purposes.

You have perhaps not heard of the old salt of the "Quebecer fleet" of timber fame who said that navigation was d—d easy. "You just keep her W.S.W. outward bound and E.N.E. coming home."

SANDWICH.—Word of the sinking in Lake Erie of the steamer Jay Gould and the foundering of the barge Commodore was brought here by the crew of the barge, four men and three women, who were picked up by the steamer Mataafa. The Gould's crew, they said, was rescued by an unidentified steamer and taken to Ashtabula, Ohio. The sinkings occurred in heavy weather off south-east shoal.

NOTICES TO MARINERS

The following notices to mariners have been issued by the Department of Marine, Ottawa, under dates given:

NOTICE TO MARINERS
No. 28 of 1918
(Atlantic No. 14)

NOVA SCOTIA

(64) Cape Breton island—East coast—Port of Sydney—Public traffic regulations.

These Regulations are issued for the guidance of Masters of Vessels entering the Port, and must be obeyed or their ships will be treated as hostile. They will remain in force until amended or cancelled by a Notice to Mariners.

These Regulations number thirty, and are divided in five sections, fully detailed in the official copy of this notice.

* * *

MARITIME PROVINCES

(65) Introduction of Summer Time—St. John and Halifax Time Balls.

Summer Time.—Summer Time, which is one hour in advance of Atlantic Standard Time, will be kept in the Maritime Provinces.

Period of Alteration.—From 2.00 a.m. on the 14th April to 31st October, 1918.

St. John and Halifax Time Balls.—All Time Signals from the St. John Observatory of the Canadian Meteorological Service will accord with the Summer Time and the Time Balls at St. John and Halifax will be dropped at 1 o'clock, corresponding to 4 hours Greenwich Mean Time.

Authority: Director of the Meterological Service, Toronto.

Departmental File: No. 45380.

PRINCE EDWARD ISLAND

(66) South coast—Bedeque bay—Miscouche gas and whistle buoy replaced by gas buoy.

Position.—2¼ miles 295° (N. 42° W. mag.) from Sea Cow head lighhouse.

Lat. N. 46° 20′ 5″, Long. W. 63° 51′ 16″

Alteration.—Miscouche fairway gas and whistle buoy has been replaced by a gas buoy.

Description.—Steel cylindrical buoy, surmounted by a pyramidal steel frame supporting the lantern.

Color.—Black and white vertical stripes.

Character of light.—White light, automatically occulted at short intervals.

Illuminant.—Acetylene, generated automatically.

Variation in 1918: 23° W.

Authority: Memo. from Commissioner of Lights.

Admiralty charts: Nos. 1942, 2034, 1651, 2516 and 2666.

Publication: St. Lawrence Pilot, Vol. 2, 1916, page 264.

Canadian List of Lights and Fog Signals. 1918: No. 738.

Departmental File: No. 42617.

PRINCE EDWARD ISLAND

(67) North coast—St. Peter harbor—Front light in operation—Correction.

Correction.—In Notice to Mariners, No. 56 (136) of 1917 it is stated that both the range lights on the western side of the entrance to St. Peter harbor were discontinued; this is an error; only the back range light was discontinued; the front light is in operation.

Authority: Departmental records.

Admiralty charts: Nos. 2034, 1651, 2516 and 2666.

Publication: St. Lawrence Pilot, Vol. 2, 1916, page 180.

Canadian List of Lights and Fog Signals, 1918: Nos. 755 and 756.

Departmental File: No. 20755M.

QUEBEC

(68) River St. Lawrence—Amendment to Regulations for the Port of Montreal.

Notice is hereby given that No. 14 of the Regulations for the Port of Montreal has been amended to read as follows:—

14. All up-coming vessels, on each occasion, before meeting downward-bound vessels at sharp turns, narrow passages, or where the navigation is intricate, shall stop, and, if necessary, come to a position of safety below the point of danger, and there remain until the channel is clear.

These directions apply to the following points:—

Cap Charles.
Cap a la Roche.
Grandmont Poulier.
Nicolet Traverse (Banc des Anglais).
Pointe au Soldat.
Bellmouth Curve.
Contrecœur Bend.
Cap St. Michel.
Pte. aux Trembles (en-haut).
Ste. Mary Current.

Note.—These directions will also apply to Ste. Croix Bar, which is below the limits of the Port of Montreal.

Authority: Departmental records.

Admiralty charts: Nos. 2778, 2779, 2780, 2782, 2783, 2786, 2787, 2788, 1127, 2830a and 2830b.

Canadian Naval Charts: Nos. 1, 2, 3, 5, 6, 8, 9, 10, 13, 15, 18, 22, 23 and 24.

Publications: Tide Tables and information connected with the Ship Channel, etc., 1918, page 6; and St. Lawrence Pilot above Quebec, 1912, pages 40, 43, 48, 69, 73, 85, 86, 91, 93, 103 and 34.

Departmental File: No. 30743.

QUEBEC
No. 30 of 1918.
(Atlantic No. 16.)

(69) Gulf of St. Lawrence—Cape Gaspe—Change in character of light.

Position.—On the southern end of Cape Gaspe.

Lat. N. 48° 45′ 15″, Long. W. 64° 9′ 35″.

Date of alteration.—On or about 20th May, 1918, without further notice.

Alteration.—The revolving white light will be replaced by a flashing white catoptric light, showing two flashes, with an interval of 6 seconds between them, every twenty-four seconds, thus:

Flash; 6 seconds interval; flash: 18 seconds interval.

For half the time of revolution, or 12 seconds, the light will be totally eclipsed: for the other half a light of 500 candle power will be visible, through which the stronger flashes will show.

Power.—Naked light 500 candles: flashes 50000 candles.

Illuminant.—Petroleum vapour, burned under an incandescent mantle.

Authority: Report from Mr. J. A. Smith, District Engineer, Quebec.

Admiralty charts: Nos. 1163, 1621 and 2516.

Publication: St. Lawrence Pilot, Vol. 1, 1916, page 98.

Canadian List of Lights and Fog Signals, 1918: No. 988.

Department File: No. 20988A.

ENGLAND.

(70) South coast—Isle of Wight—St. Catherine point light—Alteration in period.

Position.—Lat. N. 50° 34½′, Long. W. 1° 17¾′.

Alteration.—The period of this flashing light has been altered from five seconds to forty-five seconds, and the duration of the flash from two-tenths of a second to four-tenths of a second.

Authority: British Admiralty N. to M. No. 514 of 1918.

Admiralty charts: Nos. 2045, 2450, 2675b and 2675c.

Publication: Channel Pilot, Part 1, 1908, page 206.

* * *

No. 31 of 1918.

BRITISH COLUMBIA.

(71) Strait of Georgia—Saturna island—East point—Change in character of light.

Position.—On East point, Saturna Island.

Lat. N. 48° 47′ 5″, Long. W. 123° 2′ 44″.

Date of alteration.—On or about 1st July, 1918, without further notice.

Alteration.—The revolving white light will, without further notice, be replaced by a flashing white catoptric light, showing two flashes, with an interval of 6 seconds between them, every twenty-four seconds, thus:

Flash; 6 seconds interval; flash; 18 seconds interval.

For half the time of revolution, or 12 seconds, the light will be totally eclipsed; for the other half a light of 800 candle power will be visible, through which the stronger flashes will show.

Power.—Naked light 800 candles; flashes 50000 candles.

Illuminant.—Petroleum vapour, burned under an incandescent mantle.

Temporary light.—A temporary fixed white light will be shown from about 15th June to about 1st July, 1918, while the new illuminating apparatus is being installed.

Authority: Report from Agent of Dept. of Marine, Victoria.

Admiralty charts: Nos. 2840, 2689, 2517 and 2531.

Publication: British Columbia Pilot, Vol. 1, 1913, page 156.

Canadian List of Lights and Fog Signals, 1918: No. 2289.

Departmental File: No. 22289A.

* * *

ALASKA.

(72) Chatham strait — Point Crowley — Light established.

Position.—On Point Crowley.

Lat. N. 56° 7′ 10″, Long. W. 134° 15′ 45″.

From the light, the summit of Mt. Howard bears 102° 30′ (N. 73° E. mag.), the tangent of point Howard bears 162° 30′ (S. 47° E. mag.), and point Crowley breaker bears 221° (S. 11° 30′ W. mag.).

Character of light.—Flashing white every 3 seconds, flash 0.3 second duration.

Elevation.—42 feet.

Power.—130 candles.

Structure.—Small white wooden house 8 feet high.

Remarks.—The light is obscured from 170° (S. 39° 30′ E. mag.) to 343° (N. 46° 30′ W. mag.).

Variation in 1918: 29° 30′ E.

Authority: U.S. Dept. of Commerce, N. to M. No. 18 of 1918.

Admiralty charts: Nos. 2463 and 2431.

Publication: Alaska Pilot, 1908' page 252.

JAPAN.

(73) Iki channel—Kyushu—North coast—Province of Hizen—Futagami Jima—Lighthouse established.

Position.—On Futagami Jima.

Lat. N. 33° 36' 18", Long. E. 129° 32' 57".

Character.—Group flashing white light, showing a double flash every 30 seconds.

Elevation.—316 feet.

Visibility.—25 miles from all points of approach.

Power.—36000 candles.

Order.—Fourth dioptric.

Structure.—Octagonal tower.

Material.—Concrete.

Color.—White.

Height.—40 feet from base to light.

Authority: Notification No. 535 of Dept. of Communications, Japan.

* * *

No. 32 of 1918.

(Atlantic No. 17.)

NOVA SCOTIA.

(74) South coast—McNutt Island—Cape Roseway—Intended change in character of light.

Position.—On Cape Roseway.

Lat. N. 43° 37' 15", Long. W. 65° 15' 45".

Date of alterations.—About 1st July, 1918.

Alterations.—The upper fixed light in the lantern will be replaced by a flashing white light, described hereunder. The maintenance of the lower fixed white light in the lighthouse tower will be discontinued.

New character of light.—Flashing white light, showing one bright flash every ten seconds.

Elevation.—120 feet

Visibility.—17 miles.

Power.—100,000 candles.

Order.—Third dioptric.

Illuminant.—Petroleum vapour, burned under an incandescent mantle.

Lantern.—The tower will be surmounted by a new polygonal iron lantern, painted red.

Note.—In the edition of the Canadian List of Lights for 1918 Cape Roseway light is described as a flashing white light, showing one flash every ten seconds, as it was expected that the changes in the light would be made in the Spring of 1918. But the changes were postponed, and the lights at present shown are two fixed white lights, 55 feet apart vertically, as heretofore.

Authority: Departmental records.

Admiralty charts: Nos. 340, 352, 730, 1651, and 2870.

Publication: Nova Scotia Pilot, 1911, page 204.

Canadian List of Lights and Fog Signals, 1918: No. 256.

Departmental File: No. 20256A.

No. 33 of 1918.

(Inland No. 10.)

ONTARIO.

(75) Lake Ontario, east end—Snake Island bank—Change in character of buoys.

The two red barrel buoys, heretofore maintained on the southeast side of Snake Island bank, have been replaced by iron conical buoys, as follows:—

(1) Position of buoy.—Lat. N. 44° 11' 10" Long. W. 76° 31' 50".

The following sextant angles fix the position of the buoy:

Tangent of Ninemile point............ 0°

Old site of light on Snake Island bank.64° 13'

Portsmouth front range lighthouse....63 15

Colour.—Red.

Depth.—3 fathoms.

(2) Position of buoy.—Lat. N. 44° 11' 1", Long. W. 76° 32' 12".

The following sextant angles fix the position of the buoy:

Portsmouth front range lighthouse.. 0°

Fourmile point lighthouse...........120° 40'

Tangent of Ninemile point 96 50

Colour.—Red.

Depth.—3 fathoms.

Authority: Departmental records.

Admiralty chart: No. 1152.

Canadian Naval chart: No. 68.

Publication: St. Lawrence Pilot, above Quebec, 1912, page 201.

Departmental File: No. 21747A.

ONTARIO.

(76) Lake Ontario — Toronto harbour — Chart issued.

New chart.—A chart of Toronto harbour and approaches, numbered 65 of the Canadian Hydrographic Survey, has just been published by the Hydrographic Survey, Department of the Naval Service of Canada.

Copies may be obtained from the Hydrographic Survey, Department of the Naval Service, Ottawa, for fifteen cents per copy, payable in advance.

Departmental File: No. 28499.

ONTARIO.

(77) Lake Huron—Saugeen River—Breakwater damaged by ice—Front range lighthouse moved.

Position.—Lat. N. 44° 30' 6", Long. W. 81° 22' 33".

Breakwater damaged by ice.—The outer fifty feet of breakwater at the mouth of Saugeen river has been damaged by ice.

Lighthouse moved.—Saugeen front range lighthouse has been moved back 95 feet in the line of range onto the undamaged portion of the breakwater, 2255 feet from the back range light.

Authority: Report from Mr. W. H. Carson, District Engineer.

Admiralty charts: Nos. 519 and 678.

Canadian Naval charts: Nos. 92 and 100.

Publication: Sailing Directions for Canadian shores of Lake Huron, 1915, page 28.

Canadian List of Lights and Fog Signals, 1918: No. 1943.

Departmental File: No. 21943R.

* * *

No. 34 of 1918.

(Pacific No. 7.)

BRITISH COLUMBIA.

(78) Vancouver Island — East coast — Nanaimo harbour—Change in position of Middle bank lighted beacon.

New position.—In 30 feet water, 170 feet 252° (S. 47° W. mag.) from the old site, and 2050 feet 273° (S. 68° W. mag.) from Gallows point fog bell.

Lat. N. 49° 9' 58", Long. W. 123° 55' 29".

New structure.—Beacon formed of a cluster of 7 piles, painted black, surmounted by a slatwork drum, painted white, and a lantern.

Character of light.—Fixed white.

Elevation.—15 feet.

Visibility.—2 miles.

Remarks.—Middle bank has been dredged to a minimum depth of 27 feet for a distance of 130 feet immediately south of the site of the old beacon.

Variation in 1918: 25° E.

Authority: Report from Agent of Dept. of Marine, Victoria.

Admiralty charts: Nos. 573, 2512, 579 and 1917.

Publication: British Columbia Pilot, Vol. 1, 1913, page 311.

Canadian List of Lights and Fog Signals, 1918: No. 2327.

Departmental File: No. 29467.

ALASKA.

(79) Revillagigedo channel — Mary Island — Change in charactistic of light.

Position.—On northeast shore of Mary Island.

Lat. N. 55° 5' 53", Long. W. 131° 10' 57".

New characteristic of light. — Group flashing every 6 seconds, thus: Flash, 0.3 second; eclipse, 0.9 second; flash, 0.3 second; eclipse 4.5 seconds.

Power.—White light, 600 candles; red light, 180 candles.

Authority: U.S. Dept. of Commerce, N. to M. No. 19 of 1918.

Admiralty charts: Nos. 2458 and 2431.

Publication. Alaska Pilot, 1908, page 136.

* * *

ALASKA.

(80) Chatham strait—Red Bluff Bay—Light established.

Position.—On southeasterly side of southernmost island at entrance to Red Bluff Bay.

From the light, the northeast point of island bears 28° 30' (N. 1° 30' mag.), the north tangent of south entrance point bears 107° 30' (N. 77° 30' E. mag.), and the small point on the south bears 171° (S. 39° E. mag.).

Character.—Fixed white light.

Elevation.—18 feet.

Power.—60 candles.

Structure.—Small white house.

Note.—The light is obscured from 67° (N. 37° E. mag.) to 208° 30' (S. 1° 30' E. mag.).

Remarks.—The light is unwatched.

Variation in 1918: 30° E.

Authority: U.S. Dept. of Commerce, N. to M. No. 19 of 1918.

Admiralty charts: Nos. 3637, 2453, and 2431.

Publication: Alaska Pilot, 1908, page 247.

* * *

No. 35 of 1918.

(Atlantic No. 18.)

NEW BRUNSWICK.

(81) East coast—Miramichi Bay—Fox Island—Swashway range—Lighthouses being rebuilt.

Former notice.—No. 102 (263) of 1917.

Range lighthouses being rebuilt.—New range lighthouses, described hereunder, are under construction to replace the old range light towers of the Swashway range on Fox Island.

Date.—The lights will be shown from the new structures when completed about 1st July, 1918, without further notice.

(1) Front Range Light.

Position.—On east side of Fox Island, 1⅓ miles from its south end.

Lat. N. 47° 6' 48", Long. W. 64° 59' 52".

Character.—Fixed white light.

Elevation.—35 feet.

Visibility.—11 miles in the line of range.

Order.—Catoptric.

New structure.—Enclosed tower, square in plan, with sloping sides; square lantern.

Material.—Wood.

Color.—White.

Height.—33 feet, from base to top of ventilator on the lantern.

(2) Back Range Light.

Position.—1,200 feet 260° (N. 76° W. mag.) from the front range light.

Character.—Fixed white light.

Elevation.—62 feet.

Visibility.—13 miles in the line of range.

Order.—Catoptric.

New structure.—Skelton tower, square in plan, with sloping sides, surmounted by an enclosed watchroom and square lantern; wooden slats on upper portion of side of skeleton framework facing alignment.

Material.—Skeleton frame, steel; watchroom and lantern, wood.

Colour.—Skeleton frame, red; slats, watchroom and lantern, white.

Height.—64 feet, from base to top of ventilator on the lantern.

Sailing directions.—The lights in one, bearing 260° (N. 76° W. mag.), lead from the deep water of the outer bay through the Swashway, in not less than 2¾ fathoms water, to the buoys in the ship channel.

Variation in 1918: 24° W.

Authority: Report from Mr. G. S. Macdonald, Resident Engineer, St. John.

Admiralty charts: Nos. 2187, 2034, 1657 and 2516.

Publication: St. Lawrence Pilot, Vol. 2, 1916, page 274.

Canadian List of Lights and Fog Signals, 1918: Nos. 861 and 862.

Departmental File: No. 20862R.

NEW BRUNSWICK.

(82) Eastcoast Shippigan Gulley—Back range light destroyed by ice.

Position.—Near the eastern end of the west beach, Shippigan gully.

Back range light carried away.—Shippigan gully back range light has been carried away by ice.

Authority: Report from N. B. Supt. of Lights.

Admiralty charts: Nos. 1633, 1715 and 2516.

Publication: St. Lawrence Pilot, Vol. 2, 1916, page 293.

Canadian List of Lights and Fog Signals, 1918: No. 906.

Departmental File: No. 20905R.

QUEBEC

(83) Gulf of St. Lawrence—Magdalen Islands—Grand Entry Harbour—Back range light pole and pier carried away by ice—Temporary light.

Former notice.—No. 103 (277) of 1910.

Position.—On the shoal inside the entrance to Grand Entry Harbour.

Back range light carried away.—The pier on which Grand Entry Harbour back range light stood has been carried away by ice.

Temporary light.—A temporary back range light has been placed in operation.

Authority: Report from Supt. of Lights, Quebec.

Admiralty charts: Nos. 1134, 2516 and 2666.

Publication: St. Lawrence Pilot, Vol. 1, 1916, page 69.

Canadian List of Lights and Fog Signals, 1918: No. 1031.1.

Departmental File: No. 21031R.

* * *

No. 36 of 1918.

(Atlantic No. 19.)

NOVA SCOTIA.

(84) South coast—Halifax Harbour—Dartmouth—Change in position of light.

Date of change.—On or about 15th June, 1918, without further notice.

New position.—310 feet 338° (South mag.) from the old site in the tower of Dartmouth Exhibition building.

Lat. N. 44° 39' 57", Long. W. 63° 34' 36".

Character.—Fixed red light.

Elevation.—127 feet.

Visibility.—10 miles in the line of range.

Power.—3000 candles.

Order.—Catoptric.

New structure.—Skeleton tower, square in plan, with sloping sides, surmounted by enclosed watchroom and square lantern; white wooden slats on upper portion of side of steel framework facing alignment.

Material.—Skeleton frame, steel; watchroom and lantern, wood.

Colour.—Skeleton frame, black; watchroom, white with a black diamond on its south side; lantern, white.

Height.—65 feet, from base to top of ventilator on the lantern.

Variation in 1918: 22° W.

Authority: Report from J. A. Leger, District Engineer, Halifax.

Admiralty charts: Nos. 311, 2320, 2410, 729, 1651, 266 and 2670.

Publication: Nova Scotia Pilot, 1911, page 136.

Canadian List of Lights and Fog Signals, 1918: No. 336.

Departmental Files: Nos. 20336 A and C.

NOVA SCOTIA.

(85) **Cape Breton Island—East coast—South Ingonish Harbour—Outer portion of breakwater carried away—Pole light moved.**

Former notice.—No. 58 (155), of 1912.

Position.—Lat. N. 46° 38′ 16″, Long. W. 60° 23′ 15″.

Part of breakwater carried away.—The outer portion of the breakwater on the north side of the entrance to South Ingonish Hárbour has been carried away.

Light moved.—The fixed white pole light has been moved farther in on the breakwater, 40 feet from its old position.

Authority: Report from N.S. Supt. of Lights.

Admiralty charts: Nos. 2727, 1651, 2516 and 2666.

Publication: St. Lawrence Pilot, Vol. 2, 1916, page 31.

Canadian List of Lights and Fog Signals, 1918: No. 525.

Departmental File: No. 20525R.

* * *

PRINCE EDWARD ISLAND

(56) **North coast—Malpeque Bay—March Water—Approach to Kier shore pier—Stakes placed.**

Former notice.—No. 13 (36) of 1917.

Two stakes have been placed at the outer end of the dredged channel leading to Kier shore pier, east side of Malpeque Bay, as follows:

(1) Position.—On north side of dredged channel at its outer end. Lat. N. 46° 31′ 29″, Long. W. 63° 42′ 20″.

Description.—Stake, with a cask at the top.

Colour.—Black.

(2) Position.—On south side of dredged channel at its outer end.

Description.—Stake, with a cask at the top.

Colour.—Red.

Authority: Report from Agent of Dept. of Marine, Charlottetown.

Admiralty charts: Nos. 1983 and 2034.

Publication: St. Lawrence Pilot, Vol. 2, 1916, page 185.

Departmental File: No. 38639.

QUEBEC

(87) **River St. Lawrence, north channel—Mouth of Chicot River—Buoy established.**

Position.—At mouth of Chicot River, on north side of entrance to dredged channel.

Lat. 46° 7′ 53″, Long W. 73° 6′ 47″.

Description.—Wooden spar buoy.

Colour.—Red.

Authority: Departmental records.

Admiralty charts: Nos. 2784 and 8230b.

Canadian Naval charts: Nos. 7a, 22 and 23.

Publication: St. Lawrence Pilot above Quebec, 1912, page 82.

Departmental File: No. 38894.

ENGLAND.

(88) **West coast—Liverpool bay—River Mersey approach—Alterations in lighting.**

1. Alteration in position of Bar light-vessel:

New Position.—At a distance of about 5 miles north-westward of former charted position.

Lat. 53° 33′ 18″ N., Long. 3° 25′ 40″ W.

Description.—A light-vessel exhibiting a group flashing white light.

2. Light-buoys established.

(a) Position.—Lat. 53° 35′ 45″ N., Long. 3° 25′ 40″ W.

Description.—A black conical light-buoy exhibiting an occulting white light.

Remarks.—This light-buoy is known as the North light-buoy.

(b) Position.—Lat. 53° 30′ 47″ N., Long. 3° 25′ 40″ W.

Description.—A red conical light-buoy exhibiting a flashing green light.

Remarks.—This light-buoy is known as the South light-buoy.

(c) Position.—Lat. 53° 32′ 18″ N., Long. 3° 19′ 20″ W.

Description.—A black and white chequered pillar light-buoy exhibiting a revolving white light.

Authority: British Admiralty N. to M. No. 560 of 1918.

Admiralty charts: Nos. 1951, 1170b, 1826, 1825b, 1824a and 2.

Publication: W. C. England Pilot, 1910, pages 389 and 390.

THE LOVEKIN FLANGED PIPE JOINT

The Lovekin Pipe Expanding & Flanging Machine Co., Philadelphia, have recently brought out a line of machinery for making expanded pipe joints possessing considerable merit. The manufacturers claim that the joint produced by this machine possesses all the economical advantages of the screw thread joint and a positive holding power besides, which is only limited by the ultimate strength of the pipe itself.

The making of this pipe joint requires no heat, hence the possibilities of crystallization and incalculable strains arising from unequal contraction in cooling are absolutely eliminated. A flexibility of operation is gained admitting of the attachment of pipes of almost any dimensions to flanges of every character, ranging from the very thinnest sheet tubing (which it would be impossible to thread) up to the heaviest grade of hydraulic piping.

Screw joints are rapidly being replaced by more reliable and safer methods. This is owing to the many defective and unmechanical features of the screw flange. The principal reasons for the engineering profession discarding it are as follows:

First, since the cross-section area of the pipe has been reduced, by cutting a thread upon it, we have by just this same amount reduced its strength against longitudinal stresses. These longitudinal stresses are the main ones to be dealt with where great variations in temperature are experienced, and these stresses frequently exceed the internal bursting pressures which have to be considered.

Second, it is very seldom possible in practice to send the screw flange entirely home upon the pipe owing to the liability of bursting the flange by forcing it on the taper of the thread. If the flange is not entirely home, that portion of the thread back of the flange hub is left exposed. This becomes a constant source of weakness against the principal enemies of the pipe joints, viz., vibrations and corrosion. During a long period of investigation it has been found that where screwed flanges failed in the face of longitudinal strains and vibrations failure took place at the back of the flange hub.

By the Lovekin method the metal in the pipe is extruded into the recesses machined in the flange hub. This is done by the cold rolling process, which improves the metal 15 per cent., as demonstrated by actual tests.

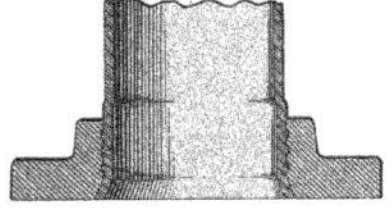

PIPE FLANGE EXPANDING MACHINE AND FLANGE CONNECTION MADE BY ITS AID.

If it were possible to upset the pipe at the point of threading to such an extent as to leave the bottom of the thread entirely beyond the original outside diameter of the pipe, we would approach an ideal connection.

This result is accomplished by the method under consideration, as will be seen by the accompanying illustrations.

THE WAR BEE

The "War Bee," recently launched from the yards of the Nova Scotia Steel and Coal Co., has the following principal dimensions: 248 ft. 9 ins. by 35 ft. and 20 ft. and will carry about 2,400 tons dead-

SHIPYARD OF THE NOVA SCOTIA STEEL & COAL CO., SHOWING THE "WAR BEE."

THE "WAR BEE," AFTER LAUNCHING AT THE YARDS OF THE NOVA SCOTIA STEEL & COAL CO.

weight. She is equipped with complete cargo-discharging gear, folding masts, and four large hatches for the rapid handling of cargo. The engines, of which an illustration appears herewith, are triple expansion, having cylinders 17, 28 and 46, by 33 in. stroke. The air, feed and bilge pumps are directly connected. The engines were built entirely in the Nova Scotia Steel and Coal Co.'s shops. Steam is furnished by two boilers 11 ft. 6 ins. diam. by 11 ft. long at 185 lbs. working pressure. Natural draft is used.

The War Bee is lighted throughout by electric light, and a separate room is provided for each officer amidships on the bridge deck.

A second ship having the same dimensions as the War Bee may be seen on the ways in the background.

REINFORCED CONCRETE VESSELS

By M. L.

The progress of construction in concrete for coasting vessels, barges and other light craft proceeds apace, and as it takes about fifteen years for concrete to attain its maximum strength and hardness there is no doubt about the value of this material for such uses. Steel and iron are found to deteriorate immediately from the day that they are put to use, unless very carefully protected by means of paint and other materials. In a reinforced concrete ship the steel takes up the whole of the tension and the cement the compression, and as it is increasing towards its maximum strength the stresses of the steel decrease, and as it is an absolute preservative of the steel, more especially when it is surrounded by water and not subjected to the air.

It will be readily understood that in constructing reinforced concrete vessels the weight of the hull must, of necessity be very much heavier than where either wood or iron is employed. The actual weight of the hull will be increased from at least 33 to 50 per cent. The whole external surface of the concrete can be made perfectly smooth, and in the case of seagoing vessels, can be coated with ordinary anti-fouling compositions, such as are employed on wooden and steel vessels. In regard to barges for river work this is not necessary. It has been

ENGINES FOR SS. "WAR BEE," 17, 28 AND 46 IN. BY 33 IN. STROKE.

found that where the ordinary lines of a vessel are adhered to there is very great expense in constructing the necessary shuttering or form work, so as to get all the correct curves longitudinally, diagonally and transversely, and it is owing to this difficulty and also the curvature of the rods, that a new method of construction, known as the Pollock method, has been adopted so as to meet and overcome these objections.

All curved work is dispensed with in their reinforced concrete system; the form work becomes of the simplest nature possible and the rods are practically straight throughout, so that no great expense is incurred. At first glance, shipowners and naval architects will not like such a departure from the preconceived designs, but when one considers the speed of 7 to 8 knots usual for coasting vessels of this size fitted with sails, and auxiliary power, or a speed up to 9 knots with full power without sails, it will be agreed that the straight line hull will cause very little, if any greater resistance, always assuming, of course, that the co-efficient of fineness is always the same, and that the lines are slightly fuller forward than aft; i.e., the centre of buoyancy slightly forward of amidships. If reinforced concrete vessels were built of this design, especially for experimental vessels, it will considerably reduce the cost and add to their strength and efficiency.

Reinforced concrete vessels cannot corrode nor rust, and their life is practically hundreds of years, whilst they are easily repaired. Not only are they fire and vermin proof, but they do not require trained shipwrights, riveters, platers, angle smiths, etc., and they entail no expense for upkeep, and their depreciation is nil. They are constructed in half to one-third of the time of steel or wooden ships, and they cost very much less than ships constructed of either of those two materials, and especially if they are constructed to standard designs, as a number can be made from the one shuttering.

THE following are the chief particulars of the standard 10,000-ton oil tankers which have been adopted by the United States Shipping Board:—Length overall, 444 ft.; length between perpendiculars, 430 ft.; beam, moulded, 59 ft.; depth, moulded, 33 ft. 3 in.; load draught, about 25 ft. 6 in.; total deadweight carrying capacity at above draught, 10,300 tons; designed speed, 10½ knots.

Eastern Canada Marine Activities

Doings of the Month at Montreal and on St. Lawrence River

June Correspondence

SIGNS of the times are reflected in the activities evident on every side, and probably nowhere is this more pronounced than at the port of Montreal. The use of the port as a passenger terminal will receive little attention this season as it has practically been decided that nothing but freight and special traffic will be handled this summer. This will be very little change from last year as the passenger movement for 1917 was exceptionally light, but the efforts this season will be concentrated on the transportation of essential freight and other necessary supplies.

A feature of present activities is the large number of arrivals that are in a sense foreign to the port, many of the vessels now coming in being those that formerly operated in other portions of the globe, notable among these are quite a number of the Blue Funnel line, which since the war have been transferred from the Liverpool-China service to that of the trans-Atlantic.

Winter activities in connection with shipbuilding are now becoming evident in the nearing to completion of many vessels laid down during the late fall or winter months. The launch of the "War Earl" for the Imperial Munitions Board, by the Canadian Vickers Co., is only one of several that will take the water during the present season. This vessel was launched early in June and her drag lines were still taut when the first portion of the keel for another vessel was placed on the slip just vacated. Nothing of a public nature marked the event, and in accordance with the accepted practice of the present times future war launchings will be carried out in a very quiet manner. During the present month the firm expect that two others will be placed in the water; these will be vessels of about 7,000 tons.

The firm of Fraser, Brace Co., anticipate the launching of two of their four 3,000-ton wooden vessels about the 22nd of this month. The hulls have been far enough advanced for launching for nearly a month but operations were somewhat delayed owing to a little misunderstanding regarding labor conditions. This, however, has been adjusted, and the men have all returned to work and the matter of a settlement is in the hands of the various officials and the Department of Labor. The chief delay in placing the boats in the water has been due to the work required in preparing the crib and the gate that is to be located at the entrance to the basin. All the machinery for the finished vessels is lying on the banks at the side of the canal adjoining the shipyard awaiting the removal of the boats from their present position, as it would be inadvisable to place the same in position owing to the greater draft that would be given to the vessels.

Following the organization of the Province of Quebec branch of the Navy League of Canada, the movement for a membership campaign has been under way, and developments are to the point where the same is to be launched. The great object of the campaign, and also the permanent aim of the organization, is to arouse public interest in the creation of a Canadian mercantile fleet, manned by Canadians. The interest already displayed has virtually assured the officials of the success of the movement.

Activities at the Montreal Dry Dock are still of abnormal character, and operations here emphasize the importance of the repair branch of the marine industry. Many vessels have received minor and other repairs since the opening of navigation, and indications point to a very active season for this company. The "Canobie," in for a new stern post and stern tube, has been completed and placed at the disposal of the owners. Work on the "Cataract"—as far as the dry-dock is concerned—is about finished, and the vessel leaves the dock on the 20th of the month, and her place will be taken by the "Calgarian," a long service lake vessel that is being replated and otherwise overhauled to fit her for ocean service. Several ships constructed for the United States Shipping Board have arrived in port during the past six weeks, some of which have received minor repairs to damages received in their trip down the lakes.

The development and expansion in the marine field has resulted in the enlargement of many plants and in the organization of others, to enable the manufacturers to take care of the greater business of the present abnormal activities. The firm of Peacock Bros., of Montreal, have recently acquired a building on Delorimier Ave. for the purpose of making their line of Weir pumps that have previously been imported from the Old Country. Rapid progress is being made on the new plant, and the machinery has been practically installed and the factory is expected to be in operation in the course of a few weeks. Recent large orders have been received by this company for service pumps and other marine specialties for many of the vessels being constructed in the different yards throughout the Dominion.

Another firm that has made extensions in order to take care of the additional business arising out of the shipbuilding industry is that of the Mason Regulator & Engineering Co. Their new factory on Dagenais street has been in operation for some time and the necessary equipment has nearly all been installed. They have recently completed a large contract for feed-water heaters and accessories for the British-American Shipbuilding Co. of Welland, Ont., and are now working on a fairly large order for the Collingwood Shipbuilding Co. for similar appliances. A recent contract received was for the auxiliary equipment in connection with six ships being built by the Canadian Vickers Co. of Montreal. In a short time they expect to start operations on the manufacture of Mason reducing valves for the United States navy.

Still another plant that has felt the pressure of the marine developments is that of the firm of Darling Bros. This company is now erecting the fourth addition to the plant in the past two years, the expansion being largely due to the increase in their ordinary business, although considerable has been as a result of government work. The present development is that of a grey iron foundry, a new departure for this concern. The plant will be equipped with all the modern conveniences for the rapid and economic handling of the work, and it is the expectation of the firm that in addition to supplying their own needs the capacity of the plant will be sufficient to enable them to cater to the outside trade.

In the interests of greater food conservation the Department of Naval Service in co-operation with the government, have fitted out the steamer "Thirty-Three" with herring and mackerel drift nets and despatched her to sea for the purpose of endeavoring to locate the herring schools off the Atlantic coast. This action has resulted from the increasing demand for herring as a food, and also to provide ample bait for the cod fishing fleets, the latter being in danger of shortage owing to the increased demand for herring as a food. In addition to acting as a scout the vessel on its first trip out succeeded in netting about ten tons of mackerel, landing the same at the port of Canso, Nova Scotia. Her work will be carried out chiefly along this coast, and any catches will be disposed of at the nearest port to which she is sailing.

The Davie Shipbuilding and Repairing Co., of Levis, P.Q., successfully launched the new steam car ferry "Canora," a vessel of 3,400 tons, on the morning of June 10. This vessel has been constructed for the Canadian Northern Pacific Railway and has been built to meet the requirements of the road in the transportation of passengers and freight between the Island of Vancouver, B.C., and the mainland. A full description of this vessel will appear in our next issue.

On a recent visit to Quebec the Hon. Mr. Ballantyne, Minister of Marine, was greatly impressed with the splendid docking facilities of the harbor and the efficient equipment of every essential character for the accommodation of shipping. It is more than possible that some action will be taken in the near future to utilize to a larger extent the existing Louise Docks and the other natural advantages of the harbor.

The Leclaire Shipbuilding Company of Sorel have acquired considerable property across the river from their present site and good progress has been made towards the establishment of a new yard to augument their present activities; existing contracts call for the construction of a number of trawlers and facilities are being provided at the new location to handle ships 250 feet in length

The port of Halifax has been chosen as the site for one of the largest shipbuilding plants yet established in the Dominion. That this will eventually culminate in an enterprise of considerable magnitude is practically assured when it is learned that the initial expenditure for plant and machinery will be in the neighborhood of $3,000,000. It is anticipated that ample facilities will be provided for the construction of steel vessels up to 10,000 tons. Among those associated with the company, which is to be known as the Halifax Shipbuilders, Limited, are James Carruthers, J. W. Norcross, and R. M. Wolvin. In connection with the attitude of the government towards the newly-organized firm it is understood that no assistance will be given to the company other that the placing of a number of contracts at fixed prices for the construction of several modern steel freighters.

In connection with further developments at the port of Halifax it is announced that the present dry-dock at the harbor, which suffered some slight damage during the explosion of last winter, has been acquired by the government and will immediately be placed in first-class condition, alterations being made and machinery installed to enable the repairs to be made to the largest vessels visiting the port.

In connection with the Halifax catastrophe it might be interesting to note that certain parties are endeavoring to have further investigation into the causes of the explosion and urges that some action be taken towards trying those involved before a jury with the purpose of placing the blame on the responsible persons.

Announcement has been made by the Hon. F. B. Carvell, Minister of Public Works, that work will be proceeded with towards the enlargement and improving of the Courtenay Bay Harbor Works at the Port of Halifax; it is also intimated that steel shipbuilding will be another activity of this company. Mr. Carvell thought that little other undertaking of an important nature would be commenced in the immediate future owing to the abnormal conditions existing through the country.

The Nova Scotia Shipbuilding and Transportation Company has recently been reorganized and is now operating as a limited company. Several small vessels of about 420 tons have been completed and sent to sea and others are still building. These will be finished in the late summer and then the keel of another large vessel will be laid for operations during the winter months .

TELL-TALE SWITCHBOARD FOR MARINE LIGHTS

THE tell-tale switchboard, illustrated herewith, is the recent product of a Canadian inventor. It is built for either fresh water or ocean-going vessels, and is particularly built to withstand salt water conditidons. All electrical equipment and fuses are of National Electric Code standard. The working parts are of cast material. The bells, buzzers and relays are all made waterproof.

This particular switchboard is approved of by both the Imperial Munitions Board and the Department of the Naval Service. It is being used successfully on numerous vessels, and is giving entire satisfaction.

The working of this switchboard is very simple and easily understood. When the mainswitch is closed, and each switch on the separate units closed, then all sailing lamps should be lighted. If, however, one of the sailing lamps fails to light, then the tell-tale lamp and buzzer, or bell, corresponding to the sailing lamp, will indicate. The operator will then know where to make repairs.

These switchboards are made in any number of units to suit the purchaser. They are constructed in four different types, viz., the simple two-wire type, the simple three-wire type, the automatic two-wire type, and the automatic three-wire type.

The great advantage with the three-wire type is that when a sailing lamp fails to light and is indicated on the switchboard the operator only has to throw over the throw-over switch on the unit indicated, when the emergency lamp in the sailing lantern will light. Should the tell-tale lamp and bell still indicate, the operator will find that both lamps in the sailing lantern are electrically damaged, or else the fuse on that unit is damaged.

The Jaynes tell-tale switchboard is protected in the allied countries, and may be purchased from the Factory Products Co., Limited, Toronto, or direct from the inventor, Leo E. Jaynes, care of Polson Iron Works, Limited, Toronto.

FORTY-SIX SHIPS TO LAUNCH THIS YEAR

Shipbuilding as a National Industry Will Be Carried on After That

OTTAWA.—Before the close of the year it is anticipated that the greater number of the forty-six vessels now being built by the Imperial Munitions Board on the Pacific and Atlantic will be ready for launching. The first will have her trial trip at Vancouver, and the whole fleet will be practically completed by September. On the Pacific twenty-seven are being built, and on the Atlantic nineteen. They are of 3,100 tons each. The manufacture of machinery for the vessels is being speeded up, and it is expected that the greater part of it will be installed by the year end.

At the completion of these contracts the Imperial Munitions Board will go out of the shipbuilding business, and the government will thereafter carry on the industry as a national enterprise.

Montreal.—The 7,000-ton cargo ship War Earl was launched from the yard of the Canadian Vickers, Limited, on Saturday morning. The vessel has a length of 380 feet and a breadth of 49 feet. Five minutes after the War Earl was launched another keel was laid down, and this ship will be completed about the first of October.

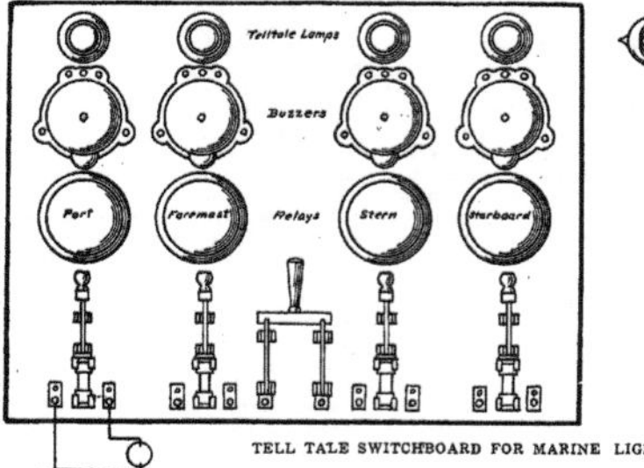

TELL TALE SWITCHBOARD FOR MARINE LIGHTS.

Work in Progress in Canadian Shipyards

A record of Work in Process of Completion—Principal Features of Specification—Approximate Launching Dates—Vessels on Order.

BRITISH COLUMBIA

YARROWS, LIMITED, ESQUIMALT, B.C.

One stern-wheel steamer, 165 ft. x 35 ft., for Indian Government service.

At present engaged chiefly on repair work.

H. VOLLMERS, NANAIMO, B.C.

One gasoline tug, 9 tons.

NEW WESTMINSTER CONSTRUCTION & ENGINEERING CO., NEW WESTMINSTER, B.C.

Four wooden steamers, Nos. 10-11-12-13, for undisclosed interests. 250 ft. x 43 ft. 6 in. x 25 ft., moulded depth, 2,800 tons d. w., 1,000 h.p., single screw engines, 12 knots, naval architect, I. Alexander. Launchings March 14, April 14, May 15 and June 15, 1918.

B.C. CONSTRUCTION & ENG. CO., POPLAR ISLAND, NEW WESTMINSTER, B.C.

Four wooden vessels, 1,800 gr. tons, 2,800 tons d.w.

NEW WESTMINSTER MARINE CO., FT. OF FURNESS ST., NEW WESTMINSTER, B.C.

STAR SHIPYARD CO., NEW WESTMINSTER, B.C.

J. A. CROLL, PORT ALBERNI, B.C.

PACIFIC CONSTRUCTION CO., PORT CCOQUITLAM, B.C.

Taken over the plant of the Coquitlam Shipbuildiug Co.

Gasoline yacht, 50 ft., June, 1918.

Three wooden freight steamers, Nos. 20-21-22, 250 ft. b. p. x 42 ft. 6 in. x 25 ft., 1,800 gr. tons, 2,800 tons d.w., trip. exp. engines, water tube boilers ,12 knots, 1,000 h.p. Two for undisclosed interests, one for builder's account. Launchings May 20, June 15, Sept., 1918.

Four new berths under construction for steel ships up to 10,000 tons.

C. E. BAINTER, PRINCE RUPERT, B.C.

Several wooden fishing steamers, for undisclosed interests.

GRAND TRUNK DRYDOCK SHIP REPAIR CO., LTD., PRINCE RUPERT, B.C.

S. A. MOULTON, PRINCE RUPERT, B.C.

Ten composite vessels for undisclosed interests.

BRITISH-AMERICAN SHIPBUILDING & ENG. CO., VANCOUVER, B.C.

Establishing plant on Kitsilano Reserve.

Twenty wooden vessels being built for undisclosed interests, 3,500 tons each.

JOHN COUGHLAN & SONS, N. VANCOUVER, B.C.

"War Camp," sister ship to "Alaska," launched in March, 1918.

One steel freight steamer, 427 ft. 9 in. x 54 ft. x 29 ft. 9 in., draft 24 ft. 2 in., speed 11½ knots, 5,730 gross tons. Launched Jan. 10, 1918, for B. Stolt Nielsen, Christiania, Norway. Both sold to the Cunard Line.

Three steel steamers, "War Chariot," "War Chief," and "War Noble," Nos. 4-5-6, 425 ft. x 54 ft., 8,800 tons cap., for the Cunard Line.

Six steel freight steamers, 425 ft. x 54 ft. x 24 ft. 2 in., 8,800 tons cap., 1 knots, for undisclosed interests. Delivery Jan., Feb., March, May and July, 1918.

One steel freight steamer, "Alaska," 427 ft. 9 in. x 54 ft. x 29 ft. 9 in., draft 24 ft. 2 in., speed 11½ knots, 5,730 gr. tons, 8,800 d.w. Launched Jan. 10, 1918, for B. Stolt Nielsen, Christiania, Norway. Sold to the Cunard Line.

THE FOUNDATION CCO., VANCOUVER, B.C.

FRASER VALLEY SHIPBUILDING CO., VANCOUVER, B.C.

GRANT, SMITH & CO., VANCOUVER, B.C.

Six wooden cargo vessels, 250 ft. x 43 ft. 6 in. x 25 ft., 3,000 tons cap., 9½ knots, for the Canadian Government.

HARRISON & LAMOND, SHIPBUILDERS, LTD., VANCOUVER, B.C.

One wooden auxiliary schooner, 225 ft. x 44 ft. x 21 ft. 4 in., 1,600 gr. tons, 2,550 tons cap., for undisclosed interests.

LYALL SHIPBUILDING CO., NORTH VANCOUVER, B.C.

"War Puget," Feb. 16, 1918, 250 ft., wood, 2,080 tons.

Ten wooden cargo ships, for undisclosed interests. Launchings Jan. 20, Mar. 2, Mar. 23, and April 11, 1918.

Building six wooden vessels for own account.

"War Caribou," April 10, 1918, 3,080tons, wood.

W. R. MANCHION, VANCOUVER, B.C.

STANDARD SHIPBUILDING CO., VANCOUVER, B.C.

Two 3,500-ton vessels for Brazilian Govt., and six for French Govt. of 4,500 tons.

Donohoe reinforced type, wood composite construction.

TAYLOR ENGINEERING CO., VANCOUVER, B.C.

Number of small vessels being constructed at total value of $300,000. 4,500-ton floating drydock, 352 ft. long.

VANCOUVER SHIPYARD, LTD., VANCOUVER, B.C.

One motor freighter, 125 ft. x 24 ft. x 12 ft., wood, 160 h.p., Bolinder's crude oil engine. To be completed Aug., for Taylor Engr. Co.

Also repairing several steamers and schooners.

THE WALLACE SHIPYARDS, LTD., N. VANCOUVER, B.C.

"War Dog," 4,500 tons, steel, May 18, 1918.

"War Power," 4,600 tons, steel, March 23, 1918.

Two steel steamers, 315 ft. x 45 ft. x 27 ft., 4,600 tons d. w., trip. exp. vertical engines ,two S. E. Scotch boilers, 10 knots, single screw, for the Canadian Government. Deliveries Dec., 1917, and Aug., 1918.

Two wooden aux. schooners, 255 ft. o. a., 225 ft. keel x 44 ft. x 21 ft. 4 in., five masts, 1,500 gr. tons, cap. 2,500 tons or 1,500,000 ft. lumber, two hatches, twin 160 h.p, Bolinders Hot-bulb engines, twin screws, for the Canadian Government.

Four wooden cargo vessels, aggregate tonnage of 17,500 tons for undisclosed interests.

WESTERN CANADA SHIPYARDS ,LTD., VANCOUVER, B.C.

"War Nootka," Jan. 4, 1918, 3,080 tons (machinery installed at Ogden Point Assembly Sheds).

Six wooden freight steamers, 250 ft. h.p., 259 ft. o. a., 43½ ft. extreme breadth, molded 42½ ft., 25 ft. molded depth., draft 22 ft., 2,800 tons d. w., for undisclosed interests. "War Nootka," launched January 4, 1918. "War Selkirk" launched March 6, 1918.

"War Tatla," "War Casco, "War Chilkat," "War Tanoo," 4 standard vessels now under construction.

CAMERON-GENOA MILLS CO., VANCOUVER, B.C.

"War Yukon," Jan. 24, 1918, 3,080 tons, wood (machinery installed at Ogden Point Assembling Sheds.

THE FOUNDATION CO. OF B. C., LTD., VICTORIA, B.C.

Building ships for undisclosed interests.

Five wooden steamers, 250 b. p. x 42 ft. 6 in. x 25 ft., 1,800 gr. tons, 2,800 tons d. w., one trip. exp. engine, 1,000 h.p., "Howden" water tube boilers, furnished and installed by owners, 10 knots, for undisclosed interests. Esplen & Son & McNaught, N.Y.C., designers.

No. 1—"War Sonchee," launched Dec. 27, 1917.

No. 2—"War Masset," launched April 11, 1918.

No. 3—"War Babine," ready for launching, awaiting propeller and rudder.

No. 4—"War Camchin," ready for launching, awaiting propeller and rudder.

No. 5—"War Nanoose," ready for launching in about five weeks.

CLARENCE HOARD, VICTORIA, B.C.

Wooden car barge for C.P.R., Jan., 1918.

VICTORIA SHIPBUILDING CCO., VICTORIA, B.C.

Constructing wooden ships for British Govt., 3,500 tons each.

NEW BRUNSWICK

C. T. WHITE & SON, LTD., ALMA, N.B. (Office at Sussex, N.B., also.)

Two schooners aux. power, 143 ft., 900 tons d. w., first to be launched in April; second in June.

JAMES X, LENTEIGNE, LOWER CARAQUET, N.B.

One schooner, 28 tons, wood.

INTERNATIONAL SHIPBUILDING CO., NEWCASTLE, N.B.

Two four-masted wooden auxiliary schooners, 155 ft. keel, 37 ft. depth, 535 net tons. One to be completed in September and the other December, 1918. For builders' account.

EUREKA SHIPBUILDING CO., NORTH HEAD, N.B.

One wooden schooner, "Mollie & Melba," 350 tons reg.

PORT COLBORNE BUILDING & REALTY CO., LTD., REXTON, N.B. (Head Office, Welland, Ont.)

Four-masted wooden schooner. Has cap. of 3 schooners per year.

GRANT & HORNE, ST. JOHN, N.B.

Two wooden cargo steamers, "War Fundy" (launched Aug., 1918) and "War Digby," 250 ft. x 42 ft. 5 in. x 25 ft. 5 in., 1,000 h.p., speed 9½ knots, 2,800 tons d.w., for undisclosed interests.

MARINE CONSTRUCTION CO., CANADA, LTD., ST. JOHN, N.B.

Four wooden auxiliary, four-masted schooners, 185 ft. o. a., 165 ft. keel x 40 ft., 1,100 d. w. tons, for builder's account. J. Murray Watts, Philadelphia, naval architect. Launching April, 1918.

One wooden schooner, 'Dorfontein," 740 reg. tons. Launching June, 1918.

PETER AND A. A. McINTYRE, ST. JOHN, N.B.

One schooner, 425 tons, 137 ft., three masts, wood, for builder's account. To be launched in June. Another to be commenced immediately.

ST. JOHN SHIPBUILDING CO., ST. JOHN, N.B. (Plant at Courtenay Bay.)

Ten five-masted auxl. schooners, oil burning engines, for undisclosed interests.

Establishing plant for steel and wood construction.

ST. MARTIN'S SHIPBUILDING CO., ST. MARTIN'S, N.B.

One wooden schooner, 450 tons, three masts. To be launched in July for builder's account.

NEWFOUNDLAND

ANGLO-NEWFOUNDLAND DEVELOPMENT CO., BOTWOOD, NFLD.

Two three-mast aux. schrs., wood, 450 tons, 150 H.P. for the builders' account.

NEWFOUNDLAND SHIPBUILDING CO., HARBOR GRACE, NFLD.

To build ships of 1,200 tons d.w. Steel ships later.

M. E. MARTIN, MORRIS ARM, NR. ST. JOHN'S, NFLD.

Two wooden schooners, 300 tons each, for builder's account.

One wooden schooner, 500 tons, for builder's account.

UNION SHIPBUILDING CO., ST. JOHN, NFLD.

ANNAPOLIS SHIPPING CO., ANNAPOLIS ROYAL, N.S.

"Hilda M. Clark," March 4, 1918, 187 ft., 640 tons reg.

Two schrs. three-masted, wood 500 tons, 170 ft. long. To be completed August and December, 1918.

NOVA SCOTIA

B. L. TUCKER, BASS RIVER, N.S.

One schooner, 350 tons, wood.

JAS. S. CREELMAN, BASS RIVER, N.S.

One three-masted schooner, wood, 170 ft. long, 500 tons. To be completed October, 1918, for builders' account.

HANKINSON SHIPBUILDING CO., BELLIVEAU'S COVE, N.S.
One schooner, 360 tons, wood.

A. A. THERIAULT, BELLIVEAU'S COVE, N.S.

Two wooden three-masted tern schooners, 150 ft. x 33 ft. x 11 ft. 11 in., 400 gr. tons, gasoline engines, for builder's account. Launchings April and June, 1918.

One wooden one-masted schr.

WESTPORT SHIPBUILDING CO., BELLIVEAU'S COVE, N.S.

One wooden schooner, 230 tons net. To be launched in October for builder's account.

ELI PUBLICOVER, BLANDFORD, N.S.

BRIDGEWATER SHIPPING Co., BRIDGEWATER, N.S.

One tern schooner, 450 tons gr., wood. Adapted for aux. power. To be completed Sept., 1918.

One tern schooner, 275 tons gr., wood. Adapted for aux. power. To be completed Nov., 1918.

Both built on owner's account and for sale.

W. A. NAUGLER, BRIDGEWATER, N.S.

One wooden three-masted schooner, "Wm. A. Naugler," 146 ft. o. a., 116 ft. keel x 32 ft. x 11 ft. 6 in., 360 gr. tons, 600 tons cap., for builder's account.

One schooner, 300 tons, wood.

H. MAC ALONEY, CANNING, N.S.

Two wooden schooners, 400 tons each. For builder's account. Launchings July and October, 1918.

S. M. FIELDS, CAPE D'OR, N.S.

One three-masted schooner, 350 tons.

Keel laid for another three-master.

CHESTER BASIN SHIPBUILDERS, LTD., CHESTER BASIN, N.S.

Two schooners, 100 tons, wood. Launchings Aug., 1917, Dec., 1917.

One four-masted schooner, 60 tons. Launching Sept., 1918.

One two-masted schooner, 100 tons. Launching June, 1918.

MORTIMER PARSONS, CHEVERIE, N.S.

One wooden three-masted schooner, 170 ft. x 38 ft. x 14 ft., 575 gr. tons, 900 tons cap., for builder's account. Launching Sept., 1918.

Two wooden three-masted schooners for builder's account.

One tern schooner, "Donald Parsons," 140 ft. keel x 34 ft. 4 in. x 12 ft. 4 in., 500 gross tons, 825 tons d.w. for Mortimer Parsons. Launched May 11, 1918.

Another vessel, same size, to be launched Nov., 1918, for builder's account.

BAY SHORE SHIPYARD, CHURCH POINT, N.S.

Building two wooden vessels for R. C. Elkin, Ltd., St. John, N.B.

MOISE BELLIVEAU, CHURCH POINT, N.S.

One schooner, 450 tons, wood.

FIDELE BOUDREAU, CHURCH POINT, N.S.

One schooner, 350 tons, wood.

J. E. GASKILL, CHURCH POINT, N.S.

Other yards at Grosses Coques, Little Brook, Meteghan and Port Wade, N.S.

Five wooden schooners from 344 to 387 reg. tons. All sister ships.

COMEAU SHIPBUILDING CO., COMEAUVILLE, N.S.

One three-masted wooden schooner, for builder's account, 450 tons.

J. W. COMEAU, COMEAUVILLE, N.S.

One schooner, 329 tons, Jan., 1918.

J. N. RAFUSE & SONS, CONQUERALL BANKS, DIGBY CO., N.S. (Other plants at Salmon River and Shelburne, N.S.)

Two wooden three-masted schooners, 120 ft. x 32 ft. x 12 ft., 700 tons cap., for builder's account. For the market. Launchings Oct. 25, 1917, and Jan. 10., 1918.

One three-masted wooden schooner, "Integra," 112 ft. x 30 ft. x 11 ft. 6 in., 600 tons cap., for J. O. Williams & Co., St. John's, Newfoundland. Launching Nov. 25, 1917.

E. F. WILLIAMS, DARMOUTH, N.S.

One schooner, 360 tons, wood.

ROBAR BROTHERS, DAYSPRING, N.S.

Schooner for Capt. Ivan Creaser, Dayspring.

MAURICE E. LEARY, DAYSPRING, N.S.

One wooden schooner, 225 tons, 135 ft. o. a. To be launched Aug. 15, 1918, for La Have Outfitting Co., La Have, N.S.

J. NEWTON PUGSLEY & CHAS. ROBERTSON, DILIGENT RIVER, N.S.

One schooner, three-masted, 475 tons.

McLEAN & McKAY, ECONOMY, N.S.

One tern schooner, wood, 400 tons net, 135 ft. keel. To be launched Sept. 1, 1918, for builders' account.

S. J. SOLEY, ROX RIVER, N.S.

One schooner, three-masted, 121 ft. keel, 300 tons. Built of native wood. To be launched Sept., 1918. For builder's account. For sale.

ALLAN & FRASER, FRASERVILLE, N.S.

One schooner, 350 tons, wood.

BERNARD W. MELANSON, GILBERT'S COVE, N.S.

One three-masted schooner, wood, 250 tons net. To be completed November, 1018, for builder's account and for sale.

AMOS BLINN, GROSSES COQUES, N.S.

One schooner, 375 tons. Jan., 1918.

One schooner, 350 tons, wood.

BINN BROS., GROSSES COQUES, N.S.

F. K. WARREN & CO., GROSSES COQUES, N.S. (Offices in Union Bank Chambers, Halifax, N.S., also.)

350-ton tern schooner, under construction Jan., 1918.

HALIFAX SHIPBUILDING CO., HALIFAX, N.S.

FAUQUIER & PORTER, HANTSPORT, N.S.

Two wooden four-masted schooners, 178 ft. x 40 ft. x 18 ft., 1,000 gross tons, 2,000 tons d. w., twin 100 h.p. oil engines. Launchings Aug. 1, 1918, and Sept. 1, 1918. Both for sale.

J. WILLARD SMITH, HILLSBURN, N.S. (Head Office, St. John, N.B.)

One tern schooner, wood, 140 ft. keel, 35 ft. beam, 450 tons net reg. tons.

HENRY COVEY, INDIAN HARBOR, N.S.

CHARLES GRIFFIN, ISAACS HARBOR, N.S.

One schooner, 40 tons, wood.

LEWIS SHIPBUILDING CO., LEWISTON, N.S.

One schooner, 670 tons. Jan., 1918.

INNOCENT COMEAU, LITTLE BROOK, N.S.

One three-masted schoner, 140 ft. x 35 ft. 6 in. x 17 ft. 6 in., 650 gr. tons, for the Weymouth Shipping Co., Weymouth, N.S., Can. Delivered Jan. 16, 1918.

J. W. RAYMOND, LITTLE BROOK, N.S.

One wooden schooner, No. 4, 140 ft. x 36 ft. x 17 ft., 575 gr. tons, for Jones Bros., Weymouth, Digby Co., N.S.

H. A. FRANK, LIVERPOOL, N.S.

Two wooden schooners, 500 tons, for builder's account.

McKEAN SHIPBUILDING CO., LIVERPOOL, N.S.

One wooden three-masted schooner, 126 ft. keel x 3 ft. x 12 ft. 9 in., 700 tons d.w., for builder's account. Launching May, 1918.

W. F. McKEAN & CO., LIVERPOOL, N.S.

One schooner 400 tons, Jan., 1918.

D. C. MULHALL, LIVERPOOL, N.S.

Two wooden schooners, one 300 gross tons, other 400 gross tons, for undisclosed interests.

N.S. SHIPBUILDING & TRANS. CO., LIVERPOOL, N.S.

One wooden tern schooner No. 2, 122 ft. x 33 ft. x 12 ft. 6 in., 450 gr. tons, 70 d.w. tons, 7 knots, for Peter Yee Wing & Co., Ltd., Sydney, Australia. R. McLeod and J. S. Gardner. Launching Feb. 1, 1918

One two-masted fishing schooner, "Sadie A. Nickle," 130 ft. o. a. x 26 ft. 10 ft. 6 in. hold, 230 d.w., two 50 h.p. oil engines, for Rafuse & Sons. Launching May 15, 1918.

One three-mast schooner, 420 tons. F-M Co. oil burning engines, for Job Bros., St. Johns, Nfld. Launching July 15.

One three-masted schooner, 270 tons, for builders' account. Launching Sept. 1.

One two-masted fishing schooner, 120 tons, for builders' account. Launching Oct. 1.

ROBIN, JONES & WHITMAN, LIVERPOOL, N.S.

One schooner, 340 tons, wood.

SOUTHERN SALVAGE CO., LIVERPOOL, N.S.

One wooden schooner, three masts, 300 gross tons, 600 tons capacity, for a West Indian firm. Launching Jan., 1918.

One wooden steamer; No. 5, 250 ft. x 43 ft. 6 in. x 25 ft., 2,900 gr. tons, 10 knots, 1,000 h.p., for undisclosed interests. Launching Fall, 1918.

One two-masted schooner, "Win-the-War," 137 ft. o. a. x 25 ft. 2 in. x 11 ft. 6 in., 187 gr. tons, for the builder's account. Launched Nov., 1917.

CONRAD & REINHARDT, LUNENBURG, N.S.

LUNENBURG MARINE RAILWAY, LUNENBURG, N.S.

FRED A. ROBAR, LUNENBURG, N.S.

SMITH & RHULAND, LUNENBURG, N.S.

Two schrs., 225 tons each. Jan., 1918.

Two schooners, 100 tons each.

"Donald Cook," 112 ft. beam, depth of hold, 11 ft., for Capt. William Cook. Launched April, 1918.

Two schooners, 168 gr. tons, for W. C. Smith.

One schooner, 150 gr. tons, for Lun. Outfitting Co.

J. B. YOUNG, LUNENBURG, N.S.

ERNST SHIPBUILDING CO., MAHONE BAY, N.S.

One two-masted schooner, 120 ft. o. a. x 25 ft. 6 in. x 10 ft. 6 in., 200 tons d.w., for Lunenburg Outfitting Co. April, 1918. Christened "Madeline Adams."

One three-masted schooner 125 ft. keel x 32 ft. x 12 ft., 600 tons, d. w. To be delivered July, 1918.

Two masted schooner, "Agnes D, McGlaston,' 130 ft. x 26 ft. 8 in. x 10 ft. 8 in. Launched Nov. 13, 1917.

Keel laid for three-masted wooden schooner, 200 tons.

J. ERNEST & SON, MAHONE BAY, N.S.

One schooner, 520 tons. Jan., 1918.

O. A. HAM, MAHONE BAY, N.S.

One schooner, "Dozle," 128 tons. Launching May 30. One motor boat about 20 tons.

McLEAN CONSTRUCTION CO., MAHONE BAY, N.S. (Leased to Montague Mahaffy, Toronto.)

One three-masted schooner, 325 tons gr. reg., 500 tons cap., West Indies freight type. Will be launched in July.

Keel laid for another three-masted schooner with cargo capacity of 425 tons.

JOHN McLEAN & SONS, MAHONE BAY, N.S.

One schooner, 95 tons, wood.

J. A. BALSOM CO., LTD., MARGARETSVILLE, N.S.

One schooner, 409 tons. Jan., 1918.

CLARE SHIPBUILDING CO., METEGHAN RIVER, N.S.

One wooden schooner for builder's account. 400 tons.

A. H. COMEAU & CO., METEGHAN, N.S.

One schooner, 400 tons, wood.

AGAPIT COMEAU, METEGHAN, N.S.

JOHN F. DEVEAU, METEGHAN, N.S.

A 300-ton schooner was launched recently for Ritcey & Co., Lunenburg, N.S., and named Charles A. Ritcey.

1 schooner, 425 tons. Jan., 1918.

1 schooner, 400 tons, wood.

J. E. GASKILL, METEGHAN, N.S.

THOMAS GERMAN, METEGHAN, N.S.

One schooner, 350 tons, wood.

R. H. HOWES CONSTRUCTION CO., METEGHAN, N.S. (Leased James Cosman's shipyards.)

Building several wooden schooners for own account.

DR. F. H. MACDONALD, METEGHAN, N.S.

One wooden four-masted schooner, "Rebecca L. Macdonald," 201 ft. o. a. x 6 ft. x 16 ft., 800 gr. tons, 1,500 tons d.w., for builder's account. Launching January 1, 1918.

One wooden fishing schooner, No. 3, 84 ft. x 17 ft. x 7 ft., 50 gr. tons, 5 h.p. Launching August, 1918.

One schooner, 544 tons, wood.

METEGHAN RAILWAY & SHIPBUILDING CO., METEGHAN, N.S.

One wooden three-masted schooner, 182 ft. x 35 ft., 450 gr. tons.

One schooner, 470 tons, wood.

CHAS. McNEIL, NEW GLASGOW, N.S.

One wooden 3-masted schooner, 103 ft. x 30 ft. x 10 ft. 9 in., 200 tons, for builder's account. Launching July, 1918.

One wooden 3-masted schooner, 142 ft. x 25 ft. 4 in. x 13 ft., 500 gr. tons, for builder's account. Launching Oct., 1918.

THE NOVA SCOTIA STEEL & COAL CO., NEW GLASGOW, N.S.

Two steel freight steamers, 257 ft. 9 in. x 35 ft. x 20 ft., 1,700 gr. tons, 2,350 tons cap., trip. exp. engines, boilers 10 ft. 6 in. x 11 ft. 6 in., 800 h.p., 8.5 knots, for undisclosed interests. Launching of one Jan., 1918.

Two steel cargo strs. Raised quarter deck type, 248 ft. 9 in. long, 1,649.68 gross tons. Triple expansion engines, 17, 28, 46 x 33 in. stroke. One, the "War Bee," for undisclosed interests to be completed June, 1918. Other for Steel Co., to be completed Oct., 1918.

Two masted schooner, launched May 16, 130 ft. o. a. x 26 ft. x 10 ft. 6 in. Designed by J. S. Gardner.

O'BRIEN BROS., NOEL, N.S.

One schooner, 325 tons, wood.

W. R. HUNTLEY & SON, PARRSBORO, N.S.

One wooden schooner, 175 ft. x 39 ft. x 17 ft., 900 gross tons, for

C. T. WHITE & SON, SUSSEX, N.S.

Two schooners, 325 tons each. Jan., 1918.

One schooner, 490 tons, wood.

One schooner, 850 tons, four masted.

WAGSTAFF & HATFIELD, PARRSBORO, N.S.

One schooner, 400 tons. Jan., 1918.

SIDNEY ST. C. JONES, PLYMPTON, N.S. (Head Office, Weymouth North, N.S.)

One tern schooner, 200 tons, net reg., 107 ft. x 28 ft. 8 in. x 10 ft. 8 in., to be launched Nov., 1918.

S. SALTER, PARRSBORO, N.S.

One 200-ton schooner, wood.

DOWLING & STODDART, PORT CLYDE, N.S.

One gas boat, 27 tons, wood.

One schooner, 300 tons, wood.

SWIME BROS., PORT CLYDE, N.S.

Several small motor trawlers, for builder's account.

G. M. COCHRANE, PORT GREVILLE, N.S.

One tern schooner, "Alfredock Hedley," 152 ft. 6 in. x 36 ft. 12 ft. 6 in., 461 tons reg., for Adam B. Mackay, of Hamilton, Ont. Launched Jan., 1918.

One 4-masted schooner, 850 tons, 155 ft. x 37 ft. x 18 ft., 2 decks, wood. To be completed Oct., 1918, for builder's account.

H. ELDERKIN & CO., PORT GREVILLE, N.S.

One wooden three-masted schooner, for undisclosed interests.

ELLIOTT GRAHAM, PORT GREVILLE, N.S.

Is building a schooner, "Khaki Lad," of about 325 tons, at Port Greville, N.S., for J. W. Kirkpatrick and others. She will be launched early in October, and is reported to have been sold to Newfoundland parties.

One schooner, 360 tons. Jan., 1918.

SMITH CANNING CO., PORT GREVILLE, N.S.

One schooner, 350 tons, wood.

WAGSTAFF & HATFIELD, PORT GREVILLE, N.S.

One schooner, 400 tons, wood.

WILLIAM CROWELL, PORT LATOUR, N.S.

J. W. RAYMOND, PORT MAITLAND, N.S.

One schooner, 375 tons. Jan., 1918.

PORT WADE SHIPBUILDING CO., PORT WADE, N.S. (Head Office, Digby.)

One 350-ton schooner, wood.

JOHN BROWN, PUBLIC LANDING, N.S.

One tow barge, 50 tons, wood.

THE CUMBERLAND SHIPBUILDING CO., PUGWASH, N.S.

Establishing plant for the construction of wooden ships.

MACKENZIE SHIPPING CO., RIVER JOHN, N.S.

One four-masted schooner, 600 tons, wood. About half built. Fitted for aux. engines.

CHARLES McLELLAN, RIVER JOHN, N.S.

One schooner, 100 tons, wood.

W. J. FOLEY, SALMON RIVER, N.S.

Building for J. N. Rafuse & Sons one wooden, 300 tons, under construction. Cap. of plant, 1,000 tons yearly.

J. N. RAFUSE & SONS, SALMON RIVER, N.S. (Other plants at Shelburne and Conquerall Banks.)

Launched recently a three-masted schooner named "Industrial," at W. J. Foley's ship yard at Salmon River, N.S., of the following dimensions: length, 113 ft, breadth, 30 ft., depth, 11½ ft., 325 tons.

ACADIAN SHIPPING CO., SAULNIERVILLE, N.S.

One schooner, 00 tons, wood.

SAULNIERVILLE SHIPBUILDING CO., LTD., SAULNIERVILLE, N.S.

J. LEWIS & SONS, SHEET HARBOR, N.S.

One schooner, wood, four-masted, 725 gr. tons. To be completed Sept., 1918. For private account.

GEO. A. COX, SHELBURNE, N.S.

One schooner, 200 tons, wood.

One schooner, 322 tons, wood.

JOSEPH McGILL SHIPBUILDING & TRANSPORTATION CO., SHELBURNE, N.S.

"Sparkling Glance," 246 tons register, for Harvey & Co., St. John's, Nfld.

One schooner, 160 tons, wood.

W. C. McKAY & SON, SHELBURNE, N.S.

One wooden schooner, 130 ft. o. a. x 26½ ft. x 11 ft. depth of hold, 140 tons register. Launched Dec. 1, 1917.

One three-masted schooner, for Messrs. Hallet, of Newfoundland.

One schooner, 620 tons, wood.

J. N. RAFUSE & SONS, SHELBURNE, N.S. (Other plants at Salmon River and Conquerall Banks.)

One wooden three-masted schooner, 120 ft. x 32 ft. x 12 ft., 700 tons cap., for builder's account. For the market. Launching Jan. 10, 1918.

One wooden three-masted schooner, 118 ft. x 33 ft. x 12 ft., 700 tons cap., for builder's account. For the market. Launching Jan. 15, 1918.

Two three-masted wooden schooners, 122 ft. 6 in. keel x 32 ft. x 12 ft., 275 net tons, for J. O. Williams & Co., St. John's, Newfoundland. One to be built at Salmon River yard and one at Ship Harbor.

THE SHELBURNE SHIPBUILDERS, LTD., SHELBURNE, N.S.

One tern schooner, "Misty Star," 145 ft. x 31 ft. x 11 ft. 6 in., 600 tons d.w., 330 tons gross, for Harvey & Co., St. John's, Newfoundland. Launching Mar. 15, 1918.

One tern schooner, 154 ft. x 32 ft. x 12 ft. 4 in., 400 gr. tons, 700 tons d.w. cap. Launching June, 1918. Not sold yet.

Two schooners, one 349 tons, other 400 tons.

EASTERN SHIPBUILDING CO., SHIP HARBOR, N.S.

One wooden schooner, 600 tons d.w., for J. N. Rafuse & Sons., Halifax, N.S., Can.

One wooden schooner, 300 tons.

STEPHEN MORASH & CO., SHIP HARBOR, N.S.

One wooden schooner, 150 ft. x 37 ft. x 13 ft., four masts, for Canadian interests. Keel laid Feb. 5, 1917.

JAMES E. PETTIS, SPENCER'S ISLAND, N.S.

One schooner, 425 tons, wood, three-masted.

CAPE BRETON SHIPBUILDING CO., SYDNEY, N.S.

One three-masted schooner.

THE TUSKET SHIPBUILDING CO., TUSKET, N.S.

AMOS H. STEVENS, TANCOOK, N.S.

ALVIN STEVENS, TANCOOK, N.S.

STANLEY MASON, TANCOOK, N.S.

ALBERT PARSONS, WALTON, N.S.

One schooner, 400 tons. Jan., 1918.

H. T. LeBLANC, WEDGEPORT, N.S.

One steam trawler, 165 ft., 600 tons, for J. N. Rafuse & Sons.

L. S. CANNING, WARDS BROOK, N.S.

One tern schooner, 350 tons, for W. C. Smith & Co., Lunenburg, N.S.

WEDGEPORT NAVIGATION & TRANSPORTATION CO., WEDGEPORT, N.S.

T. K. BENTLEY, WEST ADVOCATE, N.S.

One 200-ton steamer.

One 4-masted schooner, wood, 511 tons. To be launched September or October. For owner's account.

J. W. KIRKPATRICK, WEST ADVOCATE, N.S.

One 350-ton schooner, wood.

BOEHNER BROS., WEST LA HAVE, N.S.

Two large fishing schooners, 150 tons each.

One beam trawler, 300 tons, crude oil engines. For local interests.

BEAZLEY BROS., WEYMOUTH, N.S. (Head Office Roy Bldg., Halifax.)

One schooner, 394 tons, three masts.

E. R. GAUDET, WEYMOUTH, N.S.

One 350-ton schooner, three-masted.

E. P. RICE, WEYMOUTH, N.S.

One three-masted schooner, 350 tons.

RICE, WARREN & CO., WEYMOUTH, N.S.

One three-masted schooner, 334 tons.

FALMOUTH SHIPBUILDING & TRANSPORTATION CO., WINDSOR, N.S.

One wooden schooner, 350 gr. tons, 700 tons cap., for undisclosed interests.

NOEL SHIPBUILDING & TRANSPORTATION CO., WINDSOR, N.S.

One wooden schooner, three masts, 138 ft. x 35 ft. x 13 ft., 450 net reg. tons. Oil auxiliary engines. Launching August, 1918, for builder's account.

C. A. NICKERSON, WOODS HARBOR, N.S.

MILTON SHIPBUILDING CO., YARMOUTH, N.S.

Taken over and enlarging old plant of James Jenkins.

350-ton schooner building.

W. O. SWEENY, YARMOUTH, N.S.

100-ton fishing schooner.

YARMOUTH SHIPBUILDING CO., YARMOUTH, N.S.

One 450-ton wooden schooner, 128 ft. keel. To be completed Oct., 1918.

ONTARIO

CAN. ALLIS-CHAMBERS (Head Office, Toronto), BRIDGEBURG, ONT.

Six standard freight steamers, 3,500 tons, 261 feet long, for undisclosed interests.

COLLINGWOOD SHIPBUILDING CO., COLLINGWOOD, ONT.

Two steel cargo steamers, Nos. 51-52, 50 ft. x 43 ft. x 25 ft., 2,500 gr. tons, 2,900 tons d. w., triple expansion engines 18-30-50 x 36, 2 boilers 14 ft. x 11 ft., 10 knots, for undisclosed interests. Delivery May and Aug., 1918.

Four deep-sea trawlers, Nos. 53-58, inclusive, 125 ft. x 23 ft. 6 in. x 13 ft. 6 in., 288 gross tons, 12¾-21½-35/24 engines, 1 cylinder 13 ft. 6 in. x 10 ft. 6 in., 10 knots, 500 h.p. for undisclosed interests.

One steel freight steamer of 3,800 tons d. w. for undisclosed interests. Keel laid May 8, 1918.

"War Wizard," May 8, 1918, 3,000 tons d. w., 261 ft. x 43 ft. 6 in. x 20 ft. depth. For undisclosed interests.

"War Witch" same size, now building. To be launched ——

R. MORRILL, COLLINGWOOD, ONT.

"Windsor," Aug. 10, steam tug, 105 ft., for Ontario Gravel & Freighting Co.

CAN. CAR & FOUNDRY CO., FORT WILLIAM, ONT. (Head Office Transportation Bldg., Montreal.)

Twelve steel mine sweepers for French Government.
145 ft. o.a. steel construction, value $2,500,000.
Plant under construction.

GREAT LAKES DREDGING CO., FORT WILLIAM, ONT.

Two wooden vessels, 2700 tons d.w. cap., 260 ft. o.a., 43 ft. beam. Triple expansion engines of 1,000 h.p. To be launched November, 1918.

"War Sioux" launched May 12, 1918. Keel laid for another twenty minutes later. Launching scheduled for November.

THUNDER BAY CONTRACTING CO., FORT WILLIAM, ONT.

One wooden freight steamer, 261 ft. long, for undisclosed interests.

NATIONAL SHIPBUILDING CO., GODERICH, ONT.

DAVIS DRYDOCK CO., FT. OF BAY ST., KINGSTON, ONT.

One motor pass. boat, 64 ft. x 18 ft. x 6 ft., 52 gr. tons, wood. For ferry service, Gananoque to Clayton. Delivery June 1, 1918.

Twenty-eight lifeboats, regulation size, wood and metal for Government and private interests.

KINGSTON SHIPBUILDING CO., KINGSTON, ONT.

Several steel trawlers for undisclosed interests. First one launched Dec. 22, 1917.

SELBY & YOULDSON, KINGSTON, ONT.

GEORGIAN BAY SHIPBUILDING & WRECKING CO., MIDLAND, ONT.

Tug, 50 tons, wood.

One tug 40 tons, wood.

MIDLAND DRY DOCK CO., MIDLAND, ONT.

Three steel freight steamers, 261 ft. x 43 ft. 6 in. x 23 ft., 3,400 tons d. w., 10 knots, for undisclosed interests. Deliveries, one in July, 1918, and two others before close of navigation, 1918.

Alterations on steamers "Mariska" and "Glenlyon."

"Western Star" being reconstructed.

MIDLAND SHIPBUILDING CO., LTD., MIDLAND, ONT.

Three steel freight steamers, 261 ft. x 43 ft., 6 ft x 23 ft., 3,400 tons d. w., 10 knots for undisclosed interests. Deliveries, one in July, 1918, and two others before opening of navigation, 1919.

PORT ARTHUR SHIPBUILDING CO., PORT ARTHUR, ONT.

Six steel trawlers on order. 135 ft. o. a. x 23 ft. 4 in. x 15 ft. 1 in., 294.5 tons, trip. exp. engines, 500 h.p., single end Scotch boiler. Two-masted.

Two of these launched June 8, 1918.

"War Isis," April 3, 1918, 3,400 tons d. w. Trawler launched same day. April 3, 1918, keel laid for "War Hatha," sister ship to "War Osiris," of same construction, nearly completed.

"War Fish," Aug. 19, 1917, 4,300 tons, steel.

"War Dance," Nov. 3, 1917, 3.400 tons, steel.

Four steel screw steamers, ocean freight service—"War Osiris," "War Hathor," "War Karma," "War Horus," 3,400 tons d. w. each, 261 ft. o.a. 251 ft. b.p. in length. Triple expansion engines, 2 Scotch boilers each, approx. 1,250 h.p. For completion, two Aug. 31st, and two Nov. '20, 1918. For undisclosed interests.

MUIR BROS. DRYDOCK CO., LTD., PORT DALHOUSIE, ONT.

J. W. GEROW, ROSSPORT, ONT.

REID WRECKING CO., SARNIA, ONT.

One steel tug, 157 ft. x 32 ft. x 19 ft., trip. exp. engines, for lake or ocean service. Keel blocks laid March 5. Launching July, 1918. For Reid Wrecking Co

WEST, PEACHEY & SONS, SIMCOOE, ONT.

Number of tugs for lumbering interests.

DOMINION SHIPBUILDING CO., TORONTO, ONT.

One steel cargo vessel, 261 ft. x 43 ft. x 28 ft. 2 in., 3,500 tons d.w. cap. For Department of Marine, Ottawa.
Have orders for five other vessels same dimensions.
"Traja" launched May 15, 1918.

THE POLSON IRON WORKS, LTD., TORONTO, ONT.

Eight steel cargo steamers, Nos. 133-4-5-6-7-9-40, 216 ft. o. a., 251 ft. b. p. x 43 ft. 6 in. x 22 ft. 11 in., draft loaded 19 ft. 6 in., single decked, 2.350 approx. gross tons, 3,500 tons cap., single screw trip, exp. engine 20½-33-54 x 36, 1,250 h.p., 10 knots, two 14 ft. x 12 ft. Scotch R. T. boilers, 180 h.p., to cost $600,000 each, for undisclosed interests. Launchings throughout 1918.

Four building and four on order.

"Asp," Feb. 11, 1918, 261 ft., 3,500 tons d.w., steel.

"Tento," Oct. 22, 1917, 261 ft., 3,500 tons d.w. steel. For Norwegian interests. Transferred to British registry.

THE THOR IRON WORKS, TORONTO, ONT.

Under same management as Dominion Iron Works

TORONTO SHIPBUILDING CO., LTD., TORONTO, ONT. (Toronto Dry Dock Co., Ltd., under same management.)

Two wooden cargo steamers, 250 ft. B.P. 42 ft. 6 in., moulded breadth 25 ft., moulded depth 22 ft., draft 2,500 tons d. w. Triple exp. engines $\frac{20 \times 33 \times 54}{40}$. Howden boiler. Launchings July and Sept., 1918. For undisclosed interests.

THE BRITISH AMERICAN SHIPBUILDING CO., WELLAND, ONT.

Two steel frs.. 3,500 tons d. w., 261 ft. O.A., 43 ft. beam, 23 ft. moulded depth, Westinghouse steam turbine engines. Delivery 1918.

WELLAND SHIPBUILDING CO., WELLAND, ONT.

One steel freight steamer, "War Weasel," 261 ft. x 43 ft. 6 in. x 23 ft., 3,300 tons d. w., trip. exp. engines, 14 ft. x 12 in. Scotch boilers, 10 knots, 1,250 h. p. Launching April, 1918.

Two deep frame cargo vessels, "War Badger" and "No. 3," 261 ft. x 43 ft. 6 in. x 23 ft., 3,300 d. w. tons, geared turbines, Howden's forced draught boilers, 10 knots, 1,250 h.p. Launchings June and August, 1918.

PRINCE EDWARD ISLAND

THE CARDIGAN SHIPBUILDING PLANT, CARDIGAN, P.E.I. (Purchased Annandale Lumber Co. plant and removed to Cardigan.)

One three-masted schooner, 325 tons, to be completed November, 1918.
Two more to be built.

QUEBEC

TIDEWATER SHIPBUILDERS, LTD., THREE RIVERS, QUE. (Formerly Sorel Shipbuilding & Coal Co.)

Three steel trawlers for undisclosed interests
One steel steam barge for the Canada Steamships Lines.

R. N. LE BLANC, BONAVENTURE, QUE.

J. Z. DEGAGNE, EBOULEMENTS, QUE.

DAVIE SHIPBUILDING & REPAIRING CO., LEVIS, QUE.

Six military barges, 130 feet long.
Eight steel trawlers, 2,035 tons.
One floating crane, 350 tons.
*One steel cargo vessel of 5,000 tons cap.
Building several steel lighters and several wooden drifters.

Steel car ferry "Canora" building for C.N.R., 308 ft. x 52 ft., cap. 20 loaded cars. Speed, 14 knots.

ATLAS CEMENT CONSTRUCTION CO., LTD., MONTREAL, QUE.

One concrete cargo steamer, "Concretia," 125 ft. x 22 ft. x 13 ft. steel ribbed, hull to be from 3 in. to 5 in., thick, for the builder's account.

CANADIAN VICKERS, LTD., MONTREAL, QUE.

Two cargo steamers, 9.400 tons, steel; 1 dredge, 2,364 tons, steel; 12 trawlers, 3,050 tons, steel; 23 drifters, 3,350 tons, wood.

Six freight steamers, 24 ft. draught, 7,000 tons cap., 11 knots, for undisclosed interests. Delivery, 1917, of two for Norwegian interests "Porsanger," 394 ft. 6 in. o. a. x 40 ft. 4 in. x 30 ft., triple exp. engines, launched Nov. 29, 1917.

"War Earl" launched June 8, 1918, 7,000 tons d.w., 380 ft. x 49 ft. Keel of sister ship laid five minutes after this launching.

One steel freight steamer, 2,360 tons, cap., for undisclosed interests.

Three steel cargo vessels, 8,200 tons, to be laid down in May, Aug. and Sept., 1918, for undisclosed interests.

FRASER BRACE & CO., MONTREAL, QUE.

Four wooden cargo steamers, 3,000 tons, for undisclosed interests. Two keels laid during Oct., 1917.

HALL ENGINEERING CO., MONTREAL, QUE.

MONTREAL DRY DOCK & SHIP REPAIRING CO., MONTREAL, QUE.

Operate dock 428 ft. long 30 ft. deep.

MONTREAL SHIPBUILDERS LTD., 37 Belmont St., MONTREAL, QUE. (Associated with Atlas Construction Co.

THE QUEBEC SHIPBUILDING AND REPAIR CO., Board of Trade Bldg., MONTREAL, QUE.

2 vessels for undisclosed interests 3,000 tons each.

QUINLAN & ROBERTSON, MONTREAL, QUE. (Yards at Quebec.)

Four wooden steamers, for undisclosed interests.

LOUIS GAUNDRY, 12 St. Peter St., QUEBEC, QUE.

QUEBEC SHIPBUILDING & REPAIR CO., ST. LAURENT, QUE.

2 schrs. 1,400 tons and 1,200 tons.

One wooden four-masted auxl. schooner "Martin Connolly," 223 ft. x 42 ft. x 20 ft., 2,100 tons d. w., for undisclosed interests. Launched Oct. 28, 1917.

QUINLAN & ROBERTSON, QUEBEC, QUE.

Four wooden steamers, totalling 6,400 tons, for undisclosed interests.

CANADIAN GOVERNMENT SHIPYARDS, SOREL, QUE.

One steel vessel for undisclosed interests.

LECLAIRE SHIPBUILDING CO., SOREL, QUE.

Building 6 steel ships at value of $1,500,000.

Six trawlers, 125 ft. single screw type expansion engines, 500 I.P.H. steel. To be completed this year for private acct.

H. H. SHEPHERD, SOREL, QUE.

Rebuilding seven drifters. Particulars withheld owing to govt. restrictions.

SINCENNES-McNAUGHTON LINES. SOREL, QUE.

One tug 410 tons, wood.

CAN. GOVERNMENT SHIPYARD, SOREL, QUE.

One steel vessel for undisclosed interests.
Building steam trawlers and wooden drifters for undisclosed interests.

ST. LAWRENCE SHIPBUILDING & STEEL CO., SOREL, QUE.

THE THREE RIVERS SHIPYARDS LTD., THREE RIVERS, QUE.

Two cargo steamers, 3,100 d. w., wood, 260 ft. x 43 ft. 6 in., 1,000 H.P. engines. Names, "War Mingan" and "War Nicolet." To be completed end of navigation season for undisclosed interests.

Marine News from Every Source

Sarnia.—The MacKenzie Milne & Co., Ltd., of Sarnia, have taken up the building of metallic life boats, and are making satisfactory progress in this line.

Frederick D. Herbert, who has been elected president and general manager of the Kearfott Engineering Co., Inc., has been identified for a number of years with the maritime industry and for the last ten years was New York manager of the Terry Steam Turbine Co. He will continue to handle the marine account of the Terry Steam Turbine Co., with offices at 95 Liberty Street, New York.

The Dominion Shipbuilding Co., Toronto, is making good progress in the construction of the new shipyard at the foot of Bathurst Street. One hull is partly plated, and berths have been laid down for five other standard steel freighters. Two gantry cranes are being erected, and will be completed shortly. The machine and blacksmiths' shops are practically completed, and will soon be ready for the machinery. The plate shop and moulding loft are being built, and are in an advanced stage of construction.

LUNENBURG, N.S.—Miss Margaret Nicholson and other young ladies who have contracted to assist with the building of small boats, at Beinn Breagh, Baddeck, left Wednesday morning for Baddeck, where they will take up their new work at once. A number of other young ladies are already at Baddeck and others have signified their intention of taking up this work. Miss Nicholson is to remain for the duration of the war, and so are some of the others. Many have, however, contracted for one year the shortest period for which a contract will be signed, and will then renew the contract if they find that the work is congenial and that they can do it.

LONDON.—A fine piece of repair work has just been done by men of the Mercantile Drydock Company at Jarrow. A steamer torpedoed Sunday was brought to the Tyne the following day with a forty-foot rent in her side near the engine-room. Early Tuesday she was placed on the Admiralty pontoon and the men worked day and night until Friday, carrying out temporary repairs to enable her to be towed around to a wharf for completion of the work. This constitutes a record for the River Tyne. Many workmen worked for forty-three hours at a stretch, while others put in four consecutive shifts without a break.

Halifax.—That the work of establishing the big steel shipbuilding plant at Halifax will be commenced next week, if the engineers' plans are completed by that time, and that within three months the keels of three 10,000-ton freighters, the largest ever built in Canada, will have been laid on the building berths, was the statement made by J. W. Norcross, president of Halifax Shipyards, Limited. "Our expenditure for the shipbuilding plant," Mr. Norcross said, "will be between $3,750,000 and $4,000,000, and if the engineers have the plans ready we will start spending that money during the latter part of next week, when we will let our contracts."

TORONTO.—The car ferry Canora, which was launched at Levis, Que., is the first vessel of its kind to be equipped with a stern gate, the invention of Mr. A. Angstrom, a naval architect of this city. The ferry, which was constructed for the Canadian Northern Railway, will ply in British Columbia waters, operating between Patricia Bay and Port Mann. The stern gate is a segment of a circle. It is a new departure of the rolling type of a shutter and is arranged to completely close the stern opening when the ship is at sea. The contrivance, which closely resembles the roll-top of a desk in design, is operated by a hoisting engine situated on the cabin deck, which will raise or lower it to any distance. When up the shutter closes in on the deck in such a way that it interferes with nothing else.

Cleveland.—The steamer Charles R. Van Hise, of the Pittsburg Steamship Company, has been taken over by the United States Shipping Board. The steamer, now on Lake Superior, will be commandeered when she reaches Lake Erie, to be cut in two and reconstructed for salt water service. The Van Hise is the largest lake vessel taken for the coast. The steamer is 446 feet keel, 50 feet beam and 29.5 feet deep. Being too wide to pass through the Welland Canal she will be cut in two sections, which will be rolled over and towed through the canal on their sides. The Van Hise was selected because of her unusual dimensions and power. She is deeper than most lake boats of her length. She has new Scotch boilers and a quadruple expansion engine. Her capacity is about 7,400 tons, but it is figured she will carry at least 9,000 tons on the coast.

Montreal.—Speaking of the Pierce McLouth, a wooden ship made at Marine City, an expert says: "The Pierce McLouth is of unusually heavy build, having six-inch bottom planking, fourteen-inch floor frames, six-inch bilge ceiling and six-inch clamps. The ship is diagonally strapped with steel, has steel straps at the head of her frames, and has two plates of seven-eighths steel, thirty inches wide, on the keelson. She is built entirely of white oak, and has an enclosed pilot house, steam steering gear, electric lights, steam windlass, patent anchors, steel spars, and, in fact, everything that would add to stability and durability. During construction all her timbers were treated with creosote. The vessel is equipped with a powerful driving outfit, having a compound engine with cylinders 24 and 48 inches in diameter and 36-inch stroke. She carries a Scotch boiler, 12 feet in diameter and 14 feet long, allowing 125 pounds of working steam pressure."

CLASSIFIED ADVERTISING

INCORPORATIONS

Ottawa.—Halifax Shipyards, Limited, the Montreal company which, with a capital of $6,000,000 was granted incorporation, is authorized by the Government to engage in a variety of pursuits. The notice of incorporation appearing in the Canada Gazette states that the company may design, construct, purchase, lease or charter steamships, ships, dredges, tugs, scows, steamship lines, transportation lines, wharves, dockyards, shipbuilding yards, marine railways, telegraph and telegraph lines, etc., on lands owned or controlled by it. The company may also build and operate steamboat and railway terminals, transportation warehouses. storage and cold storage facilities, yards and stockyards; etc., on its own land. The company may construct steamship works for the manufacture of machines, railway or railway equipment and all supplies for steam boats and vessels generally. Power houses and structures for the development and utilization of water, steam, electric and other power may be built and operated. It can also carry on the business of transporting mail and freight upon land and water of the dominion and engage in all branches of towing, wrecking and salvage in Canadian waters.

MARINE

Washington.—Deliveries of steel ships to the Shipping Board in the first two weeks of June numbered 16, with a total weight tonnage of 89,162. Because of German submarine activity off the American coast, the Shipping Board ordered bonus of 25 per cent. of their monthly wages paid to all seamen employed on American merchant vessels in the coastwise, West Indian and South and Central American trade.

A Pacific Port.—Messages received here said the Pacific Steamship Company's freight and passenger steamer Ravalli was destroyed by fire en route to Alaskan points recently. When the fire was discovered the boat was beached and the passengers and crew put ashore. The fire did not stop burning until it reached the water's edge. All the baggage was saved. Most of the cargo was destroyed. The Ravalli was a vessel of 777 tons.

London.—Twelve new British companies have been registered with the object of constructing ferro-concrete vessels. They are as follows: Marine and General Concrete Construction Syndicate, Ltd., capital £600,000; Scottish Concrete Ship Co., £10,000; Blacketts Concrete Ships, Ltd., £5,000; Gloucester Ferro Concrete Shipbuilding Co., Ltd., £20,000; Stuarts Concrete Ship Co., Ltd., £15,000; Amble Ferro Concrete Co., Ltd., £25,000; Wear Concrete Building Co., Ltd., £70,000; Concrete Seacraft, Ltd., £12,500; Aberdeen Concrete Shipbuilding Co., Ltd., £40,000; Blair-Frost Ferro Concrete Co., Ltd., £60,000; Cambelltown Ferro Concrete Shipbuilding Co., capitalization not given.

10" Special Double-Suction Circulating Pump with 8 x 8 Vertical Engine.

MORRIS

is specializing on

Circulating Pumps FOR Surface Condensers

Have you our Catalog?

MORRIS MACHINE WORKS
BALDWINSVILLE, N.Y., U.S.A.

Canadian Sales Agents:
STOREY PUMP & EQUIPMENT COMPANY
TORONTO

Somewhere in Canada where Shop Foremen see this list of High Grade Machinery, these will be selected quickly as most of this Machinery cannot be duplicated anywhere else for immediate shipment.

Largest Stock of Mining Machinery in Canada.

Coal, Coke, High Speed Steel, High Speed Drills.

Send Us Your Inquiries.

Air Compressors, Alley & McLellan 600 ft. at 100 lb. pressure.
Air Compressor, Rand 800 ft..
Motors Heavy Duty 10-25-40-75-125.
Tube Mills, Fraser Chalmers, Power & Mining 22 ft.
Ball Mill Hardinge.
Tanks, Rails, Pumps, Aiken Classifiers, Crushers, Holman Drills, Oliver Filterers, Hoists, Boilers, Pumps, Transformers, Lighting Generators, Motors. 2 Ton Alco Truck, Wire Rope, 5 Ton Travelling Crane, 5 Ton Crane, Hoist Engine, Concrete Mixer.
Complete List of Mining Machinery and Lathes.

Send Inquiries for Prices:

Zenith Coal and Steel Products, Limited
1410 Royal Bank Building, Toronto, Ontario
402 McGill Building, Montreal, Quebec

IRON AND STEEL
FERRO-MANGANESE
FLUORSPAR
FIREBRICKS

MITCHELLS LIMITED
142 QUEEN STREET GLASGOW (SCOTLAND)
Cable Address:—"IRONCROWN GLASGOW"

Over 30 Years' Experience Building

ENGINES AND Propeller Wheels

H. G. TROUT CO.
King Iron Works
226 OHIO ST.
BUFFALO, N. Y.

If what you need is not advertised, consult our Buyers' Directory and write advertisers listed under proper heading.

YARROWS

Limited

Associated with YARROW & CO., GLASGOW

WORKS AT ESQUIMALT, B. C.

Telegrams and Cables "Yarrows, Victoria"

SHIPBUILDERS, ENGINEERS AND SHIPREPAIRERS

IRON AND BRASS FOUNDERS

VESSELS CONVERTED FROM COAL BURNING TO OIL FUEL BURNING SYSTEMS.
MANUFACTURE OF MANGANESE BRONZE PROPELLER BLADES A SPECIALITY.

MARINE RAILWAY, LENGTH 300 ft., CAPACITY 2500 TONS DEAD WEIGHT

Address: P.O. Box 1595 **Victoria B. C.**

Larger Vessels Docked in Government Graving Dock, Esquimalt—Lowest rates on the Pacific Coast.

LONDON PARIS CHRISTIANIA MILAN

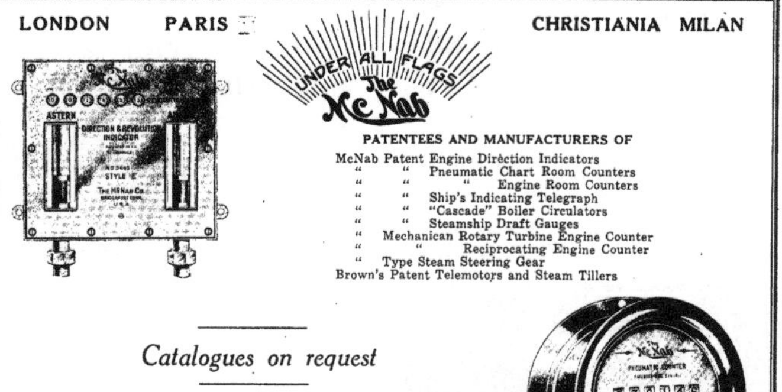

PATENTEES AND MANUFACTURERS OF

McNab Patent Engine Direction Indicators
" " Pneumatic Chart Room Counters
" " " Engine Room Counters
" " Ship's Indicating Telegraph
" " "Cascade" Boiler Circulators
" " Steamship Draft Gauges
" Mechanican Rotary Turbine Engine Counter
" " Reciprocating Engine Counter
" Type Steam Steering Gear
Brown's Patent Telemotors and Steam Tillers

Catalogues on request

THE McNAB COMPANY

Bridgeport, Conn., U.S.A.

If what you need is not advertised, consult our Buyers' Directory and write advertisers listed under proper heading.

You Are Looking

at an illustration we would like to make twice as big, because it illustrates the

PRESTON

Adjustable Bevel Ship Band Saw

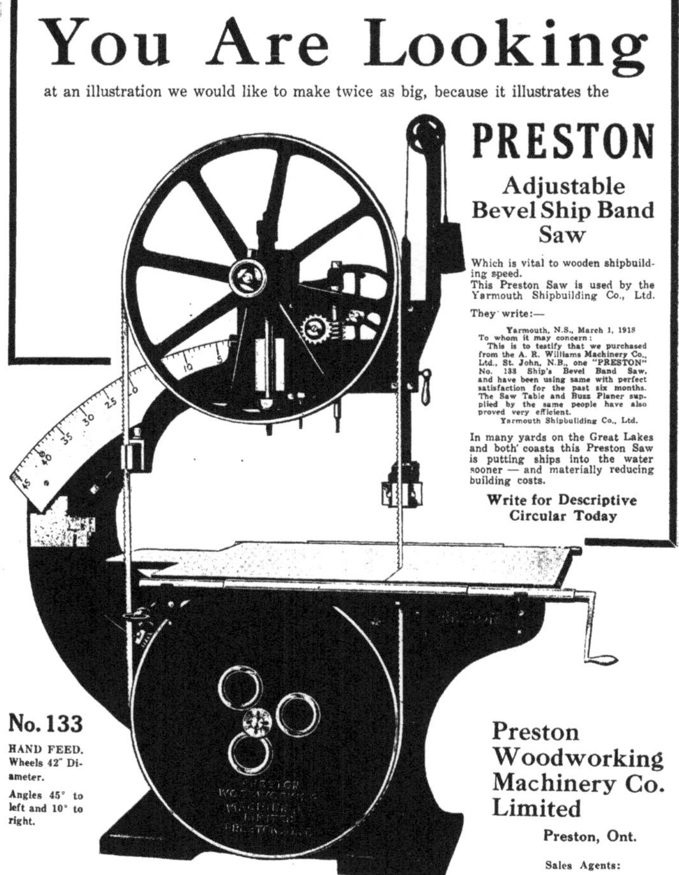

Which is vital to wooden shipbuilding speed.
This Preston Saw is used by the Yarmouth Shipbuilding Co., Ltd.

They write:—

Yarmouth, N.S., March 1, 1918
To whom it may concern:
This is to testify that we purchased from the A. R. Williams Machinery Co., Ltd., St. John, N.B., one "PRESTON" No. 133 Ship's Bevel Band Saw, and have been using same with perfect satisfaction for the past six months. The Saw Table and Buzz Planer supplied by the same people have also proved very efficient.
Yarmouth Shipbuilding Co., Ltd.

In many yards on the Great Lakes and both coasts this Preston Saw is putting ships into the water sooner — and materially reducing building costs.

Write for Descriptive Circular Today

No. 133

HAND FEED.
Wheels 42" Diameter.

Angles 45° to left and 10° to right.

Preston Woodworking Machinery Co. Limited

Preston, Ont.

Sales Agents:

A. R. Williams Machinery Co.
St. John, Vancouver and Toronto

BUYERS' DIRECTORY

ACCUMULATORS, HYDRAULIC
Smart-Turner Mach. Co., Hamilton, Ont.

AERATING RESERVOIRS
Spray Engineering Co., Boston, Mass.

AIR PORTS
Mitchell Co., The Robert, Montreal, Que.
Turnbull Elevator Mfg. Co., Toronto, Ont.

ALLOYS, BRASS AND COPPER
Dom. Copper Products Co., Ltd., Montreal, Que.
Mitchell Co., The Robert, Montreal, Que.
Mueller Mfg. Co., H., Sarnia, Ont.
United Brass & Lead, Ltd., Toronto, Ont.

ANCHORS
Beauchemin & Fils, Sorel, P.Q.
Hopkins & Co., F. H., Montreal, Que.
Kearfott Engineering Co., New York, N.Y.
McNab Co., Bridgeport, Conn.
Henry Rogers Sons & Co., Wolverhampton, Eng.
Wm. C. Wilson & Co., Toronto, Ont.

ARCHITECT, NAVAL
Watts, J. Murray, Philadelphia, Pa.

ASBESTOS GOODS
Wm. C. Wilson & Co., Toronto, Ont.

BABBITT METAL
Aikenhead Hardware, Ltd., Toronto, Ont.
Hoyt Metal Company, Toronto, Ontario.
Tallman Brass & Metal Co., Hamilton, Ont.
Wilkinson & Kompass, Hamilton, Ont.
Wm. C. Wilson & Co., Toronto, Ont.

BAROMETERS
Wilson & Co., Wm. C., Toronto, Canada.

BAR IRON
Mitchells Ltd., Glasgow, Scotland.

BARS, GRATE
Babcock & Wilcox, Ltd., Montreal, Que.

BEARINGS, BRASS
Empire Mfg. Co., London, Ont.
Mueller Mfg. Co., H., Sarnia, Ont.
Mitchell Co., The Robert, Montreal, Que.
Tallman Brass & Metal Co., Hamilton, Ont.
United Brass & Lead Co., Toronto, Ont.

BELLS, SHIPS, ENGINE ROOM, ETC.
Aikenhead Hardware, Ltd., Toronto, Ont.
Clarke & Bros. Co., Montreal, Que.
Cory & Son, Inc., Chas., New York, N.Y.
Empire Mfg. Co., London, Ont.
Mitchell Co., The Robert, Montreal, Que.
Morrison Brass Mfg. Co., James, Toronto, Ont.
Mueller Mfg. Co., H., Sarnia, Ont.
United Brass & Lead, Ltd., Toronto, Ont.

BELTING, LEATHER
Aikenhead Hardware, Ltd., Toronto, Ont.
Wm. C. Wilson & Co., Toronto, Ont.

BIBBS, COMPRESSION
Empire Mfg. Co., London, Ont.
Mitchell Co., The Robert, Montreal, Que.
Mueller Mfg. Co., H., Sarnia, Ont.
United Brass & Lead, Ltd., Toronto, Ont.

BINNACLES
Hopkins & Co., F. H., Montreal, Que.
Morrison Brass Mfg. Co., James, Toronto, Ont.

BLOCKS, CARGO, HEEL AND TACKLE
Aikenhead Hardware, Ltd., Toronto, Ont.
Hopkins & Co., F. H., Montreal, Que.
Loveridge, Ltd., Docks, Cardiff, Wales.

BLOWERS, TURBO-
Mason Regulator & Engin. Co., Montreal, Que.

BOAT CHOCKS
Corbet Fdry. & Machine Co., Owen Sound, Ont.
Marten-Freeman Co., Toronto, Ont.

BOILER COMPOUND
Wm. C. Wilson & Co., Toronto, Ont.

BOILER CIRCULATORS
Eckliff Circulator Co., Detroit, Mich.

BOILER FEED PUMPS
Can. Ingersoll-Rand Co., Ltd., Sherbrooke, Que.
Goldie & McCulloch Co., Galt, Ont.
Smart-Turner Mach. Co., Hamilton, Ont.
Williams Machinery Co., A. R., Toronto, Ont.

BOILER FITTINGS
Empire Mfg. Co., London, Ont.
Goldie & McCulloch Co., Galt, Ont.
McAvity & Sons Ltd., T., St. John, N.B.
Wager Furnace Bridge Wall Co., Inc., New York, N.Y.

BOILERS, MARINE
Babcock & Wilcox, Ltd., Montreal, Que.
Can. Fairbanks-Morse Co., Montreal, Que.
Can. Vickers, Ltd., Montreal, Que.
Collingwood Shipbuilding Co., Collingwood, Ont.
Doxford & Sons, William, Sunderland, England.
Engr. & Mach. Wks. of Can., St. Catharines, Ont.
Goldie & McCulloch, Ltd., Galt, Ont.
Hall Engineering Works, Montreal, Que.
Mason Regulator & Engin. Co., Montreal, Que.
Montreal Dry Docks & Shipbuilding Co., Montreal, Que.
National Shipbuilding Co., Goderich, Ont.
Marsh Engineering Works, Belleville, Ont.
Polson Iron Works, Toronto, Ontario.
Port Arthur Shipbuilding Co., Port Arthur, Ont.
Waterous Engine Works Co., Brantford, Ont.
Williams Machinery Co., A. R., Toronto, Ont.

BOOKS, TECHNICAL, MARINE
MacLean Publishing Co., Toronto, Ont.

BOLTS
London Bolt & Hinge Works, London, Ont.
Mitchell Co., The Robert, Montreal, Que.
United Brass & Lead, Ltd., Toronto, Ont.
Wilkinson & Kompass, Hamilton, Ont.

BROKER
Page & Jones, Mobile, Ala., U.S.A.

BRASS GOODS
Corbet Fdry. & Machine Co., Owen Sound, Ont.
Goldie & McCulloch Co., Galt, Ont.
McAvity & Sons, T., St. John, N.B.
Mitchell Co., The Robert, Montreal, Que.
Mueller Mfg. Co., H., Sarnia, Ont.
Tallman Brass & Metal Co., Hamilton, Ont.

BUCKETS, CLAMSHELL
Beatty & Sons, Welland, Ont.
Morris Crane & Hoist Co., Herbert, Niagara Falls, Ont.

BUCKETS, DUMP
Morris Crane & Hoist Co., Herbert, Niagara Falls, Ont.

BUCKETS, COALING
Beatty & Sons, Welland, Ont.
Engr. & Mach. Wks. of Can., St. Catharines, Ont.
Hopkins & Co., F. H., Montreal, Que.
Marsh Engineering Works, Belleville, Ont.
Reed & Co., Geo., Montreal, Que.

CABLE
Hopkins & Co., F. H., Montreal, Que.
McNab Co., Bridgeport, Conn.
Wm. C. Wilson & Co., Toronto, Ont.

CABLE, LEAD COVERED AND ARMORED
Standard Underground Cable Co., Hamilton, Ont.

CABLE, ACCESSORIES
Standard Underground Cable Co., Hamilton, Ont.

CAPSTANS
Advance Mach. & Welding Co., Montreal, Que.
Dake Engine Co., Grand Haven, Mich.
Hopkins & Co., F. H., Montreal, Que.
Kennedy & Sons, Wm., Owen Sound, Ont.
Wm. C. Wilson & Co., Toronto, Ont.

CALKING TOOLS, ELECTRIC
Aikenhead Hardware, Ltd., Toronto, Ont.

CALKING TOOLS, PNEUMATIC
Aikenhead Hardware, Ltd., Toronto, Ont.
Can. Ingersoll-Rand Co. Sherbrooke, Que.

CALORIFIERS
Low & Sons, Ltd., Archibald, Glasgow, Scotland

CASTINGS
Beauchemin & Fils, Sorel, P.Q.
Can. Steel Foundries, Ltd., Montreal, Que.
Collingwood Shipbuilding Co., Collingwood, Ont.
Wm. Hamilton Co., Peterboro, Ont.
Kennedy & Sons, Wm., Owen Sound, Ont.
Goldie & McCulloch Co., Galt, Ont.
Marsh Engineering Works, Belleville, Ont.
Mitchell Co., Ltd., Robt., Montreal, Que.
Mueller Mfg. Co., H., Sarnia, Ont.
Tallman Brass & Metal Co., Hamilton, Ont.
United Brass & Lead, Ltd., Toronto, Ont.
Waterous Engine Works Co., Brantford, Ont.

CASTINGS, ALLOY
Mitchell Co., The Robert, Montreal, Que.
Mueller Mfg. Co., H., Sarnia, Ont.
Tolland Mfg. Co., Montreal, Que.
Tallman Brass & Metal Co., Hamilton, Ont.
United Brass & Lead, Ltd., Toronto, Ont.

CASTINGS, ALUMINUM
Empire Mfg. Co., London, Ont.
Mitchell Co., The Robert, Montreal, Que.
Tallman Brass & Metal Co., Hamilton, Ont.
United Brass & Lead Co., Toronto, Ont.

CASTINGS, BRASS
Crouse-Hinds Co. of Canada, Ltd., Toronto, Ont.
Empire Mfg. Co., London, Ont.
Goldie & McCulloch Co., Galt, Ont.
McAvity & Sons Ltd., T., St. John, N.B.
McNab Co., Bridgeport, Conn.
Mitchell Co., Ltd., Robt., Montreal, Que.
Mueller Mfg. Co., H., Sarnia, Ont.
Tolland Mfg. Co., Montreal, Que.
Tallman Brass & Metal Co., Hamilton, Ont.
United Brass & Lead Co., Toronto, Ont.
Waterous Engine Works Co., Brantford, Ont.

CASTINGS, GREY IRON, MALLEABLE, ALUMINUM
Crouse-Hinds Co. of Canada, Ltd., Toronto, Ont.
Engr. & Mach. Wks. of Can., St. Catharines, Ont.
McAvity & Sons, T., St. John, N.B.
McNab Co., Bridgeport, Conn.
Mitchell Co., The Robert, Montreal, Que.
Mueller Mfg. Co., H., Sarnia, Ont.
Waterous Engine Works Co., Brantford, Ont.

CASTINGS, MANGANESE STEEL

CASTINGS, MANGANESE BRONZE
Tallman Brass & Metal Co., Hamilton, Ont.
United Brass & Lead Co., Toronto, Ont.

CELLAR DRAINERS
Mueller Mfg. Co., H., Sarnia, Ont.

CHAINS
Aikenhead Hardware, Ltd., Toronto, Ont.
Kearfott Engineering Co., New York, N.Y.
Hopkins & Co., F. H., Montreal, Que.
Morris Crane & Hoist Co., Herbert, Niagara Falls, Ont.
Henry Rogers, Sons & Co., Wolverhampton, Eng.
Wm. C. Wilson & Co., Toronto, Ont.

CHAIN BLOCKS AND SLINGS
Morris Crane & Hoist Co., Herbert, Niagara Falls, Ont.

CHANDLERY, SHIP
Beauchemin & Fils, Sorel, P.Q.
Hopkins & Co., F. H., Montreal, Que.
Leckie, Ltd., John, Toronto, Ont.

CLAMPS, STEAM AND WATER
Mueller Mfg. Co., H., Sarnia, Ont.

CLOCKS
American Steam Gauge & Valve Mfg. Co., Boston, Mass.
Morrison Brass Mfg. Co., James, Toronto, Ont.
Williams Machinery Co., A. R., Toronto, Ont.

CLOSETS
Mueller Mfg. Co., H., Sarnia, Ont.
United Brass & Lead, Ltd., Toronto, Ont.

COAL
Nova Scotia Steel & Coal Co., New Glasgow, N.S.

COAL HANDLING MACHINERY
Morris Crane & Hoist Co., Herbert, Niagara Falls, Ont.
Waterous Engine Works Co., Brantford, Ont.

COCKS, BILGE, DISCHARGE, INDICATOR,
McAvity & Sons Ltd., T., St. John, N.B.
Mitchell Co., The Robert, Montreal, Que.
Morrison Brass Mfg. Co., James, Toronto, Ont.

COCKS, BASIN
Empire Mfg. Co., London, Ont.
Mitchell Co., The Robert, Montreal, Que.
United Brass & Lead, Ltd., Toronto, Ont.

COMPASSES
Wm. C. Wilson & Co., Toronto, Ont.

COMPRESSORS, AIR
Can. Fairbanks-Morse Co., Montreal, Que.
Canadian Ingersoll-Rand Co., Sherbrooke, Que.
Hopkins & Co., F. H., Montreal, Que.
Smart-Turner Mach. Co., Hamilton, Ont.
Williams Machinery Co., A. R., Toronto, Ont.

COMPRESSED AIR TANKS
Scaife & Sons Co., Wm. B., Oakmont, Pa.

CONDENSERS
Goldie & McCulloch Co., Galt, Ont.
Kearfott Engineering Co., New York, N.Y.
Morris Machine Works, Baldwinsville, N.Y.
Smart-Turner Mach. Co., Hamilton, Ont.
Weir Ltd., G. & J., Cathcart, Glasgow, Scotland.
Williams Machinery Co., A. R., Toronto, Ont.

CONDULETS, MARINE
Crouse-Hinds Co. of Canada, Ltd., Toronto, Ont.

CONTRACTORS' SUPPLIES
McAvity & Sons, T., St. John, N.B.

CONSULTING ENGINEER
Watt, J. Murray, Philadelphia, Pa.

CONVEYORS, ASH, COAL
Babcock & Wilcox, Ltd., Montreal, Que.
Hopkins & Co., F. H., Montreal, Que.

COPING MACHINES
Bertram & Sons, Ltd., John, Dundas, Ont.

COPPER TUBES, SHEETS AND RODS
Dom. Copper Products Co., Montreal, Que.
Tallman Brass & Metal Co., Hamilton, Ont.

COTTON, CALKING
Wilson & Co., Wm. C., Toronto, Canada.

COVERS, CANVAS, FOR HATCHES, LIFE-BOATS, ETC.
Leckie, Ltd., John, Toronto, Ont.
Waterous Engine Works Co., Brantford, Ont.

COUNTERS, REVOLUTION
American Steam Gauge & Valve Mfg. Co., Boston, Mass.

CRANES
Aikenhead Hardware, Ltd., Toronto, Ont.
Can. Fairbanks-Morse Co., Montreal, Que.
Hopkins & Co., F. H., Montreal, Que.
Williams Machinery Co., A. R., Toronto, Ont.

CRANES, ELECTRIC
Babcock & Wilcox, Ltd., Montreal, Que.
Morris Crane & Hoist Co., Herbert, Niagara Falls, Ont.
Smart-Turner Mach. Co., Hamilton, Ont.

CRANES, GOLIATH AND PNEUMATIC
Morris Crane & Hoist Co., Herbert, Niagara Falls, Ont.

CRANES, GRANTRY, PORTABLE, JIB
Morris Crane & Hoist Co., Herbert, Niagara Falls, Ont.
Smart-Turner Mach. Co., Hamilton, Ont.

CRANES, OVERHEAD TRAVELLING
Morris Crane & Hoist Co., Herbert, Niagara Falls, Ont.

CRANK SHAFTS
Canada Foundries and Forgings, Welland, Ont.

DAVITS, BOAT
Corbet Fdry. & Machine Co., Owen Sound, Ont.
Hopkins & Co., F. H., Montreal, Que.
Marten, Freeman Co., Toronto, Ont.
Waterous Engine Works Co., Brantford, Ont.

DEAD LIGHTS, BRASS
Goldie & McCulloch Co., Galt, Ont.
Morris Crane & Hoist Co., Herbert, Niagara Falls, Ont.
Mitchell Co., Ltd., Robt., Montreal, Que.

DECK LIGHTS
Mitchell Co., The Robert, Montreal, Que.
Turnbull Elevator Mfg. Co., Toronto, Ont.

DECK PLUGS, ELECTRIC
Crouse-Hinds Co. of Canada, Ltd., Toronto, Ont.
Mitchell Co., The Robert, Montreal, Que.

DECKING FOR SHIPS
Marine Decking & Supply Co., Philadelphia, Pa.

DERRICKS
Aikenhead Hardware, Ltd., Toronto, Ont.
Dake Engine Co., Grand Haven, Mich.
Hopkins & Co., F. H., Montreal, Que.
Marsh Engineering Works, Belleville, Ont.
Morris Crane & Hoist Co., Herbert, Niagara Falls, Ont.

DISTILLERS
Kearfott Engineering Co., New York, N.Y.

DREDGES
Collingwood Shipbuilding Co., Collingwood, Ont.
Norbom Engineering Co., Philadelphia, Pa.
Polson Iron Works, Toronto.

DRILLS, AIR
Aikenhead Hardware, Ltd., Toronto, Ont.
Can. Ingersoll-Rand Co. Sherbrooke, Que.
Hopkins & Co., F. H., Montreal, Que.

DRILLS, CENTRE
Wilt Twist Drill Co., Walkerville, Ont.

DRILLS, BLACKSMITH AND BIT STOCK
Wilt Twist Drill Co., Walkerville, Ont.

DRILLS, ELECTRIC
Aikenhead Hardware, Ltd., Toronto, Ont.
Wilkinson & Kompass, Hamilton, Ont.

DRILLS, HIGH SPEED
Wilt Twist Drill Co., Walkerville, Ont.
Zenith Coal & Steel Products, Ltd., Montreal,

DRILLS, TRACK
Wilt Twist Drill Co., Walkerville, Ont.

DRILLS, RACHET AND HAND
Wilt Twist Drill Co., Walkerville, Ont.

DRILLS, TWIST
Aikenhead Hardware, Ltd., Toronto, Ont.
Williams Machinery Co., A. R., Toronto, Ont.
Wilt Twist Drill Co., Walkerville, Ont.

DRY DOCKS
Can. Vickers, Ltd., Montreal, Que.
Collingwood Shipbuilding Co., Collingwood, Ont.
Doxford & Sons, William, Sunderland, England.
Georgian Bay Shipbuilding & Wrecking Co., Midland, Ont.
National Shipbuilding Co., Goderich, Ont.
Polson Iron Works, Toronto, Ont.
Port Arthur Shipbuilding Co., Port Arthur, Ont.
Yarrows, Limited, Victoria, B.C.

ECONOMIZERS, FUEL
Babcock & Wilcox, Ltd., Montreal, Que.

EJECTORS
Empire Mfg. Co., London, Ont.
Mitchell Co., The Robert, Montreal, Que.
Morrison Brass Mfg. Co., James, Toronto, Ont.
Smart-Turner Mach. Co., Hamilton, Ont.

ELECTRIC LAMPS
Mitchell Co., The Robert, Montreal, Que.
Wm. C. Wilson & Co., Toronto, Ont.

ELECTRO-PLATING
Mitchell Co., The Robert, Montreal, Que.
Tallman Brass & Metal Co., Hamilton, Ont.
United Brass & Lead Co., Toronto, Ont.

ELECTRIC WELDING
Beauchemin & Fils, Sorel, P.Q.

ELEVATING MACHINERY
Goldie & McCulloch, Ltd., Galt, Ont.
Wm. Hamilton Co., Peterboro, Ont.
Morris Crane & Hoist Co., Herbert, Niagara Falls, Ont.
Waterous Engine Works Co., Brantford, Ont.

ELEVATORS
Turnbull Elevator Mfg. Co., Toronto, Ont.

ENGINES, HOISTING
Corbet Fdry. & Machine Co., Owen Sound, Ont.
Hopkins & Co., F. H., Montreal, Que.
Kennedy & Sons, Wm., Owen Sound, Ont.
Marsh Engineering Works, Belleville, Ont.
Port Arthur Shipbuilding Co., Port Arthur, Ont.
Williams Machinery Co., A. R., Toronto, Ont.

ENGINE, INTERNAL COMBUSTION
Doxford & Sons, William, Sunderland, England.

ENGINES, MARINE
Bolinders Co., New York, N.Y.
Can. Fairbanks-Morse Co., Montreal, Que.
Can. Vickers, Ltd., Montreal, Que.
Doxford & Sons, William, Sunderland, England.
Goldie & McCulloch, Ltd., Galt, Ont.
Hopkins & Co., F. H., Montreal, Que.
Iron Works, Ltd., Owen Sound, Ont.
Mason Regulator & Engin. Co., Montreal, Que.
National Shipbuilding Co., Goderich, Ont.
Norbom Engineering Co., Philadelphia, Pa.
Polson Iron Works, Toronto, Ont.
Port Arthur Shipbuilding Co., Port Arthur, Ont.
Trout Co., H. G., Buffalo, N.Y.
Waterous Engine Works Co., Brantford, Ont.
Williams Machinery Co., A. R., Toronto, Ont.

ENGINE STARTERS (AIR)
Scaife & Sons Co., Wm. B., Oakmont, Pa.

ENGINES, STEERING
Corbet Fdry. & Machine Co., Owen Sound, Ont.
Dake Engine Co., Grand Haven, Mich.
Kennedy & Sons, Wm., Owen Sound, Ont.
Wm. C. Wilson & Co., Toronto, Ont.

ENAMELWARE
Mueller Mfg. Co., H., Sarnia, Ont.
United Brass & Lead Co., Toronto, Ont

EVAPORATORS
Kearfott Engineering Co., New York, N.Y.
Mason Regulator & Engin. Co., Montreal, Que.
McNab Co., Bridgeport, Conn.
Weir Ltd., G. & J., Cathcart, Glasgow, Scotland.

EXTRACTORS, GREASE
American Steam Gauge & Valve Mfg. Co., Boston, Mass.

EYE BOLTS AND NUTS
Canada Foundries and Forgings, Welland, Ont.
Mitchell Co., The Robert, Montreal, Que.
United Brass & Lead, Ltd., Toronto, Ont.

FANS
Aikenhead Hardware, Ltd., Toronto, Ont.
Empire Mfg. Co., London, Ont.
Reed & Co., Geo. W., Montreal, Que.
Smart-Turner Mach. Co., Hamilton, Ont.
Williams Machinery Co., A. R., Toronto, Ont.

FENDERS, ROPE
Hopkins & Co., F. H., Montreal, Que.
Leckie, Ltd., John, Toronto, Ont.
Wilson & Co., Wm. C., Toronto, Canada.

FERRO-MANGANESE
Mitchells, Ltd., Glasgow, Scotland.

FILES
Aikenhead Hardware, Ltd., Toronto, Ont.
Williams Machinery Co., A. R., Toronto, Ont.

FIRE BRICKS
Beveridge Paper Co., Montreal, Que.
Mitchells, Ltd., Glasgow, Scotland.
Williams Machinery Co., A. R., Toronto, Ont.

FILTERS, FEED WATER
Mason Regulator & Engin. Co., Montreal, Que.

FITTINGS, MARINE
Hopkins & Co., F. H., Montreal, Que.
McAvity & Sons, T., St. John, N.B.
Mitchell Co., The Robert, Montreal, Que.
United Brass & Lead, Ltd., Toronto, Ont.

FITTINGS, MOTOR BOAT
Empire Mfg. Co., London, Ont.
Mitchell Co., The Robert, Montreal, Que.
Mueller Mfg. Co., H., Sarnia, Ont.
United Brass & Lead, Ltd., Toronto, Ont.

FIXTURES, ELECTRIC
Cory & Son, Inc., Chas., New York, N.Y.
Crouse-Hinds Co. of Canada, Ltd., Toronto, Ont.
Mitchell Co., The Robert, Montreal, Que.
Tallman Brass & Metal Co., Hamilton, Ont.

FLAG POLES, STEEL
Dennis Wire & Iron Works Co., London, Ont.

FLOW METERS
Spray Engineering Co., Boston, Mass.

FLUE CLEANERS
Wm. C. Wilson & Co., Toronto, Ont.

FORGES
Aikenhead Hardware, Ltd., Toronto, Ont.
Hopkins & Co., F. H., Montreal, Que.

FLANGING AND EXPANDING MACHINES, PIPE
Lovekin Pipe Expanding & Flanging Mach. Co., Philadelphia, Pa.

FLOODLIGHTS, ELECTRIC
Crouse-Hinds Co. of Canada, Ltd., Toronto, Ont.

FLUORSPAR
Mitchells, Ltd., Glasgow, Scotland.
Scythes & Co., Toronto, Ont.

FORGINGS, ALL KINDS
Aikenhead Hardware, Ltd., Toronto, Ont.
Collingwood Shipbuilding Co., Collingwood, Ont.
Nova Scotia Steel & Coal Co., New Glasgow, N.S.

FORGINGS, STEEL AND IRON
Canada Foundries and Forgings, Welland, Ont.

FURNACE BRIDGE WALLS
Wager Furnace Bridge Wall Co., Inc., 149 Broadway, New York, N.Y.

GAUGES, RECORDING
American Steam Gauge & Valve Mfg. Co., Boston, Mass.
Empire Mfg. Co., London, Ont.

GASKETS
Wm. C. Wilson & Co., Toronto, Ont.

GAUGES COCKS
McAvity & Sons, T., St. John, N.B.
Mitchell Co., The Robert, Montreal, Que.

GAUGE GLASSES
Wm. C. Wilson & Co., Toronto, Ont.

GAUGES, WATER, PRESSURE, COMPOUND AND VACUUM
Aikenhead Hardware, Ltd., Toronto, Ont.
Babcock & Wilcox, Ltd., Montreal, Que.
Empire Mfg. Co., London, Ont.
McAvity & Sons, T., St. John, N.B.
McNab Co., Bridgeport, Conn.
Morrison Brass Mfg. Co., James, Toronto, Ont.

GENERATORS AND CONVERTORS
Can. Fairbanks-Morse Co., Montreal, Que.

GLAZING COMPOSITION
Kuhls, H. B. Fred., 6411 3rd Ave., Brooklyn, N.Y.

GONGS
Clark & Bro., C. O., Montreal, Que.
Mitchell Co., Ltd., Robt., Montreal, Que.

GRAPHITE
Wm. C. Wilson & Co., Toronto, Ont.

GRATINGS
Corbet Fdry. & Machine Co., Owen Sound, Ont.
Canada Wire & Iron Goods Co., Hamilton, Ont.

GRINDERS, ELECTRIC
Wilkinson & Kompass, Hamilton, Ont.

GRINDERS, PNEUMATIC
Can. Ingersoll-Rand Co. Sherbrooke, Que.

GUARDS, MACHINERY
Dennis Wire & Iron Works Co., London, Ont.

GUY RODS AND ANCHORS, ELECTRIC
Crouse-Hinds Co. of Canada, Ltd., Toronto, Ont.

HAMMERS
Canada Foundries and Forgings, Ltd., Welland, Ont.

HARDWARE, MARINE
Hopkins & Co., F. H., Montreal, Que.
Mitchell Co., The Robert, Montreal, Que.

HEADLIGHTS, ELECTRIC
Crouse-Hinds Co. of Canada, Ltd., Toronto, Ont.
Hopkins & Co., F. H., Montreal, Que.
Tallman Brass & Metal Co., Hamilton, Ont.

HEATING EQUIPMENT
Empire Mfg. Co., London, Ont.
Low & Sons, Ltd., Archibald, Glasgow, Scotland

HEATERS, FEED, WATER
Babcock & Wilcox, Ltd., Montreal, Que.
Goldie & McCulloch Co., Galt, Ont.
Kearfott Engineering Co., New York, N.Y.
Mason Regulator & Engin. Co., Montreal, Que.
McNab Co., Bridgeport, Conn.
Weir Ltd., G. & J., Cathcart, Glasgow, Scotland.

HINGES
London Bolt & Hinge Works, London, Ont.
Mitchell Co., Ltd., Robt., Montreal, Que.

HOOKS
Hopkins & Co., F. H., Montreal, Que.
Morris Crane & Hoist Co., Herbert, Niagara Falls, Ont.

HOISTS, ASH
Beatty & Sons, Welland, Ont.
Marsh Engineering Works, Belleville, Ont.
St. Clair Bros., Galt, Ont.
Waterous Engine Works Co., Brantford, Ont.

HOIST BLOCKS
Morris Crane & Hoist Co., Herbert, Niagara Falls, Ont.

HOISTS, CHAIN
Aikenhead Hardware, Ltd., Toronto, Ont.
Can. Fairbanks-Morse Co., Montreal, Que.
Dake Engine Co., Grand Haven, Mich.
Hopkins & Co., F. H., Montreal, Que.
Morris Crane & Hoist Co., Herbert, Niagara Falls, Ont.
Williams Machinery Co., A. R., Toronto, Ont.

HOISTS, CARGO, MOVING, ETC.
Dake Engine Co., Grand Haven, Mich.
Hopkins & Co., F. H., Montreal, Que.
Marsh Engineering Works, Belleville, Ont.
Waterous Engine Works Co., Brantford, Ont.

HOISTING MACHINERY
Beatty & Sons, Welland, Ont.
Can. Ingersoll-Rand Co. Sherbrooke, Que.
Corbet Fdry. & Machine Co., Owen Sound, Ont.
Wm. Hamilton Co., Peterboro, Ont.
Marsh Engineering Works, Belleville, Ont.
Morris Crane & Hoist Co., Herbert, Niagara Falls, Ont.
Waterous Engine Works Co., Brantford, Ont.
Williams Machinery Co., A. R., Toronto, Ont.

HOSE
Wm. C. Wilson & Co., Toronto, Ont.

INDICATORS, ENGINE
American Steam Gauge & Valve Mfg. Co., Boston, Mass.
Cory & Son, Inc., Chas., New York, N.Y.
McNab Co., Bridgeport, Conn.

INDICATORS, SPEED
Aikenhead Hardware, Ltd., Toronto, Ont.
Cory & Son, Inc., Chas., New York, N.Y.

INJECTORS
Aikenhead Hardware, Ltd., Toronto, Ont.
Empire Mfg. Co., London, Ont.
Mitchell Co., The Robert, Montreal, Que.
Morrison Brass Mfg. Co., James, Toronto, Ont.
Williams Machinery Co., A. R., Toronto, Ont.

INGOTS
Broughton Copper Co., Ltd., Manchester, Eng.

INSULATORS, ELECTRIC
Crouse-Hinds Co. of Canada, Ltd., Toronto, Ont.

INSTRUMENTS, NAUTICAL
Leckie, Ltd., John, Toronto, Ont.

IRON AND STEEL
Mitchells Ltd., Glasgow, Scotland.

JACKS
Hopkins & Co., F. H., Montreal, Que.
Morris Crane & Hoist Co., Herbert, Niagara Falls, Ont.

JOINTS, BALL SLIP
Norbom Engineering Co., Philadelphia, Pa.

INSURANCE, MARINE
Toronto Insurance & Vessel Agency, Toronto, Ont.

JOURNAL WEDGES
Canada Foundries and Forgings, Ltd., Welland, Ont.

LATHES
Bertram & Sons, Ltd., John, Dundas, Ont.

LEATHER, LACE
Wm. C. Wilson & Co., Toronto, Ont.

LIFEBUOYS, BELTS AND PRESERVERS
Fosbery Co., Barking, Essex, Eng.
Hopkins & Co., F. H., Montreal, Que.
Wm. C. Wilson & Co., Toronto, Ont.

LIFEBOATS
Hopkins & Co., F. H., Montreal, Que.
Wm. C. Wilson & Co., Toronto, Ont.

LIFE BOAT EQUIPMENT
Eckliff Circulator Co., Detroit, Mich.
Hopkins & Co., F. H., Montreal, Que.
Leckie, Ltd., John, Toronto, Ont.
Wm. C. Wilson & Co., Toronto, Ont.

LIFE JACKETS
Fosbery Co., Barking, Essex, Eng.
Hopkins & Co., F. H., Montreal, Que.
Leckie, Ltd., John, Toronto, Ont.

LIGHTS, ALL KINDS
Can. Fairbanks-Morse Co., Montreal, Que.
Crouse-Hinds Co. of Canada, Ltd., Toronto, Ont.
Hopkins & Co., F. H., Montreal, Que.
Mitchell Co., The Robert, Montreal, Que.
Morrison Brass Mfg. Co., James, Toronto, Ont.
Tallman Brass & Metal Co., Hamilton, Ont.

LIGHTS, SIDE, PORT
Goldie & McCulloch Co., Galt, Ont.
Crouse-Hinds Co. of Canada, Ltd., Toronto, Ont.
Hopkins & Co., F. N., Montreal, Que.
McAvity & Sons Ltd., T., St. John, N.B.
Mitchell Co., The Robert, Montreal, Que.
Tallman Brass & Metal Co., Hamilton, Ont.
Turnbull Elevator Mfg. Co., Toronto, Ont.
Wilson & Co., Wm. C., Toronto, Ont.

LIGHTING SETS
Can. Fairbanks-Morse Co., Montreal.
Cory & Son, Inc., Chas., New York, N.Y.
Crouse-Hinds Co. of Canada, Ltd., Toronto, Ont.
Kearfott Engineering Co., New York, N.Y.
Williams Machinery Co., A. R., Toronto, Ont.
Wilson & Co., Wm. C., Toronto, Canada.

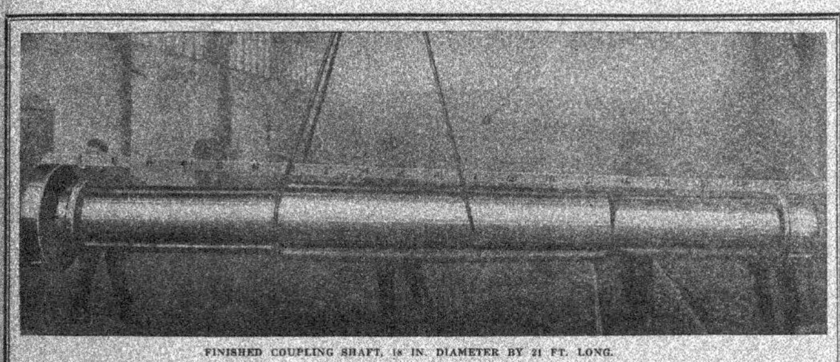

FINISHED COUPLING SHAFT, 18 IN. DIAMETER BY 21 FT. LONG.

LOADERS, WAGON AND TRUCK
Morris Crane & Hoist Co., Herbert, Niagara Falls, Ont.
Waterous Engine Works Co., Brantford, Ont.

LOCKERS, METAL
Dennis Wire & Iron Works Co., London, Ont.
Reed & Co., Geo. W., Montreal, Que.

LUBRICATORS
Empire Mfg. Co., London, Ont.
Mitchell Co., The Robert, Montreal, Que.

MACHINISTS
Corbet Fdry. & Machine Co., Owen Sound, Ont.
Wm. Hamilton Co., Peterboro, Ont.
Hyde Engineering Works, Montreal, Que.
Mitchell Co., The Robert, Montreal, Que.
United Brass & Lead Co., Toronto, Ont.

MANGANESITE, PASTE
Wm. C. Wilson & Co., Toronto, Ont.

MARINE CASTINGS
Wm. Hamilton Co., Peterboro, Ont.
Goldie & McCulloch Co., Galt, Ont.
Mitchell Co., Ltd., Robt., Montreal, Que.
McAvity & Sons, T., St. John, N.B.
Mueller Mfg. Co., H., Sarnia, Ont.
Tallman Brass & Metal Co., Hamilton, Ont.
United Brass & Lead, Ltd., Toronto, Ont.
Waterous Engine Works Co., Brantford, Ont.

MARINE SPECIALTIES
Goldie & McCulloch Co., Galt, Ont.
Hopkins & Co., F. H., Montreal, Que.
McAvity & Sons, T., St. John, N.B.
Mitchell Co., Ltd., Robt., Montreal, Que.
United Brass & Lead, Ltd., Toronto, Ont.

MOTORS
Aikenhead Hardware, Ltd., Toronto, Ont.
Can. Fairbanks-Morse Co., Montreal, Que.
Williams Machinery Co., A. R., Toronto, Ont.
Zenith Coal & Steel Products, Ltd., Montreal.

NAUTICAL INSTRUMENTS
Wm. C. Wilson & Co., Toronto, Ont.

NET LIFTERS
Dake Engine Co., Grand Haven, Mich.
Leckie, Ltd., John, Toronto, Ont.

NOZZLES, ALL KINDS
Empire Mfg. Co., London, Ont.
Mitchell Co., The Robert, Montreal, Que.
Tolland Mfg. Co., Boston, Mass.

NUTS
Mitchell Co., The Robert, Montreal, Que.
Wilkinson & Kompass, Hamilton, Ont.
United Brass & Lead, Ltd., Toronto, Ont.

OAKUM
Wm. C. Wilson & Co., Toronto, Ont.

OIL CUPS
Empire Mfg. Co., London, Ont.
Mitchell Co., Ltd., Robt., Montreal, Que.

OILCLOTHING
Wm. C. Wilson & Co., Toronto, Ont.

OIL, LINSEED
Canada Linseed Oil Mills, Montreal, Que.

OILS AND GREASE
Wm. C. Wilson & Co., Toronto, Ont.

OAKUM
Davey & Sons W O., Jersey City., N.J.
Leckie, Ltd., John, Toronto, Ont.

ORNAMENTAL IRONWORK
Canada Wire & Iron Goods Co., Hamilton, Ont.

OXY-ACETLYENE OUTFITS
Can. Fairbanks-Morse Co., Montreal, Que.
Hopkins & Co., F. H., Montreal, Que.
McAvity & Sons Ltd., T., St. John, N.B.
Williams Machinery Co., A. R., Toronto, Ont.

PAINT
Ault & Wiborg Co. of Can., Ltd., Toronto, Ont.
Wm. C. Wilson & Co., Toronto, Ont.

PAINTS, ANTI-FOULING
Kuhls, H. B. Fred., 6411 3rd Ave., Brooklyn, N.Y.

PAINTS, ANTI-CORROSIVE
Kuhls, H. B. Fred. 6411 3rd Ave., Brooklyn, N.Y.

PAINTING EQUIPMENT, AUTOMATIC
Spray Engineering Co., Boston, Mass.

PACKING, AMMONIA
France Packing Co, Philadelphia, Pa.
Wm. C. Wilson & Co., Toronto, Ont.

PACKING, MARINE
Aikenhead Hardware, Ltd., Toronto, Ont.
France Packing Co., Philadelphia, Pa.
Wm. C. Wilson & Co., Toronto, Ont.

PACKING, METALLIC
Aikenhead Hardware, Ltd., Toronto, Ont.
France Packing Co., Philadelphia, Pa.

PACKING, STEAM
Aikenhead Hardware, Ltd., Toronto, Ont.
France Packing Co., Philadelphia, Pa.
Wm. C. Wilson & Co., Toronto, Ont.

PANELBOARD AND CABINETS, ELECTRIC
Crouse-Hinds Co. of Canada, Ltd., Toronto, Ont
Preston Woodworking Machy. Co., Preston, Ont.

PATTERNS, WOOD AND METAL
Beauchemin & Fils, Sorel, P.Q.

PIPE, LAPWELD, CAST IRON, RIVETED
Empire Mfg. Co., London, Ont.
Norbom Engineering Co., Philadelphia, Pa.
United Brass & Lead, Ltd., Toronto, Ont.

PIPE RAILING, IRON AND BRASS
Dennis Wire & Iron Works Co., London, Ont.
Mitchell Co., The Robert, Montreal, Que.

PITCH
Reed & Co., Geo., Montreal, Que.

PITCH, PINE
Wilson & Co., Wm. C., Toronto, Canada.

PLANERS
Bertram & Sons, Ltd., John, Dundas, Ont.

PLANERS, STANDARD AND ROTARY
Yates Mach. Co., P. B., Hamilton, Ont.

PLATE CLAMPS
Morris Crane & Hoist Co., Herbert, Niagara Falls, Ont.

PLATE PUNCH TABLES
Norbom Engineering Co., Philadelphia, Pa.

PLUMBING EQUIPMENT
Low & Sons, Ltd., Archibald, Glasgow, Scotland.
Empire Mfg. Co., London, Ont.
Mitchell Co., The Robert, Montreal, Que.
Mueller Mfg. Co., H., Sarnia, Ont.
Tallman Brass & Metal Co., Hamilton, Ont.
Wm. C. Wilson & Co., Toronto, Ont.
United Brass & Lead, Ltd., Toronto, Ont.

PROPELLOR BLADES, BRONZE
Corbet Fdry. & Machine Co., Owen Sound, Ont.
Empire Mfg. Co., London, Ont.
Mitchell Co., The Robert, Montreal, Que.
Tallman Brass & Metal Co., Hamilton, Ont.
United Brass & Lead, Ltd., Toronto, Ont.
Yarrows, Limited, Victoria, B.C.

PROPELLER WHEELS
Goldie & McCulloch Co., Galt, Ont.
Kennedy & Sons, Wm., Owen Sound, Ont.
Trout Co., H. G., Buffalo, N.Y.

PIPING, STEAM
Babcock & Wilcox, Ltd., Montreal, Que.

PORT LIGHTS
Goldie & McCulloch Co., Galt, Ont.
Mitchell Co., Ltd., Robt., Montreal, Que.
Mueller Mfg. Co., H., Sarnia, Ont.
Turnbull Elevator Mfg. Co., Toronto, Ont.
Tallman Brass & Metal Co., Hamilton, Ont.
Wm. C. Wilson & Co., Toronto, Ont.
United Brass & Lead, Ltd., Toronto, Ont.

PROPELLER WHEELS
Beauchemin & Fils, Sorel, P.Q.
McNab Co., Bridgeport, Conn.

PULLEYS
Wm. Hamilton Co., Peterboro, Ont.
Smart-Turner Mach. Co., Hamilton, Ont.
Williams Machinery Co., A. R., Toronto, Ont.

PUMP, AIR
Can. Ingersoll-Rand Co. Sherbrooke, Que.
Smart-Turner Mach. Co., Hamilton, Ont.
Weir Ltd., G. & J., Cathcart, Glasgow, Scotland.
Williams Machinery Co., A. R., Toronto, Ont.

PUMPS
Can. Fairbanks-Morse Co., Montreal, Que.
Canada Foundries & Forgings, Welland, Ont.
Goldie & McCulloch, Ltd., Galt, Ont.
Hopkins & Co., F. H., Montreal, Que.
McAvity & Sons, T., St. John, N.B.
Williams Machinery Co., A. R., Toronto Ont.

PUMPS, CENTRIFUGAL
Can. Ingersoll-Rand Co. Sherbrooke, Que.
Wm. Hamilton Co., Peterboro, Ont.
Goldie & McCulloch Co., Galt, Ont.
Kearfott Engineering Co., New York, N.Y.
Norbom Engineering Co., Philadelphia, Pa.
Smart-Turner Mach. Co., Hamilton, Ont.
Waterous Engine Works Co., Brantford, Ont.
Williams Machinery Co., A. R., Toronto, Ont.

PUMPS, BILGE
Goldie & McCulloch Co., Galt, Ont.
Mitchell Co., Ltd., Robt., Montreal, Que.
Mueller Mfg. Co., H., Sarnia, Ont.

PUMPS, CIRCULATING
Goldie & McCulloch Co., Galt, Ont.
McNab Co., Bridgeport, Conn.
Morris Machine Works, Baldwinsville, N.Y.

PUMPS, FEED WATER
Goldie & McCulloch Co., Galt, Ont.
McNab Co., Bridgeport, Conn.
Weir Ltd., G. & J., Cathcart, Glasgow, Scotland.
Williams Machinery Co., A. R., Toronto, Ont.

PUMPS, REES ROTURBO
Goldie & McCulloch Co., Galt, Ont.

PUMPS, HAND AND POWER
Aikenhead Hardware, Ltd., Toronto, Ont.
Mitchell Co., The Robert, Montreal, Que.
Smart-Turner Mach. Co., Hamilton, Ont.
Williams Machinery Co., A. R., Toronto, Ont.

PUMPS, HIGH PRESSURE
Canadian Ingersoll-Rand Co., Sherbrooke, Que.
Smart-Turner Mach. Co., Hamilton, Ont.

PUMPS, STEAM TURBO
Wm. C. Wilson & Co., Toronto, Ont.

PUMPING MACHINES
Norbom Engineering Co., Philadelphia, Pa.

PUNCHES
Bertram & Sons, Ltd., John, Dundas, Ont.

PUNCHES, SINGLE, DOUBLE AND MULTIPLE
Norbom Engineering Co., Philadelphia, Pa.

PURIFIERS, WATER
Babcock & Wilcox, Ltd., Montreal, Que.

RADIATORS, STEAM, ELECTRIC
Empire Mfg. Co., London, Ont.
Low & Sons, Ltd., Archibald, Glasgow, Scotland.
Wm. C. Wilson & Co., Toronto, Ont.

RAILS, OVERHEAD
Morris Crane & Hoist Co., Herbert, Niagara Falls, Ont.

RANGES
Hopkins & Co., F. H., Montreal, Que.
Wm. C. Wilson & Co., Toronto, Ont.

RECEIVERS, AIR
Can. Ingersoll-Rand Co. Sherbrooke, Que.

REGULATORS, FEED WATER
American Steam Gauge & Valve Mfg. Co., Boston, Mass.

REGULATORS, PRESSURE
Mason Regulator & Engin. Co., Montreal, Que.

RELEASING GEARS
Eckliff Circulator Co., Detroit, Mich.

REPAIRS, MARINE
Corbet Foundry & Mach. Co., Owen Sound, Ont.
Can. Vickers, Ltd., Montreal, Que.
Collingwood Shipbuilding Co., Collingwood, Ont.
Engr. & Mech. Wks. of Can., St. Catharines, Ont.
Georgian Bay Shipbuilding & Wrecking Co., Midland, Ont.
Hyde Engineering Works, Montreal, Que.
Iron Works, Ltd., Owen Sound, Ont.
Kennedy & Sons, Wm., Owen Sound, Ont.
Muir Bros. Dry Dock Co., Port Dalhousie, Ont.
National Shipbuilding Co., Goderich, Ont.
Port Arthur Shipbuilding Co., Port Arthur, Ont.
Yarrow, Limited, Victoria, B.C.

RIGGING SCREWS
Hopkins & Co., F. H., Montreal, Que.

RIGGING, WIRE ROPE
Leckie, Ltd., John, Toronto, Ont.

RIVETS
London Bolt & Hinge Works, London, Ont.
Severance Mfg. Co., S., Glassport, Pa.
Wilkinson & Kompass, Hamilton, Ont.

RIVETERS, PNEUMATIC
Can. Ingersoll-Rand Co., Ltd., Sherbrooke, Que.

RODS, COPPER, BRASS, BRONZE
Standard Underground Cable Co., Hamilton, Ont.
Tallman Brass & Metal Co., Hamilton, Ont.

ROLLS, STRAIGHTENING, BENDING
Bertram & Sons, Ltd., John, Dundas, Ont.

ROOFING
Reed & Co., Geo., Montreal, Que.
Wm. C. Wilson & Co., Toronto, Ont.

ROPE BLOCKS
Aikenhead Hardware, Ltd., Toronto, Ont.
Can. Fairbanks-Morse Co., Montreal, Que.
Morris Crane & Hoist Co., Herbert, Niagara Falls, Ont.

ROPE
Hopkins & Co., F. H., Montreal, Que.
Leckie, Ltd., John, Toronto, Ont.
McNab Co., Bridgeport, Conn.
Wm. C. Wilson & Co., Toronto, Ont.

RUBBER COATS
Wm. C. Wilson & Co., Toronto, Ont.

SAW MILL MACHINERY
Wm. Hamilton Co., Peterboro, Ont.
Preston Woodworking Machy. Co., Preston, Ont.
Waterous Engine Works Co., Brantford, Ont.
Yates Machine Co., P. B., Hamilton, Ont.

SAWS, BAND
Preston Woodworking Machy. Co., Preston, Ont.

SCALES, BOILERS, ENGINES
Can. Fairbanks-Morse Co., Montreal, Que.

SCOWS
Collingwood Shipbuilding Co., Collingwood, Ont.
Polson Iron Works, Toronto, Ont.

SCREENS, WIRE
Dennis Wire & Iron Works Co., London, Ont.

SCREWS, COACH
London Bolt & Hinge Works, London, Ont.

SEAM PAINT
Kuhls, H. B. Fred., 6411 3rd Ave., Brooklyn, N.Y.

SEPARATORS, OIL, STEAM
Mason Regulator & Engin. Co., Montreal, Que.
Smart-Turner Mach. Co., Hamilton, Ont.

SHAFTING
Wm. Hamilton Co., Peterboro, Ont.
Mitchell Co., The Robert, Montreal, Que.
Wilkinson & Kompass, Hamilton, Ont.

SHAFTING, BRONZE
Empire Mfg. Co., London, Ont.

SHEARS
Bertram & Sons, Ltd., John, Dundas, Ont.
Norbom Engineering Co., Philadelphia, Pa.

SHIPBUILDING TOOLS
Aikenhead Hardware, Ltd., Toronto, Ont.
Can. Ingersoll-Rand Co. Sherbrooke, Que.

SHIPS, BUILDERS OF
Can. Vickers, Ltd., Montreal, Que.
Collingwood Shipbuilding Co., Collingwood, Ont.
Doxford & Sons, William, Sunderland, England.
Georgian Bay Shipbuilding & Wrecking Co., Midland, Ont.
National Shipbuilding Co., Goderich, Ont.
Polson Iron Works, Toronto, Ont.
Port Arthur Shipbuilding Co., Port Arthur, Ont.
Yarrows, Limited, Victoria, B.C.

SHIP BROKERS
Page & Jones, Mobile, Ala.

SHIP PLATES
Nova Scotia Steel & Coal Co., New Glasgow, N.S.

SLEDGES
Wilkinson & Kompass, Hamilton, Ont.

SLINGS
Hopkins & Co., F. H., Montreal, Que.
Morris Crane & Hoist Co., Herbert, Niagara Falls, Ont.

SPECIAL MACHINERY
Corbet Fdry. & Machine Co., Owen Sound, Ont.
Wm. Hamilton Co., Peterboro, Ont.
Miller Bros. & Sons, Ltd., Montreal, Que.
Smart-Turner Mach. Co., Hamilton, Ont.

SMOOTH-ON
Wm. C. Wilson & Co., Toronto, Ont.

SPIKES
Wm. C. Wilson & Co., Toronto, Ont.

SPRAY COOLING SYSTEMS
Spray Engineering Co., Boston, Mass.

STEAMSHIP AGENTS
Page & Jones, Mobile, Ala.

STEAM SPECIALTIES
Can. Fairbanks-Morse Co., Montreal, Que.
Corbet Fdry. & Machine Co., Owen Sound, Ont.
Empire Mfg. Co., London, Ont.
Mitchell Co., The Robert, Montreal, Que.

STEAM TRAPS
Aikenhead Hardware, Ltd., Toronto, Ont.
American Steam Gauge & Valve Mfg. Co., Boston, Mass.
Empire Mfg. Co., London, Ont.
Mason Regulator & Engin. Co., Montreal, Que.
Mitchell Co., The Robert, Montreal, Que.
Smart-Turner Mach. Co., Hamilton, Ont.

STEEL, HIGH SPEED
Hopkins & Co., F. H., Montreal, Que.
Nova Scotia Steel & Coal Co., New Glasgow, N.S.
Wilkinson & Kompass, Hamilton, Ont.
Zenith Coal & Steel Products, Ltd., Montreal.

STEEL SHELVING
Dennis Wire & Iron Works, London, Ont.

STEEL WORK, STRUCTURAL
Babcock & Wilcox, Ltd., Montreal, Que.
Corbet Fdry. & Machine Co., Owen Sound, Ont.
Wm. Hamilton Co., Peterboro, Ont.

STEERING GEARS
Corbet Fdry. & Machine Co., Owen Sound, Ont.
Hopkins & Co., F. H., Montreal, Que.
Engr. & Mach. Wks. of Can., St. Catharines, Ont.
Wm. C. Wilson & Co., Toronto, Ont.

STOCK RACKS FOR BARS, PIPING, ETC.
Mitchell Co., The Robert, Montreal, Que.
Morris Crane & Hoist Co., Herbert, Niagara Falls, Ont.

STOKERS, MECHANICAL
Babcock & Wilcox, Ltd., Montreal, Que.

SUPERHEATERS, STEAM
Babcock & Wilcox, Ltd., Montreal, Que.
Goldie & McCulloch Co., Galt, Ont.

SWITCHBOARDS, ELECTRIC
Crouse-Hinds Co. of Can., Ltd., Toronto, Ont.

TALLOW
Can. Economic Lubricant Co., Montreal, Que.
Wm. C. Wilson & Co., Toronto, Ont.

TANKS, STEEL
Corbet Foundry & Mach. Co., Owen Sound, Ont.
Goldie & McCulloch, Ltd., Galt, Ont.
Hopkins & Co., F. H., Montreal, Que.
Marsh Engineering Works, Belleville, Ont.
Port Arthur Shipbuilding Co., Port Arthur, Ont.
Reed & Co., Geo., Montreal, Que.
Scaife & Sons Co., Wm. B., Oakmont, Pa.

TANKS (AIR, GAS AND LIQUID)
Marsh Engineering Works, Belleville, Ont.
Scaife & Sons Co., Wm. B., Oakmont, Pa.

TAPPING MACHINES
Mueller Mfg. Co., H., Sarnia, Ont.

TELEGRAPHS, SHIPS
McNab Co., Bridgeport, Conn.
Cory & Sons, Inc., Chas., New York, N.Y.
Morrison Brass Mfg. Co., James, Toronto, Ont.
Wilson & Co., Wm., Toronto, Ont.

TELEPHONES, MARINE
Cory & Son, Inc., Chas., New York, N.Y.
McNab Co., Bridgeport, Conn.

TESTERS, METER
Mueller Mfg. Co., H., Sarnia, Ont.

THUMB SCREWS AND NUTS
Canada Foundries & Forgings, Welland, Ont.
Mitchell Co., The Robert, Montreal, Que.
United Brass & Lead, Ltd., Toronto, Ont.

TRACK SYSTEMS
Morris Crane & Hoist Co., Herbert, Niagara Falls, Ont.

TRAVELLING BLOCKS
Morris Crane & Hoist Co., Herbert, Niagara Falls, Ont.

TROLLEYS
Can. Fairbanks-Morse Co., Montreal, Que.
Morris Crane & Hoist Co., Herbert, Niagara Falls, Ont.

TROLLEY HOISTS
Morris Crane & Hoist Co., Herbert, Niagara Falls, Ont.

TRUCKS, HAND, ELECTRIC
Aikenhead Hardware, Ltd., Toronto, Ont.
Can. Fairbanks-Morse Co., Montreal, Que.

TUBES, BOILER
Babcock & Wilcox, Ltd., Montreal, Que.
Broughton Copper Co., Ltd., Manchester, Eng.
Goldie & McCulloch Co., Galt, Ont.

TUBES, COPPER AND BRASS
Mueller Mfg. Co., H., Sarnia, Ont.
Tallman Brass & Metal Co., Hamilton, Ont.
Standard Underground Cable Co., Hamilton, Ont.

TUGS
Polson Iron Works, Toronto, Ont.
Collingwood Shipbuilding Co., Collingwood, Ont.

TURBINES, STEAM
Goldie & McCulloch Co., Galt, Ont.

TURBINES, DIRECT-DRIVING AND GEARED
Doxford & Sons, William, Sunderland, England.

TURNBUCKLES
Canada Foundries & Forgings, Welland, Ont.
Hopkins & Co., F. H., Montreal, Que.

TURNTABLES
Morris Crane & Hoist Co., Herbert, Niagara Falls, Ont.

SPIKES, SMALL RAILROAD
Severance Mfg. Co., S., Glassport, Pa.

UNIONS, ALL KINDS
Dart Union Company, Toronto, Ont.

VALVES, AIR
Mitchell Co., The Robert, Montreal, Que.
Mueller Mfg. Co., H., Sarnia, Ont.

VALVE, DISCS
Wm. C. Wilson & Co., Toronto, Ont.

VALVES
American Steam Gauge & Valve Mfg. Co., Boston, Mass.
Babcock & Wilcox, Ltd., Montreal, Que.
Can. Fairbanks-Morse Co., Montreal, Que.
Empire Mfg. Co., London, Ont.
McAvity & Sons, Ltd., T., St. John, N.B.
Mason Regulator & Engin. Co., Montreal, Que.
McNab Co., Bridgeport, Conn.
Norbom Engineering Co., Philadelphia, Pa.
Williams Machinery Co., A. R., Toronto, Ont.
Wm. C. Wilson & Co., Toronto, Ont.
United Brass & Lead, Ltd., Toronto, Ont.

VALVES, FOOT
Aikenhead Hardware, Ltd., Toronto, Ont.
Mitchell Co., The Robert, Montreal, Que.
Smart-Turner Mach. Co., Hamilton, Ont.

VALVES, STOP, REDUCING, SAFETY CHECK, ETC.
Aikenhead Hardware, Ltd., Toronto, Ont.
American Steam Gauge & Valve Mfg. Co., Boston, Mass.
McAvity & Sons Ltd., T., St. John, N.B.
Mitchell Co., The Robert, Montreal, Que.
Morrison Brass Mfg. Co., James, Toronto, Ont.

VALVES, MIXING
Mitchell Co., The Robert, Montreal, Que.
Mueller Mfg. Co., H., Sarnia, Ont.

VALVES, REDUCING, PRESSURE
Mitchell Co., The Robert, Montreal, Que.
Mueller Mfg. Co., H., Sarnia, Ont.

VARNISHES
Aikenhead Hardware, Ltd., Toronto, Ont.
Ault & Wiborg Co. of Can., Ltd., Toronto, Ont.
Leckie, Ltd., John, Toronto, Ont.
Reed & Co., Geo., Montreal, Que.

VENTILATORS, COWL
McNab Co., Bridgeport, Conn.
Mitchell Co., The Robert, Montreal, Que.

VENTILATION EQUIPMENT
Empire Mfg. Co., London, Ont.
Hopkins & Co., F. H., Montreal, Que.
Low & Sons, Ltd., Archibald, Glasgow, Scotland

WASHERS
London Bolt & Hinge Works, London, Ont.
Wm. C. Wilson & Co., Toronto, Ont.

WATER COLUMNS
Mitchell Co., The Robert, Montreal, Que.
Morrison Brass Mfg. Co., James, Toronto, Ont.

WELDING, ELECTRIC
Hall Engineering Works, Montreal, Que.
Marine Welding Co., Buffalo, N.Y.
Beatty & Sons, M., Welland, Ont.

WATER SOFTENERS
Babcock & Wilcox, Ltd., Montreal, Que.

WATER SUPPLY SYSTEMS
Mueller Mfg. Co., H., Sarnia, Ont.

WATER HEATERS
Empire Mfg. Co., London, Ont.
Morrison Brass Mfg. Co., James, Toronto, Ont.

WHISTLES AND SYRENS
Empire Mfg. Co., London, Ont.
McAvity & Sons, T., St. John, N.B.
McNab Co., Bridgeport, Conn.
Mitchell Co., Ltd., Robt., Montreal, Que.
Morrison Brass Mfg. Co., Jas., Toronto, Ont.

WINCHES, CARGO
Aikenhead Hardware, Ltd., Toronto, Ont.
Corbet Fdry. & Machine Co., Owen Sound, Ont.
Hopkins & Co., F. H., Montreal, Que.
Marsh Engineering Works, Belleville, Ont.

WINCHES, DOCK, SHIP
Beatty & Sons, M., Welland, Ont.
Marsh Engineering Works, Belleville, Ont.
Miller Bros. & Sons, Ltd., Montreal, Que.
Morris Crane & Hoist Co., Herbert, Niagara Falls, Ont.
Wilson & Co., Wm. C., Toronto, Canada.

WINCHES, TOWING
Corbet Foundry & Mach. Co., Owen Sound, Ont.

WINCHES, TRAWL
Beatty & Sons, M., Welland, Ont.
Wm. C. Wilson & Co., Toronto, Ont.

WINDLASSES
Corbet Fdry. & Machine Co., Owen Sound, Ont.
Dake Engine Co., Grand Haven, Mich.
Hopkins & Co., F. H., Montreal, Que.
Wilson & Co., Wm. C., Toronto, Canada.

WIPER CAPS, OILER BOXES, ETC.
Mitchell Co., The Robert, Montreal, Que.
Morrison Brass Mfg. Co., James, Toronto, Ont.

WIRE, COPPER CLAD STEEL
Standard Underground Cable Co., Hamilton, Ont.

WIRE, COPPER, BRASS, BRONZE
Standard Underground Cable Co., Hamilton, Ont.
Tallman Brass & Metal Co., Hamilton, Ont.

WIRE, INSULATED
Standard Underground Cable Co., Hamilton, Ont.

WIRELESS OUTFITS
Marconi Wireless Telegraph Co., Montreal, Que.

WIRE ROPE
Zenith Coal & Steel Products, Ltd., Montreal,

WOOD WORKING MACHINERY
Aikenhead Hardware, Ltd., Toronto, Ont.
Can. Fairbanks-Morse Co., Montreal, Que.
Preston Woodworking Machy. Co., Preston, Ont.
Yates Mach. Co., P. B., Hamilton, Ont.

WOOD BORING TOOLS
Aikenhead Hardware, Ltd., Toronto, Ont.
Can. Ingersoll-Rand Co., Sherbrooke, Que.

WOODITE GAUGE GLASS WASHERS
Wm. C. Wilson & Co., Toronto, Ont.

WRENCHES
Canada Foundries & Forgings, Welland, Ont.

YACHT BROKER
Watt, J. Murray, Philadelphia, Pa.

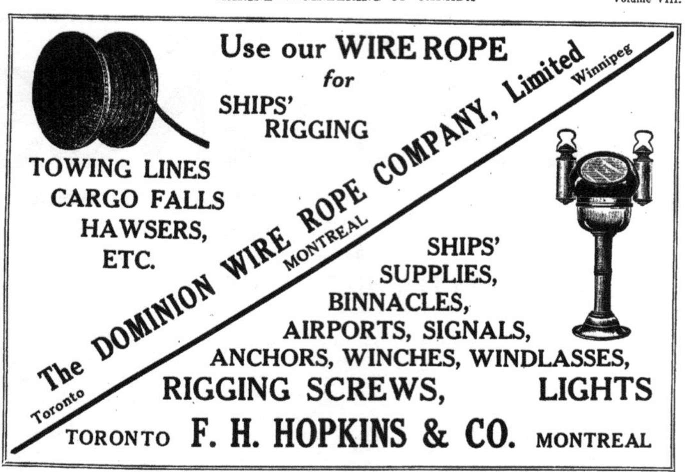

INDEX TO ADVERTISERS

Harris Heavy Pressure and its Advantages
HARRIS METAL
HEAVY PRESSURE
THE CANADA METAL CO.
TORONTO
1. A complete immunity from hot bearings is secured. HARRIS HEAVY PRESSURE having a lower co-efficient of friction than any other known metal.
2. A scored journal is impossible, and if through any failure of lubrication a bearing should run hot, HARRIS HEAVY PRESSURE, owing to its special properties, will act as a lubricant, saving the journal from injury and preventing any delay to traffic.
3. It will stand the heaviest pressures, always running cool, even under the most trying conditions.
4. It will wear from 50 to 100 per cent. longer on general machinery bearings than any other Babbitt metal.
5. It effects a saving in lubrication.
6. It preserves the journals, and materially increases their life. A journal after running a short time with HARRIS HEAVY PRESSURE attains a after running a short time with HARRIS HEAVY PRESSURE attains a
7. It is easily applied and, if properly applied, no abrasive force will remove it.
8. Its cheapness. The first cost is moderate. It gives a longer life to the bearings, resulting in a great economy, as the number of renewals is thereby considerably reduced; its specific gravity is low in comparison with other metals; does not deteriorate with re-melting; and these advantages, together with its unequalled anti-fricion properties, render it the cheapest as well as the best metal for all general machinery bearings.
ORDER A BOX FROM OUR NEAREST FACTORY
THE CANADA METAL CO., LIMITED
HAMILTON
MONTREAL
TORONTO
WINNIPEG
VANCOUVER

Raw and Boiled
Linseed Oil
"Maple-Leaf Brand"
THE most exacting requirements of Shipbuilders or Governments are met by our brand.
The Canada Linseed Oil Mills
LIMITED
Mills at Toronto and Montreal
Write the nearest Mill

SMART-TURNER MARINE PUMPS
justify by Years of Service the high standard to which they are built. LET US SHOW YOU.
The Smart-Turner Machine Co., Ltd.
HAMILTON, CANADA

MORRISON'S
MARINE SPECIALTIES
Highest Quality
plus high grade construction and modern design
are what you get when buying the Morrison line of Marine Specialties. Every article is guaranteed reliable and thoroughly factory tested.
Approved by the Dominion Marine Inspection Department, Lloyd's Survey and Munition Board.
Write for our new catalog. It consists of 60 pages with 150 cuts.
The James Morrison Brass Mf'g. Co.
LIMITED
TORONTO
CANADA
Ship's Side Cock with Locking Shield, Bronze
J.M.T. Reducing Valve
Morrison Improved Twin Marine Safety Valve
Overboard Discharge Valve, Spring Loaded
Asbestos Packed Boiler Drain Cock, Bronze
"Marine" Angle Valve, Bronze
Morrison Improved Pressure Gauge

CIRCULATES IN EVERY PROVINCE OF CANADA AND ABROAD

MARINE ENGINEERING of Canada

A monthly journal dealing with the progress and development of Merchant and Naval Marine Engineering, Shipbuilding, the building of Harbors and Docks, and containing a record of the latest and best practice throughout the Sea-going World. Published by
The MacLean Publishing Co., Limited
MONTREAL, Southam Building TORONTO 143-153 University Ave. WINNIPEG, 1207 Union Trust Bldg. LONDON, ENG., 88 Fleet St.

Vol. VIII. Publication Office, Toronto—July, 1918 No. 7

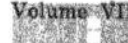

MARINE BRASS CASTINGS

ROUGH OR FINISHED

Navy Brass,
Bronze
and Gunmetal
Castings

Alloy Castings
of any size and
weight to your
Specifications

Babbitt Metal to Naval Specifications

TOLLAND MFG. COMPANY

1165 Carriere Rd. Montreal, Que., Can.

Stern Casting
for Ice Breaker
"JOHN D. HAZEN"

STEEL CASTINGS

Ship Castings are a Specialty

CANADIAN STEEL FOUNDRIES LIMITED

General Offices:
Transportation Bldg.,
Montreal

Works:
Montreal and
Welland

Canadian-Built Ocean-going Steamer "Reginolite"

The fourth ship launched on an order of five for the IMPERIAL OIL CO.

The "Reginolite" was recently launched and is here seen on her trial trip. She is built for ocean service and measures:—

Length .... 250 feet
Breadth ....43 feet 9 inches
Depth ....25 feet moulded

The trials, although carried out in stormy weather, were highly successful, the guaranteed speed being exceeded by one and one-half knots.

We also recently launched the first two of six trawlers, now being built for the Naval Service Department. Other craft are nearing completion.

We are makers of steel and wooden ships, engines, boilers, castings and forgings.

PLANT FITTED WITH MODERN APPLIANCES FOR QUICK WORK. Dry Docks and Shops Equipped to Operate Day and Night on Repairs.

The Collingwood Shipbuilding Co., LIMITED

COLLINGWOOD, ONTARIO, CANADA

LOW'S SPECIALTIES FOR SHIPS

We are Specialists in Appliances for

HEATING and PLUMBING WORK

We also make all kinds of

BRASS and SHEET METAL WORK

REQUIRED IN THE CONSTRUCTION OF SHIPS.

We have supplied many well-known vessels with all their requirements in the departments referred to.

Low's Patent Steam Radiators

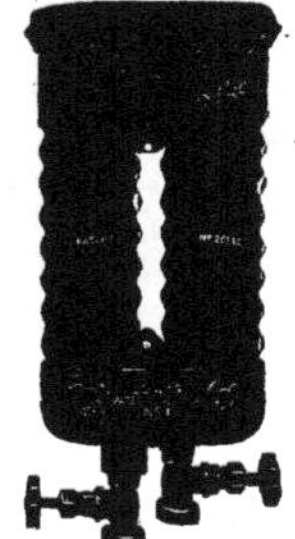

Strong enough to stand High-Pressure.

All surface in actual contact with Steam.

No Webs used to make up the surface.

Gives most heat for smallest space occupied.

Internal Tubes give accelerated circulation of air.

Low's Patent Calorifiers

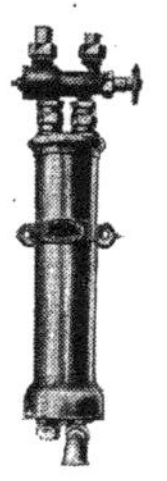

45 GALLONS OF HOT WATER PER MINUTE.

Largest output with minimum of space occupied.

PATENT VALVE PREVENTS ACCIDENTAL SCALDING.
ADOPTED BY THE ADMIRALTY.
STEAM CANNOT BE TURNED ON BEFORE THE WATER.

"Highlow" Electric Radiators

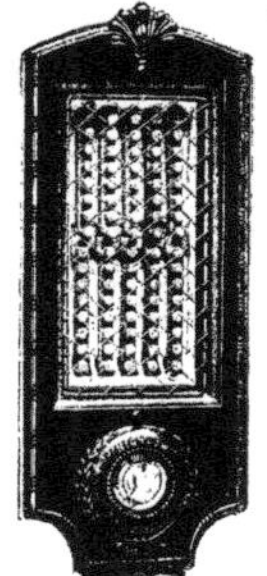

Specially designed for Ships' use. Simple in Construction.

Guaranteed for Two Years.

No Mica or other complication to get out of order.

Large quantities of usual Ships' voltages in Stock.

ARCHIBALD LOW & SONS, LTD.

MERKLAND WORKS, PARTICK, GLASGOW

LIVERPOOL AGENTS: A. J. Nevill & Co., 9 Cook St.

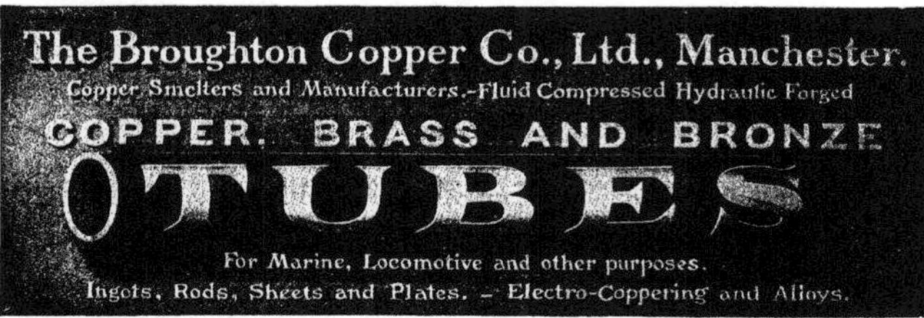

If what you need is not advertised, consult our Buyers' Directory and write advertisers listed under proper heading.

MARINE ENGINEERING
of Canada

Volume VIII. TORONTO, JULY, 1918 Number 7

War Time Effort at Toronto Ship Building Co.'s Plant

Standard Wooden Ships Being Built in Toronto to Same Design as Those Constructed on Pacific Coast

By A. G. WEBSTER
Associate Editor Marine Engineering

ALTHOUGH wooden ships of small tonnage have been built in Toronto, it is only quite recently that any substantial development has taken place in this industry. The extraordinary demand for ships due to the losses from submarines caused the Imperial Munitions Board to extend its activities into a new field. A standard type of wooden ship propelled by steam was adopted by the board and contracts let with various builders, among whom was the Toronto Shipbuilding Co. This concern which was organized last summer rented a property from the Toronto Harbor Commission, in the Ashbridge's Bay industrial district. The shipyard is alongside what is known now as the Don Ship Channel being also the Don River diversion. The officers of the company, whose names are given below, are well known in marine circles, and have shown considerable enterprise in establishing what is to all intents and purposes a new industry in Toronto.

One of the principal difficulties that confronted the company was the getting together of an organization to carry out the work. The building of wooden ships had practically become a lost art, as the industry had been declining for many years. For this reason there was a great dearth of skilled ship carpenters, a situation aggravated by the sudden demand for artisans of this type. However, this difficulty was gradually overcome and good progress is being made with the work of construction. Being practically all outside work the hard winter retarded progress to some extent.

The company secured a contract for two ships, the first keel being laid last August and the second in October. They are both standard design, identical with those being built on the Pacific coast and other yards in Canada.

The principal dimensions of this standardized vessel are:

Hull Dimensions

Length B.P.	250 ft.
Length O.A.	259 ft.
Breadth, extreme	43 ft. 6 in.
Breadth, moulded	42 ft. 6 in.
Depth, moulded	25 ft.
Depth over keel	26 ft. 11 in.
Draft for displacement	22 ft.
Approximate deadweight on 20 ft. maximum draft	2,500 tons
Approximate deadweight on 21 ft. maximum draft	2,800 tons

General Description

Vessel is of the single deck cargo type, built principally of Douglas fir from British Columbia with hold beams, wood deck houses and rails. The ship has an elliptical stern with long poop deck aft, and raised fo'c-sle forward. There will be five hatches, one deep ballast tank with longitudinal divisions. Six watertight wooden bulkheads, one bunker bulkhead, which is a non-watertight bulkhead, one screen bulkhead, and one watertight door between engine room and tunnel. Six cargo winches, one of which is to be a warping winch, will be installed. Windlass on the fo'c'sle head suited to handle anchors and full scope of chain, also arranged for warping as usual. Fore peak to be fitted for fresh water, with filling pipes, suctions, etc. After peak to be fitted in the same manner. Culinary water will be distributed from separate steel tanks of about four hundred gallons capacity each.

The vessel is to be driven by single screw, with engines abaft admidships. The vessel is to be built to Lloyd's requirements for A1 classification, and to the requirements of the British Board of Trade, as far as necessary for a cargo steamer. Long lengths of lumber are used in the keels, keelsons, planking and ceiling. Fastenings consists of treenails, screw bolts, and drift bolts of galvanized iron and black iron; galvanized bolts being used when in contact with sea water. The keel is sided 24 in., moulded 20 in., and in four lengths. The scarphs are 10 ft. long and are fastened with 1 in. galvanized bolts, ends of scarphs are spiked with three 10 in. spikes at each end. The keel has a ½ in. plate rubber, spiked with ⅝ in. rag bolts one foot apart.

Keelsons

There are three wood keelsons, the centre keelson is 24 in. by 20 in. and sister keelsons are 20 in. by 20 in. Centre keelson is in four lengths and scarphed with 10 ft. scarphs, and the sister keelsons are in four lengths with 10 ft. scarphs.

A steel rider keelson is fitted on top of the wood keelsons, as shown in cross section. The steel keelson, centre and sister keelsons, frames and keel, are fastened with through screw bolts of 1¼ in. galvanized and black iron. In addition, the wood keelsons are edge bolted with one 1¼ in. screw bolt in each space between frames. The steel keelson extends

GENERAL VIEW OF TWO WOODEN HULLS UNDER CONSTRUCTION.

VIEW SHOWING HULL FRAMING WITH STERN POST IN POSITION.

DECK VIEW LOOKING AFT, DECKING PARTLY LAID.

from the fore peak bulkhead to the after peak bulkhead, and is constructed with a wide foundation plate having a built-up box girder 12 in. wide and 24 in. deep, riveted to centre of foundation plate. The stem is 18 in. side and mould 26 in. The cut water at stem is reduced to 5 in., gradually to widen out to conform to the curve of the forefoot as it widens out to meet the keel.

Framing and Rudder

The apron sides 24 in. by 28 in. The apron is connected to the stem with 1½ in. galvanized drift bolts, spaced 12 in. centre and staggered. The propeller post is moulded 30 in., and generally sided 24 in. In way of shaft tube it is swelled out to 32 in. The rudder post is 24 in. square tenoned into the keel. Steel castings will be secured to the rudder post for the fitting and hanging of the rudder. The rudder will be of steel plate to Lloyds' requirements. The pintles will be of steel. brass bushed and working in cast steel gudgeons. The shaft logs are 36 in. square and 11 ft. long for housing the 21 in. diameter stern tube. The stuffing box for shaft is to be secured to end of log and flange of stern tube bolted through stern post.

The frames are double to side 12 in. and to mould 24 in. at keel 18 in. at long floor futtock, 16 in. at the turn of the bilge, 14 in. in way of the top of thick ceiling, 13 in. in way of hold beams, 11 in. at deck and 7 in. at upper deck. Double frames are fastened to one another with fir treenails.

Ceiling and Decking

The ceiling at flat of floor is 10 in. thick, thick bilge ceiling 14 in. by 14 in. upper strakes 10 in. by 12 in., and remainder under main deck 8 in. by 12 in. The ceiling in long poop and forecastle is 6 in. thick. The ceiling is fastened with 1⅛ in. iron bolts. The main deck beams are 36 in. spaced scantlings. All hatch beams 14 in. sided and 14 in. moulded. Beams in way of tank bulkheads to side 14 in. The poop, bridge and forecastle deck beams are spaced 3 in., mould 12 in., and side 12 in.

Main deck planking is 4 in. by 4 in., finishing about 3¾ in. of clear, straight edge grain fir. The main deck is fastened with 7-16 in. by 8 in. galvanized spikes, two in each strake in each beam. The upper deck is fastened with ⅜ in. by 7 in. spikes, two in each strake in each beam. Decking in lengths 20 ft. and 40 ft. averaging 30 ft.

Outside Planking and Ceiling

The first garboard is 18 in. x 10 in. with 5 ft. scarphs. Second garboard 16 in. x 9 in. and third garboard 14 in. x 8 in. Bottom planking 12 in. x 6 in., bilge planking 10 in. x 6 in., side planking 8 in. x 6 in. up to the guard. Top side planking 8 in. x 5 in. Garboard strakes are fastened with four ⅞ in. galvanized bolts, spaced about 3 in. centres; bolts being 32 in. and 20 in. long. The bottom and bilge planking is fastened with two treenails and two galvanized spikes in each frame, and galvanized butt bolts. Bottom ceiling is 12 in. x 10 in., in long lengths, having a limber strake fitted adjacent to the sister keelsons, as shown on midship section. In way of the deep water tank, the limber strake is edge and face bolted with 1⅛ in. black drift bolts. All outside planking is caulked with best handpicked oakum. Caulking in the proportion of one strand of oakum to each inch thickness of plank. Outside plank seams up to 10 ft. waterline are cemented. Balance of outside seams puttied. Decks are caulked with the best handpicked oakum, in the proportion of one strand of oakum to each inch thickness of plank.

Miscellaneous

Machinery spaces are to be lined with ⅛ in. wrought iron, properly secured in the way of all parts of woodwork exposed to risk of fire. In general terms. the stokehold will require to be completely sheathed and such parts of the engine room as required. The casing around engine and boilers in the 'tween decks is built of solid 4 in. lumber, well fastened with 1 in. drift bolts. Bunker spaces to be lined with 1¼ in. ceiling laid on grounds 3 in. thick to form air spaces. This ceiling is covered with ⅛ in. black iron sheets. All spaces intended for use as bunkers to be so fitted with air spaces and sheeting. Ventilation of bunkers to be very closely considered and carried out.

Steering engine to be arranged directly over the rudder, with all necessary pulleys, chains, quadrants, etc., as required to give the desired leads, and to be fitted with suitable hand gear. Control rods to be laid from the steering engine to the steering wheel. The shaft tunnel is formed aft of lumber 10 in. x 6 in. on sides and 10 in x 4 in. on top. Beams are 6 in. x 6 in., spaces 3 ft. apart. Bulwarks are 4 in. in planking. The ship is fitted with four bulkheads, namely collision bulkhead, aft peak bulkhead and bulkheads at fore and aft ends of machinery space. The bulkheads are formed of double sheeting 3 in., lumber laid diagonally and fitted between courses with pointed, canvas, caulked and made watertight. The ship will be fitted with two wood masts, having double derrick, each derrick to lift 5 tons.

The masts will be of sufficient length to permit the installation of a wireless outfit, The ship will be supplied with two class A lifeboats to British Board of Trade requirements. A dinghy and service boat will also be supplied. The anchors will be stockless, self tripping type. The crew will be housed forward under the fo'c'sle deck. Accommodation for the engineers will be fitted alongside starboard engine casing and for the petty officers on the port side of casing. The bridge house will be built forward end of poop deck and fitted up to accommodate saloon, officers, wireless cabin. The wheel house will be built on top of saloon house and arranged to accommodate the captain and wireless room. Steering wheel will be on the bridge together with compasses and telegraphs.

Main Engine Details

The main engine which is being installed in the standard wooden ships is a striple expansion, surface condensing, engine, the three cylinders working on separate cranks placed at angles of 120 degrees. The h.p. cylinder is 20 in.; I. P. 33 in. and l.p. 54 in. diameter, with a common stroke of 40 in. The average working horse power is above 1,000 at sea at about 65 to 70 revs. per min. The cylinders are made of close grained grey iron. The h.p. cylinder has a piston valve and the other two have double ported slide valves. The piston rods, cross heads and connecting rods are of open hearth steel to Lloyds' requirements. The slide valve motion is of the Stephenson link type with double eccentrics. A direct steam driven reversing engine will be provided and also a hand reversing gear. Columns and bed plant are cast iron. The condenser is cast iron with rolled brass tube plate, brass tubes and has 1,560 sq. ft. of cooling surface.

The crank shaft is of built up section in three parts with solid forged couplings. Crank webs, crank shaft and thrust shaft are of steel. The propeller shaft is steel with coupling forged on fitted to propeller with taper. The shaft is 12 in. diameter; a continuous brass liner is fitted on the tail shaft, running in the lignum vitae bearing in stern tube.

The propeller is cast iron, a right hand screw of solid section with four blades 14 ft. 6 in., diameter and 15 ft. 3 in. pitch. The propeller is taper bored to fit shaft, with through key and secured by nut, having left hand thread. The stern tube is cast iron fitted with forged steel nut on left side. The main exhaust pipe is 17 in. diameter. The discharge pipe from circulating pump to condenser is 8 in. diameter. The air pump discharge pipe is 6 in. diameter.

Boiler Installation

The boiler installation consists of two three-element Howden patent combination sectional water tube boilers complete with mountings, smokeboxes and air heater of Howden type arranged for working under the Howden system of forced draft.

Each boiler is made up of three elements having a total heating surface of 2,800 sq. ft., while the superheater has 117 sq. ft. of heating surface. The grate area for .75 in. forced draught is 45.8 sq. ft., the fire bars being 5 ft. long. The evaporation per hour from feed at 120 degs. Fah., is 13,750 lbs., thus the boiler will give a total evaporation of 13,750 lbs. per lb. of fuel per hour from combined heat absorbing surface of 2,800 sq. ft. and 45.8 sq. ft., .75 in. forced draught grate area, with coal of not less than 12,000 B.t.u. per lb.

The boiler is supported in steel framework in such a manner as to allow free expansion. The units are encased with brick and have doors at the side for getting at the tubes for cleaning or renewing. The feed and blow off enter the lower drum at the front head into an arrangement of internal pipes. Provision is made in combustion chamber and smoke box door for removing soot from the tubes belonging to the boiler and air heater. The forced draught is of Howden latest and improved type, as are also the fire doors which, when opened, automatically shut off the draught. The space occupied by the boiler including brickwork is as follows:—Width, 12 ft. 6⅜ in.; depth 12 ft. 8 in.; and height 12 ft. 9 in. The boilers are tested by water pressure to 360 lbs. per sq. in. for the working pressure of 180 lbs. per sq. in.

Auxiliary Machinery

The auxiliary machinery comprises pumps for the various services and deck machinery. The independent boiler feed pump is of the Weir long stroke simplex type, 10 x 6 x 12 in., capable of discharging 20,000 lbs. of water per hour at normal speed against a boiler pressure of 180 lbs. Only one independent pump is required if the main engine is standard type, as it is supplied with ram feed pumps driven off the air pump levers. For general service a vertical duplex, Worthington type, pump 6 x 4 x6 in., is used. The general service includes boiler feed, circulating, deck, fire and sanitary pumps; a duplex ballast pump 7½ x 9 x 10 in. will be installed for low pressure service, capable of pumping about 100 tons of water per hour.

The auxiliary condenser is of the surface atmospheric type, having approximately 400 sq. ft. of cooling surface. The feed water heater is of the exhaust surface type of sufficient heating surface to deal with 25,000 lbs. of feed water per hour from a temperature corresponding to vacuum of 25 in. to atmospheric boiling point. The gravity type feed water filter will be capable of dealing with about 25,000 lbs. of feed water per hour. An evaporator will be supplied, capable of evaporating 15 tons of sea water per day.

The deck machinery will consist of five 2½ ton, 8 x 10 in. horizontal type steam winches, having extended ends for warping. Each ship will also have one 8 x 8 inch double gear windlass with independent gipsy.

(1) BEGINNING TO ERECT FRAMING. (2) INTERIOR OF HULL SHOWING STEEL KEELSON. (3) LOOKING TOWARDS THE BOW, INSIDE. (4) LOOKING TOWARDS THE STERN, INSIDE.

STERN VIEW OF WAR ONTARIO READY FOR LAUNCHING.

SHOWING PLATFORM INSIDE HULL FOR BUILDING CEILING.

The accompanying illustrations show the freighters at various stages of building and are interesting in that they show the method of construction. At the time of writing the planking, ceiling and decking on No. 1 hull have been completed, while on No. 2 hull the outside planking only has been finished. With favorable weather good progress is being made and No. 1 hull will probably be launched some time in July. C. L. Hays, the superintendent, was formerly in charge of No. 1 section of the Welland ship canal.

The officers of the Toronto Shipbuilding Co. are as follows:—C. S. Boone, president; J. E. Russell, vice-president and general manager; C. A. Boone, 2nd vice-president; J. M. Russell, secretary; John J. Manley, treasurer; C. L. Hays, engineer and superintendent; J. J. Whalen, master boat builder.

All the main machinery is supplied to the ship builders by the Imperial Munitions Board, the orders for equipment being distributed direct by the board. The main propelling engines for these two vessels are being built by the Canadian Bridge Co., Walkerville, Ont., and the boilers by Polson Iron Works, Toronto. The auxiliary and deck machinery are being supplied by the Imperial Munitions Board.

War Ontario Launched

The first hull was launched on June 29 and named "War Ontario," the christening ceremony being performed by Lady Hearst, wife of the Premier of Ontario.

Some delay occurred in releasing the boat from the ways, as owing to a slight sinking of the ground when she was about to take the water the ways sank down and held her. It was therefore, necessary to raise them with jacks to give them a proper slope. At the end of twenty-four minutes the trouble was adjusted and the "War Ontario" glided into the channel gracefully.

No speeches were delivered, and at the conclusion of the ceremony Lady Hearst was presented with a boquet of flowers by Miss Edna Russell. Lady Hearst presented John J. Whelan, master shipbuilder of the Russell Co., with a handsome engraved solid gold watch. Loud cheers were afterwards given for Lady Hearst and Mr. Russell.

Among those present on the platform were Mayor Church, Ald. Geo. Ramsden, Ald. Sam Ryding, A. H. Jeffery, Strachan Johnson, Dr. Jameson, John Laxton, Geo. Wright, E. J. Boland, Richard McWilliams, director of the Russell Company; Revs. O'Brien and Bushnell, C. L. Hays, Capt. Sullivan, M. J. Haney, A. T. Bradley, Capt. Manly, J. Sing, Mr. Lambert, representing Col. Gears, chairman of the Imperial Munitions Board, and representatives of Lloyd's.

U.S. NAVY HONORS CHIEF BOATSWAIN'S MATE

Secretary Daniels announces that the Navy Department has awarded the medal of honor and a gratuity of $100 to John Mackenzie, chief boatswain's mate, United States Naval Reserve Force, for extraordinary heroism which resulted in saving from possible destruction the U.S.S. Remlik, a converted yacht now on patrol service in European waters.

The case is unique in that it has to do with one of the latest engines of war. As is well known, United States destroyers and other submarine fighters carry depth charges containing a large amount of high explosives which are dropped in the path of enemy submarines and explode under water. These have proved effective weapons in the destruction of U-boats, and they are safe enough when the safety pins are affixed; but when they get beyond control and the safety pin comes out they are a source of serious danger to the vessels carrying them. It will be recalled that the men on the U. S. S. Manley who lost their lives in the collision of that destroyer with a British vessel were killed by the explosion of one of these bombs.

In a heavy gale on the morning of December 17, 1917, a depth charge on the Remlik broke loose from its position on the stern. The box went overboard, but the charge was hurled in the opposite direction and went bouncing about the deck. As it weighed hundreds of pounds, it was impossible for anyone to lift the bomb and carry it to safty. It was even dangerous for anyone to go to that part of the ship as the seas were washing over the stern. As the officers and crew watched the bomb some one shouted, 'The pin's come out." Realizing the danger Mackenzie, exclaiming, "Watch me; I'll get it," dashed down the deck and flung himself upon the charging cylinder. Three times he almost had his arms about the bomb, but each time it tore from him, once almost crushing him. The fourth time he got a firm grip on it and heaved it upright on one flat end. Then he sat on it and held it down. The charge might have broken loose again and exploded at any moment, blowing Mackenzie to bits; but he held on firmly until lines could be run to him and man and depth bomb safely lashed. Soon afterwards the ship was headed up into the sea and the charge carried to a place of safety.

The commanding officer of the Remlik, in his report recommending that the medal of honor be conferred on Mackenzie, says:

"Mackenzie, in acting as he did, exposed his life and prevented a serious accident to the ship and probable loss of the ship and entire crew. Had this depth charge exploded on the quarter deck with the sea and wind that existed at the time there is no doubt that the ship would have been lost."

TRIAL TRIP OF S.S. "WAR POWER" AT VANCOUVER

The War Power, a product of the Wallace Shipyards Ltd., had a very successful trial trip last week. The run was made under easy steam to Nanaimo, where the ship was bunkered. Leaving Nanaimo the official trial commenced and a speed of 10 knots was easily maintained, while over the measured course between Point Atkinson and Hollyburn Wharf 12 knots was recorded.

The War Power, which is now loading at Vancouver and will proceed to sea as soon as finished, is the second deep sea boat turned out by the Wallace yards, while a third is expected to be launched this month.

High Pressure Air Compressor Design and Application*

By Joseph M. Ford

Compressed air at high pressure is becoming an increasingly important medium in modern engineering practice and in naval warfare. The accompanying paper deals with the machines which produce high pressure air, leaving on one side the question of power transmission by compressed air. In passing, the principal advantages of this system are referred to, and will be seen to be of considerable value. They consist of the facility with which energy can be stored; the unlimited rate at which accumulated energy may be converted into useful work; and, apart from loss of efficiency, leakage cannot cause any awkward consequences.

TESTING MULTI-STAGE COMPRESSORS

ALTHOUGH extremely interesting results can be obtained from scientific tests of air-compressors, the manufacturer is usually content with tests for the efficiency of the machine. A purchaser requires to know the power required per cubic foot of air actually delivered, since this is the basis of comparison for compressors. The problem, then, is to determine exactly how much air is being delivered. The two methods usually adopted are: the "pumping-up" method and the orifice test. In the former the compressor is run at its full pressure, pumping into a reservoir, and from this latter the air is led to a receiver of known capacity. The details of the method are well known, being exactly the same in principle as used on tests of ordinary compressors for pneumatic tools. The orifice test is now being largely adopted, on account of the possibilities of error with the "pumping-up" method due to inaccurate temperature readings, oil and water in the bottles, etc. The orifice test method is also well known in connection with ordinary compressor work and need not be described.

Useful information as to the condition of piston rings and valves may be obtained from indicator cards, which also show to some extent the efficiency of the cooling. If, on applying Brauer's construction to the diagram, the compression line is found to be almost isothermal, it does not necessarily indicate that the cooling is almost perfect, as is sometimes stated. The line may indeed appear to be the isothermal, due almost certainly to leaky piston rings or suction valves. Lines giving an exponent value in PVn of less than 1.26 should be looked upon with suspicion.

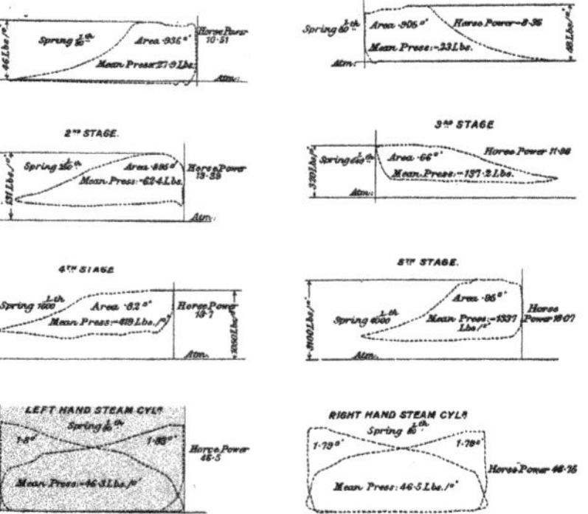

FIG. 32a. FIG. 32b.

If any stage pressure is very much above that estimated, the trouble will likely be due to the valves of the next higher stage leaking, or to its own delivery valve not being tight. Judicious use of the cylinder drain cocks, and the behavior of the stage pressure gauges immediately on shutting down, enables conclusions to be drawn as to the source of trouble without the necessity of taking indicator cards.

Maintenance of Compressors

The usual practices of engine maintenance apply to compressors, but unfortunately the care and attention given to these machines is frequently less than that necessary. Particularly is this the case in some Diesel engine plants, where the compressor, until it gives trouble, is looked upon as an engine detail. The lubricating system, above all, must be kept in the highest state of efficiency, as failure in, say, the high-pressure stage, where these is usually considerable ring friction, generally leads to a cracked

FIG. 30. "BROTHERHOOD" THREE-STAGE COMPRESSOR, 600 CU. FT. CAPACITY.

*Conclusion of paper read before the Greenock (Scotland) Association of Shipbuilders and Engineers. See issues of May 31 and 28 for previous instalments.

A complete set of indicator diagrams from this machine is given in Figs. 32 A and B. It may be mentioned that the shape of the suction lines in the cards from the fourth and fifth stages results from the fact that the cylinder volumes are fairly large compared with the cubical contents of the intercoolers, the lowest point on the line indicating the commencement of transfer. The comparatively small clearance volume in the first stage cylinders, shown by the rapid drop in pressure on the back stroke, is also worthy of note, the reasons for the small volume having previously been discussed.

Fig. 33 shows a four-stage direct-coupled machine by the same builders, capable of charging a 45 cub. ft. reservoir to 2,000 lb. per sq. inch in 80 minutes when running at 250 r.p.m., whilst the five-stage direct-coupled machine shown in Fig. 34, charges a 40 cub. ft. receiver to 3,000 lb. per sq. inch in one hour at 350 r.p.m.

A point to which attention may be directed is that the intercoolers on these machines are generally of the submerged type, that is, the cooler consists of a copper or steel pipe coiled round the cylinders in the tank. This tank construction simplifies very considerably the cylinder castings, and the cooling effect is probably enhanced due to the whole of the connections being submerged, but of course this method is not always admissible. The cooling is still further improved by water injection, the water in the case of a steam-driven machine being obtained by condensation, and therefore quite free from foreign matter.

The piston packing in these compressors is of the fibre-ring type, the high-pressure gland fibres being fitted in removable boxes, which can be repacked at the bench—an obvious advantage. All valves, which are mainly of the thimble type, have removable seatings, so that a replacement is readily effected.

An unusual feature is the provision of an indicator to show that the circulating water is flowing through the casing, which in many cases is very convenient, as it is not always easy to arrange for the water to be discharged into a funnel. The importance of maintaining the circulation will be readily understood from a study of the temperature charts previously given.

Diesel Engine Compressors

Special attention has been devoted to the design of air-compressors for Diesel engines; product of one firm ranging from small hand-driven machines for charging starting bottles in case of emergency, to those used on some of the largest machines yet constructed. A patented design of three-stage compressor for 1,000 lb. per sq. in. working pressure, which has given very satisfactory results from marine work, is shown in Figs. 35 to 38, a special feature of this machine being the arrangement of the second or intermediate stage. By suitably proportioning the cylinder and intercooler dimensions it is possible to dispense entirely with valves in this stage, which means that the intercooler volume is part of the intermediate pressure cylinder clearance.

The principle will perhaps be better understood by reference to Figs. 39 and 40, which shows the "Reavell" patent V class duplex compressor. The air is delivered from the low-pressure stage down through the intercooler to the second stage cylinder, and on the down stroke it is compressed, and again passes up through the same intercooler on its way to the high-pressure cylinder—the high-pressure suction pipe being in direct connection with the low-pressure delivery passage. The only modification to the low-pressure valves which this entails is that the valve cover must be arranged to withstand the intermediate pressure delivery pressure instead of the low-pressure as is usual.

FIG. 34. "BROTHERHOOD" FIVE-STAGE COMPRESSOR.

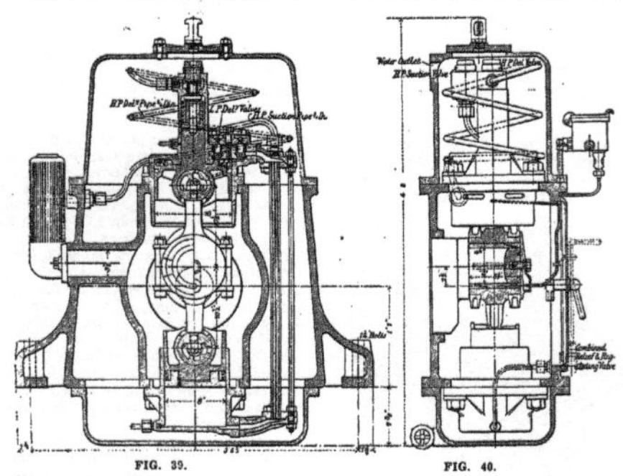
FIG. 39. FIG. 40.

Since the air passes through the same intercooler twice, only one of these is required, instead of two as with other designs, and apart from the simplicity of construction resulting therefrom, the efficiency is considerably increased owing to the fact that the second stage of compression is effected in the cylinder and cooler itself—the reasons for improved efficiency having already been dealt with. The marine compressor under consideration is exactly the same in principle, only a single intercooler being provided.

In the remarks on cylinder clearance volume, it was pointed out that in a Diesel engine compressor it is sometimes advantageous to have considerable cylinder clearances, and this is borne out by a study of this machine. The clearance volume in the low-pressure stage is, of course, reduced as far as possible, the method of securing the piston to the rod being such that the clearance space is the minimum. The intermediate-pressure clearance, however, is very large, being, in fact, the whole volume of the intercooler tubes and connecting pipe; whilst the high-pressure clearance is also very large, being the volume of the coiled aftercooler in the jacket. Apart from the features which have already been dealt with in the notes on clearance, this arrangement allows the high-pressure valves to be located outside the water jacket—a distinct advantage—but, further, the air is cooled to a large extent after compression before it cooled to a large extent after compression before it reaches the delivery valve, which cooling undoubtedly lengthens the life of the cylinder, which may not be the limit of the damage.

When the circulating water is chalky or muddy, the cooling system must be periodically cleaned if the highest efficiency is to be maintained. Valves should be examined from time to time for cracks. It is of the utmost importance that relief valves should be in perfect order, for these are in every way as important as a boiler safety valve, and under no conditions whatever should the valves be wedged or otherwise put out of operation to stop a leak. They should be tested for blowing off at frequent intervals. If stage pressures are rather higher than normal, it means probably, as previously explained, that valves are leaking.

FIG. 31. "BROTHERHOOD" FIVE-STAGE COMPRESSOR. 10,000 CU. FT. CAPACITY.

Provided the pressures are not excessive, no trouble will result from running in this condition, but matters should be put right at the first opportunity. In justice to Diesel engine air-compressors, it must be stated that the frequency with which the engine log book records "compressor trouble" is due in many cases to negligence. There is still a big scope for improvement in the design of multi-stage compressors.

FIG. 33. "BROTHERHOOD" FOUR-STAGE COMPRESSOR.

Modern Machines

Through the courtesy of various makers it has been possible to reproduce several photographs and drawings of some of the latest high-pressure air-compressors, and

FIG. 35. FIG. 36.

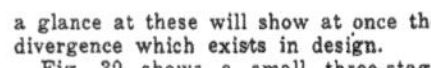

a glance at these will show at once the divergence which exists in design.

Fig. 30 shows a small three-stage compressor, chain-driven, and having a capacity of 600 cubic feet of free air per hour, compressed to 3 000 lb. per square inch. As has been already explained, in small machines the tendency is to reduce the number of stages of compression for a given final pressure, and this statement is borne out by a comparison between this machine and that shown in Fig. 31.

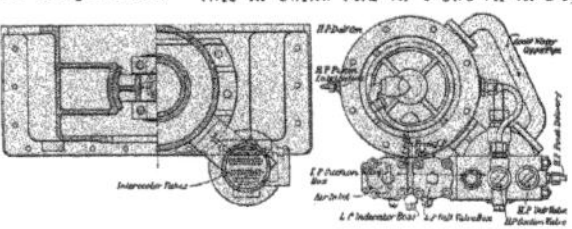

FIG. 37. FIG. 38.

Although the working pressure of this latter is the same, viz., 3,000 lb. per sq. inch, the capacity is 10,000 cub. ft. free air per hour, or more than 16 times as great as that of the small machine; and in this case the compression is divided into five stages. The machine is steam-driven, two double-acting steam cylinders being at the ends and the air cylinders in the centre, while the air piston rods are coupled side by side to the horizontal bridle which is attached at its extremities to the steam line crossheads.

lated oil in the latter, special provision is made for draining.

Another lever-driven compressor as used on engines of continental design, is shown in Figs. 48 to 53, and, as in the previous machines, there are no valves in the second stage. In this case the intermediate-pressure cylinder is located below the crosshead, and a patented feature is the utilization of the end of this cylinder as an auxiliary low-pressure cylinder. Normally the valves in the cover of this cylinder are inoperative, but when there is considerable manoeuvring, and in consequence a big demand for air, such as obtains on entering port, they can be brought into action, either automatically or by hand, when, as will be seen from the sections, this cylinder delivers air into the top of the intercooler at the same time as the main low-pressure cylinder, the whole then passing to the intermediate cylinder in the usual way. By this simple means, the capacity of the compressor can temporarily be increased by about 30 per cent., which means that it will not be necessary to keep such a large auxiliary machine running as would otherwise be the case. As would be expected, very special attention has to be paid to the proportioning of such a machine, on account of the alteration to the distribution of the total work which results from such a device. In this particular machine the high-pressure valves are arranged directly in the cylinder head, access to them being obtained from the engine upper platform through the small covers on the bonnet.

Fig. 55 shows a well-designed machine of a type similar to those used for torpedo work, for a delivery pressure of 3,500 lb. per sq. inch. The capacity at this pressure and 340 r.p.m. is 30 cub. ft. of free air per hour. The machine is a four-stage steam-driven type and is a self-contained unit, the inverted steam cylinder, as will be seen, being placed beneath the air cylinders. Apart from the very compact arrangement which is thus obtained, the drive is direct, so that the stresses in the connecting rods, for example, are much less than they would be with the "side-by-side" arrangement. Another noteworthy feature in the design is the provision of a separate cylinder for each stage of compression, instead of having recourse to differential pistons, which latter arrangement, although possessing many advantages, is decidedly inferior from a thermal point

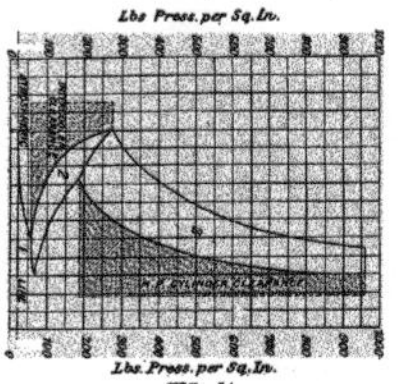

Lbs. Press. per Sq. In.

Lbs. Press. per Sq. In.

FIG. 54.

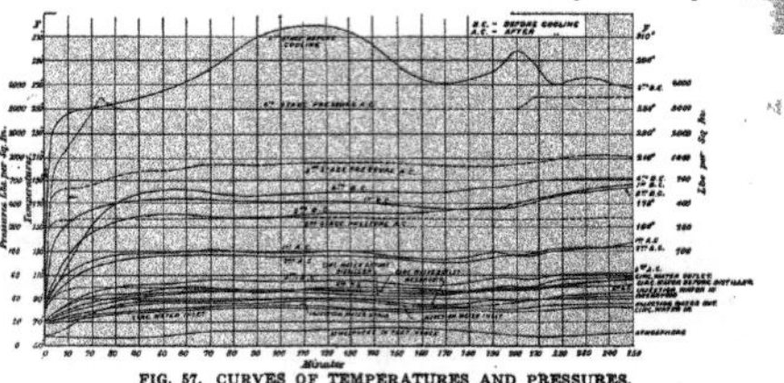
FIG. 57. CURVES OF TEMPERATURES AND PRESSURES.

FIG. 55. "GENERAL" FOUR-STAGE COMPRESSOR.

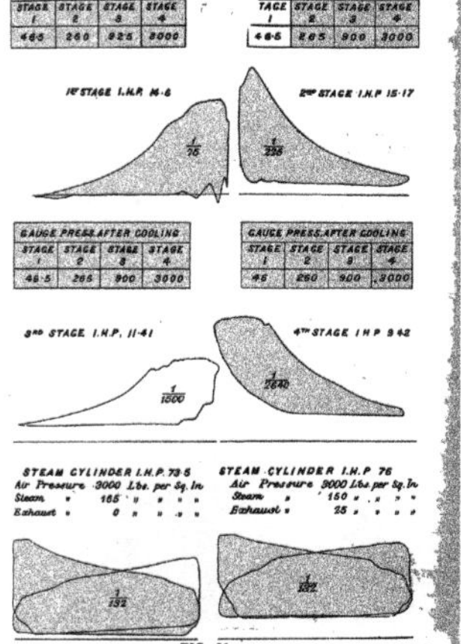

FIG. 56.

latter and reduces carbonization troubles. The theoretical indicator diagram which results from these large clearances and "valveless intermediate" construction is given in Fig. 54, annexed, and applies to all machines of this type.

As far as the actual construction of the machine is concerned, attention may be called to the very effective cylinder jacketing, especially at the breech end of the low-pressure cylinder; also the simplicity of the high-pressure cylinder casting, with its absence of valve pockets, should be noted. It might be noted that one small point which is open to criticism is the method of making the air connection between the jacket and the liner —viz., by the forced-in sleeves, for should a liner by any chance become scored or cracked, it might prove a little difficult to remove the sleeves so as to permit the withdrawal of the liner. This compressor is driven from the forward end of the engine crank shaft, and the stage pressures adopted in this case are: low pressure 60 lb. per sq. inch, and intermediate pressure 295 lb. per sq. inch, when working at its full capacity against a final delivery pressure of 1,000 lb. per sq. inch.

FIG. 42.

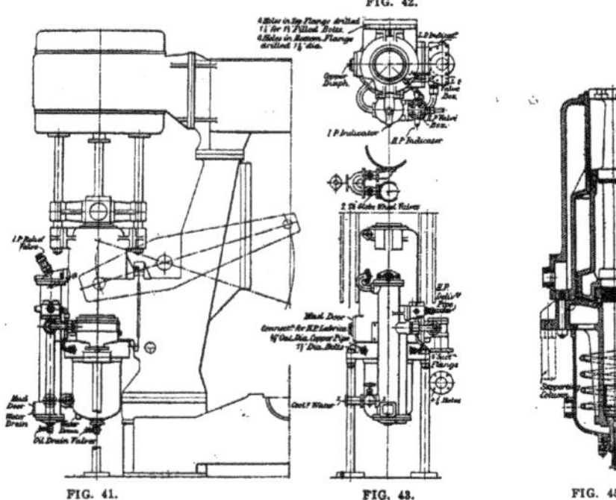

FIG. 41. FIG. 43.

FIG. 46.

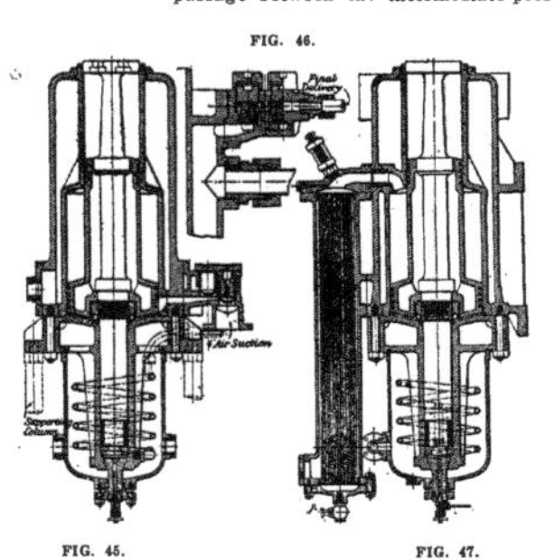

FIG. 45. FIG. 47.

In modern high-power marine Diesel engines of the two-cycle slow-running type, it has become usual, for various reasons, to drive the scavenging pumps by means of levers, in accordance with the well-known marine practice, and in many cases the compressor drive can conveniently be arranged in conjunction therewith.

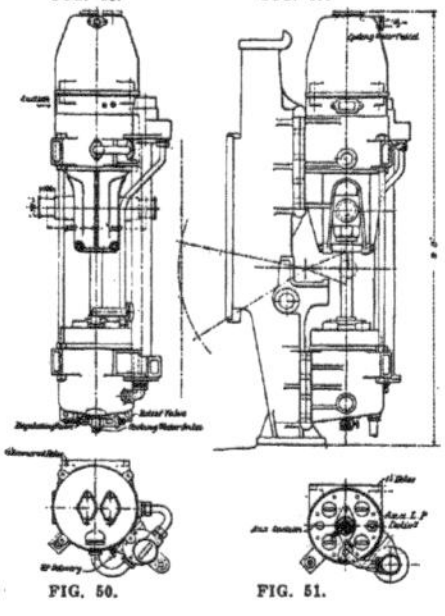

FIG. 48. FIG. 49.

FIG. 50. FIG. 51.

FIG. 52. FIG. 53.

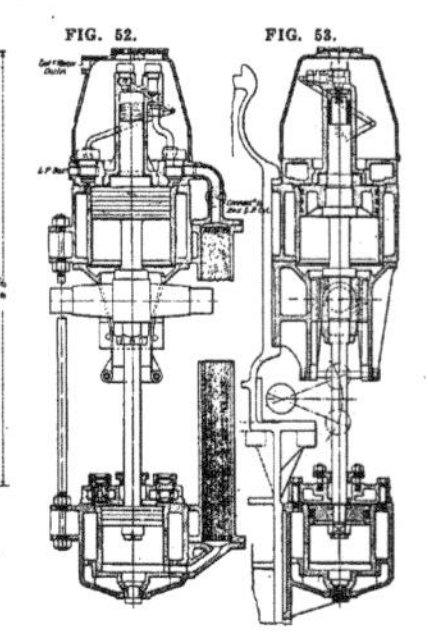

Figs. 41 to 44 show the compressor arrangement which has been successfully adopted on a large British marine engine, whilst sections of the machine itself are given in Figs. 45 to 47. This compressor also embodies the Reavell "Valveless intermediate" patents, the direct passage between the intermediate-pressure cylinder and the intercooler being clearly shown in Fig. 46. The simple construction of the intercooler is worth noticing, the tubes, which are usually of copper, being expanded into the steel tube plates. When the thickness of the latter, gauge of the tubes, and method of expanding are carefully chosen, this system leaves little to be desired, and, further, it permits of a greater cooling surface being obtained, for the same overall dimensions, than would be the case were the usual ferrule system adopted. However, the tubes must be of the highest quality material and must be thoroughly tested before use, for should a tube failure occur, it would be more difficult to repair than in the case of a cooler with ferrules.

The valve arrangements are seen in Figs. 45 and 47, the low-pressure suction and delivery valves being interchangeable, whilst the high-pressure valves, which have removable seats, are also interchangeable. This, of course, reduces the number of spares which have to be carried, in addition to obviating the possibility of mistakes when reassembling after an overhaul.

To minimize the danger, should the high-pressure cooling pipe or any other part under pressure give out, three large bursting diaphragms are provided on the water jacket (two being on the high-pressure bonnet, and one on the opposite side of the low-pressure valve box); these are ruptured, thus giving free egress to the water and air, when the pressure in the jacket exceeds a pre-determined figure. It may be remarked that the arrangement of this compressor, with the high-pressure cylinder at the bottom, is unusual, and to avoid difficulties which might otherwise result from accumu-

AIR-COMPRESSOR TRIAL, JANUARY 11, 1915; ENGINE No. 539. BY THE GENERAL ENGINE AND BOILER CO. NEW CROSS, LONDON.

Gauge Pressures after Cooling. Stage 1	Stage 2	Stage 3	Stage 4	Boiler Pressure.	Back Pressure.	Counter. Revs.	Time, Mins.	Air Inlet Atmosphere.	First Stage before Cooling.	First Stage After	Second Stage Before.	Second Stage After.	Third Stage Before.	Third Stage After	Fourth Stage Before.	Fourth Stage After.	Circulating Water Inlet	Circulating water before Distiller.	Circulating Water Outlet.	Injection Water Inlet.	Injection Water Drain	Pressure.
Pump Up.																						
								deg. F.	deg. F.	deg. F.	deg. F.	deg. F.	deg. F.	deg. F.	deg. F.	deg. F.	deg. F.	deg. F.	deg. F.	deg. F.	deg. F.	deg. F.
42	248	500	500	155	6	1,020	3-0															
46	252	580	1,000	150	7.5	2,380	7-0	57	156	109	250	101	122	83	95	78	68	81	82	78	78	60
46	260	580	1,500	150	8	3,680	11-0															
46	262	581	2,000	150	8	4,850	14-25	56	162	116	250	104	135	88	130	81	74	82	86	78	78	73
48	270	600	2,500	150	9	5,980	17-45															
49	280	640	3,000	155	10	7,040	20-45	56	165	124	150	116	150	91	152	82	76	86	93	78	78	79
Three Hours' Run.																						
49	280	700	3,500	127	10	8,030	23-40															
49	270	800	3,000	140	9.5	14,050	40-0	58	172	128	260	127	157	101	178	92	75	92	94	87	78	87
48	280	826	3,000	140	9.5	18,080	53-0	58	172	128	260	125	157	101	178	92	80	96	99	88	80	85
48	280	910	3,000	135	9	23,600	70-0	61	173	131	310	129	160	106	180	97	81	97	99	87	81	88
48	280	902	3,000	135	9	27,900	83-0	62	172	127	300	125	152	101	175	93	78	93	95	88	84	87
48	285	910	3,000	135	9	33,050	98-0	62	172	128	320	127	162	104	177	97	80	96	96	90	82	88
48	285	925	3,000	137	9	38,170	113-0	63	171	128	320	128	163	104	178	96	81	96	98	90	84	88
48	285	925	3,000	137	9	42,800	128-0	64	171	128	320	129	163	105	180	95	81	97	99	90	81	85
46	285	925	3,000	150	9	53,781	161-0	62	167	120	270	122	163	101	179	92	81	93	99	84	79	87
47	285	920	3,000	125	9	58,664	176-0	62	166	124	270	123	161	101	180	94	82	95	95	85	81	88
47	280	910	3,000	140	9	63,701	191-0	60	168	129	280	130	170	109	185	100	80	99	100	90	83	89
47	275	900	3,000	135	9	67,360	202-0	60	168	127	300	128	163	104	180	98	78	97	98	88	81	89
50	285	920	3,500	150	10	72,306	217-0	62	175	131	270	134	173	110	189	101	79	102	105	95	85	92
55	320	1,050	3,500	155	10	77,276	232-0	64	182	135	275	136	178	112	192	101	81	105	108	97	86	94
58	330	1,040	3,500	195	10	82,256	247-0	66	183	138	265	135	177	113	191	102	82	106	109	98	87	95

Capacity of Machine—30 Cub. Ft. of Air per Hour at 3,000 lb. per Sq. In.

	Average Revolutions per Minute	Circulating Water lb. per Min.	Injection Water. lb. per Min.	Average Steam Pressure lb. per sq. in.	Average Exhaust Pressure lb. per sq. in.
Pump up	388	244	0.6	151	8
Three hours' run	381	248	0.56	146	9

of view, on account of the reduced cooling surface per unit of heat generated. In addition to the effective cooling resulting from the arrangement adopted, intercoolers are provided, and, further, a distiller is fitted which provides pure water for injection into the air cylinderse.

Figs. 56 and 57 and the table show the results of an exhaustive trial carried out by the builders on one of these machines. The curves reveal many points of interest, whilst an idea of the distribution of work through the machine can be obtained from the cards, which show a mechanical efficiency of about 68 per cent.—a very satisfactory result considering the large amount of piston friction and the two connecting rods, etc. The card from the fourth stage would seem to indicate a large cylinder clearance, but it may be pointed out that sometimes the indicator itself adds quite a large percentage clearance to a small high-pressure cylinder—especially when connections are long, due to the necessity of passing through a water jacket. Regarding the intercooling effect as shown by the curves, it might be thought that improvement could be effected. In practice, however, a long continuous run at full pressure is very seldom required of compressors of the charging type—therefore the intercoolers need not be of such dimensions as those required for a continuous duty machine. Again, the question of economy is of less importance.

With the increasing number of applications of high-pressure air, and especially with the growth of the Diesel engine industry, more attention is being paid to the design of high-pressure machines, and in view of the scarcity of trustworthy literature on the subject, the foregoing brief outline of some of the points which arise may serve a useful purpose.

FITTING OIL ENGINES IN OLD SHIPS

By M. M.

A suggestion was made some time ago that all existing sailing ships should be fitted with oil engines, whereby the number of voyages made within a given time could be doubled. One objection that has been put forward to this suggestion is that the majority of the vessels that would be equipped are old ones, and would not be able to withstand the vibration of the motors. This object is now, however, supported by experience, as many old sailing vessels have been fitted with this auxiliary means of motive power. For instance, a vessel built in 1883, which was originally built as a tug, was afterwards converted into a seagoing lighter, and is now a schooner, with auxiliary power. A still more striking example is the brigantine "Tyne," built in 1867, which has recently been fitted with two 50 h.p. hot-bulb engines. It is said that motors have been installed in wooden ships in which some of the timbers were half-rotted through, yet they were put into service and gave satisfactory results. These examples effectually dispose of the objection that the vibration of the motors will prove disastrous. A much more reasonable objection at the present time is that the auxiliaries, being slow compared with full-powered craft, while being visible from long distances owing to their tall masts and sails are an easy prey for submarines. This drawback, however, can be overcome by sending such vessels to remote parts of the globe, where they can relieve full-powered ships for service in the danger zones. In order to effect the conversion of auxiliaries, sailing ships not now so equipped, would mean the suspension of a certain proportion of the work on new full-powered ships, but this policy would be amply justified in view of the more rapid increase of tonnage thereby brought about.

THERMIT

IN the process of melting aluminum by means of the thermit process the slag represents three-quarters of the contents of the crucible and, for the present at any rate, has to be thrown away as useless. Attempts have been made to convert it into a refractory but without success. Thermit products are restricted to the rarer kinds of metals and alloys which cannot be produced readily by ordinary fuel methods. Prior to the war the various compositions necessary for working the process came from Germany, but now that the company has been brought under entire British control, means have been found for producing everything that was formerly produced from materials of German origin, from materials obtainable in this country. This result has necessarily involved a great deal of research, but it has been successfully carried out, and there are now some thirty metals and alloys dealt with in this way. As has been known for some time, the Germans have made use of thermit in connection with incendiary bombs dropped from aircraft, and generally thermit is now a recognized military store. The thermit process for melting metal cannot, of course, be recommended on the score of economy, but where it is necessary to do so quickly in circumstances in which time is more important than cost, then it has an important place in industry. The use of the process for ship repairs and similar work during the war has been invaluable, and it is now a common thing for ships to carry a thermit outfit on voyages.

Most Desirable Size and Speeds of Cargo Ships for Economy

The following article gives the methods by which the author arrives at the most desirable size and speed of any cargo steamer for any desired duty, and the compromises necessary to harmonize the, at times, conflicting interests of owner, designer and builder.

FROM correspondence which recently appeared in a British technical journal on the subject of the best size and speed of general cargo steamers, it appeared to the present writer that there was a desire for some further definition of these qualities, for the different trades upon which vessels are engaged. The conditions induced by the war have made the subject of national importance, as the new merchant fleet which will be brought into being should be such that the best results will be obtained with the minimum expenditure of material, labor, and time. The ultimate object of all interested in shipping is that the vessels engaged shall transport the maximum amount of cargo in relation to their initial cost, and at the cheapest possible freight rate. The maxima and minima of these requirements are not coincident, and it is therefore necessary to compromise between them. The initial cost of construction of vessels has varied greatly during different periods of the past, and this has made it difficult for owners having different sizes of ships built at different periods to estimate their relative efficiencies on a common basis, i.e., an unsuitable ship built in, say, 1908 might, on account of the very low initial cost, be able successfully to compete against another of better design built in 1913. With the object of combining the interests of the designer, the owner, and the builder, and of eliminating a number of difficult corrections, the writer has endeavored to formulate a method by which the most economical dimensions can be arrived at, for any length of voyage, condition of loading, or speed.

General.—The efficient part of a vessel's working life is when transporting cargo from one port to another, and the period spent in loading and discharging, although necessary, can only be considered as lost time. The efficiency of a vessel transporting cargo at a minimum speed (which will be defined later) can therefore be expressed at:

$$E = \frac{T - t}{T}$$

where T is a convenient total period of time, and t is the proportion of T which is required for loading, discharging, docking, &c.

Size of Vessel.—There are many inducements towards building large vessels, the most important of which are: (1) Reduced initial cost in relation to the deadweight; (2) Reduced horsepower, crew, and amount of coal required in relation to the deadweight; (3) Greater seaworthiness. The problem cannot be considered sectionally, however, and an owner is sometimes prevented from utilizing these advantages fully on account of—(4) The length of voyage; (5) The increased length of time required to load and discharge a large ship (which time might for convenience be termed "Detention"); (6) The unsuitability of certain harbors, loading berths, and dry docks; (7) Greater loss in case of mishap or disaster. The lost time suggested under item 5 is of considerable magnitude, as the number of hatches or derrick systems which can be arranged in vessels of equal proportions will vary as $\sqrt[3]{\text{deadweight}}$, or, with holds of equal length, the work to be done at each hatch will vary as B, D, H, where B and D are the breadth and depth respectively. H (see Fig. 6 overleaf) is the average distance through which the cargo has to be transported.

It will be seen from this and the figures given in Table I that the average work to be done at each hatch in a 250-ft. vessel is about 16,000 ft.-tons as compared with 250,000 ft.-tons in a 570-ft. ship.

Items 1 and 2 enumerated at the top of this column represent savings in cost of transportation, whereas item 5 represents loss of invested capital; it follows that the most suitable size of vessel is that which gives the best economic adjustment between these opposing factors. (See Diagram A above).

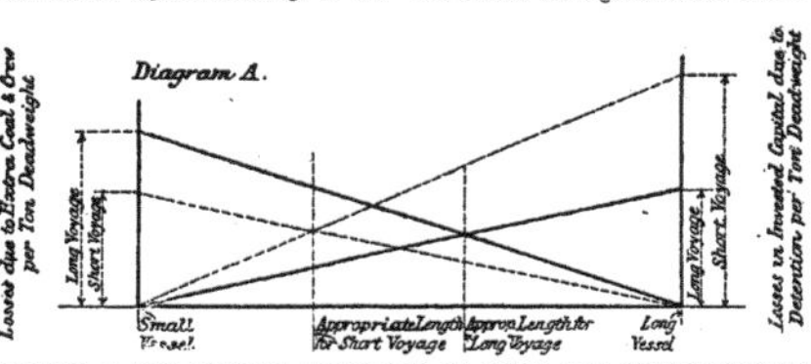

DIAGRAM A.—THE ECONOMIC ADJUSTMENT OF VESSEL SIZE BETWEEN SAVINGS IN COST OF TRANSPORTATION AND LOSS OF INVESTED CAPITAL.

A relatively long voyage or rapid system of loading and discharging will increase the value of E, while an increase in dimensions or an increase of speed for a fixed length of voyage and fixed derrick speed will reduce it, and from these facts it may be presumed that length (or dimensions) may vary in some proportion to E.

Method of Calculating Results.—The problem is to find the dimensions which will give the best compromise between items 1 and 2 and item 5, and in order to obtain information covering a wide field, it was decided to calculate the possible working values of 25 vessels, for voyages of 1,000, 4,000 and 8,000 nautical miles, for two speeds of loading and discharging, and for service speeds of 8 knots, 9½ knots, 11 knots, 12½ knots, and 14 knots.

The designs were divided into five groups of 250 ft., 330 ft., 410 ft., 490 ft., and 570 ft., and each group was subdivided to give the speeds already mentioned.

The arrangement of each size of vessel is shown on Figs. 1 to 5 on the next page; the length of machinery space is for a speed of 11 knots. In the vessels above this speed it will be increased in length, but it was assumed that this increase would not affect the number of hatches or derrick systems.

Each vessel is of the poop, bridge, and forecastle type, these erections extending over about 50 per cent. of the vessel's length. The ratio of length to depth is 13½ times, this being the maximum which Lloyd's will allow with normal scantlings, and it may also be said that this ratio gives the maximum length for derrick arrangements and minimum draught in relation to the deadweight. The breadth has in each case been made equal to $\frac{\text{length}}{10} + 12$ ft.

The number of winches and derrick systems were arranged to give the maximum number of receiving stations on the quay side, and it was assumed in the calculations that the derricks would be worked on the "yardarm" system if two winches were fitted abreast, and by "swinging" if only single winches were arranged. It has been assumed that the bulkheads are placed so that the work to be done will be equally divided between the winches or derrick systems.

*Paper read at the spring meetings of the Fifty-Ninth session of the Institution of Naval Architects, March 20, 1918.

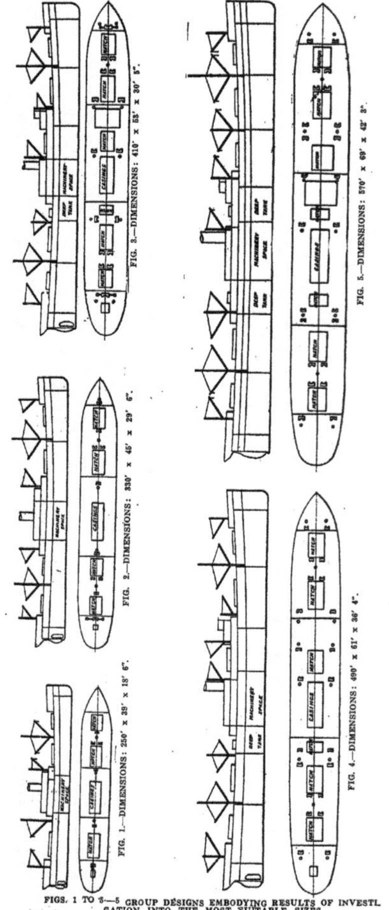

FIG. 1.—DIMENSIONS: 250' x 39' x 18' 6".

FIG. 2.—DIMENSIONS: 330' x 45' x 29' 6".

FIG. 3.—DIMENSIONS: 410' x 53' x 30' 5".

FIG. 4.—DIMENSIONS: 490' x 61' x 36' 4".

FIG. 5.—DIMENSIONS: 570' x 69' x 42' 3".

FIGS. 1 TO 5—5 GROUP DESIGNS EMBODYING RESULTS OF INVESTI-GATION INTO THE MOST SUITABLE SIZES FOR CARGO STEAMERS.

d) The coal per voyage has been es-
ıated from a curve, the ordinates of
ich are 2 lb. per horse-power per
ır at 800 horse-power, 1.7 lbs. at
00 horse-power, and 1.6 lbs. where the
·se-power is 4,000 or more. The coal
port use has been assumed as pro-
:tional to the weight of cargo and the
tance H. It was also assumed that
sels making an 8,000-mile voyage

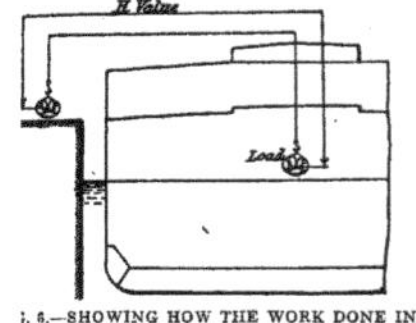

:. 6.—SHOWING HOW THE WORK DONE IN DISCHARGING CARGO VARIES WITH THE BEAM OF THE SHIP.

uld be rebunkered at a port inter-
diate between the terminal ports.

(e) The net tonnage is that obtainable Board of Trade regulations, and in-des the present allowances for pro-ling space.

(f) The money values are not the pre-it ones, but as they are comparatively rect between each size and speed of p, and between initial cost of vessel and cost of transportation, they have no influence on the results of the relative efficiencies. Notwithstanding this statement, however, the effect of increased labor and other rates has been investigated, and the results are given later.

(g) Line No. 20 in Table I. gives the "tons carried" in relation to the initial cost of vessel, and line 30 gives the estimated freight rates. Cross curves of these have been drawn for each stated condition, and Figs. 7 and 9 illustrate these curves for three of the conditions.

Results.—It will be seen from these curves that the ratio of "tons carried" defines the appropriate length most prominently, but, as cost of coal, labor, etc., must also be considered, it was thought necessary to fix the appropriate length line indicated, by interpolation, between the highest part of the "tons carried" curve and the lowest part of the "freight rate" curve. To complete the main investigation, the appropriate lengths of each condition and distance have been plotted on a base efficiency E, and these results are indicated on Fig. 10. The spots on this figure for the lengths appropriate to 8,000 miles and 45 per cent. derrick capacity are greater proportionately than those for 4,000 miles and maximum derrick capacity, the explanation of this difference being that the vessels on the longer voyage have intermediate coaling stations. The difference due to this fact does not, however, affect the results to any appreciable extent.

Figs. 7 and 9 represent the extreme limits of efficiency for the calculated conditions, their relation being $\frac{8,000}{1,000} \times \frac{1}{0.45} = 17.8$ times the loading speed, or length of voyage, and it is thought that from these and the curves on Fig. 10, lengths appropriate to best results for any known or average length of voyage, condition of loading or speed, can be obtained, for vessels engaged in the carriage of cargo only and having no assistance from the revenue obtained by carriage of passengers.

To illustrate the effect of increased cost of coal, labor, and tonnage dues over the originally calculated amounts, additional lines have been indicated on Fig. 8 to show the effect of increasing the cost of these items by 20 per cent., and it will be seen that although this would tend to increase the appropriate length, such increase would not be of great magnitude. Some calculations have been made to find the effect of vessels making return voyages in ballast trim, and it has been found that if the ballast run is estimated as an extended length of voyage, and not as lost time, in calculating the E value, a correct result will be obtained.

Fig. 11 gives the relative "tons carried" and "freight rate" in terms of speed for the appropriate lengths, and these have been plotted as a percentage

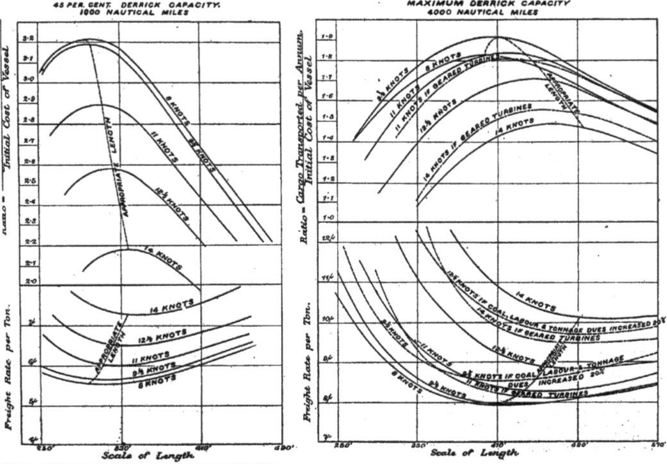

FIG. 7.—RATIO OF FREIGHT RATE TO SHIP LENGTH FOR 45 PER CENT. DERRICK CAPACITY, 1,000 NAUTICAL MILES

FIG. 8.—RATIO OF FREIGHT RATE TO SHIP'S LENGTH FOR MAXIMUM DERRICK CAPACITY, 4,000 NAUTICAL MILES.

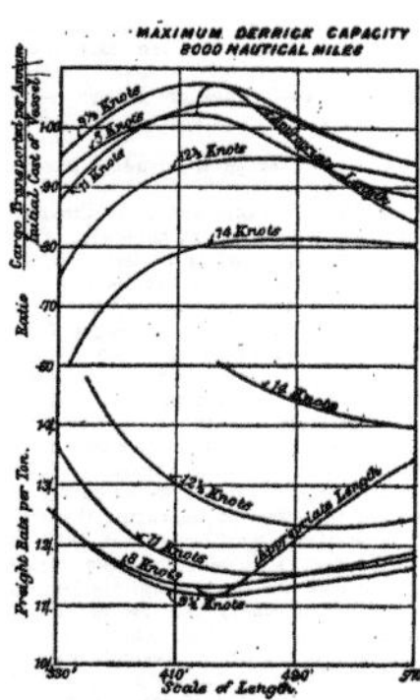

FIG. 9.—MAXIMUM DERRICK CAPACITY, 8,0000 NAUTICAL MILES.

of the best and lowest results. It will be seen that the most economical speeds are from 9 knots to 9½ knots, but it is understood that higher speeds are necessary when cargoes of a perishable nature are carried.

It has not been found convenient to illustrate the curves for each calculated condition of loading, but Table II. is given to show the effect of altered discharging speed on the "tons carried" and "freight rate."

Other Types and Proportions.—The appropriate lengths shown on Fig. 10 will be applicable to other types of cargo vessels, provided that in comparing the dimensions of one vessel with another the characteristics and proportions are similar.

Stowage Capacity.—Stowage capacity will often influence a designer in arriving at the most suitable size of a vessel, and it is found that the stowage rate per ton of cargo for fixed speed is reduced as length is increased; it may, therefore, in some cases be found profitable to choose a slightly shorter vessel than the curves indicate.

Improved Types of Machinery.—The effect of geared turbine installations with double reduction gears has also been investigated from the particulars given on Fig. 12, and the results are plotted for speeds of 11 knots and 14 knots on Fig. 8. It appears that these improved propulsive elements with reduced coal consumption would not appreciably alter the appropriate length of vessel. They, however, affect the earning power considerably, and on account of the present relatively high initial cost of this type of engine for low powers, geared turbines are evidently unsuitable for speeds of less than about 10½ knots.

Conclusions.—It appears from these investigations and the results of actual ships tabulated hereafter that the most suitable length of a "tramp" trader would be between 380 ft. and 400 ft. with a speed of 9 knots to 9½ knots. The maximum efficiency E of the most up-to-date cargo lines does not exceed 0.60, which, with a speed of 12½ knots would have a length of about 450 ft. and a corresponding maximum draught of about 28 ft. In view of this maximum draught, it appears to be an open question whether the proposed deepening of trade routes should be considered prior to legislation being completed for improvements in our harbors, such as increased berthing and loading facilities at quays, increased storage capacity, and the removal of many port restrictions which have caused detention to our ships in the past.

The curves of gross deadweight given on Fig. 13 will be found useful for comparing one vessel with another.

The writer desires to thank the directors of Messrs. Scotts' Shipbuilding and Engineering Company for their permission to publish these results. He also thanks the friends who have advised and assisted him in the development of these notes.

Appendix

To show the possibile use of, and the method of applying the curves on Fig. 10, the following illustrations are given:

No. 1.—A cargo tramp, 400 ft. in length, and having five hatches, each of which is worked by derricks at one end,

APPROPRIATE LENGTH IN TERMS OF EFFICIENCY.

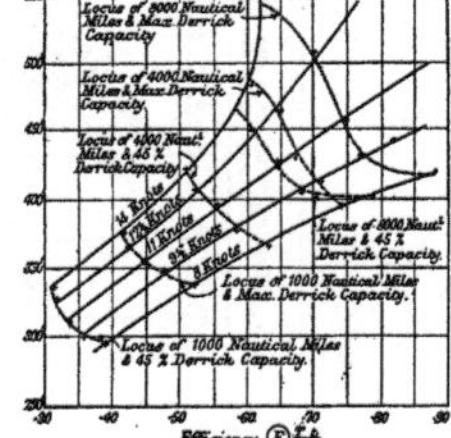

FIG. 10.—APPROPRIATE LENGTH IN TERMS OF EFFICIENCY.

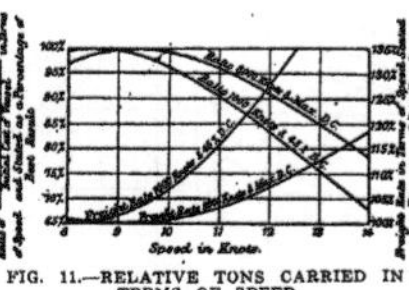

FIG. 11.—RELATIVE TONS CARRIED IN TERMS OF SPEED.

The detail curves for the condition nearest to this vessel's efficiency show that if the vessel had been 23 ft. shorter the "tons carried" would have been increased by 1 per cent. and the "freight rate" reduced by ½ per cent.

No. 2.—Proposed vessel for service between Bombay and Glasgow, or neighboring ports, a distance of 6,500 nautical miles. Speed on service, 11 knots. Full cargoes can be obtained on all voyages, and it can be handled at an average of 150 tons per derrick system per day. Swinging derricks only to be arranged for, and the cargo can be handled over both sides of vessel at Eastern ports, but will be worked to and from a quay wall in Britain. The vessel will bunker before embarking and also at an intermediate port on each voyage. What is the most suitable length?

		Ft.		Ft.		Ft.
Lengths		360		400		440
		Tons		Tons		Tons
Gross deadweight per Fig. 13		5,750		7,750		9,900
Less coal for half-voyage	530		620		710	
Less stores, fresh water, etc.	100	630	120	740	140	850
Net cargo to be handled		5,120		7,010		9,050
Number of derricks on each side ...		8		9		10
Estimated time per trip:—						
		Days		Days		Days
Time to load at Bombay....		4.27		5.20		6.03
Time to discharge at Glasgow		8.54		10.40		12.06
Time to coal on route?		2.00		2.00		2.00
Time for shifts in harbor, etc., and all other delays ..		3.00		3.00		4.00
Total lost time		17.81		20.60		24.09
Total time at sea		24.60		24.60		24.60
Total time per trip		42.41		45.20		48.69
Efficiency E		.058		0.545		0.50
		Ft.		Ft.		Ft.
Lengths per Fig. 10 for these efficiencies		402		389		376

Column 2 gives the best agreement between the trial and the resulting lengths, and the most suitable length would be about 395 ft. long.

No. 3.—Vessel 600 ft. long engaged on a voyage of 5,000 nautical miles, and having a service speed of 12 knots. Thirteen derricks are arranged on each side, and these load or discharge from lighters on each side at each terminal port, each at the rate of 200 tons per day. Is this the most suitable length?

	Tons
Gross deadweight (from Fig. 13)........	23,500
Less coal, stores, fresh water, etc.......	1,900
Net cargo to be handled	21,600
	Days
Time required to load and discharge.... $\frac{21{,}600 \times 2}{26 \times 200}$	8.3

*Estimated from the 400-ft. ship in relation to the deadweight.

Time required for coaling and shifting..	3.0
Average time lost for repairs, etc., per voyage	2.7
Total lost time per voyage	14.0
Total time sailing	17.4
Total time per trip	31.4
Efficiency E $\frac{17.4}{31.4}$ .... =	0.555
	Ft.
Length corresponding to this efficiency from Fig. 10 .... =	412

It is evident that the efficiency will improve considerably with this reduced length, and on this account a length of 460 ft. having 10 winches each side might be tried.

The particulars would then be:

	Tons
Gross deadweight (Fig. 13)	10,700
Less coal, stores, fresh water, etc.	1,200
Net cargo to be handled	9,500
	Days
Time required to load and discharge .... $\frac{9{,}500 \times 2}{20 \times 200}$	4.75
Time required for coaling and shifting....	2.45
Average time lost for repairs, etc., per voyage	2.0
Total lost time per voyage	9.20
Total time sailing	17.4
Total time per trip	26.6
E value	0.654
	Ft.
Corresponding length	460

This smaller vessel would transport 15 per cent. additional cargo in relation to the initial cost, and the freight rate would be reduced by about 4 per cent.

The writer is not aware of any existing vessel of the basis particulars in this latter case, but such a length, speed, and length of voyage has been suggested by Sir John Biles in his paper to the Institution in the year 1900.

Table III. gives the actual working qualities of seven vessels, and indicates how their lengths as built compare with the calculated lengths.

UNSINKABLE SHIPS

By Mark Meredith

No vessel can be made other than relatively safe. The modern steamer is enormously safer than the steamer of twenty or thirty years ago, when the penetration of the single plate, which constituted the skin, meant that the inrush of water could find its way all fore and aft, and to port and starboard. But the unsinkable ship has yet to be built.

That is not to say that naval architects could not design and that shipbuilders and engineers could not construct something in the form of a vessel which would be very nearly unsinkable—that is, which would only sink after it had been destroyed afloat and converted into a mass of unrelated parts. The design of such a craft might be based on one or other of two ideas, both of which have been strongly advocated many times, but neither of them has yet been proved to be practicable.

They both failed because they did not take cognisance of the fact that a ship must do more than float. She must also engage in trading enterprises, otherwise she is only a curiosity of design and workmanship, representing the soundness of a theory, but also the absolute uselessness of that theory for any practical purpose. And although a good deal is being said now about protection from enemy mines and submarines, much of it is valueless because it does not lead to any result.

The first idea is extreme subdivision. Let a ship be subdivided into such a large number of watertight compartments that, no matter where she is struck, the number of those torn open will not be sufficient to disturb the buoyancy of the vessel. And let there be longitudinal as well as cross bulkheads, let all the bulkheads be carried well above the waterline, and let there be no doors in the bulkheads. This, combined with a double skin all fore and aft, the two skins well apart, and the space between cut up into watertight compartments, would, so the advocates of this system say, ensure buoyancy in spite of any imaginable accident. So would the other idea of arranging in or about a ship so much buoyant material as would keep her afloat. The second idea may be dismissed in a very few words. It would be totally impossible to fit in or about a modern steamer as much buoyant material as would keep her afloat once her own buoyancy was abolished, or even partially abolished. A steamer would have to carry a cargo of cork, and even then her own weight would probably be more than the cargo would support. If, on the other hand, the buoyant materials were airtight cases or compartments, the idea is simply that of subdivision under another name.

Subdivision presents difficulties so great that, so far as can be seen at present, it is impracticable, and it has even been argued that subdivision might actually constitute a danger—especially longitudinal subdivision.

A ship cannot be a successful economic unit if she is handicapped by her design or by the number and extent of the safety devices which she carries. Absolute safety is not reconcilable with even moderate efficiency. One of the most recent plans for making ships immune from sinking by mine or torpedo includes the suggestion that all their engines and boilers should be in duplicate. That in itself rules out the whole plan as impracticable. A ship can no more carry about a deadweight of useless machinery than it can carry at some distance from each of its sides a screen of anti-torpedo netting or steel plates. With regard to all safety devices there are two questions that must be asked. If it is a matter of internal design, would it seriously impair the efficiency of the vessel for trading purposes, either by misusing space or reducing speed relative to horsepower? If it is a matter of outside protection, would it reduce speed or spoil the vessel's steering and manoeuvring caapcity? If the answer to either is yes, the plan will not do.

A SHIPBUILDING branch of the Indian Munitions Board has recently been established at Calcutta, and it is believed that a start will soon be made with the construction of small craft that might be useful in Mesopotamia, says the "Iron and Coal Trades Review." So far as Bombay is concerned, nearly a century ago most of the famous East Indiamen were built there by many generations of the famous Parsee family of Wadia, and it was not until the introduction of steam-propelled iron vessels that the industry came to an end. A great effort, however, is to be made to revive it in more than one centre.

NEULAND MAGNETIC REDUCTION GEAR

THIS magnetic gear has been developed particularly for use in ships where the high speed of the turbine must be reduced to match the most efficient propeller speed. It can be designed for a great range of reduction ratios, the apparent limits being about 6 to 1 and 40 to 1, it being practicable to obtain the higher limit with but a single reduction.

This gear is characteristic of a mechanical gear, in that it has a fixed ratio of reduction for which it has been designed, and in this respect it is like a synchronous motor which when supplied with an alternating current of fixed frequency will rotate at a corresponding speed; in the gear it is the speed of the driver, in place of the frequency, which determines the speed of the secondary or driven element. It resembles the synchronous motor also in its ability to carry a certain maximum load before falling out of step with the driver. In design, therefore, a suitable overload margin must be allowed above the normal operating load.

Experimental Arrangement

In order to corroborate the working of this principle an experimental machine, shown in the accompanying engraving, was built, and to save time a stator and other parts on hand were used in its construction. The machine has a field with six poles, thirty-three bars in the secondary, and twenty-seven teeth on the stator. The reduction ratio is, therefore, 5½ to 1. The inner stator indicated at B in Fig. 2 was omitted from this machine.

Operation

The operation of this gear is indicated on the accompanying chart, Fig. 3, in which the curves indicate the maximum load the machine will carry with corresponding amper turns per pole. Curve 1 represents the maximum torque possible for which the machine is designed; curve 2 represents the torque delivered to the propeller shaft or low speed side of the machine, and curve 3 the corresponding torque on the high speed side. It will be noted that the first part of the curve resembles the torque characteristics of a series motor, while the latter part is much like a saturation curve and indicates that by increasing the pole and stator sections the increased force beyond the bend of the curve would be made available for increasing the flux density at the air gap and, therefore, adding materially to the torque of the secondary.

The principle of the gear can best be understood by reference to Fig. 2. The driven or primary field element A is rotated by the turbine shaft. This field element, which consists of two poles, is energized by some suitable source of current. Surrounding it are two stationary cages or stators B and C, the stator B consisting of laminated bars of sheet iron mounted on non-magnetic supporting rods, while the stator C is built up of laminations, assembled and supported, as is usually the custom.

The rods and teeth are magnetically energized, as they come within the field of the primary element, so that as the latter rotates it sets up a wave of magnetism that sweeps around the stators. Between the stators is the driven member D, also made up of bars of laminated iron.

FIG. 1—ASSEMBLY OF NEULAND MAGNETIC SPEED REDUCTION GEAR.

It will be observed that there are 64 teeth on each stator, while the secondary or driven member has 66, so that at only two points are the teeth of the stators and secondary in alignment. With the parts at rest and under no load, this

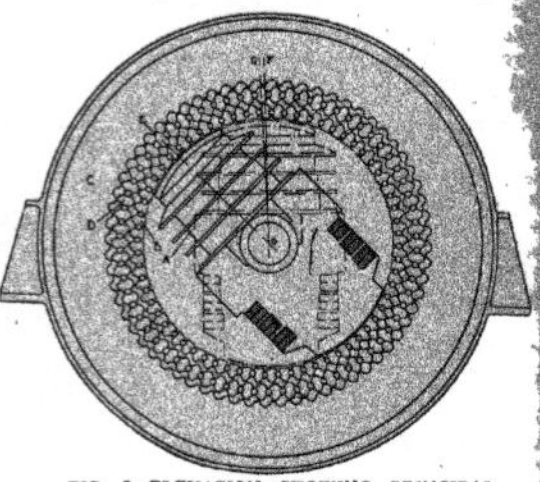

FIG. 2—ELEVATION SHOWING PRINCIPAL ELECTRICAL FEATURES OF GEAR.

alignment will be formed in the field of greatest magnetic flux—in other words, directly in line with the poles of the driving member, as shown at E in the diagram.

If the driven member were held stationary while the driving member is rotated to the position shown in broken lines, the teeth of the stator and the secondary will be out of alignment and the magnetic flux will set up a torque between the teeth, which will only be satisfied when the secondary is permitted to rotate far enough to bring its teeth into coincidence with the teeth of the stator, this point corresponding to position of greatest magnetic flux.

In other words, while the driver is moving through the angle A O F the driven member moves through the angle G O F. In a complete rotation of the driver there will be a movement of the secondary through the space of two teeth, and there will be a reduction in speed of 33 to 1. Of course, the driven member will lag behind the driver by an amount depending upon the load it has to carry.

Reduction Ratio

It will be noted that the reduction ratio is equal to the number of secondary bars divided by the number of poles in the primary, while the number of stator teeth in each cage must be smaller than the number of secondary teeth by the number of poles in the primary. It is also evident that if the number of teeth in the secondary is greater than that in the stator the secondary will rotate against instead of with the driver.

The torque on the secondary is not exerted by the primary, as in a slipping magnetic clutch, but is obtained by the pull of the stator teeth upon the secondary bars, the primary merely furnishing the magnetic force.

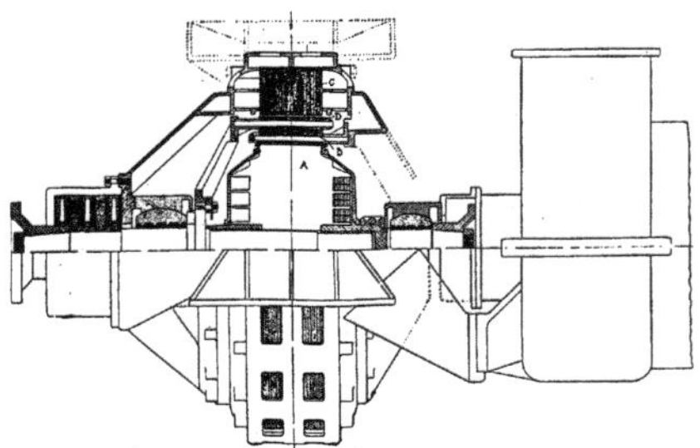

FIG. 4.—LONGITUDINAL SECTION SHOWING DETAILS OF GEAR.

The secondary and stationary teeth form as many alignments as there are poles, and if the driver is provided with more than two poles the ratio between the secondary and the stator teeth must be such as to provide the same number of alignments.

Full Size Design

Based on the tests of the experimental machine, designs have been prepared for a full size unit, which is shown in Figs. 2 and 4. This machine wil have the following characteristics: Normal horsepower, 1,500; overload margin, with normal field, 60 per cent.; speeds, 3,600-109; reduction ratio, 33-1. The design differs from the experimental machine, in that an additional stationary cage is inserted between the field and secondary, the purpose of which is to increase the torque on the secondary. Since the torque on the inner and outer circumference is the same, a torque practically double in value is produced on the secondary without increasing the core length, and, in spite of the fact that the field ampere turns must be increased to supply the necessary force for the added gap, a substantial reduction in weight and an increased efficiency result.

The gear has two poles, sixty-six secondary bars and sixty-four stationary bars and teeth. The field has been especially developed to fill the requirements of high speed, balance and simple machining. It is circular in section, and can practicallv be finished on the lathe and boring mill.

Field Winding

The field copper, wound directly in circular grooves turned in the core, comprises a bare strip insulated from the core by substantial bakelite plates and between turns by a strip of fish paper. The insulating plates are of such thickness as to wedge the strip in the groove under tension, so that once wound and held by a bronze wire hand the considerable centrifugal forces will be unable to displace it and thus cause unbalancing. The separating flanges are for the purpose of forming a plurality of grooves to prevent a cumulative strain on the outer flange and to make possible a strip conductor of reasonable width. Each groove contains two layers separated by a bakelite plate and wound in opposite directions so as to bring all leads to the outside for easy connection.

Particular attention has been paid to the design of the cages. Their arrangement combines the greatest possible section in the laminated bars with maximum strength of the non-magnetic supporting rods. A non-magnetic ring having axial holes drilled through it, into which the rods are pressed, serves as the supporting member for the laminated bars, each of which is forced between adjacent rods and insulated from them by a fibre tube on the rods one-sixteenth inch in thickness.

The rings in turn are screwed to the spider of the secondary and the stationary end brackets of the stator respectively. A narrow ring, insulated from the rods, ties each of the cages on the other side, and thus makes them thoroughly rigid. Screwed to the spider of the secondary cage is the main or driven shaft, which extends through the bore in the field, so that the field is partly supported by and rotates around the main shaft on bearings provided on either end. The two main bearings, one at the end of the main shaft and the other on a stub extension of the field sleeve, support the weight of the primary and secondary and thereby make the gear unit independent of the turbine and propeller shaft bearings.

Mechanical Details

Oiling of the main, and particularly of the internal field bearings, is by pressure circulation, the oil being introduced into a circular groove in the centre of the left hand main bearing through a hole; from the groove into the centre hole of the main shaft through several evenly spaced radial holes; out at the right end of the shaft and back through axial grooves in the bearing surfaces into a circular pocket of the spider and through holes back into the reservoir of the main bearing.

For the sake of simplicity and unit system of assembly the turbine is screwed to arms extendin g from the gear bracket and the thrust bearing is made a part of the gear. An extension of the main shaft carries the thrust flanges, while the thrust collars are supported in a housing, so that the thrust is transferred to a specially rigid bracket, to the legs of the frame and from there to the supporting beams of the ship.

In order to reduce the windage losses to a minimum the high velocity field member is entirely enclosed by aluminum plates on either side screwed to the field and a thin strip fastened to the plates and enclosing the outer circumference, so that externally the whole resembles a smooth drum and offers as little resistance to rapid rotation.

The weight of this unit, including the thrust bearing, is estimated at 24,000 lbs., which, it will be noted, is less than 50 per cent. the weight of a mechanical gear for similar duty.

The losses are made up of the exciting loss, which in this design is normally 4.5 k.w., or about 1/3 of 1 per cent. the iron loss and the windage and friction losses, and the total should not exceed 2 per cent., making the efficiency 98 per cent.

There are many ways of figuring the size of a feed pipe but the best practice for an engineer to follow is to install a pipe of the same size as the connection provided in the boiler. A feed main should be increased in proportion to the size and number of feed connections and it is well to make an additional allowance of 25 per cent. for friction, etc.

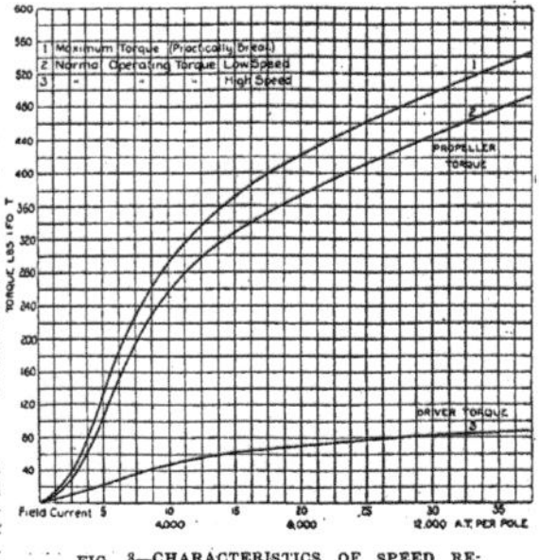

FIG. 3—CHARACTERISTICS OF SPEED REDUCTION GEARS.

MARINE FOUNDRY PRODUCTS OF MESSRS. YARROWS, LTD.

DURING the past few months Messrs. Yarrows, Ltd., Victoria, B.C., have been making a large number of interesting castings.

They have an order on hand for making thirty cast iron propellers 14 ft. 6 in. dia., weighing aboubt 6½ tons apiece. Seventeen of these have been completed, and the accompanying illustration (photograph No. 1) shows one of these propellers being shipped for installation on one of the Imperial Munitions Board vessels. These propellers are swept up in loam, and as many as four propellers have been cast successfully off the same mould.

In addition to these propellers, the firm has turned out several large marine engine castings consisting of H.P., LP. and L.P. cylinders and bedplates for engines 24 x 38 x 62 x 42 in. stroke.

One of the illustrations shows a large L.P. cylinder, the casting of which weighed 8½ tons. The H.P. and LP. for this same engine weighed 4½ tons and 6 tons respectively, the bedplate being cast in three sections, the total weight being 17 tons.

Another illustration shows one of the H.P. cylinder castings which Messrs. Yarrows, Ltd., are making for the United States Emergency Fleet Corporation. These castings are being turned out at the rate of one a month, and weigh approximately six tons.

Another illustration shows a large cast iron nitric retort made in one piece, and weighing 8 tons, this retort being 12 ft. high and 6 ft. ¼ in. inside diameter.

CAST IRON RETORT, MADE BY YARROWS, LTD.

All kinds of marine work is undertaken by the firm and a specialty is made of manganese bronze propellers, these being urned out weighing as much as 8,000 lbs.

Four manganese bronze blades were turned out by Messrs. Yarrows, Ltd., in the short time of five days, each blade weighing about 3,500 lbs. This performance was very creditable as no previous preparations had been made, the job being for an emergency breakdown.

MARINE CASTINGS MADE BY YARROWS, LTD., AT THEIR FOUNDRY AT ESQUIMAULT, B.C. LEFT: H.P. CYLINDER; RIGHT: L.P. CYLINDER; UPPER: PROPELLER

The MacLean Publishing Company
LIMITED
(ESTABLISHED 1888)

JOHN BAYNE MACLEAN - - - - - - President
H. T. HUNTER - - - - - - - - - Vice-President
H. V. TYRRELL - - - - - - - - - General Manager

PUBLISHERS

MARINE ENGINEERING
of Canada

A monthly journal dealing with the progress and development of Merchant and Naval Marine Engineering, Shipbuilding, the building of Harbors and Docks, and containing a record of the latest and best practice throughout the Sea-going World.

B. G. NEWTON, Manager.
J. M. WILSON, Editor. A. R. KENNEDY, Asst. Editor.
Associate Editors:
A. G. WEBSTER J. H. RODGERS (Montreal) W. F. SUTHERLAND

OFFICES

CANADA—
Montreal—Southam Building, 128 Bleury St., Telephone Main 1004.
Toronto—143-153 University Ave., Telephone Main 7324.
Winnipeg—1207 Union Trust Building. Telephone Main 3449.
Eastern Representative, E. M. Pattison.
Ontario Representative, S. S. Moore.
Toronto and Hamilton Representative, J. N. Robinson.

UNITED STATES—
New York—R. B. Huestis, 111 Broadway, New York, Telephone 8971 Rector.
Chicago—A. H. Byrne, 900 Lytton Bldg., 14 E. Jackson Street. Phone Harrison 1147.
Boston—C. L. Morton, Room 733, Old South Bldg., Telephone Main 1024.

GREAT BRITAIN—
London—The MacLean Company of Great Britain, Limited, 88 Fleet Street, E.C. E. J. Dodd, Director, Telephone Central 12960. Address: Atabek, London, England.

SUBSCRIPTION RATE

Canada, $1.00; United States, $1.50; Great Britain, Australia and other colonies, 4s. 6d. per year; other countries, $1.50. Advertising rates on request.

Subscribers, who are not receiving their paper regularly, will confer a favor by telling us. We should be notified at once of any change in address giving both old and new.

Office of Publication, 143-153 University Avenue, Toronto, Ontario.

Vol. VIII JULY, 1918 No. 7

PRINCIPAL CONTENTS

OUR NEED FOR SHIPS

THE argument has several times been advanced in these columns that Canada must consider the establishment of a shipbuilding industry on a permanent basis if we are to place our products, and particularly our manufactures, in those markets where they can be disposed of to the best advantage. Foreign nations may seek out agricultural products with their own ships but they will want to bring manufactured goods to our shores, not to take them away.

This argument is appealing in the United States in a somewhat different way. In an address before the Illinois Manufacturers' Association, Edward N. Hurley, chairman of the United States Shipping Board, declared that the United States would be in such a favorable position as regards tonnage after the war that it was up to the manufacturers to increase their output so that the ships might be operated to advantage.

If the United States has plenty of ships after the war and Canada has but few it does not take a great deal of figuring to arrive at a conclusion as to which country will be in the best position to develop foreign trade.

COAL AND COLD

ADVICES from all coal producing centres in Canada and the United States reveal a similarity of opinion regarding the coal situation which demands all the attention and action which it is possible to give.

Some time ago it was stated that no coal would go to Montreal from Nova Scotia this summer for the simple reason there are no ships available. While there may be some small amount taken up by rail, the great bulk of the fuel supply must be water-borne during the shipping season.

It is now stated that the Nova Scotia coal output continues to decrease. Commenting on this the Sydney correspondent of MARINE ENGINEER writes:

"For the first six months of 1918 the production has fallen off from the record of the first six months of 1917 by about 330,000 tons, and it is only too probable that by the end of the year the total outputs will be less than those of 1917 by about 330,000 tons, and it is only too probable that by the end of the year the total outputs will be less than those of 1917 by almost half a million tons. At the same time it is hoped the rate of decline in the last half of 1918 will not be so rapid as it was during the last six months. That is the best that can be hoped for."

The problem of coal production is not the least of the problems that face the Allied leaders, and if the coal production declines to a point where it restricts the output of munitions and the transport of troops the gravity of the situation will appear in its true light. According to our authority quoted above things are approaching such a point that if activities vital to the success of the war are touched the private consumer will have to freeze, if need be, because the needs of the army and navy come before the requirements of people at this time.

ALL FOR NATURE AND NATURE FOR ALL

BEFORE another winter passes the possibilities attending a real coal famine will be brought home to many of us. It may be that personally we will not be physically inconvenienced but indirectly there will be many things to remind us that conditions are different from what they used to be, and careful thought will impress on us the truth that they never will the same again.

Political crises may come and go, questions involving taxes, tariffs, coinage, and social considerations generally may be decided and re-decided to meet the whims and fancies of parties in power so that ill-effects can be re-adjusted and inconveniences removed. But—interference with and misuse of natural resources bring their own reward. Nature out-Shylocks Shylock because she always gets her pound of flesh even though she may have to wait a thousand years for it.

Indiscretions of humanity in utilizing nature's resources have cost mankind untold sums and it is not putting it too strongly to say that the future conditions of existence of the human race are being decided now, and the decision is in two stages. First, is or is not man going to so conserve naturally recurring sources of power so that the human race will be independent of limited supplies of fuel? Second, are these sources of power to pass into the control of a select circle of individuals or be retained by the State as a national asset for the use of the people?

Suppose some clever, ambitious individual devised a scheme for shutting off sunlight and put up arbitrary rates for its supply irrespective of what his plant investment cost—in other words, got all it would stand. How long would the people stand for it? A reasonable amount of light and heat is necessary for every individual and as long as it can be obtained from any source whatever, it is a moral obligation of all responsible governments to see that it is supplied.

Fo'c'stle Days
or
Reminiscences of a Wind Jammer
By Capt. Geo. S. Laing

The steamer and the motorship have deprived the seafarer of much of the old-time romance attaching to a sailor's life. The advent of wireless has also increased the safety of it. Deprived of these features, a sailor is immediately dependent on his own resources, and Captain Laing gives us a true and seaman-like reflection of sea life under canvas. The present feverish activity in constructing self-propelled craft affords a suitable background for incidents in a career in which gaffs, booms, tar and canvas held sway before being displaced by mechanical unloaders, derrick masts, fuel oil, coal and wire rope.

Part VI.

Oases for Shipwrecked Seamen

Sometimes an Antarctic whaler will land a crew on the larger islets of the Southern Ocean, and the Governments of several maritime nations have put stores and built rough shelter on these outposts, to succor any forlorn sailors who may be cast up on their shores. Clothing, bedding and food are coopered up in good casks and a coat of tar and sand put over all. Then with roped tarpaulins and a rough boarding the material is left to fill its mission of mercy and cheer if so needed. Small stores, such as matches, needles, twines, nails, etc., are not forgotten, and conspicuous finger-posts on different parts of the rocky beaches give directions and valuable information as to local peculiarities. Any unlawful use of such stores would be very properly treated as a criminal offence, and where vessels have had need to use the stuff the nation to which they belong think it an honor to replenish the stock, generally through the medium of a gunboat or small cruiser from Cape Town or Freemantle.

Wild Ocean Scenery

The "Maggie Dorrit's" boys were beginning to be quite seasoned to heavy weather, and oilskins and long leather sea boots became the standing regalia. For two days we might be running at a speed of 11 miles an hour under a main topgallant sail, then a day or so under reefed upper topsails and foresail; and sometimes that had to be rolled up on the jack-stay and marled down with extra gaskets, whilst the two lower topsails with a dead fore-topmast-staysail were the only "rags" left on the barque as she cut her way along the lower fringe of the globe, chased by watery hills that smoked and roared their "freedom of the seas."

Very little "heaving to" is done in this district, as the sea-room is more than ample and the gales mostly favorable. The grandeur of heavy winds and seas must now be told. When a vessel is "running heavy" before a high-wind and sea there are three places from which one may view the sights and feel the thrills in comparative safety. One place of vantage is the poop, another is the fo'c'sle head, and the third is from an upper topsail yard, seated in the bunt and hanging on to the tie.

We were running now for all she was worth, and the poop ladder was scaled. The wind blew 50 miles an hour, and that velocity was raised to 60 in the squalls that came along each hour or so. Such squalls even in the glare of noonday are clothed in celestial scud sufficient to cause a temporary darkness. This phenomena adds to their fiendish howling and whistling as the invisible force tears through sheave holes and makes crowbars out of chain sheets. The shrouds and backstays seem to emit a metallic sound not unlike a cat's concert answering a street bagpipe artist, whose love of national medicine has affected his "expression."

One cannot face the wind and speak, and it is only by means of the rail or lifeline that you can hand-over-hand yourself to the wheel-box gratings, where the helmsman is "grinding up salt water for the captain's ducks," which is Jack's nonsensical phrase for steering. What a thing of life and beauty a sailing ship is! The sea was literally lashed into steaming froth or spindrift, which the gale at times picked up and hurled on board, as if it was grape-shot out of a cannon's mouth.

Down went her stern till the next big wave towered thirty feet above the taffrail. Then came an irresistible thud as the rolling water mound picked the frail craft up on its gurgling bosom and practically threw her ahead. When the wave was right amidships the vessel shivered heavily, for her bow and stern were, for the moment, without support. Meantime the crest water fell inboard at her waist, tumbling in cascades over both rails and filling the well deck fore and aft. Thus the "Maggie Dorrit" staggered on, blown, shoved and batter-rammed to the eastward, hoping that Australia's Great Bight would soon act as a breakwater and hand out more gentle samples of meteorology.

From the fo'c'sle head, which was as dry as a bone when wind and sea were aft, a magnificent spectacle was also to be seen. The shelter of the reefed foresail made it very comfortable here, and as she pointed her jib-boom heavenwards and with a return motion fell head down into the next valley the horizon beyond was hid from view for the moment and a wall of sloping green and white water lay in our path. Our speed now was over 12 miles an hour, taking 8 from the wind and 4 from the scend of the sea. A good helmsman was now the most important man in the crew, for a vessel in a heavy seaway behaves to a great extent in accordance with how she is sculled or guided.

To "broach to" at a time like this would mean perhaps losing our spars or foundering through the hatches giving in under-sea poundage. A peculiarity of "running" before wind and sea is noticed in relation to rolling. For perhaps ten minutes the vessel will act like a half-tide rock in a ground swell, practically wallowing under water from the break of the poop to the foremast, and rolling, laboring and moaning as if in her death struggles. Then all of a sudden she will clear herself by means of scuppers and ports and sit on her keel as steady as if she was on the blocks of a dry dock. Only for half a minute perhaps, but in such a seething turmoil this is a picture that arouses admiration, as well as appreciation, from those who may be watching a "slant" to get from one end of the ship to the other without being floated over the topgallant rail into Davy Jones' locker, whose maw is never full. Then suddenly as if she lamented her good behavior, the centre of gravity will renew its overtures, and equilibrium gone mad will make our ship's waist into a mill race.

Scene From Aloft

From the fore upper topsail yard the gale scenery was perhaps more inspiring than the view obtained from the deck. The rolling motion was more pronounced of course, and looking down on the ship as she plowed along can never be forgotten. The sea appeared to be flush with both rails and the roar of the wind through the network of masting and tackles was a song from the wilds. When looking aft at the canvas on the mainmast one noticed the graceful cut of the topsails, showing that the geometry of a good sailmaker is essential for equal distribution of wind stresses. The roach in the foot was as true as a rainbow, while the perpendicular seams, reef bands, clew clothes and earing squares were as nicely depicted as a blueprint could show them.

The tumbling of the seas on deck modulate the screeching of the wind god, so that shrill pitches and muffled sounds play their part. Then a little sunshine and a patch or two of "dungaree" sky puts the sea picture in the footlights, but not for long, as a black squall with clouds low and threatening to cover the trucks can be seen rising over the mizzen top.

Looking down from squared yards one can also see the model of the ship. The "Maggie Dorrit's" greatest beam was just a little abaft the foremast, and from that point of greatest breadth she gracefully tapered away to the stern. Where a vessel lacks this point of naval architecture she invariably carries dangerous water on her quarters. She will, however, carry more cargo, but again she will be slower.

Look at our favorite albatross hanging on to nothing just above the mizzen truck. Legend has it that these birds are the restless souls of drowned seamen and less beautiful reveries have been given philosophic thought. One thing is certain, the lonely waters on which we now sail would be very desolate without the pinioned crew.

PART VI.

A TYPICAL MIDDLE WATCH

WE came on deck at midnight and mustered around the quarter-deck captain. After the names were called and the order given to "Relieve the wheel and lookout" those who could generally made for the galley. One of the apprentices had to keep his two hours on the lea side of the poop handy to the officer, and the duties of this unfortunate were many. He could be called upon to light the binnacle lamp every half-hour that the helmsman might see his course. Then the standard compass half-way up the mizzen-mast frequently needed attention. This was quite an acrobatic job as one had to go out on a foot rope stretched between the swifter shroud and the mast to get at the standard. At the time I write of, these lamps as well as side-lights generally burnt colza oil when the ship was clear of traffic and colza has really to be persuaded to burn. The boy who kept the lee poop had also to strike the bells every half-hour, which was answered by the forward lookout.

Then the officer might require a bucket of sea water drawn to take the temperature, with the result that you were sent forward with an order to the fo'c'stle head man to "keep his eyes skinned for bergs." On the way alon a chunk of salt water flops over in her waist and knocks you down on the slippery teak decks. As you pass the galley door the tail-end of some tobacco juice enters your eyes which are already heavy with brine and broken sleep. And so the time wears on.

Just as four bells comes in sight a heavy squall looms up from aft and the mate whistles for the spare hands to "Stand by the top gallant halyards." When this squall hits her the wheelsman is cautioned to "Keep her dead before it." Everything aloft cracks like fire kindling and just as the officer looks astern to see if the squall is thinning, the old man pops his bald pate up the companion scuttle, gives a snort of satisfaction and turns below again to rest. The main brace bumkins are just outside of the old man's port and when a squall hits the "Maggie Dorrit" he knows it even in his sleep.

More Stomach Trouble

To square up old acounts with the steward a moonless middle watch was picked out for chicken stealing. The coop was lashed on top of our half-deck and Fid, the oldest apprentice (or ancient mariner) acted as master plunderer. The fowl had to be killed, plucked, cooked and eaten within the watch. As the bird was being caught the other three boys made various noises in the lamp locker and boatswain's stores, both these rooms being situated at the after-part of the half-deck house. Tilbury and myself were promised a share of the "grand feed" as we turned out at 4 a.m. but we haven't got it yet for here is what happened:—

Fid was scared after he had killed the bird and thought that chopping its head and feet off was sufficient preparation for cooking, so into the oven, feathers and entrails too, went the hen. Water and grease were poured over for basting, but soon the smell of feathers became pungent, although every one agreed that the wind being aft would prevent the mate from getting a scent.

The smoke, however, soon cleared the whole watch from their comfort in the galley and Fid became nearly demented. Then some joker said "Why not boil it" and in a jiffy the "grand feed" was flopped into the coffee pot which stood its watch on the railed top of the stove. With the smoke nuisance gone, some hearts throbbed easier, but old "Rope-yarns," the mate, with his sea voice, thundered the order along to "lower away the main top-gallant halyards." The chain splice in the sheet had carried away and the sail was in danger of being shook to threads while the yard might have sprung.

Repairing the Damage

The things that were said about that top-gallant yard and sheet would hold a printing machine stock still. After taking the wind out of the sail with clew, spilling and leech lines the work aloft went slow. Extra rovins had to be passed round the jack-stay while the sheet had to be respliced in the top and rove through its wandering course of sheave-holes, lead blocks and fair-leaders to the deck. With a stolen chicken in the pot, this ordinary seamanship took extraordinary time to execute. The old barque even aggravated matters by rolling almost continuously as if she enjoyed the situation.

Our cook, alias Dr., generally came to the galley at four-thirty each morning, but the same devil that broke the top-gallant sheet also broke the cook's slumbers, for even with the wind aft and the mainsail furled, words that are not within hail of the most coveted dictionary came floating up to the hard-working fellows on the mainmast.

It was eight bells before the sail was set again and Tilbury and I were watering at the mouth for a taste of chicken, having even dreamt of a princely banquet. As all hands tailed on to the halyards some one whispered, "Was it good?" while another more despondent and knowing creature answered "yes, the smell was."

Of course every living soul on board was innocent for the time being, but in broad daylight as the old man took his chronometer sights it was very apparent that more than a top gallant sheet had been broken during the night. After breakfast was over the four boys were called aft, not however before we had made up our minds to all share in the blame and, of course, punishment. The court martial was staged below in the cabin and as we shuffled aft in Indian-file the grins of the steersman and watch officer would have baffled an actor sworn to solemnity.

The Verdict

At the bottom of the companion steps the flunkey stood guard with a visage as controlled as that of a veteran butler in the Royal household. The old man's voice could be heard in the cathedral gloominess and it said plainly: "Send them right in, Steward." As we lined up athwart the saloon deck the steward carried in the circumstantial evidence on a platter. It would appear that the old skipper could not contain himself, for he still lingered in his stateroom out of sight. Every second that we stood waiting for our judge seemed like an hour and if the captain hadn't eventually come out of his cabin the chances are that we would all have burst into uncontrollable fits of giggling.

There was the chicken, feathers and all, dyed with coffee, and swollen to the dimensions of a poisoned pup. It looked like a ball of oakum or a plum duff that had been in a melee on liberty night in Iquique.

The longer the captain called on composure to enshroud him the worse his dilemma became and it was quite evident that he was even short of breath in rushing out of his cabin and hurriedly shouting at us, "Give the horrid mess to the pigs, and after this let my private property alone or I'll ropesend the whole bunch of you." As we climbed up on deck Fid was was heard to whisper: "I wish he had let us take it to the half-deck." Our pig-pen was jammed under the main-stay at the mast-coating of the foremast and thence we trudged in processional manner, the butt of the whole crew. For the rest of the trip the different brands of "Dutchmen" in the fo'c'stle poked fun at the apprentices by asking varied queries concerning "dot domd schickken."

St. Paul and Amsterdam Rocks

The old barque wallowed along past the 78th meridian of east longitude with the above mentioned rocks away to the northward. The novelty of sighting the Australian land in a fortnight became a strong, fascinating hope to boys who were on their first lone trail. We tried to guess what had happened in our far-off homes in the last three months, and the receiving of mail from that magnetic quarter alone was sufficient to build up a youngster's curiosity to breaking strain. Having only sighted one small island off Brazils since leaving Hamburg, Germany, excepting, of course, the south coast of England, the reader can perhaps imagine what deep-water sailing means.

Tilbury wondered what sort of life he would have in the land of the kangaroo sheep and gold. The wash of the ocean water on our decks had caused a green slime to almost approach the status of grass and in a reasonably good day we were put to scrubbing it off with coir brooms and ashes from the galley range. The birds were still with us, from the giant albatross to the tiny lightening juggler or petrel. These wee fellows flit about in a very exciting manner just a few inches above the roaring waves, while the larger pinioned seamen soar at an altitude more in keeping with the topsail yards.

Antics of the Southern Ocean Birds

We were constantly noticing something new about our feathered friends. Although they never came to rest on our spars, it was quite a sight to watch them having a rest in a heavy sea. They would sit down on the brow of an oncoming wave and just as the crest was about to break on them they would rise a few feet for a moment, letting the roller part of the water crest pass and then slide down the back of the sea mountain and float up the next incline, an antic in parallel with a switch-back railroad.

All our clothes and bedding were damp from the long continued heavy weather and the "kites" or royals and light staysails had been in the gaskets for nearly a month. Everyone tried to look forward to the day when the old man would haul the Maggie Dorrit's head to the northward where fine weather would be found under the shadow of Tasmania.

Bird Strategy

A day came when the wind and sea took a siesta, and our birds ventured alongside, the gurgling of displacement waters being absent. The dead swell brought the cape pigeons within reach of our hands as we threw small cubes of fat pork to them. The larger ones were shy and kept astern or abeam perhaps a hundred yards away. As time went on the big birds became infuriated at seeing the little fellows get all the pickings. Their resentment was shown by a shrill cry and a circling around from port to starboard. Then they came to rest and held council together and subsequently an albatross got under way with his side-wheeler motion and when in the air he took his bearings. Methods had assuredly been analysed and procedure moulded at their meeting, for intelligence was very apparent in what followed.

The big bird's move was interpreted, he meant to fly low and in a parallel line with the ship's side close in, trusting to pick up a little food in the hurried manoeuvre. On he came with nervousness written in his eyes. All the small fry scurried for safety, but the albatross had too much way on and failure resulted in his first attempt. To try and stop just where the food was, he took in his hinged wings and stuck his web feet into the water, but still he forged ahead past the goal. Another trial, however, bore its reward, and throttling his flight much earlier he landed in the right place. With a gobble or two everything for eating was soon picked up and the largest sea bird known went on his way shouting to his more timid companions that the feat was easily done.

Our Deck-house is Smashed in by a Sea

We were soon rolling along again before a moderate gale and high sea. Our half-deck was like a little island and to get in and out of it without a mishap was quite an undertaking. This was our last really big blow "running the easting down" to Brisbane, but it left its memories. After the barque behaving remarkably well when the wind was at its height we little dreamt of any mishap for the elements, both wind and sea were actually falling. Tilbury and the writer were fast asleep in the "gravy-eye" part of the morning watch, i.e. 4 to 6 a.m., and who can beat a sea apprentice at sleeping soundly? when suddenly a sense of being rolled down a greased rainbow into the mighty deep was experienced. The old barque had shut her weather eye for a second and allowed a granite-topped sea to mount the main sheer-pole and drop squarely on the side of our wooden house.

Human atoms never moved more quickly than Tilbury and myself as we emerged from the gaping wound in the side of our home, followed by mess kits, miscellaneous clothing, and sundry flotsam. Our seamanship was far enough advanced to tell us to fly or swim for the poop ladders, and likely enough our movements were a composition of both the aforementioned arts. Finally we arrived on the poop, our knee caps rattling like aspen leaves and our teeth chattering with the cold douche from the South Indian ocean. And what a humane greeting we received from old "Rope-yarns," the mate: "Lift the sail-locker hatch and get under cover you confound-

THE MAGGIE DORRIT DISCHARGING IN PORT ADELAIDE, AUSTRALIA.

ed Jonahs; that's an answer to your whistling for wind on the equator, and will do you good." Under our breath we thanked Providence for safeguarding men of the mate's stamp, who must suffer with enlargement of the heart and its affections.

None of us had an interest in trying to save our belongings and trusted childlike to the watch on deck amusing themselves along that philanthropic line of action. The steward, who thought the ship was foundering, was seized with a reactionary style of behaviour and threw his hoarding methods to the winds, for he actually sent us a dry shift of clothing from that most wonderous branch of Petticoat Lane—the deep water slop-chest. The whole catastrophe was caused through the wind falling much quicker than the sea could act in harmony, hence the barque's speed was too slow and her usual lively motions too ungraceful for Father Neptune's onslaught.

When the old man heard of the escapade he shook with guffaws and said, "That'll be square for the chicken business and make them hardy sailors." For at least another two weeks the boys had the pleasure of using the sail-locker as a home, till "Chips," the Russian Finn carpenter, put our half-deck in shipshape form again.

(To Be Continued.)

FISHING SCHOONER LAUNCHED

A two-masted fishing schooner of 130 feet over all, 26 ft. beam, 10 ft. 6 in. hold, was launched on May 16 from the yards of the Nova Scotia Shipbuilding and Transporting Co., Ltd. Capt. Roland Knickle, of Lunenburg, will be the master. The vessel is built after a model made by J. S. Gardner, and has brought very favorable comment from those who are in a position to know. The name of the craft is Sadie A. Knickle.

This is the third vessel launched by this company within nine months. They already have another three-masted schooner under construction of 420 tons. They will also lay the keels of two more vessels on their own account.

SEAMANSHIP AND NAVIGATION

Conducted by "The Skipper" *

Articles of direct interest to mariners, discussions of seamanship, navigation rules, and allied topics. Inquiries, from readers are invited and will be replied to promptly through these columns, unless otherwise requested.

THE MARINERS COMPASS—III.

IN practical navigation there is no more important duty than finding the deviation of the compass and keeping a record of same. Each watch officer should find the deviation twice in his four-hour spell, that is, provided the ship's course remains the same throughout the watch. If, however, the vessel's head is altered, then a deviation for each new course is essential.

Deviation records thus found are carefully entered into a chart-room book made for the purpose, and such books if kept in proper order go a long way in absolving an unfortunate mate or master who may be under the fire of a marine court of enquiry. What are the circumstances that may tend to change the amount of deviation on any point of the compass? The ship changing her position, the ship heeling or listing, nature of different cargoes, extra heavy seas smashing on board, or steering a long time on one course; also be careful to keep mechanical tools away from the binnacle.

The finest compass adjuster can only give you a card of local deviations, and see that the correcting magnets are in a position to give the compass a maximum of equilibrium under the changing conditions that the vessel will eventually find herself in. Adjusters have therefore a very important vocation, but it is up to the navigator to constantly figure out the deviations on each course as he goes along.

At sea when out of sight of land, the most common methods of keeping tab on the deviation are computed by compass bearings of the sun, moon and stars in conjunction with the ship's apparent time, declination of the celestial bodies ,and an approximate latitude. The Burdwood or Davis Tables are carried on all British Empire ships for this purpose, and the nautical almanac known as "Brown's" must also be in possession of every officer. When the true bearing of the heavenly object is found, then the difference between that bearing and the bearing by compass is the error. By applying the variation from the chart to the error the deviation is secured.

Another method of finding the deviation is in connection with chart work. As you sail or steam along the coast take the compass bearing of two conspicuous objects when in line. Then put your parallel rulers over the two objects on the chart, work this direction to the nearest magnetic compass rose and read off the magnetic bearing. The difference between the deck bearing and the chart bearing is your deviation for the point on which the ship is heading. See illustration for this method.

There are times when the heavenly bodies are hid indefinitely through overcast weather. One cannot always find range lights or conspicuous objects in line either. When the land has been "made" under such unfavorable circumstances and the captain wishes to know the deviations on several points which he will be bound to use during the coming night, the necessary corrections can be approximately found thus.

Swing the vessel under very easy helm and steady her properly on the four cardinal points, N.S.E.W., and also on the four quadrantal points, N.E., S.E., S.W., N.W. When steadied on these 8 equi-distant courses take the compass

*Editor's Note.—This department is conducted by a qualified ship's master, and will deal with questions involving seamanship and navigation. The attention of young men desiring to obtain certificates as mates and masters is particularly directed to this department.

10 IN. COMPASS OF LORD KELVIN TYPE WITH AZIMUTH MIRROR AND SHADOW PIN.

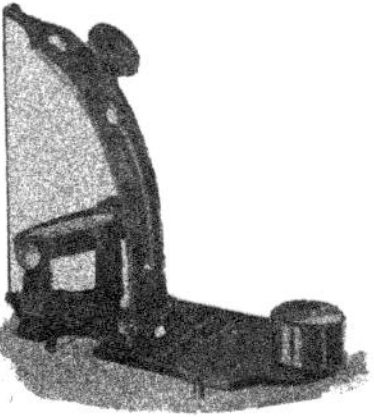

AZIMUTH MIRROR FOR USE WITH STANDARD COMPASS BOWLS

CANADIAN MADE PELORUS USED ALSO FOR COURSE CORRECTING AND SETTING

bearing of any point of land. After seeing the ship put on her original course, work out a set of deviations or construct a course in the following manner.

Assume that the 8 compass bearings of the same object read thus:—

TABLE 1.

Ship's Head by Compass	Bearing of Object by Compass	Magnetic Bearing Found	Deviations Found	Examples in Translation
1. N.	N 57° E	N 53° E	4° W	
2. NE	N 35° E	N 53° E	18° E	Mag. Bear. N 53° E
3. East	N 31° E	N 53° E	22° E	Comp. " N 39° E
4. SE	N 39° E	N 53° E	14° E	Dev. 14° E on SE
5. S	N 50° E	N 53° E	3° E	Mag. Bear. N 53° E
6. SW	N 64° E	N 53° E	11° W	Comp. " N 64° E
7. West	N 75° E	N 53° E	22° W	Dev. 11° W on SW
8. N.W.	N 74° E	N 53° E	21° W	
	425 ÷ 8			
Mean or Magnetic Bearing	N 53° E			

The difference between the two sets of bearings gives the deviations and the rule for naming them is similar to that for naming the error in azimuths and amplitudes. If the magnetic is to the right of the compass bearing the deviation is easterly. If the magnetic is to the left of the compass bearing the deviations is westerly.

Try the above problem for example of that rule. First compass bearing is N57E. The corresponding magnetic bearing, N53°E, is to the left of N57E, hence the deviation is W. I have made it still plainer in the drawings herewith. Some of the finest opportunities of making up reliable deviation cards or curves are to be had when the vessel is at anchor for a few hours awaiting a berth or pratique. When the wind, tide, or current fails to make the craft head on all the necessary points, a kick or two with the propellor will give the desired result, although much can be done with a ship's rudder even when the anchor is down and swinging is necessary. In connection with the method shown here in figures remember that the vessel should not be less than five miles from the object and if she is ten or twelve the results will be all the more accurate as radius of swing will be negligible.

As a general rule all compass work of this nature is concerned, as far as bearings and deviations go, with the "Standard" compass, which is another name for the navigating compass. The steering compass must of course be used for all purposes if it is the only one on board, and some small craft only carry the one. As these articles are intended for all classes of seamen as well as beginners, there is no special text book followed, the aim being to introduce the different problems, explain them in a practical and sailor fashion and thus make the student an advanced scholar when a particular navigational course is taken up. The following method may not be found in your text book but it will help greatly in getting used to the juggling which has to be done in course correcting and such like work. We will take the most common, or say the oldest, form of deviation card first and work out the deviations in a manner adopted on board some vessels.

TABLE 2.

Ship's Head by Compass	Deviation by Adjuster or Captain	Extended to Magnetic
North	18° E	N 18° E
NxE	13° E	N 24° E
NNE	8° E	N 30° E
NExN	2° E	N 36° E
NE	4° W	N 41° E
NExE	9° W	N 47° E
ENE	13° W	N 55° E
ExN	16½° W	N 62° E
East	20° W	N 70° E
SxS	23° W	N 79° E
ESE	25° W	N 88° E
SExE	28° W	S 84° E
SE	30° W	S 75° E
SExS	30° W	S 64° E
SSE	28° W	S 50° E
SxE	26° W	S 37° E
South	23° W	S 23° E

The two left-hand columns are what might be handed to you, while the extended column is your own figuring; found by applying easterly deviation to the right and westerly to the left. The question involved now is, what must one steer by compass to make so and so magnetic. After laying off your course on the chart you find it reads N. 36° E. magnetic, look at the card, bring your finger down the extended column till you come to that course, then run over the card thwartships and find course to steer, in this case N.E. by N. or N.34° E. But it is not always so simple as your course may lay between the set figures. Say that your magnetic course was 7. 70° E, to find the course to steer by compass we have no such number in the extended column, but it must lay between S 75° E and S 64 E, and as the deviations are the same for these two directions, just take a middle course, run your finger thwartships and call the desired direction or course S E ½ S or S 39° E. Other cases can be worked out by proportion or rule of three. The deviations shown herewith are rather large, but as long as the principle is understood that is the main thing. In practice the writer has had to contend with deviations amounting to 12°, 15° and 20°, but the vessel in question was an old rattle trap of a steamboat in the Baltic trade where a list or heel of two or three points through deck loads was common and the type of compass the worst and cheapest on the market. With a Lord Kelvin make of compass a vessel seldom carries more than 5° or 6°. of deviation and ships that stay pretty much in a regular trade can get the disturbance down much finer by making the compass, its records and its failings a constant but self-satisfying study.

Another kind of card or record has to be understood. It appears in this form:

TABLE 3.

Ship's Head Magnetic	Deviation by Adjuster or Captain	Extended to Compass
South	14° W	S 14° W
SxW	10° W	S 21° W
SSW	5° W	S 27° W
SWxS	0°	S 34° W
SW	5° E	S 40° W
SWxW	9° E	S 47° W
WSW	14° E	S 53° W
WxS	18° E	S 61° W
West	23° E	S 67° W
WxN	27° E	S 74° W
WNW	30° E	S 82° W
NWxW	32° E	N 88° W
NW	34° E	N 79° W
NWxN	32° E	N 66° W
NNW	30° E	N 52° W
NxW	27° E	N 38° W

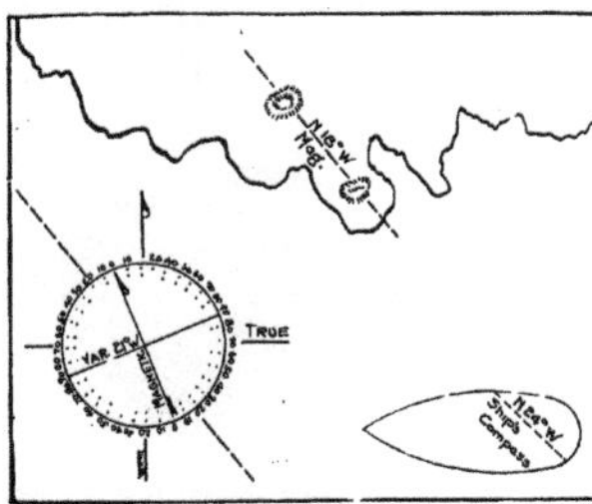

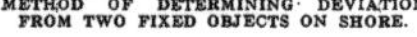
METHOD OF DETERMINING DEVIATION FROM TWO FIXED OBJECTS ON SHORE.

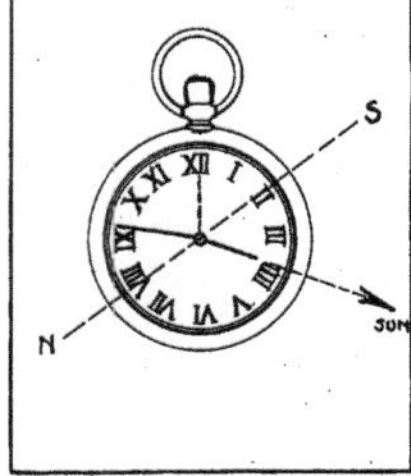

METHOD OF DETERMINING DIRECTION BY THE USE OF A WATCH OR CHRONOMETER

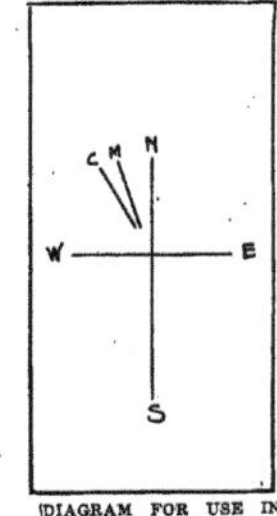

DIAGRAM FOR USE IN DETERMINING COMPASS ERROR AND DEVIATION.

Remember that the var. and dev. combined make the compass "error." Add them if names are the same, but subtract less from greater if names are unlike, in which case the "error" takes name of greater.

This comes to your hand with the first two columns, the third or extended column you figure out for yourself. Then to find the course to steer: Should the magnetic course from the chart be S 11° W, simply pick out S x W in magnetic column and draw your finger thwartships to the extended figures and the course to give the helmsman is found, in this case S 21° W. Say that your magnetic course from the chart was W x S ½ S, i.e., S 73° W. By working in between WxS and W S W we could assume a 16° easterly deviation and its application to S 73° W would make your helmsman's course S 57° W, which is in keeping with the proportional values of the extended column. If you care for stricter methods by finer interpolation, the answers in the aggregate will just about even up with the practical ship methods shown here.

Look back at last month's article if you fail to grasp anything that has been spoken of. The higher problems of amplitude and azimuth touched on at the beginning of this instalment and worked through the agency of almanacs and tables will be thoroughly gone into in due course. Meantime there is a considerable amount of interesting and practical work to be dealt with, such as logs, leads, moorings, anchor work, and "dead reckoning."

Explanation of Diagrams

The compass shown is a 10 in. Lord Kelvin type with an azimuth mirror on top and a shadow pin shipped. By the use of prisms the reflection of any heavenly body can be brought into the radius of the graduated card in the bowl and thus the bearing found for such work as this article covers. The attached shadow pin is the "good old reliable" that was used before fancy prism and mirror outfits were on the market. This brass pin can only throw a shadow and give a bearing when the sun is bright, whereas the optical instrument can be used when the sun, moon, or star is a mere blur. When the highest reading of the sextant is taken at local noon on board, the shadow pin, if the sun is bright, will cast a true north and south bearing on the compass and thus provide the navigator with the "error." By applying the variation to this error the deviation is found. The writer was ten years afloat before he saw an azimuth mirror, and all kinds of work was performed with the ordinary pin. Of course lunar or stellar work was impossible, the sun problems were the only ones suitable for the pin.

In the middle cut you see another design of azimuth mirror for shipping on top of the standard compass bowl. These handy outfits can be used for bearings of terrestrial objects as well as celestial, and in this manner they are used in coasting work.

The instrument in the box is a Canadian made pelorus, also used in work connected with course correcting and setting. This instrument is becoming very popular in the new merchant marine of Canada and the U.S.

On the chart question you will find that the two mountains when in line bear N 18° W magnetic. Suppose then that they bore N 24° W by the ship's compass when they were in line, what is the deviation? Answer 6°. How do you name it? By looking at the right hand figure and consulting the dummy compass with the lines representing the compass and magnetic bearings shown. The magnetic bearing is to the right of the compass bearing so that fact names the 6° as easterly deviation. If you used the true bearing of the mountains in a line, in conjunction with the bearing from the ship, the result would give you the whole compass "error." Then by applying the variation from the chart to the error you would get the magnetic bearing and finally the deviation as before.

The watch device or diagram shows a handy way of finding th true north or south directions when ordinary circumstances are absent and one is adrift in a forest or in the air or on the water. Cases of emergency come to most people whose business is fighting the elements.

Provided you can see the sun, just take out your watch and point the hour hand directly at it, as you see in the figure.

Then the arc that lies between the hour hand as it points to the sun and the figure XII. on the watch when bisected will give the true north and south points of the compass. Practice this at different times of the day and the simple results will soon verify the usefulness of the drawing,

Our aviators sometimes use this method of finding approximate courses when gunfire has disabled their compasses.

When used to the north of the Cancer tropics, in Canada for instance, the observer will always be facing to the southward. On the other side of the sun's declination the game is in acordance with the observer's latitude.

To be continued

NOTICE TO MARINERS

The following notices to mariners have been issued by the Department of Marine, Ottawa, under date given:

No. 37 of 1918
(Inland No. 11.)
ONTARIO.

(89) River St. Lawrence—Thousand slands—Middle channel—Color of beacons.

Previous notice.—No. 101 (360) of 1915.

The following beacons marking the Canadian middle channel, Thousand Islands, including the drums surmounting them, will hereafter be white throughout:— —

Name	Lat. N.	Long. W
Wood Island beacon...	44° 21′ 41″	75° 59′ 43″
Camelot Isalnd becon..	44° 18′ 8″	76° 6′ 55″
Punts beacon	44° 17′ 58″	76° 7′ 56″

N. to M. No. 37 (89) 6-6-18

Authority.—Departmental records.

Admiralty charts: Nos. 2789(i), 259(b) and 1152.

Publication: St. Lawrence Pilot above Quebec. 1912, pages 175, 176 and 177.

Departmental File: No. 18286.

ONTARIO

(90) River St. Lawrence—Thousand Islands—Middle channel—Bass rock beacon destroyed by ice.

Previous notice.—No. 101 (360) of 1915.

Position.—On Bass rock island.

Lat. N. 44° 17′ 13″, Long. W. 76° 9′ 24″.

Beacon carried away.—Bass rock beacon has been carried away by ice.

Re-establishment. — It will be re-established within a few days, without further notice. N. to M. No. 37, (90) 6-6-618.

Authority: Departmental records.

Admiralty charts: Nos. 1789(i), 259(b) and 1152.

Publication: St. Lawrence Pilot above Quebec, 1912, page 183.

Departmental File: No. 18286.

ONTARIO

(91) Lake Erie, western end—Wreck westward of Colchester reef.

Position of wreck.—The wreck of the whaleback steamer "Henry Cort" lies sunk at a point 3.82 miles 275° 20′ (N. 82° 40′ W. mag.) from Colchester reef lighthouse; 8.25 miles 117° 10′ (S. 60° 50′ E. mag.) from Detroit river lighthouse; and 5.47 miles 11° 20′ (N. 13° 20′ E. mag.) from Middle Sister Island.

Lat. N. 41° 56′ 17″, Long. W. 82° 58′ 36″.

The wreck lies north and south, about 80 0feet southward of the charted course for down bound vessels, and is entirely submerged with about 8 feet water over the deck.

Buoys marking wreck.—The wreck is marked by two striped spar buoys, one at the bow and one at the stern, and by a gas buoy showing a green light occulting at 3 seconds intervals, moored abreast of the middle of the wreck.

N to M No. 37 (91) 6-6-618.

Variation in 1918: 2° W.

Authority: U. S. H. O. N. to M. No. 18 of 1918; and Departmental records.

Admiralty charts: Nos 490, 332 and 678.

Publication: U.S.H.O. Publication No 108D, 1907, page 86.

Departmental File: No. 15379.

N. 38 of 1918.
(Atlantic No. 20.)
NEW BRUNSWICK.

(92) North coast—Chaleur Bay—Bathurst harbor—Change in position of back range light.

Former notices.—No. 47 (150) of 1915; and No. 91 (237) of 1917.

New position of back range light.—On the shoal inside Carron point, 1500 feet 208° (S. 52° 45′ W. mag.) from the front range light on Carron point.

Character.—White light, automatically occulted at short intervals.

Elevation.—38 feet.

Visibility.—11 miles.

Illuminant.—Acetylene, compressed in acetone.

New structure.—White wooden pole set in a square cribwork pier with battered sides lens lantern on top of pole; small white shed at base of pole.

Height.—35 feet from top of pier to top of lantern.

Remarks.—The light is unwatched.

Sailing directions.—The range lights in one mark the axis of the dredged channel over the bar at the entrance to Bathurst harbor, from the outer red conical buoy to the turn in the channel at Alston point. N. to M. No. 38 (92) 7-6-18.

Variation in 1918: 24° 45′ W.

Authority: Records, Chief Engineers' Office, Dept. of Marine.

Admiralty charts: Nos. 1715 and 2516.

Publication: St. Lawrence Pilot Vol. 2, 1916, page 307.

Canadian List of Lights and Fog Signals, 1918: No. 931.

Departmental Files: Nos. 20931A and 20930C.

QUEBEC

(93) River St. Lawrence—Langelois point light not discontinued.

Position.—On south shore of River St. Lawrence,, about ½ mile northeastward of Langelois point.

Lat. N. 46° 35′ 6″, Long. W. 71° 59′ 24″.

Light not discontinued.—Langelois point light is in operation; its maintenance has not been discontinued; and Notice to Mariners No. 116 (301) of 1917 is therefore cancelled.

Character of light.—Fixed white.

N. to M. No. 38 (93) 7-6-18.

Authority: Departmental records.

Admiralty charts: Nos. 2779, 2830(a) and 797.

Canadian Naval charts: Nos. 16 and 24.

Publication: St. Lawrence Pilot above Quebec, 1912, page 39.

Canadian List of Lights and Fog Signals, 1918: No. 1273.

Departmental Files: Nos. 21273K and R.

No. 39 of 1918.
(Atlantic No. 21.)
EAST COAST OF CANADA

(94) Regulations with regard to Vessels' Lights.

The following extract from "Defence of Canada Order, 1917," revised to 15th May, 1918, is published for the information of mariners. This extract cover sthe Regulations with regard to Vessels' Lights.

"22A. The Masters of all vessels (other than those employed exclusively in Lake or River Service) shall comply with the following orders regarding ships' lights:—

(1) Anchor Lights.—No electrically-lit lanterns shall be employed by any vessel as anchor lights. The normal brilliancy of all other anchor lanterns shall be reduced by fifty per cent.

(2) Masthead Lights.—No masthead light of a brilliancy exceeding two and one-half candle power is to be exhibited. The reflectors are to be removed from the lanterns. Masthead lights are never to be used unless the Master considers it absolutely necessary.

(3) Side Lights.—No side light of a brilliancy exceeding eight candle power shall be exhibited. In clear weather, and when specially ordered five candle power lamps are to be exhibited. Oil side lamps are only to be exhibited if electric lights are not available. Reflectors are to be removed from the lanterns.

(4) Stern Lights.—No stern light is to be exhibited execept to avoid danger of collision, and such light is to be extinguished as soon as the danger is past. Such light shall be of two and one-half candle power.

(5) Other Lights.—No lights visible from outboard, either aloft, on deck or below, except those required by the Regulations for the Prevention of Collisions at Sea, and such as may be necessary for authorized signalling purposes, shall be used on any vessel. This shall apply to all vessels whether under way or at anchor.

(6) The above orders shall apply to vessels of every description, other than H.M. Ships, within the waters on the East Coast of Canada and extending up the St. Lawrence River as far as the port of Quebec.

(7) Vessels carrying volatile oil or spirits in bulk shall exhibit (in lieu of oil lamps) electrically-lit lanterns not exceeding in brilliancy fifty per cent. of the brilliancy of the normal oil lamps."

Authority: Dept. of the Naval Service.
Departmental File: No. 30545.

No. 40 of 1918
(Inland No. 12)

UNITED STATES OF AMERICA

(95) Lake Ontario, east end—Stony island—Gas buoy established—Point Peninsula—Gas buoy discontinued.

(1) Position of Stony island gas buoy No. 2.—Off northeast end of Stony island.

Description.—Conical buoy, with skeleton superstructure.

Character of light.—Occulting white light, visible 10 seconds and eclipsed 10 seconds alternately.

Power.—120 candles.

(2) Gas buoy discontinued.—Point Peninsula gas buoy No. 1 has been permanently discontinued. N. to M. No. 40 (95) 11-6-18.

Authority: U.S. Dept. of Commerce N. to M. No. 22 of 1918.

Admiralty charts: Nos. 1152 and 797.

Publication: St. Lawrence Pilot above Quebec, 1912, pages 207 and 204.

UNITED STATES OF AMERICA

(96) Lake Huron, south end—Change in position of Lake Huron light-vessel—Changes in buoyage.

(1) New position of light-vessel.—1.79 miles 11° 30' (N. 14° 50' E. mag.) from Fort Gratiot lighthouse, and 480 feet east of Point Edward range line.

From the light-vessel, Point Edward front range light bears 183° (S. 6° 20' W. mag.), and the north tangent of Windemere hotel bears 247° 30' (S. 70° 50' W. mag.)

(2) New position of Lake Huron cut buoy No. 2.—2340 feet southerly from light-vessel, and 480 feet east of Point Edward range line.

From the buoy, Point Edward front range light bears 183° 30' (S. 6° 50" W. mag.), Fort Gratiot lighthouse bears 195° (S. 18° 20' W. mag.), and the north tangent of Windemere hotel bears 268° 30' (N. 88° 10' W. mag.).

Depth.—19 feet.

(3) Position of Lake Huron cut buoy No. 1. New buoy.—Abreast Lake Huron cut buoy No. 2, and 480 feet west of Point Edward range line.

From the buoy, Point Edward front range light bears 177° 30' (S. 0° 50' W. mag.), Fort Gratiot lighthouse bears 188° (S. 11° 20' W. mag.), and the north tangent of Windemere hotel bears 268° 30' (N. 88' 10' W. mag.)

Description.—Black spar buoy.

Depth.—19 feet. N. to M. No 40 (96) 11-6-18.

Variation in 1918: 3° 20' W.

Authority: U.S. Dept. of Commerce N. to M. Nos. 20 and 22 of 1918.

Admiralty charts: Nos. 330, 519 and 678.

Canadian Naval chart: No. 90.

Publication: U. S. H. O. Publication No. 108C, 1907, page 42.

UNITED STATES OF AMERICA

(97) Lake Huron—Pointe aux Barques—Gas and whistle buoy replaced by gas and bell buoy.

Position.—In about 6 fathoms, about 2 miles 56° (N. 60° E. mag.) from Pointe aux Barques lighthouse.

Description of gas and bell buoy.—Cylindrical buoy, with skeleton superstructure.

Color.—Black.

Character of light.—Flashing white light every 3 seconds, flash 0.3 second duration.

Elevation.—12 feet.

Power.—70 candles.

N. to M. No. 40 (97) 11-6-18.

Anuthority: U.S. Dept. of Commerce N. to M. No. 22 of 1918.

Admiralty charts: Nos. 519 and 678.

Publication: U. S. H. O. Publication No. 108C, 1907, page 47.

UNITED STATES OF AMERICA

(98) River St. Mary—Vidal shoals channel—Changes made in aids to navigation.

Aids are to temporarily mark limits of dangerous shoal areas under improvement.

(1) Vidal shoals channel gas buoy No. 1, moved temporarily to a point about 330 yards 105° (S. 72° E. mag.) from the intersection of the lines of Canadian canal upper entrance range and Vidal shoal channel range, and marks the northerly edge of shoal on south line of Vidal shoals channel.

(2) Vidal shoals channel gas buoy No. 2, to be temporarily established at a point about 175 yards 241° 30' (S. 64° 30' W. mag.) from the intersection of above-named range lines and marks the southerly limits of shoals. The gas buoy is conical, with skeleton superstructure, and shows a group flashing red light, 2 flashes every 15 seconds, thus: flash 2 seconds, eclipse 2 seconds, flash 2 seconds, eclipse 9 seconds, of 35 candle-power, 11 feet above water.

(3) Vidal shoals junction gas buoy, H. S., discontinued. N to M. No. 40 (98) 11-6-18.

Variation in 1918: 3° W.

Authority: U.S. Dept. of Commerce N. to M. No. 22 of 1918.

Admiralty charts: Nos. 324, 329 and 678.

Publication: U. S. H. O. Publication No. 108A, 1906, page 28.

Canadian List of Lights and Fog Signals, 1918: Pages 136 and 137.

Departmental File. No. 26150.

No. 41 of 1918
(Atlantic No. 22)

PRINCE EDWARD ISLAND

(99) East coast—Boughton river—Dredged cut west of Annandale wharf—Stakes.

Previous notice.—No. 65 (216) of 1916, last paragraph.

Stakes—The dredged cut through the shoal west of Annandale wharf is no wmarked by 8 stakes, 4 on the north side and 4 on the south side. The stakes on the north side of the cut have bushes attached to their tops.

N. to M. No. 41 (99) 14-6-18.

Authority: Report from Agent of Dept. of Marine, Charlottetown.

Admiralty charts: Nos. 2005 and 2034.

Publication: St. Lawrence Pilot, Vol. 2, 1916, page 167.

Departmental File: No. 19684.

PRINCE EDWARD ISLAND

(100) East coast—St. Mary Bay—Two new buoys established.

(1) Position.—2100 feet 182° 20' (S. 26° W. mag.) from Marsh point, the point on the south side of Panmure island ½ mile west of Panmure island wharf.

Lat. N 46° 7' 40", Long. W. 62° 29' 45".

Description.—Wooden spar buoy.

Color.—Black.

Depth.—12 feet. N. to M. No. 41 (100) 16-6-18.

(2) Position.—1400 feet 332° 20' (N. 4° W. mag.) from Hicken point, the point on the south side of St. Mary bay west of St. Mary bay wharf.

Lat. N. 46° 7' 37", Long. W. 62° 29' 0".

Description.—Wooden spar buoy.

Color.—Red.

Depth.—12 feet.

Variation in 1918: 23° 40' E.

Authority: Report from Agent of Dept. of Marine, Charlottetown.

Admiralty charts: Nos. 2029 and 2034.

Publication: St. Lawrence Pilot, Vol. 2, 1916, page 171.

Departmental File: No. 30924.

PRINCE EDWARD ISLAND

(101) South coast—Northumberland strait—Bedeque bay — Summerside — Front range lighthouse moved.

New position.—Summerside front range lighthouse on the railway wharf has been moved 36 feet westward in the line of range. It now stands on the west side of the wharf, 344 feet from its outer end.

Lat. N. 46° 23' 18", Long. 63° 46' 41".

N. to M. No. 41 (101) 14-6-18.

Authority: Report from Agent of Dept. of Marine, Charlottetown.

Admiralty charts: Nos. 1912, 2034, 1651, 2516 and 2666.

Jublication: St. Lawrence Pilot, Vol. 2, page 265.

Canadian List of Lights and Fog Signals, 1918: No. 736.

Departmental File: No. 20736R.

QUEBEC.

(102) River St. Lawrence below Montreal—Vicinity of Longue Pointe—Reserved harbor area buoyed.

Two special and temporary gas buoys have been placed to mark a reserved harbor area in the vicinity of Longue Pointe:—

(1) Position of more northerly buoy.—1150 feet 82° (S. 83° E. mag.) from Tetreauville front range light.

Lat. N. 45° 35' 54", Long. W. 73° 30' 21".

Color.—Red.

Character of light.—Fixed red light.

(2) Position of more southerly buoy.—600 feet 62° (N. 77° E. mag.) from the spire of Longue Pointe church.

Lat. N. 45° 35' 11", Long. W. 73° 30' 14".

Color.—Red.

Character of light.—Fixed red light.

Prohibited area.—No vessel is allowed inside or westward of the line between these two gas buoys.

N. to M. No. 41 (102) 14-6-18.

Variation in 1918: 15° W.

Authority: Departmental records.

Admiralty charts: Nos. 1127, 2788 and 2830b.

Canadian Naval Charts: Nos. 1, 2 and 22.

Publication: St. Lawrence Pilot above Quebec, 1912, pages 96 and 97.

Departmental File: No. 25577.

WALES

(103) South coast—Cardiff roads and approach—Light buoys established.

(a) Position.—At a distance of 16 cables, 295° (N. 49° W. mag.) from Monkstone lighthouse, in the position formerly occupied by Middle Cariff buoy which it replaces, Monkstone lighthouse Lat. 51° 24¾' N., Long. 3° 06' W.

Description.—A black conical light-buay exhibiting a flashing white light every ten seconds.

(b) Position.—At a distance of 12¾ cables, 234° (C. 70° W. mag.), from Monkstone lighthouse.

Description.—A black conical light-buoy exing a flashing red light every ten seconds.

N. to M. No. 41 (103) 14-6-18.

Variation in 1918: 16° W.

Authority: British Admiralty N. to M. No. 608 of 1918.

Admiralty charts: Nos. 1182, 2682 and 1179.

Publication: W. C. England Pilot, 1910, pages 209 and 211.

A. JOHNSTON,
Deputy Minister.

Department of Marine,
Ottawa, Canada, June 14th, 1918.

OWING to their great strength and lasting properties, New South Wales hardwoods are particularly suitable for shipbuilding. Many of the coastal rivers have bad sand bars, and the continual bumping when the vessels are crossing is a severe trial; but, nevertheless, many coasters have been running continuously for periods of from 20 to 30 years, and are still fit for service. For use in connection with the harbor ferry service of Sydney preference is given to wooden ships, as they are better able to withstand the continual strain of bumping much better than steel. These vessels are from 200 to 300 ft. long. On account of their hardness New South Wales timbers are not so liable to damage by marine insects as soft woods, and frequently hulls are planked to the waterline with hardwoods and the tops finished with soft woods. Considerable quantities of timber suitable for keels have been shipped from New South Wales to the Pacific Coast of North America during the last few years and the trade is growing.—L. E.

In a railroad station eating-house out in a lonely spot in the West they pass a basket of sandwiches and you are to help yourself. It happened that Big Bill got a sandwich without any meat, and he yelled out:

"Say, Jack, shuffle them again, I got the joker."

Eastern Canada Marine Activities

Doings of the Month at Montreal and on St. Lawrence River

July Correspondence

MARINE activities in and about Montreal continue with unabated interest. The summer months have seen some remarkable advancement in the progress that has been made by the different companies engaged in the construction and repair of vessels. It is expected that almost double the tonnage that has already been launched will take the water before the winter sets in.

Two of the standard Imperial Munitions Board 3,000-ton wooden vessels, the War Erie and the War Huron, under construction at the yard of the Fraser, Brace Co. were successfully floated from their building site in the yard basin on the afternoon of July 4, and the following morning were towed out into the Lachine Canal adjoining the shipyard. In order to obtain sufficient depth of water in the basin it was necessary to pump the water from the canal into the basin, a specially constructed concrete retaining slab being located at the entrance to maintain the water in the basin. Both vessels were afloat about four in the afternoon but divers were required for the removal of some of the supporting blocks before the water was allowed to return to the canal level. The floating level for these two boats was about three feet above that of the canal. Additional excavation has since been made to prepare the basin for the launching of the other two vessels now being constructed on the upper level. Work is progressing rapidly on the War Niagara and the War Ottawa and the launching dates for these two vessels have been fixed for Saturday July 27 and August 10 respectively. The two recently floated are now being fitted with the engines and machinery, after which the superstructure and the rigging will be completed. Nothing is definitely known yet as to further work at this yard as no announcement has been made as to additional ship contracts.

Another War Vessel

A feature of marine progress on the River St. Lawrence was the official inspection given to the steamship Porsanger, recently constructed by the Canadian Vickers Company. This steel vessel with a displacement of upwards of 7,000 tons, is the largest of its kind yet built in this country. She was launched early this year and for the past several months has been fitting out at the yards of the builders and the recent inspection marks the final event before being placed in service. While the vessel was originally ordered and built for private interests she will, for the duration of the war operate as a cargo steamer for the British government.

The Concrete Vessel

The initial trial trip of the first power propelled concrete vessel built in Eastern Canada was made during the present week. The Concretia, which was built by the Montreal Shipbuilders, Limited, was launched last fall, but owing to unavoidable circumstances the work of installing the machinery and the completion of the superstructure was delayed until the spring. While no official statement has been given out regarding the results of her trial trip—which was made up the Lachine Canal and the St. Lawrence to Prescott and return—it was announced by one of the officers of the company, that the behaviour of the boat under actual operating conditions was entirely satisfactory in every particular, and proved to her builders the possibilities of this type of vessel for regular service.

Less Grain Passing

General cargo traffic through the Lachine Canal during the month of June was considerably less than for the corresponding period of 1917. The most notable decrease was in the amount of grain passage, this year's tonnage for June being only about 43 per cent. of that for the same month last year. Wheat shows the most surprising figures, the movement being 357,031 bushels, against 1,489,094 bushels for June of last year. Oats also showed a decrease of about 710,016 bushels, the amount coming down this June being only 204,082 bushels. The only grains showing increases were barley and flaxseed, the former 631,140 bushels against 273,960 bushels, and in the case of flaxseed this year's passage was 34,000 bushels, while none of this grain was transported through the canal in the month of June, 1917. Considerably more coal was carried this year, and the general produce showed increases. The great falling off in passenger traffic was due to the opening of this service from two to four weeks later this year.

A Great Programme

The magnitude of the shipbuilding programme has created considerable interest among marine officials in connection with the need that will eventually arise for capable engineers to operate the large number of vessels that are now being built for lake and ocean service. In order to meet the coming situation the Montreal lodge for the International Union of Steam and Operating Engineers has taken the initiative, in appointing a committee to interview the Hon. Mr. Ballantyne with the object of adopting some method to offset the existing shortage by establishing some kind of training school along the lines now maintained by the United States government. No action has as yet been taken in this connection but officials of the recently organized Navy League anticipate giving their co-operation to the organization of any scheme that will advance the interests of this branch of the service. The committee that has been appointed by the local Union No. 593, to take the matter up with the Canadian authorities is as follows: S. J. Maguire, president; A. Munro, J. Dillon, T. M. Clark, T. Hunt, S. Sains and A. Hamlin, the latter president of Local No. 588 of the Steam and Operating Engineers.

A Great Success

The hopes of the officers of the Quebec branch of the Navy League of Canada were more than realized when, at the conclusion of the recent whirlwind campaign, the enrollment totaled the aggregate number of 22,500. While some little attention has been given to acquiring members in other sections, the recent campaign was confined to Montreal and the surrounding districts, and from the satisfactory results that have already been attained, the League anticipate a good response from the remainder of the Province. Mr. Glass, the secretary of the league in Montreal, realizes that it will be more difficult to arouse interest in some of the inland provinces, but thinks on the whole that the permanent nature of the Navy League, and its great object will appeal to all those who have the interests of Canada at heart.

Anxious Moments

In view of the fact that the Hon. Mr. Ballantyne has been favorably impressed with the advantageous facilities of the Quebec docks and harbor, and the intimation that increased use would probably be made of these in the near future, the announcement by certain officials of the Imperial Government Purchasing Commission, that additional charges would be entailed by Utilizing Quebec as a shipping terminal, has created some adverse feeling in the Quebec Board of Trade, and steps are being taken to rectify the wrong impression that seems to prevail regarding these conditions. It appears that the actual haulage distance from the West is less by about 200 miles when transporting freight from Winnipeg to Quebec than it would be taking the same to Montreal.

Exceptional progress has been made on the two wooden vessels recently launched by theh firm of Quinlan and Robertson at their yard at Quebec, and during the last several weeks the installation of the machinery has proceeded rapidly, particularly on the second vessel where the experience gained on the first boat has enabled the mechanical staff to work to better advantage in placing the equipment. The trial trip of the first vessel the War Mohawk, is looked for early in August.

Very High Tide Water

Throughout the tidal section of the St.

Lawrence River the country experienced an eventful period during the colse of the first week in July, when an exceptional high tide flooded many sections of the adjoining country. Much of the lower town in the City of Quebec was seriously flooded when the water rose to a height of 21 feet, forcing the passage over the protecting walls provided for a high tide of 18 feet, the previous high record. The flooding was repeated at the following tide, but from then rapidly subsided to normal, but in the city and country districts considerable damage was done. The river was so rough that operation of the various ferries and river boats was continued under great difficulties.

Development at St. John

Announcement has been made that the improvements and extensions that are to be made at the Port of St. John, N.B., will result in this place having one of the largest dry docks on the Continent. The dry dock when completed will be 1,150 feet in length and 125 feet wide at the bottom and capable of accomodating vessels with draft as great as 40 ft., which includes the largest vessels now in service by any country in the world. In addition to the construction of the dry dock, the St. John Dry Dock and Shipbuilding Company will engage in the building of steel vessels of 10,000 tons displacement. Operations will commence almost immediately, and it is anticipated that the entire work will be finished in about three years. The controlling interest will be held by the Canadian Dredging Co. of Midland, Ont., the general manager in all probability being D. S. Pratt of Midland now in a similar position with the Canadian Dredging Co.

A Great Undertaking

If all expectations are realized the new shipbuilding plant to be constructed by the Halifax Shipbuilders Limited will compare favorably with any of those yet built or contemplated on this side of the Atlantic. While nothing definite has been given out it is believed that everything possible will be done to expedite the work of erection to enable a start to be made, within three months, on the three 10,000-ton steel vessels now under contract. The shipyard site will have a water frontage of 2,500 feet, near the vicinity of the recent explosion. While the essential undertaking is the building of vessels, it is intended eventually to enlarge the existing dry dock to accommodate much larger vessels.

METAL CUTTING BAND SAW

THE band saw machine illustrated herewith is a recent product of the Napier Saw Works, Springfield, Mass., and is adapted to the rapid cutting of soft steel, tool steel and soft metals of all kinds.

A gravity feed operated by the balanced weight of the tilting head is used. A spring at the back of the machine is in such a position and of such tension that the feeding weight is constant from the start to the finish of any cut which can be made on the machine.

BAND SAW MACHINE FOR CUTTING METALS.

The base is of cabinet type with lubricating tank enclosed. All bearings are extra large. The bearings in the clutch pulley, and in the driving and idle wheels, which carry the tension of the saw, are heavy and rigid.

The cutting band is lower than the returning band thus making it possible to cut off any length required. The cutting band is supported by truss arms, each carrying a pair of guides specially designed.

By means of the roll guides the band saw is directed so that its travel between the truss arms is absolutely vertical, thus as- suring straight cutting. These guides once set are in perfect adjustment for any saw which may be placed on the machine. This is a feature not found on any other machine.

A rotary pump of simple design provides amply cutting compound for the saw. The teeth of the band saw are downward in the cut which insures that the cutting compound flows at once to the point of the saw tooth where it is needed.

The manufacturers recommend a saw speed of 120 lin. ft. per min. for soft steel, 100 lin. ft. per min. for tool steel and 150 lin. ft. per min. for soft metals. A one-horsepower motor is required for the operation of the saw which occupies a floor space 4 ft. x 6 ft.

WOULD EXCHANGE IDEAS IN TRADE

The Canadian Industrial Reconstruction Association is planning a general Dominion-wide interchange of views and contentions next year. The executive council has discussed tentatively arrangements for delegations to travel throughout Canada as a means of bringing the East and the West to a closer understanding on questions affecting the general welfare of the Dominion.

Sir John Willison, the president, states that the idea of encouraging the various Canadian interests to understand each other is one of the chief objects of the association. "That is how we can hope to successfully negotiate the trying post-bellum period," said he. The delegations that will visit different parts of the Dominion will consist of farmers, manufacturers, business men, financiers, etc. They will be restricted to no special interest or section of the Canadian business and industrial community.

Sir John Willison states that plans for various university fellowships are going ahead, and that when the coming academic year opens these will take definite shape.

Lieut.-Col. Ibbetson Leonard, of London, Ont., has been awarded the D.S.O. He is a member of the firm of E. Leonard & Sons, boiler and engine makers.

Work in Progress in Canadian Shipyards

A record of Work in Process of Completion—Principal Features of Specification—Approximate Launching Dates—Vessels on Order.

BRITISH COLUMBIA

YARROWS, LIMITED, ESQUIMALT, B.C.
One stern-wheel steamer, 165 ft. x 35 ft., for Indian Government service.
At present engaged chiefly on repair work.

H. VOLLMERS, NANAIMO, B.C.
One gasoline tug, 9 tons.

NEW WESTMINSTER CONSTRUCTION & ENGINEERING CO., NEW WESTMINSTER, B.C.
Four wooden steamers, Nos. 10-11-12-13, for undisclosed interests. 250 ft. x 43 ft. 6 in. x 25 ft., moulded depth, 2,800 tons d. w., 1,000 h.p., single screw engines, 12 knots, naval architect, I. Alexander. Launchings March 14, April 14, May 15 and June 15, 1918.

B.C. CONSTRUCTION & ENG. CO., POPLAR ISLAND, NEW WESTMINSTER, B.C.
Four wooden vessels, 1,800 gr. tons, 2,800 tons d.w.

NEW WESTMINSTER MARINE CO., FT. OF FURNESS ST., NEW WESTMINSTER, B.C.

STAR SHIPYARD CO., NEW WESTMINSTER, B.C.

J. A. CROLL, PORT ALBERNI, B.C.

PACIFIC CONSTRUCTION CO., PORT CCOQUITLAM, B.C.
Taken over the plant of the Coquitlam Shipbuildiug Co.
Gasoline yacht, 50 ft., June, 1918.
Three wooden freight steamers, Nos. 20-21-22, 250 ft. b. p. x 42 ft. 6 in. x 25 ft., 1,800 gr. tons, 2,800 tons d.w., trip. exp. engines, water tube boilers ,12 knots, 1,000 h.p. Two for undisclosed interests, one for builder's account. Launchings May 20, June 15, Sept., 1918.
Four new berths under construction for steel ships up to 10,000 tons.

C. E. BAINTER, PRINCE RUPERT, B.C.
Several wooden fishing steamers, for undisclosed interests.

GRAND TRUNK DRYDOCK SHIP REPAIR CO., LTD., PRINCE RUPERT, B.C.

S. A. MOULTON, PRINCE RUPERT, B.C.
Ten composite vessels for undisclosed interests.

BRITISH-AMERICAN SHIPBUILDING & ENG. CO., VANCOUVER, B.C.
Establishing plant on Kitsilano Reserve.
Twenty wooden vessels being built for undisclosed interests, 3,500 tons each.

JOHN COUGHLAN & SONS, N. VANCOUVER, B.C.
"War Camp," sister ship to "Alaska," launched in March, 1918.
One steel freight steamer, 427 ft. 9 in. x 54 ft. x 29 ft. 9 in., draft 24 ft. 2 in., speed 11½ knots, 5,730 gross tons. Launched Jan. 10, 1918, for B. Stolt Nielsen, Christiania, Norway. Both sold to the Cunard Line.
Three steel steamers, "War Chariot," "War Chief," and "War Noble," Nos. 4-5-6, 425 ft. x 54 ft., 8,800 tons cap., for the Cunard Line.
Six steel freight steamers, 425 ft. x 54 ft. x 24 ft. 2 in., 8,800 tons cap., 1 knots, for undisclosed interests. Delivery Jan., Feb., March, May and July, 1918.
One steel freight steamer, "Alaska," 427 ft. 9 in. x 54 ft. x 29 ft. 9 in., draft 24 ft. 2 in., speed 11½ knots, 5,730 gr. tons, 8,800 d.w. Launched Jan. 10, 1918, for B. Stolt Nielsen, Christiania, Norway. Sold to the Cunard Line.

THE FOUNDATION CCO., VANCOUVER, B.C.

FRASER VALLEY SHIPBUILDING CO., VANCOUVER, B.C.

GRANT, SMITH & CO., VANCOUVER, B.C.
Six wooden cargo vessels, 250 ft. x 43 ft. 6 in. x 25 ft., 3,000 tons cap., 9½ knots, for the Canadian Government.

HARRISON & LAMOND, SHIPBUILDERS, LTD., VANCOUVER, B.C.
One wooden auxiliary schooner, 225 ft. x 44 ft. x 21 ft. 4 in., 1,600 gr. tons, 2,550 tons cap., for undisclosed interests.

LYALL SHIPBUILDING CO., NORTH VANCOUVER, B.C.
"War Puget," Feb. 16, 1918, 250 ft., wood, 2,080 tons.
Ten wooden cargo ships, for undisclosed interests. Launchings Jan. 20, Mar. 2, Mar. 23, and April 11, 1918.
Building six wooden vessels for own account.
"War Caribou," April 10, 1918, 3,080tons, wood.

W. R. MANCHION, VANCOUVER, B.C.

STANDARD SHIPBUILDING CO., VANCOUVER, B.C.
Two 3,500-ton vessels for Brazilian Govt., and six for French Govt. of 4,500 tons.
Donohoe reinforced type, wood composite construction.

TAYLOR ENGINEERING CO., VANCOUVER, B.C.
Number of small vessels being constructed at total value of $300,000.
4,500-ton floating drydock, 352 ft. long.

VANCOUVER SHIPYARD, LTD., VANCOUVER, B.C.
One motor freighter, 125 ft. x 24 ft. x 12 ft., wood, 160 h.p., Bolinder's crude oil engine. To be completed Aug., for Taylor Engr. Co.
Also repairing several steamers and schooners.

THE WALLACE SHIPYARDS, LTD., N. VANCOUVER, B.C.
"War Dog," 4,500 tons, steel, May 18, 1918.
"War Power," 4,600 tons, steel, March 23, 1918.
Two steel steamers, 315 ft. x 45 ft. x 27 ft., 4,600 tons d. w., trip. exp. vertical engines ,two S. E. Scotch boilers, 10 knots, single screw, for the Canadian Government. Deliveries Dec., 1917, and Aug., 1918.
Two wooden aux. schooners, 255 ft. o. a., 225 ft. keel x 44 ft. x 21 ft. 4 in., five masts, 1,500 gr. tons, cap. 2,500 tons or 1,500,000 ft. lumber, two hatches, twin 160 h.p. Bolinders Hot-bulb engines, twin screws, for the Canadian Government.
Four wooden cargo vessels, aggregate tonnage of 17,500 tons for undisclosed interests.

WESTERN CANADA SHIPYARDS ,LTD., VANCOUVER, B.C.
"War Nootka," Jan. 4, 1918, 3,080 tons (machinery installed at Ogden Point Assembly Sheds).
Six wooden freight steamers, 250 ft. h.p., 259 ft. o. a., 43½ ft. extreme breadth, molded 42½ ft., 25 ft. molded depth., draft 22 ft., 2,800 tons d. w., for undisclosed interests. "War Nootka," launched January 4, 1918. "War Selkirk" launched March 6, 1918.
"War Tatla," "War Casco, "War Chilkat," "War Tanoo," 4 standard vessels now under construction.

CAMERON-GENOA MILLS CO., VANCOUVER, B.C.
"War Yukon," Jan. 24, 1918, 3,080 tons, wood (machinery installed at Ogden Point Assembling Sheds.

THE FOUNDATION CO. OF B. C., LTD., VICTORIA, B.C.
Building ships for undisclosed interests.
Five wooden steamers, 250 b. p. x 42 ft. 6 in. x 25 ft., 1,800 gr. tons, 2,800 tons d. w., one trip. exp. engine, 1,000 h.p., "Howden" water tube boilers, furnished and installed by owners, 10 knots, for undisclosed interests. Esplen & Son & McNaught, N.Y.C., designers.
No. 1—"War Sonchee," launched Dec. 27, 1917.
No. 2—"War Masset," launched April 11, 1918.
No. 3—"War Babine," ready for launching, awaiting propeller and rudder.
No. 4—"War Camchin," ready for launching, awaiting propeller and rudder.
No. 5—"War Nanoose," ready for launching in about five weeks.

CLARENCE HOARD, VICTORIA, B.C.
Wooden car barge for C.P.R., Jan., 1918.

VICTORIA SHIPBUILDING CCO., VICTORIA, B.C.
Constructing wooden ships for British Govt., 3,500 tons each.

NEW BRUNSWICK

C. T. WHITE & SON, LTD., ALMA, N.B. (Office at Sussex, N.B., also.)
Two schooners aux. power, 143 ft., 900 tons d. w., first to be launched in April; second in June.

JAMES X, LENTEIGNE, LOWER CARAQUET, N.B.
One schooner, 28 tons, wood.

INTERNATIONAL SHIPBUILDING CO., NEWCASTLE, N.B.
Two four-masted wooden auxiliary schooners, 155 ft. keel, 37 ft. depth, 535 net tons. One to be completed in September and the other December, 1918. For builders' account.

EUREKA SHIPBUILDING CO., NORTH HEAD, N.B.
One wooden schooner, "Mollie & Melba," 350 tons reg.

PORT COLBORNE BUILDING & REALTY CO., LTD., REXTON, N.B. (Head Office, Welland, Ont.)
Four-masted wooden schooner. Has cap. of 3 schooners per year.

GRANT & HORNE, ST. JOHN, N.B.
Two wooden cargo steamers, "War Fundy" (launched Aug., 1918) and "War Digby," 250 ft. x 42 ft. 5 in. x 25 ft. 5 in., 1,000 h.p., speed 9½ knots, 2,800 tons d.w., for undisclosed interests.

MARINE CONSTRUCTION CO., CANADA, LTD., ST. JOHN, N.B.
Four wooden auxiliary, four-masted schooners, 185 ft. o. a., 165 ft. keel x 40 ft., 1,100 d. w. tons, for builder's account. J. Murray Watts, Philadelphia, naval architect. Launching April, 1918.
One wooden schooner, "Dorfontein," 740 reg. tons. Launching June, 1918.

PETER AND A. A. McINTYRE, ST. JOHN, N.B.
One schooner, 425 tons, 137 ft., three masts, wood, for builder's account. To be launched in June. Another to be commenced immediately.

ST. JOHN SHIPBUILDING CO., ST. JOHN, N.B. (Plant at Courtenay Bay.)
Ten five-masted auxl. schooners, oil burning engines, for undisclosed interests.
Establishing plant for steel and wood construction.

ST. MARTIN'S SHIPBUILDING CO., ST. MARTIN'S, N.B.
One wooden schooner, 450 tons, three masts. To be launched in July for builder's account.

NEWFOUNDLAND

ANGLO-NEWFOUNDLAND DEVELOPMENT CO., BOTWOOD, NFLD.
Two three-mast aux. schrs., wood, 450 tons, 150 H.P. for the builders' account.

NEWFOUNDLAND SHIPBUILDING CO., HARBOR GRACE, NFLD.
To build ships of 1,200 tons d.w. Steel ships later.

M. E. MARTIN, MORRIS ARM, NR. ST. JOHN'S, NFLD.
Two wooden schooners, 300 tons each, for builder's account.
One wooden schooner, 500 tons, for builder's account.

UNION SHIPBUILDING CO., ST. JOHN, NFLD.

ANNAPOLIS SHIPPING CO., ANNAPOLIS ROYAL, N.S.
"Hilda M. Clark," March 4, 1918, 187 ft., 640 tons reg.
Two schrs. three-masted, wood 500 tons, 170 ft. long. To be completed August and December, 1918.

NOVA SCOTIA

B. L. TUCKER, BASS RIVER, N.S.
One schooner, 350 tons, wood.

JAS. S. CREELMAN, BASS RIVER, N.S.
One three-masted schooner, wood, 170 ft. long, 500 tons. To be completed October, 1918, for builders' account.

HANKINSON SHIPBUILDING CO., BELLIVEAU'S COVE, N.S.
One schooner, 360 tons, wood.

A. A. THERIAULT, BELLIVEAU'S COVE, N.S.

Two wooden three-masted tern schooners, 150 ft. x 33 ft. x 11 ft. 11 in., 400 gr. tons, gasoline engines, for builder's account. Launchings April and June, 1918.
One wooden one-masted schr.

WESTPORT SHIPBUILDING CO., BELLIVEAU'S COVE, N.S.

One wooden schooner, 230 tons net. To be launched in October for builder's account.

ELI PUBLICOVER, BLANDFORD, N.S.

BRIDGEWATER SHIPPING Co., BRIDGEWATER, N.S.

One tern schooner, 450 tons gr., wood. Adapted for aux. power. To be completed Sept., 1918.

One tern schooner, 275 tons gr., wood. Adapted for aux. power. To be completed Nov., 1918.

Both built on owner's account and for sale.

W. A. NAUGLER, BRIDGEWATER, N.S.

One wooden three-masted schooner, "Wm. A. Naugler," 146 ft. o. a., 116 ft. keel x 32 ft. x 11 ft. 6 in., 360 gr. tons, 600 tons cap., for builder's account.

One schooner, 300 tons, wood.

H. MAC ALONEY, CANNING, N.S.

Two wooden schooners, 400 tons each. For builder's account. Launchings July and October, 1918.

S. M. FIELDS, CAPE D'OR, N.S.

One three-masted schooner, 350 tons.
Keel laid for another three-master.

CHESTER BASIN SHIPBUILDERS, LTD., CHESTER BASIN, N.S.

Two schooners, 100 tons, wood. Launchings Aug., 1917, Dec., 1917.
One four-masted schooner, 60 tons. Launching Sept., 1918.
One two-masted schooner, 100 tons. Launching June, 1918.

MORTIMER PARSONS, CHEVERIE, N.S.
One wooden three-masted schooner, 170 ft. x 38 ft. x 14 ft., 575 gr. tons, 900 tons cap., for builder's account. Launching Sept., 1918.
Two wooden three-masted schooners for builder's account.
One tern schooner, "Donald Parsons," 140 ft. keel x 34 ft. 4 in. x 12 ft. 4 in., 500 gross tons, 825 tons d.w. for Mortimer Parsons. Launched May 11, 1918.
Another vessel, same size, to be launched Nov., 1918, for builder's account.

BAY SHORE SHIPYARD, CHURCH POINT, N.S.

Building two wooden vessels for R. C. Elkin, Ltd., St. John, N.B.

MOISE BELLIVEAU, CHURCH POINT, N.S.
One schooner, 450 tons, wood.

FIDELE BOUDREAU, CHURCH POINT, N.S.

One schooner, 350 tons, wood.

J. E. GASKILL, CHURCH POINT, N.S.
Other yards at Grosses Coques, Little Brook, Meteghan and Port Wade, N.S.
Five wooden schooners from 344 to 387 reg. tons. All sister ships.

COMEAU SHIPBUILDING CO., COMEAUVILLE, N.S.

One three-masted wooden schooner, for builder's account, 450 tons.

J. W. COMEAU, COMEAUVILLE, N.S.

One schooner, 329 tons, Jan., 1918.

J. N. RAFUSE & SONS, CONQUERALL BANKS, DIGBY CO., N.S. (Other plants at Salmon River and Shelburne, N.S.)
Two wooden three-masted schooners, 120 ft. x 32 ft. x 12 ft., 700 tons cap., for builder's account. For the market. Launchings Oct. 25, 1917, and Jan. 10., 1918.
One three-masted wooden schooner, "Integra," 112 ft. x 30 ft. x 11 ft. 6 in., 600 tons cap., for J. O. Williams & Co., St. John's, Newfoundland. Launching Nov. 25, 1917.

E. F. WILLIAMS, DARMOUTH, N.S.

One schooner, 360 tons, wood.

ROBAR BROTHERS, DAYSPRING, N.S.
Schooner for Capt. Ivan Creaser, Dayspring.

MAURICE E. LEARY, DAYSPRING, N.S.
One wooden schooner, 225 tons, 135 ft. o. a. To be launched Aug. 15, 1918, for La Have Outfitting Co., La Have, N.S.

J. NEWTON PUGSLEY & CHAS. ROBERTSON, DILIGENT RIVER, N.S.

One schooner, three-masted, 475 tons.

McLEAN & McKAY, ECONOMY, N.S.
One tern schooner, wood, 400 tons net, 135 ft. keel. To be launched Sept. 1, 1918, for builders' account.

S. J. SOLEY, ROX RIVER, N.S.

One schooner, three-masted, 121 ft. keel, 300 tons. Built of native wood. To be launched Sept., 1918. For builder's account. For sale.

ALLAN & FRASER, FRASERVILLE, N.S.

One schooner, 350 tons, wood.

BERNARD W. MELANSON, GILBERT'S COVE, N.S.

One three-masted schooner, wood, 250 tons net. To be completed November, 1018, for builder's account and for sale.

AMOS BLINN, GROSSES COQUES, N.S.

One schooner, 375 tons. Jan., 1918.
One schooner, 350 tons, wood.

BINN BROS., GROSSES COQUES, N.S.

F. K. WARREN & CO., GROSSES COQUES, N.S. (Offices in Union Bank Chambers, Halifax, N.S., also.)
350-ton tern schooner, under construction Jan., 1918.

HALIFAX SHIPBUILDING CO., HALIFAX, N.S.

FAUQUIER & PORTER, HANTSPORT, N.S.
Two wooden four-masted schooners, 178 ft. x 40 ft. x 18 ft., 1,000 gross tons, 2,000 tons d. w., twin 100 h.p. oil engines. Launchings Aug. 1, 1918, and Sept. 1, 1918. Both for sale.

J. WILLARD SMITH, HILLSBURN, N.S. (Head Office, St. John N.B.)

One tern schooner, wood, 140 ft. keel, 35 ft. beam, 450 tons net reg. tons.

HENRY COVEY, INDIAN HARBOR, N.S.

CHARLES GRIFFIN, ISAACS HARBOR, N.S.

One schooner, 40 tons, wood.

LEWIS SHIPBUILDING CO., LEWISTON, N.S.
One schooner, 670 tons. Jan., 1918.

INNOCENT COMEAU, LITTLE BROOK, N.S.

One three-masted schoner, 140 ft. x 35 ft. 6 in. x 17 ft. 6 in., 650 gr. tons, for the Weymouth Shipping Co., Weymouth, N.S., Can. Delivered Jan. 18, 1918.

J. W. RAYMOND, LITTLE BROOK, N.S.

One wooden schooner, No. 4, 140 ft. x 36 ft. x 17 ft., 575 gr. tons, for Jones Bros., Weymouth, Digby Co., N.S.

H. A. FRANK, LIVERPOOL, N.S.

Two wooden schooners, 500 tons, for builder's account.

McKEAN SHIPBUILDING CO., LIVERPOOL, N.S.

One wooden three-masted schooner, 126 ft. keel x 3 ft. x 12 ft. 9 in., 700 tons d.w., for builder's account. Launching May, 1918.

W. F. McKEAN & CO., LIVERPOOL, N.S.
One schooner 400 tons, Jan., 1918.

D. C. MULHALL, LIVERPOOL, N.S.

Two wooden schooners, one 300 gross tons, other 400 gross tons, for undisclosed interests.

N.S. SHIPBUILDING & TRANS. CO., LIVERPOOL, N.S.

One wooden tern schooner No. 2, 122 ft. x 33 ft. x 12 ft. 6 in., 450 gr. tons, 70 d.w. tons, 7 knots, for Peter Yee Wing & Co., Ltd., Sydney, Australia. R. McLeod and J. S. Gardner. Launching Feb. 1, 1918
One two-masted fishing schooner, "Sadie A. Nickle," 130 ft. o. a. x 26 ft. 10 ft. 6 in. hold, 230 d.w., two 50 h.p. oil engines, for Rafuse & Sons. Launching May 15, 1918.
One three-mast schooner, 420 tons. F-M Co. oil burning engines, for Job Bros., St. Johns, Nfld. Launching July 15.
One three-masted schooner, 270 tons, for builders' account. Launching Sept. 1.

One two-masted fishing schooner, 120 tons, for builders' account. Launching Oct. 1.

ROBIN, JONES & WHITMAN, LIVERPOOL, N.S.

One schooner, 340 tons, wood.

SOUTHERN SALVAGE CO., LIVERPOOL, N.S.

One wooden schooner, three masts, 300 gross tons, 600 tons capacity, for a West Indian firm. Launching Jan., 1918.

One wooden steamer, No. 5, 250 ft. x 43 ft. 6 in. x 25 ft., 2,900 gr. tons, 10 knots, 1,000 h.p., for undisclosed interests. Launching Fall, 1918.

One two-masted schooner, "Win-the-War," 137 ft. o. a. x 25 ft. 2 in. x 11 ft. 6 in., 187 gr. tons, for the builder's account. Launched Nov., 1917.

CONRAD & REINHARDT, LUNENBURG, N.S.

LUNENBURG MARINE RAILWAY, LUNENBURG, N.S.

FRED A. ROBAR, LUNENBURG, N.S.

SMITH & RHULAND, LUNENBURG, N.S.

Two schrs., 225 tons each. Jan., 1918.
Two schooners, 100 tons each.

"Donald Cook," 112 ft. beam, depth of hold, 11 ft., for Capt. William Cook. Launched April, 1918.
Two schooners, 168 gr. tons, for W. C. Smith.
One schooner, 150 gr. tons, for Lun. Outfitting Co.

J. B. YOUNG, LUNENBURG, N.S.

ERNST SHIPBUILDING CO., MAHONE BAY, N.S.

One two-masted schooner, 120 ft. o. a. x 25 ft. 6 in. x 10 ft. 6 in., 200 tons d.w., for Lunenburg Outfitting Co. April, 1918. Christened "Madeline Adams."

One three-masted schooner 125 ft. keel x 32 ft. x 12 ft., 600 tons, d. w. To be delivered July, 1918.
Two masted schooner, "Agnes D. McGlaston," 130 ft. x 26 ft. 8 in. x 10 ft. 8 in. Launched Nov. 13, 1917.
Keel laid for three-masted wooden schooner, 200 tons.

J. ERNEST & SON, MAHONE BAY, N.S.

One schooner, 520 tons. Jan., 1918.

O. A. HAM, MAHONE BAY, N.S.

One schooner, "Dozle," 128 tons. Launching May 30. One motor boat about 20 tons.

McLEAN CONSTRUCTION CO., MAHONE BAY, N.S. (Leased to Montague Mahaffy, Toronto.)
One three-masted schooner, 325 tons gr. reg., 500 tons cap., West Indies freight type. Will be launched in July.

Keel laid for another three-masted schooner with cargo capacity of 425 tons.

JOHN McLEAN & SONS, MAHONE BAY, N.S.

One schooner, 95 tons, wood.

J. A. BALSOM CO., LTD., MARGARETSVILLE, N.S.

One schooner, 409 tons. Jan., 1918.

CLARE SHIPBUILDING CO., METEGHAN RIVER, N.S.

One wooden schooner for builder's account. 400 tons.

A. H. COMEAU & CO., METEGHAN, N.S.

One schooner, 400 tons, wood.

AGAPIT COMEAU, METEGHAN, N.S.

JOHN F. DEVEAU, METEGHAN, N.S.

A 300-ton schooner was launched recently for Ritcey & Co., Lunenburg, N.S., and named Charles A. Ritcey.
1 schooner, 425 tons. Jan., 1918.
1 schooner, 400 tons, wood.

COLLINGWOOD SHIPBUILDING CO., COLLINGWOOD, ONT.

Two steel cargo steamers, Nos. 51-52, 50 ft. x 43 ft. x 25 ft., 2,500 gr. tons, 2,900 tons d. w., triple expansion engines 18-30-50 x 36, 2 boilers 14 ft. x 11 ft., 10 knots, for undisclosed interests. Delivery May and Aug., 1918.

Four deep-sea trawlers, Nos. 53-58, inclusive, 125 ft. x 23 ft. 6 in. x 13 ft. 6 in., 288 gross tons, 12¾-21½-35/24 engines, 1 cylinder 13 ft. 6 in. x 10 ft. 6 in., 10 knots, 500 h.p. for undisclosed interests.

One steel freight steamer of 3,800 tons d. w. for undisclosed interests. Keel laid May 8, 1918.

"War Wizard," May 8, 1918, 3,000 tons d. w., 261 ft. x 43 ft. 6 in. x 20 ft. depth. For undisclosed interests.

"War Witch" same size, now building. To be launched ——

R. MORRILL, COLLINGWOOD, ONT.

"Windsor," Aug. 10, steam tug, 105 ft., for Ontario Gravel & Freighting Co.

CAN. CAR & FOUNDRY CO., FORT WILLIAM, ONT. (Head Office Transportation Bldg., Montreal.)

Twelve steel mine sweepers for French Government.
145 ft. o.a. steel construction, value $2,500,000.
Plant under construction.

GREAT LAKES DREDGING CO., FORT WILLIAM, ONT.

Two wooden vessels, 2700 tons d.w. cap., 260 ft. o.a., 43 ft. beam. Triple expansion engines of 1,000 h.p. To be launched November, 1918.

"War Sioux" launched May 12, 1918. Keel laid for another twenty minutes later. Launching scheduled for November.

THUNDER BAY CONTRACTING CO., FORT WILLIAM, ONT.

One wooden freight steamer, 261 ft. long, for undisclosed interests.

NATIONAL SHIPBUILDING CO., GODERICH, ONT.

DAVIS DRYDOCK CO., FT. OF BAY ST., KINGSTON, ONT.

One motor pass. boat, 64 ft. x 18 ft. x 6 ft., 52 gr. tons, wood. For ferry service, Gananoque to Clayton. Delivery June 1, 1918.

Twenty-eight lifeboats, regulation size, wood and metal for Government and private interests.

KINGSTON SHIPBUILDING CO., KINGSTON, ONT.

Several steel trawlers for undisclosed interests. First one launched Dec. 22, 1917.

SELBY & YOULDSON, KINGSTON, ONT.

GEORGIAN BAY SHIPBUILDING & WRECKING CO., MIDLAND, ONT.

Tug, 50 tons, wood.

One tug 40 tons, wood.

MIDLAND DRY DOCK CO., MIDLAND, ONT.

Three steel freight steamers, 261 ft. x 43 ft. 6 in. x 23 ft., 3,400 tons d. w., 10 knots, for undisclosed interests. Deliveries, one in July, 1918, and two others before close of navigation, 1918.

Alterations on steamers "Mariska" and "Glenlyon."

"Western Star" being reconstructed.

MIDLAND SHIPBUILDING CO., LTD., MIDLAND, ONT.

Three steel freight steamers, 261 ft. x 43 ft., 6 ft x 23 ft., 3,400 tons d. w., 10 knots for undisclosed interests. Deliveries, one in July, 1918, and two others before opening of navigation, 1919.

PORT ARTHUR SHIPBUILDING CO., PORT ARTHUR, ONT.

Six steel trawlers on order. 135 ft. o. a. x 23 ft. 4 in. x 15 ft. 1 in., 294.5 tons, trip. exp. engines, 500 h.p., single end Scotch boiler. Two-masted.

Two of these launched June 8, 1918.

"War Isis," April 3, 1918, 3,400 tons d. w. Trawler launched same day. April 3, 1918, keel laid for "War Hatha," sister ship to "War Osiris," of same construction, nearly completed.

"War Fish," Aug. 19, 1917, 4,300 tons, steel.

"War Dance," Nov. 3, 1917, 3.400 tons, steel.

Four steel screw steamers, ocean freight service—"War Osiris," "War Hathor," "War Karma," "War Horus," 3,400 tons d. w. each, 261 ft. o.a. 251 ft. b.p. in length. Triple expansion engines, 2 Scotch boilers each, approx. 1,250 h.p. For completion, two Aug. 31st, and two Nov. '20 1918. For undisclosed interests.

MUIR BROS. DRYDOCK CO., LTD., PORT DALHOUSIE, ONT.

J. W. GEROW, ROSSPORT, ONT.

REID WRECKING CO., SARNIA, ONT.

One steel tug, 157 ft. x 32 ft. x 19 ft., trip. exp. engines, for lake or ocean service. Keel blocks laid March 5. Launching July, 1918. For Reid Wrecking Co.

WEST, PEACHEY & SONS, SIMCOOE, ONT.

Number of tugs for lumbering interests.

DOMINION SHIPBUILDING CO., TORONTO, ONT.

One steel cargo vessel, 261 ft. x 43 ft. x 28 ft. 2 in., 3,500 tons d.w. cap. For Department of Marine, Ottawa.
Have orders for five other vessels same dimensions.
"Traja" launched May 15, 1918.

THE POLSON IRON WORKS, LTD., TORONTO, ONT.

Eight steel cargo steamers, Nos. 133-4-5-6-7-9-40, 216 ft. o. a., 251 ft. b. p. x 43 ft. 6 in. x 22 ft. 11 in., draft loaded 19 ft. 6 in., single decked, 2,350 approx. gross tons, 3,500 tons cap., single screw trip, exp. engine 20½-33-54 x 36, 1,250 h.p., 10 knots, two 14 ft. x 12 ft. Scotch R. T. boilers, 180 h.p., to cost $600,000 each, for undisclosed interests. Launchings throughout 1918.

Four building and four on order.

"Asp," Feb. 11, 1918, 261 ft., 3,500 tons d.w., steel.

"Tento," Oct. 22, 1917, 261 ft., 3,500 tons d.w. steel. For Norwegian interests. Transferred to British registry.

THE THOR IRON WORKS, TORONTO, ONT.

Under same management as Dominion Iron Works.

TORONTO SHIPBUILDING CO., LTD., TORONTO, ONT. (Toronto Dry Dock Co., Ltd., under same management.)

Two wooden cargo steamers, 250 ft. B.P. 42 ft. 6 in., moulded breadth 25 ft., moulded depth 22 ft., draft 2,500 tons d. w. Triple exp. engines $\frac{20 \times 33 \times 54}{40}$. Howden boiler. Launchings July and Sept., 1918. For undisclosed interests.

THE BRITISH AMERICAN SHIPBUILDING CO., WELLAND, ONT.

Two steel frs., 3,500 tons d. w., 261 ft. O.A., 43 ft. beam, 23 ft. moulded depth, Westinghouse steam turbine engines. Delivery 1918.

WELLAND SHIPBUILDING CO., WELLAND, ONT.

One steel freight steamer, "War Weasel," 261 ft. x 43 ft. 6 in. x 23 ft., 3,300 tons d. w., trip. exp. engines, 14 ft. x 12 in. Scotch boilers, 10 knots, 1,250 h. p. Launching April, 1918.

Two deep frame cargo vessels, "War Badger" and "No. 3," 261 ft. x 43 ft. 6 in. x 23 ft., 3,300 d. w. tons, geared turbines, Howden's forced draught boilers, 10 knots, 1,250 h.p. Launchings June and August, 1918.

PRINCE EDWARD ISLAND

THE CARDIGAN SHIPBUILDING PLANT, CARDIGAN, P.E.I. (Purchased Annandale Lumber Co. plant and removed to Cardigan.)

One three-masted schooner, 325 tons, to be completed November, 1918.
Two more to be built.

QUEBEC

TIDEWATER SHIPBUILDERS, LTD., THREE RIVERS, QUE.

(Formerly Sorel Shipbuilding & Coal Co.)
Three steel trawlers for undisclosed interests.
One steel steam barge for the Canada Steamships Lines.

R. N. LE BLANC, BONAVENTURE, QUE.

J. Z. DEGAGNE, EBOULEMENTS, QUE.

DAVIE SHIPBUILDING & REPAIRING CO., LEVIS, QUE.

Six military barges, 130 feet long.
Eight steel trawlers, 2,035 tons.
One floating crane, 350 tons.
*One steel cargo vessel of 5,000 tons cap.
Building several steel lighters and several wooden drifters.

Steel car ferry "Canora" building for C.N.R., 308 ft. x 52 ft., cap. 20 loaded cars. Speed, 14 knots.

ATLAS CEMENT CONSTRUCTION CO., LTD., MONTREAL, QUE.

One concrete cargo steamer, "Concretia," 125 ft. x 22 ft. x 13 ft., steel ribbed, hull to be from 3 in. to 5 in., thick, for the builder's account.

CANADIAN VICKERS, LTD., MONTREAL, QUE.

Two cargo steamers, 9.400 tons, steel; 1 dredge, 2,364 tons, steel; 12 trawlers, 3,050 tons, steel; 23 drifters, 3,350 tons, wood.

Six freight steamers; 24 ft. draught, 7,000 tons cap., 11 knots, for undisclosed interests. Delivery, 1917, of two for Norwegian interests. "Porsanger," 394 ft. 6 in. o. a. x 40 ft. 4 in. x 30 ft., triple exp. engines, launched Nov. 29, 1917.

"War Earl" launched June 8, 1918, 7,000 tons d.w., 380 ft. x 49 ft. Keel of sister ship laid five minutes after this launching.

One steel freight steamer, 2,360 tons, cap., for undisclosed interests.

Three steel cargo vessels, 8,200 tons, to be laid down in May, Aug. and Sept., 1918, for undisclosed interests.

FRASER BRACE & CO., MONTREAL, QUE.

Four wooden cargo steamers, 3,000 tons, for undisclosed interests. Two keels laid during Oct., 1917.

HALL ENGINEERING CO., MONTREAL, QUE.

MONTREAL DRY DOCK & SHIP REPAIRING CO., MONTREAL, QUE.

Operate dock 428 ft. long 30 ft. deep.

MONTREAL SHIPBUILDERS LTD., 37 Belmont St., MONTREAL, QUE. (Associated with Atlas Construction Co.

THE QUEBEC SHIPBUILDING AND REPAIR CO., Board of Trade Bldg., MONTREAL, QUE.

2 vessels for undisclosed interests 3,000 tons each.

QUINLAN & ROBERTSON, MONTREAL, QUE. (Yards at Quebec.)

Four wooden steamers, for undisclosed interests.

LOUIS GAUNDRY, 12 St. Peter St., QUEBEC, QUE.

QUEBEC SHIPBUILDING & REPAIR CO., ST. LAURENT, QUE.

2 schrs. 1,400 tons and 1,200 tons.

One wooden four-masted auxl. schooner "Martin Connolly," 223 ft. x 42 ft. x 20 ft., 2,100 tons d. w., for undisclosed interests. Launched Oct. 28, 1917.

QUINLAN & ROBERTSON, QUEBEC, QUE.

Four wooden steamers, totalling 6,400 tons, for undisclosed interests.

CANADIAN GOVERNMENT SHIPYARDS, SOREL, QUE.

One steel vessel for undisclosed interests.

LECLAIRE SHIPBUILDING CO., SOREL, QUE.

Building 6 steel ships at value of $1,500,000.
Six trawlers, 125 ft. single screw type expansion engines, 500 I.P.H. steel. To be completed this year for private acct.

H. H. SHEPHERD, SOREL, QUE.

Rebuilding seven drifters. Particulars withheld owing to govt. restrictions.

SINCENNES-McNAUGHTON LINES. SOREL, QUE.

One tug 410 tons, wood.

CAN. GOVERNMENT SHIPYARD, SOREL, QUE.

One steel vessel for undisclosed interests.
Building steam trawlers and wooden drifters for undisclosed interests.

ST. LAWRENCE SHIPBUILDING & STEEL CO., SOREL, QUE.

THE THREE RIVERS SHIPYARDS LTD., THREE RIVERS, QUE.

Two cargo steamers, 3,100 d. w., wood, 260 ft. x 43 ft. 6 in., 1,000 H.P. engines. Names, "War Mingan" and "War Nicolet." To be completed end of navigation season for undisclosed interests.

Marine News from Every Source

Halifax.—John Schwartz, master, is advertising the schooner, Leba J. Schwartz, for sale. She is 95 tons register and was built in 1912.

Sault Ste. Marie.—Reports for the June traffic through the locks show that the iron ore shipments were 9,876,913 tons, the largest on record for June, exceeding by more than 300,000 tons the previous high record of 9,518,351 tons in June, 1917.

Montreal.—A 7,000-ton cargo steamer, the War Duchess, a sister ship to the War Earl, was launched at the shipyard of the Canadian Vickers, Limited, in this city. There was no public function. Col. W. I. Gear released the new vessel at 12.15 promptly and without a hitch.

Dominion incorporation has been granted to the Dominion Shipbuilding Co. Ltd., to construct, hire, purchase, worth and charter steamships, etc., to carry on business as iron founders, mechanical engineers, makers of munitions, tools, machinery, etc. Capital stock is $3,000,000, and the head office is in Toronto.

Halifax.—The contract for building the steel shipyards here has been awarded by Halifax Shipyards, Limited, to the Bedford Construction Co. (formerly Cavicchi & Pegano,) incorporated in Nova Scotia, it was announced by officials of the company, who predicted that the first of the three shipbuilding berths will be completed within four months' time.

Niagara Falls.—It is interesting to note that in the colossal enterprise in the building of the Chippawa-Queenston Hydro-Electric Power Canal, involving the expenditure of millions of dollars, the employment of a thousand men and the most modern and efficient machinery in the world, during the twelve months the great public work has been in operation, only two men have been killed.

Port Arthur—At 9.20 p.m., July 20, the steamer "War Hathor" was successfully launched at the shipyards of the Port Arthur Shipbuilding Company, Port Arthur, Ontario, Miss Hazel Whalen, daughter of the president of the company acting as sponsor. This seamer is being built on account of the Imperial Munitions Board for ocean freight service, and is almost identical to the steamer "War Isis."

Quebec.—The new Canadian Northern car ferry Canora, which has been built here by the Davie Shipbuilding & Repairs Co., was launched at the company's dry-dock at Lauzon. The new steamer, which will ply between Victoria, B.C., Patricia Bay and Port Mann, is 308 feet long, 52 feet beam, has a capacity of 20 loaded cars and a speed of 14 knots an hour. The Canora is a coal burner and cost between $400,000 and $500,000.

A New York despatch says that a syndicate of Canadian and American interests will construct ten steel ships of 8,800 tons each at Prince Rupert at a cost of approximately $16,000,000. The stetel is to be fabricated in Pittsburg, and assembled in Prince Rupert, at the floating dry dock of the Grand Trunk Pacific Railway. It is stated that the new company has leased the dry dock for five years for shipbuilding purposes.

New Westminster, B.C.—The work of installing engines and boilers and otherwise fitting the War Edensaw for sea is proceeding satisfactorily at the Tenth street wharf, and if the necessary equipment comes forward when wanted, the builders, the New Westminster Construction & Engineering Co., Ltd., expect to have this work completed within thirty days. Part of the main engines are already on board and the boilers, which are of the sectional type, are partially installed. Long ere the Edensaw is completed the next hull, that of the War Kitimat, will be alongside for completion.

THE S.S. Niagara, owned by the Union Steamship Company, of New Zealand, recently had an interesting round voyage from Vancouver to Australasia. She is a "combination" steamer, with two reciprocating engines driving the wing shafts and exhausing into a low-pressure turbine in the centre shaft. On the run to Canada one blade of the central propeller was lost, and the propeller was removed at Vancouver and an order given for a new one to be fitted on the vessel's next arrival at the port. The Niagara left with only her reciprocating engines working, and it was expected that she would have no difficulty in keeping up to her normal speed.

New Telephone Cable, Windsor to Detroit.—A new submarine telephone cable connecting Windsor with Detroit, and providing ample trans-river facilities for many years to come, has just been completed at a cost of $5,300. The cable is 3,000 feet in length, and weighs about fifteen tons. One hundred and two additional circuits, including "phantom" circuits, are provided. The cable was laid from a specially designed windlass reel placed on a powerful tug. A portion of it was placed in a trench on the river bottom in order to have it out of the way of dragging anchors. There are now three separate submarine cables connecting Windsor and Detroit.

London.—One of the surprises of the war is the work being carried on by the salvage department of the Admiralty. It has saved 407 vessels. New types of salvage machinery have come into use in the operations, and have been developed to an extent that would have been impossible in peace time. One illustration of this is the actual raising of ships from deep water by means of pontoons. It had always been held that a vessel, sunk in deep water, of an actual displacement or dead-weight exceeding 1,600 tons, must be regarded as a total loss. This theory has been proved unsound. Many big liners have been raised and repaired.

Fredericton.—The story is being told here of how two small steamships, which were laid up for twenty years because no purchasers could be found at any price, have just been sold for $1,260,000, and the case is said to be the most remarkable on record as showing the extraordinary advance in the value of vessel property. The two freighters, Port Caroline and Port Denison, were already old when at the end of the Boer war the owners decided to lay them up. They were out of commission until 1915. Then they were reclassed and offered for sale. Nobody would buy them. Now word has reached Boston that their lucky owners have disposed of them for more than a million and a quarter in cash.

Washington.—There is already a controversy on over the relative merits of concrete and steel ships, and in some places it has already settled down to a mild squabble between concrete and steel interests. The latter claim the building of concrete boats is being discouraged by the authorities. One report has it that "the steel interests" have the ear of the Shipping Board, and have used their access to it to discourage an extensive concrete programme. This seems a fantastic explanation, because there is not steel enough for the ships ordered, and concrete is advanced, not as a substitute for steel ships but as an additional source of ships over and beyond the limitations imposed by the available steel.

Washington.—The Fourth of July was celebrated by the splash of American ships which Charles M. Schwab, director-general of the Emergency Fleet Corporation, said would reecho in the ears of the German emperor. From one minute after twelve o'clock, when the first ship was launched at Superior, Wis., until late in the day, cargo carriers and other types of vessels were sent overboard in every part of the country to help build the ocean bridge for the allied fighting forces in Europe. The offices of the Emergency Fleet Corporation were kept open late into the night to receive official reports of the launchings. Incomplete reports up to shortly after midnight gave the number of vessels sent overboard as fifty-five, aggregating 299,364 tons dead weight. They represent 36 steel ships and 19 wooden.

Vancouver.—One of the chief objects of the visit of W. P. Hinton, general manager of the Grand Trunk Pacific, to the Pacific coast centres this month was to accelerate the construction of steel ships at the Prince Rupert shipyards. That city, in view of its natural advantages, supplemented by developments of recent years, has not received its share of attention in the shipbuilding activities. Hon. T. D. Pattulo, Minister of Lands, has been in correspondence with the G. T. P. management for a considerable period with regard to this question, and now Mr. Hinton expects that negotiations in progress will shape up at an early date, which will give the G. T. P. coast terminal what is desired, and that shipbuilding may be started without delay.

Ottawa.—Of 46 wooden steamships being constructed in Canada by the Im-

perial Munitions Board the great proportion will be ready to take the high seas before the present year ends. The first of them is to have her final trial trip at Vancouver next week, and a whole fleet will be ready by September. The work of installing machinery and equipment in these first launched will naturally require more time than when the operations are systematized by experience, but the work is being rushed as much as possible of the whole fleet. Twenty-seven are being built on the Pacific coast and 19 in the East. Each will have a capacity of 3,100 tons when they are completed. The Munitions Board will go out of business so far as shipbuilding is concerned. The industry hereafter will be directed and developed by the Government through the Department of Marine and Fisheries.

Ottawa.—A railway board judgment just issued reaffirms the right of railways to give reduced rates to points where a railway has to meet water competition. The application under consideration was that of the Board of Trade of Sidney, B.C., which wanted the board to rule that coast terminal rates should apply to shipments to Sidney via Great Northern and Canadian Northern car ferry services from Vancouver City. Sidney is located on Vancouver Island, and is eighteen miles nearer Vancouver than Victoria, but the rates to Victoria are lower than those that apply to Sidney. The board finds that as Victoria is a port for ocean vessels and there is actual competition via the Panama Canal and other water routes the railway company is justified in maintaining the lower rates to Victoria without making them applicable to intermediate non-competitive points like Sidney.

Port Stanley.—A fall of two feet in the level of Lake Erie has been reported by residents of Port Stanley. The fall is sufficient to cause much trouble in the unloading of boats, as it brings the decks a considerable distance below the docks. This makes a steep grade up which the merchandise must be taken. The Bessemer experienced this difficulty at Port Stanley and great care had to be exercised in unloading the coal cars. There is a variation in the height of the water level in Lake Erie every year, but the water line this season is reported to be much lower than it has been for years. No satisfactory explanation of the cause has been advanced. There is a periodic rise and fall in all the Great Lakes, but according to past experiences the water should be still rising and should not commence to recede until next year. If the phenomenon was temporary, a strong wind might have been the cause, but in this event it is not likely that the water would remain at the low level for more than a day. Rainfall in the north country affects the lakes to a great extent, and if the fall there has been much less than normal it may account for the lower level in Lake Erie.

DOMINION STEEL OUTPUT

The annual output of the Dominion Steel Corporation to March 31, 1918, compared with previous years, was as follows:

	31st March, 1917.	31st March, 1918.
Pig iron	346,926	332,231
Steel ingots.....	377,079	374,332
Blooms and billets for sale	144,051	139,557
Rails	17,495	17,108
Wire rods for sale	67,492	73,650
Bars	5,259	1,542
*Wire	35,142	27,165
Nails	20,175	16,347

*This includes wire used in the manufacture of nails shown in next line.

The output of ingots, which is the best measure of the Steel Company's production, remains practically at the standard attained during the two previous years, beyond which no considerable advance is possible until the new works now under construction become effective.

The total production from all collieries for the past five years was as follows:

Year ending:	Gross tons.
1918	3,781,615
1917	4,279,772
1916	5,261,198
1915	4,550,612
1914	5,047,683

The report adds: "Your directors again have to express their regret that, despite every effort on their part to prevent it, there has been a further serious diminution in the colliery output. As in the previous year, this falling off is to be attributed chiefly to the serious depletion of the working force through the operation of the Military Service Act."

INDIAN FIRM WANTS STEEL FOR CARS

A VERY important firm in Calcutta, India, who have contracts with the Indian government for works of national importance, wish to secure in Canada the following material for building cars in their works at Calcutta:

3,075 bars mild steel channels, 9" x 2.184" x 25 lb. per ft. x 19' 6".
2,050 bars mild steel channels, 9" x 2.814" x 25 lb. per ft. x 27' 6".
515 bars mild steel angles, 4" x 5" x ½" x 21' 9".
2,565 mild steel plates, 14' 0" x 2' 8½" x 3/16".
515 mild steel plates, 17' 0" x 3' 6" x 5/16".
587 mild steel plates, 12' 0" x 2' 9" x ⅜".
1,025 mild steel plates, 10' 0" x 2' 6" x ⅜".
268 mild steel plates, 20' 0" x 4' 0" x ½".
256 mild steel plates, 19' 4" x 2' 0" x ⅝".

All the above to be to "M.C.B." specifications but minimum tensile strength to be 58,500 pounds per square inch.

1,025 bars refined iron flat, 3" x 1" x 17' 1".
103 bars refined iron flat, 4" x 2" x 12' 9".
256 bars refined iron flat, 2¾" x 1¼" x 12' 9".
103 bars refined iron flat, 2" x 1" x 14' 3".
515 bars refined iron flat, 3¼" x 1" x 16' 0".
256 bars refined iron round, 3⅛" dia. x 21' 5".
340 bars refined iron round, 6" dia. x 16' 1".
1,070 bars refined iron round, 3" dia. x 16' 1".

Must have tensile steel strength of 89,660 to 100,500 per square inch.

1,025 bars high tensile steel, 1⅛" dia. 11' 0".
1,025 bars high tensile steel, 3¼" dia. 17' 0".

Must have tensile strength of 89,660 to 100,500 per square inch.

503 bars double refined iron round, 2½" x 20' 1".
256 bars double refined iron round, 4" x 12' 0".
103 bars double refined iron round, 2½" x 22' 4".
206 bars double refined iron round, 1⅛" x 17' 8".
2,050 bars double refined iron rd., 1" x 13' 6".
245 bars double refined iron flat, 2½" x 1⅛" x 15' 7".

All to standard specification for double refined iron.

7,175 all mild steel round, ⅝" dia. x 15' 0".
13,735 all mild steel round, ½" dia. x 15' 0".
15,375 all mild steel round, ⅜" dia. x 15' 0".

All to "M.C.B." specifications.

515 bars flat spring steel, 3" x ½" x 15' 1".
26,140 bars flat mild steel, 1¼" x 3/16" x ⅛" x 8' 0".

"M.C.B." specifications.

13,650 full twisted coal chain, 3/16" diameter.
882,000 hexagon nuts, ⅜", British thread.
63,000 hexagon nuts, ⅝", British thread.
126,000 hexagon nuts, ¾", British thread.
4,200 hexagon nuts, 1½", British thread.
14,700 split pins, 3/16" x 1¾".
11,550 split pins, ¼" x 2¼".

Incorporation has been granted to the British Colonies Transportation Co., Ltd., to build, buy, sell, equip and operate steamships, sailing ships, etc. Capital is placed at $750,000. Place of business will be St. John, N.B.

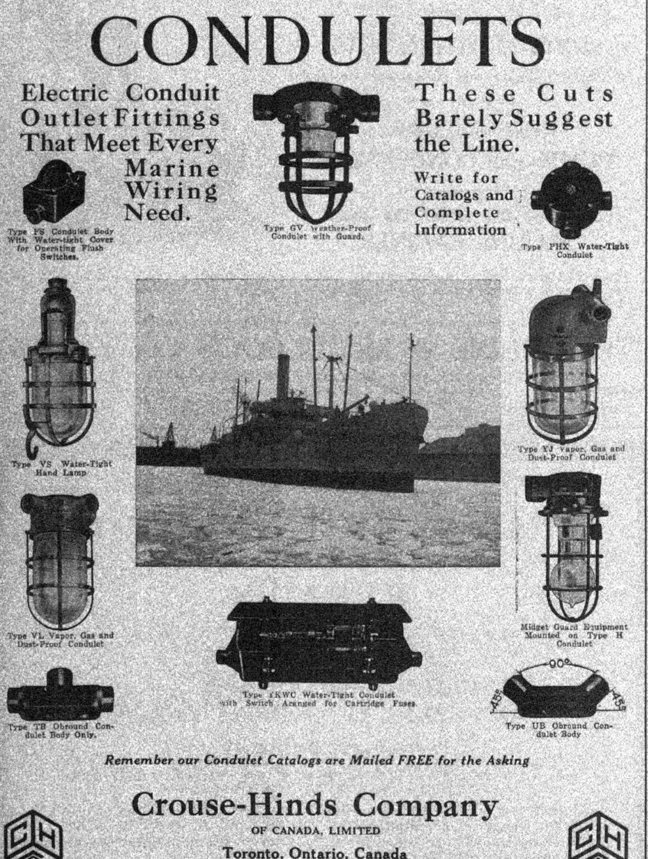
CONDULETS
Electric Conduit Outlet Fittings That Meet Every Marine Wiring Need.
These Cuts Barely Suggest the Line.
Write for Catalogs and Complete Information
Type FS Condulet Body With Water-tight Cover for Operating Flush Switches.
Type GV Weather-Proof Condulet with Guard.
Type PHX Water-Tight Condulet
Type VS Water-Tight Hand Lamp
Type YJ Vapor, Gas and Dust-Proof Condulet
Type VL Vapor, Gas and Dust-Proof Condulet
Midget Guard Equipment Mounted on Type H Condulet
Type TB Obround Condulet Body Only.
Type YKWC Water-Tight Condulet with Switch Aranged for Cartridge Fuses.
90°
45°
45°
Type UB Obround Condulet Body
Remember our Condulet Catalogs are Mailed FREE for the Asking
Crouse-Hinds Company
OF CANADA, LIMITED
Toronto, Ontario, Canada

Steel Castings

from ¼ of a pound to 30,000 pounds

SHIP CASTINGS

Steel Propeller Wheels — A Specialty — Steel Stockless Anchors

BEAUCHEMIN & FILS, LIMITED

SOREL, CANADA

We Do Contract Work for Ship Repair and Fitting-Out.

Quick Service a Specialty

CONDENSERS	WINDLASSES
PUMPS	WINCHES
FEED WATER HEATERS	STEERING ENGINES
	PROPELLERS
EVAPORATORS	MARINE ENGINES
FANS	BOILERS
AUXILIARY TURBINES	SHAFTING
	VALVES

CONDENSER AND BOILER TUBES

Immediate Delivery

Kearfott Engineering Co., Inc.

Frederick D. Herbert, President

95 Liberty Street, New York City Telephone Cortland 3415

MARINE Castings

Brass, Gunmetal, Manganese Bronze, Delta Metal, Nickel Alloys, Aluminum, etc.

MARINE AND LOCOMOTIVE ENGINE BEARINGS

MACHINE WORK AND ELECTRO PLATING. METAL PATTERN MAKING.

United Brass & Lead, Ltd., Toronto, Ont.

BERTRAM MACHINE TOOLS

For Structural, Bridge and Shipbuilding Plants

Modern in design and built for heavy service, our line embraces a varied equipment of Punches, Shears, Bending and Straightening Rolls, Coping Machines, Rotary and Plate Planers.

The assistance and advice of our engineers are yours for the asking.

Double Punch and Shear.
Capacity—
Shears 8-in. by 1½-in. plate.
Punches 2½-in. hole in 1¼-in. plate.

The John Bertram & Sons Co., Limited

DUNDAS, ONTARIO, CANADA

MONTREAL 723 Drummond Bldg. TORONTO 1002 C.P.R. Bldg. VANCOUVER 609 Bank of Ottawa Bldg.

WINNIPEG 1205 McArthur Bldg.

YARROWS

Limited

Associated with YARROW & CO., GLASGOW

WORKS AT ESQUIMALT, B. C.

SHIPBUILDERS, ENGINEERS AND SHIP REPAIRERS

IRON AND BRASS FOUNDERS

VESSELS CONVERTED FROM COAL BURNING TO OIL FUEL BURNING SYSTEMS.
MANUFACTURE OF MANGANESE BRONZE PROPELLER BLADES A SPECIALITY.

MARINE RAILWAY, LENGTH 300 ft., CAPACITY 2500 TONS DEAD WEIGHT

Telegrams and Cables "Yarrows, Victoria"

Address P.O. Box 1595 Victoria B. C.

Larger Vessels Docked in Government Graving Dock, Esquimalt—Lowest rates on the Pacific Coast.

LONDON PARIS ST. JOHN, N.B. MILAN

THE McNAB COMPANY

Bridgeport, Conn., U.S.A.

PATENTEES AND MANUFACTURERS OF

McNab Patent Engine Direction Indicators
" " Pneumatic Chart Room Counters
" " Pneumatic Engine Room Counters
" " Ship's Indicating Telegraph

McNab Patent "Cascade" Boiler Circulators
" " Steamship Draft Gauges
" Mechanical Rotary Turbine Engine Counter
" " Reciprocating Engine Counter
" Type Steam Steering Gear

Brown's Patent Telemotors and Steam Tillers

EMERGENCY FLEET CORPORATION TECHNICAL ORDER No. 30 SPECIFIES THE INSTALLATION OF McNAB PATENT ENGINE DIRECTION INDICATORS AND McNAB PATENT PNEUMATIC CHART ROOM COUNTERS ON ALL STEEL SHIPS UNDER CONSTRUCTION FOR THEM.

Catalogues on request.

BUYERS' DIRECTORY

ACCUMULATORS, HYDRAULIC
- Smart-Turner Mach. Co., Hamilton, Ont.

AERATING RESERVOIRS
- Spray Engineering Co., Boston, Mass.

AIR PORTS
- Mitchell Co., The Robert, Montreal, Que.
- Turnbull Elevator Mfg. Co., Toronto, Ont.

ALLOYS, BRASS AND COPPER
- Dom. Copper Products Co., Ltd., Montreal, Que.
- Mitchell Co., The Robert, Montreal, Que.
- Mueller Mfg. Co., H., Sarnia, Ont.
- Tallman Brass & Metal Co., Hamilton, Ont.
- United Brass & Lead, Ltd., Toronto, Ont.

ANCHORS
- Beauchemin & Fils, Sorel, P.Q.
- Hopkins & Co., F. H., Montreal, Que.
- Kearfott Engineering Co., New York, N.Y.
- McNab Co., Bridgeport, Conn.
- Henry Rogers Sons & Co., Wolverhampton, Eng.
- Wm. C. Wilson & Co., Toronto, Ont.

ARCHITECT, NAVAL
- Watts, J. Murray, Philadelphia, Pa.

ASBESTOS GOODS
- Wm. C. Wilson & Co., Toronto, Ont.

BABBITT METAL
- Aikenhead Hardware, Ltd., Toronto, Ont.
- Hoyt Metal Company, Toronto, Ontario.
- Tallman Brass & Metal Co., Hamilton, Ont.
- Wilkinson & Kompass, Hamilton, Ont.
- Wm. C. Wilson & Co., Toronto, Ont.

BAROMETERS
- Wilson & Co., Wm. C., Toronto, Canada.

BAR IRON
- Mitchells Ltd., Glasgow, Scotland.

BARS, GRATE
- Babcock & Wilcox, Ltd., Montreal, Que.

BEARINGS, BRASS
- Empire Mfg. Co., London, Ont.
- Mueller Mfg. Co., H., Sarnia, Ont.
- Mitchell Co., The Robert, Montreal, Que.
- Tallman Brass & Metal Co., Hamilton, Ont.
- United Brass & Lead Co., Toronto, Ont.

BELLS, SHIPS, ENGINE ROOM, ETC.
- Aikenhead Hardware, Ltd., Toronto, Ont.
- Clarke & Bros. Co., Montreal, Que.
- Cory & Son, Inc., Chas., New York, N.Y.
- Empire Mfg. Co., London, Ont.
- Mitchell Co., The Robert, Montreal, Que.
- Morrison Brass Mfg. Co., James, Toronto, Ont.
- Mueller Mfg. Co., H., Sarnia, Ont.
- Tallman Brass & Metal Co., Hamilton, Ont.
- United Brass & Lead, Ltd., Toronto, Ont.

BELTING, LEATHER
- Aikenhead Hardware, Ltd., Toronto, Ont.
- Wm. C. Wilson & Co., Toronto, Ont.

BIBBS, COMPRESSION
- Empire Mfg. Co., London, Ont.
- Mitchell Co., The Robert, Montreal, Que.
- Mueller Mfg. Co., H., Sarnia, Ont.
- United Brass & Lead, Ltd., Toronto, Ont.

BINNACLES
- Hopkins & Co., F. H., Montreal, Que.
- Morrison Brass Mfg. Co., James, Toronto, Ont.

BLOCKS, CARGO, HEEL AND TACKLE
- Aikenhead Hardware, Ltd., Toronto, Ont.
- Hopkins & Co., F. H., Montreal, Que.
- Leveridge, Ltd., Docks, Cardiff, Wales.

BLOWERS, TURBO-
- Mason Regulator & Engin. Co., Montreal, Que.

BOAT CHOCKS
- Corbet Fdry. & Machine Co., Owen Sound, Ont.
- Marten-Freeman Co., Toronto, Ont.

BOILER COMPOUND
- Wm. C. Wilson & Co., Toronto, Ont.

BOILER CIRCULATORS
- Eckliff Circulator Co., Detroit, Mich.

BOILER FEED PUMPS
- Can. Ingersoll-Rand Co., Ltd., Sherbrooke, Que.
- Goldie & McCulloch Co., Galt, Ont.
- Smart-Turner Mach. Co., Hamilton, Ont.
- Williams Machinery Co., A. R., Toronto, Ont.

BOILER FITTINGS
- Empire Mfg. Co., London, Ont.
- Goldie & McCulloch Co., Galt, Ont.
- McAvity & Sons Ltd., T., St. John, N.B.
- Wager Furnace Bridge Wall Co., Inc., New York, N.Y.

BOILERS, MARINE
- Babcock & Wilcox, Ltd., Montreal, Que.
- Can. Fairbanks-Morse Co., Montreal, Que.
- Can. Vickers, Ltd., Montreal, Que.
- Collingwood Shipbuilding Co., Collingwood, Ont.
- Doxford & Sons, William, Sunderland, England.
- Engr. & Mach. Wks. of Can., St. Catharines, Ont.
- Goldie & McCulloch, Ltd., Galt, Ont.
- Hall Engineering Works, Montreal, Que.
- Mason Regulator & Engin. Co., Montreal, Que.
- Montreal Dry Docks & Shipbuilding Co., Montreal, Que.
- National Shipbuilding Co., Goderich, Ont.
- Marsh Engineering Works, Belleville, Ont.
- Polson Iron Works, Toronto, Ontario.
- Port Arthur Shipbuilding Co., Port Arthur, Ont.
- Waterous Engine Works Co., Brantford, Ont.
- Williams Machinery Co., A. R., Toronto, Ont.

BOOKS, TECHNICAL, MARINE
- MacLean Publishing Co., Toronto, Ont.

BOLTS
- London Bolt & Hinge Works, London, Ont.
- Mitchell Co., The Robert, Montreal, Que.
- United Brass & Lead, Ltd., Toronto, Ont.
- Wilkinson & Kompass, Hamilton, Ont.

BROKER
- Page & Jones, Mobile, Ala., U.S.A.

BRASS GOODS
- Corbet Fdry. & Machine Co., Owen Sound, Ont.
- Goldie & McCulloch Co., Galt, Ont.
- McAvity & Sons, T., St. John, N.B.
- Mitchell Co., The Robert, Montreal, Que.
- Mueller Mfg. Co., H., Sarnia, Ont.
- Tallman Brass & Metal Co., Hamilton, Ont.

BUCKETS, CLAMSHELL
- Beatty & Sons, Welland, Ont.
- Morris Crane & Hoist Co., Herbert, Niagara Falls, Ont.

BUCKETS, DUMP
- Morris Crane & Hoist Co., Herbert, Niagara Falls, Ont.

BUCKETS, COALING
- Beatty & Sons, Welland, Ont.
- Engr. & Mach. Wks. of Can., St. Catharines, Ont.
- Hopkins & Co., F. H., Montreal, Que.
- Marsh Engineering Works, Belleville, Ont.
- Reed & Co., Geo., Montreal, Que.

BUSHINGS, BRONZE
- Oberdorfer Brass Co., M. L., Syracuse, N.Y.

CABLE
- Hopkins & Co., F. H., Montreal, Que.
- McNab Co., Bridgeport, Conn.
- Wm. C. Wilson & Co., Toronto, Ont.

CABLE, LEAD COVERED AND ARMORED
- Standard Underground Cable Co., Hamilton, Ont.

CABLE, ACCESSORIES
- Standard Underground Cable Co., Hamilton, Ont.

CAPSTANS
- Advance Mach. & Welding Co., Montreal, Que.
- Dake Engine Co., Grand Haven, Mich.
- Hopkins & Co., F. H., Montreal, Que.
- Kennedy & Sons, Wm., Owen Sound, Ont.
- Wm. C. Wilson & Co., Toronto, Ont.

CALKING TOOLS, ELECTRIC
- Aikenhead Hardware, Ltd., Toronto, Ont.

CALKING TOOLS, PNEUMATIC
- Aikenhead Hardware, Ltd., Toronto, Ont.
- Can. Ingersoll-Rand Co. Sherbrooke, Que.

CALORIFIERS
- Low & Sons, Ltd., Archibald, Glasgow, Scotland

CASTINGS
- Beauchemin & Fils, Sorel, P.Q.
- Can. Steel Foundries, Ltd., Montreal, Que.
- Collingwood Shipbuilding Co., Collingwood, Ont.
- Wm. Hamilton Co., Peterboro, Ont.
- Kennedy & Sons, Wm., Owen Sound, Ont.
- Goldie & McCulloch Co., Galt, Ont.
- Marsh Engineering Works, Belleville, Ont.
- Mitchell Co., Ltd., Robt., Montreal, Que.
- Mueller Mfg. Co., H., Sarnia, Ont.
- Tallman Brass & Metal Co., Hamilton, Ont.
- United Brass & Lead, Ltd., Toronto, Ont.
- Waterous Engine Works Co., Brantford, Ont.

CASTINGS, ALLOY
- Mitchell Co., The Robert, Montreal, Que.
- Mueller Mfg. Co., H., Sarnia, Ont.
- Oberdorfer Brass Co., M. L., Syracuse, N.Y.
- Tolland Mfg. Co., Montreal, Que.
- Tallman Brass & Metal Co., Hamilton, Ont.
- United Brass & Lead, Ltd., Toronto, Ont.

CASTINGS, ALUMINUM
- Empire Mfg. Co., London, Ont.
- Mitchell Co., The Robert, Montreal, Que.
- Tallman Brass & Metal Co., Hamilton, Ont.
- United Brass & Lead Co., Toronto, Ont.

CASTINGS, BRASS
- Crouse-Hinds Co. of Canada, Ltd., Toronto, Ont.
- Empire Mfg. Co., London, Ont.
- Goldie & McCulloch Co., Galt, Ont.
- McAvity & Sons Ltd., T., St. John, N.B.
- McNab Co., Bridgeport, Conn.
- Mitchell Co., Ltd., Robt., Montreal, Que.
- Mueller Mfg. Co., H., Sarnia, Ont.
- Oberdorfer Brass Co., M. L., Syracuse, N.Y.
- Tolland Mfg. Co., Montreal, Que.
- Tallman Brass & Metal Co., Hamilton, Ont.
- United Brass & Lead Co., Toronto, Ont.
- Waterous Engine Works Co., Brantford, Ont.

CASTINGS, GREY IRON, MALLEABLE, ALUMINUM
- Crouse-Hinds Co. of Canada, Ltd., Toronto, Ont.
- Engr. & Mach. Wks. of Can., St. Catharines, Ont.
- McAvity & Sons, T., St. John, N.B.
- McNab Co., Bridgeport, Conn.
- Mitchell Co., The Robert, Montreal, Que.
- Mueller Mfg. Co., H., Sarnia, Ont.
- Waterous Engine Works Co., Brantford, Ont.

CASTINGS, MANGANESE STEEL

CASTINGS, MANGANESE BRONZE
- Oberdorfer Brass Co., M. L., Syracuse, N.Y.
- Tallman Brass & Metal Co., Hamilton, Ont.
- United Brass & Lead Co., Toronto, Ont.

CELLAR DRAINERS
- Mueller Mfg. Co., H., Sarnia, Ont.

CHAINS
- Aikenhead Hardware, Ltd., Toronto, Ont.
- Kearfott Engineering Co., New York, N.Y.
- Hopkins & Co., F. H., Montreal, Que.
- Morris Crane & Hoist Co., Herbert, Niagara Falls, Ont.
- Henry Rogers, Sons & Co., Wolverhampton, Eng.
- Wm. C. Wilson & Co., Toronto, Ont.

CHAIN BLOCKS AND SLINGS
- Morris Crane & Hoist Co., Herbert, Niagara Falls, Ont.

CHANDLERY, SHIP
- Beauchemin & Fils, Sorel, P.Q.
- Hopkins & Co., F. H., Montreal, Que.
- Leckie, Ltd., John, Toronto, Ont.

CLAMPS, STEAM AND WATER
- Mueller Mfg. Co., H., Sarnia, Ont.

CLOCKS
- American Steam Gauge & Valve Mfg. Co., Boston, Mass.
- Morrison Brass Mfg. Co., James, Toronto, Ont.
- Williams Machinery Co., A. R., Toronto, Ont.

CLOSETS
- Mueller Mfg. Co., H., Sarnia, Ont.
- United Brass & Lead, Ltd., Toronto, Ont.

COAL
- Nova Scotia Steel & Coal Co., New Glasgow, N.S.

COAL HANDLING MACHINERY
- Morris Crane & Hoist Co., Herbert, Niagara Falls, Ont.
- Waterous Engine Works Co., Brantford, Ont.

COCKS, BILGE, DISCHARGE, INDICATOR,
- McAvity & Sons Ltd., T., St. John, N.B.
- Mitchell Co., The Robert, Montreal, Que.
- Morrison Brass Mfg. Co., James, Toronto, Ont.

COCKS, BASIN
- Empire Mfg. Co., London, Ont.
- Mitchell Co., The Robert, Montreal, Que.
- United Brass & Lead, Ltd., Toronto, Ont.

COMPASSES
- Wm. C. Wilson & Co., Toronto, Ont.

COMPRESSORS, AIR
- Can. Fairbanks-Morse Co., Montreal, Que.
- Canadian Ingersoll-Rand Co., Sherbrooke, Que.
- Hopkins & Co., F. H., Montreal, Que.
- Smart-Turner Mach. Co., Hamilton, Ont.
- Williams Machinery Co., A. R., Toronto, Ont.

COMPRESSED AIR TANKS
- MacKinnon Steel Co., Sherbrooke, Que.
- Scaife & Sons Co., Wm. B., Oakmont, Pa.

CONDENSERS
- Goldie & McCulloch Co., Galt, Ont.
- Kearfott Engineering Co., New York, N.Y.
- Morris Machine Works, Baldwinsville, N.Y.
- Smart-Turner Mach. Co., Hamilton, Ont.
- Weir Ltd., G. & J., Cathcart, Glasgow, Scotland.
- Williams Machinery Co., A. R., Toronto, Ont.

CONDULETS, MARINE
- Crouse-Hinds Co. of Canada, Ltd., Toronto, Ont.

CONTRACTORS' SUPPLIES
- McAvity & Sons, T., St. John, N.B.

CONSULTING ENGINEER
- Watt, J. Murray, Philadelphia, Pa.

CONVEYORS, ASH, COAL
- Babcock & Wilcox, Ltd., Montreal, Que.
- Hopkins & Co., F. H., Montreal, Que.

COPING MACHINES
- Bertram & Sons, Ltd., John, Dundas, Ont.

COPPER TUBES, SHEETS AND RODS
- Dom. Copper Products Co., Montreal, Que.
- Tallman Brass & Metal Co., Hamilton, Ont.

COTTON, CALKING
- Wilson & Co., Wm. C, Toronto, Canada.

COVERS, CANVAS, FOR HATCHES, LIFE-BOATS, ETC.
- Leckie, Ltd., John, Toronto, Ont.
- Waterous Engine Works Co., Brantford, Ont.

COUNTERS, REVOLUTION
- American Steam Gauge & Valve Mfg. Co., Boston, Mass.

CRANES
- Aikenhead Hardware, Ltd., Toronto, Ont.
- Can. Fairbanks-Morse Co., Montreal, Que.
- Hopkins & Co., F. H., Montreal, Que.
- Williams Machinery Co., A. R., Toronto, Ont.

CRANES, ELECTRIC
- Babcock & Wilcox, Ltd., Montreal, Que.
- Morris Crane & Hoist Co., Herbert, Niagara Falls, Ont.
- Smart-Turner Mach. Co., Hamilton, Ont.

CRANES, GOLIATH AND PNEUMATIC
- Morris Crane & Hoist Co., Herbert, Niagara Falls, Ont.

CRANES, GRANTRY, PORTABLE, JIB
- Morris Crane & Hoist Co., Herbert, Niagara Falls, Ont.
- Smart-Turner Mach. Co., Hamilton, Ont.

CRANES, OVERHEAD TRAVELLING
- Morris Crane & Hoist Co., Herbert, Niagara Falls, Ont.

CRANK SHAFTS
- Canada Foundries and Forgings, Welland, Ont.

DAVITS, BOAT
Corbet Fdry. & Machine Co., Owen Sound, Ont.
Hopkins & Co., F. H., Montreal, Que.
Marten, Freeman Co., Toronto, Ont.
Waterous Engine Works Co., Brantford, Ont.

DEAD LIGHTS, BRASS
Goldie & McCulloch Co., Galt, Ont.
Morris Crane & Hoist Co., Herbert, Niagara Falls, Ont.
Mitchell Co., Ltd., Robt., Montreal, Que.

DECK LIGHTS
Mitchell Co., The Robert, Montreal, Que.
Turnbull Elevator Mfg. Co., Toronto, Ont.

DECK PLUGS, ELECTRIC
Crouse-Hinds Co. of Canada, Ltd., Toronto, Ont.
Mitchell Co., The Robert, Montreal, Que.

DECKING FOR SHIPS
Marine Decking & Supply Co., Philadelphia, Pa.

DERRICKS
Aikenhead Hardware, Ltd., Toronto, Ont.
Dake Engine Co., Grand Haven, Mich.
Hopkins & Co., F. H., Montreal, Que.
Marsh Engineering Works, Belleville, Ont.
Morris Crane & Hoist Co., Herbert, Niagara Falls, Ont.

DISTILLERS
Kearfott Engineering Co., New York, N.Y.

DREDGES
Collingwood Shipbuilding Co., Collingwood, Ont.
Norbom Engineering Co., Philadelphia, Pa.
Polson Iron Works, Toronto.

DRILLS, AIR
Aikenhead Hardware, Ltd., Toronto, Ont.
Can. Ingersoll-Rand Co. Sherbrooke, Que.
Hopkins & Co., F. H., Montreal, Que.

DRILLS, CENTRE
Wilt Twist Drill Co., Walkerville, Ont.

DRILLS, BLACKSMITH AND BIT STOCK
Wilt Twist Drill Co., Walkerville, Ont.

DRILLS, ELECTRIC
Aikenhead Hardware, Ltd., Toronto, Ont.
Wilkinson & Kompass, Hamilton, Ont.

DRILLS, HIGH SPEED
Wilt Twist Drill Co., Walkerville, Ont.
Zenith Coal & Steel Products, Ltd., Montreal,

DRILLS, TRACK
Wilt Twist Drill Co., Walkerville, Ont.

DRILLS, RACHET AND HAND
Wilt Twist Drill Co., Walkerville, Ont.

DRILLS, TWIST
Aikenhead Hardware, Ltd., Toronto, Ont.
Williams Machinery Co., A. R., Toronto, Ont.
Wilt Twist Drill Co., Walkerville, Ont.

DRY DOCKS
Can. Vickers, Ltd., Montreal, Que.
Collingwood Shipbuilding Co., Collingwood, Ont.
Doxford & Sons, William, Sunderland, England.
Georgian Bay Shipbuilding & Wrecking Co., Midland, Ont.
National Shipbuilding Co., Goderich, Ont.
Polson Iron Works, Toronto, Ont.
Port Arthur Shipbuilding Co., Port Arthur, Ont.
Yarrows, Limited, Victoria, B.C.

ECONOMIZERS, FUEL
Babcock & Wilcox, Ltd., Montreal, Que.

EJECTORS
Empire Mfg. Co., London, Ont.
Mitchell Co., The Robert, Montreal, Que.
Morrison Brass Mfg. Co., James, Toronto, Ont.
Smart-Turner Mach. Co., Hamilton, Ont.

ELECTRIC LAMPS
Mitchell Co., The Robert, Montreal, Que.
Wm. C. Wilson & Co., Toronto, Ont.

ELECTRO-PLATING
Mitchell Co., The Robert, Montreal, Que.
Tallman Brass & Metal Co., Hamilton, Ont.
United Brass & Lead Co., Toronto, Ont.

ELECTRIC WELDING
Beauchemin & Fils, Sorel, P.Q.

ELEVATING MACHINERY
Goldie & McCulloch, Ltd., Galt, Ont.
Wm. Hamilton Co., Peterboro, Ont.
Morris Crane & Hoist Co., Herbert, Niagara Falls, Ont.
Waterous Engine Works Co., Brantford, Ont.

ELEVATORS
Turnbull Elevator Mfg. Co., Toronto, Ont.

ENGINES, HOISTING
Corbet Fdry. & Machine Co., Owen Sound, Ont.
Hopkins & Co., F. H., Montreal, Que.
Kennedy & Sons, Wm., Owen Sound, Ont.
Marsh Engineering Works, Belleville, Ont.
Port Arthur Shipbuilding Co., Port Arthur, Ont.
Williams Machinery Co., A. R., Toronto, Ont.

ENGINE, INTERNAL COMBUSTION
Doxford & Sons, William, Sunderland, England.

ENGINES, MARINE
Bolinders Co., New York, N.Y.
Can. Fairbanks-Morse Co., Montreal, Que.
Can. Vickers, Ltd., Montreal, Que.
Doxford & Sons, William, Sunderland, England.
Goldie & McCulloch, Ltd., Galt, Ont.
Hopkins & Co., F. H., Montreal, Que.
Iron Works, Ltd., Owen Sound, Ont.
Mason Regulator & Engin. Co., Montreal, Que.
National Shipbuilding Co., Goderich, Ont.
Norbom Engineering Co., Philadelphia, Pa.
Polson Iron Works, Toronto, Ont.
Port Arthur Shipbuilding Co., Port Arthur, Ont.
Trout Co., H. G., Buffalo, N.Y.
Waterous Engine Works Co., Brantford, Ont.
Williams Machinery Co., A. R., Toronto, Ont.

ENGINE STARTERS (AIR)
Scaife & Sons Co., Wm. B., Oakmont, Pa.

ENGINES, STEERING
Corbet Fdry. & Machine Co., Owen Sound, Ont.
Dake Engine Co., Grand Haven, Mich.
Kennedy & Sons, Wm., Owen Sound, Ont.
Wm. C. Wilson & Co., Toronto, Ont.

ENAMELWARE
Mueller Mfg. Co., H., Sarnia, Ont.
United Brass & Lead Co., Toronto, Ont

EVAPORATORS
Kearfott Engineering Co., New York, N.Y.
Mason Regulator & Engin. Co., Montreal, Que.
McNab Co., Bridgeport, Conn.
Weir Ltd., G. & J., Cathcart, Glasgow, Scotland.

EXTRACTORS, GREASE.
American Steam Gauge & Valve Mfg. Co., Boston, Mass.

EYE BOLTS AND NUTS
Canada Foundries and Forgings, Welland, Ont.
Mitchell Co., The Robert, Montreal, Que.
United Brass & Lead, Ltd., Toronto, Ont.

FANS
Aikenhead Hardware Ltd., Toronto, Ont.
Empire Mfg. Co., London, Ont.
Reed & Co., Geo. W., Montreal, Que.
Smart-Turner Mach. Co., Hamilton, Ont.
Williams Machinery Co., A. R., Toronto, Ont.

FENDERS, ROPE
Hopkins & Co., F. H., Montreal, Que.
Leckie, Ltd., John, Toronto, Ont.
Wilson & Co., Wm. C., Toronto, Canada.

FERRO-MANGANESE
Mitchells, Ltd., Glasgow, Scotland.

FILES
Aikenhead Hardware, Ltd., Toronto, Ont.
Williams Machinery Co., A. R., Toronto, Ont.

FIRE BRICKS
Beveridge Paper Co., Montreal, Que.
Mitchells, Ltd., Glasgow, Scotland.
Williams Machinery Co., A. R., Toronto, Ont.

FILTERS, FEED WATER
MacKinnon Steel Co., Sherbrooke, Que.
Mason Regulator & Engin. Co., Montreal, Que.

FITTINGS, MARINE
Hopkins & Co., F. H., Montreal, Que.
McAvity & Sons, T., St. John, N.B.
Mitchell Co., The Robert, Montreal, Que.
United Brass & Lead, Ltd., Toronto, Ont.

FITTINGS, MOTOR BOAT
Empire Mfg. Co., London, Ont.
Mitchell Co., The Robert, Montreal, Que.
Mueller Mfg. Co., H., Sarnia, Ont.
United Brass & Lead, Ltd., Toronto, Ont.

FIXTURES, ELECTRIC
Cory & Son, Inc., Chas., New York, N.Y.
Crouse-Hinds Co. of Canada, Ltd., Toronto, Ont.
Mitchell Co., The Robert, Montreal, Que.
Tallman Brass & Metal Co., Hamilton, Ont.

FLAG POLES, STEEL
Dennis Wire & Iron Works Co., London, Ont.

FLOW METERS
Spray Engineering Co., Boston, Mass.

FLUE CLEANERS
Wm. C. Wilson & Co., Toronto, Ont.

FORGES
Aikenhead Hardware, Ltd., Toronto, Ont.
Hopkins & Co., F. H., Montreal, Que.

FLANGING AND EXPANDING MACHINES, PIPE
Lovekin Pipe Expanding & Flanging Mach. Co., Philadelphia, Pa.

FLOODLIGHTS, ELECTRIC
Crouse-Hinds Co. of Canada, Ltd., Toronto, Ont.

FLUORSPAR
Mitchells, Ltd., Glasgow, Scotland.
Scythes & Co., Toronto, Ont.

FORGINGS, ALL KINDS
Aikenhead Hardware, Ltd., Toronto, Ont.
Collingwood Shipbuilding Co., Collingwood, Ont.
Nova Scotia Steel & Coal Co., New Glasgow, N.S.

FORGINGS, STEEL AND IRON
Canada Foundries and Forgings, Welland, Ont.

FURNACE BRIDGE WALLS
Wager Furnace Bridge Wall Co., Inc., 149 Broadway, New York, N.Y.

GAUGES, RECORDING
American Steam Gauge & Valve Mfg. Co., Boston, Mass.
Empire Mfg. Co., London, Ont.

GASKETS
Wm. C. Wilson & Co., Toronto, Ont.

GAUGES COCKS
McAvity & Sons, T., St. John; N.B.
Mitchell Co., The Robert, Montreal, Que.

GAUGE GLASSES
Wm. C. Wilson & Co., Toronto, Ont.

GAUGES, WATER, PRESSURE, COMPOUND AND VACUUM
Aikenhead Hardware, Ltd., Toronto, Ont.
Babcock & Wilcox, Ltd., Montreal, Que.
Empire Mfg. Co., London, Ont.
McAvity & Sons, T., St. John, N.B.
McNab Co., Bridgeport, Conn.
Morrison Brass Mfg. Co., James, Toronto, Ont.

GENERATORS AND CONVERTORS
Can. Fairbanks-Morse Co., Montreal, Que.

GLAZING COMPOSITION
Kuhls, H. B. Fred., 6411 3rd Ave., Brooklyn, N.Y.

GONGS
Clark & Bro., C. O., Montreal, Que.
Mitchell Co., Ltd., Robt., Montreal, Que.

GRAPHITE
Wm. C. Wilson & Co., Toronto, Ont.

GRATINGS
Corbet Fdry. & Machine Co., Owen Sound, Ont.
Canada Wire & Iron Goods Co., Hamilton, Ont.
MacKinnon Steel Co., Sherbrooke, Que.

GRINDERS, ELECTRIC
Wilkinson & Kompass, Hamilton, Ont.

GRINDERS, PNEUMATIC
Can. Ingersoll-Rand Co. Sherbrooke, Que.

GUARDS, MACHINERY
Dennis Wire & Iron Works Co., London, Ont.

GUY RODS AND ANCHORS, ELECTRIC
Crouse-Hinds Co. of Canada, Ltd., Toronto, Ont.

HAMMERS
Canada Foundries and Forgings, Ltd., Welland, Ont.

HARDWARE, MARINE
Hopkins & Co., F. H., Montreal, Que.
Mitchell Co., The Robert, Montreal, Que.

HEADLIGHTS, ELECTRIC
Crouse-Hinds Co. of Canada, Ltd., Toronto, Ont.
Hopkins & Co., F. H., Montreal, Que.
Tallman Brass & Metal Co., Hamilton, Ont.

HEATING EQUIPMENT
Empire Mfg. Co., London, Ont.
Low & Sons, Ltd., Archibald, Glasgow, Scotland

HEATERS, FEED, WATER
Babcock & Wilcox, Ltd., Montreal, Que.
Goldie & McCulloch Co., Galt, Ont.
Kearfott Engineering Co., New York, N.Y.
Mason Regulator & Engin. Co., Montreal, Que.
McNab Co., Bridgeport, Conn.
Weir Ltd., G. & J., Cathcart, Glasgow, Scotland.

HINGES
London Bolt & Hinge Works, London, Ont.
Mitchell Co., Ltd., Robt., Montreal, Que.

HOOKS
Hopkins & Co., F. H., Montreal, Que.
Morris Crane & Hoist Co., Herbert, Niagara Falls, Ont.

HOISTS, ASH
Beatty & Sons, Welland, Ont.
Marsh Engineering Works, Belleville, Ont.
St. Clair Bros., Galt, Ont.
Waterous Engine Works Co., Brantford, Ont.

HOIST BLOCKS
Morris Crane & Hoist Co., Herbert, Niagara Falls, Ont.

HOISTS, CHAIN
Aikenhead Hardware, Ltd., Toronto, Ont.
Can. Fairbanks-Morse Co., Montreal, Que.
Dake Engine Co., Grand Haven, Mich.
Hopkins & Co., F. H., Montreal, Que.
Morris Crane & Hoist Co., Herbert, Niagara Falls, Ont.
Williams Machinery Co., A. R., Toronto, Ont.

HOISTS, CARGO, MOVING, ETC.
Dake Engine Co., Grand Haven, Mich.
Hopkins & Co., F. H., Montreal, Que.
Marsh Engineering Works, Belleville, Ont.
Waterous Engine Works Co., Brantford, Ont.

HOISTING MACHINERY
Beatty & Sons, Welland, Ont.
Can. Ingersoll-Rand Co. Sherbrooke, Que.
Corbet Fdry. & Machine Co., Owen Sound, Ont.
Wm. Hamilton Co., Peterboro, Ont.
Marsh Engineering Works, Belleville, Ont.
Morris Crane & Hoist Co., Herbert, Niagara Falls, Ont.
Waterous Engine Works Co., Brantford, Ont.
Williams Machinery Co., A. R., Toronto, Ont.

HOSE
Wm. C. Wilson & Co., Toronto, Ont.

INDICATORS, ENGINE
American Steam Gauge & Valve Mfg. Co., Boston, Mass.
Cory & Son, Inc., Chas., New York, N.Y.
McNab Co., Bridgeport, Conn.

INDICATORS, SPEED
Aikenhead Hardware, Ltd., Toronto, Ont.
Cory & Son, Inc., Chas., New York, N.Y.

INJECTORS
Aikenhead Hardware, Ltd., Toronto, Ont.
Empire Mfg. Co., London, Ont.
Mitchell Co., The Robert, Montreal, Que.
Morrison Brass Mfg. Co., James, Toronto, Ont.
Williams Machinery Co., A. R., Toronto, Ont.

INGOTS
Broughton Copper Co., Ltd., Manchester, Eng.

INSULATORS, ELECTRIC
Crouse-Hinds Co. of Canada, Ltd., Toronto, Ont.

INSTRUMENTS, NAUTICAL
Leckie, Ltd., John, Toronto, Ont.

IRON AND STEEL
Mitchells Ltd., Glasgow, Scotland.

JACKS
Hopkins & Co., F. H., Montreal, Que.
Morris Crane & Hoist Co., Herbert, Niagara Falls, Ont.

JOINTS, BALL SLIP
Norbom Engineering Co., Philadelphia, Pa.

INSURANCE, MARINE
Toronto Insurance & Vessel Agency, Toronto, Ont.

JOURNAL WEDGES
Canada Foundries and Forgings, Ltd., Welland, Ont.

JUTE PACKING
Stratford Oakum Co., Geo., Jersey City, N.J.

LATHES
Bertram & Sons, Ltd., John, Dundas, Ont.

LEATHER, LACE
Wm. C. Wilson & Co., Toronto, Ont.

LIFEBUOYS, BELTS AND PRESERVERS
Fosbery Co., Barking, Essex, Eng.
Hopkins & Co., F. H., Montreal, Que.
Wm. C. Wilson & Co., Toronto, Ont.

LIFEBOATS
Hopkins & Co., F. H., Montreal, Que.
Wm. C. Wilson & Co., Toronto, Ont.

LIFE BOAT EQUIPMENT
Eckliff Circulator Co., Detroit, Mich.
Hopkins & Co., F. H., Montreal, Que.
Leckie, Ltd., John, Toronto, Ont.
Wm. C. Wilson & Co., Toronto, Ont.

LIFE JACKETS
Fosbery Co., Barking, Essex, Eng.
Hopkins & Co., F. H., Montreal, Que.
Leckie, Ltd., John, Toronto, Ont.

LIGHTS, ALL KINDS
Can. Fairbanks-Morse Co., Montreal, Que.
Crouse-Hinds Co. of Canada, Ltd., Toronto, Ont.
Hopkins & Co., F. H., Montreal, Que.
Mitchell Co., The Robert, Montreal, Que.
Morrison Brass Mfg. Co., James, Toronto, Ont.
Tallman Brass & Metal Co., Hamilton, Ont.

LIGHTS, SIDE, PORT
Goldie & McCulloch Co., Galt, Ont.
Crouse-Hinds Co. of Canada, Ltd., Toronto, Ont.
Hopkins & Co., F. N., Montreal, Que.
McAvity & Sons Ltd., T., St. John, N.B.
Mitchell Co., The Robert, Montreal, Que.
Tallman Brass & Metal Co., Hamilton, Ont.
Turnbull Elevator Mfg. Co., Toronto, Ont.
Wilson & Co., Wm. C., Toronto, Ont.

LIGHTING SETS
Can. Fairbanks-Morse Co., Montreal.
Cory & Son, Inc., Chas., New York, N.Y.
Crouse-Hinds Co. of Canada, Ltd., Toronto, Ont.
Goldie & McCulloch Co., Ltd., Galt, Ont.
Kearfott Engineering Co., New York, N.Y.
Williams Machinery Co., A. R., Toronto, Ont
Wilson & Co., Wm. C., Toronto, Canada.

LOADERS, WAGON AND TRUCK
Morris Crane & Hoist Co., Herbert, Niagara Falls, Ont.
Waterous Engine Works Co., Brantford, Ont.

LOCKERS, METAL
Dennis Wire & Iron Works Co., London, Ont.
Reed & Co., Geo. W., Montreal, Que.

LUBRICATORS
Empire Mfg. Co., London, Ont.
Mitchell Co., The Robert, Montreal, Que.

MACHINISTS
Corbet Fdry. & Machine Co., Owen Sound, Ont.
Wm. Hamilton Co., Peterboro, Ont.
Hyde Engineering Works, Montreal, Que.
Mitchell Co., The Robert, Montreal, Que.
United Brass & Lead Co., Toronto, Ont.

MANGANESITE, PASTE
Wm. C. Wilson & Co., Toronto, Ont.

MARINE CASTINGS
Wm. Hamilton Co., Peterboro, Ont.
Goldie & McCulloch Co., Galt, Ont.
Mitchell Co., Ltd., Robt., Montreal, Que.
McAvity & Sons, T., St. John, N.B.
Mueller Mfg. Co., H., Sarnia, Ont.
Tallman Brass & Metal Co., Hamilton, Ont.
United Brass & Lead, Ltd., Toronto, Ont.
Waterous Engine Works Co., Brantford, Ont.

MARINE SPECIALTIES
Goldie & McCulloch Co., Galt, Ont.
Hopkins & Co., F. H., Montreal, Que.
McAvity & Sons, T., St. John, N.B.
Mitchell Co., Ltd., Robt., Montreal, Que.
United Brass & Lead, Ltd., Toronto, Ont.

MOTORS
Aikenhead Hardware, Ltd., Toronto, Ont.
Can. Fairbanks-Morse Co., Montreal, Que.
Williams Machinery Co., A. R., Toronto, Ont.
Zenith Coal & Steel Products, Ltd., Montreal.

NAUTICAL INSTRUMENTS
Wm. C. Wilson & Co., Toronto, Ont.

NET LIFTERS
Dake Engine Co., Grand Haven, Mich.
Leckie, Ltd., John, Toronto, Ont.

NOZZLES, ALL KINDS
Empire Mfg. Co., London, Ont.
Mitchell Co., The Robert, Montreal, Que.
Tolland Mfg. Co., Boston, Mass.

NUTS
Mitchell Co., The Robert, Montreal, Que.
Wilkinson & Kompass, Hamilton, Ont.
United Brass & Lead, Ltd., Toronto, Ont.

OIL CUPS
Empire Mfg. Co., London, Ont.
Mitchell Co., Ltd., Robt., Montreal, Que.

OILCLOTHING
Wm. C. Wilson & Co., Toronto, Ont.

OIL, LINSEED
Canada Linseed Oil Mills, Montreal, Que.

OILS AND GREASE
Wm. C. Wilson & Co., Toronto, Ont.

OAKUM
Davey & Sons W O., Jersey City., N.J.
Leckie, Ltd., John, Toronto, Ont.
Stratford Oakum Co., Geo., Jersey City, N.J.
Wm. C. Wilson & Co., Toronto, Ont.

ORNAMENTAL IRONWORK
Canada Wire & Iron Goods Co., Hamilton, Ont.

OXY-ACETLYENE OUTFITS
Can. Fairbanks-Morse Co., Montreal, Que.
Hopkins & Co., F. H., Montreal, Que.
McAvity & Sons Ltd., T., St. John, N.B.
Williams Machinery Co., A. R., Toronto, Ont.

PAINT
Ault & Wiborg Co. of Can., Ltd., Toronto, Ont.
Wm. C. Wilson & Co., Toronto, Ont.

PAINTS, ANTI-FOULING
Kuhls, H. B. Fred., 6411 3rd Ave., Brooklyn, N.Y.

PAINTS, ANTI-CORROSIVE
Kuhls, H. B. Fred. 6411 3rd Ave., Brooklyn, N.Y.

PAINTING EQUIPMENT; AUTOMATIC
Spray Engineering Co., Boston, Mass.

PACKING, AMMONIA
France Packing Co., Philadelphia, Pa.
Wm. C. Wilson & Co., Toronto, Ont.

PACKING, MARINE
Aikenhead Hardware, Ltd., Toronto, Ont.
France Packing Co., Philadelphia, Pa.
Wm. C. Wilson & Co., Toronto, Ont.

PACKING, METALLIC
Aikenhead Hardware, Ltd., Toronto, Ont.
France Packing Co., Philadelphia, Pa.

PACKING, STEAM
Aikenhead Hardware, Ltd., Toronto, Ont.
France Packing Co., Philadelphia, Pa.
Wm. C. Wilson & Co., Toronto, Ont.

PANELBOARD AND CABINETS, ELECTRIC
Crouse-Hinds Co. of Canada, Ltd., Toronto, Ont
Preston Woodworking Machy. Co., Preston, Ont.

PATTERNS, WOOD AND METAL
Beauchemin & Fils, Sorel, P.Q.

PIPE, LAPWELD, CAST IRON, RIVETED
Empire Mfg. Co., London, Ont.
Norbom Engineering Co., Philadelphia, Pa.
United Brass & Lead, Ltd., Toronto, Ont.

PIPE RAILING, IRON AND BRASS
Dennis Wire & Iron Works Co., London, Ont.
Mitchell Co., The Robert, Montreal, Que.

PITCH
Reed & Co., Geo., Montreal, Que.

PITCH, PINE
Wilson & Co., Wm. C., Toronto, Canada.

PLANERS
Bertram & Sons, Ltd., John, Dundas, Ont.

PLANERS, STANDARD AND ROTARY
Yates Mach. Co., P. B., Hamilton, Ont.

PLATE CLAMPS
Morris Crane & Hoist Co., Herbert, Niagara Falls, Ont.

PLATE PUNCH TABLES
Norbom Engineering Co., Philadelphia, Pa.

PLUMBING EQUIPMENT
Low & Sons, Ltd., Archibald, Glasgow, Scotland.
Empire Mfg. Co., London, Ont.

Mitchell Co., The Robert, Montreal, Que.
Mueller Mfg. Co., H., Sarnia, Ont.
Tallman Brass & Metal Co., Hamilton, Ont.
Wm. C. Wilson & Co., Toronto, Ont.
United Brass & Lead, Ltd., Toronto, Ont.

PROPELLOR BLADES, BRONZE
Corbet Fdry. & Machine Co., Owen Sound, Ont.
Empire Mfg. Co., London, Ont.
Mitchell Co., The Robert, Montreal, Que.
Tallman Brass & Metal Co., Hamilton, Ont.
United Brass & Lead, Ltd., Toronto, Ont.
Yarrows, Limited, Victoria, B.C.

PROPELLER WHEELS
Goldie & McCulloch Co., Galt, Ont.
Kennedy & Sons, Wm., Owen Sound, Ont.
Trout Co., H. G., Buffalo, N.Y.

PIPING, STEAM
Babcock & Wilcox, Ltd., Montreal, Que.

PORT LIGHTS
Goldie & McCulloch Co., Galt, Ont.
Mitchell Co., Ltd., Robt., Montreal, Que.
Mueller Mfg. Co., H., Sarnia, Ont.
Turnbull Elevator Mfg. Co., Toronto, Ont.
Tallman Brass & Metal Co., Hamilton, Ont.
Wm. C. Wilson & Co., Toronto, Ont.
United Brass & Lead, Ltd., Toronto, Ont.

PROPELLER WHEELS
Beauchemin & Fils, Sorel, P.Q.
McNab Co., Bridgeport, Conn.

PULLEYS
Wm. Hamilton Co., Peterboro, Ont.
Smart-Turner Mach. Co., Hamilton, Ont.
Williams Machinery Co., A. R., Toronto, Ont.

PUMP, AIR
Can. Ingersoll-Rand Co. Sherbrooke, Que.
Smart-Turner Mach. Co., Hamilton, Ont.
Weir Ltd., G. & J., Cathcart, Glasgow, Scotland.
Williams Machinery Co., A. R., Toronto, Ont.

PUMPS
Can. Fairbanks-Morse Co., Montreal, Que.
Canada Foundries & Forgings, Welland, Ont.
Goldie & McCulloch, Ltd., Galt, Ont.
Hopkins & Co., F. H., Montreal, Que.
McAvity & Sons, T., St. John, N.B.
Oberdorfer Brass Co., M. L., Syracuse, N.Y.
Williams Machinery Co., A. R., Toronto Ont.

PUMPS, CENTRIFUGAL
Can. Ingersoll-Rand Co. Sherbrooke, Que.
Wm. Hamilton Co., Peterboro, Ont.
Goldie & McCulloch Co., Galt, Ont.
Kearfott Engineering Co., New York, N.Y.
Norbom Engineering Co., Philadelphia, Pa.
Smart-Turner Mach. Co., Hamilton, Ont.
Waterous Engine Works Co., Brantford, Ont.
Williams Machinery Co., A. R., Toronto, Ont.

PUMPS, BILGE
Goldie & McCulloch Co., Galt, Ont.
Mitchell Co., Ltd., Robt., Montreal, Que.
Mueller Mfg. Co., H., Sarnia, Ont.

PUMPS, CIRCULATING
Goldie & McCulloch Co., Galt, Ont.
McNab Co., Bridgeport, Conn.
Morris Machine Works, Baldwinsville, N.Y.

PUMPS, FEED WATER
Goldie & McCulloch Co., Galt, Ont.
McNab Co., Bridgeport, Conn.
Weir Ltd., G. & J., Cathcart, Glasgow, Scotland.
Williams Machinery Co., A. R., Toronto, Ont.

PUMPS, LIFT AND FORCE
Low & Sons, A., Glasgow, Scotland.

PUMPS, REES ROTURBO
Goldie & McCulloch Co., Galt, Ont.

PUMPS, HAND AND POWER
Aikenhead Hardware, Ltd., Toronto, Ont.
Mitchell Co., The Robert, Montreal, Que.
Smart-Turner Mach. Co., Hamilton, Ont.
Williams Machinery Co., A. R., Toronto, Ont.

PUMPS, HIGH PRESSURE
Canadian Ingersoll-Rand Co., Sherbrooke, Que.
Smart-Turner Mach. Co., Hamilton, Ont.

PUMPS, STEAM TURBO
Wm. C. Wilson & Co., Toronto, Ont.

PUMPING MACHINES
Norbom Engineering Co., Philadelphia, Pa.

PUNCHES
Bertram & Sons, Ltd., John, Dundas, Ont.

PUNCHES, SINGLE, DOUBLE AND MULTIPLE
Norbom Engineering Co., Philadelphia, Pa.

PURIFIERS, WATER
Babcock & Wilcox, Ltd., Montreal, Que.

RADIATORS, STEAM, ELECTRIC
Empire Mfg. Co., London, Ont.
Low & Sons, Ltd., Archibald, Glasgow, Scotland.
Wm. C. Wilson & Co., Toronto, Ont.

RAILS, OVERHEAD
Morris Crane & Hoist Co., Herbert, Niagara Falls, Ont.

RANGES
Hopkins & Co., F. H., Montreal, Que.
Wm. C. Wilson & Co., Toronto, Ont.

RECEIVERS, AIR
Can. Ingersoll-Rand Co. Sherbrooke, Que.
MacKinnon Steel Co., Sherbrooke, Que.

REGULATORS, FEED WATER
American Steam Gauge & Valve Mfg. Co. Boston, Mass.

REGULATORS, PRESSURE
Mason Regulator & Engin. Co., Montreal, Que.

RELEASING GEARS
Eckliff Circulator Co., Detroit, Mich.

REPAIRS, MARINE
Corbet Foundry & Mach. Co., Owen Sound, Ont
Can. Vickers, Ltd., Montreal, Que.
Collingwood Shipbuilding Co., Collingwood, Ont.
Engr. & Mach. Wks. of Can., St. Catharines, Ont.
Georgian Bay Shipbuilding & Wrecking Co. Midland, Ont.
Hyde Engineering Works, Montreal, Que.
Iron Works, Ltd., Owen Sound, Ont.
Kennedy & Sons, Wm., Owen Sound, Ont.
Muir Bros. Dry Dock Co., Port Dalhousie, Ont.
National Shipbuilding Co., Goderich, Ont.
Port Arthur Shipbuilding Co., Port Arthur, Ont.
Yarrow, Limited, Victoria, B.C.

RIGGING SCREWS
Hopkins & Co., F. H., Montreal, Que.

RIGGING, WIRE ROPE
Leckie, Ltd., John, Toronto, Ont.

RIVETS
London Bolt & Hinge Works, London, Ont.
Severance Mfg. Co., S., Glassport, Pa.
Wilkinson & Kompass, Hamilton, Ont.

RIVETERS, PNEUMATIC
Can. Ingersoll-Rand Co., Ltd., Sherbrooke, Que.

RODS, COPPER, BRASS, BRONZE
Standard Underground Cable Co., Hamilton, Ont
Tallman Brass & Metal Co., Hamilton, Ont.

ROLLS, STRAIGHTENING, BENDING
Bertram & Sons, Ltd., John, Dundas, Ont.

ROOFING
Reed & Co., Geo., Montreal, Que.
Wm. C. Wilson & Co., Toronto, Ont.

ROPE BLOCKS
Aikenhead Hardware, Ltd., Toronto, Ont.
Can. Fairbanks-Morse Co., Montreal, Que.
Morris Crane & Hoist Co., Herbert, Niagara Falls, Ont.

ROPE
Hopkins & Co., F. H., Montreal, Que.
Leckie, Ltd., John, Toronto, Ont.
McNab Co., Bridgeport, Conn.
Stratford Oakum Co., Geo., Jersey City, N.J.
Wm. C. Wilson & Co., Toronto, Ont.

RUBBER COATS
Wm. C. Wilson & Co., Toronto, Ont.

SAW MILL MACHINERY
Wm. Hamilton Co., Peterboro, Ont.
Preston Woodworking Machy. Co., Preston, Ont.
Waterous Engine Works Co., Brantford, Ont.
Yates Machine Co., P. B., Hamilton, Ont.

SAWS, BAND
Preston Woodworking Machy. Co., Preston, Ont.

SCALES, BOILERS, ENGINES
Can. Fairbanks-Morse Co., Montreal, Que.

SCOWS
Collingwood Shipbuilding Co., Collingwood, Ont.
Polson Iron Works, Toronto, Ont.

SCREENS, WIRE
Dennis Wire & Iron Works Co., London, Ont.

SCREWS, COACH
London Bolt & Hinge Works, London, Ont.

SEAM PAINT
Kuhls, H. B. Fred., 6411 3rd Ave., Brooklyn, N.Y.

SEPARATORS, OIL, STEAM
Mason Regulator & Engin. Co., Montreal, Que.
Smart-Turner Mach. Co., Hamilton, Ont.

SHAFTING
Wm. Hamilton Co., Peterboro, Ont.
Mitchell Co., The Robert, Montreal, Que.
Wilkinson & Kompass, Hamilton, Ont.

SHAFTING, BRONZE
Empire Mfg. Co., London, Ont.
Tallman Brass & Metal Co., Hamilton, Ont.

SHEARS
Bertram & Sons, Ltd., John, Dundas, Ont.
Norbom Engineering Co., Philadelphia, Pa.

SHIPBUILDING TOOLS
Aikenhead Hardware, Ltd., Toronto, Ont.
Can. Ingersoll-Rand Co. Sherbrooke, Que.

SHIPS, BUILDERS OF
Can. Vickers, Ltd., Montreal, Que.
Collingwood Shipbuilding Co., Collingwood, Ont.
Doxford & Sons, William, Sunderland, England.
Georgian Bay Shipbuilding & Wrecking Co., Midland, Ont.
National Shipbuilding Co., Goderich, Ont.
Polson Iron Works, Toronto, Ont.
Port Arthur Shipbuilding Co., Port Arthur, Ont.
Yarrows, Limited, Victoria, B.C.

SHIP BROKERS
Page & Jones, Mobile, Ala.

SHIP PLATES
Nova Scotia Steel & Coal Co., New Glasgow, N.S.

SLEDGES
Wilkinson & Kompass, Hamilton, Ont.

SLINGS
Hopkins & Co., F. H., Montreal, Que.
Morris Crane & Hoist Co., Herbert, Niagara Falls, Ont.

SPECIAL MACHINERY
Corbet Fdry. & Machine Co., Owen Sound, Ont.
Wm. Hamilton Co., Peterboro, Ont.
Miller Bros. & Sons, Ltd., Montreal, Que.
Smart-Turner Mach. Co., Hamilton, Ont.

SMOOTH-ON
Wm. C. Wilson & Co., Toronto, Ont.

SPIKES
Wm. C. Wilson & Co., Toronto, Ont

SPRAY COOLING SYSTEMS
Spray Engineering Co., Boston, Mass.

STEAMSHIP AGENTS
Page & Jones, Mobile, Ala.

STEAM SPECIALTIES
Can. Fairbanks-Morse Co., Montreal, Que.
Corbet Fdry. & Machine Co., Owen Sound, Ont.
Empire Mfg. Co., London, Ont.
Mitchell Co., The Robert, Montreal, Que.

STEAM TRAPS
Aikenhead Hardware, Ltd., Toronto, Ont.
American Steam Gauge & Valve Mfg. Co., Boston, Mass.
Empire Mfg. Co., London, Ont.
Mason Regulator & Engin. Co., Montreal, Que.
Mitchell Co., The Robert, Montreal, Que.
Smart-Turner Mach. Co., Hamilton, Ont.

STEEL, HIGH SPEED
Hopkins & Co., F. H., Montreal, Que.
Nova Scotia Steel & Coal Co., New Glasgow, N.S.
Wilkinson & Kompass, Hamilton, Ont.
Zenith Coal & Steel Products, Ltd., Montreal.

STEEL SHELVING
Dennis Wire & Iron Works, London, Ont.

STEEL WORK, STRUCTURAL
Babcock & Wilcox, Ltd., Montreal, Que.
Corbet Fdry. & Machine Co., Owen Sound, Ont.
Wm. Hamilton Co., Peterboro, Ont.
MacKinnon Steel Co., Sherbrooke, Que.

STEERING GEARS
Corbet Fdry. & Machine Co., Owen Sound, Ont.
Hopkins & Co., F. H., Montreal, Que.
Engr. & Mach. Wks. of Can., St. Catharines, Ont.
Wm. C. Wilson & Co., Toronto, Ont.

STOCK RACKS FOR BARS, PIPING, ETC.
Mitchell Co., The Robert, Montreal, Que.
Morris Crane & Hoist Co., Herbert, Niagara Falls, Ont.

STOKERS, MECHANICAL
Babcock & Wilcox, Ltd., Montreal, Que.

SUPERHEATERS, STEAM
Babcock & Wilcox, Ltd., Montreal, Que.
Goldie & McCulloch Co., Galt, Ont.

SWITCHBOARDS, ELECTRIC
Crouse-Hinds Co. of Can., Ltd., Toronto, Ont.

TALLOW
Can. Economic Lubricant Co., Montreal, Que.
Wm. C. Wilson & Co., Toronto, Ont.

TANKS, STEEL
Corbet Foundry & Mach. Co., Owen Sound, Ont.
Goldie & McCulloch, Ltd., Galt, Ont.
Hopkins & Co., F. H., Montreal, Que.
MacKinnon Steel Co., Sherbrooke, Que.
Marsh Engineering Works, Belleville, Ont.
Port Arthur Shipbuilding Co., Port Arthur, Ont.
Reed & Co., Geo., Montreal, Que.
Scaife & Sons Co., Wm. B., Oakmont, Pa.

TANKS (AIR, GAS AND LIQUID)
MacKinnon Steel Co., Sherbrooke, Que.
Marsh Engineering Works, Belleville, Ont.
Scaife & Sons Co., Wm. B., Oakmont, Pa.

TAPPING MACHINES
Mueller Mfg. Co., H., Sarnia, Ont.

TELEGRAPHS, SHIPS
McNab Co., Bridgeport, Conn.
Cory & Sons, Inc., Chas., New York, N.Y.
Morrison Brass Mfg. Co., James, Toronto, Ont.
Wilson & Co., Wm., Toronto, Ont.

TELEPHONES, MARINE
Cory & Son, Inc., Chas., New York, N.Y.
McNab Co., Bridgeport, Conn.

TESTERS, METER
Mueller Mfg. Co., H., Sarnia, Ont.

THUMB SCREWS AND NUTS
Canada Foundries & Forgings, Welland, Ont.
Mitchell Co., The Robert, Montreal, Que.
United Brass & Lead, Ltd., Toronto, Ont.

TRACK SYSTEMS
Morris Crane & Hoist Co., Herbert, Niagara Falls, Ont.

TRAVELLING BLOCKS
Morris Crane & Hoist Co., Herbert, Niagara Falls, Ont.

TROLLEYS
Can. Fairbanks-Morse Co., Montreal, Que.
Morris Crane & Hoist Co., Herbert, Niagara Falls, Ont.

TROLLEY HOISTS
Morris Crane & Hoist Co., Herbert, Niagara Falls, Ont.

TRUCKS, HAND, ELECTRIC
Aikenhead Hardware, Ltd., Toronto, Ont.
Can. Fairbanks-Morse Co., Montreal, Que.

TUBES, BOILER
Babcock & Wilcox, Ltd., Montreal, Que.
Broughton Copper Co., Ltd., Manchester, Eng.
Goldie & McCulloch Co., Galt, Ont.

TUBES, COPPER AND BRASS
Mueller Mfg. Co., H., Sarnia, Ont.
Tallman Brass & Metal Co., Hamilton, Ont.
Standard Underground Cable Co., Hamilton, Ont.

TUGS
Polson Iron Works, Toronto, Ont.
Collingwood Shipbuilding Co., Collingwood, Ont.

TURBINES, STEAM
Goldie & McCulloch Co., Galt, Ont.

TURBINES, DIRECT-DRIVING AND GEARED
Doxford & Sons, William, Sunderland, England.

TURNBUCKLES
Canada Foundries & Forgings, Welland, Ont.
Hopkins & Co., F. H., Montreal, Que.

TURNTABLES
Morris Crane & Hoist Co., Herbert, Niagara Falls, Ont.

SPIKES, SMALL RAILROAD
Severance Mfg. Co., S., Glassport, Pa.

UNIONS, ALL KINDS
Dart Union Company, Toronto, Ont.

VALVES, AIR
Mitchell Co., The Robert, Montreal, Que.
Mueller Mfg. Co., H., Sarnia, Ont.

VALVE, DISCS
Wm. C. Wilson & Co., Toronto, Ont.

VALVES
American Steam Gauge & Valve Mfg. Co., Boston, Mass.
Babcock & Wilcox, Ltd., Montreal, Que.
Can. Fairbanks-Morse Co., Montreal, Que.
Empire Mfg. Co., London, Ont.
McAvity & Sons, Ltd., T., St. John, N.B.
Mason Regulator & Engin. Co., Montreal, Que.
McNab Co., Bridgeport, Conn.
Norbom Engineering Co., Philadelphia, Pa.
Williams Machinery Co., A. R., Toronto, Ont.
Wm. C. Wilson & Co., Toronto, Ont.
United Brass & Lead, Ltd., Toronto, Ont.

VALVES, FOOT
Aikenhead Hardware, Ltd., Toronto, Ont.
Mitchell Co., The Robert, Montreal, Que.
Smart-Turner Mach. Co., Hamilton, Ont.

VALVES, STOP, REDUCING, SAFETY CHECK, ETC.
Aikenhead Hardware, Ltd., Toronto, Ont.
American Steam Gauge & Valve Mfg. Co., Boston, Mass.
McAvity & Sons Ltd., T., St. John, N.B.
Mitchell Co., The Robert, Montreal, Que.
Morrison Brass Mfg. Co., James, Toronto, Ont.

VALVES, MIXING
Mitchell Co., The Robert, Montreal, Que.
Mueller Mfg. Co., H., Sarnia, Ont.

VALVES, REDUCING, PRESSURE
Mitchell Co., The Robert, Montreal, Que.
Mueller Mfg. Co., H., Sarnia, Ont.

VALVES, STORM
Low & Sons, A., Glasgow, Scotland.

VARNISHES
Aikenhead Hardware, Ltd., Toronto, Ont.
Ault & Wiborg Co. of Can., Ltd., Toronto, Ont.
Leckie, Ltd., John, Toronto, Ont.
Reed & Co., Geo., Montreal, Que.

VENTILATORS, COWL
McNab Co., Bridgeport, Conn.
Mitchell Co., The Robert, Montreal, Que.

VENTILATION EQUIPMENT
Empire Mfg. Co., London, Ont.
Hopkins & Co., F. H., Montreal, Que.
Low & Sons, Ltd., Archibald, Glasgow, Scotland

WASHERS
London Bolt & Hinge Works, London, Ont.
Wm. C. Wilson & Co., Toronto, Ont.

WATER COLUMNS
Mitchell Co., The Robert, Montreal, Que.
Morrison Brass Mfg. Co., James, Toronto, Ont.

WELDING, ELECTRIC
Hall Engineering Works, Montreal, Que.
Marine Welding Co., Buffalo, N.Y.
Beatty & Sons, M., Welland, Ont.

WATER SOFTENERS
Babcock & Wilcox, Ltd., Montreal, Que.

WATER SUPPLY SYSTEMS
Mueller Mfg. Co., H., Sarnia, Ont.

WATER HEATERS
Empire Mfg. Co., London, Ont.
Morrison Brass Mfg. Co., James, Toronto, Ont.

WHISTLES AND SYRENS
Empire Mfg. Co., London, Ont.
McAvity & Sons, T., St. John, N.B.
McNab Co., Bridgeport, Conn.
Mitchell Co., Ltd., Robt., Montreal, Que.
Morrison Brass Mfg. Co., Jas., Toronto, Ont.

WINCHES, CARGO
Aikenhead Hardware, Ltd., Toronto, Ont.
Corbet Fdry. & Machine Co., Owen Sound, Ont.
Hopkins & Co., F. H., Montreal, Que.
Marsh Engineering Works, Belleville, Ont.

WINCHES, DOCK, SHIP
Beatty & Sons, M., Welland, Ont.
Marsh Engineering Works, Belleville, Ont.
Miller Bros. & Sons, Ltd., Montreal, Que.
Morris Crane & Hoist Co., Herbert, Niagara Falls, Ont.
Wilson & Co., Wm. C., Toronto, Canada.

WINCHES, TOWING
Corbet Foundry & Mach. Co., Owen Sound, Ont.

WINCHES, TRAWL
Beatty & Sons, M., Welland, Ont.
Wm. C. Wilson & Co., Toronto, Ont.

WINDLASSES
Corbet Fdry. & Machine Co., Owen Sound, Ont.
Doke Engine Co., Grand Haven, Mich.
Hopkins & Co., F. H., Montreal, Que.
Wilson & Co., Wm. C., Toronto, Canada.

WIPER CAPS, OILER BOXES, ETC.
Mitchell Co., The Robert, Montreal, Que.
Morrison Brass Mfg. Co., James, Toronto, Ont.

WIRE, COPPER CLAD STEEL
Standard Underground Cable Co., Hamilton, Ont.

WIRE, COPPER, BRASS, BRONZE
Standard Underground Cable Co., Hamilton, Ont.
Tallman Brass & Metal Co., Hamilton, Ont.

WIRE, INSULATED
Standard Underground Cable Co., Hamilton, Ont.

WIRELESS OUTFITS
Marconi Wireless Telegraph Co., Montreal, Que.

WIRE ROPE
Zenith Coal & Steel Products, Ltd., Montreal.

WOOD WORKING MACHINERY
Aikenhead Hardware, Ltd., Toronto, Ont.
Can. Fairbanks-Morse Co., Montreal, Que.
Preston Woodworking Machy. Co., Preston, Ont.
Yates Mach. Co., P. B., Hamilton, Ont.

WOOD BORING TOOLS
Aikenhead Hardware, Ltd., Toronto, Ont.
Can. Ingersoll-Rand Co., Sherbrooke, Que.

WOODITE GAUGE GLASS WASHERS
Wm. C. Wilson & Co., Toronto, Ont.

WRENCHES
Canada Foundries & Forgings, Welland, Ont.

YACHT BROKER
Watt, J. Murray, Philadelphia, Pa.

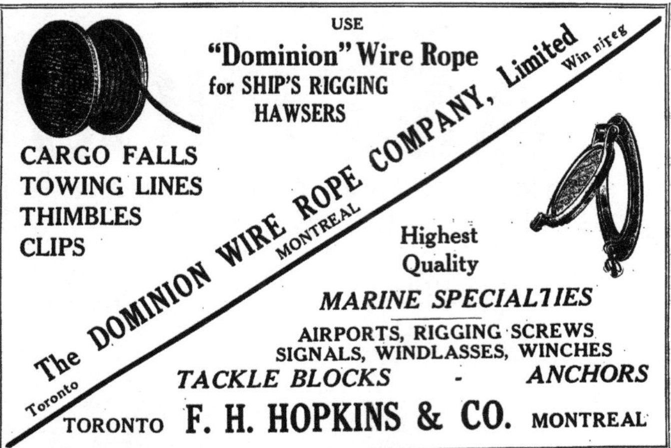

INDEX TO ADVERTISERS

CIRCULATES IN EVERY PROVINCE OF CANADA AND ABROAD

MARINE ENGINEERING of Canada

A monthly journal dealing with the progress and development of Merchant and Naval Marine Engineering, Shipbuilding, the building of Harbors and Docks, and containing a record of the latest and best practice throughout the Sea-going World. Published by
The MacLean Publishing Co., Limited
MONTREAL, Southam Building TORONTO 143-153 University Ave. WINNIPEG, 1207 Union Trust Bldg. LONDON, ENG., 88 Fleet St.

Vol. VIII. Publication Office, Toronto—August, 1918 No. 8

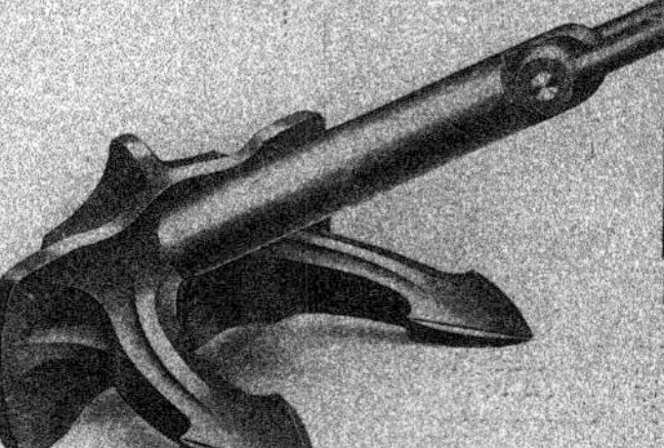

MARINE ENGINEERING of Canada

Volume VIII. TORONTO, AUGUST, 1918 Number 7

Heat Treatment of Large Forgings: Methods and Apparatus*

By H. H. Ashdown **

So many factors enter into the proper methods of heat treating steel that, in spite of the modern equipment used and the many works bearing on the subject, each particular article is a problem in itself and must be approached in that light. The present article deals with the various methods in use and their resultant effects on the quality of the steel treated.

SO much has already been said and written concerning the heat treatment of steels, and one finds in all works, where any form of treatment is practised, so many authorities on this subject, that the lay mind would think that perfection in every possible way had been achieved. Further, with the perfect system of pyrometric control now installed in most of our steel works, the greater part of that period has dealt with them in a commercial sense. His proposal therefore is to treat this matter in as practical a way as possible, dealing with the subject from a commercial or works point of view rather than approaching it as a subject of research. Having thus mapped out the course of the paper, it has been decided to exclude, as far as possible, technical terms relative to structural conditions, and in the subsequent discussion to evade such technical references, the nomenclature of metallography being, generally speaking, more confusing than helpful to the practical man.

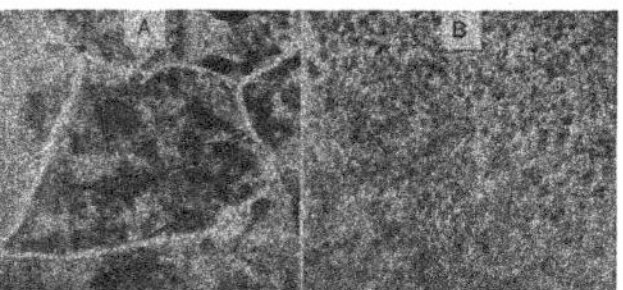

FIG. 1. TENSILE AND BENDING TESTS OF STEEL FORGINGS WITH PHOTO MICROGRAPHS, MAGNIFIED 100 DIAMETERS.
A. Forged and oil treated tensile, 51.5; elongation 16.6; reduction 5.7; bending 10°.
B. After correct heat treatment. Tensile 50.6; elongation 16.6; reduction 24.1; bending 180°.

impression gained is the impossibility of failures to occur due to incorrect treatment.

The heat treatment of steel may cover a very wide field, but the heat treatment of steel forgings suggests at once the impression of handling materials both in quantity and in the mass. To give a broad sense of comparison, one might compare the treatment of a piece of 1-inch round steel with a forging weighing some 60 tons, or a furnace full of forgings in the aggregate weighing 200 tons—this being on a parallel with the vastly different conditions ruling when casting ingots of a few cwts. in weight and those weighing 100 tons. From experimental evidence and casual observation it is quite evident that there is still much wanting in this direction, in spite of our greatly advanced knowledge and high attainments in research on this subject of recent years.

The author has been closely associated with the investigation, manufacture, and treatment of steels in all their phases for the past twenty-five years, and for

In order to introduce and prove the value of metallography in the works, it is expedient to submit photomicrographs together with some form of practical illustration, an example of which is shown in Fig. 1, which shows at sight the fundamental principles underlying a moderately careful treatment, and then to compare their relative comparative strengths with that of bricks and mortar and concrete. It will, of course, be appreciated that these protomicrographs are strictly comparative, each being of 100 magnifications. At the outset it is suggested that the greater percentage of failures occurring is not so much due to the lack of necessary scientific knowledge, but from the want of knowing how to apply that knowledge in works practice, and also in a large measure from the lack of observation generally, when forgings are subjected to heat treatment.

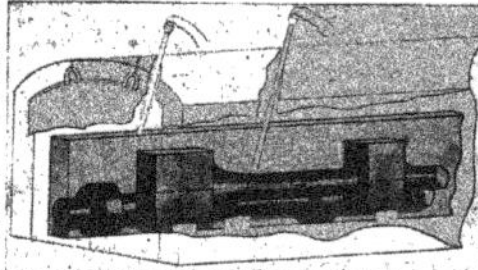

FIG. 2. MISCELLANEOUS FORGINGS IN THE FURNACE ILLUSTRATING THE USE OF PYROMETERS.

The rough outline sketch of a furnace about 8 feet by 10 feet by 100 feet long, Fig. 2, contains miscellaneous forgings such as are now comparatively common in big commercial steel works. These furnaces are periodically charged up with forgings of all descriptions, say, for example, heavy gun forgings, marine shafting, turbine shafting, and the like, possibly amounting to 150 tons in the aggregate. In the first place they are heated to a predetermined temperature. Annealing temperatures vary greatly, from anything between 650 deg. C (1,200 deg. F.), and 927 deg. C. (1,700 deg. F.), and the results obtained from the forging naturally vary accordingly. These temperatures are questioned at times, and one is told that there is little need to trouble, as the forgings as a rule pass test. That may be so, but unfortunately one hears from time to time of forgings, machine parts of all descriptions, breaking down prematurely, due to fatigue. Fatigue compares very favorably with a medical practitioner's "influenza"; it temporarily covers a very wide field of inherent troubles. The cause of the so-called fatigue could often be located at

*A paper read before the Institute of Mechanical Engineers.

**Sir W. G. Armstrong Whitworth Co., Newcastle-on-Tyne, England.

the manufacturer's works before such forgings were put into service.

In Figs. 3 and 4, are seen various types of structures from similar steels in the so-called annealed condition. The author prefers not to suggest anything that might be embraced in already stringent specifications, but this condition of things is doubtless a matter that must receive closer attention in the immediate future. The temperature of 650 deg. C. (1,200 deg. F.) is, as all know, well under the change-point of steels in general, and unless that critical temperature is exceeded comparatively little change is effected in the structural condition. A forging which has been finished at a high temperature, or parts which have been raised to a forging heat and have received little or no subsequent work, on cooling retain a very coarse structure.

To make this point more clear, an example of two operations in the forging of a marine shaft, Fig. 5, is submitted By withdrawing the already cogged billet A from the furnace, the amount of work shown in B is performed. Portion B_1 is finished, but in order to heat up portion B_2 for the subsequent forging operation, portion of B_1 must also be included in the furnace, and this naturally becomes subjected to a prolonged forging heat, subsequently receiving little or no work upon it. The structural condition of this portion of such a shaft may not be unlike that of an unannealed steel casting, an example of the structure bedescription came before the author's naing shown in Fig. 6. A bad case of this description came before the author's notice some little time ago. The forging in question was a double intermediate shaft which actually cracked in the lathe during the turning operation; and, within the writer's knowledge, this is by no means an isolated case. The shaft in question was cut through this defective area, and tests taken in the immediate neighborhood gave the results shown in the illustration.

The shaft was then subjected to a prescribed heat treatment, and the results of the test, together with the microstructure, are given side by side for comparison in Fig. 8. This regrettable feature applies not only to large forgings, but equally so the whole way down the scale, even to drop stampings. It is not the intention to name any particular forgings of which examples are given, nor their source of manufacture; suffice it to be said that it has been my privilege to visit many large steel works both at home and abroad.

By way of an example of a comparatively small forging, a special type of heavy bolt was made by heating to forging temperature for their full length pieces of round bars and "upsetting" a portion to form the head as shown in Fig. 7. These forgings were subsequently normalized (this in the the author's opinion is a very vague term), at 677 deg. C. (1,250 deg. F.); but when such bolts were subjected to a shock test, it was by no means uncommon for the head to shell off. The reason for this will be readily understood when one examines

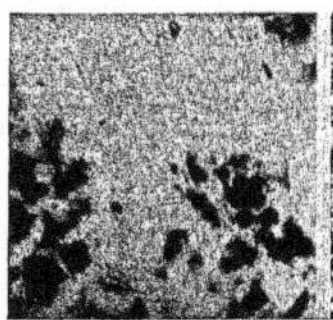

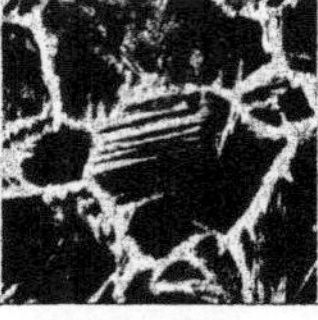
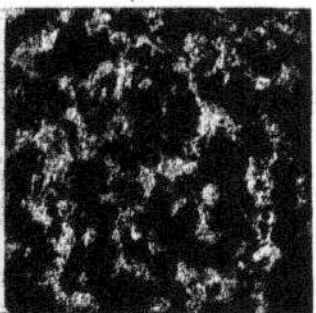

Lightly annealed — Drastically overheated — Drastically overheated — Well annealed

HEAVY STEEL FORGINGS. MEDIUM CARBON STEEL.

FIG. 3. PHOTO MICROGRAPHS MAGNIFIED 100 DIAMETERS SHOWING THE EFFECTS OF HEAT TREATMENT ON STEEL FORGINGS.

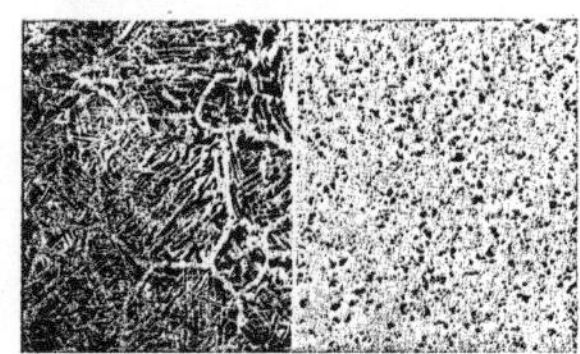

FIG. 4. BAD AND GOOD. ANNEALING OF LOW CARBON STEEL, MAGNIFICATION 100 DIAMETERS.

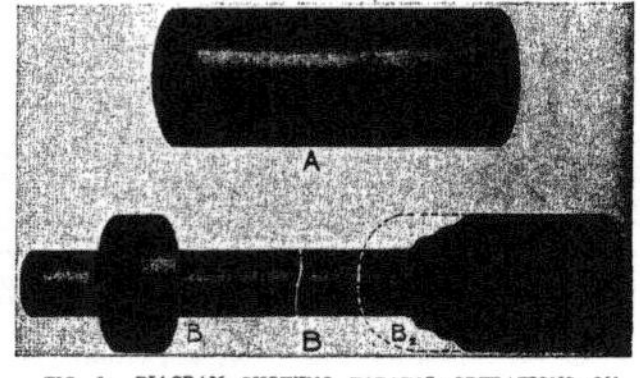

FIG. 5. DIAGRAM SHOWING FORGING OPERATIONS ON MARINE SHAFT.

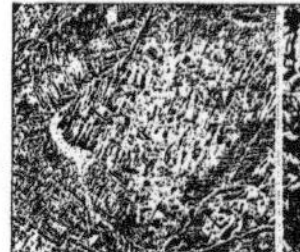

FIG. 6. ANNEALED STEEL CASTING. MAGNIFICATION 25 DIAMETERS

FIG. 8. MAGNIFICATION 100 DIAMETERS
C—Drastically overheated. Yield 16.8; tensile 31.5; elongation 16.5; reduction 24; bend 70°.
D—Subsequently normalized. Yield 18.5; tensile 32.8; elongation 29.8; reduction 52.3; bend 180°.

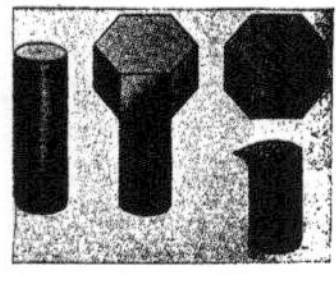

FIG. 7. HEAVY BOLT BROKEN UNDER SHOCK TEST

what had taken place in the forging operation; the only portion which had received any work was the material forming the head, and where this adjoined the body one had two entirely different structures.

Although Fig. 9 was not taken from this particular type of forging, it will show the great contrast in adjoining structures when forgings are finished at high temperatures and are not correctly subsequently heat-treated. Now, after these forgings received a proper heat treatment, a uniformity in structure was obtained together with the disappearance of any further shelling of the head when subjected to the same repeated shock stresses.

FIG. 9. ETCHED PIECE OF STEEL. Actual size. Photo micograph on junction line. Magnification 100 diameters.

A misconception more or less common in heat treatment of steels is that full advantage can be taken of the initial forging heat, and that forgings or stampings finished at a high temperature will, on cooling, regain a fine structural condition. This is entirely wrong, and if it is desired to obtain the best normal structural condition, the heat in such forgings must be allowed first to fall below its recalescence temperature and then reheated somewhat above its calescence temperature and preferably cooled in air.

With regard to high-temperature annealing and high-temperature quenching for hardening, it is generally known that as soon as one has reasonably passed the calescence temperature of the particular steel under observation, a refining in the structure will result therefrom, and also that, practically speaking, the maximum hardness of the steel can be effected by quenching from this temperature. It is also known that after exceeding reasonable limits of this temperature, in conjunction with the time factor, the coarse structure redevelops, and develops more rapidly the higher the temperature. A most important point often overlooked is the comparatively long time taken to raise forgings of appreciable size or in quantity to these elevated temperatures, and it is during this unnecessary period, which is costly in both fuel and labor, that the forgings are caused to deteriorate rapidly. Some of the big industrial institutions are favorable to high-temperature quenching, and it would be interesting to learn their reasons for this. The author has never found it necessary to exceed, either for annealing or quenching, when treating forgings in the mass, a temperature of 843 deg. C. (1,550 deg. F.), and has had no reason to regret the results obtained.

Reference has already been made to the lack of observation during the process of annealing. Let us return once again to the annealing furnace, and give it a few moments' attention whilst the heat being gradually taken up by the forgings. What does one notice? Naturally, the smaller or projecting sections become heated long before the larger or depressed portions. Assuming now, for the sake of argument, that the predetermined temperature is to be 870 deg. C. (1,600 deg. F.), it is easy to understand that if in a short time the small sections have attained this heat, it will be a considerable period before the larger sections acquire this same temeprature, and, whilst the process of uniformity is being obtained, the smaller sections have been most unduly heated. Such points as these are often the fruitful cause of many failures, and they, unfortunately, are points which are frequently overlooked, and it is here where a good furnaceman becomes invaluable. The author has found it an exceedingly good plan first to get a good initial soaking heat through the forgings, just under the "change-point," say about 704 deg. C. (1,300 deg. F), where, if at all, little harm is done by a long heating, and then gradually to raise the whole of the forgings to the predetermined temperature.

Pyrometers

Pyrometers are of great value and service in indicating temperatures, but in inexperienced or careless hands, they are more harmful than helpful, and so far as recorders are concerned, little value should be placed on them without knowing that careful observation had also been kept on the furnace. The record simply indicates the temperature of the area in which the pyrometer is located, Fig. 2, 10, 11, and 12, and a great diversity of temperature may often be seen in large furnaces within a few feet of this position. In the author's opinion, the greatest value of a recorder is the moral effect it has on the furnacemen. It certainly causes them to be more attentive than they otherwise might be to their fire control.

The following instance relative to the deception of pyrometers is given: Let us imagine the pyrometer being in contact with the shaft end of the forging, Fig. 12; one might then record the desired temperature for many hours, and yet the body of the forging could still be several hundred degrees below that indicated. Indeed, the author has seen comparatively small engine parts, of which, within a length of three feet, after heating and quenching, one end has been in its softest possible condition whilst the other end has been practically unfileable. When it is fully realized that pyrometers simply indicate a local temperature, and that uniformity of temperature is entirely dependent on careful observation, many of the so-called mysteries, together with their consequent troubles, will disappear.

Although only one form of pyrometer has been illustrated, the Fery radiating pyrometer, Siemens water pyrometer, and Brearley sentinel pyrometer, Fig. 13, have each their own spheres of usefulness. In the latter case, for instance, the sentinels can be placed in positions quite inaccessible to other forms of pyrometers and again can be laid in rotating bodies, their complete fusion denoting their prescribed temperatures.

When it is generally realized that forgings, after their structures have once been refined by a correct heat treatment and are quickly cooled, give better mechanical and physical tests generally, the fallacy of sand annealing will soon disappear. These remarks apply equally so to those which have been quenched and have received a subsequent light annealing or tempering to conform to specified tests. It is sometimes an advantage to bury in sand forgings made from steels akin to those exhibiting air-hardening properties, with a view to rendering them soft for machining; but if such forgings after machining are to be assembled in the machine in this annealed state, it is then better to cool quickly and reheat slightly below the change-point.

If it is intended to anneal forgings in sand with the object of softening for machining, they should be fully covered before they have fallen below the critical temperature, and for comparatively small forgings hot sand should be employed. For want of a little knowledge on this subject much apparent slackness is practised, and this in a large measure accounts often for much of the so-called hard and soft spots which cause endless trouble and expense in the machine shop.

Oil Quenching

Attention has already been called to the want of uniformity in heating for annealing, and this applies equally to the case of forgings being heated for quenching, in order both to improve their structural condition adn also to effect a certain standard of hardness. It might appear somewhat elementary to embrace such remarks as these, but when one finds that such important essentials are commonly overlooked, it is well to make it quite clear that all forgings, whatever their size, must be quenched only when uniformly heated above the change-point of the steel.

Excessive heating of forgings before quenching is not only productive of a bad structural condition, but also renders the cooling medium unnecessarily hot, and thus effects a comparatively slow cooling of the forging through its critical range, allowing partial precipitation of the ferrite to take place. It will be generally admitted that the quicker a forging can be cooled until the critical range has been passed the better will be its structural condition, and consequently its physical properties will be of a correspondingly higher order. This, in the author's opinion, is a matter deserving of closer attention than it would appear to have received in the past.

This point leads up to the liquid used as a quenching medium. The author trusts he is not exceeding the privilege conferred on him if he mentions the subject of gun forgings. Of recent years marked developments have been seen in the size and weight of these forgings until they have assumed ponderous masses and require for their quenching tanks of oil from 8 to 10 feet in diameter and approaching 100 feet in depth. If further developments are to take place, the quenching of these colossal masses will have to be seriously considered if we are to obtain the best results, and not only compete with, but lead, competitive nations.

Those who have experimented by water-quenching materials in the mass, know that greatly superior results can be obtained, and also that the structural condition is all that can be desired. The author believes he is correct in stating that this form of quenching, if not in use in this country, is in operation in foreign countries, and of the forgings so treated which have come before him, both for microscopical examination and general testing, to say the least, the results have been highly commendable. He is disposed to think that steel manufacturers would welcome the opportunity of the choice of quenching mediums both from the point of view of producing a better article, and also by dispensing with a costly fluid, and one which is attendant with great risks when large masses of steel are being hardened. It is possible that slight modifications in the analysis of forgings so treated may have to be made, and at the same time it must also be remembered that quenching mediums which are highly suited for treating large and heavy forgings are not necessarily the best for light forgings or those of thin section, but such matters as these may be safely left to the discretion of experienced manufacturers.

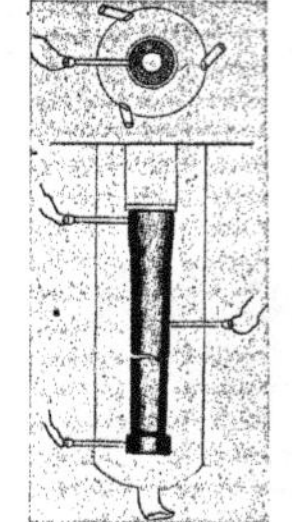

FIG. 11. ILLUSTRATING THE USE OF PYROMETER.

Specifications

Many specifications covering the composition and treatment of materials leave much to be desired, and suggest anything but progress. They bind the manufacturer in every possible direction, and thus restrict him in the production of an article much superior to that to which he is now limited. An example is cited of specifications covering certain high-class forgings. They are specified to give after heat-treatment, say, 30 to 35 tons tensile with a combined elongation and reduction of area of, say, 80. If a manufacturer produces a better article giving 37 tons together with a sum of 90 representing the combined elongation and reduction, this forging would be rejected, and would have to be re-annealed to conform with the maximum stress. Suffice it to say this forging after the re- treatment, although conforming to specification, would be much inferior than when in its original heat-treated condition.

The author hopes that engineers and the steel manufacturers' customer will correct him if his interpretation of specifications is incorrect. He looks upon specifications as a working basis, drawn up between the engineer and the steel manufacturer when entering into contracts, to be used as a guidance for both parties and with reasonable discretion by the inspecting engineer. If specifications were so framed that they were to be enforced to the letter, one could see little use for the intelligent engineer as an inspector.

Research

As research investigation should be of great value and assistance to the steel manufacturer in guiding him out of his occasional troubles, it is felt that this paper would be greatly wanting in a most essential detail if reference were not made to this very important subject. In the author's opinion the great value of research work is often lost owing to its want of application. The research chemist may be a highly qualified man, and his conclusions as to the cause of failure within reasonable limits of being correct, but generally speaking, he is not in a position to apply his reasoning into works practice, and his reports being often of a highly technical character are to some extent beyond the technical range of our good practical men. It would therefore appear that a gap yet exists between the works and the laboratory, and requires to be bridged across in such a manner that the full value of research investigation can be infused into works practice.

It is often stated that the fine structural conditions and excellent tests obtained by treating pieces in the laboratory cannot be obtained when treating masses in the works. This idea is totally wrong; by reasonably careful control of furnace conditions, and where quenching is necessary, regulating these conditions also, equally good results can be obtained. To substantiate this statement, in Fig. 14, is shown the structural condition of a forging weighing about 30 tons, which had been subjected to a prolonged heating before quenching. This was subsequently given a refining treatment, requenched and tempered, with the result shown in the adjoining photograph. This is by no means an exceptional case, and it is questioned if a much finer structure can be produced in the laboratory when dealing with a 1-inch diameter bar.

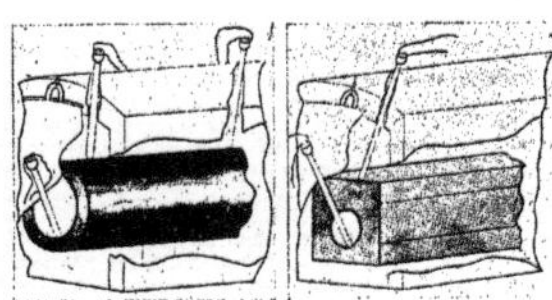

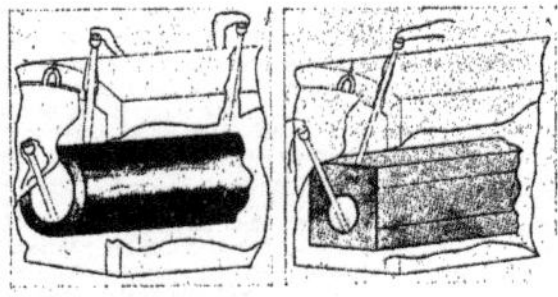

FIG. 10. FURNACE DIAGRAMS ILLUSTRATING THE USE OF PYROMETER.

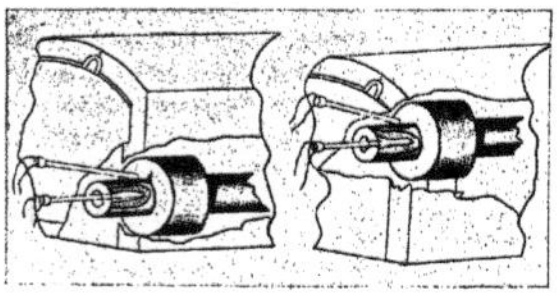

FIG. 12. PYROMETER IN CONTACT WITH THE SHAFT END OF A FORGING

Ghosts

Ghosts in steel forgings, as is gener-

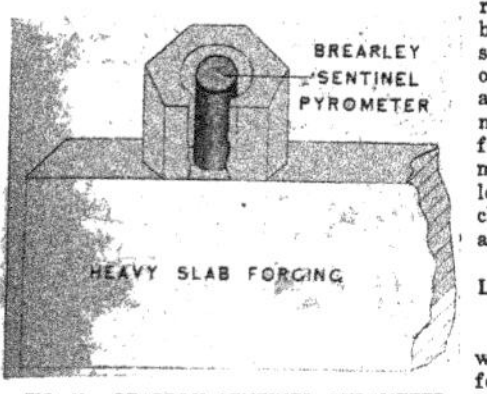

FIG. 13. BEARDLY SENTINEL PYROMETER APPLIED TO HEAVY SLAB FORGING.

ally known, are caused by the presence of non-metallic enclosures; and whatever form of heat treatment these forgings may subsequently receive, it is practically impossible to remove these ghost-marked areas. It is, however, within the writer's experience, possible to reduce the size of the ghost-marks, but the treatment necessary is both drastic and costly, and possibly the remedy may prove little better than the evil. This is a subject for which there is yet much room for investigation; and although in some cases their presence need not be looked upon with any degree of alarm, there is no question that a steel free from ghosts is a much superior article. In any case, prevention is always better than cure, and it is quite possible for the steel-maker to cast ingots of 100 tons in weight with the major portion entirely free from ghosts.

It has been endeavored to indicate briefly the principles underlying successful heat treatment on a commercial scale, and although in a small degree the value of science in the works has been discounted, no one is more anxious than the writer to see it take a prominent place. This end, however, can only be attained by creating in those who have the handling of this class of work a personal interest, and this will be best achieved by occasional practical demonstrations.

The great value of correct and careful heat-treatment does not yet appear to be appreciated by many whom it should seriously concern. When it is generally realized that forgings made from the best steel produced can be hopelessly spoiled, due either to insufficient heating or by overheating, and that forgings such as are now under review may be worth many thousands of dollars, and again from this same cause premature fracture may take place with a consequent great loss of life, this subject may be given closer attention and considered more as a crude form of craft.

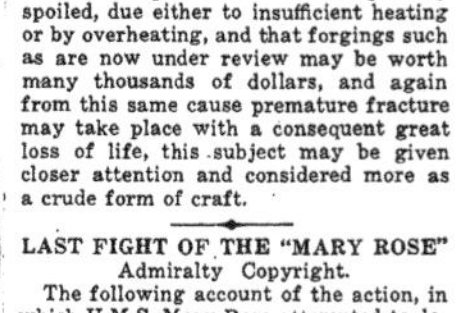

Drastically Heated Before Quenching. Refined and Nominally Heated Before Quenching.

FIG. 14. 30-TON FORGING SHOWING EFFECTS OF QUENCHING AND TEMPERING.

LAST FIGHT OF THE "MARY ROSE"

Admiralty Copyright.

The following account of the action, in which H.M.S. Mary Rose attempted to defend the convoy committed to her charge, compiled from official sources, has been placed at our disposal:—

H.M.S. Mary Rose left a Norwegian port in charge of a west-bound convoy of merchant ships in the afternoon of Oct. 16. At dawn on Oct. 17 flashes of gunfire were sighted astern. The captain of the Mary Rose, Lieutenant-Commander Charles Fox, who was on the bridge at the time, remarked that he supposed it was a submarine shelling the convoy, and promptly turned his ship to investigate; all hands were called to action stations.

The Mary Rose had increased to full speed, and in a short time three light cruisers were sighted coming towards them at high speed out of the morning mist. The Mary Rose promptly challenged, and receiving no reply, opened fire with every gun that would bear, at a range of about four miles. The German light cruisers appeared to have been nonplussed by this determined single-handed onslaught, as they did not return the fire until the range had closed to three miles. They then opened fire, and the Mary Rose held gallantly on through a barrage of bursting shells until only a mile separated her from the enemy. Up to this point the German markmanship was poor, but as the British destroyer turned to bring her torpedo tubes to bear, a salvo struck her, bursting in the engine-room and leaving her disabled, a log on the water. All guns, with the exception of the after one, were out of action, and their crews killed or wounded; but the after gun continued in action under the direction of Sub-Lieutenant Marsh, R.N.V.R., as long as the gun would bear. The captain came down from the wrecked bridge and passed aft, encouraging and cheering his deafened men. He stopped beside the wrecked remains of the midship gun and shouted to the survivors of its crew: "God bless my heart, lads, get her going again; we're not done yet!"

The enemy were now pouring a concentrated fire into the motionless vessel. One of the boilers, struck by a shell, exploded, and through the inferno of escaping steam, smoke, and the vapour of bursting shell, came that familiar, cheery voice: "We're not done yet."

As the German light cruisers sped past, two able seamen (Able Seaman French and Able Seaman Bailey), who alone survived among the torpedo tubes' crews, on their own initiative laid and fired the remaining torpedo. Able Seaman French was killed immediately, and Able Seaman Bailey badly wounded. Realizing that the enemy had passed ahead, and that the 4-in. gun could no longer be brought to bear on them, the captain went below and set about destroying the cyphers. The First Lieutenant (Lieutenant Bavin), seeing one of the light cruisers returning towards them, called the gunner (Mr. Handcock), and bade him sink the ship. The captain then came on deck and gave the order "Abandon ship." All the boats had been shattered by shell fire at their davits, but the survivors launched a Carley raft and paddled clear of the ship.

The German light cruiser detailed to administer the coup-de-grace then approached to within 300 yards and poured a succession of salvoes into the already riddled hull. The Mary Rose sank at 7.15 a.m. with colors flying. The captain, first lieutenant, and gunner were lost with the ship, but the handful of survivors, in charge of Sub-Lieutenant J. R. D. Freeman, R.N., on the Carley raft, fell in some hours later with a lifeboat belonging to one of the ships of the convoy. Sailing and rowing, they made the Norwegian coast some 48 hours later, and were tended with the utmost kindness by the Norwegian authorities. All survivors unite in testifying to the cheerful courage of the senior surviving officer, Sub-Lieutenant Freeman, throughout the last phase of this ordeal. Able Seaman Bailey, who, despite severe shrapnel wounds in the leg, persisted in taking his turn at the oar, is also specially mentioned for an invincible light-heartedness throughout.

Unhappily, there is no record of what was in the mind of the captain of the Mary Rose when he made that single-handed dash in the face of such preposterous odds. The convoy which was in his charge lay ahead of him, and, as he apparently supposed was being attacked by the gunfire of a hostile submarine. When, on rushing to the scene, he realized that it was to meet not a submarine, but three of Germany's newest and fastest light cruisers, it is conceivable that the original intention of rescue was not supplanted in his mind by considerations of higher strategy. He held on unflinchingly, and he died, leaving to the annals of the service an episode not less glorious than that in which Sir Richard Grenville perished.

Heavy Oil Engine Developed in Europe

Automatic Starting, Stopping and Reversing Render This Engine Easily Operated—Care Taken in Designing Auxiliary Details Makes For Efficiency at Sea

THE extent to which continental engineers in Europe have contributed to the development of the internal combustion marine engine has been viewed for some time with some alarm by British firms who are at present employed exclusively on war work.

Many of the ships now building in Canada and the United States will be motor-engined, and the fact that many of these engines are built under license from one or other European pioneer firm imparts considerable interest to the engine shown in the accompanying engravings. This particular type of engine is known as the Kromhout, and the engine described is said to be the largest of its type. It is built by the firm of Goedkoof, of Amsterdam, and is known as the 4M6 type, operating on the two-cycle system, having four cylinders of 16¾-in. bore by 18¾-in. stroke, and it develops its full power at about 240 revolutions per minute.

Operation

The starting and reversing of the engine is effected by means of compressed air. The cam shaft, marked E on the drawing, is carried on the front of the engine, two cams being fitted in way of each cylinder for operating the air starting—and consequently reversing—valves, according to the direction into which the engine is required to run. In addition, the cam shaft has four cams for operating the four feed pumps, which are mounted on the front of the engine, between the No. 2 and No. 3 cylinders, at a height convenient for observation and overhaul.

The cam shaft is operated by the hand wheel A, by means of which the shaft can be moved in a fore and aft direction either with the engine stationary or with it running, the cams being so arranged that the movement of the cam shaft, with the engine stationary, is bound to admit compressed air for starting purposes to the cylinder in the best position for commencing the rotation of the engine in the required direction, either to go ahead or astern. The engine has, therefore, no dead centre, but will start in any position ahead or astern simply by the operation of the hand wheel A in the required direction.

Air is admitted to the requisite cylinder through the valves marked G, the jackets of which are water cooled, and immediately above the valves marked G a small horizontal cone valve is fitted for indicating purposes.

The hand lever C mounted on No. 4 cylinder operates all four compression relief valves. The hand lever B, shown between the third and fourth cylinders, is for the purpose of regulating the stroke of the fuel feed pumps, and, consequently, the revolutions of the engine, the shaft operated by the lever B being also connected to the governor F, so that the strokes of the fuel feed pumps are adjusted as required by the governor in the event of the load being suddenly taken off for any reason. The hand wheel D operates a friction clutch used for disconnecting the propeller shaft from the engine when the signal "stop" is telegraphed down from the navigating bridge; the governor then automatically comes into action and reduces the stroke of the fuel feed pumps as necessary, so that the engine runs at about 100 revolutions per minute, at which speed the cylinder vaporizers remain at a proper temperature to keep the engine running an indefinite time, the quantity of scavenging air being automatically adjusted by the governor immediately the hand wheel D is moved to the neutral or stop position.

If, with the engine turning "ahead," the telegraph would ring down "astern," there is no need for the clutch to be operated, and the hand wheel D is not touched; the engineer simply operates hand wheel A, the engine comes to a rest and then commences to rotate in the opposite direction. This operation can be effected at full speed as well as reduced speeds.

Governor Has Several Functions

The governor F is a piece of mechanism with several functions. It is, of course, primarily designed to control the revolutions per minute of the engine. It does so in two ways. When the load is taken off it reduces the pump stroke, and hence the revolutions to the minimum of 100 mentioned above. It, at the same time, throttles the scavenging air in or-

SIDE ELEVATION OF COMPLETE ENGINE

der to maintain the required temperature in the combustion chamber. It further has one particular advantage, and that is, when the engine at sea is running at the normal full revolutions of 240 per minute, if the propeller, in a heavy sea, heaves out of the water, thus taking the load off the engine, the governor instantly prevents the revolutions from increasing above 240, and thus overcomes any tendency to racing. At the same time, in a seaway, the revolutions are maintained at full speed by means of the lever B.

At the after end of the engine, and behind the control wheel D, are placed the bilge and circulating pumps, each being in duplicate. They are operated by the shaft marked I, which is driven off the crank shaft by a silent chain, the pump shaft running at 150 revolutions when the main engine is turning at 240. The thrust block K is of the ordinary marine type.

Lubrication is effected by means of two pump feed lubricators marked H, mounted between the first and second and third and fourth cylinders at a height convenient for the engineer to see the quantity of oil in the reservoirs. The lubricators are operated by the cam shaft, and are provided with sight feeds and supply oil to the cylinder and all bearings, except the thrust block bearings, which are provided with separate wick feed lubricators.

The lubricators are of a special type, and possess the distinct feature of having no valves at all in the delivery pipes. The big end bearings and main bearings are all lined with white metal, and all main bearings can be removed and replaced or interchanged without dismantling the engine.

Detail Features

The engine possesses several small features which are intended to make for efficiency in service at sea. Among these may be mentioned the control wheel marked J, which enables the quantity of scavenging air to be adjusted by the engineer to his liking in relation to the revolutions per minute. Then, again, at the back of the engine a hand control valve is fitted to each circulating water supply pipe, so that the quantity of circulating water per cylinder can be adjusted as may be necessary, provision being made by means of safety valves on the circulating water pumps to prevent any damage happening to the pumps should all four control valves be closed down inadvertently at the same time.

The scavenging air is drawn into the crank case through specially designed valves, which are silent in their action, and are provided with gauze guards to prevent any foreign matter being drawn into the crank case. The compressed air on the down stroke of the piston passes up through the piston to the cylinder, and in this way a draught of cold air passes the gudgeon pin and helps to cool it.

Another detail which adds to the convenient use of the engine at sea is the receptacle for methylated spirit required for the preliminary heating of the lamps. A deep cup, with a long handle, is provided, which can be conveniently filled without danger of spilling, put into position and lit up without any chance of the spirit overflowing if the ship is rolling heavily. The builders of the engine point out, however, that its special feature is the vaporizing arrangements, by the use of which blow-lamps are required for starting purposes only. They explain that once the engine is running it will continue to do so at any revolutions between 100 and 240 for any length of time without the blow-lamps having to be relit, and without any danger of its failing to pick up the full load immediately, even if it be running at its lowest number of revolutions for several hours. This result is obtained without any outside mechanism beyond the fact that the quantity of scavenging air is always proportionate to the revolutions per minute. They point out, too, that the engine requires no water injection under any circumstances whatsoever.

Auxiliary Set

Included in the equipment of the 4 M 6 type engine is an auxiliary set, the horse-power of which varies according to the requirements of the owners—that is to say, the work that the auxiliary engine is required to do. The auxiliary engine is supplied primarily to drive the air compressor for the compressed air required for starting and manœuvring the main engines, but owners sometimes require the auxiliary engine also to drive a dynamo and supplement the bilge and ballast pumps.

With a 4 M 6 type engine, an 8 horse-power engine would be sufficient for driving the air compressor only, but as in the type of vessel to which engines of this power are supplied electric light is almost always fitted, the normal horse-power of the auxiliary engine is 12, so that it may drive the dynamo, as well as an auxiliary set of bilge and ballast pumps, and in the case of vessels working in hot climates, ventilating fans, etc.

The auxiliary engine is generally constructed on the same principle as the main engine, but is not, of course, provided with reversing gear. It uses the same oil, and, in common with other types of Kromhout engines, when started, runs continuously without the starting blow-lamps having to be relit under any conditions of load. We are indebted to "The Engineer" for data and illustration used in connection with this article.

FAST WORK ON THIS VESSEL

In Eleven Hours Engines and Boilers Were Placed in Position

BELFAST, Ireland.—Workman, Clark & Company, local shipbuilders, have achieved a world's record in completing an 8,000-ton standard ship in fifteen days after she was launched. The vessel was launched at 9 o'clock in the morning; by 8 o'clock the same evening all her engines and boilers were in position.

Cleveland, Ohio—The grain movement from Duluth and Fort William is expected to start about September 15, and as the movement of grain from Lake Michigan ports is increasing, the demand for tonnage may cut into the ore trade.

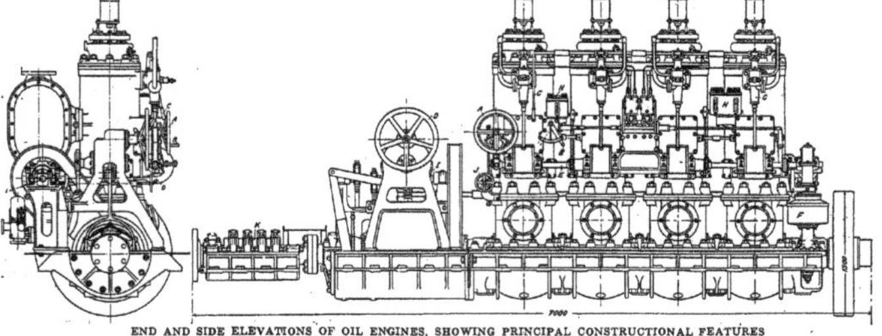

END AND SIDE ELEVATIONS OF OIL ENGINES, SHOWING PRINCIPAL CONSTRUCTIONAL FEATURES

Marine Boiler Corrosion; Causes and Prevention

Performance Depends to a Certain Extent Upon Initial Treatment of Boiler—Sea Water One of Principal Causes of Corrosion—Lime Useful in Treatment

THE subject of corrosion of marine boilers, along with causes and prevention, were the subject of a paper prepared by the late D. E. Rees (member) before he lost his life in warfare last year. The paper was read before the Institute of Marine Engineers last December, and gave rise to a valuable and interesting discussion thereon, which is appended hereto.

Initial Treatment

The life of a marine boiler depends to a certain extent on the care and attention bestowed on it during the first few months of its existence, provided that the plates have been properly treated at the building of the boiler—that is, all mill scale removed and the plates kept dry and free from atmospheric corrosion. The mill scale, if not removed, has a very irritant action on the clean metal of the boiler, and is most hurtful to the plates if allowed to remain on them. When new boilers are put under the varying conditions and temperatures found on ship board the greatest care and attention is needed for the first period of working if the boilers are to be kept free from corrosion; the new metal of the boiler at this stage being most open to attack from all sources of corrosive agencies. It is advisable on the first voyage to put into the boilers before closing them up 5 lbs. of lime per 1,000 I.H.P. and afterwards for a period of about six days 2 lbs. of lime per day per 1,000 I.H.P. should be passed through the hotwell as milk of lime, and 1 lb. of lime per day per 1,000 I.H.P. during the remainder of the voyage. At the first opportunity the boilers should be opened up, and if a thin coating of lime is found on the internal surfaces the use of lime with the feed water may be gradually reduced, but if such is not the case and the boiler water shows any signs of black or red oxides, then the use of lime with the feed water should be continued. Daily tests of the boiler water with a chlorine testing bottle should also be made, and should the water contain a greater amount than 100 grs. per gallon the lime should be increased in quantity. If, on testing, the water was found to be acid, carbonate of soda should be added to the feed water until the water becomes a safe alkaline again.

Having thus given the boilers a good start in life, constant attention on the part of those in charge is required if the boilers are to be kept in anything like a good condition. As the conditions under which different sets of boilers work vary so considerably, and the management of them so often changes hands, it is a matter of great difficulty to make any fixed laws or rules governing their treatment; and it is also a well-known fact that seldom do two different sets of boilers require the same treatment. However, if the following methods of dealing with the corrosive agencies are attended to they will help to prolong the life of any marine boiler considerably.

One of the principal causes of corrosion is sea water, it being practically impossible to entirely exclude it from the boilers even under the most favorable conditions. One of the constituents of sea water is chloride of magnesium ($Mg\ Cl_2$), and under certain conditions sea water in contact with heated copper, brass iron or steel surfaces becomes acid by the conversion of the chloride of magnesium into hydrochloric acid ($H\ Cl$) and magnesia (Mg). The action is as follows:—Where chloride of magnesium is present it attacks the iron, and when water containing it is used as the boiler feed serious corrosion is produced, this being due to the high temperature, and under high pressure this salt is decomposed, as shown by the following equation:—$Mg\ Cl_2 + 2H_2O = Mg\ (OH) + 2HCl$.

The magnesium hydrate is precipitated and the HCl dissolves in the water. This acid dissolves a certain quantity of iron from the boiler furnaces, forming chloride of iron ($Fe\ Cl_2$), as soon as the latter is formed it is decomposed by the magnesia already liberated, precipitating oxide of iron and re-forming chloride of magnesium, as will be seen from the equations $H\ Cl_2 + Fe = Fe\ Cl_2 + H$, and $Fe\ Cl + Mg = Mg\ Cl + Fe$ the H in the first equation being given off into the water. This oxide of iron deposited is ferrous oxide, and is black in color, and remains so unless air is allowed to get into the boiler, when it becomes ferric oxide and changes in color from black to red or brown. Thus, if corrosion is to be prevented, sea water must be kept out of the boilers; and this can only be done by keeping condensers tight, evaporators free from priming, and a supply of fresh water carried in engine-room tanks for use in case of any emergency. The carrying of fresh water is, unfortunately, a condition which seldom exists on the majority of steamers today, as by keeping the fresh water in engine-room tanks the cargo-carrying capacity is diminished, so that, from the shipowners' point of view, the boilers must be sacrificed to cargo.

For the rendering of sea water non-corrosive about 8 lbs. of quicklime, or 45 lbs. of soda crystals per ton, would be required, and such quantities are, of course, prohibitive when there is any considerable leakage of sea water with the boiler feed. The use of lime has the further disadvantage that out of the 8 lbs. mentioned above about 5¾ lbs. would be deposited as scale. Sea water in itself has sufficient sulphate of lime to produce 3¼ lbs. of scale for every ton of water, so that if the proper proportion of lime be added the total quantity of scale which will be deposited is 9 lbs. Although this scale, when evenly coated over the surfaces of the boiler, may form a protection against the corrosive action of the chloride of magnesium, it is, at the best, only an expensive and unsatisfactory remedy, as it increases the consumption of fuel and may also damage the boiler by overheating when accumulated.

Evaporator Operation

Another point which demands careful attention, and is not sufficiently recognized by marine engineers, is the working of evaporators for the production of the make-up feed water. Evaporators are, of course, very essential for the making up the loss of feed water due to leakages at glands and joints, but they should be blown down before the brine becomes too concentrated, otherwise the magnesium chloride will be decomposed and give off hydrochloric acid, which will pass over into the boilers with the distilled water, thus producing acidity in the fresh feed water. The action of the acid formed in this way differs from that formed in the boiler by the decomposition of sea water, inasmuch as it is not destroyed immediately afterwards by re-uniting with the magnesia, but is carried in with the feed water and held in solution throughout the entire volume of the boiler water. It is thus in a position to attack all parts which are not protected, also the parts where circulation is poor. As this action goes on the acid is becoming more concentrated, and will continue in its harmful action until prevented from entering the boiler. It is, therefore, essential that all evaporators should be fitted with baffle plates, have their vapour pipes led well above the dome, salinometer cocks fitted, and the density taken at frequent intervals. The density should not be allowed to exceed $\frac{3\frac{1}{2}}{32}$, and when this point is reached the evaporator should be blown down and re-filled.

As will be seen from the above, the use of soda and lime should receive careful consideration, and the quantities added to the feed water should be governed by the condition of working in each case such as the density of the water, the rate of working and the time which must elapse before the next cleaning of the boilers. Experience points out that the use of lime is more generally satisfactory on a voyage, soda being used in cases where the boiler water is acid through the density of the brine in the evaporator being excessive, and where no vegetable oil has been allowed to enter the boiler.

Lime Treatment

As before stated, sea-water cannot absolutely be prevented from getting into the boilers, therefore it is an excellent plan to continually use a small quantity of milk of lime to neutralise this sea water. One pound of lime per day per 1,000 I.H.P., dissolved in fresh water in the following way may suffice, and the lime thus used should be ordinary unslaked lime. It should be finely powdered and kept in a dry place, the milk of lime being made by mixing 1 lb. of this lime in a gallon of fresh water, the solution to be strained through gauze wire before use in order to get rid of any lumps or solid impurities.

Carbonate of soda is effective in changing the sulphate of lime found in the sea water into sulphate of soda, which is soluble, therefore harmless. Carbonate of lime is also formed and is easily got rid of by blowing out. Formerly practically all corrosion in boilers was attributed to galvanic action, and a certain amount of corrosion is still attributed to this cause. If two separate plates of metal, one more corrodible than the other, are taken and partially immersed in a corrosive liquid and connected in the atmosphere by a wire, a current of electricity is generated which originates from the more corrodible plate and passes by way of the liquid to the less corrodible plate, then through the wire and back to the more corrodible plate. If the wire be removed and the plates placed so that they complete their own circuit the current will still exist. Therefore, if in one construction there be two metals of different electrical potential exposed to corrosive influence, the electro-positive metal will be attacked and a current will pass between the component metals. Now, in boilers containing sea water this action takes place between the different metals used in the construction of the boiler. To counteract the galvanic action, zinc plates placed in perfect metallic contact with the plates of the boiler should be used, as the position of zinc in the list of electro-positives causes the zinc to be attacked before the iron, and, therefore, the boiler plates would be left in good condition. This action, of course only takes place when sea water is allowed to enter the boilers, and zinc plates are only fitted to minimize the action of the objectionable sea water. Therefore, if sea water is kept out of the boilers the zinc will not readily act, and there will be little need of its use in such large quantities, thus lessening a very expensive item in the working of marine boilers. To afford an efficient protection by the use of zinc there must be perfect metallic contact between the zinc slabs and the iron or steel plates of the boiler. The practice of hanging the zinc slabs by wire or hooks from stays or tubes is an absolutely useless one as the zinc is not serving the purpose for which it is intended; and although the zinc may show signs of deterioration, yet in such a case this would be from attack by the sea water, and not through any electric couple.

Lubrication Effects

Another cause of corrosion is the presence of animal or vegetable oil in the feed water. By using compounds containing such oils as lubricants in steam cylinders the exhaust steam carries them over to the condensers, and the fatty acids liberated from these oils by decomposition will cause pitting wherever the sludgy deposit can find a resting place in the boiler. Only a minimum quantity of the highest grade of hydro-carbon oil should be used in steam cylinders, and in lubricating piston and valve rods the same precaution should be observed. Apart from the evil effects of acidity the hydro-carbon deposited upon the heating surfaces of a boiler is most harmful, as a thin film of this deposit forms a very bad conductor, thus preventing the heat passing through to the water and so causing the heating surfaces to burn, blister and crack.

Under the care of careful and competent engineers a marine engine can be operated without a particle of internal lubrication, only a very small amount of oil being used for the purpose of swabbing the rods. Should it, however, be necessary to resort to internal lubrication, various samples of oils should be thoroughly analysed by vaporising a small quantity of the oil over a Bunsen burner and carefully noting the temperatures at which vaporisation takes place, the remaining deposits being tested for grit, acids and other impurities. Thus by taking this deposit at various temperatures before the point of vaporisation is reached and testing same, a good resulting oil could be found, for it is in this remaining deposit that the evil effects of cheaper oils lie. Feed water containing oil should be purified on its way to the boiler by passing it through some efficient filter, of which there are several types of varying degrees of efficiency on the market. By the use of a filter a good deal of oil is prevented from entering the boilers and carrying on its corrosive action. Where vegetable oils have been allowed to enter the feed water the use of soda to neutralise the acids is sometimes attended with trouble in the form of a soapy scum on the surface of the water in the boiler, and this is frequently carried into the H.P. cylinder bv priming. In such cases lime alonc should be used to neutralise the acids from the oil.

Effect of Air

Air has been a well-known cause of corrosion for many years, and many instances of rapid corrosion have been proved to have been caused by the feed pumps drawing air from the hotwell and the feed being delivered to the boiler at a level considerably below the water line. Corrosion arising from the presence of air takes place in the following manner:—Small beads or bubbles of air are expelled from the water on boiling and attach themselves tenaciously to the heating surfaces. The oxygen at once attacks the plates, and the action results in iron rust, and produces a thin crust or excrescence. This, when washed away by the circulation or dislodged by expansion and contraction, leaves beneath it a small hole or pit.

Pitting, when once started progresses very rapidly, as the indentations form ideal resting places for the bubbles of air and at the same time present fresh surfaces for attack. Fresh water at 32° F. absorbs oxygen to the extent of 4.9 per cent. of its own weight, at 50° F. it absorbs 3.8 per cent, and at 68° it absorbs 3.1 per cent., whilst salt water absorbs more air than fresh water. It will thus be seen that, as far as possible, air must be prevented from entering the boiler with the feed water, and the best method of doing this is to have an automatically-controlled independent feed pump working in conjunction with a direct-contact feed heater of the "Weir" or other approved type, where the air is liberated from the water in its passage through the heater. Care should also be taken to ensure that the feed pump glands are kept tight, to guard against air entering in this way.

In the majority of cases it is impossible to entirely prevent air from entering the boilers, at the same time a good deal of the air can be kept from attacking the plates if the internal feed pipes are led into the steam space and slots cut in them so that the water can spray out of the pipe; most of the air will then be given out to the steam, thus preventing it going down with the water to the heating surface and having a chance of attacking the plates.

The design of a boiler also enters into the prevention of corrosion by air, for if a boiler is designed to give us good a circulation as possible, then the air bubbles would have a lesser chance of finding resting places. Another practice which is very detrimental to boilers is that of filling them up from dock or river waters, and especially from rivers that have their courses leading through coal-mining districts, as water that is pumped out of coal mines contains, among other constituents, iron sulphate ($Fe SO_4$), and this oxidises the iron when brought into contact with the atmosphere, the action forming iron oxide (FeO_3) and leaving sulphuric acid, this being brought about by the hydrogen amalgamating with the SO_4 and producing H_2SO_4, which, of course, is anything but a desirable constituent of boiler feed water.

The Chairman: I think you will agree with me that, having heard the paper read, the loss of Mr. Rees is a decided loss to the Institute, for he was evidently a very thoughtful man, and the paper he has written should certainly be of great assistance to our sea-going members. Mr. Rees, of course, cannot reply to any criticisms which are made by the members, so what has to be done is to have a more or less debating class amongst ourselves, and for you to offer your opinions. I know there are other gentlemen present who will take exception to anything you say, and by doing so we shall get a discussion.

Mr. F. M. Timpson: I thoroughly agree

with the chairman's remarks as to the loss of Mr. Rees. He has certainly shown himself to have been an engineer who has studied his profession very closely, and he has given in his paper a good many items, some of which are perhaps new to us. At any rate, they are of very great general value. The class of paper which Mr. Rees has put before us is looked forward to by our members. To the paper we cannot offer a criticism; we can only concur as to what has been said as regards the style of it. I may say, in respect of some remarks, especially in regard to evaporators, that the working of this apparatus is not always given the attention that it needs. In one part of the paper Mr. Rees speaks of not being able to get the evaporators down is very important. It is recognised as such by some of the eminent makers. Some have introduced apparatus which keeps the evaporator from an undesirable density. The various formulae given for the chemical solutions are very useful, and I regret that the writer was not spared to be with us.

Mr. John B. Harvey: The late Mr. Rees was the author of a very valuable paper on "Corrosion of Marine Boilers," and I quite agree with that gentleman that new boilers, if not thoroughly well attended to and treated in a proper manner when first placed in commission, will be productive of endless trouble before they are very old. Lime or soda, when placed in a boiler, should be thoroughly dissolved, and not put in all at once; it should be allowed to drip into the feed water continuously from a tank. There is no doubt that evaporators do not receive the attention they require, often being allowed to prime through, not being blown down at the necessary intervals. A very little priming will be the cause of a great deal of harm, and will soon be the means of creating much scale on the heating surfaces of the boilers as well as causing corrosion. The author, in his paper, mentions internal feed pipes. In quite a number of marine boilers these are led across the tops of the tubes with a branch leading down between each nest of tubes, with the result that the feed water is forced down towards the furnace, and as it is not possible to get rid of all the air from the feed water before it enters the boiler the remaining air is carried down on to the top and sides of the furnace, which will be the means of starting corrosion sooner than anything. In my opinion the internal feed pipes should be led across the tops of the tubes with pipes slotted on the side facing the combustion chamber, so that the feed will meet the water as it is circulating away from the combustion chamber in the direction of the front end of the boiler. If all internal feed pipes were fitted in this manner it would prevent a great deal of the corrosion which occurs on the furnace, as the air would then have a better chance of being liberated into the steam space before being carried down on to the heating surfaces.

Mr. S. G. Martlew: I beg to thank the late member for his paper, which no doubt will be thought over a great deal by many engineers and lead to discussion in mess-rooms and other places where they congregate, as most of us are unable to get to the Institute in these stirring days. He speaks of the use of lime to treat the boiler water, but does not refer to putting the lime into a tank and passing the feed through it. By circulating the water via a lime-tank and suitable filters prior to entering the boiler, objectionable influences are greatly reduced in potency; in fact, the whole of these, including globular and emulsified oil, may be caught externally if electrolytic action is arranged to take place in addition to the mechanical and chemical safeguards. Removal of corrosive or scale-forming matter in the feed should preferably be effected outside, and not within the boiler, otherwise the steam generator is turned into a chemical works, with injurious results arising mainly from the reactions and the settlements of impurities upon the heating surfaces. An old and inexpensive method in starting up new boilers is to first limewash the inside, and (or use a little sea-water for a very short time as feed), by using the surfacing cock, a thin scale, say 1-32 inch thick, will adhere to the interior structure. This dispenses with the considerable amount of lime referred to in the author's opening paragraph, thus minimising deposit and loss of efficiency. I notice that there seems lack of balance in two of the equations employed to represent chemical reactions, due possibly to inadvertent errors in the MSS. The first one should, I think, read:—$MgCl_2+2H_2O=Mg(OH)_2+2HCl$; and the third one:—$2FeCl_2+Mg_2=2MgCl_2+Fe_2$, or else $FeCl_2+MgO=FeO+MgCl_2$. At the end of the paper may be added: $FeSO_4+2H_2O=H_2SO_4+FeO_2+H_2$, the nascent hydrogen, H, being a corrosive factor as well as the sulphuric acid. The presence of oil obviously being detrimental to boilers, for internal lubrication when working engines in or out of port and manoeuvring generally, I have often succeeded in preventing slide-valves, etc., from squeaking or grunting by pouring a little water in by the grease cups. the only addition throughout the entire voyages being a small quantity of best cylinder-oil for rod-swabbing. Where zinc-plates and their supporting iron studs or hangers touch, both metals should be scraped bright at the surfaces of contact to ensure good results. I have found that by enclosing the lower part of each plate with a perforated guard-tray, as the zinc disintegrates, any large pieces are thereby stopped from falling upon the heating surfaces, thus obviating corrosion, over-heating and waste of an expensive metal. By emptying, cleaning, drying and closing up steam generators which are to be out of service, they may be kept from wasting, and in the case of water-tube types, when in and out of service, special measures against corrosion are sometimes necessary. The author does not refer to methods taken to repair damage done by corrosion. Upon surveying the boilers of one vessel I found deep pits and scars on the combustion chamber tops and sides, and having scraped these down to the bare metal, filled one lot with Portland cement and the other with zinc powder-paste. Care was observed when steaming to prevent the entry of air or other injurious fluids or solids, and upon subsequent examination the old defects were barely perceptible and no new ones arose while I remained as chief. Seeing no mention of external corrosion in this paper, but having met with some serious instances, I feel bound to state that this must particularly be guarded against. Leaky joints, either above or on the shell, causing insidious percolation beneath the lagging, deposit and cooling of ashes next to unprotected front plates, wash of impure bilge water against the bottom, and steam or water leaking at tube or stay ends—all these are responsible for wastage on the outsides of marine boilers.

Mr. A. J. McLeod: The paper written by the late Mr. Rees touches upon a subject which is of vital importance to the marine engineer, and it is to be regretted he was not spared to read the paper himself before us.

The author appears to have followed the practice advocated in a pamphlet published some years ago and quoted verbatim in several text books, notably Jamieson's "Steam and Steam Engines." The conclusions given represent good practice and have long been accepted by marine engineers.

In establishing new boilers, almost always a considerable amount of "bleeding" takes place. or, in other words, porous oxide is formed excessively, and the most rapid method of dealing with this is to follow the directions given in the paper supplemented by judicious "blowing down." The writer speaks from actual experience with new boilers, and first tried these methods some ten years ago at the time the above-mentioned pamphlet was published; the results fully confirming the efficacy of the methods advocated.

Ten years, however, represents a long stride in engineering progress, and after a boiler has become established with a protective coat it is doubtless better practice to introduce lime to the extra-feed water on the principle of concentration up to the point of deposition in the manner used by various water softening processes now on the market, and thus pass into the boiler softened water. It doesn't seem quite right to introduce lime into a boiler and afterwards laboriously chip it off again; the external application of lime. while being equally effective has not this objection.

The danger of hydrochloric acid given off by the evaporator is rather overstated, this would only occur when the density reached saturation point. Twenty-five ounces per gallon, or 5-32, and most engineers know how hopeless it would be to attempt to make up feed with the evaporator at such high densities. However, should any (HCl.) hydrochloric acid be formed it would be an aqueos vapour, which when diluted with the boiler, would form an extremely weak solution, which could readily be neutralised by soda.

Galvanic action is really a weak acid-uous solution setting up an electrolytic action between the dissimilar metals used in the boiler's construction, e.g., ferrous and non-ferrous. A boiler in this condition closely resembles a "plating bath," the zinc plates which we introduce being the anodes and the boiler itself the cathode. As in the plating bath unless a good metallic contact is established between the anode or electro-positive element and the cathode or electro-negative element, no deposition from the anode to the cathode would take place but both would become corroded in proportion to their electro-chemical equivalents; similarly both boiler plates and zinc plates would be corroded unless proper metallic contact be established between the zinc and iron.

Corrosion in the very nature of things will occur in and about boilers, and undoubtedly the greatest enemy of all is free oxygen. Oxygen will attack every known element in greater or less degree so therefore no steamship should be without adequate air-extracting devices attached to the feed-water systems so as to de-aciate the water before it enters the boiler. Only constant care and vigilance will keep in hand and restrict the many corrosive influences met with in marine boilers.

Mr. Wm. McLaren: We all regret that our member, Mr. Rees, has passed away as a victim of the submarine, and at the moment we can only say that he died for his country. It is difficult to criticise a paper written by one who has passed away. It is left to us to make the best of the means at our disposal. The author has evidently given some study to the care of boilers. There is no one amongst us but has his own particular way of treating boilers, and, as Mr. Rees admits, no two boilers can be run alike. I came out of a boiler last night, it was a boiler of the Lancashire type. It has been giving trouble, the feed water having been passed through a water softener, and when the furnace rings were examined one would think that either they had been scoured or some acid had been acting on the plates. They were cleaned and painted with Portland cement wash. The boiler has been at work for the past seven months; there has been some sign of pitting showing ever since the water softener was put into use about eight years ago. The other parts of the boiler where pitting is has been in patches, "not regular," all on the heating surface, changing about, first one place and then another has been attacked. There is not any signs to be seen about the shell plates. The boiler was left to cool from steam. Scale was there, but it was not difficult to get rid of it. The boiler water had been drawn from an artesian well 370 ft. deep and run in a water softener, also town's water on occasion when either the softener was being cleaned or well pump under repair. Mr. Martlew remarked upon the use of cement. If you wash the heating surfaces as well as the shell surfaces of a boiler which is coated with Portland cement you will find, after the boiler has been run for a time, that this acts as a preventative to pitting. Would it not be an advantage to test marine boilers on land, in the boiler works, before they are fitted into the ship? Would it not be an advantage for the life of the boiler and also for the life of the engines to supply steam to it in the boiler works, to keep it going for a few weeks on end, with all the conditions that it would be put to when it was fitted into the ship? That, I think, is a question for the metallurgist. Some boiler tubes go for the life of the ship and boiler, say 10 or 20 years. Some only go for a few months. There is nothing harder on the marine engineer, and it is always a cry of "lubricator"—that is, the engine slide valves grunting —when it is the blessed boiler giving off some sort of scale or deposit that goes pumping into your engine slide valve casings. You hear of this grunting, and that is from the boiler tubes and plates "at it," hence the poor engine has had to suffer under this scale. I say, give the boiler some work to do, at least before it is fitted into the ship. Whether, when under such test the boiler supplies steam to an engine, or whether it is merely under test for blowing off, it should be supplied either with normal town water or you can put it under water such as would be used in seafaring conditions. It is like starting from the sea water to the evaporator. I quite agree with the author that the boiler should be treated early with sea water. Certainly we have the different rivers from which we have to take water. Until you are trading with these particular places you never have the chance of finding out the conditions under which your boiler is working. One point I have neglected to mention. Why should you stick to these calculations or quantities of I.H.P. when we ought to take as a basis the amount of water evaporated?

Mr. J. Shanks: In reading Mr. Rees' paper I have been very much impressed with its excellence; it is one of very great value to sea-going engineers, and shows that Mr. Rees had given diligent thought to the management of marine boilers and if everyone aplied themselves to the recommendations made in the paper, I feel sure we should hear little or nothing about corrosion. Mr. Rees emphasizes the importance of giving boilers a good start in life, and all marine engineers know that this is the great consideration. Too often however this is neglected. In the leading lines of steamers I think I may say that all the precautions recommended are adopted in regard to the cleaning of the boilers, washing the heating surfaces with cement wash, filling them with fresh water and fitting zinc plates in metallic contact with the vital surfaces before the vessel starts on her first voyage. When these precautions are taken there is little fear, provided the boilers are in the care of an intelligent engineer with an up-to-date plant on board, of trouble accruing from corrosion. I would like, however, to emphasise to sea-going engineers the importance of keeping sea water out of the boiler from the first day of its life. Many still believe that the introduction of sea water into a new boiler creates a protective scale on the heating surfaces; nothing could be more deceptive. In regard to the management, or rather arrangement of evaporators, Mr. Rees recommends that the vapor pipe should be led well above the dome; this, in my opinion, is important. I know of one well-known line of steamers where the vapor pipe is led to such a height as to prevent the possibility of water being carried on to the condenser, and the evaporator is practically being worked under the pressure in the condenser, and the results are most satisfactory. The judicious introduction of soda or lime with the feed water is to be recommended, but first of all the engineer must know the character of the water in the boilers and the feed water he is putting in. In regard to lubricating oils, this opens out a wide question, but there is no reason, with our present knowledge on the subject, why oils should be used for internal lubrication of cylinders which may leave a residuum which is detrimental to the boilers. In any case a modern ship should be fitted with an efficient feed water filter. Our hon. secretary, I think, should be thanked for bringing this paper before us, and the Institute will treasure the memory of Mr. Rees, who has given his life for his country, and leaves behind him such a valuable contribution to the transactions. I feel sure that the members who have read it will benefit.

The Hon. Secretary: The presence of mill scale is a feature in connection with the care and treatment of boilers which it is well to have emphasized, as has been done in the paper submitted for our consideration. The mill scale if left on the plates has a very injurious effect, fortunately it is comparatively seldom that this scale survives the various operations involved in the building of a boiler, still, it is sometimes found, and ought to be removed immediately it is noticed. All boilers ought to be examined periodically, preferably every voyage, if of three or four months' duration, so that any changes taking place in the internal parts may be carefully observed and noted for guidance in regard to treatment from voyage to voyage. The periodical examination of boilers should start as soon as the trial trip is finished, as the beginning of their working days requires to be watched and safeguarded from the possible presence of germs, which may develop into troubles later on.

The worst case of mill scale corrosion that came within my experience was in the boilers of a new steamer that had made one voyage. In this instance circumstances prevented an examination of the boilers prior to sailing from the home port. The scale was on the inside shell plates and the corrosive action had so developed that in places, when the scale was removed, the depth of the pitting was quite a quarter of an inch.

The MacLean Publishing Company

LIMITED

(ESTABLISHED 1888)

JOHN BAYNE MACLEAN - - - - President
H. T. HUNTER - - - - - - Vice-President
H. V. TYRRELL - - - - - - General Manager

PUBLISHERS OF

MARINE ENGINEERING of Canada

A monthly journal dealing with the progress and development of Merchant and Naval Marine Engineering, Shipbuilding, the building of Harbors and Docks, and containing a record of the latest and best practice throughout the Sea-going World.

B. G. NEWTON, Manager. A. R. KENNEDY, Editor.

Associate Editors:

J. H. RODGERS (Montreal). W. F. SUTHERLAND.

OFFICES

CANADA—

Montreal—Southam Building, 128 Bleury St., Telephone Main 1004.
Toronto—143-153 University Ave., Telephone Main 7324.
Winnipeg—1207 Union Trust Building, Telephone Main 3449.
Eastern Representative—E. M. Pattison.
Ontario Representative—S. S. Moore.
Toronto and Hamilton Representative—J. N. Robinson.

UNITED STATES—

New York—R. B. Huestis, 111 Broadway, New York, Telephone 8971 Rector
Chicago—A. H. Byrne, 900 Lytton Bldg., 14 E. Jackson Street, 'Phone Harrison 1147
Boston—C. L. Morton, Room 733, Old South Bldg., Telephone Main 1024

GREAT BRITAIN—

London—The MacLean Company of Great Britain, Limited, 88 Fleet Street, E.C., E. J. Dodd, Director, Telephone Central 1960. Address: Atabek, London, England.

SUBSCRIPTION RATE

Canada, $1.00; United States, $1.50; Great Britain, Australia and other colonies, 4s. 2d. per year; other countries, $1.50. Advertising rates on request.

Subscribers who are not receiving their paper regularly will confer a favor by telling us. We should be notified at once of any change in address, giving both old and new.

Office of Publication, 143-153 University Avenue, Toronto, Ontario.

Vol. VIII. AUGUST 1918 No. 8

PRINCIPAL CONTENTS

Merchant Marine Should Be Recognized

THE week from September 1 to 7 is to be set aside in the Dominion as Sailors' Week, with the special purpose of raising money for the support of the dependents of those who have lost their lives while serving on the British Merchant Marine.

With the object there can be no quarrel. There is not even room for a good argument concerning the wonderful service performed on the trade routes by the gallant men of the Merchant Marine.

But with the system that allows the dependents of these men to be made the objects of a week's pity and hat-in-hand giving there is every fault to be found.

Why should there be no pensions for those left behind by these men? Is their service not as worthy of recognition as that of any other branch? Is the work they perform less honorable and less vital to the existence of the nation?

Not a bit of it. The great bulk of the hardships on the seas in this war have been endured by the men of the Merchant Marine. They have seen more of the submarine than any other branch of the service, and they have been instrumental in carrying men, supplies, ammunition—in fact everything that has enabled the Allied armies to carry on the war in the different fields.

In the face of the submarines, of the floating mine and the sunken mine, the men of the Merchant Marine went ahead, and with a courage that was wonderful and a tenacity of purpose that was traditional, turned an adverse balance in 1913 into a trade balance in 1917.

If the men of the Merchant Marine had shirked, this work would not have been done. The men in the front could not have been fed. The supplies and munitions could not have been sent across. The commercial supremacy of Britain could not have been maintained.

The service of the Merchant Marine is not a mean service. It is deserving of real and substantial recognition. It is not enough to depend upon Sailors' Week," or tag days or voluntary giving. It should be recognized and rewarded on a straight and decent basis. There should be pensions and allowances for dependents of those who lost their lives in the service of the Merchant Marine, and it is high time the matter was put on the permanent basis that it deserves.

Letting Trade Routes Slide

SIR JOSEPH MACLAY, controller of shipping in Britain, has brought forward a fact that should not be passed lightly over. Speaking of the fact that Britain has provided the protection for U. S. forces crossing the Atlantic, he said:—

"But I might add, since the fact may not be well known, that we are only able to face these new responsibilities by sacrificing for the time not only British, but Imperial interests. Ships, which under normal circumstances are engaged in the trades between the British Isles and the Far East, Australasia, and India, have had to be withdrawn from service, and we have been compelled to sacrifice to a large extent communication between the Mother Country and the Dominions and the Southern Seas."

This statement was not made as a complaint; Sir Joseph was not urging that Britain was making too great sacrifices—rather it was simply a statement of absolute fact with which the people should be made acquainted.

Britain is letting her foreign trade slide; she is taking off boat after boat from trade routes that have been, and will be, matters of commercial life and death to her. The Island country is sparing nothing—Britain is in it to the neck, and will stay in it to the neck till it's all over and cleared up.

At the same time it is well to look forward to the day when trade routes will have to be won back, and when commerce and industry will have to ply again. War not only breaks down the enemy, but it weakens the people waging it.

When it is considered that a condensing engine or turbine using about 10 lbs. of steam per h.p. hour only converts into useful work about twenty per cent. of the heat value of the fuel while a large portion of the remainder goes to waste in the condenser cooling water, it is readily realized that the advantages of putting the exhaust steam to useful work more than outweigh the two to one advantage in economy that the central station possesses.

How the Paul Jones Lost Her Crew

Being a Narrative of the Days When the Captain and Mate Used Poor Whiskey and Big Fists to Gather up a Crew—Policeman Was Found in the Lot—Chased Into Service on the Trim Clipper "Paul Jones"

By E. J. PATRICK

THE harbor of Port Philip, in Australia, was crowded with the shipping of all nations in the early months of 1886. Every berth alongside the piers of Port Melbourne and Williamstown was occupied by sailing ships from all parts of the world. As a result many ships had to lie out in the bay and discharge or load their cargoes into lighters alongside. As the double process meant a stay of anything from six weeks to twelve, plenty of opportunity was given to us boys to make the acquaintance of fellow seafarers. We were lying three-quarters of a mile from the pier at Port Melbourne, and it was the work of the writer, with two other boys, to pull to and from the quay several times a day. Within a week of our arrival we had pretty well memorized the name, nationality, and character of every ship in sight.

The "Mystery Ship," as she was known, was a large, full-rigged ship lying out in the bay about half a mile farther out than ourselves. All sorts of rumors were current about her. She was evidently loaded down all ready for departure; it was known that she had come in from New York, unloaded, and filled up again, and yet she still lay at anchor, sails made fast, and ready to take the pilot aboard and get towed outside the Heads. Rumor had it that her skipper was a real "hard case," and that the mate, a fellow standing 6 feet 3 inches in his bare feet, was an even tougher case than the "old man."

Rumor in this case, was a proved fact, and we soon learned that his beautiful model of a Yankee clipper, painted white from stem to stern, was a veritable "floating hell" for the sailor. Every mother's son of her crew, with the exception of the skipper, mate, cook and carpenter, had deserted her at the first chance after she dropped anchor. Three of the fellows, two able seamen and a boy or ordinary seaman, had actually swam ashore after dark, risking the chance of meeting their end through a shark, rather than that of spending another term at sea in this beautiful "devil."

The mate, in true Yankee style, obtained implicit lightning-like obedience to his slightest behest by means of belaying pin or his muscular strength. The 103-day passage from New York to Melbourne had been an experience that few of the crowd would face again. True, in running away, they forfeited their wages for the period, but that was compensated for by the fact that the month's pay from Melbourne was nearly twice that of New York. Even with a good ship, a crew was hard to obtain, hence the "Paul Jones," in all her stately sea dress, and outward attractiveness to the sailor's eye failed completely to gather up even a nucleus of a crew. I came to make the acquaintance of the youth of the trio who had swam ashore. He was a tall, strongly-built lad, and was lying low and keeping out of the way, until the "—— —— hooker" had cleared out, as he strongly expressed it.

Those veritable land sharks, the "boarding masters," had been offered quite princely sums per head for every man they could obtain to make up a crew. More than one poor druken unconscious mortal was dumped into the "Paul Jones's" boat after dark and taken on board, there to remain a prisoner until the complement was on board. This process of getting a crew is called "Shanghai-ing," and is resorted to only by rough-and-tough ships. The very nature of the transaction makes haste a necessity, for if one "Shanghaied" sailor escaped, the game would be up, and the authorities become wise. As in this case, discrimination is often thrown to the winds, and any strong, able-bodied male, who can be netted into the low-class bar-room, and induced to take part in the drinking and hilarity, is likely to awake with the swish of the briny as the accompaniment to his waking thoughts, and a very mixed memory as to his last conscious hours.

No prettier sight on God's ocean can be seen or heard than the early morning preparations on a fine big sailing ship, for her voyage, may be across two hemispheres. The tuneful chanty of the men as they heave at the windlass or capstan, bringing up the "mud hook," and the slanting rays of the morning sun bidding them bon voyage, whilst the saucy, diminutive-looking tow-boat, leading ahead with the big hawser attached, all goes to make up a picture that, once seen, can never be displaced for its sea charm.

But on a certain dark night, with no chanties to mark the event, no fuss or racket, the "Paul Jones" lifted her anchor and, unseen and unheard, made for the open sea. The word ran round our ship like a fire-call in the early morning, that the "Paul Jones" had gone. I looked to the accustomed spot, and sure enough it was so. What tragedy and crime was represented in her departure; what kind of a crowd had she got, and had any of our pals involuntarily joined the gang? We should never know, perhaps. Poor beggars, I thought, there's a tough time ahead of them. This was Tuesday morning, and if anyone had suggested that on Saturday morning following we should see the "Paul Jones" crew pass under our stern on their way back to Port Melbourne pier, we should have thought him a maniac, and yet this is what actually happened.

She was hardly outside the Heads, with some of her square sails just being shaken out, and the tow-boat still standing by to take back the pilot when heavy smoke began pouring out of the forward hatch. Where there's smoke there's fire, and in this case fire of such a fierceness and rapidity that it only took a few minutes to see that the ship was doomed. Fire-fighting appliances on a sailing ship are usually of a neglible quantity, and so it proved in this case. The tow-boat was still within reach, and she quickly came alongside prepared to take off the crew. And what a crew! Never shall be dimmed the sight of those mixed human samples as they stood and talked on the deck of the tug, as they passed close by us. The first news of the burning had come by wire from the signal station at Port Philip Heads, and now we looked on the ragged remnants of the ship. Of the twenty-two men, only four were sailors. There was one fellow with a long white blouse, who looked like a painter. Another fellow was quite a "toff" in his well cut suit. Still another bore evidence of being well acquainted with a bullock team and general farm work in the back blocks, and to crown all, there was a policeman, who had actually been roped in and put to sleep like the rest, and dumped aboard.

It would be a safe bet that not one of those individuals would again take his drinks by the sad sea waves, unless he knew well his ground. The "Paul Jones" got her crew all right, but evidently there was some one who struck a better way of getting ashore again than by swimming.

ACCORDING to the *Revue de Chimie Industrielle*, the best composition for dressing the hulls of sea-going vessels is made up as follows: Six parts of tallow, melted hot, to which the following are added, and the mixture constantly stirred: six parts carbonate of lime, two parts of powdered quicklime, four of charcoal, four of sulphur, four of felspar, two of dried zinc white, thirty-three of spirit varnish, twenty-three of coal-tar naphtha, sixteen of 92 per cent. spirits of wine, and ten of ordinary varnish. To this mixture is added the coloring matter desired, in a dry and powdered form. The following additional ingredients are said to increase the efficiency of the composition: Tallow, six parts; copper, ten; iron, three; brown hematite, twenty-five.

Fo'c'stle Days
or
Reminiscences of a Wind Jammer

By Capt. Geo. S. Laing

The steamer and the motorship have deprived the seafarer of much of the old-time romance attaching to a sailor's life. The advent of wireless has also increased the safety of it. Deprived of these features, a sailor is immediately dependent on his own resources, and Captain Laing gives us a true and seaman-like reflection of sea life under canvas. The present feverish activity in constructing self-propelled craft affords a suitable background for incidents in a career in which gaffs, booms, tar and canvas held sway before being displaced by mechanical unloaders, derrick masts, fuel oil, coal and wire rope.

PART VII.

AT last our craft was hauled to the north-east, our longitude being about 140° east of Greenwich. This manoeuvre meant two big items on the voyage, the coming of finer weather and the hopes of arriving in Australia in two or three weeks. As we had only sighted South Trinidad off Brazils since we left the Lizards and Wolf—the famous guardians of the English Channel—the sight of land in itself was no mean picture to look forward to. The reader must imagine what it would be for him or her not to lay eyes on anything but sea and sky for four months. It was quite a test in many ways and one became impatient at the last of the trip. Even the same faces focused on one's mental mirror without the least deviation made us long to see a stranger.

The smashing-in of the half-deck was soon forgotten, the whole affair was only a small incident to seasoned sea-dogs, although naturally it loomed up in a first voyager's mind.

Approaching the Largest Island on Our Sphere—Australia

We were approximately 1,200 miles from the anchorage in Moreton Bay at the mouth of the Brisbane river, and it looked as if we were going to "break off." This meant going round Tasmania instead of through Bass' Straits. The men in the forecastle were looking out their shore togs, and a general wash up and tidying of things was in evidence.

By all that we could read between the lines it appeared that the "Maggie Dorrit" would be minus 8 A.B.s, 1 apprentice, plus the second mate, cook and sailmaker. It had leaked out that the second mate was well acquainted with the Queensland Kanaka schooners that plied their semi-slavery voyages amongst the ocean islands, giving trinkets and offering contracts to the negro islanders for work on the Australian sugar plantations. At the time I write of, thirty years ago, running away or deserting the ship in these parts was the most common thing to do. A few of the incentives that caused desertion were, the gold stories, free life in the bush or sheep farming, the lure of the coasting traffic with its big money and good grub, and a sailor's love of rum and "sheoke." Sheoke was the term given to colonial beer.

An exceptionally clear night with a southerly wind gave us another enchanting view of the celestial beauties of the Australian heavens. The southern cross stood out with its chief actors resembling vivid eyes of molten metal. Its famous pointers (a) Centauri and (b) Centauri gave the key to other constellations and Canopus, Ackernar and Fomalhaut in the offing. Spica shows to advantage in these parts and sailors always remember it as having the shape of a spanker—the fore and aft lower sail on a barque's mizenmast. To look aloft and watch the mast heads describe semicircles amongst such glittering jewellery was an inspiration for any one who admires Nature's pictures.

USING THE SEXTANT.

Our feathered sailors gradually grew less in number as we sailed north, land having no attraction for their unwearied feet and wings. How different an instinct to the migratory types of birds whom Nature has endowed with sense of direction and distance that they may the more readily find a haven of rest after their sojourn over the ocean deserts. The sight of two coasters one evening, just as the sun lowered himself down in peek-a-bo clouds on the western waters was sufficient to cause every man and boy on board to look at the strange far-off vessels as if they had been approaching pirates with the black flag industry a foregone conclusion.

Making Ready for Port

Heaving the hand-log became more of a pleasure now and when the watches were relieved there was quite a demand amongst the crew to know how fast the old "Maggie Dorrit" was sailing. Jack's mind was working like yeast, old shellbacks that were usually stolid became playful like kittens and one could note the anticipation bubbling out of their eyes as visions of upcountry life and beachcomberism passed through their weatherbeaten countenances. The majority of them had danced, drank and gambled in the land of the Kangaroo before, but this time they were to be wiser and not land in jail, to be thrown back on the first outward-bounder that swung into the stream with a full cargo.

Fid the ancient mariner, had been working up Mercator sailings from each day's noon position, and thus keeping the non-navigating part of the crew abreast of the situation. Our albatross family had disappeared entirely and we had come into coast-colored waters moreover we could smell the land. This would appear an overdrawn expression but it is a literal truth. After coming in from the trackless ocean the coastal atmosphere of any islands or continent is very materially changed and a sailor's nostrils are susceptible and as keen at the scent as a hound's might be.

We were all kept busy with sand and canvas on the paint and teakwood. A coat of white lead was put on the bulwarks, whilst the teak and greenheart had to be dressed up with copal varnish and very reverently swabbed with fresh water—even if there wasn't enough on board for ordinary clothes washing. The Old Country ships were greatly admired in Australian ports and skippers and mates vied with each other in bringing into port a clean and tid ycraft glistening with fresh paint and varnish. When the time came the sails were also "harbor-stowed" and ropes "flemish-coiled" or "jagged up" in the rigging. Double sheets were either "becketed" or unrove and stowed below.

Everyone of the crew seemed to work harder and quicker than at any other stage of the passage. The getting near port business was certainly a tonic of wonderful potency.

Land on the Port Beam

The coast line soon came in sight, a faint bluish strip to the westward. Fid announced to Tilbury "the land of your adoption, but don't attempt desertion till the Maggie Dorrit is on the eve of departure, otherwise you may be caught." Tilbury promised to take this advice and as we will see later, he did.

Standing in for the land with a fair wind, after being off the earth for four months, acted as an invigoration both to mind and muscle, almost to a point of exuberance. To throw the thoughts back to leaving Europe was like picturing some funereal happening, but rolling gently into a Queensland estuary had a picnic effect. A few shore birds ventured out to welcome us, and as is usual with such messengers, they perched on the "trucks," a point of vantage for all orators and acrobats. As the land rose and gave more definite outlines, revealing in detail a handsome topography, the desertion spirit entered more souls than the parties already mentioned. There can be no more exciting moment in a youngster's existence than his maiden entrance to a foreign strand some 15,000 miles from his real home. Utopian dreams flash across his mind, fag ends of Marryat, Stevenson and Bullen yarns unravel themselves as he scales the ratlines of the topmast shouting "Aye, aye, sir" to the gruff order, "break the buntline stops, you boys." On the Maggie Dorrit we used wooden pegs for buntline stops on such sails as topsails and courses and the withdrawal from the strand of these pegs was preferable to breaking them from the deck as is usual with lighter stops of sail twine. When entering port a boy was handy in each "top" for such light but important duties.

Shoal Water Symbols

The chain cables had been dragged from their lockers in the fore-hold and bent on to the anchors off Bass' Straits, but now the mud hooks were launched out-board in their shank-painter garb ready to be "cock-billed" into "ringstopper" position, which makes them await the final "let go."

Our cook actually donned a white cap and the galley underwent a wonderful change in appearance, whilst the steward laid cocoanut runners and coir mats around the saloon and cabins. We passed a few islands and ran our number up at the mizen gaff end under the old "red duster" of the merchant fleet, a flag by the way, that is better understood today than it was in Maggie Dorrit times. Those international flag signals were interpreted by the lighthouse people and flashed by telegram to Brisbane.

The brass on the wheel, binnacle, skylight and companionway simply glistened after its bathbrick and colza dressing, and a finish with whitening. The varnished teakwood was as well preserved as on the day of the barque's launching back in 1860. Able seamen, eight in number, covering six nationalities, flitted around like flunkeys whose alacrity centres on tips, and there was a reason for it all. Would those adventurers not be cooking their own "damper" (bush made bread) and swinging a handkerchief "swag" over their shoulder ere the week was out? Ye wanderlust gods, but Australia had a pull.

The old barque shot into Moreton Bay with the Pile lighthouse over the lee whisker-boom and the scenery was grand. There the terraced levels of wooded country, and there the open spaces of cultivated or cleared lands stood out from aloft. Our old man was stumping the weather side of the poop as if he was verifying the great work of Cook or Van Diemen. His bamboo silk suit raised him at a glance as the autocrat of the vessel, while his "as you go," "luff a little," "meet her, damn you," was rattled off to the quartermaster with a knowing style that meant "I savvy" all about your desertion dreams and actions."

A large, powerful tug bore down on us from the river, the mouth of which was hid in timber land. The man in the weather chains was yelling out his soundings in sea drawl—"by the mark seventeen," "deep sixteen," then as the water shoaled such phrases as "by the deep eight," "a quarter less eight," and "a half seven," etc. "Deep sixteen" means 96 feet of water, and "a half seven" means 45 feet of water.

The carpenter stood, hammer in hand, at the cat-head, having previously seen that the necessary amount of cable was "ranged" on the fore part of the windlass and the norman pin shipped upright to prevent "riding."

Then the reduction of white wings became the main ordeal and the rattling of blocks, squeaking of gins, with modulations from patent ball-bearing sheaves mixed in with "shantying" made on the whole a noise that must be heard to be understood. Finally the helm is put "hard down" and the barque rounds to while the head sails are lowered as the mizen stay-sail and spanker foot are "flat-sheeted" to aid in the graceful manoeuvre. The square sails hanging in their gear, but aback, are sufficient to stop the ship's way and as the leadsman announces "by the mark ten, up and down, sir," the old man sings out to "stand by to let go." With a fur-

CAPT. COOK PARLEYING WITH THE NATIVES ON THE EAST COAST OF AUSTRALIA, 1770.

ther yell from the leadsman of "she's going astern, sir," the final command bawled from the poop could have been heard a mile away, "let go—give her the 45 shackle outside the hawse-pipe." In due course the mate bawls back "anchor's got her, sir," and that order is timed as the end of the passage proper, the vessel's movements being then subject to tugs, pilots and other inshore control such as depth on bar, stage of tide, wharf accommodation and so on.

The distance covered on the trip was something over 15,000 miles and time on passage 115 days. How would that suit some of the people who think five days on the Atlantic equals a short term in jail? But we must not digress.

Every mother's son laid aloft to harbor-stow the sails, putting a tight "skin" on each white wing and passing the gaskets after the manner of a "seizing" instead of the winding fashion used at sea, when furling.

We had orders to stay at anchor for the night and the change in routine was very welcome, especially for the boys who were exempted from anchor watch by some fluke at the expense of the carpenter and sailmaker. After all the rolling and diving of the passage the Maggie Dorrit rode to her anchor in this peaceful bay as if she was a stationary model laid down on a mirror. Not a move, no sheets straining or banging with the weather lurch, no sea water on deck rushing madly over hatch-combings and mast-coats and filling one's long boots when brace hauling at the main fyfe-rail. No officer shouting superfluous commands because he felt sleepy enough to drown himself, but lacked the energy to go over the rail. Instead, we were to act the role of farmers—turn in for the whole night.

The tug had gone to some lightship but would return at daybreak. Some of us made large entries in our diaries while others began their home correspondence. The transition from boundless sea room to a land-locked anchorage brought to the eyes as well as the mind that peculiar feast of fascination that one fails to thoroughly describe. As the shades of evening fell over the ship, and anchor lights were displayed, the cry of a cockatoo from the woods and the whiff of a camp fire were as gun shots to the fact that we were really "in port."

Poor Tilbury swore by every serving mallet and marline spike in the boatswain's locker that no ship would ever carry him out of this country. There he stood gazing over the topgallant rail almost ready to drop on his knees like a Mussulman and salaam to the dark coastline. His hatred of the sea was pitiful.

Next morning at 5 a.m. we hove short and had breakfast. The screw tug "Wallaby" took our 11-inch coir hawser and aided us in breaking out the anchor by pulling us ahead a little till the anchor was "away." Just as the hook showed its stock the tug was given more rope and with a graceful swing the old craft headed in for the Brisbane river. At eight sharp the ensign, house-flag and quarantine colors flew from their respective masts, mizen, main and fore. The big job was the rigging in of the jibboom as "standing booms" or a bowsprit and jibboom in one piece, were not known on craft like the Maggie Dorrit.

Towing Upstream

The Brisbane river is semi-tropical in its floral environment and the eucalyptus or gum trees seem to fortify the air whilst other fragrant smells of sun-gloried foliage strike the new arrival as pleasant and worth coming for.

Of course the panorama from aloft was sublime. The "Wallaby" could have towed two vessels of our size as she belonged to the great fleet of ocean tugs that were built on the Tay and Clyde for foreign service and had come out to Australia under her own steam. Such towboats were to be found in the Hoogli, Chittagong and Rangoon and also Melbourne, Sydney, Newcastle and other British Empire outposts.

As we passed some clearings the women and children waved a welcome and at other places the more stolid Chinese coolies or Kanaka negroes worked away mechanically, heeding no one. Suddenly an awkward bend in the river came in view and our jibboom almost touched the overhanging trees on the bank as the old barque swept round on starboard helm, the "Wallaby" meantime appearing broad on the port bow till the channel took a straight course,

Our old man was able to converse intimately with the pilot on many local topics as he had spent a few years on the coral-studded waters of the Great Barrier Reef in connection with the pearl fisheries, and it was quite interesting to be sent on the poop for a few minutes and be within earshot of their conversation which seemed to range mostly on old acquaintances who had led hard lives under a Botany Bay sort of career, with booze, bush life, gold craze and coast schooners all in the setting.

A slowdown for a few minutes allowed the port authorities to hail our captain as regards bill of health, cargo and store manifests, and an enquiry as to whether we had lost any men overboard on the passage. At last we could see the bridge at Brisbane and the shipping at the wharves. Uppermost in our minds was the arrival of mail from home. A full rigged ship, two steamboats and a Kanaka schooner comprised the port fleet at the dock moorings, whilst a four-master with painted ports rode light in midstream.

By the time the Maggie Dorrit got tied up to D. and L. Brown's wharf, the letters had arrived aft and were distributed by an officer. The mail steamers took six weeks to make the passage from England, but the mail lost no value on that account and to boys who had never been more than a mile from an apple tree till their sojourn on the barque, the perusal of letters from parent and chums went high. Half a dozen epistles were opened at once and a bite of news taken from each, whilst newspaper cuttings or photos were scanned only to be rolled up unceremoniously with torn envelopes and tar-fingered note paper, when the mate's voice denoted murder if we did not get about some work or other that the crew were handling. The sad part of the letter business was the outstanding fact that the A. B.s received no missives from any one. Likely some of them were under false names, and for conscription motives which obtained in their respective countries, such beautiful cords as filial attachment had been cruelly severed, and the wound cauterized with time, rum and consequent abandonment.

OUR TUG GETTING UP STEAM.

The customs officials and hatch surveyors hobnobbed with the old man and the agent's clerk, every one agreeing that whatever might be debatable it was quite in order to lower the gauge in the square-face which stood sentry near the cigar box on the saloon table. Old "rope-yarns," the mate, eyed the bottle through an open skylight, but he wasn't in the running for free drinks.

First Day Happenings in Brisbane

The stevedore told the mate what purchases to rig up from the topmast heads for cargo output. Our main yard had to be cockbilled and a dolly timber lashed to the inshore rail for handling the transfer whip which landed the miscellaneous cargo on the wharf. As first voyagers we were paying more attention to a crowd of larrikins than to second mate and bosun orders, and occasionally we would trip over gins, grease pots and other outfits that lay around the main hatch.

But the greatest sight of this eventful day was the arrival alongside of the ship's butcher, with his load of fresh beef, vegetables and fruits. We would have tackled the stuff raw. In our pathos we must have looked like a restaurant cat which had lost its favorite patrons and had been hounded off the premises in a starving condition.

A paper reporter came aboard and had a talk to us. He mentioned that the so-and-so had beaten us by seven days on the Hamburg-Brisbane passage, but a smell from the galley aroused our senses far more than the announcement of defeat humiliated them. Dusk found us at supper and the fresh mess was such a treat that the Queensland mosquitoes were allowed to draw blood from our faces and arms without much disputation till it was time to "clear up the wreck," a sailor's phrase for putting things in order after a meal.

It soon leaked out that the old man had word from his home that a little stranger had arrived when we were "running the easting down" and the father was so pleased that he gave each apprentice in the half-deck a shilling.

Cook, Columbus, Magellan—any old discoverer that you care to mention—wasn't in it with pride and glee, when Fid, our senior boy, towed us up through Queen street, one of the main thoroughfares of Brisbane. Sea youngsters had a minus quantity in ceremony amongst their own class and when we passed within hail of some other brass-bounders the informal salute ran something like the following:

"Hullo, there, what ship?"

"The 'Gladys' of Bristol."

"Are you out from home?"

"No, we're from 'Frisco."

"What kind of grub do you get?"

"Oh, rotten, but we have had a rice cargo since leaving home and naturally we got swollen up like nigger babies."

"Is your 'old man' any good?"

"Good! No, he's a Welshman with a sixty-fourth in the ship and he is going to buy a farm if he can sell out the slop-chest."

"Does he crack on?" (carry sail.)

"No, he's a bare pole man, carries sail in the locker, but the old girl (ship) is fast, which covers up his real actions."

With some further remarks concerning the whereabouts of a cheap eating-house where the wily Chink served up savory dishes to hungry growing boys, the confab came to an end, but not before the youths of each craft had invited the others to visit them on board and compare further notes.

Wooden Shipbuilding Record.—Canada now holds the record for speed in wooden shipbuilding on the American Continent. The feat was accomplished by the Quinlan and Robertson Shipbuilding Co. of Quebec in the construction of the "War Seneca," a wooden steamer. This vessel was launched on June 13 and a dock steam trial was made last Friday—just 48 days later—by turning over the main engine and auxiliaries under steam. This is six days better than the previous best made in the United States, which is said to have been 54 days. This boat was built for the Imperial Munitions Board.

SEAMANSHIP AND NAVIGATION

Conducted by "The Skipper" *

Articles of direct interest to mariners, discussions of seamanship, navigation rules, and allied topics. Inquiries from readers are invited and will be replied to promptly through these columns, unless otherwise requested.

THESE three avenues of dead reckoning can never be too rigidly followed up in practice. The fact that they are frequently neglected to some extent, until disaster appears, only illustrates further their vast importance. Whether it is on the ocean or the coast or the lake and river, the apparatus for ascertaining the ship's speed and the depth of water should be used at regular times even when the weather is clear and the ship's exact position known. Only by this means can the hand or mechanical logs or the hand and mechanical leads be of real use when fogs, mists, or blinding snow and rain shuts out the view to those on the bridge or poop.

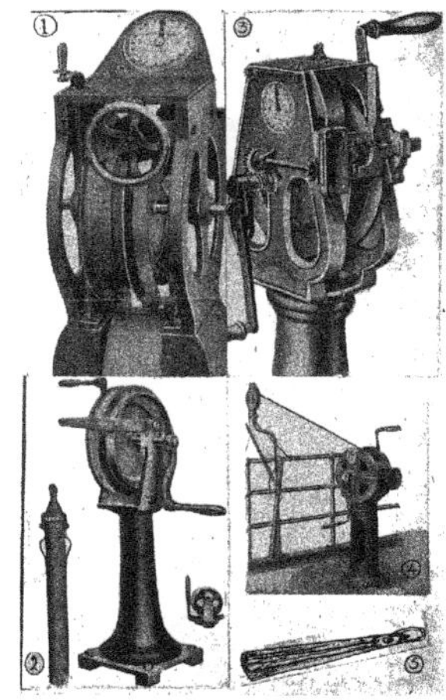

VARIOUS TYPES OF SOUNDING APPARATUS.

The Hand Log

Log lines of this nature are generally made of material similar to signal halyard stuff with three or four strands. The log chips or log bags which are supposed to hold the line stationary on the surface of the water, and duplicate the work of a tape line, are set out in the illustration. The former is made of wood with ballasted sector as a keel, while the latter, which the writer has always used, is made of canvas.

The whole outfit is kept on a reel which finds a place in beckets somewhere around the taff-rail. A hand log line is marked on the principle that the length of each knot on the line must be in the same proportion to a nautical mile as the number of seconds of the time glass is to an hour. As a rule two glasses are made, the use of either one being a matter of choice, as long as the line is adapted accordingly.

Say that a coil of log line stuff is thrown at you with the command to "make a new hand log," the procedure should be something like this:

As all new ropes have an elasticity a new log line should be treated in

*Editor's Note.—This department is conducted by a qualified ship's master, and will deal with questions involving seamanship and navigation. The attention of young men desiring to obtain certificates as mates and masters is particularly directed to this department.

some manner which will cause it to lose that property. That can be done in two ways, either by stretching the line by subjecting it to pressure when dry, thereby making shrinkage by water or further stretching when in use almost impossible. Or another method is to wet the line, thus shrinking it and marking it when in this condition. Naturally the marks will be short spaced, but will come about true when eventually used

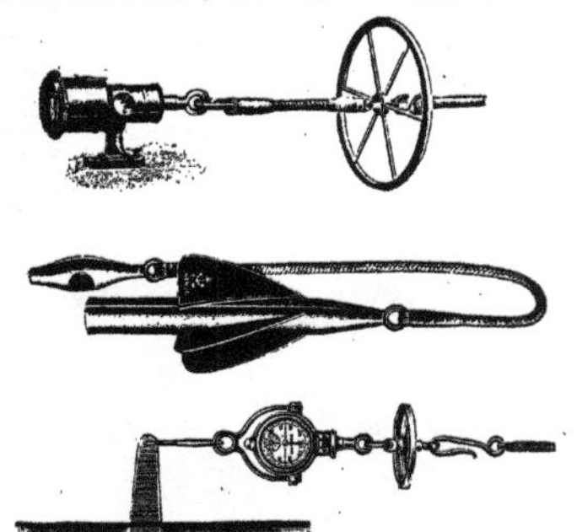

VARIOUS TYPES OF MECHANICAL LOGS AND INDICATING DEVICES.

through strain and then remain that way. To check both ways and get a reliable line remeasure now and again, especially after making a new one. Further, the results from heaving the hand log depend greatly on the two operators, there is a third party required but he simply holds the reel.

If the man holding the glass "turns" when he is told to, and shouts "stop" when the sand has just disappeared, or if a watch is used and the measurement is exact, the results must be good, provided of course that the quartermaster or mate who is paying out the line does not allow "bights" to fall in the water nor allow the reel to undo itself.

Here is the method of working out the spaces between the marks or knots. For a 14-second line:

House seconds 3,600 : interval seconds, 14:: 6,080 feet in mile × 14 ÷ 85,120 ÷ 3,600 = 23 feet 7 inches.

An approximate method is sometimes used. Add a cipher to the seconds and divide by six, that gives the feet, any remainder multiplied by two gives the inches.

140÷6=23 and doubling the remainder makes 4 inches. So between the two methods a mean would be about 23 feet 6 inches.

A 28-second log line would have its marks spaced at 46 feet 8 inches.

Now measure off as "stray line" from 10 to 15 fathoms and put a piece of white rag into the strand to indicate the beginning of the marked line. This stray line lets the log chip get clear of the rudder or propeller eddies, so the white rag as it passed over the taff-rail is the beginning of the real measurement and the signal to "turn" if a glass is used or to shout "time" if a watch is used. The arrow in illustration denotes the "toggle" on the chip bridle. After chalking on a deck seam the length between knots mark the line in one of the ways herein mentioned:

At 1 knot, put small strip of leather.
At 2 knots, put two strips of leather.
At 3 knots, put a single knot, with piece of cord.
At 4 knots, put two knots on the cord.
At 5 knots, put a single knot.
At 6 knots, three knots on the cord, and so on. A more simplified way of marking a log line is by putting a piece of cord in the required position with the requisite number of knots tied in it.

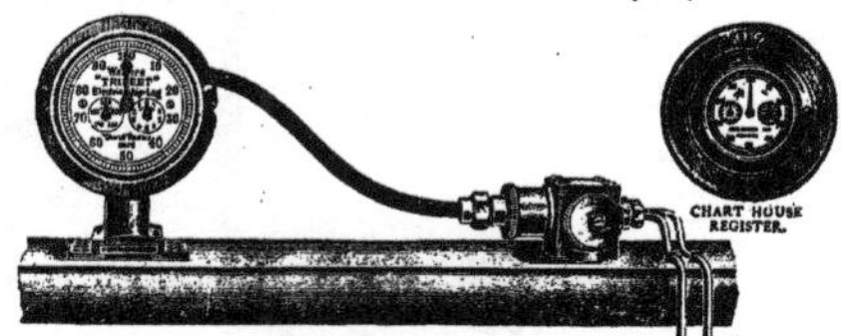

MECHANICAL LOG MOUNTED ON RAIL.

Unless special circumstances dictate otherwise, the hand log is "hove" to leeward, and although mechanical logs are the order of the day it is a seaman's business to understand how to make and how to use the old non-mechanical articles.

Even a fair guess at a ship's speed can be made by throwing a piece of wood overboard at the stem and timing its arrival at the taff-rail by watch. Say that your ship is 500 feet long, and the wood takes 30 seconds to pass from end to end, then the measurement must come out at approximately 10 miles per hour.

One can also measure the velocity of a tide when at anchor with the hand log, and to some extent the trend or drift of a current can also be found in a roadstead by noting the direction in which the log "chip" or bag takes when thrown clear of the ship and the line paid out.

If the vessel is moving slowly one can drop the deep sea lead on the bottom and pay out line. This gives an accurate knowledge almost in regard to the direction of an oblique current. Leads by the way are generally "hove" to windward and the words "watch there watch" are shouted first by the man who actually drops the lead and in turn the men who may be ranged along the weather rail with a few fathoms of slack line in their hands.

Mechanical Logs

In regard to mechanical logs which you see in the illustration, the best results are had when no one is allowed to "monkey" with them. As a rule they only require oiling, say, twice a day. Always close the oil holes with the loose sleeve for that purpose, otherwise the log is bound to get grit in it from the smoke-stack. Rate your patent log by observation distances, by coastal work, such as lighthouse to lighthouse and also remember that in a heavy head sea the mechanical log may over-register, whilst in a following or even quarterly sea the machine may underestimate the ship's actual speed. Any one who, as a boy, has pulled in hand logs every two hours, knows what a help he gets when the bag or chip is on top of a sea and then how the line nearly cuts his fingers off when these creators of friction are in between two mounds of water. These same forces work on the mechanical logs too.

Mechanical logs are subject to disturbance in many ways. For instance, the least knock which the rotator may get from floating debris may cause false registration or again rope shakings or pieces of waste that get thrown overboard entwine themselves on the line and away goes the accuracy of the mechanism. For such reasons one must give close scrutiny to such apparatus.

After pulling in a patent log line the taff-rail end must at once be paid out into the sea as the rotator puts thousands of kinks into the line and only in this way can the line be pulled in clear and straight.

The use of a trustworthy log is perhaps best appreciated in thick weather when one is trying to crawl up a coast line on elbow courses and where the log and lead readings prompt the changes of the ship's head. As I have pointed out before, the confidence required in such predicaments can only be had through using these dead reckoning instruments in fine clear weather, thus verifying their work and getting to know them sufficiently to be able to trust them when they are the only means of registering a ship's position. For instance, take the Atlantic passage on which you have several chances of dropping the lead on one or other of the Newfoundland banks and of course measuring the distance between casts with the log. What happens? Some skippers never dream of such a thing as verifying the ship's position in the aforementioned way, while others fairly jump with joy at the opportunity. Who is the best man when it comes on thick?

As the mechanical log line is the medium through which the mechanism revolves, the make of this line is of a plaited nature, i.e., no strands, whilst the hand log line is made of strand material as it measures by marks and not by rotary motion. If using a sand glass with the hand log rate it frequently with a watch.

In the illustrations a chart house register is shown which is electrically connected at the taff-rail.

Hand Leads

These are of various weights but generally a 7-lb. lead is considered useful around harbors whilst a 14-lb. one does for river or other inshore work and a 28-pounder for deep sea use.

The hand lead line measures depths up to 20 fathoms or 120 feet. There are nine "marks" and 11 "deeps" on the line. Here are the "marks":

At 2 fathoms, 2 leather fingers.
At 3 fathoms, 3 leather fingers.
At 5 fathoms, white rag, linen.
At 7 fathoms, red rag, bunting.
At 10 fathoms, piece leather with hole in it.
At 13 fathoms, blue rag, flannel or serge.
At 15 fathoms, white rag, linen.
At 17 fathoms, red rag, bunting.
At 20 fathoms, cord with two knots in it.

The measurements of 1, 4, 6, 8, 9, 11, 12, 14, 16, 18 and 19 are the "deeps" and are calculated by judgment from any of the "marks." An eye splice in the end of the lead line is put over the end of the lead and brought up to a close contact with the becket which is worked into the hole in the top as you see it in figure five of the illustrations. A line marked carefully at each foot is very necessary for harbor or dock work in places where the skipper and mates are thrown on their own initiative, and these places are legion when you start hiking round the sphere in a Canadian tramp steamer or auxiliary schooner.

Deep Sea Lead

This line is marked the same as the hand line to 20 fathoms, after which every tenth fathom is marked by an additional knot and every intervening fifth fathom by a single knot.

There is 1 knot at 25 fathoms, 3 knots at 30 fathoms, 1 knot at 35 fathoms, 4 knots at 40 fathoms, and so on. At 100 fathoms the mark is a piece of leather with two holes in it. This line may be

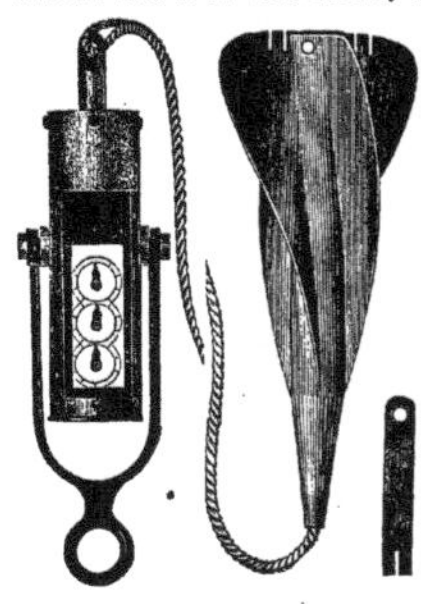

MECHANICAL LOG AND INDICATING DEVICE

marked up to 120 fathoms and of course can only be used at such depths from a vessel absolutely stopped. Herein lies the great advantage of

Mechanical Leads

These wonderful sounding machines can be used with good results in all moderate speeds, say up to 12 miles an hour. It is almost useless to try and explain the ins and outs of mechanical leads as each maker hands out lucid instructions with his machine.

In figure 1 the winding in is done by an electric motor, whilst the clock device indicates the length of line actually out. The depth is found of course when the lead is hove on board and the tube removed and read against a graduated scale. On the lead striking the ground this high class instrument rings a bell on the machine and the man at the brake arrests the motion of the drum, in a very slow manner, however.

In Figure 2 you see a very simple but efficient machine which is being used on our Canadian merchant fleet and is made

In Figures 3 and 4 other types are shown and it can be said of them all that "practice makes perfect" in their use. A clumsy or nervous creature in a sounding crew will wreck the whole outfit.

Even although your ship is travelling at 10 or 12 miles an hour a simple outfit such as figure 2 will almost speak to you and it is only a matter of getting a few drills in. Of course this machine has no electric bells to warn one that the bottom has been touched, but the line will suddenly slacken and wobble when the lead touches. At that moment the man at the brake brings the line up *gently* and then the winding-in process begins.

Another anxious moment comes when the ground instrument is dangling between the water and the taff-rail. The little "fair-lead" or pulley gives the line an overhang but be very careful that the winding crew go slow at the finish and thus reduce the chances of bumping the lead against the ship's side, for in many patent leads a glass tube is part of the recording apparatus.

Courts of enquiry into strandings are constantly dealing with men's certificates for not using the lead, and yet for some unexplained want of common sense, navigators of all classes stand before these nautical tribunals and bite the dust. Make your crews use the lead until they are heard to remark in their sleep the old landsman's warble:

I'd rather be courting the old man's daughter
Than having my hooks in this cold water.
By the mark seven.
Take a spell at this lead
For I'm d——d near dead.
By the deep eleven.

The scientific knowledge used in the making of mechanical leads is either the law that the pressure of water on a body increases with the depth or that the volume of air decreases in the same proportion as pressure upon it increases. Glass tubes and scales are called into use to register the depth.

Lookout

In saying something under this important heading, it must be acknowledged that a great number of sailors thought that "being on the lookout" simply meant being in a stated place—forecastle head, lee bridge, crowsnest on the foremast. But it means really scanning the horizon when in these places of trust. Whilst the piracy of our enemies with their torpedo murder have made a ship's lookout on a par with a soldier's sentry duties, that is where it should be all the time, peace or war.

Hundreds of vessels have gone into broken water and sat down on a reef when there was really time for a smart lookout man to shout "breakers ahead, hard over," or some such warning.

Repeated proof of ships' crews adrift in their lifeboats through the parent vessel catching fire or colliding with ice being passed unheeded by vessels close to, brings home to us what a lookout really means. It means life or death as well as protection of property.

(To be continued.)

Eastern Canada Marine Activities

Doings of the Month at Montreal and on St. Lawrence River

(August Correspondence)

GOOD progress has been made in the midsummer activities in this and surrounding districts, but shipbuilding has experienced some slight delay owing to the delay in the delivery of needed equipment. This has been particularly noticeable in connection with steel plates, but in general the yards report very satisfactory operations. At the yard of Fraser, Brace Co., rapid work is being made on the fitting out of those boats recently launched. Fitting out of the S. S. War Earl and the S.S. War Duchess, two 7,000-ton steel cargo steamers, recently launched at the yard of Canadian Vickers, is nearing completion and these vessels will be placed in commission in the near future. At this yard there are now under construction, two 7.000-ton, one 4,300-ton, and one 8,100-ton steamers for the Imperial and Canadian Governments.

New Shipping Master

Under the new civic service regulations Capt. Joseph Osteus Gray has been appointed shipping master for the port of Montreal. This position demands the services of a semi-technical man and the long experience of Capt. Gray with marine work generally, particularly in connection with shipbuilding and and operation, will fit him remarkably well for this important position. No salary is attached to this position but the normal fees of the shipping master of the port approximate $3,500 per year. Capt. Gray is well known in Montreal as the founder of a pioneer school of navigation and seamanship, training young men for the merchant marine. The school was organized last October with the object of meeting the urgent need for technical men in this service. Since 1911 Capt. Gray has acted as superintendent of wharves and shipping in the port of Montreal. Capt. Gray has had a very interesting career, starting as an apprentice at the age of 14, and had the record of being the youngest man to receive his master's certificate on passing the requirements of the nautical college at Liverpool. Capt. Gray has had a wide experience in every branch of the service and is well qualified for the duties of his new position.

The announcement that Jas. Carruthers has severed his connection with the Halifax Shipyards has come as a surprise to the shipping world. Mr. Carruthers has not only disassociated himself with the Board of Directors but has also withdrawn his stock application. The reason for this action has been the objection of Mr. Carruthers to the policy adopted by the company regarding the organization of a certain subsidiary company known as the Maritime Wrecking Company.

New Developments

Officials of the Wm. Lyall Shipbuilding Company have recently announced the closing of a contract with representatives of the French Government, for eighteen 1,500-ton wooden vessels. These boats will be constructed along similar lines to those now being built at the Vancover yards. The new vessels will be of the schooner rigged type, fitted with auxiliary engines.

Interest is again being shown in the damming of the south channel of the St. Lawrence River at the Longue Sault Rapids. It is understood that this most recent application will be made by the St. Lawrence River Power Company, which is at present developing 100.000 electrical horse power at Messina Springs. N.Y.; this company is stated as a subsidiary corporation of the Aluminum Company of America.

Activities At Quebec

After considerable agitation the officials of the Port of Quebec have succeeded in getting the Imperial Shipping Board to recognize the advantages of the port for a shipping terminus. It has been announced that the loading of two ships a week will be allowed until the closing of navigation.

In addition to recent contracts for two 5,100-ton steel steamers, the Davie Shipbuilding Co. has secured a contract to build twelve wooden vessels, 200 feet in length, for the French Government. Great activity is noted at the Davie yard. Repairs are now in progress on the ocean steamer Celtic Prince, also to the machinery on the Government steamer Montcalm and other vessels. Completion of the twelve wooden vessels are expected in a year's time.

At a recent sitting of the Royal Commission, appointed to inquire into the labor troubles at the various provincial shipyards, an incident arose in connection with certain statements which resulted in the resignation of J. M. Walsh, a representative of the Internationl Union. The action of Mr. Walsh has somewhat disorganized the work of the commission.

Following the final official test of the new Quebec bridge the structure has been formally taken over by the Government. This last test comprised the resting of two mammoth trains of engines and fifty-five loaded cars each, upon the center span. These trains were run on to the bridge from opposite sides, stopped in the center for a few minutes and simultaneously started shorewards. High speed tests were also given.

CARGO STEAMER LAUNCHED AT VICKERS'

Marked Activity in Connection With Marine Work at Big Montreal Yards

The successful launch of the S. S. "Samnanger" took place at the works of Canadian Vickers, Limited, recently. This makes the third launch from Canadian Vickers yards since the present open season of navigation.

The dimensions of the "Samnaner," which is a 7,000 tonner, are as follows:

Length, 380 feet; breath, 49 feet; depth, 30 ft.

The "Samnanger" is a sister ship to the "Porsanger," which was recently delivered by Canadian Vickers, Limited, to Messrs. Furness, Withy & Co. of Montreal, who are acting as managers on behalf of the British Government.

It is expected that the "Samnanger" will be completed within two or three weeks.

The vessel was launched by Captain H. Jonassen, of Bergen, Norway. As in previous cases there was no ceremony.

The rapid production now going on at these works was strikingly evidenced during the launch. On the dock their latest ship, the "War Earl," was being painted after having run her steam trials last Thursday, while the sister ship, the "War Duchess," was lying in the basin with all machinery on board, getting ready for trials to take place in about two or three weeks.

There is, therefore, every indication that at the end of the present month Canadian Vickers' yard alone will have completed and handed over four 7,000-ton cargo steamers, while on the berths there will be five other vessels, several of them in an advanced stage of completion.

THE SUCCESSFUL men are they who have worked while their neighbors' minds were vacant or occupied with passing trivialities, who have been acting while others have been wrestling with indecision. They are the men who have tried to read all that has been written about their craft; who have learned from the masters and fellow-craftsmen of experience, and profited thereby; who have gone about with their eyes open, noting the good points of other men's work, and considered how they might do it better. Thus they have carried themselves above mediocrity, and in striving to do things the best they could, have educated themselves in the truest manner.—Santa Fe Magazine.

PROGRESS IN NEW EQUIPMENT

There is Here Provided in Compact Form a Monthly Compendium of Shipbuilding and Marine Engineering Auxiliary Product Achievement

IMPROVED FLOOR TYPE BORING, MILLING AND DRILLING MACHINE

THIS machine is driven from a motor mounted on top of the column. The drive is direct connected to main drive shaft, there being no belting whatever involved in this design. The spindle drive is controlled by a pair of friction cone clutches, located at back of saddle, and accessible for adjustment. This arrangement provides a reversal of spindle for back facing and tapping. The driving pinion for the spindle meshes with a large diameter gear face cut directly on the face plate. This location of spindle drive prevents spindle torsion, and as a result eliminates one of the most frequent causes of chatter when milling. The front end of spindle slides through an adjustable bearing carried in the spindle sleeve, but the spindle does not rotate in this bearing. The rotating motion is taken in another adjustable bearing, and on the external diameter of the spindle sleeve. The advantage of this design is a provision of take-up for wear on the sliding spindle bearing.

A very prominent feature of this tool, and one which is exclusive in this design, is the sensitive and powerful concentric screw feed of spindle, accomplished by means of a differential train of gears. The only thing that limits the length of spindle feed when it is traversed by this method is the factor of practicability, as there is no mechanical limit of feed, such as is encountered with a rack and pinion or an auxiliary screw feed. This method of feeding permits continuous traverse of spindle without resetting. The feed is applied between main bearings, requiring no overhanging support at end of saddle.

The spindle is traversed by a long bronze nut which engages square thread on the spindle, and which has a bearing only on the sides of the spindle thread. It will be seen that this arrangement provides a very long bearing, and as the two rotate together at the same rate of speed, except when feed is engaged, the possibility of wear is very remote. However, in case of wear, there is an adjustment to take it up. The end thrust in either direction is taken on ball bearings.

The thrust of spindle when milling is taken in a most rigid manner, directly on the main saddle casting, and is entirely independent of the end thrust of spindle for boring. The principle of carrying feed and speed gear trains in the saddle as one unit lends to the machine a facility for operation not readily surpassed.

There are twelve changes of speed, and twelve changes of feed embodied in this design. All feeds are at the same rate per revolution of spindle, whether applied to spindle, saddle or column traverse, and no two of them can be engaged at the same time. It is notable in connection with this speed and feed arrangement that any one of the twelve feeds can be applied to any one of the twelve spindle speeds, making in reality one hundred and forty-four actual rates of feed. Power rapid traverse, independent of the regular feeds, is provided for spindle, saddle and column in every direction. With one lever, machine can be instantly started and stopped, or reversed, independent of main drive or motor.

The gear shifts are all of the sliding transmission type, and are tightly and neatly enclosed, a feature that adds not only to the life and appearance of the machine, but also provides that safety

BORING MILL ADAPTED FOR MANY OPERATIONS.

FRONT VIEW OF GEAR BOX.

for the operator so essential to State law requirements and rapid production of work.

All traversing gears are located consistent with the most approved method of design, that is, between the ways, and close to the guiding side.

GEAR BOX WITH COVER REMOVED SHOWING MECHANISM.

The gears and shafts are made of chrome nickel steel, specially heat treated. The spindle is made of high carbon hammered crucible steel and accurately ground to secure correct alignment.

The oiling of the saddle parts is accomplished by the syphon system, which insures a continuous supply of clean oil to the bearings. The counterweight for the saddle operates inside the column, out of the way of the operator, thus assuring safety in accordance with State laws.

When this machine is correctly aligned in position with the rigid outer support column, and the wide, unyielding floor plate, it possesses to a remarkable indispensable tool in factories and shipyards handling a variety of massive castings which must be machined accurately and quickly.

After several years of experience, designing and experimenting in a very practical way, the manufacturers of this machine are thoroughly convinced of its high efficiency under severe service. One of its greatest advantages is its universal range of adaptability. It may be used to bore, mill, drill, tap, spline, and for oil grooving or rotary planing at one setting. When swivel table is used, the different sides of work may be finished without resetting. The design of this machine is such that its actual manipulation requires a relatively low proportion of the operator's time, creating a wider opportunity, and a greater incentive for him to increase production.

This machine, of improved design, is built by the Landis Tool Company, of Waynesboro, Pa. It has been designed especially to meet the requirements in a general way, of shipyards, navy yards, turbine works, etc., and will handle a wide range of heavy machine work. Briefly it combines the necessary durability and simplicity of operation to insure accuracy and quantity of all work usually machined on floor type boring machines.

ANGLE BENDING MACHINES

THE accompanying illustrations show several types of power bending machines developed by Kane & Roach, Syracuse, N.Y., for the bending of angles and other structural shapes.

In working out the design of these machines, the bending rolls were placed as close to the floor as possible in order to facilitate the placing of heavy work in the machine. These machines may be equipped for either individual motor or belt drive with reversing pulleys as shown in the illustrations. In cases where belt drive is employed the shifters are so constructed that the operator of the machine can manipulate either shifter to provide for running the machine forward or back, as required. In handling certain classes of work it is necessary to reverse the machine several times to provide for running the work back and forth, and in such cases a conveniently located operating lever is particularly important. It will be noticed that the bending rolls are almost in contact with each other, which results in several desirable features. In the first place, having the rolls close together practically avoids the flat or straight spot at either end of a piece that is being bent, this not only saving time in bending the ends of the work to the required curvature, but also saving material through avoiding the necessity of cropping off the extreme ends of the work which can not be bent to the desired form.

Another advantage of having the rolls close together is that the material can be started into the roll to be bent without having to start the bend by hand. This is the means of making a very important reduction in the cost of bending work. A sliding gag under the adjusting screw

BELT DRIVEN ANGLE BENDING MACHINE.

MOTOR DRIVEN ANGLE BENDING MACHINE SHOWING ANGLE IN ROLLS

BENDING MACHINES FOR LIGHT ANGLES.

provides for holding down the top roll, and this gag can be slid out from under the screw to enable the top roll to be raised out of the way when it is desired to put work in the machine or take it out after a bend has been completed. With this arrangement the screw adjustment does not have to be altered. When one piece has been removed and another piece has been put into position on the lower rolls the upper roll is dropped into place and the gag slid back under the adjusting screw, after which the machine is ready to continue its operation.

Side rolls are provided at each side of the bending rolls, and these may be swiveled to any angle according to the radius of the curve to which work is being bent. These side rolls can also be adjusted in or out to meet the requirements of different classes of work. They effectually prevent side twisting or buckling of the work during the bending operation and enable the piece to come out perfectly true. The rolls are made of high carbon steel, and the outside sections slide in or out to provide for handling any width of work that comes within the capacity of the machine. The shafts carrying these rolls are exceedingly heavy, and all the front bearings that carry the roller shafts are bushed with bronze. The driving gears are made of steel with cut teeth. With the exception of differences in size the design of all three of these machines is the same. Each machine is provided with a scale and pointer to show just where to set the rolls for bending work to any desired radius of curvature, which is an important feature in simplifying the setting of the machine for any given job.

All three rolls are positively driven by means of gears, thus avoiding any slip.

On the No. 14 angle bending machine light angles up to 2½ in. x 2½ in. x ¼ in. can be bent. To provide for bending larger angles, the manufacturers have brought out a heavy-duty machine that is built in three sizes. The No. 22 machines handles angle irons up to 3 in. x 3 in. x ⅜ in.; the No. 23 machine handles work up to 4 in. x 4 in. x ½ in.; and the No. 26 machine handles work up to 6 in. x 6 in. x ⅝ in. In addition to bending angles, these machines can be employed for bending equivalent sizes of I-beams, channels, T-irons, rounds, squares, tubular stock, flat stock, and other shapes by simply making "filling in" collars or rolls to suit any particular kind of work. It is also possible to bend either one angle or two angles at a time, according to the quantity of work which is being handled.

GLOBE VALVE

The Empire Manufacturing Co., London, have brought out a new line of globe valves which possess very desirable features. The hexagons are unusually heavy and have ample metal for a full long pipe thread. The bonnet is made of a uniformly heavy casting in which ample provision has been made for long threads on the valve stem and packing gland. High pressure steam packing is used and the construction of the valve is such that it may readily be packed under pressure. When the valve is open an unusually full and open waterway is provided owing to the ample diameter of the valve seating.

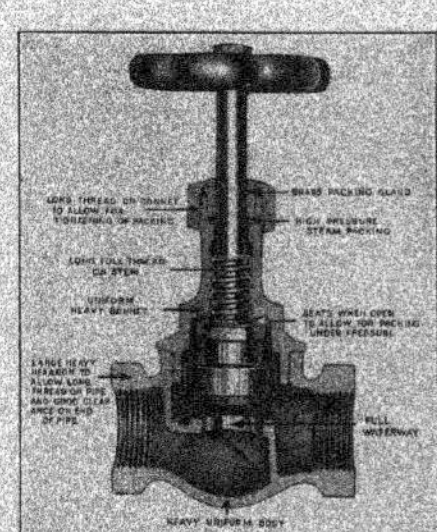

IMPROVED GLOBE VALVE

STEAMER WEXFORD PROBABLY LOCATED

Speculation Revived Concerning Fate of Vessel Lost in Big Storm

Goderich.—What is considered the first authentic information of the whereabouts of the steamer Wexford, lost in the great November storm of 1915, was brought to port by the captain of the steamer Mariska. On his course from Chicago to Goderich he sighted and passed within twenty feet of two spars, both at approximately the distance apart the spacing of the Wexford's spars would indicate. Both of these were seen distinctly in the fall of the water between seas, one shorter than the other, with the after spar slightly bent. His familiarity with the vessel when trading on the upper lakes strengthens his conclusion that this can be none other than the Wexford. The location is 15 miles northwest by north of Point Clark, and 16¾ miles northwest of Kincardine.

The last seen of this vessel was on the fateful Sunday when the Kaministiquia, which had left Goderich that morning, had met the Wexford about the middle of the afternoon, then on her course for Goderich. All that had previously been found was a lifeboat and several bodies which came ashore near Grand Bend, thirty miles below here. From the alleged location of the vessel it is apparent that she headed into the northerly storm, but had not made many miles before foundering. The vessel's spars now indicate that she was finally heading down the lake.

TURNS OUT FIRST RIVETLESS SHIP

Electrical Welding Has Been Used In Putting the Plates Together

Building of a steel ship without rivets has been effected in a shipyard on the south coast of England, and its construction may mark a new era in the shipbuilding industry. A process of electrical welding was used for joining the plates, in place of the usual riveting and caulking. By means of an electric arc, the joints are submitted to intense heat, and the plates are fused together. The process is not entirely new, as auxiliary work has been done in the past by electric welding. During the last year, developments have been made which have permitted of the extension of this method in ship construction. A saving of between twenty and twenty-five per cent. is saved in both time and material, judging from experimental work done on the new vessel just launched.

The general adoption of electrical welding in shipbuilding would permit a material speeding-up of production. The electric process is particularly economical in the assembling of bulkheads, deck structures and other interior work. The United States is keeping in touch with the developments in this work in Great Britain, and arrangements are under way for the construction of several 10,000-ton standard ships by the same process. These large vessels will contain about 2½ per cent. of the number of rivets originally intended, while the British boat was absolutely rivetless.

Making Big Cranes.—The Dominion Bridge Company Limited, Montreal, has been awarded the contract for building two large shipbuilding cranes of the travelling gantry type with 7-ton man trolleys, for the Tidewater Shipbuilders, Limited, Three Rivers, P.Q.

ASSOCIATION AND PERSONAL

A Monthly Record of Current Association News and of Individuals Who Have Been More or Less Prominent in Marine Circles

J. Murray Watts, naval architect and engineer, of Philadelphia, has been commissioned captain in the 57h Engineers (Inland Waterways). While Captain Watts is overseas the business carried on

CAPTAIN J. MURRAY WATTS.

in his name will be handled by the engineering firm of Cornell and Matthews, naval architects and engineers, room 712 Bulletin Bldg., N.E. corner Juniper and Filbert Sts., Philadelphia.

George H. Madgett, who formerly was affiliated with the Hamilton Bridge Co., Hamilton, Ont., now is connected with the Standard Steel Construction Co., Welland, Ont., in an important capacity.

James Carruthers has resigned from the presidency of the Halifax Shipyards Co. He is reported to object to the formation of a subsidiary concern which he contends may deprive the shareholders of the parent company of profits it might make.

W. E. Burris, a former Perth, Ont., boy, has been promoted to the position of manager of the Victoria, B.C., plant of the Foundation Company, which is building warships for the United States Navy. Their plant at Tacoma, Wash., is the largest shipbuilding plant in the United States.

LICENSED PILOTS

ST. LAWRENCE RIVER

Captain Walter Collins, 43 Main Street, Kingston, Ont.; Captain M. McDonald, River Hotel, Kingston, Ont.; Captain Charles J. Martin, 13 Balaclava Street, Kingston, Ont.; Captain T. J. Murphy, 11 William Street, Kingston, Ont.

ST. LAWRENCE RIVER, BAY OF QUINTE, AND MURRAY CANAL

Captain James Murray, 106 Clergy Street, Kingston, Ont.; Capt. James H. Martin, 259 Johnston Street, Kingston, Ont.; John Corkery, 17 Rideau Street, Kingston, Ont.; Captain Daniel H Mills, 272 University Avenue, Kingston, Ont.

MONTREAL PILOTS' ASSOCIATION

President—Alberic Angers, Montreal.
Secretary—C. B. Hamelin, Champlain, Que.

ASSOCIATIONS

DOMINION MARINE ASSOCIATION

President—A. A. Wright, Toronto. Secretary—Francis King, Kingston, Ont.

GREAT LAKES AND ST. LAWRENCE RIVER RATE COMMITTEE

Chairman—W. F. Herman, Cleveland, Ohio. Secretary—Jas. Morrison, Montreal.

INTERNATIONAL WATER LINES PASSENGER ASSOCIATION.

President—O. H. Taylor, New York.
Secretary—M. R. Nelson, 1184 Broadway, New York.

SHIPPING FEDERATION OF CANADA.

President—Andrew A. Allan, Montreal; Manager and Secretary—T. Robb, 218 Board of Trade, Montreal; Treasurer, J. R. Binning, Montreal.

SHIPMASTERS' ASSOCIATION OF CANADA.

Secretary—Captain E. Wells, 45 St. John Street, Halifax, N.S.

GRAND COUNCIL N.A.M.E. OFFICERS.

A. R. Milne, Kingston, Ont., Grand President.
J. E. Belanger, Bienville, Levis, Grand Vice-President.
Neil J. Morrison, P.O. Box 238, St. John, N.B., Grand Secretary-Treasurer.
J. W. McLeod, Owen Sound, Ont., Grand Conductor.
Lemuel Winchester, Charlottetown, P.E.I., Grand Doorkeeper.
Alf. Charbonneau, Sorel, Que., and J. Scott, Halifax, N.S., Grand Auditors.

Capt. Joseph Osteers Gray has been appointed shipping master for the port

CAPTAIN JOSEPH O. GRAY

of Montreal. Capt. Gray is well known in Montreal as the founder of a pioneer school of navigation and seamanship. Capt. Gray has had a wide experience in every branch of the service and is well qualified for the duties of his new position.

Brockville, Ont.—The steam barge M. H. Wormington, which struck a rock near Sister Light, in the St. Lawrence River three weeks ago, has been refloated and taken to Montreal. Pontoons were used to raise the vessel, and much of her cargo of coal had to be lightered before the craft could be extricated. Her stern lay in 25 feet of water, while the bow rested on the shoal. Wrecking operations were conducted by the owners, the Vincennes-MacNaughton Co., Montreal.

1918 Directory of Subordinate Councils, National Association of Marine Engineers.

Name.	No.	President.	Address.	Secretary.	Address.
Toronto,	1	Arch. McLaren,	324 Shaw Street	E. A. Prince,	49 Eaton Ave.
St. John,	2	W. L. Hurder,	209 Douglas Avenue	G. T. G. Blewett,	36 Murray St.
Collingwood,	3	John Osburn,	Collingwood, Ont.	Robert McQuade,	Collingwood, Ont.
Kingston,	4	Joseph W. Kennedy,	395 Johnston Street	James Gillie,	101 Clergy St.
Montreal,	5	Eugene Hamelin,	Jeanne Mance Street	M. Lazure	120 Ribard St.
Victoria,	6	John E. Jeffcott,	Esquimault, B.C.	Peter Gordon,	808 Blanchard St.
Vancouver,	7	Isaac N. Kendall	319 11th St. E., Vanc.	E. Read	232 13th St. W.
Levis,	8	Michael Latulippe,	Lauzon, Levis, Que.	Arthur Jolin,	Lauzon, W.
Sorel,	9	Nap. Beaudoin,	Sorel, Que.	Alf. Charbonneau,	Box 204, Sorel, Que.
Owen Sound,	10	John W. McLeod	570 4th Ave.	R. J. McLeod,	682 8th St.
Windsor,	11	Alex. McDonald,	28 Crawford Ave.	Neil Maitland,	[illegible] London St. W.
Midland,	12	Geo. McDonald	Midland, Ont.	A. E. House,	Box 333
Halifax,	13	Robert Blair	29 Parrsboro Street	Chas. E. Pearce.	Portland St., Dartmouth, N.S.
Sault Ste. Marie,	14	Charles H. Innes,	27 Euclid Road	Wm. Hindmarch	34 Euclid Rd.
Charlottetown,	15	J. A. Rowe	176 King Street	Chas. Cumming.	27 Euston St.
Twin City,	16	H. W. Cross,	436 Ambrose St	J. W. Farquharson,	169 College St.
				A. H. Archand,	Champlain, Que.

Work in Progress in Canadian Shipyards

A record of Work in Process of Completion—Principal Features of Specification—Approximate Launching Dates—Vessels on Order.

BRITISH COLUMBIA

YARROWS, LIMITED, ESQUIMALT, B.C.
One stern-wheel steamer, 165 ft. x 35 ft., for Indian Government service.
At present engaged chiefly on repair work.

H. VOLLMERS, NANAIMO, B.C.
One gasoline tug, 9 tons.

NEW WESTMINSTER CONSTRUCTION & ENGINEERING CO., NEW WESTMINSTER, B.C.
Four wooden steamers, Nos. 10-11-12-13, for undisclosed interests. 250 ft. x 43 ft. 6 in. x 25 ft., moulded depth, 2,800 tons d. w., 1,000 h.p., single screw engines, 12 knots, naval architect, I. Alexander. Launchings March 14, April 14, May 15 and June 15, 1918.

B.C. CONSTRUCTION & ENG. CO., POPLAR ISLAND, NEW WESTMINSTER, B.C.
Four wooden vessels, 1,800 gr. tons, 2,800 tons d.w.

NEW WESTMINSTER MARINE CO., FT. OF FURNESS ST., NEW WESTMINSTER, B.C.

STAR SHIPYARD CO., NEW WESTMINSTER, B.C.

J. A. CROLL, PORT ALBERNI, B.C.

PACIFIC CONSTRUCTION CO., PORT CCOQUITLAM, B.C.
Taken over the plant of the Coquitlam Shipbuildiug Co.
Gasoline yacht, 50 ft., June, 1918.
Three wooden freight steamers, Nos. 20-21-22, 250 ft. b. p. x 42 ft. 6 in. x 25 ft., 1,800 gr. tons, 2,800 tons d.w., trip. exp. engines, water tube boilers ,12 knots, 1,000 h.p. Two for undisclosed interests, one for builder's account. Launchings May 20, June 15, Sept., 1918.
Four new berths under construction for steel ships up to 10,000 tons.

C. E. BAINTER, PRINCE RUPERT, B.C.
Several wooden fishing steamers, for undisclosed interests.

GRAND TRUNK DRYDOCK SHIP REPAIR CO., LTD., PRINCE RUPERT, B.C.

S. A. MOULTON, PRINCE RUPERT, B.C.
Ten composite vessels for undisclosed interests.

BRITISH-AMERICAN SHIPBUILDING & ENG. CO., VANCOUVER, B.C.
Establishing plant on Kitsilano Reserve.
Twenty wooden vessels being built for undisclosed interests, 3,500 tons each.

JOHN COUGHLAN & SONS, N. VANCOUVER, B.C.
"War Camp," sister ship to "Alaska," launched in March, 1918.
One steel freight steamer, 427 ft. 9 in. x 54 ft. x 29 ft. 9 in., draft 24 ft. 2 in., speed 11½ knots, 5,730 gross tons. Launched Jan. 10. 1918, for B. Stolt Nielsen, Christiania, Norway. Both sold to the Cunard Line.
Three steel steamers, "War Chariot," "War Chief," and "War Noble," Nos. 4-5-6, 425 ft. x 54 ft., 8,800 tons cap., for the Cunard Line.
Six steel freight steamers, 425 ft. x 54 ft. x 24 ft. 2 in., 8,800 tons cap., 1 knots, for undisclosed interests. Delivery Jan., Feb., March, May and July, 1918.
One steel freight steamer, "Alaska," 427 ft. 9 in. x 54 ft. x 29 ft. 9 in., draft 24 ft. 2 in., speed 11½ knots, 5,730 gr. tons, 8,800 d.w. Launched Jan. 10, 1918, for B. Stolt Nielsen, Christiania, Norway. Sold to the Cunard Line.

THE FOUNDATION CCO., VANCOUVER, B.C.

FRASER VALLEY SHIPBUILDING CO., VANCOUVER, B.C.

GRANT, SMITH & CO., VANCOUVER, B.C.
Six wooden cargo vessels, 250 ft. x 43 ft. 6 in. x 25 ft., 3,000 tons cap., 9½ knots, for the Canadian Government.

HARRISON & LAMOND, SHIPBUILDERS, LTD., VANCOUVER, B.C.
One wooden auxiliary schooner, 225 ft. x 44 ft. x 21 ft. 4 in., 1,600 gr. tons, 2,550 tons cap., for undisclosed interests.

LYALL SHIPBUILDING CO., NORTH VANCOUVER, B.C.
"War Puget," Feb. 16, 1918, 250 ft., wood, 2,080 tons.
Ten wooden cargo ships, for undisclosed interests. Launchings Jan. 20, Mar. 2, Mar. 23, and April 11, 1918.
Building six wooden vessels for own account.
"War Caribou," April 10, 1918, 3,080tons, wood.

W. R. MANCHION, VANCOUVER, B.C.

STANDARD SHIPBUILDING CO., VANCOUVER, B.C.
Two 3,500-ton vessels for Brazilian Govt., and six for French Govt. of 4,500 tons.
Donohoe reinforced type, wood composite construction.

TAYLOR ENGINEERING CO., VANCOUVER, B.C.
Number of small vessels being constructed at total value of $300,000.
4,500-ton floating drydock, 352 ft. long.

VANCOUVER SHIPYARD, LTD., VANCOUVER, B.C.
One motor freighter, 125 ft. x 24 ft. x 12 ft., wood, 160 h.p., Bolinder's crude oil engine. To be completed Aug., for Taylor Engr. Co.
Also repairing several steamers and schooners.

THE WALLACE SHIPYARDS, LTD., N. VANCOUVER, B.C.
"War Dog," 4,500 tons, steel, May 18, 1918.
"War Power," 4,600 tons, steel, March 23, 1918.
Two steel steamers, 315 ft. x 45 ft. x 27 ft., 4,600 tons d. w., trip. exp. vertical engines ,two S. E. Scotch boilers, 10 knots, single screw, for the Canadian Government. Deliveries Dec., 1917, and Aug., 1918.
Two wooden aux. schooners, 255 ft. o. a., 225 ft. keel x 44 ft. x 21 ft. 4 in., five masts, 1,500 gr. tons, cap. 2,500 tons or 1,500,000 ft. lumber, two hatches, twin 160 h.p. Bolinders Hot-bulb engines, twin screws, for the Canadian Government.
Four wooden cargo vessels, aggregate tonnage of 17,500 tons for undisclosed interests.

WESTERN CANADA SHIPYARDS ,LTD., VANCOUVER, B.C.
"War Nootka," Jan. 4, 1918, 3,080 tons (machinery installed at Ogden Point Assembly Sheds).
Six wooden freight steamers, 250 ft. h.p., 259 ft. o. a., 43½ ft. extreme breadth, molded 42⅓ ft., 25 ft. molded depth., draft 22 ft., 2,800 tons d. w., for undisclosed interests. "War Nootka," launched January 4, 1918. "War Selkirk" launched March 6, 1918.
"War Tatla," "War Casco, "War Chilkat," "War Tanoo," 4 standard vessels now under construction.

CAMERON-GENOA MILLS CO., VANCOUVER, B.C.
"War Yukon," Jan. 24, 1918, 3,080 tons, wood (machinery installed at Ogden Point Assembling Sheds.

THE FOUNDATION CO. OF B. C., LTD., VICTORIA, B.C.
Building ships for undisclosed interests.
Five wooden steamers, 250 b. p. x 42 ft. 6 in. x 25 ft., 1,800 gr. tons, 2,800 tons d. w., one trip. exp. engine, 1,000 h.p., "Howden" water tube boilers, furnished and installed by owners, 10 knots, for undisclosed interests. Esplen & Son & McNaught, N.Y.C., designers.
No. 1—"War Sonchee," launched Dec. 27, 1917.
No. 2—"War Masset," launched April 11, 1918.
No. 3—"War Babine," ready for launching, awaiting propeller and rudder.
No. 4—"War Camchin," ready for launching, awaiting propeller and rudder.
No. 5—"War Nanoose," ready for launching in about five weeks.

CLARENCE HOARD, VICTORIA, B.C.
Wooden car barge for C.P.R., Jan., 1918

VICTORIA SHIPBUILDING CCO., VICTORIA, B.C.
Constructing wooden ships for British Govt., 3,500 tons each.

NEW BRUNSWICK

C. T. WHITE & SON, LTD., ALMA, N.B. (Office at Sussex, N.B., also.)
Two schooners aux. power, 143 ft., 900 tons d. w., first to be launched in April; second in June.

JAMES X, LENTEIGNE, LOWER CARAQUET, N.B.
One schooner, 28 tons, wood.

INTERNATIONAL SHIPBUILDING CO., NEWCASTLE, N.B.
Two four-masted wooden auxiliary schooners, 155 ft. keel, 37 ft. depth, 535 net tons. One to be completed in September and the other December, 1918. For builders' account.

EUREKA SHIPBUILDING CO., NORTH HEAD, N.B.
One wooden schooner, "Mollie & Melba," 350 tons reg.

PORT COLBORNE BUILDING & REALTY CO., LTD., REXTON, N.B. (Head Office, Welland, Ont.)
Four-masted wooden schooner. Has cap. of 3 schooners per year.

GRANT & HORNE, ST. JOHN, N.B.
Two wooden cargo steamers, "War Fundy" (launched Aug., 1918) and "War Digby," 250 ft. x 42 ft. 5 in. x 25 ft. 5 in., 1,000 h.p., speed 9½ knots, 2,800 tons d.w., for undisclosed interests.

MARINE CONSTRUCTION CO., CANADA, LTD., ST. JOHN, N.B.
Four wooden auxiliary, four-masted schooners, 185 ft. o. a., 165 ft. keel x 40 ft., 1,100 d. w. tons, for builder's account. J. Murray Watts, Philadelphia, naval architect. Launching April, 1918.
One wooden schooner, "Dorfontein," 740 reg. tons. Launching June, 1918.

PETER AND A. A. McINTYRE, ST. JOHN, N.B.
One schooner, 425 tons, 137 ft., three masts, wood, for builder's account. To be launched in June. Another to be commenced immediately.

ST. JOHN SHIPBUILDING CO., ST. JOHN, N.B. (Plant at Courtenay Bay.)
Ten five-masted auxl. schooners, oil burning engines, for undisclosed interests.
Establishing plant for steel and wood construction.

ST. MARTIN'S SHIPBUILDING CO., ST. MARTIN'S, N.B.
One wooden schooner, 450 tons, three masts. To be launched in July for builder's account.

NEWFOUNDLAND

ANGLO-NEWFOUNDLAND DEVELOPMENT CO., BOTWOOD, NFLD.
Two three-mast aux. schrs., wood, 450 tons, 150 H.P. for the builders' account.

NEWFOUNDLAND SHIPBUILDING CO., HARBOR GRACE, NFLD.
To build ships of 1,200 tons d.w. Steel ships later.

M. E. MARTIN, MORRIS ARM, NR. ST. JOHN'S, NFLD.
Two wooden schooners, 300 tons each, for builder's account.
One wooden schooner. 500 tons, for builder's account.

UNION SHIPBUILDING CO., ST. JOHN, NFLD.

ANNAPOLIS SHIPPING CO., ANNAPOLIS ROYAL, N.S.
"Hilda M. Clark," March 4, 1918, 187 ft., 640 tons reg.
Two schrs. three-masted, wood 500 tons, 170 ft. long. To be completed August and December, 1918.

NOVA SCOTIA

B. L. TUCKER, BASS RIVER, N.S.
One schooner, 350 tons, wood.

JAS. S. CREELMAN, BASS RIVER, N.S.
One three-masted schooner, wood, 170 ft. long, 500 tons. To be completed October, 1918, for builders' account.

HANKINSON SHIPBUILDING CO., BELLIVEAU'S COVE, N.S.
One schooner, 380 tons, wood.

A. A. THERIAULT, BELLIVEAU'S COVE, N.S.
Two wooden three-masted tern schooners, 150 ft. x 33 ft. x 11 ft. 11 in., 400 gr. tons, gasoline engines, for builder's account. Launchings April and June, 1918.
One wooden one-masted schr.

WESTPORT SHIPBUILDING CO., BELLIVEAU'S COVE, N.S.
One wooden schooner, 230 tons net. To be launched in October for builder's account.

ELI PUBLICOVER, BLANDFORD, N.S.

BRIDGEWATER SHIPPING Co., BRIDGEWATER, N.S.
One tern schooner, 450 tons gr., wood. Adapted for aux. power. To be completed Sept., 1918.
One tern schooner, 275 tons gr., wood. Adapted for aux. power. To be completed Nov., 1918.
Both built on owner's account and for sale.

W. A. NAUGLER, BRIDGEWATER, N.S.
One wooden three-masted schooner, "Wm. A. Naugler," 146 ft. o. a., 116 ft. keel x 32 ft. x 11 ft. 6 in., 360 gr. tons, 600 tons cap., for builder's account.
One schooner, 300 tons, wood.

H. MAC ALONEY, CANNING, N.S.
Two wooden schooners, 400 tons each. For builder's account. Launchings July and October, 1918.

S. M. FIELDS, CAPE D'OR, N.S.
One three-masted schooner, 350 tons.
Keel laid for another three-master.

CHESTER BASIN SHIPBUILDERS, LTD., CHESTER BASIN, N.S.
Two schooners, 100 tons, wood. Launchings Aug., 1917, Dec, 1917.
One four-masted schooner, 60 tons. Launching Sept., 1918.
One two-masted schooner, 100 tons. Launching June, 1918.

MORTIMER PARSONS, CHEVERIE, N.S.
One wooden three-masted schooner, 170 ft. x 38 ft. x 14 ft., 575 gr. tons, 900 tons cap., for builder's account. Launching Sept., 1918.
Two wooden three-masted schooners for builder's account.
One tern schooner, "Donald Parsons," 140 ft. keel x 34 ft. 4 in. x 12 ft. 4 in., 500 gross tons, 825 tons d.w. for Mortimer Parsons. Launched May 11, 1918.
Another vessel, same size, to be launched Nov., 1918, for builder's account.

BAY SHORE SHIPYARD, CHURCH POINT, N.S.
Building two wooden vessels for R. C. Elkin, Ltd., St. John, N.B.

MOISE BELLIVEAU, CHURCH POINT, N.S.
One schooner, 450 tons, wood.

FIDELE BOUDREAU, CHURCH POINT, N.S.
One schooner, 350 tons, wood.

J. E. GASKILL, CHURCH POINT, N.S.
Other yards at Grosses Coques, Little Brook, Meteghan and Port Wade, N.S.
Five wooden schooners from 344 to 387 reg. tons. All sister ships.

COMEAU SHIPBUILDING CO., COMEAUVILLE, N.S.
One three-masted wooden schooner, for builder's account, 450 tons.

J. W. COMEAU, COMEAUVILLE, N.S.
One schooner, 329 tons, Jan., 1918.

J. N. RAFUSE & SONS, CONQUERALL BANKS, DIGBY CO., N.S. (Other plants at Salmon River and Shelburne, N.S.)
Two wooden three-masted schooners, 120 ft. x 32 ft. x 12 ft., 700 tons cap., for builder's account. For the market. Launchings Oct. 25, 1917, and Jan. 10., 1918.
One three-masted wooden schooner, "Integra," 112 ft. x 30 ft. x 11 ft. 6 in., 600 tons cap., for J. O. Williams & Co., St. John's, Newfoundland. Launching Nov. 25, 1917.

E. F. WILLIAMS, DARMOUTH, N.S.
One schooner, 360 tons, wood.

ROBAR BROTHERS, DAYSPRING, N.S.
Schooner for Capt. Ivan Creaser, Dayspring.

MAURICE E. LEARY, DAYSPRING, N.S.
One wooden schooner, 225 tons, 135 ft. o. a. To be launched Aug. 15, 1918, for La Have Outfitting Co., La Have, N.S.

J. NEWTON PUGSLEY & CHAS. ROBERTSON, DILIGENT RIVER, N.S.
One schooner, three-masted, 475 tons.

McLEAN & McKAY, ECONOMY, N.S.
One tern schooner, wood, 400 tons net, 135 ft. keel. To be launched Sept. 1, 1918, for builders' account.

S. J. SOLEY, ROX RIVER, N.S.
One schooner, three-masted, 121 ft. keel, 300 tons. Built of native wood. To be launched Sept., 1918. For builder's account. For sale.

ALLAN & FRASER, FRASERVILLE, N.S.
One schooner, 350 tons, wood.

BERNARD W. MELANSON, GILBERT'S COVE, N.S.
One three-masted schooner, wood, 250 tons net. To be completed November, 1018, for builder's account and for sale.

AMOS BLINN, GROSSES COQUES, N.S.
One schooner, 375 tons. Jan., 1918.
One schooner, 350 tons, wood.

BINN BROS., GROSSES COQUES, N.S.

F. K. WARREN & CO., GROSSES COQUES, N.S. (Offices in Union Bank Chambers, Halifax, N.S., also.)
350-ton tern schooner, under construction Jan., 1918.

HALIFAX SHIPBUILDING CO., HALIFAX, N.S.

FAUQUIER & PORTER, HANTSPORT, N.S.
Two wooden four-masted schooners, 178 ft. x 40 ft. x 18 ft., 1,000 gross tons, 2,000 tons d. w., twin 100 h.p. oil engines. Launchings Aug. 1, 1918, and Sept. 1, 1918. Both for sale.

J. WILLARD SMITH, HILLSBURN, N.S. (Head Office, St. John, N.B.)
One tern schooner, wood, 140 ft. keel, 35 ft. beam, 450 tons net reg. tons.

HENRY COVEY, INDIAN HARBOR, N.S.

CHARLES GRIFFIN, ISAACS HARBOR, N.S.
One schooner, 40 tons, wood.

LEWIS SHIPBUILDING CO., LEWISTON, N.S.
One schooner, 670 tons. Jan., 1918.

INNOCENT COMEAU, LITTLE BROOK, N.S.
One three-masted schoner, 140 ft. x 35 ft. 6 in. x 17 ft. 6 in., 650 gr. tons, for the Weymouth Shipping Co., Weymouth, N.S., Can. Delivered Jan. 18, 1918.

J. W. RAYMOND, LITTLE BROOK, N.S.
One wooden schooner, No. 4, 140 ft. x 36 ft. x 17 ft., 575 gr. tons, for Jones Bros., Weymouth, Digby Co., N.S.

H. A. FRANK, LIVERPOOL, N.S.
Two wooden schooners, 500 tons, for builder's account:

McKEAN SHIPBUILDING CO., LIVERPOOL, N.S.
One wooden three-masted schooner, 126 ft. keel x 3 ft. x 12 ft. 9 in., 700 tons d.w., for builder's account. Launching May, 1918.

W. F. McKEAN & CO., LIVERPOOL, N.S.
One schooner 400 tons, Jan., 1918.

D. C. MULHALL, LIVERPOOL, N.S.
Two wooden schooners, one 300 gross tons, other 400 gross tons, for undisclosed interests.

N.S. SHIPBUILDING & TRANS. CO., LIVERPOOL, N.S.
One wooden tern schooner No. 2, 122 ft. x 33 ft. x 12 ft. 6 in., 450 gr. tons, 70 d.w. tons, 7 knots, for Peter Yee Wing & Co., Ltd., Sydney, Australia. R. McLeod and J. S. Gardner. Launching Feb. 1, 1918
One two-masted fishing schooner, "Sadie A. Nickle," 130 ft. o. a. x 26 ft. 10 ft. 6 in. hold, 250 d.w., two 50 h.p. oil engines, for Rafuse & Sons. Launching May 15, 1918.
One three-mast schooner, 420 tons. F-M Co. oil burning engines, for Job Bros., St. Johns, Nfld. Launching July 15.
One three-masted schooner, 270 tons, for builders' account. Launching Sept. 1.
One two-masted fishing schooner, 120 tons, for builders' acount. Launching Oct. 1.

ROBIN, JONES & WHITMAN, LIVERPOOL, N.S.
One schooner, 340 tons, wood.

SOUTHERN SALVAGE CO., LIVERPOOL, N.S.
One wooden schooner, three masts, 300 gross tons, 600 tons capacity, for a West Indian firm. Launching Jan., 1918.
One wooden steamer, No. 5, 250 ft. x 43 ft. 6 in. x 25 ft., 2,900 gr. tons, 10 knots, 1,000 h.p., for undisclosed interests. Launching Fall, 1918.
One two-masted schooner, "Win-the-War," 137 ft. o. a. x 25 ft. 2 in. x 11 ft. 6 in., 187 gr. tons, for the builder's account. Launched Nov., 1917.

CONRAD & REINHARDT, LUNENBURG, N.S.

LUNENBURG MARINE RAILWAY, LUNENBURG, N.S.

FRED A. ROBAR, LUNENBURG, N.S.

SMITH & RHULAND, LUNENBURG, N.S.
Two schrs., 225 tons each. Jan., 1918.
Two schooners, 100 tons each.
"Donald Cook," 112 ft. beam, depth of hold, 11 ft., for Capt. William Cook. Launched April, 1918.
Two schooners, 168 gr. tons, for W. C. Smith.
One schooner, 150 gr. tons, for Lun. Outfitting Co.

J. B. YOUNG, LUNENBURG, N.S.

ERNST SHIPBUILDING CO., MAHONE BAY, N.S.
One two-masted schooner, 120 ft. o. a. x 25 ft. 6 in. x 10 ft. 6 in., 200 tons d.w., for Lunenburg Outfitting Co. April, 1918. Christened "Madeline Adams."
One three-masted schooner 125 ft. keel x 32 ft. x 12 ft., 600 tons, d. w. To be delivered July, 1918.
Two masted schooner, "Agnes D. McGlaston," 130 ft. x 26 ft. 8 in. x 10 ft. 8 in. Launched Nov. 13, 1917.
Keel laid for three-masted wooden schooner, 200 tons.

J. ERNEST & SON, MAHONE BAY, N.S.
One schooner, 520 tons. Jan., 1918.

O. A. HAM, MAHONE BAY, N.S.
One schooner, "Dozie," 128 tons. Launching May 30. One motor boat about 20 tons.

McLEAN CONSTRUCTION CO., MAHONE BAY, N.S. (Leased to Montague Mahaffy, Toronto.)
One three-masted schooner, 325 tons gr. reg., 500 tons cap., West Indies freight type. Will be launched in July.
Keel laid for another three-masted schooner with cargo capacity of 425 tons.

JOHN McLEAN & SONS, MAHONE BAY, N.S.
One schooner, 95 tons, wood.

J. A. BALSOM CO., LTD., MARGARETSVILLE, N.S.
One schooner, 409 tons. Jan., 1918.

CLARE SHIPBUILDING CO., METEGHAN RIVER, N.S.
One wooden schooner for builder's account. 400 tons.

A. H. COMEAU & CO., METEGHAN, N.S.
One schooner, 400 tons, wood.

AGAPIT COMEAU, METEGHAN, N.S.

JOHN F. DEVEAU, METEGHAN, N.S.
A 300-ton schooner was launched recently for Ritcey & Co., Lunenburg, N.S., and named Charles A. Ritcey.
1 schooner, 425 tons. Jan., 1918.
1 schooner, 400 tons, wood.

J. E. GASKILL, METEGHAN, N.S.

THOMAS GERMAN, METEGHAN, N.S.

One schooner, 350 tons, wood.

R. H. HOWES CONSTRUCTION CO., METEGHAN, N.S. (Leased James Cosman's shipyards.)

Building several wooden schooners for own account.

DR. F. H. MACDONALD, METEGHAN, N.S.

One wooden four-masted schooner, "Rebecca L. Macdonald," 201 ft. o. a. x 6 ft. x 16 ft., 800 gr. tons, 1,500 tons d.w., for builder's account. Launching January 1, 1918.

One wooden fishing schooner, No. 3, 84 ft. x 17 ft. x 7 ft., 50 gr. tons, 5 h.p. Launching August, 1918.

One schooner, 544 tons, wood.

METEGHAN RAILWAY & SHIPBUILDING CO., METEGHAN, N.S.

One wooden three-masted schooner, 182 ft. x 35 ft., 450 gr. tons.

One schooner, 470 tons, wood.

CHAS. McNEIL, NEW GLASGOW, N.S.

One wooden 3-masted schooner, 103 ft. x 30 ft. x 10 ft. 9 in., 200 tons, for builder's account. Launching July, 1918.

One wooden 3-masted schooner, 142 ft. x 25 ft. 4 in. x 13 ft., 500 gr. tons, for builder's account. Launching Oct., 1918.

THE NOVA SCOTIA STEEL & COAL CO., NEW GLASGOW, N.S.

Two steel freight steamers, 257 ft. 9 in. x 35 ft. x 20 ft., 1,700 gr. tons, 2,350 tons cap., trip. exp. engines, boilers 10 ft. 6 in. x 11 ft. 6 in., 800 h.p., 8.5 knots, for undisclosed interests. Launching of one Jan. 1918.

Two steel cargo strs. Raised quarter deck type, 248 ft. 9 in. long, 1,649.68 gross tons. Triple expansion engines, 17, 28, 46 x 33 in. stroke. One, the "War Bee," for undisclosed interests to be completed June, 1918. Other for Steel Co., to be completed Oct., 1918.

Two masted schooner, launched May 16, 130 ft. o. a. x 26 ft. x 10 ft. 6 in. Designed by J. S. Gardner.

O'BRIEN BROS., NOEL, N.S.

One schooner, 325 tons, wood.

W. R. HUNTLEY & SON, PARRSBORO, N.S.

One wooden schooner, 175 ft. x 39 ft. x 17 ft., 900 gross tons, for

C. T. WHITE & SON, SUSSEX, N.S.

Two schooners, 325 tons each. Jan., 1918.

One schooner, 490 tons, wood.

One schooner, 850 tons, four masted.

WAGSTAFF & HATFIELD, PARRSBORO, N.S.

One schooner, 400 tons. Jan., 1918.

SIDNEY ST. C. JONES, PLYMPTON, N.S. (Head Office, Weymouth North, N.S.)

One tern schooner, 200 tons, net reg., 107 ft. x 28 ft. 8 in. x 10 ft. 8 in., to be launched Nov., 1918.

S. SALTER, PARRSBORO, N.S.

One 200-ton schooner, wood.

DOWLING & STODDART, PORT CLYDE, N.S.

One gas boat, 27 tons, wood.

One schooner, 300 tons, wood.

SWIME BROS., PORT CLYDE, N.S.

Several small motor trawlers, for builder's account.

G. M. COCHRANE, PORT GREVILLE, N.S.

One tern schooner, "Alfredock Hedley," 152 ft. 6 in. x 36 ft. 12 ft. 6 in., 461 tons reg., for Adam B. Mackay, of Hamilton, Ont. Launched Jan., 1918.

One 4-masted schooner, 850 tons, 155 ft. x 37 ft. x 18 ft., 2 decks, wood. To be completed Oct., 1918, for builder's account.

H. ELDERKIN & CO., PORT GREVILLE, N.S.

One wooden three-masted schooner, for undisclosed interests.

ELLIOTT GRAHAM, PORT GREVILLE, N.S.

Is building a schooner, "Khaki Lad," of about 325 tons, at Port Greville, N.S., for J. W. Kirkpatrick and others. She will be launched early in October, and is reported to have been sold to Newfoundland parties.

One schooner, 360 tons. Jan., 1918.

SMITH CANNING CO., PORT GREVILLE, N.S.

One schooner, 350 tons, wood.

WAGSTAFF & HATFIELD, PORT GREVILLE, N.S.

One schooner, 400 tons, wood.

WILLIAM CROWELL, PORT LATOUR, N.S.

J. W. RAYMOND, PORT MAITLAND, N.S.

One schooner, 375 tons. Jan., 1918.

PORT WADE SHIPBUILDING CO., PORT WADE, N.S. (Head Office, Digby.)

One 350-ton schooner, wood.

JOHN BROWN, PUBLIC LANDING, N.S.

One tow barge, 50 tons, wood.

THE CUMBERLAND SHIPBUILDING CO., PUGWASH, N.S.

Establishing plant for the construction of wooden ships.

MACKENZIE SHIPPING CO., RIVER JOHN, N.S.

One four-masted schooner, 600 tons, wood. About half built. Fitted for aux. engines.

CHARLES McLELLAN, RIVER JOHN, N.S.

One schooner, 100 tons, wood.

W. J. FOLEY, SALMON RIVER, N.S.

Building for J. N. Rafuse & Sons one wooden, 300 tons, under construction. Cap. of plant, 1,000 tons yearly.

J. N. RAFUSE & SONS, SALMON RIVER, N.S. (Other plants at Shelburne and Conquerall Banks.)

Launched recently a three-masted schooner named "Industrial," at W. J. Foley's ship yard at Salmon River, N.S., of the following dimensions: length, 113 ft. breadth, 30 ft., depth, 11½ ft., 325 tons.

ACADIAN SHIPPING CO., SAULNIERVILLE, N.S.

One schooner, 80 tons, wood.

SAULNIERVILLE SHIPBUILDING CO., LTD., SAULNIERVILLE, N.S.

J. LEWIS & SONS, SHEET HARBOR, N.S.

One schooner, wood, four-masted, 725 gr. tons. To be completed Sept., 1918. For private account.

GEO. A. COX, SHELBURNE, N.S.

One schooner, 200 tons, wood.

One schooner, 322 tons, wood.

JOSEPH McGILL SHIPBUILDING & TRANSPORTATION CO., SHELBURNE, N.S.

"Sparkling Glance." 246 tons register, for Harvey & Co., St. John's, Nfld.

One schooner, 160 tons, wood.

W. C. McKAY & SON, SHELBURNE, N.S.

One wooden schooner, 130 ft. o. a. x 26½ ft. x 11 ft. depth of hold, 140 tons register. Launched Dec. 1, 1917.

One three-masted schooner, for Messrs. Hallet, of Newfoundland.

One schooner, 620 tons, wood.

J. N. RAFUSE & SONS, SHELBURNE, N.S. (Other plants at Salmon River and Conquerall Banks.)

One wooden three-masted schooner, 120 ft. x 32 ft. x 12 ft., 700 tons cap., for builder's account. For the market. Launching Jan. 10, 1918.

One wooden three-masted schooner, 118 ft. x 33 ft. x 12 ft., 700 tons cap., for builder's account. For the market. Launching Jan. 15, 1918.

Two three-masted wooden schooners, 122 ft. 6 in. keel x 32 ft. x 12 ft., 275 net tons, for J. O. Williams & Co., St. John's, Newfoundland. One to be built at Salmon River yard and one at Ship Harbor.

THE SHELBURNE SHIPBUILDERS, LTD., SHELBURNE, N.S.

One tern schooner, "Misty Star," 145 ft. x 31 ft. x 11 ft. 6 in., 600 tons d.w., 330 tons gross, for Harvey & Co., St. John's, Newfoundland. Launching Mar. 15, 1918.

One tern schooner, 154 ft. x 32 ft. x 12 ft. 4 in., 400 gr. tons, 700 tons d.w. cap. Launching June, 1918. Not sold yet.

Two schooners, one 349 tons, other 400 tons.

EASTERN SHIPBUILDING CO., SHIP HARBOR, N.S.

One wooden schooner, 600 tons d.w.; for J. N. Rafuse & Sons., Halifax, N.S., Can.

One wooden schooner, 300 tons.

STEPHEN MORASH & CO., SHIP HARBOR, N.S.

One wooden schooner, 150 ft. x 37 ft. x 13 ft., four masts, for Canadian interests. Keel laid Feb. 5, 1917.

JAMES E. PETTIS, SPENCER'S ISLAND, N.S.

One schooner, 425 tons, wood, three-masted.

CAPE BRETON SHIPBUILDING CO., SYDNEY, N.S.

One three-masted schooner.

THE TUSKET SHIPBUILDING CO., TUSKET, N.S.

AMOS H. STEVENS, TANCOOK, N.S.

ALVIN STEVENS, TANCOOK, N.S.

STANLEY MASON, TANCOOK, N.S.

ALBERT PARSONS, WALTON, N.S.

One schooner, 400 tons. Jan., 1918.

H. T. LeBLANC, WEDGEPORT, N.S.

One steam trawler, 165 ft., 500 tons, for J. N. Rafuse & Sons.

L. S. CANNING, WARDS BROOK, N.S.

One tern schooner, 350 tons, for W. C. Smith & Co., Lunenburg, N.S.

WEDGEPORT NAVIGATION & TRANSPORTATION CO., WEDGEPORT, N.S.

T. K. BENTLEY, WEST ADVOCATE, N.S.

One 200-ton steamer.

One 4-masted schooner, wood, 511 tons. To be launched September or October. For owner's account.

J. W. KIRKPATRICK, WEST ADVOCATE, N.S.

One 350-ton schooner, wood.

BOEHNER BROS., WEST LA HAVE, N.S.

Two large fishing schooners, 150 tons each.

One beam trawler, 300 tons, crude oil engines. For local interests.

BEAZLEY BROS., WEYMOUTH, N.S. (Head Office Roy Bldg., Halifax.)

One schooner, 394 tons, three masts.

E. R. GAUDET, WEYMOUTH, N.S.

One 350-ton schooner, three-masted.

E. P. RICE, WEYMOUTH, N.S.

One three-masted schooner, 350 tons.

RICE, WARREN & CO., WEYMOUTH, N.S.

One three-masted schooner, 334 tons.

FALMOUTH SHIPBUILDING & TRANSPORTATION CO., WINDSOR, N.S.

One wooden schooner, 350 gr. tons, 700 tons cap., for undisclosed interests.

NOEL SHIPBUILDING & TRANSPORTATION CO., WINDSOR, N.S.

One wooden schooner, three masts, 138 ft. x 35 ft. x 13 ft., 450 net reg. tons. Oil auxiliary engines. Launching August, 1918, for builder's account.

C. A. NICKERSON, WOODS HARBOR, N.S.

MILTON SHIPBUILDING CO., YARMOUTH, N.S.

Taken over and enlarging old plant of James Jenkins.

350-ton schooner building.

W. O. SWEENY, YARMOUTH, N.S.

100-ton fishing schooner.

YARMOUTH SHIPBUILDING CO., YARMOUTH, N.S.

One 450-ton wooden schooner, 128 ft. keel. To be completed Oct., 1918.

ONTARIO

CAN. ALLIS-CHAMBERS (Head Office, Toronto), BRIDGEBURG, ONT.

Six standard freight steamers, 3,500 tons, 261 feet long, for undisclosed interests.

COLLINGWOOD SHIPBUILDING CO., COLLINGWOOD, ONT.

Two steel cargo steamers, Nos. 51-52, 50 ft. x 43 ft. x 25 ft., 2,500 gr. tons, 2,900 tons d. w., triple expansion engines 18-30-50 x 36, 2 boilers 14 ft. x 11 ft., 10 knots, for undisclosed interests. Delivery May and Aug., 1918.

Four deep-sea trawlers, Nos. 53-58, inclusive, 125 ft. x 23 ft. 6 in. x 13 ft. 6 in., 288 gross tons, 12¾-21½-35/24 engines, 1 cylinder 13 ft. 6 in. x 10 ft. 6 in., 10 knots, 500 h.p. for undisclosed interests.

One steel freight steamer of 3,800 tons d. w. for undisclosed interests. Keel laid May 8, 1918.

"War Wizard," May 8, 1918, 3,000 tons d. w., 261 ft. x 43 ft. 6 in. x 20 ft. depth. For undisclosed interests.

"War Witch" same size, now building. To be launched ——

R. MORRILL, COLLINGWOOD, ONT.

"Windsor," Aug. 10, steam tug, 105 ft., for Ontario Gravel & Freighting Co.

CAN. CAR & FOUNDRY CO., FORT WILLIAM, ONT. (Head Office Transportation Bldg., Montreal.)

Twelve steel mine sweepers for French Government.
145 ft. o.a. steel construction, value $2,500,000.
Plant under construction.

GREAT LAKES DREDGING CO., FORT WILLIAM, ONT.

Two wooden vessels, 2700 tons d.w. cap., 260 ft. o.a., 43 ft. beam. Triple expansion engines of 1,000 h.p. To be launched November, 1918.

"War Sioux" launched May 12, 1918. Keel laid for another twenty minutes later. Launching scheduled for November.

THUNDER BAY CONTRACTING CO., FORT WILLIAM, ONT.

One wooden freight steamer, 261 ft. long, for undisclosed interests.

NATIONAL SHIPBUILDING CO., GODERICH, ONT.

DAVIS DRYDOCK CO., FT. OF BAY ST., KINGSTON, ONT.

One motor pass. boat, 64 ft. x 18 ft. x 6 ft., 52 gr. tons, wood. For ferry service, Gananoque to Clayton. Delivery June 1, 1918.

Twenty-eight lifeboats, regulation size, wood and metal for Government and private interests.

KINGSTON SHIPBUILDING CO., KINGSTON, ONT.

Several steel trawlers for undisclosed interests. First one launched Dec. 22, 1917.

SELBY & YOULDSON, KINGSTON, ONT.

GEORGIAN BAY SHIPBUILDING & WRECKING CO., MIDLAND, ONT.

Tug, 50 tons, wood.

One tug 40 tons, wood.

MIDLAND DRY DOCK CO., MIDLAND, ONT.

Three steel freight steamers, 261 ft. x 43 ft. 6 in. x 23 ft., 3,400 tons d. w., 10 knots, for undisclosed interests. Deliveries, one in July, 1918, and two others before close of navigation, 1918.

Alterations on steamers "Mariska" and "Glenlyon."

"Western Star" being reconstructed.

MIDLAND SHIPBUILDING CO., LTD., MIDLAND, ONT.

Three steel freight steamers, 261 ft. x 43 ft., 6 ft x 23 ft., 3,400 tons d. w., 10 knots for undisclosed interests. Deliveries, one in July, 1918, and two others before opening of navigation, 1919.

PORT ARTHUR SHIPBUILDING CO., PORT ARTHUR, ONT.

Six steel trawlers on order. 135 ft. o. a. x 23 ft. 4 in. x 15 ft. 1 in., 294.5 tons, trip. exp. engines, 500 h.p., single end Scotch boiler. Two-masted.

Two of these launched June 8, 1918.

"War Isis," April 3, 1918, 3,400 tons d. w. Trawler launched same day. April 3, 1918, keel laid for "War Hatha," sister ship to "War Osiris," of same construction, nearly completed.

"War Fish," Aug. 19, 1917, 4,300 tons, steel.

"War Dance," Nov. 3, 1917, 3.400 tons, steel.

Four steel screw steamers, ocean freight servicel—"War Osiris," "War Hathor," "War Karma," "War Horus," 3,400 tons d. w. each, 261 ft. o.a. 251 ft. b.p. in length. Triple expansion engines, 2 Scotch boilers each, approx. 1,250 h.p. For completion, two Aug. 31st, and two Nov. '20. 1918. For undisclosed interests.

MUIR BROS. DRYDOCK CO., LTD., PORT DALHOUSIE, ONT.

J. W. GEROW, ROSSPORT, ONT.

REID WRECKING CO., SARNIA, ONT.

One steel tug, 157 ft. x 32 ft. x 19 ft., trip. exp. engines, for lake or ocean service. Keel blocks laid March 5. Launching July, 1918. For Reid Wrecking Co.

WEST, PEACHEY & SONS, SIMCOOE, ONT.

Number of tugs for lumbering interests.

DOMINION SHIPBUILDING CO., TORONTO, ONT.

One steel cargo vessel, 261 ft. x 43 ft. x 28 ft. 2 in., 3,500 tons d.w. cap. For Department of Marine, Ottawa.
Have orders for five other vessels same dimensions.
"Traja" launched May 15, 1918.

THE POLSON IRON WORKS, LTD., TORONTO, ONT.

Eight steel cargo steamers, Nos. 133-4-5-6-7-9-40, 216 ft. o. a., 251 ft. b. p. x 43 ft. 6 in. x 22 ft. 11 in., draft loaded 19 ft. 6 in., single decked, 2,350 approx. gross tons, 3,500 tons cap., single screw trip. exp. engine 20½-33-54 x 36, 1,250 h.p., 10 knots, two 14 ft. x 12 ft. Scotch R. T. boilers, 180 h.p., to cost $600,000 each, for undisclosed interests. Launchings throughout 1918.

Four building and four on order.

"Asp," Feb. 11, 1918, 261 ft., 3,500 tons d.w., steel.

"Tento," Oct. 22, 1917, 261 ft., 3,500 tons d.w. steel. For Norwegian interests. Transferred to British registry.

THE THOR IRON WORKS, TORONTO, ONT.

Under same management as Dominion Iron Works

TORONTO SHIPBUILDING CO., LTD., TORONTO, ONT. (Toronto Dry Dock Co., Ltd., under same management.)

Two wooden cargo steamers, 250 ft. B.P. 42 ft. 6 in., moulded breadth 25 ft., moulded depth 22 ft., draft 2,500 tons d. w. Triple exp. engines $\frac{20\times33\times54}{40}$. Howden boiler. Launchings July and Sept., 1918. For undisclosed interests.

THE BRITISH AMERICAN SHIPBUILDING CO., WELLAND, ONT.

Two steel frs., 3,500 tons d. w., 261 ft. O.A., 43 ft. beam, 23 ft. moulded depth, Westinghouse steam turbine engines. Delivery 1918.

WELLAND SHIPBUILDING CO., WELLAND, ONT.

One steel freight steamer, "War Weasel," 261 ft. x 43 ft. 6 in. x 23 ft., 3,300 tons d. w., trip. exp. engines, 14 ft. x 12 in. Scotch boilers, 10 knots, 1,250 h. p. Launching April, 1918.

Two deep frame cargo vessels, "War Badger" and "No. 3," 261 ft. x 43 ft. 6 in. x 23 ft., 3,300 d. w. tons, geared turbines, Howden's forced draught boilers, 10 knots, 1,250 h.p. Launchings June and August. 1918.

PRINCE EDWARD ISLAND

THE CARDIGAN SHIPBUILDING PLANT, CARDIGAN, P.E.I. (Purchased Annandale Lumber Co. plant and removed to Cardigan.)

One three-masted schooner, 325 tons, to be completed November, 1918.
Two more to be built.

QUEBEC

TIDEWATER SHIPBUILDERS, LTD., THREE RIVERS, QUE. (Formerly Sorel Shipbuilding & Coal Co.)

Three steel trawlers for undisclosed interests
One steel steam barge for the Canada Steamships Lines.

R. N. LE BLANC, BONAVENTURE, QUE.

J. Z. DEGAGNE, EBOULEMENTS, QUE.

DAVIE SHIPBUILDING & REPAIRING CO., LEVIS, QUE.

Six military barges, 130 feet long.
Eight steel trawlers, 2,035 tons.
One floating crane, 350 tons.
*One steel cargo vessel of 5,000 tons cap.
Building several steel lighters and several wooden drifters.

Steel car ferry "Canora" building for C.N.R., 308 ft. x 52 ft., cap. 20 loaded cars. Speed, 14 knots.

ATLAS CEMENT CONSTRUCTION CO., LTD., MONTREAL, QUE.

One concrete cargo steamer, "Concretia," 125 ft. x 22 ft. x 13 ft., steel ribbed, hull to be from 3 in. to 5 in., thick, for the builder's account.

CANADIAN VICKERS, LTD., MONTREAL, QUE.

Two cargo steamers, 9:400 tons, steel; 1 dredge, 2,364 tons, steel; 12 trawlers, 3,050 tons, steel; 23 drifters, 3,360 tons, wood.

Six freight steamers, 24 ft. draught, 7,000 tons cap., 11 knots, for undisclosed interests. Delivery, 1917, of two for Norwegian interests. "Porsanger," 394 ft. 6 in. o. a. x 40 ft. 4 in. x 30 ft., triple exp. engines, launched Nov. 29, 1917.

"War Earl" launched June 8. 1918, 7,000 tons d.w., 380 ft. x 49 ft. Keel of sister ship laid five minutes after this launching.

One steel freight steamer, 2,360 tons, cap., for undisclosed interests.
Three steel cargo vessels, 8,200 tons, to be laid down in May, Aug. and Sept., 1918, for undisclosed interests.

FRASER BRACE & CO., MONTREAL, QUE.

Four wooden cargo steamers, 3,000 tons, for undisclosed interests. Two keels laid during Oct., 1917.

HALL ENGINEERING CO., MONTREAL, QUE.

MONTREAL DRY DOCK & SHIP REPAIRING CO., MONTREAL, QUE.

Operate dock 428 ft. long 30 ft. deep.

MONTREAL SHIPBUILDERS LTD., 37 Belmont St., MONTREAL, QUE. (Associated with Atlas Construction Co.

THE QUEBEC SHIPBUILDING AND REPAIR CO., Board of Trade Bldg., MONTREAL, QUE.

2 vessels for undisclosed interests 3,000 tons each.

QUINLAN & ROBERTSON, MONTREAL, QUE. (Yards at Quebec.)

Four wooden steamers, for undisclosed interests.

LOUIS GAUNDRY, 12 St. Peter St., QUEBEC, QUE.

QUEBEC SHIPBUILDING & REPAIR CO., ST. LAURENT, QUE.

2 schrs. 1,400 tons and 1,200 tons.
One wooden four-masted auxl. schooner "Martin Connolly," 223 ft. x 42 ft. x 20 ft., 2,100 tons d. w., for undisclosed interests. Launched Oct. 28, 1917.

QUINLAN & ROBERTSON, QUEBEC, QUE.

Four wooden steamers, totalling 6,400 tons, for undisclosed interests.

CANADIAN GOVERNMENT SHIPYARDS, SOREL, QUE.

One steel vessel for undisclosed interests.

LECLAIRE SHIPBUILDING CO., SOREL, QUE.

Building 6 steel ships at value of $1,500,000.
Six trawlers, 125 ft. single screw type expansion engines, 500 I.P.H. steel. To be completed this year for private acct.

H. H. SHEPHERD, SOREL, QUE.

Rebuilding seven drifters. Particulars withheld owing to govt. restrictions.

SINCENNES-McNAUGHTON LINES. SOREL, QUE.

One tug 410 tons, wood.

CAN. GOVERNMENT SHIPYARD, SOREL, QUE.

One steel vessel for undisclosed interests.
Building steam trawlers and wooden drifters for undisclosed interests.

ST. LAWRENCE SHIPBUILDING & STEEL CO., SOREL, QUE.

THE THREE RIVERS SHIPYARDS LTD., THREE RIVERS, QUE.

Two cargo steamers, 3,100 d. w., wood, 260 ft. x 43 ft. 6 in., 1,000 H.P. engines. Names, "War Mingan" and "War Nicolet." To be completed end of navigation season for undisclosed interests.

Aikenhead's

"Toledo"

Pipe Threading Tools

The top illustration shows the Toledo Adjustable Ratchet Threading Tool No. 1A. It is adjustable for variations in sizes of fittings. Threads of standard taper or straight lock nut threads may be cut with it, or it can be had reversed in all its parts for cutting left-hand threads. One man can easily thread 1" to 2" pipe with this Toledo tool.

The third illustration from the top of page shows the Toledo Geared Adjustable Threading Tool No. 2, which is furnished complete with Ratchet and Driving Cross. With this light, compact tool one man can easily thread 2½" to 4" pipe.

Photographed Advantages

The upper photograph shows a man threading a 4" bend **in position** with the Toledo Geared Adjustable Threading Tool No. 2. With nothing but an ordinary threading tool to do the work, this whole pipe line would have had to be taken down.

In the lower photograph a man is shown threading a pipe near the ceiling of an engine room. The position was awkward and the platform shaky, but the job was easily handled with a Toledo Adjustable Ratchet Threading Tool No. 1A.

Engine Room Equipment

As well as stocking complete lines of all high-grade threading tools, we carry a full assortment of engine room equipment. Your enquiries are solicited.

Write to-day

Aikenhead Hardware Limited

17, 19, 21 Temperance Street, Toronto

Quebec.—The "War Gaspe," the third wooden vessel to leave the shipyards of Quinlan and Robertson, here, since last spring, was successfully launched July 27.

St. John, N.B.—Two wooden vessels were launched on August 7. The new vessels are the Celina K. Goldman, 477 tons, built at St. Martins, and the Vincent A. White, 460 tons, launched at Alma.

Levis, Que.—A syndicate of several shipbuilding companies of the Dominion may shortly form a syndicate and erect a $1,500,000 plant. Dussault & Hutchison, Commercial street, Levis, Que., are interested.

Prince Rupert, B. C.—A contract has been closed for the construction of five steel ships of 8,500 tons each, and preparations for their building are already underway at the shipbuilding plant of the Grand Trunk Pacific.

Fort William, Ont.—The first trawler built by the Canadian Car and Foundry Co. for the French Government was launched recently. The vessel was christened "Navrin," and the ceremony was performed by Miss Parks.

Platinum Found in British Columbia. A platinum and placer gold strike has taken place near Athalmer on Tobey Creek, a tributary of the Columbia river. The strike is extensive and the deposits both of platinum and gold are stated to be rich.

Ottawa, Ont.—The chief press censor announced recently that the steamer Siberian Prince, which went ashore at Lawrencetown, 15 miles east of Halifax, during a heavy fog, on July 2, has been floated and is now safely docked. The damage done was slight.

Sarnia, Ont.—Officials of the Northern Navigation Co. have received word that the steamer Huronic, which blew out a cylinder-head in Lake Superior, had reached Port Arthur and made repairs, and would leave on schedule time. The accident happened about 45 miles from Port Arthur.

Paper Export Regulations.—Orders for export, except for Canada, may be placed in such quantities, size and weight as export requirements demand, without regard to domestic regulations, the pulp and paper section of the War Industries Board announced last Friday. This ruling applies to orders now on the books and to future orders.

Production of Anthracite in U.S.—Production of anthracite coal during the week ending August 3 established a new record for this year, with 1,75049,0 tons. In announcing this the fuel administration gave warning that this one week will not save the situation, and urged miners and operators to increased endeavor.

New Shipbuilding Records in July in U.S.—With the launching of 123 vessels totalling 631,944 deadweight tons, and the delivery of 41 others of 235,025 deadweight tons, new world shipbuilding records were established in July by American shipyards, the Shipping Board announced. The July launchings alone were greater than those of any single year in the past.

Level of Lakes in July.—Lakes Superior and Erie attained a higher level in July than in June, according to the monthly summary prepared by the United States Lake Survey Office, Detroit. Lakes Michigan and Huron were the only lakes maintaining a level in July higher than the average for that month in the last ten years.

Three Rivers, Que. — The National Shipbuilding Corporation, of Boston, has orders for between eight and ten millions of dollars' worth of wood ships on foreign account, and has decided to have them built in this city, where it has just secured control of the yard of the Three Rivers Shipbuilding Co. The plant of this company having about 15 ways will be utilized for these orders.

Toronto.—Considerable significance is attached locally to the fact that the Dominion Shipbuilding Company has doubled its authorized capitalization. It is known that this company has been very busy since its establishment in Toronto, and it has already successfully completed seevral orders.

Goderich, Ont.—The by-law to grant exemption from taxes and other concessions to the National Shipbuilding Co. for ten years was carried by an overwhelming majority, the vote being 511 for the by-law and 47 against. The company will erect a new boiler shop, to employ at least 30 men, in addition to the plant they already have.

Marine Rates Advanced in New York.—Marine underwriters have advanced war risks on sailing vessels both for coastwise and trans-ocean routes because of the continued activity of U-boats in coastal waters. Rates jumped to 3 per cent. and in some cases to 4 per cent. for sailings between American ports, while trans-ocean rates were advanced to 10 per cent. by some underwriters.

Vancouver, B.C. — The Vancouver Board of Trade is making efforts to have Japanese freight shipped direct to ports in British Columbia instead of through United States ports, whence the bulk of it is transhipped to this province. The stringent United States regulations regarding shipments to Canada make it almost impossible to secure certain necessary foodstuffs which arrive from the Orient.

Schooner Launched at Meteghan, N.S.—The first vessel built by the Howes Construction Co. of Boston and New York at their yard at Meteghan river, was floated on July 25 after several attempts had proved abortive. The new three-master is 150 feet long, 34 ft. 6 in. beam, 14.9 ft. depth of hold, with a gross tonnage of 485 and net 400 tons.

Victoria, B.C.—Contracts have been actually placed with one of the well known shipbuilding companies for the construction of twenty wooden vessels in Victoria, and the Provincial Government has granted an option for a few days on a site situated on the Songhese Reserve. This is the substance of an announcement made by Hon. John Hart, Minister of Finance, recently.

Fort William, Ont.—The envy of five sister ships in various stages of construction in adjacent berths, the Wauarin, first of the French mine sweepers constructed at the plant of the Canadian Car and Foundry Company, moved to her maiden dip on July 29. Within 30 minutes after she started towards the slip, the keel of a seventh mine sweeper was laid in her place. The Wauarin will have her boilers installed at the docks, her guns placed fore and aft, her upperworks finished and go through a series of target tests on Thunder Bay all before she starts on her way down the lakes for the seaboard. It is understood that a second launching was to be made on August 7 and others soon after.

Port Greville, N. B.—The four-masted schooner Fredie E., on July 25, was launched from the shipyard of H. Elderkin & Company. Fitted with gasoline equipment for hoisting purposes and modern in every particular, she is a vessel of 669 tons register, 199 feet long, 38 9 wide and 19 feet deep. From Port Greville the Fredie E. goes to St. John to load lumber for Durban, South Africa. Construction on another schooner of about 275 tons is to be started at once, so as to have her ready for the water before the close of navigation this fall. This vessel will be built on the co-operative plan, the workmen taking shares at first cost and sharing in any profit realized from her sale.

EMPIRE
Largest Brass Manufacturers in British Empire
Quality
First
MARINE BRASS GOODS
A-810 Square Head Steam Cock
A-2512 Steam Whistle with Valve
A-860 Standard Gate Valve Solid Wedge; Rising Stem
All goods are factory tested and can be depended upon to give reliable service. They are heavy in weight and made to formula in our own plant.
Back of the Empire "E" (trade mark) stands the guarantee for quality and service of the largest brass manufacturers in the British Empire.
A few of our products:
Steam Traps
Brass Castings
Steam Whistles
Pressure Gauges
Brass Valves
Compresion Bibbs
Ship Lavatories and
Water Closets
Steam Stops
Lubricators
Brass Fittings
Basin Cocks
E
Quality Mark
NAVAL CASTINGS
Send us your specifications for Brass, Bronze and Gun Metal Naval Castings.
Secure a copy of our finely illustrated catalogue - contains the full Empire line
E
Quality Mark
Empire Manufacturing Company
LIMITED
LONDON and TORONTO

Will Build More Ships at Vancouver.—A despatch from London, England, states that a number of prominent shipbuilders, including Sir William Beardmore, called on Hon. C. C. Ballantyne, Canadian Minister of Marine, with reference to shipbuilding in Canada. The result of the interview has been the granting of permission for a company to undertake the construction of twenty wooden vessels at Vancouver for Norwegian registry. Numerous applications have been received to build steel ships in Canada for either Allied or neutral registry, but these have been consistently refused owing to the fact that the Dominion government has launched a shipbuilding programme of its own, for which it is difficult to secure the necessary ship plates.

"War Gaspe" Launched at Quebec.—Christened the "War Gaspe," the third three thousand-ton wooden steamer built at Limoilou shipyards of Messrs. Quinlan and Robertson, was launched recently. The launching was witnessed by a very large number of citizens who had congregated in the shipyard and surrounding frontages. The War Gaspe, built for the Imperial Munitions Board, is the finest vessel in construction and lines yet put out of the Limoilou shipyards. The War Gaspe, like her sister ships previously launched, measures 261 feet in length, and has a cargo capacity of 3,100 tons, and will be fitted up with 1,000 horse-power triple-expansion engines and steam generated by the latest type of water tube boilers. She is built of British Columbia fir, reinforced with steel and will carry no sail, being a standard steamer fitted out with Marconi wireless apparatus.

Large Drydock in St. John.—Canada is to have the largest drydock on this side of the Atlantic, and one of the largest in the world. The St. John Drydock and Shipbuilding Co. has already secured the contract for its construction from the Department of Public Works, Ottawa. The contract calls for the construction of a drydock which will be of the first class and the largest on this side of the Atlantic, and will, when completed, be capable of accommodating the greatest ships of any navy in the world, or the largest mercantile vessels now built or contemplated. The length of the new dock will be 1,150 feet, and its width at the bottom 125 feet, with forty feet of water over the sill. The control of theh enterprise is in the hands of the Canada Dredging Co., of Midland, Ontario, a strong and wealthy corporation, which has acquired control of the stock of the St. John Drydock and Shipbuilding Co. The men who are associated in the new company are, according to the "Journal of Commerce," James Playfair, Midland, president and general manager of the Great Lakes Transportation Co, and president of the Midland Iron and Steel Co.; Hon. W. H. Richardson, Kingston; D. L. White, Jr., Midland; W. J. Sheppard, Waubaushene; W. E. Phin, Hamilton; D. S. Pratt, Midland; George Y. Chown, Kingston; R. Hobson, Hamilton; Col. Thomas A. Duff, Toronto; J. A. Paisley, Cleveland, and J. B. Craven, New York. The general manager of the company will probably be D. S. Pratt of Midland, Ont., who occupies a similar position at present with the Canadian Dredging Co.

Vancouver, B. C.—Difficulties beeween masters and mates and the steamship and tugboat companies operating on the Pacific Coast came up for investigation under the Industrial Disputes Investigation Act two weeks ago, the commission of enquiry consisting of Messrs. W. E. Burns, chairman; J. H. McVety, for the Canadian Merchant Service Guild; and E. A. James for the companies. The men demand recognition of their organization and higher wages. Nineteen or twenty of the companies, it is understood, have accepted the wage schedule issued by the guild, but the C. P. R., the G. T. P. and the Union Steamship Company claim they have no dispute with the men and refuse to deal with the guild in any way.

She is named the Richard B. Silver, and is a fine specimen of craftsmanship, having been built to fulfil the requirements of the British Lloyd's rating, with whom she is classed A1, 12 years with a star, this being the highest possible rating ever awarded to vessels, unless built of Burmah teak wood.

She has been purchased by H. W. Adams and other of Lunenburg, N.S., and will proceed to Liverpool, N.S., to load for Buenos Ayres under the command of Capt. A. H. Zinck. The vessel is equipped with stockless anchors, has a Fairbanks-Morse type Z engine aboard, is fastened with galvanized bolts throughout, and is more than ordinarily heavily rigged. Mr. George B. Morecroft, the manager, is to be congratulated upon adding such a vessel to those already issued from St. Mary's Bay yards, and he has a second well under way, being built from exactly the same model. He will proceed to lay the keel of a third on the stocks recently occupied by the Richard B. Silver.

THE PORT OF CALCUTTA

By Mark Meredith

A report on the progress of the port of Calcutta is of greater interest to the English people than that of any other port of India, because of Calcutta's historical associations, and also because Calcutta is, with the exception of London, the largest city of the British Empire, in point of population. It is among the twelve largest cities of the world and is the centre of English trade in India.

The sturdy English pioneers of old Calcutta founded an empire in a fit of absence of mind, as the late Sir John Seely so aptly phrased it, but it was with a very determined mind that they set out to make Calcutta a world-centre of trade and a real English emporium for Hindustan and its neighbors. The hold of British merchants in other parts of India has relaxed. Bombay is far behind Calcutta in the solidity and permanence of its European institutions and great trading and mercantile establishments. Whatever may be said by the small band of Europeans in Bombay as to urbs prima in Indis, Calcutta is still the heart and pulse of the British in India, and we trust the British community of the "City of Palaces" will be true to the great traditions of the past. Charnock and the long list of great English names associated with Calcutta, should never be disgraced by any wavering on the part of the British in the delta of the Ganges during the difficult and troublous times ahead. The future historians, we hope, will be able to record as fine a list of great names of Britishers as is now associated with Calcutta's stirring past.

Few great cities in the British Empire can boast of so grand a river frontage as Calcutta. The Hughli, commercially speaking, is the most important channel by which the River Ganges enters the Bay of Bengal. It assumes its distinctive name about 120 miles from the sea and from Calcutta to the sea, a distance of about 80 miles, the river is a record of engineering improvement and success. The tide on the river runs rapidly, the headwave of the advancing tide sometimes exceeding seven feet in height and is felt as far up as Calcutta where disaster to small craft is of frequent occurrence.

The old East India Company took special measures to assist the difficult navigation of the Hughli river on which Calcutta stands, and thus when the port of Calcutta was presently founded it came at once under Government control. In 1870 a Port Trust was formed and during the past forty years it has constantly been busy building up the port and extending its accommodation to keep pace with trade. In 1870 there were only six screw-pile jetties, six cranes and six sheds for the accommodation of overseas trade. The vast accommodation now available consists chiefly of riverside jetties and of the Kidderpore docks. The latter were constructed in 1884-5 at a cost of approximately two million pounds.

Great schemes for extending the port have been framed in recent years, and in order to make the new arrangements thoroughly in keeping with the most up-to-date methods of dock construction and port management a special committee visited British and European continental ports in 1913. The constant congestion of the port led the provincial government in the same year to appoint a committee to consider the whole question of future port development and especially the formulation of a well-defined and carefully thought out policy. This committee approved an important new dock extension scheme just undertaken by the port commissioners and declared that "it will ensure an ample margin for the expansion of trade which is likely to take place in any period that can reasonably be foreseen," and that the future expansion of seaborne trade should take place in the neighborhod of the docks. Elaborate improvements of the railway approaches to the port, the development of road communication and the improvement of the river were the other principal directions to which they directed activities.

ONE OF MANY INSTALLATIONS OF MARTEN-FREEMAN COMPENSATING DAVITS

FOOL PROOF—RELIABLE—FAST OPERATING

TIME IS LIFE

WHAT GUARANTEE HAVE YOU THAT IT WILL BE POSSIBLE TO PUT ALL YOUR LIFEBOATS OVERSIDE IN THE FEW MINUTES AVAILABLE?

THE MARTEN-FREEMAN COMPANY, LIMITED

301 MANNING CHAMBERS, TORONTO, ONTARIO.

OBERDORFER

BRONZE GEARED PUMPS

"A Pump for Every Motor"

THE gas engine operator knows that positive lubrication and cooling is impossible unless the circulation pump is reliable and regular.

As an evidence of the worth of Oberdorfer Bronze Geared Pumps, they are forming a part of many of the leading motors in service, and are recognized for their positive pressure, lifting power without priming, and adaptability to varying speeds and conditions.

Made in ten sizes and six styles. Write for our latest Pump Booklet. It will interest you.

M. L. Oberdorfer Brass Company

806 East Water Street, Syracuse, N.Y., U.S.A.

Georgian Bay Shipbuilding & Wrecking Co., Ltd.

Modern Marine Railway. Capacity 1,000 tons.

Specialists in the Construction of Wooden Ships

Complete equipment, skilled workmen. Satisfactory production guaranteed. Repairs and overhauling of all kinds given immediate attention.

You want your work done thoroughly. Consult us. Our many years of practical experience at your service.

MIDLAND, ONTARIO

With Exceptional Facilities for Placing

Fire and Marine Insurance

In all Underwriting Markets

Agencies: TORONTO, MONTREAL, WINNIPEG, VANCOUVER, PORT ARTHUR.

Protect Your Plant!

War-time conditions, spy work and enemy activities make it necessary to safeguard all plants, munitions works, factories, shipyards, etc. Daily the newspapers tell of fires, explosions and other destructive happenings that are obviously not accidental.

DENNIS CHAIN LINK FENCE

is the only safeguard. It is built of massive wire, stands seven feet high and bristles with vicious barbed wire. Built for the express purpose of establishing a "safety zone" about industrial plants to keep out intruders, thieves, firebugs, cranks and incendiaries. A fact-giving illustrated folder is yours for the asking.

Toronto, 36 Lombard Street.

THE DENNIS WIRE & IRON WORKS CO., LIMITED

LONDON, CANADA

Halifax Montreal Ottawa Winnipeg Vancouver

NITER CAKE FOR PICKLING BRASS

By Mark Meredith.

Niter cake, as a substitute for sulphuric acid in pickling brass, is finding extensive use in Great Britain. The quantity so used in that country now runs into several hundred tons per week. Difficulty in obtaining raw material, coupled with the increased demand for sulphuric acid for other purposes, has resulted in an extended application of niter cake solutions in place of dilute sulphuric acid for pickling annealed brass.

Niter cake is essentially crude acid sodium sulphate, and while the latter in the pure anhydrous state contains theoretically 40.8 per cent. of sulphuric acid, the free acid found in niter cake may vary from 3 to 30 per cent. For pickning, the niter cake solution should show 3 to 5 per cent. sulphuric acid on titration; there is no advantage to be gained in using solutions of higher acid concentration.

While niter cake is a variable product, annealed brass from a pickling point of view may be even more so. Given clean work, niter cake solutions replace dilute sulphuric acid quite efficiently, but with dirty work the difference is much more marked, and niter cake solutions, even under the most favorable conditions, may fail to pickle the work satisfactorily. It is little consolation for the manufacturer to know that his troubles lie in his annealing furnaces. Much may be done to overcome troubles as follows:

The niter cake solution should be as hot as possible. Its acid content should be tested frequently and maintained at 3 to 5 per cent. by the addition of niter cake. The hot annealed products may be quenched in water, whereby much scale is mechanically loosened prior to pickling.

The hot annealed work may be placed direct into the niter-cake solution, and the pickle can thus be maintained at a high temperature without auxiliary steam heating.

Electrochemical aid might be sought by using a low voltage current, making the lead lining of the vat the cathode and the work to be pickled the anode.

The difficulties met with in the successful use of niter cake solution can only be overcome in a satisfactory manner by paying close attention to the conditions governing the annealing so as to obtain the annealed brass as clean as possible, and by using the niter cake solution under conditions which will most strongly stimulate its pickling activity.

SURFACE HARDENING BRONZE AND COPPER

In occasional instances it is desirable to harden the surface of copper and bronze, such things as dies being preferably so dealt with to enable them to withstand wear. This hardening is best done by alloying tin into the surface of the metal after the articles are finished, the process being as follows:—Metal which is machined all over—and not mere castings having the "skin" left on—is taken and all grease and dirt removed by well scouring with caustic potash or soda, rinsing in clean water and drying in non-resinous sawdust. The articles are then heated to a red heat and thinly coated with pure tin, and the heat continued for some minutes, when all excess of tin is wiped off with a piece of tow and the heat maintained until the tin left is absorbed, after which the articles are allowed to cool in the air. Or the articles can be thoroughly cleaned and have the surface thinly coated with tin in the usual way, using zinc chloride as a flux. The work must then be thoroughly washed in hot water to removed any excess of chloride of zinc, dried quickly, and then heated to redness in a muffle in which a reducing atmosphere is maintained, the process being complete as soon as the tin is absorbed. As the metals are somewhat tender at red heat, they should be kept on an iron plate or fire-clay tile during the period of heating. It is also possible to secure this kind of surface alloying with a blow-pipe flame, because absorption takes place at a red heat, as most persons using a copper soldering bit find out to their cost o occasion, and a "burnt" tinned solderin bit wants some hard work to file it u clean and fit for use for its proper pur pose. The same principle is involved i "pyro-plating" on steel, in this case silve and gold leaf being used, and partly al loyed with the steel by heat in a clos muffle.

BROKEN GAUGE GLASSES

By M. M.

The following hints concerning broke gauge glasses cannot be too forcibly im pressed upon engineers. It is not only matter of knowing how, but also bein able to practically apply the knowledg effectively and expeditiously. The cock on the boiler should be immediately shu and the glass replaced without dela The new glass should be free from flaw or scratches, with ends ground squar or fire-finished and of the correct lengt If too long it would restrict the passag of steam to the glass; if too short, th packing may work over the edges of th glass. In many makes of gauges th correct length of the glass is stamped o the framing.

Before replacing it, the whole of th old packing should be removed, and th screws of the adjusting glands mad easily workable. The packing should b placed on the glass, and while screwin up the bottom gland the glass should b kept in contact with the metal of th lower cock. Care should be taken not t screw up the upper gland hard befor the lower one, as this may lift the glass and perhaps allow the lower packing t choke the orifice.

The glands should at first be screwe up hand-tight, after which the steam an drain cocks should be opened a smal amount to heat the glass gradually, whe after a short interval the water coc should be gradually opened, and then th drain cock closed, the steam and wate cocks being then fully opened gradually and the glands adjusted as required. Th glass should now be tested by closing th steam cock and opening the drain, whe water should rush out freely. The wate cock should then be closed and the stean cock opened, when steam should rush ou freely. When the drain is closed and th water cock open, careful note should b made of how the water rises in the glass It should, if the fitting is properly made rise smartly to the water level of th other glass.

Welland, Ont.—It is reported that th cost of reinforcing the Welland Rive bridge is estimated at $5,000. Plans ar now being prepared.

Quality

Service

FOR COOL BEARINGS

We Manufacture:

Phosphor Bronze Tail Shaft Liners, Pump Liners, Stuffing Boxes, Stern Tube Bushings and Brass Castings of every description.

Tallman Brass & Metal Limited

HAMILTON, ONT.

DART UNIONS

Never Leak, Rust nor Corrode At the Joint

Heavy malleable pipe ends with BRONZE AGAINST BRONZE *at the joint.*

They last, and their upkeep cost over a period of years doesn't amount to a single cent.

It is worth your while to use Dart Unions.

OBTAINABLE AT EVERY PORT

Manufactured and Guaranteed by

DART UNION COMPANY, Limited

TORONTO, CANADA

WILKINSON & KOMPASS
TORONTO HAMILTON WINNIPEG
IRON AND STEEL
HEAVY HARDWARE
MILL SUPPLIES
AUTOMOBILE ACCESSORIES
WE SHIP PROMPTLY

Castings

Honor Roll
The MacLean Publishing Company Ltd.
Publishers of
MacLean's Magazine
Farmers' Magazine
The Financial Post
Hardware and Metal
Canadian Grocer
Dry Goods Review
Men's Wear Review
Bookseller and Stationer
Canadian Machinery
Power House
Canadian Foundryman
Marine Engineering
Sanitary Engineer
Printer and Publisher
D.W.G. Davies
John Dring
Harold Fogarty
A.O. Thompson
E.W. Earle
H. Rose
M.A. Sanderson
G.E. Pearson
G.A. Mundie
Desmond Hemingway
Victor Bercot
G.I. Cook
Hector Pearson
C.W. Catlow
H.S. Jones
Geo. Aubrey
Arthur Batty
Harold Cross
A.W. Hardy
E.A. Humphries
L. T. Snashall
Sidney Metcalf
F.H. Thomas
T. McGillicuddy
Gordon Bennett
Walter Pearson
Roy. Jarratt
A. Macdonald
Ernest. Madigan
L. Makepeace
Harold. Gildner
Edwin. Mackie
C. Denham
F. Black
A.W. MacIntyre
J.W. Zimmerman
C.F. Breckon
P.W. Gowans
C. Hawkins
H.L. Southall
Andrew O'Malley

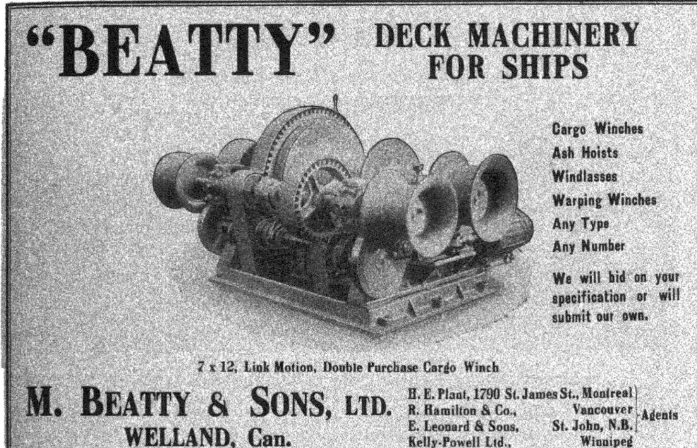
"BEATTY" DECK MACHINERY FOR SHIPS
Cargo Winches
Ash Hoists
Windlasses
Warping Winches
Any Type
Any Number
We will bid on your specification or will submit our own.
7 x 12, Link Motion, Double Purchase Cargo Winch
M. BEATTY & SONS, LTD.
WELLAND, Can.
H. E. Plant, 1790 St. James St., Montreal
R. Hamilton & Co., Vancouver
E. Leonard & Sons, St. John, N.B.
Kelly-Powell Ltd., Winnipeg
Agents

Dollars or Eyes

Which?

What Are Your Men's Eyes Worth?

Stoco Safety Goggles cost $12.00 a dozen—with interesting discounts for quantity orders. At the lowest possible estimate it is several hundred times cheaper to provide every man in your employ with a pair of Stoco Safety Goggles than to have one of them lose an eye.

If you go shopping for eye protection you may find some one who will sell you more goggles per dollar than we will, but you cannot buy more protection or greater durability per dollar anywhere. Stoco Safety Goggles will make every pair of eyes in your factory absolutely safe for an indefinite period—we have never heard of one of our Protection Glasses wearing out.

The Stoco Safety Goggle is practically indestructible. It has no soldered joints to come apart, no screws to work loose and fall out. When lenses become pitted they may be readily changed by anyone without the aid of special tools. Being constructed entirely of german silver, the Stoco Safety Goggle will not rust or corrode from the action of perspiration and may be sterilized daily in any manner desired without damage.

Eyes are safe behind the Stoco Safety Goggle. The lenses are ground from optical glass especially selected for its strength. The one-piece front prevents particles entering between the eyes—an ever present danger with goggles lacking this exclusive feature. Perforated metal side shields provide effective protection against rivet heads, metal or stone chips and other missiles.

We shall be glad to send interested responsible people a sample of the Stoco Safety Goggle on receipt of request on letterhead or accompanied by business card.

STANDARD OPTICAL COMPANY
GENEVA NEW YORK

BUYERS' DIRECTORY

ACCUMULATORS, HYDRAULIC
- Smart-Turner Mach. Co., Hamilton, Ont.

AERATING RESERVOIRS
- Spray Engineering Co., Boston, Mass.

AIR PORTS
- Mitchell Co., The Robert, Montreal, Que.
- Turnbull Elevator Mfg. Co., Toronto, Ont.

ALLOYS, BRASS AND COPPER
- Dom. Copper Products Co., Ltd., Montreal, Que.
- Mitchell Co., The Robert, Montreal, Que.
- Mueller Mfg. Co., H., Sarnia, Ont.
- Tallman Brass & Metal Co., Hamilton, Ont.
- United Brass & Lead, Ltd., Toronto, Ont.

ANCHORS
- Beauchemin & Fils, Sorel, P.Q.
- Hopkins & Co., F. H., Montreal, Que.
- Kearfott Engineering Co., New York, N.Y.
- McNab Co., Bridgeport, Conn.
- Henry Rogers Sons & Co., Wolverhampton, Eng.
- Wm. C. Wilson & Co., Toronto, Ont.

ARCHITECT, NAVAL
- Watts, J. Murray, Philadelphia, Pa.

ASBESTOS GOODS
- Wm. C. Wilson & Co., Toronto, Ont.

BABBITT METAL
- Aikenhead Hardware, Ltd., Toronto, Ont.
- Hoyt Metal Company, Toronto, Ontario.
- Tallman Brass & Metal Co., Hamilton, Ont.
- Wilkinson & Kompass, Hamilton, Ont.
- Wm. C. Wilson & Co., Toronto, Ont.

BAROMETERS
- Wilson & Co., Wm. C., Toronto, Canada.

BAR IRON
- Mitchells Ltd., Glasgow, Scotland.

BARS, GRATE
- Babcock & Wilcox, Ltd., Montreal, Que.

BEARINGS, BRASS
- Empire Mfg. Co., London, Ont.
- Mueller Mfg. Co., H., Sarnia, Ont.
- Mitchell Co., The Robert, Montreal, Que.
- Tallman Brass & Metal Co., Hamilton, Ont.
- United Brass & Lead Co., Toronto, Ont.

BELLS, SHIPS, ENGINE ROOM, ETC.
- Aikenhead Hardware, Ltd., Toronto, Ont.
- Clarke & Bros. Co., Montreal, Que.
- Cory & Son, Inc., Chas., New York, N.Y.
- Empire Mfg. Co., London, Ont.
- Mitchell Co., The Robert, Montreal, Que.
- Morrison Brass Mfg. Co., James, Toronto, Ont.
- Mueller Mfg. Co., H., Sarnia, Ont.
- Tallman Brass & Metal Co., Hamilton, Ont.
- United Brass & Lead, Ltd., Toronto, Ont.

BELTING, LEATHER
- Aikenhead Hardware, Ltd., Toronto, Ont.
- Wm. C. Wilson & Co., Toronto, Ont.

BIBBS, COMPRESSION
- Empire Mfg. Co., London, Ont.
- Mitchell Co., The Robert, Montreal, Que.
- Mueller Mfg. Co., H., Sarnia, Ont.
- United Brass & Lead, Ltd., Toronto, Ont.

BINNACLES
- Hopkins & Co., F. H., Montreal, Que.
- Morrison Brass Mfg. Co., James, Toronto, Ont.

BLOCKS, CARGO, HEEL AND TACKLE
- Aikenhead Hardware, Ltd., Toronto, Ont.
- Hopkins & Co., F. H., Montreal, Que.
- Loveridge, Ltd., Docks, Cardiff, Wales.

BLOWERS, TURBO-
- Mason Regulator & Engin. Co., Montreal, Que.

BOAT CHOCKS
- Corbet Fdry. & Machine Co., Owen Sound, Ont.
- Marten-Freeman Co., Toronto, Ont.

BOILER COMPOUND
- Wm. C. Wilson & Co., Toronto, Ont.

BOILER CIRCULATORS
- Eckliff Circulator Co., Detroit, Mich.

BOILER FEED PUMPS
- Can. Ingersoll-Rand Co., Ltd., Sherbrooke, Que.
- Goldie & McCulloch Co., Galt, Ont.
- Morris Mach. Works, Baldwinsville, N.Y.
- Smart-Turner Mach. Co., Hamilton, Ont.
- Williams Machinery Co., A. R., Toronto, Ont.

BOILER FITTINGS
- Empire Mfg. Co., London, Ont.
- Goldie & McCulloch Co., Galt, Ont.
- McAvity & Sons Ltd., T., St. John, N.B.
- Wager Furnace Bridge Wall Co., Inc., New York, N.Y.

BOILERS, MARINE
- Babcock & Wilcox, Ltd., Montreal, Que.
- Can. Fairbanks-Morse Co., Montreal, Que.
- Can. Vickers, Ltd., Montreal, Que.
- Collingwood Shipbuilding Co., Collingwood, Ont.
- Doxford & Sons, William, Sunderland, England.
- Engr. & Mach. Wks. of Can., St. Catharines, Ont.
- Goldie & McCulloch, Ltd., Galt, Ont.
- Hall Engineering Works, Montreal, Que.
- Mason Regulator & Engin. Co., Montreal, Que.
- Montreal Dry Docks & Shipbuilding Co., Montreal, Que.
- National Shipbuilding Co., Goderich, Ont.
- Marsh Engineering Works, Belleville, Ont.
- Polson Iron Works, Toronto, Ontario.
- Port Arthur Shipbuilding Co., Port Arthur, Ont.
- Waterous Engine Works Co., Brantford, Ont.
- Williams Machinery Co., A. R., Toronto, Ont.

BOOKS, TECHNICAL, MARINE
- MacLean Publishing Co., Toronto, Ont.

BOLTS
- London Bolt & Hinge Works, London, Ont.
- Mitchell Co., The Robert, Montreal, Que.
- United Brass & Lead, Ltd., Toronto, Ont.
- Wilkinson & Kompass, Hamilton, Ont.

BROKER
- Page & Jones, Mobile, Ala., U.S.A.

BRASS GOODS
- Corbet Fdry. & Machine Co., Owen Sound, Ont.
- Goldie & McCulloch Co., Galt, Ont.
- McAvity & Sons, T., St. John, N.B.
- Mitchell Co., The Robert, Montreal, Que.
- Mueller Mfg. Co., H., Sarnia, Ont.
- Tallman Brass & Metal Co., Hamilton, Ont.

BUCKETS, CLAMSHELL
- Beatty & Sons, Welland, Ont.
- Morris Crane & Hoist Co., Herbert, Niagara Falls, Ont.

BUCKETS, DUMP
- Morris Crane & Hoist Co., Herbert, Niagara Falls, Ont.

BUCKETS, COALING
- Beatty & Sons, Welland, Ont.
- Engr. & Mach. Wks. of Can., St. Catharines, Ont.
- Hopkins & Co., F. H., Montreal, Que.
- Marsh Engineering Works, Belleville, Ont.
- Reed & Co., Geo., Montreal, Que.

BUSHINGS, BRONZE
- Oberdorfer Brass Co., M. L., Syracuse, N.Y.

CABLE
- Hopkins & Co., F. H., Montreal, Que.
- McNab Co., Bridgeport, Conn.
- Wm. C. Wilson & Co., Toronto, Ont.

CABLE, LEAD COVERED AND ARMORED
- Standard Underground Cable Co., Hamilton, Ont.

CABLE, ACCESSORIES
- Standard Underground Cable Co., Hamilton, Ont.

CAPSTANS
- Advance Mach. & Welding Co., Montreal, Que.
- Dake Engine Co., Grand Haven, Mich.
- Hopkins & Co., F. H., Montreal, Que.
- Kennedy & Sons, Wm., Owen Sound, Ont.
- Wm. C. Wilson & Co., Toronto, Ont.

CALKING TOOLS, ELECTRIC
- Aikenhead Hardware, Ltd., Toronto, Ont.

CALKING TOOLS, PNEUMATIC
- Aikenhead Hardware, Ltd., Toronto, Ont.
- Can. Ingersoll-Rand Co. Sherbrooke, Que.

CALORIFIERS
- Low & Sons, Ltd., Archibald, Glasgow, Scotland

CASTINGS
- Beauchemin & Fils, Sorel, P.Q.
- Can. Steel Foundries, Ltd., Montreal, Que.
- Collingwood Shipbuilding Co., Collingwood, Ont.
- Wm. Hamilton Co., Peterboro, Ont.
- Kennedy & Sons, Wm., Owen Sound, Ont.
- Goldie & McCulloch Co., Galt, Ont.
- Marsh Engineering Works, Belleville, Ont.
- Mitchell Co., Ltd., Robt., Montreal, Que.
- Mueller Mfg. Co., H., Sarnia, Ont.
- Tallman Brass & Metal Co., Hamilton, Ont.
- United Brass & Lead, Ltd., Toronto, Ont.
- Waterous Engine Works Co., Brantford, Ont.

CASTINGS, ALLOY
- Mitchell Co., The Robert, Montreal, Que.
- Mueller Mfg. Co., H., Sarnia, Ont.
- Oberdorfer Brass Co., M. L., Syracuse, N.Y.
- Tolland Mfg. Co., Montreal, Que.
- Tallman Brass & Metal Co., Hamilton, Ont.
- United Brass & Lead, Ltd., Toronto, Ont.

CASTINGS, ALUMINUM
- Empire Mfg. Co., London, Ont.
- Mitchell Co., The Robert, Montreal, Que.
- Tallman Brass & Metal Co., Hamilton, Ont.
- United Brass & Lead Co., Toronto, Ont.

CASTINGS, BRASS
- Crouse-Hinds Co. of Canada, Ltd., Toronto, Ont.
- Empire Mfg. Co., London, Ont.
- Goldie & McCulloch Co., Galt, Ont.
- McAvity & Sons Ltd., T., St. John, N.B.
- McNab Co., Bridgeport, Conn.
- Mitchell Co., Ltd., Robt., Montreal, Que.
- Mueller Mfg. Co., H., Sarnia, Ont.
- Oberdorfer Brass Co., M. L., Syracuse, N.Y.
- Tolland Mfg. Co., Montreal, Que.
- Tallman Brass & Metal Co., Hamilton, Ont.
- United Brass & Lead Co., Toronto, Ont.
- Waterous Engine Works Co., Brantford, Ont.

CASTINGS, GREY IRON, MALLEABLE, ALUMINUM
- Crouse-Hinds Co. of Canada, Ltd., Toronto, Ont.
- Engr. & Mach. Wks. of Can., St. Catharines, Ont.
- McAvity & Sons, T., St. John, N.B.
- McNab Co., Bridgeport, Conn.
- Mitchell Co., The Robert, Montreal, Que.
- Mueller Mfg. Co., H., Sarnia, Ont.
- Waterous Engine Works Co., Brantford, Ont.

CASTINGS, MANGANESE STEEL

CASTINGS, MANGANESE BRONZE
- Oberdorfer Brass Co., M. L., Syracuse, N.Y.
- Tallman Brass & Metal Co., Hamilton, Ont.
- United Brass & Lead Co., Toronto, Ont.

CELLAR DRAINERS
- Mueller Mfg. Co., H., Sarnia, Ont.

CHAINS
- Aikenhead Hardware, Ltd., Toronto, Ont.
- Kearfott Engineering Co., New York, N.Y.
- Hopkins & Co., F. H., Montreal, Que.
- Morris Crane & Hoist Co., Herbert, Niagara Falls, Ont.
- Henry Rogers, Sons & Co., Wolverhampton, Eng.
- Wm. C. Wilson & Co., Toronto, Ont.

CHAIN BLOCKS AND SLINGS
- Morris Crane & Hoist Co., Herbert, Niagara Falls, Ont.

CHANDLERY, SHIP
- Beauchemin & Fils, Sorel, P.Q.
- Hopkins & Co., F. H., Montreal, Que.
- Leckie, Ltd., John, Toronto, Ont.

CLAMPS, STEAM AND WATER
- Mueller Mfg. Co., H., Sarnia, Ont.

CLOCKS
- American Steam Gauge & Valve Mfg. Co., Boston, Mass.
- Morrison Brass Mfg. Co., James, Toronto, Ont.
- Williams Machinery Co., A. R., Toronto, Ont.

CLOSETS
- Mueller Mfg. Co., H., Sarnia, Ont.
- United Brass & Lead, Ltd., Toronto, Ont.

COAL
- Nova Scotia Steel & Coal Co., New Glasgow, N.S.

COAL HANDLING MACHINERY
- Morris Crane & Hoist Co., Herbert, Niagara Falls, Ont.
- Waterous Engine Works Co., Brantford, Ont.

COCKS, BILGE, DISCHARGE, INDICATOR,
- McAvity & Sons Ltd., T., St. John, N.B.
- Mitchell Co., The Robert, Montreal, Que.
- Morrison Brass Mfg. Co., James, Toronto, Ont.

COCKS, BASIN
- Empire Mfg. Co., London, Ont.
- Mitchell Co., The Robert, Montreal, Que.
- United Brass & Lead, Ltd., Toronto, Ont.

COMPASSES
- Wm. C. Wilson & Co., Toronto, Ont.

COMPRESSORS, AIR
- Can. Fairbanks-Morse Co., Montreal, Que.
- Canadian Ingersoll-Rand Co., Sherbrooke, Que.
- Hopkins & Co., F. H., Montreal, Que.
- Smart-Turner Mach. Co., Hamilton, Ont.
- Williams Machinery Co., A. R., Toronto, Ont.

CONDENSERS
- Goldie & McCulloch Co., Galt, Ont.
- Kearfott Engineering Co., New York, N.Y.
- Morris Machine Works, Baldwinsville, N.Y.
- Smart-Turner Mach. Co., Hamilton, Ont.
- Weir Ltd., G. & J., Cathcart, Glasgow, Scotland.
- Williams Machinery Co., A. R., Toronto, Ont.

CONDULETS, MARINE
- Crouse-Hinds Co. of Canada, Ltd., Toronto, Ont.

CONTRACTORS' SUPPLIES
- McAvity & Sons, T., St. John, N.B.

CONSULTING ENGINEER
- Watt, J. Murray, Philadelphia, Pa.

CONVEYORS, ASH, COAL
- Babcock & Wilcox, Ltd., Montreal, Que.
- Hopkins & Co., F. H., Montreal, Que.

COPING MACHINES
- Bertram & Sons, Ltd., John, Dundas, Ont.

COPPER TUBES, SHEETS AND RODS
- Dom. Copper Products Co., Montreal, Que.
- Tallman Brass & Metal Co., Hamilton, Ont.

COTTON, CALKING
- Wilson & Co., Wm. C., Toronto, Canada.

COVERS, CANVAS, FOR HATCHES, LIFE-BOATS, ETC.
- Leckie, Ltd., John, Toronto, Ont.
- Waterous Engine Works Co., Brantford, Ont.

COUNTERS, REVOLUTION
- American Steam Gauge & Valve Mfg. Co., Boston, Mass.

CRANES
- Aikenhead Hardware, Ltd., Toronto, Ont.
- Can. Fairbanks-Morse Co., Montreal, Que.
- Hopkins & Co., F. H., Montreal, Que.
- Williams Machinery Co., A. R., Toronto, Ont.

CRANES, ELECTRIC
- Babcock & Wilcox, Ltd., Montreal, Que.
- Morris Crane & Hoist Co., Herbert, Niagara Falls, Ont.
- Smart-Turner Mach. Co., Hamilton, Ont.

CRANES, GOLIATH AND PNEUMATIC
- Morris Crane & Hoist Co., Herbert, Niagara Falls, Ont.

CRANES, GRANTRY, PORTABLE, JIB
- Morris Crane & Hoist Co., Herbert, Niagara Falls, Ont.
- Smart-Turner Mach. Co., Hamilton, Ont.

CRANES, OVERHEAD TRAVELLING
- Morris Crane & Hoist Co., Herbert, Niagara Falls, Ont.

CRANK SHAFTS
- Canada Foundries and Forgings, Welland, Ont.

DAVITS, BOAT
- Corbet Fdry. & Machine Co., Owen Sound, Ont.
- Hopkins & Co., F. H., Montreal, Que.
- Marten, Freeman Co., Toronto, Ont.
- Waterous Engine Works Co., Brantford, Ont.

DEAD LIGHTS, BRASS
Goldie & McCulloch Co., Galt, Ont.
Morris Crane & Hoist Co., Herbert, Niagara Falls, Ont.
Mitchell Co., Ltd., Robt., Montreal, Que.
DECK LIGHTS
Mitchell Co., The Robert, Montreal, Que.
Turnbull Elevator Mfg. Co., Toronto, Ont.
DECK PLUGS, ELECTRIC
Crouse-Hinds Co. of Canada, Ltd., Toronto, Ont.
Mitchell Co., The Robert, Montreal, Que.
DECKING FOR SHIPS
Marine Decking & Supply Co., Philadelphia, Pa.
DERRICKS
Aikenhead Hardware, Ltd., Toronto, Ont.
Dake Engine Co., Grand Haven, Mich.
Hopkins & Co., F. H., Montreal, Que.
Marsh Engineering Works, Belleville, Ont.
Morris Crane & Hoist Co., Herbert, Niagara Falls, Ont.
DISTILLERS
Kearfott Engineering Co., New York, N.Y.
DREDGES
Collingwood Shipbuilding Co., Collingwood, Ont.
Morris Mach. Works, Baldwinsville, N.Y.
Norbom Engineering Co., Philadelphia, Pa.
Polson Iron Works, Toronto.
DRILLS, AIR
Aikenhead Hardware, Ltd., Toronto, Ont.
Can. Ingersoll-Rand Co. Sherbrooke, Que.
Hopkins & Co., F. H., Montreal, Que.
DRILLS, CENTRE
Wilt Twist Drill Co., Walkerville, Ont.
DRILLS, BLACKSMITH AND BIT STOCK
Wilt Twist Drill Co., Walkerville, Ont.
DRILLS, ELECTRIC
Aikenhead Hardware, Ltd., Toronto, Ont.
Wilkinson & Kompass, Hamilton, Ont.
DRILLS, HIGH SPEED
Wilt Twist Drill Co., Walkerville, Ont.
Zenith Coal & Steel Products, Ltd., Montreal,
DRILLS, TRACK
Wilt Twist Drill Co., Walkerville, Ont.
DRILLS, RACHET AND HAND
Wilt Twist Drill Co., Walkerville, Ont.
DRILLS, TWIST
Aikenhead Hardware, Ltd., Toronto, Ont.
Williams Machinery Co., A. R., Toronto, Ont.
Wilt Twist Drill Co., Walkerville, Ont.
DRY DOCKS
Can. Vickers, Ltd., Montreal, Que.
Collingwood Shipbuilding Co., Collingwood, Ont.
Doxford & Sons, William, Sunderland, England.
Georgian Bay Shipbuilding & Wrecking Co., Midland, Ont.
National Shipbuilding Co., Goderich, Ont.
Polson Iron Works, Toronto, Ont.
Port Arthur Shipbuilding Co., Port Arthur, Ont.
Yarrows, Limited, Victoria, B.C.
ECONOMIZERS, FUEL
Babcock & Wilcox, Ltd., Montreal, Que.
EJECTORS
Empire Mfg. Co., London, Ont.
Mitchell Co., The Robert, Montreal, Que.
Morrison Brass Mfg. Co., James, Toronto, Ont.
Smart-Turner Mach. Co., Hamilton, Ont.
ELECTRIC LAMPS
Mitchell Co., The Robert, Montreal, Que.
Wm. C. Wilson & Co., Toronto, Ont.
ELECTRO-PLATING
Mitchell Co., The Robert, Montreal, Que.
Tallman Brass & Metal Co., Hamilton, Ont.
United Brass & Lead Co., Toronto, Ont.
ELECTRIC WELDING
Beauchemin & Fils, Sorel, P.Q.
ELEVATING MACHINERY
Goldie & McCulloch, Ltd., Galt, Ont.
Wm. Hamilton Co., Peterboro, Ont.
Morris Crane & Hoist Co., Herbert, Niagara Falls, Ont.
Waterous Engine Works Co., Brantford, Ont.
ELEVATORS
Turnbull Elevator Mfg. Co., Toronto, Ont.
ENGINES, HOISTING
Corbet Fdry. & Machine Co., Owen Sound, Ont.
Hopkins & Co., F. H., Montreal, Que.
Kennedy & Sons, Wm., Owen Sound, Ont.
Marsh Engineering Works, Belleville, Ont.
Port Arthur Shipbuilding Co., Port Arthur, Ont.
Williams Machinery Co., A. R., Toronto, Ont.
ENGINE, INTERNAL COMBUSTION
Doxford & Sons, William, Sunderland, England.
ENGINES, MARINE
Bolinders Co., New York, N.Y.
Can. Fairbanks-Morse Co., Montreal, Que.
Can. Vickers, Ltd., Montreal, Que.
Doxford & Sons, William, Sunderland, England.
Goldie & McCulloch, Ltd., Galt, Ont.
Hopkins & Co., F. H., Montreal, Que.
Iron Works, Ltd., Owen Sound, Ont.
Mason Regulator & Engin. Co., Montreal, Que.
Morris Mach. Works, Baldwinsville, N.Y.
National Shipbuilding Co., Goderich, Ont.
Norbom Engineering Co., Philadelphia, Pa.
Polson Iron Works, Toronto, Ont.
Port Arthur Shipbuilding Co., Port Arthur, Ont.
Trout Co., H. G., Buffalo, N.Y.
Waterous Engine Works Co., Brantford, Ont.
Williams Machinery Co., A. R., Toronto, Ont.
ENGINE STARTERS (AIR)
Scaife & Sons Co., Wm. B., Oakmont, Pa.
ENGINES, STEERING
Corbet Fdry. & Machine Co., Owen Sound, Ont.
Dake Engine Co., Grand Haven, Mich.
Kennedy & Sons, Wm., Owen Sound, Ont.
Wm. C. Wilson & Co., Toronto, Ont.
ENAMELWARE
Mueller Mfg. Co., H., Sarnia, Ont.
United Brass & Lead Co., Toronto, Ont
EVAPORATORS
Kearfott Engineering Co., New York, N.Y.
Mason Regulator & Engin. Co., Montreal, Que.
McNab Co., Bridgeport, Conn.
Weir Ltd., G. & J., Cathcart, Glasgow, Scotland.
EXTRACTORS, GREASE
American Steam Gauge & Valve Mfg. Co., Boston, Mass.
EYE BOLTS AND NUTS
Canada Foundries and Forgings, Welland, Ont.
Mitchell Co., The Robert, Montreal, Que.
United Brass & Lead, Ltd., Toronto, Ont.
FANS
Aikenhead Hardware Ltd., Toronto, Ont.
Empire Mfg. Co., London, Ont.
Reed & Co., Geo. W., Montreal, Que.
Smart-Turner Mach. Co., Hamilton, Ont.
Williams Machinery Co., A. R., Toronto, Ont.
FENDERS, ROPE
Hopkins & Co., F. H., Montreal, Que.
Leckie, Ltd., John, Toronto, Ont.
Wilson & Co., Wm. C., Toronto, Canada.
FERRO-MANGANESE
Mitchells, Ltd., Glasgow, Scotland.
FILES
Aikenhead Hardware, Ltd., Toronto, Ont.
Williams Machinery Co., A. R., Toronto, Ont.
FIRE BRICKS
Beveridge Paper Co., Montreal, Que.
Mitchells, Ltd., Glasgow, Scotland.
Williams Machinery Co., A. R., Toronto, Ont.
FILTERS, FEED WATER
MacKinnon Steel Co., Sherbrooke, Que.
Mason Regulator & Engin. Co., Montreal, Que.
FITTINGS, MARINE
Hopkins & Co., F. H., Montreal, Que.
McAvity & Sons, T., St. John, N.B.
Mitchell Co., The Robert, Montreal, Que.
United Brass & Lead, Ltd., Toronto, Ont.
FITTINGS, MOTOR BOAT
Empire Mfg. Co., London, Ont.
Mitchell Co., The Robert, Montreal, Que.
Mueller Mfg. Co., H., Sarnia, Ont.
United Brass & Lead, Ltd., Toronto, Ont.
FIXTURES, ELECTRIC
Cory & Son, Inc., Chas., New York, N.Y.
Crouse-Hinds Co. of Canada, Ltd., Toronto, Ont.
Mitchell Co., The Robert, Montreal, Que.
Tallman Brass & Metal Co., Hamilton, Ont.
FLAG POLES, STEEL
Dennis Wire & Iron Works Co., London, Ont.
FLOW METERS
Spray Engineering Co., Boston, Mass.
FLUE CLEANERS
Wm. C. Wilson & Co., Toronto, Ont.
FORGES
Aikenhead Hardware, Ltd., Toronto, Ont.
Hopkins & Co., F. H., Montreal, Que.
FLANGING AND EXPANDING MACHINES, PIPE
Lovekin Pipe Expanding & Flanging Mach. Co., Philadelphia, Pa.
FLOODLIGHTS, ELECTRIC
Crouse-Hinds Co. of Canada, Ltd., Toronto, Ont.
FLUORSPAR
Mitchells, Ltd., Glasgow, Scotland.
Scythes & Co., Toronto, Ont.
FORGINGS, ALL KINDS
Aikenhead Hardware, Ltd., Toronto, Ont.
Collingwood Shipbuilding Co., Collingwood, Ont.
Nova Scotia Steel & Coal Co., New Glasgow, N.S.
FORGINGS, STEEL AND IRON
Canada Foundries and Forgings, Welland, Ont.
FURNACE BRIDGE WALLS
Wager Furnace Bridge Wall Co., Inc., 149 Broadway, New York, N.Y.
GAUGES, RECORDING
American Steam Gauge & Valve Mfg. Co., Boston, Mass.
Empire Mfg. Co., London, Ont.
GASKETS
Wm. C. Wilson & Co., Toronto, Ont.
GAUGES COCKS
McAvity & Sons, T., St. John, N.B.
Mitchell Co., The Robert, Montreal, Que.
GAUGE GLASSES
Wm. C. Wilson & Co., Toronto, Ont.
GAUGES, WATER, PRESSURE, COMPOUND AND VACUUM
Aikenhead Hardware, Ltd., Toronto, Ont.
Babcock & Wilcox, Ltd., Montreal, Que.
Empire Mfg. Co., London, Ont.
McAvity & Sons, T., St. John, N.B.
McNab Co., Bridgeport, Conn.
Morrison Brass Mfg. Co., James, Toronto, Ont.
GENERATORS AND CONVERTORS
Can. Fairbanks-Morse Co., Montreal, Que.
GLAZING COMPOSITION
Kuhls, H. B. Fred., 6411 3rd Ave., Brooklyn, N.Y.
GOGGLES
Standard Optical Co., Geneva, N.Y.
GONGS
Clark & Bro., C. O., Montreal, Que.
Mitchell Co., Ltd., Robt., Montreal, Que.
GRAPHITE
Wm. C. Wilson & Co., Toronto, Ont.
GRATINGS
Can. Welding Works, Montreal, Que.
Corbet Fdry. & Machine Co., Owen Sound, Ont.
Canada Wire & Iron Goods Co., Hamilton, Ont.
MacKinnon Steel Co., Sherbrooke, Que.
GRINDERS, ELECTRIC
Wilkinson & Kompass, Hamilton, Ont.
GRINDERS, PNEUMATIC
Can. Ingersoll-Rand Co. Sherbrooke, Que.
GUARDS, MACHINERY
Dennis Wire & Iron Works Co., London, Ont.
GUY RODS AND ANCHORS, ELECTRIC
Crouse-Hinds Co. of Canada, Ltd., Toronto, Ont.
HAMMERS
Canada Foundries and Forgings, Ltd., Welland, Ont.
HARDWARE, MARINE
Hopkins & Co., F. H., Montreal, Que.
Mitchell Co., The Robert, Montreal, Que.
HEADLIGHTS, ELECTRIC
Crouse-Hinds Co. of Canada, Ltd., Toronto, Ont.
Hopkins & Co., F. H., Montreal, Que.
Tallman Brass & Metal Co., Hamilton, Ont.
HEATING EQUIPMENT
Empire Mfg. Co., London, Ont.
Low & Sons, Ltd., Archibald, Glasgow, Scotland
HEATERS, FEED, WATER
Babcock & Wilcox, Ltd., Montreal, Que.
Goldie & McCulloch Co., Galt, Ont.
Kearfott Engineering Co., New York, N.Y.
Mason Regulator & Engin. Co., Montreal, Que.
McNab Co., Bridgeport, Conn.
Weir Ltd., G. & J., Cathcart, Glasgow, Scotland.
HINGES
London Bolt & Hinge Works, London, Ont.
Mitchell Co., Ltd., Robt., Montreal, Que.
HOOKS
Hopkins & Co., F. H., Montreal, Que.
Morris Crane & Hoist Co., Herbert, Niagara Falls, Ont.
HOISTS, ASH
Beatty & Sons, Welland, Ont.
Marsh Engineering Works, Belleville, Ont.
St. Clair Bros., Galt, Ont.
Waterous Engine Works Co., Brantford, Ont.
HOIST BLOCKS
Morris Crane & Hoist Co., Herbert, Niagara Falls, Ont.
HOISTS, CHAIN
Aikenhead Hardware, Ltd., Toronto, Ont.
Can. Fairbanks-Morse Co., Montreal, Que.
Dake Engine Co., Grand Haven, Mich.
Hopkins & Co., F. H., Montreal, Que.
Morris Crane & Hoist Co., Herbert, Niagara Falls, Ont.
Williams Machinery Co., A. R., Toronto, Ont.
HOISTS, CARGO, MOVING, ETC.
Dake Engine Co., Grand Haven, Mich.
Hopkins & Co., F. H., Montreal, Que.
Marsh Engineering Works, Belleville, Ont.
Waterous Engine Works Co., Brantford, Ont.
HOISTING MACHINERY
Beatty & Sons, Welland, Ont.
Can. Ingersoll-Rand Co. Sherbrooke, Que.
Corbet Fdry. & Machine Co., Owen Sound, Ont.
Wm. Hamilton Co., Peterboro, Ont.
Marsh Engineering Works, Belleville, Ont.
Morris Crane & Hoist Co., Herbert, Niagara Falls, Ont.
Waterous Engine Works Co., Brantford, Ont.
Williams Machinery Co., A. R., Toronto, Ont.
HOSE
Wm. C. Wilson & Co., Toronto, Ont.
INDICATORS, ENGINE
American Steam Gauge & Valve Mfg. Co., Boston, Mass.
Cory & Son, Inc., Chas., New York, N.Y.
McNab Co., Bridgeport, Conn.
INDICATORS, SPEED
Aikenhead Hardware, Ltd., Toronto, Ont.
Cory & Son, Inc., Chas., New York, N.Y.
INJECTORS
Aikenhead Hardware, Ltd., Toronto, Ont.
Empire Mfg. Co., London, Ont.
Mitchell Co., The Robert, Montreal, Que.
Morrison Brass Mfg. Co., James, Toronto, Ont.
Williams Machinery Co., A. R., Toronto, Ont.
INGOTS
Broughton Copper Co., Ltd., Manchester, Eng.
INSULATORS, ELECTRIC
Crouse-Hinds Co. of Canada, Ltd., Toronto, Ont.
INSTRUMENTS, NAUTICAL
Leckie, Ltd., John, Toronto, Ont.
IRON AND STEEL
Mitchells Ltd., Glasgow, Scotland.
JACKS
Hopkins & Co., F. H., Montreal, Que.
Morris Crane & Hoist Co., Herbert, Niagara Falls, Ont.
JOINTS, BALL SLIP
Norbom Engineering Co., Philadelphia, Pa.
INSURANCE, MARINE
Toronto Insurance & Vessel Agency, Toronto, Ont.
JOURNAL WEDGES
Canada Foundries and Forgings, Ltd., Welland, Ont.
JUTE PACKING
Stratford Oakum Co., Geo., Jersey City, N.J.
LATHES
Bertram & Sons, Ltd., John, Dundas, Ont.
LEATHER, LACE
Wm. C. Wilson & Co., Toronto, Ont.
LENSES FOR GOGGLES
Standard Optical Co., Geneva, N.Y.
LIFEBUOYS, BELTS AND PRESERVERS
Fosbery Co., Barking, Essex, Eng.
Hopkins & Co., F. H., Montreal, Que.
Wm. C. Wilson & Co., Toronto, Ont.
LIFEBOATS
Hopkins & Co., F. H., Montreal, Que.
Wm. C. Wilson & Co., Toronto, Ont.
LIFE BOAT EQUIPMENT
Eckliff Circulator Co., Detroit, Mich.
Hopkins & Co., F. H., Montreal, Que.
Leckie, Ltd., John, Toronto, Ont.
Wm. C. Wilson & Co., Toronto, Ont.
LIFE JACKETS
Fosbery Co., Barking, Essex, Eng.
Hopkins & Co., F. H., Montreal, Que.
Leckie, Ltd., John, Toronto, Ont.
LIGHTS, ALL KINDS
Can. Fairbanks-Morse Co., Montreal, Que.
Crouse-Hinds Co. of Canada, Ltd., Toronto, Ont.
Hopkins & Co., F. H., Montreal, Que.
Mitchell Co., The Robert, Montreal, Que.
Morrison Brass Mfg. Co., James, Toronto, Ont.
Tallman Brass & Metal Co., Hamilton, Ont.
LIGHTS, SIDE, PORT
Goldie & McCulloch Co., Galt, Ont.
Crouse-Hinds Co. of Canada, Ltd., Toronto, Ont.
Hopkins & Co., F. N., Montreal, Que.
McAvity & Sons Ltd., T., St. John, N.B.
Mitchell Co., The Robert, Montreal, Que.
Tallman Brass & Metal Co., Hamilton, Ont.
Turnbull Elevator Mfg. Co., Toronto, Ont.
Wilson & Co., Wm. C., Toronto, Ont.
LIGHTING SETS
Can. Fairbanks-Morse Co., Montreal.
Cory & Son, Inc., Chas., New York, N.Y.
Crouse-Hinds Co. of Canada, Ltd., Toronto, Ont.
Goldie & McCulloch Co., Ltd., Galt, Ont.
Kearfott Engineering Co., New York, N.Y.
Williams Machinery Co., A. R., Toronto, Ont
Wilson & Co., Wm. C., Toronto, Canada.

LOADERS, WAGON AND TRUCK
Morris Crane & Hoist Co., Herbert, Niagara Falls, Ont.
Waterous Engine Works Co., Brantford, Ont.
LOCKERS, METAL
Dennis Wire & Iron Works Co., London, Ont.
Reed & Co., Geo. W., Montreal, Que.
LUBRICATORS
Empire Mfg. Co., London, Ont.
Mitchell Co., The Robert, Montreal, Que.
MACHINISTS
Corbet Fdry. & Machine Co., Owen Sound, Ont.
Wm. Hamilton Co., Peterboro, Ont.
Hyde Engineering Works, Montreal, Que.
Mitchell Co., The Robert, Montreal, Que.
United Brass & Lead Co., Toronto, Ont.
MANGANESITE, PASTE
Wm. C. Wilson & Co., Toronto, Ont.
MARINE CASTINGS
Wm. Hamilton Co., Peterboro, Ont.
Goldie & McCulloch Co., Galt, Ont.
Mitchell Co., Ltd., Robt., Montreal, Que.
McAvity & Sons, T., St. John, N.B.
Mueller Mfg. Co., H., Sarnia, Ont.
Tallman Brass & Metal Co., Hamilton, Ont.
United Brass & Lead, Ltd., Toronto, Ont.
Waterous Engine Works Co., Brantford, Ont.
MARINE SPECIALTIES
Goldie & McCulloch Co., Galt, Ont.
Hopkins & Co., F. H., Montreal, Que.
McAvity & Sons, T., St. John, N.B.
Mitchell Co., Ltd., Robt., Montreal, Que.
United Brass & Lead, Ltd., Toronto, Ont.
MOTORS
Aikenhead Hardware, Ltd., Toronto, Ont.
Can. Fairbanks-Morse Co., Montreal, Que.
Williams Machinery Co., A. R., Toronto, Ont.
Zenith Coal & Steel Products, Ltd., Montreal,
NAUTICAL INSTRUMENTS
Wm. C. Wilson & Co., Toronto, Ont.
NET LIFTERS
Dake Engine Co., Grand Haven, Mich.
Leckie, Ltd., John, Toronto, Ont.
NOZZLES, ALL KINDS
Empire Mfg. Co., London, Ont.
Mitchell Co., The Robert, Montreal, Que.
Tolland Mfg. Co., Boston, Mass.
NUTS
Mitchell Co., The Robert, Montreal, Que.
Wilkinson & Kompass, Hamilton, Ont.
United Brass & Lead, Ltd., Toronto, Ont.
OIL CUPS
Empire Mfg. Co., London, Ont.
Mitchell Co., Ltd., Robt., Montreal, Que.
OILCLOTHING
Wm. C. Wilson & Co., Toronto, Ont.
OIL, LINSEED
Canada Linseed Oil Mills, Montreal, Que.
OILS AND GREASE
Wm. C. Wilson & Co., Toronto, Ont.
OAKUM
Davey & Sons W O., Jersey City., N.J.
Leckie, Ltd., John, Toronto, Ont.
Stratford Oakum Co., Geo., Jersey City, N.J.
Wm. C. Wilson & Co., Toronto, Ont.
ORNAMENTAL IRONWORK
Canada Wire & Iron Goods Co., Hamilton, Ont.
Can. Welding Works, Montreal, Que.
OXY-ACETLYENE OUTFITS
Can. Fairbanks-Morse Co., Montreal, Que.
Can. Welding Works, Montreal, Que.
Hopkins & Co., F. H., Montreal, Que.
McAvity & Sons Ltd., T., St. John, N.B.
Williams Machinery Co., A. R., Toronto, Ont.
PAINT
Ault & Wiborg Co. of Can., Ltd., Toronto, Ont.
Wm. C. Wilson & Co., Toronto, Ont.
PAINTS, ANTI-FOULING
Kuhls, H. B. Fred., 6411 3rd Ave., Brooklyn, N.Y.
PAINTS, ANTI-CORROSIVE
Kuhls, H. B. Fred. 6411 3rd Ave., Brooklyn, N.Y.
PAINTING EQUIPMENT, AUTOMATIC
Spray Engineering Co., Boston, Mass.
PACKING, AMMONIA
France Packing Co., Philadelphia, Pa.
Wm. C. Wilson & Co., Toronto, Ont.
PACKING, MARINE
Aikenhead Hardware, Ltd., Toronto, Ont.
France Packing Co., Philadelphia, Pa.
Wm. C. Wilson & Co., Toronto, Ont.
PACKING, METALLIC
Aikenhead Hardware, Ltd., Toronto, Ont.
France Packing Co., Philadelphia, Pa.
PACKING, STEAM
Aikenhead Hardware, Ltd., Toronto, Ont.
France Packing Co., Philadelphia, Pa.
Wm. C. Wilson & Co., Toronto, Ont.
PANELBOARD AND CABINETS, ELECTRIC
Crouse-Hinds Co. of Canada, Ltd., Toronto, Ont
Preston Woodworking Machy. Co., Preston, Ont.
PATTERNS, WOOD AND METAL
Beauchemin & Fils, Sorel, P.Q.
PIPE, LAPWELD, CAST IRON, RIVETED
Empire Mfg. Co., London, Ont.
Norbom Engineering Co., Philadelphia, Pa.
United Brass & Lead, Ltd., Toronto, Ont.
PIPE RAILING, IRON AND BRASS
Dennis Wire & Iron Works Co., London, Ont.
Mitchell Co., The Robert, Montreal, Que.
PITCH
Reed & Co., Geo., Montreal, Que.
PITCH, PINE
Wilson & Co., Wm. C., Toronto, Canada.
PLANERS
Bertram & Sons, Ltd., John, Dundas, Ont.
PLANERS, STANDARD AND ROTARY
Yates Mach. Co., P. B., Hamilton, Ont.
PLATE CLAMPS
Morris Crane & Hoist Co., Herbert, Niagara Falls, Ont.
PLATE PUNCH TABLES
Norbom Engineering Co., Philadelphia, Pa.
PLUMBING EQUIPMENT
Low & Sons, Ltd., Archibald, Glasgow, Scotland.
Empire Mfg. Co., London, Ont.
Mitchell Co., The Robert, Montreal, Que.
Mueller Mfg. Co., H., Sarnia, Ont.
Tallman Brass & Metal Co., Hamilton, Ont.
Wm. C. Wilson & Co., Toronto, Ont.
United Brass & Lead, Ltd., Toronto, Ont.
PROPELLOR BLADES, BRONZE
Corbet Fdry. & Machine Co., Owen Sound, Ont.
Empire Mfg. Co., London, Ont.
Mitchell Co., The Robert, Montreal, Que.
Tallman Brass & Metal Co., Hamilton, Ont.
United Brass & Lead, Ltd., Toronto, Ont.
Yarrows, Limited, Victoria, B.C.
PROPELLER WHEELS
Goldie & McCulloch Co., Galt, Ont.
Kennedy & Sons, Wm., Owen Sound, Ont.
Trout Co., H. G., Buffalo, N.Y.
PIPING, STEAM
Babcock & Wilcox, Ltd., Montreal, Que.
PORT LIGHTS
Goldie & McCulloch Co., Galt, Ont.
Mitchell Co., Ltd., Robt., Montreal, Que.
Mueller Mfg. Co., H., Sarnia, Ont.
Turnbull Elevator Mfg. Co., Toronto, Ont.
Tallman Brass & Metal Co., Hamilton, Ont.
Wm. C. Wilson & Co., Toronto, Ont.
United Brass & Lead, Ltd., Toronto, Ont.
PROPELLER WHEELS
Beauchemin & Fils, Sorel, P.Q.
McNab Co., Bridgeport, Conn.
PULLEYS
Wm. Hamilton Co., Peterboro, Ont.
Smart-Turner Mach. Co., Hamilton, Ont.
Williams Machinery Co., A. R., Toronto, Ont.
PUMP, AIR
Can. Ingersoll-Rand Co. Sherbrooke, Que.
Smart-Turner Mach. Co., Hamilton, Ont.
Weir Ltd., G. & J., Cathcart, Glasgow, Scotland.
Williams Machinery Co., A. R., Toronto, Ont.
PUMPS
Can. Fairbanks-Morse Co., Montreal, Que.
Canada Foundries & Forgings, Welland, Ont.
Goldie & McCulloch, Ltd., Galt, Ont.
Hopkins & Co., F. H., Montreal, Que.
McAvity & Sons, T., St. John, N.B.
Oberdorfer Brass Co., M. L., Syracuse, N.Y.
Williams Machinery Co., A. R., Toronto Ont.
PUMPS, CENTRIFUGAL
Can. Ingersoll-Rand Co. Sherbrooke, Que.
Wm. Hamilton Co., Peterboro, Ont.
Goldie & McCulloch Co., Galt, Ont.
Kearfott Engineering Co., New York, N.Y.
Morris Mach. Works, Baldwinsville, N.Y.
Norbom Engineering Co., Philadelphia, Pa.
Smart-Turner Mach. Co., Hamilton, Ont.
Waterous Engine Works Co., Brantford, Ont.
Williams Machinery Co., A. R., Toronto, Ont.
PUMPS, BILGE
Goldie & McCulloch Co., Galt, Ont.
Mitchell Co., Ltd., Robt., Montreal, Que.
Morris Mach. Works, Baldwinsville, N.Y.
Mueller Mfg. Co., H., Sarnia, Ont.
PUMPS, CIRCULATING
Goldie & McCulloch Co., Galt, Ont.
McNab Co., Bridgeport, Conn.
Morris Machine Works, Baldwinsville, N.Y.
PUMPS, FEED WATER
Goldie & McCulloch Co., Galt, Ont.
McNab Co., Bridgeport, Conn.
Morris Mach. Works, Baldwinsville, N.Y.
Weir Ltd., G. & J., Cathcart, Glasgow, Scotland.
Williams Machinery Co., A. R., Toronto, Ont.
PUMPS, LIFT AND FORCE
Low & Sons, A., Glasgow, Scotland.
PUMPS, REES ROTURBO
Goldie & McCulloch Co., Galt, Ont.
PUMPS, HAND AND POWER
Aikenhead Hardware, Ltd., Toronto, Ont.
Mitchell Co., The Robert, Montreal, Que.
Smart-Turner Mach. Co., Hamilton, Ont.
Williams Machinery Co., A. R., Toronto, Ont.
PUMPS, HIGH PRESSURE
Canadian Ingersoll-Rand Co., Sherbrooke, Que.
Morris Mach. Works, Baldwinsville, N.Y.
Smart-Turner Mach. Co., Hamilton, Ont.
PUMPS, STEAM TURBINE
Morris Mach. Works, Baldwinsville, N.Y.
Wm. C. Wilson & Co., Toronto, Ont.
PUMPING MACHINES
Norbom Engineering Co., Philadelphia, Pa.
PUNCHES
Bertram & Sons, Ltd., John, Dundas, Ont.
PUNCHES, SINGLE, DOUBLE AND MULTIPLE
Norbom Engineering Co., Philadelphia, Pa.
PURIFIERS, WATER
Babcock & Wilcox, Ltd., Montreal, Que.
RADIATORS, STEAM, ELECTRIC
Empire Mfg. Co., London, Ont.
Low & Sons, Ltd., Archibald, Glasgow, Scotland.
Wm. C. Wilson & Co., Toronto, Ont.
RAILS, OVERHEAD
Morris Crane & Hoist Co., Herbert, Niagara Falls, Ont.
RANGES
Hopkins & Co., F. H., Montreal, Que.
Wm. C. Wilson & Co., Toronto, Ont.
RECEIVERS, AIR
Can. Ingersoll-Rand Co. Sherbrooke, Que.
Can. Welding Works, Montreal, Que.
MacKinnon Steel Co., Sherbrooke, Que.
REGULATORS, FEED WATER
American Steam Gauge & Valve Mfg. Co., Boston, Mass.
REGULATORS, PRESSURE
Mason Regulator & Engin. Co., Montreal, Que.
RELEASING GEARS
Eckliff Circulator Co., Detroit, Mich.
REPAIRS, MARINE
Corbet Foundry & Mach. Co., Owen Sound, Ont
Can. Vickers, Ltd., Montreal, Que.
Collingwood Shipbuilding Co., Collingwood, Ont.
Engr. & Mach. Wks. of Can., St. Catharines, Ont.
Georgian Bay Shipbuilding & Wrecking Co., Midland, Ont.
Hyde Engineering Works, Montreal, Que.
Iron Works, Ltd., Owen Sound, Ont.
Kennedy & Sons, Wm., Owen Sound, Ont.
Muir Bros. Dry Dock Co., Port Dalhousie, Ont.
National Shipbuilding Co., Goderich, Ont.
Port Arthur Shipbuilding Co., Port Arthur, Ont.
Yarrow, Limited, Victoria, B.C.
RIGGING SCREWS
Hopkins & Co., F. H., Montreal, Que.
RIGGING, WIRE ROPE
Leckie, Ltd., John, Toronto, Ont.
RIVETS
London Bolt & Hinge Works, London, Ont.
Severance Mfg. Co., S., Glassport, Pa.
Wilkinson & Kompass, Hamilton, Ont.
RIVETERS, PNEUMATIC
Can. Ingersoll-Rand Co., Ltd., Sherbrooke, Que.
RODS, COPPER, BRASS, BRONZE
Standard Underground Cable Co., Hamilton, Ont.
Tallman Brass & Metal Co., Hamilton, Ont.
ROLLS, STRAIGHTENING, BENDING
Bertram & Sons, Ltd., John, Dundas, Ont.
ROOFING
Reed & Co., Geo., Montreal, Que.
Wm. C. Wilson & Co., Toronto, Ont.
ROPE BLOCKS
Aikenhead Hardware, Ltd., Toronto, Ont.
Can. Fairbanks-Morse Co., Montreal, Que.
Morris Crane & Hoist Co., Herbert, Niagara Falls, Ont.
ROPE
Hopkins & Co., F. H., Montreal, Que.
Leckie, Ltd., John, Toronto, Ont.
McNab Co., Bridgeport, Conn.
Stratford Oakum Co., Geo., Jersey City, N.J.
Wm. C. Wilson & Co., Toronto, Ont.
RUBBER COATS
Wm. C. Wilson & Co., Toronto, Ont.
SAW MILL MACHINERY
Wm. Hamilton Co., Peterboro, Ont.
Preston Woodworking Machy. Co., Preston, Ont.
Waterous Engine Works Co., Brantford, Ont.
Yates Machine Co., P. B., Hamilton, Ont.
SAWS, BAND
Preston Woodworking Machy. Co., Preston, Ont.
SCALES, BOILERS, ENGINES
Can. Fairbanks-Morse Co., Montreal, Que.
SCOWS
Collingwood Shipbuilding Co., Collingwood, Ont.
Polson Iron Works, Toronto, Ont.
SCREENS, WIRE
Dennis Wire & Iron Works Co., London, Ont.
SCREWS, COACH
London Bolt & Hinge Works, London, Ont.
SEAM PAINT
Kuhls, H. B. Fred., 6411 3rd Ave., Brooklyn, N.Y.
SEPARATORS, OIL, STEAM
Mason Regulator & Engin. Co., Montreal, Que.
Smart-Turner Mach. Co., Hamilton, Ont.
SHAFTING
Wm. Hamilton Co., Peterboro, Ont.
Mitchell Co., The Robert, Montreal, Que.
Wilkinson & Kompass, Hamilton, Ont.
SHAFTING, BRONZE
Empire Mfg. Co., London, Ont.
Tallman Brass & Metal Co., Hamilton, Ont.
SHEARS
Bertram & Sons, Ltd., John, Dundas, Ont.
Norbom Engineering Co., Philadelphia, Pa.
SHIPBUILDING TOOLS
Aikenhead Hardware, Ltd., Toronto, Ont.
Can. Ingersoll-Rand Co. Sherbrooke, Que.
SHIPS, BUILDERS OF
Can. Vickers, Ltd., Montreal, Que.
Collingwood Shipbuilding Co., Collingwood, Ont.
Doxford & Sons, William, Sunderland, England.
Georgian Bay Shipbuilding & Wrecking Co., Midland, Ont.
National Shipbuilding Co., Goderich, Ont.
Polson Iron Works, Toronto, Ont.
Port Arthur Shipbuilding Co., Port Arthur, Ont.
Yarrows, Limited, Victoria, B.C.
SHIP BROKERS
Page & Jones, Mobile, Ala.
SHIP PLATES
Nova Scotia Steel & Coal Co., New Glasgow, N.S.
SLEDGES
Wilkinson & Kompass, Hamilton, Ont.
SLINGS
Hopkins & Co., F. H., Montreal, Que.
Morris Crane & Hoist Co., Herbert, Niagara Falls, Ont.
SPECIAL MACHINERY
Corbet Fdry. & Machine Co., Owen Sound, Ont.
Wm. Hamilton Co., Peterboro, Ont.
Miller Bros. & Sons, Ltd., Montreal, Que.
Smart-Turner Mach. Co., Hamilton, Ont.
SMOOTH-ON
Wm. C. Wilson & Co., Toronto, Ont.
SPIKES
Wm. C. Wilson & Co., Toronto, Ont.
SPRAY COOLING SYSTEMS
Spray Engineering Co., Boston, Mass.
STEAMSHIP AGENTS
Page & Jones, Mobile, Ala.
STEAM SPECIALTIES
Can. Fairbanks-Morse Co., Montreal, Que.
Corbet Fdry. & Machine Co., Owen Sound, Ont.
Empire Mfg. Co., London, Ont.
Mitchell Co., The Robert, Montreal, Que.
STEAM TRAPS
Aikenhead Hardware, Ltd., Toronto, Ont.
American Steam Gauge & Valve Mfg. Co., Boston, Mass.
Empire Mfg. Co., London, Ont.
Mason Regulator & Engin. Co., Montreal, Que.
Mitchell Co., The Robert, Montreal, Que.
Smart-Turner Mach. Co., Hamilton, Ont.
STEEL, HIGH SPEED
Hopkins & Co., F. H., Montreal, Que.
Nova Scotia Steel & Coal Co., New Glasgow, N.S.
Wilkinson & Kompass, Hamilton, Ont.
Zenith Coal & Steel Products, Ltd., Montreal.

STEEL SHELVING
Dennis Wire & Iron Works, London, Ont.

STEEL WORK, STRUCTURAL
Babcock & Wilcox, Ltd., Montreal, Que.
Can. Welding Works, Montreal, Que.
Corbet Fdry. & Machine Co., Owen Sound, Ont.
Wm. Hamilton Co., Peterboro, Ont.
MacKinnon Steel Co., Sherbrooke, Que.

STEERING GEARS
Corbet Fdry. & Machine Co., Owen Sound, Ont.
Hopkins & Co., F. H., Montreal, Que.
Engr. & Mach. Wks. of Can., St. Catharines, Ont.
Wm. C. Wilson & Co., Toronto, Ont.

STOCK RACKS FOR BARS, PIPING, ETC.
Mitchell Co., The Robert, Montreal, Que.
Morris Crane & Hoist Co., Herbert, Niagara Falls, Ont.

STOKERS, MECHANICAL
Babcock & Wilcox, Ltd., Montreal, Que.

SUPERHEATERS, STEAM
Babcock & Wilcox, Ltd., Montreal, Que.
Goldie & McCulloch Co., Galt, Ont.

SWITCHBOARDS, ELECTRIC
Crouse-Hinds Co. of Can., Ltd., Toronto, Ont.

TALLOW
Can. Economic Lubricant Co., Montreal, Que.
Wm. C. Wilson & Co., Toronto, Ont.

TANKS, STEEL
Can. Welding Works, Montreal, Que.
Corbet Foundry & Mach. Co., Owen Sound, Ont.
Goldie & McCulloch, Ltd., Galt, Ont.
Hopkins & Co., F. H., Montreal, Que.
MacKinnon Steel Co., Sherbrooke, Que.
Marsh Engineering Works, Belleville, Ont.
Port Arthur Shipbuilding Co., Port Arthur, Ont.
Reed & Co., Geo., Montreal, Que.
Scaife & Sons Co., Wm. B., Oakmont, Pa.

TANKS (AIR, GAS AND LIQUID)
Can. Welding Works, Montreal, Que.
MacKinnon Steel Co., Sherbrooke, Que.
Marsh Engineering Works, Belleville, Ont.
Scaife & Sons Co., Wm. B., Oakmont, Pa.

TAPPING MACHINES
Mueller Mfg. Co., H., Sarnia, Ont.

TELEGRAPHS, SHIPS
McNab Co., Bridgeport, Conn.
Cory & Sons, Inc., Chas., New York, N.Y.
Morrison Brass Mfg. Co., James, Toronto, Ont.
Wilson & Co., Wm., Toronto, Ont.

TELEPHONES, MARINE
Cory & Son, Inc., Chas., New York, N.Y.
McNab Co., Bridgeport, Conn.

TESTERS, METER
Mueller Mfg. Co., H., Sarnia, Ont.

THUMB SCREWS AND NUTS
Canada Foundries & Forgings, Welland, Ont.
Mitchell Co., The Robert, Montreal, Que.
United Brass & Lead, Ltd., Toronto, Ont.

TRACK SYSTEMS
Morris Crane & Hoist Co., Herbert, Niagara Falls, Ont.

TRAVELLING BLOCKS
Morris Crane & Hoist Co., Herbert, Niagara Falls, Ont.

TROLLEYS
Can. Fairbanks-Morse Co., Montreal, Que.
Morris Crane & Hoist Co., Herbert, Niagara Falls, Ont.

TROLLEY HOISTS
Morris Crane & Hoist Co., Herbert, Niagara Falls, Ont.

TRUCKS, HAND, ELECTRIC
Aikenhead Hardware, Ltd., Toronto, Ont.
Can. Fairbanks-Morse Co., Montreal, Que.

TUBES, BOILER
Babcock & Wilcox, Ltd., Montreal, Que.
Broughton Copper Co., Ltd., Manchester, Eng.
Goldie & McCulloch Co., Galt, Ont.

TUBES, COPPER AND BRASS
Mueller Mfg. Co., H., Sarnia, Ont.
Tallman Brass & Metal Co., Hamilton, Ont.
Standard Underground Cable Co., Hamilton, Ont.

TUGS
Polson Iron Works, Toronto, Ont.
Collingwood Shipbuilding Co., Collingwood, Ont.

TURBINES, STEAM
Goldie & McCulloch Co., Galt, Ont.

TURBINES, DIRECT-DRIVING AND GEARED
Doxford & Sons, William, Sunderland, England.

TURNBUCKLES
Canada Foundries & Forgings, Welland, Ont.
Hopkins & Co., F. H., Montreal, Que.

TURNTABLES
Morris Crane & Hoist Co., Herbert, Niagara Falls, Ont.

SPIKES, SMALL RAILROAD
Severance Mfg. Co., S., Glassport, Pa.

UNIONS, ALL KINDS
Dart Union Company, Toronto, Ont.

VALVES, AIR
Mitchell Co., The Robert, Montreal, Que.
Mueller Mfg. Co., H., Sarnia, Ont.

VALVE, DISCS
Wm. C. Wilson & Co., Toronto, Ont.

VALVES
American Steam Gauge & Valve Mfg. Co., Boston, Mass.
Babcock & Wilcox, Ltd., Montreal, Que.
Can. Fairbanks-Morse Co., Montreal, Que.
Empire Mfg. Co., London, Ont.
McAvity & Sons, Ltd., T., St. John, N.B.
Mason Regulator & Engin. Co., Montreal, Que.
McNab Co., Bridgeport, Conn.
Norbom Engineering Co., Philadelphia, Pa.
Williams Machinery Co., A. R., Toronto, Ont.
Wm. C. Wilson & Co., Toronto, Ont.
United Brass & Lead, Ltd., Toronto, Ont.

VALVES, FOOT
Aikenhead Hardware, Ltd., Toronto, Ont.
Mitchell Co., The Robert, Montreal, Que.
Smart-Turner Mach. Co., Hamilton, Ont.

VALVES, STOP, REDUCING, SAFETY CHECK, ETC.
Aikenhead Hardware, Ltd., Toronto, Ont.
American Steam Gauge & Valve Mfg. Co., Boston, Mass.
McAvity & Sons Ltd., T., St. John, N.B.
Mitchell Co., The Robert, Montreal, Que.
Morrison Brass Mfg. Co., James, Toronto, Ont.

VALVES, MIXING
Mitchell Co., The Robert, Montreal, Que.
Mueller Mfg. Co., H., Sarnia, Ont.

VALVES, REDUCING, PRESSURE
Mitchell Co., The Robert, Montreal, Que.
Mueller Mfg. Co., H., Sarnia, Ont.

VALVES, STORM
Low & Sons, A., Glasgow, Scotland.

VARNISHES
Aikenhead Hardware, Ltd., Toronto, Ont.
Ault & Wiborg Co. of Can., Ltd., Toronto, Ont.
Leckie, Ltd., John, Toronto, Ont.
Reed & Co., Geo., Montreal, Que.

VENTILATORS, COWL
Can. Welding Works, Montreal, Que.
McNab Co., Bridgeport, Conn.
Mitchell Co., The Robert, Montreal, Que.

VENTILATION EQUIPMENT
Empire Mfg. Co., London, Ont.
Hopkins & Co., F. H., Montreal, Que.
Low & Sons, Ltd., Archibald, Glasgow, Scotland

WASHERS
London Bolt & Hinge Works, London, Ont.
Wm. C. Wilson & Co., Toronto, Ont.

WATER COLUMNS
Mitchell Co., The Robert, Montreal, Que.
Morrison Brass Mfg. Co., James, Toronto, Ont.

WELDING, ELECTRIC
Can. Welding Works, Montreal, Que.
Hall Engineering Works, Montreal, Que.
Marine Welding Co., Buffalo, N.Y.
Beatty & Sons, M., Welland, Ont.

WATER SOFTENERS
Babcock & Wilcox, Ltd., Montreal, Que.

WATER SUPPLY SYSTEMS
Mueller Mfg. Co., H., Sarnia, Ont.

WATER HEATERS
Empire Mfg. Co., London, Ont.
Morrison Brass Mfg. Co., James, Toronto, Ont.

WHISTLES AND SYRENS
Empire Mfg. Co., London, Ont.
McAvity & Sons, T., St. John, N.B.
McNab Co., Bridgeport, Conn.
Mitchell Co., Ltd., Robt., Montreal, Que.
Morrison Brass Mfg. Co., Jas., Toronto, Ont.

WINCHES, CARGO
Aikenhead Hardware, Ltd., Toronto, Ont.
Corbet Fdry. & Machine Co., Owen Sound, Ont.
Hopkins & Co., F. H., Montreal, Que.
Marsh Engineering Works, Belleville, Ont.

WINCHES, DOCK, SHIP
Beatty & Sons, M., Welland, Ont.
Marsh Engineering Works, Belleville, Ont.
Miller Bros. & Sons, Ltd., Montreal, Que.
Morris Crane & Hoist Co., Herbert, Niagara Falls, Ont.
Wilson & Co., Wm. C., Toronto, Canada.

WINCHES, TOWING
Corbet Foundry & Mach. Co., Owen Sound, Ont.

WINCHES, TRAWL
Beatty & Sons, M., Welland, Ont.
Wm. C. Wilson & Co., Toronto, Ont.

WINDLASSES
Corbet Fdry. & Machine Co., Owen Sound, Ont.
Dake Engine Co., Grand Haven, Mich.
Hopkins & Co., F. H., Montreal, Que.
Wilson & Co., Wm. C., Toronto, Canada.

WIPER CAPS, OILER BOXES, ETC.
Mitchell Co., The Robert, Montreal, Que.
Morrison Brass Mfg. Co., James, Toronto, Ont.

WIRE, COPPER CLAD STEEL
Standard Underground Cable Co., Hamilton, Ont.

WIRE, COPPER, BRASS, BRONZE
Standard Underground Cable Co., Hamilton, Ont.
Tallman Brass & Metal Co., Hamilton, Ont.

WIRE, INSULATED
Standard Underground Cable Co., Hamilton, Ont.

WIRELESS OUTFITS
Marconi Wireless Telegraph Co., Montreal, Que.

WIRE ROPE
Zenith Coal & Steel Products, Ltd., Montreal.

WOOD WORKING MACHINERY
Aikenhead Hardware, Ltd., Toronto, Ont.
Can. Fairbanks-Morse Co., Montreal, Que.
Preston Woodworking Machy. Co., Preston, Ont.
Yates Mach. Co., P. B., Hamilton, Ont.

WOOD BORING TOOLS
Aikenhead Hardware, Ltd., Toronto, Ont.
Can. Ingersoll-Rand Co. Sherbrooke, Que.

WOODITE GAUGE GLASS WASHERS
Wm. C. Wilson & Co., Toronto, Ont.

WRENCHES
Canada Foundries & Forgings, Welland, Ont.

YACHT BROKER
Watt, J. Murray, Philadelphia, Pa.

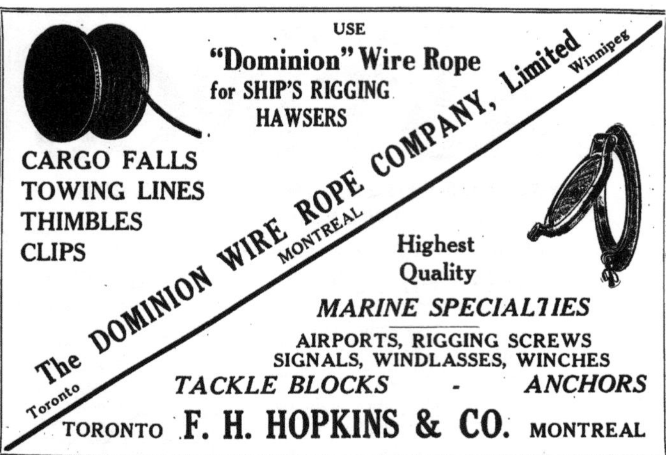

INDEX TO ADVERTISERS

CIRCULATES IN EVERY PROVINCE OF CANADA AND ABROAD

MARINE ENGINEERING of Canada

A monthly journal dealing with the progress and development of Merchant and Naval Marine Engineering, Shipbuilding, the building of Harbors and Docks, and containing a record of the latest and best practice throughout the Sea-going World. Published by The MacLean Publishing Co., Limited

MONTREAL, Southam Building TORONTO 143-153 University Ave. WINNIPEG, 1207 Union Trust Bldg. LONDON, ENG., 88 Fleet St.

Vol. VIII. Publication Office, Toronto—September, 1918 No. 9

If what you need is not advertised, consult our Buyers' Directory and write advertisers listed under proper heading.

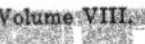

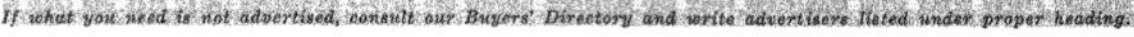

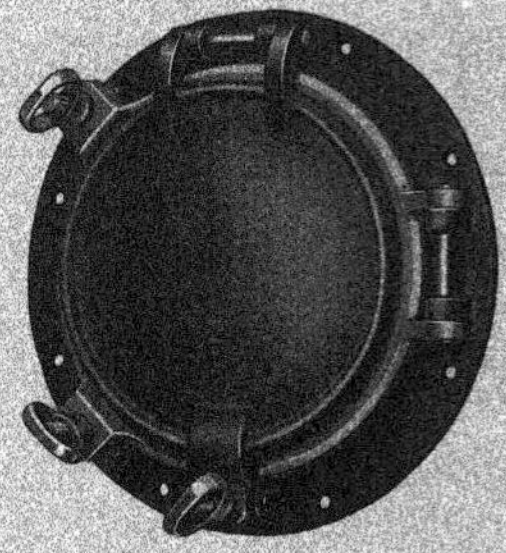

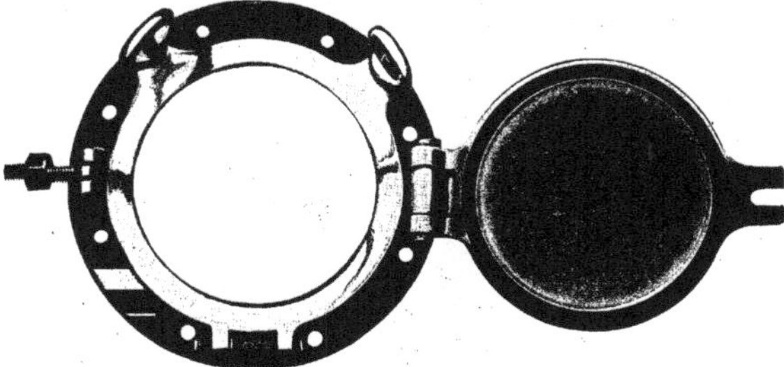

If interested, tear out this page and place with letters to be answered.

MARINE ENGINEERING
of Canada

Volume VIII. TORONTO, SEPTEMBER, 1918 Number 9

Can Canada Supply The Crews For Her Merchant Marine?

By T. H. FENNER, Associate Editor.

NOW that Canada has definitely embarked on a programme of ship construction, with the idea of creating a national mercantile marine, the question of furnishing the necessary personnel at once becomes of interest.

It must be borne in mind that a great many of the men who are now joining the mercantile marine and naval reserves are doing so out of the desire to help their country in a time of stress, and very few of these men are likely to continue a seafaring life after the present crisis is past. Where, then, is the steady supply of the right kind of men to be found to maintain our growing fleet and conduct it in safety over the seven seas, to develop Canada's foreign trade to the highest possible degree.

It is well to first look into the subject and see the type of man that is required in the various departments of a modern ship, and in studying this question we will consider the ordinary cargo carrier, or tramp, of say 5,000 tons d.w., as this is the type of ship that carries the bulk of the world's trade and constitutes the backbone of the mercantile marine. Furthermore, the men who man this type are very often they who, later on in life, are found on the bridge or in the engine rooms of the fast passenger liner, for the qualities called for in the successful navigation of the tramp are equally useful in any other class of vessel.

The vessel under consideration will generally carry 5,000 tons of cargo, with an indicated horse power of 1,200 to 1,500, on a coal consumption of 20 to 25 tons per 24 hours, the speed averaging from 9 to 10 knots when vessel is clean. It is not uncommon for these vessels to be away from their home port from 1½ to 3 years, a point to be borne in mind when considering the men who work them. Her crew will consist of from 28 to 34 all told, depending on the owners. This crew is composed of (not counting in the captain, who is a being apart) 3 officers, 1 boatswain, 1 carpenter, 1 steward, 6 sailors, and a boy, in the deck department. The engine room staff consists of four engineers, storekeeper, donkey man, and one greaser, six firemen, and two trimmers. There will also be two boys in the steward's department, one of these exclusively for the engineers' service.

In the past, the deck officer, which term includes the 1st, 2nd, and 3rd mates, graduated to the bridge in either one of two ways. He "came in through the hawse pipe," which is the nautical way of saying that he had sailed as boy and ordinary seaman in sailing ships for four years, then passing for a second officer's certificate. This type of ship's officer is rapidly disappearing. He was a first class seaman, generally, and possessed of many admirable qualities, but as a rule a man of little or no education outside of his immediate needs.

The Other Way

The alternative, and what is now the more usual way, is for the prospective officer to serve an apprenticeship of four years, in steam or sail, before passing for second officer. In some cases, the training takes place on special ships, where the boy gets a general as well as a nautical education, and where the parents pay a more or less heavy fee for the privilege. There are also ships maintained by the government for boys who have not the means to serve on the private training ships, and there are again training establishments, afloat and ashore, maintained by the Seamen's Orphanage Society. All this tends to produce a type of ship's officer who is not only a good navigator, a good seaman, with a fairly extensive knowledge of ship's business, but also an officer with a considerable amount of polish. When it is remembered that the ship's captain and officers very often are the only representatives of their country in a foreign port, the advantage of having good appearance and manners is obvious.

The Engineer Officers

The engineer officers are of another type, in some ways. First of all they are not purely seafaring men, although many of them become so by remaining the greater part of their lives at sea. But the important difference is this: the deck officer's training is intimately connected with ships and seagoing matters from the very first, whereas the engineer officer, in many cases, is absolutely ignorant of ships till he makes his first voyage to sea. The British B. O. T. requires that before a candidate can sit for a second class engineer's certificate he must prove that he has served an apprenticeship of 5 years at the making and repairing of marine engines and boilers, three years at least of which must be served in the shop. The rest may be served in drawing office or at pattern making. He must then serve for 12 months in an ocean going steamer in charge of a watch, after which he is eligible for examination. It is usual for the apprentice to follow, during his five years, a systematic course of evening classes devoted to the theory of mechanical engineering, so that when he steps aboard his ship with the rank of fourth engineer to make his initial voyage, he is generally a fairly good mechanic, with a good grounding in technical knowledge, and a woeful ignorance of the seagoing part of marine engineering, which, however, he soon acquires more or less painfully.

The Duties of the Men

Having outlined the type of man required in each of the principal departments, let us look further into the nature of their duties and their prospects, with a view to finding out what there is about a career in the mercantile marine to attract the youth of Canada. It must not be forgotten that in the British Islands a large percentage of the youthful population have a natural inclination to go to sea, with a fine disregard for future prospects, or for any consideration but the adventures to be met with This is not to be wondered at when one considers how much the life of Great Britain is bound up with ships and shipping.

The "Old Man"

To begin at the top with the captain, or, as he is usually called, the "Old Man." Before obtaining an appointment as captain, he has been sailing, in steam and sail, or steam only, as the case may be, for probably 15 years. He has risen slowly through the various grades of 3rd mate, second mate, and first mate, acquiring skill and experience and much diverse knowledge till at last he stands upon the bridge, supreme arbiter of his little world. To him is entrusted a ship and cargo valued at thousands of pounds, and also many human lives. On his ability to make a safe and speedy passage from port to port, to handle the cargo, or get the best out of longshoremen, depends very often the profit or loss of a voyage, and in some cases the chartering of the vessel from some outlying port is in the skipper's hands. He must possess quite an amount of knowledge connected with ships' business, be an authority on the Mercantile Marine Act, do duty as physician and surgeon, and generally look after the welfare of his little kingdom. If he is a wise and firm disciplinarian he will have a smart and happy ship. If he is weak, his weakness will extend right down to the trimmer in the bunkers. He is a being apart from every other on board, even to the extent of not signing the ship's articles of agreement as a seaman,

TYPICAL HIGH-CLASS CARGO STEAMER

which is done by every other member of the crew. The average salary for a captain of the class of vessel we are considering was, in pre-war days, from £25 to £30 per month, or from $125 to $150 per month, all found, with frequently a bonus for good voyages, or freedom from accident. This salary has increased since the war, and it is highly probable that it will remain at a higher rate after the war. I am taking figures at the British rates, because the bulk of these freight carriers are British, and Canadian ships after the war will have to compete with British and foreign ships for their trade, and with the exception of the U. S. A., other nations' ships are manned cheaper than British.

Everything Found

The salary itself is not so small as would appear to the average Canadian reader at the first glance. The captain of a ship is supplied with everything necessary to his comfort, and he need not spend any of his salary for his living expenses. Then his salary can be augmented in various ways, legitimate, and otherwise, though it is the exception for a ship's captain to resort to the latter. He holds an honorable and definite position in the eyes of his fellow men, and very often retires from the sea when comparatively young, to a business career on shore, very often as ship's surveyor, average adjustor, ship store dealer, etc., or marine superintendent of a steamship line.

The chief officer's duties consist in assisting in the navigation of the vessel, and in being the general executive of the ship. To him the bosun reports daily, and receives his directions as to the work to be performed around the decks, standing rigging, etc. The upkeep of the ship above decks and in the cargo holds is under his general supervision, and the skipper's orders are addressed directly to him for carrying out. The chief officer always holds a master's certificate and must be capable of stepping into the master's position at any minute.

The Second Officer

The second officer is very often called the navigating officer of the vessel, chiefly because he takes the sights at noon with the skipper and works out the ship's position. This is, of course, worked out by the captain and 1st officer also, and results checked. Besides keeping his watch and navigating duties the 2nd mate has certain parts of the standing and running gear in his charge, and in port is in charge of the loading and unloading of the after holds. It may be said here that in most steamers three mates are carried, and the watches are divided between them, making 4 hours on duty and 8 off. During the 8 hours off there are usually odds and ends to attend to, occupying some of the time. The second officer quite often holds a master's certificate, and always a first officer's.

The third officer is the junior. He usually takes the bridge from 8 to 12, morning and evening, when the skipper is usually around, and one or other of the senior officers within hail. To him is relegated the duty of looking after the ship's signal flags and various other minor necessaries, while in port he takes charge of the fore hold. He usually holds a second officer's certificate, and this is the first step on his way to a command, after his probating period is over. Officers' salaries vary in different companies, but the average peace time rate was from £14 to £18 per month for 1st officer, from £10 to £14 for second officer, and from £8 to £20 for third officer. Canadian rates of pay in the coastal service approximated to these rates as about 1 1-3 or 1½ to 1.

The Engineering Department

Turning now to the engineer's department, which every engineer knows is the most important, we find a little society, strictly self-contained, and very jealous of its prerogatives. It must be explained that with the introduction of steam and the accompanying invasion of the engineer a feeling of resentment was born in the breast of the old time sailor against "these interlopers," as he called the engineer. The engineer, on his side, felt that he was a much superior being to the old "shell backs," and it is only within very recent years that they have begun to learn that they are mutually dependent on each other, and to get together accordingly. Many amusing, and many unpleasant, experiences resulted therefrom. The writer remembers well, on one occasion, where the chief engineer and the captain had rather strained relations. The chief ordered all the steam heaters shut off the saloon and officers' and engineers' rooms. This was in New York in the month of December, and it was far from being a joke. The chief explained to the captain that the heaters were taking too much water from the donkey boiler, and if they were continued one of the winches would have to be stopped. The skipper perforce had to take his word for it, and the worst part of the whole thing lay in the fact that the skipper could go ashore all day and be comfortable, but the officers and engineers all suffered to gratify the chief's personal spite.

TRIPLE EXPANSION MARINE ENGINE IN ERECTING SHOP.

Next To The Captain

The chief engineer ranks next to the captain, and is as supreme below decks as the captain is above. He is consulted by the captain on all questions of coaling, tanking fresh water, speed at which to make the run from port to port, and many other details connected with the commercial running of the vessel. He has direct responsibility over all the machinery in the vessel, on deck as well as in the engine room. On him depends, in the last resort, the ability of the ship to perform her duties, for the most skilful navigation avails nothing if the motive power is idle. On him and his staff depends the ability to load and discharge cargo rapidly, which means short stays in port. This is dependent on the winches being in shape to stand up to their work. He must know to within a very few tons how much coal it will take to get from port to port and how much is the minimum amount he can have in his bunkers. He must be able to advise the captain as to the most economical speed to make a voyage, and he must set his differential expansion gear so as to get the most possible from his engines. He must exercise a watchful care over the operation of the boilers and engines, determine the limiting amount of density to be allowed in the boilers, and see that his orders are carried out.

He keeps a daily log of engine room happenings, and certain standard items, such as number of revolutions, steam pressure, coal burnt, percentage of ash, oil used, etc., and at the end of each run he compiles an abstract, accompanied usually by a set of indicator diagrams, and forwards it to the owners, where it is closely gone over and compared with other voyages. He must be prepared for any emergency that may arise, and must be a good enough mechanic to make any repairs with the limited tools at his command, consisting usually of a hand forge, and tools, with hand hammers, chisels, and files. He must also thoroughly understand the principles underlying the design and construction of his boilers, engines and auxiliary machinery, so that he can detect and remedy in time any defect that may appear. A breakdown 1,000 miles from anywhere is not a pleasant contingency, and the chief feels, as the immortal "McAndrews" of Kipling felt,

"All mine the fault, all mine, O Lord,
And no one else but me,
The fault that leaves ten thousand tons
A log upon the sea."

It takes years to make a competent marine engineer, and a man has usually eight or ten years' sea experience on top of his apprenticeship before he attains a position as chief.

The Second Engineer

The second engineer is the executive officer below, as the chief officer is on deck. He is the man who directs the junior engineers and firemen, greasers, etc., and does the actual supervision of all the work. He keeps the watch from 4 to 8 morning and evening, and this enables him to be around during the forenoon to look things over and give his orders to the juniors. Before a port is reached he presents the chief with a list of necessary work, and he and the chief consult together and decide just what shall be done, and it is then left for the second to carry out. On the ability of the second depends the chief's hope of happiness, as with a competent man in this position the chief can lay back and refrain from any active interference in the daily routine. This is better for both chief and second as it increases the prestige of both. The second must always hold a first class certificate, so that he can take charge if necessary.

The Third Engineer

The third engineer, who must hold a second class

certificate, is an important man in the engine room. He takes the most unpleasant watch in the ship, that from 12 to 4, morning and evening, known to all mariners as the "gravy-eyed" watch. This is because he never gets a real sleep while at sea, and if the run is very long, he is usually heavy eyed and weary. Besides his engine room duties he has the task of looking after all the winches on either the forward or after deck, and in between ports, he must work during his off duty hours to have them in good order for the next port. In port he does all the overhauling of the main engine below the cylinders, that is, connecting rods, eccentrics and valve gear, and pump link gear. As soon as the second finds he is up to his job he has more or less a free hand.

Then The Fourth

The fourth engineer is usually just out of his apprenticeship, and putting in his qualifying time for a second engineer's certificate. He keeps the chief's watch, 8 to 12, morning and evening, so that he always has a senior engineer within call. To him fall all the most unpleasant jobs in the engine room, such as bilge diving, and all work on the boiler tops, pumps, etc., while he also has a set of winches and the windlass to look after. If he is quick to learn, and strikes a ship carrying a decent second, he will be happy; if not, his time as fourth engineer will be the reverse of pleasant.

Engineers are better paid than the deck officers, the chief's salary being nearly as high as the captain's, and the junior's accordingly.

The life is an attractive one, as it is possible in a few years, to visit every important sea port in the world, and broaden one's mind by seeing different peoples and countries. During the ship's stay in port there is ample opportunity to see some of the country, after working hours, and given a good ship and shipmates, the time at sea is also pleasant, especially in tropical waters.

A Good Training

The outstanding point, as far as the engineer is concerned, is that after spending some time at sea, he can come back to shore life and pursue his profession on shore, and he will find that the time at sea has given him quickness, confidence, and the ability to handle men. It is an experience that is invaluable and unique. On the other hand, if he elects to remain at sea and become chief engineer, he will usually try to get to better ships and, if possible, become chief of a first class liner, which is an enviable position.

The question of training is a big one, and one that will have to be worked out gradually. In Great Britain they are busy now discussing ways and means of educating boys to man their ships after the war.

In Canada we have a large number of youths who enter machine shops and in two or three years become machinists, and to many of these the prospect of being anything more than machinists is remote. They are mostly intelligent, and have a certain amount of education. Now, if opportunities were afforded these boys to go through the different departments of a marine engine building shop, coupled with judicious technical education, they would have an opportunity of entering the mercantile marine as engineers, with an infinitely better prospect than remaining a machinist. It is essential that they should go through a good course of fitting and erecting, marking off, etc., and should be developed as mechanics, because it must always be remembered that there is no machine shop round the corner at sea, and if the services of a machine shop are called on in a foreign port, there will be some lengthy explanations to make to the owner when the ship gets home.

A Profession—Not a Trade

It is important to the owner to have a class of first rate men to draw from, if he is to run his ships at the highest efficiency, and after the war this will have to be done if the Canadian ship is to compete successfully with other nations. One way to secure the right class of men is to instil into the youths that they are not being brought up to a trade, but to a profession. Give them every opportunity to gain the technical knowledge necessary for this object. Make conditions on the vessels as comfortable and attractive as possible, the emoluments as high as is commercially practicable, and get the cooperation of the Government in making their examinations strict enough to ensure the elimination of all but the most competent.

This will all help in interesting the youth with some ambition, and after a few of the boys have made voyages to distant lands and got the taste of salt water, their tales will prove a powerful help in drawing others.

It would seem that the best time to attract the youth of the country to a seafaring career is at the time when they are leaving school to enter life. Thousands of boys leave school every year to drift into occupations that can lead nowhere. If the attractions of a seafarer's life and its prospects were to be laid before these boys then, many recruits for the deck departments could be found. In fairness to the boys they should be given the opportunity of entering the ship owner's employ as apprentices, and proper facilities given them while at sea to study the art of navigation, so that at the end of four years they would be ready to pass their examinations and become junior officers.

Now is the time for ship owners to get together and determine the means of tackling this subject, or we will find ourselves with a merchant fleet, but without the personnel to run it.

Developing the Marine Specialties Trade

Bawden Machine Company Are Turning Out a Large Line of Marine Goods—There is Need of Making This Class of Goods Proof Against All Sorts of Hard Usage and Abuse

AN interesting line of marine specialities, consisting of steering engines, ship windlasses, and winches, is manufactured by the Bawden Machine Company, Toronto. All these machines are built with a large amount of strength and rigidity to meet the severe conditions prevailing in their sphere of usefulness.

The winches are made in sizes varying from 7 inch cylinder diameter by 12 inch stroke, to 10 inch cylinder diameter by 12 inch stroke. The frames are made of a one piece casting, which makes the most rigid construction, besides eliminating possible trouble from bolts loosening up during severe working conditions. Anybody who has had experience of the way winches are run by stevedores will appreciate this point. This method also facilitates the aligning of cylinders and crank shafts. The gears are made either of the helical or plain spur gear type, and in either single or compound gear arrangement. The clutch controls are arranged so that the act of throwing in the compound gear will release the single and vice versa. Extended drum shafts can be fitted by removing the warping drums from the shaft ends, and fitting flanged couplings in their place, with outboard bearings at the end of the extension shafts. The crosshead and crank pin bearings are of hammered babbit metal in cast iron boxes, and the crank pin end is of the marine box type. The crank pins are made a pressed fit in the discs, and then rivetted over and discs are fitted with a shrunk ring. These winches can be fitted with a steam controlled reversing gear if required, by arranging the eccentric sheaves at right angles to the crank, with a butterfly control valve admitting steam to either end of the cylinder as required. The whole arrangement makes a well balanced and rugged machine, well fitted for the duties it has to perform. The illustration shows the general arrangement.

STEERING COLUMNS AND WHEELS

Steam Windlasses

The windlasses are manufactured in approximately the same sizes as the winches, except that the stroke is shorter, being 10 in. as against 12 in. on the winches. The larger sizes are fitted with cast steel gears, cast iron being used in the smaller sizes. Particular attention has been paid in the building of these windlasses to making the various levers for steam control, brakes, and clutches easily accessible to the operator as can be seen from the illustration. For facilitating quicker warping, which is necessary very often when moving a vessel into a berth, the second motion shaft can be run independently, and by a convenient arrangement of clutches, the engine can be run without moving the gears. Provision has been made for moving the windlass by hand, either by means of a walking beam, or from the second motion shaft. The walking beam with levers shipped in position can be seen in the illustration. The wildcats on main shaft are mounted on bushings, enabling them to work independently of each other as required. The steam reversing feature is also embodied in the windlass. This is a distinct improvement over the hand reversing gear, eliminating as it does two eccentric sheaves, with straps rods and link motions, with their attendant maintenance troubles, and costs. The ordinary Stephenson link motion on a winch or windlass is subjected to much rougher usage than on almost any other kind of engine, and wears correspondingly quickly. In the Bawden Machine Company's design, the reversing is effected by means of a simple butterfly valve, in conjunction with a single eccentric set without lap or lead on the valve. As in the winches, babbited bearings are used, and the construction throughout is of the strongest.

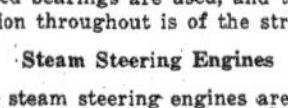

CORNER OF STEERING ENGINE SHOP

Steam Steering Engines

The steam steering engines are manufactured in sizes varying from 6 in. diam. by 6 in. stroke to 9 in. diameter by 9 in.

STEERING ENGINE WITH OUTSIDE CHAIN DRUM

FRONT VIEW OF WINDLASS

stroke. The same general characteristics such as rigidity of construction, and attention to details, appear in the steering engines as appear in the other machines described. They are made in a variety of designs to suit the individual requirement of their various purchasers. The Imperial Munitions Board among others have purchased several of these units for use in vessels being built to their order in various parts of Canada. The two chief types are those with the underslung chain drum and guide sheaves, and those with the shaft continued out from the bed of the engine, with the chain drum mounted some distance from the engine, and supported by an outboard bearing. The hunting gear for controlling the steam valves is of the usual type, driven through right angled shafts with bevel gears from the main worm wheel, and the vertical spindle is provided with the usual arrangement of safety stops to prevent the wheel being turned too far in either of the extreme positions. The piston type of steam valve is used throughout in the construction of these engines, for both cylinders and control.

A corner of the erecting shop is shown in the illustration, with several of the underslung type of engine in course of erection. Special attention is given to the fitting of the various parts, in each design of engine, and they have already made good reputation for speed of control and good working abilities. Another illustration shows a portion of the shop devoted to the building of the steering columns for the ships' wheelhouses, and a good view is shown of the various parts, as well as the completed columns, with the wheels mounted in place. It is a good sign to see Canadian firms keeping pace with the demands for new equipment that the shipbuilding industry will require in ever larger quantities.

SURVIVE U-BOAT ATTACKS

To the Nova Scotia Shipbuilding and Transportation Co., Ltd., of Liverpool, Nova Scotia, has come during this year the record of having two tern schooners built by the above company, with J. S. Gardiner as master builder, survive the attacks of the submarine peril.

CARGO WINCH.

Early in the spring the "Maid of Harlech" was set upon by the underseas craft, which inflicted considerable damage upon it and being deserted by the crew, was towed into Oran, Algiers, where the French Consul immediately corresponded with the builders for necessary materials for repairing the vessel. This was at once attended to by this company which forwarded the required supplies, and the vessel was re-commissioned.

Now comes the tern schooner "Bianca," launched here from the same yard in June, 1917, also meeting a similar fate, to be towed into the port of Halifax by the American fishing vessel "Commonwealth," returning as it were from the depths of the sea.

That excellent material and workmanship was put into the construction of both these vessels is now completely established, and the company have every reason to be proud of its record. Since last November they have launched four other ships, and have two now under construction, into which the like quality of labor and material have been, and are being put.

8" x 8" SHIPS WINDLASS.

Causes of Failure in Boiler Plates

Effect of Grain Growth—Alteration of Crystalline Structure by Mechanical Deformation—Some Remedies

By WALTER ROSENHAIN and D. HANSEN.

THE occasional cases of failures in boiler plates met with in practice have formed the subject of several papers and discussions before the Iron and Steel Institute in recent years. A number of such cases have been investigated by the authors, and an account of one which offers features of particular importance which do not appear to have been previously noticed was read before the Iron and Steel Institutes in May. These are of special importance because it may be found that they afford a clue to the cause of failure in other cases, particularly in boiler plates of the largest dimensions.

The failure occurred in the last stage of the manufacture of the plate. The size and dimensions of the plate are illustrated in Fig. 1. The plate has a thickness of 1¾ in. and measures 4 ft. 4 in. in width by 11 ft. in length. It was manufactured under a stringent specification, but cracked during the straightening of the edges after the bending operations had been completed. Inquiry showed that the bending operations had been carried out in stages in the cold, the plate being subjected to intermediate annealings between the various stages. The position of the crack which formed in the plate is indicated in the diagram.

The material of the plate was first submitted to chemical analysis, mechanical tests, and general microscopic examination. The results obtained were as follows:

CHEMICAL ANALYSIS

	Per Cent.		Per Cent.
Carbon	0.16	Manganese	0.628
Silicon	0.079	Nickel	0.10
Sulphur	0.030	Chromium	nil
Phosphorous . .	0.048		

There is nothing abnormal in this composition, which represents a mild steel of high quality.

Tensile tests were taken from the outside and inside of the plate as received, with the results in Table 1, columns 1 and 2:

Here again there is nothing abnormal, except perhaps a slight indication of an unusual condition of the steel in the comparatively large difference between elastic limit and yield stress. It was thought that possibly this peculiarity might arise from the existence of internal stresses in the material, and in order to remove these as far as possible without changing the structural condition of the steel, a portion of the plate was annealed at 550° C. for 30 minutes. The results of tensile tests of a plate in this condition are given in the third column of Table 1. It will be seen that the difference between elastic limit and yield stress is still comparatively large.

In order, further, to test this point, and also to ascertain how far the tensile tests obtained on the material, as received and after annealing at 550° C., correspond to the best properties which the material is capable of attaining, a sample of the plate was normalized by

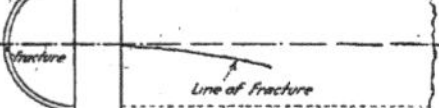

FIG. 1—WHERE THE FRACTURE OCCURRED.

heating to 900° C. followed by cooling in air. The results of tensile tests made on the sample thus treated are given in column 4 of Table 1. Here it will be seen that the elastic limit has come very much closer to the yield stress, while the yield stress itself has been raised. The

whether the properties of the steel were really as satisfactory as the tensile tests would indicate. For this purpose an impact test has been used, for although it is recogized that the conditions under which failure occurs in boiler plate possess no apparent resemblance to those of an impact test, yet experience has repeatedly shown that materials which give a low figure under an impact test are liable to fail under apparently static conditions.

The form of impact test employed is that known as the international notched bar impact test, made with a modification of the Charpy impact testing machine, and on specimens measuring 10 mm. by 10 mm. in section by 53.3 mm. in length, having in the middle a rounded notch with a radius of two-thirds of a millimeter. On the material as received this test gave a mean figure of 0.75

TABLE 1—PHYSICAL TESTS OF THE FAILED PLATE

Particulars.	Plate as Received		Plate Annealed 550° C.	Plate Normalized 900° C.
	Outside 1	Inside 2	Outside 3	Outside 4
Diameter, in.	0.375	0.375	0.375	0.375
Cross sectional area, sq. in.	0.1105	0.1105	0.1105	0.1105
Elastic limit, tons per sq. in.	14.2	11.3	15.4	18.3
Yield stress, tons per sq. in.	18.3	16.1	18.7	19.15
Ultimate stress, tons per sq. in.	26.88	27.24	27.61	27.94
Modules, lb. per sq. in.................	29.8×10^6	29.8×10^6	30.4×10^6	30.2×10^6
Extension per cent. on 1.3 in.*	31.6	33.1	34.5	42.2
Reduction of area per cent.	59.6	60.7	59.1	62.5

*A gage length of 1.3 in. is chosen to give a ratio of gage length to diameter equal to 3.5.

FIGS. 2 AND 3—STRUCTURE IN TRANSVERSE AND LONGITUDINAL SECTION OF THE ORIGINAL METAL. 50 DIAMETERS. A CONSIDERABLE AMOUNT OF BANDING IS PRESENT.

ultimate stress has only been slightly affected, but, on the other hand the elongation has been markedly improved.

Since the tensile tests showed little or no departure from the normal in the material of this plate it became desirable to apply other tests in order to ascertain

kgm. per sq. cm., the actual values obtained being: 0.84, 0.88, 0.66, 1.08, 0.86, 1.20. These figures are of course very abnormally low, a reasonable value for a boiler plate of this kind being from 8 to 11 kgm. per sq. cm. It was thought that possibly this low value might be

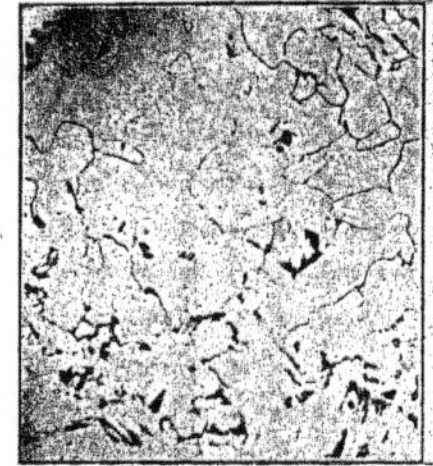

FIG. 4 (LEFT)—LARGE FERRITE CRYSTALS IN THE CARBONLESS BENDS OF THE STRUCTURE AFTER FURTHER ETCHING; MAGNIFICATION IS 150 DIAM.

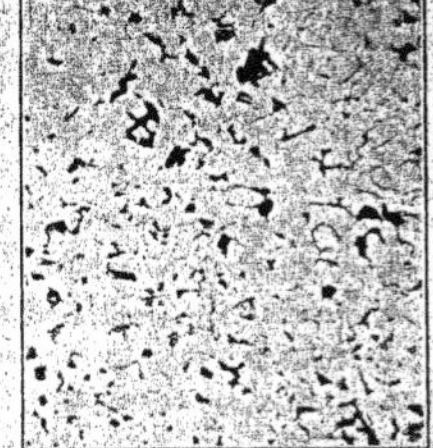

FIG. 5 (CENTRE)—SAME MATERIAL AFTER NORMALIZING; MAG. 150 DIAM.

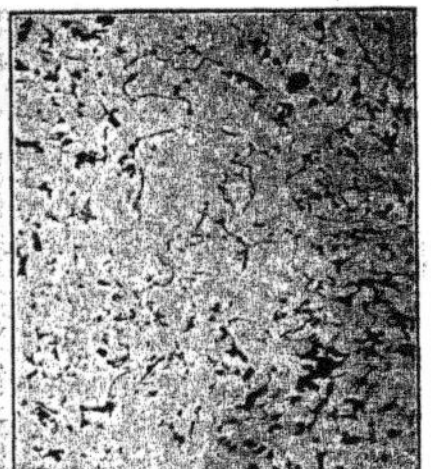

FIG. 6 (RIGHT)—STRUCTURE OF ONE OF CARBONLESS AREAS OF THE SPECIMEN WHICH HAS BEEN HAMMERED IN THE COLD AND THEN ANNEALED AT 650° C. MAG. 150.

due to cold work which the plate had received, leaving it in a work-hardened and, possibly, internally strained condition. The impact tests were therefore repeated on specimens of the plate which had been annealed for thirty minutes at 550° C., in the same way as had been done with the tensile test pieces. The mean result of six impact tests made on the steel in this condition gives a value of 2.90 kgm. per sq. cm., the actual figures obtained being as follows: 2.10, 3.86, 2.64, 3.36, 3.52, 1.92.

It will be seen that this very low temperature annealing, by removing cold work and internal stress has improved the impact behaviour of the material quite appreciably, but that, even when thus treated, it is still very far below the normal value for steel of this grade. This is indicated by the impact figures given on samples of the plate after normalizing at 900° C., when values of 10.78 and 11.72—mean, 11.25 kgm. per sq. cm. were obtained.

It is evident from these figures that the steel of the fractured plate is in an abnormally bad condition, presumably as the result of some treatment—thermal or mechanical, or both—which it has received during manufacture, and it became necessary to discover, if possible, the cause of this abnormality.

The general microscopic examination of the steel showed at first sight nothing abnormal. The structure in general transverse and longitudinal section is shown under a magnification of 50 diameters in photomicrographs Figs. 2 and 3. It will be seen that the scale of the structure, so far as ferrite-pearlite distribution is concerned, appears to be satisfactory, but there is a considerable amount of banding present, although this amount is not in itself abnormal for a plate of such large size.

More careful examination of the structure, however, particularly after it had been etched in such a way as to develop the ferrite boundaries, revealed a striking pecularity. This takes the form of relatively very large ferrite crystals in the carbonless bands of the structure. These are illustrated, under a magnification of 150 diameters, in Fig. 4. The corresponding grain size of the same material, after normalizing, is shown in Fig. 5 under the same magnification. It should be noted, however, that the normalized structure shown in Fig. 5 has been obtained not by treating a small laboratory sample but from a comparatively large piece of the plate about a foot square which had been subjected to the heat treatment described. The most careful study of the steel in both conditions revealed no other difference between the "as received" and normalized conditions. The inference is thus indicated that the abnormal impact behaviour of the steel as received may be due to the development of coarse crystals in the carbonless bands which occur in this material, and the possibility is suggested that the failure of this plate may be connected with the phenomenon of grain growth which has in recent years been discovered in the case of iron and very low carbon steel.

The subject of grain growth is of fundamental importance in connection with the further investigation of this plate, and it is referred to it in greater detail at this point.

Phenomena which are now recognized as coming under the general title of grain growth were discovered and described by Stead[1] and Charpy[2]. A considerable advance in our knowledge of the subject was, however, made by Sau-

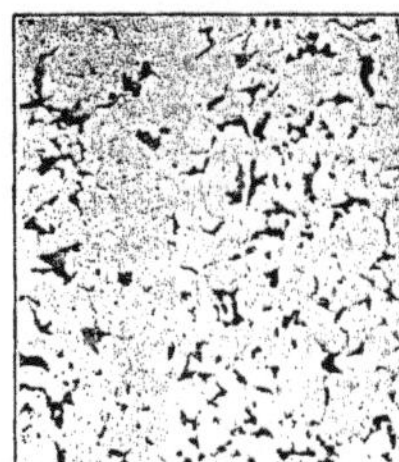

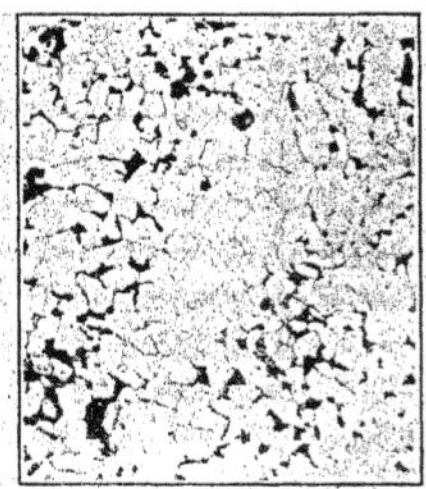

FIGS. 7, 8, 9—INCROSTRUCTURES AT 150 DIAMETERS OF THE STEEL, CORRESPONDING TO THE VARIOUS FORMS OF HEAT TREATMENT OUTLINED IN TABLE 3.

veur[3], who made the well-known experiment of straining by compression a conical piece of nearly pure iron, and subsequently annealing the piece thus treated at a temperature below the lowest critical point. On cutting a section and etching it, a band of very large ferrite crystals was found at one point, and this

FIG. 10—MICROSTRUCTURE OF PLATE NO. 2 USED EXPERIMENTALLY. VERY LITTLE BANDING PRESENT, THE PLATE BEING ONLY ½-INCH THICK

led to the view that there is a critical amount of plastic deformation which, for a given annealing temperature, below the critical range produces very rapid grain growth.

The subject has been more fully investigated by Chappell,[4] and has also been dealt with in America by Sherry.[5] The latter author has shown that grain growth occurs, not only in comparatively pure iron, but in any region existing in a mass of mild steel from which pearlite is absent or nearly absent—in the carbonless bands such as those met with in boiler plates, provided, of course, that the necessary treatment, consisting of plastic deformation of the right intensity followed by annealing at a correspondingly low temperature, has been applied.

In view of the results obtained by the authors just referred to, the observations made on the boiler plate which forms the subject of this paper at once suggested that the development of coarse ferrite crystals in the carbonless bands of the plate was the result of grain growth following upon deformation in the cold and subsequent low temperature annealing. When it is borne in mind that this plate was bent cold and then annealed several times in succession, it will be seen that the conditions likely to produce grain growth in carbon-free areas had been present.

The authors, however, were not satisfied with a general inference of this kind, but endeavored experimentally to reproduce the conditions under which the steel had developed the coarse and relatively brittle structure which it possessed when received. For this purpose two series of experiments were undertaken. In both series the material was first normalized in order to destroy the previously existing coarse crystals and to bring the material into the condition in which it gives a satisfactorily high impact figure. Deformation was then applied to the material in two ways; in one case, in the cold (by hammering), and in the second case at a temperature between 600° and 700° C., or below the critical range. Specimens treated in both ways were then annealed at 650° C. for 30 min. The microstructure was examined both before and after this last annealing, and impact tests were taken on the material at each stage.

The resulting structure in one of the carbonless areas of the specimen which has been hammered in the cold and subsequently annealed at 650° C. is shown in Fig. 6 under a magnification of 150 diameters. Comparison with Fig. 5 shows at once that considerable grain growth has taken place, although the resulting grains are not quite so large or well developed as those in Fig. 4. The sample which has been hammered between 600° and 700° C. gives a very similar structure, and the impact figure in this case is brought down to 1.56 kgm. per sq. cm.

In order to test the matter further another series of experiments was undertaken in which varying amounts of mechanical deformation were applied in the cold followed by annealing at 650° C. In order, however, to prove that it was not the annealing process alone which resulted in the reduction of the impact figure, the normalized sample was also annealed at 650° C. without previous mechanical deformation. The results obtained by impact tests on specimens thus treated are given in Table 2:

TABLE 2—TESTS ON BOILER PLATE NO. 1

Treatment	Resistance to Impact. Kilogrammeter per Square Centimeter
Normalized at 900 deg. C.	10.46
Normalized at 900 deg. C.	8.92
Normalized; annealed at 650 deg. C.	9.04
Normalized; severely deformed; annealed, 650 deg. C.	11.7
Normalized; reduced 12.4 per cent.; annealed, 650 deg. C.	10.66
Normalized; reduced 7.1 per cent.; annealed, 650 deg. C.	8.44
Normalized; reduced 6.9 per cent.; annealed, 650 deg. C.	10.04
Normalized; reduced 4.9 per cent.; annealed, 650 deg. C.	8.14
Normalized; reduced 3 per cent.; annealed, 650 deg. C.	6.34

In this table the amount of mechanical deformation is measured by percentage reduction of thickness produced by pressing in the cold in a powerful press.

The results given in Table 2 are instructive. It will be seen that large amounts of reduction actually improve the impact strength slightly, but with decreasing amounts of mechanical deformation followed by low temperature annealing the impact strength is very much reduced, although the lowest value obtained in this way, 6.34 kgm. per sq. cm., is still very much better than that

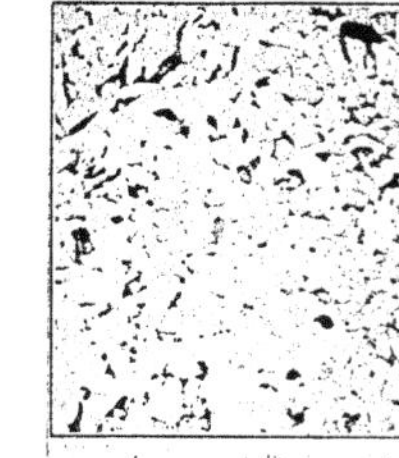

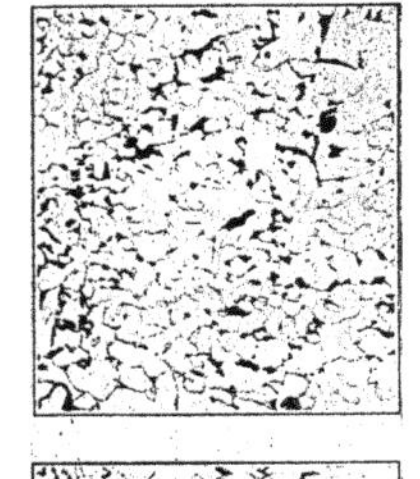

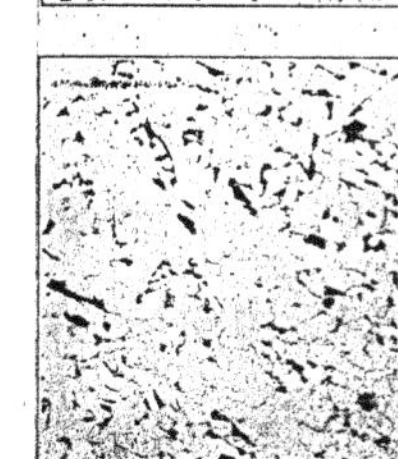

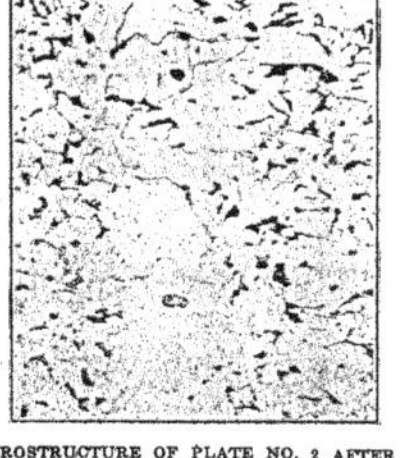

FIG. 11, 12 (UPPER), 13 AND 14 (LOWER), MICROSTRUCTURE OF PLATE NO. 2, AFTER TREATMENT REFERRED TO IN TABLE 2.

found in the plate in its condition as received, or that described in the hammered sample given above. There is nothing to suggest, however, that hammering, as distinct from such deformation as occurs in cold bending, has any specific effect. It should further be borne in mind that when a thick plate is bent in the cold, a considerable range of plastic deformation is produced, ranging from a maximum at the surface of the plate to zero at the neutral axis. Somewhere within this range the critical deformation, corresponding to the annealing temperature employed, is likely to occur.

The microstructure corresponding to the various forms of treatment referred to in Table 2 are illustrated in Figs. 7' 8 and 9, at a magnification of 150 diameters. Fig. 7 refers to the last specimen mentioned in the table having the lowest impact figure and correspondingly showing the largest development of grain growth in the carbonless bands. Fig. 8 refers to the material as normalized and annealed at 650° C. without intermediate deformation. It will be seen that here there is no appreciable difference in grain size between the carbonless band and the adjacent steel. Finally, Fig. 9 refers to the material which has been severely deformed and subsequently annealed at 650° C., giving a high impact figure. Here it will be seen that the grain has been very much refined even in the carbonless areas, and this corresponds in a striking manner with the very high impact figure, 11.7.

When the evidence above described is carefully considered it will be seen to afford a considerable degree of proof of the view that the brittleness, as evidenced by the very low impact figures and actual failure in manufacture which has been found in the plate under discussion, arises from the existence of coarse ferrite crystals due to grain growth in the carbonless bands of the steel, and that this grain-growth is the result of a moderate amount of deformation in the cold, followed by low temperature annealing. It is further evident that normalizing the material, or indeed merely heating it to a temperature above the critical range, is sufficient entirely to obliterate this grain growth and all its evil effects.

It will be seen that this conclusion indicates that the presence of carbonless bands, which is regarded as a normal feature and has not hitherto been considered a serious source of danger or weakness in a boiler plate, may become the cause of failure if associated with a suitable combination of mechanical deformation and low temperature annealing. If carbonless bands are to be regarded as a normal feature in boiler plates—and in existing practice this is probably inevitable—and if deformation in the cold, such as bending, etc., is otherwise a desirable practice, it seems that subsequent normalizing is necessary, or certainly desirable. as a safeguard against dangers of the kind described here.

In order further to test the view which has been advanced above, the authors have endeavored to carry out similar experiments and tests on other samples of boiler plate, but the other samples at their disposal came in every case from plates of much smaller size and thickness, with the result that the banding, where it existed to a marked extent, was on a much smaller scale. Experiments on these plates were, however, made in order that the results might be regarded as a check on the observations already described. In the case of a plate half an inch thick, which may be referred to as No. 2, the chemical analysis was as follows:

	Per Cent.		Per Cent.
Carbon	0.123	Phosphorous	0.057
Silicon	0.014	Manganese	0.49
Sulphur	0.03		

which again indicates a steel of satisfactory composition. The general microstructure of this plate in the condition as received is shown in Fig. 10. A certain amount of banding is present, but not on the scale found in the first plate described. A piece of this plate was

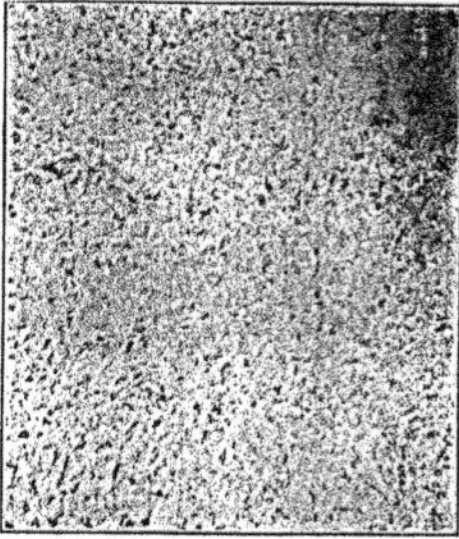

FIG. 15—MICROSTRUCTURE OF ANOTHER ½-INCH PLATE. THERE IS AN ABSENCE OF MARKED BANDING.

normalized at 950° C., and portions were subsequently treated as follows:

Hammered cold and annealed at 650 deg. C.

Hammered between 600 and 700 deg. C., and annealed at 650 deg. C.

Annealed at 650 deg. C. without previous mechanical treatment.

Impact tests have subsequently been carried out on the samples thus treated, with the results given in Table 3:

TABLE 3—TESTS OF BOILER PLATE NO. 2

Treatment	Energy to Fracture. Kilogrammeters
As normalized at 950 deg. C.	11.06
Normalized at 950 deg. C.; hammered cold and annealed at 650 deg. C....	5.52
Normalized, 950 deg. C.; hammered between 600 deg. C. and 700 deg. C., and annealed at 650 deg. C.	7.18
Normalized at 950 deg. C., annealed at 650 deg. C., without mechanical treatment	10.44

It will be seen that the normalized material again gives a high value, and that this value is not appreciably diminished by a further anealing at 650° C. On the other hand, cold hammering followed by annealing at 650° C. lowers the impact figure to one half of the normal value, while hammering between 600 and 700° C. reduces it considerably but to a lesser extent. The corresponding microstructures are illustrated in Figs. 11, 12, 13 and 14. Fig. 11 shows the material as normalized, Fig. 12 shows it after normalizing and reannealed at 650° C. without mechanical treatment, Fig. 13 shows the effects of cold hammering followed by annealing at 650° C., and Fig. 14 shows the effect of hammering between 600° and 700° C., followed by annealing at 650° C.

It will be seen that, in general terms, the results obtained with this material are of the same kind as those found in the first plate but, probably owing to the smaller scale of the banding originally existing in this steel the results are not quite so intense in character. It may be mentioned that this plate also had failed in practice, but not during manufacture, and in a manner which is not necessarily related to the phenomenon of grain growth. The experiments on this plate serve to confirm the observations made with the first example, but they indicate that in plates of a smaller thickness the effects are not likely to be so serious as in the larger plates.

The results obtained with plates Nos. 2 and 3 thus confirm the view that the low impact figures found in the first plate, and to a lesser extent in plate No. 2, are associated with the coarse crystal structure in the carbonless bands, and that these are the result of grain growth produced by slight deformation and subsequent low temperature annealing; also that normalizing in every case completely removes this source of weakness.

[1]Stead, *Journal of the Iron and Steel Institute*, 1898, No. I, p. 145; *ibid.*, No. II. p. 137.

[2]Charpy, *Comptes Rendus*, vol. cli.

[3]Sauveur, *Proceedings of the International Congress for Testing Materials, Sixth Congress*, 1912, vol. xi.

[4]Chappell, *Journal of the Iron and Steel Institute*, 1914, No. I, p. 460.

[5]Sherry, *Faraday Society*, December, 1916.

Allied Tungsten Pool.—An international agreement for the pooling of all available tungsten among the United States and the allies, reached through the Inter-allied Munitions Council at Paris, was announced to-day by chairman Baruch of the War Industries Board. Its terms will be made public if the arrangement is finally approved by President Wilson. Tungsten is produced largely in Colorado, but some comes from South America and Sweden and other European countries. Distribution of the supply probably will be directed by the War Industries Board.

Niagara Falls, Ont.—Such good progress has been made on the 13½-foot pipe line in Victoria Park for the Hydro that it was announced water will be turned into the pipe on October 23. This will add fifty thousand horse power to the output of the Ontario Power Company, of the Hydro system, and take care of power shortage experienced last Winter. This city will also tap the big pipe with a two-foot main for additional water supply.

The "War Taurus" Takes the Water at Polson's

Steel Vessel Being Turned Out to the Order of the Imperial Munitions Board—Launching Was a Complete Success in Every Way—Work Held Up For Some Time by Strike

A FURTHER useful contribution to Canada's war effort was made on Thursday, September 19th, when the steel steamer "War Taurus" was successfully launched from the yards of the Polson Iron Works, Toronto. The launch passed off without the slightest sign of a hitch to mar the event, the vessel starting to move with the firing of the gun, and being moored at her berth almost before the disturbance incidental to her striking the water had died away. The "War Taurus" is one of six similar vessels which are being built for the Imperial Munitions Board. The work of building the vessel has been greatly hindered by a strike of the marine steam fitters and helpers, and it has been necessary to launch her without fitting the sea connections, which will cause further delay in the completion of the hull, ready for sea. The strike was occasioned by the refusal of the company to recognize the steam fitters' union and run a closed shop, which the company has never done in its thirty-five years' existence. Without discussing here the merits of the case it is certainly a lamentable state of affairs that when ships are so badly needed the building should be held up for weeks by labor troubles.

THE LAUNCH OF THE "WAR TAURUS"

The "War Taurus" is a steel vessel of 261 feet in length by forty-three feet six inches beam, and a moulded depth of twenty-three feet. She has been constructed under the classification of the British Corporation and her propelling machinery consists of triple expansion reciprocating engines, the cylinder dimensions being H.P. 20½ inches dia. I.P. 33 in. and L.P. 54 in. with a stroke of 36 capable of developing 1,250 horsepower. She has two boilers of the Scotch marine type, 14 feet diameter by 12 feet long, working at 180 lbs. gauge pressure.

Haste Not Everything

Discussing the shipbuilding situation with Mr. Frank E. Wall, the works manager and chief engineer of the Polson Company, the subject of making records in ship building was brought up. Mr. Wall pointed out that building and completing a vessel in the shortest possible number of days was not so desirable an accomplishment as the general public might be led to believe. He pointed out that many of these ships, after a short trial trip, had to be re-docked and considerable time spent on them, before they could undertake a long voyage, but this phase was never referred to in the published reports of the records achieved. He held it was more desirable to take a reasonable length of time for the building and completion of the ship, so that when she was completed she was ready to start on a voyage of any duration without the necessity of dry docking

There are several more vessels on the stocks in the Polson Company's yard, and it is hoped to launch three more during the early days of October. It is interesting to know, in this connection, that with the exception of some of the auxiliary machinery practically everything else is built from the raw material in the company's shops.

STRINGERLESS SHIPS

By R. C.

Until quite recent times it was considered absolutely essential to the stiffening of the side plating of a ship that longitudinal side stringers should be fitted between the bilge brackets and the beam knees, this being in accordance with Lloyd's rules. Many vessels are now, however, being built without side stringers, compensation being provided by slightly increasing the depth of the beam knees, these modifications being sufficient to satisfy the rules issued by the classification societies. Many advantages follow the elimination of the side stringers. In the first place there is a considerable saving in the cost of construction of the vessel; moreover, there is additional cargo space, and the ship is very convenient to discharge, there being no shelves for the lodgment of such cargoes as grain or coal. Further, the increased depth of the bilge brackets either avoids or reduces the unsupported span of the frame between the bilge brackets and the beam knees, which admits of a reduction in the size of frame so that it is possible to utilize bulb angle frames instead of the built sections provided for by the rules. It is open to question, however, whether it is structurally safe to dispense entirely with side stringers, although the fact that many builders are at least convinced of the advantage which result from this method of construction.

THE "WAR TAURUS" ON THE WAYS

MODERN CONDENSER PRACTICE*

By D. D. Pendleton

FOR high duty work, 28-inch vacuum or better, the type of condenser to be selected should be thoroughly studied, taking into consideration the engineering, physical and commercial conditions. The question of water is of prime importance and should be studied as to : Quantity, kind, clean, dirty, acid, etc.; pumping head; source, natural or artificial; average temperature. With these points settled, the commercial vacuum which is practical to maintain may be determined, always bearing in mind that the average log of a properly designed, rated and operated surface condenser will show better vacuum than a jet machine and in turn better than a barometer machine, due principally to the conditions as to air leaks and exhaust piping losses.

Surface Condensers

The surface machine also allows the condensate to be reclaimed as pure, distilled water for boiler feed, thus requiring only a moderate amount of make-up water to be treated in cases of unsuitable boiler feed supply. There is no limit to the size of surface equipments—they are now being made in sizes from 100 square feet up to 70,000 square feet.

The barometric condenser is limited in size by the expensive exhaust piping. The low level jet is also limited as to size. This is due to the removal or tail pump and air pump limitations and to the power necessary for driving the pumps. Units of 7,500 to 10,000 kilowatts seem to be the practical limit for these pumps, although there have been installed in certain cases twin jets of 20,000 kilowatts capacity.

A 20,000 kilowatt turbine consuming 12½ pounds per kilowatt hour or a total of 250,000 of steam would require for surface work 31,250 gallons of water per minute at 15-degrees difference, with 70-degree water and 28-inch vacuum, against a 10-foot head. This would require approximately 100 horse-power, the air and condensate pump of the hydraulic type requiring 140 horsepower additional or a total of 240 horsepower with one set of auxiliaries.

The jet at a 5-degree difference would require 20,000 gallons per minute against a 40-foot head if no booster pump were required, or approximately 400 horsepower. The pump would require about 250 horsepower more, making a total of 650 horsepower.

Conditions Influencing Choice

After considering the local conditions, the choice of a condenser should be made upon the power requirements of operating cost, always considering the points mentioned in the preceding paragraph. Assuming a case of 40,000 pounds of steam per hour, 28-inch vacuum with 70-degree water, the power requirements are as follows: Surface condensers at 15 degree difference would require 5,000 gallons per minute. Since the discharge should be sealed, a working head of 10 feet should be ample, under ordinary circumstances, requiring a water horsepower of 12½. At an efficiency of 75 per cent. on the centrifugal pump, a pumping horsepower of 16 2/3 would be required. Using the wet and dry system the air pump would require approximately six horsepower, the condensate pump about 2½ horsepower, making a total of approximately 25 horsepower. The low type jet owing to the low efficiency of the removal pump would require more power.

At a temperature difference of 5 degrees, however, less water, or 3,100 gallons per minute would be required. A head of 36 feet being the discharge head from the condenser, plus friction would make approximately 27.9 water horsepower. The efficiency of the removal pump being about only 50 per cent, it would require a total of 55.8 horsepower. The air pump, however, would be approximately twice as large as for the surface condenser and would require 12 horsepower, giving a total of 68 horsepower for the jet system. The barometric type would require same amount of water, but the water must be lifted to the condenser head using the total static head plus suction and lift on the pump, less the suction effect of vacuum. This usually can be considered a net head of say 40 feet, making a total of 31 water horsepower; and at an efficiency of say 70 per cent. on the supply pump, a total of 44½ horsepower would be required.

The same air pump would be used as in the case of the jet, requiring 12 horsepower, making a total of 56½ horsepower. The hydraulic or other types of air pumps and the steam jet would materially increase the air removal power requirements.

Summing the case up we have: Surface, 25 horsepower; barometric, 65½ horsepower—an increase over the surface of 126 per cent. low type jet, 68 horsepower plus an increase over the surface of 172 per cent. The relative costs would be: One hundred per cent. for the low type jet; 100 per cent. for barometric, not including the exhaust piping; and about 200 per cent. for surface equipment.

The rating of the surface of a condenser depends principally upon the amount of heat that can be dissipated by each square foot of surface, this being determined by the heat transfer per degree difference between the mean circulating water temperature and the temperature of the vacuum to be attempted.

Condenser Tubes

The tubes in surface condensers have been a source of much thought and experiment. The material used should be pure electrolytic copper, high grade spelter, having not over 0.07 per cent. lead, 0.03 per cent. iron and .05 per cent. cadmium, and pure tin. No scrap should be used. The most common mixtures are: Pure copper, muntz metal (60 per cent. copper, 40 per cent. zinc); admiralty (70 per cent copper, 29 per cent. zinc and 1 per cent. tin). The standard tubes usually are No. 16 or 18 B. w. g. in thickness and ⅝ inch, ¾ or 1 inch outside diameter, depending upon conditions.

The quality of water available and the temperature usually determine the vacuum which is practical to maintain. This quantity can be varied for the same vacuum by varying the assumed terminal temperature difference, which will directly vary the surface for any given rate of heat transfer. A terminal temperature difference of 5, 10, 15, or 20 degrees can be used. A ratio of heat transfer from 300 to 500 may be employed. For standard commercial work, a terminal temperature difference of 15 degrees is usually conceded the best practice and a heat transfer of 300 to 350 is conceded the most conservative.

Water Velocities

Eminently good engineers have advocated water velocities for surface condensers as high as 10 feet per second, stating emphatically that 8 feet is conceded practical and standard. From tests of large condensers, a water velocity of 4.7 feet per second seems to be the average; for the purpose of discussion, say 5 feet. Now suppose that this velocity be increased to 8 or 10 feet; the power requirements for such conditions would be excessive. All talk of 8 to 10 feet per second seems to be absurd, in view of the fact that there are very few commercial and practical plants operating under such conditions.

Water quantities are usually figured at a 5-degree terminal difference as standard practice, although results have easily been maintained at the vacuum temperature or a 100-per cent. water efficiency.

The removal or tail pump, as already stated, requires the larger amount of power—usually on river conditions the water has to be supplied to a cold well by a booster pump, drawn into the condenser by suction and expelled against the vacuum plus any friction and outside static head. The booster pump power has not been considered in the example of power requirements for the types already given.

Air Pump Requirements

The air pump is required to remove air coming in with the water as well as air entering through leaks in the exhaust or other piping and air entering with the steam. Here also the water quantities are figured as in the low type jet machines. The water supply can be more easily figured for it can be pumped directly to the barometric head without booster or auxiliary pumps. The air pump requirements are the same as in the low type jet.

After fixing the type of condenser to be used, the next step should be a study of the auxiliaries for equipment. These can be divided into two kinds: Motor driven and steam turbine or steam engine driven. It is very easy to analyze the power requirements of the motor-driven air, circulating and hotwell pumps for the electrical energy is taken directly from the switchboard. Steam turbine or engine drives, however, bring up the question of the return of heat to the feed water or heat balance and this so often complicates matters that we lose sight of the true power consumed. Roughly, 90 per cent. of the steam of the auxiliaries will be returned as heat to the feed water provided there is enough water of a suitable temperature to absorb this heat. The amount of feed water available for absorbing this heat varies with the load on the main unit. The result is that at times of light load there is a grave danger of showing an excess of exhaust steam. This militates against the use of steam-driven auxiliaries to some extent and had led to the development of combined drives using turbines and motors.

In the case of steam jets, this device takes boiler steam, compresses the non-condensible vapors from the condenser and discharges the whole into a feed water heater working at atmospheric pressure. All of the heat except that lost in radiation and leakage is returned to the feed water and for that reason it would seem that the amount of steam required made little difference in the general economy of the whole plant. However, as pointed out previously, care must be taken to make certain that the available steam for heating the feed water is not excessive. One way of analyzing the steam consumption of such an apparatus as a steam jet air pump is as follows: Suppose the pump takes 3 per cent. of the main turbine steam and that the main unit is rated at 5,000 kilowatts or say 60,000 pounds of steam per hour, the steam jet, therefore, requires 1,800 pounds of steam per hour. If this steam were expanding through the same range, that is from boiler pressure down, in the main turbine, it would produce about 150 kilowatts per hour. This then, in a sense, is the power consumption of the air pump and is to be compared with the power that some other type of pump would require for the same condition.

Booster Pump

The efficiency of the booster pump for the low type jet and the supply pump for the barometric can be assumed as standard, consistent with good centrifugal pump design at a given head and quantity. However, the efficiency of the removal pump for the jet, owing to the slow entering water velocities in the impeller when pumping from a vacuum, is seldom over 50 per cent. The barometric, therefore, as far as power requirements are concerned is more efficient than the jet and less efficient than the surface. No condensate is reclaimed for boiler feed purposes in either case.

Where water supply is a serious problem some artificial method of cooling a given supply of circulating water is employed. This is done in several ways: Cooling towers of the natural draft enclosed type or of the natural draft open or rack type, forced draft, and combined forced and natural draft. In standard cooling tower practice it has been determined that with the air temperature at 75 degrees Fahr. and at 70 per cent. relative humidity, the most economical vacuum is 27 inches, based on a 30-inch barometer, or a vacuum having a temperature of 115 degrees Fahr.

Practical Cooling Limits

Rating all condensers on a 5-degree difference—surface, jet or barometric—thus requiring the minimum amount of water, it will be necesary to cool the water from 110 degrees Fahr. Commercially, under the given air conditions, it has been found practical to cool to only 85 degrees Fahr. or sometimes to 80 degrees Fahr., giving a 25 or 30-degree cooling range, which with steam at 1,000 B. t. u. per pound would require 25 and 33.3 pounds of water per pound of steam, respectively. The type of equipment to select would depend upon existing local conditions, such as space available and power consumption.

The natural draft open open type of cooler requires more ground space and can be placed only where the vapors and driftage will do no harm to surrounding buildings or property. This type requires a pumping head of approximately 30 feet and no other power. The natual draft enclosed chimney type takes less ground space, the chimney or stack is usually about 75 feet high.

Area of the Cooling Pond

It is impossible to state generally the area of a pond necessary for a given quantity of water, but generally speaking, at a working pressure on the spray nozzles of 6 to 7 pounds, a pond of 3 square feet per gallon of water sprayed will have to be used. The area of a natural cooling pond is obviously very large as the heat dissipated per square foot depends upon the relative humidity, the temperature of the water and the velocity of the air currents. These are naturally, on account of varied conditions, very discordant.

In any water cooling device 90 per cent. of the cooling effect is through evaporation and it is, therefore, vitally necessary that sufficient unsaturated air be given free access to the proper quantity of water. The air in all cases should go off the tower or pond in a saturated condition.

TESTING GUNS FOR AIRCRAFT

The Browning machine gun has successfully undergone a test to determine its value for use with aircraft. This is one of three types of machine guns with which the rate of fire can be so synchronized with the revolutions of the propeller of a tractor airplane that the gun can be fired by the pilot of a combat plane through the revolving blades. Firing in that fashion, it is necessary to aim the machine gun by steering the plane directly at the target. The direction of the plane gives direction to the fire and the pilot can fire the machine gun while controlling the plane.

Connected With Engine

Airplane propellers revolve at from 800 to 2,000 revolutions per minute. The machine gun is connected with the airplane engine by a mechanical or hydraulic device, and impulses from the crank shaft are transmitted to the machine gun. The rate of fire of the machine gun is constant and its fire is synchronized with the revolving propeller blades by "wasting" a certain percentage of the impulses it receives from the airplane engine and by having the remaining impulses trip or pull the trigger so that the gun fires just at the fraction of the second when the propeller blades are clear of the line of fire.

The pilot operates the gun by means of a lever which controls the circuit and allows the impulses to trip the trigger.

Severe Test Given Gun

The test given the Browning gun was severe. A gun was mounted on the frame of an American combat plane and connected with the airplane engine. The test was conducted on the ground and in place of the propeller a metal disk was attached to the crank shaft. The Browning gun was then required to register hits on the metal disk as it revolved at varying speeds from 400 to 2,000 revolutions per minute. The slightest "hang fire" or delay in action on the part of the gun would have been shown by the failure of the bullets to hit precisely on the spot on the disk representing the centre of the zone of fire. The gun functioned perfectly.

The Browning gun to be used with aircraft is the heavy type with the water jacket removed.

Will Also Use Marlin Gun

Besides the Browning, the United States will also employ the Marlin aircraft gun as a synchronized weapon. Several thousand of these have been manufactured and the gun is in quantity production.

The British and French use the Vickers as a synchronized machine gun.

The Lewis aircraft machine gun is used by the British, French, and American forces, but for a different purpose. In a two-seated combat plane, fixed machine guns are mounted forward to be operated by the pilot and flexible guns are mounted to be operated by the observer in the rear seat of the plane. The observer operates Lewis guns on flexible mounts, firing to right or left of the plane.

It is of vital importance to have absolute reliability of function in a synchronized machine gun on tractor airplanes.

EXPERIENCE with large gas engines appears to show that nickel steel alloys which have a small coefficient of heat expansion and are fairly strong under impactive loads should be suitable for the cylinders. The cylinder and water jacket liner of harder material is fitted. Cast steel is unsuitable for such cylinders—is usually cast in one piece, and a separate Practical Engineer, London.

Port Colborne, Ont.—Although the main swing bridge here is out of use at present, owing to the steamer "Malton" having run into it, the accident is not expected to hold up navigation.

Vancouver Firms Pool Engine and Boiler Orders

One Order For Twelve Boilers and Twenty-four Engines Amounting to Well Over Half Million—Big Chance For Developing Carrying Trade Between Canada and Eastern Countries

BUSINESS amounting to between a half and three-quarters of a million dollars was placed with Eastern Canada concerns for contractors working on French merchant marine orders at the Pacific Coast. J. A. McCulloch, of Vancouver, acting for three firms at the coast, has been in Ontario for the past few days in connection with this work. His trip East had to do with a new form of buying that has been found satisfactory, viz., the pooling of orders and the appointing of one purchasing agent to place them all.

Mr. McCulloch has been closely connected with the munitions and shipbuilding business for some time in the West. His first experience was in the munitions business at Winnipeg, he having gone to the coast two years ago, and since then has followed the business of ship construction very closely from an engineering standpoint.

A Busy Place Now

"Shipbuilding has made the Pacific Coast a busy place," remarked Mr. McCulloch to this paper. "It looks right now as though the work in hand would guarantee that we would be well engaged for a year or eighteen months yet if nothing more turns up." Mr. McCulloch represented on his buying trip the Northern Construction Co., of Vancouver, the New Westminster Engineering and Construction Co., and the Pacific Construction Co. of Coquitlam. These companies have French orders now, 12 vessels in all. They are 1,500 ton French cargo boats, 205 feet long, 40 ft. beam, twin vertical, surface condensing engines of 275 indicated horse power. Mr. McCulloch's special business was the placing of the orders for the 12 boilers and the 24 engines needed for this work.

Can Buy Better

"We thought before, and we know now for sure that we can do better buying in this way than by each of the concerns sending a man down here to look for shop capacity to turn out the boilers and engines that will be required. If a man came into the East now with an order for a boiler and a couple of engines he would get a very scant hearing and delivery would be absolutely a matter of convenience to the shop handling the business. When a person can go to the makers of boilers and engines and say, 'Here is an order for twelve boilers and 24 engines,' we find that we can in this way secure the undivided attention of the shop and deliveries are better. This allows us to proceed at once with the construction of the vessels, and rush them along to the stage where they will be ready for the fittings to be placed in them. We know that a shipment will be coming along every month or so and this will fit in very nicely with our plans at the coast. Building ships is like making munitions. You don't want stuff piled up. You want to get a nice even flow of the necessary material to your yards just the same as keeping shells going through the various operations in a machine shop.

Who Gets the Work?

The contracts were placed by Mr. McCulloch as follows:

The Allis-Chalmers got 10 engines, of the type mentioned above, while Goldie-McCulloch of Galt will furnish 14 of the same type, 275 indicated horse power.

The first order for boilers was placed with the International Engineering Works of Amherst, N.S., the remainder having not been settled when Mr. McCulloch was preparing to leave, but this contract he expected to place before going west. In all the business was between a half and three-quarters of a million. Delivery will be made as far as possible to coincide with the progress of construction at the coast. The contractors with the French government rely on the fact that the work is for that government's war efforts to secure priority ratings that will enable them to get the material for the mechanical equipment needed.

A Great Work

"We have at the coast now six wooden and two steel yards, and there must be in the neighborhood of four thousand men working in these plants. It is estimated that a million a month is put into circulation through the operation of the shipbuilding plants. In the building of wooden vessels we are well situated at the coast in regard to timber. We can get practically everything that we need with the exception of some of the lignumvitae and other materials used for fitting."

In regard to the building of engines at the coast Mr. McCulloch considered that on the smaller types they could compete with the East, but on the larger ones it would be difficult. In fact this work has not been undertaken. The high freight rates worked against them to some extent. "But there is one thing of which I am tolerable certain," remarked Mr. McCulloch, "and that is that the Pacific coast could have had an engine building industry had they gone at it right at the start when there was plenty of work offering to give the industry the necessary work to keep going. The building of boilers could also have been established there as a good industry now. There was some uncertainty about the contracts at the start, when they did not appear to be of such a nature as to warrant much of an outlay. But the work that has been secured there lately both for wood and steel makes it certain that the boiler and engine proposition could have been established.

Looking to the Future

"What of the future? Will the shipbuilding industry be permanent in British Columbia?" asked MARINE ENGINEERING.

"That all depends," remarked the westerner, "on whether they decide to go in and take advantage of the situation as it exists at the present time, and as it will undoubtedly exist for some time after the conclusion of the war. There is a great Pacific coast trade that is looking for some person to develop it. Some of the men at the coast claim it is the biggest opportunity that has ever opened, but it is going to take money and courage to develop it. The carrying trade from the Pacific coast of Canada to Japan and China is large and it is going to be larger. There are loads for bottoms both ways. If the Canadians don't get in and handle this trade it is going to be attended to by the people of Japan and China. That is the common belief of many of the men at the coast who pay a very great deal of attention to such matters, and there is very good reason for what they say. So far there has not been any movement made to meet this situation, but it should be handled by private interest. It is hardly a matter for direct Government action.

The Labor Situation

"How about the labor situation?"

"About the same as the East is," was Mr. McCulloch's opinion. "There have been too many strikes there on the coast. I regard the finding of Senator Robertson as a very fair and just one, although some of the labor men do not seem to see it that way. We have had no trouble in securing men for the work. Some of the men are beginning to see that high rate of pay generally brings everything along with it. It generally works out that way. Where wages run high the prices of commodities will not be far distant in the advance."

Prices for Material

In regard to prices for staple articles that go into the construction of steel ships, it is apparent that there is an advance of some size over prices here or at mills. In the matter of ship plate, it is sold from warehouse at the coast at 12 cents per pound. The Gov-

Eastern Canada Marine Activities

Doings of the Month at Montreal and on St. Lawrence River

(September Correspondence)

WHEN the season's work has been finished it will undoubtedly be found that the achievements of Canada in the matter of shipbuilding will compare very favorably with her facilities to compete in this industry. Many of the yards have had exceptional results in the building of both the wooden and the steel vessels and the general operations will demonstrate her ability to take her share of the construction work of the future. The problem of acquiring material has often affected the situation in many yards but on the whole the delivery has been very satisfactory. Plates for the steel shipbuilders have been coming through from the States in good volume and no complaints of a serious character are heard from local builders.

Fourth Boat Launched

On Saturday noon, Sept. 21, the last of the four 3,000-ton cargo wooden vessels under construction by the Fraser Brace Co., of Montreal, for the Imperial Munitions Board was successfully towed out of the basin into the Lachine Canal and placed alongside of the other three launched earlier in the season. Some little trouble was experienced in the launching of the last boat, owing to miscalculations in the cutting of the ropes or the knocking out of the trigger posts. The ceremony was set for the Wednesday, but it was Friday morning before the vessel was completely in the water, the bow remaining fast on the bank after several attempts to remove her. No serious damage was done to the hull as ample water was kept under the vessel. Pumps were kept operating night and day filling the basin to its maximum depth, this being approximately five feet higher than the normal level of the canal. The installation of the machinery in the War Huron and the War Erie, the first two launched, is progressing rapidly, so much so, that the first named is now about ready for her initial trial trip, if not already taken. The War Ottawa, the third ship launched, is being rapidly fitted out both as to the machinery and the superstructure, some delay having been experienced in the latter owing to the inability to obtain some of the necessary lumber. The contractors for the vessels are doing all the work on the boats with the exception of the mechanical equipment, this being installed by the Imperial Munitions Board, who have a staff of engineers and workmen on the site. Immediately the basin has been cleared of the water now in it work will be commenced in clearing up the yard for the start on six smaller wooden vessels, 206 feet in length, for the French government, operations on these to continue throughout the winter. It is anticipated that the experience gained in the building of the larger vessels will enable the builders to make much better progress on these smaller ships, the completion of which is looked for early next year. Owing to the depth of the basin into which the large boats were launched it was necessary to place a coffer-dam at the entrance with sufficient height to permit of an additional depth of from three to five feet of water in the basin, to float those constructed on the floor, and also to launch those erected on the bank. In the case of the smaller vessels the coffer-dam will still be required, but the extra depth of water will not be necessary owing to the lighter draft.

War Niagara Inspected

Following the launch of the War Niagara, the 3,000-ton wooden vessel lately constructed by Fraser Brace Co., she was taken to the Montreal Dry Dock for inspection to see if any damage had resulted from the slight mishap during the launching proceedings. It was found however that everything was in first class condition and she will be refloated immediately and returned to the site to undergo the completion of her superstructure and the installation of her machinery. Operations at the Montreal Dry Dock this season have been heavier than in any other year of its history owing to the abnormal activities in the shipping industry and the heavy demand for all classes of boats, many obsolete vessels having been placed in dry dock and repaired or overhauled for temporary service.

Progress at Vicker's Plant

In reading of records in shipbuilding, launching and fitting out, reported frequently from Great Britain and the United States, the achievement in and about Montreal should not be overlooked, as the work being performed here may come as a surprise to Canadians in general, and particularly to those not closely associated with the shipbuilding industry. During the past season remarkable activity has continued at the yard of the Canadian Vickers Limited, and before navigation closes this fall it is anticipated that nearly 74,000 tons of shipping will have been launched. Early last month there was launched at the yards the 7,400-ton steel cargo steamer "Samnanger," and following the launching the installation of the machinery and boilers, all of which were constructed at the works, was begun the same day. Nine days later the engines were run under steam, and six days after that the ship was complete, that is, fifteen days after the empty hull was launched. This equals the record recently reported from England of the completion of an 8,000-ton cargo steamer fifteen days after launching. The "War Earl," "War Duchess," and the "War Joy," are other vessels completed or nearing completion during the present season by this company, all of them over 7,000 tons, while four more cargo steamers are in course of construction with two others ready to lay down. Deliveries of cargo vessels from this plant since the opening of navigation this year totals around 30,000 tons. Of the nine vessels included in the year's programme, three will form units of the Canadian government's service. Considering the small number of men available in Canada for this industry the showing is exceptionally good.

Labor Troubles Settled

The Royal Commission on Shipyards, which has been in session since the beginning of August, inquiring into the labor disputes that have arisen in several of the yards along the St. Lawrence has recently completed its work in this connection, and as a result an agreement has been signed between the owners and the workmen, which assures for the duration of the European war uninterrupted ship construction work in the different yards where strikes were threatened. There is also a prospect that additional agreements will shortly be negotiated with other yards throughout Canada where trouble of this character has developed. The essential features of the agreement is the settlement of a nine-hour day and the sliding rate of wages regulated by the cost of living as shown by the official figures published by the Department of Labor.

---

ernment at the present time recognize the price of 7½ cents at mills, while for points around here warehouses sell it at 10 cents per pound. Prices for all articles needed in brass trade are around the 40c mark. Mr. McCulloch believes that it is possible to establish a steel industry at the coast, and states that capacity could be secured the year around. "It is a fact," though," admitted Mr. McCulloch, "that Fernie coke has not secured the coast market, not because it is not good enough, but because the men who want it cannot depend on deliveries. There are fine ridges of iron ore near the coast. It will take capital and work to do it, but it is not impossible."

Mr. McCulloch is an Ontario boy, being born at Cornwall. e received his university course in Toronto, and has been working as a mechanical engineer in the West for some years, spending some time in Winnipeg with the Manitoba Bridge, after which he went to the coast.

The MacLean Publishing Company
LIMITED
(ESTABLISHED 1888)

JOHN BAYNE MACLEAN - - - - - President
H. T. HUNTER - - - - - Vice-President
H. V. TYRRELL - - - - - General Manager

PUBLISHERS OF

MARINE ENGINEERING
of Canada

A monthly journal dealing with the progress and development of Merchant and Naval Marine Engineering, Shipbuilding, the building of Harbors and Docks, and containing a record of the latest and best practice throughout the Sea-going World.

B. G. NEWTON, Manager. A. R. KENNEDY, Editor.

Associate Editors:
J. H. RODGERS (Montreal). W. F. SUTHERLAND.

OFFICES

CANADA—
Montreal—Southam Building, 128 Bleury St., Telephone Main 1004.
Toronto—143-153 University Ave., Telephone Main 7324.
Winnipeg—1207 Union Trust Building, Telephone Main 3449.
Eastern Representative—E. M. Pattison.
Ontario Representative—S. S. Moore.
Toronto and Hamilton Representative—J. N. Robinson.

UNITED STATES—
New York—R. B. Huestis, 111 Broadway, New York, Telephone 8971 Rector
Chicago—A. H. Byrne, 900 Lytton Bldg., 14 E. Jackson Street, 'Phone Harrison 1147
Boston—C. L. Morton, Room 733, Old South Bldg., Telephone Main 1024

GREAT BRITAIN—
London—The MacLean Company of Great Britain, Limited, 88 Fleet Street, E.C., E. J. Dodd, Director, Telephone Central 1960. Address: Atabek, London, England.

SUBSCRIPTION RATE

Canada, $1.00; United States, $1.50; Great Britain, Australia and other colonies, 4s. 2d. per year; other countries, $1.50. Advertising rates on request.

Subscribers who are not receiving their paper regularly will confer a favor by telling us. We should be notified at once of any change in address, giving both old and new.

Office of Publication, 143-153 University Avenue, Toronto, Ontario.

Vol. VIII. SEPTEMBER, 1918 No. 9

PRINCIPAL CONTENTS

Radical Change in Ship Construction Methods

ANOTHER application of the electric welding outfit is in welding ship hulls. A ship has been recently built in Great Britain in which rivetting has been largely dispensed with, and the joints made by electro-welding.

This opens up a field of wide possibilities once the difficulties of initial cost of electrodes and plant are overcome. In point of strength there should be no difficulty for the strength of a rivetted joint seldom exceeds 70% of the solid, and good welding can be at least as strong. For the outer skin of the ship, where plates overlap each other, this method of making a joint should be easier and better than rivetting and it dispenses with caulking. For plates on tank tops, it should be a quick and serviceable method, but for joining framing and intercostals, it would hardly seem as good as rivetting, the extra stiffening of the rivet heads being lost. In many of the large passenger ships of recent building, the rivet heads in the upper courses of plates, as high as the hurricane deck, have been left full as this materially decreases vibration by stiffening the plate.

For watertight bulkheads this method is very suitable, and wherever it can be employed it eliminates the marking off, boring and rivetting of plates which means an immense saving of labor, and it also saves the rough usage of the plates occasioned in drifting holes into alignment. It will be interesting to watch the further development of this latest improvement on our old methods.

The Real Meaning of Success

J. J. WARREN, managing director of the Consolidated Mining and Smelting Co. of Canada, at Rossland, B.C., has announced that the company will give a $500 scholarship to the "son of any employee of the company working at day labor, who heads his class in the matriculation examinations for applied science in the British Columbia University."

The stipulation that the boy shall be the son of a man working at day labor is not a form of patronage. It faces a condition that actually exists, viz., that the son of a day laborer as a general thing stands a very poor chance of getting a course in applied science at a university. Too often the force of circumstances that made the father a day laborer is operating to do the same thing to the son.

The company that gets close to its employees—that sees in industrial life something more than clock punching and dividend notices—that wants the sons of its day laborers to have a chance to occupy better positions—that company is going to succeed in the higher meaning of the word success.

For after all success cannot be measured entirely by the expansion of plants, the paying of dividends or the declaring of bonuses.

AN extraordinary story of a boy's enterprise was related at London Sessions, when Robert Scott (17), engineer, pleaded guilty to receiving stolen tools. Mr. St. John Hutchinson said accused was undoubtedly a lad of exceptional ability with a future before him. He left the Hugh Myddelton School when fourteen years of age, and entered prosecutor's employ, showing such engineering skill that at fifteen he was earning 35s. a week. While still that age he and another boy took a back room and commenced making screws and nuts. The partner, reaching eighteen, had to join the army, but Scott continued working, and was so successful that he was able to take premises at £130 a year and employ eleven men and eight women. He had saved £400 and spent it on machinery, and a further sum saved had been invested in war stock. He now held large contracts with the Government relating to aeroplane work. Sir R. Wallace, K.C., said that all must regret that the lad had used his marvellous gifts in such a way. He would give him another chance and bind him over to come up for judgment if called upon.—Birmingham Post.

* * *

AMERICAN officers in Canada are well pleased with the way Canadians are handling their orders. They started out to make this place the granary of the Empire, but they simply can't keep from sticking a machine shop in between the farms.

REWARDING INITIATIVE IS TO SECURE MORE OF IT

Frank E. Wall, General Manager of the Polson Iron Works, Has Filled Many Responsible Positions

"THERE'S what I mean by initiative." The speaker indicated a small gripping device on his desk. "Of our sixteen hundred men," he continued, "I told you that about fifty study at night. One of these studious boys made that and brought it to me."

The speaker was Mr. Frank E. Wall, general manager of the Polson Iron Works, Limited, Toronto.

The gripping device was a most practical-looking metal chuck. In size and outward appearance it was not unlike an old-fashioned wooden potato-masher. But that five-inch projection—corresponding to the rounded handle of the potato-masher—was square.

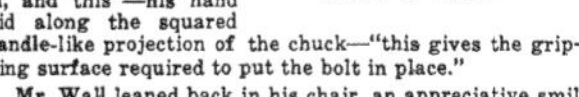

FRANK E. WALL

"It will save labor and time hitherto required for squaring ends of stay-bolts red hot under the hammer," Mr. Wall explained. "You know how a stay-bolt is ordinarily put in: the end squared to provide a gripping surface. Then, when the bolt is in place, the squared end is cut off, leaving about one inch protruding, which is riveted tight to the boiler plate. Well, this chuck grips the staybolt without it having been squared, and this"—his hand slid along the squared handle-like projection of the chuck—"this gives the gripping surface required to put the bolt in place."

Mr. Wall leaned back in his chair, an appreciative smile prefacing his next remark.

"I am sending a personal letter to the young man who made that chuck and with my letter a tidy cheque. It may be an incentive to other men in our plant."

"And to other men in other plants," was the unspoken thought of MARINE ENGINEERING.

Came Early to America

In London, England, thirty-five years ago, Frank E. Wall was born. A year later the Statue of Liberty greeted him and his parents and, in due time, the public schools of the Republic taught him the three R's. High school and University beckoned, but his chosen work called.

Consequently, before his twentieth year, young Wall had completed his apprenticeship and was well advanced in a course of study under the private tutelage of a man who, in the words of his one-time pupil, "is the peer of the best naval engineer that ever put ships on paper."

The year 1905 found Mr. Wall in the United States Navy yards at Norfolk, Va., filling his first situation of importance. From then on he held responsible positions, one an appointment to the engineering staff of the Public Utilities Commission of New York City. The subways, then in course of construction, presented engineering problems that Wall helped to solve.

Along in 1915 the Mobile Shipbuilding Corporation had scouts out for a man capable of designing and supervising the erection of a new shipbuilding plant. Wall's record marked him as a likely man for the job. But Wall himself didn't look older than his thirty-two years. He moved to Mobile, Alabama. The more he heard of the Mobile plant the more he wanted to build it. And he did build it—so thoroughly well that Uncle Sam at war wanted him.

So 1917 found Frank E. Wall with the United States Shipping Board as a supervisor of steel ship construction. And here he remained until the Spring of 1918, when he accepted the position he has since filled with credit to himself and to Polson's.

"Not a book on shipbuilding or mechanics—not a paper worth reading—gets by; study them all," said Mr. Wall. And that from a young man in a big position is a fine appreciation to the technical press.

A photograph? Mr. Wall was persuaded. He found many of work and one of particular interest that had nothing to do with work. It was a snapshot of Mrs. Frank E. Wall and their little family of four.

"I thank her for my success—for eighty per cent. of it," Mr. Wall said seriously in conclusion.

Bill Haywood in Jail Now

BIG Bill Haywood, leader of the I.W.W. in United States, has been sent to prison for twenty years, and a stiff fine has been imposed as well. Other lesser lights have gone down for lesser terms. The peculiar thing is that Haywood and those associated with him have been able to get away with their rubbish as long as they have and remain outside the prison gates. Of course, Haywood has been jailed a good many times. It is no new sensation to him to look at current events from behind the bars. But this time it looks as though Bill were in for a good long look, and by the time twenty years turn over there won't be much rumpus left in the carcass of said Bill.

Haywood is well known in Western Canada. In fact the Crow's Nest Pass district in the south of British Columbia, the towns along the foothills of the Rockies in Alberta, and all the district where socialism runs rampant, were the real stamping ground for Haywood. He's not altogether a pleasant chap to gaze upon. He's big, well developed, but has one optic that is badly damaged. But to meet the man off the stump is to get a shock. He is not the lawless tub-thumper any more. His voice is mild if anything, and he betrays nothing of the bravado or the apostle of blood and thunder that has made him famous. He accepts arrest in the very best spirit. In fact it would almost seem that he did so with a feeling of pity for the police, the law makers, and all and sundry who had anything to do with skidding him toward the cooler.

But allowing for all that, he's a better resident in jail than outside it. The one question that has never been satisfactorily answered is, "How did you get away with it all, Bill?" For certain it is that had the small fry of the land taken it upon themselves to make such a stink as Bill stirred up, the law would have had them as soon as the sheriff had time to pull on his knee boots and fasten his office tag on the lapel of his coat.

* * *

IT'S an awful sensation for a man to go away for a couple of week's holidays and then come back and find that business has gone along splendidly during his absence.

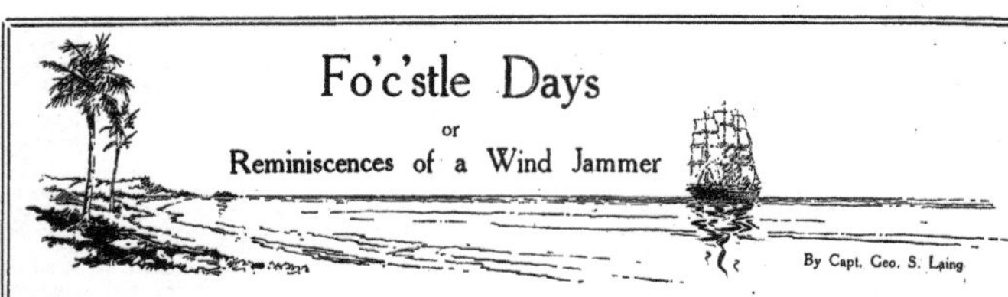

The steamer and the motorship have deprived the seafarer of much of the old-time romance attaching to a sailor's life. The advent of wireless has also increased the safety of it. Deprived of these features, a sailor is immediately dependent on his own resources, and Captain Laing gives us a true and seaman-like reflection of sea life under canvas. The present feverish activity in constructing self-propelled craft affords a suitable background for incidents in a career in which gaffs, booms, tar and canvas held sway before being displaced by mechanical unloaders, derrick masts, fuel oil, coal and wire rope.

Part VIII

AUSTRALIA'S FAMOUS 12c MEALS

SOME of our purchases on the first night ashore in Brisbane consisted of an album of local views, a copy of the paper which had written up our passage and a "burst" at a favorite "feed shop" kept, of course, by a Chinaman. At the time I write of the Australian people had found it necessary to prevent any further immigration of the almond-eyed race for fear the country would be spoilt for our own British stock. The rabbit also, which had been imported, had already become a nuisance, if not a danger, and a price was on its head too.

When the Chink saw four new sea apprentices enter, some of the kitchen help were called in as detectives, for, "rough-housing" was not unusual. None of the "Maggie Dorrit's" boys were that way inclined except Ffd, the other three being too green and timid on their first voyage abroad.

As we made our way back to D. and L. Brown's wharf, a few jocular remarks were thrown out for our ears by the passerby. The Australian folks of all classes had always a liking for young sailors. Those of them who had experienced the passage from Britain in sail could never forget to pass such words along as:

"Where are you boys from?"

"Have a good pasage out?"

Then with a final salutation the well-wishers would turn on their heels and throw back in glee such deck orders as:

"Heave the log."

"Light the binnacle."

Perhaps a faint "any mulligatawny left?" would also pierce the ear. This query had reference to a soup that was well known on all immigrant vessels, and one must remember that the great bulk of the colonials had spent from three to four months on the trip to the land of their adoption, and amongst the features of such a tedious passage the cramped accomodation and the daily menu were "institutions" of miraculous power, when contrasted with man's mode of living ashore.

Our Moral Welfare

Another expression of comradeship was shown to sea-urchins by inviting them to people's homes on the least shadow of acquaintance, in any of the Australian harbors. Say that a parent was strolling around the docks on a Sunday with some members of the household, and he came across a ship hailing from his native sea-port at home. That was sufficient, the boys of the half-deck were made welcome at this party's house. Many others were ever ready to welcome the sea apprentice to places of wholesome amusement, such as musical evenings, Saturday afternoon games and Sunday meetings. At the time this reminiscence speaks of, a grea number of Australians were deserter from vessels—apprentices and other who had made good in their new spher As soon as any of that class came t the coast from ventures in the interio they made straight for the shipping t renew old associations and in some case to offer enticements.

FLORA OF AUSTRALIA

Without A Crew

Before our barque was half discharge the fo'c'stle hands had gone. Even ou second mate had deserted, likewise th cook and sailmaker. Tilbury, howeve hung on till the day before we saile but he got his belongings ashore piece meal, whilst our joint financial help wa given at a farewell ceremony at th favorite feed shop. We had to be ver foxy to insure a safe desertion for ou parson's son, whose hatred of a sea lif was akin to madness.

A cable told us that our craft wa expected at Newcastle, N.S.W where a cargo of black diamond awaited us for Coquimbo, Chil S.A. We had seen two aborigine very old men, with their lan tresses and withered features. much-travelled individual had als interested us in his yarns abou the peculiar creatures that Austra lia can boast of, such as duck-bille platypus, laughing jackass bir marsupial mammals, opossum rat cockatoos, etc. Our much-inform ed friend had carried his swa from Melbourne to the Cape Yor country in the tropics and, o course, had originally been "brass-bounder" like ourselves.

We were soon assembled o shore to shake a good-bye flippe with Tilbury. From our scar drawings we had been able t throw seven shilings in the war derer's way, and at midnight h started for the interior with hi swag and black thorn regalia.

coasters, steam and sail, the shipping consisted of barques and full rigged ships, mostly loading coal for Chili, Peru, California and the Philippine Isrands. A particular fleet of Sydney colliers were known as the "60 milers" as they had settled down to that short run much the same as a section of the "Geordie brigs" stayed, year out and year in; on the route between Newcastle and London in our home waters, and with the same cargo as we have under discussion—coal. In passing it might be emphasized that whether on land or sea one of the greatest problems of mankind is to transport coal.

Ocular demonstration of horse power was in evidence at our ballast berth, and showed how hard it is to break away from primitive methods. This equine study was one of circus regularity and utility. In the discharge of our earth ballast, a horse dragged the main purchase through a shoe or snatch block on the dock, thus raising a 500 lb. basket of material at a time from the hold. At a certain spot the noble animal stopped and the load was transferred to a yard-arm jig, which in turn swung it ashore. On the limit of swing being reached a "tripping line" tightened from the bottom of the basket and shot the ballast out in the air to drop on a hill whose base had to be kept a certain distance from the dock side.

Like all other sailor towns Newcastle was full of "free and easys," a term given to saloons that have a singing and dancing hall off the bar. Two of these varnished hells were known as the Clarendon and the Black Diamond and I am sure that some of my readers have heard yarns concerning them. Our "old man" told us to keep clear of these places, altho' at our age the Chink feed shops had the heavy pull. Captains and mates frequently gave good advice about these liquor palaces, and then steered straight for them on the hopes of having a stormy spree all to themselves.

Boys Describe Their Craft

Over a score of "brass-bounders" strolled about the streets and parks of this famous coal port of the southern hemisphere in the evenings and we soon got a store of information, which was more or less exaggerated, concerning items vital to boy life at sea.

Some boys would have us believe that their "packet" was as comfortable as the "Valhalla" or the "Sunbeam," two world famed yachts. Then again some imaginative apprentice would describe his "hooker" as a real slave dhow, where every officer kept a belaying pin up his sleeve, and knuckle dusters were worn after sunset.

Of course, the normal boy would call a spade a spade, and many of us were far better treated than the men who were our superiors, had been in their time. The fact was, that life on "windjammers" had certain drawbacks which were inseparable from it. The danger of going aloft, or working in the waist at the braces when seas were tumbling over the weather rail, the unhomelike surroundings and disciplined rules were all irritable to the boy who had come to sea believing in the assurance of utopian voyages that took in a few tropical islands and on which feasting and laying around in a long-sleeved chair became the principle manœuvres.

Every one must "toe the line" according to graded position. Howling gales, buffeting seas and the constant strain of fighting these elements prevent ships' officers from acquiring the highly polished, though maybe hollow, manners of a Spanish nobleman, whose greatest worry in any one day of his life might be the removal of a geranium petal from his immaculate and artistic summer vest. So it can easily be imagined that quite a few sea-apprentices thought themselves martyrs to an unconquerable profession, when in reality they were spoilt mother's sons being licked into shape.

Let it be known, however, that from a point of view of food, living apartments and remuneration, life on windjammers was a disgrace to the British Empire. Now that this war has opened up so many channels of common sense and avenues of clear vision, Jack of the merchant fleet is a little better thought of and God knows it wasn't before time. With the three essentials raised, the animal comforts of the sailor are progressing, but it will take very strong representation to keep his status up after the war. In the meantime we have men in high places stating that the merchant navy will henceforth and for ever be a national fleet, a twin sister to the powder puffs of the gigantic royal fleet. Amen—so be it.

(To be continued).

DEVELOP FLUORSPAR MINE IN B.C.

"The Consolidated Mining & Smelting Co. have started development of the fluorspar deposit which they bonded on the north fork of the Kettle river," writesA. E. Haggen. "This is one of the most important mineral discoveries recently made in the province of British Columbia. It is used in the manufacture of hydrofluoric acid, of which a large amount is consumed in the electrolytic lead refinery at Trail, and as a flux in silver-lead and copper ores. With the establishment of the iron and steel and glass-making industries in British Columbia the existence of fluorspar will prove of great value. In the smelting of iron and steel fluorspar and phosphorus contents, and increases the tensile strength of the metal. Other uses are in the manufacture of spiegeleisen, foundry work, cupola, furnaces, manufacture of enamels, glazes, fireproof ware, apochromatic lenses, gems and carbon electrodes for flaming arc lamps, so that the discovery of this deposit is not only of value in existing metallurgical industries, but in the future industrial development of the province. A good deal of the mineral has already been packed out on horses, but a road is being built to connect the deposit with the railway, a distance of about eight miles.

SEAMANSHIP AND NAVIGATION

Conducted by "The Skipper"*

Articles of direct interest to mariners, discussions of seamanship, navigation rules, and allied topics. Inquiries from readers are invited and will be replied to promptly through these columns, unless otherwise requested.

BEFORE proceeding to work out the numerous problems of dead reckoning and observation it becomes imperative that the sailor student know something about the motions of the mighty waters. To handle a vessel in a seamanlike manner without acquiring such knowledge is simply an impossibility.

Tides

A practical definition of this word is the breathing of the fluid part of our globe, or, the alternate elevation and depression of the sea.

These movements are caused principally through the attractive power of the sun and moon. The latter has twice the drawing power of the former luminary as it is much nearer the earth, thus at full and change of the moon the highest tides occur, or within two days of these periods.

That high tides should occur simultaneously on opposite sides of the globe is accounted for by the fact that the attraction of the sea immediately under the moon is greater than that upon the solid globe taken as a whole, thus the water involved rises. On the other side of the globe the reverse is the case, the waters lacking rigidity like the solid earth, are less strongly attracted, consequently a bulge away from the moon takes place. Is that clear?

The sun's action on the waters as far as attraction is concerned is similar to that of the moon with the difference however, that it only manages to augment or modify the stronger effect of the moon, according to the relative position of the celestial bodies to that of the globe.

Without splitting scientific hairs it can easily be noticed that such governing laws of motion are influenced by contour of continents and other tributary causes such as indentation of coasts, proximity of islands to mainlands, the ebb of rivers, coast gales, etc.

Look upon tides then as vertical motions, against horizontal motions of water which come under the category of currents. The combined pull of sun and moon cause spring tides, which are high, whilst neap tides or low tides are the result of sun and moon pulling at right angles to one another.

These tide waves have their birth and unobstructed action on the broad expanses of ocean and hence do not occur in landlocked waters such as our Great Lakes, which, however, answer to nature's call by sending their surplus waters back to Father Ocean (their original source) through the medium of ebb motion—back to sea levels.

How is it that we find abnormal sea movements in the Bay of Fundy, Canada, and the Bristol Channel, England? Tides in these places rise and fall from 30 to 50 feet and more at certain times. This is excessive, but not when you take into account the fact that in both cases the tide becomes bottled owing to geographical outline. The tides that enter these estuaries have no alternative but to rise abnormally as their motion has not only been influenced by the coast line but it has been virtually stopped altogether.

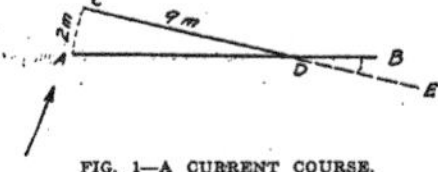

FIG. 1—A CURRENT COURSE.

Thus it can be seen that the motions of the tide are not perceptible on the oceans where their movements are constant and free, but these same waters cause an approaching and receding motion on the coasts which is termed flood tide and ebb tide respectively. Further on in our studies we will take up these matters in relation to chart work, as the navigator has to find how the tides affect the depth of water in harbor and river mouths. Very few harbors allow a ship to have a roving commission, stage of tide and vessel's draught have to be constantly watched, if "grounding" or "running ashore" are to be avoided.

Here and there on the charts, especially at big ports, the time of high water at full and change of the moon is given in roman numerals.

Some very practical illustrations of tide effects are given to the observant sailor. For instance the "falls" and "rips" that can be seen off the coasts. At first sight these water flurries are apt to be taken as indications of a reef or other submerged danger, but in reality the eddy patches only denote the effect of the tide on an uneven bottom, perhaps changing from sand and shells to rock and kelp, or again inshore ledges giving place to deeper beds. In these and in other such cases the disturbance on the surface resembles the turbulent motion seen at sluice valves, sewage outlets and gate and lock operations where force is at work under water.

Cross seas are also a coast product of the tides caused by wind and tide acting in different directions on the waters and setting up a short but dangerous sea, especially where small craft with little freeboard are concerned. The writer's most startling experiences with tides were gained in such grand "schools" as the Pentland Firth in the north of Scotland, also in the Hebrides, Norwegian coast, and in tropical places, such as Sunda Straits. In these spots and, of course, in many others, the becalmed windjammer or the famous 8 knot steam tramp with a foul bottom almost became helpless in the strong tides which sometimes attain a velocity of 6 to 10 miles an hour. If a low powered steamer breaks her sheer frequently in such a tide the chance of grounding or collision becomes a possibility.

The only good point that aids vessels of a helpless nature in getting beyond the power of such coast confined tides is the fact that every six hours the tide runs in the opposite direction. In this respect it is often advantageous to await the precise moment at which one enters a narrow channel, for if the tide is favorable to the ship's course such places can nearly always be negotiated and cleared on one tide. Tides of less than five miles an hour are the most common and are just taken as they come, due allowance, however, has to be made in relation to courses and speeds and the effect of the local tide on such factors.

As the moon is the chief factor in forming the tides and it takes approximately 28 days for it to make a circuit of the earth, it can be deduced that at fortnightly periods the tides will have their greatest movements—springs. The intermediate periods will also create the sluggish tides called neaps.

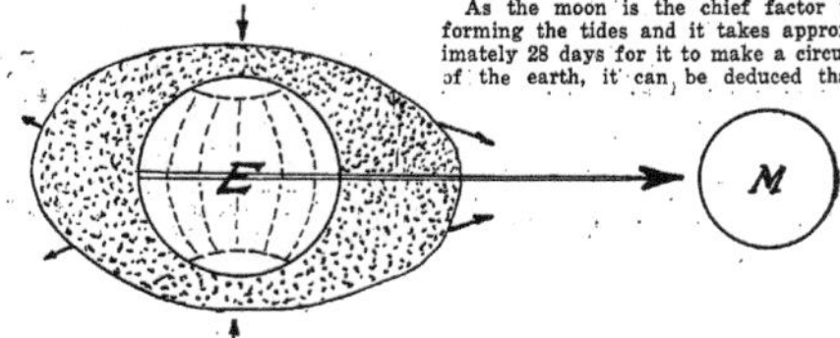

MOON'S ATTRACTION ON WATERS OF THE GLOBE

Currents

The science of the sea is still an infant and the line of demarcation between tides and currents is a flexible

*Editor's Note.—This department is conducted by a qualified ship's master, and will deal with questions involving seamanship and navigation. The attention of young men desiring to obtain certificates as mates and masters is particularly directed to this department.

one. It can be said, however, that a broad difference between the two motions is very apparent. In the case of tides we have the earth's satellite (moon) exerting the main force by pulling the waters of the earth towards it as if by a magnet. But in the case of currents we find that other agents enter the field of action. Among them, the winds, temperatures, and axial movements of the sphere can be mentioned. Salinity in its relation to specific gravity, evaporation, especially within the tropical boundaries, and rain or precipitation also figure in the wonderful machinery that gives impetus to ocean currents. In a general way solar energy in the form of heat is perhaps the big stick in currents of the ocean, as it is in the currents of the air—wind.

The water of seas and oceans is constantly on the move, stagnation is unknown. True, very deep parts of the ocean move slower than the surface parts, but nevertheless a universal commotion is ever at work and those who have studied physical geography from all angles are at one on this point. From a sailor's viewpoint the currents that affect his craft are, of course, on the surface, altho' a few well defined streams are known to have sub-surface motion—the strong parts of the Gulf Stream, Labrador, Agulhas and other great ocean rivers. Quite a few fathoms down these large motions are known to have a velocity.

An important current, however, is the purely surface motion called a "drift." This movement is set up with wind and naturally follows such steady winds as "trades" and "monsoons." Again, the drift motion answers the command of gales even of short duration and this must ever be borne in mind whether off Sable Island or the Burlings.

Beginners must remember that in sea parlance a current is named after the compass point **toward** which it flows. This the opposite rule to naming a wind. For instance, a S.W. current moves to the S.W. and, of course, would carry your ship there, whereas a S.W. wind moves to the N.E. Clinch that difference before going any further. Some teachers put it this way, currents set to, and winds blow **from.** We speak then of a current setting to the E.N.E. with a **drift** of 4 miles an hour. The **set** is the direction, the **drift** is the speed.

How The Great Currents Act

The laws that govern the well defined currents already mentioned seem to act in the following manner. All heated or super-heated waters within the tropics are impelled towards the poles. All polar or cold waters trend towards the earth's centre—the equator. This will cause the waters of Cancer—the northern tropic—to flow northerly, and the waters of the southern tropic—Capricorn—to flow southerly. Each of the polar waters will act in opposite ways and so feed the movements.

Thus we find that the Labrador current with its berg studded bosom sets ever to the south to melt its glacier born craft in the warm waters of the Gulf Stream. This current as you will find flows northerly along the Florida coast and branches off to the N.E. at Cape Hatteras. While its velocity decreases and its river dimensions broaden out till it is hardly recognisable, it is due entirely to the warm waters of this equator born stream that the British Isles has such an equable climate.

In the chart shown herewith the student will find that apart from the deflections caused by coasts or islands, and also remembering that the earth's rotary motion and winds have an effect on ocean movements, the main bodies of water act in the manner just mentioned. In many cases where it is not even dreamt of, the vast difference between the ship's position by observation and her position by dead reckoning is entirely due to currents and for this reason a chart of prevailing winds and currents should always be consulted. Only by such procedure can finer navigation be expected.

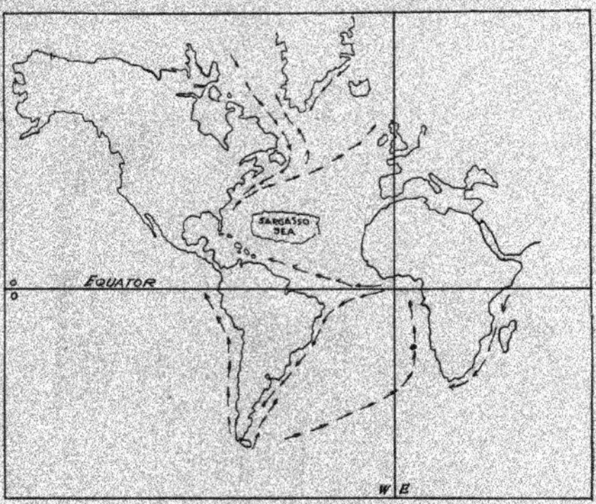

COURSE OF OCEAN CURRENTS

Unwelcome Antics Of The Sea

Father Ocean has a few abnormal movements, but we just have to put up with them. Damage and not service is the result of such motions, and they are frequently related to volcanic eruptions with subsequent changes on the sea floors. I have in mind the famous Krakatoa (Sunday Straits, Java) incident, where a large part of the island was literally thrown into the sea by volcanic action, causing a huge upheaval of water to race wildly on the nearby town of Anjer and wipe the place off the map. After visiting this spot as a boy and hearing of its peculiar record, I was more than interested some years later when within five hundred miles of the same island we steamed through a patch of calm water the color of milk. Some terrific movement on the sea floor must have happened minus the wave effect.

Figure 1 illustrates a current course. The student may be able to realize more fully in chart work how important this kind of work is. Suppose your ship at A, and you wish to steer to B. The course would be east true, but a current sets N.N.E. with a drift of 2 miles per hour. Ship's speed is 9 miles per hour.

Roughly, such a current would carry the vessel to port so that you, at least, know that you must keep her a little to the starboard hand.

Draw your line from A to B. From A lay a off an hour's current and mark that C. From C cut A B with an hour's run (9 miles), and call that point D. Join C to D and extend to E. Then B D E angle is the amount to hang her up to the starboard hand, in this case the southward (12°). With the parallel rulers this can easily be found at the nearest compass rose on the chart. S 78° E would be the course instead of East.

You will note that the parallelogram of forces enters the problem. Of course, a current right ahead or right astern simply reduces or augments your speed. An oblique force is different however.

Figure 2 gives an idea of how the waters of our globe follow the moon on one side and bulge out on the other, representing the high tides at full and change of the celestial body in question.

The chart illustrates a few of the ocean's main currents. All cold water is bound south, and all warm water bound north in our hemisphere. The opposite condition obtains in the southern half of the world. In the north Atlantic the Gulf Stream and Labrador motions are depicted. In the part south of the equator, the Brazil current is hiking to Cape Horn, whilst Cape Horn's icy waters take a N.E. course, and a coast course off Peru and Chili.

(To be continued)

PROGRESS IN NEW EQUIPMENT

There is Here Provided in Compact Form a Monthly Compendium of Shipbuilding and Marine Engineering Auxiliary Product Achievement

RADIAL DRILLING MACHINE

A wall radial drilling machine, as made by the Lynd Farquhar Co. of Boston, Mass., is described and illustrated herewith. This is a well designed and carefully built machine, and special care has been given to locate the entire control of the machine within easy reach of the operator.

The arm is constructed of extra heavy channels, accurately planed top and bottom, with substantial supporters at each end, and is supported from outer end to top of wall bracket by heavy steel brace bars.

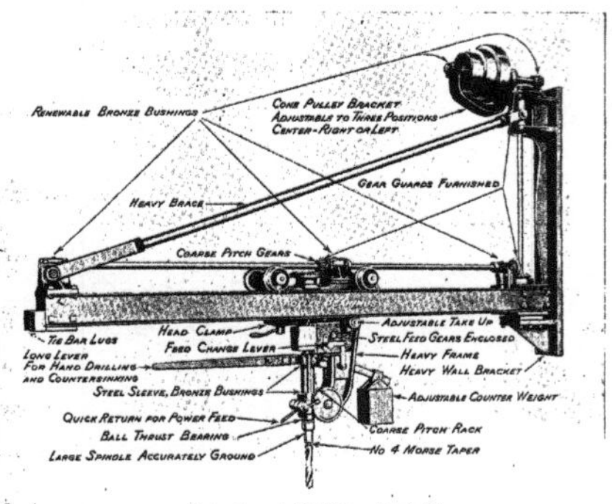

WALL RADIAL DRILLING MACHINE

The wall bracket is heavily ribbed, planed on back and where bolted to the wall is 10 in. wide, 6 ft. 10½ in. high. Bracket at top of machine with bevel gear housing can be located at three positions for convenience in connecting belt drive from countershaft.

The motor application is made by mounting 5 to 7½ H.P. variable speed motor on suitable brackets, which will be furnished at extra cost, in place of bracket that carries bevel gear housings.

The head is exceedingly rigid, mounted on four flanged wheels fitted with roller bearings, and moves with extreme ease from end to end of arm. All gears are accurately cut from the solid, feed gears being of steel. The bearings are bronze bushed and renewable. The wheels, carrying heads, are fitted with high grade roller bearings.

A hand lever feed, nicely counterbalanced by adjustable weight for light drilling and countersinking, can be adjusted to remain in any position. The geared power feed has two changes, .015 to .025 per revolution of the spindle. Can be changed while drill is in operation, and provides a good range of feed for drills up to 2½ in. diameter. An automatic release of power feeds to spindle is provided at extreme traverse of spindle to prevent damage to feed gears.

The spindle is of high carbon steel, 2½ in. diameter in bearings, is accurately ground, runs in long bronze bushings, (renewable) within a steel sleeve; has 7 in. traverse; No. 4 Morse taper hole; is fitted with high grade heavy duty ball thrust bearing. Upper end of spindle 2⅜ in. dia., and slides through heavy steel driving sleeve to which is keyed the main driving gear.

An adjustable eccentric wheel on under side of arm keeps the carriage in proper adjustment along the channels. A clamp lever conveniently located, clamps head rigidly to the arm. Tie bar lugs are provided at extreme end of arm to receive tie bar in event it should be found desirable in extra heavy drilling. The countershaft is self oiling, tight and loose pulleys 16 in. diameter, 4½ in. face, and should run 350 R.P.M.

CABINET SURFACER

A cabinet surfacer built by the Oliver Machinery Co. of Grand Rapids, Mich., is illustrated herewith. This machine has a number of improved features.

The demand for an accurate smooth planer capable of a wide range of work both in soft and hard woods is well met in the design of this planer, its correct design, compactness, feeding power and general efficiency making it readily adaptable to the requirements of pattern makers, aeroplane manufacturers, government shops, engineering works, shipbuilders and railroad car shops.

The machine is made in two widths to plane either 24 in. or 30 in. wide and the bed will lower to receive and plane material up to 8 inches thick. The standard feeds furnished are 14, 18, 24 and 31 feet per minute.

The sides are of cored form, the girts are heavily ribbed machine jointed and bolted. Ample material, properly distributed, eliminates strains and vibration.

To assure perfect balance, hence smooth planing, the cylinder is of the circular type, made of a crucible steel forging of uniform texture and belted at both ends. It carries two thin high speed air hardening steel knives securely clamped against hard steel chip breakers, whose lips are shaped to repel shavings and chips. The entire cylinder is nicely ground to perfect size. The journals are long, of unusually large diameter and machine ground.

The cylinder pulleys are two in number, and their faces are grooved spirally, preventing air pockets under the belts and augmenting the belt power through its close contact with the pulleys.

The cylinder bearings are of the self-oiling side-clamping type with large oil chambers. They are readily adjusted to hold the cylinder firmly in place for smooth planing. Readjustment of caps can be made instantly.

The back pressure bar follows the cylinder and is held by adjustable screws for regulating the hold-down pressure on the lumber as it leaves the cut. The throat between the front and rear pressure bars is only 2¼ inches for service in planing smoothly and without end clipping on very short stock.

The chip breaking pressure bar before the cut holds the lumber firmly to the bed as it is planed, and prevents the chips from tearing its surface. It is attached to the main frame in such position that it raises concentrically with the cylinder, thus preventing the bar raising into the knives. A steel spring shoe is secured to the lower part of the bar, which not only holds the stock firmly on the bed but yields to any ordinary inequalities in rough stock. An adjustable weight regulates the pressure.

The bed is of suitable dimensions, gibbed properly, and is raised in a wide central slide that is deep to give the bed the proper rigidity. A guide is located at each side to prevent material from leading away from the bed and striking the frame or gearing. The bed is supported on heavy square thread screws with ball bearings, reducing friction and permitting the raising of the bed without exertion. The bevel gears used in elevating the bed are machine cut and are protected from dirt by a wide shelf that covers the inner space below the table and keeps the shavings within the frame making it very easy to clean around the machine. Directly underneath the cylinder is a centre plate adjustable to wear, removable from the bed when worn. It may be redressed and refitted or discarded for a new one as the occasion demands.

SINGLE CYLINDER FOUR ROLL DOUBLE BELTED CABINET SURAFACER.

There are four feed rolls, one pair back of the cylinder to feed the material free of the knives and assist in carrying it through the machine. They are large in diameter, made of hammered steel and are supported in self-oiling boxes. The upper feed-in roll is corrugated for gripping the stock firmly. All the rolls are driven by a train of heavy gearing, three pitch, the teeth machine cut, insuring a smooth and positive feed. The feed-in rolls are weighted and the delivery rolls have spring tension. The upper delivery roll has a cover over it and is provided with a scraper which keeps the rolls free from shavings.

Sectional feed rolls may be provided when so desired. It consists of sections 1⅛ inches wide and 4 inches diameter. While giving a horizontal drive as positive as a solid roll, they yield vertically independently of each other to the extent of 3/16 inch. Each section is composed of an outer ring enclosing four sections or seats placed radially to the centre of the roll shaft and each carrying a helical spring.

The sectional chip breaker consists of 1⅛ inch sections pivoted in connection with section weights, on two horizontal bars attached to two plates. They yield concentric to the cylinder independently of each other.

SECTIONAL FEED ROLL SHOWING POSITIVE DRIVE AND STRONG CONSTRUCTION.

The gearing that transmits the power to the rolls is all machine cut and perfectly finished throughout. Such gears as revolve on steel studs are provided with oiling devices that are right. Grease cups mounted on each stud furnish oil through channels in the studs to the inner surface of the hubs, producing effective lubrication. The studs are large and the bearings for shafts that pass through the machine are fitted with dust-proof spring oilers. The master driving pinion is hammered crucible steel and machine cut.

A grinding attachment is provided which will quickly and accurately grind the knives of planers and jointers without the trouble of taking the knives off the machine. The attachment can be quickly mounted on the machine or removed therefrom. The motor is mounted in grinder head; current is taken from an ordinary lamp socket.

METHOD FOR THE COMBUSTION OF BROWN COAL

By M. M.

Mr. E. H. Miller has introduced a process for the economical combustion of inferior qualities of coal in boiler furnaces, which gives a greater thermal return than when such coals are burned in air in the usual way. The principle of the process is to apply the conditions that exist in the water gas producer to the boiler furnace, and use exhaust steam for gas production. To accomplish this the producer forms part of the boiler setting, and is placed immediately under the present fire grate area. The ash pit and grate are dispensed with, and in their place is a chamber which is used for the combustion of the gas. By means of this chamber, which is situated immediately under the tubes in the tubular boiler, complete combustion of the coal is assured. The use of exhaust steam obtained from auxiliaries for gas generation permits of the latent heat of the steam being made available which represents nearly five-sixths of the original heat used to form the steam. The water gas, when generated, is at high temperature, and is mixed with air pre-heated by passing through the flues, and ignited. As the combustion takes place almost immediately, after the gas is generated, conditions are such that permit of obtaining a high percentage of the thermal contents of the fuel as effective heat for evaporation.

Experiments have been made using Morwell brown coal, containing over 20 per cent of water, and excellent results have been obtained. Complete control of the combustion is possible, the combustion being absolutely smoke-less. Morwell brown coal lends itself particularly to the preparation of water gas, the high percentage of the water in the coal being in no way detrimental to the formation of the gas. Such coal, when burnt in chain grate or under feed stokers, usually begins to burn when about to be discharged to the ashpit, the high percentage of water having lowered the temperature of the combustion zone long before combustion is complete. The advantage of generating water gas instead of producer gas (carbon monoxide) is that about twice the quantity of coal can be gasified on the same grate area when making water gas than when making producer gas; also advantage can be of the latent heat exhaust steam.

ASSOCIATION AND PERSONAL

A Monthly Record of Current Association News and of Individuals Who Have Been More or Less Prominent in Marine Circles

Arthur Irish has joined the staff of C. E. Disher & Co., brokers, Vancouver, and is calling on the city jobbing trade.

H. J. Main has been appointed Assistant Superintendent, Farnham Division, C.P.R., in place of W. J. Pickrell, promoted.

J. A. Cook has been appointed Assistant Superintendent, Smiths Falls Division, C.P.R., in place of H. J. Main, transferred.

H. J. Main has been appointed Assistant Superintendent, Farnham Division of the C.P.R., in place of W. J. Pickrell, promoted.

Capt. Geo. S. Laing of Toronto, under the auspices of the Ontario Division of the Navy League, will entertain local audiences throughout the season with his illustrated lecture on "Our Merchant Fleet In Peace or War."

W. L. Macdonald, broker, Rogers Bldg., Vancouver, has been appointed official adjuster of disputes between employees and employers in the shipbuilding trade. His decision is binding on both parties.

The Shepard Electric Crane & Hoist Co., of Montour Falls, N.Y., have appointed A. J. Barnes as export manager, with headquarters at Montour Falls, N.Y. Mr. Barnes will also continue to be director of publicity for the company.

Lieut. Edward Boynton has received his discharge from the naval air service after being overseas for a number of years. He has taken a position with J. A. M. Taylor, Toronto, and will call on the trade in Toronto and other industrial centres. Previous to going overseas Lieut. Boynton was with the Canadian Fairbanks-Morse Co. He was in many of the opertaions around Gallipoli and at Saloniki. At the former place he took part in 24 out of 31 raids that were conducted there, and for this received decoration.

Mr. William J. Langton, superintendent of the Dominion Transport Company, in Toronto, has been appointed general manager of the company with headquarters in Montreal. He will assume his new duties the first of next month, and will be succeeded here by Mr. Richard Walker, agent at Ottawa. Mr. Langton was born in Toronto and joined the office staff of the company thirty years ago, occupying his present position for the past thirteen years. He is president of the Transportation Club and is also connected with the Rotary Club, the Masons and other fraternal organizations.

George Chahoon, jun., president of the Laurentide Co., will presently join the ranks of the industrial experts in the war service of the United States Government. It has been known among Mr. Chahoon's friends for some time that he had volunteered his services for any work to which Washington might assign him in connection with the war. Announcement was made recently that he was to be associated with the chemical branch of the War Welfare Work of the American Government, with headquarters at Baltimore. Mr. Chahoon's departure will involve no change in the organization of the Laurentide Co. His services are being loaned to Washington for the period of the war, but he will remain president of the company as heretofore.

Ottawa, Ont.—Architect John A. Pearson, Centre Block, Parliament Hill, Ottawa, will receive tenders until noon of September 27 for the iron stairs, ladders, etc., required for the central heating plant of the new Parliament Buildings.

Buckley Bay, Queen Charlotte Id., B.C.—A wireless plant with aerial towers 200 feet high will be erected here for the Marconi Wireless Telegraph Co. of Canada, Limited. For further information enquire of the manager, L. S. Hankins, 172 William street, Montreal.

LICENSED PILOTS

ST. LAWRENCE RIVER

Captain Walter Collins, 43 Main Street, Kingston, Ont.; Captain M. McDonald, River Hotel, Kingston, Ont.; Captain Charles J. Martin, 13 Balaclava Street, Kingston, Ont.; Captain T. J. Murphy, 11 William Street, Kingston, Ont.

ST. LAWRENCE RIVER, BAY OF QUINTE, AND MURRA YCANAL

Captain James Murray, 106 Clergy Street, Kingston, Ont.; Capt. James H. Martin, 259 Johnston Street, Kingston, Ont.; John Corkery, 17 Rideau Street, Kingston, Ont.; Captain Daniel H. Mills, 272 University Avenue, Kingston, Ont.

MONTREAL PILOTS' ASSOCIATION

President—Alberic Angers, Montreal.

Secretary—C. B. Hamelin, Champlain, Que.

ASSOCIATIONS

DOMINION MARINE ASSOCIATION

President—A. A. Wright, Toronto. Secretary—Francis King, Kingston, Ont.

GREAT LAKES AND ST. LAWRENCE RIVER RATE COMMITTEE

Chairman—W. F. Herman, Cleveland, Ohio. Secretary—Jas. Morrison, Montreal.

INTERNATIONAL WATER LINES

President—O. H. Taylor, New York.

Secretary—M. R. Nelson, 1184 Broadway, New York.

SHIPPING FEDERATION OF CANADA

President..Andrew A. Allan, Montreal; Manager and Secretary—T. Robb, 218 Board of Trade, Montreal; Terasurer—J. R. Binning, Montreal.

SHIPMASTERS' ASSOCIATION OF CANADA

Secretary—Captain E. Wells, 46 St. John Street, Halifax, N.S.

GRAND COUNCIL N.A.M.E. OFFICERS

A. R. Milne, Kingston, Ont., Grand President.

J. E. Belanger, Bienville, Levis, Grand Vice-President.

Neil J. Morrison, P.O. Box 228, St. John, N.B., Grand Secretary-Treasurer.

J. W. McLeod, Owen Sound, Ont., Grand Conductor.

Lemuel Winchester, Charlottetown, P.E.I., Grand Doorkeeper.

Alf. Charbonneau, Sorel, Que., and J. Scott, Halifax, N.S., Grand Auditors.

1918 Directory of Subordinate Councils, National Association of Marine Engineers.

Name	No.	President	Address	Secretary	Address
Toronto	1	Arch. McLaren	324 Shaw Street	E. A. Prince	49 Eaton Ave.
St. John	2	W. L. Hurder	209 Douglas Avenue	G. T. G. Blewett	36 Murray St.
Collingwood	3	John Osburn	Collingwood, Ont.	Robert McQuade	Collingwood, Ont.
Kingston	4	Joseph W. Kennedy	395 Johnston Street	James Gillie	101 Clergy St.
Montreal	5	Eugene Hamelin	Jeanne Mance Street	M. Lazure	120 Ribard St.
Victoria	6	John E. Jeffcott	Esquimalt, B.C.	Peter Gordon	808 Blanchard St.
Vancouver	7	Isaac N. Kendall	319 11th St. E., Vanc.	E. Read	232 13th St. W.
Levis	8	Michael Latulippe	Lauzon, Levis, Que.	Arthur John	Lauzon, W.
Sorel	9	Nap. Beaudoin	Sorel, Que.	Alf. Charbonneau	Box 204, Sorel, Que.
Owen Sound	10	John W. McLeod	570 4th Ave.	R. J. McLeod	662 8th St.
Windsor	11	Alex. McDonald	28 Crawford Ave.	Neil Maitland	London St. W.
Midland	12	Geo. McDonald	Midland, Ont.	A. E. House	Box 333
Halifax	13	Robert Blair	29 Parrsboro Street	Chas. E. Pearce	Portland St., Dartmouth, N.S.
Sault Ste. Marie	14	Charles H. Innes	27 Euclid Road	Wm. Hindmarch	34 Euclid Rd.
Charlottetown	15	J. A. Rowe	176 King Street	Chas. Cumming	27 Euston St.
Twin City	16	H. W. Cross	436 Ambrose Street	J. W. Farquharson	169 College St.
				A. H. Archand	Champlain, Que.

Work in Progress in Canadian Shipyards

A record of Work in Process of Completion—Principal Features of Specification—Approximate Launching Dates—Vessels on Order.

BRITISH COLUMBIA

YARROWS, LIMITED, ESQUIMALT, B.C.

One stern-wheel steamer, 165 ft. x 35 ft., for Indian Government service.

At present engaged chiefly on repair work.

H. VOLLMERS, NANAIMO, B.C.

One gasoline tug, 9 tons.

NEW WESTMINSTER CONSTRUCTION & ENGINEERING CO., NEW WESTMINSTER, B.C.

Four wooden steamers, Nos. 10-11-12-13, for undisclosed interests. 250 ft. x 43 ft. 6 in. x 25 ft., moulded depth, 2,800 tons d. w., 1,000 h.p., single screw engines, 12 knots, naval architect, I. Alexander. Launchings March 14, April 14, May 15 and June 15, 1918.

B.C. CONSTRUCTION & ENG. CO., POPLAR ISLAND, NEW WESTMINSTER, B.C.

Four wooden vessels, 1,800 gr. tons, 2,800 tons d.w.

NEW WESTMINSTER MARINE CO., FT. OF FURNESS ST., NEW WESTMINSTER, B.C.

STAR SHIPYARD CO., NEW WESTMINSTER, B.C.

J. A. CROLL, PORT ALBERNI, B.C.

PACIFIC CONSTRUCTION CO., PORT CCOQUITLAM, B.C.

Taken over the plant of the Coquitlam Shipbuildiug Co.

Gasoline yacht, 50 ft., June, 1918.

Three wooden freight steamers, Nos. 20-21-22, 250 ft. b. p. x 42 ft. 6 in. x 25 ft., 1,800 gr. tons, 2,800 tons d.w., trip. exp. engines, water tube boilers ,12 knots, 1,000 h.p. Two for undisclosed interests, one for builder's account. Launchings May 20, June 15, Sept., 1918.

Four new berths under construction for steel ships up to 10,000 tons.

C. E. BAINTER, PRINCE RUPERT, B.C.

Several wooden fishing steamers, for undisclosed interests.

GRAND TRUNK DRYDOCK SHIP REPAIR CO., LTD., PRINCE RUPERT, B.C.

S. A. MOULTON, PRINCE RUPERT, B.C.

Ten composite vessels for undisclosed interests.

BRITISH-AMERICAN SHIPBUILDING & ENG. CO., VANCOUVER, B.C.

Establishing plant on Kitsilano Reserve.

Twenty wooden vessels being built for undisclosed interests, 3,500 tons each.

JOHN COUGHLAN & SONS, N. VANCOUVER, B.C.

"War Camp," sister ship to "Alaska," launched in March, 1918.

One steel freight steamer, 427 ft. 9 in. x 54 ft. x 29 ft. 9 in., draft 24 ft. 2 in., speed 11½ knots, 5,730 gross tons. Launched Jan. 10, 1918, for B. Stolt Nielsen, Christiania, Norway. Both sold to the Cunard Line.

Three steel steamers, "War Chariot," "War Chief," and "War Noble," Nos. 4-5-6, 425 ft. x 54 ft., 8,800 tons cap., for the Cunard Line.

Six steel freight steamers, 425 ft. x 54 ft. x 24 ft. 2 in., 8,800 tons cap., 1 knots, for undisclosed interests. Delivery Jan., Feb., March, May and July, 1918.

One steel freight steamer, "Alaska," 427 ft. 9 in. x 54 ft. x 29 ft. 9 in., draft 24 ft. 2 in., speed 11½ knots, 5,730 gr. tons, 8,800 d.w. Launched Jan. 10, 1918, for B. Stolt Nielsen, Christiania, Norway. Sold to the Cunard Line.

THE FOUNDATION CCO., VANCOUVER, B.C.

FRASER VALLEY SHIPBUILDING CO., VANCOUVER, B.C.

GRANT, SMITH & CO., VANCOUVER, B.C.

Six wooden cargo vessels, 250 ft. x 43 ft. 6 in. x 25 ft., 3,000 tons cap., 9½ knots, for the Canadian Government.

HARRISON & LAMOND, SHIPBUILDERS, LTD., VANCOUVER, B.C.

One wooden auxiliary schooner, 225 ft. x 44 ft. x 21 ft. 4 in., 1,600 gr. tons, 2,550 tons cap., for undisclosed interests.

LYALL SHIPBUILDING CO., NORTH VANCOUVER, B.C.

"War Puget," Feb. 16, 1918, 250 ft., wood, 2,080 tons.

Ten wooden cargo ships, for undisclosed interests. Launchings Jan. 20, Mar. 2, Mar. 23, and April 11, 1918.

Building six wooden vessels for own account.

"War Caribou," April 10, 1918, 3,080tons, wood.

W. R. MANCHION, VANCOUVER, B.C.

STANDARD SHIPBUILDING CO., VANCOUVER, B.C.

Two 3,500-ton vessels for Brazilian Govt., and six for French Govt. of 4,500 tons.

Donohoe reinforced type, wood composite construction.

TAYLOR ENGINEERING CO., VANCOUVER, B.C.

Number of small vessels being constructed at total value of $300,000.

4,500-ton floating drydock, 352 ft. long.

VANCOUVER SHIPYARD, LTD., VANCOUVER, B.C.

One motor freighter, 125 ft. x 24 ft. x 12 ft., wood, 160 h.p., Bolinder's crude oil engine. To be completed Aug., for Taylor Engr. Co.

Also repairing several steamers and schooners.

THE WALLACE SHIPYARDS, LTD., N. VANCOUVER, B.C.

"War Dog," 4,500 tons, steel, May 18, 1918.

"War Power," 4,600 tons, steel, March 23, 1918.

Two steel steamers, 315 ft. x 45 ft. x 27 ft., 4,600 tons d. w., trip. exp. vertical engines ,two S. E. Scotch boilers, 10 knots, single screw, for the Canadian Government. Deliveries Dec., 1917, and Aug., 1918.

Two wooden aux. schooners, 255 ft. o. a., 225 ft. keel x 44 ft. x 21 ft. 4 in., five masts, 1,500 gr. tons, cap. 2,500 tons or 1,500,000 ft. lumber, two hatches, twin 160 h.p. Bolinders Hot-bulb engines, twin screws, for the Canadian Government.

Four wooden cargo vessels, aggregate tonnage of 17,500 tons for undisclosed interests.

WESTERN CANADA SHIPYARDS ,LTD., VANCOUVER, B.C.

"War Nootka," Jan. 4, 1918, 3,080 tons (machinery installed at Ogden Point Assembly Sheds).

Six wooden freight steamers, 250 ft. h.p., 259 ft. o. a., 43½ ft. extreme breadth, molded 42½ ft., 25 ft. molded depth., draft 22 ft., 2,800 tons d. w., for undisclosed interests. "War Nootka," launched January 4, 1918. "War Selkirk" launched March 6, 1918.

"War Tatia," "War Casco, "War Chilkat," "War Tanoo," 4 standard vessels now under construction.

CAMERON-GENOA MILLS CO., VANCOUVER, B.C.

"War Yukon," Jan. 24, 1918, 3,080 tons, wood (machinery installed at Ogden Point Assembling Sheds.

THE FOUNDATION CO. OF B. C., LTD., VICTORIA, B.C.

Building ships for undisclosed interests.

Five wooden steamers, 250 b. p. x 42 ft. 6 in. x 25 ft., 1,800 gr. tons, 2,800 tons d. w., one trip. exp. engine, 1,000 h.p., "Howden" water tube boilers, furnished and installed by owners, 10 knots, for undisclosed interests. Esplen & Son & McNaught, N.Y.C., designers.

No. 1—"War Sonchee," launched Dec. 27, 1917.

No. 2—"War Masset," launched April 11, 1918.

No. 3—"War Babine," ready for launching, awaiting propeller and rudder.

No. 4—"War Camchin," ready for launching, awaiting propeller and rudder.

No. 5—"War Nanoose," ready for launching in about five weeks.

CLARENCE HOARD, VICTORIA, B.C.

Wooden car barge for C.P.R., Jan., 1918.

VICTORIA SHIPBUILDING CCO., VICTORIA, B.C.

Constructing wooden ships for British Govt., 3,500 tons each.

NEW BRUNSWICK

C. T. WHITE & SON, LTD., ALMA, N.B. (Office at Sussex, N.B., also.)

Two schooners aux. power, 143 ft., 900 tons d. w., first to be launched in April; second in June.

JAMES X. LENTEIGNE, LOWER CARAQUET, N.B.

One schooner, 28 tons, wood.

INTERNATIONAL SHIPBUILDING CO., NEWCASTLE, N.B.

Two four-masted wooden auxiliary schooners, 155 ft. keel, 37 ft. depth, 535 net tons. One to be completed in September and the other December, 1918. For builders' account.

EUREKA SHIPBUILDING CO., NORTH HEAD, N.B.

One wooden schooner, "Mollie & Melba," 350 tons reg.

PORT COLBORNE BUILDING & REALTY CO., LTD., REXTON, N.B. (Head Office, Welland, Ont.)

Four-masted wooden schooner. Has cap. of 3 schooners per year.

GRANT & HORNE, ST. JOHN, N.B.

Two wooden cargo steamers, "War Fundy" (launched Aug., 1918) and "War Digby," 250 ft. x 42 ft. 5 in. x 25 ft. 5 in., 1,000 h.p., speed 9½ knots, 2,800 tons d.w., for undisclosed interests.

MARINE CONSTRUCTION CO., CANADA, LTD., ST. JOHN, N.B.

Four wooden auxiliary, four-masted schooners, 185 ft. o. a., 165 ft. keel x 40 ft., 1,100 d. w. tons, for builder's account. J. Murray Watts, Philadelphia, naval architect. Launching April, 1918.

One wooden schooner, "Dorfontein," 740 reg. tons. Launching June, 1918.

PETER AND A. A. McINTYRE, ST. JOHN, N.B.

One schooner, 425 tons, 137 ft., three masts, wood, for builder's account. To be launched in June. Another to be commenced immediately.

ST. JOHN SHIPBUILDING CO., ST. JOHN, N.B. (Plant at Courtenay Bay.)

Ten five-masted auxl. schooners, oil burning engines, for undisclosed interests.

Establishing plant for steel and wood construction.

ST. MARTIN'S SHIPBUILDING CO., ST. MARTIN'S, N.B.

One wooden schooner, 450 tons, three masts. To be launched in July for builder's account.

NEWFOUNDLAND

ANGLO-NEWFOUNDLAND DEVELOPMENT CO., BOTWOOD, NFLD.

Two three-mast aux. schrs., wood, 450 tons, 150 H.P. for the builders' account.

NEWFOUNDLAND SHIPBUILDING CO., HARBOR GRACE, NFLD.

To build ships of 1,200 tons d.w. Steel ships later.

M. E. MARTIN, MORRIS ARM, NR. ST. JOHN'S, NFLD.

Two wooden schooners, 300 tons each, for builder's account.

One wooden schooner, 500 tons, for builder's account.

UNION SHIPBUILDING CO., ST. JOHN, NFLD.

ANNAPOLIS SHIPPING CO., ANNAPOLIS ROYAL, N.S.

"Hilda M. Clark," March 4, 1918, 187 ft., 640 tons reg.

Two schrs. three-masted, wood 500 tons, 170 ft. long. To be completed August and December, 1918.

NOVA SCOTIA

B. L. TUCKER, BASS RIVER, N.S.

One schooner, 350 tons, wood.

JAS. S. CREELMAN, BASS RIVER, N.S.

One three-masted schooner, wood, 170 ft. long, 500 tons. To be completed October, 1918, for builders' account.

HANKINSON SHIPBUILDING CO., BELLIVEAU'S COVE, N.S.
One schooner, 360 tons, wood.

A. A. THERIAULT, BELLIVEAU'S COVE, N.S.
Two wooden three-masted tern schooners, 150 ft. x 33 ft. x 11 ft, 11 in., 400 gr. tons, gasoline engines, for builder's account. Launchings April and June, 1918.
One wooden one-masted schr.

WESTPORT SHIPBUILDING CO., BELLIVEAU'S COVE, N.S.
One wooden schooner, 230 tons net. To be launched in October for builder's account.

ELI PUBLICOVER, BLANDFORD, N.S.

BRIDGEWATER SHIPPING Co., BRIDGEWATER, N.S.
One tern schooner, 450 tons gr., wood. Adapted for aux. power. To be completed Sept., 1918.
One tern schooner, 275 tons gr., wood. Adapted for aux. power. To be completed Nov., 1918.
Both built on owner's account and for sale.

W. A. NAUGLER, BRIDGEWATER, N.S.
One wooden three-masted schooner, "Wm. A. Naugler," 146 ft. o. a., 116 ft. keel x 32 ft. x 11 ft. 6 in., 360 gr. tons, 600 tons cap., for builder's account.
One schooner, 300 tons, wood.

H. MAC ALONEY, CANNING, N.S.
Two wooden schooners, 400 tons each. For builder's account. Launchings July and October, 1918.

S. M. FIELDS, CAPE D'OR, N.S.
One three-masted schooner, 350 tons.
Keel laid for another three-master.

CHESTER BASIN SHIPBUILDERS, LTD., CHESTER BASIN, N.S.
Two schooners, 100 tons, wood. Launchings Aug., 1917, Dec., 1917.
One four-masted schooner, 60 tons. Launching Sept., 1918.
One two-masted schooner, 100 tons. Launching June, 1918.

MORTIMER PARSONS, CHEVERIE, N.S.
One wooden three-masted schooner, 170 ft. x 38 ft. x 14 ft., 575 gr. tons, 900 tons cap., for builder's account. Launching Sept., 1918.
Two wooden three-masted schooners for builder's account.
One tern schooner, "Donald Parsons," 140 ft. keel x 34 ft. 4 in. x 12 ft. 4 in., 500 gross tons, 825 tons d.w. for Mortimer Parsons. Launched May 11, 1918.
Another vessel, same size, to be launched Nov., 1918, for builder's account.

BAY SHORE SHIPYARD, CHURCH POINT, N.S.
Building two wooden vessels for R. C. Elkin, Ltd., St. John, N.B.

MOISE BELLIVEAU, CHURCH POINT, N.S.
One schooner, 450 tons, wood.

FIDELE BOUDREAU, CHURCH POINT, N.S.
One schooner, 350 tons, wood.

J. E. GASKILL, CHURCH POINT, N.S.
Other yards at Grosses Coques, Little Brook, Meteghan and Port Wade, N.S.
Five wooden schooners from 344 to 387 reg. tons. All sister ships.

COMEAU SHIPBUILDING CO., COMEAUVILLE, N.S.
One three-masted wooden schooner, for builder's account, 450 tons.

J. W. COMEAU, COMEAUVILLE, N.S.
One schooner, 329 tons, Jan., 1918.

J. N. RAFUSE & SONS, CONQUERALL BANKS, DIGBY CO., N.S. (Other plants at Salmon River and Shelburne, N.S.)
Two wooden three-masted schooners, 120 ft. x 32 ft. x 12 ft., 700 tons cap., for builder's account. For the market. Launchings Oct. 25, 1917, and Jan. 10., 1918.
One three-masted wooden schooner, "Integra," 112 ft. x 30 ft. x 11 ft. 6 in., 600 tons cap., for J. O. Williams & Co., St. John's, Newfoundland. Launching Nov. 25, 1917.

E. F. WILLIAMS, DARMOUTH, N.S.
One schooner, 360 tons, wood.

ROBAR BROTHERS, DAYSPRING, N.S.
Schooner for Capt. Ivan Creaser, Dayspring.

MAURICE E. LEARY, DAYSPRING, N.S.
One wooden schooner, 225 tons, 135 ft. o. a. To be launched Aug. 15, 1918, for La Have Outfitting Co., La Have, N.S.

J. NEWTON PUGSLEY & CHAS. ROBERTSON, DILIGENT RIVER, N.S.
One schooner, three-masted, 475 tons.

McLEAN & McKAY, ECONOMY, N.S.
One tern schooner, wood, 400 tons net, 135 ft. keel. To be launched Sept. 1, 1918, for builders' account.

S. J. SOLEY, ROX RIVER, N.S.
One schooner, three-masted, 121 ft. keel, 300 tons. Built of native wood. To be launched Sept., 1918. For builder's account. For sale.

ALLAN & FRASER, FRASERVILLE, N.S.
One schooner, 350 tons, wood.

BERNARD W. MELANSON, GILBERT'S COVE, N.S.
One three-masted schooner, wood, 250 tons net. To be completed November, 1018, for builder's account and for sale.

AMOS BLINN, GROSSES COQUES, N.S.
One schooner, 375 tons. Jan., 1918.
One schooner, 350 tons, wood.

BINN BROS., GROSSES COQUES, N.S.

F. K. WARREN & CO., GROSSES COQUES, N.S. (Offices in Union Bank Chambers, Halifax, N.S., also.)
350-ton tern schooner, under construction Jan., 1918.

HALIFAX SHIPBUILDING CO., HALIFAX, N.S.

FAUQUIER & PORTER, HANTSPORT, N.S.
Two wooden four-masted schooners, 178 ft. x 40 ft. x 18 ft., 1,000 gross tons, 2,000 tons d. w., twin 100 h.p. oil engines. Launchings Aug. 1, 1918, and Sept. 1, 1918. Both for sale.

J. WILLARD SMITH, HILLSBURN, N.S. (Head Office, St. John, N.B.)
One tern schooner, wood, 140 ft. keel, 35 ft. beam, 450 tons net reg. tons.

HENRY COVEY, INDIAN HARBOR, N.S.

CHARLES GRIFFIN, ISAACS HARBOR, N.S.
One schooner, 40 tons, wood.

LEWIS SHIPBUILDING CO., LEWISTON, N.S.
One schooner, 670 tons. Jan., 1918.

INNOCENT COMEAU, LITTLE BROOK, N.S.
One three-masted schoner, 140 ft. x 35 ft. 6 in. x 17 ft. 6 in., 650 gr. tons, for the Weymouth Shipping Co., Weymouth, N.S., Can. Delivered Jan. 18, 1918.

J. W. RAYMOND, LITTLE BROOK, N.S.
One wooden schooner, No. 4, 140 ft. x 36 ft. x 17 ft., 575 gr. tons, for Jones Bros., Weymouth, Digby Co., N.S.

H. A. FRANK, LIVERPOOL, N.S.
Two wooden schooners, 500 tons, for builder's account.

McKEAN SHIPBUILDING CO., LIVERPOOL, N.S.
One wooden three-masted schooner, 126 ft. keel x 3 ft. x 12 ft. 9 in., 700 tons d.w., for builder's account. Launching May, 1918.

W. F. McKEAN & CO., LIVERPOOL, N.S.
One schooner 400 tons, Jan., 1918.

D. C. MULHALL, LIVERPOOL, N.S.
Two wooden schooners, one 300 gross tons, other 400 gross tons, for undisclosed interests.

N.S. SHIPBUILDING & TRANS. CO., LIVERPOOL, N.S.
One wooden tern schooner No. 2, 122 ft. x 33 ft. x 12 ft. 6 in., 450 gr. tons, 70 d.w. tons, 7 knots, for Peter Yee Wing & Co., Ltd., Sydney, Australia. R. McLeod and J. S. Gardner. Launching Feb. 1, 1918
One two-masted fishing schooner, "Sadie A. Nickle," 130 ft. o. a. x 26 ft. 10 ft. 6 in. hold, 250 d.w., two 50 h.p. oil engines, for Rafuse & Sons. Launching May 15, 1918.
One three-mast schooner, 420 tons. F-M Co. oil burning engines, for Job Bros., St. Johns, Nfld. Launching July 15.
One three-masted schooner, 270 tons, for builders' account. Launching Sept. 1.
One two-masted fishing schooner, 120 tons, for builders' account. Launching Oct. 1.

ROBIN, JONES & WHITMAN, LIVERPOOL, N.S.
One schooner, 340 tons, wood.

SOUTHERN SALVAGE CO., LIVERPOOL, N.S.
One wooden schooner, three masts, 300 gross tons, 600 tons capacity, for a West Indian firm. Launching Jan., 1918.
One wooden steamer, No. 5, 250 ft. x 43 ft. 6 in. x 25 ft., 2,900 gr. tons, 10 knots, 1,000 h.p., for undisclosed interests. Launching Fall, 1918.
One two-masted schooner, "Win-the-War," 137 ft. o. a. x 25 ft. 2 in. x 11 ft. 6 in., 187 gr. tons, for the builder's account. Launched Nov., 1917.

CONRAD & REINHARDT, LUNENBURG, N.S.

LUNENBURG MARINE RAILWAY, LUNENBURG, N.S.

FRED A. ROBAR, LUNENBURG, N.S.

SMITH & RHULAND, LUNENBURG, N.S.
Two schrs., 225 tons each. Jan., 1918.
Two schooners, 100 tons each.
"Donald Cook," 112 ft. beam, depth of hold, 11 ft., for Capt. William Cook. Launched April, 1918.
Two schooners, 168 gr. tons, for W. C. Smith.
One schooner, 150 gr. tons, for Lun. Outfitting Co.

J. B. YOUNG, LUNENBURG, N.S.

ERNST SHIPBUILDING CO., MAHONE BAY, N.S.
One two-masted schooner, 120 ft. o. a. x 25 ft. 6 in. x 10 ft. 6 in., 200 tons d.w., for Lunenburg Outfitting Co. April, 1918. Christened "Madeline Adams."
One three-masted schooner 125 ft. keel x 32 ft. x 12 ft., 600 tons, d. w. To be delivered July, 1918.
Two masted schooner, "Agnes D. McGlaston," 130 ft. x 26 ft. 8 in. x 10 ft. 8 in. Launched Nov. 13, 1917.
Keel laid for three-masted wooden schooner, 200 tons.

J. ERNEST & SON, MAHONE BAY, N.S.
One schooner, 520 tons. Jan., 1918.

O. A. HAM, MAHONE BAY, N.S.
One schooner, "Dozle," 128 tons. Launching May 30. One motor boat about 20 tons.

McLEAN CONSTRUCTION CO., MAHONE BAY, N.S. (Leased to Montague Mahaffy, Toronto.)
One three-masted schooner, 325 tons gr. reg., 500 tons cap., West Indies freight type. Will be launched in July.
Keel laid for another three-masted schooner with cargo capacity of 425 tons.

JOHN McLEAN & SONS, MAHONE BAY, N.S.
One schooner, 95 tons, wood.

J. A. BALSOM CO., LTD., MARGARETSVILLE, N.S.
One schooner, 409 tons. Jan., 1918.

CLARE SHIPBUILDING CO., METEGHAN RIVER, N.S.
One wooden schooner for builder's account. 400 tons.

A. H. COMEAU & CO., METEGHAN, N.S.
One schooner, 400 tons, wood.

AGAPIT COMEAU, METEGHAN, N.S.

JOHN F. DEVEAU, METEGHAN, N.S.
A 300-ton schooner was launched recently for Ritcey & Co., Lunenburg, N.S., and named Charles A. Ritcey.
1 schooner, 425 tons. Jan., 1918.
1 schooner, 400 tons, wood.

Fairbanks Scales

For more than 90 Years

The World's Standard for Accurate Weighing

The Canadian Fairbanks-Morse Co., Ltd.

Halifax, St. John, Quebec, Montreal, Ottawa, Toronto, Hamilton, Windsor, Winnipeg, Saskatoon, Calgary, Vancouver, Victoria.

J. E. GASKILL, METEGHAN, N.S.

THOMAS GERMAN, METEGHAN, N.S.

One schooner, 350 tons, wood.

R. H. HOWES CONSTRUCTION CO., METEGHAN, N.S. (Leased James Cosman's shipyards.)

Building several wooden schooners for own account.

DR. F. H. MACDONALD, METEGHAN, N.S.

One wooden four-masted schooner, "Rebecca L. Macdonald," 201 ft. o. a. x 6 ft. x 16 ft., 800 gr. tons, 1,500 tons d.w., for builder's account. Launching January 1, 1918.

One wooden fishing schooner, No. 8, 84 ft. x 17 ft. x 7 ft., 50 gr. tons, 5 h.p. Launching August, 1918.

One schooner, 544 tons, wood.

METEGHAN RAILWAY & SHIPBUILDING CO., METEGHAN, N.S.

One wooden three-masted schooner, 182 ft. x 35 ft., 450 gr. tons.

One schooner, 470 tons, wood.

CHAS. McNEIL, NEW GLASGOW, N.S.

One wooden 3-masted schooner, 108 ft. x 30 ft. x 10 ft. 9 in., 200 tons, for builder's account. Launching July, 1918.

One wooden 3-masted schooner, 142 ft. x 25 ft. 4 in. x 13 ft., 500 gr. tons, for builder's account. Launching Oct., 1918.

THE NOVA SCOTIA STEEL & COAL CO., NEW GLASGOW, N.S.

Two steel freight steamers, 257 ft. 9 in. x 35 ft. x 20 ft., 1,700 gr. tons, 2,350 tons cap., trip. exp. engines, boilers 10 ft. 6 in. x 11 ft. 6 in., 800 h.p., 8.5 knots, for undisclosed interests. Launching of one Jan, 1918.

Two steel cargo strs. Raised quarter deck type, 248 ft. 9 in. long, 1,649.68 gross tons. Triple expansion engines, 17, 28, 46 x 33 in. stroke. One, the "War Bee," for undisclosed interests to be completed June, 1918. Other for Steel Co., to be completed Oct., 1918.

Two masted schooner, launched May 16, 130 ft. o. a. x 26 ft. x 10 ft. 6 in. Designed by J. S. Gardner.

O'BRIEN BROS., NOEL, N.S.

One schooner, 325 tons, wood.

W. R. HUNTLEY & SON, PARRSBORO, N.S.

One wooden schooner, 175 ft. x 39 ft. x 17 ft., 900 gross tons, for

C. T. WHITE & SON, SUSSEX, N.S.

Two schooners, 325 tons each. Jan., 1918.

One schooner, 490 tons, wood.

One schooner, 850 tons, four masted.

WAGSTAFF & HATFIELD, PARRSBORO, N.S.

One schooner, 400 tons. Jan., 1918.

SIDNEY ST. C. JONES, PLYMPTON, N.S. (Head Office, Weymouth North, N.S.)

One tern schooner, 200 tons, net reg., 107 ft. x 28 ft. 8 in. x 10 ft. 3 in., to be launched Nov., 1918.

S. SALTER, PARRSBORO, N.S.

One 200-ton schooner, wood.

DOWLING & STODDART, PORT CLYDE, N.S.

One gas boat, 27 tons, wood.

One schooner, 300 tons, wood.

SWIME BROS., PORT CLYDE, N.S.

Several small motor trawlers, for builder's account.

G. M. COCHRANE, PORT GREVILLE, N.S.

One tern schooner, "Alfredock Hedley," 152 ft. 6 in. x 36 ft. 12 ft. 6 in., 461 tons reg., for Adam B. Mackay, of Hamilton, Ont. Launched Jan., 1918.

One 4-masted schooner, 850 tons, 155 ft. x 37 ft. x 18 ft., 2 decks, wood. To be completed Oct., 1918, for builder's account.

H. ELDERKIN & CO., PORT GREVILLE, N.S.

One wooden three-masted schooner, for undisclosed interests.

ELLIOTT GRAHAM, PORT GREVILLE, N.S.

Is building a schooner, "Khaki Lad," of about 325 tons, at Port Greville, N.S., for J. W. Kirkpatrick and others. She will be launched early in October, and is reported to have been sold to Newfoundland parties.

One schooner, 360 tons. Jan., 1918.

SMITH CANNING CO., PORT GREVILLE, N.S.

One schooner, 350 tons, wood.

WAGSTAFF & HATFIELD, PORT GREVILLE, N.S.

One schooner, 400 tons, wood.

WILLIAM CROWELL, PORT LATOUR, N.S.

J. W. RAYMOND, PORT MAITLAND, N.S.

One schooner, 375 tons. Jan., 1918.

PORT WADE SHIPBUILDING CO., PORT WADE, N.S. (Head Office, Digby.)

One 350-ton schooner, wood.

JOHN BROWN, PUBLIC LANDING, N.S.

One tow barge, 50 tons, wood.

THE CUMBERLAND SHIPBUILDING CO., PUGWASH, N.S.

Establishing plant for the construction of wooden ships.

MACKENZIE SHIPPING CO., RIVER JOHN, N.S.

One four-masted schooner, 600 tons, wood. About half built. Fitted for aux. engines.

CHARLES McLELLAN, RIVER JOHN, N.S.

One schooner, 100 tons, wood.

W. J. FOLEY, SALMON RIVER, N.S.

Building for J. N. Rafuse & Sons one wooden, 300 tons, under construction. Cap. of plant, 1,000 tons yearly.

J. N. RAFUSE & SONS, SALMON RIVER, N.S. (Other plants at Shelburne and Conquerall Banks.)

Launched recently a three-masted schooner named "Industrial," at W. J. Foley's ship yard at Salmon River, N.S., of the following dimensions: length, 113 ft., breadth, 30 ft., depth, 11½ ft., 325 tons.

ACADIAN SHIPPING CO., SAULNIERVILLE, N.S.

One schooner, 60 tons, wood.

SAULNIERVILLE SHIPBUILDING CO., LTD., SAULNIERVILLE, N.S.

J. LEWIS & SONS, SHEET HARBOR, N.S.

One schooner, wood, four-masted, 725 gr. tons. To be completed Sept., 1918. For private account.

GEO. A. COX, SHELBURNE, N.S.

One schooner, 200 tons, wood.

One schooner, 322 tons, wood.

JOSEPH McGILL SHIPBUILDING & TRANSPORTATION CO., SHELBURNE, N.S.

"Sparkling Glance," 246 tons register, for Harvey & Co., St. John's, Nfld.

One schooner, 160 tons, wood.

W. C. McKAY & SON, SHELBURNE, N.S.

One wooden schooner, 130 ft. o. a. x 26½ ft. x 11 ft. depth of hold, 140 tons register. Launched Dec. 1, 1917.

One three-masted schooner, for Messrs. Hallet, of Newfoundland.

One schooner, 620 tons, wood.

J. N. RAFUSE & SONS, SHELBURNE, N.S. (Other plants at Salmon River and Conquerall Banks.)

One wooden three-masted schooner, 120 ft. x 32 ft. x 12 ft., 700 tons cap., for builder's account. For the market. Launching Jan. 10, 1918.

One wooden three-masted schooner, 118 ft. x 33 ft. x 12 ft., 700 tons cap., for builder's account. For the market. Launching Jan. 15, 1918.

Two three-masted wooden schooners, 122 ft. 6 in. keel x 32 ft. x 12 ft., 275 net tons, for J. O. Williams & Co., St. John's, Newfoundland. One to be built at Salmon River yard and one at Ship Harbor.

THE SHELBURNE SHIPBUILDERS, LTD., SHELBURNE, N.S.

One tern schooner, "Misty Star," 145 ft. x 31 ft. x 11 ft. 6 in., 600 tons d.w., 330 tons gross, for Harvey & Co., St. John's, Newfoundland. Launching Mar. 15, 1918.

One tern schooner, 154 ft. x 32 ft. x 12 ft. 4 in., 400 gr. tons, 700 tons d.w. cap. Launching June, 1918. Not sold yet.

Two schooners, one 349 tons, other 400 tons.

EASTERN SHIPBUILDING CO., SHIP HARBOR, N.S.

One wooden schooner, 600 tons d.w., for J. N. Rafuse & Sons., Halifax, N.S., Can.

One wooden schooner, 300 tons.

STEPHEN MORASH & CO., SHIP HARBOR, N.S.

One wooden schooner, 150 ft. x 37 ft. x 13 ft., four masts, for Canadian interests. Keel laid Feb. 5, 1917.

JAMES E. PETTIS, SPENCER'S ISLAND, N.S.

One schooner, 425 tons, wood, three-masted.

CAPE BRETON SHIPBUILDING CO., SYDNEY, N.S.

One three-masted schooner.

THE TUSKET SHIPBUILDING CO., TUSKET, N.S.

AMOS H. STEVENS, TANCOOK, N.S.

ALVIN STEVENS, TANCOOK, N.S.

STANLEY MASON, TANCOOK, N.S.

ALBERT PARSONS, WALTON, N.S.

One schooner, 400 tons. Jan., 1918.

H. T. LeBLANC, WEDGEPORT, N.S.

One steam trawler, 165 ft., 600 tons, for J. N. Rafuse & Sons.

L. S. CANNING, WARDS BROOK, N.S.

One tern schooner, 350 tons, for W. C. Smith & Co., Lunenburg, N.S.

WEDGEPORT NAVIGATION & TRANSPORTATION CO., WEDGEPORT, N.S.

T. K. BENTLEY, WEST ADVOCATE, N.S.

One 200-ton steamer.

One 4-masted schooner, wood, 511 tons. To be launched September or October. For owner's account.

J. W. KIRKPATRICK, WEST ADVOCATE, N.S.

One 350-ton schooner, wood.

BOEHNER BROS., WEST LA HAVE, N.S.

Two large fishing schooners, 150 tons each.

One beam trawler, 300 tons, crude oil engines. For local interests.

BEAZLEY BROS., WEYMOUTH, N.S. (Head Office Roy Bldg., Halifax.)

One schooner, 394 tons, three masts.

E. R. GAUDET, WEYMOUTH, N.S.

One 350-ton schooner, three-masted.

E. P. RICE, WEYMOUTH, N.S.

One three-masted schooner, 350 tons.

RICE, WARREN & CO., WEYMOUTH, N.S.

One three-masted schooner, 334 tons.

FALMOUTH SHIPBUILDING & TRANSPORTATION CO., WINDSOR, N.S.

One wooden schooner, 350 gr. tons, 700 tons cap., for undisclosed interests.

NOEL SHIPBUILDING & TRANSPORTATION CO., WINDSOR, N.S.

One wooden schooner, three masts, 138 ft. x 35 ft. x 13 ft., 450 net reg. tons. Oil auxiliary engines. Launching August, 1918, for builder's account.

C. A. NICKERSON, WOODS HARBOR, N.S.

MILTON SHIPBUILDING CO., YARMOUTH, N.S.

Taken over and enlarging old plant of James Jenkins.

350-ton schooner building.

W. O. SWEENY, YARMOUTH, N.S.

100-ton fishing schooner.

YARMOUTH SHIPBUILDING CO., YARMOUTH, N.S.

One 450-ton wooden schooner, 128 ft. keel. To be completed Oct., 1918.

ONTARIO

CAN. ALLIS-CHAMBERS (Head Office, Toronto), BRIDGEBURG, ONT.

Six standard freight steamers, 3,500 tons, 261 feet long, for undisclosed interests.

COLLINGWOOD SHIPBUILDING CO., COLLINGWOOD, ONT.

Two steel cargo steamers, Nos. 51-52, 50 ft. x 43 ft. x 25 ft., 2,500 gr. tons, 2,900 tons d. w., triple expansion engines 18-30-50 x 36, 2 boilers 14 ft. x 11 ft., 10 knots, for undisclosed interests. Delivery May and Aug., 1918.

Four deep-sea trawlers, Nos. 53-58, inclusive, 125 ft. x 23 ft. 6 in. x 13 ft. 6 in., 288 gross tons, 12¾-21½-35/24 engines, 1 cylinder 13 ft. 6 in. x 10 ft. 6 in., 10 knots, 500 h.p. for undisclosed interests.

One steel freight steamer of 3,800 tons d. w. for undisclosed interests. Keel laid May 8, 1918.

"War Wizard," May 8, 1918, 3,000 tons d. w., 261 ft. x 43 ft. 6 in. x 20 ft. depth. For undisclosed interests.

"War Witch" same size, now building. To be launched ——

R. MORRILL, COLLINGWOOD, ONT.

"Windsor," Aug. 10, steam tug, 105 ft., for Ontario Gravel & Freighting Co.

CAN. CAR & FOUNDRY CO., FORT WILLIAM, ONT. (Head Office Transportation Bldg., Montreal.)

Twelve steel mine sweepers for French Government.
145 ft. o.a. steel construction, value $2,500,000.
Plant under construction.

GREAT LAKES DREDGING CO., FORT WILLIAM, ONT.

Two wooden vessels, 2700 tons d.w. cap., 260 ft. o.a., 43 ft. beam. Triple expansion engines of 1,000 h.p. To be launched November, 1918.

"War Sioux" launched May 12, 1918. Keel laid for another twenty minutes later. Launching scheduled for November.

THUNDER BAY CONTRACTING CO., FORT WILLIAM, ONT.

One wooden freight steamer, 261 ft. long, for undisclosed interests.

NATIONAL SHIPBUILDING CO., GODERICH, ONT.

DAVIS DRYDOCK CO., FT. OF BAY ST., KINGSTON, ONT.

One motor pass. boat, 64 ft. x 18 ft. x 6 ft., 52 gr. tons, wood. For ferry service, Gananoque to Clayton. Delivery June 1, 1918.

Twenty-eight lifeboats, regulation size, wood and metal for Government and private interests.

KINGSTON SHIPBUILDING CO., KINGSTON, ONT.

Several steel trawlers for undisclosed interests. First one launched Dec. 22, 1917.

SELBY & YOULDSON, KINGSTON, ONT.

GEORGIAN BAY SHIPBUILDING & WRECKING CO., MIDLAND, ONT.

Tug, 50 tons, wood.

One tug 40 tons, wood.

MIDLAND DRY DOCK CO., MIDLAND, ONT.

Three steel freight steamers, 261 ft. x 43 ft. 6 in. x 23 ft., 3,400 tons d. w., 10 knots, for undisclosed interests. Deliveries, one in July, 1918, and two others before close of navigation, 1918.

Alterations on steamers "Mariska" and "Glenlyon."

"Western Star" being reconstructed.

MIDLAND SHIPBUILDING CO., LTD., MIDLAND, ONT.

Three steel freight steamers, 261 ft. x 43 ft., 6 ft x 23 ft., 3,400 tons d. w., 10 knots for undisclosed interests. Deliveries, one in July, 1918, and two others before opening of navigation, 1919.

PORT ARTHUR SHIPBUILDING CO., PORT ARTHUR, ONT.

Six steel trawlers on order. 135 ft. o. a. x 23 ft. 4 in. x 15 ft. 1 in., 294.5 tons, trip. exp. engines, 500 h.p., single end Scotch boiler. Two-masted.

Two of these launched June 8, 1918.

"War Isis," April 3, 1918, 3,400 tons d. w. Trawler launched same day. April 3, 1918, keel laid for "War Hatha," sister ship to "War Osiris," of same construction, nearly completed.

"War Fish," Aug. 19, 1917, 4,300 tons, steel.

"War Dance," Nov. 3, 1917, 3,400 tons, steel.

Four steel screw steamers, ocean freight servicel—"War Osiris," "War Hathor," "War Karma," "War Horus," 3,400 tons d. w. each, 261 ft. o.a. 251 ft. b.p. in length. Triple expansion engines, 2 Scotch boilers each, approx. 1,250 h.p. For completion, two Aug. 31st, and two Nov. '20 1918. For undisclosed interests.

MUIR BROS. DRYDOCK CO., LTD., PORT DALHOUSIE, ONT.

J. W. GEROW, ROSSPORT, ONT.

REID WRECKING CO., SARNIA, ONT.

One steel tug, 157 ft. x 32 ft. x 19 ft., trip. exp. engines, for lake or ocean service. Keel blocks laid March 5. Launching July, 1918. For Reid Wrecking Co.

WEST, PEACHEY & SONS, SIMCOOE, ONT.

Number of tugs for lumbering interests.

DOMINION SHIPBUILDING CO., TORONTO, ONT.

One steel cargo vessel, 261 ft. x 43 ft. x 28 ft. 2 in., 3,500 tons d.w. cap. For Department of Marine, Ottawa.
Have orders for five other vessels same dimensions.
"Traja" launched May 15, 1918.

THE POLSON IRON WORKS, LTD., TORONTO, ONT.

Eight steel cargo steamers, Nos. 133-4-5-6-7-9-40, 216 ft. o. a., 251 ft. b. p. x 43 ft. 6 in. x 22 ft. 11 in., draft loaded 19 ft. 6 in., single decked, 2,350 approx. gross tons, 3,500 tons cap., single screw trip, exp. engine 20½-33-54 x 36, 1,250 h.p., 10 knots, two 14 ft. x 12 ft. Scotch R. T. boilers, 180 h.p., to cost $600,000 each, for undisclosed interests. Launchings throughout 1918.

Four building and four on order.

"Asp," Feb. 11, 1918, 261 ft., 3,500 tons d.w., steel.

"Tento," Oct. 22, 1917, 261 ft., 3,500 tons d.w. steel. For Norwegian interests. Transferred to British registry.

THE THOR IRON WORKS, TORONTO, ONT.

Under same management as Dominion Iron Works

TORONTO SHIPBUILDING CO., LTD., TORONTO, ONT. (Toronto Dry Dock Co., Ltd., under same management.)

Two wooden cargo steamers, 250 ft. B.P. 42 ft. 6 in., moulded breadth 25 ft., moulded depth 22 ft., draft 2,500 tons d. w. Triple exp. engines $\frac{20\times 33\times 54}{40}$. Howden boiler. Launchings July and Sept., 1918. For undisclosed interests.

THE BRITISH AMERICAN SHIPBUILDING CO., WELLAND, ONT.

Two steel frs., 3,500 tons d. w., 261 ft. O.A., 43 ft. beam, 28 ft. moulded depth, Westinghouse steam turbine engines. Delivery 1918.

WELLAND SHIPBUILDING CO., WELLAND, ONT.

One steel freight steamer, "War Weasel," 261 ft. x 43 ft. 6 in. x 23 ft., 3,300 tons d. w., trip. exp. engines, 14 ft. x 12 in. Scotch boilers, 10 knots, 1,250 h. p. Launching April, 1918.

Two deep frame cargo vessels, "War Badger" and "No. 3," 261 ft. x 43 ft. 6 in. x 23 ft., 3,300 d. w. tons, geared turbines, Howden's forced draught boilers, 10 knots, 1,250 h.p. Launchings June and August, 1918.

PRINCE EDWARD ISLAND

THE CARDIGAN SHIPBUILDING PLANT, CARDIGAN, P.E.I. (Purchased Annandale Lumber Co. plant and removed to Cardigan.)

One three-masted schooner, 325 tons, to be completed November, 1918.
Two more to be built.

QUEBEC

TIDEWATER SHIPBUILDERS, LTD., THREE RIVERS, QUE.

(Formerly Sorel Shipbuilding & Coal Co.)
Three steel trawlers for undisclosed interests.
One steel steam barge for the Canada Steamships Lines.

R. N. LE BLANC, BONAVENTURE, QUE.

J. Z. DEGAGNE, EBOULEMENTS, QUE.

DAVIE SHIPBUILDING & REPAIRING CO., LEVIS, QUE.

Six military barges, 130 feet long.
Eight steel trawlers, 2,035 tons.
One floating crane, 350 tons.
*One steel cargo vessel of 5,000 tons cap.
Building several steel lighters and several wooden drifters.

Steel car ferry "Canora" building for C.N.R., 308 ft. x 52 ft., cap. 20 loaded cars. Speed, 14 knots.

ATLAS CEMENT CONSTRUCTION CO., LTD., MONTREAL, QUE.

One concrete cargo steamer, "Concretia," 125 ft. x 22 ft. x 13 ft., steel ribbed, hull to be from 3 in. to 5 in., thick, for the builder's account.

CANADIAN VICKERS, LTD., MONTREAL, QUE.

Two cargo steamers, 9,400 tons, steel; 1 dredge, 2,364 tons, steel; 12 trawlers, 3,050 tons, steel; 23 drifters, 3,350 tons, wood.

Six freight steamers, 24 ft. draught, 7,000 tons cap., 11 knots, for undisclosed interests. Delivery, 1917, of two for Norwegian interests. "Porsanger," 394 ft. 6 in. o. a. x 40 ft. 4 in. x 30 ft., triple exp. engines, launched Nov. 29, 1917.

"War Earl" launched June 8, 1918, 7,000 tons d.w., 380 ft. x 49 ft. Keel of sister ship laid five minutes after this launching.

One steel freight steamer, 2,360 tons, cap., for undisclosed interests.

Three steel cargo vessels, 8,200 tons, to be laid down in May, Aug. and Sept., 1918, for undisclosed interests.

FRASER BRACE & CO., MONTREAL, QUE.

Four wooden cargo steamers, 3,000 tons, for undisclosed interests. Two keels laid during Oct., 1917.

HALL ENGINEERING CO., MONTREAL, QUE.

MONTREAL DRY DOCK & SHIP REPAIRING CO., MONTREAL, QUE.

Operate dock 428 ft. long 30 ft. deep.

MONTREAL SHIPBUILDERS LTD., 37 Belmont St., MONTREAL, QUE. (Associated with Atlas Construction Co.

THE QUEBEC SHIPBUILDING AND REPAIR CO., Board of Trade Bldg., MONTREAL, QUE.

2 vessels for undisclosed interests 3,000 tons each.

QUINLAN & ROBERTSON, MONTREAL, QUE. (Yards at Quebec.)

Four wooden steamers, for undisclosed interests.

LOUIS GAUNDRY, 12 St. Peter St., QUEBEC, QUE.

QUEBEC SHIPBUILDING & REPAIR CO., ST. LAURENT, QUE.

2 schrs. 1,400 tons and 1,200 tons.
One wooden four-masted auxl. schooner "Martin Connolly," 223 ft. x 42 ft. x 20 ft., 2,100 tons d. w., for undisclosed interests. Launched Oct. 28, 1917.

QUINLAN & ROBERTSON, QUEBEC, QUE.

Four wooden steamers, totalling 6,400 tons, for undisclosed interests.

CANADIAN GOVERNMENT SHIPYARDS, SOREL, QUE.

One steel vessel for undisclosed interests.

LECLAIRE SHIPBUILDING CO., SOREL, QUE.

Building 6 steel ships at value of $1,500,000.
Six trawlers, 125 ft. single screw type expansion engines, 500 I.P.H. steel. To be completed this year for private acct.

H. H. SHEPHERD, SOREL, QUE.

Rebuilding seven drifters. Particulars withheld owing to govt. restrictions.

SINCENNES-McNAUGHTON LINES, SOREL, QUE.

One tug 410 tons, wood.

CAN. GOVERNMENT SHIPYARD, SOREL, QUE.

One steel vessel for undisclosed interests.
Building steam trawlers and wooden drifters for undisclosed interests.

ST. LAWRENCE SHIPBUILDING & STEEL CO., SOREL, QUE.

THE THREE RIVERS SHIPYARDS LTD., THREE RIVERS, QUE.

Two cargo steamers, 3,100 d. w., wood, 260 ft. x 43 ft. 6 in., 1,000 H.P. engines. Names, "War Mingan" and "War Nicolet." To be completed end of navigation season for undisclosed interests.

Canada's Leading Tool House

Aikenhead's

Quality Tools for All Purposes

You want to, you need to, and you should conserve time and labor on every ship you build—and in your boiler plant.

You can save days—and dollars wherever holes are drilled by employing

Portable Electric Drills

Have your men take Van Dorn Drills to your work. Out in the open, in rain, Van Dorn Drills will stay on the job as long as your most earnest worker weathers the storm.

And they will drill every material from wood to armour plate, cutting neat and cutting clean. Note their

Capacity in Solid Steel

DAOOO	3-16 in.
DAOO	¼ in.
DAOOX	5-16 in.
DAO	⅜ in.
DA1	½ in.
DA1X	⅝ in.
DA2	⅞ in.
DA2X	1 in.

Van Dorn Portable Electric Drills will drill 50 per cent. more holes, at one-third the cost for power, than the best pneumatic drill of similar rated capacity. Not a claim without foundation, but a fact that we would like your permission to prove again by demonstration in your plant.

Say we may—and ask for Catalog.

AIKENHEAD HARDWARE LIMITED

17 TEMPERANCE STREET, TORONTO

Marine News From Every Source

Port Coquitlam, B.C.—Building of 1,500-ton ships will begin as soon as the Pacific Construction Company's two new ways, now under construction, are ready.

North Vancouver, B.C.—Preparations for the construction of a long dock fronting the shipyards of the Lyall Shipbuilding Company are under way. This dock will facilitate the work of installing boilers and engines in ships built in the Lyall yards.

Contract Closed.—Fraser & Chalmers of Canada, Limited, of Montreal, Que., have been awarded a contract by the Corp. of the Town of Pointe Claire, Pointe Claire, Que., for a 1,500-gallon centrifugal pump, direct connected to a Sterling gasoline engine.

Three Rivers, Que.—The construction of a $10,000 dredge is contemplated by the Tidewater Shipbuilding Company. J. J. Collins, C.E., supervisor, Three Rivers.

Three Rivers, Que.—The Boston Shipbuilding Company, Limited, will soon have four vessels ready for launching. This company has a contract for eight new ships, according to reports.

Halifax, N. S.—Rapid progress is being made in the preliminary work in connection with the developments at the site of the Halifax Shipyards. Considerable excavation has been done for general construction work, but the main effort is concentrated on clearing up the ground and grading for the keels of four large vessels. The capacity of the yard when completed will be for four steel hulls, 500 feet in length.

London.—Another world's record has been made at a Belfast shipbuilding yard by the completion of a standard ship in five working days after the launching of the vessel. The boat took the ways on August 22, and the work of putting in the machinery was started the same day. Steam was gotten up on the 26th, trials were completed yesterday, and she was handed over to the owners this morning.

Halifax.—Satisfactory progress is being made in connection with the work on the Halifax shipyards, states the managing director, Roy M. Wolvin, according to an item in The Canadian Engineer. The graving dock is working to capacity day and night, and the number of men employed has recently been doubled. On the completion of the plant the company will proceed to build the 10,000-ton vessels contracted for by the Dominion Government. The largest ocean freighters now being built in Canada are 8,400 tons, constructed at the Vickers yard in Montreal.

Toronto.—Tenders will be received by registered post only, addressed to the chairman, Board of Control, City Hall, Toronto, up to 12 o'clock noon on Tuesday, September 10, 1918, for the construction and delivery of hand-operated travelling crane. Specifications and forms of tender may be obtained at the Works Department, Room 12, City Hall.

Port Arthur, Ont.—The Port Arthur Shipbuilding Company has been allocated the construction of two more full canal sized freighters by the Dominion Government, delivery to be made in the spring of 1919. The boats will each have a capacity of 3,400 tons. The Dominion Government has placed contracts for seven new boats in various shipyards, and of these Port Arthur has obtained two.

London.—The first concrete ship built at the Barrow shipyards was successfully launched, being the first of the 10,000-ton barges now under construction for the department of the controller of merchant shipbuilding. The Barrow shipyards did not begin operations until the beginning of 1918. Many shipbuilding experts were present at the launching, including a representative of the admiralty. A thousand-ton concrete vessel also was launched at the new shipyard at Barnstaple. Like that launched at the Barrow yards, its construction occupied four months. The site of the Barnstaple shipyard was marsh land last March.

Ottawa.—Canada has again smashed all records for speed in wooden shipbuilding. The latest feat has lowered this country's own mark by about sixty per cent. The "War Camchin," a wooden ship of 3,100 tons, built for the Imperial Munitions Board, was launched by the Foundation Company ship yards at Victoria, B.C., on August 31. Installation of machinery commenced at the Imperial Munitions Board's installation plant at Victoria on Sept. 3rd. The installation work was completed in twelve and one-half working days. On September 17th a successful seagoing trip was held. This vessel will commence loading cargo within seven days of the trial trip.

Vancouver, B.C.—On August the 22nd at 7 p.m. the wooden steamer War Tanoo, the sixth and last hull to be built by the Western Canada Shipyards for the Imperial Munitions Board, was launched at the False Creek yard under the supervision of W. Clark, superintendent of the yard. Mrs. J. Sims, wife of one of the officials of the firm of Grant, Smith, McDonell & Co., christened the vessel. In tow of two tugs the steamer War Chilcat passed through the False Creek bridge at 6.45 o'clock the night before. The vessel left at 6 o'clock and struck a flood tide, making the passage through both bridges without accident. She was towed over to the Ogden Point assembly plant, where she will be outfitted for sea. The yard has completed its work well within the contract time.

Victoria.—Nine thousand people witnessed the keel laying ceremonies of the Foundation Company's shipyards here recently, the commencement of a contract for 20 ships for the French government being celebrated in the city by a half-holiday. Premier Oliver and S. F. Tolmie, M.P., officiated at the laying of the first keel. At the second the principals were Hon. John Hart and Mayor Todd. Subsequently the War Nanoose, the last of the vessels being constructed by the Imperial Munitions Board, was launched, Miss M. Oliver, daughter of the provincial premier, acting as sponsor. Two bands were in attendance one being the shipyard band of the Portland yards of the Foundation Company, which was brought here by Bayley Hopkins, manager for the Pacific Northwest.

St. John N.B.—The certificate of the master, Captain Charles E. Dagwell, has been suspended during the war as a result of the inquiry held here on Saturday regarding the circumstances surrounding the loss of the schooner Dornfontein, which was attacked by gun-fire and destroyed by burning by a German submarine off the mouth of the Bay of Funday on August 21st. Chief emphasis is laid on the fact that the captain had handed over to the commander of the enemy submarine his secret sailing orders, making no attempt to destroy them, notwithstanding the peremptory orders he had received. From his conduct the court found that the captain had attached only secondary importance to the document, that he had made light of his duties and responsibility, had been guilty of unheard of neglect, and that he had been gravely negligent, but not with criminal intent. Although the crew was of mixed nationality and two of them spoke German, the court was unable to connect the disaster with any prearranged signals or notification to the enemy.

ARGENTINE NAVIGATION—ITS ORIGIN AND GROWTH

By R. E.

It used to be a common saying in Australia that New Zealand and the "Union Line" meant pretty much the same thing—the country and the company were so closely interwoven that the one was regarded as the complement of the other, and much the same thing may be said, with certain differences, of Argentine and the Argentine Navigation Company, for along 3,000 miles of river Argentina may be said, without exaggeration, to have its origin in Nicholas Michanovitch, for until he appeared on the scene the river transport of the river and immense territories of Argentina was infinitesimal and of deep sea transport she had none.

It was in 1864 that Nicholas Michanovitch, then a youth of seventeen landed at Monte Video, without friends, without money, and unable to speak the Spanish tongue. The war with Paraguay soon began, and young Michanovitch, taking "anything that offered." obtained employment on store ships, in a comparatively short space of time becoming master of a small coaster at Buenos Aires. Then followed, in 1875, the hiring of a couple of tugs, and the opening of a combined office and store—a single room—and on these small, but, as they proved, secure foundations, Nicholas Michanovitch began to build. The tugs, which at first, he was able to hire, he soon became able to buy, and, as time went on, to add to their number, and when we remember that at Buenos Aires in those days there was no "tying up" ships having to lie out in the river and discharge into lighters, it followed that, with the trade of the port steadily increasing, lighters and tugs became in still greater demand, and the man who could supply them prospered correspondingly. Michanovitch gradually extended his activities, placed an order in England for a cargo steamer, and this vessel was the first of her class to enter the port of Buenos Aires, as it then was—in 1880. An opposition tug and lighterage company was then bought out by Michanovitch, who, by 1889, owned over 30 tugs and lighters, and, by the end of 1889 was the possessor of a fleet of over 100 vessels, and now the rapidly developing trade of Buenos Aires and the Plate ports carried with it a corresponding increase, in the, by this time, extensive business of which Michanovitch was the master-mind. Year by year business expanded, and in 1909 Michanovitch formed the Argentine Navigation Company (Nicholas Michanovitch, Limited) which, on the outbreak of the present war, owned a fleet of over 300 vessels, and was described by the Chairman at the R.M.S.P. meeting, in language by no means overdrawn, as "an important concern, with an extensive organization for linking up the South American ports by means of vessels, river craft, tugs, lighters, etc."

From the beginning the Argentine Navigation Company prospered exceedingly. An extract from the chairman's report of December, 1910, reads: "Argentine has over three thousand miles of river available for navigation, and nearly two thousand miles of sea coast. On all this vast extent of waterway the fleet of this company is constantly plying, serving the ports, the townships, and the settlements. Our service across the estuary of the River Plate between the capitals of Argentine and Uruguay, is of invaluable public utility by connecting those two countries. The steamers undertaking these nightly runs are handsome modern boats, equipped with all that can be desired for the comfort of passengers. Another service of upwards of one thousand miles unites Argentine with Paraguay, giving a fast service between the two capitals. Even beyond Asuncion the service continues to the distant town of Concepcion (as far north, almost, as the latitude of Rio), and, at the same time combines a service on the Alto Parana. Fifty ports and townships are in this manner linked up with the metropolis of Buenos Aires and put in touch with Europe. Still another service navigates the waters of the Uruguay to meet the requirements on both the Argentine side and that of the neighbouring republic."

GERMANY SEEKS IRON ORE FROM THE SWEDES

London.—That Germany is contemplating the probable loss of the iron deposits on the Lorraine front, from which she draws so large a proportion of the raw material for her munitions, is indicated by the report of a reliable authority in Berlin, who says that Germany already is making approaches to Sweden with the view of drawing on the iron ore in that country.

It is pointed out that while Sweden is neutral, she is within her rights in disposing of any of her products to any of the belligerents, just as it is legitimate for the allied powers to inform any and every neutral that they shall get no assistance from the allied nations in the way of foodstuffs and raw materials if the neutrals assist Germany with supplies for the manufacture of munitions.

THIS VESSEL STUCK GOING TO THE BASIN

MONTREAL.—Fraser, Brace Co., had a launching recently. This is the last of 4 of the 3,000 ton wooden boats being built by this firm for the I.M.B. By some slight or unknown reason the vessel failed to move freely and the stern swung too freely, resulting in the disarrangement of the supporting timbers, so that the keel at the bow cut into the skids and prevented it taking the water; she is now lying at an angle of about 20 deg. in both directions with her bow uppermost. Mr. Underhill, the Supt., expects to have her in the basin soon. Work is now under way to haul her from her inclined position. As soon as the basin is clear and the old timbers removed, work will be commenced on six smaller wooden vessels, 206 feet in length, for the French Government.

METHODS OF METAL FUSION

By O. C.

DURING the progress of the world-war the scientific welder has risen to the heights of fame, for he has grasped with both hands the opportunities given him to make good in a profession which hitherto had never come fully into its own. The making of permanent metallic joints had been practised in one form or another from the remote days when the ancient craftsman pursued his task by hammering pieces of metal together which had been previously raised to welding heat. The success of the method depended largely on the skill of the mechanic, particularly as to the exactness of his observations of temperature and the conditions of the surfaces of the metal to be joined.

All methods of welding are more or less valuable, each in its own place, from the soft soldering generally used for joining together pieces of thin metal, to the much higher form known by the name of the oxy-acetylene process, which is the last word in this industry. The tensile strength of soft solder is extremely low; it has a much lower strength than the metals to be joined and is principally used for tin-plates, zinc, lead and sometimes brass or copper, wherever no serious strain is encountered. Riveting, brazing welding by water-gas, electric welding and thermit welding are all more or less in practice in various directions, whilst all have their respective uses. These systems are not in general use to anything like the extent of the oxy-acteylene method.

Processes

Thermit-welding is a process specially applicable for the joining of iron and mild steel of great thickness, and has been much used for the welding of rails and the repair of large steel castings. Perhaps its greatest use has been in connection with the welding of tramway joints; but there can be little doubt that this process is destined to be much more largely utilized than heretofore.

The term "autogenous" is usually applied to welding joints by the fusion of metal in action by the flames of the blowpipe. Generally speaking, however, it is more particularly applied to the oxy-acetylene process—most commonly employed in the use of oxy-acetylene welding, which is of comparative recent date. The first blowpipes working with acetylene under pressure were made in 1901 by MM. Fouche and Picard, but since then the process has been subject to extremely rapid strides of progress and its great success has given rise to birth of other processes.

In starting to weld, the operator has one main object—to succeed in uniting metal in such a manner that its properties are not weakened or destroyed. In constructional work certain joints can only be satisfactorily made by autogenous welding, from the industrial point of view, such as in the case of certain branches to tubes, pieces for machines, aeroplanes and other special devices.

Limitations

Considered in the light of a process whose chief function is to restore to use machinery which has broken down under the stress or strain of use or in accident there is nothing in sight to approach it either in efficiency or in economy. It is a process which, used by capable well-trained workmen with a good knowledge of the principles involved, has no rival for the perfect repair of broken metal. There are cases where the application of autogenous welding does not give results, as in the joining of a thick portion to a thin piece where it is impossible to avoid the effects of expansion; but, obviously, no expert would ever set out to achieve results which he knew from the first were unattainable. For the repair of articles in cast iron, bronze and alloys of aluminum it is often indispensable to have the pre-heating of the article followed, after welding, by a slow-cooling in order to avoid the effects of expansion and contraction.

The successful welder should naturally understand the physical and chemical properties of the products which he uses, and which can leave such effects that the strength and solidity of the joint he makes are seriously compromised. It is necessary that he should know the value of the different metals and alloys which he may have to join and the melting point of each. It is of prime importance that he study fully the questions of expansion and contraction. If the temperature of a metallic body is raised progressively throughout the mass, or is lowered in the same way, the phenomena of expansion generally has no bad effects such as breaking or deformation, since its action is uniform. This is not so when the heat is applied at one point or part of the body. The metal tries to expand at this place, and since no force can stop it, it breaks or deforms from that which opposes it. It is necessary to obviate these effects, and in autogenous welding there are numerous devices at the disposal of the experienced operator such as allow for free play in preparing the work of warming the whole piece simultaneously, heating opposing parts, artificial breaking, and the sprinkling of common water in parts to create a force of contraction to counterbalance that of expansion. The method of treatment varies with each metal, and each piece according to shape and dimensions. The effects of expansion are now more or less due to the inferior character of the metal.

It is important to note that the tenacity of various metals, however strong, becomes practically nil when raised to a temperature which is still below the melting point. Breaks on cooling are much more common in these matters than in others, and one is able to avoid them on knowing that at a certain temperature their tenacity diminishes very rapidly. The texture of the metal considerably influences the mechanical properties.

Material Used

Another point which cannot safely be overlooked is the necessity of adding special metals in order to produce the perfect weld. The principal quality which one should require of the metal to be added is that it should be as pure as possible. Metal of the same nature could be added when there is no risk of lowering the quality of the parts to be joined, but this is not generally the case, and the addition to the line of welding a metal which is purer and better improves the holding power of the joint.

Autogenous welding can be applied advantageously in multitudes of repairs in iron and steel. It is often the only process by which fractured, deformed or worn-out parts can be re-created. The application of the process to the manufacture of material for war purposes and to the reconstruction of derelict plants during these last years of abnormal activity in Britain has been on an abnormal scale, and it has nowhere as yet been possible to form even an approximate idea of the extent to which the process has been tried. One may, however, glimpse something of the utility of autogenous welding when it is learned that one firm alone is reported to have dealt with something like ten thousand fractured machines per annum. Great strides have been made by British engineers in this industry, some branches of which Germany had made particularly her own, principally in the method of early apprenticing her youth and specially training them for a long period for the work involved.

There is nothing mechanical or slipshod about the process. It demands the very highest skill and a special course of study if the operator is ever to become a master craftsman whose work can be depended upon. Such is the nature of the business that the repaired part may be made to appear to the eye a perfect model of the original machine and yet prove wholly illusory in operation and fail to stand up to its work. Thus, it need hardly be said, it is not necessarily due to design or a desire to deceive. It simply means that the operator, in his ignorance, has failed to take account of one or other of the vital factors essential to complete success. Firms of the highest repute in the trade, knowing how deceptive outside appearances may prove, enforce the most rigid tests, and no repaired unit is allowed to leave the works until it emerges successfully from the ordeal. It not infrequently happens that the weld proves stronger than the rest of the machine, so much so that ordinary cutting tools altogether fail to make an impression upon the superfluous metal.

As the art of welding becomes better known, so its uses extend. To-day thousands of welders are engaged in helping to turn out wonderful machines and devices, from which perhaps other yet more wonderful things may be evolved.

Quebec.—News reached this city from Three Rivers to the effect that the Three Rivers Shipbuilding Company has sold its yards to the Boston Shipping Company. The new company is said to have a contract for the construction of eight new vessels, and is calling for 1,000 men in addition to the 1,000 already employed by the Three Rivers Company. Four vessels are nearing completion at Three Rivers and will shortly be launched.

Montreal.—An enviable record was equalled when the 7,400-ton steamer "Sammanger" was completed fifteen days after being launched. The installation of engines and boilers was commenced immediately after launching, and in six days a steam trial of the engines was held. Nine days after the hull was complete, thus equalling the record made recently in England, when an 8,000-ton steamer was completed fifteen days after the launch.

Large Shipbuilding Contract.—John E. Russel, vice-president and general manager of the Toronto Shipbuilding Co., has announced the closing of a contract to build ten wooden ships for French interests. Mr. Russell stated that the bulk of these vessels are guaranteed for delivery in July and August, 1919, and as the present capacity of the plant is fully engaged, more land was necessary. Negotiations for the acquiring of this were practically complete, and the plant will be kept running on full capacity all the winter.

Collingwood.—The new steamer War Witch has been successfully launched from the yard of the Collingwood Shipbuilding Co. This is a sister ship to the War Wizard, launched four months ago. The War Witch was built to the order of the Imperial Munitions Board for ocean-going service, and is fitted accordingly. The leading dimensions are as follows: Length over all, 261 feet, beam 43½ feet, and moulded depth 20 feet. She is fitted with triple expansion engines and steam is supplied by two Scotch boilers, 14 ft. dia. by 11 ft. long. The War Witch will be completed and fitted out for sea without delay.

Montreal.—The royal commission on shipyards, which has been in session daily since August 6, last, inquiring into labor disputes which have arisen in shipyards in Montreal, Quebec, Levis, Three Rivers and Sorel, saw, as one important result of their work, the signing of an agreement which will assure uninterrupted construction work for the duration of the war in four yards where strikes were threatened. Other shipyards are expected to follow suit. The agreement is to date from September 1, and continue in effect for the duration of the war. Its main conditions are a nine-hour day's work instead of ten hours, as hitherto, time and a half for all overtime, and in certain cases double time. After February 1st the scale of pay to be revised in accordance with any increase or decrease in cost of living as shown by the labor department, and disputes to be finally referred to a board of conciliation with no halt to ship construction between whiles. Firms signing this agreement are: Fraser, Brace and Company, Ltd., Montreal; Davie Shipbuilding and Repairing Company, Ltd., Lauzon, Levis; The Quebec Shipbuilding and Repair Company, Ltd., Quebec; Quinlan and Robertson, Ltd., Quebec.

New Westminster.—No time is to be lost in the construction of the five wooden steamers for the French government to be built at the Poplar Island yard by the New Westminster Construction & Engineering, Ltd. The keel timbers are already coming in. All five vessels are to be completed in ten months. Four berths will be used, and the frames of the fifth vessel will be cut, so that practically speaking the yard will be working on all five simultaneously. The five steamers ordered by the French government are of 1,500 tons deadweight carrying capacity, and it is understood they are to be used as general cargo carriers in the coasting trade and between England and France. They will be equipped with steam engines, and are of the twin screw type. The plans have been drawn by Tams Lemoyne & Co. of New York, regarded as the leading marine architects on the continent. Orders for some of the engines have already been placed with the Goldie McCullough Co. of Galt, Ont., and with Allis-Chalmers Co. The local yard have a working agreement with the Western Canada yard and the Coquitlam yard, and the engines for all twelve vessels distributed among the three plants will be ordered together.

If interested, tear out this page and place with letters to be answered.

WILKINSON & KOMPASS
TORONTO HAMILTON WINNIPEG
IRON AND STEEL
HEAVY HARDWARE
MILL SUPPLIES
AUTOMOBILE ACCESSORIES
WE SHIP PROMPTLY

Castings

Mason No. 55 Style Pump Pressure Regulator

Reilly Multi-Screen Feed Water Filter

Mason No. 126 Style Marine Reducing Valve

Reilly Navy Type Feed Water Heater

Reilly Marine Evaporator Submerged Type

MASON

The value of a ship is judged by its service. That service depends upon its equipment. Equipment of the Mason quality is a guarantee of reliable service. The service of our staff of expert engineers has also been of value to others; can they be of service to you? We are steam room specialists. Our line is complete. Let us get in touch with you.

We manufacture Reilly Evaporators, Reilly Distillers, Reilly Feed Heaters, Reilly Feed Water Filters, and furnish Mason Reducing Valves for Marine Service.

We are the sole Canadian manufacturers of these auxiliaries, and make a specialty of prompt deliveries.

Our new shop gives us ample facilities for quick service.

The
Mason Regulator & Engineering Company, Limited

135-153 Dagenais St.
MONTREAL

506 Kent Building
TORONTO

ELASTIC
TRADE MARK
SEAM COMPOSITION

YARROWS

Limited

Associated with YARROW & CO., GLASGOW

WORKS AT ESQUIMALT B. C.

Telegrams and Cables "Yarrows, Victoria"

SHIPBUILDERS, ENGINEERS AND SHIP REPAIRERS

IRON AND BRASS FOUNDERS

VESSELS CONVERTED FROM COAL BURNING TO OIL FUEL BURNING SYSTEMS.
MANUFACTURE OF MANGANESE BRONZE PROPELLER BLADES A SPECIALITY.

MARINE RAILWAY, LENGTH 300 ft., CAPACITY 2500 TONS DEAD WEIGHT

Address: P.O. Box 1595 Victoria B. C.

Larger Vessels Docked in Government Graving Dock, Esquimalt—Lowest rates on the Pacific Coast.

LONDON PARIS ST. JOHN, N.B. MILAN

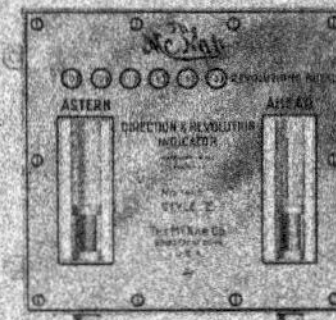

THE McNAB COMPANY

Bridgeport, Conn., U.S.A.

PATENTEES AND MANUFACTURERS OF

McNab Patent Engine Direction Indicators
" " Pneumatic Chart Room Counters
" " Pneumatic Engine Room Counters
" " Ship's Indicating Telegraph

McNab Patent "Cascade" Boiler Circulators
" " Steamship Draft Gauges
" Mechanical Rotary Turbine Engine Counter
" " Reciprocating Engine Counter
" Type Steam Steering Gear

Brown's Patent Telemotors and Steam Tillers

EMERGENCY FLEET CORPORATION TECHNICAL ORDER No. 30 SPECIFIES THE INSTALLATION OF McNAB PATENT ENGINE DIRECTION INDICATORS AND McNAB PATENT PNEUMATIC CHART ROOM COUNTERS ON ALL STEEL SHIPS UNDER CONSTRUCTION FOR THEM.

Catalogues on request.

CANADA
FOUNDRIES & FORGINGS
LIMITED

DROP FORGED STEEL

LARGE

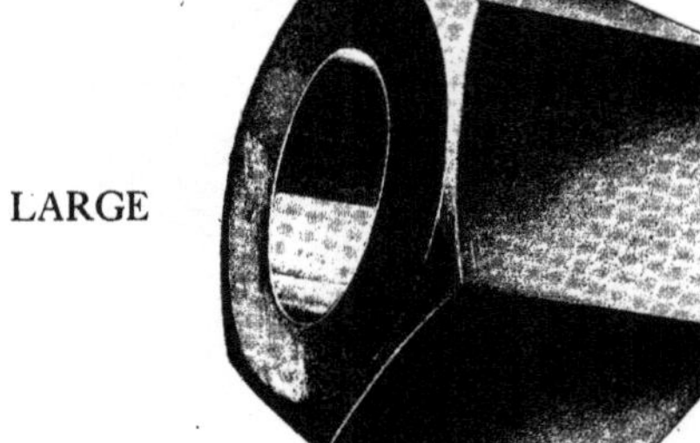

SIZES

MARINE BOILER STAY NUTS

FORGED FROM THE SOLID BAR TO LLOYD'S SPECIFICATIONS.
EVERY NUT NEEDS A WRENCH.

FORGED STEEL WRENCHES

EVERY NUT NEEDS A WRENCH.
PRICE ON APPLICATION.

SMITHERIES AT WELLAND, ONT.

BUYERS' DIRECTORY

ACCUMULATORS, HYDRAULIC
Smart-Turner Mach. Co., Hamilton, Ont.

AERATING RESERVOIRS
Spray Engineering Co., Boston, Mass.

AIR PORTS
Mitchell Co., The Robert, Montreal, Que.
Turnbull Elevator Mfg. Co., Toronto, Ont.

ALLOYS, BRASS AND COPPER
Dom. Copper Products Co., Ltd., Montreal, Que.
Mitchell Co., The Robert, Montreal, Que.
Mueller Mfg. Co., H., Sarnia, Ont.
Tallman Brass & Metal Co., Hamilton, Ont.
United Brass & Lead, Ltd., Toronto, Ont.

ANCHORS
Beauchemin & Fils, Sorel, P.Q.
Hopkins & Co., F. H., Montreal, Que.
Kearfott Engineering Co., New York, N.Y.
McNab Co., Bridgeport, Conn.
Henry Rogers Sons & Co., Wolverhampton, Eng.
Wm. C. Wilson & Co., Toronto, Ont.

ARCHITECT, NAVAL
Watts, J. Murray, Philadelphia, Pa.

ASBESTOS GOODS
Wm. C. Wilson & Co., Toronto, Ont.

BABBITT METAL
Aikenhead Hardware, Ltd., Toronto, Ont.
Hoyt Metal Company, Toronto, Ontario.
Tallman Brass & Metal Co., Hamilton, Ont.
Wilkinson & Kompass, Hamilton, Ont.
Wm. C. Wilson & Co., Toronto, Ont.

BAROMETERS
Wilson & Co., Wm. C., Toronto, Canada.

BAR IRON
Mitchells Ltd., Glasgow, Scotland.

BARS, GRATE
Babcock & Wilcox, Ltd., Montreal, Que.

BEARINGS, BRASS
Empire Mfg. Co., London, Ont.
Mueller Mfg. Co., H., Sarnia, Ont.
Mitchell Co., The Robert, Montreal, Que.
Tallman Brass & Metal Co., Hamilton, Ont.
United Brass & Lead Co., Toronto, Ont.

BELLS, SHIPS, ENGINE ROOM, ETC.
Aikenhead Hardware, Ltd., Toronto, Ont.
Clarke & Bros. Co., Montreal, Que.
Cory & Son, Inc., Chas., New York, N.Y.
Empire Mfg. Co., London, Ont.
Mitchell Co., The Robert, Montreal, Que.
Morrison Brass Mfg. Co., James, Toronto, Ont.
Mueller Mfg. Co., H., Sarnia, Ont.
Tallman Brass & Metal Co., Hamilton, Ont.
United Brass & Lead, Ltd., Toronto, Ont.

BELTING, LEATHER
Aikenhead Hardware, Ltd., Toronto, Ont.
Wm. C. Wilson & Co., Toronto, Ont.

BIBBS, COMPRESSION
Empire Mfg. Co., London, Ont.
Mitchell Co., The Robert, Montreal, Que.
Mueller Mfg. Co., H., Sarnia, Ont.
United Brass & Lead, Ltd., Toronto, Ont.

BINNACLES
Hopkins & Co., F. H., Montreal, Que.
Morrison Brass Mfg. Co., James, Toronto, Ont.

BLOCKS, CARGO, HEEL AND TACKLE
Aikenhead Hardware, Ltd., Toronto, Ont.
Hopkins & Co., F. H., Montreal, Que.
Loveridge, Ltd., Docks, Cardiff, Wales.

BLOWERS, TURBO.
Mason Regulator & Engin. Co., Montreal, Que.

BOAT CHOCKS
Corbet Fdry. & Machine Co., Owen Sound, Ont.
Marten-Freeman Co., Toronto, Ont.

BOILER COMPOUND
Wm. C. Wilson & Co., Toronto, Ont.

BOILER CIRCULATORS
Eckliff Circulator Co., Detroit, Mich.

BOILER FEED PUMPS
Can. Ingersoll-Rand Co., Ltd., Sherbrooke, Que.
Darling Bros., Ltd., Montreal, Que.
Goldie & McCulloch Co., Galt, Ont.
Morris Mach. Works, Baldwinsville, N.Y.
Smart-Turner Mach. Co., Hamilton, Ont.
Williams Machinery Co., A. R., Toronto, Ont.

BOILER FITTINGS
Empire Mfg. Co., London, Ont.
Goldie & McCulloch Co., Galt, Ont.
McAvity & Sons Ltd., T., St. John, N.B.
Wager Furnace Bridge Wall Co., Inc., New York, N.Y.

BOILERS, MARINE
Babcock & Wilcox, Ltd., Montreal, Que.
Can. Fairbanks-Morse Co., Montreal, Que.
Can. Vickers, Ltd., Montreal, Que.
Collingwood Shipbuilding Co., Collingwood, Ont.
Doxford & Sons, William, Sunderland, England.
Engr. & Mach. Wks. of Can., St. Catharines, Ont.
Goldie & McCulloch, Ltd., Galt, Ont.
Hall Engineering Works, Montreal, Que.
Mason Regulator & Engin. Co., Montreal, Que.
Montreal Dry Docks & Shipbuilding Co., Montreal, Que.
National Shipbuilding Co., Goderich, Ont.
Marsh Engineering Works, Belleville, Ont.
Polson Iron Works, Toronto, Ontario.
Port Arthur Shipbuilding Co., Port Arthur, Ont.
Waterous Engine Works Co., Brantford, Ont.
Williams Machinery Co., A. R., Toronto, Ont.

BOOKS, TECHNICAL, MARINE
MacLean Publishing Co., Toronto, Ont.

BOLTS
London Bolt & Hinge Works, London, Ont.
Mitchell Co., The Robert, Montreal, Que.
United Brass & Lead, Ltd., Toronto, Ont.
Wilkinson & Kompass, Hamilton, Ont.

BROKER
Page & Jones, Mobile, Ala., U.S.A.

BRASS GOODS
Corbet Fdry. & Machine Co., Owen Sound, Ont.
Goldie & McCulloch Co., Galt, Ont.
McAvity & Sons, T., St. John, N.B.
Mitchell Co., The Robert, Montreal, Que.
Mueller Mfg. Co., H., Sarnia, Ont.
Tallman Brass & Metal Co., Hamilton, Ont.

BUCKETS, CLAMSHELL
Beatty & Sons, Welland, Ont.
Morris Crane & Hoist Co., Herbert, Niagara Falls, Ont.

BUCKETS, DUMP
Morris Crane & Hoist Co., Herbert, Niagara Falls, Ont.

BUCKETS, COALING
Beatty & Sons, Welland, Ont.
Engr. & Mach. Wks. of Can., St. Catharines, Ont.
Hopkins & Co., F. H., Montreal, Que.
Marsh Engineering Works, Belleville, Ont.
Reed & Co., Geo., Montreal, Que.

BUSHINGS, BRONZE
Oberdorfer Brass Co., M. L., Syracuse, N.Y.

CABLE
Hopkins & Co., F. H., Montreal, Que.
McNab Co., Bridgeport, Conn.
Wm. C. Wilson & Co., Toronto, Ont.

CABLE, LEAD COVERED AND ARMORED
Standard Underground Cable Co., Hamilton, Ont.

CABLE, ACCESSORIES
Darling Bros., Ltd., Montreal, Que.
Standard Underground Cable Co., Hamilton, Ont.

CAPSTANS
Advance Mach. & Welding Co., Montreal, Que.
Dake Engine Co., Grand Haven, Mich.
Hopkins & Co., F. H., Montreal, Que.
Kennedy & Sons, Wm., Owen Sound, Ont.
Wm. C. Wilson & Co., Toronto, Ont.

CALKING TOOLS, ELECTRIC
Aikenhead Hardware, Ltd., Toronto, Ont.

CALKING TOOLS, PNEUMATIC
Aikenhead Hardware, Ltd., Toronto, Ont.
Can. Ingersoll-Rand Co. Sherbrooke, Que.

CALORIFIERS
Low & Sons, Ltd., Archibald, Glasgow, Scotland

CASTINGS
Beauchemin & Fils, Sorel, P.Q.
Can. Steel Foundries, Ltd., Montreal, Que.
Collingwood Shipbuilding Co., Collingwood, Ont.
Wm. Hamilton Co., Peterboro, Ont.
Kennedy & Sons, Wm., Owen Sound, Ont.
Goldie & McCulloch Co., Galt, Ont.
Marsh Engineering Works, Belleville, Ont.
Mitchell Co., Ltd., Robt., Montreal, Que.
Mueller Mfg. Co., H., Sarnia, Ont.
Tallman Brass & Metal Co., Hamilton, Ont.
United Brass & Lead, Ltd., Toronto, Ont.
Waterous Engine Works Co., Brantford, Ont.

CASTINGS, ALLOY
Mitchell Co., The Robert, Montreal, Que.
Mueller Mfg. Co., H., Sarnia, Ont.
Oberdorfer Brass Co., M. L., Syracuse, N.Y.
Tolland Mfg. Co., Montreal, Que.
Tallman Brass & Metal Co., Hamilton, Ont.
United Brass & Lead, Ltd., Toronto, Ont.

CASTINGS, ALUMINUM
Empire Mfg. Co., London, Ont.
Mitchell Co., The Robert, Montreal, Que.
Tallman Brass & Metal Co., Hamilton, Ont.
United Brass & Lead Co., Toronto, Ont.

CASTINGS, BRASS
Crouse-Hinds Co. of Canada, Ltd., Toronto, Ont.
Empire Mfg. Co., London, Ont.
Goldie & McCulloch Co., Galt, Ont.
McAvity & Sons Ltd., T., St. John, N.B.
McNab Co., Bridgeport, Conn.
Mitchell Co., Ltd., Robt., Montreal, Que.
Mueller Mfg. Co., H., Sarnia, Ont.
Oberdorfer Brass Co., M. L., Syracuse, N.Y.
Tolland Mfg. Co., Montreal, Que.
Tallman Brass & Metal Co., Hamilton, Ont.
United Brass & Lead Co., Toronto, Ont.
Waterous Engine Works Co., Brantford, Ont.

CASTINGS, GREY IRON, MALLEABLE, ALUMINUM
Crouse-Hinds Co. of Canada, Ltd., Toronto, Ont.
Darling Bros., Ltd., Montreal, Que.
Engr. & Mach. Wks. of Can., St. Catharines, Ont.
McAvity & Sons, T., St. John, N.B.
McNab Co., Bridgeport, Conn.
Mitchell Co., The Robert, Montreal, Que.
Mueller Mfg. Co., H., Sarnia, Ont.
Waterous Engine Works Co., Brantford, Ont.

CASTINGS, MANGANESE STEEL

CASTINGS, MANGANESE BRONZE
Oberdorfer Brass Co., M. L., Syracuse, N.Y.
Tallman Brass & Metal Co., Hamilton, Ont.
United Brass & Lead Co., Toronto, Ont.

CELLAR DRAINERS
Mueller Mfg. Co., H., Sarnia, Ont.

CHAINS
Aikenhead Hardware, Ltd., Toronto, Ont.
Kearfott Engineering Co., New York, N.Y.
Hopkins & Co., F. H., Montreal, Que.
Morris Crane & Hoist Co., Herbert, Niagara Falls, Ont.
Henry Rogers, Sons & Co., Wolverhampton, Eng.
Wm. C. Wilson & Co., Toronto, Ont.

CHAIN BLOCKS AND SLINGS
Morris Crane & Hoist Co., Herbert, Niagara Falls, Ont.

CHANDLERY, SHIP
Beauchemin & Fils, Sorel, P.Q.
Hopkins & Co., F. H., Montreal, Que.
Leckie, Ltd., John, Toronto, Ont.

CLAMPS, STEAM AND WATER
Mueller Mfg. Co., H., Sarnia, Ont.

CLOCKS
American Steam Gauge & Valve Mfg. Co., Boston, Mass.
Morrison Brass Mfg. Co., James, Toronto, Ont.
Williams Machinery Co., A. R., Toronto, Ont.

CLOSETS
Mueller Mfg. Co., H., Sarnia, Ont.
United Brass & Lead, Ltd., Toronto, Ont.

COAL
Nova Scotia Steel & Coal Co., New Glasgow, N.S.

COAL, HANDLING MACHINERY
Morris Crane & Hoist Co., Herbert, Niagara Falls, Ont.
Waterous Engine Works Co., Brantford, Ont.

COCKS, BILGE, DISCHARGE, INDICATOR,
McAvity & Sons Ltd., T., St. John, N.B.
Mitchell Co., The Robert, Montreal, Que.
Morrison Brass Mfg. Co., James, Toronto, Ont.

COCKS, BASIN
Empire Mfg. Co., London, Ont.
Mitchell Co., The Robert, Montreal, Que.
United Brass & Lead, Ltd., Toronto, Ont.

COMPASSES
Wm. C. Wilson & Co., Toronto, Ont.

COMPRESSORS, AIR
Can. Fairbanks-Morse Co., Montreal, Que.
Canadian Ingersoll-Rand Co., Sherbrooke, Que.
Darling Bros., Ltd., Montreal, Que.
Hopkins & Co., F. H., Montreal, Que.
Smart-Turner Mach. Co., Hamilton, Ont.
Williams Machinery Co., A. R., Toronto, Ont.

CONDENSERS
Darling Bros., Ltd., Montreal, Que.
Goldie & McCulloch Co., Galt, Ont.
Kearfott Engineering Co., New York, N.Y.
Morris Machine Works, Baldwinsville, N.Y.
Smart-Turner Mach. Co., Hamilton, Ont.
Weir, Ltd., G. & J., Cathcart, Glasgow, Scotland.
Williams Machinery Co., A. R., Toronto, Ont.

CONDULETS, MARINE
Crouse-Hinds Co. of Canada, Ltd., Toronto, Ont.

CONTRACTORS' SUPPLIES
McAvity & Sons, T., St. John, N.B.

CONSULTING ENGINEER
Watt, J. Murray, Philadelphia, Pa.

CONVEYORS, ASH, COAL
Babcock & Wilcox, Ltd., Montreal, Que.
Hopkins & Co., F. H., Montreal, Que.

COPING MACHINES
Bertram & Sons, Ltd., John, Dundas, Ont.

COPPER TUBES, SHEETS AND RODS
Dom. Copper Products Co., Montreal, Que.
Tallman Brass & Metal Co., Hamilton, Ont.

COTTON, CALKING
Wilson & Co., Wm. C., Toronto, Canada.

COVERS, CANVAS, FOR HATCHES, LIFE-BOATS, ETC.
Leckie, Ltd., John, Toronto, Ont.
Waterous Engine Works Co., Brantford, Ont.

COUNTERS, REVOLUTION
American Steam Gauge & Valve Mfg. Co., Boston, Mass.

CRANES
Aikenhead Hardware, Ltd., Toronto, Ont.
Can. Fairbanks-Morse Co., Montreal, Que.
Hopkins & Co., F. H., Montreal, Que.
Williams Machinery Co., A. R., Toronto, Ont.

CRANES, ELECTRIC
Babcock & Wilcox, Ltd., Montreal, Que.
Morris Crane & Hoist Co., Herbert, Niagara Falls, Ont.
Smart-Turner Mach. Co., Hamilton, Ont.

CRANES, GOLIATH AND PNEUMATIC
Morris Crane & Hoist Co., Herbert, Niagara Falls, Ont.

CRANES, GRANTRY, PORTABLE, JIB
Morris Crane & Hoist Co., Herbert, Niagara Falls, Ont.
Smart-Turner Mach. Co., Hamilton, Ont.

CRANES, OVERHEAD TRAVELLING
Morris Crane & Hoist Co., Herbert, Niagara Falls, Ont.

CRANK SHAFTS
Canada Foundries and Forgings, Welland, Ont.

DAVITS, BOAT
Corbet Fdry. & Machine Co., Owen Sound, Ont.
Hopkins & Co., F. H., Montreal, Que.
Marten, Freeman Co., Toronto, Ont.
Waterous Engine Works Co., Brantford, Ont.

DEAD LIGHTS, BRASS
Goldie & McCulloch Co., Galt, Ont.
Morris Crane & Hoist Co., Herbert, Niagara Falls, Ont.
Mitchell Co., Ltd., Robt., Montreal, Que.

DECK LIGHTS
Mitchell Co., The Robert, Montreal, Que.
Turnbull Elevator Mfg. Co., Toronto, Ont.

DECK PLUGS, ELECTRIC
Crouse-Hinds Co. of Canada, Ltd., Toronto, Ont.
Mitchell Co., The Robert, Montreal, Que.

DECKING FOR SHIPS
Marine Decking & Supply Co., Philadelphia, Pa.

DERRICKS
Aikenhead Hardware, Ltd., Toronto, Ont.
Dake Engine Co., Grand Haven, Mich.
Hopkins & Co., F. H., Montreal, Que.
Marsh Engineering Works, Belleville, Ont.
Morris Crane & Hoist Co., Herbert, Niagara Falls, Ont.

DISTILLERS
Kearfott Engineering Co., New York, N.Y.

DREDGES
Collingwood Shipbuilding Co., Collingwood, Ont.
Morris Mach. Works, Baldwinsville, N.Y.
Norbom Engineering Co., Philadelphia, Pa.
Polson Iron Works, Toronto.

DRILLS, AIR
Aikenhead Hardware, Ltd., Toronto, Ont.
Can. Ingersoll-Rand Co. Sherbrooke, Que.
Hopkins & Co., F. H., Montreal, Que.

DRILLS, CENTRE
Wilt Twist Drill Co., Walkerville, Ont.

DRILLS, BLACKSMITH AND BIT STOCK
Wilt Twist Drill Co., Walkerville, Ont.

DRILLS, ELECTRIC
Aikenhead Hardware, Ltd., Toronto, Ont.
Wilkinson & Kompass, Hamilton, Ont.

DRILLS, HIGH SPEED
Wilt Twist Drill Co., Walkerville, Ont.
Zenith Coal & Steel Products, Ltd., Montreal.

DRILLS, TRACK
Wilt Twist Drill Co., Walkerville, Ont.

DRILLS, RACHET AND HAND
Wilt Twist Drill Co., Walkerville, Ont.

DRILLS, TWIST
Aikenhead Hardware, Ltd., Toronto, Ont.
Williams Machinery Co., A. R., Toronto, Ont.
Wilt Twist Drill Co., Walkerville, Ont.

DRY DOCKS
Can. Vickers, Ltd., Montreal, Que.
Collingwood Shipbuilding Co., Collingwood, Ont.
Doxford & Sons, William, Sunderland, England.
Georgian Bay Shipbuilding & Wrecking Co., Midland, Ont.
National Shipbuilding Co., Goderich, Ont.
Polson Iron Works, Toronto, Ont.
Port Arthur Shipbuilding Co., Port Arthur, Ont.
Yarrows, Limited, Victoria, B.C.

ECONOMIZERS, FUEL
Babcock & Wilcox, Ltd., Montreal, Que.

EJECTORS
Darling Bros., Ltd., Montreal, Que.
Empire Mfg. Co., London, Ont.
Mitchell Co., The Robert, Montreal, Que.
Morrison Brass Mfg. Co., James, Toronto, Ont.
Smart-Turner Mach. Co., Hamilton, Ont.

ELECTRIC LAMPS
Mitchell Co., The Robert, Montreal, Que.
Wm. C. Wilson & Co., Toronto, Ont.

ELECTRO-PLATING
Mitchell Co., The Robert, Montreal, Que.
Tallman Brass & Metal Co., Hamilton, Ont.
United Brass & Lead Co., Toronto, Ont.

ELECTRIC WELDING
Beauchemin & Fils, Sorel, P.Q.

ELEVATING MACHINERY
Darling Bros., Ltd., Montreal, Que.
Goldie & McCulloch, Ltd., Galt, Ont.
Wm. Hamilton Co., Peterboro, Ont.
Morris Crane & Hoist Co., Herbert, Niagara Falls, Ont.
Waterous Engine Works Co., Brantford, Ont.

ELEVATORS
Darling Bros., Ltd., Montreal, Que.
Turnbull Elevator Mfg. Co., Toronto, Ont.

ENGINES, HOISTING
Corbet Fdry. & Machine Co., Owen Sound, Ont.
Hopkins & Co., F. H., Montreal, Que.
Kennedy & Sons, Wm., Owen Sound, Ont.
Marsh Engineering Works, Belleville, Ont.
Port Arthur Shipbuilding Co., Port Arthur, Ont.
Williams Machinery Co., A. R., Toronto, Ont.

ENGINE, INTERNAL COMBUSTION
Doxford & Sons, William, Sunderland, England.

ENGINES, MARINE
Bolinders Co., New York, N.Y.
Can. Fairbanks-Morse Co., Montreal, Que.
Can. Vickers, Ltd., Montreal, Que.
Doxford & Sons, William, Sunderland, England.
Goldie & McCulloch, Ltd., Galt, Ont.
Hopkins & Co., F. H., Montreal, Que.
Iron Works, Ltd., Owen Sound, Ont.
Mason Regulator & Engin. Co., Montreal, Que.
Morris Mach. Works, Baldwinsville, N.Y.
National Shipbuilding Co., Goderich, Ont.
Norbom Engineering Co., Philadelphia, Pa.
Polson Iron Works, Toronto, Ont.
Port Arthur Shipbuilding Co., Port Arthur, Ont.
Trout Co., H. G., Buffalo, N.Y.
Waterous Engine Works Co., Brantford, Ont.
Williams Machinery Co., A. R., Toronto, Ont.

ENGINE STARTERS (AIR)
Scaife & Sons Co., Wm. B., Oakmont, Pa.

ENGINES, STEERING
Corbet Fdry. & Machine Co., Owen Sound, Ont.
Dake Engine Co., Grand Haven, Mich.
Kennedy & Sons, Wm., Owen Sound, Ont.
Wm. C. Wilson & Co., Toronto, Ont.

ENAMELWARE
Mueller Mfg. Co., H., Sarnia, Ont.
United Brass & Lead Co., Toronto, Ont.

EVAPORATORS
Kearfott Engineering Co., New York, N.Y.
Mason Regulator & Engin. Co., Montreal, Que.
McNab Co., Bridgeport, Conn.
Weir Ltd., G. & J., Cathcart, Glasgow, Scotland.

EXTRACTORS, GREASE
American Steam Gauge & Valve Mfg. Co., Boston, Mass.
Darling Bros., Ltd., Montreal, Que.

EYE BOLTS AND NUTS
Canada Foundries and Forgings, Welland, Ont.
Mitchell Co., The Robert, Montreal, Que.
United Brass & Lead, Ltd., Toronto, Ont.

FANS
Aikenhead Hardware Ltd., Toronto, Ont.
Empire Mfg. Co., London, Ont.
Reed & Co., Geo. W., Montreal, Que.
Smart-Turner Mach. Co., Hamilton, Ont.
Williams Machinery Co., A. R., Toronto, Ont.

FENDERS, ROPE
Hopkins & Co., F. H., Montreal, Que.
Leckie, Ltd., John, Toronto, Ont.
Wilson & Co., Wm. C., Toronto, Canada.

FERRO-MANGANESE
Mitchells, Ltd., Glasgow, Scotland.

FILES
Aikenhead Hardware, Ltd., Toronto, Ont.
Williams Machinery Co., A. R., Toronto, Ont.

FIRE BRICKS
Beveridge Paper Co., Montreal, Que.
Mitchells, Ltd., Glasgow, Scotland.
Williams Machinery Co., A. R., Toronto, Ont.

FILTERS, FEED WATER
Darling Bros., Ltd., Montreal, Que.
MacKinnon Steel Co., Sherbrooke, Que.
Mason Regulator & Engin. Co., Montreal, Que.

FITTINGS, MARINE
Hopkins & Co., F. H., Montreal, Que.
McAvity & Sons, T., St. John, N.B.
Mitchell Co., The Robert, Montreal, Que.
United Brass & Lead, Ltd., Toronto, Ont.

FITTINGS, MOTOR BOAT
Empire Mfg. Co., London, Ont.
Mitchell Co., The Robert, Montreal, Que.
Mueller Mfg. Co., H., Sarnia, Ont.
United Brass & Lead, Ltd., Toronto, Ont.

FIXTURES, ELECTRIC
Cory & Son, Inc., Chas., New York, N.Y.
Crouse-Hinds Co. of Canada, Ltd., Toronto, Ont.
Mitchell Co., The Robert, Montreal, Que.
Tallman Brass & Metal Co., Hamilton, Ont.

FLAG POLES, STEEL
Dennis Wire & Iron Works Co., London, Ont.

FLOW METERS
Spray Engineering Co., Boston, Mass.

FLUE CLEANERS
Wm. C. Wilson & Co., Toronto, Ont.

FORGES
Aikenhead Hardware, Ltd., Toronto, Ont.
Hopkins & Co., F. H., Montreal, Que.

FLANGING AND EXPANDING MACHINES, PIPE
Lovekin Pipe Expanding & Flanging Mach. Co., Philadelphia, Pa.

FLOODLIGHTS, ELECTRIC
Crouse-Hinds Co. of Canada, Ltd., Toronto, Ont.

FLUORSPAR
Mitchells, Ltd., Glasgow, Scotland.
Scythes & Co., Toronto, Ont.

FORGINGS, ALL KINDS
Aikenhead Hardware, Ltd., Toronto, Ont.
Collingwood Shipbuilding Co., Collingwood, Ont.
Nova Scotia Steel & Coal Co., New Glasgow, N.S.

FORGINGS, STEEL AND IRON
Canada Foundries and Forgings, Welland, Ont.

FURNACE BRIDGE WALLS
Wager Furnace Bridge Wall Co., Inc., 169 Broadway, New York, N.Y.

GAUGES, RECORDING
American Steam Gauge & Valve Mfg. Co., Boston, Mass.
Empire Mfg. Co., London, Ont.

GASKETS
Wm. C. Wilson & Co., Toronto, Ont.

GAUGES COCKS
McAvity & Sons, T., St. John, N.B.
Mitchell Co., The Robert, Montreal, Que.

GAUGE GLASSES
Wm. C. Wilson & Co., Toronto, Ont.

GAUGES, WATER, PRESSURE, COMPOUND AND VACUUM
Aikenhead Hardware, Ltd., Toronto, Ont.
Babcock & Wilcox, Ltd., Montreal, Que.
Empire Mfg. Co., London, Ont.
McAvity & Sons, T., St. John, N.B.
McNab Co., Bridgeport, Conn.
Morrison Brass Mfg. Co., James, Toronto, Ont.

GENERATORS AND CONVERTORS
Can. Fairbanks-Morse Co., Montreal, Que.

GLAZING COMPOSITION
Kuhls, H. B. Fred., 6411 3rd Ave., Brooklyn, N.Y.

GOGGLES
Standard Optical Co., Geneva, N.Y.

GONGS
Clark & Bro., C. O., Montreal, Que.
Mitchell Co., Ltd., Robt., Montreal, Que.

GRAPHITE
Wm. C. Wilson & Co., Toronto, Ont.

GRATINGS
Can. Welding Works, Montreal, Que.
Corbet Fdry. & Machine Co., Owen Sound, Ont.
Canada Wire & Iron Goods Co., Hamilton, Ont.
MacKinnon Steel Co., Sherbrooke, Que.

GRINDERS, ELECTRIC
Wilkinson & Kompass, Hamilton, Ont.

GRINDERS, PNEUMATIC
Can. Ingersoll-Rand Co. Sherbrooke, Que.

GUARDS, MACHINERY
Dennis Wire & Iron Works Co., London, Ont.

GUY RODS AND ANCHORS, ELECTRIC
Crouse-Hinds Co. of Canada, Ltd., Toronto, Ont.

HAMMERS
Canada Foundries and Forgings, Ltd., Welland, Ont.

HARDWARE, MARINE
Hopkins & Co., F. H., Montreal, Que.
Mitchell Co., The Robert, Montreal, Que.

HEADLIGHTS, ELECTRIC
Crouse-Hinds Co. of Canada, Ltd., Toronto, Ont.
Hopkins & Co., F. H., Montreal, Que.
Tallman Brass & Metal Co., Hamilton, Ont.

HEATING EQUIPMENT
Darling Bros., Ltd., Montreal, Que.
Empire Mfg. Co., London, Ont.
Low & Sons, Ltd., Archibald, Glasgow, Scotland

HEATERS, FEED, WATER
Darling Bros., Ltd., Montreal, Que.
Babcock & Wilcox, Ltd., Montreal, Que.
Goldie & McCulloch Co., Galt, Ont.
Kearfott Engineering Co., New York, N.Y.
Mason Regulator & Engin. Co., Montreal, Que.
McNab Co., Bridgeport, Conn.
Weir Ltd., G. & J., Cathcart, Glasgow, Scotland.

HINGES
London Bolt & Hinge Works, London, Ont.
Mitchell Co., Ltd., Robt., Montreal, Que.

HOOKS
Hopkins & Co., F. H., Montreal, Que.
Morris Crane & Hoist Co., Herbert, Niagara Falls, Ont.

HOISTS, ASH
Beatty & Sons, Welland, Ont.
Marsh Engineering Works, Belleville, Ont.
St. Clair Bros., Galt, Ont.
Waterous Engine Works Co., Brantford, Ont.

HOIST BLOCKS
Morris Crane & Hoist Co., Herbert, Niagara Falls, Ont.

HOISTS, CHAIN
Aikenhead Hardware, Ltd., Toronto, Ont.
Can. Fairbanks-Morse Co., Montreal, Que.
Dake Engine Co., Grand Haven, Mich.
Hopkins & Co., F. H., Montreal, Que.
Morris Crane & Hoist Co., Herbert, Niagara Falls, Ont.
Williams Machinery Co., A. R., Toronto, Ont.

HOISTS, CARGO, MOVING, ETC.
Dake Engine Co., Grand Haven, Mich.
Hopkins & Co., F. H., Montreal, Que.
Marsh Engineering Works, Belleville, Ont.
Waterous Engine Works Co., Brantford, Ont.

HOISTING MACHINERY
Beatty & Sons, Welland, Ont.
Can. Ingersoll-Rand Co. Sherbrooke, Que.
Corbet Fdry. & Machine Co., Owen Sound, Ont.
Wm. Hamilton Co., Peterboro, Ont.
Marsh Engineering Works, Belleville, Ont.
Morris Crane & Hoist Co., Herbert, Niagara Falls, Ont.
Waterous Engine Works Co., Brantford, Ont.
Williams Machinery Co., A. R., Toronto, Ont.

HOSE
Wm. C. Wilson & Co., Toronto, Ont.

INDICATORS, ENGINE
American Steam Gauge & Valve Mfg. Co., Boston, Mass.
Cory & Son, Inc., Chas., New York, N.Y.
McNab Co., Bridgeport, Conn.

INDICATORS, SPEED
Aikenhead Hardware, Ltd., Toronto, Ont.
Cory & Son, Inc., Chas., New York, N.Y.

INJECTORS
Aikenhead Hardware, Ltd., Toronto, Ont.
Empire Mfg. Co., London, Ont.
Mitchell Co., The Robert, Montreal, Que.
Morrison Brass Mfg. Co., James, Toronto, Ont.
Williams Machinery Co., A. R., Toronto, Ont.

INGOTS
Broughton Copper Co., Ltd., Manchester, Eng.

INSULATORS, ELECTRIC
Crouse-Hinds Co. of Canada, Ltd., Toronto, Ont.

INSTRUMENTS, NAUTICAL
Leckie, Ltd., John, Toronto, Ont.

IRON AND STEEL
Mitchells Ltd., Glasgow, Scotland.

JACKS
Hopkins & Co., F. H., Montreal, Que.
Morris Crane & Hoist Co., Herbert, Niagara Falls, Ont.

JOINTS, BALL SLIP
Norbom Engineering Co., Philadelphia, Pa.

INSURANCE, MARINE
Toronto Insurance & Vessel Agency, Toronto, Ont.

JOURNAL WEDGES
Canada Foundries and Forgings, Ltd., Welland, Ont.

JUTE PACKING
Stratford Oakum Co., Geo., Jersey City, N.J.

LATHES
Bertram & Sons, Ltd., John, Dundas, Ont.

LEATHER, LACE
Wm. C. Wilson & Co., Toronto, Ont.

LENSES FOR GOGGLES
Standard Optical Co., Geneva, N.Y.

LIFEBUOYS, BELTS AND PRESERVERS
Fosbery Co., Barking, Essex, Eng.
Hopkins & Co., F. H., Montreal, Que.
Wm. C. Wilson & Co., Toronto, Ont.

LIFEBOATS
Hopkins & Co., F. H., Montreal, Que.
Wm. C. Wilson & Co., Toronto, Ont.

LIFE BOAT EQUIPMENT
Eckliff Circulator Co., Detroit, Mich.
Hopkins & Co., F. H., Montreal, Que.
Leckie, Ltd., John, Toronto, Ont.
Wm. C. Wilson & Co., Toronto, Ont.

LIFE JACKETS
Fosbery Co., Barking, Essex, Eng.
Hopkins & Co., F. H., Montreal, Que.
Leckie, Ltd., John, Toronto, Ont.

LIGHTS, ALL KINDS
Can. Fairbanks-Morse Co., Montreal, Que.
Crouse-Hinds Co. of Canada, Ltd., Toronto, Ont.
Hopkins & Co., F. H., Montreal, Que.
Mitchell Co., The Robert, Montreal, Que.
Morrison Brass Mfg. Co., James, Toronto, Ont.
Tallman Brass & Metal Co., Hamilton, Ont.

LIGHTS, SIDE, PORT
Goldie & McCulloch Co., Galt, Ont.
Crouse-Hinds Co. of Canada, Ltd., Toronto, Ont.
Hopkins & Co., F. N., Montreal, Que.
McAvity & Sons Ltd., T., St. John, N.B.
Mitchell Co., The Robert, Montreal, Que.
Tallman Brass & Metal Co., Hamilton, Ont.
Turnbull Elevator Mfg. Co., Toronto, Ont.
Wilson & Co., Wm. C., Toronto, Ont.

LIGHTING SETS
Can. Fairbanks-Morse Co., Montreal.
Cory & Son, Inc., Chas., New York, N.Y.
Crouse-Hinds Co. of Canada, Ltd., Toronto, Ont.
Goldie & McCulloch Co., Ltd., Galt, Ont.
Kearfott Engineering Co., New York, N.Y.
Williams Machinery Co., A. R., Toronto, Ont.
Wilson & Co., Wm. C., Toronto, Canada.

LOADERS, WAGON AND TRUCK
Morris Crane & Hoist Co., Herbert, Niagara Falls, Ont.
Waterous Engine Works Co., Brantford, Ont.

LOCKERS, METAL
Dennis Wire & Iron Works Co., London, Ont.
Reed & Co., Geo. W., Montreal, Que.

LUBRICATORS
Empire Mfg. Co., London, Ont.
Mitchell Co., The Robert, Montreal, Que.

MACHINISTS
Corbet Fdry. & Machine Co., Owen Sound, Ont.
Wm. Hamilton Co., Peterboro, Ont.
Hyde Engineering Works, Montreal, Que.
Mitchell Co., The Robert, Montreal, Que.
United Brass & Lead Co., Toronto, Ont.

MANGANESITE, PASTE
Wm. C. Wilson & Co., Toronto, Ont.

MARINE CASTINGS
Wm. Hamilton Co., Peterboro, Ont.
Goldie & McCulloch Co., Galt, Ont.
Mitchell Co., Ltd., Robt., Montreal, Que.
McAvity & Sons, T., St. John, N.B.
Mueller Mfg. Co., H., Sarnia, Ont.
Tallman Brass & Metal Co., Hamilton, Ont.
United Brass & Lead, Ltd., Toronto, Ont.
Waterous Engine Works Co., Brantford, Ont.

MARINE SPECIALTIES
Goldie & McCulloch Co., Galt, Ont.
Hopkins & Co., F. H., Montreal, Que.
McAvity & Sons, T., St. John, N.B.
Mitchell Co., Ltd., Robt., Montreal, Que.
United Brass & Lead, Ltd., Toronto, Ont.

MOTORS
Aikenhead Hardware, Ltd., Toronto, Ont.
Can. Fairbanks-Morse Co., Montreal, Que.
Williams Machinery Co., A. R., Toronto, Ont.
Zenith Coal & Steel Products, Ltd., Montreal,

NAUTICAL INSTRUMENTS
Wm. C. Wilson & Co., Toronto, Ont.

NET LIFTERS
Dake Engine Co., Grand Haven, Mich.
Leckie, Ltd., John, Toronto, Ont.

NOZZLES, ALL KINDS
Empire Mfg. Co., London, Ont.
Mitchell Co., The Robert, Montreal, Que.
Tolland Mfg. Co., Boston, Mass.

NUTS
Mitchell Co., The Robert, Montreal, Que.
Wilkinson & Kompass, Hamilton, Ont.
United Brass & Lead, Ltd., Toronto, Ont.

OIL CUPS
Empire Mfg. Co., London, Ont.
Mitchell Co., Ltd., Robt., Montreal, Que.

OILCLOTHING
Wm. C. Wilson & Co., Toronto, Ont.

OIL, LINSEED
Canada Linseed Oil Mills, Montreal, Que.

OILS AND GREASE
Wm. C. Wilson & Co., Toronto, Ont.

OAKUM
Davey & Sons W. O., Jersey City., N.J.
Leckie, Ltd., John, Toronto, Ont.
Stratford Oakum Co., Geo., Jersey City, N.J.
Wm. C. Wilson & Co., Toronto, Ont.

ORNAMENTAL IRONWORK
Canada Wire & Iron Goods Co., Hamilton, Ont.
Can. Welding Works, Montreal, Que.

OPTICAL GOODS
Standard Optical Co., Geneva, N.Y.

OXY-ACETLYENE OUTFITS
Can. Fairbanks-Morse Co., Montreal, Que.
Can. Welding Works, Montreal, Que.
Hopkins & Co., F. H., Montreal, Que.
McAvity & Sons Ltd., T., St. John, N.B.
Williams Machinery Co., A. R., Toronto, Ont.

PAINT
Ault & Wiborg Co. of Can., Ltd., Toronto, Ont.
Wm. C. Wilson & Co., Toronto, Ont.

PAINTS, ANTI-FOULING
Kuhls, H. B. Fred., 6431 3rd Ave., Brooklyn, N.Y.

PAINTS, ANTI-CORROSIVE
Kuhls, H. B. Fred., 6414 3rd Ave., Brooklyn, N.Y.

PAINTING EQUIPMENT, AUTOMATIC
Spray Engineering Co., Boston, Mass.

PACKING, AMMONIA
France Packing Co., Philadelphia, Pa.
Wm. C. Wilson & Co., Toronto, Ont.

PACKING, MARINE
Aikenhead Hardware, Ltd., Toronto, Ont.
France Packing Co., Philadelphia, Pa.
Wm. C. Wilson & Co., Toronto, Ont.

PACKING, METALLIC
Aikenhead Hardware, Ltd., Toronto, Ont.
France Packing Co., Philadelphia, Pa.

PACKING, STEAM
Aikenhead Hardware, Ltd., Toronto, Ont.
France Packing Co., Philadelphia, Pa.
Wm. C. Wilson & Co., Toronto, Ont.

PANELBOARD AND CABINETS, ELECTRIC
Crouse-Hinds Co. of Canada, Ltd., Toronto, Ont.
Preston Woodworking Machy. Co., Preston, Ont.

PATTERNS, WOOD AND METAL
Beauchemin & Fils, Sorel, P.Q.

PIPE, LAPWELD, CAST IRON, RIVETED
Empire Mfg. Co., London, Ont.
Norbom Engineering Co., Philadelphia, Pa.
United Brass & Lead, Ltd., Toronto, Ont.

PIPE RAILING, IRON AND BRASS
Dennis Wire & Iron Works Co., London, Ont.
Mitchell Co., The Robert, Montreal, Que.

PITCH
Reed & Co., Geo., Montreal, Que.

PITCH, PINE
Wilson & Co., Wm. C., Toronto, Canada.

PLANERS
Bertram & Sons, Ltd., John, Dundas, Ont.

PLANERS, STANDARD AND ROTARY
Yates Mach. Co., P. B., Hamilton, Ont.

PLATE CLAMPS
Morris Crane & Hoist Co., Herbert, Niagara Falls, Ont.

PLATE PUNCH TABLES
Norbom Engineering Co., Philadelphia, Pa.

PLUMBING EQUIPMENT
Low & Sons, Ltd., Archibald, Glasgow, Scotland.
Empire Mfg. Co., London, Ont.
Mitchell Co., The Robert, Montreal, Que.
Mueller Mfg. Co., H., Sarnia, Ont.
Tallman Brass & Metal Co., Hamilton, Ont.
Wm. C. Wilson & Co., Toronto, Ont.
United Brass & Lead, Ltd., Toronto, Ont.

PROPELLOR BLADES, BRONZE
Corbet Fdry. & Machine Co., Owen Sound, Ont.
Empire Mfg. Co., London, Ont.
Mitchell Co., The Robert, Montreal, Que.
Tallman Brass & Metal Co., Hamilton, Ont.
United Brass & Lead, Ltd., Toronto, Ont.
Yarrows, Limited, Victoria, B.C.

PROPELLER WHEELS
Goldie & McCulloch Co., Galt, Ont.
Kennedy & Sons, Wm., Owen Sound, Ont.
Trout Co., H. G., Buffalo, N.Y.

PIPING, STEAM
Babcock & Wilcox, Ltd, Montreal, Que.

PORT LIGHTS
Goldie & McCulloch Co., Galt, Ont.
Mitchell Co., Ltd., Robt., Montreal, Que.
Mueller Mfg. Co., H., Sarnia, Ont.
Turnbull Elevator Mfg. Co., Toronto, Ont.
Tallman Brass & Metal Co., Hamilton, Ont.
Wm. C. Wilson & Co., Toronto, Ont.
United Brass & Lead, Ltd., Toronto, Ont.

PROPELLER WHEELS
Beauchemin & Fils, Sorel, P.Q.
McNab Co., Bridgeport, Conn.

PULLEYS
Wm. Hamilton Co., Peterboro, Ont.
Smart-Turner Mach. Co., Hamilton, Ont.
Williams Machinery Co., A. R., Toronto, Ont.

PUMP, AIR
Can. Ingersoll-Rand Co. Sherbrooke, Que.
Smart-Turner Mach. Co., Hamilton, Ont.
Weir Ltd., G. & J., Cathcart, Glasgow, Scotland.
Williams Machinery Co., A. R., Toronto, Ont.

PUMPS
Bawden Machine Co., Toronto.
Can. Fairbanks-Morse Co., Montreal, Que.
Canada Foundries & Forgings, Welland, Ont.
Darling Bros., Ltd., Montreal, Que.
Goldie & McCulloch, Ltd., Galt, Ont.
Hopkins & Co., F. H., Montreal, Que.
McAvity & Sons, T., St. John, N.B.
Oberdorfer Brass Co., M. L., Syracuse, N.Y.
Williams Machinery Co., A. R., Toronto Ont.

PUMPS, CENTRIFUGAL
Can. Ingersoll-Rand Co. Sherbrooke, Que.
Darling Bros., Ltd., Montreal, Que.
Wm. Hamilton Co., Peterboro, Ont.
Goldie & McCulloch Co., Galt, Ont.
Kearfott Engineering Co., New York, N.Y.
Morris Mach. Works, Baldwinsville, N.Y.
Norbom Engineering Co., Philadelphia, Pa.
Smart-Turner Mach. Co., Hamilton, Ont.
Waterous Engine Works Co., Brantford, Ont.
Williams Machinery Co., A. R., Toronto, Ont.

PUMPS, BILGE
Darling Bros., Ltd, Montreal, Que.
Goldie & McCulloch Co., Galt, Ont.
Mitchell Co., Ltd., Robt., Montreal, Que.
Morris Mach. Works, Baldwinsville, N.Y.
Mueller Mfg. Co., H., Sarnia, Ont.

PUMPS, CIRCULATING
Darling Bros., Ltd., Montreal, Que.
Goldie & McCulloch Co., Galt, Ont.
McNab Co., Bridgeport, Conn.
Morris Machine Works, Baldwinsville, N.Y.

PUMPS, FEED WATER
Darling Bros., Ltd., Montreal, Que.
Goldie & McCulloch Co., Galt, Ont.
McNab Co., Bridgeport, Conn.
Morris Mach. Works, Baldwinsville, N.Y.
Weir Ltd., G. & J., Cathcart, Glasgow, Scotland.
Williams Machinery Co., A. R., Toronto, Ont.

PUMPS, LIFT AND FORCE
Darling Bros., Ltd., Montreal; Que.
Low & Sons, A., Glasgow, Scotland.

PUMPS, REES ROTURBO
Goldie & McCulloch Co., Galt, Ont.

PUMPS, HAND AND POWER
Aikenhead Hardware, Ltd., Toronto, Ont.
Darling Bros., Ltd., Montreal, Que.
Mitchell Co., The Robert, Montreal, Que.
Smart-Turner Mach. Co., Hamilton, Ont.
Williams Machinery Co., A. R., Toronto, Ont.

PUMPS, HIGH PRESSURE
Canadian Ingersoll-Rand Co., Sherbrooke, Que.
Darling Bros., Ltd., Montreal, Que.
Morris Mach. Works, Baldwinsville, N.Y.
Smart-Turner Mach. Co., Hamilton, Ont.

PUMPS, STEAM TURBINE
Darling Bros., Ltd., Montreal, Que.
Morris Mach. Works, Baldwinsville, N.Y.
Wm. C. Wilson & Co., Toronto, Ont.

SHIP BROKERS
- Page & Jones, Mobile, Ala.

SHIP PLATES
- Nova Scotia Steel & Coal Co., New Glasgow, N.S.

SLEDGES
- Wilkinson & Kompass, Hamilton, Ont.

SLINGS
- Hopkins & Co., F. H., Montreal, Que.
- Morris Crane & Hoist Co., Herbert, Niagara Falls, Ont.

SPECIAL MACHINERY
- Corbet Fdry. & Machine Co., Owen Sound, Ont.
- Wm. Hamilton Co., Peterboro, Ont.
- Miller Bros. & Sons, Ltd., Montreal, Que.
- Smart-Turner Mach. Co., Hamilton, Ont.

SMOOTH-ON
- Wm. C. Wilson & Co., Toronto, Ont.

SPIKES
- Wm. C. Wilson & Co., Toronto, Ont.

SPRAY COOLING SYSTEMS
- Spray Engineering Co., Boston, Mass.

STEAMSHIP AGENTS
- Darling Bros., Ltd., Montreal, Que.
- Page & Jones, Mobile, Ala.

STEAM SPECIALTIES
- Can. Fairbanks-Morse Co., Montreal, Que.
- Corbet Fdry. & Machine Co., Owen Sound, Ont.
- Empire Mfg. Co., London, Ont.
- Mitchell Co., The Robert, Montreal, Que.

STEAM TRAPS
- Aikenhead Hardware, Ltd., Toronto, Ont.
- American Steam Gauge & Valve Mfg. Co., Boston, Mass.
- Darling Bros., Ltd., Montreal, Que.
- Empire Mfg. Co., London, Ont.
- Mason Regulator & Engin. Co., Montreal, Que.
- Mitchell Co., The Robert, Montreal, Que.
- Smart-Turner Mach. Co., Hamilton, Ont.

STEEL, HIGH SPEED
- Hopkins & Co., F. H., Montreal, Que.
- Nova Scotia Steel & Coal Co., New Glasgow, N.S.
- Wilkinson & Kompass, Hamilton, Ont.
- Zenith Coal & Steel Products, Ltd., Montreal.

STEEL SHELVING
- Dennis Wire & Iron Works, London, Ont.

STEEL WORK, STRUCTURAL
- Babcock & Wilcox, Ltd., Montreal, Que.
- Can. Welding Works, Montreal, Que.
- Corbet Fdry. & Machine Co., Owen Sound, Ont.
- Wm. Hamilton Co., Peterboro, Ont.
- MacKinnon Steel Co., Sherbrooke, Que.

STEERING GEARS
- Corbet Fdry. & Machine Co., Owen Sound, Ont.
- Hopkins & Co., F. H., Montreal, Que.
- Engr. & Mach. Wks. of Can., St. Catharines, Ont.
- Wm. C. Wilson & Co., Toronto, Ont.

STOCK RACKS FOR BARS, PIPING, ETC.
- Mitchell Co., The Robert, Montreal, Que.
- Morris Crane & Hoist Co., Herbert, Niagara Falls, Ont.

STOKERS, MECHANICAL
- Babcock & Wilcox, Ltd., Montreal, Que.

SUPERHEATERS, STEAM
- Babcock & Wilcox, Ltd., Montreal, Que.
- Goldie & McCulloch Co., Galt, Ont.

SWITCHBOARDS, ELECTRIC
- Crouse-Hinds Co. of Can., Ltd., Toronto, Ont.

TALLOW
- Can. Economic Lubricant Co., Montreal, Que.
- Wm. C. Wilson & Co., Toronto, Ont.

TANKS, STEEL
- Can. Welding Works, Montreal, Que.
- Corbet Foundry & Mach. Co., Owen Sound, Ont.
- Goldie & McCulloch, Ltd., Galt, Ont.
- Hopkins & Co., F. H., Montreal, Que.
- MacKinnon Steel Co., Sherbrooke, Que.
- Marsh Engineering Works, Belleville, Ont.
- Port Arthur Shipbuilding Co., Port Arthur, Ont.
- Reed & Co., Geo., Montreal, Que.
- Scaife & Sons Co., Wm. B., Oakmont, Pa.

TANKS (AIR, GAS AND LIQUID)
- Can. Welding Works, Montreal, Que.
- MacKinnon Steel Co., Sherbrooke, Que.
- Marsh Engineering Works, Belleville, Ont.
- Scaife & Sons Co., Wm. B., Oakmont, Pa.

TAPPING MACHINES
- Mueller Mfg. Co., H., Sarnia, Ont.

TELEGRAPHS, SHIPS
- McNab Co., Bridgeport, Conn.
- Cory & Sons, Inc., Chas., New York, N.Y.
- Morrison Brass Mfg. Co., James, Toronto, Ont.
- Wilson & Co., Wm., Toronto, Ont.

TELEPHONES, MARINE
- Cory & Son, Inc., Chas., New York, N.Y.
- McNab Co., Bridgeport, Conn.

TESTERS, METER
- Mueller Mfg. Co., H., Sarnia, Ont.

THUMB SCREWS AND NUTS
- Canada Foundries & Forgings, Welland, Ont.
- Mitchell Co., The Robert, Montreal, Que.
- United Brass & Lead, Ltd., Toronto, Ont.

TRACK SYSTEMS
- Morris Crane & Hoist Co., Herbert, Niagara Falls, Ont.

TRAVELLING BLOCKS
- Morris Crane & Hoist Co., Herbert, Niagara Falls, Ont.

TROLLEYS
- Can. Fairbanks-Morse Co., Montreal, Que.
- Morris Crane & Hoist Co., Herbert, Niagara Falls, Ont.

TROLLEY HOISTS
- Morris Crane & Hoist Co., Herbert, Niagara Falls, Ont.

TRUCKS, HAND, ELECTRIC
- Aikenhead Hardware, Ltd., Toronto, Ont.
- Can. Fairbanks-Morse Co., Montreal, Que.

TUBES, BOILER
- Babcock & Wilcox, Ltd., Montreal, Que.
- Broughton Copper Co., Ltd., Manchester, Eng.
- Goldie & McCulloch Co., Galt, Ont.

TUBES, COPPER AND BRASS
- Mueller Mfg. Co., H., Sarnia, Ont.
- Tallman Brass & Metal Co., Hamilton, Ont.
- Standard Underground Cable Co., Hamilton, Ont.

TUGS
- Polson Iron Works, Toronto, Ont.
- Collingwood Shipbuilding Co., Collingwood, Ont.

TURBINES, STEAM
- Goldie & McCulloch Co., Galt, Ont.

TURBINES, DIRECT-DRIVING AND GEARED
- Doxford & Sons, William, Sunderland, England.

TURNBUCKLES
- Canada Foundries & Forgings, Welland, Ont.
- Hopkins & Co., F. H., Montreal, Que.

TURNTABLES
- Morris Crane & Hoist Co., Herbert, Niagara Falls, Ont.

SPIKES, SMALL RAILROAD
- Severance Mfg. Co., S., Glassport, Pa.

UNIONS, ALL KINDS
- Dart Union Company, Toronto, Ont.

VALVES, AIR
- Mitchell Co., The Robert, Montreal, Que.
- Mueller Mfg. Co., H., Sarnia, Ont.

VALVE, DISCS
- Wm. C. Wilson & Co., Toronto, Ont.

VALVES
- American Steam Gauge & Valve Mfg. Co., Boston, Mass.
- Babcock & Wilcox, Ltd., Montreal, Que.
- Can. Fairbanks-Morse Co., Montreal, Que.
- Darling Bros., Ltd., Montreal, Que.
- Empire Mfg. Co., London, Ont.
- McAvity & Sons, Ltd., T., St. John, N.B.
- Mason Regulator & Engin. Co., Montreal, Que.
- McNab Co., Bridgeport, Conn.
- Norbom Engineering Co., Philadelphia, Pa.
- Williams Machinery Co., A. R., Toronto, Ont.
- Wm. C. Wilson & Co., Toronto, Ont.
- United Brass & Lead, Ltd., Toronto, Ont.

VALVES, FOOT
- Aikenhead Hardware, Ltd., Toronto, Ont.
- Mitchell Co., The Robert, Montreal, Que.
- Smart-Turner Mach. Co., Hamilton, Ont.

VALVES, STOP, REDUCING, SAFETY CHECK, ETC.
- Aikenhead Hardware, Ltd., Toronto, Ont.
- American Steam Gauge & Valve Mfg. Co., Boston, Mass.
- Darling Bros., Ltd., Montreal, Que.
- McAvity & Sons Ltd., T., St. John, N.B.
- Mitchell Co., The Robert, Montreal, Que.
- Morrison Brass Mfg. Co., James, Toronto, Ont.

VALVES, MIXING
- Darling Bros., Ltd., Montreal, Que.
- Mitchell Co., The Robert, Montreal, Que.
- Mueller Mfg. Co., H., Sarnia, Ont.

VALVES, REDUCING, PRESSURE
- Mitchell Co., The Robert, Montreal, Que.
- Mueller Mfg. Co., H., Sarnia, Ont.

VALVES, STORM
- Low & Sons, A., Glasgow, Scotland.

VARNISHES
- Aikenhead Hardware, Ltd., Toronto, Ont.
- Ault & Wiborg Co. of Can., Ltd., Toronto, Ont.
- Leckie, Ltd., John, Toronto, Ont.
- Reed & Co., Geo., Montreal, Que.

VENTILATORS, COWL
- Can. Welding Works, Montreal, Que.
- McNab Co., Bridgeport, Conn.
- Mitchell Co., The Robert, Montreal, Que.

VENTILATION EQUIPMENT
- Empire Mfg. Co., London, Ont.
- Hopkins & Co., F. H., Montreal, Que.
- Low & Sons, Ltd., Archibald, Glasgow, Scotland

WASHERS
- London Bolt & Hinge Works, London, Ont.
- Wm. C. Wilson & Co., Toronto, Ont.

WATER COLUMNS
- Darling Bros., Ltd., Montreal, Que.
- Mitchell Co., The Robert, Montreal, Que.
- Morrison Brass Mfg. Co., James, Toronto, Ont.

WELDING, ELECTRIC
- Can. Welding Works, Montreal, Que.
- Hall Engineering Works, Montreal, Que.
- Marine Welding Co., Buffalo, N.Y.
- Beatty & Sons, M., Welland, Ont.

WATER SOFTENERS
- Babcock & Wilcox, Ltd., Montreal, Que.

WATER SUPPLY SYSTEMS
- Mueller Mfg. Co., H., Sarnia, Ont.

WATER HEATERS
- Empire Mfg. Co., London, Ont.
- Morrison Brass Mfg. Co., James, Toronto, Ont.

WHISTLES AND SYRENS
- Empire Mfg. Co., London, Ont.
- McAvity & Sons, T., St. John, N.B.
- McNab Co., Bridgeport, Conn.
- Mitchell Co., Ltd., Robt., Montreal, Que.
- Morrison Brass Mfg. Co., Jas., Toronto, Ont.

WINCHES, CARGO
- Aikenhead Hardware, Ltd., Toronto, Ont.
- Corbet Fdry. & Machine Co., Owen Sound, Ont.
- Hopkins & Co., F. H., Montreal, Que.
- Marsh Engineering Works, Belleville, Ont.

WINCHES, DOCK, SHIP
- Beatty & Sons, M., Welland, Ont.
- Marsh Engineering Works, Belleville, Ont.
- Miller Bros. & Sons, Ltd., Montreal, Que.
- Morris Crane & Hoist Co., Herbert, Niagara Falls, Ont.
- Wilson & Co., Wm. C., Toronto, Canada.

WINCHES, TOWING
- Corbet Foundry & Mach. Co., Owen Sound, Ont.

WINCHES, TRAWL
- Beatty & Sons, M., Welland, Ont.
- Wm. C. Wilson & Co., Toronto, Ont.

WINDLASSES
- Corbet Fdry. & Machine Co., Owen Sound, Ont.
- Dake Engine Co., Grand Haven, Mich.
- Hopkins & Co., F. H., Montreal, Que.
- Wilson & Co., Wm. C., Toronto, Canada.

WIPER CAPS, OILER BOXES, ETC.
- Mitchell Co., The Robert, Montreal, Que.
- Morrison Brass Mfg. Co., James, Toronto, Ont.

WIRE, COPPER CLAD STEEL
- Standard Underground Cable Co., Hamilton, Ont.

WIRE, COPPER, BRASS, BRONZE
- Standard Underground Cable Co., Hamilton, Ont.
- Tallman Brass & Metal Co., Hamilton, Ont.

WIRE, INSULATED
- Standard Underground Cable Co., Hamilton, Ont.

WIRELESS OUTFITS
- Marconi Wireless Telegraph Co., Montreal, Que.

WIRE ROPE
- Zenith Coal & Steel Products, Ltd., Montreal.

WOOD WORKING MACHINERY
- Aikenhead Hardware, Ltd., Toronto, Ont.
- Can. Fairbanks-Morse Co., Montreal, Que.
- Preston Woodworking Machy. Co., Preston, Ont.
- Yates Mach. Co., P. B., Hamilton, Ont.

WOOD BORING TOOLS
- Aikenhead Hardware, Ltd., Toronto, Ont.
- Can. Ingersoll-Rand Co. Sherbrooke, Que.

WOODITE GAUGE GLASS WASHERS
- Wm. C. Wilson & Co., Toronto, Ont.

WRENCHES
- Canada Foundries & Forgings, Welland, Ont.

YACHT BROKER
- Watt, J. Murray, Philadelphia, Pa.

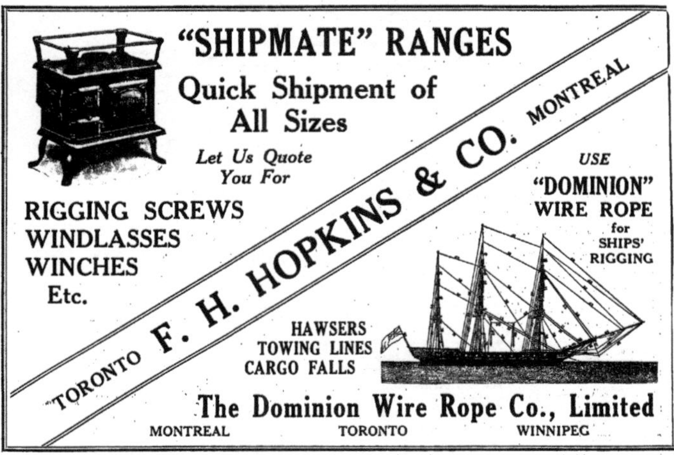

INDEX TO ADVERTISERS

CIRCULATES IN EVERY PROVINCE OF CANADA AND ABROAD

MARINE ENGINEERING of Canada

A monthly journal dealing with the progress and development of Merchant and Naval Marine Engineering, Shipbuilding, the building of Harbors and Docks, and containing a record of the latest and best practice throughout the Sea-going World. Published by
The MacLean Publishing Co., Limited
MONTREAL, Southam Building TORONTO 143-153 University Ave. WINNIPEG, 1207 Union Trust Bldg. LONDON, ENG., 88 Fleet St.

Vol. VIII. Publication Office, Toronto—October, 1918 No. 10

BUY
VICTORY BONDS
TO THE UTMOST LIMIT OF YOUR ABILITY

Canadian Steel Foundries, Limited, Montreal

Canadian-Built Ocean-going Steamer "Reginolite"

The fourth ship launched on an order of five for the IMPERIAL OIL CO.

The "Reginolite" was recently launched and is here seen on her trial trip. She is built for ocean service and measures:—

Length ....250 feet
Breadth ....43 feet 9 inches
Depth ....25 feet moulded

The trials, although carried out in stormy weather, were highly successful, the guaranteed speed being exceeded by one and one-half knots.

We also recently launched the first two of six trawlers, now being built for the Naval Service Department. Other craft are nearing completion.

We are makers of steel and wooden ships, engines, boilers, castings and forgings.

PLANT FITTED WITH MODERN APPLIANCES FOR QUICK WORK. Dry Docks and Shops Equipped to Operate Day and Night on Repairs.

The Collingwood Shipbuilding Co.,
LIMITED
COLLINGWOOD, ONTARIO, CANADA

If what you need is not advertised, consult our Buyers' Directory and write advertisers listed under proper heading.

MARINE ENGINEERING of Canada

Volume VIII. TORONTO, OCTOBER, 1918 Number 10

Will the Marine Steam Engine Survive the Marine Oil Engine

Development in Marine Propulsion—Different Types of Marine Motors—The Future Marine Engine

By T. H. FENNER, Associate Editor

DEVELOPMENT in the science of marine propulsion has been comparatively slow, each stage being pursued till nothing more was possible along that particular line, when a distinct departure from previous practice was made and the same perfecting process gone through. Beginning with the paddle engine, various styles of engines were adapted to the driving of paddle wheels, and it was not till 1836, or practically thirty years after the first steamboat had been tried, that John Ericsson and Francis P. Smith, each working independently, introduced the screw propeller. The engines used for driving paddle wheels were mostly of the side lever or grasshopper type, the only difference between these two being the position of the fulcrum pin of the lever. In the side lever the fulcrum was in the centre of the lever, one end of the lever being connected to the piston rod, and the other to the connecting rod, making a well balanced arrangement of the moving parts. In the grasshopper engine of the fulcrum pin was placed at one end of the lever, with the cylinder at the other, the connecting rod working from a pin between the two. The side lever was the one in general use, though for river work the beam engine came into some considerable favor, and can still be seen to-day in use at most of the seaports of this continent. Other types of engine used were the oscillating cylinder, in which the cylinders were supported on trunnions, and the piston rod connected directly to the crank, thus doing away with the necessity of a connecting rod; and in modern paddle-wheel boats, diagonal engines are fitted, the cylinders being placed at an angle of 45° to the crank shaft, leaving space underneath for condenser, air pumps, etc. The paddle wheel was retained in use for ocean going vessels till about the year 1862, when the Cunard Line steamer "Scotia," of 4,000 h.p., was built, she being about the last steamer to be fitted with paddle wheels, for the trans-Atlantic trade. For coastal and cross channel service, the paddle steamer remained in favor up till very recent date, and some very fine and fast vessels were built, with speeds running up to 21 and 22 knots per hour.

The Screw Propeller Introduced

The screw propeller, from 1836 to 1850, was in the experimental stage, and owing to the shaft being so low down in the ship, a modification of the engine was necessary. The screw propeller had to turn considerably faster than the paddle, and to obtain the necessary speed, geared engines were resorted to. These were sometimes of the oscillating type, with the driving shaft above the cylinder, and large gears connecting with the small pinions on the propeller shaft. Beam engines were also adapted to gearing, and were more or less successful.

TRIPLE EXPANSION MARINE ENGINE IN ERECTING SHOP.

With the introduction of higher pressure, the engines could be made to run at higher speeds, so that the gearing was dispensed with. The horizontal engine came into use, but owing to the narrow beam, various devices had to be used to enable the cylinder to be kept close to the crank. The trunk engine was one of these, in which the piston was extended to work through the cylinder at each end, and the connecting rod attached to an eye-bolt on the trunk, much in the same way as an automobile rod is connected to-day. Another device was the return connecting rod, in which the cylinder was set on one side of the shaft, and the connecting rod on the other. The piston had two rods instead of one, one passing over the shaft, and the other under. The end of these rods was connected to a cross-head, sliding in guides, and the connecting rod was connected in the centre of this cross-head. The inverted diagonal engine was then tried, and this led to the adoption of the inverted direct-acting engine, which has held its place ever since.

With the introduction of higher boiler pressure, and a greater understanding of the thermo dynamics of the steam engine, the way was opened for another radical departure, the two stage expansion. The first steamer to be fitted for the trans-Atlantic trade was the steamer "Holland," in 1870, and the pressure she carried was 60 pounds per inch. The economies effected by the use of the compound surface condensing engine were sufficient to lead to more general use, and the boiler was gradually improved to carry higher pressures, until when 100 pounds per square inch was reached the triple expansion engine came on the scene. This was about 1881, and from then till 1900 the only development made was to increase the number of expansions to four, in the quadruple expansion engine, and to increase steam pressure to 200 pounds and over. Of course, many details were improved, and various forms of valve gear introduced such as the Joy and Bremme, while the slide valve was replaced by the piston valve on the high and intermediate pressure cylinders. The size of the units in use led to steam and hydraulic gears for reversing, and steam steering gear came into its own. However, it is safe to say that from the introduction of the triple expansion engine, in 1881, to the invention and practical application of the turbine by Parsons, in 1900, the evolution of the marine engine was entirely one of detail.

The modern triple expansion engine is capable of developing one horse power on a consumption of 1.5 pounds coal per hour, and no improvement over that figure seems at all likely. The use of superheated steam is limited by the ability of the packing and working surfaces to withstand the high temperatures incidental to it, and it seemed as though marine engineering had approached as near perfection as possible, when the steam turbine made its advent.

A Distinct Change

Here was a radical change from anything previously used in marine engineering. Every engine hitherto had been of the reciprocating type, with necessarily a large number of moving parts, subject to wear, and offering possibilities of breakdown. The turbine, being a rotary engine, had but one moving part, the rotor itself. It was capable of running at high speed and with a range of temperatures much greater than was possible in the multi-cylinder reciprocating engine, therefore, being a much more efficient heat engine. It was found, however, that at low speeds its efficiency dropped considerably, and therefore, it was not a suitable motor for ships of less than 16 to 17 knots, which practically confined its use to the high-class passenger trade, and naval uses. In these spheres the turbine proved itself considerably more economical than the reciprocating engine of the same power, and has been largely adopted. The introduction of the geared turbine and turbo-electric machinery has increased the advantage, and by enabling the turbine to be run at its most efficient speed while the propellor is also operating at the speed best suited to its capabilities, the combined efficiency of the two is brought to the highest possible percentage.

The Internal Combustion Engine

Having apparently exhausted the possibilities of the steam engine, a new factor has been coming into the field during the last few years, namely, the internal combustion engine. Considering the phenomenal success of this type of motor in the automobile and stationary field, it was to be expected that it would sooner or later invade the marine industry. However, its progress has been slow, for reasons that will be seen later. As a heat engine, the internal combustion engine is capable of much higher thermal efficiency than the reciprocating or rotary steam engine, and it furthermore eliminates the fuel losses incidental to the burning of fuel in a separate boiler. They are roughly dividable into two classes, first those in which the fuel and air are drawn into the cylinder by the piston, and then compressed by the return stroke of the piston, ignition being caused by an electrical spark at the moment of greatest compression. These engines operate on gasoline chiefly. The second class are those of the hot bulb and Diesel type, in which the air alone is compressed, the fuel being injected at the compression pressure by a separate air compressor, and ignition caused by the temperature of the compressed air in the cylinder. These engines are operated at high pressures, up to 500 pounds per square inch, and the fuel is heavy oil. This is the type of engine that is being used as a marine motor for ocean going vessels in considerable numbers, and is thought by many to be a strong rival of the steam engine.

How They Compare

We have thus three distinct types of marine engine, all in actual use, and each possessing certain characteristic advantages. The oldest, and the one in most general use, is the triple expansion and quadruple expansion reciprocating engine. These engines have been fitted into thousands of vessels of all sizes and have run long voyages with a freedom from accident and a small maintenance cost; which has made their position an exceedingly strong one. The majority of the steam vessels afloat are fitted with this type of engine. On this account principles and methods of construction are so well understood that the cost of production in normal times is low, and in the event of repairs spare parts are easily procured in the home port. They are economical in steam consumption and capable of running without a stop for weeks at a time. The writer has steamed continuously for forty-three days without a stop for any purpose. They are capable of very long life, it being quite usual for the engines of a ship to outlast the hull. The skilled attendance required is not excessive and is readily available, at least in the older countries. They are capable of being handled, in case of a breakdown, as a one or two cylinder engine. The crank shaft is usually of the built-up type, with a coupling between each crank, and is interchangeable. The chances of a total breakdown are remote in the single screw steamer, and in the twin screw almost unheard of. Of course an accident to the propellers themselves could put a vessel out of commission but this could be so with any type of engine. We may say that in the modern multi-expansion engine we have an economical, reliable, and proven prime mover, long past the experimental stage and suitable to almost all types of vessels, from coasting boats to battleships.

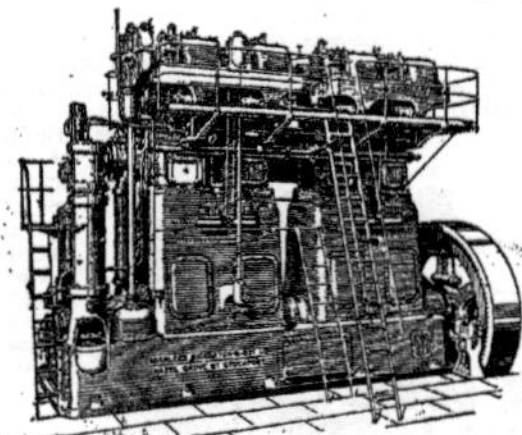

500 B.H.P. MIRRILEES DIESEL ENGINE

Against this engine, what does the steam turbine have to. claim? Less weight per horse power, absence of complicated moving parts, absence of vibration, especially desirable in passenger ships, and small head room required, a very important consideration in fighting ships, as it allows of the protective deck being carried much lower. At high speeds it is more economical, but for the ordinary class of cargo vessel it has not been found to possess any advantage over the reciprocating type. This is due to the fact that in order to get the highest efficiency from the steam it is necessary to run the turbine at a high velocity, and at the low powers required by the slow cargo vessel, this cannot be well arranged. However, the use of the high speed turbine, with gears interposed between turbine and propellor shaft has made it possible to run the turbine at the highest efficient speed, and at the same time use a propeller of a pitch and diameter and correct number of revolutions for the particular hull to which it is fitted. It is claimed that with this type of machinery, savings of 15 to 20 per cent. can be made in fuel consumption over reciprocating engines of the same power. There is also a saving in weight amounting to about 10 per cent., and in space to nearly 25 per cent.

A further improvement coming into use with turbine machinery is the turbo electric combination. In this case the turbine drives a generator, which in turn drives electric motors connected to the propeller shafts. This arrangement permits of ease of manoeuvring through easily controlled speed variation, reversing, stopping, and starting. The overall efficiency of dynamos, motors and turbines is high. A further combination has been used of turbo generators with mechanical gearing between the motors and shafts. The combination of high pressure reciprocating engines exhausting into low pressure turbines has also been adopted with marked success, notably in the White Star liners "Olympic" and "Laurentic." For passenger ships of large size and intermediate speed this form of engine seems to be very well adapted. For the purpose of comparison between the three representative types of steam engines we have the particulars of three sister ships built for the United States government. These are the "Jupiter," "Cyclops" and "Neptune." They are of 20,000 tons displacement and 12,000 tons deadweight capacity. The principal dimensions are, length 548 ft x 65 ft. beam, x 39¼ ft. deep. The "Jupiter" is fitted with turbo-electric machinery, the "Cyclops" with reciprocating engines, and the "Neptune" with Parsons geared turbines. The weight of the propelling machinery alone compares as follows:

	Tons
Jupiter, with turbo-electric drive	156
Neptune, with geared turbine "	150
Cyclops, with reciprocating engine drive	280

It is seen therefore that the weight of the reciprocating machinery is nearly double that of the turbine, and that between geared drive and electro drive there is practically no difference as regards weight. From the standpoint of economy we have the results as follows:

Steam consumption at maximum speed.

	Pounds per shaft h.p.
Jupiter	11.1
Neptune	13.4
Cyclops	14 0

These steam consumptions are at maximum speed. When we get down to lower speeds the reciprocating engine shows more favorably in steam consumption. Take for instance four vessels of U. S. navy: the "Florida" and "Utah" are equipped with Parsons turbines, the "Delaware," with reciprocating engines, and the "New Mexico" with electric drive. The water rate per effective horse power per hour at different speeds are as follows:

	12 knots	19 knots	21 knots
New Mexico (turbo electric)	17 3	15.0	16.4
Delaware (reciprocating)	22.0	18.7	21.0
Utah (Parsons turbines)	28.7	20.3	21.0
Florida (Parsons turbines)	31 8	24.0	23 0

From this it will be seen that the vessel with the electric drive is the most economical at all speeds, while at the lower speeds the reciprocating engine is more economical than the straight turbine. Probably one of the best examples of a direct connected turbine installation is the "Mauretania," of 68,000 h.p., and 26 knots. Her consumption is about 11.5 lbs. water per shaft horse power, and 15 lbs. coal per s.h.p. per hour.

The Oil Engine

The engines we have been considering up to now have been steam driven, and that implies boilers and a large crew to attend them. In ships where oil is used as fuel in the boiler furnaces the crew is cut down, but wherever steam is generated in a vessel separate from the machine where it is to do its work there are unavoidable losses. The oil engine makes its stand on the fact that it is the most efficient heat engine yet evolved; that it is self contained. the combustion of the fuel and the utilizing of the energy being in one chamber; elimination of boilers and saving of space, besides less attendance. Truly a fairly good list to its credit. We find in actual performance that this engine can develop a shaft horse power on about .4 lb. oil fuel per hour against about .75 lb. per hour when burnt under a boiler with highly efficient turbines. The cost of a Diesel installation about equals the cost of a geared turbine installation. Progress in the construction of the Diesel engine has been retarded by the enormous demands occasioned by the war on standard engine builders, but notwithstanding these handicaps many ships have been fitted with these motors. One of the difficulties has been, and is, that this type of engine is a very expensive one to manufacture, it being estimated by Professor Lueke, of Columbia, that the cost is never less than $60 per h.p. However, the low cost of operating should offset this, and the larger cruising radius per ton of fuel is also a considerable advantage. For units up to 1,000 h.p. there seems to be a considerable field for the internal combustion engine of the Diesel, semi-Diesel, or hot bulb engine, and the well known Bolinder engine forms a good example of the latter. As an earnest that the development of the oil engine is proceding apace, comes the news from the Clyde of the trial trips of the motor vessel, "Glenapp," the largest vessel yet fitted in Great Britain with this type of engine. The Glenapp is of 10,000 tons d.w. capacity and is equipped with two six-cylinder Burmeister and Wain four-cycle Diesel engines of 3,300 b.h.p. each. This is quite a large installation. These large units have been made possible by the overcoming of the trouble experienced in reversing, as these large engines are reversed as easily as a steam engine. The saving in space is not so large compared to the reciprocating engine as it that of the turbine, but the oil engine, owing to its much lower consumption, requires only about one half the amount of reserve fuel on board that a vessel using oil-burning boilers would require. On the other hand, carrying the same amount, she would require her fuel tanks filling only once for every twice of the other vessel, thus making a large saving in port dues, pilotage, and delays, etc., while these savings, compared to the vessel burning coal amount to about 10 to 1. The oil engine for marine purposes is long past the experimental stage, and while it is capable of much improvement it has established itself as a competitor very firmly. After the war, when manufacturers of engines get time to take it up, the development of the oil engine will proceed with vigor.

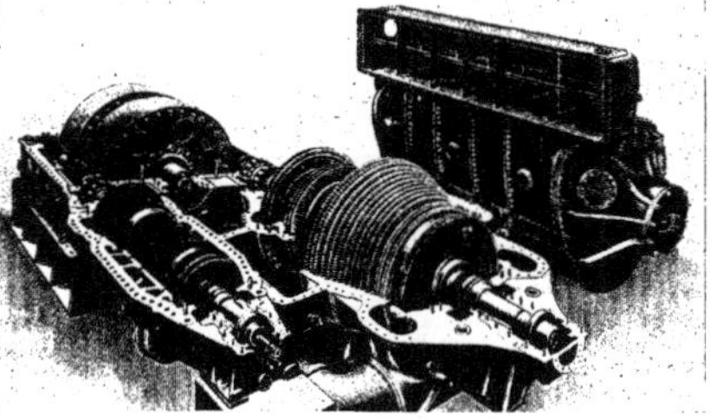

COMPOUND TURBINE TOP CASE REMOVED, SHOWING GEARS

From the foregoing it would seem as though the future of marine propulsion would be divided between two classes of motor. For the large, high speed and high powered vessel of the navy and merchant service the turbine, using steam generated from oil fuel, and driving an electric generator, which furnishes energy to the motors on propeller shafts, seems to be the ideal power. It combines the highest efficiency of the turbine with the highest propeller efficiency. It is easily handled and has a higher efficiency at any speed than the direct connected turbine or reciprocating engine. Its actual fuel saving over the reciprocating type is from 20 to 25 per cent.

For the cargo steamer of moderate speed and power ranging from 1,500 h.p. to 10,000 h.p. it is probable that the internal combustion engine will gradually replace the reciprocating engine. The owner of a cargo steamer, either on a regular route or tramping, usually has to figure very closely between a profit and a loss. This does not apply to the present time, but undoubtedly will in the years after the war. To him the savings of the oil engine installation should appeal. It gives him probably 10% more cargo space and reduces his wages bill by the elimination of firemen and trimmers. He makes an actual saving in fuel cost over coal, and the cost of placing it on board is less. In the case of steamers on long continuous voyages there is considerable expense incurred at coaling ports, which can be mostly saved by the larger cruising radius of the oil engined vessel. Lastly there is the elimination of delays lasting from 3 to 24 hours or more, occasioned by the stokehold crew being ashore on a jamboree at sailing time. The reciprocating engine will die slowly, but disappear it will.

THE DANGER OF SEDIMENT IN MARINE BOILERS

THE troubles and dangers incidental to the accumulation of sediment in marine boilers were fully explained and preventive measures indicated in a paper read before the Institute of Marine Engineers by Mr. W. R. Austin (member).

"Engineers in charge of multitubular boilers will agree that sediment in such boilers is most objectionable, and anyone called to survey them looks with suspicion when he sees it present in abnormal quantity on its removal from the boilers.

It not infrequently happens that furnaces become distorted without the cause or the occasion being known, and observations lead to the belief that not a few accidents of this kind are attributable to the presence of sediment. As enquiries have revealed the fact that many watch keeping engineers are but imperfectly aware of the dangers arising from its presence in the boilers, these lines are written to point out where the risk of accident generally arises.

If a vessel's boilers were always filled by hose from the shore the quantity of sediment found at the time of cleaning would be very much less. It is, however, not always possible to obtain such water owing to the absence of facilities, while occasionally one hears of an objection on account of the expense. "Needs must when the de'il drives." The ship must sail, so the engineer is obliged to fill his boilers from the dock cr river in which his ship is then lying.

At such a time the water may appear quite clear as viewed from the deck, but appearances are deceptive. As an example, some double bottom tanks were filled with water from the St. Lawrence River as the ship was proceeding to sea. The water appeared to be clear, yet when the tanks were opened on this side several inches of silt were found on the tank bottoms. It must have come from the river, as when previously filled the tanks had contained sea water.

Blow-off cocks are placed so low down on the ship's bilge that in many docks they are in the immediate vicinity of the mud of the dock bottom, and boilers filled in this way always get with the water a certain amount of sediment. The engineer may have used every means in cleaning the boilers to ensure safety in working, but by filling them as indicated he nullifies to a great extent all his care, and admits a most insidious enemy. Once in the boilers sediment becomes a source of danger while they remain under steam. It settles down over the surface of the boiler bottom and its removal by blowing off from the bottom is not very effective, only that portion being ejected which lies within the stream lines of the flowing water close to the mouth of the blow-off pipe.

The first occasion I would mention where risk arises from the presence of sediment is in the act of raising steam. In this operation time is essential. The danger from racking strains caused by unequal expansion is so generally recognized that very few new boilers are now installed which have not the means of circulating the water within when raising steam. The danger to the shell is one which looms largely in the eye of the engineer, and ofttimes it is feared to the exclusion of another which he has overlooked. He circulates to save the boiler shell from undue strain, but in doing so brings into existence conditions which are detrimental to the furnaces. The movements due to the circulation of the water within the boiler disturb the sediment. It is thrown into circulation and flows with the water currents, whose course is liable to change with any local change of temperature. It thus becomes a menace owing to the variable temperature of the furnaces at such a time.

A specific instance of this is shown by the practice of many engineers who, when raising steam, light up the lower furnaces of a set of boilers before the wing furnaces; by this course avoiding as far as it is possible any inequality of temperature in the structure of the boiler shell. If the centre furnace only is lit when the circulation of the water begins, the water rises by convection from the furnace C as shown by the full lines on the sketch, and returns to the lower parts of the boiler via the wing furnaces. The comparatively cool surfaces of the nests of tubes in each of these furnaces assist in smoothing the path of least resistance for the falling currents of water. The sediment is carried by the water and a part of it falling between the tubes is deposited on the wing furnace crowns A and B. It lies there, and when these furnaces are lit up and generate heat, all the elements necessary to collapse of the crown plates are present.

In some cases a furnace so injured may remain unnoticed until it is subsequently examined. Then, on a search being made no trace of the cause of collapse can be found, as the sediment was thrown off when the plate became distorted.

To prevent injury in these circumstances, uniformity of temperature in each furnace is required, and it is suggested that the circulation of the water within the boilers should not begin until all the furnaces are lit and uniformly burning, as thereby the risk of sediment settling on the furnace crown plates would be minimized.

The next occasion on which sediment becomes a menace arises after the ship has put to sea. If rough weather be then experienced those who at such a time have blown through a gauge glass must have noticed that the water after blowing through is much discolored. It resembles in many instances coffee "grounds." This is an indication of the state of the water in the boiler. The

speeds are required and often called for. The engineer in the stokehold to save time and suit the steam production to the requirements of the engines frequently shuts off. This method of regulating the steam supply, in view of what has been said, is a most reprehensible one and should be strictly prohibited. A uniform heat should be maintained in each furnace by regulating the supply of oil fuel from each burner as may be required for the lessened need of the engines. This course is absolutely essential where single burners are fitted. In cases where twin burners are fitted to each furnace then one of these may be shut off on all furnaces in a case of urgency; but the better course is to regulate the valves which control the fuel supply to the burners. By doing so uniformity of steam production in each furnace is maintained. Evaporation will be reduced, and the risk avoided which is always attendant on shutting off the entire oil fuel supply from any individual furnace.

A third instance where sediment may give trouble occurs when a vessel after rough weather arrives at an anchorage and the boiler fires are banked. The rolling of the ship has ceased, but the sediment is still in suspension in the water. If all the stop valves are then closed the tendency is to stop the circulation and check the rise of steam from the heating surfaces. With banked fires the furnace temperature has fallen considerably, the upward movements of the steam grow more sluggish, and one can conceive a point being reached when steam may be generated on a furnace crown plate and remain there because of the insufficiency of heat necessary to throw it off, backed by the lack of movement in the water. In such circumstances the sediment in suspension falls on the crown plate and becomes a menace. Instances have occurred where loud reports were heard in boilers lying under banked fires, and after the report there was heard gurgling up to the steam space a considerable volume of steam. One can only speculate on what caused the report. It may have been due to a sudden inrush of water on a part of the crown plate which was slightly overheated. As the crown plate was not distorted one naturally asks what could have disturbed the sediment. The probability is that owing to the conditions of temperature existing at the moment, steam may have been generated but had not left the surface of the plate, and over this had fallen a slight coating of sediment arresting it and causing the steam to be isolated and insulated. As it absorbs more heat it ultimately bursts the enveloping sediment, thus admitting the water to the overheated surface and causing the report. The writer has frequently heard such reports when boilers were under banked fires. The boiler pressure was 90 lbs. and in no instance was damage found in the furnace at that pressure. With higher pressures and greater density and temperature of steam accidents have occurred while lying under banked fires, and for this reason the danger is pointed out.

As a means of avoiding the risk of accident while lying under banked fires, steam should be used for a considerable time after banking has taken place. By doing so steam generated is drawn to the steam space and the water is kept in circulation until such time as the sediment has settled on the boiler bottom. Fires should always be banked at the front. If banked at the back, a large part of the surface of the furnace crown plate is exposed to the cold air and thereby downward currents are induced in the water within the boilers which may lead to a deposit of sediment on the crown plate.

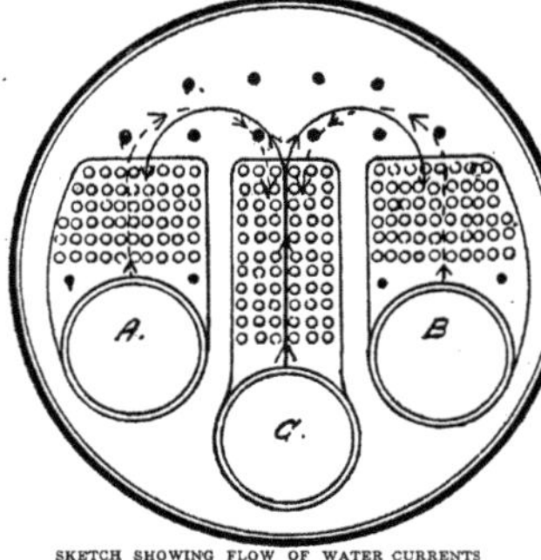

SKETCH SHOWING FLOW OF WATER CURRENTS

It is worthy of note that distortion occurs more frequently to furnaces of extreme than to those of small diameter. As the plate thickness is determined in the patent types of furnace by the diameter, and the rule used gives an ample factor of safety having been fixed after exhaustive hydraulic tests, it follows that the question of strength does not here arise. Furnaces of all diameters are equally liable to collapse if solid saline matter is found abnormally thick on the plate. That furnaces of extreme diameter suffer more than those of small diameter when no scale is present, appears to be induced by their form. They have a broader back and falling sediment will settle and remain with greater certainty, until disturbed, on such a crown plate than it will on one of small diameter. Given similar working conditions in boilers fitted respectively with small and extreme diameters of furnace, any other deleterious matter present would affect the smaller diameter equally with the extreme diameter, and the predominance of accidents with furnaces of extreme type is a proof that the cause of accident is a movable one and produced by a change of temperature at the sources of heat. This causes a deposit of sediment on the crown plate, and on a restoration of normal working conditions the plate becomes overheated.

As already stated a knowledge of the action of water within the boilers when at work is not generally known, nor are the dangers fully appreciated. The subject might well form part of the viva voce examination held by the Board of Trade when granting engineers' certificates. This would lead to a general consideration of the subject and bear fruit (if one may say so) in reducing the number of such accidents.

Furnace troubles are like a disease, contracted by defective arrangements and lack of care. It is present and demands attention, but prevention is the better way. Exclude the sediment from the boilers and the trouble will not arise. It is even now entirely a question of ways and means and the exercising of reasonable care.

In existing vessels it is suggested that a test cock should be fitted at the blow off cock level. By using this, water drawn off in a glass would show if the water outside the ship was such as could be used for filling the boilers. If mud is present then the means of filtering would be improvised as the water was being pumped into the boilers through the upper manhole.

In new vessels the following suggestions if adopted would eliminate the risk arising from the presence of sediment.

The blow-off pipe should only be used for blowing off and not for filling the boilers. This can be done by fitting a non-return valve on the blow-off pipe permitting the water to be ejected from the boiler, but preventing its admission through this pipe.

Water for feeding or filling the boilers in part or when the water is muddy should be filtered. This can be attained by a filter (with a locked by-pass for use at sea) being fitted between the donkey pump sea cock and its suction valve chest. If more than one pump can feed the boilers each should draw from a tee piece on the bottom of the filter.

From the boiler circulating valve on the donkey pump suction chest the circulating pipe should be connected to the blow-off pipe between the non-return valve above mentioned and the blow-off valve on the boiler.

Launching Held at the Dominion Yards

St. Mihiel Has Deadweight Capacity of 4,300 Tons—Progress That Has Been Made in Building a Splendid Industrial Plant on Reclaimed Land

IN the closing days of September the S.S. "St. Mihiel" was launched from the yard of the Dominion Shipbuilding Co., Toronto, being the first of two similar ships to be built there, together with six other vessels of 3,550 tons, and five trawlers.

The St. Mihiel is a steel cargo steamer of the following dimensions. Length o.a. 261 feet. Breadth moulded 43 feet 6 inches, and depth moulded 28 feet two inches. She is subdivided by five bulkheads, four of which are watertight, and one non-watertight, and has a deadweight capacity of 4,300 tons. Built to Lloyd's classification 100A1. The propelling equipment consists of a set of inverted reciprocating triple expansion engines, with cylinders of the following dimensions: H.P. 20 inch diameter, I. P. 33½ inches and L.P. 55 inches, the stroke being 40 inches. Steam is supplied by two Scotch marine boilers, 14 ft. 6 in. dia. by 12 ft. 0 in. long, built by John Inglis Co. The indicated horse power is estimated at 1,400, with a working steam pressure of 180 lbs. per sq. inch. The vessel will carry her 4,300 tons deadweight at an average speed of 10 knots on a consumption of 22½ tons of coal per twenty-four hours. For cargo handling purposes she is equipped with six reversible single drum steam wrenches, working at four hatchways. There is also a steam windlass on the foc's'le lead for raising anchor and warping purposes. The winches, windlass and steam steering gear are made by the Bawden Machine Co., Toronto, the winches being the 7 in. x 7 in. x 12 in. size. This is the type of vessel known as the Fredericstadt Bulk Carrier.

THE ST. MIHIEL READY TO LAUNCH

The Dominion Shipbuilding Co.

The yard from which this vessel was launched is a striking example of triumphing over obstacles which is a distinguishing characteristic of the Canadian. The Dominion Shipbuilding Co. was formed after the destruction by fire of the Thor Iron Works, which were situated at the foot of Bathurst street and fronting on Toronto Bay. This portion of the water front was included in the Toronto Harbor Commission reclamation scheme, so it was decided to lay out the new plant on the reclaimed land. As soon as the retaining wall was built at the edge of the reclaimed area, and the filling in carried out to some extent, the company started driving the piles down to bed rock to form the foundation for their building berths, which together with the shops were to cover 15½ acres. Work was carried on with such vigor that to-day there are five fully equipped building berths, four of which are occupied by ships in various stages of construction, the fifth having just been vacated by the St. Mihiel. All this has been accomplished, and the main buildings erected for the shops in the course of a few months on land which was practically not in existence the 5th of December last year.

The Buildings and Building Berths

There are at present erected and in operation, the main building, containing angle and plate furnaces, the binding slabs, angle iron smith's shop, and the punch shop. The machinery in these shops is sufficient to allow of handling about 75 to 100 tons steel in a working day of nine hours, all on the ground floor.

THE ST. MIHIEL TAKING THE WATER

The power house containing two Sullivan air compressors is in this building on the ground floor, also the warehouse and stock room, the joiners' and carpenters' shop, and the mould loft. On the second floor the rooms are all large and well lighted, and some idea of their size can be got when it is said the building is 485 feet long by 210 feet wide in way of punch shop, and 110 feet in way of power house, etc. The building is of steel frame construction with concrete walls reinforced, and the roofing is of corrugated iron over the punch shop, the mould loft being covered with a built-up roof.

100-TON SHEER LEGS

The plate and angle furnaces are of the oil burning type and are capable of handling the longest frames and plates required, made and installed by Toronto Incinerator Co.

Foundations are in and steel framing erected for a second building, which will contain an electrical shop, pipe shop, machine shop and forge, and pattern shop. This building when completed will be 425 feet long by 110 feet wide, of the same construction as the punch shop. As soon as this building is completed, or sooner if conditions permit, two other buildings containing foundry and boiler shops respectively will be erected. It will be see from the foregoing account that not only are ships being built and launched, but the yard is being practically built up at the same time, which in these strenuous days is quite a noteworthy achievement.

The machinery already installed for handling steel at the ship's side consists of three gantry cranes, and for installing engines and boilers a 100 ton shear legs is installed. There are at present on the stocks four ships, one of the St. Mihiel type, and three of a slightly smaller size, known as the improved Cunard Type Bulk Carrier, having a deadweight capacity of 3,550 tons. The second vessel of the St. Mihiel type will probably be launched in October, and it is hoped to launch some of the smaller ships in the near future. The steel necessary for the construction of these vessels is all either in the yard or on the way. There will therefore be no delay in the construction through waiting on material. Contracts have been placed for the engines and boilers of five ships and three trawlers. The capacity of the yard when completed will be about 12 ships per year of full canal size, that is 261 feet long by 43 feet six inches beam, and about 4,300 tons deadweight. To ensure this capacity the layout of the yard has been planned so as to eliminate all unnecessary handling of material, enabling each department to go ahead with its particular part without being delayed by the preceding one.

PLATE PUNCHING MACHINE

Mr. Dahlgren, the general manager, has been connected with the shipbuilding business since 1893, his early years being spent with the Superior Shipbuilding Co., Superior, Wis. He was with this company in various capacities from 1893 to 1900, and from 1902 to 1912. The intervening two years was spent with the Union Iron Works, San Francisco. Coming to Canada in 1912 he was first assistant superintendent, becoming later superintendent and then general superintendent of the Western Dry Dock and Shipbuilding Co., Port Arthur, remaining with them till 1916. He then came to the Thor Iron Works, Toronto, as superintendent and in 1917 was appointed manager of this company. On the formation of the Dominion Shipbuilding Co. in August, 1917, he was appointed general manager, and for a time while the transition of the two companies was taking place he held both positions.

The Dominion Shipbuilding Co. with the up-to-date equipment installed should occupy a prominent place in the development of Canada's shipbuilding development.

YARD AS IT WILL LOOK WHEN COMPLETED

BOILER AND OTHER REPAIRS BY ELECTRIC WELDING*

WELDING is one of the oldest branches of the working of metals. In some respects it is a lost art, as there are good grounds to believe that the ancients were able to weld some of the bronze alloys. In the following remarks the author proposes to confine himself to the welding of iron and steel, unless otherwise stated. A weld is the intimate union of two pieces of metal, produced when the pieces have been raised to welding heat, by pressure or hammering, and the welding state of a metal only exists within a limited range of temperature, being something like 100° for iron and steel, but varies with the metal. As a rule, good iron will stand a higher temperature than steel, although certain steel, such as blistered or good shear will stand a high temperature. In the smith's fire steel can, and should be forged with a lighter tool than iron, the blows being in rapid succession. In the ideal weld the two surfaces to be united are brought to the plastic heat together, neither at too high or too low a temperature, when the point of juncture should be as strong relative to its section as any other portion. From the foregoing remarks, however, it will be appreciated that much depends upon the skill and experience of the operator, and it is recognised in ordinary engineering practice that an allowance has to be made for inevitable human frailties.

The first process of electric arc welding to be employed in a commercial sense was that of De Bernardos, which was used in Messrs. Lloyd and Lloyd's Works, over twenty years ago, in the welding of flanges and branches to iron and steel pipes. In the De Bernardos process a carbon is employed, an arc being drawn between the carbon and the job, a portion of which is brought to welding heat, and the added metal is heated in the flame of the arc. In the early Bernardos process the work was made the negative pole and the carbon the positive, but latterly the poles were reversed, thus doing away with the dangers of carbonisation of the metal caused by the natural flow of carbon particles from the positive of the negative. The Bernardos process is still largely employed in this country. Slavianoff substituted a metal electrode for the carbon electrode of the Bernados process, although Bernardos as far back as 1885 had the idea of using a hollow carbon filled with the adding metal. In the carbon electrode of the Bernardos does not seem to have anticipated, his difficulty being that, like many other great inventors, he was in advance of the means and appliances of his time. The names of many investigators and workers in our own and other countries during the eighties and nineties of last century could be honorably mentioned, each doing their little bit to advance what is practically a new trade. Among them Charles Lewis Coffin, of Detroit, U.S.A.; Mark Wesley Dewey, of New York, U.S.A.; Pommée, of Altona, near Hamburg; W. P. Thompson, of Liverpool; Thos. Odlum, of Virginia, U.S.A.; Francis Todd, of Newcastle-on-Tyne, and Joseph Fouilloud, of Paris.

*Read before the members of the Institute of Marine Engineers, on March 12th, by Mr. R. S. Kennedy, Member of Council, I.M.E., M.I.C.E., Member N.E.G. Inst. of E. and S.

We have already referred in the Bernardos process to the arrangement of the poles of the electric arc. Now it is generally agreed that the province of the engineer is to utilise the forces and methods of Nature for the benefit of mankind, and Nature in this case has provided that the positive pole of the electric arc shall be much better than the negative pole. We consequently arrange in electric arc welding that the positive pole shall be on the bigger mass, which in 999 cases out of 1,000 is the job, and the negative pole on the smaller mass of metal, which in modern electric arc welding is the metallic pencil of the adding material. By working with Nature we thus provide favorable conditions for the first essential of a good weld, namely, that the pieces to be united shall be brought to a welding heat at the same time. You will note that we have only provided favorable conditions; the actual carrying out of this requirement rests with the skill of the operator. This consideration of the difference in temperature of the two poles of the electric arc makes it at once apparent why direct current is more suitable than alternating for arc welding. On the other hand, alternating current is quite suitable, and probably better than direct current for what is known as resistance welding or for spot welding.

The author's Company* were the first to employ the metallic electrode in this country on a commercial scale—namely, early in 1910, although about a year previously Mr. Copeman, of the Furness Lines, had carried out a few experimental jobs to his own vessels. Since 1910 the annual output of the British Arc Welding Company has increased at least 100 times, and during the present war its services have been utilised in directions which would not have been permitted under peace conditions. In making this statement, however, the author wishes to acknowledge assistance received from kindly and constructive criticism from the Board of Trade and Lloyd's Register in pre-war days,but everything has now been speeded up. In particular, the tests of electric-welded specimens carried out to the instructions of the Board of Trade in 1909 and 1910 were of great value.

These tests were made not only with the object of getting at the tensile strength of the weld, but of finding out if the process of welding affected the neighboring material. Numbers of specimens were tested, and some of these were annealed, but it was found that annealing made no difference to the results, and the material immediately adjacent to the weld behaved in a normal manner. These tests gave a tensile strength of about 17 to 18 tons per square inch, but since then improvements in the materials and methods have increased the tensile strength of weld in boiler steel to about 27 tons per square inch. In practice, however, the author would not recommend that a tensile strength of more than 20 tons per square inch be worked to, this giving a sufficient margin for possible small defects in workmanship. It might here be remarked that in no single case has the author known an electric weld to give way suddenly; failure has always been preceded by a small crack, which has gradually developed.

*The British Arc Welding Co., Ltd.

Electric arc welding is primarily a form of autogenous welding—that is to say, the metals to be united are heated to such a temperature that they will fuse together on contact without the application of external pressure. It is, however, found in practice that the application of even the moderate amount of pressure produced by a hand hammer increases the tensile strength and tenacity of the weld some 5 per cent. It is, however, essential that this work should be put into the material when it is at welding heat or, at any rate, above the black heat. It may here be remarked that it is often said that the value of metal added in this fashion is analogous to the ball of iron obtained in the puddling furnace. This, however, is not the case, and the better results are probably due to the fact that the iron wire used is of the very best material, with preferably a small percentage of manganese. This iron wire has been very heavily worked in the process of manufacture, and subsequently annealed, and as used by the author's firm shows a tensile strength of 28 tons with an elongation of 50 per cent. Somewhat similar results are obtained in another field with cast iron, which has several times been re-melted. The whole question of the amount of work put into the material of a weld is very fascinating, and there is no doubt that the capacity of a weld for taking up rapidly alternating strains for a long period, and for absorbing sudden shocks, very much depends upon this factor.

Returning to our blacksmith, whether under the spreading chestnut tree or in in the more prosaic conditions of the modern smithy, we find that they all employ some kind of flux, usually sand or borax. This flux surrounds the heated iron or steel and protects it against the impurities of the fuel, removing at the same time the coating of scales. Some impure wrought irons flux themselves, but with steel other mixtures are used. The flux, as its name indicates, also increases the fluidity of the heated metal.

In electric arc welding with a metallic electrode one great advantage is that, with the exception of the atmosphere, we have no impurities to guard except such as are introduced in the materials. The source of heat is pure, and we have to see that the job is properly cleaned and the metallic electrode of suitable material. Still, to provide against oxidisation and also to increase the fluidity of the metal a flux is necessary to good work in arc welding, and the heated metal is protected from oxidisation by an inert gas given off by the flux. The most convenient method of applying the flux is to coat evenly the metallic electrode, thus providing a constant and uniform supply.

Electric arc welding is a process of building up, and consists of adding metal to an existing structure. For this type of welding the electric arc has one great advantage in its high temperature. This is the highest known, and thus by the application of a small number of calories a part of the job, say, about ⅛ in. diameter, is almost instantaneously raised to welding heat, and the drop of adding metal from the pencil, also at welding heat is united to it, and the process of building up is continued till the required section is reached. The small quantity of heat required does not cause any undue expansion of the job in hand, and contraction troubles are reduced to a minimum. It is quite a common practice to weld over a riveted seam, although in this case it is necessary that rivets in the area dealt with should be completely welded over, and not left half covered. After welding a seam it is necessary to caulk the landing edge for some 6 ins. at each end of welded portion. Cracks in furnaces, end plates, combustion chambers, etc., are dealt with by cutting out the defective portion, leaving a V-shaped opening, which is filled in with the welding material. Work can be carried out directly overhead, or in any position that is accessible to the welding pencil, and where the operator can see what he is doing. As the work is one requiring constant attention on the part of the operator, it is advisable, in order to get the best job, to make it as accessible as possible, and that the operator should be reasonably comfortable.

In common with all hand welding, a good job depends on the conscientious work of the man. The author's firm have always trained their own welders, and keep them in constant employment. A full report is made of each job, and the name of the welder recorded, and the whole object of the training is to inculcate a sense of responsibility.

The materials at present dealt with on a commercial scale are wrought iron and steel and cast steel, and occasionally cast iron. The range of temperature of the welding heat is the determining factor in the adaptability of a substance for welding. Much successful work has been done with cast iron, notably with castings of considerable

FIG. 1—(TOP) BEFORE WELDING. FIG. 2—(BOTTOM) AFTER WELDING.

age, which have not been subjected to corrosive action, and with the good mixtures of more modern times. It is probable that there is a welding temperature of cast iron, but the range of this temperature is very small, something of the nature of 10°.

The voltage across the metallic arc is about 22 to 25, and the writer adds an equal steadying resistance which makes the voltage at the terminals of the dynamo about 45. A substantial resistance is employed which is put in circuit by an automatic switch, when the welder breaks his arc, thus keeping the load on the machine constant. The amperes actually employed are about 175, but in practice a 200-ampere machine is necessary, while the author's firm use machines designed for 250 amperes. In the big passenger liners it is the practice to weld from the ship's dynamo, suitable welding and substitutional resistance being provided. By a special winding of the dynamo, known as separate excitation, the machine can be steadied under varying loads, but even in this case the author still prefers to retain the substitutional resistances in addition.

The design of the portable machinery for generating electricity presents many interesting problems. Plant is designed to meet the varying conditions, and consists of wagons generating their own electricity, portable petrol driven generating sets, self-propelled or dumb barges with steam-driven or paraffin sets, steam turbine plants, and last, but not least, the motor generator sets. This last plant is of great service in a port like London, where the docks are well served with electric power mains at a constant voltage. The design of the dynamo is a matter for the electrical engineer, but the conditions of working are trying, and it is advisable to have ample commutator surface and good ventilation, as in urgent marine repairs it is possible that a machine may be asked to run almost continuously for two or three weeks.

The preparation of a job for electric welding is a matter of considerable importance, as the presence of impurities is likely to be detrimental to the weld. In dealing with the external or fire surfaces of a boiler it is usually sufficient to use an ordinary chipping hammer, and then thoroughly wire-brush the metal to be dealt with; but some superintending engineers prefer to have a light chipping taken over the surface, which is, of course, the ideal preparation. In marine work, however, the time available is often so short that as a general rule the former method is adopted. When, however, it comes to dealing with the water surfaces of a boiler greater care is necessary, especially if zinc plates have been freely used. The welder, if a properly-trained man, would at once recognize this difficulty and apply the only remedy, which is to chip down till pure metal is reached.

Arc welding being a building up process, cracks are dealt with by veeing out at the line of fracture, the vee being made wide enough to ensure that the welder can reach with his pencil to the bottom on either side with a certainty of striking his arc at any required position. As the welder is a highly skilled man, it is usual for the boilermakers to prepare the work to instructions, and the welder himself puts in the finishing touches. The welding in of new backs to combustion chambers or tube plates, or work of that kind is dealt with in precisely similar manner, although here certain allowances have to be made for the work drawing together as the welder proceeds. It should be mentioned that in dealing with cracks it is absolutely essential that the whole of the fractured portion be cut away till a solid chipping is obtained, and then go a bit deeper to be on the safe side. If welding is carried out over a partially cut away fracture it is certain that sooner or later it will work to the surface. One of the most unsatisfactory matters we have to deal with is the welding of a crack in the original weld of a furnace, as it is most difficult to say where the defective weld ends, and a further defective portion some short distance along may work back into the part dealt with.

As in all engineering matters, it is better to know the worst and deal with it. The writer recalls an incident in our early days—about 1910—when we were called in to weld a crack, apparently about one inch long, in the back of a combustion chamber of a Swedish vessel. Our man started to cut out the crack when with a loud report the chamber back split right across, showing a fracture a full sixteenth open. This caused great alarm at first, and we were charged with using undue vigor, but on veeing out the fracture for welding it was found that the back was grooved right across on the water side, so we were exonerated. It is a merciful dispensation of Providence that such defects develop mainly when the boiler is cold or under banked fires, and it is generally recognized that a boiler is never safer than when warmed up and steaming steadily. Owing to its higher temperature the electric arc is more suitable for dealing with the heavier sections than the oxy-acetylene or oxy-coal gas, while, on the other hand, for thickness of 3-16 in. and under one or other of the gas systems is preferable.

The author has been asked to summarise as briefly as possible the conclusions reached in the very able papers recently read by Commander E. P. Jessop and Naval Constructor H. G. Knox, both of the U. S. navy. The principal welding consisted of the repairing of the cylinders of some eighteen German vessels, where large pieces had been broken from the upper portions. The method of repair consisted of the welding in by the electric arc or oxy-acetylene gas of a new piece in cast steel or cast iron to replace the portion broken away. In arc welding the old and useful device of tapping short steel studs into the cast iron was used to enable the added steel (in this case) to make a surer weld. The electric arc welding repairs were carried out with the cylinders in place, while with the oxy-acetylene process it was necessary to remove the cylinders so that the joints for welding could be laid in a horizontal position, and also that the cylinders could be heated. Commander Jessop quite truly points out that the great difficulty found in the arc welding of the cast iron surface was to get the first layer of the adding steel material to adhere; and that this layer was always added before the patch was put in position for welding. In the oxy-acetylene jobs, as before remarked, the cylinders were secured in place, and, the joints being horizontal, both sides of the joint were made fluid, and cast iron sticks melted into the bath thus formed. Both methods appear to have given excellent results, and the repairs are certainly the largest of their nature that have yet been carried out, and reflect the greatest credit on all concerned. It would not be wise, however, to generalise on the treatment of cast iron from these results. You will remember that we have before remarked that with good mixtures of cast iron one can with fair certainty make a good weld. It must be remembered that these were high class vessels, and that in all probability the very best metal would be used in their cylinders and liners, and certainly in superheater jobs the H.P. cylinders and liners would be of a very special mixture, which so far as the author's knowledge is concerned has only been made in this country during the last five or six years. He trusts that we may hear further on this point, but his present information is that these vessels were superheater jobs.

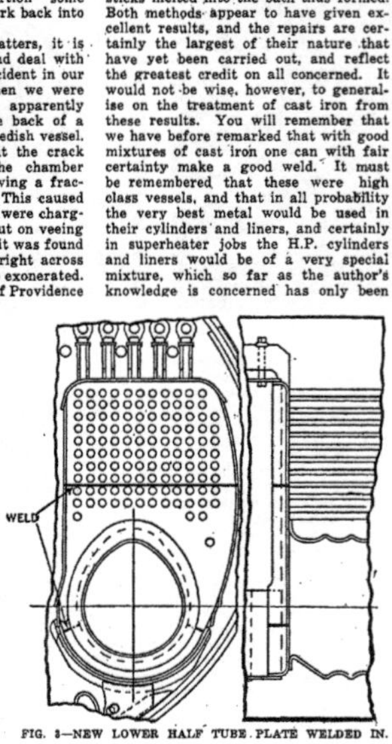

WELD

FIG. 3—NEW LOWER HALF TUBE PLATE WELDED IN.

The author claims that arc welding, where carried out by skilled operators with suitable materials, is absolutely reliable, and can point to some 20,000 jobs, some of a very big nature, while the percentage of even partial failures would, at any rate, be on the right side of the decimal point. These partial failures would be mainly accounted for where the work was carried out under unfavorable conditions, and often in the nature of a forlorn hope. Great difficulties are met with in hurried repairs to the lower portions of the hulls of vessels in dry dock, where water is constantly dripping from the leaky portion, and owing to the cement inside it is often impossible to stop it in the time available. It must, however, be remembered that metal added by the heat of the electric arc or other methods has not been subjected to the same amount of work as a rolled steel plate or forging. It is, therefore, not so well adapted to take up work suddenly applied, and one would not recommend it for a position of responsibility where such conditions arise. This, however, is a condition generally recognised by engineers with all welds.

The question of the resistance of welds to rapidly alternating stresses and shocks is somewhat obscure. Some year or two ago the author's firm were asked to weld the broken piston rod of a 10 cwt. steam hammer which had already been twice welded in the fire. This was carried out, and is now running satisfactorily. It is not permitted at the present time to refer specially to work carried out, but outside of the boiler repairs, repairs to hulls include the welding of broken stern frames, "A" frames for twin screws and the welding in of a new piece of stem is quite an everyday occurrence.

Boiler repairs are of infinite variety, and include the welding up of cracks to any extent, the welding in of new plates, thickening up of corroded surfaces, and building up of landing edges and defective rivets. Leaky stays and tubes have been welded in position with excellent results, and in cases of trouble with stays with loose washers it is excellent practice to build up from the solid plate to form the washer, which can then be faced off with a special tool.

Before proceeding to show the few slides which it has been found possible to prepare in these strenuous times, the author wishes to refer to a few gentlemen who have been of great assistance in the development of arc welding in this country early in 1910.

The superintendent engineer of Geo. Thompson & Co. had carried out the first electric welding repair of any size, being the welding up of a number of cracks in the Purves furnaces of the ss. Moravian. This was closely followed up with large repairs to the circumferential seams of the boilers of the Port of London authority's hoppers Nos. 3 and 4. One of our vice-presidents was early in the field, and it was due to his insistence that resistance plant was designed to weld from the ordinary electric-lighting sets of the larger vessels. Generally, however, it was found that the process supplied a long-felt want, and the author's task consisted mainly in seeing that none but fully-trained welders were allowed to undertake any welding repairs. The author's father, Mr. John Kennedy, and his Hamburg colleague, Mr. Bartlett, were the prime movers in introducing the process of the metallic electrode to this country, and the former was a tower of strength when in the early days it was necessary to overdraw at the bank, while Mr. Halket and Mr. Thom were indefatigable in assisting and advising in early experiments.

As referred to above, a number of slides were shown by Mr. Kennedy, illustrating in a general way the type of work carried out and a few of the repairs are shown in the accompanying illustrations.

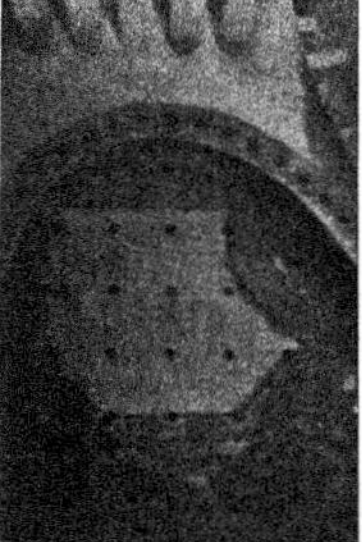

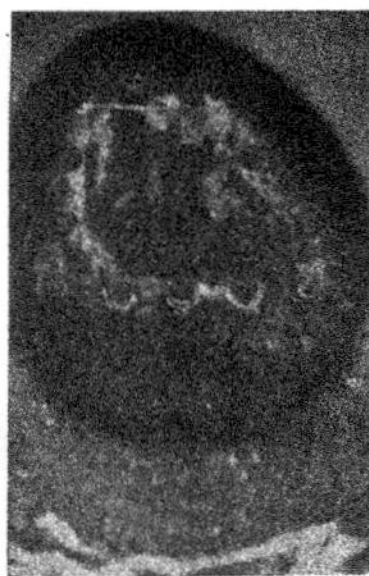

FIG. 4—MARINE BOILER COMBUSTION CHAMBER BACK

One of the slides was of the first motor wagon plant used for electric arc welding. The chassis was originally built by J. and E. Hall, of Dartford, to W. A. Steven's patents as a petrol electric motor-'bus, and was the forerunner of the present Tilling-Stevens petrol electric motor-'buses. This machine ran experimentally between Roehampton and Brighton, but was bought by the author's company and the electrical equipment converted to arc welding purposes, still retaining the electric road drive. The same principle was adopted by the War Office for portable searchlights. The chassis is driven by two motors, which engage the driving wheels through a worm drive: in the later machines the driving is from one electric motor, which drives a cardan shaft, and ordinary differential gear to the driving wheels. When the machine arrives at the job the current is switched from the road drive to the welding circuit, so that the same engine and dynamo answer both purposes.

Figs. 1 and 2 show repairs carried out in January, 1912, to one of the Canadian Pacific liners in Liverpool. Fig. 1 shows the defective portion of the flanging of the front end plate cut out ready for welding, and Fig. 2 the completed repairs.

Fig. 3 shows a repair carried out in May, 1912, to one of the Atlantic transport liners at Tilbury. Two tube plates were thus dealt with, and in a number of other furnaces smaller portions were cut out and new pieces welded in. This repair was the most difficult that had been up to this time attempted, as it was necessary to weld the new lower half to the existing half tube plate perforated with holes for the tubes. It was, however, satisfactorily carried out, and has never given any trouble. It may be of interest to mention that nine tube plates have just been similarly dealt with at Cardiff for the same owners.

Fig. 4 shows a repair carried out in November, 1911, to the back plate of the combustion chamber of a marine boiler. The first view shows the defective portion of the back plate cut away, the second, the new portion of plate in position for welding, and the third, the completed job. It will be noted that the plate has been cut through the line of stays, being the method recommended by the surveyors, which is undoubtedly preferable.

ENGINE BUILDING ON PACIFIC COAST

In a recent issue of this paper an article appeared entitled, "Vancouver Firms Pool Engine and Boiler Orders."

In the course of this article, Mr. McCulloch was quoted as stating with regard to the building of engines, that on the smaller types they could compete, but on the larger ones it would be difficult. In fact it had not been undertaken.

We are informed by Mr. A. F. Menzies, who is now in Ottawa representing the Wallace Shipyards, of Vancouver, that the construction of large marine engines and boilers has been undertaken and well carried out, and is proceeding very successfully. The following list of engines already built and in course of construction by this company remove any doubt on the matter:

S.S. War Dog, 1,350 I.H.P., now at sea.

S.S. War Power, 1,650 I.H.P., now at sea.

S.S. War Storm, 1,650 I.H.P., under construction.

S.S. War Cayuse, 1,000 I.H.P., now at sea.

S.S. War Atlas, 1,000 I.H.P., ready for trial trip.

One steamer, 1,000 I.H.P., under construction.

Vessel No. 100, 1,800 I.H.P., under construction.

Vessel No. 106, 1,800 I.H.P., under construction.

Two engines of 2,500 I.H.P.,, plans in hand.

The next size to be undertaken will be of 3,000 I.H.P.

The boilers for the above vessels are being constructed at the Vulcan Iron Works, Vancouver, and this firm are also building boilers for the vessels under construction at the yard of the J. J. Coughlan & Sons, whose boiler shop was recently destroyed by fire.

The foundry of the Wallace Shipyards Ltd. has furnished all the iron and brass castings for the above vessels, excepting the first two. The foundry has also turned out a number of propellers for the Imperial Munitions Board, and one large manganese bronze propeller for a coast steamer. From the foregoing it will be seen that engine and boiler building on the west coast is an accomplished fact, and we are pleased to correct the impression given by our previous article.

Incorporation has been granted at Ottawa to the Maple Leaf Shipping Co. Limited, and the incorporators are A. C. McMaster, A. McIntosh, H. C. Perkins, B. A. Farley, and M. I. Bergetts. The objects of the company are to build, construct, hire, purchase and charter steamships and other vessels of any class, etc. The head office is in Toronto.

Incorporation has been granted to the Nova Scotia Transportation Co. Ltd to build, construct, hire, and purchase steamships andj other vessels and to carry on business as ship owners, warehouse men, etc. The head office is in Toronto.

Vancouver, B.C.—Erection of three wireless telegraph stations at a cost of approximately $45,000 is contemplated by Lord Rhondda's estate. V. Lloyd-Owen, 850 Hastings St. W., Vancouver, is the architect.

The "Chicora" Has a Real Career and History

It Lived and Served For Over Half a Century—When First Built She Represented Positively the Last Word in the Business of Ship Construction

By T. H. FENNER, Associate Editor

AN interesting link with a war of by-gone days is furnished by the old-time privateer, "Chicora," which, after a useful career, more than half a century, is now enjoying a well-earned rest at her wharf in Toronto Bay. The "Chicora" was built at Messrs. Laird's yard at Birkenhead, England, in 1864, shortly after her more famous predecessor, the "Alabama." Though built for the same service, she never really entered it, as she was interrupted in her passage and interned in Halifax.

When she first floated down the Mersey she represented the last word in shipbuilding, and it speaks volumes for the builders, both of hull and engines, that she is to-day quite ready for putting back into service. Looking at her as she lies at the wharf, more or less neglected, one can still imagine how she must have looked stealing out of the Mersey on her blockade running errand. Built for speed, with a hull like a yacht, and two funnels with a decided rake to them, she must have fulfilled the old sea story writer's description of a "long, low, rakish looking craft." Needless to say, she is of the side wheel paddle type with feathering floats, and her engines, built by Messrs. Fawcett, Preston & Co. of Liverpool, are of the oscillating cylinder variety. She has two cylinders set directly under the crank shaft, of 50 in. dia. x 60 in. stroke, condensing. The air and feed pumps, etc., are worked from a crank in the centre of the crank shaft, between the two driving cranks. Her hull dimensions are 230 feet long by 46 feet 10 inches over all beam, and a depth of 10 feet nine inches. She can make 15 knots at her best, and in the year 1864 must have been reckoned a flyer.

Fortunately, or unfortunately, according to the individual viewpoint, she never succeeded in accomplishing any blockade running, and after one or two attempts, was taken off the service and laid up in Halifax.

Taken To Lower Lakes

About the year 1868 she was brought from Halifax to Buffalo by N. Milloy & Co. of Niagara, for freight and passenger traffic on the Great Lakes. She had to be cut for passing through the canals between Montreal and her destination, and she was put together in Buffalo and performed useful service on Lake Erie till 1870, when she once more was called on for military purposes. This was the occasion of the Riel rebellion, and the expedition under Col., afterwards Sir Garnet, Wolsely, was to be carried from Collingwood to the shores of Thunder Bay, on Lake Superior, and the "Chicora" was eminently suitable for this work. However, she met with obstacles even in this. As no armed troops could go through U. S. territory by rail, it was thought that the water transportation would overcome that difficulty. However, the only lock at the Soo that the "Chicora" could pass through was on the American side, so she was not allowed to carry her troops through. Finally, another vessel was secured on Lake Superior, the troops landed, and the "Chicora" returned to Collingwood to embark a further contingent. Delivering these, she returned again, this time embarking Col. Wolsely and the balance of the expedition. On this occasion the troops were landed and the "Chicora" passed through the locks empty, re-embarking her passengers the other side and proceeding on her journey to Prince Arthur's Landing. Her side was found too long to be placed alongside the landing, and accordingly, she had to be anchored and the troops and cargo landed by small boats.

THE "CHICORA" AS SHE IS TO-DAY

Further Service on Upper Lakes

After the trooping business was carrie through the "Chicora" was placed in service, carrying freight and passengers up the western shores of Lake Superior. Here, however, her general characteristics were unsuitable for the traffic, having, as she did, too much passenger accommodation in proportion to her carrying capacity, and her high speed making the expense of running her overbalance the money she could earn. She was, accordingly, withdrawn from service and laid up till the summer of 1874 when she had a brief period of her former glory.

Becomes a Royal Yacht

Being unquestionably the fastest and finest vessel on the lakes at that period she was selected to be the private yacht of the then Governor-General, Lord Dufferin, who was about to make a tour of the Upper Lakes during the months of July and August. For this auspicious event she was thoroughly overhauled, and her cabin accommodation was altered so as to comprise the various sleeping and reception rooms required by her distinguished passengers. Resplendent in new paint, glittering brasswork, with the Governor-General's standard floating at the mast head, and all gaily dressed in bunting, she looked the part when the Vice-Regal party stepped on board. To make it all complete the captain and crew were dressed in full naval uniform. She well fulfilled her part and after completing her charter retired gracefully to rest for another period, being laid up at the dock in Collingwood.

A New Career

Lying idly at the wharf, she attracted the attention of the late Barlow Cumberland, who conceived the idea of bringing her down to Lake Ontario to run on the Toronto-Niagara route. Knowing she was of little use to her owners where she was, he considered that she could be bought cheap and run profitably, provided he could get her down to Lake Ontario. This presented some considerable problems and was considered by many to be impossible. There were several reasons for this: She had been brought to Buffalo in halves and put together there, and guards fitted to her hull there which made the hull too wide to go through the smallest lock on the Welland, which was a bare 26 feet wide. Furthermore, if she could be got through the lock the only place she could be joined together was at Muir's dock at Port Dalhousie, and this was above the last lock leading into Lake Ontario. This lock was only 200 feet long, and the "Chicora" was 230 feet long. These obstacles seemed to be insuperable, but Mr. Cumberland was not easily put off. By careful examination, he found that the guards could be removed with small expense, and when the paddles had been removed there would be ¾ of an inch to spare in the narrow Welland lock. So far so good, but there still remained the question of putting her together. Here again luck was on their side. The Government intended to empty the five mile stretch between the last lock and St. Catharines in order to put gates on the lower lock at St. Catharines. It was also discovered that the high level gates at the Dalhousie lock were of the same depth as the low level gates, and the lock extended for 33 feet beyond them. Here then was an opportunity to get her through. After the close of navigation the "Chicora," having been brought through the canal in halves to Port Dalhousie, she was put together in Muir's dry dock, and then, the high level gates being opened, she was floated into the lock. This made practically a lock five miles long and the process of lowering it naturally took a few days. However, the water gradually fell till it was at the same level as Lake Ontario, when the lower gates were opened and the "Chicora" floated out on to Lake Ontario and the last stage of the voyage to Toronto commenced. Arrangements had been made to keep a tug in readiness to tow her down, this being the only tug not laid up for the season. When in sight of the entrance to Toronto Bay, disaster nearly befell both the tug and her charge. Something happened to the tug's engine and she stopped. The wind was blowing on shore, and no assistance was available. The anxiety of those on board can be readily imagined, but just as things looked their blackest a cheering toot from the tug boat's whistle was heard and the tow line tightened up, as gathering way, they headed up for the entrance. A short time sufficed to lay the "Chicora" at her wharf, where, during the winter she was prepared for her long and useful career on Lake Ontario. The following season she was put on the run, which she continued in for 35 years, between Niagara, Lewiston, Queenston and Toronto. For the three years previous to 1917 she was on the Toronto-Olcott, N. Y., service, and since then has been laid up.

The "Chicora" still has the original engines in her, but her boilers have been renewed. The hull is composed of all the original framework, but she had practically a new bottom fitted in 1902. The upper plating is the same that was put in her 54 years ago.

WOODEN SHIPBUILDING

It has fallen to the Oregon District to have the distinction of starting the first of the all-wood 5,000-ton steamers for the Emergency Fleet Corporation, authorization having been given October 8, by Chas. E. Piez, general manager of the Emergency Fleet Corporation, for theh enlarged type of vessel to be laid down. Some time may be required to work out copies of plans and specifications covering this vessel, so that contracts will probably be awarded soon after an additional appropriation asked of Congress is made available. Numerous designs have been offered the Emergency Fleet Corporation for 5,000-ton wooden vessels, but the Columbia River type, as this is designated, is the first of size other than a composite ship to be both approved and ordered built.

About the time of the delivery of the first Hough ships which are 3,500-ton carriers, it was foreseen by builders that the time was not far distant when a larger vessel of native wood would find strong favor because of the advantage in carrying 1,500 tons more deadweight with virtually the same maintenance charges, so this 5,000-ton type was started. The plans and specifications were drawn under the direction of officers of the Wood Ship Division of the Oregon District, on the occasion of the visit in July of Mr. Piez and Director General Schwab. They brought up the subject of a larger wooden carrier, and were surprised when informed that such a vessel had already been designed.

On instruction of Mr. Piez the wood shipbuilders met with the district officers and went over the plans carefully and when they were sent to New York for inspection by Lloyd's and the American Bureau of Shipping, they had the backing of all the plants in the state. It was September 17 that the two classification societies approved of the big vessel, and in turn the plans were taken to Philadelphia and laid before the experts of the Emergency Fleet Corporation.

The ship will have a length of 344 ft 5 inches overall, and 315 feet between perependiculars, with a beam moulded of 48 feet, and depth, moulded, of 34 feet 6 inches, the vessel having a full shelter deck and two t'ween decks. The material for these ships will be of unusual length, and in some features such as keel, keelson, stem and stern construction will call for pieces of large dimension.

In undertaking this new work two outstanding features are important, they being that the yards can lay down the vessels and get out all other material without expanding their present equipment and machinery under order for the Ferris and Hough types, which include main engines of 1,400 horsepower to be placed in the larger vessels, thereby saving expense for the builder and the Government. Experience may bring about an increase in the size of the main machinery, but for the present it is not intended to add more power than the other vessels have had installed.

The construction of this new type will be restricted to Pacific Coast yards, and the larger portion of that to the plants in Oregon and Washington.

MARINE

TORONTO.—The "War Hydra" was launched from the yards of Polson Iron Works on Tuesday last week. The launch was originally scheduled for Saturday, but unforeseen delays occasioned the postponement till Tuesday. The "War Hydra" is a sister ship of the "War Taurus" launched from the same yard last month. She is a steel, single screw vessel of 261 feet long, by 43 feet 6 inches beam, by 23 feet moulded depth, built to the classification of the British Corporation. Her propelling machinery consists of triple-expansion engines of 1,250 h.p., the cylinders being 20½ x 33 x 54, with a 36-inch stroke. She has two boilers, Scotch marine type, 14 feet diameter by 12 feet long.

Toronto.—The steamer Cayuga of the Niagara Division of the Canada Steamships Lines lately made her final trip of the season between Toronto, Queenston and Lewiston. With the navigation season on this division of the company closed, it is interesting to note that the Cayuga, which opened and closed the season, made 303 trips, covering almost 25,000 miles. Apart from losing a wheel while entering the Yonge street slip a few months ago, the Cayuga had her best season, despite war conditions and its effects on excursion travel. With the exception of the hospital patients and a few transport troops, which are still at the Niagara Camp, the Cayuga on her 2 o'clock trip brought over the last batch of troops and a number of motor transports.

OTTAWA.—Delay in Canada's shipbuilding program is being occasioned by the shortage of steel supplies for yards engaged in this work. Hon. C. C. Ballantyne laid down a program in the spring, contemplating a large umber of vessels, for which contracts were let and shipbuilding yards erected. Arrangements had been made with the U. S. for about 80,000 tons of plates up to July 1st, but owing to the tremendous demands in the States, deliveries to Canada have fallen considerably short of expectations. Every effort is being made to hasten deliveries, and improvement is expected in the next few months.

Vancouver.—A banquet was tendered by the Wallace Shipyards recently to Capt. J. Nelson Craven, on the eve of his return to the Old Country. Capt. Craven has been supervising the building of the four vessels for Messrs. Chambers & Co., of Liverpool. He spoke very highly of the quality of the vessels, and said that they would be excellent ambassadors of B. C. shipbuilding. Capt. Craven was presented with a silver traveller's comfort, and congratulatory speeches were made by the principal guests.

The MacLean Publishing Company
LIMITED
(ESTABLISHED 1888)

JOHN BAYNE MACLEAN - - - - President
H. T. HUNTER - - - - - Vice-President
H. V. TYRRELL - - - - - - General Manager

PUBLISHERS OF

MARINE ENGINEERING of Canada

A monthly journal dealing with the progress and development of Merchant and Naval Marine Engineering, Shipbuilding, the building of Harbors and Docks, and containing a record of the latest and best practice throughout the Sea-going World.

B. G. NEWTON, Manager. A. R. KENNEDY, Editor.

Associate Editors:

J. H. RODGERS (Montreal). W. F. SUTHERLAND.

OFFICES

CANADA—
Montreal—Southam Building, 128 Bleury St., Telephone Main 1004.
Toronto—143-153 University Ave., Telephone Main 7324.
Winnipeg—1207 Union Trust Building, Telephone Main 3449.
Eastern Representative—E. M. Pattison.
Ontario Representative—S. S. Moore.
Toronto and Hamilton Representative—J. N. Robinson.

UNITED STATES—
New York—R. B. Huestis, 111 Broadway, New York, Telephone 8971 Rector
Chicago—A. H. Byrne, 900 Lytton Bldg., 14 E. Jackson Street, 'Phone Harrison 1147
Boston—C. L. Morton, Room 733, Old South Bldg., Telephone Main 1024

GREAT BRITAIN—
London—The MacLean Company of Great Britain, Limited, 88 Fleet Street, E.C., E. J. Dodd, Director, Telephone Central 1960. Address: Atabek, London, England.

SUBSCRIPTION RATE

Canada, $1.00; United States, $1.50; Great Britain, Australia and other colonies, 4s. 2d. per year; other countries, $1.50. Advertising rates on request.

Subscribers who are not receiving their paper regularly will confer a favor by telling us. We should be notified at once of any change in address, giving both old and new.

Office of Publication, 143-153 University Avenue, Toronto, Ontario.

Vol. VIII. OCTOBER, 1918 No. 10

PRINCIPAL CONTENTS

Public and Private Methods

WHY is it that government institutions seem so often to lack in the fine precision of detail that makes private operations in the same line a success?

For instance, right now such a condition exists at Vancouver. In the Vancouver *World* of recent date the following appears:—

"Vessels are unloaded, re-loaded and get away from Vancouver just as quickly as anywhere on the Pacific Coast."

That is, some vessels are.

When a C. P. R. boat comes in she finds one empty shed to receive her freight and another full shed to be emptied into her hold. Loading is going on at one end of the ship and unloading at the other.

Also the freight does not accumulate in the shed. There are cars, all the cars needed, to take it away and so prevent congestion. Also switching engines to move the cars.

And likewise at the Great Northern dock.

But not so, not by any means so, at the other docks and at the government dock in particular. There the freight piles up and piles up in great mountains so that it takes a gang nearly twice as big to work a hatch as the gang working a hatch in a C. P. R. steamer.

And moreover, if C. P. R. or Great Northern boats are in the longshoremen find it better to go to them than go out to the government dock, where they are a long way from home and not a solitary restaurant is working at night when a man wants some hot coffee to see him through.

Now that condition is too often indicative of the difference between public and private ownership. The private company has to study actual competitive conditions and meet them. The government can afford to take the position too often of knowing that a deficit is not going to wreck their business. The Vancouver case looks like one that will stand some explanation.

Protection Needed For the Public

THERE are controllers for fuel and for food. There are officials to see that people don't eat too much, and there are others to see that they don't burn too much coal.

The country is sadly in need just now of a controller or some such official whose special duty it will be to see that the general public don't get soaked when they come to spend their money.

The people who need protection right now are the people who spend their money and get in return an article that is a fraud and a scandal. People need protection against shoes that are sold as leather, and which in reality have only the merest touch of leather to cover up a lot of substitute. They won't stand up against wear. And yet they are sold in the open market and at a good price.

People need protection in the purchase of clothes. The average purchaser is not an expert in this business. He cannot readily detect shoddy. And yet he puts up his good money for the clothing of himself and his family, only to find that he has paid a good price for something that is scandalously distant from what he was led to believe.

"Get the money" seems to be the big word in too many business concerns now. The idea of "service" is being crowded so near to the back door that it's not going to take a great deal of coaxing for it to depart entirely from the premises.

"Get the money." The people have it now. "Get the money." Never mind what follows. "Get the money." There's a new bunch of suckers coming on the market every Monday morning. "Get the money." Don't worry about the man who is spending it—he's supposed to have his eyes open.

There are men and firms who are doing business in an honorable way. Against such the public needs no protection. But in this age of money-grabbing shysters there certainly is need of ample protection against the gang of exploiters that are foisting their shoddy rubbish on the market.

Tin is coming down to where it will soon be on speaking terms with the rest of us. Not long ago it had reached the dizzy height where a pound was worth well nigh $1.50. Now it can't quite get its chin up to the dollar mark even with one toe on the ground. About the time tin hit the $1.50 mark there should have been a couple of lynchings.

MAN'S OWN EXPERIENCE IS WORTH A LOT TO HIM

H. L. Treat is Production Superintendent for T. McAvity & Sons, Ltd., of St. John, N.B.—His ideas are worth considering.

THE shipbuilding industry is pulling men east and west. It is also pulling them from the south. That's how it happened that Harry Lee Treat is connected with the T. McAvity & Sons, Ltd., of St. John, N.B. He has made a success of the munitions business, and was early in it at a time when a great deal that is now accepted fact was in the experimental stage.

Mr. Treat graduated from the Middletown High School, in which town he was born. He started as an apprentice with Messrs. Pratt & Whitney Co., of Hartford, Conn. Soon after he completed his training he developed a special aptitude for tool and experimental work, holding a series of positions with the Royal Typewriter, Noiseless Typewriter, and Bryant Electric Co. He also spent some time designing tools and fixtures with the Singer Mfg. Co.

HARRY LEE TREAT

In the Munitions Business

Mr. Treat was called on by the Ansonia Mfg. Co. when they started the manufacture of munitions. There was little or no precedent to go by in those early days, neither had the technical papers started in with their splendid work of explaining the making of shells and fuses. It was a case where individual initiative alone could solve the problem. To this work of laying out and designing special machinery and tools, Mr. Treat devoted his best energy and training, and he succeeded.

It was about two years ago, early in 1916, that Mr. Treat came to Canada, joining the McAvity Co. staff as production superintendent, which position he still holds. He has always taken a keen interest in sports, holding city championships several times as a bowler. Baseball owns him as an ardent supporter, and basketball is also a favorite. Much of his interest in sport has been in connection with the Y.M.C.A. in the various centres in which he has lived.

The McAvity people have consistently developed along sane and legitimate lines. They followed the increased demand with increased facilities for turning out the work.

Mr. Treat has very pronounced views of his own regarding the chances of the mechanic for advancement to the better positions in the industrial world. "A young man," said the St. John man, "has simply got to gather his own experience. That sort of training is, after all, the most valuable. He must, as he goes along, sift out the worthless—for after all, one is bound to meet with considerable of that. After a man gets his start and experience he can decide what line is his channel. Let him take the work that interests him, and then go ahead and stick to it. Jumping around from one line to another seldom gets a man anywhere. A man is never educated," continued Mr. Treat, "in fact life is a perpetual education to the man who will keep his mind open, and never allow himself to get to the point where he thinks he can learn no more. Keep in touch with the technical and trade papers. There are excellent opportunities presented in them. If mechanics would use the staffs of the technical papers more they would find many problems easy of solution. It is a good thing for a mechanic to get a problem of his own under discussion in these papers. In that way he can bring to bear on it the experience and opinions of other mechanics who may have encountered and mastered the same point. In brief," concluded Mr. Treat, "the man who succeeds can never quit picking up new points. They are coming out all the time, and there are lots of good ones in the list."

Better Play It Safe This Time

MECHANICS are these days receiving exceptional wages. Some of them, most we hope, are salting away neat little sums for the rainy season. Among the thrifty, Victory Bonds are regarded as one of the premier investments, and the new issue will undoubtedly be very largely taken by this class. But occasionally one hears of cases in which men have not been careful—when, in fact, they have been reckless as regards their own futures and those depending upon them. An actual instance that stands out is that of the mechanic who was a bear to work and who was making something like $3,000 a year—probably more than the superintendent of his shop was receiving. He was, in fact, a good workman. This man purchased a car—not a flivver—a car. One night while out driving with his wife, he met with an accident. The car turned turtle; he was killed and his wife was seriously injured. The after records revealed that the man owned the car—an $1,800 machine. It was paid for. In the home there was a piano on which two payments had been made; there was a life insurance policy for $500 and a chattel mortgage of $400. The machine had been smashed—it was worthless. This was the record of a man who had been making exceptional money—and this is the record of what he did with it—practically nothing to place on the credit side against disaster. There are many men who will say it is nobody's business what they do with their money. Probably not; but to a good many others the good old moral will appeal: Buy Victory Bonds and give the clouds a silver lining.

We're Busy Dodgin' Germs

OH folks imagine nowadays that every place they turn, some person's waitin' for to hoist on them a Spanish germ. They see them settin' on the road, and campin' on the trees, and scatterin' forty different ways whene'er they hear a sneeze.

They're sprayin' dope on dollar bills that camp inside the bank, they're killin' germs that venture there, the fat ones and the lank.

You see a man come in the car, there's murder in his eye, to see if any germ-stuffed jay is comin' on too nigh—he sizes up the line what's there, and if he hears a sneeze, he trembles from his stomach up and wabbles at the knees.

And when he goes to get some grub he grabs the battin' card, and gazes at the things thereon and ponders long and hard.

He's sure the soup is full of germs, on fish they'll camp, 'tis true, and on the liver and the rice will dwell ten million Flu—the waiter, too, his eye looks bad, there's death upon his paw, there's torture written on his chest and sickness on his jaw.

That waiter should be run right in and planted in the coup, this thumb with sixteen kinds of germs has gamboled in the soup.

Oh, there ain't much fun in livin' now. I'd rather be a worm, what camps inside some lonely spot what's free from any germ, than hoofin' round the streets these days a-scared of folks like you, what's tryin' to fasten on to me big hunks of Spanish Flu.—ARK.

Fo'c'stle Days

or

Reminiscences of a Wind Jammer

By Capt. Geo. S. Laing

The steamer and the motorship have deprived the seafarer of much of the old-time romance attaching to a sailor's life. The advent of wireless has also increased the safety of it. Deprived of these features, a sailor is immediately dependent on his own resources, and Captain Laing gives us a true and seaman-like reflection of sea life under canvas. The present feverish activity in constructing self-propelled craft affords a suitable background for incidents in a career in which gaffs, booms, tar and canvas held sway before being displaced by mechanical unloaders, derrick masts, fuel oil, coal and wire rope.

Part IX.

The Ballast Mountains of Newcastle

After discharging half of our ballast we were towed down to the coal berths, where a few car loads were dumped into the hold. Then we towed upstream again to the berth at the Dyke. This mode of procedure ensured sufficient stiffening in the vessel's bottom compatible with safety in a sailing ship. The ballast mountains of this great export harbor were in themselves monuments of the vast sailing-ship trade, altho' sailing craft of to-day are generally fitted with a cellular bottom for water ballast which does away with the old system entirely. Looking from aloft over the Dyke area of Newcastle one could see thousands of tons of cobble stone which had come all the way from Rio de Janeiro, 9,000 miles, as ships' ballast for that stormy route. Then there were sand and gravel hills that had come in the same manner from South Africa. Other accumulations were there too that had given stability to the white-winged fleets bound for Newcastle from all directions.

Southerly Buster

In towing down to the loading berth the second time we were almost capsized in a squall peculiar to the Australian coast and known as a "southerly buster," an ugly storm to handle a light ship in. As a rule they are wind tigers of short duration but quite a number of vessels parted their moorings alongside the quays, and did minor damage with bowsprits or taffrails by fouling other craft. During the squall the air was charged with sand and gravel and coal screenings and the effect was anything but pleasant.

A letter arrived from Tilbury, wrapped in evasion and told of his thankfulness to be clear of the "horrid sea life." This epistle was delivered by the postman. Had it come through the agent's and captain's medium our accomplice work might have leaked out.

When under the coal hoists a dealer from Sydney came on board with skins for sale, and each boy bought for half-a-guinea a nice rug. The choice comprised wallaroo, wallaby and opossum furs nicely sewn on green scalloped cloth. No matter how far off the home going might be, sea apprentices gathered up curios for that eventful time.

A photographer in this port also took the picture of each vessel and after mounting it in a lifebuoy frame sold his product amongst the officers and boys. This was a very catching device, the circular lifebuoy being painted with white enamel and the house and national flags appearing, one on each side, with the vessel in the centre. We made quite a few friends on shore, some of them being introductions from Brisbane folks, but soon we must leave for Coquimbo, our Chilian port of destination. This lonely run across the southern Pacific meant at least 6,600 miles and was in many ways an extension of our "running the easting down" along the contracted hoops that encircle the broad oceans which roll and thunder beneath the southern firmament.

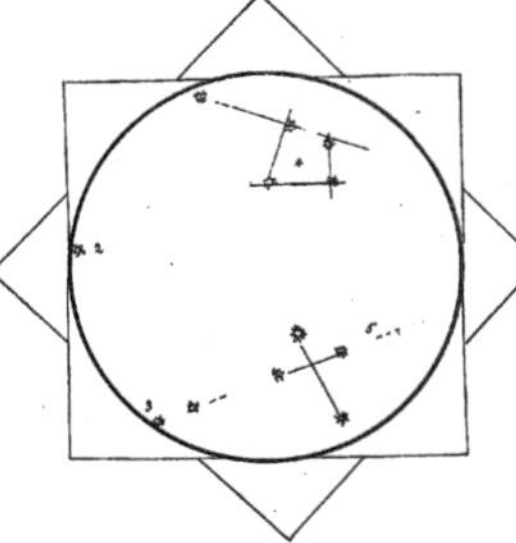

1—SPICA ON THE GAP END OF THE SPANKER. 2—ANTARIS OR THE SCORPION. 3—SOUTHERN CROSS POINTERS. 4—THE SPANKER—AFTER THE SAIL OF THAT NAME. 5—THE SOUTHERN CROSS.

As regards long distances in an east or west direction in high latitudes you may notice that a Mercator projection shows courses as curves with the sweep towards the adjacent pole. In such localities as we were sailing in the degrees of longitude were much smaller than they would be on or near the equator. The earth being a spheroid and the meridians of longitude all converging at the poles, it can be readily seen that a degree on the parallel of 40°, 50° or 60° latitude will differ in each case. Thus it means that while a degree of longitude is 60 miles on the equator it is only about 30 miles in 60° N. or S. It is therefore a common affair for an old wind-jammer to make in one day's run as much as ten degrees of longitude in "running the easting down." In low latitudes, i.e., near the equator, it would take a fast mail boat to alter her longitude at that daily rate.

This Australian-Chilian trip was sometimes attended with fire through spontaneous combustion, and in a few cases iron vessels had sailed for thousands of miles with a smouldering volcano under the hatches. In other cases ships had been blown up or abandoned when the fire proved the master. With coal cargoes on such long voyages the danger is ever present, and vessels from Shields or Cardiff, Old Country ports, were even subject to this horrible condition more than the craft from Australia to Chili.

Sailing Day

It came to sailing day and our Blue Peter was run up to the fore truck announcing in sailor parlance that anyone with a debt to collect had better come along. The 3rd mate was promoted to 2nd mate and a cook from the bush had been procured. No sailmaker could be found, but a crew of fair seamen from the local jail and four others of the "Murrumbidgee" type were brought on board the "Maggie Dorrit" after she swung into mid-stream. This rough bunch of globe trotters could not be trusted to act rationally alongside of a dock, hence our movement into the river. A tug soon brought the old man off and then towed us out to sea, where there are no back doors or sheoke nets. As we passed out through the turbulent waters of the Knobbies the flag was dipped to the shore station, and a few caps and handkerchiefs fluttered from fast receding figures—some of whom were known to us, while others simply waved a bon voyage as a matter of an established custom. To many persons a sheoke net will bear description. It is not likely that such things are used now, as a great reform has come to us in recent years.

A Sheoke Net

Like all other young countries Australia was flooded with vile drink, and sheoke was the native beer. The drunkenness amongst ships' crews from cabin boy upwards was so accompanied with drowning accidents through falling off accomodation ladders and gang-planks that the port authorities in Newcastle ordered all vessels to spread a net under their gangways, to catch, if possible, the inebriated salt, should he roll off the ladder. This affair was more or less of a joke as all preventive measures are in connection with evils that need absolute annihilation. Some sailors would have gone to sleep in the sheoke nets if police and ships' watchmen had not moved them on to a more appropriate place. Other stupified creatures would spin incoherent yarns about losing shillings and half-crowns in the net and wondering why they couldn't find the money again. The net by the way had a three inch mesh.

Every sailing ship that entered Australian harbors from Great Britain or continental ports such as Hamburg or Rotterdam generally carried a consignment of whiskey, gins and rums, all cheap immature drinks that made fortunes for the distillers and brokers, and beastly work for the police and coroners.

Fiery stuff from Scotland that cost 10c a bottle to put it in a ship's hold would bring $1.50 to $3 in the land of the kangaroo. Hollands gin which could be distilled from reeking vegetable garbage and similar German potions from the same source, but called kummel, polluted the country.

In this connection a great many sailors who deserted in Australia spoiled their good chances, at the same time a great number of run-away apprentices of thirty years ago are big men in the Commonwealth to-day.

Waking Up The New Crew

After getting an offing of four or five miles the towboat signalled to haul the towrope in. The four members of the fo'c'stle who had come on board through the medium of that glorified kaiser—the boarding-house master—were lying around the fo'c'stle in various attitudes of recumbent leisure. This privilege was sometimes given to such seamen until the towboat had gone, as they have been known to jump overboard in their delirium as long as the towboat was ahead, knowing that there was still a chance to be taken ashore again.

All the fore and aft canvas was set and a final salute was dipped to the tug while she in turn gave three long toots on the steam whistle and a short one. Old "ropeyarns" was sent forward to "let these sundowners know where they are." One out of the three could stand up, but before long a new face joined us at the topsail halyards, and later on the other two appeared altho' one was crawling. He, of course, wanted to go aloft or out on the boom, where he should get killed, so the mates had a lively time.

None of the new rascals that were in such a hazy condition knew the ship's name or where we were bound to altho' they had "touched the pen" in signing on. As to reading aloud the ship's "articles" to those fellows, all that they generally remembered about was the spectre of a bald headed clerk mumbling a long-winded discourse jammed full of "saids." "aforesaids" and notwithstandings," but bawling out in clear English "substitutes at the master's option." The clause was in relation to food and it was stretched in its interpretation until its literal meaning fell overboard.

The Beach-comber

In the days I am writing of the forward crew of a ship were looked upon as necessary evils with the consequence that a great number of them descended in the social scale till they were out of sight. While it is lofty minded to remember that human clay can to some extent mould its own destiny through intelligent clamor, the faults were mostly on the shipowner's side. Miserable pay, miserable quarters and miserable food were the three deterrents that made seamen forget they had a soul. The finished article accruing from such a low plane was called the beach-comber.

To this human wreck every foreign strand became a paradise, pro tem, where he ate little but drank much. He acted as a jackanapes to the lowest caste of the natives on whose shore he happened to be sprawling. To all fools the boomerang comes back and what may be high living to-day turns into hell's gridirons to-morrow. And so it was with the beach-comber, who after making a passage on a ship and regaining his health, would deliberately swim ashore from the anchorage under cover of darkness and squirm into the fandango huts of Callao, Iquique or Pisagua. He was the guest of the evening and equally as much at home on the other side of the Andes, when the ship struck Rio de Janeiro or Monte Video.

Our heathen sailor could curse in Spanish or Hindustani till the cocoanut palms or mango trees shook their fruit into his grasp. In other words he would act as the abject slave of his king, the boarding house master (marine species). This "king" would use the beach-comber to entice new members to leave their vessels. As fresh arrivals came ashore "dead beefs" were shipped outward bound on vessels weighing their anchors. In this way there was a business movement of supply and demand.

The only time a beach-comber was clean and tidy was when the port of destination was sighted ahead. He seldom saw any land over the taffrail, for on leaving a port he was non compos mentis.

Many of these misguided creatures were good sailormen when their only danger was the whack from a Cape Horn sea or a crack on the nose from a reef point, but when in harbor there were only two safe places for them, and these places were the jail and the hospital. Even then, if such institutions were not wary our amphibious clown might be found down at the mole bumming coin or tobacco from the various boats' crews or joining the members of a liberty day in their yells and riotous behaviour.

To all southern water navigators the stars shown herewith were well known celestial pictures.

WAR HORUS TAKES WATER AT PT. ARTHUR

Busy Spot on Orders for the Imperial Munitions Board

At 3 p.m. October 5, the Port Arthur Shipbuilding Company, Limited, Port Arthur, Ontario, launched the steamer "War Horus," built to the order of the Imperial Munitions Board.

This vessel is a steel screw, single deck, general freight-carrying ship with a straight stem and semi-elliptical stern with poop, bridge and forecastle, built on the transverse system with inner bottoms throughout. General dimensions are as follows:

Length over all	261 ft.
Length B. P.	251 ft.
Molded breadth	43 ft. 6 in.
Molded depth	23 ft.
Gross tonnage	2,240
Deadweight tonnage	3,400

Cargo will be handled by four steel derrick posts fitted with eight booms, each boom served by 7 x 12 reversible double drum steam winch.

Propelling machinery consists of a triple expansion surface condensing engine, having cylinders 20½ ft.-34½ ft.-55 ft. x 40 ft. Steam will be supplied by two Scotch boilers 15 ft. in diameter by 11 ft. long, with a working pressure of 190 lbs. and developing about 1500 I.H.P.

SEAMANSHIP AND NAVIGATION

Conducted by "The Skipper"*

Articles of direct interest to mariners, discussions of seamanship, navigation rules, and allied topics. Inquiries from readers are invited and will be replied to promptly through these columns, unless otherwise requested.

THE SCIENCE OF THE SEA—Part II

Rollers

This name is given to a freak wave that has no crest but rolls into the beach of some West African ports, and also strikes the islands of Ascension and St. Helena at odd times. It may be known at other spots too. The roller is very dangerous to ships anchored in a roadstead and is also a fiend of destruction to all harbor works and facilities. Wriggling inshore like a switch-back railway the roller seldom breaks in foam till it hits the sand. Its origin is still a mystery.

Bores in Rivers

The bore movement should perhaps belong to the part of the article discussing tides, as it is an inshore agent, but here it is; for its antics must be chronicled. Every navigator and sailor should know of it.

Some half a dozen rivers on the globe are visited, especially at spring tides, with the bore. The writer has experiences of it on the Hoogli, India, and the Seine, France. On the Seine the kick of the bore is felt at Rouen and when it is due the harbor master's deputy visits each ship and gives instruction as to extra moorings, otherwise the craft moored alongside the quays would all break adrift. In some cases cargo planks and gangways are pulled up till the danger is over.

The bore seems to be caused by a fight between coast and river waters. When the flood or incoming tide advances on the coast some rivers refuse to be overpowered and keep on emptying their ebb water into the sea. The resistance goes on till a balance of power on the ocean side smashes through like a raging lion and carries a wave of victory 15 to 20 feet high far inland. This is called the bore, and to certain craft it is a real danger, as it rushes along the banks of the river. Even after the wave has subsided the kick proceeds and frequently parts ships' moorings. If steaming or at single anchor the dangers are not great.

When anchored at the double moorings below Calcutta, one can glance over the rail at the river ebbing, say at four miles an hour, and almost like the shot of a gun the next look at the Hoogli's muddy waters discloses the fact that the bore in the lower reaches of the river has suddenly made the famous tributary run in the opposite direction—flood movement.

Sailors must grapple with the waters as the farmer does with the land, and no man in charge of a ship's deck should be ignorant of the many moves and dodges of the seas, rivers and oceans.

Winds

Wind as well as water enters into the seaman's profession so it will be quite in order to write them up together.

Air and wind are synonymous—wind being simply air in motion. The same agents that give motion to the waters also give motion to the air. Thus we have the ice of polar seas, the heat of the equator, the earth's rotation and the distribution of sea and land all determining the wind circulation of our globe.

As an instance of the laws that govern winds let us examine the "trades." Here we have a N.E. wind north of the equator and a S.E. wind south of the equator. Temperature plays its part in this way. The superheated air in the Tropic of Cancer expands and rises into the upper atmosphere, thus inviting heavier and cooler air to take its place. This air or wind must come from the north. Then with the earth revolving from the west to east on its axis, a deflection to the eastward is the result and you have the N. E. trade wind. The same laws govern the Tropic of Capricorn and produce S. E. wind.

These famous winds are divided by the doldrums—a belt of meteorological samples made up of calms, squalls and torrential rains. The doldrum area has a north and south motion according to the sun's declination but in the main it hangs around between 5° on either side of the equator, a preference for the northern hemisphere being reported and apparently substantiated.

Another example of wind action is the land and sea breezes experienced on coasts, especially tropical coasts. This is what happens, the land takes

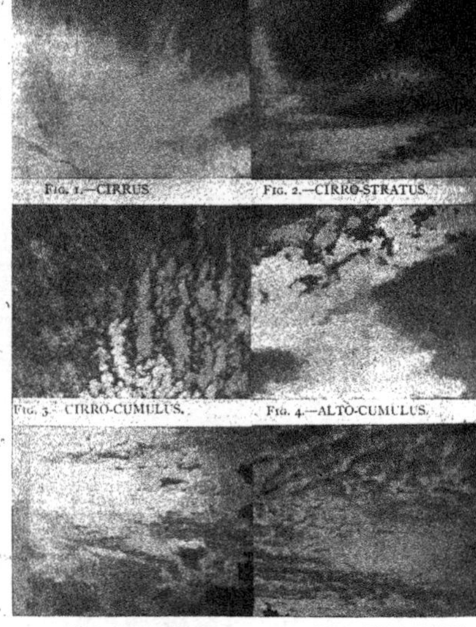

FIG. 1.—CIRRUS
FIG. 2.—CIRRO-STRATUS
FIG. 3.—CIRRO-CUMULUS
FIG. 4.—ALTO-CUMULUS

FIG. 5—ALTO-STRATUS FIG. 6—STRATO-CUMULUS

*Editor's Note.—This department is conducted by a qualified ship's master, and will deal with questions involving seamanship and navigation. The attention of young men desiring to obtain certificates as mates and masters is particularly directed to this department.

the sun's heat quicker than the nearby sea and thus the air ashore rises thereby inviting a cool breeze from the sea. At night when the sun has gone, the land loses its heat very quickly and the air or wind rushes out over the water—creating the "land breeze."

It is easily understood that the winds of the southern hemisphere are more trustworthy and continuous than those of the northern hemisphere, simply because the southern half of the world is not broken up so unevenly with continents, but is largely made up of ocean. Maury's "brave west winds," or "roaring forties," blow right round the sphere in the southern latitudes between the parallels of 40° and 50°.

Variable winds are those which appear to blow without any regularity as to time, place or direction. Such winds answer to local causes, as for example configuration of nearby land, especially in regard to mountain ranges, and the vagaries of moisture and temperature.

Monsoon winds are periodic according to season—summer or winter, and the laws that govern them are similar to those already mentioned in connection with land and sea breezes on tropical coasts. The latter, however, are diurnal motions while the monsoon changes his tune semi-annually.

Monsoon winds blow towards the relatively warm land in the summer season, and towards the relatively warm sea in the winter season. Thus India has a wet summer as the S.W. monsoon blows from the equator, say from May to October. From November to March the wind blows seaward, or from N.E., emphasizing the fact again that air circulation on the surface of the globe is mainly dependent on temperature. The word monsoon is derived from an Arabic root, which reminds us also that our friend the Arab has a long pennant in the history of sailoring.

To show how winds are affected by surface conditions such as a mountain range, take the Magellan Straits, which could be aptly called the Cape Horn canal. The meteorological differences so apparent at the Atlantic and Pacific entrances to this narrow channel are attributed to the Andes mountains that push their glacial toes into Terra del Fuego, and the famous squalls called willy-waws are a product emanating from the aerial strife that goes on in this lonely but grand school for seamen.

Plateaux and deserts have also a big say in the air circulation of our globe, as do lakes in the interior of a country. It must be remembered also that the direction of the wind changes with altitude, and its velocity in the upper atmosphere is frequently different to the speed of the same wind on the ground. Our brave airmen have secured a mass of data along this line of knowledge, which will go far in advancing certain theories that otherwise have remained undeveloped for centuries.

Although the marine world of to-day is almost entirely mechanical, the steamboat mate and master must study the winds almost as much as their brother in windjammers, who depend on the piping breezes to make a smart passage. A man who studies the winds knows when to expect fog or rains; he also knows how to manœuvre when pack or growler ice draws a line across his bows. The berg in all his majesty can also be more readily avoided when the man on the bridge understands something of wind and its pressures. Then leeway is a big factor in all craft.

The law of storms which includes the tropical cyclones will be a subject in itself, but what has been covered in this article will, it is hoped, act as an incentive to young seamen to take up the study of air motion—wind.

A scale for wind tabulation known as the "Beaufort's Scale" has long been laid down in books treating on meteorology, but its value to the rising generation of seamen has deteriorated owing to the passing era of canvas as a means of propulsion.

To give an approximate idea of wind forces and velocities the calm is, of course, the index position or reading with ciphers denoting absence of motion. A gentle breeze blows at a velocity of say 10 miles an hour and has almost a force of half a pound to the square foot.

A moderate wind may have a rate of 15 miles an hour and presses on a surface at about one pound to the square foot.

The moderate gale runs along at 30 miles an hour and registers five pounds to the square foot.

A gale that would make the ordinary tramp steamer perhaps "heave to" would be blowing at 50 or 60 miles an hour and might have a pressure of 15 pounds to the square foot.

In hurricanes with a velocity of over 70 miles an hour the seaman is entirely at the mercy of the elements—both air and water—and a ship may or may not be under proper command.

When studying "windology" from your ship's deck, you will at once become familiar with aerial shipping — the clouds. These moisture carriers are, of course, sailing at different altitudes according to their cargoes, but the sailor's most noted cloud is called "scud," as it is the lowest and moves like a well driven smoke over his frail craft. Our scientists call scud by a more dignified name cirrus.

It will be noted that clouds in the upper regions may be heading in an opposite direction to the wind and lower clouds. What does this prove? Simply that air as well as water has counter currents or motions. Where this counter movement is well defined it gives a sailor, especially if his craft is in the temperate zones, a remindor that either a change of wind in direction and probably force, or a spell of broken weather is in the vicinity. If, however, this phenomenon is seen in the trade winds it simply denotes the upper air acting as a feeder by returning to its old source to be used over again when nature changes its height and temperature. And again let it be emphasized that temperature, or heat and cold, or elasticity and congelation, whichever you like, commands all movements of air and water, both horizontally and perpendicularly.

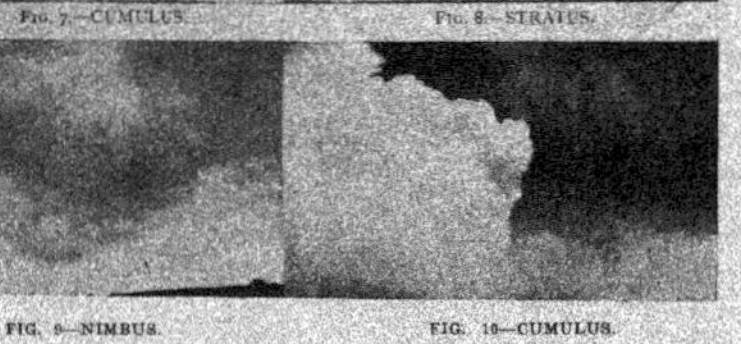

Fig. 7—Cumulus. Fig. 8—Stratus.

Fig. 9—Nimbus. Fig. 10—Cumulus.

In some later article on tending ships in tide ways, storms or anchorages, the student will more easily undestand why his vessel acts this way or that way, if he will meantime study the motions of the sea, wind and clouds.

The chart shows the different cloud systems.

Fig. 1. If low down it sometimes brings rain as with a rising southerly gale.

Fig. 2 and 8. Generally a fine weather sky.

Fig. 3 and 5. Settled for the day, but a change coming.

Fig. 4, 6 and 7. Fine cumulus (high) moderate and stormy respectively.

Fig. 9 and 10. Rain and squalls.

PROGRESS IN NEW EQUIPMENT

There is Here Provided in Compact Form a Monthly Compendium of Shipbuilding and Marine Engineering Auxiliary Product Achievement

STRAIGHT EDGING AND JOINTING MACHINE

A recently designed wood working tool seen at Toronto Exhibition apart from its mechanical merit deserves attention on account of its strictly Canadian origin and design. In designing the straight edging and jointing machine shown here the Canada Machinery Corporation, Galt, have produced a machine whose excellent mechanical design renders possible accurate and quantity production.

Former machines designed for similar work were made with an overhead movable arm supporting the saw arbor with the feed chains travelling in the saw table and grooved to clear the saw blade. Many disadvantages resulted from the overhead suspension of the saw arbor, cuttings were extremely hard to dispose of properly and short stock was difficult to saw without contrivances apt to get out of order. The placing of the saw arbor below the table and the making of the machine into an under-cutting type at once removes these difficulties and provides a machine which is capable of sawing pieces as short as 7 in. and which permits of the feeding of one piece immediately after another so that the cutting of the saw is continuous.

STRAIGHT EDGING AND JOINTING MACHINE.

To saw pieces as short as 7 in. in the older types of machines a smaller blade was necessarily required and, as is well known from experience, a small saw requires a much higher speed and more power to equal the performance of a moderately sized saw and permits of a better cutting angle for the saw teeth.

The drive from the countershaft is extremely powerful, an 8x8 in. pulley being provided for driving the saw mandrel. In consideration of the fact that an adequately supported and well proportioned mandrel is essential to good work the mandrel has been made 1 15-16 in. diameter and is supported in three long 7 in. bearings. An adjustable and thrust is provided for the taking up of end play resulting from wear. The mandrel or saw arbor runs as quietly and as smoothly as a shaper spindle.

In the mounting of the saw blade on the arbor considerable attention has been given to the securing of a strong and rigid connection. 7 in. collars are provided on each side of the blade permitting the use of thin saws and the saving of stock. The saw arbor is carried through these collars and the assembly is securely fastened by a 1 1-8 in. nut on the end. Coned bushings are provided which automatically center the saw in the correct position on the arbor. The provision which has been made for removing saws is an especially commendable feature. Saws can be removed and replaced with the utmost facility, the time taken being no longer than that required to remove blade from an ordinary rip saw. A movable section of the table, on being actuated by a crank, carries with it to one side one of the feed chains and exposes the whole saw mounting to view.

The feed consists of two travelling chains with serrated surface, one on either side of the saw blade. These chains are supported in long ways provided with adjustment for wear and are cleaned and oiled at each revolution. The truth of cut depends entirely on the travel of the chains whose vertical height is adjustable to suit either rough or smooth lumber.

The stock is held to the feed chains by heavy feed rolls swing suspended and under the action of heavy feed springs. The feed mechanism is driven by a combination of cone pulleys and reduction gears from the saw arbor. Four changes of feed, 50, 75, 115, and 175 ft. per minute, are available through four step cone pulleys actuated by a handwheel. A handle is provided for instantly disengaging the feed.

The production may be estimated from machines in use which are ripping 1 inch kiln dried maple at a rate of 115 ft. per minute and 2 in. similar stock at 75 ft. per minute.

The machine when used with a planer tooth saw will take a cut so straight and smooth that the stock can be glued up without further dressing, and one machine can break out as much stock for table tops, dressers, etc., as 4 or 5 hand-feed rip saws. When stock is being edged by hand theh operator must keep his blade within the stock, thus wasting lumber, whereas the 6 x 11 saw will dress along the edge without gouging.

In the design of this machine the operator's safety has been the first consideration and under no circumstance is it possible for the operator's hands to come into proximity to the saw blade, and it is impossible once the stock is engaged in the feed mechanism for it to be thrown back on the operator. The efficiency and comfort of the operator is also provided for by the sawdust hood which, on account of the saw being driven from below, carries away all dust and trimmings.

The capacity of the machine is such that it will take stock 29 in. wide and in thicknesses up to 3 in. with a 14 in. blade.

The table and feed rolls are raised and lowered by sq. thread screws equipped with ball bearings to take end thrust. The fence is instantly adjustable and is of the self-locking type and is provided with a graduated scale on the front bracket.

A workman of average intelligence will in a few days become quite proficient in the operation of this machine.

In one plant two of these saws are in operation On each saw is a boy feeding in stock and a girl taking stock away. This work was formerly done by experienced men, using band rip saws and hand-feed circular saws.

NEW HORIZONTAL BORING MILL

The demand for an ever-increasing supply of large machine tools by the different government departments and private manufacturers has been met by machine tool makers in a very satisfactory manner. A type of machine called for very largely is the horizontal boring mill of the floor type. In this connection we illustrate this month a mill manufactured by the Giddings & Lewis Mfg. Co., of Fond du Lac, Wis., and described by them as No. 4 floor type boring, drilling and milling machine.

The machine has been designed with a view to adaptability, and the operations that can be performed on it include boring, drilling, milling, tapping, threading, facing, turning and slotting. The accompanying cut shows clearly the general arrangement of the machine. Following are some of the salient points in the makeup of the machine.

Floor Plate on Runway

The large floor plate is firmly bolted and doweled to the runway, both of which are supplied with generous-sized surfaces resting on the foundation. They are cross-ribbed and have metal properly distributed, thus making them strong enough to remain in perfect alignment after being properly installed. The "T" slots of the floor plate are conveniently positioned to receive the anchor bolts securing the work and they are machined out of the solid metal. The floor plate itself is machined all over, affording convenience in gauging and aligning the work.

The Column

Great care has been taken to furnish the necessary strength for this important part. It is well ribbed, and has metal correctly distributed to withstand all unusual strains to which it is subjected. The base or bottom surface is supplied with an unusual spread in all directions where it rides the generous-sized ways of the bed. Lost motion between these parts is eliminated by means of long taper and square-locked gibs.

The End Support

While not subjected to the major strains such as the column, this part is made unusually strong in order to perform its functions when used as the outer support for long boring bars, etc. When necessary to use, and it is possible, this end support should be placed as close to the work as convenient. This is accomplished by sliding the piece over the top of the floor plate, it being guided by a long key. It is entirely independent from the rest of the machine and can be easily removed when it is desired to place extra large work on the floor plate. All of the adjustments on this unit are made by hand in the most convenient and accurate manner possible.

The Headstock

The headstock is of boxed design and unusual strength, furnishing perfect support to all of the moving parts. It has very large vertical and horizontal dimensions on the face of the column. Rigidity is obtained by means of two taper square-locked gibs. The face or outside surface is supplied with a cover which is removed to adjust the spindle sleeve bearing and other inside adjustments. Located on the top and within reach of the operator are oil reservoirs which supply lubrication for all of the bearings. From the bottom of this unit is suspended the operator's platform, which is designed to ride with the headstock at any position. When the headstock is at its lowest extreme the platform supports telescope and allow it to ride the ways of the bed. This platform places the operator directly before and convenient to all controlling mechanism and makes it possible to actively control the machine at all times, thus keeping it continually in the cut and making maximum efficiency possible.

The Spindle and Sleeve

The centre of the heat-treated, hammered high carbon steel spindle is positioned unusually close to the face of the column, eliminating an undesirable overhang. It is ground to exact size and the front end is bored to receive the Morse taper shank, tang, and drift key, while the back end is equipped with a ball thrust bearing, through which the feed is transmitted by the use of an extra long ram, carrying the rack, which disengages with the pinion at both extremes. The power is transmitted to the spindle from the sleeve by means of two long spindled keys, diametrically opposite. At each end of the sleeve a take-up collet is provided and by its proper adjustment the slide of the spindle is made snug, thereby allowing for precision alignment. The sleeve, like the spindle, is made of heat-treated, hammered, high carbon steel, and is of unusual length and strength. It is rigidly supported in two generous-sized adjustable bronze bearings placed far apart. Each of these bearings is independently adjusted for taking up wear, thus eliminating all lost motion. The front face of this sleeve is prepared to receive milling cutters and attachments.

To the spindle sleeve are secured two large driving bull gears. The front or face gear, being the larger, receives its power through the sliding back gear shaft. The power is transmitted through the variable speed unit located at the upper right-hand corner of the headstock. This selective gear speed unit contains heat-treated steel clash gears, cut with stub teeth to provide additional strength and ease of operation. The high speed shafts in this unit are ball seated and all heavy-duty shafts have generous-sized phosphor bronze bearings.

NEW HORIZONTAL BORING MILL

NAVAL ARCHITECTURE COURSE OPEN TO WOMEN STUDENTS

The Committee on Public Information, division on women's war work, issues the following:

A special short course in naval architecture, of not less than six weeks, is offered by six coeducational universities and technical schools in the United States.

The course is given at the suggestion of the United States Civil Service Commission in order to increase the supply of ship draftsmen so needed by the Government at the present time.

It is open to senior students in technical courses or graduates of technical schools. After six weeks of intensive training the graduates will be eligible for the lowest grade of ship-drafting position under the Navy Department. The plan is to develop them in the Government drafting rooms.

Universities and schools are urged by the commission to open this course to women, since it is largely to the women that the Government must look to supply the increasing demand for ship draftsmen.

The schools which now offer such a course in naval architecture are: Massachusetts Institute of Technology, University of Michigan, Pennsylvania State College, University of California, University of Washington, and University of Texas.

Mr. A. R. Dufresne, assistant chief engineer of the Department of Public Works, has resigned to accept the position of manager of the St. John Dry Dock and Shipbuilding Company of St. John, N.B. This company is under contract with the government to construct a dry dock of the largest class, also a breakwater, and to dredge an extensive channel and basin in St. John harbor.

THE STEERING OF SHIPS

By M. M.

All ships must possess the power to manœuvre, but exactly to what extent will depend on the type of the vessel and the use for which it is intended. Although all vessels possess the power to manœuvre it can hardly be said that the majority of ships are really easy to handle. It is true they are handled and handled effectively, but nevertheless captains often wish that they had more control over their vessels than is given them even by twin propellers and the ordinary rudder.

It will not be without interest to examine what takes place when helm is given to a ship. As the rudder at first goes over, the ship for the moment continues on her course and there is a sudden concentration of water between the rudder and the deadwood aft. This sets up an increase of pressure on both the rudder and the deadwood which pushes away the stern of the ship in the opposite direction to which the rudder is turning. The ship also moves bodily outwards. The instantaneous effect therefore is to move the ship along a course which is curved in the opposite way to that in which the ship is required to turn finally. In a short time the ship takes up a definite, but not really steady swing. This swing is helped by the pressure on the bow, the excess pressure on the deadwood aft being reduced. Shortly after this the vessel settles down to a steady swing, the pressures on bow and the rudder turning her, but the pressure on the deadwood aft is now on the opposite side to what it was originally, with the result that it retards the turning of the vessel. Equilibrium must eventually be established when the middle line of the ship takes up a definite angle to the direction in which the centre of gravity of the ship is travelling. This angle is called the drift angle. The distance between the original course of the vessel and the position of the ship when she is moving in exactly the opposite direction to her original one is called the tactical diameter of the vessel. If this is to be small the deadwood aft should be well cut away.

When the ship settles down on her turning circle, about the centre of which she rotates, there is some point—usually well forward of amidships—on the vessel which only has a motion along the middle line, every other point on the vessel really moving in some other direction. This point is called the pivoting point; and the resistance of the various parts under water to turning depend on their distance from this pivoting point. Since the pivoting point is forward of amidships, it follows that the aft deadwood is more effective in reducing turning than the forward deadwood.

When the rudder is first put over, the centre of pressure on it is below the centre of pressure of the force opposing the lateral motion of the ship, and in consequence the vessel at first heels towards the centre of the turning circle. When steady motion is established, centrifugal force acts on the vessel through a point generally above the water line and certainly above the centre of lateral resistance. This force is more powerful than the pressure on the rudder. with the result that the vessel heels outwards. Although this is very generally true, it would be possible to conceive of a case where the pressure on the rudder was so great and relatively high, and the centre of gravity of the ship, through which the centrifugal force acts, so low, that the ship might heel inwards on the turning circle instead of outwards.

It is, of course, well known that wind will affect the steering of a ship. If she is moving with the wind on the beam, the centre of pressure of the wind force on the above-water portion may be forward of abaft the centre of lateral resistance of the under-water portion. In any case, helm will have to be carried one way or another to correct the tendency of the wind to turn the ship. This will always decrease the speed of the vessel. In one particular case, it so happened that the centre of pressure of wind was abaft the centre of lateral resistance, the deadwood aft was cut away, bringing the latter point further forward, making matters worse, so that a good deal of helm had to be carried with a beam-wind.

It is generally understood that wind can affect the speed of a ship a good deal. If the wind is directly ahead, it will retard the motion of a ship considerably by direct pressure, although it will not affect the helm. If it is on either bow, it will not only retard the speed on account of its direct pressure, but also by the fact that helm will have to be carried to keep the vessel straight. With wind directly on the beam, helm will always practically be carried, and the speed of the ship will be retarded on this account, although the wind pressure has no direct effect.

Rudders are divided into several classes. The most common form is the ordinary merchantile rudder in which the whole area of the rudder is abaft the axis of rotation. For many years the most common type of rudder in war vessels has been the balanced rudder. This takes several different forms. It may be completely balanced and supported by the rudder head and a bottom pintle, or it may be completely balanced and also completely underhung and supported from two points on the rudder stock. There is another form of rudder described as semi-balanced, in which a small portion only of the rudder area is forward of the axis, the rudder being pivoted on the rudderhead and on one or more pintles, the portion of the rudder below the bottom pintle being completely underhung.

The ordinary mercantile form of rudder in general use because it is easily handled, although it is not so economical in form as some of the other types speeds of merchant vessels being generally small, does not make the rudder unmanageable in size. The steering gear for it has to be larger and heavier than the more effective rudder of the balanced or semi-balanced type, all of its area being abaft the axis, the twisting forces acting on it are much greater than with the latter type. For vessels with cruiser sterns — which includes practically all war vessels—the balanced type of rudder becomes almost a necessity, although in the last few years certain merchant vessels fitted with cruiser sterns have still been given the ordinary merchant type of rudder, and it is doubtful if there is any reason to depart from this form in general practice. If particularly rapid manoeuvring is required there may be some reason for it.

There is not a very accurate way of working up the strength of rudders from first principles, as the forces acting on them have never been very accuratel[y] determined. Formulæ are used for thi[s] purpose in certain cases which are admittedly comparative. For the majorit[y] of merchant vessels the necessary rudde[r] sizes are all given in the rules of th[e] registration societies. It can hardly b[e] said that a rudder is particularly effective in controlling a ship, in fact, i[f] specially delicate manoeuvring is required in a vessel, twin screws mus[t] always be fitted to assist the rudde[r].

JACK OF THE ROYAL NAVY

By "The Skipper."

Who leads a strenuous disciplined life
At home, abroad, on every sea?
Who drills with boats and cutlass knife
That we may never bow the knee
To enemies who fain would say
"By powers divine we lead the way?"
Jack of the Royal Navy.

Who fights below in hulls of steel
'Neath lee of armour plate,
And waits perchance that sickening heel,
That "list" which means his fate,
With superstructure all aflare
And screws still running in the air?
Jack of the Royal Navy.

When luckier ships return to port
And chronicle der tag,
Who calls their own part naval sport
And looks aloft at Britain's flag,
Whispering in his loved one's ears
"To the ship that sank you owe the cheers?"
Jack of the Royal Navy.

JACK OF THE MERCHANT NAVY

By "The Skipper."

Who echoes from the stok'hol' floor
The chief's last order "give her hell!"
Whips "diamonds" through the furnace door
With powerful will, yet knowing well
That "Bill," the underwater brute,
Is now approaching inch by inch
With aim "divine"—a pirate's loot,
One Prussian gloat, "Here is a cinch,"
"A ship of wounded British skunks,"
We'll kill them easily in their bunks,"
"Just aim right for that blazing cross."
She's gone—a Berlin prize—an Allied loss?
Firemen of the Merchant Navy.

Who cruises round the homeland coast
Or hikes for "Gib" or Montreal
In floating coffins which are lost
Through "Bill's" presumptuous, heathenish gall
Who peers through salted gale and fog
From fo'c'sle head or bridge or mast?
Who steers or sounds or reads the log
To bring our Empire through the blast?
Who takes the U-boats' bloody way
With calm, determined counter-stroke?
And bravely waits until the day
When sway of pirates will be broke?
A.B.'s of the Merchant Navy.

Fort William, Ont.—A contract fo[r] twenty-five steel vessels for the Ameri[can] Government has been secured b[y] the Canadian Car and Foundry Co.

Work in Progress in Canadian Shipyards

A record of Work in Process of Completion—Principal Features of Specification—Approximate Launching Dates—Vessels on Order.

BRITISH COLUMBIA

YARROWS, LIMITED, ESQUIMALT, B.C.

One stern-wheel steamer, 165 ft. x 35 ft., for Indian Government service.

At present engaged chiefly on repair work.

H. VOLLMERS, NANAIMO, B.C.

One gasoline tug, 9 tons.

NEW WESTMINSTER CONSTRUCTION & ENGINEERING CO., NEW WESTMINSTER, B.C.

Four wooden steamers, Nos. 10-11-12-13, for undisclosed interests. 250 ft. x 43 ft. 6 in. x 25 ft., moulded depth, 2,800 tons d. w., 1,000 h.p., single screw engines, 12 knots, naval architect, I. Alexander. Launchings March 14, April 14, May 15 and June 15, 1918.

B.C. CONSTRUCTION & ENG. CO., POPLAR ISLAND, NEW WESTMINSTER, B.C.

Four wooden vessels, 1,800 gr. tons, 2,800 tons d.w.

NEW WESTMINSTER MARINE CO., FT. OF FURNESS ST., NEW WESTMINSTER, B.C.

STAR SHIPYARD CO., NEW WESTMINSTER, B.C.

J. A. CROLL, PORT ALBERNI, B.C.

PACIFIC CONSTRUCTION CO., PORT CCOQUITLAM, B.C.

Taken over the plant of the Coquitlam Shipbuildiug Co.

Gasoline yacht, 50 ft., June, 1918.

Three wooden freight steamers, Nos. 20-21-22, 250 ft. b. p. x 42 ft. 6 in. x 25 ft., 1,800 gr. tons, 2,800 tons d.w., trip. exp. engines, water tube boilers ,12 knots, 1,000 h.p. Two for undisclosed interests, one for builder's account. Launchings May 20, June 15, Sept., 1918.

Four new berths under construction for steel ships up to 10,000 tons.

C. E. BAINTER, PRINCE RUPERT, B.C.

Several wooden fishing steamers, for undisclosed interests.

GRAND TRUNK DRYDOCK SHIP REPAIR CO., LTD., PRINCE RUPERT, B.C.

S. A. MOULTON, PRINCE RUPERT, B.C.

Ten composite vessels for undisclosed interests.

BRITISH-AMERICAN SHIPBUILDING & ENG. CO., VANCOUVER, B.C.

Establishing plant on Kitsilano Reserve.

Twenty wooden vessels, being built for undisclosed interests, 3,500 tons each.

JOHN COUGHLAN & SONS, N. VANCOUVER, B.C.

"War Camp," sister ship to "Alaska," launched in March, 1918.

One steel freight steamer, 427 ft. 9 in. x 54 ft. x 29 ft. 9 in., draft 24 ft. 2 in., speed 11½ knots, 5,730 gross tons. Launched Jan. 10, 1918, for B. Stolt Nielsen, Christiania, Norway. Both sold to the Cunard Line.

Three steel steamers, "War Chariot," "War Chief," and "War Noble." Nos. 4-5-6, 425 ft. x 54 ft., 8,800 tons cap., for the Cunard Line.

Six steel freight steamers, 425 ft. x 54 ft. x 24 ft. 2 in., 8,800 tons cap., 1 knots, for undisclosed interests. Delivery Jan., Feb., March, May and July, 1918.

One steel freight steamer, "Alaska," 427 ft. 9 in. x 54 ft. x 29 ft. 9 in., draft 24 ft. 2 in., speed 11½ knots, 5,730 gr. tons, 8,800 d.w. Launched Jan. 10, 1918, for B. Stolt Nielsen, Christiania, Norway. Sold to the Cunard Line.

THE FOUNDATION CCO., VANCOUVER, B.C.

FRASER VALLEY SHIPBUILDING CO., VANCOUVER, B.C.

GRANT, SMITH & CO., VANCOUVER, B.C.

Six wooden cargo vessels, 250 ft. x 43 ft. 6 in. x 25 ft., 3,000 tons cap., 9½ knots, for the Canadian Government.

HARRISON & LAMOND, SHIPBUILDERS, LTD., VANCOUVER, B.C.

One wooden auxiliary schooner, 225 ft. x 44 ft. x 21 ft. 4 in., 1,600 gr. tons, 2,550 tons cap., for undisclosed interests.

LYALL SHIPBUILDING CO., NORTH VANCOUVER, B.C.

"War Puget," Feb. 16, 1918, 250 ft., wood, 2,080 tons.

Ten wooden cargo ships, for undisclosed interests. Launchings Jan. 20, Mar. 2, Mar. 23, and April 11, 1918.

Building six wooden vessels for own account.

"War Caribou," April 10, 1918, 3,080tons, wood.

W. R. MANCHION, VANCOUVER, B.C.

STANDARD SHIPBUILDING CO., VANCOUVER, B.C.

Two 3,500-ton vessels for Brazilian Govt., and six for French Govt. of 4,500 tons.

Donohoe reinforced type, wood composite construction.

TAYLOR ENGINEERING CO., VANCOUVER, B.C.

Number of small vessels being constructed at total value of $300,000. 4,500-ton floating drydock, 352 ft. long.

VANCOUVER SHIPYARD, LTD., VANCOUVER, B.C.

One motor freighter, 125 ft. x 24 ft. x 12 ft., wood, 160 h.p., Bolinder's crude oil engine. To be completed Aug., for Taylor Engr. Co.

Also repairing several steamers and schooners.

THE WALLACE SHIPYARDS, LTD., N. VANCOUVER, B.C.

"War Dog," 4,500 tons, steel, May 18, 1918.

"War Power," 4,600 tons, steel, March 23, 1918.

Two steel steamers, 315 ft. x 45 ft. x 27 ft., 4,600 tons d. w., trip. exp. vertical engines ,two S. E. Scotch boilers, 10 knots, single screw, for the Canadian Government. Deliveries Dec., 1917, and Aug., 1918.

Two wooden aux. schooners, 255 ft. o. a., 225 ft. keel x 44 ft. x 21 ft. 4 in., five masts, 1,500 gr. tons, cap. 2,500 tons or 1,500,000 ft. lumber, two hatches, twin 160 h.p, Bolinders Hot-bulb engines, twin screws, for the Canadian Government.

Four wooden cargo vessels, aggregate tonnage of 17,500 tons for undisclosed interests.

WESTERN CANADA SHIPYARDS ,LTD., VANCOUVER, B.C.

"War Nootka," Jan. 4, 1918, 3,080 tons (machinery installed at Ogden Point Assembly Sheds).

Six wooden freight steamers, 250 ft. h.p., 259 ft. o. a., 43½ ft. extreme breadth, molded 42½ ft., 25 ft. molded depth., draft 22 ft., 2,800 tons d. w., for undisclosed interests. "War Nootka," launched January 4, 1918. "War Selkirk" launched March 6, 1918.

"War Tatla," "War Casco, "War Chilkat," "War Tanoo," 4 standard vessels now under construction.

CAMERON-GENOA MILLS CO., VANCOUVER, B.C.

"War Yukon," Jan. 24, 1918, 3,080 tons, wood (machinery installed at Ogden Point Assembling Sheds.

THE FOUNDATION CO. OF B. C., LTD., VICTORIA, B.C.

Building ships for undisclosed interests.

Five wooden steamers, 250 b. p. x 42 ft. 6 in. x 25 ft., 1,800 gr. tons, 2,800 tons d. w., one trip. exp. engine, 1,000 h.p., "Howden" water tube boilers, furnished and installed by owners, 10 knots, for undisclosed interests. Esplen & Son & McNaught, N.Y.C., designers.

No. 1—"War Sonchee," launched Dec. 27, 1917.

No. 2—"War Masset," launched April 11, 1918.

No. 3—"War Babine," ready for launching, awaiting propeller and rudder.

No. 4—"War Camchin," ready for launching, awaiting propeller and rudder.

No. 5—"War Nanoose," ready for launching in about five weeks.

CLARENCE HOARD, VICTORIA, B.C.

Wooden car barge for C.P.R., Jan., 1918.

VICTORIA SHIPBUILDING CCO., VICTORIA, B.C.

Constructing wooden ships for British Govt., 3,500 tons each.

NEW BRUNSWICK

C. T. WHITE & SON, LTD., ALMA, N.B. (Office at Sussex, N.B., also.)

Two schooners aux. power, 143 ft., 900 tons d. w., first to be launched in April; second in June.

JAMES X, LENTEIGNE, LOWER CARAQUET, N.B.

One schooner, 28 tons, wood.

INTERNATIONAL SHIPBUILDING CO., NEWCASTLE, N.B.

Two four-masted wooden auxiliary schooners, 155 ft. keel, 37 ft. depth, 535 net tons. One to be completed in September and the other December, 1918. For builders' account.

EUREKA SHIPBUILDING CO., NORTH HEAD, N.B.

One wooden schooner, "Mollie & Melba," 350 tons reg.

PORT COLBORNE BUILDING & REALTY CO., LTD., REXTON, N.B. (Head Office, Welland, Ont.)

Four-masted wooden schooner. Has cap. of 3 schooners per year.

GRANT & HORNE, ST. JOHN, N.B.

Two wooden cargo steamers, "War Fundy" (launched Aug., 1918) and "War Digby," 250 ft. x 42 ft. 5 in. x 25 ft. 5 in., 1,000 h.p., speed 9½ knots, 2,800 tons d.w., for undisclosed interests.

MARINE CONSTRUCTION CO., CANADA, LTD., ST. JOHN, N.B.

Four wooden auxiliary, four-masted schooners, 185 ft. o. a., 165 ft. keel x 40 ft., 1,100 d. w. tons, for builder's account. J. Murray Watts, Philadelphia, naval architect. Launching April, 1918.

One wooden schooner, "Dorfontein," 740 reg. tons. Launching June, 1918.

PETER AND A. A. McINTYRE, ST. JOHN, N.B.

One schooner, 425 tons, 137 ft., three masts, wood, for builder's account. To be launched in June. Another to be commenced immediately.

ST. JOHN SHIPBUILDING CO., ST. JOHN, N.B. (Plant at Courtenay Bay.)

Ten five-masted auxl. schooners, oil burning engines, for undisclosed interests.

Establishing plant for steel and wood construction.

ST. MARTIN'S SHIPBUILDING CO., ST. MARTIN'S, N.B.

One wooden schooner, 450 tons, three masts. To be launched in July for builder's account.

NEWFOUNDLAND

ANGLO-NEWFOUNDLAND DEVELOPMENT CO., BOTWOOD, NFLD.

Two three-mast aux. schrs., wood, 450 tons, 150 H.P. for the builders' account.

NEWFOUNDLAND SHIPBUILDING CO., HARBOR GRACE, NFLD.

To build ships of 1,200 tons d.w. Steel ships later.

M. E. MARTIN, MORRIS ARM, NR. ST. JOHN'S, NFLD.

Two wooden schooners, 300 tons each, for builder's account.

One wooden schooner, 500 tons, for builder's account.

UNION SHIPBUILDING CO., ST. JOHN, NFLD.

ANNAPOLIS SHIPPING CO., ANNAPOLIS ROYAL, N.S.

"Hilda M. Clark," March 4, 1918, 187 ft., 640 tons reg.

Two schrs. three-masted, wood 500 tons, 170 ft. long. To be completed August and December, 1918.

NOVA SCOTIA

B. L. TUCKER, BASS RIVER, N.S.

One schooner, 350 tons, wood.

JAS. S. CREELMAN, BASS RIVER, N.S.

One three-masted schooner, wood, 170 ft. long, 500 tons. To be completed October, 1918, for builders' account.

HANKINSON SHIPBUILDING CO., BELLIVEAU'S COVE, N.S.
One schooner, 360 tons, wood.

A. A. THERIAULT, BELLIVEAU'S COVE, N.S.

Two wooden three-masted tern schooners, 150 ft. x 33 ft. x 11 ft. 11 in., 400 gr. tons, gasoline engines, for builder's account. Launchings April and June, 1918.

One wooden one-masted schr.

WESTPORT SHIPBUILDING CO., BELLIVEAU'S COVE, N.S.

One wooden schooner, 230 tons net. To be launched in October for builder's account.

ELI PUBLICOVER, BLANDFORD, N.S.

BRIDGEWATER SHIPPING Co., BRIDGEWATER, N.S.

One tern schooner, 450 tons gr., wood. Adapted for aux. power. To be completed Sept., 1918.

One tern schooner, 275 tons gr., wood. Adapted for aux. power. To be completed Nov., 1918.

Both built on owner's account and for sale.

W. A. NAUGLER, BRIDGEWATER, N.S.

One wooden three-masted schooner, "Wm. A. Naugler," 146 ft. o. a., 116 ft. keel x 32 ft. x 11 ft. 6 in., 360 gr. tons, 600 tons cap., for builder's account.

One schooner, 300 tons, wood.

H. MAC ALONEY, CANNING, N.S.

Two wooden schooners, 400 tons each. For builder's account. Launchings July and October, 1918.

S. M. FIELDS, CAPE D'OR, N.S.

One three-masted schooner, 350 tons.
Keel laid for another three-master.

CHESTER BASIN SHIPBUILDERS, LTD., CHESTER BASIN, N.S.

Two schooners, 100 tons, wood. Launchings Aug., 1917, Dec., 1917.
One four-masted schooner, 60 tons. Launching Sept., 1918.
One two-masted schooner, 100 tons. Launching June, 1918.

MORTIMER PARSONS, CHEVERIE, N.S.

One wooden three-masted schooner, 170 ft. x 38 ft. x 14 ft., 575 gr. tons, 900 tons cap., for builder's account. Launching Sept., 1918.

Two wooden three-masted schooners for builder's account.

One tern schooner, "Donald Parsons," 140 ft. keel x 34 ft. 4 in. x 12 ft. 4 in., 500 gross tons, 825 tons d.w. for Mortimer Parsons. Launched May 11, 1918.

Another vessel, same size, to be launched Nov., 1918, for builder's account.

BAY SHORE SHIPYARD, CHURCH POINT, N.S.

Building two wooden vessels for R. C. Elkin, Ltd., St. John, N.B.

MOISE BELLIVEAU, CHURCH POINT, N.S.

One schooner, 450 tons, wood.

FIDELE BOUDREAU, CHURCH POINT, N.S.

One schooner, 350 tons, wood.

J. E. GASKILL, CHURCH POINT, N.S.

Other yards at Grosses Coques, Little Brook, Meteghan and Port Wade, N.S.

Five wooden schooners from 344 to 387 reg. tons. All sister ships.

COMEAU SHIPBUILDING CO., COMEAUVILLE, N.S.

One three-masted wooden schooner, for builder's account, 450 tons.

J. W. COMEAU, COMEAUVILLE, N.S.

One schooner, 329 tons, Jan., 1918.

J. N. RAFUSE & SONS, CONQUERALL BANKS, DIGBY CO., N.S. (Other plants at Salmon River and Shelburne, N.S.)

Two wooden three-masted schooners, 120 ft. x 32 ft. x 12 ft., 700 tons cap., for builder's account. For the market. Launchings Oct. 25, 1917, and Jan. 10., 1918.

One three-masted wooden schooner, "Integra," 112 ft. x 30 ft. x 11 ft. 6 in., 600 tons cap., for J. O. Williams & Co., St. John's, Newfoundland. Launching Nov. 25, 1917.

E. F. WILLIAMS, DARMOUTH, N.S.

One schooner, 360 tons, wood.

ROBAR BROTHERS, DAYSPRING, N.S.

Schooner for Capt. Ivan Creaser, Dayspring.

MAURICE E. LEARY, DAYSPRING, N.S.

One wooden schooner, 225 tons, 135 ft. o. a. To be launched Aug. 15, 1918, for La Have Outfitting Co., La Have, N.S.

J. NEWTON PUGSLEY & CHAS. ROBERTSON, DILIGENT RIVER, N.S.

One schooner, three-masted, 475 tons.

McLEAN & McKAY, ECONOMY, N.S.

One tern schooner, wood, 400 tons net, 135 ft. keel. To be launched Sept. 1, 1918, for builders' account.

S. J. SOLEY, ROX RIVER, N.S.

One schooner, three-masted, 121 ft. keel, 300 tons. Built of native wood. To be launched Sept., 1918. For builder's account. For sale.

ALLAN & FRASER, FRASERVILLE, N.S.

One schooner, 350 tons, wood.

BERNARD W. MELANSON, GILBERT'S COVE, N.S.

One three-masted schooner, wood, 250 tons net. To be completed November, 1018, for builder's account and for sale.

AMOS BLINN, GROSSES COQUES, N.S.

One schooner, 375 tons. Jan., 1918.
One schooner, 350 tons, wood.

BINN BROS., GROSSES COQUES, N.S.

F. K. WARREN & CO., GROSSES COQUES, N.S. (Offices in Union Bank Chambers, Halifax, N.S., also.)

350-ton tern schooner, under construction Jan., 1918.

HALIFAX SHIPBUILDING CO., HALIFAX, N.S.

FAUQUIER & PORTER, HANTSPORT, N.S.

Two wooden four-masted schooners, 178 ft. x 40 ft. x 18 ft., 1,000 gross tons, 2,000 tons d. w., twin 100 h.p. oil engines. Launchings Aug. 1, 1918, and Sept. 1, 1918. Both for sale.

J. WILLARD SMITH, HILLSBURN, N.S. (Head Office, St. John, N.B.)

One tern schooner, wood, 140 ft. keel, 35 ft. beam, 450 tons net reg. tons.

HENRY COVEY, INDIAN HARBOR, N.S.

CHARLES GRIFFIN, ISAACS HARBOR, N.S.

One schooner, 40 tons, wood.

LEWIS SHIPBUILDING CO., LEWISTON, N.S.

One schooner, 670 tons. Jan., 1918.

INNOCENT COMEAU, LITTLE BROOK, N.S.

One three-masted schoner, 140 ft. x 35 ft. 6 in. x 17 ft. 6 in., 650 gr. tons, for the Weymouth Shipping Co., Weymouth, N.S., Can. Delivered Jan. 16, 1918.

J. W. RAYMOND, LITTLE BROOK, N.S.

One wooden schooner, No. 4, 140 ft. x 36 ft. x 17 ft., 575 gr. tons, for Jones Bros., Weymouth, Digby Co., N.S.

H. A. FRANK, LIVERPOOL, N.S.

Two wooden schooners, 500 tons, for builder's account.

McKEAN SHIPBUILDING CO., LIVERPOOL, N.S.

One wooden three-masted schooner, 126 ft. keel x 3 ft. x 12 ft. 9 in., 700 tons d.w., for builder's account. Launching May, 1918.

W. F. McKEAN & CO., LIVERPOOL, N.S.

One schooner 400 tons, Jan., 1918.

D. C. MULHALL, LIVERPOOL, N.S.

Two wooden schooners, one 300 gross tons, other 400 gross tons, for undisclosed interests.

N.S. SHIPBUILDING & TRANS. CO., LIVERPOOL, N.S.

One wooden tern schooner No. 2, 122 ft. x 33 ft. x 12 ft. 6 in., 450 gr. tons, 70 d.w. tons, 7 knots, for Peter Yee Wing & Co., Ltd., Sydney, Australia. R. McLeod and J. S. Gardner. Launching Feb. 1, 1918

One two-masted fishing schooner, "Sadie A. Nickle," 130 ft. o. a. x 26 ft. 10 ft. 6 in. hold, 230 d.w., two 50 h.p. oil engines, for Rafuse & Sons. Launching May 15, 1918.

One three-mast schooner, 420 tons. F-M Co. oil burning engines, for Job Bros., St. Johns, Nfld. Launching July 15.

One three-masted schooner, 270 tons, for builders' account. Launching Sept. 1.

One two-masted fishing schooner, 120 tons, for builders' account. Launching Oct. 1.

ROBIN, JONES & WHITMAN, LIVERPOOL, N.S.

One schooner, 340 tons, wood.

SOUTHERN SALVAGE CO., LIVERPOOL, N.S.

One wooden schooner, three masts, 300 gross tons, 600 tons capacity, for a West Indian firm. Launching Jan., 1918.

One wooden steamer, No. 5, 250 ft. x 43 ft. 6 in. x 25 ft., 2,900 gr. tons, 10 knots, 1,000 h.p., for undisclosed interests. Launching Fall, 1918.

One two-masted schooner, "Win-the-War," 137 ft. o. a. x 25 ft. 2 in. x 11 ft. 6 in., 187 gr. tons, for the builder's account. Launched Nov., 1917.

CONRAD & REINHARDT, LUNENBURG, N.S.

LUNENBURG MARINE RAILWAY, LUNENBURG, N.S.

FRED A. ROBAR, LUNENBURG, N.S.

SMITH & RHULAND, LUNENBURG, N.S.

Two schrs., 225 tons each. Jan., 1918.
Two schooners, 100 tons each.
"Donald Cook," 112 ft. beam, depth of hold, 11 ft., for Capt. William Cook. Launched April, 1918.
Two schooners, 168 gr. tons, for W. C. Smith.
One schooner, 150 gr. tons, for Lun. Outfitting Co.

J. B. YOUNG, LUNENBURG, N.S.

ERNST SHIPBUILDING CO., MAHONE BAY, N.S.

One two-masted schooner, 120 ft. o. a. x 25 ft. 6 in. x 10 ft. 6 in., 200 tons d.w., for Lunenburg Outfitting Co. April, 1918. Christened "Madeline Adams."

One three-masted schooner 125 ft. keel x 32 ft. x 12 ft., 600 tons, d. w. To be delivered July, 1918.

Two masted schooner, "Agnes D. McGlaston," 130 ft. x 26 ft. 8 in. x 10 ft. 8 in. Launched Nov. 13, 1917.

Keel laid for three-masted wooden schooner, 200 tons.

J. ERNEST & SON, MAHONE BAY, N.S.

One schooner, 520 tons. Jan., 1918.

O. A. HAM, MAHONE BAY, N.S.

One schooner, "Dozle," 128 tons. Launching May 30. One motor boat about 20 tons.

McLEAN CONSTRUCTION CO., MAHONE BAY, N.S. (Leased to Montague Mahaffy, Toronto.)

One three-masted schooner, 325 tons gr. reg., 500 tons cap., West Indies freight type. Will be launched in July.

Keel laid for another three-masted schooner with cargo capacity of 425 tons.

JOHN McLEAN & SONS, MAHONE BAY, N.S.

One schooner, 95 tons, wood.

J. A. BALSOM CO., LTD., MARGARETSVILLE, N.S.

One schooner, 409 tons. Jan., 1918.

CLARE SHIPBUILDING CO., METEGHAN RIVER, N.S.

One wooden schooner for builder's account. 400 tons.

A. H. COMEAU & CO., METEGHAN, N.S.

One schooner, 400 tons, wood.

AGAPIT COMEAU, METEGHAN, N.S.

JOHN F. DEVEAU, METEGHAN, N.S.

A 300-ton schooner was launched recently for Ritcey & Co., Lunenburg, N.S., and named Charles A. Ritcey.
1 schooner, 425 tons. Jan., 1918.
1 schooner, 400 tons, wood.

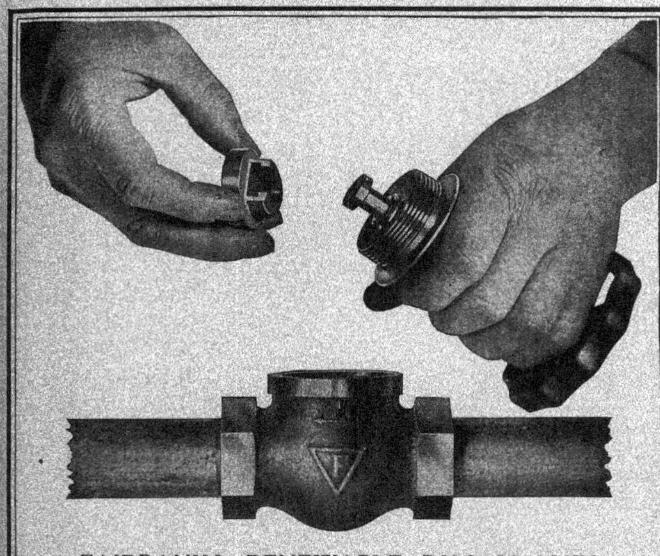

FAIRBANKS RENEWABLE DISC VALVES

WOULD REDUCE YOUR VALVE EXPENSE EVEN THOUGH THEY COST DOUBLE THE PRESENT PRICE. THE BAKELITE DISC GIVES MAXIMUM SERVICE UNDER ALL CONDITIONS OF STEAM, AIR, AND WATER. WHEN THE DISC DOES WEAR OUT, IT CAN BE REPLACED IN LESS THAN ONE MINUTE WITH ONLY ONE TOOL—A WRENCH TO REMOVE THE BONNET. SPECIFY FAIRBANKS VALVES ON YOUR ORDER.

THE CANADIAN FAIRBANKS-MORSE COMPANY, LIMITED

CANADA'S DEPARTMENTAL HOUSE FOR MECHANICAL GOODS

DEPARTMENTS

SCALE, VALVE, STEAM GOODS, ELECTRICAL, AUTOMOBILE SUPPLY, TRANSMISSION, OIL ENGINE, MACHINERY, PUMP, CONTRACTORS, RAILWAY, MACHINE SHOP SUPPLY

SALES OFFICES

HALIFAX, ST. JOHN, QUEBEC, MONTREAL, OTTAWA, TORONTO, HAMILTON, WINDSOR, WINNIPEG, SASKATOON, CALGARY, VANCOUVER, VICTORIA

J. E. GASKILL, METEGHAN, N.S.

THOMAS GERMAN, METEGHAN, N.S.

One schooner, 350 tons, wood.

R. H. HOWES CONSTRUCTION CO., METEGHAN, N.S. (Leased James Cosman's shipyards.)

Building several wooden schooners for own account.

DR. F. H. MACDONALD, METEGHAN, N.S.

One wooden four-masted schooner, "Rebecca L. Macdonald," 201 ft. o. a. x 6 ft. x 16 ft., 800 gr. tons, 1,500 tons d.w., for builder's account. Launching January 1, 1918.

One wooden fishing schooner, No. 3, 84 ft. x 17 ft. x 7 ft., 50 gr. tons, 5 h.p. Launching August, 1918.

One schooner, 544 tons, wood.

METEGHAN RAILWAY & SHIPBUILDING CO., METEGHAN, N.S.

One wooden three-masted schooner, 182 ft. x 35 ft., 450 gr. tons.

One schooner, 470 tons, wood.

CHAS. McNEIL, NEW GLASGOW, N.S.

One wooden 3-masted schooner, 103 ft. x 30 ft. x 10 ft. 9 in., 200 tons, for builder's account. Launching July, 1918.

One wooden 3-masted schooner, 142 ft. x 25 ft. 4 in. x 13 ft., 500 gr. tons, for builder's account. Launching Oct., 1918.

THE NOVA SCOTIA STEEL & COAL CO., NEW GLASGOW, N.S.

Two steel freight steamers, 257 ft. 9 in. x 35 ft. x 20 ft., 1,700 gr. tons, 2,350 tons cap., trip. exp. engines, boilers 10 ft. 6 in. x 11 ft. 6 in., 800 h.p., 8.5 knots, for undisclosed interests. Launching of one Jan, 1918.

Two steel cargo strs. Raised quarter deck type, 248 ft. 9 in. long, 1,649.68 gross tons. Triple expansion engines, 17, 28, 46 x 33 in. stroke. One, the "War Bee," for undisclosed interests to be completed June, 1918. Other for Steel Co., to be completed Oct., 1918.

Two masted schooner, launched May 16, 130 ft. o. a. x 26 ft. x 10 ft. 6 in. Designed by J. S. Gardner.

O'BRIEN BROS., NOEL, N.S.

One schooner, 325 tons, wood.

W. R. HUNTLEY & SON, PARRSBORO, N.S.

One wooden schooner, 175 ft. x 39 ft. x 17 ft., 900 gross tons, for

C. T. WHITE & SON, SUSSEX, N.S.

Two schooners, 325 tons each. Jan., 1918.

One schooner, 490 tons, wood.

One schooner, 850 tons, four masted.

WAGSTAFF & HATFIELD, PARRSBORO, N.S.

One schooner, 400 tons. Jan., 1918.

SIDNEY ST. C. JONES, PLYMPTON, N.S. (Head Office, Weymouth North, N.S.)

One tern schooner, 200 tons, net reg., 107 ft. x 28 ft. 8 in. x 10 ft. 3 in., to be launched Nov., 1918.

S. SALTER, PARRSBORO, N.S.

One 200-ton schooner, wood.

DOWLING & STODDART, PORT CLYDE, N.S.

One gas boat, 27 tons, wood.

One schooner, 300 tons, wood.

SWIME BROS., PORT CLYDE, N.S.

Several small motor trawlers, for builder's account.

G. M. COCHRANE, PORT GREVILLE, N.S.

One tern schooner, "Alfredock Hedley," 152 ft. 6 in. x 36 ft. 12 ft. 6 in., 461 tons reg., for Adam B. Mackay, of Hamilton, Ont. Launched Jan., 1918.

One 4-masted schooner, 850 tons, 155 ft. x 37 ft. x 18 ft., 2 decks, wood. To be completed Oct., 1918, for builder's account.

H. ELDERKIN & CO., PORT GREVILLE, N.S.

One wooden three-masted schooner, for undisclosed interests.

ELLIOTT GRAHAM, PORT GREVILLE, N.S.

Is building a schooner, "Khaki Lad," of about 325 tons, at Port Greville, N.S., for J. W. Kirkpatrick and others. She will be launched early in October, and is reported to have been sold to Newfoundland parties.

One schooner, 360 tons. Jan., 1918.

SMITH CANNING CO., PORT GREVILLE, N.S.

One schooner, 350 tons, wood.

WAGSTAFF & HATFIELD, PORT GREVILLE, N.S.

One schooner, 400 tons, wood.

WILLIAM CROWELL, PORT LATOUR, N.S.

J. W. RAYMOND, PORT MAITLAND, N.S.

One schooner, 375 tons. Jan., 1918.

PORT WADE SHIPBUILDING CO., PORT WADE, N.S. (Head Office, Digby.)

One 350-ton schooner, wood.

JOHN BROWN, PUBLIC LANDING, N.S.

One tow barge, 50 tons, wood.

THE CUMBERLAND SHIPBUILDING CO., PUGWASH, N.S.

Establishing plant for the construction of wooden ships.

MACKENZIE SHIPPING CO., RIVER JOHN, N.S.

One four-masted schooner, 600 tons, wood. About half built. Fitted for aux. engines.

CHARLES McLELLAN, RIVER JOHN, N.S.

One schooner, 100 tons, wood.

W. J. FOLEY, SALMON RIVER, N.S.

Building for J. N. Rafuse & Sons one wooden, 300 tons, under construction. Cap. of plant, 1,000 tons yearly.

J. N. RAFUSE & SONS, SALMON RIVER, N.S. (Other plants at Shelburne and Conquerall Banks.)

Launched recently a three-masted schooner named "Industrial," at W. J. Foley's ship yard at Salmon River, N.S., of the following dimensions: length, 113 ft. breadth, 30 ft., depth, 11½ ft., 325 tons.

ACADIAN SHIPPING CO., SAULNIERVILLE, N.S.

One schooner, 90 tons, wood.

SAULNIERVILLE SHIPBUILDING CO., LTD., SAULNIERVILLE, N.S.

J. LEWIS & SONS, SHEET HARBOR, N.S.

One schooner, wood, four-masted, 725 gr. tons. To be completed Sept., 1918. For private account.

GEO. A. COX, SHELBURNE, N.S.

One schooner, 200 tons, wood.

One schooner, 322 tons, wood.

JOSEPH McGILL SHIPBUILDING & TRANSPORTATION CO., SHELBURNE, N.S.

"Sparkling Glance," 246 tons register, for Harvey & Co., St. John's, Nfld.

One schooner, 160 tons, wood.

W. C. McKAY & SON, SHELBURNE, N.S.

One wooden schooner, 130 ft. o. a. x 26½ ft. x 11 ft. depth of hold, 140 tons register. Launched Dec. 1, 1917.

One three-masted schooner, for Messrs. Hallet, of Newfoundland.

One schooner, 620 tons, wood.

J. N. RAFUSE & SONS, SHELBURNE, N.S. (Other plants at Salmon River and Conquerall Banks.)

One wooden three-masted schooner, 120 ft. x 32 ft. x 12 ft., 700 tons cap., for builder's account. For the market. Launching Jan. 10, 1918.

One wooden three-masted schooner, 118 ft. x 33 ft. x 12 ft., 700 tons cap., for builder's account. For the market. Launching Jan. 15, 1918.

Two three-masted wooden schooners, 122 ft. 6 in. keel x 32 ft. x 12 ft., 275 net tons, for J. O. Williams & Co., St. John's, Newfoundland. One to be built at Salmon River yard and one at Ship Harbor.

THE SHELBURNE SHIPBUILDERS, LTD., SHELBURNE, N.S.

One tern schooner, "Misty Star," 145 ft. x 31 ft. x 11 ft. 6 in., 600 tons d.w., 330 tons gross, for Harvey & Co., St. John's, Newfoundland. Launching Mar. 15, 1918.

One tern schooner, 154 ft. x 32 ft. x 12 ft. 4 in., 400 gr. tons, 700 tons d.w. cap. Launching June, 1918. Not sold yet.

Two schooners, one 349 tons, other 400 tons.

EASTERN SHIPBUILDING CO., SHIP HARBOR, N.S.

One wooden schooner, 600 tons d.w., for J. N. Rafuse & Sons, Halifax, N.S., Can.

One wooden schooner, 300 tons.

STEPHEN MORASH & CO., SHIP HARBOR, N.S.

One wooden schooner, 150 ft. x 37 ft. x 13 ft., four masts, for Canadian interests. Keel laid Feb. 5, 1917.

JAMES E. PETTIS, SPENCER'S ISLAND, N.S.

One schooner, 425 tons, wood, three-masted.

CAPE BRETON SHIPBUILDING CO., SYDNEY, N.S.

One three-masted schooner.

THE TUSKET SHIPBUILDING CO., TUSKET, N.S.

AMOS H. STEVENS, TANCOOK, N.S.

ALVIN STEVENS, TANCOOK, N.S.

STANLEY MASON, TANCOOK, N.S.

ALBERT PARSONS, WALTON, N.S.

One schooner, 400 tons. Jan., 1918.

H. T. LeBLANC, WEDGEPORT, N.S.

One steam trawler, 165 ft., 600 tons, for J. N. Rafuse & Sons.

L. S. CANNING, WARDS BROOK, N.S.

One tern schooner, 350 tons, for W. C. Smith & Co., Lunenburg, N.S.

WEDGEPORT NAVIGATION & TRANSPORTATION CO., WEDGEPORT, N.S.

T. K. BENTLEY, WEST ADVOCATE, N.S.

One 200-ton steamer.

One 4-masted schooner, wood, 511 tons. To be launched September or October. For owner's account.

J. W. KIRKPATRICK, WEST ADVOCATE, N.S.

One 350-ton schooner, wood.

BOEHNER BROS., WEST LA HAVE, N.S.

Two large fishing schooners, 150 tons each.

One beam trawler, 300 tons, crude oil engines. For local interests.

BEAZLEY BROS., WEYMOUTH, N.S. (Head Office Roy Bldg. Halifax.)

One schooner, 394 tons, three masts.

E. R. GAUDET, WEYMOUTH, N.S.

One 350-ton schooner, three-masted.

E. P. RICE, WEYMOUTH, N.S.

One three-masted schooner, 350 tons.

RICE, WARREN & CO., WEYMOUTH, N.S.

One three-masted schooner, 334 tons.

FALMOUTH SHIPBUILDING & TRANSPORTATION CO., WINDSOR, N.S.

One wooden schooner, 350 gr. tons, 700 tons cap., for undisclosed interests.

NOEL SHIPBUILDING & TRANSPORTATION CO., WINDSOR, N.S.

One wooden schooner, three masts, 138 ft. x 35 ft. x 13 ft., 450 net reg. tons. Oil auxiliary engines. Launching August, 1918, for builder's account.

C. A. NICKERSON, WOODS HARBOR, N.S.

MILTON SHIPBUILDING CO., YARMOUTH, N.S.

Taken over and enlarging old plant of James Jenkins.

350-ton schooner building.

W. O. SWEENY, YARMOUTH, N.S.

100-ton fishing schooner.

YARMOUTH SHIPBUILDING CO., YARMOUTH, N.S.

One 450-ton wooden schooner, 128 ft. keel. To be completed Oct., 1918.

ONTARIO

CAN. ALLIS-CHAMBERS (Head Office, Toronto), BRIDGEBURG, ONT.

Six standard freight steamers, 3,500 tons, 261 feet long, for undisclosed interests.

COLLINGWOOD SHIPBUILDING CO., COLLINGWOOD, ONT.

Two steel cargo steamers, Nos. 51-52, 50 ft. x 43 ft. x 25 ft., 2,500 gr. tons, 2,900 tons d. w., triple expansion engines 18-30-50 x 36, 2 boilers 14 ft. x 11 ft., 10 knots, for undisclosed interests. Delivery May and Aug., 1918.

Four deep-sea trawlers, Nos. 53-58, inclusive, 125 ft. x 23 ft. 6 in. x 13 ft. 6 in., 288 gross tons, 12¾-21½-35/24 engines, 1 cylinder 13 ft. 6 in. x 10 ft. 6 in., 10 knots, 500 h.p. for undisclosed interests.

One steel freight steamer of 3,800 tons d. w. for undisclosed interests. Keel laid May 8, 1918.

"War Wizard," May 8, 1918, 3,000 tons d. w., 261 ft. x 43 ft. 6 in. x 20 ft. depth. For undisclosed interests.

"War Witch" same size, now building. To be launched ——

R. MORRILL, COLLINGWOOD, ONT.

"Windsor," Aug. 10, steam tug, 105 ft., for Ontario Gravel & Freighting Co.

CAN. CAR & FOUNDRY CO., FORT WILLIAM, ONT. (Head Office Transportation Bldg., Montreal.)

Twelve steel mine sweepers for French Government.
145 ft. o.a. steel construction, value $2,500,000.
Plant under construction.

GREAT LAKES DREDGING CO., FORT WILLIAM, ONT.

Two wooden vessels, 2700 tons d.w. cap., 260 ft. o.a., 43 ft. beam. Triple expansion engines of 1,000 h.p. To be launched November, 1918.

"War Sioux" launched May 12, 1918. Keel laid for another twenty minutes later. Launching scheduled for November.

THUNDER BAY CONTRACTING CO., FORT WILLIAM, ONT.

One wooden freight steamer, 261 ft. long, for undisclosed interests.

NATIONAL SHIPBUILDING CO., GODERICH, ONT.

DAVIS DRYDOCK CO., FT. OF BAY ST., KINGSTON, ONT.

One motor pass. boat, 64 ft. x 18 ft. x 6 ft., 52 gr. tons, wood. For ferry service, Gananoque to Clayton. Delivery June 1, 1918.

Twenty-eight lifeboats, regulation size, wood and metal for Government and private interests.

KINGSTON SHIPBUILDING CO., KINGSTON, ONT.

Several steel trawlers for undisclosed interests. First one launched Dec. 22, 1917.

SELBY & YOULDSON, KINGSTON, ONT.

GEORGIAN BAY SHIPBUILDING & WRECKING CO., MIDLAND, ONT.

Tug, 50 tons, wood.

One tug 40 tons, wood.

MIDLAND DRY DOCK CO., MIDLAND, ONT.

Three steel freight steamers, 261 ft. x 43 ft. 6 in. x 23 ft., 3,400 tons d. w., 10 knots, for undisclosed interests. Deliveries, one in July, 1918, and two others before close of navigation, 1918.

Alterations on steamers "Mariska" and "Glenlyon."

"Western Star" being reconstructed.

MIDLAND SHIPBUILDING CO., LTD., MIDLAND, ONT.

Three steel freight steamers, 261 ft. x 43 ft., 6 ft x 23 ft., 3,400 tons d. w., 10 knots for undisclosed interests. Deliveries, one in July, 1918, and two others before opening of navigation, 1919.

PORT ARTHUR SHIPBUILDING CO., PORT ARTHUR, ONT.

Six steel trawlers on order. 135 ft. o. a. x 23 ft. 4 in. x 15 ft. 1 in., 294.5 tons, trip. exp. engines, 500 h.p., single end Scotch boiler. Two-masted.

Two of these launched June 8, 1918.

"War Isis," April 3, 1918, 3,400 tons d. w. Trawler launched same day. April 3, 1918, keel laid for "War Hatha," sister ship to "War Osiris," of same construction, nearly completed.

"War Fish," Aug. 19, 1917, 4,300 tons, steel.

"War Dance," Nov. 3, 1917, 3,400 tons, steel.

Four steel screw steamers, ocean freight services—"War Osiris," "War Hathor," "War Karma," "War Horus," 3,400 tons d. w. each. 261 ft. o.a. 251 ft. b.p. in length. Triple expansion engines, 2 Scotch boilers each, approx. 1,250 h.p. For completion, two Aug. 31st, and two Nov. '20. 1918. For undisclosed interests.

MUIR BROS. DRYDOCK CO., LTD., PORT DALHOUSIE, ONT.

J. W. GEROW, ROSSPORT, ONT.

REID WRECKING CO., SARNIA, ONT.

One steel tug, 157 ft. x 32 ft. x 19 ft., trip. exp. engines, for lake or ocean service. Keel blocks laid March 5. Launching July, 1918. For Reid Wrecking Co

WEST, PEACHEY & SONS, SIMCOOE, ONT.

Number of tugs for lumbering interests.

DOMINION SHIPBUILDING CO., TORONTO, ONT.

One steel cargo vessel, 261 ft. x 43 ft. x 28 ft. 2 in., 3,500 tons d.w. cap. For Department of Marine, Ottawa.
Have orders for five other vessels same dimensions.
"Traja" launched May 15, 1918.

THE POLSON IRON WORKS, LTD., TORONTO, ONT.

Eight steel cargo steamers, Nos. 133-4-5-6-7-9-40, 216 ft. o. a., 251 ft. b. p. x 43 ft. 6 in. x 22 ft. 11 in., draft loaded 19 ft. 6 in., single decked. 2,350 approx. gross tons, 3,500 tons cap., single screw trip, exp. engine 20½-33-54 x 36, 1,250 h.p., 10 knots, two 14 ft. x 12 ft. Scotch R. T. boilers, 180 h.p., to cost $600,000 each, for undisclosed interests. Launchings throughout 1918.

Four building and four on order.

"Asp," Feb. 11, 1918, 261 ft., 3,500 tons d.w., steel.

"Tento," Oct. 22, 1917, 261 ft., 3,500 tons d.w. steel. For Norwegian interests. Transferred to British registry.

THE THOR IRON WORKS, TORONTO, ONT.

Under same management as Dominion Iron Works

TORONTO SHIPBUILDING CO., LTD., TORONTO, ONT. (Toronto Dry Dock Co., Ltd., under same management.)

Two wooden cargo steamers, 250 ft. B.P. 42 ft. 6 in., moulded breadth 26 ft., moulded depth 22 ft., draft 2,500 tons d. w. Triple exp. engines $\frac{20 \times 33 \times 54}{40}$. Howden boiler. Launchings July and Sept., 1918. For undisclosed interests.

THE BRITISH AMERICAN SHIPBUILDING CO., WELLAND, ONT.

Two steel frs., 3,500 tons d. w., 261 ft. O.A., 43 ft. beam, 23 ft. moulded depth, Westinghouse steam turbine engines. Delivery 1918.

WELLAND SHIPBUILDING CO., WELLAND, ONT.

One steel freight steamer, "War Weasel," 261 ft. x 43 ft. 6 in. x 23 ft., 3,300 tons d. w., trip. exp. engines, 14 ft. x 12 in. Scotch boilers, 10 knots, 1,250 h. p. Launching April, 1918.

Two deep frame cargo vessels, "War Badger" and "No. 3," 261 ft. x 43 ft. 6 in. x 23 ft., 3,300 d. w. tons, geared turbines, Howden's forced draught boilers, 10 knots, 1,250 h.p. Launchings June and August, 1918.

PRINCE EDWARD ISLAND

THE CARDIGAN SHIPBUILDING PLANT, CARDIGAN, P.E.I. (Purchased Annandale Lumber Co. plant and removed to Cardigan.)

One three-masted schooner, 325 tons, to be completed November, 1918.
Two more to be built.

QUEBEC

TIDEWATER SHIPBUILDERS, LTD., THREE RIVERS, QUE.

(Formerly Sorel Shipbuilding & Coal Co.)

Three steel trawlers for undisclosed interests
One steel steam barge for the Canada Steamships Lines.

R. N. LE BLANC, BONAVENTURE, QUE.

J. Z. DEGAGNE, EBOULEMENTS, QUE.

DAVIE SHIPBUILDING & REPAIRING CO., LEVIS, QUE.

Six military barges, 130 feet long.
Eight steel trawlers, 2,035 tons.
One floating crane, 350 tons.
*One steel cargo vessel of 5,000 tons cap.
Building several steel lighters and several wooden drifters.

Steel car ferry "Canora" building for C.N.R., 308 ft. x 52 ft., cap. 20 loaded cars. Speed, 14 knots.

ATLAS CEMENT CONSTRUCTION CO., LTD., MONTREAL, QUE.

One concrete cargo steamer, "Concretia," 125 ft. x 22 ft. x 13 ft., steel ribbed, hull to be from 3 in. to 5 in., thick, for the builder's account.

CANADIAN VICKERS, LTD., MONTREAL, QUE.

Two cargo steamers, 9,400 tons, steel; 1 dredge, 2,364 tons, steel; 12 trawlers, 3,050 tons, steel; 23 drifters, 3,350 tons, wood.

Six freight steamers, 24 ft. draught, 7,000 tons cap., 11 knots, for undisclosed interests. Delivery, 1917, of two for Norwegian interests "Porsanger," 394 ft. 6 in. o. a. x 40 ft. 4 in. x 30 ft., triple exp. engines, launched Nov. 29, 1917.

"War Earl" launched June 8, 1918, 7,000 tons d.w., 380 ft. x 49 ft. Keel of sister ship laid five minutes after this launching.

One steel freight steamer, 2,360 tons, cap., for undisclosed interests.

Three steel cargo vessels, 8,200 tons, to be laid down in May, Aug. and Sept., 1918, for undisclosed interests.

FRASER BRACE & CO., MONTREAL, QUE.

Four wooden cargo steamers, 3,000 tons, for undisclosed interests. Two keels laid during Oct., 1917.

HALL ENGINEERING CO., MONTREAL, QUE.

MONTREAL DRY DOCK & SHIP REPAIRING CO., MONTREAL, QUE.

Operate dock 428 ft. long 30 ft. deep.

MONTREAL SHIPBUILDERS LTD., 37 Belmont St., MONTREAL, QUE. (Associated with Atlas Construction Co.

THE QUEBEC SHIPBUILDING AND REPAIR CO., Board of Trade Bldg., MONTREAL, QUE.

2 vessels for undisclosed interests 3,000 tons each.

QUINLAN & ROBERTSON, MONTREAL, QUE. (Yards at Quebec.)

Four wooden steamers, for undisclosed interests.

LOUIS GAUNDRY, 12 St. Peter St., QUEBEC, QUE.

QUEBEC SHIPBUILDING & REPAIR CO., ST. LAURENT, QUE.

2 schrs. 1,400 tons and 1,200 tons.
One wooden four-masted auxl. schooner "Martin Connolly," 228 ft. x 42 ft. x 20 ft., 2,100 tons d. w., for undisclosed interests. Launched Oct. 28, 1917.

QUINLAN & ROBERTSON, QUEBEC, QUE.

Four wooden steamers, totalling 6,400 tons, for undisclosed interests.

CANADIAN GOVERNMENT SHIPYARDS, SOREL, QUE.

One steel vessel for undisclosed interests.

LECLAIRE SHIPBUILDING CO., SOREL, QUE.

Building 6 steel ships at value of $1,500,000.
Six trawlers, 125 ft. single screw type expansion engines, 500 I.P.H. steel. To be completed this year for private acct.

H. H. SHEPHERD, SOREL, QUE.

Rebuilding seven drifters. Particulars withheld owing to govt. restrictions.

SINCENNES-McNAUGHTON LINES, SOREL, QUE.

One tug 410 tons, wood.

CAN. GOVERNMENT SHIPYARD, SOREL, QUE.

One steel vessel for undisclosed interests.
Building steam trawlers and wooden drifters for undisclosed interests.

ST. LAWRENCE SHIPBUILDING & STEEL CO., SOREL, QUE.

THE THREE RIVERS SHIPYARDS LTD., THREE RIVERS, QUE.

Two cargo steamers, 3,100 d. w., wood, 260 ft. x 43 ft. 6 in., 1,000 H.P. engines. Names, "War Mingan" and "War Nicolet." To be completed end of navigation season for undisclosed interests.

Aikenhead's Automatic Wahlstrom Chuck

is a time-saver on every chucking operation. Ordinarily it takes minutes to make each tool change, but with this Wahlstrom Chuck it takes but **two seconds.** The tool is perfectly centered automatically while spindle is running.

The one keyless, colletless chuck that never slips, because the jaws close on the entire shank of the tool in a grip that becomes firmer as resistance increases. You can use this better chuck profitably.

See the Broken Tang

See the Broken Tang

Van Dorn

Portable Electric Drills

give truly exceptional service inside or out in all weathers. Operate on either direct or alternating current—on alternating, any frequency from 20 to 60 or 80 to 125 cycles (single or split phase).

Capacity in Solid Steel

DAOOO	3-16 in.	DAI	½ in.
DAOO	¼ in.	DA1X	⅝ in.
DAOOX	5-16 in.	DA2	⅞ in.
DAO	⅜ in.	DA2X	1 in.

Bolt Stocks and Dies

In our complete stock are makes you've known favorably for years—that of the American Tap & Die Co., for instance. Other makes of the highest grade, cutting exact size or over size as required.

U-2 Universal is a large size portable electric tool room Grinder

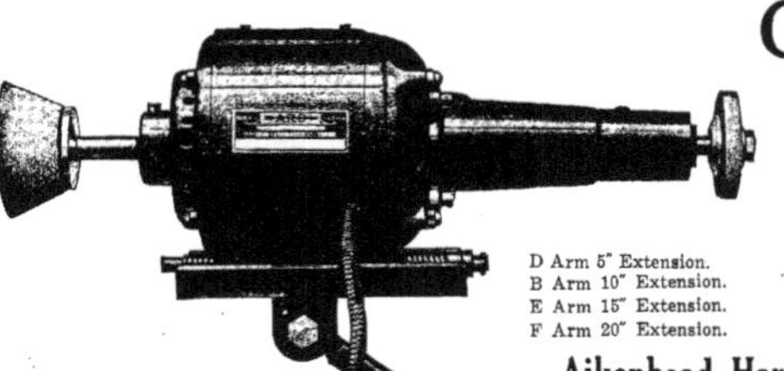

Fitted with "D" arm as illustration shows. This Grinder attached to your lathe, shaper or planer will prove a money-saver, money-making tool.

The "D" arm attachment, directly connected to armature, which runs at 10,000 r.p.m., will swing wheels up to 5" in diameter. And, arms of greater length can be substituted whenever desired.

D Arm 5" Extension.
B Arm 10" Extension.
E Arm 15" Extension.
F Arm 20" Extension.

FULL 1-3 H.P. MOTOR

Aikenhead Hardware, Limited
17, 19, 21 Temperance Street
TORONTO, CANADA

Esquimalt, B.C.—The Dominion Government are considering the building of a new dry dock.

Barrington, N. S.—The old shipbuilding industry of this port may be revived in the near future, as local financial interests are understood to be considering establishing a shipyard.

Nanaimo, B.C.—Local capital is behind a company formed here to purchase an eight-acre site and construct a three-ways shipbuilding slip to fill a contract for six steamers expected from an allied Government.

Fort William, Ont.—The Canadian Car and Foundry Co. have recently completed the construction of twelve mine sweepers at their yards here. One of these vessels, having completed her steam trials, is now on her way down the lakes to Montreal.

Victoria, B.C.—The yard recently operated by the Cameron-Genoa Mills Shipbuilders, Ltd., at Point Ellice, has been taken over by the Foundation Co. of this city. An order for 20 vessels for the French Government has been secured and three keels are now laid.

Victoria.—H. B. Pickering, manager of the Foundation Company here, has interviewed various engineering concerns in Vancouver relative to the construction of engine boilers and auxiliary machinery for part of the twenty vessels being built here to the order of the French Government. The Foundation Company will give Vancouver and Victoria firms the first opportunity to build this equipment.

Halifax.—Speaking at Halifax, Mr. Ballantyne, Minister of Marine, said that one of his objects in visiting Halifax was to consult with Admiral Story, the naval defences of Halifax and the Atlantic coast. While naturally unable to go into details, Mr. Ballantyne said he could assure the citizens that the Naval Department was fully alive to the situation.

Cleveland.—The Montreal Transportation Co. have sold to Cuban sugar interests the steamer Paipoonge and the barge Thunder Bay. The boats will be taken to the coast this fall, after being cut to pass through the Welland Canal. The Paipoonge was formerly the Corona, and the Thunder Bay the Malta, belonged to the Pittsburgh Steamship Co.

St. Johns.—Speaking at St. Johns the Minister of Marine said that the $5,000,000 mill at Sydney, N.S., now under construction, would be rolling plates next July. Its capacity would be 250,000 tons. Referring to the probability of a shipbuilding plant at St. Johns he said the Government would be willing to give any company that started contracts on the same basis as any other yard in Canada.

Annapolis Royal, N.S.—A new dry dock capable of handling vessels up to 5,000 tons is reported to be under construction here. A twin schooner, 1,100 tons, was launched in September by the Annapolis Shipbuilding Co., and another of 1,000 tons is expected to be launched in December.

A Double Event In Vancouver.—A notable record in shipbuilding annals on the West Coast was made when two vessels, totalling 13,600 tons d.w., were launched on the same day and within one hour and a half, from the yards of the Coughlan Shipbuilding Co. and Wallace Shipbuilding Co., respectively. The "War Noble," launched by the Coughland yard, is a steel freighter of 8,800 tons capacity. She is 427 feet long, 54 feet beam, and 29 feet 9 inches deep. She was launched just 63 days after laying the keel, which constitutes a Canadian record. The "War Storm," launched by the Wallace Shipbuilding Co., is a 4,800 ton steamer, 315 feet long, 45 feet beam, and 27 feet deep. Her engines are of the triple expansion reciprocating type, with cylinders 24 ft. 36 in., and 63 in. diameter by 45 in. stroke, and are to indicate 1,700 H.P. Steam will be supplied by two Scotch boilers, 15 ft. 6 in. diameter by 11 ft. 3 in. long. The engines were built by the Wallace Company and the boilers by the Vulcan Iron Works of Vancouver. The contract price for the "War Noble" was $1,452,000, or $165 per ton, while that for the "War Storm" was $984,000, or $205 per ton.

Ottawa.—Sealed tenders addressed to the undersigned and endorsed "Tender for Protective Works at Steveston, Fraser River, B.C.," will be received until 12 o'clock noon, Tuesday, November 5, 1918, for the construction of protective works to existing jetty at Steveston, at the mouth of Fraser River, District of New Westminster, British Columbia.

Plans and forms of contract can be seen and specification and forms of tender obtained at this department, at the offices of the District Engineers at New Westminster, B.C., Victoria, B.C.; at the Postoffices, Vancouver, B C., and Stevestin, B.C.

Tenders will not be considered unless made on printed forms supplied by the department, and in accorance with conditions contained therein.

Each tender must be accompanied by an accepted cheque on a chartered bank payable to the order of the Minister of Public Works, equal to 10 per cent. of the amount of the tender. War Loan Bonds of the Dominion will also be accepted as security or war bonds and cheques if required to make up an odd amount.

Note—Blue prints can be obtained at this department by depositing an accepted bank cheque for the sum of $20 payable to the order of the Minister of Public Works, which will be returned if the intending bidder submits a regular bid.

By order,
R. C. DESROCHERS,
Secretary.

Department of Public Works, Ottawa, October 2, 1918.

WASHINGTON.—The concrete ship program is well under way. All five of the Government owned yards are now ready to lay keels, and most of the yards are more than half completed. The yards are located at Wilmington, N.C., Jacksonville, Mobile, San Francisco, and San Diego, Cal. The present program provides for the building of 38 tankers and freight steamers of 7,800 tons d.w., three freighters of 3,500 tons and one of 3,000 tons deadweight. In addition to these vessels the Emergency Fleet Corporation have plans for building twenty-one 500-ton concrete barges, for inland waters, all scheduled to be delivered before December 1st this year.

CATALOGUES

Bolinder Marine Oil Engine.—This is the latest publication describing the Bolinder engines. The make-up is very attractive and the contents interesting and instructive. Detailed descriptions of the various models of engines are each accompanied by a photographic engraving, and a line drawing showing over all dimensions. Following these are illustrations of a large number of vessels of diverse types fitted with the Bolinder engine, with a short introductory to each type.

Besides being of great interest to those considering the use of oil engines the publication is to be commended for its artistic composition, the colored engravings being particularly good specimens of this art.

THE SIGNPOSTS OF THE OCEAN

By Mark Meredith

No one who has ever come into contact with the sea can go for a very long time without wishing to know something of the science of navigation, which inspires the sailor with light-hearted confidence as he sails over the pathless sea. There is little of the wonderful in the fact that an express train reaches its destination, but on the sea there are no railway lines, and nothin but a wide expanse of sea and sky to guide the mariner on his way.

Perhaps the most important man on board ship—even in these days, is the man at the wheel, for he is the guardian of the compass—or perhaps better, its servant. Unless otherwise directed, his eyes may hardly ever leave it. The ship is to be steered on a particular course shown by the compass, and that course must not be forgotten or mistaken; his attention must be concentrated on it the whole of the time and must never be relaxed. Only experience and vigilance assure the ship keeping to the right course. The "compass true" lies at the root of the sailor's confidence; and as he leaves behind him every tower, every hill and every lighthouse, to journey on an unmarked plain of waters, he knows that his compass will never fail him on the darkest night or in the foggiest weather.

But it is not enough to know the direction in which the ship is going, for the distance travelled must also be measured. If we sail from Britain westwards we know we shall come to America, but to avoid striking its shores anywhere we must know when they are being approached and where a safe harbor can be found, and so from hour to hour the speed of the ship is estimated by means of the log. This is a coil of string wound upon a reel, to the end of which is tied the log, a small piece of wood shaped and balanced to float steadily in the water; the log is thrown overboard, and noting the time that a measured length of string takes to run off the reel provides the data for the computation of the speed of the ship. The measures distances on the log-line are marked by knots; the time bears the same proportion to the hour as the knot does to the nautical mile, and hence we speak of the ship having a speed of so many knots; this estimated speed is written down in the "log book," which is the sailor's journal, for here are recorded the force and direction of the wind, the state of the weather, the course steered by compass, the nautical events of each hour of the voyage, all ships or land seen, and any changes which occur from time to time. Each day at noon the distance run in the previous twenty-four hours is added up and applied in the direction that has been steered by compass; and thus is ascertained the progress made in the voyage. This method of navigation is called "dead reckoning."

Navigation by dead reckoning is comparatively simple and easy, and may often be relied upon for short distances, but there are causes which make it untrustworthy. In different parts of the world are currently or varying strength, some running as fast as seven miles an hour. A captain may be steering west on his way to America, with fine weather and smooth water, and enter a current of sea water, many miles in width, running rapidly to the northward ;there is nothing in the appearance of the sea to tell him he is in such a current, and if he has only his dead reckoning to trust to, he may be wrecked on the coast of Newfoundland at a time when he thinks he is 500 miles away from the nearest land and 1,500 miles from New York. Or a strong southerly wind acting always on the side of a vessel will drift her bodily to the northward, although the steersman may keep her head pointing west. Or a ship may be so drifted about and twisted and turned by stormy weather and changeable winds that the dead reckoning may be entirely muddled up and lost. Then reliance must be placed on that branch of navigation called "nautical astronomy." Here, use is made of the sextant and a chronometer, and by taking observations of the sun, the position of the ship can be determined. Let us see exactly what this means. There are no marks on the sea to show where the ship is, so the chart is ruled with imaginary lines crossing each other, and dividing the blank surface of the sea into little spaces; if it is possible to find which square the ship is in it is possible to determine the progress made. The lines running east and west mark the degrees of latitude, and those running north and south the degrees of longitude; only when a sailor knows his latitude and longitude can he determine the position of his ship. If a ball of clay is held in the hand it is impossible to measure the position of any point on it, because there is no starting point on it to measure from; but if you run a long needle through the middle of it you at once get two starting points—namely, where the needle goes in and where it comes out. Such a ball is our earth, and the north and the south poles are its only natural fixed points; the circumference of the earth half way between the poles is called the equator, the lines of latitude are those parallel to the equator, and we count 90 degrees of latitude from the equator to the pole. The equator is obviously the most natural line to start from and sailors of all nationalities count their latitude from it. The lines of longitude are at right angles to those of latitude, but there seems no natural reason for beginning to count the longitude from any one place in particular rather than another, so the French start theirs from Paris, we start from Greenwich, and the Russians from Petrograd Greenwich was chosen as it is the site of the national observatory and the centre of all astronomical calculations.

To determine the position of the ship let us first consider how the latitude is found when out at sea. The sun at noon in the northern hemisphere is always due south; if we sail away towards it we find that it gets higher and higher every day at noon, until we get near the equator, when at twelve o'clock it is exactly overhead, and there is no shadow whatever, or, if we sail northwards the sun sinks lower and lower, until in the polar regions the sun is sometimes only on the horizon at noon. The sextant is an instrument for measuring angles, and when the captain comes on deck at noon it is to measure the height of the sun above the horizon. He begins to observe a few minutes before twelve o'clock and watches it slowly rising, until at last it stops and slowly begins to descend; the highest altitude that has been observed gives the latitude by a very short calculation, in which the chief element to be considered is the position of the sun itself in reference to the equator. Only twice in the year is the sun exactly above the equator, and that is at the equinoxes. In the summer it comes much nearer to us, and in winter it goes much further off. Its distance north or south of the equator is called the declination, and the path it appears to follow, crossing and recrossing the equator is called the ecliptic; therefore, if we did not calculate the declination with the altitude we should only get our distance from some point on the ecliptic, instead of our latitude, which is our distance from the equator. There is annually published at the Greenwich Royal Observatory a "nautical almanac," in which all the movements of the sun, moon, planets, and principal stars have been calculated beforehand. From this book the sailor gets the sun's declination, corrects it for the moment of noon at which he takes his observation, and so is able to derive the latitude. It often happens that the sun is obscured by clouds at noon, but we can get the latitude at night on the same principle from a medidian altitude of the moon, of a planet, or of a star. There is also a method of "double altitude," which consists in estimating from two altitudes of a heavenly body not on the meridian, what must be its altitude when it is on the meridian.

By one or other of these methods the seaman finds his latitude, but to know the position of his ship he must also know his longitude, and to find this is not so easy a matter. As the sun rises in the east, it is daylight earlier the further east you go, and the greater the difference of longitude the greater the difference of time between any two places. The two questions of time and longitude are so intimately connected that they may almost be said to be identical. The globe is divided into 360 degrees of longitude, which is all passed over by the sun in the course of 24 hours—that is at the rate of 15 degrees every hour. If, therefore, we know that there is one hour's difference of time between London and Copenhagen, it follows that Copenhagen must be 15 degrees, or 900 miles distant. If then a ship is in the Atlantic at exactly 12 o'clock,

the captain knows that he is 15 degrees west of Greenwich, and that this is his longitude. Therefore in order to find our longitude at sea we require to know two things: First, the time exactly to a second, on board the ship, and secondly, the time at the same instant and as accurately as Greenwich. Great accuracy is indispensable, because if one hour of time corresponds to 15 degrees of longitude a mistake of one minute will make a difference of 15 miles in the position of the ship.

The exact time on board ship is found by an observation of the sun usually taken at eight or nine in the morning, when it is rising rapidly. This observation, with somewhat intricate mathematic calculations, gives accurately the time which would be roughly shown on a sun-dial. It is then only necessary to know with the same accuracy, the time at Greenwich at the instant the observation was taken, and by comparing these two times we at once get the longitude. This Greenwich time may be found in various ways, but the simplest is to carry on board the ship a clock which was set to London time before the ship sailed. Such clocks are specially made to go correctly and evenly in all climates, and by ingenious contrivances they are compensated for changes of temperature. They are called chronometers. A large ship carries several of them, so that, by comparing one with the other a more accurate Greenwich time may be arrived at. They are hung in swinging cradles, so that they may not feel the rolling of the ship; even the cradles are supported from a foundation of tow or wool, that they may not suffer from any shock or jar. The chronometer-room is placed in the middle of the ship, and is kept as far as possible at a uniform temperature. Chronometers are made to go two days without being wound up, but it is usual to wind them up daily in case of accidents. In a man-of-war the sentry at the captain's cabin is not relieved in the morning till it is reported to him that the chronometers have been wound up, and if he is thus retained at his post beyond his proper term of duty he takes good care to make it known that the chronometers have not been wound up. The chronometers are all compared one with another every day, and a register is kept of their performances, from which it is at once seen which are going most steadily, and can most surely be relied upon. When a ship makes a long voyage the Greenwich times shown by the chronometers is carefully tested when she arrives in a port whose longitude is known. Thus the chronometer is of vast importance in marine work.

When the latitude and longitude have been ascertained by trustworthy observation that latitude and longitude are accepted as true, however much they may differ from the results afforded by the dead reckoning and the position of the ship is marked on the chart according to these observations. The deviation may be due to bad steering, or to the effect of a current, and in iron ships the difference may be due to deflection of the compass over and above its known deviation. This deviation is carefully ascertained for each point of the compass before the ship leaves port, and a table of deviations is supplied to the ship, whatever she is built of, but a change in the cargo of the ship may affect the compass or a different distribution of any iron on board. It is well therefore every day to check the compass bearing of the sun with its true bearing.

The foregoing are the methods generally in use among navigators when far out at sea in blue water, but the use of the lead and line, when nearing land are more commonly used in shallow water. Many parts of the sea have been so carefully surveyed that not only is the depth of water known and marked on the chart, but the nature of the bottom has also been recorded, whether it is of rock, sand, shells, clay or mud. The English Channels have been most elaborately surveyed, and in foggy weather, when no stars, moon or sun is visible, the seamen can confidently grope his way up the English Channel, trusting to the lead alone. There is a cavity filled with tallow at the end of the leaden plummet, the lead line is marked at every two or three fathoms, and when the depth has been ascertained by the line, the lead is drawn up to the surface and the nature of the bottom known by the marks on the tallow. Three or four casts of the lead will determine the position of the ship. If the first cast of the lead shows 80 fathoms of water and a sandy bottom the ship runs a mile east, and then gets 28 fathoms, with sand and shells on the bottom; the ship runs a further mile, and a third cast of the lead shows 27 fathoms, with a bottom of broken shells, the probability is that there is only one place in the Channel where such a result can be obtained, if there is any doubt a fourth cast of the lead would show the ship's position unmistakably.

But for ocean voyages it is to the heavenly bodies alone that the seamen can look for guidance. Amidst the ceaseless change of winds, and waves, and weather and stars, which to the landsman are mere beautiful objects in the night landscape, to a sailor they bring boundless confidence and security, for they are the signposts of certain accuracy.

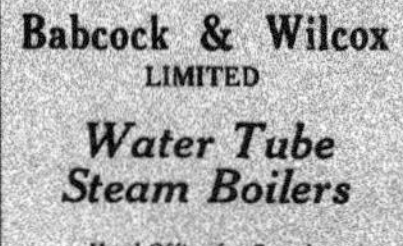

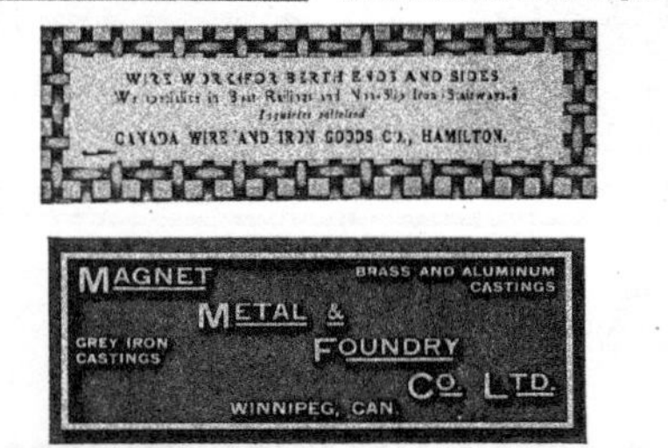

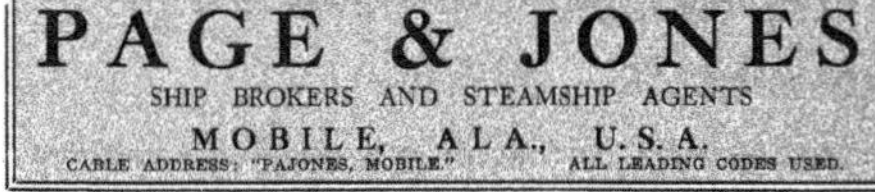

We Are Able to Meet Shipbuilders'

immediate needs because of our modern manufacturing facilities.

There are no long waits for delivery, and this is a big consideration in these strenuous times of delays.

We can give speedy delivery on:—

VERTICAL DUPLEX MARINE PUMPS
MARINE REDUCING VALVES
DEXTER VALVE RE-SEATING MACHINES*
ROCHESTER LUBRICATORS
COMBINED AIR AND CIRCULATING PUMPS

DARLING BROS, LIMITED, 120 Prince Street
MONTREAL

Vancouver Calgary Winnipeg Toronto Halifax

**These are a real asset to any marine engineer*

SHIPS BELLS

Made from Pure Bell Metal
Complete with Attachments

C. O. CLARK & BROS.

1510 ST. PATRICK STREET MONTREAL, QUE.

MILLER BROS. & SONS
LIMITED

120 Dalhousie St. Montreal

GREY IRON CASTINGS
SHIPS WINCHES

MARINE MACHINERY

Steam Pumps
Power Pumps
Pump Strainers
Hoisting Engines

Extra Heavy Flanges
Anchor Windlass
Steering Engine

6 x 6 Anchor Windlass
For French Mine Sweepers and Trawlers

THE BAWDEN MACHINE CO., LIMITED, 163 Sterling Road, TORONTO

LET US QUOTE

On Your Requirements in

Sheet Metal Stampings For Ships' Equipment

A history of continuous development since 1861 in the sheet metal stamping industry is a guarantee of our ability to give you first-class service.

The Pedlar Plant at Oshawa is one of the largest sheet metal plants in the British Empire. In equipment and personnel it is second to none. Our batteries of heavy presses and hammers are capable of handling all classes of work. We are at present engaged on important contracts for

SHIPS' VENTILATING COWLS
METAL RAFTS and METAL LIFEBOATS

In normal times we produce a great deal of automobile body stampings, our work in this line including the bodies of many widely-known makes of car.

Oshawa, where our plant is located, being on the main line of three transcontinental railway systems, offers also unexcelled facilities for prompt shipment to any point.

Write us for quotations on the stampings you need

The Pedlar People Limited

(Established 1861)

HEAD OFFICE AND FACTORY : OSHAWA, ONT.

Branches at Montreal, Ottawa, Toronto, London, Winnipeg, Vancouver

If what you need is not advertised, consult our Buyers' Directory and write advert isers listed under proper heading.

THE WAGER FURNACE BRIDGE WALL
A preferred and valuable feature in marine and stationary boilers—endorsed by governments; steamship, freight and passenger steamer companies; railroads; stationary plants; others.
WAGER FURNACE BRIDGE WALL CO., Inc.
OF NEW YORK SINGER BUILDING
Philadelphia, Detroit, Seattle, Portland
San Francisco Vancouver, B. C.

Castings

WILKINSON & KOMPASS
TORONTO HAMILTON WINNIPEG
IRON AND STEEL
HEAVY HARDWARE
MILL SUPPLIES
AUTOMOBILE ACCESSORIES
WE SHIP PROMPTLY

Mason No. 55 Style Pump Pressure Regulator
Reilly Multi-Screen Feed Water Filter
Mason No. 126 Style Marine Reducing Valve
MASON
The value of a ship is judged by its service. That service depends upon its equipment. Equipment of the Mason quality is a guarantee of reliable service. The service of our staff of expert engineers has also been of value to others; can they be of service to you? We are steam room specialists. Our line is complete. Let us get in touch with you.
We manufacture Reilly Evaporators, Reilly Distillers, Reilly Feed Heaters, Reilly Feed Water Filters, and furnish Mason Reducing Valves for Marine Service.
We are the sole Canadian manufacturers of these auxiliaries, and make a specialty of prompt deliveries.
Our new shop gives us ample facilities for quick service.
Reilly Navy Type Feed Water Heater
The
Mason Regulator & Engineering Company, Limited
135-153 Dagenais St. MONTREAL
506 Kent Building TORONTO
Reilly Marine Evaporator Submerged Type

Fast, Efficient Service

We are prepared to supply the shipbuilding trade promptly with good quality Grey Iron and Brass Castings. Any size--any quantity.

MARINE BOILERS AND ENGINES
PARTS AND FITTINGS

Get in touch with us. Enquiries and orders given prompt attention.

Waterous

BRANTFORD, ONTARIO, CANADA

Serve by Saving!

GET RID of your old, cumbersome and **dangerous** blocks and tackles. Don't keep a hundred years behind the times. Equip your vessels with the simplest, fastest and **safest** life-boat-handling devices known:

J-H Lifeboat Windlasses & Rapid Releasing Hooks

With J-H Windlasses two men can raise a loaded lifeboat, and one man can safely control its descent. J-H Windlasses are equipped with steel cable—no ropes to kink, rot or burn.

J-H Releasing Hooks **insure the instant release of both ends of lifeboat. J-H Hooks are used on lifeboats of the United States Emergency Fleet, and hundreds of others. Write to-day for illustrated pamphlet.**

ECKLIFF CIRCULATORS

are guaranteed to create proper circulation in Scotch boilers. That's a guarantee of higher efficiency and lower expense. Write to-day for folder.

ECKLIFF CIRCULATOR CO.

62 Shelby Street - Detroit, Michigan

280

Over 6,000,000 Tons

of new ships under construction will have high pressure piping flanged by the **LOVEKIN METHOD.**

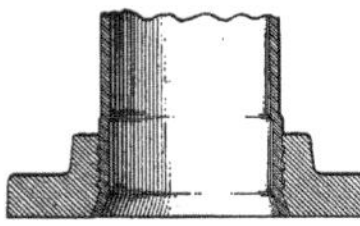

The Lovekin Flange

Our machines not only reduce labor and material costs and produce the strongest and most uniform joint obtainable—but SPEED-UP PRODUCTION as well.

Used by many large plants and all navy yards. Write us for list and further information.

LOVEKIN PIPE EXPANDING AND FLANGING MACHINE CO.

521 Phila. Bank Bldg. - PHILADELPHIA, PA.

If any advertisement interests you, tear it out now and place with letters to be answered.

If what you need is not advertised, consult our Buyers' Directory and write advertisers listed under proper heading.

BUYERS' DIRECTORY

ACCUMULATORS, HYDRAULIC
- Smart-Turner Mach. Co., Hamilton, Ont.

AERATING RESERVOIRS
- Spray Engineering Co., Boston, Mass.

AIR PORTS
- Mitchell Co., The Robert, Montreal, Que.
- Turnbull Elevator Mfg. Co., Toronto, Ont.

ALLOYS, BRASS AND COPPER
- Dom. Copper Products Co., Ltd., Montreal, Que.
- Mitchell Co., The Robert, Montreal, Que.
- Mueller Mfg. Co. H., Sarnia, Ont.
- Tallman Brass & Metal Co., Hamilton, Ont.
- United Brass & Lead, Ltd., Toronto, Ont.

ANCHORS
- Beauchemin & Fils, Sorel, P.Q.
- Hopkins & Co., F. H., Montreal, Que.
- Kearfott Engineering Co., New York, N.Y.
- McNab Co., Bridgeport, Conn.
- Henry Rogers Sons & Co., Wolverhampton, Eng.
- Wm. C. Wilson & Co., Toronto, Ont.

ARCHITECT, NAVAL
- Watts, J. Murray, Philadelphia, Pa.

ASBESTOS GOODS
- Wm. C. Wilson & Co., Toronto, Ont.

BABBITT METAL
- Aikenhead Hardware, Ltd., Toronto, Ont.
- Hoyt Metal Company, Toronto, Ontario.
- Tallman Brass & Metal Co., Hamilton, Ont.
- Wilkinson & Kompass, Hamilton, Ont.
- Wm. C. Wilson & Co., Toronto, Ont.

BAROMETERS
- Wilson & Co., Wm. C., Toronto, Canada.

BAR IRON
- Mitchells Ltd., Glasgow, Scotland.

BARS, GRATE
- Babcock & Wilcox, Ltd., Montreal, Que.

BEARINGS, BRASS
- Empire Mfg. Co., London, Ont.
- Mueller Mfg. Co., H., Sarnia, Ont.
- Mitchell Co., The Robert, Montreal, Que.
- Tallman Brass & Metal Co., Hamilton, Ont.
- United Brass & Lead Co., Toronto, Ont.

BELLS, SHIPS, ENGINE ROOM, ETC.
- Aikenhead Hardware, Ltd., Toronto, Ont.
- Clarke & Bros. Co., Montreal, Que.
- Cory & Son, Inc., Chas. New York, N.Y.
- Empire Mfg. Co., London, Ont.
- Mitchell Co., The Robert, Montreal, Que.
- Morrison Brass Mfg. Co., James, Toronto, Ont.
- Mueller Mfg. Co., H., Sarnia, Ont.
- Tallman Brass & Metal Co., Hamilton, Ont.
- United Brass & Lead, Ltd., Toronto, Ont.

BELTING, LEATHER
- Aikenhead Hardware, Ltd., Toronto, Ont.
- Wm. C. Wilson & Co., Toronto, Ont.

BIBBS, COMPRESSION
- Empire Mfg. Co., London, Ont.
- Mitchell Co., The Robert, Montreal, Que.
- Mueller Mfg. Co., H., Sarnia, Ont.
- United Brass & Lead, Ltd., Toronto, Ont.

BINNACLES
- Hopkins & Co., F. H., Montreal, Que.
- Morrison Brass Mfg. Co., James, Toronto, Ont.

BLOCKS, CARGO, HEEL AND TACKLE
- Aikenhead Hardware, Ltd., Toronto, Ont.
- Hopkins & Co., F. H., Montreal, Que.
- Loveridge, Ltd., Docks, Cardiff, Wales.

BLOWERS, TURBO-
- Mason Regulator & Engin. Co., Montreal, Que.

BOAT CHOCKS
- Corbet Fdry. & Machine Co., Owen Sound, Ont.
- Marten-Freeman Co., Toronto, Ont.

BOILER COMPOUND
- Wm. C. Wilson & Co., Toronto, Ont.

BOILER CIRCULATORS
- Eckliff Circulator Co., Detroit, Mich.

BOILER FEED PUMPS
- Can. Ingersoll-Rand Co., Ltd., Sherbrooke, Que.
- Darling Bros., Ltd., Montreal, Que.
- Goldie & McCulloch Co., Galt, Ont.
- Morris Mach. Works, Baldwinsville, N.Y.
- Smart-Turner Mach. Co., Hamilton, Ont.
- Williams Machinery Co., A. R., Toronto, Ont.

BOILER FITTINGS
- Empire Mfg. Co., London, Ont.
- Goldie & McCulloch Co., Galt, Ont.
- McAvity & Sons Ltd., T., St. John, N.B.
- Wager Furnace Bridge Wall Co., Inc., New York, N.Y.

BOILER MOUNTINGS
- Jenkins Bros., Montreal.

BOILERS, MARINE
- Babcock & Wilcox, Ltd., Montreal, Que.
- Can. Fairbanks-Morse Co., Montreal, Que.
- Can. Vickers, Ltd., Montreal, Que.
- Collingwood Shipbuilding Co., Collingwood, Ont.
- Doxford & Sons, William, Sunderland, England.
- Engr. & Mach. Wks. of Can., St. Catharines, Ont.
- Goldie & McCulloch, Ltd., Galt, Ont.
- Hall Engineering Works, Montreal, Que.
- Mason Regulator & Engin. Co., Montreal, Que.
- Montreal Dry Docks & Shipbuilding Co., Montreal, Que.
- National Shipbuilding Co., Goderich, Ont.
- Marsh Engineering Works, Belleville, Ont.
- Polson Iron Works, Toronto, Ontario.
- Port Arthur Shipbuilding Co., Port Arthur, Ont.
- Waterous Engine Works Co., Brantford, Ont.
- Williams Machinery Co., A. R., Toronto, Ont.

BOOKS, TECHNICAL, MARINE
- MacLean Publishing Co., Toronto, Ont.

BOLTS
- London Bolt & Hinge Works, London, Ont.
- Mitchell Co., The Robert, Montreal, Que.
- United Brass & Lead, Ltd., Toronto, Ont.
- Wilkinson & Kompass, Hamilton, Ont.

BROKER
- Page & Jones, Mobile, Ala., U.S.A.

BRASS GOODS
- Corbet Fdry. & Machine Co., Owen Sound, Ont.
- Goldie & McCulloch Co., Galt, Ont.
- McAvity & Sons, T., St. John, N.B.
- Mitchell Co., The Robert, Montreal, Que.
- Mueller Mfg. Co., H., Sarnia, Ont.
- Tallman Brass & Metal Co., Hamilton, Ont.

BUCKETS, CLAMSHELL
- Beatty & Sons, Welland, Ont.
- Morris Crane & Hoist Co., Herbert, Niagara Falls, Ont.

BUCKETS, DUMP
- Morris Crane & Hoist Co., Herbert, Niagara Falls, Ont.

BUCKETS, COALING
- Beatty & Sons, Welland, Ont.
- Engr. & Mach. Wks. of Can., St. Catharines, Ont.
- Hopkins & Co., F. H., Montreal, Que.
- Marsh Engineering Works, Belleville, Ont.
- Reed & Co., Geo., Montreal, Que.

BUSHINGS, BRONZE
- Oberdorfer Brass Co., M. L., Syracuse, N.Y.

CABLE
- Hopkins & Co., F. H., Montreal, Que.
- McNab Co., Bridgeport, Conn.
- Wm. C. Wilson & Co., Toronto, Ont.

CABLE, LEAD COVERED AND ARMORED
- Standard Underground Cable Co., Hamilton, Ont.

CABLE, ACCESSORIES
- Darling Bros., Ltd., Montreal, Que.
- Standard Underground Cable Co., Hamilton, Ont.

CAPSTANS
- Advance Mach. & Welding Co., Montreal, Que.
- Dake Engine Co., Grand Haven, Mich.
- Hopkins & Co., F. H., Montreal, Que.
- Kennedy & Sons, Wm., Owen Sound, Ont.
- Wm. C. Wilson & Co., Toronto, Ont.

CALKING TOOLS, ELECTRIC
- Aikenhead Hardware, Ltd., Toronto, Ont.

CALKING TOOLS, PNEUMATIC
- Aikenhead Hardware, Ltd., Toronto, Ont.
- Can. Ingersoll-Rand Co. Sherbrooke, Que.

CALORIFIERS
- Low & Sons, Ltd., Archibald, Glasgow, Scotland

CASTINGS
- Beauchemin & Fils, Sorel, P.Q.
- Can. Steel Foundries, Ltd., Montreal, Que.
- Collingwood Shipbuilding Co., Collingwood, Ont.
- Wm. Hamilton Co., Peterboro, Ont.
- Kennedy & Sons, Wm., Owen Sound, Ont.
- Goldie & McCulloch Co., Galt, Ont.
- Marsh Engineering Works, Belleville, Ont.
- Mitchell Co., Ltd., Robt., Montreal, Que.
- Mueller Mfg. Co., H., Sarnia, Ont.
- Tallman Brass & Metal Co., Hamilton, Ont.
- United Brass & Lead, Ltd., Toronto, Ont.
- Waterous Engine Works Co., Brantford, Ont.

CASTINGS, ALLOY
- Magnet Metal & Foundry Co., Ltd., Winnipeg.
- Mitchell Co., The Robert, Montreal, Que.
- Mueller Mfg. Co., H., Sarnia, Ont.
- Oberdorfer Brass Co., M. L., Syracuse, N.Y.
- Tolland Mfg. Co., Montreal, Que.
- Tallman Brass & Metal Co., Hamilton, Ont.
- United Brass & Lead, Ltd., Toronto, Ont.

CASTINGS, ALUMINUM
- Empire Mfg. Co., London, Ont.
- Magnet Metal & Foundry Co., Ltd., Winnipeg.
- Mitchell Co., The Robert, Montreal, Que.
- Tallman Brass & Metal Co., Hamilton, Ont.
- United Brass & Lead Co., Toronto, Ont.

CASTINGS, BRASS
- Crouse-Hinds Co. of Canada, Ltd., Toronto, Ont.
- Empire Mfg. Co., London, Ont.
- Goldie & McCulloch Co., Galt, Ont.
- McAvity & Sons Ltd., T., St. John, N.B.
- McNab Co., Bridgeport, Conn.
- Mitchell Co., Ltd., Robt., Montreal, Que.
- Mueller Mfg. Co., H., Sarnia, Ont.
- Oberdorfer Brass Co., M. L., Syracuse, N.Y.
- Tolland Mfg. Co., Montreal, Que.
- Tallman Brass & Metal Co., Hamilton, Ont.
- United Brass & Lead Co., Toronto, Ont.
- Waterous Engine Works Co., Brantford, Ont.

CASTINGS, GREY IRON, MALLEABLE, ALUMINUM
- Crouse-Hinds Co. of Canada, Ltd., Toronto, Ont.
- Darling Bros., Ltd., Montreal, Que.
- Engr. & Mach. Wks. of Can., St. Catharines, Ont.
- Magnet Metal & Foundry Co., Ltd., Winnipeg.
- McAvity & Sons, T., St. John, N.B.
- McNab Co., Bridgeport, Conn.
- Mitchell Co., The Robert, Montreal, Que.
- Mueller Mfg. Co., H., Sarnia, Ont.
- Waterous Engine Works Co., Brantford, Ont.

CASTINGS, MANGANESE STEEL

CASTINGS, MANGANESE BRONZE
- Oberdorfer Brass Co., M. L., Syracuse, N.Y.
- Tallman Brass & Metal Co., Hamilton, Ont.
- United Brass & Lead Co., Toronto, Ont.

CELLAR DRAINERS
- Mueller Mfg. Co., H., Sarnia, Ont.

CHAINS
- Aikenhead Hardware, Ltd., Toronto, Ont.
- Kearfott Engineering Co., New York, N.Y.
- Hopkins & Co., F. H., Montreal, Que.
- Morris Crane & Hoist Co., Herbert, Niagara Falls, Ont.
- Henry Rogers, Sons & Co., Wolverhampton, Eng.
- Wm. C. Wilson & Co., Toronto, Ont.

CHAIN BLOCKS AND SLINGS
- Morris Crane & Hoist Co., Herbert, Niagara Falls, Ont.

CHANDLERY, SHIP
- Beauchemin & Fils, Sorel, P.Q.
- Hopkins & Co., F. H., Montreal, Que.
- Leckie, Ltd., John, Toronto, Ont.

CLAMPS, STEAM AND WATER
- Mueller Mfg. Co., H., Sarnia, Ont.

CLOCKS
- American Steam Gauge & Valve Mfg. Co., Boston, Mass.
- Morrison Brass Mfg. Co., James, Toronto, Ont.
- Williams Machinery Co., A. R., Toronto, Ont.

CLOSETS
- Mueller Mfg. Co., H., Sarnia, Ont.
- United Brass & Lead, Ltd., Toronto, Ont.

COAL
- Nova Scotia Steel & Coal Co., New Glasgow, N.S.

COAL HANDLING MACHINERY
- Morris Crane & Hoist Co., Herbert, Niagara Falls, Ont.
- Waterous Engine Works Co., Brantford, Ont.

COCKS, BILGE, DISCHARGE, INDICATOR,
- McAvity & Sons Ltd., T., St. John, N.B.
- Mitchell Co., The Robert, Montreal, Que.
- Morrison Brass Mfg. Co., James, Toronto, Ont.

COCKS, BASIN
- Empire Mfg. Co., London, Ont.
- Mitchell Co., The Robert, Montreal, Que.
- United Brass & Lead, Ltd., Toronto, Ont.

COMPASSES
- Wm. C. Wilson & Co., Toronto, Ont.

COMPRESSORS, AIR
- Can. Fairbanks-Morse Co., Montreal, Que.
- Canadian Ingersoll-Rand Co., Sherbrooke, Que.
- Darling Bros., Ltd., Montreal, Que.
- Hopkins & Co., F. H., Montreal, Que.
- Smart-Turner Mach. Co., Hamilton, Ont.
- Williams Machinery Co., A. R., Toronto, Ont.

CONDENSERS
- Darling Bros., Ltd., Montreal, Que.
- Goldie & McCulloch Co., Galt, Ont.
- Kearfott Engineering Co., New York, N.Y.
- Morris Machine Works, Baldwinsville, N.Y.
- Smart-Turner Mach. Co., Hamilton, Ont.
- Weir Ltd., G. & J., Cathcart, Glasgow, Scotland.
- Williams Machinery Co., A. R., Toronto, Ont.

CONDULETS, MARINE
- Crouse-Hinds Co. of Canada, Ltd., Toronto, Ont.

CONTRACTORS' SUPPLIES
- McAvity & Sons, T., St. John, N.B.

CONSULTING ENGINEER
- Watt, J. Murray, Philadelphia, Pa.

CONVEYORS, ASH, COAL
- Babcock & Wilcox, Ltd., Montreal, Que.
- Hopkins & Co., F. H., Montreal, Que.

COPING MACHINES
- Bertram & Sons, Ltd., John, Dundas, Ont.

COPPER TUBES, SHEETS AND RODS
- Dom. Copper Products Co., Montreal, Que.
- Tallman Brass & Metal Co., Hamilton, Ont.

COTTON, CALKING
- Wilson & Co., Wm. C., Toronto, Canada.

COVERS, CANVAS, FOR HATCHES, LIFE-BOATS, ETC.
- Leckie, Ltd., John, Toronto, Ont.
- Waterous Engine Works Co., Brantford, Ont.

COUNTERS, REVOLUTION
- American Steam Gauge & Valve Mfg. Co., Boston, Mass.

COWLS, SHIPS' VENTILATORS
- Pedlar People, Ltd., Oshawa, Ont.

COUPLINGS, AIR HOSE
- Cleveland Pneumatic Tool Co. of Canada, Toronto.

CRANES
- Aikenhead Hardware, Ltd., Toronto, Ont.
- Can. Fairbanks-Morse Co., Montreal, Que.
- Hopkins & Co., F. H., Montreal, Que.
- Williams Machinery Co., A. R., Toronto, Ont.

CRANES, ELECTRIC
- Babcock & Wilcox, Ltd., Montreal, Que.
- Morris Crane & Hoist Co., Herbert, Niagara Falls, Ont.
- Smart-Turner Mach. Co., [illegible]

CRANES, GOLIATH AND PNEUMATIC
- Morris Crane & Hoist Co., Herbert, Niagara Falls, Ont.

CRANES, GRANTRY, PORTABLE, JIB
- Morris Crane & Hoist Co., Herbert, Niagara Falls, Ont.
- Smart-Turner Mach. Co., Hamilton, Ont.

CRANES, OVERHEAD TRAVELLING
- Morris Crane & Hoist Co., Herbert, Niagara Falls, Ont.

CRANK SHAFTS
- Canada Foundries and Forgings, Welland, Ont.

DAVITS, BOAT
Corbet Fdry. & Machine Co., Owen Sound, Ont.
Hopkins & Co., F. H., Montreal, Que.
Marten, Freeman Co., Toronto, Ont.
Waterous Engine Works Co., Brantford, Ont.

DEAD LIGHTS, BRASS
Goldie & McCulloch Co., Galt, Ont.
Morris Crane & Hoist Co., Herbert, Niagara Falls, Ont.
Mitchell Co., Ltd., Robt., Montreal, Que.

DECK LIGHTS
Mitchell Co., The Robert, Montreal, Que.
Turnbull Elevator Mfg. Co., Toronto, Ont.

DECK PLUGS, ELECTRIC
Crouse-Hinds Co. of Canada, Ltd., Toronto, Ont.
Mitchell Co., The Robert, Montreal, Que.

DECKING FOR SHIPS
Marine Decking & Supply Co., Philadelphia, Pa.

DERRICKS
Aikenhead Hardware, Ltd., Toronto, Ont.
Dake Engine Co., Grand Haven, Mich.
Hopkins & Co., F. H., Montreal, Que.
Marsh Engineering Works, Belleville, Ont.
Morris Crane & Hoist Co., Herbert, Niagara Falls, Ont.

DISTILLERS
Kearfott Engineering Co., New York, N.Y.

DREDGES
Collingwood Shipbuilding Co., Collingwood, Ont.
Morris Mach. Works, Baldwinsville, N.Y.
Norbom Engineering Co., Philadelphia, Pa.
Polson Iron Works, Toronto.

DRILLS, AIR
Aikenhead Hardware, Ltd., Toronto, Ont.
Can. Ingersoll-Rand Co. Sherbrooke, Que.
Hopkins & Co., F. H., Montreal, Que.

DRILLS, CENTRE
Wilt Twist Drill Co., Walkerville, Ont.

DRILLS, BLACKSMITH AND BIT STOCK
Wilt Twist Drill Co., Walkerville, Ont.

DRILLS, ELECTRIC
Aikenhead Hardware, Ltd., Toronto, Ont.
Wilkinson & Kompass, Hamilton, Ont.

DRILLS, HIGH SPEED
Wilt Twist Drill Co., Walkerville, Ont.
Zenith Coal & Steel Products, Ltd., Montreal.

DRILLS, TRACK
Wilt Twist Drill Co., Walkerville, Ont.

DRILLS, RACHET AND HAND
Wilt Twist Drill Co., Walkerville, Ont.

DRILLS, TWIST
Aikenhead Hardware, Ltd., Toronto, Ont.
Williams Machinery Co., A. R., Toronto, Ont.
Wilt Twist Drill Co., Walkerville, Ont.

DRY DOCKS
Can. Vickers, Ltd., Montreal, Que.
Collingwood Shipbuilding Co., Collingwood, Ont.
Doxford & Sons, William, Sunderland, England.
Georgian Bay Shipbuilding & Wrecking Co., Midland, Ont.
National Shipbuilding Co., Goderich, Ont.
Polson Iron Works, Toronto, Ont.
Port Arthur Shipbuilding Co., Port Arthur, Ont.
Yarrows, Limited, Victoria, B.C.

ECONOMIZERS, FUEL
Babcock & Wilcox, Ltd., Montreal, Que.

EJECTORS
Darling Bros., Ltd., Montreal, Que.
Empire Mfg. Co., London, Ont.
Mitchell Co., The Robert, Montreal, Que.
Morrison Brass Mfg. Co., James, Toronto, Ont.
Smart-Turner Mach. Co., Hamilton, Ont.

ELECTRIC LAMPS
Mitchell Co., The Robert, Montreal, Que.
Wm. C. Wilson & Co., Toronto, Ont.

ELECTRO-PLATING
Mitchell Co., The Robert, Montreal, Que.
Tallman Brass & Metal Co., Hamilton, Ont.
United Brass & Lead Co., Toronto, Ont.

ELECTRIC WELDING
Beauchemin & Fils, Sorel, P.Q.

ELEVATING MACHINERY
Darling Bros., Ltd., Montreal, Que.
Goldie & McCulloch, Ltd., Galt, Ont.
Wm. Hamilton Co., Peterboro, Ont.
Morris Crane & Hoist Co., Herbert, Niagara Falls, Ont.
Waterous Engine Works Co., Brantford, Ont.

ELEVATORS
Darling Bros., Ltd., Montreal, Que.
Turnbull Elevator Mfg. Co., Toronto, Ont.

ENGINES, HOISTING
Corbet Fdry. & Machine Co., Owen Sound, Ont.
Hopkins & Co., F. H., Montreal, Que.
Kennedy & Sons, Wm., Owen Sound, Ont.
Marsh Engineering Works, Belleville, Ont.
Port Arthur Shipbuilding Co., Port Arthur, Ont.
Williams Machinery Co., A. R., Toronto, Ont.

ENGINE, INTERNAL COMBUSTION
Doxford & Sons, William, Sunderland, England.

ENGINES, MARINE
Bolinders Co., New York, N.Y.
Can. Fairbanks-Morse Co., Montreal, Que.
Can. Vickers, Ltd., Montreal, Que.
Doxford & Sons, William, Sunderland, England.
Goldie & McCulloch, Ltd., Galt, Ont.
Hopkins & Co., F. H., Montreal, Que.
Iron Works, Ltd., Owen Sound, Ont.
Mason Regulator & Engin. Co., Montreal, Que.
Morris Mach. Works, Baldwinsville, N.Y.
National Shipbuilding Co., Goderich, Ont.
Norbom Engineering Co., Philadelphia, Pa.
Polson Iron Works, Toronto, Ont.
Port Arthur Shipbuilding Co., Port Arthur, Ont.
Trout Co., H. G., Buffalo, N.Y.
Waterous Engine Works Co., Brantford, Ont.
Williams Machinery Co., A. R., Toronto, Ont.

ENGINE STARTERS (AIR)
Scaife & Sons Co., Wm. B., Oakmont, Pa.

ENGINES, STEERING
Corbet Fdry. & Machine Co., Owen Sound, Ont.
Dake Engine Co., Grand Haven, Mich.
Kennedy & Sons, Wm., Owen Sound, Ont.
Wm. C. Wilson & Co., Toronto, Ont.

ENAMELWARE
Mueller Mfg. Co., H., Sarnia, Ont.
United Brass & Lead Co., Toronto, Ont.

EVAPORATORS
Kearfott Engineering Co., New York, N.Y.
Mason Regulator & Engin. Co., Montreal, Que.
McNab Co., Bridgeport, Conn.
Weir Ltd., G. & J., Cathcart, Glasgow, Scotland.

EXTRACTORS, GREASE
American Steam Gauge & Valve Mfg. Co., Boston, Mass.
Darling Bros., Ltd., Montreal, Que.

EYE BOLTS AND NUTS
Canada Foundries and Forgings, Welland, Ont.
Mitchell Co., The Robert, Montreal, Que.
United Brass & Lead, Ltd., Toronto, Ont.

FANS
Aikenhead Hardware Ltd., Toronto, Ont.
Empire Mfg. Co., London, Ont.
Reed & Co., Geo. W., Montreal, Que.
Smart-Turner Mach. Co., Hamilton, Ont.
Williams Machinery Co., A. R., Toronto, Ont.

FENDERS, ROPE
Hopkins & Co., F. H., Montreal, Que.
Leckie, Ltd., John, Toronto, Ont.
Wilson & Co., Wm. C., Toronto, Canada.

FERRO-MANGANESE
Mitchells, Ltd., Glasgow, Scotland.

FILES
Aikenhead Hardware Ltd., Toronto, Ont.
Williams Machinery Co., A. R., Toronto, Ont.

FIRE BRICKS
Beveridge Paper Co., Montreal, Que.
Mitchells, Ltd., Glasgow, Scotland.
Williams Machinery Co., A. R., Toronto, Ont.

FILTERS, FEED WATER
Darling Bros., Ltd., Montreal, Que.
MacKinnon Steel Co., Sherbrooke, Que.
Mason Regulator & Engin. Co., Montreal, Que.

FITTINGS, MARINE
Hopkins & Co., F. H., Montreal, Que.
McAvity & Sons, T., St. John, N.B.
Mitchell Co., The Robert, Montreal, Que.
United Brass & Lead, Ltd., Toronto, Ont.

FITTINGS, MOTOR BOAT
Empire Mfg. Co., London, Ont.
Mitchell Co., The Robert, Montreal, Que.
Mueller Mfg. Co., H., Sarnia, Ont.
United Brass & Lead, Ltd., Toronto, Ont.

FIXTURES, ELECTRIC
Cory & Son, Inc., Chas., New York, N.Y.
Crouse-Hinds Co. of Canada, Ltd., Toronto, Ont.
Mitchell Co., The Robert, Montreal, Que.
Harvey Hubbell Co. of Canada, Toronto, Can.
Tallman Brass & Metal Co., Hamilton, Ont.

FLAG POLES, STEEL
Dennis Wire & Iron Works Co., London, Ont.

FLOW METERS
Spray Engineering Co., Boston, Mass.

FLUE CLEANERS
Wm. C. Wilson & Co., Toronto, Ont.

FORGES
Aikenhead Hardware, Ltd., Toronto, Ont.
Hopkins & Co., F. H., Montreal, Que.

FLANGING AND EXPANDING MACHINES, PIPE
Lovekin Pipe Expanding & Flanging Mach. Co., Philadelphia, Pa.

FLOODLIGHTS, ELECTRIC
Crouse-Hinds Co. of Canada, Ltd., Toronto, Ont.

FLUORSPAR
Mitchells, Ltd., Glasgow, Scotland.
Scythes & Co., Toronto, Ont.

FORGINGS, ALL KINDS
Aikenhead Hardware, Ltd., Toronto, Ont.
Collingwood Shipbuilding Co., Collingwood, Ont.
Nova Scotia Steel & Coal Co., New Glasgow, N.S.

FORGINGS, STEEL AND IRON
Canada Foundries and Forgings, Welland, Ont.

FURNACE BRIDGE WALLS
Wager Furnace Bridge Wall Co., Inc., 149 Broadway, New York, N.Y.

GAUGES, RECORDING
American Steam Gauge & Valve Mfg. Co., Boston, Mass.
Empire Mfg. Co., London, Ont.

GASKETS
Wm. C. Wilson & Co., Toronto, Ont.

GAUGES COCKS
McAvity & Sons, T., St. John, N.B.
Mitchell Co., The Robert, Montreal, Que.

GAUGE GLASSES
Wm. C. Wilson & Co., Toronto, Ont.

GAUGES, WATER, PRESSURE, COMPOUND AND VACUUM
Aikenhead Hardware, Ltd., Toronto, Ont.
Babcock & Wilcox, Ltd., Montreal, Que.
Empire Mfg. Co., London, Ont.
Jenkins Bros., Montreal.
McAvity & Sons, T., St. John, N.B.
McNab Co., Bridgeport, Conn.
Morrison Brass Mfg. Co., James, Toronto, Ont.

GUARDS, LAMPS
Cleveland Pneumatic Tool Co. of Can., Toronto.

GENERATORS AND CONVERTORS
Can. Fairbanks-Morse Co., Montreal, Que.

GLAZING COMPOSITION
Kuhls, H. B. Fred., 6411 3rd Ave., Brooklyn, N.Y.

GOGGLES
Standard Optical Co., Geneva, N.Y.

GONGS
Clark & Bro., C. O., Montreal, Que.
Mitchell Co., Ltd., Robt., Montreal, Que.

GRAPHITE
Wm. C. Wilson & Co., Toronto, Ont.

GRATINGS
Can. Welding Works, Montreal, Que.
Corbet Fdry. & Machine Co., Owen Sound, Ont.
Canada Wire & Iron Goods Co., Hamilton, Ont.
MacKinnon Steel Co., Sherbrooke, Que.

GRINDERS, ELECTRIC
Wilkinson & Kompass, Hamilton, Ont.

GRINDERS, PNEUMATIC
Can. Ingersoll-Rand Co. Sherbrooke, Que.

GUARDS, MACHINERY
Dennis Wire & Iron Works Co., London, Ont.

GUY RODS AND ANCHORS, ELECTRIC
Crouse-Hinds Co. of Canada, Ltd., Toronto, Ont.

HAMMERS
Canada Foundries and Forgings, Ltd., Welland, Ont.

HARDWARE, MARINE
Hopkins & Co., F. H., Montreal, Que.
Mitchell Co., The Robert, Montreal, Que.

HEADLIGHTS, ELECTRIC
Crouse-Hinds Co. of Canada, Ltd., Toronto, Ont.
Hopkins & Co., F. H., Montreal, Que.
Tallman Brass & Metal Co., Hamilton, Ont.

HEATING EQUIPMENT
Darling Bros., Ltd., Montreal, Que.
Empire Mfg. Co., London, Ont.
Low & Sons, Ltd., Archibald, Glasgow, Scotland

HEATERS, FEED, WATER
Darling Bros., Ltd., Montreal, Que.
Babcock & Wilcox, Ltd., Montreal, Que.
Goldie & McCulloch Co., Galt, Ont.
Kearfott Engineering Co., New York, N.Y.
Mason Regulator & Engin. Co., Montreal, Que.
McNab Co., Bridgeport, Conn.
Weir Ltd., G. & J., Cathcart, Glasgow, Scotland.

HINGES
London Bolt & Hinge Works, London, Ont.
Mitchell Co., Ltd., Robt., Montreal, Que.

HOOKS
Hopkins & Co., F. H., Montreal, Que.
Morris Crane & Hoist Co., Herbert, Niagara Falls, Ont.

HOISTS, ASH
Beatty & Sons, Welland, Ont.
Marsh Engineering Works, Belleville, Ont.
St. Clair Bros., Galt, Ont.
Waterous Engine Works Co., Brantford, Ont.

HOIST BLOCKS
Morris Crane & Hoist Co., Herbert, Niagara Falls, Ont.

HOISTS, CHAIN
Aikenhead Hardware, Ltd., Toronto, Ont.
Can. Fairbanks-Morse Co., Montreal, Que.
Dake Engine Co., Grand Haven, Mich.
Hopkins & Co., F. H., Montreal, Que.
Morris Crane & Hoist Co., Herbert, Niagara Falls, Ont.
Williams Machinery Co., A. R., Toronto, Ont.

HOISTS, CARGO, MOVING, ETC.
Dake Engine Co., Grand Haven, Mich.
Hopkins & Co., F. H., Montreal, Que.
Marsh Engineering Works, Belleville, Ont.
Waterous Engine Works Co., Brantford, Ont.

HOISTING MACHINERY
Beatty & Sons, Welland, Ont.
Can. Ingersoll-Rand Co. Sherbrooke, Que.
Corbet Fdry. & Machine Co., Owen Sound, Ont.
Wm. Hamilton Co., Peterboro, Ont.
Marsh Engineering Works, Belleville, Ont.
Morris Crane & Hoist Co., Herbert, Niagara Falls, Ont.
Waterous Engine Works Co., Brantford, Ont.
Williams Machinery Co., A. R., Toronto, Ont.

HOSE
Wm. C. Wilson & Co., Toronto, Ont.

INDICATORS, ENGINE
American Steam Gauge & Valve Mfg. Co., Boston, Mass.
Cory & Son, Inc., Chas., New York, N.Y.
McNab Co., Bridgeport, Conn.

INDICATORS, SPEED
Aikenhead Hardware, Ltd., Toronto, Ont.
Cory & Son, Inc, Chas., New York, N.Y.

INJECTORS
Aikenhead Hardware, Ltd, Toronto, Ont.
Empire Mfg. Co., London, Ont.
Mitchell Co., The Robert, Montreal, Que.
Morrison Brass Mfg. Co., James, Toronto, Ont.
Williams Machinery Co., A. R., Toronto, Ont.

INGOTS
Broughton Copper Co., Ltd, Manchester, Eng.

INSULATORS, ELECTRIC
Crouse-Hinds Co. of Canada, Ltd, Toronto, Ont.

INSTRUMENTS, NAUTICAL
Leckie, Ltd., John, Toronto, Ont.

IRON AND STEEL
Mitchells Ltd., Glasgow, Scotland.

JACKS
Hopkins & Co., F. H., Montreal, Que.
Morris Crane & Hoist Co., Herbert, Niagara Falls, Ont.

JOINTS, BALL SLIP
Norbom Engineering Co., Philadelphia, Pa.

INSURANCE, MARINE
Toronto Insurance & Vessel Agency, Toronto, Ont.

JOURNAL WEDGES
Canada Foundries and Forgings, Ltd., Welland, Ont.

JUTE PACKING
Stratford Oakum Co., Geo., Jersey City, N.J.

LATHES
Bertram & Sons, Ltd., John, Dundas, Ont.

LEATHER, LACE
Wm. C. Wilson & Co., Toronto, Ont.

LENSES FOR GOGGLES
Standard Optical Co., Geneva, N.Y.

LIFEBUOYS, BELTS AND PRESERVERS
Fosbery Co., Barking, Essex, Eng.
Hopkins & Co., F. H., Montreal, Que.
Wm. C. Wilson & Co., Toronto, Ont.

LIFEBOATS
Hopkins & Co., F. H., Montreal, Que.
Wm. C. Wilson & Co., Toronto, Ont.

LIFE BOAT EQUIPMENT
Eckliff Circulator Co., Detroit, Mich.
Hopkins & Co., F. H., Montreal, Que.
Leckie, Ltd., John, Toronto, Ont.
Wm. C. Wilson & Co., Toronto, Ont.

LIFE JACKETS
Fosbery Co., Barking, Essex, Eng.
Hopkins & Co., F. H., Montreal, Que.
Leckie, Ltd., John, Toronto, Ont.

LIGHTS, ALL KINDS
Can. Fairbanks-Morse Co., Montreal, Que.
Crouse-Hinds Co. of Canada, Ltd., Toronto, Ont.
Hopkins & Co., F. H., Montreal, Que.
Mitchell Co., The Robert, Montreal, Que.
Morrison Brass Mfg. Co., James, Toronto, Ont.
Tallman Brass & Metal Co., Hamilton, Ont.

LIGHTS, SIDE, PORT
Goldie & McCulloch Co., Galt, Ont.
Crouse-Hinds Co. of Canada, Ltd., Toronto, Ont.
Hopkins & Co., F. H., Montreal, Que.
McAvity & Sons Ltd., T., St. John, N.B.
Mitchell Co., The Robert, Montreal, Que.
Tallman Brass & Metal Co., Hamilton, Ont.
Turnbull Elevator Mfg. Co., Toronto, Ont.
Wilson & Co., Wm. C., Toronto, Ont.

LIGHTING SETS
Can. Fairbanks-Morse Co., Montreal.
Cory & Son, Inc., Chas., New York, N.Y.
Crouse-Hinds Co. of Canada, Ltd., Toronto, Ont.
Goldie & McCulloch Co., Ltd., Galt, Ont.
Kearfoit Engineering Co., New York, N.Y.
Williams Machinery Co., A. R., Toronto, Ont.
Wilson & Co., Wm. C., Toronto, Canada.

LOADERS, WAGON AND TRUCK
Morris Crane & Hoist Co., Herbert, Niagara Falls, Ont.
Waterous Engine Works Co., Brantford, Ont.

LOCKERS, METAL
Dennis Wire & Iron Works Co., London, Ont.
Reed & Co., Geo. W., Montreal, Que.

LUBRICATORS
Empire Mfg. Co., London, Ont.
Mitchell Co., The Robert, Montreal, Que.

MACHINERY-METAL AND WOOD BORING
Cleveland Pneumatic Tool Co. of Can., Toronto.

MACHINISTS
Corbet Fdry. & Machine Co., Owen Sound, Ont.
Wm. Hamilton Co., Peterboro, Ont.
Hyde Engineering Works, Montreal, Que.
Mitchell Co., The Robert, Montreal, Que.
United Brass & Lead Co., Toronto, Ont.

MANGANESITE, PASTE
Wm. C. Wilson & Co., Toronto, Ont.

MARINE CASTINGS
Wm. Hamilton Co., Peterboro, Ont.
Goldie & McCulloch Co., Galt, Ont.
Mitchell Co., Ltd., Robt., Montreal, Que.
McAvity & Sons, T., St. John, N.B.
Mueller Mfg. Co., H., Sarnia, Ont.
Tallman Brass & Metal Co., Hamilton, Ont.
United Brass & Lead, Ltd., Toronto, Ont.
Waterous Engine Works Co., Brantford, Ont.

MARINE SPECIALTIES
Goldie & McCulloch Co., Galt, Ont.
Hopkins & Co., F. H., Montreal, Que.
McAvity & Sons, T., St. John, N.B.
Mitchell Co., Ltd., Robt., Montreal, Que.
United Brass & Lead, Ltd., Toronto, Ont.

MOTORS
Aikenhead Hardware, Ltd., Toronto, Ont.
Can. Fairbanks-Morse Co., Montreal, Que.
Williams Machinery Co., A. R., Toronto, Ont.
Zenith Coal & Steel Products, Ltd., Montreal.

NAUTICAL INSTRUMENTS
Wm. C. Wilson & Co., Toronto, Ont.

NET LIFTERS
Dake Engine Co., Grand Haven, Mich.
Leckie, Ltd., John, Toronto, Ont.

NOZZLES, ALL KINDS
Empire Mfg. Co., London, Ont.
Mitchell Co., The Robert, Montreal, Que.
Tolland Mfg. Co., Boston, Mass.

NUTS
Mitchell Co., The Robert, Montreal, Que.
Wilkinson & Kompass, Hamilton, Ont.
United Brass & Lead, Ltd., Toronto, Ont.

OIL CUPS
Empire Mfg. Co., London, Ont.
Mitchell Co., Ltd., Robt., Montreal, Que.

OILCLOTHING
Wm. C. Wilson & Co., Toronto, Ont.

OIL, LINSEED
Canada Linseed Oil Mills, Montreal, Que.

OILS AND GREASE
Wm. C. Wilson & Co., Toronto, Ont.

OAKUM
Davey & Sons W. O., Jersey City, N.J.
Leckie, Ltd., John, Toronto, Ont.
Stratford Oakum Co., Geo., Jersey City, N.J.
Wm. C. Wilson & Co., Toronto, Ont.

ORNAMENTAL IRONWORK
Canada Wire & Iron Goods Co., Hamilton, Ont.
Can. Welding Works, Montreal, Que.

OPTICAL GOODS
Standard Optical Co., Geneva, N.Y.

OXY-ACETLYENE OUTFITS
Can. Fairbanks-Morse Co., Montreal, Que.
Can. Welding Works, Montreal, Que.
Hopkins & Co., F. H., Montreal, Que.
McAvity & Sons Ltd., T., St. John, N.B.
Williams Machinery Co., A. R., Toronto, Ont.

PAINT
Ault & Wiborg Co. of Can., Ltd., Toronto, Ont.
Wm. C. Wilson & Co., Toronto, Ont.

PAINTS, ANTI-FOULING
Kuhls, H. B. Fred., 6411 3rd Ave., Brooklyn, N.Y.

PAINTS, ANTI-CORROSIVE
Kuhls, H. B. Fred, 6411 3rd Ave., Brooklyn, N.Y.

PAINTING EQUIPMENT, AUTOMATIC
Spray Engineering Co., Boston, Mass.

PACKING, AMMONIA
France Packing Co., Philadelphia, Pa.
Wm. C. Wilson & Co., Toronto, Ont.

PACKING, MARINE
Aikenhead Hardware, Ltd., Toronto, Ont.
France Packing Co., Philadelphia, Pa.
Wm. C. Wilson & Co., Toronto, Ont.

PACKING, METALLIC
Aikenhead Hardware, Ltd., Toronto, Ont.
France Packing Co., Philadelphia, Pa.

PACKING, STEAM
Aikenhead Hardware, Ltd., Toronto, Ont.
France Packing Co., Philadelphia, Pa.
Wm. C. Wilson & Co., Toronto, Ont.

PANELBOARD AND CABINETS, ELECTRIC
Crouse-Hinds Co. of Canada, Ltd., Toronto, Ont.
Preston Woodworking Machy. Co., Preston, Ont.

PATTERNS, WOOD AND METAL
Beauchemin & Fils, Sorel, P.Q.

PIPE, LAPWELD, CAST IRON, RIVETED
Empire Mfg. Co., London, Ont.
Norbom Engineering Co., Philadelphia, Pa.
United Brass & Lead, Ltd., Toronto, Ont.

PIPE RAILING, IRON AND BRASS
Dennis Wire & Iron Works Co., London, Ont.
Mitchell Co., The Robert, Montreal, Que.

PITCH
Reed & Co., Geo., Montreal, Que.

PITCH, PINE
Wilson & Co., Wm. C., Toronto, Canada.

PLANERS
Bertram & Sons, Ltd., John, Dundas, Ont.

PLANERS, STANDARD AND ROTARY
Yates Mach. Co., P. B., Hamilton, Ont.

PLATE CLAMPS
Morris Crane & Hoist Co., Herbert, Niagara Falls, Ont.

PLATE PUNCH TABLES
Norbom Engineering Co., Philadelphia, Pa.

PLUGS, ATTACHMENT
Harvey Hubbel Co. of Canada, Toronto, Can.

PLUMBING EQUIPMENT
Low & Sons, Ltd., Archibald, Glasgow, Scotland.
Empire Mfg. Co., London, Ont.
Mitchell Co., The Robert, Montreal, Que.
Mueller Mfg. Co., H., Sarnia, Ont.
Tallman Brass & Metal Co., Hamilton, Ont.
Wm. C. Wilson & Co., Toronto, Ont.
United Brass & Lead, Ltd., Toronto, Ont.

PROPELLOR BLADES, BRONZE
Corbet Fdry. & Machine Co., Owen Sound, Ont.
Empire Mfg. Co., London, Ont.
Mitchell Co., The Robert, Montreal, Que.
Tallman Brass & Metal Co., Hamilton, Ont.
United Brass & Lead, Ltd., Toronto, Ont.
Yarrows, Limited, Victoria, B.C.

PROPELLER WHEELS
Goldie & McCulloch Co., Galt, Ont.
Kennedy & Sons, Wm., Owen Sound, Ont.
Trout Co., H. G., Buffalo, N.Y.

PIPING, STEAM
Babcock & Wilcox, Ltd., Montreal, Que.

PORT LIGHTS
Goldie & McCulloch Co., Galt, Ont.
Mitchell Co., Ltd., Robt., Montreal, Que.
Mueller Mfg. Co., H., Sarnia, Ont.
Turnbull Elevator Mfg. Co., Toronto, Ont.
Tallman Brass & Metal Co., Hamilton, Ont.
Wm. C. Wilson & Co., Toronto, Ont.
United Brass & Lead, Ltd., Toronto, Ont.

PROPELLER WHEELS
Beauchemin & Fils, Sorel, P.Q.
McNab Co., Bridgeport, Conn.

PULLEYS
Wm. Hamilton Co., Peterboro, Ont.
Smart-Turner Mach. Co., Hamilton, Ont.
Williams Machinery Co., A. R., Toronto, Ont.

PUMP, AIR
Can. Ingersoll-Rand Co. Sherbrooke, Que.
Smart-Turner Mach. Co., Hamilton, Ont.
Weir Ltd., G. & J., Cathcart, Glasgow, Scotland.
Williams Machinery Co., A. R., Toronto, Ont.

PUMPS
Bawden Machine Co., Toronto.
Can. Fairbanks-Morse Co., Montreal, Que.
Canada Foundries & Forgings, Welland, Ont.
Darling Bros., Ltd., Montreal, Que.
Goldie & McCulloch, Ltd., Galt, Ont.
Hopkins & Co., F. H., Montreal, Que.
McAvity & Sons, T., St. John, N.B.
Oberdorfer Brass Co., M. L., Syracuse, N.Y.
Williams Machinery Co., A. R., Toronto Ont.

PUMPS, CENTRIFUGAL
Can. Ingersoll-Rand Co. Sherbrooke, Que.
Darling Bros., Ltd., Montreal, Que.
Wm. Hamilton Co., Peterboro, Ont.
Goldie & McCulloch Co., Galt, Ont.
Kearfott Engineering Co., New York, N.Y.
Morris Mach. Works, Baldwinsville, N.Y.
Norbom Engineering Co., Philadelphia, Pa.
Smart-Turner Mach. Co., Hamilton, Ont.
Waterous Engine Works Co., Brantford, Ont.
Williams Machinery Co., A. R., Toronto, Ont.

PUMPS, BILGE
Darling Bros., Ltd., Montreal, Que.
Goldie & McCulloch Co., Galt, Ont.
Mitchell Co., Ltd., Robt., Montreal, Que.
Morris Mach. Works, Baldwinsville, N.Y.
Mueller Mfg. Co., H., Sarnia, Ont.

PUMPS, CIRCULATING
Darling Bros., Ltd., Montreal, Que.
Goldie & McCulloch Co., Galt, Ont.
McNab Co., Bridgeport, Conn.
Morris Machine Works, Baldwinsville, N.Y.

PUMPS, FEED WATER
Darling Bros., Ltd., Montreal, Que.
Goldie & McCulloch Co., Galt, Ont.
McNab Co., Bridgeport, Conn.
Morris Mach. Works, Baldwinsville, N.Y.
Weir Ltd., G. & J., Cathcart, Glasgow, Scotland.
Williams Machinery Co., A. R., Toronto, Ont.

PUMPS, LIFT AND FORCE
Darling Bros., Ltd., Montreal, Que.
Low & Sons, A., Glasgow, Scotland.

PUMPS, REES ROTURBO
Goldie & McCulloch Co., Galt, Ont.

PUMPS, HAND AND POWER
Aikenhead Hardware, Ltd., Toronto, Ont.
Darling Bros., Ltd., Montreal, Que.
Mitchell Co., The Robert, Montreal, Que.
Smart-Turner Mach. Co., Hamilton, Ont.
Williams Machinery Co., A. R., Toronto, Ont.

PUMPS, HIGH PRESSURE
Canadian Ingersoll-Rand Co., Sherbrooke, Que.
Darling Bros., Ltd., Montreal, Que.
Morris Mach. Works, Baldwinsville, N.Y.
Smart-Turner Mach. Co., Hamilton, Ont.

PUMPS, STEAM TURBINE
Darling Bros., Ltd., Montreal, Que.
Morris Mach. Works, Baldwinsville, N.Y.
Wm. C. Wilson & Co., Toronto, Ont.

PUMPING MACHINES
Norbom Engineering Co., Philadelphia, Pa.

PUNCHES
Bertram & Sons, Ltd., John, Dundas, Ont.

PUNCHES, SINGLE, DOUBLE AND MULTIPLE
Norbom Engineering Co., Philadelphia, Pa.

PURIFIERS, WATER
Babcock & Wilcox, Ltd., Montreal, Que.

RADIATORS, STEAM, ELECTRIC
Empire Mfg. Co., London, Ont.
Low & Sons, Ltd., Archibald, Glasgow, Scotland.
Wm. C. Wilson & Co., Toronto, Ont.

RAILS, OVERHEAD
Morris Crane & Hoist Co., Herbert, Niagara Falls, Ont.

RANGES
Hopkins & Co., F. H., Montreal, Que.
Wm. C. Wilson & Co., Toronto, Ont.

RECEIVERS, AIR
Can. Ingersoll-Rand Co. Sherbrooke, Que.
Can. Welding Works, Montreal, Que.
Darling Bros., Ltd., Montreal, Que.
MacKinnon Steel Co., Sherbrooke, Que.

REGULATORS, FEED WATER
American Steam Gauge & Valve Mfg. Co., Boston, Mass.

REGULATORS, PRESSURE
Mason Regulator & Engin. Co., Montreal, Que.

RELEASING GEARS
Eckliff Circulator Co., Detroit, Mich.

REPAIRS, MARINE
Corbet Foundry & Mach. Co., Owen Sound, Ont.
Can. Vickers, Ltd., Montreal, Que.
Collingwood Shipbuilding Co., Collingwood, Ont.
Engr. & Mach. Wks. of Can., St. Catharines, Ont.
Georgian Bay Shipbuilding & Wrecking Co., Midland, Ont.
Hyde Engineering Works, Montreal, Que.
Iron Works, Ltd., Owen Sound, Ont.
Kennedy & Sons, Wm., Owen Sound, Ont.
Muir Bros. Dry Dock Co., Port Dalhousie, Ont.
National Shipbuilding Co., Goderich, Ont.
Port Arthur Shipbuilding Co., Port Arthur, Ont.
Yarrow, Limited, Victoria, B.C.

RIGGING SCREWS
Hopkins & Co., F. H., Montreal, Que.

RIGGING, WIRE ROPE
Leckie, Ltd., John, Toronto, Ont.

RIVETS
London Bolt & Hinge Works, London, Ont.
Severance Mfg. Co., S., Glassport, Pa.
Wilkinson & Kompass, Hamilton, Ont.

RIVETERS, PNEUMATIC
Can. Ingersoll-Rand Co., Ltd., Sherbrooke, Que.
Cleveland Pneumatic Tool Co. of Can., Toronto.

RODS, COPPER, BRASS, BRONZE
Standard Underground Cable Co., Hamilton, Ont.
Tallman Brass & Metal Co., Hamilton, Ont.

ROLLS, STRAIGHTENING, BENDING
Bertram & Sons, Ltd., John, Dundas, Ont.

ROOFING
Reed & Co., Geo., Montreal, Que.
Wm. C. Wilson & Co., Toronto, Ont.

ROPE BLOCKS
Aikenhead Hardware, Ltd., Toronto, Ont.
Can. Fairbanks-Morse Co., Montreal, Que.
Morris Crane & Hoist Co., Herbem, Niagara Falls, Ont.

ROPE
Hopkins & Co., F. H., Montreal, Que.
Leckie, Ltd., John, Toronto, Ont.
McNab Co., Bridgeport, Conn.
Stratford Oakum Co., Geo., Jersey City, N.J.
Wm. C. Wilson & Co., Toronto, Ont.

RUBBER COATS
Wm. C. Wilson & Co., Toronto, Ont.

SAW MILL MACHINERY
Wm. Hamilton Co., Peterboro, Ont.
Preston Woodworking Machy. Co., Preston, Ont.
Waterous Engine Works Co., Brantford, Ont.
Yates Machine Co., P. B., Hamilton, Ont.

SAWS, BAND
Preston Woodworking Machy. Co., Preston, Ont.

SCALES, BOILERS, ENGINES
Can. Fairbanks-Morse Co., Montreal, Que.

SCOWS
Collingwood Shipbuilding Co., Collingwood, Ont.
Polson Iron Works, Toronto, Ont.

SCREENS, WIRE
Dennis Wire & Iron Works Co., London, Ont.

SCREWS, COACH
London Bolt & Hinge Works, London, Ont.

SEAM PAINT
Kuhls, H. B. Fred., 6411 3rd Ave., Brooklyn, N.Y.

SEPARATORS, OIL, STEAM
Darling Bros., Ltd., Montreal, Que.
Mason Regulator & Engin. Co., Montreal, Que.
Smart-Turner Mach. Co., Hamilton, Ont.

SHAFTING
Wm. Hamilton Co., Peterboro, Ont.
Mitchell Co., The Robert, Montreal, Que.
Wilkinson & Kompass, Hamilton, Ont.

SHAFTING, BRONZE
Empire Mfg. Co., London, Ont.
Tallman Brass & Metal Co., Hamilton, Ont.

SHEARS
Bertram & Sons, Ltd., John, Dundas, Ont.
Norbom Engineering Co., Philadelphia, Pa.

SHIPBUILDING TOOLS
Aikenhead Hardware, Ltd., Toronto, Ont.
Can. Ingersoll-Rand Co. Sherbrooke, Que.

SHIPS, BUILDERS OF
Can. Vickers, Ltd., Montreal, Que.
Collingwood Shipbuilding Co., Collingwood, Ont.
Doxford & Sons, William, Sunderland, England.
Georgian Bay Shipbuilding & Wrecking Co., Midland, Ont.
National Shipbuilding Co., Goderich, Ont.
Polson Iron Works, Toronto, Ont.
Port Arthur Shipbuilding Co., Port Arthur, Ont.
Yarrows, Limited, Victoria, B.C.

The "Spraco" *Paint Gun*

It Paints 3 to 12 Times Faster

than paint can be brushed on. Moreover, in the hands of an unskilled man, Spraco Pneumatic Painting Equipment produces finely finished surfaces, free from streaks and brush marks.

Easily it covers rough surfaces and those extremely difficult or impossible to reach with a hand brush.

Spraco Pneumatic Painting Equipment will save you 50 per cent. or more on labor and 15 per cent. or more on paint. Shipbuilding and industrial plants using Spraco Equipment will corroborate this statement.

Say you are interested.

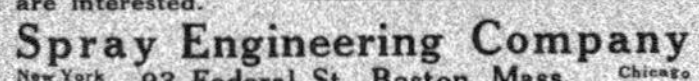

Spray Engineering Company

New York **93 Federal St., Boston, Mass.** Chicago

Cable Address—SPRACO BOSTON Code—WESTERN UNION

Representatives in Canada:

RUDEL-BELNAP MACHINERY CO.

95 McGill Street, Montreal, Quebec, Canada

The "Little David" Caulker on the deck of the "War Ottawa."

Pneumatic caulking has revolutionized this part of shipbuilding; in the building of the "War Ottawa" the "Little David" caulker broke all records. The tool gives the oakum the correct twist as it enters the seam, and is approved by both the British and American Lloyd's.

Any of our branches will give your further information.

Canadian Ingersoll-Rand Company, Limited

General Offices: Montreal, Quebec

Branch Offices:

SYDNEY, SHERBROOKE, MONTREAL, TORONTO, COBALT, WINNIPEG, NELSON, VANCOUVER

SHIP BROKERS
Page & Jones, Mobile, Ala.
SHIP PLATES
Nova Scotia Steel & Coal Co., New Glasgow, N.S.
SLEDGES
Wilkinson & Kompass, Hamilton, Ont.
SLINGS
Hopkins & Co., F. H., Montreal, Que.
Morris Crane & Hoist Co., Herbert, Niagara Falls, Ont.
SPECIAL MACHINERY
Corbet Fdry. & Machine Co., Owen Sound, Ont.
Wm. Hamilton Co., Peterboro, Ont.
Miller Bros. & Sons, Ltd., Montreal, Que.
Smart-Turner Mach. Co., Hamilton, Ont.
SOCKETS, PULL KEY AND KEYLESS
Harvey Hubbell Co. of Can., Toronto, Can.
SMOOTH-ON
Wm. C. Wilson & Co., Toronto, Ont.
SPIKES
Wm. C. Wilson & Co., Toronto, Ont.
SPRAY COOLING SYSTEMS
Spray Engineering Co., Boston, Mass.
STAMPINGS, SHEET METAL
Pedlar People, Ltd., Oshawa, Ont.
STEAMSHIP AGENTS
Darling Bros., Ltd., Montreal, Que.
Page & Jones, Mobile, Ala.
STEAM SPECIALTIES
Can. Fairbanks-Morse Co., Montreal, Que.
Corbet Fdry. & Machine Co., Owen Sound, Ont.
Empire Mfg. Co., London, Ont.
Mitchell Co., The Robert, Montreal, Que.
STEAM TRAPS
Aikenhead Hardware, Ltd., Toronto, Ont.
American Steam Gauge & Valve Mfg. Co., Boston, Mass.
Darling Bros., Ltd., Montreal, Que.
Empire Mfg. Co., London, Ont.
Mason Regulator & Engin. Co., Montreal, Que.
Mitchell Co., The Robert, Montreal, Que.
Smart-Turner Mach. Co., Hamilton, Ont.
STEEL, HIGH SPEED
Hopkins & Co., F. H., Montreal, Que.
Nova Scotia Steel & Coal Co., New Glasgow, N.S.
Wilkinson & Kompass, Hamilton, Ont.
Zenith Coal & Steel Products, Ltd., Montreal.
STEEL SHELVING
Dennis Wire & Iron Works, London, Ont.
STEEL WORK, STRUCTURAL
Babcock & Wilcox, Ltd., Montreal, Que.
Can. Welding Works, Montreal, Que.
Corbet Fdry. & Machine Co., Owen Sound, Ont.
Wm. Hamilton Co., Peterboro, Ont.
MacKinnon Steel Co., Sherbrooke, Que.
STEERING GEARS
Corbet Fdry. & Machine Co., Owen Sound, Ont.
Hopkins & Co., F. H., Montreal, Que.
Engr. & Mach. Wks. of Can., St. Catharines, Ont.
Wm. C. Wilson & Co., Toronto, Ont.
STOCK RACKS FOR BARS, PIPING, ETC.
Mitchell Co., The Robert, Montreal, Que.
Morris Crane & Hoist Co., Herbert, Niagara Falls, Ont.
STOKERS, MECHANICAL
Babcock & Wilcox, Ltd., Montreal, Que.
SUPERHEATERS, STEAM
Babcock & Wilcox, Ltd., Montreal, Que.
Goldie & McCulloch Co., Galt, Ont.
SWITCHBOARDS, ELECTRIC
Crouse-Hinds Co. of Can., Ltd., Toronto, Ont.
TALLOW
Can. Economic Lubricant Co., Montreal, Que.
Wm. C. Wilson & Co., Toronto, Ont.
TANKS, STEEL
Can. Welding Works, Montreal, Que.
Corbet Foundry & Mach. Co., Owen Sound, Ont.
Goldie & McCulloch, Ltd., Galt, Ont.
Hopkins & Co., F. H., Montreal, Que.
MacKinnon Steel Co., Sherbrooke, Que.
Marsh Engineering Works, Belleville, Ont.
Port Arthur Shipbuilding Co., Port Arthur, Ont.
Reed & Co., Geo., Montreal, Que.
Scaife & Sons Co., Wm. B., Oakmont, Pa.
TANKS (AIR, GAS AND LIQUID)
Can. Welding Works, Montreal, Que.
MacKinnon Steel Co., Sherbrooke, Que.
Marsh Engineering Works, Belleville, Ont.
Scaife & Sons Co., Wm. B., Oakmont, Pa.
TAPPING MACHINES
Mueller Mfg. Co., H., Sarnia, Ont.
TELEGRAPHS, SHIPS
McNab Co., Bridgeport, Conn.
Cory & Sons, Inc., Chas., New York, N.Y.
Morrison Brass Mfg. Co., James, Toronto, Ont.
Wilson & Co., Wm., Toronto, Ont.
TELEPHONES, MARINE
Cory & Son, Inc., Chas., New York, N.Y.
McNab Co., Bridgeport, Conn.
TESTERS, METER
Mueller Mfg. Co., H., Sarnia, Ont.
THUMB SCREWS AND NUTS
Canada Foundries & Forgings, Welland, Ont.
Mitchell Co., The Robert, Montreal, Que.
United Brass & Lead, Ltd., Toronto, Ont.
TRACK SYSTEMS
Morris Crane & Hoist Co., Herbert, Niagara Falls, Ont.
TRAVELLING BLOCKS
Morris Crane & Hoist Co., Herbert, Niagara Falls, Ont.
TROLLEYS
Can. Fairbanks-Morse Co., Montreal, Que.
Morris Crane & Hoist Co., Herbert, Niagara Falls, Ont.
TROLLEY HOISTS
Morris Crane & Hoist Co., Herbert, Niagara Falls, Ont.
TRUCKS, HAND, ELECTRIC
Aikenhead Hardware, Ltd., Toronto, Ont.
Can. Fairbanks-Morse Co., Montreal, Que.
TUBES, BOILER
Babcock & Wilcox, Ltd., Montreal, Que.
Broughton Copper Co., Ltd., Manchester, Eng.
Goldie & McCulloch Co., Galt, Ont.
TUBES, COPPER AND BRASS
Mueller Mfg. Co., H., Sarnia, Ont.
Tallman Brass & Metal Co., Hamilton, Ont.
Standard Underground Cable Co., Hamilton, Ont.
TUGS
Polson Iron Works, Toronto, Ont.
Collingwood Shipbuilding Co., Collingwood, Ont.
TURBINES, STEAM
Goldie & McCulloch Co., Galt, Ont.
TURBINES, DIRECT-DRIVING AND GEARED
Doxford & Sons, William, Sunderland, England.
TURNBUCKLES
Canada Foundries & Forgings, Welland, Ont.
Hopkins & Co., F. H., Montreal, Que.
TURNTABLES
Morris Crane & Hoist Co., Herbert, Niagara Falls, Ont.
SPIKES, SMALL RAILROAD
Severance Mfg. Co., S., Glassport, Pa.
UNIONS, ALL KINDS
Dart Union Company, Toronto, Ont.
VALVES, AIR
Mitchell Co., The Robert, Montreal, Que.
Mueller Mfg. Co., H., Sarnia, Ont.
VALVE, DISCS
Wm. C. Wilson & Co., Toronto, Ont.
VALVES
American Steam Gauge & Valve Mfg. Co., Boston, Mass.
Babcock & Wilcox, Ltd., Montreal, Que.
Can. Fairbanks-Morse Co., Montreal, Que.
Darling Bros., Ltd., Montreal, Que.
Empire Mfg. Co., London, Ont.
McAvity & Sons, Ltd., T., St. John, N.B.
Mason Regulator & Engin. Co., Montreal, Que.
McNab Co., Bridgeport, Conn.
Norbom Engineering Co., Philadelphia, Pa.
Williams Machinery Co., A. R., Toronto, Ont.
Wm. C. Wilson & Co., Toronto, Ont.
United Brass & Lead, Ltd., Toronto, Ont.
VALVES, FOOT
Aikenhead Hardware, Ltd., Toronto, Ont.
Mitchell Co., The Robert, Montreal, Que.
Smart-Turner Mach. Co., Hamilton, Ont.
VALVES, STOP, REDUCING, SAFETY CHECK, DISCHARGE, SUCTION
Aikenhead Hardware, Ltd., Toronto, Ont.
American Steam Gauge & Valve Mfg. Co., Boston, Mass.
Darling Bros., Ltd., Montreal, Que.
McAvity & Sons Ltd., T., St. John, N.B.
Mitchell Co., The Robert, Montreal, Que.
Morrison Brass Mfg. Co., James, Toronto, Ont.
VALVES, MIXING
Darling Bros., Ltd., Montreal, Que.
Mitchell Co., The Robert, Montreal, Que.
Mueller Mfg. Co., H., Sarnia, Ont.
VALVES, REDUCING, PRESSURE
Mitchell Co., The Robert, Montreal, Que.
Mueller Mfg. Co., H., Sarnia, Ont.
VALVES, STORM
Low & Sons, A., Glasgow, Scotland.
VARNISHES
Aikenhead Hardware, Ltd., Toronto, Ont.
Ault & Wiborg Co. of Can., Ltd., Toronto, Ont.
Leckie, Ltd., John, Toronto, Ont.
Reed & Co., Geo., Montreal, Que.
VENTILATORS, COWL
Can. Welding Works, Montreal, Que.
McNab Co., Bridgeport, Conn.
Mitchell Co., The Robert, Montreal, Que.
VENTILATION EQUIPMENT
Empire Mfg. Co., London, Ont.
Hopkins & Co., F. H., Montreal, Que.
Low & Sons, Ltd., Archibald, Glasgow, Scotland
WASHERS
London Bolt & Hinge Works, London, Ont.
Wm. C. Wilson & Co., Toronto, Ont.
WATER COLUMNS
Darling Bros., Ltd., Montreal, Que.
Mitchell Co., The Robert, Montreal, Que.
Morrison Brass Mfg. Co., James, Toronto, Ont.
WELDING, ELECTRIC
Can. Welding Works, Montreal, Que.
Hall Engineering Works, Montreal, Que.
Marine Welding Co., Buffalo, N.Y.
Beatty & Sons, M., Welland, Ont.
WATER SOFTENERS
Babcock & Wilcox, Ltd., Montreal, Que.
WATER SUPPLY SYSTEMS
Mueller Mfg. Co., H., Sarnia, Ont.
WATER HEATERS
Empire Mfg. Co., London, Ont.
Morrison Brass Mfg. Co., James, Toronto, Ont.
WHISTLES AND SYRENS
Empire Mfg. Co., London, Ont.
McAvity & Sons, T., St. John, N.B.
McNab Co., Bridgeport, Conn.
Mitchell Co., Ltd., Robt., Montreal, Que.
Morrison Brass Mfg. Co., Jas., Toronto, Ont.
WINCHES, CARGO
Aikenhead Hardware, Ltd., Toronto, Ont.
Corbet Fdry. & Machine Co., Owen Sound, Ont.
Hopkins & Co., F. H., Montreal, Que.
Marsh Engineering Works, Belleville, Ont.
WINCHES, DOCK, SHIP
Beatty & Sons, M., Welland, Ont.
Marsh Engineering Works, Belleville, Ont.
Miller Bros. & Sons, Ltd., Montreal, Que.
Morris Crane & Hoist Co., Herbert, Niagara Falls, Ont.
Wilson & Co., Wm. C., Toronto, Canada.
WINCHES, TOWING
Corbet Foundry & Mach. Co., Owen Sound, Ont.
WINCHES, TRAWL
Beatty & Sons, M., Welland, Ont.
Wm. C. Wilson & Co., Toronto, Ont.
WINDLASSES
Corbet Fdry. & Machine Co., Owen Sound, Ont.
Dake Engine Co., Grand Haven, Mich.
Hopkins & Co., F. H., Montreal, Que.
Wilson & Co., Wm. C., Toronto, Canada.
WIPER CAPS, OILER BOXES, ETC.
Mitchell Co., The Robert, Montreal, Que.
Morrison Brass Mfg. Co., James, Toronto, Ont.
WIRE, COPPER CLAD STEEL
Standard Underground Cable Co., Hamilton, Ont.
WIRE, COPPER, BRASS, BRONZE
Standard Underground Cable Co., Hamilton, Ont.
Tallman Brass & Metal Co., Hamilton, Ont.
WIRE, INSULATED
Standard Underground Cable Co., Hamilton, Ont.
WIRELESS OUTFITS
Marconi Wireless Telegraph Co., Montreal, Que.
WIRE ROPE
Zenith Coal & Steel Products, Ltd., Montreal.
WOOD WORKING MACHINERY
Aikenhead Hardware, Ltd., Toronto, Ont.
Can. Fairbanks-Morse Co., Montreal, Que.
Preston Woodworking Machy. Co., Preston, Ont.
Yates Mach. Co., P. B., Hamilton, Ont.
WOOD BORING TOOLS
Aikenhead Hardware, Ltd., Toronto, Ont.
Can. Ingersoll-Rand Co., Sherbrooke, Que.
WOODITE GAUGE GLASS WASHERS
Wm. C. Wilson & Co., Toronto, Ont.
WRENCHES
Canada Foundries & Forgings, Welland, Ont.
YACHT BROKER
Watt, J. Murray, Philadelphia, Pa.

Your
Workers
will smile
behind
STOCO
SAFETY GOGGLES
Industrial efficiency, which is really another way of saying maximum production, is absolutely dependent on the state of mind of the employee. If you can replace the strained, worried look of the man who must be constantly alert to dodge the eye hazard of his work with the assured smile of the man who knows his eyes are safe, you'll see his production begin to jump. In other words, don't expect 100% productive effort from the man who can only devote 50% of his attention to his work.
STOCO Safety Goggles are comfortable and completely protective. If you are interested in that kind of eye protection ask us on your letterhead for a sample.
STANDARD OPTICAL CO.
GENEVA, N.Y.

INDEX TO ADVERTISERS

Harris Heavy Pressure and its Advantages

1. **A complete immunity from hot bearings is secured, HARRIS HEAVY PRESSURE** having a lower co-efficient of friction than any other known metal.
2. **A scored journal is impossible**, and if through any failure of lubrication a bearing should run hot, **HARRIS HEAVY PRESSURE**, owing to its special properties, will act as a lubricant, saving the journal from injury and preventing any delay to traffic
3. It will stand the **heaviest pressures**, always running cool, even under the most trying conditions.
4. It will **wear** from 50 to 100 per cent. longer on general machinery bearings than any other Babbitt metal.
5. It effects a **saving in lubrication.**
6. It **preserves the journals**, and materially **increases their life.** A journal after running a short time with **HARRIS HEAVY PRESSURE** attains a perfectly smooth and highly polished surface.
7. It is easily applied and, if properly applied, no abrasive force will remove it.
8. **Its cheapness.** The first cost is moderate. It gives a longer life to the bearings, resulting in a great economy, as the number of renewals is thereby considerably reduced; its specific gravity is low in comparison with other metals; does not deteriorate with re-melting; and these advantages, together with its unequalled anti-fricion properties, render it the **cheapest** as well as the **best metal** for all general machinery bearings.

ORDER A BOX FROM OUR NEAREST FACTORY

THE CANADA METAL CO., LIMITED

HAMILTON
MONTREAL

TORONTO

WINNIPEG
VANCOUVER

TANKS

We make steel tanks of all kinds, for compressed air, gas or liquids; also ornamental iron work, iron stairs, gratings and railings, structural steel work, ventilators, etc.

Canadian Welding Works, Ltd.
MONTREAL, QUE.

A Pair of Winners

For High-Class Marine Service

Send us your inquiries for Simplex and Duplex Vertical Pumps, also Horizontal Pumps.

The Smart-Turner Machine Co.
LIMITED
Hamilton - Canada

CIRCULATES IN EVERY PROVINCE OF CANADA AND ABROAD

MARINE ENGINEERING of Canada

A monthly journal dealing with the progress and development of Merchant and Naval Marine Engineering, Shipbuilding, the building of Harbors and Docks, and containing a record of the latest and best practice throughout the Sea-going World. Published by
The MacLean Publishing Co., Limited
MONTREAL, Southam Building TORONTO 143-153 University Ave. WINNIPEG, 1207 Union Trust Bldg. LONDON, ENG., 88 Fleet St.

Vol. VIII. Publication Office, Toronto—November, 1918 No. 11

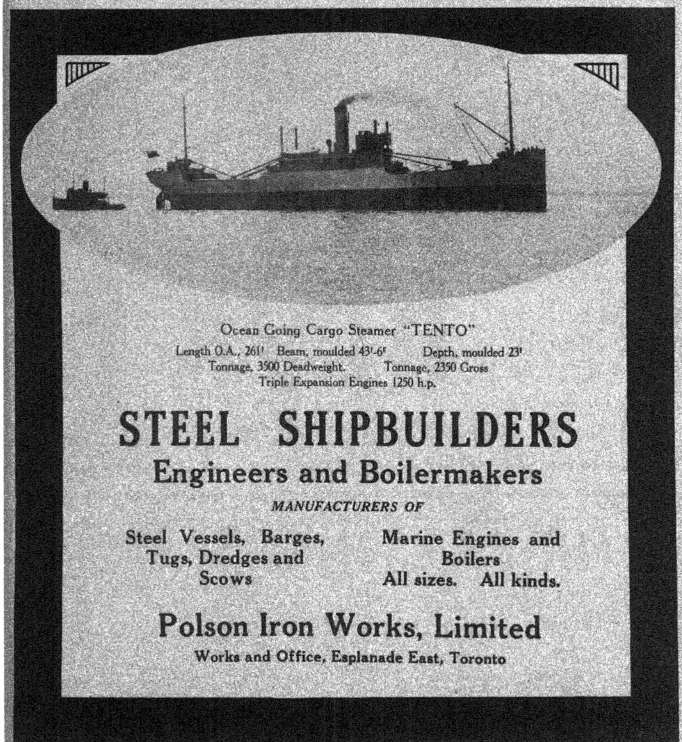

If what you need is not advertised, consult our Buyers' Directory and write advertisers listed under proper heading.

Planning for War —and Peace

In the field of manufacture, as on the field of battle, American strategy has been a big factor in the winning of the war.

Our fighting men at the front, and our fighting men back of the front, down to the humblest of the workers in shop and shipyard, have given heroic demonstrations of their courage, initiative, resourcefulness and spirit of self-sacrifice.

Our war plans have made good. But what about our Peace plans?

Now that the Unconditional Surrender of the Hun looms as an imminent possibility, with its accompanying cessation of war orders and the resumption of intense competition, what does it mean to us industrially? Is it a threat or a promise?

We believe it is a promise—but only to those who continue to seek diligently *now* for better methods of management and production.

Looking over our own record of the war plans in the making and carrying out of which the Keller organization has been privileged to co-operate, it is gratifying to realize that while we have been working for the success of the Allied cause, we have also been working for the post-war success of ourselves and our customers.

By striving steadily to increase our production of Keller-Made Master-Built Pneumatic Tools, and to improve the tools themselves so that they can be depended upon to increase the production of the Shipyards, Shops and Foundries using them, we have not only "done our bit" to render certain the outcome of the great Battle for Democracy, but have at the same time and to the same extent helped to pave the way for the winning of the great Battle for World Trade that is to follow.

Preventable inefficiency is Treason during the war. After the war it will be Business Suicide.

William H. Keller *President*

KELLER PNEUMATIC TOOL COMPANY, GRAND HAVEN, MICH., U. S. A.

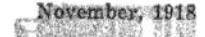

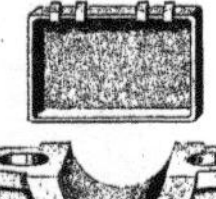

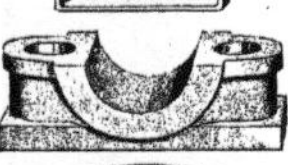

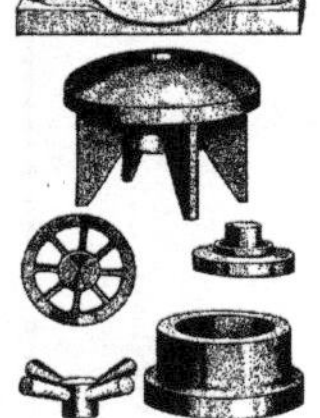

We Are Equipped To Deliver On Short Notice Ship Castings In All Sizes from a Few Ounces To Many Hundred Pounds

TO impress upon your attention the wide range we are equipped to manufacture, we illustrate here a group of small miscellaneous ship castings. This is in order to draw a comparison. At the same time our Foundry daily turns out Castings of one thousand pounds and more in weight.

We are at the present time supplying Government contractors with their brass and bronze castings for shipbuilding purposes.

An interesting booklet entitled "MUELLER Castings" is yours for the asking. Illustrated therein will be found a widely diversified collection of castings that we have made on special order for numerous customers. The castings shown in this advertisement will present a fair idea of what we are making in the line of small ship castings. Send us your specifications. They will be treated as confidential, and even if we can't get together in a business sense we may be able to impart some information on your problem that will be of assistance to you.

H. MUELLER MANUFACTURING CO., LTD.

SARNIA, CANADA

"Victory Loan, 1918, H. Mueller Mfg. Co., Ltd., first in Canada to exceed quota and receive Honor Flag."

LOW'S SPECIALITIES FOR SHIPS

We are Specialists in Appliances for

HEATING and VENTILATION
FITTINGS for PLUMBING WORK

Also all kinds of

BRASS and SHEET METAL WORK

REQUIRED IN THE CONSTRUCTION OF SHIPS.

We have supplied many well-known vessels with all their requirements in the departments referred to.

Low's Patent "Unifix" Coupling Joints

TIGHT AT ANY PRESSURE
FOR ALL TIME

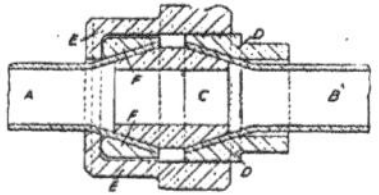

No jointing required.
All metal to metal.
No brazing required.

Pipes merely coned out and jointed together

Low's Patent Storm Valves

Fitted with Indicating Deck Plates. Approved by the Board of Trade

New design to meet recent Board of Trade requirements. No sluice or other valve required. Minimum of space occupied.

ARCHIBALD LOW & SONS, LTD.
MERKLAND WORKS, PARTICK, GLASGOW

LIVERPOOL AGENTS: A. J. Nevill & Co., 9 Cook St.
N.E. COAST AGENTS: Ryder, Mumme & Co., Milburn House, Newcastle-on-Tyne
LONDON OFFICE: 58 Fenchurch St., E.C.3

Agents for Canada: A. J. Widston & Co., 17 John St., Montreal; also at 2 Rector St., New York

If what you need is not advertised, consult our Buyers' Directory and write advertisers listed under proper heading.

MARINE ENGINEERING
of Canada

Volume VIII. TORONTO, NOVEMBER, 1918 Number 11

Feeding and Circulating the Water in Boilers

Earlier Forms of Feeding—Flow of Water in Circular Boilers—Flow of Water in Water Tube Boilers—From a Paper Read Before the Institute of Marine Engineers, London

By JOHN WATSON

IT has been vividly expressed by leading men in the scientific and technical world that it is incumbent upon all to use every means possible for improving and economizing in every direction, and probably most engineers have been evolving in their minds various schemes to attain this end both in leisure hours and in the larger spheres of experience and work.

Circular Boilers

In order that we may follow clearly the line of thought I propose to lay before you this evening, I should like to show you (what the older engineers will know well, but which may not be so familiar to the younger men) some of the methods which have been adopted for introducing the feed into boilers, and I am selecting the older fashioned circular boiler for general illustration, as most of us have had our training in this and the locomotive and old Navy type.

As far as I am aware the earliest boiler had no internal pipe at all, the spigot on the check valve merely entering into the boiler.

A later form had the spigot lengthened somewhat, and opened out on the top or bottom of the spigot to prevent the water from impinging so directly on the walls of the boiler or tubes.

A still later form had a short internal pipe carried up to within a few inches of the water level, and the water delivered there as shown in No. 2B.

In each of these methods (the water being delivered in bulk) it was found that corrosion of the plates and tubes alongside was very rapid.

Even quite recently a boiler was fitted with the feed as shown at No. 3B.

Then the pipe was carried across the top of the tubes and perforated with small holes either on one or both sides. This, of course, spread over the surface of the boiling water the cold feed in the form of a "spray," instead of the full volume in one local spot. The improvement of this form lies in the fact that, instead of the large volume of comparatively cold feed being delivered in one local spot, the feed, separated into a large number of small streams, is spread over a large surface of the water. This, however, still presents itself to the writer's mind as forming a "cold blanket" on the hot boiler water.

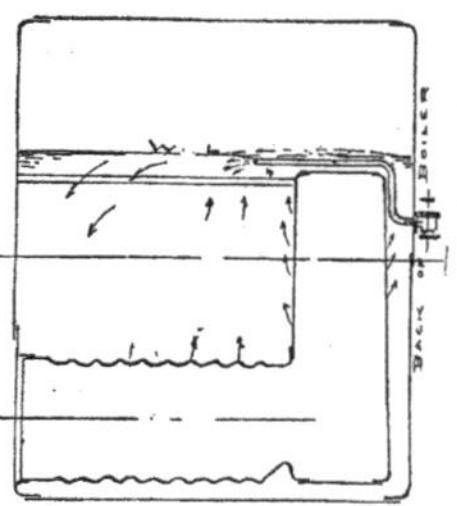

FEED PIPE ON BACK OF BOILER
INTERNAL PIPE—OPEN END

We may assume it is an axiom that there is always a right and wrong way in everything and a correct sequence in all our work if we can only find it out and understand it, because Nature's laws are fixed and immovable and insistent.

If we take a small quantity of cold water and pour it in a kettle in which the water is boiling the water immediately ceases to boil, and this fact is what has just been mentioned, and will be referred to as the "blanket" or "blanketing."

Experiments were made many years ago (if my memory serves me right, by Messrs. Vickers), in which it was demonstrated beyond doubt that the water at the back of a single ended boiler, when under full steam, was higher than at the front or tube end.

Now this is what appears to the writer should be expected.

It is perhaps not easy to make the effect quite clear in a diagram, but we hope the following explanation will make the principles clear. The fiercest heat will naturally be at the combustion chamber and along the tops of the furnaces, and, in a lesser degree, off the tubes. The furnaces being so close together, the water in the whole of the centre part of the boiler between X and X will be rising, but rising most rapidly at the back end and, in a lesser degree (a large proportion of the heat having passed into the boiler water) at the front; therefore, the circulation will be taking place from the centre of the boiler towards the wings, and from the back of the boiler towards the front. Then, as this water is also very hot, having already passed over the hot surfaces, it is unable by the natural law to disturb the water at the bottom (because the water is colder at the bottom), but simply flows over and bypasses this colder water (in the shaded portion of 5B) and rises again to the surface.

It would appear there will probably be a difference of about 1,700° F. between the temperature of the gases in the furnace and on combustion chamber plates and the temperature of the gases at the front end of the boiler where the gases pass into the uptake. This figure of 1,700° is estimated, assuming the temperature in the furnace and combustion chamber at about 2,500° F., 500° F. at

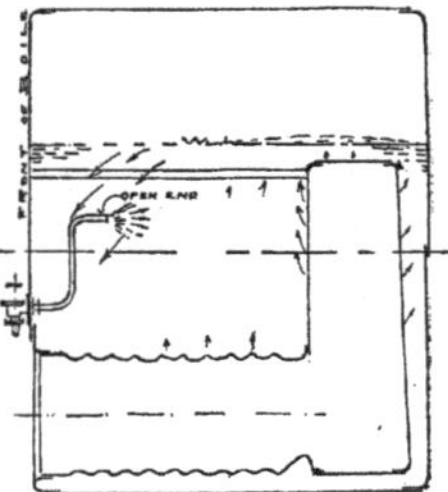

FEED PIPE ON FRONT OF BOILER
INTERNAL PIPE—OPEN END

base of funnel and 800° F. at the outlet of tubes (front end of boiler opening to uptake)—*i.e.*, 2,500°—800°=1,700° F. These figures, while only approximate (one would like to have actual figures) should, we think, be sufficient to indicate that the flow of circulation will probably be as shown in sketch No. 5B.

If, then, this is the natural line of circulation in this boiler, the feed water should be introduced in some way following the lines of flow and facilitating them and on no account going contrariwise.

It therefore does not appear to be sound to introduce the cold water into the hottest part of the boiler, but rather that it should be introduced in some position where the water would naturally be cooler and where the temperature would more nearly approximate to the temperature of the incoming feed.

Water Tube Boilers

It will be convenient now to examine the probable flow and circulation in a water-tube boiler. Water-tube boilers have been so varied in type and design, and generally with such a very small volume of water, rapid evaporation and generation of steam, that "Feed" and "Circulation" have been vital factors.

In the early type fitted in the *Propontis*, after an investigation due to an accident, the boiler design generally was not unfavorably criticized in the report, but the "circulation was deficient."

Take the case, however, of a modern water tube-boiler, in the simplest form, which has become more or less familiar to engineers, known as the "three drum type," with practically straight tubes connecting the steam and two lower water drums, and which has proved of such efficient service in Navy vessels, and also some other services. In the extremely simple form of the latest design there appears to be no question now in regard to circulation, even although the large downcomers—formerly considered absolutely essential for circulation—are no longer fitted.

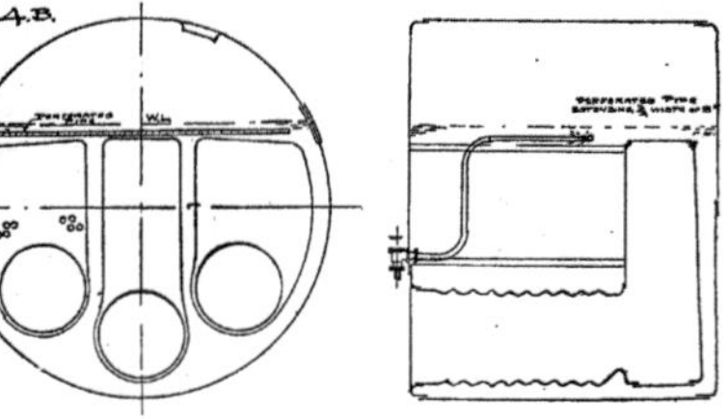

No. 4 B.—Internal Pipe carried across the top of tubes. Perforated in one or both sides.

Following the natural sequence already indicated in regard to the circular boiler, the circulation will be *up* the tubes next to and nearest to the fire in the combustion chamber, and *down* the outer row or rows furthest away from the fire.

In the case of an ordinary W.T. boiler working under forced draught there will probably be a difference of 2,250° F. between the temperature of the gases impinging on the tubes next to the fire, and the temperature of the gases (having passed through the stack of tubes, the heat having passed into the boiler water), impinging on the outer row of tubes remote from the fire.

This estimate assumes the temperature of the gases in the combustion chamber to be about 3,000° F., the temperature at the base of the funnel about 500° F., and just outside the outer row of tubes 750° F., *i.e.*, 3,000°—750°=2,250° F. This difference of 2,250° appears sufficient to give every confidence that the natural circulation of the boiler will be as indicated in sketch No. 7B., which is intended to show the natural stream lines of circulation.

If the feed is delivered in the top (or steam drum), where the temperature of the boiling water corresponds to the steam pressure, it must necessarily act (like the cold water in the kettle) as a blanket on the boiling water.

The writer some years ago evolved a method of heating the feed water for a boiler distribution in the boiler, but through stress of other matters it was allowed to lapse.

Recently the matter has again been brought into prominence by reason of the large units into which boiler power has now been divided, and the ideas have been further developed with a view to practical application.

The subject of heating the feed water in the Scotch boiler engaged the attention of the Institute of Marine Engineers at a comparatively recent date, and I do not propose to enlarge further upon the general principles involved, but to lay before you the results of the investigation referred to above, in the hope that it may be of interest and utility.

A great number of devices have been developed for heating the feed water by means inside, as well as outside, the boiler, but these have frequently been found too cumbersome, costly, or complicated to meet with general acceptance, especially for vessels on long voyages.

It is generally accepted that boilers in which the circulation is good, *i.e.*, in which the water flows freely and rapidly over the heating surfaces, absorbing the heat, are the most efficient.

Further, as is well known by engineers, boilers in which any quantity of water lies dead, and in which the circulation is not freely cyclical and positive, will not be economically efficient, and such a condition has also a deleterious effect on the boiler, producing undue strains, causing distortion, leaky seams, deterioration of material, and cracked plates.

Looking now broadly at the subject of "feeding" the boilers, Nature provides us with an analogy, which demonstrates for us the importance of correct "feeding" in the marvellous provision which is made in the animal kingdom, so that the "feed" shall pass through *preparatory processes* before it is ready for assimilation into the vital energizing parts.

With these fundamental points before us, and bearing in mind that the general principle in regard to circulation of the water in the boiler has frequently been discussed and demonstrated, and that the hot water must, by reason of the change which has taken place from being cold, *rise to the surface*, it will be readily perceived that the manner of introducing the feed into the modern boiler with a large heating surface and high pressure is of paramount importance if the highest duty is to be obtained, and *we should not be satisfied with anything less than the highest duty.*

In the special devices referred to, the effort has been made to utilize the streams of natural circulation of the boiler, first, to mingle with the incoming feed in an enlarged internal feed pipe, and then in the "outlet" nozzle, to again aid the circulating streams by inducing them through the "gills," so that the feed water will be thoroughly mixed with the boiler water *before it actually touches on any part of the boiler elements.*

Taking now the water-tube boiler for illustration first:—

Figure 1 shows the inlet connection in the lower drum of a water-tube boiler with upper (or steam drum) and lower (or water drum), and the inlet for the hot boiler water is shown at A, the feed nozzle at B, which is designed so as to draw in the hot water with the feed, and the helical or suitable shaped vane marked C. gives a turning mixing effect to the now united mass of flowing water with a view to thorough mixing. After passing this vane, the first nozzle comes into action and a certain portion of the mixed water delivered into the boiler water. The remaining portion of the water then passes through another simple injector (similarly designed to that where the feed enters the internal pipe), and the same process is repeated through a series of injectors and nozzles, as shown in the arrangement Fig. 2. It will thus be seen that along the whole length of the boiler there will be a continual stream of water—representing the "feed," plus, say, twice its volume—being directed toward the tubes in the hottest part of the boiler.

The illustration Fig. 2 shows the arrangement where the feed is delivered into the lower drum, and the upper part of the illustration where the feed is delivered into the steam (or upper) drum —the hot water inlet orifices being arranged on the top side in the lower drum and on the under side in the upper (or steam) drum.

The outlet nozzles are especially designed in series to give an easy flow of the water through them and to draw the hot water through the "gills," so as to further mix and heat the feed.

It will easily be seen that from the position where this internal pipe is placed in the lower drum, with its hot inlet orifices on top, and the suitably placed "gills," there will be a considerable effect in *inducing the down current through the outer tubes* (those further away from the fires), and in rapidly circulating the currents *up* the tubes nearest the fire.

It does not appear as if there could be any doubt in general principle that the correct place to deliver the "feed" is the lower (or water) drum, where the water will naturally be cooler, and where the fresh supply of water can be *delivered at as nearly as possible the same temperature as the surrounding water,* and this is one of the principal features of the arrangements under consideration. When the feed is thus delivered into the lower drums, either the check valve on *each* lower drum should be capable of delivering the full supply feed (in case of one check valve on either side breaking down or getting out of order), or there should be a levelling pipe between the lower

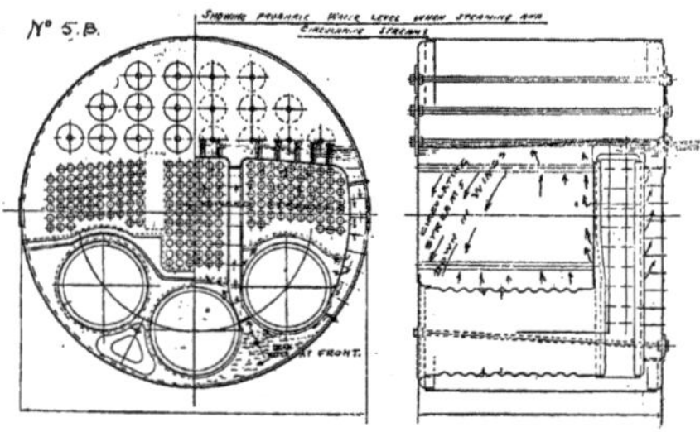

VIEW SHOWING DIFFERENCE IN WATER LEVEL

drums to ensure the maintenance of an equal head of water in the stack of tubes on each side.

In regard to the possibility of air getting in and pocketed no fear need be entertained on this point so far as the tubes are concerned, in view of their approximation to the vertical (any air bubbles running up on the cool side of the tube), provided no feed division plate is fitted in the drum, and if concern is felt as to the "bead" around the tubes a slight notch, rounded, could be made if neces-

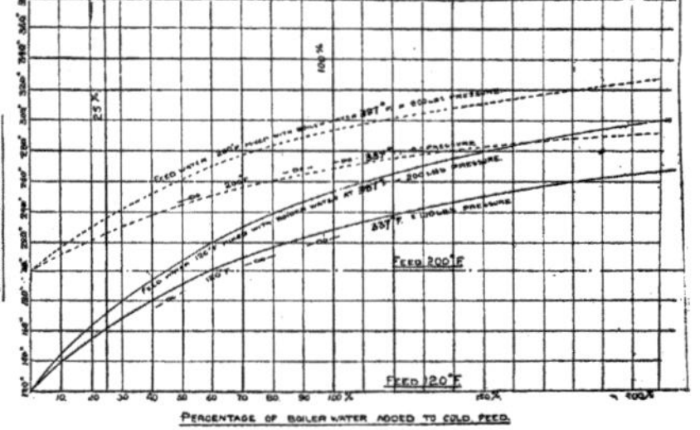

sary, to allow the last air particles to escape.

Should it be necessary, however, for other reasons to deliver the "feed" into the upper drum, as shown on Fig. 2. (in the upper part), then the fittings shown will overcome the objections (inasmuch as the "feed" is heated in the internal pipe) to the ordinary spray pipe, and assists the circulation, but cannot, for the reasons already shown and by reason of "blanketing" the hottest water and steam rising from the tubes be quite so efficient.

The outlook nozzles are shown clearly in Figs. 1a and 2a.

The arrangements as applied to the ordinary type of circular boiler is shown in Fig. 3, ffrom which it will be observed that the application of the general principle of the apparatus is the same, the hottest water being taken from the upper surface, delivered in the more or less "dead" water at the bottom, and flowing up between the furnaces into the upward hot streams rising from the furnaces and tubes.

The idea of delivering the feed into this "dead" water has been frequently advocated, and in some patented arrangements has been done, but we have every confidence that the simple apparatus now being laid before you will make its adoption and utility more generally possible and desirable. It appears obvious that heat will pass more readily into the feed water by *direct contact with hot water* than by injection of steam, *by reason of their being more readily mixed.* We find that 1 lb. of feed water being delivered into the internal feed pipe at 120°, and mixed with only ¼ lb. (an addition of 25 per cent.) of the hottest water in the boiler, at, say, 387°, would produce water of a mean temperature of 173.6° to be delivered into the boiler; while if it had been mixed with an *equal* quantity of hot water in the enlarged internal pipe (an addition of 100 per cent. hot water instead of 25 per cent.), the mean temperature (if thoroughly mixed) would be 237°.

The proportions of the mixed feed which might be attained in the internal pipe are shown in diagram No. 4, and, examining this, you will find that with "feed" water being delivered at the boiler at 200°, and being mixed with twice the quantity of the hottest water in the boiler (corresponding to 200 lbs. per square inch), the water as actually delivered into the boiler would be about 323°; and this we estimate could be attained for a feed supply pipe with check valve 2½ in. bore by an internal pipe enlarged to only about 4½ in. dia., with fittings, as shown.

It will be observed here that with feed at 120° into a boiler working at 100 lbs. pressure per square inch, the advantage gained by mixing the feed inlet, as proposed, with twice the quantity of the hot boiler water, would be to raise the temperature of the water delivered at the outlet nozzles by 180°, while with feed at 200° the outlet would be raised by 68°.

From this it will be evident the advantage that would be gained in installations where no external feed heater is used.

Fig. No. 6B shows the special internal pipe when fitted near the working water levels of the boiler.

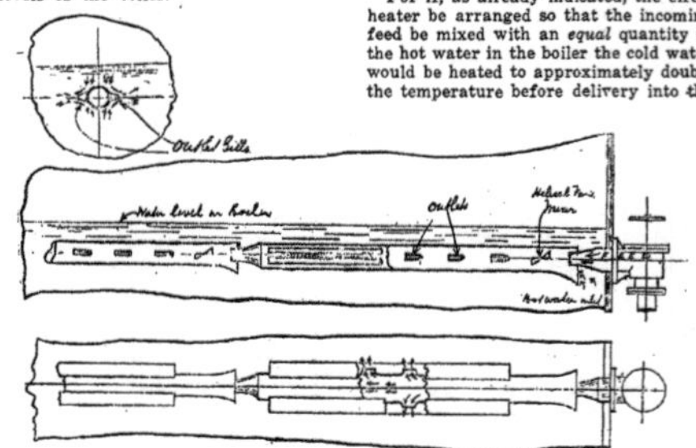

FIG. 2—ARRANGEMENT IN WATER TUBE BOILER

The table shows the effect of "mixing" in some specific cases, and it should be borne in mind that this heating of the "feed" takes place in the internal feed pipe before the feed can reach any of the surfaces of the boiler itself.

With surface feed heaters there is, of course, heat loss in passing through the tubes and frictional loss in driving the water through them, producing additional "head" on the pumps.

With any of the ordinary feed heaters it has generally been found impracticable to heat the feed to more than 200° F. at the pumps, and this temperature in long lengths of feed pipes will be considerably reduced before reaching the boilers—probably down to 190° or 185°, so that, even where these are fitted, there appears ample scope for the devices under consideration to be of very considerable advantage in safe and economical feeding.

In vessels with a large number of auxiliaries and in turbine vessels it may probably be desirable to use a feed heater for utilizing any surplus exhaust steam not required by turbines or evaporators, with the fitting described, to complete the heating of the feed and promote circulation.

In many vessels, however, we have every confidence that the feed can be effectively dealt with in the boiler itself by the means which we have endeavored to lay before you, without any further encumbrances in the way of external feed heaters.

It will readily be seen that while it may be necessary to obtain a high degree of economy throughout by utilizing exhaust steam for feed heating purposes, *i.e.*, condensing this steam by imparting its heat in this process to the feed, yet, even under these conditions, it is often necessary to put a considerable quantity of cold water into a boiler. The circo-feed heater under these conditions cannot fail to be of distinct value in reducing the bad effect of the cold water on the boiler.

For if, as already indicated, the circo-heater be arranged so that the incoming feed be mixed with an *equal* quantity of the hot water in the boiler the cold water would be heated to approximately double the temperature before delivery into the boiler, and it is manifest that the circo-heater has only to be enlarged at a small extra cost so that the feed mixes with *twice* the quantity or more, that a greater benefit may be derived.

We therefore consider that whether a feed heater be adopted for heating the feed for economical purposes or none, the fitting of a circo-heater ensures that *whatever the temperature of the feed*, its temperature is raised considerably (from, say, 25 per cent. to 150 per cent., according to the size of internal pipe and nozzle, temperature of the feed, and the temperature corresponding to the press of the boiler) *before the feed water is actually mingled with the body of the water in the boiler.*

You will also observe that there are no valves, springs or other working fittings to be adjusted or to get out of order, and that if properly designed and fitted with suitable materials it should not require any attention, except on occasional overhaul to ensure pipes and nozzles are clear.

Referring now to the opening sentence, it will be observed that no reference has been made to any economy to be secured by the use of this special appliance. On carefully considering the matter, it will be perceived that there can be no direct economy, for the simple reason that the fitting is self-contained in the boiler, and consequently no heat is saved which would otherwise be lost.

But while this is true we may fairly expect an indirect economy, due to steadier steaming under all conditions—a direct economy in repairs and, in addition, lengthening the life of the boiler, with a greater confidence in its safety. These are surely valuable assets to the owner and his clients, as well as to the sea-going engineer.

It is, generally speaking, only owners and superintendents who have the opportunity of fitting and testing proposals of this nature, and the desire to test depends upon how sound the proposals are, how clearly they have been stated and the principle made clear; and whether they appear to offer something which, if attained, is worth having and will provide an adequate return for the outlay.

How far these conditions have been fulfilled remains to be seen, but the effort has been made to attain an efficient natural circulation of the water in the

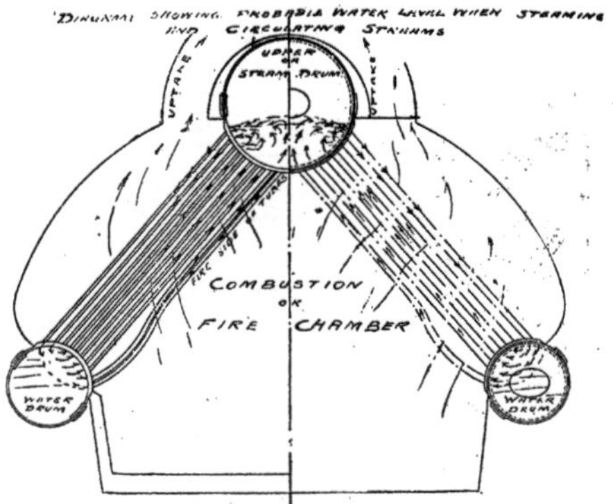

CIRCULATION IN WATER TUBE BOILER

boiler and at the same time heating the cold feed before delivering into the boiler *in the simplest possible manner*. Further, how far the devices would achieve the purpose in view would require actual demonstration in practice, but we should be pleased to have the frankest criticism with a view to confirming that the proposals meet the conditions just now stated.

THE SHIPBUILDING SITUATION

The recent statement on the merchant shipbuilding situation by Lord Pirrie, the Controller-General of Merchant Shipbuilding, was of very great interest as a statement of the present position of affairs, but it was also noteworthy in that it stated clearly that the First Lord of the Admiralty, Sir Eric Geddes, had definitely promised to release a number of the workers at present employed on warship construction for service in the construction of merchant ships, says *Engineering*, London. In his speech on the Navy Estimates in July of this year Sir Eric Geddes pointed out that upon this country had fallen almost exclusively the task of meeting and defeating the submarine peril, with the result that a very large proportion of the shipyard labor of the country was exclusively employed upon the construction of vessels for this service. Sir Eric Geddes further pointed out that as the American programme of anti-submarine construction materialized the situation as regards this form of shipbuilding effort in our yards would be eased and the labor would thus become available for replacing our own merchant ship losses. That this satisfactory condition of affairs has now been reached is a natural inference from the statement of Lord Pirrie, and it is to be hoped that the transfer of workers will take place at the earliest possible moment.

It is to be noted that among the shipyards which will be affected by the reduction of naval construction are some of the largest and most important in the country, and the class of labor employed in them contains a higher proportion of highly-skilled men than is usual in yards building cargo vessels exclusively. The yards affected are largely equipped, and the staff and other workers organized, for the rapid production of the highest class of merchant tonnage rather than purely cargo vessels, and it therefore becomes a question of importance whether the machinery and man-power now becoming available would be more usefully employed in producing standard cargo vessels or in replacing a number of the larger vessels of the liner type of which so many have been lost during the war. Of the number of such vessels lost no record is available, but the large number of important ships of which the loss has been officially reported pointed with certainty to the fact that after the war there will be a very serious shortage of this class of vessel. In view of the intense competition which is certain to follow the cessation of hostilities and the severe handicap under which British shipping companies will of necessity be placed owing to the loss of so many of their ships, it is worth while considering whether the construction of a number of such vessels, predicted in July by Sir Eric Geddes, could not now be sanctioned. The fact that the plant and machinery now becoming available would be working on its normal quality of product and therefore presumably at its maximum efficiency is a point to be borne in mind in deciding its sphere of greatest usefulness. It is neither desirable nor necessary that such vessels if constructed should be fitted out for their post-war trade until the end of the war, where such fitting-out involves elaborate passenger accommodation, etc., but the urgent need for vessels of the liner type will be at once admitted.

Another class for which there will be an urgent demand and for the construction of which a high degree of skill is required is that of the high-speed cross-Channel steamer. Most of those which survive war service will be quite unfit to resume their normal duties and their replacement will be an urgent necessity as traffic across Channel will be very heavy. This type of vessel is also very largely the product of the shipyards and labor which produces the liner class of vessel and is therefore suitable for construction by the plant and labor now to be set free from naval work.

The ultimate decision as to the types of vessel most urgently required and consequently the work upon which the extra building facilities will be employed must, of course, be dependent upon facts known only to the shipping authorities, but it is to be hoped that in making their decision due weight will be given to the need for the replacing of special types of merchant vessel and to the most efficient use of the facilities available.

HUNS BUILDING PEACE SHIPS

That the great maritime nations lately at war with Germany, and bitterly mindful of the injuries which they have suffered at German hands, should possibly have influence in determining the scope and conditions of Germany's maritime activities, is left entirely out of account (by the Germans), says the *Morning Post*, in a leading article. That is evident from the amount of tonnage that is being turned out of the German yards month by month; and the record is certainly a tribute to German resources after four years of war.

All the big lines are building huge vessels, on a scale that suggests that in the German shipyards there is no lack of either material or labor.

For instance, it is known that the Hamburg-Amerika Line is now building at Hamburg one vessel of 56,000 tons, another of 30,000 tons, and three others of over 20,000 tons. The North German Lloyd Stettin, is building two vessels of 35,000 tons, and more than a dozen others of 12,000 tons or over; and a similar record of activity comes from other shipyards.

It is estimated that at the present time very nearly a million tons of new shipping of the most efficient type is under construction in Germany; so that, with the restoration after the war of the shipping now interned in neutral ports, or in the hands of enemy powers, Germany would enter on the competition for overseas trade in a decidedly advantageous position.

The Admiralty returns of submarine losses during July show that the world tonnage was decreased by over 300,000 tons in that period; and of those losses more than half were British. As July is comparatively a good month, the leeway that world-shipping will have to make up may be gauged.

Surely no other persuasion is required to convince the Allies that no peace can be tolerable which does not ensure the satisfaction of their claims against Germany's mercantile marine. Whatever other indemnities may be exacted or foregone, it is impossible that the Allies should not require indemnity to the full limit of Germany's capacity to pay, for the destruction which she has wrought upon the world's merchant shipping.

THE FUTURE OF BRITISH SHIPPING

By M. M.

The future of British shipping after the war is somewhat obscure. There will, no doubt, be an immense amount of work to be done in re-provisioning Europe and in providing industry with the raw materials, of which it is so woefully short. Moreover, though a vast amount of tonnage now used as auxiliary to the Naval forces, will be freed soon after peace is attained, there will be immense demands on the shipping of the work in repatriating the armies of Britain and of the United States. The building programmes now under execution will not be sufficient to cope with all demands for transport, and accordingly there has recently been a great spurt in shipping companies' shares. A table has recently been published showing the movement of the securities of four important concerns.

Company	Price 7th Sept., 1918	Lowest Price 1918	1914
Cunard Line	5⅝	3⅞	1⅜
Furness, Withy & Co..	4 3-16	2⅞	1¼
P. & O. stock (deferred)	450	327	260
Royal Mail S. P. Co...	155	120	78

Collingwood.—The two steel vessels, "Thunder Bay" and "Paipoonge," have been cut in two at Collingwood Shipbuilding Co., for passing down to salt water. The two parts of the "Thunder Bay" have started their long trip, and the "Paipoonge" will start as soon as the tugs return from the "Thunder Bay" tow.

Fort William.—The last of the mine sweepers built by the Canadian Car and Foundry Company for the French government is expected to be ready to sail for France on Saturday. With her will go the remainder of the French marines long familiar to Fort William, and who have been very popular visitors on account of their excellent discipline and never failing good nature.

Description of U. S. Reinforced Concrete Ships

The 3,500-Ton Concrete Ship is of Same Dimensions as Wooden Ship—Strength of Hull—The Metacentric Height, Stability and Period—Local Strength of Hull Members—Hull Construction

IN a special report to the chairman of the Shipping Board, R. J. Wigg, chief engineer, Department of Concrete Ship Construction, Emergency Fleet Corporation, sums up the conclusions of the Concrete Ship Department as to the advisability of constructing concrete ships, and from this we make the following quotations from *International Marine Engineering*:— (1) The reinforced concrete ship can be built structurally equal to any steel ship. (2) The available information assures, with all the certainty possible, short of actual experience under service conditions, that the concrete ship will be durable for several years, assuring satisfactory service throughout the probable duration of the present war. (3) The cost of the reinforced concrete ship complete will vary between 100 dols. (£20 16s. 8d.) and 125 dols. (£26 0s. 10d.) per ton deadweight depending upon the number of ships built and the conditions of construction. The cost of the hull only will be between 30 dols. (£6 5s). and 40 dols. (£8 6s. 8d.) per ton deadweight. (4) The construction of concrete hulls will not interfere with the present programme for the construction of steel and wood hulls in so far as labor or materials are concerned. (5) The Concrete Ship Department has completed the detailed plans for a 3,500-ton concrete ship, so that construction of such ships can start immediately. (6) It is estimated that beween 150 and 200 3,500-ton concrete hulls can, if construction is started immediately, be completed by December 31, 1918, totalling approximately 600,000 tons. (7) It is further estimated if the construction of yards is begun immediately and the construction of hulls of 7,500 tons each commences by June, 1918, that 250 can be completed by August, 1919.

The 3,500-ton Concrete Ship

The standard concrete ship of 3,500 tons deadweight carrying capacity is of the same size, dimensions and form as the 3,500-ton standard wood ship except that the sheer line amidships has been slightly altered and no outer keel is fitted. The arrangement follows closely that of the wood ship, including the number and location of bulkheads. The propelling machinery design for the wood ship has been provided for without essential change in this vessel. The principal dimensions of the concrete ship are as follows:

	Ft.	In.
Length overall	281	10
Length between perpendiculars	268	0
Beam, over shell	46	0
Depth, amidships, on side	28	3
Draught	23	6
Full load displacement	6,175 tons	

The comparative weights of similar ships built of concrete, wood and steel are given in Table I.

Crete, Wood and Steel Vessels

The comparative weights of similar ships built of concrete, wood and steel are given in Table I.

TABLE I—Comparative Weights—Concrete, Wood and Steel Vessels.

	Concrete	Wood	Steel
Hull	2,500	2,300	1,160
Fittings, outfit and equip'nt.	191	191	180
Propelling machinery	206	206	200
Margin	75	80	60
Ship (light)	2,972	2,777	1,600
Reserve feed	80	80	80

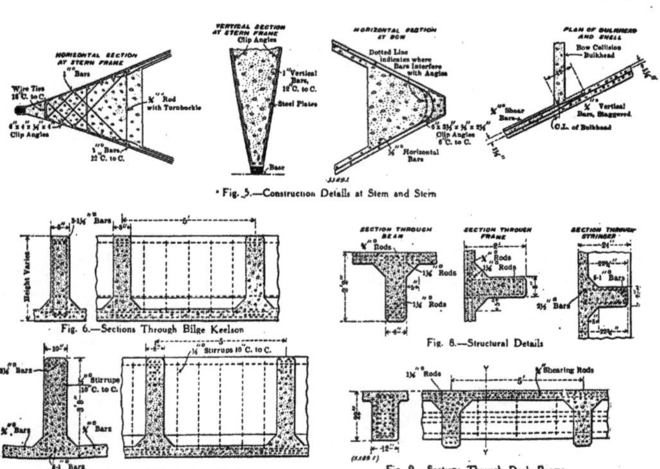

Fig. 5.—Construction Details at Stem and Stern

Fig. 6.—Sections Through Bilge Keelson

Fig. 7.—Sections Through Center Keelson

Fig. 8.—Structural Details

Fig. 9.—Sections Through Deck Beams

Ordnance	23	23	23
Fuel	300	300	300
Stores	40	40	40
Cargo	2,760	2,680	3,057
Total deadweight	3,203	3,123	3,500
Full-load displacement	6,175	5,900	5,100
Percentage deadweight to full-load displacement	52	53	68.6

TABLE II—Stresses in Government 3,500-ton Concrete Ship.

Condition	Maximum Bending Moment, Ft.-tons	Maximum Tons per Sq. In. Fibre Stress in Deck Reinforcement	Maximum Tons per Sq. In. Fibre Stress in Keel Reinforcement	Lb. per Sq. In. Fibre Stress in Concrete
Ship without cargo, hogging	25,175	5.53	2.80*	728
Ship fully loaded, hogging	37,000	5.63	2.95*	766
Ship without cargo, sagging	14,400	1.28	2.63†	270
Ship light with enough cargo in forward hole to trim, sagging	11,960	1.07	2.19†	210
Ship fully loaded, sagging	9,400	0.84	1.72†	70

*Keel. †Deck.

Strength of Hull

The strength of the ship as a girder supported on the crest of a wave amidships, hogging, and also on the crest of two waves, one at each end, sagging, was calculated for five conditions. The same basic conditions were assumed as to length, depth and form of waves and the same method of procedure that is standard practice in calculating the strength of steel ships was followed throughout. The maximum bending moments and fibre stresses in the steel reinforcements and concrete in the various conditions are given in Table II.

Metacentric Height, Stability and Period of Roll

The metacentric heights in the light (ship without cargo) and full-load conditions are, respectively, 2.15 ft. and 2.2 ft. The best practice at the present time places these values between the limits of 1 ft. and 3 ft. for vessels of this type and size. The maximum righting arm occurs at 51.5 deg. and 46.5 deg. for the vessel light and fully loaded, respectively, the extreme ranges being 89 deg. and 8.5 deg. respectively. The freeboard amidships at the side is 4 ft. 9 in. While an investigation of the period of roll is a laborious operation of doubtful value and seldom attempted in the design of steel vessels, the Concrete Ship Department, nevertheless, considered it safe to say that the concentration of relatively great weight in the decks and shell in the case of the concrete vessel would aid materially in increasing the period of roll anticipated.

Good practice in steel merchant ships for a vessel of this type gives a maximum stress in the outer fibre of from 5 tons to 8 tons per square inch figured on the same basis as given above for the concrete ship. In addition to the stress due to the ship acting as a girder

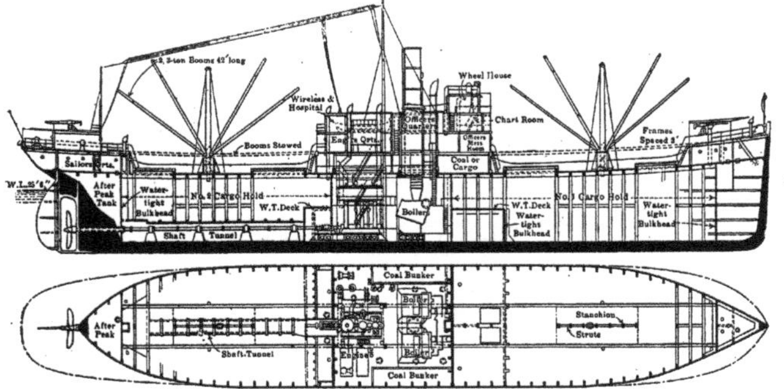

Fig. 1.—Inboard Profile and Hold Plan of 3,500-Ton Concrete Ship

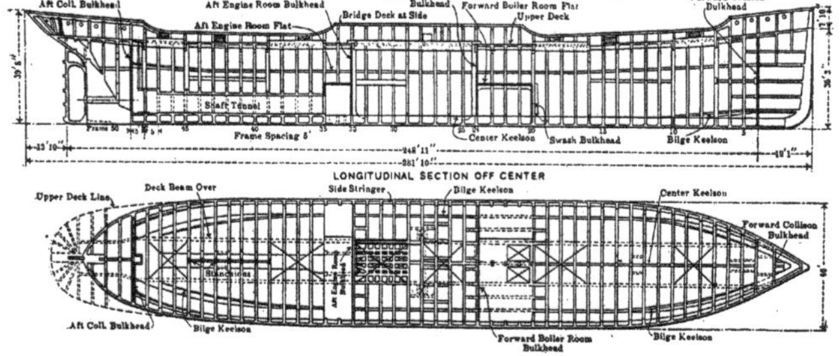

FIG. 2—FRAME PLAN.

PLAN BELOW UPPER DECK

there is the local stress between frames, where the plating must act as a beam for that space. This action is seldom taken account of in steel ships, but has been given full consideration in the design of the 268-ft. concrete ship.

Local Strength of Hull Members

Although the transverse strength of vessels is not usually investigated with any degree of accuracy, except in the case of naval vessels, as the scantlings are taken from the books of the classification societies, nevertheless, in the case of the concrete vessel, the complete transverse section was figured by the Concrete Ship Department with numerous cases of loading and heeling for every frame. The strength of the transverse frames was investigated for a large number of conditions of loading and for various immersions of the vessel. The transverse frames, Fig. 4, are designed to stand the outside water pressure—water to the gunwales— with minimum cargo load, for maximum cargo load and a sagging draught of 15 ft. 6 in. and for listed positions with loading light and heavy. All of these conditions of loading are to be met by the frames with stresses not to exceed 1,500 lb. per square inch in the steel. The results of the analysis agree with those of Dr. Bruhn published in the Transactions of the Institute of Naval Architects, 1901 and 1904. From Dr. Bruhn's analysis it is clear that the design of the concrete frame is so made that the frame is working at the above-noted safe stresses under loadings that would develop double the steel stresses in a standard steel frame of similar dimensions. In steel ships, this local action between frames would increase the stresses in the bottom plating of the ship acting as a girder from 5 tons to 10 tons, and in many cases to a much higher figure. In this concrete ship design the tensile strength in the reinforcement, due to local bending, amounts to about 3 tons per square foot, bringing the total up to slightly less than 8 tons. The bulkheads have been designed to carry a head of water on either side up to the deck. The collision bulkheads fore and aft were designed for 1,500 lb. per square inch in the concrete and 16,000 lb. per square inch in the steel. The steel stress in the engine-room bulkheads was advanced to 20,000 lb. per square inch. In the design of bulkheads for steel ships it is common practice to allow a unit stress in the steel of 22,000 lb. to 23,000 lb. per square inch. The deck is designed to carry 5 ft. of water or equivalent, which is in excess of loading on the decks of standard steel ships being built by the Emergency Fleet Corporation.

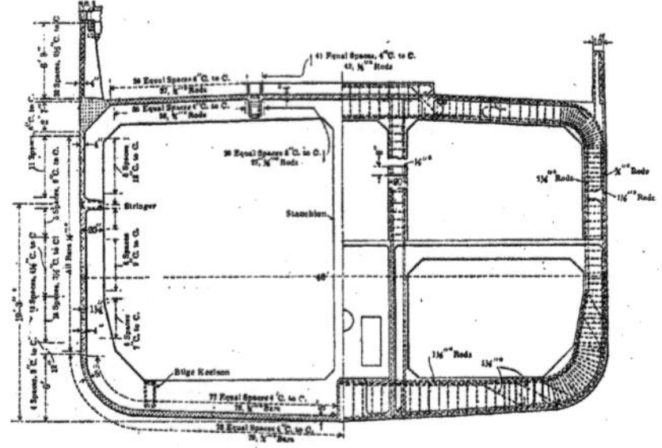
MIDSHIP SECTION SHOWING SHELL REINFORCEMENT.

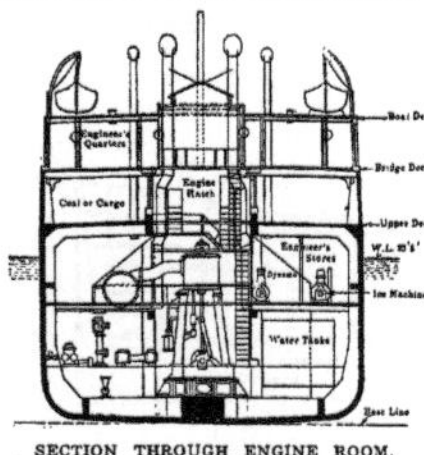

SECTION THROUGH ENGINE ROOM, LOOKING FORWARD.

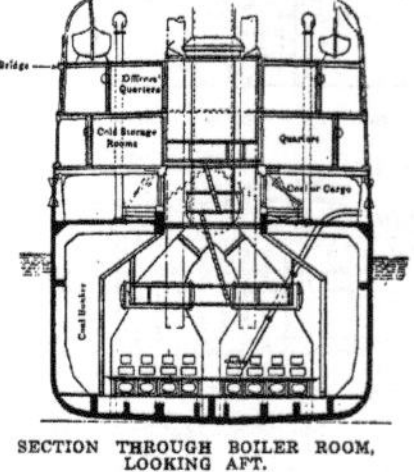

SECTION THROUGH BOILER ROOM, LOOKING AFT.

Hull Construction

As shown by Fig. 2, the parallel middle body of the hull extends 35 per cent. of the length. There is a dead rise of 9 in. The hull is subdivided into five watertight compartments by four transverse watertight bulkheads, Fig. 2, and the bottom is strengthened by a centre keelson and two bilge keelsons, Figs. 6 and 7, while the sides of the vessel are strengthened by side stringers at about the middle depth of the hull.

The keelsons consist of concrete girders reinforced with rods in the upper and lower sections and are tied together with frequently spaced stirrups. The keelsons are worked into the transverse frames.

The frames, spaced 5 ft. apart, are also reinforced concrete girders, Fig. 8, and are continued at the main deck to form the deck beams supporting the main deck, which is made of reinforced concrete slabs 4 in. thick.

The shell of the vessel is 5 in. thick on the bottom and up to 6 ft. above the base line, while the remainder is 4 in. thick. The shell is reinforced at the outer and inner edges with ⅛-inch to ¾-in. square bars parallel to the water line and spaced from 4 in. to 12 in. between centres. The horizontal reinforcing rods are located 1¼ in. from the face of the concrete shell. Vertical ¾-in. shear bars are fitted between the horizontal rods extending clear around the shell and deck.

Between the hatches in the cargo holds the deck beams, Fig. 9, are supported by reinforced concrete stanchions. The foundations for the machinery consist of steel plate girders seated on extra heavy framing consisting of longitudinal reinforced girders.

As previously stated, the hull is subdivided by four transverse watertight bulkheads, two of which are the collision bulkheads and two enclosing the machinery space. The bulkheads of concrete slabs reinforced with transverse and longitudinal steel and stiffened by vertical stringers and beams which are tied into the main frames and deck beams.

Vancouver, B.C.—The Pacific Construction Co. are planning to build a big machine shop and warehouse on Industrial Island for the purpose of outfitting vessels built at the company's plant at Coquitlam. The company will launch the wooden steamer "Antonias" for New York interests, and she will be fitted out at False Creek. Another vessel will be ready for launching almost immediately after. This company have also contracts for two 1,500-ton vessels for the French government, and three 2,200-ton vessels for the Belgian government, all wooden ships.

The Training of Engineering Apprentices

Necessity of Training—Standard of Elementary Education Required—The Apprentice From the Employer's Point of View—A Suggested Course of Training—An Apprentice Club in Scotch Engineering Works

By T. H. FENNER, Associate Editor

THE happy turn of the world's wheel that brought in sight the end of war, has concurrently turned men's minds to the vista of peace, with its obligations and rewards. With reference to this an important announcement was made recently by a minister of the Government, concerning the efforts the Canadian Government is about to make, or is making, to provide a merchant fleet. Thirty-one steamers, varying from 4,000 to 10,000 tons, are to be built in Canada, and presumably all the boilers, engines, etc., will also be built in Canada. This is an auspicious event, coming, as it does, at a period when the emergency work of the new Canadian industry was about all over, and the immediate future a matter of doubt. There is every prospect that we shall have a permanent shipbuilding and marine engineering industry, and now is the time to look to securing the supply of men to keep it going. The professions intimately connected with shipbuilding are of the naval architect and the marine engineer. The trades connected with it are legion, some of them highly skilled, some less so. The naval architect is purely a highly technical personage, connected only with the more complicated features of ship design, stability, etc. With him we are not concerned in this article. Neither need we here consider the plumber, steamfitter, electrician, plater, riveter, joiner, carpenter, etc. These are all trades that are in much request in ship construction, but they are trades, and can never be professions. The marine engineer, unlike these, has it entirely within his own power to remain a tradesman or become a professional man. I say, entirely within himself, but that is not strictly correct. It could be more truly put if I say, according to the opportunities afforded him in his apprenticeship days, coupled with his own ambition. Canada has not in the past resorted very much to the apprenticeship system to produce tradesmen, and it may be said that she has not produced tradesmen to any great extent. If marine engine building is to grow to a real industry, then marine engine builders must be trained to enable them to build engines in competition with nations who have a supply of expert tradesmen always on hand. This training must be got systematically, over a sufficient number of years, and under conditions that will enable the neophyte to register a steady progress. During this time his technical education can be looked after and arranged according to capacity shown, so that at the end of his apprenticeship it is quite plain as to whether a tradesman or an engineer has been developed. The question of the best way of achieving these results has been the theme of much discussion, especially in Great Britain, where the apprenticeship system is very fully developed. There has been much difference of opinion expressed among the men at the head of the marine engineering profession, and it must be remembered that these men are the products of the apprenticeship system, sometimes coupled with a university course, but just as often with technical education obtained in evening classes. In this country the big railway shops have developed a system of training apprentices on railway work, and giving them some technical education at the same time, which is a step in the right direction, but naturally, these boys are developed chiefly for railway practice, to become locomotive experts, qualified to be superintendents of motive power, etc.

Standard of Education Necessary

In the first place, before a boy is admitted to an apprenticeship at all, some recognized standard of education should be required. A boy who has attained the age of 16, which he should have before entering the engineering profession, without having a thorough grounding in arithmetic, including algebra, up to quadratic equations, and a good grounding in geometry, coupled with the ability to speak and write good English, is handicapped in the beginning. He is not handicapped in so far as becoming a tradesman goes, but he has a lot of leeway to make up if he is going to become an engineer. The reader may think that a good deal of stress is laid on the terms tradesman and engineer, but this is really necessary, as there is a tendency in Canada to include in the term of engineer everyone from the man in charge of a fried potato cart to the designer of the Quebec bridge. A marine engineer, in the strict sense of the term, is a man who can design, build, and operate a marine power plant, and in virtue of these attainments he is a mechanical engineer. A man who, as is often the case here, graduates to a position in charge of engines through the stoke hold, greasing, etc., is not an engineer. He may be competent to exercise supervision over a set of engines in the same sense that a locomotive driver does, but of the principles and underlying science that those engines are constructed from, he is unaware. This is not his fault, as he has never had the opportunity to acquire such knowledge, and too often, not the elementary education necessary to grasp the opportunity had it presented itself. The same man, if taken from his engine room and put in an erecting or fitting shop, would be of no use except as a laborer, as he has never learnt the trade. Yet he rejoices in the name of a marine engineer.

The Apprentice From the Employer's Standpoint

In discussing the question of apprentices, it is often considered merely from the view of the future prospects of the boy, the firm he is apprenticed to being considered only as a means to an end. Too often the boy himself gets the view, which is bad for him and for the firm. While most employers are only too glad to help apprentices along and offer them opportunities for acquiring the technical side of their work, they cannot be expected to form a secondary education body, which pays for the privilege of teaching. In direct opposition to this idea, many engineering firms in the United Kingdom only took apprentice engineers on payment of a heavy premium, and in return undertook to educate them in their profession. Their other apprentices were apprenticed to fitting, turning, pattern making, as the case might be, but were distinctly tradesmen. This is the supply that the country had to draw from for its working forces. Some of these boys, by industrious application in their own time to technical classes, became qualified for better positions, and some of them rise to high positions. However, it was strictly their own effort. It must be always borne in mind that an apprentice is a source of expense to his employer for the first two years, no matter how good he may be, or how anxious to learn. When he has been three years, the employer begins to get some return from him. After his five years are completed, he usually leaves to get experience of some other shop, or more money, so that any expense put into his training by the employer does not come back to them direct. However, they get the benefit of a continuous supply of labor, which is necessary for carrying on business. There is another benefit in that every apprentice who leaves a shop and becomes a successful engineer, is a perpetual advertisement to that shop, and bearing, as he usually does, an affectionate remembrance of the place where he learnt his business, will reciprocate by placing business their way whenever possible. It may be accepted that, in the larger sense, it is well worth while for the employer to make an effort to encourage his apprentices to acquire knowledge, and to pick out the best according to ability shown, to become eligible for staff positions. A standardized system for all shops would go a long way to help in this effort.

Suggested Course of Training

As a general rule, marine engineering apprentices have always in view, spending some period of their lives at sea, and qualifying for the certificates of the ship lines, works managers, consulting engineers, are all positions open to the marine engineer who is ambitious. It is, therefore, necessary that the apprenticeship period should cover enough time in the shop itself to allow a thorough grasp of the elements of the trade to be obtained, as well as to have the necessary shop service to qualify for the Board of Trade examinations. The Canadian Government examinations are modelled on those of the Board of Trade, and the Board of Trade certificate is valid all over the Empire. It is on the following this plan does not become an expert tradesman, but after all, why should he? As long as he acquires the knowledge of how a thing should be done, and the best way of doing it, that is what he will require to round out the technical knowledge required from his university course. Most of the boys following this course of training are destined for staff appointments, and that is what their training fits them for. To come back to our average boy, what is the best course to pursue to benefit himself, his employer, and produce the best

FIG. 1—READING ROOM AND LIBRARY

FIG. 2—A CORNER OF THE STUDY

Board of Trade. Probably 75 per cent. of them do actually go to sea for a varying period, according to how the life strikes them, or to their ability to pass the examinations. A considerable number choose this as their life's work, and remain at sea, rising in their profession till they are chief engineer of a large ship. There is no finer class of man afloat or ashore than the marine engineer, and he is generally a well-informed man technically, besides possessing marked mechanical ability and a profound knowledge of boilers, engines, pumps, under hard working conditions, coupled with a perfect knowledge of question of technical instruction during the period of apprenticeship that the differences of opinion are felt. The opponents of the evening technical class method hold that a boy of 16 to 21, who has worked hard all day, is in no fit condition to receive instruction in difficult technical subjects at night. They hold that technical instruction should be imparted during the day. The upholders of the evening class claim that to give technical education during the ordinary working hours, means disorganizing the work of the shop, and prevents the boys getting the class of work they would wish. This is because the best men for the engineering trade and profession? If the boy comes to his apprenticeship with the educational acquirements referred to earlier in this article, the necessity for some of the classes attended in the first two years is eliminated. In fact, classes to teach elementary mathematics should not be required in an evening school, as every boy should be kept at school till he has received that instruction. Therefore, let the first year in the shop be devoted to, say three months in the tool stores, getting familiar with the various tools in use before going out in the shops to use them, and the remainder of the time in

FIG. 3—RECREATION ROOM

FIG. 4—GYMNASIUM

their construction. Other men, after a few years at sea, enter the service of Lloyd's or the Board of Trade, as surveyors, a position they are well qualified for.

Superintendent engineers of steam- jobs cannot be left standing while the boys attend classes. For those boys whose parents have the means, the ideal way would be to attend the university courses while in session, and devote the remaining time to the shop. The boy the fitting shop. The evening class should be confined to acquiring the elementary course in machine construction and drawing. In a shop employing enough apprentices there will be some one man in charge of them, and looking after their welfare. If not, the boy's immediate foreman may be depended on to observe his progress. By this time he will have become sufficiently acquainted with the use of his tools to be entrusted with small jobs, and what he has learned at his drawing class will give him an intelligent interest in what he is doing. The second year should be still in the fitting shop, and the evening class work extended to take in trigonometry in addition to the second year machine construction and drawing. At the end of the second year his capabilities as reported by the foreman, and the results achieved in his evening class, should be such as to entitle him to a period on the marking off table, where his knowledge of reading drawings can be put to practical account. This marking off table work is excellent training. After three to six months of this work he should be moved to the erecting shop, where he will combine his knowledge of fitting, with his ability to understand a drawing, in the fitting up of the completed article. His evening classes during this year should include the study of physics relating to the heat engine. At the end of the third year an examination should be held by the firm, covering all the shop experience and technical education received up to date. If the candidate shows sufficiently well in this, he should be admitted to the drawing office, and during the next two years should be given the opportunity to attend advanced technical courses in the day time, either held on the firm's premises, or at the local university or technical college. During this time he will have occasion to visit ships under construction, and under repair, and his knowledge will enable him to grasp the essence of what he sees. At the end of his apprenticeship, if he elects to stay awhile with the firm, they will have a useful man. If, as is most likely, he moves to another firm, or takes a few years' seagoing experience, he has the necessary equipment to become a success. For the boy who does not qualify there still remains a chance. He should pursue his technical studies in the evening, and his shop work will be according to what he shows himself good for. At the end of his fourth year he may be given another opportunity to enter the drawing office. That would still give him a year, and a very useful year. If he does not succeed then, he has still the technical schools to go to. At the end of his apprenticeship he will be, in most cases, an excellent tradesman, with a fair technical education. He may go to sea and become a first class man. He may elect to remain at his trade ashore and work up to a general foreman or superintendent. The fact of him not passing the examination does not condemn him, but merely ensures that the boys of real ability will get their chance. At the least, he becomes a tradesman capable of earning a comfortable living, and by his skill and ability in this direction, helping on the industry of the country.

In connection with this subject the accompanying illustrations show what is being done in Great Britain to encourage the boys to be interested in their work. The recreation rooms shown here are in the works of Scott's Shipbuilding and Engineering Co., Greenoch. The fee for membership in the club is one shilling or 25 cents per annum. It is open every evening except Sundays, and the proceeds of the subscriptions are presented annually to the local infirmary in the name of the club. One of the staff is present every evening to assist any boys who wish to study their homework connected with their technical classes, and there is also a physical instructor in the gymnasium. There is a corner containing lathes, etc., and any boy who wants to pursue a hobby is allowed the use of these tools and scrap pieces from yard and foundry are furnished them to practise on.

EXPERIENCES OF A BRITISH NAVAL CONVOY OFFICER

An officer of the Royal Navy, who is at present Captain of a commisioned escort ship for convoys, has put together some interesting notes with regard to his voyages. The following are some of his experiences as described by himself:

"At 7 a.m., we sighted two life-saving vessels in good order, with masts up and sails lowered. One vessel was painted grey, with "No. 5" on her bow, and a black rudder; the other craft was camouflaged. We passed close by them in case they contained any occupants, but both were empty. Two days later, at 7.15 a.m., we passed some hatches and other wreckage, and five hours later received information that a submarine was outside the mouth of the river for which we were making. Next morning we passed a water-logged three-masted schooner awash, with her sails hanging over her sides, and at 6.30 a.m. took in an "S.O.S." signal from a ship which was being attacked by the submarine we heard of yesterday At 8.20 a.m., we stopped off the entrance to the river and shipped the pilot. After gathering speed for 10 knots, the starboard anchor caused a diversion by releasing itself, the cable running out to a clinch. A very thorough medical inspection was conducted off the Quarantine Station. We then passed an island which had been a marsh last fall, but which is now a flourishing shipyard.

"On this voyage we had a leading fireman in the ship who had seen a variety of active service during the war. An electrician in civil life, he had joined the Canadian Army at the beginning of the war, and had fought with them at Ypres, where he was wounded in the right wrist and ankle. Discharged from the Army, he joined the Navy and was sent to Mesopotamia, where he was in the vessel which tried to run supplies through to Kut-el-Amara. He was wounded by shrapnel and splinters on the Tigris, caught dysentery, and was invalided to England. After his discharge from hospital he joined the Merchant Service and went to the White Sea. He was afterwards in a vessel which was torpedoed 350 miles from land, and then joined us. He told me he had lost three brothers in the war.

"Being short of complement for our homeward journey, we embarked three of the gun's crew from a ship which had recently been sunk with the loss of 10 hands. The gunlayer told me that he was at the gun when a torpedo struck the ship on her starboard side. He thought a second torpedo struck her, as he heard two explosions with an interval of 25 seconds between them. The bulkhead was fractured before the stokehold, and the gasolene from the tank, running into the stokehold, quickly ignited, converting the ship into a roaring furnace almost at once. One lifeboat capsized, as the ship was still going ahead, and four hands out of the 14 in her were pinned under the boat and drowned. The rest scrambled on to her as she detached herself and drifted away, but the wind and tide were driving against each other, and soon these men found the flaming vessel drifting down on them. When only 10 yards distant, she broke in half, and the oil, burning furiously, began to spread over the sea. The men, who expected to be burned alive, jumped from the boat and began to swim for their lives. The good swimmers got away too far to return and were all drowned, but the weak swimmers soon tired, and returned to the boat. Using their hands as paddles, these men kept the boat in the channel which separated two sheets of burning oil. As the halves of the ship drifted apart this channel grew wider, and eventually both parts of the burning vessel went down. The men remained for six hours on the upturned boat before being rescued, but the floating oil continued to burn throughout the following day.

"This gunlayer had won the Distinguished Service Medal in the Mediterranean. He had also been torpedoed when his ship was 100 miles from land. On this occasion he had escaped as the ship sank, owing to the wheat cargo in the after-hold swelling, when the imprisoned air burst the hatches and flung him clear of the ship's propellers."

London.—The Admiralty announces that the output of world tonnage in the last quarter exceeded the losses from all causes by nearly half a million gross tons.

The United Kingdom built new shipping to the amount of 411,395 tons; the other allies and neutrals 972,735 tons.

The tonnage of merchant vessels completed in the United Kingdom and entered into service in October was 136,100.

London.—Speaking in the House of Commons to-day, Right Hon. Thomas James Macnamara, Parliamentary Secretary to the Admiralty, stated that 8,946,000 tons of British merchant ships had been lost during the war up to September 30 last by enemy action. Of this number 5,443,000 tons had been replaced by new construction, and by the purchase of ships abroad and the utilization of captured enemy ships.

Interesting Repairs to Damaged Machinery

When Brazil Broke Relations With Germany She Confiscated Interned Vessels—Their Crews Had Displayed Great Ingenuity in Doing Damage—Report of Repairs Sent to Institute of Marine Engineers

By R. N. DUNCAN

The Institute of Marine Engineers, London, received lately the following communication from one of their members resident in Brazil, describing the damage done to their vessels by the German crews, and the methods pursued in repairing them. It is interesting to note that British marine engineers are holding positions of trust in Brazil, and that they were employed in the congenial task of circumventing Germany's efforts by putting their own vessels into commission in the service of the Allies.

This accompanies a letter I have written to Mr. Geo. Adams, my old chief when sailing in the Shaw, Saville and Albion Co., who will no doubt give you a sight of the same, for therein I promised to send some particulars of work accomplished here that I am sure will be of interest to the Institute. Possibly you will have particulars from other sources of similar work, and this can supplement the same.

DAMAGE TO S.S. CABADELLO

When Brazil broke relations with the enemy of civilization and decided to confiscate the vessels interned in these ports, it was more or less known that the machinery of the same had been considerably damaged by their respective crews, and I may say a good deal of diabolical ingenuity was employed in some cases to this end. At the time mentioned I was employed as chief engineer to a large Frigorifica Company in the interior, and as everything was running smoothly in the job, and I had an old shipmate as second, I felt my marine engineering experience could be well employed in the repairs, so that the boats could rapidly be brought into commission again. The president of the Frigorifica Company, who is also president of the Paulista Railway Company, Conselheiro Antonio Prado, quickly put the wheels in motion, and the service was soon organized. The Government appointed Dr. A. Gomes de Mattos, of Rio de Janeiro, who is surveyor to the British Corporation there, as fiscal over the work. I may here say Dr. Mattos is an engineer of considerable experience, having served an apprenticeship with the old firm of Humphrey, Tennant and Co. on the Thames.

The heavy repairs were carried out by the Paulista Railway Company, at the works in Jundiahy, situated some 140 kilometres from Santos, and an excellent job they made of them. The railway company, although Brazilian, the chiefs of the works are British, and their names are Mr. Alfred Williams and Mr. Adam Gray. I enclose you their photographs; Mr. Williams served his apprenticeship with the London, Brighton and South Coast Railway at Brighton, and it was 45 years ago on the 2nd of this month that he arrived in Santos. This gentleman is locomotive superintendent for the Paulista Company. Mr. Adam Gray served his apprenticeship with Dubs and Co., of Glasgow, and has been in this country 30 years; he is chief of the Paulista Company's works at Rio Claro, where are situated the carriage and waggon shops and all the service for the narrow gauge.

The new names given to steamers that have been repaired in the Port of Santos are as follows: Cabadello, Macao, Palmares, Maranguape, Therezina, *Maceio, Baependy, Alfenas, Aracaju, Pelotas.

(* Since reported to be torpedoed.—J. A.)

The principal damage was to the cylinders, from which great pieces were broken out, as you will see from the photographs. In the two first boats most of the steel crossheads had disappeared (these crossheads weighed

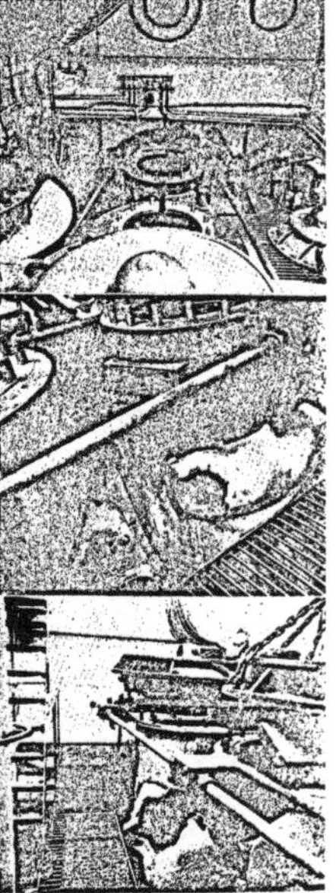

DAMAGE TO S.S. BACPENDY

some 380 kilos finished). In one case the crosshead pins had been cut half-way through, close up to the fillet, top half on one side and bottom half on the other, with the oxy-acetylene flame, then carefully soldered over and filled up. Several of the main piston valves and slide valves had vanished, likewise the latter jobs, to save time, the patch was fitted, then marked all round and returned to the shops, when the piece was bolted down on to the sole plate of a locomotive wheel lathe, and with a boring bar between the centres that was made for the purpose the patch was turned to the exact radius. After being

DAMAGED CYLINDERS S.S. ALFENAS

the main intermediate stop valves, and most of the vitals of the main auxiliary boiler feed pumps. Regarding the repairs to the cylinders, these were chipped out to a good formation and pieces fitted in with chain studding at the junction between cylinder and patch. In the first repairs after the patch had been fitted the boring bar was used just to clean up the surface overall, but in fitted in its place finally it was found there was nothing to be done except to run over the surface with a pneumatic emery wheel. On examination of the repair after the trial trip it was found that the bearing was quite equal to the rest of the cylinder face. I am not sending photos of all the jobs, but what are sent I am sure will be sufficient to let the members see what we had to contend with, and after completion with the lagging in place, an outsider would perhaps be a little bit sceptical in the matter unless the engine was stripped for a close examination.

SHOWING DAMAGED CROSSHEAD

S.S. Cabadello.—"A"—This is a view showing a piece broken out of the H.P. cylinder and liner at the top port. The main intermediate stop valve had gone and the intermediate piston valve. "B"—Pieces broken out of exhaust between H.P. and I.P. cylinders. Expansion glands between cylinders on both port and starboard sides were missing. Repairs were effected by chipping out to a good formation, fitting a piece of heavy boiler plate inside with a heavy wrought iron ring shrunk on over the flanges. Note the steel drift on the intermediate side showing how the pieces were split off. "C"—Similar damage to connection between I.P. and L.P. cylinders. Repairs were effected in a similar manner. "D" and "Da"—Two views taken in the shop showing the way the steel crosshead was treated, doubtless with the idea that it would fail at sea, and possibly with a loss of life. When suspicion was aroused the piece was taken to the shop and supported at each end; a load (hydraulic) of 150 tons was applied at the centre before it showed signs of giving. A fitter was then told to drive his chisel into the place, when a piece of the weld flew out.

S.S. Macao.—"A"—View showing H.P. cylinder on top of L.P. cover, doubtless ready for dumping overboard into the mud at the bottom of the river, where it would be difficult to locate. Note the broken feet of the cylinder and hole through exhaust passage. The cylinder had ultimately been dropped on the L.P. cover, breaking it. The main intermediate stop valve had gone and the I.P. valve. "B"—Another view of the above showing where the cylinder had been removed from; candles were placed in the exhaust passage to make the photo clear. "C"—A piece of the broken liner from

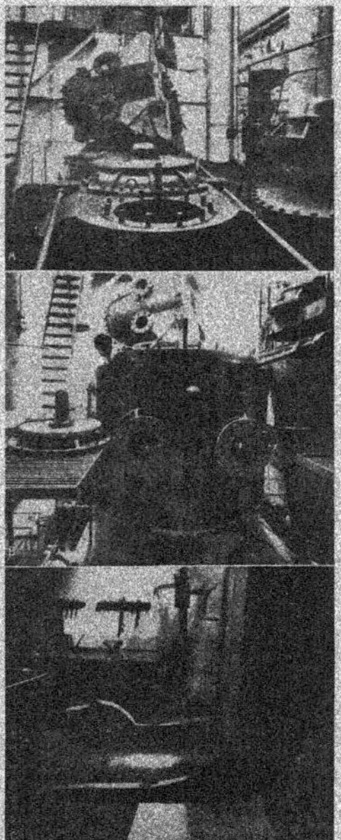

DAMAGE TO S.S. MACAO

the H.P. cylinder on the bottom platform, where it was found.

S.S. **Baependy.**—"A"—View looking aft from the H.P. cylinder. The photographs of this boat were taken after ten days' work chipping and filling up the places requiring patches. Three piston valves had disappeared and the L.P. "D" valve broken, found on board. "B"—Showing piece out of the L.P. cylinder. Note where the port opening has been broken away; this required a considerable amount of time to fit a new piece. "C"—View looking for'ard showing pieces out of the first and second I.P. cylinders. "D"—Showing some of the broken pieces found on board, including the L.P. valve. The planing machine that the new valve was faced on had to be altered to suit the work.

S.S. **Alfenas.**—"A"—This is a sister ship to the Baependy; the damage, as you will see, is somewhat greater. The photograph gives a better view of the four cylinders, and was taken before commencing the repairs. "B"—View of the L.P. cylinder showing a large piece broken out of the wall through into the valve casing.

The work of repairing these latter two steamers was executed in the space of two months. After completion, and before going out on the trial trip, the shaft was uncoupled and the engines run for three hours at about 120 revolutions a minute.

Enclosed is also photo of the new L.P. valve made for the "Aracaju."

Now I do not know that I can add anything further except that Mr. Williams was good enough to have prepared for me fifteen small blue prints showing how the repairs were effected on the "Baependy" and the "Alfenas," and these are enclosed. Should any member of the Institute desire further information I shall be only too happy to reply.

To Mr. Alfred Williams is due the credit for the manner in which he organized their portion of the service, and in this he was ably assisted by Mr. Adam Gray, and also by Mr. Storch, of the Paulista Railway Company. It is perhaps reasonable that those at this end desire you to feel that, although far away, the desire is ever with us to do our "bit."

Windsor.—Intimation is made by parties in close touch with Charles Miller, of Toronto, and the Windsor Ferry Co., Limited, that plans are proceeding for the construction of a new type of ferry boat, to be built with end-on landings, much the same style as used at Burrard Inlet, plying between Vancouver, B.C., and North Vancouver, and also between New York and Jersey City. The upper deck will be fitted for passengers, while the lower deck will be used exclusively for autos and other vehicles.

Quebec.—One of the results of the armistice has been the issuing of instructions by the Imperial Munitions Board to discontinue the fitting of guns and the quarters for gun-screws on the wooden ships now under construction.

CANADIAN ROBERT DOLLAR MAY PUT LINE ON PACIFIC

Special to MARINE ENGINEERING

OTTAWA, November 21.—The recent meeting here with shipping and export interests to consider the question of procuring tonnage to meet the requirements of Canadian trade was called by Hon. A. K. Maclean, and there was a large and representative gathering from different parts of the Dominion.

The manufacturers pointed out that large quantities of goods were now available for export, particularly to South Africa, Australia, and South America. Steel products are the classes of goods most frequently mentioned, but there has been a greater accumulation in many lines than the public has any idea of. Although many of the large plants had been diverted to munitions of one kind or another, the manufacturers realized that it would be unwise to drop their foreign connection altogether in many other lines, and a production of those was continued. The necessity now is to find bottoms to maintain regular sailings and take care of the trade to and from South Africa, Australia, New Zealand, the West Indies, and the various South American countries.

The manufacturers want something definite on which they can reply. They want to be sure they will have a boat on a certain date, with others at regular intervals. There were formerly six steamers running from Canada to New Zealand, for example, but there has not been one since 1916. It has been the same with the West Indies, where the boats also were British registered. The government will ask for the return of all such ships at once.

All the C. P. R. boats were requisitioned. The Empress of Russia and Asia have been carrying U. S. troops. They will be put back at once. The Empress of Japan and the Mounteagle were taken for the Siberian expedition, but that adventure will not be so extensive as was anticipated and they will be put back on the old route as soon as possible.

An interesting development on the Pacific is that the Canadian Robert Dollar Company proposes to inaugurate and maintain regular monthly sailings from Vancouver to Shanghai, Hong Kong, Manila and Singapore. For the present the boats used will be the Bessie Dollar, the Melville Dollar and the Harold Dollar. To enable them to maintain regular sailings and regulate their cargoes the Robert Dollar Company have erected large lumber mills at Vancouver. They accept all the regular cargo available and fill up any spare space with the products of their own mills.

They are now erecting at Shanghai large warehouses at a cost of over half a million dollars. "And if you could see the warehouses which can be built in China for half a million dollars you would be amazed," said Mr. Robert Dollar when he gave me the particulars. "They will be maintained to store such products as we may carry, pending the time of sending them up into the interior. Do you realize that one-seventh of the world's population is packed away in the Yang-Tse Valley? This will give us access to it." If developments warrant it, the Robert Dollar Company is prepared to put a fleet of ten big modern freighters on the Pacific service.

RAMMING OF THE "SHAW"

The United States destroyer Shaw was rammed by the Cunard line steamship Aquitania, and was cut in two and sunk on October 9, it was learned to-day upon the arrival of the "Melita" here.

The survivors of the Melita, a Canadian Pacific steamship, said the Shaw was one of a number of destroyers convoying a big fleet carrying American troops to an English port.

The fleet was steaming on a zig-zag course. At 5.45 a.m. the steering gear of the Shaw jammed while she was running at a 35-knot speed, and the destroyer failed to answer her helm. Her course was directly across the bow of the oncoming Aquitania, a unit in the fleet. The Aquitania's bow struck the Shaw like a monster knife, the survivors said, passing through her just forward of the bridge, and not seeming to lose any headway.

The forward part of the Shaw, in which some of the crew were asleep, remained afloat 20 minutes. Some of the men jumped overboard and were picked up injured.

The collision penetrated one of the Shaw's oil tanks and the aft section of the destroyer burst into flames below the deck under the bridge.

Some of those who perished, including one of the officers, were burned to death and others were drowned when the forward part sank.

The aft section was towed into Portland, where 26 of the men suffering with burns and from immersion were removed to a hospital.

New York.—Ocean passenger traffic to Europe will be resumed by early summer, New York steamship agents declared here to-day. The need of the European population for food, raw materials, for industry and reconstruction will absorb all tonnage for some time, it was stated. There is no immediate prospect of the government control of shipping being removed. Although some tourist traffic may be under way in several months, an official of a large American line stated that there was no chance of it assuming pre-war proportions. He said it was not merely a question of providing steamship accommodations at this end, but return journeys would have to be considered, as all accommodations from the other side would be needed for troops. Steamship offices here have been swamped for several days with telegrams and inquiries regarding the resumption of ocean travel.

Installation of Hot Bulb Engine in Motor Ship

Detailed Description of Installation in Swedish Motor Craft—Engine Designed With Water Injection—Unusually Large Oil Storage Capacity—This Makes it Necessary to Carry Large Supply of Water

IT is not often, unfortunately, in these days of war restrictions that we are able to give so detailed a description of a motor ship installation as is shown in the accompanying illustrations of a Swedish craft. The vessel in question, which is constructed of wood, has been built to the highest class of the Bureau Veritas and also to the rules enforced by the Swedish Government. She measures 152 ft. overall, with a beam of 28 ft. and a depth of 12 ft. 3 ins., her draught being about 10 ft. when carrying the normal full deadweight of 650 tons.

In appearance and design, the "Calcium," as she is called, is very similar to many steam cargo boats of similar tonnage. Two hatchways are provided, each measuring 22 ft. by 12 ft., while the hold has a capacity of 25,000 cubic ft. An electric winch, with a lifting capacity of 3 tons, is fixed on deck near each mast for working the cargo; the anchor windlass is also driven by an electric motor.

The captain, officers and engineers are accommodated aft under the poop, quarters for the men being arranged in the forecastle. All the living quarters and the engine-room are lit and heated by electricity.

A three-cylinder Avance hot-bulb engine of 270 b.h.p. is located right aft. Contrary to what is now the almost universal practice, this engine is provided with water injection, for which large tanks of fresh water have to be carried.

When examining the plans of the installation, one is at once struck by the enormous size of the exhaust pipes and water-cooled silencing arrangements, the exhaust main being considerably larger in diameter than the bore of the cylinders, while the water-cooled trunks into which the exhaust pipes from the cylinders discharge are unusually bulky. The exhaust is discharged into a short funnel through a silencer.

650-TON SWEDISH SHIP "CALCIUM."

The engine is directly reversible and, of course, starts on compressed air, but a clutch is provided for disconnecting the propeller shaft, as is customary in this type of motor. An unusually large storage capacity for fuel oil has been provided, and this feature involves the carrying of an equal quantity of fresh water for the injection. There are no fewer than five fuel oil tanks, the four under the poop at the forward end of the engine-room having a capacity of 1,320 gallons each, while the fifth, which is aft, holds 930 gallons, making a total of 6210 gallons. The consumption is about 20 gallons an hour, and the speed, loaded, 8 knots, which figures give the vessel a radius, when loaded, of 2,500 nautical miles.

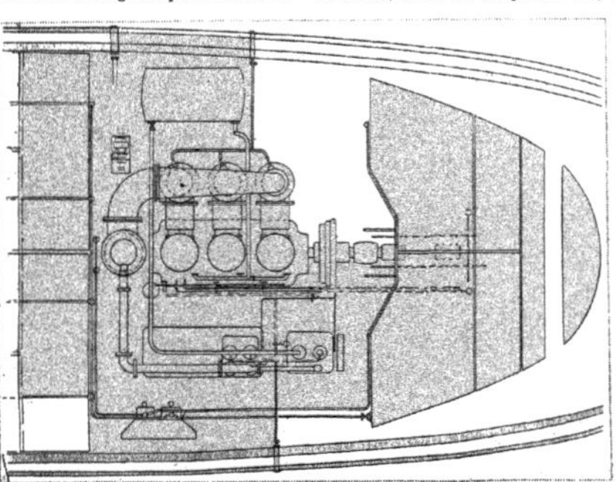

PLAN OF INSTALLATION OF 270 B.H.P. AVANCE ENGINE.

The fresh-water tanks have the same capacity as those for fuel oil.

On the right of the 'midship section, which is a view looking towards the stern, is a running and measuring tank for the fuel oil, while below it are two pumps—one for oil being worked by hand, while the other, for water, is driven by an electric motor. It will be noted that the oil tanks, being all high up, the oil will feed by gravity to the running tank, whereas two of the water tanks rest on the engine-room, hence the water has to be pumped from them when about half empty.

Two auxiliary oil engines are installed in the engine-room; one single-cylinder motor which drives an air compressor, and a two-cylinder engine direct coupled to the dynamo that provides current for lighting, heating and power. There is also a centrifugal pump driven by an electric motor.

Abaft the after fuel oil and fresh-water tanks are two tanks for lubricating oil, each with a capacity of 600 gallons.

The piping arrangements appear to be very complete, but as these are clearly shown in the illustrations there is no need to describe them here. As would be expected, the pumps are interconnected so that either the bilge pump on the engine or the centrifugal pump can be used for pumping out the bilges.

**Motor Ship and Motor Boat.*

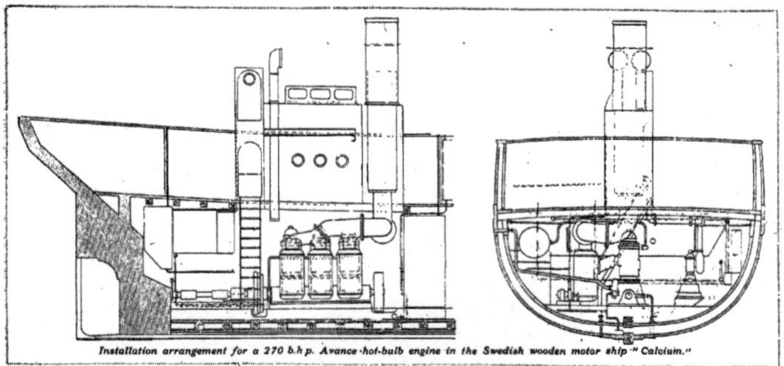

Installation arrangement for a 270 b.h.p. Avance hot-bulb engine in the Swedish wooden motor ship "Calcium."

LONGITUDINAL AND TRANSVERSE CROSS SECTION

TEMPORARY REPAIR OF A BROKEN STOP VALVE

By T. H. F.

AN interesting repair of a bad break came under my observation when sailing as 3rd engineer of "S.S.———." We were bound from Cardiff to Nagasaki, Japan, though as we were carrying Welsh coal for the Japanese navy, and Japan then being at war with Russia, our ostensible destination was Shanghai for orders. However, we had come down the Red Sea with its usual discomforts, across the Indian Ocean, and were making up for Colombo to coal. We sighted the harbour about 7 a.m., and about 7.30 the first officer gave orders to the bo'sun to get steam on the windlass, preparatory to anchoring. Now the bo-sun had sailed in steamers so long that he was, in his own mind, something of an engineer. Therefore, instead of notifying the engineer on watch that steam was wanted on deck, he undertook to put steam on himself. The two valves controlling steam to the fore and aft deck were situated in the fidley, at the bridge deck, and easy of access to anyone. The gallant bo'sun, whose engineering knowledge really was all comprised in the ability to turn a wheel to the right or left, which would open or shut a valve, naturally never thought of the fact that the steam had not been on the deck pipes since leaving Suez, so he stepped into the fidley, took the wheel in his tarry hands, gave it a mighty twist, and then things happened. It was here that the chain of events reached me personally. The messroom steward had just entered my room to call me, and I was sitting on the edge of my bunk, rubbing the sleep from my eyes, when a terrific report, followed by the roar of escaping steam burst on my ears. I jumped from the bunk into my engine-room slippers, and in pyjamas and slippers, made for the engine-room. After the first shock my ear could still distinguish the beat of the engines, and there was no sign of steam in the engine-room. Going on the top grating I opened the door leading to the boiler tops, but though the noise was plainer here, there was no steam. As I turned back again, the second came up from below, and made toward the door, beckoning me to follow. He went to the auxiliary stop valve on the port boiler and motioned me to the starboard boiler. These were shut off, and quietness reigned once more. This sounds simple, but any of my readers who have gone on boiler tops in tropical weather clad in pyjamas and slippers will appreciate the pleasure of it. As we came out onto the top grating again, I asked the second what was damaged, and elicited at first a very complete and highly ornamental biography of our bo'sun, bo'suns in general, deck officers and skippers, winding up with the announcement that the donkey boiler stop valve had been blown off. This was the case.

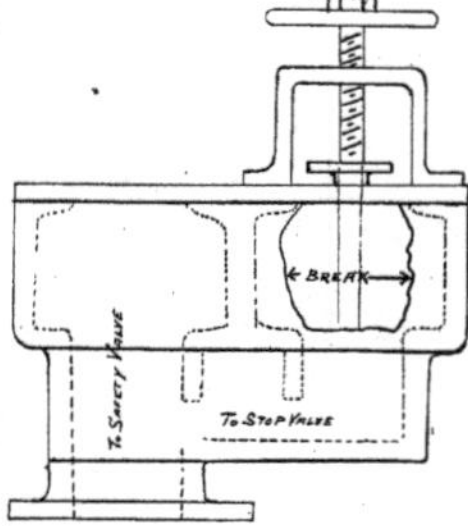

FIG. 1—SHOWING BREAK IN STOP VALVE WALL.

Steam for the winches and windlass was taken from the main boilers, through a reducing valve, and from the donkey boiler. The pipe from the reducing valve joined the donkey boiler pipe close to the valves controlling the deck steam. The donkey boiler was situated on the main deck, and from the donkey boiler stop valve to the bridge deck where controlling valves were situated, was about 30 feet in length, with a rise of about 3 feet. Now it was the practice to leave steam when at sea, open on the main boilers to the reducing valve, and up to the fidley, so that if steam was needed on deck, it could be opened without anyone having to go on the boiler tops. As no one but an engineer was supposed to touch the valves on deck, there was no danger. When the bo'sun threw open the valve a slug of water started off, and the first obstacle it struck was the donkey boiler stop valve, which surrendered. The stop valve was contained in the same casting as the safety valve, which made it more awkward. The first thing to be done was to blank the pipe off between the donkey boiler and the supply from the reducing valve, so that steam could be put on deck to handle the anchor in Colombo. This little job being done, a hurried visit to the messroom was paid, where the dried up remains of breakfast were bolted just as the stand-by rang for the pilot. This was the beginning of a perfect day, the remainder being spent between tallying coal and doing little jobs in the engine-room, such as packing the H.P. gland, etc., with the sweat forming pools wherever one stood. The delights of a coaling port ending like anything else, we were under way again about 8 p.m. with the next stop Singapore.

Next morning after breakfast, the chief, second and myself went forth to inspect the damage, and see what could be done. It was essential to have the donkey boiler at work in port, so that the main boilers could be cleaned. The side of the stop valve wall had been blown clean out as shewn by the sketch, but the lower port and safety valves were intact. The cover was good, and the valve itself not damaged. The first thing suggesting itself was to make a patch out of some plate we had, ¾-inch thick, but upon going further into the matter found this would be a job of some considerable magnitude, and other means were looked for. It was finally decided not to attempt a patch, but to use a temporary expedient that would serve till we arrived at a home port, where we could get a new casting. Looking through the stores we found a copper bend 90°, with flanges each end. This was the right size, 4 inches, if I remember rightly. The whole casting was taken from the donkey boiler top and lowered down the fidley, and taken to the engine-room vice bench. The broken wall of the stop valve was drilled all round and cut off with a chisel, leaving a flat face. The seat of the stop valve was drawn out, and holes drilled for studs to fit the copper flange. Of course, this took time, as it had to be done by the engineers in the course of their watch off. However, it was finally completed, and the casting replaced. The bend being bolted on to the casting its vertical flange came in approximately the same position as the stop valve flange had been, and the pipe was connected up with a blank flange in between. The method of working was as follows; On arriving in port the steam for the winches would be used from the main boilers during the remainder of the day of arrival. When cargo working for the day was finished, steam was shut off from the main boilers, the blank flange was removed and the donkey boiler lit up, steam being ready for the morning. When getting ready for leaving the modus operandi was reversed. The main boilers would be, of course, lit up and steaming and when cargo was finished the donkey boiler was blown down, the blank flange put in again, and steam opened up from the main boilers. By this means there were no delays, and the time of putting in and removing the blank was about 10 minutes. As we called and worked cargo at Moji, Hong Kong, Moulmein, Fiume, Aggiamarina, Tripoli, Tunis, Tarbis, Holmis, before we arrived at Antwerp, some months later, where our casting was awaiting us, we got quite used to the blank flange drill, and somewhat missed it after the new casting was placed.

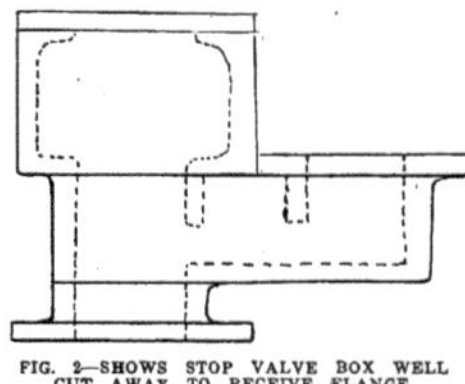

FIG. 2—SHOWS STOP VALVE BOX WELL CUT AWAY TO RECEIVE FLANGE.

REMARKABLE ESCAPE FROM SUBMARINE

One of the most remarkable experiences of the war is that of a British Stoker Petty Officer who escaped in a miraculous manner from one of our submarines which had sunk some time ago in home waters from an accidental cause. Although the Petty Officer was fighting for his life, he showed a wonderful example of indomitable courage and perseverance, and of refusal to acknowledge defeat. Alone, in almost complete darkness, with the gradually rising water, receiving electric shocks, and, towards the end, suffering from the effects of chlorine gas and a badly crushed hand, yet, in spite of continual disappointments he worked on for nearly two hours, keeping quite cool to the last, and at the seventh attempt at opening the hatch succeeded in escaping.

It was about 10.30 in the morning when the mishap to the submarine occurred, and it at once became apparent that she was taking in a great deal of water. The Stoker Petty Officer's first impulse was to close the lower conning tower hatch, but this he could not do, as some men had been ordered up the conning tower, so he went aft to see if all the men were out of the engine-room. He met one man coming forward and ordered him to put on a lifebelt and to "keep his head" till he had a chance of getting up the conning tower hatch. Having satisfied himself there was no one left aft, he made his way to the conning tower hatch with the intention of closing it, but, before he reached it, water was pouring in in a mighty volume, and that meant that his chances of closing the hatch were perfectly hopeless.

With the weight of water the vessel began to dip forward, and his only hope of escaping drowning was to shut himself in the engine-room. But before he closed the doors he shouted again to see if there was anyone who was still alive. Getting no answer, he reluctantly closed the doors against the rising water.

At this time the engine-room was in complete darkness save for the glimmer from one pilot lamp. The effect of the salt water on the electric batteries was to generate chlorine gas, and the air was becoming overpowering. The water had short-circuited the electric current, so that practically everything he touched gave him a shock. Moreover the room was oppressively hot.

He tried to think of a means of escape, and conceived the idea of opening the hatch and floating to the surface, but on trying to open the hatch he found that the tremendous pressure of the water outside prevented him moving it. He had always accepted the theory that the pressure inside a sunken air-locked vessel could be greater than the pressure outside. So to increase the pressure inside he opened a valve and admitted more water. When he considered the pressure was sufficient to blow him out he opened the hatch, but it instantly closed to again, as he had in sufficient pressure. With his shoulder, and exerting all his strength, he lifted the hatch, but again, with the weight of the water, it slammed to, crushing his fingers. With difficulty he released them, and once more opened the valve and admitted water until the engine-room was flooded right up to the coaming of the hatch. The air in this confined space was under tremendous pressure, greater than that of the water outside, so he was able to open the hatch and rise rapidly to the surface, where he was picked up by a destroyer.

London, Eng.—Some interesting figures showing the success of theh convoy system are given in a despatch from England. Since adopting the system 26,000,000 tons of food and 35,000,000 tons of munitions have been convoyed, and the percentage of foodstuffs lost has been reduced from nearly 10 per cent. to 1 per cent. This summer, out of 307 ships carrying the Argentine wheat crop, only one was lost. The grand total of merchant ships convoyed was 85,772, and of those only 433 were lost.

London.—The output of shipping in the United Kingdom for October was 136,000 tons, which is 15,000 tons less than the September losses. The world's output exceeded, for the quarter ending September, the world's losses by 500,000 tons. At the rate of the October output it would take Britain two and one-half years to make good her own losses, which for the whole period of the submarine war were 9,000,000 tons, of which 550,000 have been replaced by building and otherwise.

The MacLean Publishing Company
LIMITED
(ESTABLISHED 1888)

JOHN BAYNE MACLEAN - - - - - President
H. T. HUNTER - - - - - - Vice-President
H. V. TYRRELL - - - - - - General Manager

PUBLISHERS OF

MARINE ENGINEERING of Canada

A monthly journal dealing with the progress and development of Merchant and Naval Marine Engineering, Shipbuilding, the building of Harbors and Docks, and containing a record of the latest and best practice throughout the Sea-going World.

B. G. NEWTON, Manager. A. R. KENNEDY, Editor.

Associate Editors:

T. H. FENNER J. H. RODGERS (Montreal) W. F. SUTHERLAND

OFFICES

CANADA—
Montreal—Southam Building, 128 Bleury St., Telephone Main 1004.
Toronto—143-153 University Ave., Telephone Main 7324.
Winnipeg—1207 Union Trust Building, Telephone Main 3449.
Eastern Representative—H. V. Tresidder.
Ontario Representative—S. S. Moore.
Toronto and Hamilton Representative—J. N. Robinson.

UNITED STATES—
New York—A. R. Lowe, 111 Broadway, New York, Telephone 8971 Rector
Chicago—A. H. Byrne, 900 Lytton Bldg., 14 E. Jackson Street, 'Phone Harrison 1147
Boston—C. L. Morton, Room 733, Old South Bldg., Telephone Main 1024

GREAT BRITAIN—
London—The MacLean Company of Great Britain, Limited, 88 Fleet Street, E.C., E. J. Dodd, Director, Telephone Central 1960. Address: Atabek, London, England.

SUBSCRIPTION RATE

Canada, $1.00; United States, $1.50; Great Britain, Australia and other colonies, 4s. 2d. per year; other countries, $1.50. Advertising rates on request.

Subscribers who are not receiving their paper regularly will confer a favor by telling us. We should be notified at once of any change in address, giving both old and new.

Office of Publication, 143-153 University Avenue, Toronto, Ontario.

Vol. VIII. NOVEMBER, 1918 No. 11

PRINCIPAL CONTENTS

Progress of the Oil Engine

THE gradually increasing number of motor ships of various sizes is worthy of note. The oil engine for marine purposes is strongly entrenched in various parts of the world, and Great Britain, at the time of writing, holds the record for the largest ship built fitted with this type of engine. The British built motor has a good record all over the world, and now that the engineering shops are relieved from the stress of war work, a further effort is likely to be made in this class of engine. Sweden is a strong competitor, and the United States are going in for the manufacture of oil engines on quite a large scale. The indications are that a large business will be done in the near future by manufacturers in this line.

Government Control in Britain

WITH the approach of peace, the position of the British mercantile marine is causing considerable anxiety to shipowners in the Old Country. Before the war Great Britain owned half of the entire shipping of the world, and among them a splendid fleet of liners which were the express messengers between the outlying posts of Empire. In their war efforts, the people of Great Britain gave freely of their ships, and their shipbuilding yards were mostly given over to fighting ship construction of various sorts. On Great Britain fell the heaviest shipping losses, and especially on the fleet of superb liners which were her pride. Now that the United States, Canada, France and Japan are all building ships, British shipowners are anxious for the Government to relax its control over them. They feel that, left to themselves, they can once more make Britain supreme on the wave, but if the strangling effects of Government control continue they will be forced out of their proud position forever. It would seem to be the irony of fate, that the country which threw its greatest commercial asset freely into the hazard of war for the benefit of the rest of the world should be rewarded by having her maritime trade taken over by the ships of the nations she saved.

German Money a Real Boomerang

THERE'S one man in United States who has broken all records for buying bonds. The way he tosses out millions for the Liberty Loan would make Carnegie or Rockefeller gasp and reach for the railing to keep from being swept off the deck.

His name is A. Mitchell Palmer. Already he has bought $60,000,000 worth, and he's still going strong. If you want to know how it's done, here's the explanation in Mr. Mitchell's own words:—

"Possibly I have some little right to be a Liberty bond salesman, if there is any merit in the maxim 'practise what you preach,' because they tell me I am the biggest buyer of Liberty bonds in America. I have got something like $60,000,000 worth, and it is a poor day when I don't subscribe for $1,000,000 more. For me it is a pleasant task because I buy Liberty bonds with the Kaiser's own money."

It so happens that Mitchell is Alien Property Custodian of United States. Here's how he operates:

"Why some few weeks ago, out in a Western city, a school-teacher who was a German-born woman, died, and in her will she bequeathed $10,000 to von Hindenburg. I got that. I invested it in Liberty bonds and the proceeds were used to buy ammunition, and now Pershing's boys are trying to deliver the legacy to von Hindenburg over in Germany.

"We have made every dollar of German money in America fight the Germans. Great iron and steel mills, which were wont to send their profits out of America back to Germany, are now sending their profits to the Treasury of the United States and their product into war munitions to destroy their owners.

"Great woolen mills over in New Jersey, which were wont to send large dividends back to Berlin, are now sending those dividends to Washington, and working every loom and spindle to make those Army suits for the boys with Pershing in France.

"Great metal, mining, and mineral companies all over the United States, owned with German money, are working night and day, three shifts to the day, to produce material, not for the German over here to plant his industry in our midst as a sort of spy system against us, but for the United States, which he sought to destroy."

It is well to realize that war is not play. It's plain hell, and a nation must handle it in that way. Love taps don't fizz on the German either on the Western front or in this country. Politicians are gradually coming to the stage where they know the German hasn't got a vote just now, and they're licked if they stop to reckon about the votes he may have in years to come.

CONTRACTS TO BE FOR MORE VESSELS

Outline Given By Minister of Marine Regarding The Future Work

In pursuance of the Government's efforts to establish a Canadian mercantile fleet, the Minister of Marine has now given out contracts for the construction of 39 vessels. These are all steel steamers, varying in tonnage from 3,400 tons to 8,100 tons deadweight capacity. They are of the one deck and two deck type. It is also the intention of the Government to construct a larger class of vessel, reaching a deadweight capacity of 10,500 tons. The vessels will all be built to the highest class of Lloyd's or the British Corporation, and to the requirements of the British Board of Trade, and Canadian Steamship Inspection Board.

There has been some delay in making a start with these vessels, but it has been entirely due to the fact that the building berths were occupied by vessels building to the order of the Imperial Munitions Board. However, these ships are in many cases vacating the ways, and the Government vessels will be laid down as fast as there is available building space. The contracts have been distributed to yards in all parts of the country, though of course, the size that can be built by the lake shipyards is limited to the capacity of the canals.

It is expected that Messrs. Canadian Vickers, Montreal, will be the first yard to launch any of these ships, as they expect to put two vessels, one of 4,300 tons, and one of 8,100 tons, into the water during November. They will be named respectively Canadian Voyageur and Canadian Pioneer, and will have as sponsors Sir Robert and Lady Borden. If all goes well after launching, it is probable that these two ships will leave the St. Lawrence before the close of navigation. During the winter, if deliveries of steel continue satisfactorily, work will have been far enough advanced to have seven or eight vessels ready for service early in the spring, while the whole of the 175,000 tons comprised in the 31 vessels, will be sailing before the end of next year.

Such is the demand for shipping that if it was desirable the Government could dispose of the vessels contracted for at a handsome profit, but such is not their intention. Being now in possession of a national railway system of considerable dimensions, the Minister stated it was the intention of the Government to keep these vessels for the Canadian people, and work them in conjunction with the national railway system. In this connection the railways will feed the ships on their eastern voyages, while the ships will feed the railways on the return voyage. The management of the steamers will be under D. B. Hanna and his staff, and will not be subject to any interference outside of the management itself.

This combination of rail and vessel transportation has worked out very successfully under private management, and if the same success can be achieved with a national venture, it will go a long way to encourage the partisans of Government ownership. However, we must wait and see.

Acting upon the government suggestion to conserve steel, the Austin Co., Cleveland, have brought out a series of designs for wood construction covering their ten standard types. For the duration of the war and as long as the steel shortage exists the conservation of steel will be of vital importance. Wood is meeting this need and has the advantage of being more readily available. The wood truss is not a new form of construction, and dates back for many years, but in its application to the standard type of building the Austin engineers have redesigned the wood truss and incorporated it into the standard method of construction. They have duplicated each of the ten standards in the wood construction with a great saving in steel and the retention of the advantages of speed permanency and practical adaptability.

In a recent publication sent out by this firm the various standard types carried out in wood construction are fully illustrated.

New Grain Elevator.—A permit has been issued to the Campbell Milling Co. for a reinforced concrete elevator to cost $130,000. The site is at the corner of Cawthra Avenue and Junction Road.

SYNOPSIS OF GOVERNMENT CONTRACTS

No.	Firm.	Location.	Tonnage.	Type.	Approximate Date of Launching	Speed.
1	Canadian Vickers, Ltd.	Montreal	4,300	Single Deck, Poop, Bridge and Forecastle	November, 1918	11 knots.
2	Canadian Vickers, Ltd.	Montreal	8,100	Two Deck, Poop, Bridge and Forecastle	November, 1918	11 knots.
3	Collingwood Shipbuilding Co.	Collingwood	3,750 Lake Type	Single Deck, Poop, Bridge and Forecastle	November, 1918	9 knots.
4	Wallace Shipyards, Ltd.	Vancouver	4,300	Single Deck, Poop, Bridge and Forecastle	February, 1919	11 knots.
5	Collingwood Shipbuilding Co.	Collingwood	3,750 Lake Type	Single Deck, Poop, Bridge and Forecastle	May, 1919	9 knots.
6	Collingwood Shipbuilding Co.	Collingwood	Lake Type	Single Deck, Poop, Bridge and Forecastle	May, 1919	9 knots.
7	Collingwood Shipbuilding Co.	Collingwood	Lake Type	Single Deck, Poop, Bridge and Forecastle	May, 1919	9 knots.
8	Tidewater Shipbuilders, Ltd.	Three Rivers	5,100	Single Deck, Poop, Bridge and Forecastle	June, 1919	11 knots.
9	Tidewater Shipbuilders, Ltd.	Three Rivers	5,100	Single Deck, Poop, Bridge and Forecastle	July, 1919	11 knots.
10	Tidewater Shipbuilders, Ltd.	Three Rivers	5,100	Single Deck, Poop, Bridge and Forecastle	September, 1919	11 knots.
11	Tidewater Shipbuilders, Ltd.	Three Rivers	5,100	Single Deck, Poop, Bridge and Forecastle	October, 1919	11 knots.
12	Davie Shipbuilding & Repairing Co.	Lauzon, Levis	5,100	Single Deck, Poop, Bridge and Forecastle	July, 1919	11 knots.
13	Davie Shipbuilding & Repairing Co.	Lauzon, Levis	5,100	Single Deck, Poop, Bridge and Forecastle	August, 1919	11 knots.
14	Port Arthur Shipbuilding Co.	Port Arthur	3,400 Lake Type	Single Deck, Poop, Bridge and Forecastle	May, 1919	9 knots.
15	Port Arthur Shipbuilding Co.	Port Arthur	Lake Type	Single Deck, Poop, Bridge and Forecastle	May, 1919	9 knots.
16	Halifax Shipbuilders, Ltd.	Halifax	8,100	Two Deck, Poop, Bridge and Forecastle	August, 1919	10 knots.
17	Halifax Shipbuilders, Ltd.	Halifax	8,100	Two Deck, Poop, Bridge and Forecastle	September, 1919	10 knots.
18	Canadian Vickers, Ltd.	Montreal	4,300	Single Deck, Poop, Bridge and Forecastle	April, 1919	11 knots.
19	Canadian Vickers, Ltd.	Montreal	8,100	Two Deck, Poop, Bridge and Forecastle	December, 1918	11 knots.
20	Canadian Vickers, Ltd.	Montreal	8,100	Two Deck, Poop, Bridge and Forecastle	May, 1919	11 knots.
21	Canadian Vickers, Ltd.	Montreal	8,100	Two Deck, Poop, Bridge and Forecastle	May, 1919	11 knots.
22	Canadian Vickers, Ltd.	Montreal	8,100	Two Deck, Poop, Bridge and Forecastle	June, 1919	11 knots.
23	Canadian Vickers, Ltd.	Montreal	8,100	Two Deck, Poop, Bridge and Forecastle	June, 1919	11 knots.
24	Victoria Machinery Depot Co.	Victoria	8,100	Two Deck, Poop, Bridge and Forecastle	November, 1919	11 knots.
25	Victoria Machinery Depot Co.	Victoria	8,100	Two Deck, Poop, Bridge and Forecastle	December, 1919	11 knots.
26	Port Arthur Shipbuilding Co.	Port Arthur	3,400 Lake Type	Single Deck, Poop, Bridge and Forecastle	June, 1919	9 knots.
27	Port Arthur Shipbuilding Co.	Port Arthur	Lake Type	Single Deck, Poop, Bridge and Forecastle	June, 1919	9 knots.
28	Wallace Shipyards, Ltd.	Vancouver	4,300	Single Deck, Poop, Bridge and Forecastle	April, 1919	11 knots.
29	Wallace Shipyards, Ltd.	Vancouver	5,100	Single Deck, Poop, Bridge and Forecastle	July, 1919	11 knots.
30	Wallace Shipyards, Ltd.	Vancouver	5,100	Single Deck, Poop, Bridge and Forecastle	September, 1919	11 knots.
31	Kingston Shipbuilding Co.	Kingston	3,750 Lake Type	Single Deck, Poop, Bridge and Forecastle	November, 1919	9 knots.
32	Port Arthur Shipbuilding Co.	Port Arthur	4,300	Single Deck, Poop, Bridge and Forecastle	October, 1919	11 knots.
33	Port Arthur Shipbuilding Co.	Port Arthur	4,300	Single Deck, Poop, Bridge and Forecastle	November, 1919	11 knots.
34	I. Coughlan & Sons	Vancouver	8,100	Two Deck, Poop, Bridge and Forecastle	July, 1919	11 knots.
35	I. Coughlan & Sons	Vancouver	8,100	Two Deck, Poop, Bridge and Forecastle	August, 1919	11 knots.
36	I. Coughlan & Sons	Vancouver	8,100	Two Deck, Poop, Bridge and Forecastle	September, 1919	11 knots.
37	I. Coughlan & Sons	Vancouver	8,100	Two Deck, Poop, Bridge and Forecastle	October, 1919	11 knots.
38	Halifax Shipbuilders, Ltd.	Halifax	10,500	Three Deck, Poop, Bridge and Forecastle	December, 1919	12 knots.
39	Halifax Shipbuilders, Ltd.	Halifax	10,500	Three Deck, Poop, Bridge and Forecastle	January, 1920	12 knots.

The steamer and the motorship have deprived the seafarer of much of the old-time romance attaching to a sailor's life. The advent of wireless has also increased the safety of it. Deprived of these features, a sailor is immediately dependent on his own resources, and Captain Laing gives us a true and seaman-like reflection of sea life under canvas. The present feverish activity in constructing self-propelled craft affords a suitable background for incidents in a career in which gaffs, booms, tar and canvas held sway before being displaced by mechanical unloaders, derrick masts, fuel oil, coal and wire rope.

Part IX.

Off To The "West Coast"

THERE was only one "west coast" in the universe to a deepwater sailor and that was the Chilian and Peruvian coasts in South America. Sometimes the phrase took in the whole shore line from Cape Horn to Portland, Oregon, or Seattle, Washington. Our great Canadian ports of the Pacific are only being placed on the international map to-day. May their export magnitude and marine status grow till such names as Prince Rupert, Victoria and Vancouver are as well known abroad as Callao, Iquique and Valparaiso, for these three ports of the south know what marine activity means to a nation.

After clearing the land we hauled to the southward to go around New Zealand. One new feature of our trip was the food and the cook. Both those gigantic items were in the ascendancy and let me hasten to say that this marine chef had a real colonial streak of humor in him. It was even noticed that on going aft he kept his pipe in his mouth till he got to the poop ladder. Now an ordinary sea-cook takes his pipe out of his jaws when he gets abreast of the main-mast as do all members of the crew.

Officers only have the right to smoke on a windjammer's quarter deck, although later on we must chronicle the smoking behaviour of the ancient mariner (Fid) which beeame appalling in its shameful laxity.

Our famous birds soon joined us and seemed to welcome us back to their ocean haunts with appropriate screams and aerial gymnastics. We had great dreams of getting fattened up on the trip, in fact that was the motive behind the culinary arrangements and the explanation was a logical one. On the "west coast," all sailing-ship cargoes were worked in and out by the crew, so the five or six weeks trip was used up in making us produce flesh and muscle in readiness for the manual labor of discharging into barges our 1,500 tons of coal and loading a similar weight of saltpetre or nitrate of potassium, which is generally the homeward cargo from Chili. Copper-ore, manganese-ore, guano and grain figure in west coast exports, but in a much smaller proportion.

We were simply deluged with dry hash for breakfast and skouse for supper, until we imagined ourselves part and parcel of an Australian feed shop with all meals gratis. What a difference to the long trip from Europe, on which the pound and pint system of rations became maddening to young growing boys.

Westerly gales were again our portion and the "Magpie Dorrit" was rolling off her 200 to 250 miles for two or three days at a time. Then a couple of days of meteorological samples in which the runs might only approximate 50 or 80 miles, and so the trip went on. When in such high latitudes and away from tropical weather a great deal of detail work was impossible as the decks were constantly awash and everything below the lower-mast doublings was saturated with salt and spray. The crew, however, when not reefing, furling or hauling on the halyards and braces, were kept busy in the daytime under the shelter of the fo'c'stle head, making various kinds of sennit for chafing mats, service for topmast rigging, etc. Coir and manilla mats were also made for the cabins, similar to the mats that one sees in houses ashore. Both the thrum and plait methods were used.

The Somnolent Habits of the Ancient Mariner—Fid.

The first heavy gale that made us reduce to lower topsails brought out Fid's sleeping qualities for the second time. All hands had been furling the upper topsails and reefing them ready for a short hoist if the chance came along. As the watch below had been called out in the middle of their sleep, the old man ordered the steward to hand round a peg of whisky. When this was done the seamen were counted for fear of any extra bold salt trying to double bank on the wet goods. Then the boys were offered something in lieu of grog such as a plug of tobacco or a piece of plum-duff, according to the captain's free-will.

However, no one could account for the absence of Fid and no one had seen him. It was a very dirty night, no moon, sleet falling and everything as black as bog-oak. Of course the orders were given to look around the mast coats and fyfe rails, under the long boat, in the galley, on top of the same place and anywhere that a crouching half drowned seaman might have hung on to when washed from his feet.

"Had anyone seen him aloft?" No one had seen him on the topsail yards. "Must have fallen off a yard-arm into the sea!" The old barque had been shipping fairly heavy dollops of ocean water as she flew along in her madness before the blast, but she had not filled herself flush with the top-gallant rail for over an hour, and that was reckoned the only possible time that a man or boy could have floated into Davy Jones' locker.

Some one had the audacity to look into the half-deck, but there was nothing there but the usual truck rolling from side to side in deep-sea glory. It was impossible to sleep in the lamplocker, as it was full of oils, paint cans, and narrow shelves to accomodate the lamps, but the boatswain's locker had a little floor space so in the hunt this door was opened. The 2nd mate stepped inside for a minute, but backed out again muttering that a bag of wedges, some rope shakings and balls of sennit covered the floor. No Fid to be found. The whole thing looked a little blue when suddenly the old man called for a life line to be stretched on the poop to enable him to get a little exercise without being damaged with the lurch of the vessel.

Continued on page 50

SEAMANSHIP AND NAVIGATION

Conducted by "The Skipper"*

Articles of direct interest to mariners, discussions of seamanship, navigation rules, and allied topics. Inquiries from readers are invited and will be replied to promptly through these columns, unless otherwise requested.

TO all aspirants of the quarter deck or bridge the study of a vessel's movements under various conditions must be of vital interest. The last two articles having dealt with the motions of the sea and wind, we will now endeavor to master the behaviour of the craft and understand her antics. Whilst each vessel may have little peculiarities, which are only found out by those who handle it, all ships or boats—great and small—act in the manner laid down herewith.

The Rudder or Helm

As phraseology of the rudder is sometimes expressed in misleading language, owing to this continent having a few inland-water craft using the cross chain method of connecting rudder quadrants and steering engine gypsies, the writer wishes it to be understood throughout all his writings that the meaning of "Port the helm" is that the steering wheel, the rudder itself, and the ship's head all go to starboard, and that "Starboard the helm" means that the steering wheel, the rudder itself, and the ship's head all go to port.

Now, to beginners this apparent muddle is very trying, but as an international institution, there it stands. When the terms were first coined all craft were manoeuvred with hand tillers, a short spar shipped into the rudder head and looking forward. Lifeboats still use them, and sundry pleasure craft with hurricane decks or roomy cock-pits are fitted with tillers.

You will easily notice that when a man steering with a tiller was ordered to "Port the helm," his most natural move was to put the tiller over towards the port rail. By doing this, however, the ship's head and the rudder itself went to starboard. The trouble followed when the steering wheel was introduced, as the operation of that mechanical medium was not in accord with the movement of the tiller, although the rudder and ship's head acted as before. It is clear now that these phrases, when once explained, will give little trouble if properly understood.

In steering phraseology it is also well to remember that "Porting the helm" and "Altering your ship's head to starboard," are synonymous, while "Starboarding the helm" and "Altering your ship's head to port," are also synonymous.

Screw Propeller Action

As a general rule all single screw vessels have right-hand engines—that is, the propeller "throws" to starboard when going ahead and, of course, to port when going astern. How does this effect the ship's head?

When a steamer is going ahead her single screw tends to cant her head to port or left. When a steamer is going astern her single screw tends to cant her head to starboard or right.

Then, how would you turn a steamer of this type short round in a small bay or dock entrance?

Answer: Full speed ahead with helm aport. As soon as she gathers headway she will cant to starboard, then helm amidships and full speed astern, which will still keep her canting to the right. Keep on the manoeuvre till she is round to where you want her. You may require three or four "full aheads on port helm" and three or four "full asterns with helm amidships," but she is bound to come.

Twin Screw Ships

How do twin screws work? The propellers both revolve in an outboard fashion. That is, the starboard screw works in the same way as it does in a single propeller ship and the port screw works in the opposite way or counter clockwise. Thus you will realize that the ship's head is perfectly balanced and should not cant to one side or another as far as propeller action is concerned.

The manoeuvring of a twin screw ship is much easier, however, than a single screw vessel, for the screws can be used more in conjunction with the rudder Here is an illustration: Say that you are steaming up a river in a twin screw craft and you wish her to make a quick turn to the port hand. What would you do to facilitate the circling movement? Of course, you would "starboard the helm" and reverse the port engine, but keep the other engine going ahead. What does reversing the port engine do? Helps to cant the ship's head to the port side. The starboard engine is already doing that and so is the rudder, so you have the three agents working in unison to make the ship swivel. The same method of action can be used in turning the ship's head to either side, whether she is going ahead or astern. Just study the action and effect of the two propellers and the ordeal will appear in its simplicity, and remember that the man who thinks a problem out for himself and understands it, is far ahead of the automaton who can do a thing, but can't give a logical reason for his actions.

That young seamen may more readily interest themselves in the movements of their craft, it may be well to give an approximate definition of the word steering.

Steering

Steering is the art that neutralizes all forces that may act on your ship in regard to her propulsion in a straight line—as for instance, a course: Graceful swinging round a bend of a tortuous river under easy helm, or canting your ship's head quickly under "hard over" conditions, are two phases of the art; whilst keeping a vessel on a prescribed point or degree of the compass is the most common demand.

What forces are at work in relation to the rudder? The forces at work are: water friction on submerged part of hull, wind pressure on upper part of hull and superstructure, and the poundage of seas on any part of the ship. Then as sub-agents in the movement we have draught, heel, speed, depth of water under vessel and minor components, local and temporary, such as propellers, tides, etc. It is essential to mention these things, as only by a study of them can one expect to handle his vessel in a seamanlike manner.

Architectural Lines

A witticism on board ship is "different ships, different long-splices," meaning, of course, that all vessels do not act or behave in the same manner while confronted with the same conditions. In regards to a vessel's movements you will find that a revenue cutter or a pleasure yacht with fine lines will act quicker and easier in steering manoeuvres than an ordinary tramp steamer with a box-car bow and heavy quarters. What inference do you draw from this? That the displacement water of the ship has a big say. In what respect? In the manner of its release. A ship built for speed has her greatest breadth or beam about the foremast and tapers from that forward point towards the stern. Thus the greatest resistance is overcome well forward and diminishes gently under the ship's bottom. Is this the case in a good cargo carrier? No, sir. Where you demand large tonnage capacity you must sacrifice speed and have clumsy lines of mould. Thus it must happen that the most of cargo ships are box built and their displacement water does not leave them so easily—in fact they, what we term, "carry" it. This is exemplified at its most demonstrative point in shallow water, and any thoughtful person can at once connect the bearing which architectural lines has to steering and a ship's movements.

Why can't you put a loaded steamer at full speed in the fourteen-mile cut of the Welland, or for that matter in the Suez or Panama canals? Because she won't steer. Because the ship's bottom and the canal bottom are too close to each other and the agitation of the displacement water would cause the ship to "break her sheer," and very likely she would cant athwart the waterway. Although the canals were as straight as an arrow, the same thing would happen.

(Continued on page 52)

*Editor's Note.—This department is conducted by a qualified ship's master, and will deal with questions involving seamanship and navigation. The attention of young men desiring to obtain certificates as mates and masters is particularly directed to this department.

PROGRESS IN NEW EQUIPMENT

There is Here Provided in Compact Form a Monthly Compendium of Shipbuilding and Marine Engineering Auxiliary Product Achievement

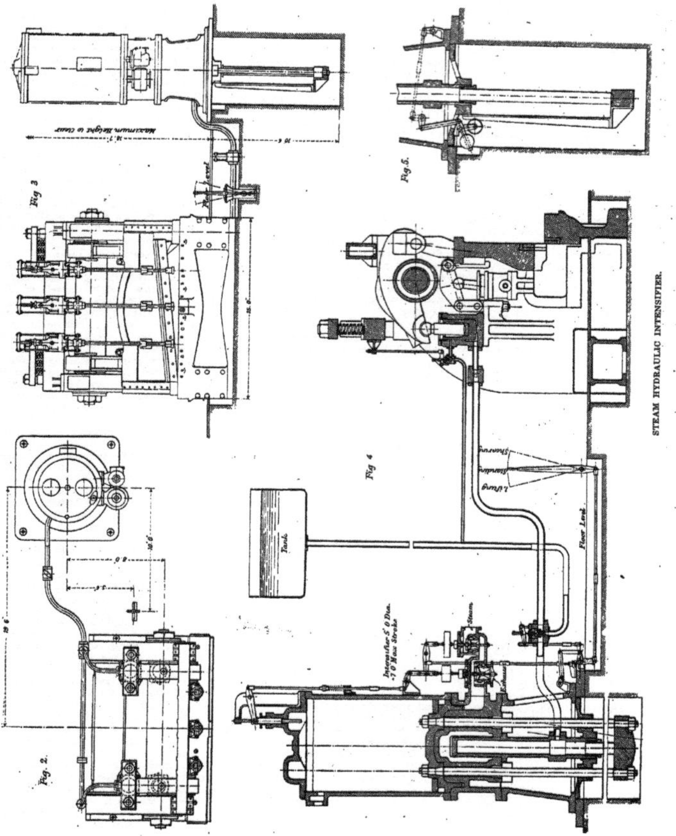

STEAM HYDRAULIC INTENSIFIER.

RESPECTIVE VIEW OF SHEARS.

STEAM HYDRAULIC PLATE SHEARS

An excellent example of up-to-date shipyard machinery is afforded by the steam hydraulic plate shears made by Messrs. Duncan Stewart and Co., Ltd., engineers, Glasgow, described and illustrated herewith, of which there are two more in course of manufacture. Two of the sets are for cutting plates up to 1¾ ins. thick and 11 feet long. The shears are quick in action and will make twenty cutting strokes per minute when cutting plates two-thirds of the maximum thickness. We are indebted to the courtesy of "Engineering" for the particulars.

The perspective view given will show the general arrangement of the shears, and the great strength of construction can be plainly seen. A line drawing of the steam hydraulic intensifier is also given, which makes the general idea of this part of the machine apparent. This intensifier is of the inverted type fitted with the rapid action automatic controlling and safety gears, which, in addition to cutting off automatically the steam to the intensifier at a point corresponding to the position of the handing lever; also prevents the handing lever from being moved further in the steaming direction than is necessary to steam the intensifier at the particular position the piston occupies in the cylinder. The shears have a gap of 36 inches, enabling six feet wide plates to be split down the middle. The side frames are massive steel castings tied together at the top by a large diameter forged steel shaft on which the cast steel rocking levers oscillate. They are further tied together at the bottom by the cast iron knife holder, and a heavy cast iron box section girder bolted in between the standards. The hydraulic cylinders are of cast steel, the rams are of chilled iron, ground on the working surfaces. Ball bearings are fitted in the top surfaces of the bolster to facilitate the easy manipulation of the plates. To the shears are fitted three cramping rams to hold the plates in position during cutting. In addition to the shears Messrs. Duncan Stewart and Co., Limited, have in hand a large number of forge presses, slab shears, billet shears and rolling mills to the designs of Mr. T. E. Holmes (late of Messrs. Davy Brothers, Sheffield), who has joined this firm. To enable them to meet the increasing demands they have nearing completion a large and well equipped engineering works.

THE TOD-ATTWOOD RAM PUMP

The vital part of a steam power plant is the boiler feed pump, and especially a marine steam plant, where the pump presents the only means of feeding the boiler. One of the latest developments in marine feed pump design is furnished by the Tod-Attwood vertical simplex pump, a description of which is given here.

A glance at the accompanying sectional elevation will at once reveal several departures from conventional design, all of which tend to make a smooth working and easily accessible pump.

The pump here described is the 11-inch by 8-inch by 12-inch size, which is designed to supply 1,250 h.p., at a piston speed of thirty feet per minute.

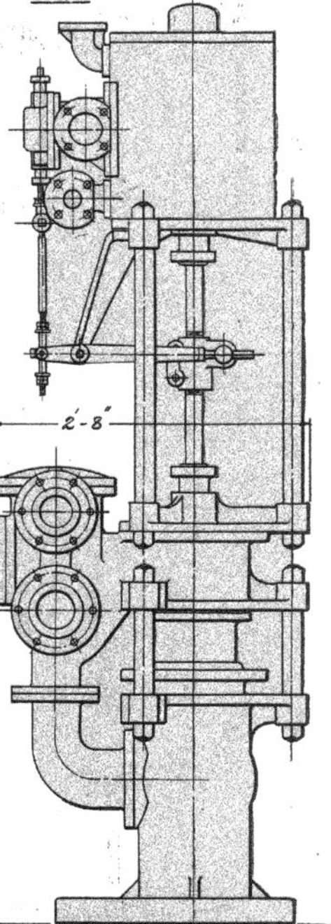

OUTSIDE VIEW SHOWING CONNECTIONS

This would require fifteen double strokes per minute, which is a very reasonable speed, conducive to long life. The steam piston is of the conical, or marine type, with two Ramsbottom rings. The valve arrangement for admitting steam is a combination of slide and piston valve, the slide valve being operated from a lever on the pump rod, and the motion of the slide giving steam to either end of the piston valve, which controls the admission to the cylinder. Unlike the majority of this style of valve, it is devoid of complications, the whole range of operations being taken care of by three main ports. The main valve chest is bolted on to the steam cylinder, and the slide valve chest is bolted on to the main chest, making access to either as convenient as possible. The ports in the cylinder bore are cut in such a manner that the piston automatically cushions itself at the end of the stroke, and so well has this feature been designed, that the suction connection can be broken with the pump under full working pressure, without any knocking at either end of the stroke. This is a remarkable claim, but it has been proved to the satisfaction of Lloyd's representatives.

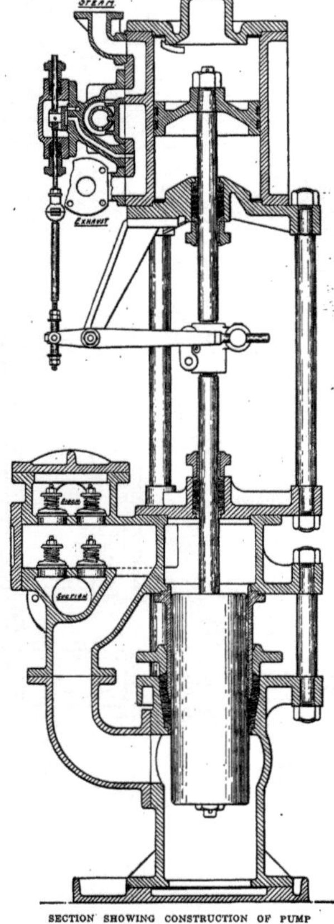

SECTION SHOWING CONSTRUCTION OF PUMP

At the water end of the pump, ample provision has been made to give easy access to suction and discharge valves. A hand hole door is provided on the suction valve box, and there is ample clearance between the top of valve studs and the discharge valve seats to enable a man to pass his hand through comfortably.

The ram is of gunmetal, as also is the cylinder liner, and the liner is arranged so as to be easily removable. There is only one packing gland which is arranged with ample working clearance round it, and the ram being water borne at all times, the wear on both liner and ram is but slight. The lower part of the chamber is made separate from the upper part, so that in the event of a break in either there is no need of a complete new water end. The steam end is supported on three rigid stanchions secured by nuts each end. All glands are bushed with gunmetal bushings, and the workmanship throughout is of the best. The pump has earned of the best. The pump has earned the seal of approval of Lloyd's surveyors here, and is at present under examination by the Bureau Veritas. The pump is manufactured by the G. H. Tod Co., Ltd., Toronto, who also manufacture a line of stationary pumps embodying some of the features of the pump described. The overall dimensions of the pump shown are 7 feet 9 inches high by 2 feet 8 inches wide. The floor space is 24 inches by 32 inches.

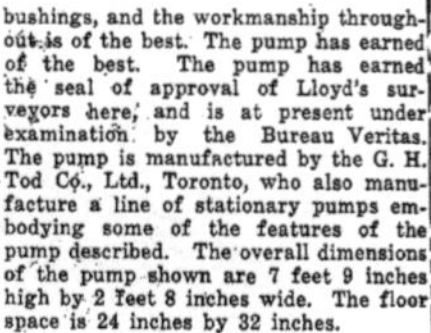

THE "OLYMPIC'S" FINE RECORD

The end of the war has made it possible to tell how Captain Bertie Hayes of the British transport Olympic, which sailed November 15 for Liverpool, won the Distinguished Service Order for sinking two U-boats on May 12, 1918, in the English Channel off Portsmouth. For this exploit $10,000 was given by the Admiralty to the Captain, officers and crew. Both submarines were destroyed at daylight, and survivors were picked up from them by the American destroyer Davis, which was about a mile away on the port quarter.

Rescued a Boat Crew

One submarine came up on the starboard bow of the Olympic as the mist cleared away and was rammed immediately. As the bow of the 50,000-ton ship was cutting her in two another U-boat appeared on the port quarter, half a mile away, and was sunk by a shot from the six-inch stern gun. Twenty-seven of her officers and men were rescued by the Davis, with five from the U-boat that was rammed. The Olympic sank another U-boat in the Mediterranean in June, 1916, by gunfire, but as there were no survivors it was not counted as official by the Admiralty.

Captain Hayes has commanded the Olympic since the end of October, 1914, when she figured in the unsuccessful attempt to tow the battleship Audacious after she had been mined. The Olympic has carried nearly 300,000 troops since then to Mudros, Alexandria, and from Canada and the United States to England and France without accident. She has had many narrow escapes from torpedoes, some missing her by barely five feet.

Cleveland.—The lake coal carriers are enjoying a breathing spell between the end of coal transportation movement and the moving of the grain. Space at the rate of 20,000,000 bushels per week for the next three weeks will be required, and this will mean the use of about 160 vessels.

Halifax.—The Portuguese Government have had Mr. Freiburger, of New York, in Halifax lately endeavoring to buy some ships for them. The Blanche H. Collins was purchased and has been transferred to the Portuguese flag, and Mr. Freiburger is now waiting permission to transfer the Kathleen Crowe.

2
h
L
B

NE

STA

J. A.

PACI

Take

Gasol

Three

6 in. x

tube boil

for builde.

Four new

C. E. BAINTER, PRI

Several wooden fishing steam

GRAND TRUNK DRYDOCK SHIP REPAIR CO., LTD., PRINCE RUPERT, B.C.

S. A. MOULTON, PRINCE RUPERT, B.C.

Ten composite vessels for undisclosed interests.

BRITISH-AMERICAN SHIPBUILDING & ENG. CO., VANCOUVER, B.C.

Establishing plant on Kitsilano Reserve.

Twenty wooden vessels being built for undisclosed interests. 3,500 tons each.

JOHN COUGHLAN & SONS, N. VANCOUVER, B.C.

"War Camp," sister ship to "Alaska," launched in March, 1918.

One steel freight steamer, 427 ft. 9 in. x 54 ft. x 29 ft. 9 in., draft 24 ft. 2 in., speed 11½ knots, 5,730 gross tons. Launched Jan. 10, 1918, for B. Stolt Nielsen, Christiania, Norway. Both sold to the Cunard Line.

Three steel steamers, "War Chariot," "War Chief," and "War Noble," Nos. 4-5-6, 425 ft. x 54 ft., 8,800 tons cap., for the Cunard Line.

Six steel freight steamers, 425 ft. x 54 ft. x 24 ft. 2 in., 8,800 tons cap., 1 knots, for undisclosed interests. Delivery Jan., Feb., March, May and July, 1918.

One steel freight steamer, "Alaska," 427 ft. 9 in. x 54 ft. x 29 ft. 9 in., draft 24 ft. 2 in., speed 11½ knots, 5,730 gr. tons, 8,800 d.w. Launched Jan. 10, 1918, for B. Stolt Nielsen, Christiania, Norway. Sold to the Cunard Line.

THE FOUNDATION CCO., VANCOUVER, B.C.

FRASER VALLEY SHIPBUILDING CO., VANCOUVER, B.C.

GRANT, SMITH & CO., VANCOUVER, B.C.

Six wooden cargo vessels, 250 ft. x 43 ft. 6 in. x 25 ft., 3,000 tons cap., 9½ knots, for the Canadian Government.

HARRISON & LAMOND, SHIPBUILDERS, LTD., VANCOUVER, B.C.

One wooden auxiliary schooner, 225 ft. x 44 ft. x 21 ft. 4 in., 1,600 gr. tons, 2,550 tons cap., for undisclosed interests.

LYALL SHIPBUILDING CO., NORTH VANCOUVER, B.C.

"War Puget," Feb. 16, 1918, 250 ft., wood, 2,080 tons.

Ten wooden cargo ships, for undisclosed interests. Launchings Jan. 20, Mar. 2, Mar. 23, and April 11, 1918.

Building six wooden vessels for own account.

"War Caribou," April 10, 1918, 3,080tons, wood.

W. R. MANCHION, VANCOUVER, B.C.

STANDARD SHIPBUILDING CO., VANCOUVER, B.C.

Two 3,500-ton vessels for Brazilian Govt., and six for French Govt. of 4,500 tons.

Donohoe reinforced type, wood composite construction.

TAYLOR ENGINEERING CO., VANCOUVER, B.C.

Number of small vessels being constructed at total value of $300,000.

4,500-ton floating drydock, 352 ft. long.

VANCOUVER SHIPYARD, LTD., VANCOUVER, B.C.

One motor freighter, 125 ft. x 24 ft. x 12 ft., wood, 160 h.p., Bolinder's crude oil engine. To be completed Aug., for Taylor Engr. Co.

Also repairing several steamers and schooners.

THE WALLACE SHIPYARDS, LTD., N. VANCOUVER, B.C.

"War Dog," 4,500 tons, steel, May 18, 1918.

"War Power," 4,600 tons, steel, March 23, 1918.

Two steel steamers, 315 ft. x 45 ft. x 27 ft., 4,600 tons d. w., trip. exp. vertical engines ,two S. E. Scotch boilers, 10 knots, single screw,

Constructing

NEW BRUNSWICK

C. T. WHITE & SON, LTD., ALMA, N.B. (Office at Sussex, N.B., also.)

Two schooners aux. power, 143 ft., 900 tons d. w., first to be launched in April; second in June.

JAMES X, LENTEIGNE, LOWER CARAQUET, N.B.

One schooner, 28 tons, wood.

INTERNATIONAL SHIPBUILDING CO., NEWCASTLE, N.B.

Two four-masted wooden auxiliary schooners, 155 ft. keel, 37 ft. depth, 535 net tons. One to be completed in September and the other December, 1918. For builders' account.

EUREKA SHIPBUILDING CO., NORTH HEAD, N.B.

One wooden schooner, "Mollie & Melba," 350 tons reg.

PORT COLBORNE BUILDING & REALTY CO., LTD., REXTON, N.B. (Head Office, Welland, Ont.)

Four-masted wooden schooner. Has cap. of 3 schooners per year.

GRANT & HORNE, ST. JOHN, N.B.

Two wooden cargo steamers, "War Fundy" (launched Aug., 1918) and "War Digby," 250 ft. x 42 ft. 5 in. x 25 ft. 5 in., 1,000 h.p., speed 9½ knots, 2,800 tons d.w., for undisclosed interests.

MARINE CONSTRUCTION CO., CANADA, LTD., ST. JOHN, N.B.

Four wooden auxiliary, four-masted schooners, 185 ft. o. a., 165 ft. keel x 40 ft., 1,100 d. w. tons, for builder's account. J. Murray Watts, Philadelphia, naval architect. Launching April, 1918.

One wooden schooner, 'Dorfontein," 740 reg. tons. Launching June, 1918.

ST. JOHN SHIPBUILDING CO, ST. JOHN, N.B. (Plant at Courtenay Bay.)

Ten five-masted auxl. schooners, oil burning engines, for undisclosed interests.

Establishing plant for steel and wood construction.

ST. MARTIN'S SHIPBUILDING CO, ST. MARTIN'S, N.B.

One wooden schooner, 450 tons, three masts. To be launched in July for builder's account.

PETER & A. A. McINTYRE, ST. JOHN, N.B.

One 900-ton schooner under construction.

NEWFOUNDLAND

ANGLO-NEWFOUNDLAND DEVELOPMENT CO., BOTWOOD, NFLD.

Two three-mast aux. schrs,. wood, 450 tons, 150 H.P. for the builders' account.

NEWFOUNDLAND SHIPBUILDING CO., HARBOR GRACE, NFLD.

To build ships of 1,200 tons d.w. Steel ships later.

M. E. MARTIN, MORRIS ARM, NR. ST. JOHN'S, NFLD.

Two wooden schooners, 300 tons each, for builder's account.

One wooden schooner, 500 tons, for builder's account.

UNION SHIPBUILDING CO., ST. JOHN, NFLD.

ANNAPOLIS SHIPPING CO., ANNAPOLIS ROYAL, N.S.

"Hilda M. Clark," March 4, 1918, 187 ft., 640 tons reg.

Two schrs. three-masted, wood 500 tons, 170 ft. long. To be completed August and December, 1918.

J. E. GASKILL, CHURCH POINT, N.S.

Other yards at Grosses Coques, Little Brook, Meteghan and Port Wade, N.S.

Five wooden schooners from 344 to 387 reg. tons. All sister ships.

COMEAU SHIPBUILDING CO., COMEAUVILLE, N.S.

One three-masted wooden schooner, for builder's account, 450 tons.

J. W. COMEAU, COMEAUVILLE, N.S.

One schooner, 329 tons, Jan., 1918.

J. N. RAFUSE & SONS, CONQUERALL BANKS, DIGBY CO., N.S. (Other plants at Salmon River and Shelburne, N.S.)

Two wooden three-masted schooners, 120 ft. x 32 ft. x 12 ft., 700 tons cap., for builder's account. For the market. Launchings Oct. 25, 1917, and Jan. 10., 1918.

One three-masted wooden schooner, "Integra," 112 ft. x 30 ft. x 11 ft. 6 in., 600 tons cap., for J. O. Williams & Co., St. John's, Newfoundland. Launching Nov. 25, 1917.

E. F. WILLIAMS, DARMOUTH, N.S.

One schooner, 360 tons, wood.

ROBAR BROTHERS, DAYSPRING, N.S.

Schooner for Capt. Ivan Cresser, Dayspring.

MAURICE E. LEARY, DAYSPRING, N.S.

One wooden schooner, 225 tons, 135 ft. o. a. To be launched Aug. 15, 1918, for La Have Outfitting Co., La Have, N.S.

J. NEWTON PUGSLEY & CHAS. ROBERTSON, DILIGENT RIVER, N.S.

One schooner, three-masted, 475 tons.

McLEAN & McKAY, ECONOMY, N.S.

One tern schooner, wood, 400 tons net, 135 ft. keel. To be launched Sept. 1, 1918, for builders' account.

S. J. SOLEY, ROX RIVER, N.S.

One schooner, three-masted, 121 ft. keel, 300 tons. Built of native wood. To be launched Sept., 1918. For builder's account. For sale.

ALLAN & FRASER, FRASERVILLE, N.S.

One schooner, 350 tons, wood.

BERNARD W. MELANSON, GILBERT'S COVE, N.S.

One three-masted schooner, wood, 250 tons net. To be completed November, 1018, for builder's account and for sale.

AMOS BLINN, GROSSES COQUES, N.S.

One schooner, 375 tons. Jan., 1918.
One schooner, 350 tons, wood.

BINN BROS., GROSSES COQUES, N.S.

F. K. WARREN & CO., GROSSES COQUES, N.S. (Offices in Union Bank Chambers, Halifax, N.S., also.)

350-ton tern schooner, under construction Jan., 1918.

HALIFAX SHIPBUILDING CO., HALIFAX, N.S.

J. WILLARD SMITH, HILLSBURN, N.S. (Head Office, St. John, N.B.)

One tern schooner, wood, 140 ft. keel, 35 ft. beam, 450 tons net reg. tons.

for

ity,

gr.
Fall,

2 in.
Nov.,

in. x 10 ft. 6 in., 200
April, 1918. Christened

[illegible] schooner 125 ft. keel x 32 ft. x 12 ft., 600 tons, d. w. To be delivered July, 1918.

Two masted schooner, "Agnes D. McGlaston," 180 ft. x 26 ft. 8 in. x 10 ft. 8 in. Launched Nov. 13, 1917.

Keel laid for three-masted wooden schooner, 200 tons.

J. ERNEST & SON, MAHONE BAY, N.S.

One schooner, 520 tons. Jan., 1918.

O. A. HAM, MAHONE BAY, N.S.

One schooner, "Dozle," 128 tons. Launching May 30. One motor boat about 20 tons.

McLEAN CONSTRUCTION CO., MAHONE BAY, N.S. (Leased to Montague Mahaffy, Toronto.)

One three-masted schooner, 325 tons gr. reg., 500 tons cap., West Indies freight type. Will be launched in July.

Keel laid for another three-masted schooner with cargo capacity of 425 tons.

JOHN McLEAN & SONS, MAHONE BAY, N.S.

One schooner, 95 tons, wood.

J. A. BALSOM CO., LTD., MARGARETSVILLE, N.S.

One schooner, 409 tons. Jan., 1918.

CLARE SHIPBUILDING CO., METEGHAN RIVER, N.S.

One wooden schooner for builder's account. 400 tons.

A. H. COMEAU & CO., METEGHAN, N.S.

One schooner, 400 tons, wood.

AGAPIT COMEAU, METEGHAN, N.S.

JOHN F. DEVEAU, METEGHAN, N.S.

A 300-ton schooner was launched recently for Ritcey & Co., Lunenburg, N.S., and named Charles A. Ritcey.
1 schooner, 425 tons. Jan., 1918.
1 schooner, 400 tons, wood.

MILTON SHIPBUILDING CO., YARMOUTH, N.S.

One wooden sailing schooner, 800 tons d.w. Keel laid April 15th, 1918, to be launched December 17th and completed December 31st. Built for sale.

J. E. GASKILL, CHURCH POINT, N.S.

One wooden schooner, 750 tons d.w. Keel laid March, 1918, to be launched December 15th. Built for J. E. Corkill.

THERIAULT SHIPBUILDING CO.

One wooden vessel, 500 tons d.w. Keel laid September 18th, to be launched in May, 1919. Built for owners account.

SIDNEY ST. C. JONES, WEYMOUTH, N.S.

One tern schooner, "Westway," 450 tons d.w. Keel laid April 18, to be launcheed December 30, 1918, and completed in January, 1919. Built for sale. 130 feet by 28 feet 8 inches beam by 10 feet 6 inches deep.

THE HANKINSON SHIPPING CO., BELLEVEAU'S COVE, N.S.

One vessel, 750 tons d.w. "Charles Doucet," fitted for auxiliary engine. Keel laid April 1, to be launched December 17. One vessel 750 tons d.w. Keel laid Aug., 1918, to be launched April, 1919, and completed. Built for owners account.

FAUQUIER & PORTER, HANTSPORT, N.S.

One vessel 1,800 tons d.w. to be fitted with two 100 h.p. crude oil engines (Canadian Fairbanks-Morse), to be launched Dec. 17, 1918, and completed January, 1919. Built for sale. Propose to lay keels for two 500-ton, 3-mast schooners as soon as this schooner is launched.

N.S. SHIPBUILDING & TRANSPORTATION CO., LIVERPOOL, N.S.

"Protea," 400 tons d.w., three-masted schooner with provision for Morse semi-Diesel engines. Keel laid Nov. 18. To be completed May account and sold to Pearson & Walker, Cape Town, S.A.

"Gordon T. Tibbs," 400 tons d.w., three-masted schooner. Keel laid Aug. 30. To be launched Dec. 15 and completed Jan. 19. Built for buildeer's account and sold to G. Tibbs & Sons, St. John's, Nfld.

One beam trawler, 900 tons d.w., 400 i.h.p., with two 200 Fairbanks-Morse semi-Diesel engines. Keel laid Nov. 19. To be completed May next. Built for Rafuse Grey et al Le Havre. Repairing schooner Karmoe for N.Y. owners.

H. ELDERKIN & CO., PORT GREVILLE, N.S.

Two vessels, 825 and 500 tons respectively. Keel laid April and August. To be launched December and January. Built for builder's account.

J. E. GASKILL, METEGHAN, N.S.

THOMAS GERMAN, METEGHAN, N.S.

One schooner, 350 tons, wood.

R. H. HOWES CONSTRUCTION CO., METEGHAN, N.S. (Leased James Cosman's shipyards.)

Building several wooden schooners for own account.

DR. F. H. MACDONALD, METEGHAN, N.S.

One wooden four-masted schooner, "Rebecca L. Macdonald," 201 ft. o. a. x 6 ft. x 16 ft., 800 gr. tons, 1,500 tons d.w., for builder's account. Launching January 1, 1918.

One wooden fishing schooner, No. 8, 84 ft. x 17 ft. x 7 ft., 50 gr. tons, 5 h.p. Launching August, 1918.

One schooner, 544 tons, wood.

METEGHAN RAILWAY & SHIPBUILDING CO., METEGHAN, N.S.

One wooden three-masted schooner, 182 ft. x 35 ft., 450 gr. tons.

One schooner, 470 tons, wood.

CHAS. McNEIL, NEW GLASGOW, N.S.

One wooden 3-masted schooner, 103 ft. x 30 ft. x 10 ft. 9 in., 200 tons, for builder's account. Launching July, 1918.

One wooden 3-masted schooner, 142 ft. x 25 ft. 4 in. x 13 ft., 500 gr. tons, for builder's account. Launching Oct., 1918.

THE NOVA SCOTIA STEEL & COAL CO., NEW GLASGOW, N.S.

Two steel freight steamers, 257 ft. 9 in. x 35 ft. x 20 ft., 1,700 gr. tons, 2,350 tons cap., trip. exp. engines, boilers 10 ft. 6 in. x 11 ft. 6 in., 800 h.p., 8.5 knots, for undisclosed interests. Launching of one Jan., 1918.

Two steel cargo strs. Raised quarter deck type, 248 ft. 9 in. long, 1,649.68 gross tons. Triple expansion engines, 17, 28, 46 x 33 in. stroke. One, the "War Bee," for undisclosed interests to be completed June, 1918. Other for Steel Co., to be completed Oct., 1918.

Two masted schooner, launched May 16, 130 ft. o. a. x 26 ft. x 10 ft. 6 in. Designed by J. S. Gardner.

O'BRIEN BROS., NOEL, N.S.

One schooner, 325 tons, wood.

W. R. HUNTLEY & SON, PARRSBORO, N.S.

One wooden schooner, 175 ft. x 39 ft. x 17 ft., 900 gross tons, for

C. T. WHITE & SON, SUSSEX, N.S.

Two schooners, 325 tons each. Jan., 1918.

One schooner, 490 tons, wood.

One schooner, 850 tons, four masted.

WAGSTAFF & HATFIELD, PARRSBORO, N.S.

One schooner, 400 tons. Jan., 1918.

S. SALTER, PARRSBORO, N.S.

One 200-ton schooner, wood.

DOWLING & STODDART, PORT CLYDE, N.S.

One gas boat, 27 tons, wood.

One schooner, 300 tons, wood.

SWIME BROS., PORT CLYDE, N.S.

Several small motor trawlers, for builder's account.

G. M. COCHRANE, PORT GREVILLE, N.S.

One tern schooner, "Alfredock Hedley," 152 ft. 6 in. x 36 ft. 12 ft. 6 in., 461 tons reg., for Adam B. Mackay, of Hamilton, Ont. Launched Jan., 1918.

One 4-masted schooner, 850 tons, 155 ft. x 37 ft. x 18 ft., 2 decks, wood. To be completed Oct., 1918, for builder's account.

ELLIOTT GRAHAM, PORT GREVILLE, N.S.

Is building a schooner, "Khaki Lad," of about 325 tons, at Port Greville, N.S., for J. W. Kirkpatrick and others. She will be launched early in October, and is reported to have been sold to Newfoundland parties.

One schooner, 360 tons. Jan., 1918.

SMITH CANNING CO., PORT GREVILLE, N.S.

One schooner, 350 tons, wood.

WAGSTAFF & HATFIELD, PORT GREVILLE, N.S.

One schooner, 400 tons, wood.

WILLIAM CROWELL, PORT LATOUR, N.S.

J. W. RAYMOND, PORT MAITLAND, N.S.

One schooner, 375 tons. Jan., 1918

PORT WADE SHIPBUILDING CO., PORT WADE, N.S. (Head Office, Digby.)

One 350-ton schooner, wood.

JOHN BROWN, PUBLIC LANDING, N.S.

One tow barge, 50 tons, wood.

THE CUMBERLAND SHIPBUILDING CO., PUGWASH, N.S.

Establishing plant for the construction of wooden ships.

MACKENZIE SHIPPING CO., RIVER JOHN, N.S.

One four-masted schooner, 600 tons, wood. About half built. Fitted for aux. engines.

CHARLES McLELLAN, RIVER JOHN, N.S.

One schooner, 100 tons, wood.

W. J. FOLEY, SALMON RIVER, N.S.

Building for J. N. Rafuse & Sons one wooden, 300 tons, under construction. Cap. of plant, 1,000 tons yearly.

J. N. RAFUSE & SONS, SALMON RIVER, N.S. (Other plants at Shelburne and Conquerall Banks.)

Launched recently a three-masted schooner named "Industrial," at W. J. Foley's ship yard at Salmon River, N.S., of the following dimensions: length, 113 ft, breadth, 30 ft., depth, 11½ ft., 325 tons.

ACADIAN SHIPPING CO., SAULNIERVILLE, N.S.

One schooner, 00 tons, wood.

SAULNIERVILLE SHIPBUILDING CO., LTD., SAULNIERVILLE, N.S.

J. LEWIS & SONS, SHEET HARBOR, N.S.

One schooner, wood, four-masted, 725 gr. tons. To be completed Sept., 1918. For private account.

GEO. A. COX, SHELBURNE, N.S.

One schooner, 200 tons, wood.

One schooner, 322 tons, wood.

JOSEPH McGILL SHIPBUILDING & TRANSPORTATION CO., SHELBURNE, N.S.

"Sparkling Glance," 246 tons register, for Harvey & Co., St. John's, Nfld.

One schooner, 160 tons, wood.

W. C. McKAY & SON, SHELBURNE, N.S.

One wooden schooner, 130 ft. o. a. x 26½ ft. x 11 ft. depth of hold, 140 tons register. Launched Dec. 1, 1917.

One three-masted schooner, for Messrs. Hallet, of Newfoundland.

One schooner, 620 tons, wood.

J. N. RAFUSE & SONS, SHELBURNE, N.S. (Other plants at Salmon River and Conquerall Banks.)

One wooden three-masted schooner, 120 ft. x 32 ft. x 12 ft., 700 tons cap., for builder's account. For the market. Launching Jan. 10, 1918.

One wooden three-masted schooner, 118 ft. x 33 ft. x 12 ft., 700 tons cap., for builder's account. For the market. Launching Jan. 15, 1918.

Two three-masted wooden schooners, 122 ft. 6 in. keel x 32 ft. x 12 ft., 275 net tons, for J. O. Williams & Co., St. John's, Newfoundland. One to be built at Salmon River yard and one at Ship Harbor.

THE SHELBURNE SHIPBUILDERS, LTD., SHELBURNE, N.S.

One tern schooner, "Misty Star," 145 ft. x 31 ft. x 11 ft. 6 in., 600 tons d.w., 330 tons gross, for Harvey & Co., St. John's, Newfoundland. Launching Mar. 15, 1918.

One tern schooner, 154 ft. x 32 ft. x 12 ft. 4 in., 400 gr. tons, 700 tons d.w. cap. Launching June, 1918. Not sold yet.

Two schooners, one 349 tons, other 400 tons.

EASTERN SHIPBUILDING CO., SHIP HARBOR, N.S.

One wooden schooner, 600 tons d.w., for J. N. Rafuse & Sons., Halifax, N.S., Can.

One wooden schooner, 300 tons.

STEPHEN MORASH & CO., SHIP HARBOR, N.S.

One wooden schooner, 150 ft. x 37 ft. x 13 ft., four masts, for Canadian interests. Keel laid Feb. 5, 1917.

JAMES E. PETTIS, SPENCER'S ISLAND, N.S.

One schooner, 425 tons, wood, three-masted.

CAPE BRETON SHIPBUILDING CO., SYDNEY, N.S.

THE TUSKET SHIPBUILDING CO., TUSKET, N.S.

AMOS H. STEVENS, TANCOOK, N.S.

ALVIN STEVENS, TANCOOK, N.S.

STANLEY MASON, TANCOOK, N.S.

ALBERT PARSONS, WALTON, N.S.

One schooner, 400 tons. Jan., 1918.

H. T. LeBLANC, WEDGEPORT. N.S.

One steam trawler, 165 ft., 600 tons, for J. N. Rafuse & Sons.

L. S. CANNING, WARDS BROOK, N.S.

One tern schooner, 350 tons, for W. C. Smith & Co., Lunenburg, N.S.

WEDGEPORT NAVIGATION & TRANSPORTATION CO., WEDGEPORT, N.S.

T. K. BENTLEY, WEST ADVOCATE, N.S.

One 200-ton steamer.

One 4-masted schooner, wood, 511 tons. To be launched September or October. For owner's account.

J. W. KIRKPATRICK, WEST ADVOCATE, N.S.

One 350-ton schooner, wood.

BOEHNER BROS., WEST LA HAVE, N.S.

Two large fishing schooners, 150 tons each.

One beam trawler, 300 tons, crude oil engines. For local interests.

BEAZLEY BROS., WEYMOUTH, N.S. (Head Office Roy Bldg., Halifax.)

One schooner, 394 tons, three masts.

E. R. GAUDET, WEYMOUTH, N.S.

One 350-ton schooner, three-masted.

E. P. RICE, WEYMOUTH, N.S.

One three-masted schooner, 350 tons.

RICE, WARREN & CO., WEYMOUTH, N.S.

One three-masted schooner, 334 tons.

FALMOUTH SHIPBUILDING & TRANSPORTATION CO., WINDSOR, N.S.

One wooden schooner, 350 gr. tons, 700 tons cap., for undisclosed interests.

NOEL SHIPBUILDING & TRANSPORTATION CO., WINDSOR, N.S.

One wooden schooner, three masts, 138 ft. x 35 ft. x 13 ft., 450 net reg. tons. Oil auxiliary engines. Launching August, 1918, for builder's account.

C. A. NICKERSON, WOODS HARBOR, N.S.

MILTON SHIPBUILDING CO., YARMOUTH, N.S.

Taken over and enlarging old plant of James Jenkins.

350-ton schooner building.

W. O. SWEENY, YARMOUTH, N.S.

100-ton fishing schooner.

YARMOUTH SHIPBUILDING CO., YARMOUTH, N.S.

One 450-ton wooden schooner, 128 ft. keel. To be completed Oct., 1918.

ONTARIO

CAN. ALLIS-CHAMBERS (Head Office, Toronto), BRIDGEBURG, ONT.

Six standard freight steamers, 3,500 tons, 261 feet long, for undisclosed interests.

COLLINGWOOD SHIPBUILDING CO., COLLINGWOOD, ONT.

Two steel cargo steamers, Nos. 51-52, 50 ft. x 43 ft. x 25 ft., 2,500 gr. tons, 2,900 tons d. w., triple expansion engines 18-30-50 x 36, 2 boilers 14 ft. x 11 ft., 10 knots, for undisclosed interests. Delivery May and Aug., 1918.

Four deep-sea trawlers, Nos. 53-58, inclusive, 125 ft. x 23 ft. 6 in. x 13 ft. 6 in., 288 gross tons, 12¾-21½-35/24 engines, 1 cylinder 13 ft. 6 in. x 10 ft. 6 in., 10 knots, 500 h.p. for undisclosed interests.

One steel freight steamer of 3,800 tons d. w. for undisclosed interests. Keel laid May 8, 1918.

"War Wizard," May 8, 1918, 3,000 tons d. w., 261 ft. x 43 ft. 6 in. x 20 ft. depth. For undisclosed interests.

"War Witch" same size, now building. To be launched ——

R. MORRILL, COLLINGWOOD, ONT.

"Windsor," Aug. 10, steam tug, 105 ft., for Ontario Gravel & Freighting Co.

CAN. CAR & FOUNDRY CO., FORT WILLIAM, ONT. (Head Office Transportation Bldg., Montreal.)

Twelve steel mine sweepers for French Government.
145 ft. o.a. steel construction, value $2,500,000.
Plant under construction.

GREAT LAKES DREDGING CO., FORT WILLIAM, ONT.

Two wooden vessels, 2700 tons d.w. cap., 260 ft. o.a., 43 ft. beam. Triple expansion engines of 1,000 h.p. To be launched November, 1918.

"War Sioux" launched May 12, 1918. Keel laid for another twenty minutes later. Launching scheduled for November.

THUNDER BAY CONTRACTING CO., FORT WILLIAM, ONT.

One wooden freight steamer, 261 ft. long, for undisclosed interests.

NATIONAL SHIPBUILDING CO., GODERICH, ONT.

KINGSTON SHIPBUILDING CO., KINGSTON, ONT.

Several steel trawlers for undisclosed interests. First one launched Dec. 22, 1917.

SELBY & YOULDSON, KINGSTON, ONT.

GEORGIAN BAY SHIPBUILDING & WRECKING CO., MIDLAND, ONT.

Tug, 50 tons, wood.

One tug 40 tons, wood.

MIDLAND DRY DOCK CO., MIDLAND, ONT.

Three steel freight steamers, 261 ft. x 43 ft. 6 in. x 23 ft., 3,400 tons d. w., 10 knots, for undisclosed interests. Deliveries, one in July, 1918, and two others before close of navigation, 1918.

Alterations on steamers "Mariska" and "Glenlyon."

"Western Star" being reconstructed.

PORT ARTHUR SHIPBUILDING CO., PORT ARTHUR, ONT.

Six steel trawlers on order. 135 ft. o. a. x 23 ft. 4 in. x 15 ft. 1 in., 294.5 tons, trip. exp. engines, 500 h.p., single end Scotch boiler. Two-masted.

MUIR BROS. DRYDOCK CO., LTD., PORT DALHOUSIE, ONT.

J. W. GEROW, ROSSPORT, ONT.

REID WRECKING CO., SARNIA, ONT.

One steel tug, 157 ft. x 32 ft. x 19 ft., trip. exp. engines, for lake or ocean service. Keel blocks laid March 5. Launching July, 1918. For Reid Wrecking Co.

"Tento," Oct. 22, 1917, 261 ft., 3,500 tons d.w. steel. For Norwegian interests. Transferred to British registry.

POLSON IRON WORKS, TORONTO

Six steel cargo steamers, Nos. 145 to 150 inclusive, for the Imperial Munitions Board. Length, 261 feet overall, 43 feet 6 inches beam and 23 feet deep, 2,350 approximate gross tons, 3,500 d.w. capacity; single screw triple expansion engine, 20½ by 33 by 54 by 36-inch stroke. 1,250 h.p., two 14 feet by 12 feet Scotch boilers 180 lbs. per sq. inch w.p.

DAVIS DRY DOCK CO., KINGSTON, ONT.

Building wooden and metallic lifeboats, general repairs.

DOMINION SHIPBUILDING CO., TORONTO, ONT.

"Le Quesnoy," 4,300 tons gross, 1,400 h.p. Triple expansion engines, 2 Scotch marine boilers, to be launched Nov. 23. Built to builder's account, Canadian registry. Also three 3,500-ton steamers on stocks in course of construction and three others same type going through shops. All ships classed.

"War Fiend," 3,500 tons d.w., triple expansion engines, 1,300 h.p. Scotch boilers, Launched Oct. 24, 1918. To be completed Nov. Built for the Imperial Munitions Board.

PORT ARTHUR SHIPBUILDING CO., PORT ARTHUR, ONTARIO

"War Karma," 3,400 tons deadweight. Built for the Imperial Munitions Board. Launched October 26. Will be completed last week of Nov. Triple expansion engines, 20½ by 34 by 56 by 40-inch stroke. 2 Scotch boilers, 15 feet by 11 feet, working pressure 190 lbs. per sq. inch.

TORONTO SHIPBUILDING CO., LTD., TORONTO, ONT. (Toronto Dry Dock Co., Ltd., under same management.)

Two wooden cargo steamers, 250 ft. B.P. 42 ft. 6 in., moulded breadth 25 ft., moulded depth 22 ft., draft 2,500 tons d. w. Triple exp. engines $\frac{20 \times 33 \times 54}{40}$. Howden boiler. Launchings July and Sept., 1918. For undisclosed interests.

THE BRITISH AMERICAN SHIPBUILDING CO., WELLAND, ONT.

Two steel frs., 3,500 tons d. w., 261 ft. O.A., 43 ft. beam, 28 ft. moulded depth, Westinghouse steam turbine engines. Delivery 1918.

WELLAND SHIPBUILDING CO., WELLAND, ONT.

One steel freight steamer, "War Weasel," 261 ft. x 43 ft. 6 in. x 23 ft., 3,300 tons d. w., trip. exp. engines, 14 ft. x 12 in. Scotch boilers, 10 knots, 1,250 h. p. Launching April, 1918.

Two deep frame cargo vessels, "War Badger" and "No. 3," 261 ft. x 43 ft. 6 in. x 23 ft., 3,300 d. w. tons, geared turbines, Howden's forced draught boilers, 10 knots, 1,250 h.p. Launchings June and August, 1918.

PRINCE EDWARD ISLAND

THE CARDIGAN SHIPBUILDING PLANT, CARDIGAN, P.E.I. (Purchased Annandale Lumber Co. plant and removed to Cardigan.)

One three-masted schooner, 325 tons, to be completed November, 1918.

QUEBEC

TIDEWATER SHIPBUILDERS, LTD., THREE RIVERS, QUE.

R. N. LE BLANC, BONAVENTURE, QUE.

J. Z. DEGAGNE, EBOULEMENTS, QUE.

DAVIE SHIPBUILDING & REPAIRING CO., LEVIS, QUE.

Six military barges, 130 feet long.
Eight steel trawlers, 2,035 tons.
One floating crane, 350 tons.
*One steel cargo vessel of 5,000 tons cap.
Building several steel lighters and several wooden drifters.

Steel car ferry "Canora" building for C.N.R., 308 ft. x 52 ft., cap. 20 loaded cars. Speed, 14 knots.

ATLAS CEMENT CONSTRUCTION CO., LTD., MONTREAL, QUE.

One concrete cargo steamer, "Concretia," 125 ft. x 22 ft. x 13 ft., steel ribbed, hull to be from 3 in. to 5 in. thick, for the builder's account.

CANADIAN VICKERS, LTD., MONTREAL, QUE.

Two cargo steamers, 9.400 tons, steel; 1 dredge, 2,364 tons, steel; 12 trawlers, 3,050 tons, steel; 23 drifters, 3,350 tons, wood.

Six freight steamers, 24 ft. draught, 7,000 tons cap., 11 knots, for undisclosed interests. Delivery, 1917, of two for Norwegian interests. "Porsanger," 394 ft. 6 in. o. a. x 40 ft. 4 in. x 30 ft., triple exp. engines, launched Nov. 29, 1917.

"War Earl" launched June 8, 1918, 7,000 tons d.w., 380 ft. x 49 ft. Keel of sister ship laid five minutes after this launching.

One steel freight steamer, 2,360 tons, cap., for undisclosed interests.

Three steel cargo vessels, 8,200 tons, to be laid down in May, Aug. and Sept., 1918, for undisclosed interests.

FRASER BRACE & CO., MONTREAL, QUE.

Four wooden cargo steamers, 3,000 tons, for undisclosed interests. Two keels laid during Oct., 1917.

HALL ENGINEERING CO., MONTREAL, QUE.

MONTREAL DRY DOCK & SHIP REPAIRING CO., MONTREAL, QUE.

Operate dock 428 ft. long 30 ft. deep.

MONTREAL SHIPBUILDERS LTD., 37 Belmont St., MONTREAL, QUE. (Associated with Atlas Construction Co.

THE QUEBEC SHIPBUILDING AND REPAIR CO., Board of Trade Bldg., MONTREAL, QUE.

2 vessels for undisclosed interests 3,000 tons each.

QUINLAN & ROBERTSON, MONTREAL, QUE. (Yards at Quebec.)

Four wooden steamers, for undisclosed interests.

LOUIS GAUNDRY, 12 St. Peter St., QUEBEC, QUE.

QUEBEC SHIPBUILDING & REPAIR CO., ST. LAURENT, QUE.

2 schrs. 1,400 tons and 1,200 tons.
One wooden four-masted auxl. schooner "Martin Connolly," 228 ft. x 42 ft. x 20 ft., 2,100 tons d. w., for undisclosed interests. Launched Oct. 28, 1917.

QUINLAN & ROBERTSON, QUEBEC, QUE.

Four wooden steamers, totalling 6,400 tons, for undisclosed interests.

CANADIAN GOVERNMENT SHIPYARDS, SOREL, QUE.

One steel vessel for undisclosed interests.

H. H. SHEPHERD, SOREL, QUE.

Rebuilding seven drifters. Particulars withheld owing to govt. restrictions.

SINCENNES-McNAUGHTON LINES, SOREL, QUE.

One tug 410 tons, wood.

CAN. GOVERNMENT SHIPYARD, SOREL, QUE.

One steel vessel for undisclosed interests.
Building steam trawlers and wooden drifters for undisclosed interests.

LECLAIRE SHIPBUILDING CO., SOREL, QUE.

Six steel trawlers 125 feet long with single screw triple expansion engines, 500 i.h.p. Two of these launched in November. One more to be launched before ice comes in. Six wooden auxiliary schooners fitted with twin screw "Skandia" engines of 120 h.p. each. These will be all in frame at the end of November and four ready for launching as soon as ice is clear next spring.

NATIONAL SHIPBUILDING CORPORATION, THREE RIVERS, QUE.

"War Mingan" and "War Radnor," launched Oct. 15 and No. 2. Built for the Imperial Munitions Board and will be operated by Anning Bros., Cardiff, Wales. Now building 10 self-propelling barges for France.

ADMIRALTY COURT REPORT

FOLLOWING is a report of the judgment of the Wreck Commissioner following an investigation held into the shelling and subsequent burning of the schooner "Dornfontein," off Brier Island, Bay of Fundy, on August 2nd last, by a German submarine. The secret orders given the captain by the Admiralty were delivered up to the submarine commander, and especial stress was laid on this circumstance:

The master, Charles Ephraim Dagwell, holding Board of Trade certificate as Master, No. 99,236, obtained in 1879, deposed that he had joined the Dornfontein when on the stocks about May 20th; that she was a wooden vessel of about 595 tons register, carrying a crew of nine, of mixed nationalities, Norwegian, Swede, Dane and Russian Fin, including two officers, the mate holding a Danish certificate. He left St. John on July 28th after having the day before received clearing from Customs, as well as secret instructions from the transport officer, Captain Mulcahy. He left at 1 o'clock on Sunday, anchoring off Partridge Island to adjust compass, leaving there on Wednesday, the 31st July, about 9 a.m., with a light wind; but the flood tide carried the vessel back, and on Thursday there was some wind, and on Friday morning the vessel was off Brier Island, steering west, about six miles off. He did not know any of the crew before and had no reason to suspect anyone on board.

Until noon of Friday the vessel had been making from 4 to 4½ knots, all sails set. She was bound to Durban, Africa, with lumber. At about 11 a.m. Friday the man at the wheel reported a vessel to the south, but he could not make her out after scanning with the glasses. He knew that submarines were near the coast, but he kept on the course W. to S.W., the bearing of the vessel approaching being about South on the port bow. It was then nearly calm. We kept looking at the strange vessel several times, but she did not appear to come any closer. We had seen submarines before, but could not detect anything strange about the construction of the vessel in sight. He was on deck all the morning, and the men were working about the decks. He looked occasionally until dinner time, then the strange vessel had been in sight about one hour, still approaching; but at noon he did not detect anything suspicious. Whilst at dinner a shot was fired. He had not taken any observations, but reckoned that at noon the Dornfontein was 16 miles off Brier Island. He knew that there was a submarine when the shot was fired, and came on deck and ordered the helm up. Another shot was then fired, a piece of the shell piercing the spanker. The vessel was then hove up, signals were hoisted on the submarine, which came up fast after the first shot. After the boats were lowered he went below and got the papers, which were in a tin box. The letter of instructions had not been opened. He had all papers in his pocket when clearing in St. John; also certificate, later placing everything in the box excepting the last named document. He did not remember that instructions were given him to destroy the secret orders in case of meeting with the enemy. He later saw the papers in the hands of the commander of the U-boat. Four hours elapsed before the Dornfontein was stripped of provisions, etc., after which they were called from below and saw his vessel in flames.

He noticed the submarine had two guns, but did not notice any other peculiarities. The Huns were decent to them. The ship was in flames at 5 o'clock, when they left the submarine to row ashore, where they landed the next morning. There was no special lookout on the Dornfontein. He kept his certificate in his pocket, not as a matter of precaution, but as he forgot to take it out. This was the second ship lost through enemy action. He knew the papers given by the Naval authorities were to be destroyed, but forgot to do so. He had conferred with Captain Mulcahy, the transport officer, and had read the document in his office, but did not remember signing the document. (This was disproved by the fact that the transport officer showed a copy duly signed by Captain Dagwell, which was acknowledged by him. He did not know that the Swedish member of his crew could speak German until he heard him on board the U-boat. When his other vessel, the Sunlight, was torpedoed, he saved his certificate, all other papers being lost.

Charles Olsen, mate, sworn, averred that he had been three years in this country; coming from Denmark, and holding a Danish certificate. He could speak a little German, having learned it at Newcastle. He joined the vessel about a week before she sailed. He heard of the submarine's presence when the shot was fired, and whilst at dinner was not told by the master that there were special instructions from the Naval Department, but was told to keep a sharp look-out for everything, although the captain did not tell him of the appearance of a vessel. Upon coming on deck the master ordered him to keep the ship off, which he did to the extent of 5 or 6 points. He saw the captain had papers in his hand, as well as a box, and a German officer took the papers when they reached the submarine, and he spoke in German to the crew. The master was with the U-boat commander in the conning tower and he did not hear any conversaton. He heard the commander tell his crew to go on board the Dornfontein and break the windlass and the engine. He thought the Swede was acting funny, like a man half drunk.

The remainder of the evidence being a corroboration of the master's evidence, the Court adjourned until four o'clock to give finding.

Finding

The Court having carefully
the evidence obtained from the o
witnesses available, viz.: mast
mate, finds that on the master
evidence some contradiction with
to his knowledge of the contents
documents containing the sailing
saying at one time that he w
aware that his instructions were
stroy such instructions upon the
ance of an enemy ship, or when
was imminent, and his subseque
mission of having read such instr
and signed the form upon whic
were printed and written, the c
being before the Court, thereby
the Court the impression that h
sidered those papers of seconda
portance.

He had placed these instructio
box which held other ship's pape
retained in his pocket his certific
which, after clearing at the
House in St. John, he had no furt
until he again reached a British

His plea is that his certifica
forgotten in his pocket, while he
away in a box the document of
importance, which the Court assu
was obliged to consult frequently.
ever, on his own admission, he d
remember what those instructions
therefore showing that he made li
his duties and of his responsibiliti
his duties to his country or his fl
his responsibility to his owne
handing over those orders to the e
although he claims that he was co
collected. An interval of five m
elapsed from the time he obeyed t
boat's signals to bring his paper
rowing away from the ship's side.

In military and naval circles, c
wartime, such neglect would bring
the individual the odium of dislc
with the possible verdict advising
tal punishment. In civil life, sinc
has begun, many persons have re
a long term of imprisonment with
fines for utterances made on the
of the moment, and which did not
with them the importance of th
heard of neglect, to follow and e
such peremptory orders as C
Dagwell had received.

He had been torpedoed before, l
heard that submarines were fre
ing and had created havoc on the
and yet in the face of his former e
ence and his knowledge of existin
ditions that danger was lurking
Western Atlantic, where many v
had already been sacrificed, he d
even give special orders to his o
or crew to be vigilant in keeping
ordinary lookout.

An object was seen by him at 1
on the 2nd of August. With the
he watched the object, but could n
fine it. It was still in sight at
He nevertheless went to his lunc
according to the mate's evidenc
not, whilst both were at table, m
what he had seen. A shot was
which drew his attention, on the

The NEW "EMCO" GLOBE VALVE

AND THE PLACE WHERE IT'S MADE

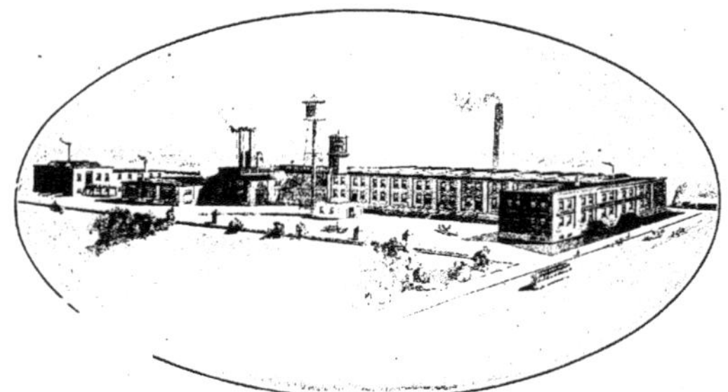

Just take a look at our

NEW EMCO

Globe Valve and note carefully a few of its good features. The long, full threads on spindle, the uniform thickness of metal.

It can be packed when open, is fitted with metal packing gland and is complete in every detail.

Every valve is packed with high-grade packing before it leaves the factory.

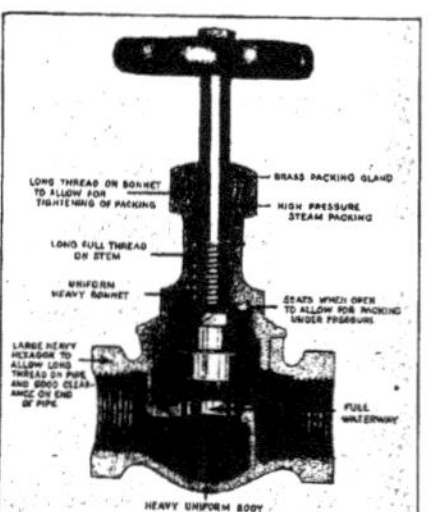

This is only one of the many new lines we have made recently to fill some long-felt wants for high-grade valves.

We have one of the most modernly equipped brass manufacturing plants under the British flag.

We can make prompt shipments from stock. Let us quote you our prices on your next brass goods requirements.

When writing us or ordering from your jobber, ask for the EMCO GLOBE VALVE. It's a winner and were proud of it.

EMPIRE MANUFACTURING COMPANY, LIMITED

LONDON TORONTO

ing of which he came on deck. A second shot was fired, at an interval of a couple of minutes, and yet on the hearing and seeing this second shot, no thought was given to the secret orders he possessed. Before the second shot was fired he had ordered the helm up, with the intention of running away, but brought his ship to the wind when the second shot struck the water a few yards from him.

The Court is of opinion that the master had ample time to reflect and destroy the document had he attached any importance to it, and the only conclusion which can be arrived at is, that he was gravely negligent, but not with criminal intent.

Whilst it has been ascertained that the crew was of mixed nationality, that two of its members spoke German, yet the Court has failed to connect this disaster with any preconceived, prearranged signals, or notification to the enemy.

In view of the fact that no data has been obtained pointing to criminal intention on the part of the master, or his crew, but finding only a total disregard of the importance of his instructions, the Court feels that in this instance a suspension of certificate will be a fit punishment to meet this neglect.

Therefore, the Court hereby suspend Captain Charles Ephraim Dagwell, Board of Trade Certificate No. 99236, for the duration of the war, until such a time when ships will be permitted to sail from any port or ports without special admiralty or governmental restrictions other than those which regulate the departure of ships in normal times, and trust that this finding will prove a deterrent to such master, in whose mind may lurk an idea that orders, instructions, issued by established authority, are of no or little importance, and that the non-fulfillment of any such orders cannot be overlooked with impunity.

As this inquiry concerns only the personal conduct of the master, other members of the crew are hereby exonerated.

Read in open court this 14th day of September, 1918.

(Sgd.) L. A. DEMERS,
Dominion Wreck Commissioner.

Concurred in
(Sgd.) A. J. MULCAHY, Lt. R.N.C.V.R.,
Naval Transport Officer.
(Sgd.) JAMES HAYES,
Assessor.

FO'C'STLE DAYS

(Continued from page 288)

This life-line was hung up in the boatswain's locker and to get it the second mate again entered that store room, yes, and for the second time he stood on Fid's body, which covered almost the entire floor. He was snoring this time almost as loud as the gale outside, and had never been aloft at all, but peacefully sleeping in the evil-smelling junk room amongst the tar and grease pots. A few marline spikes hung by their lanyards and point downwards, were banging against the wooden bulkhead just above the culprit's face. After being bullied by the mates and joked at by the seamen, Fid came into the half-deck to say that some one might have called him, but there was no fighting this time, he simply made a lurch at a plug of tobacco and chewed away in contentment.

Sighting another vessel was a rare occasion on the Australian-Chilian trip. Every skipper who left Newcastle, New South Wales, seemed to place his ship on a great circle of his own. The birds were our only solace, the clouds, winds and waves our only study. Could man or boy be further away from an ideal home? I trow not, and yet the ocean life, for youngsters at any rate, had a most fascinating and magnetic call. This lure of the elements and the romance of travel, taken together, will forever fill the quiver of Father Neptune with hosts of honorable sons.

Soon we were to cross the 180th meridian, or the north and south line that is directly opposite to the index or Greenwich meridian. In other words we had sailed half-way round the old sphere on the "water, water everywhere, and not a drop to drink." Think what the lonely ocean must be to the crazed shipwrecked sailor who abandons the parent ship in one of the small boats and has used up his drinking water and maggoty biscuits. This brings out a point that cannot escape an old sailor; the fact that such inseparable hardships were always a part of "Jack's" life, and yet it took this horrible submarine policy of the Hun to place British Empire seamen on a plane with ordinary mortals.

Skin-friction Versus Speed

It was noticed that our barge took more driving to maintain speeds over 10 miles an hour. Even with two topgallant sails and a half gale whistling through the shrouds, plus a lumpy sea shoving at her stern, the "Magpie Dorrit" showed signs of a foul bottom. Under her light displacement mark she was covered with barnacles as big as a pen knife and a six months' crop of sea grass waved on her bilge.

How to overcome this skin-friction has long been a thorn in the flesh of captains and owners of iron or steel ships. The wooden vessel has the best of it on this tack, as copper sheathing almost eliminates the awful drawback. The smoothness of the yellow metal jacket which is put over the wooden ship's bottom and perhaps the poisonous verdigris, act against the invitation of either sea grass or shell fish attachment. Later on our voyage, and at a time when tropical calms make it possible, the patent scrubber will be put in use to try and scrape some of the large barnacles off.

Our new second mate treated me with a little indulgence and frequently rang the poop bell when I happened to have my head under my wing at the lee side of the companion-way. As Tilbury had gone the night watches seemed very lonely, but now and again the officer would unbend and relate his apprentice experiences, which to my first-voyage ears sounded like Captain Kettle's yarns. Thus the wearisome watches of the southern trip went on. But some nights the time was taken up with real actions instead of yarns and dreams. Handling a foresail or main-sail, or reefing topsails and hitching all running gear safely to the pin rails in stormy weather makes a big hole in four hours, and knocks sleepy boys into wide-awake men.

Beachcombers Again

Two of our beachcombers boasted of being ready to swim ashore from the anchorage in Coquimbo, and no one appeared to be very sorry if they kept their promise. The devil himself would need the help of a dozen imps and elfs to do anything with such human clay. These ocean bums had raked the west coast of South America from Talcahuano to Callao more than once, and Singapore and Hong Kong jails had already been graced with their presence.

In looking back at this kind of sea monster the writer can recollect a few who apparently had a germ of salvation in them, provided it was nurtured by tactful superiors. On the other hand the vast majority of these homeless creatures, even though they may prove themselves fairly good seamen on the voyage, take a high temperature the minute a ship enters a harbor. As soon as the anchor is down, or the warps fast to a quay, "John Beachcomber, Esq.," must go ashore, and in some instances long before the captain has interviewed his agent and consul our brain-wracked gentleman rope-hauler is the chief attraction in one of the low drinking and dancing dens which are almost in reach of a ship's jib-boom in too many harbors of the world. A week of this life, knowing not whether it is mid-day or mid-night and the generous boarding-house master sits on his prey in the stern-sheets of a small boat which is pulled out to some loaded ship about to depart. After being hauled up and thrown into a lower bunk in a strange and dark fo'c'stle, he sleeps like a log, only to be roughly handled by the mate's shouting "shake a leg there, you're afloat again, what's your number," or similar parley.

Our "benevolent" friend, the boarding-house master, was a recognized snake in the grass, in fact he entered the lower political fields through his bribery and double-facedness in handling poor Jack. He has not vanished yet, but in most marine communities his rotten make-up is well under the glare of law and order. If a tribute must be given to such a one, it would not be exaggeration to say that in refinement or intelligence the average sailor's boarding-house master would rank with a hangman and in sordid dealings he might be on a par with a German U-boat skipper—the kind that slap their lager filled paunches and giggle at the sight of women and children being swallowed by the sea.

(To be continued).

SEAMANSHIP AND NAVIGATION

(Continued from page 289)

Now, against this experience take such winding and narrow passages as we find in the Norwegian Fiords, Hebrides of Scotland or the Magellan Straits down south. What happens when you let a steamer plug full speed through these wonderful channels? Does she break her sheer or run aground? No, sir. Why? Simply because the water is deep and her displacement does not become master of the steering proposition.

Did it never occur to you how comical it appears to see a 25,000 H.P. steamer towing through some shallow waters? It's because she can't steer. Then again, the law courts often reflect the peculiar conditions of a collision in some narrow and shallow waterway, between two craft, where, apparently, no one can be actually blamed for the mishap. The writer remembers having to tie up in the Suez Canal to the bank to allow a war ship the right of way. Just as she got abreast of our tramp what happened? The fighting ship took an awful liking for the humble merchantman and scraped along her side like a drifting ship, as helpless as a log. Her unwieldy tonnage and deep draught in such a narrow and shallow waterway made her irresponsible. Even merchant ships stopped at Port Suez and Port Said to hook on an extension rudder, but a war ship's rudder is not in such a handy place for a temporary adjustment, as it is practically underneath her.

Effects of Wind and Sea

A ship's movements in regard to the elements are noteworthy: Left to her own free will any craft will present her broadside to the wind and sea, and as Jack puts it, "Roll her guts out." When under way, however, she changes her tactics, and immediately desires to fly into the teeth of the wind and sea, as the greatest pressure is on her lee bow. You will find in practice then, that your ship will steady on her course, with a beam wind and sea, with the rudder angled "a port" if the wind is on your port side, but angled "a starboard" if wind and sea are on your right or starboard hand. In other words, your ship will carry "weather helm" or, as coasting men call it, "up stick." In this respect remember also that "lee helm" or "down stick" always means bringing a craft "to the wind," which is synonymous with the almost extinct word "luff," a grand old wind-jammer expression.

Running Dead Before It

Let us notice now the behaviour of a ship when running before the wind and sea. In this predicament the helmsman finds that she steadies on her course with the rudder amidships. What are a ship's tendencies when steaming or sailing in this manner? A set determination to "gripe" and "yaw," which means that owing to the quartermaster's slowness in watching his vessel's head, she brings the wind and sea first on one quarter and then on another. A vessel is most stubborn to settle down to quiet steering after she has been excited into this kind of behaviour, and it is only right to mention that a good steersman or, as they are called on ocean craft, quartermasters, are very valuable men, indeed. In sail, for instance, a bad helmsman may at any time in a stiff breeze be the means of "springing" a gaff boom or yard, or "carrying away" a sheet, whilst in a tramp steamer a bad steersman may fill the well-decks up flush with the rail through his inefficiency.

Head On

In steaming into a head sea and wind a loaded ship steers pretty good, but if in ballast her behaviour becomes more erratic as the wind and sea increase. As in running dead before it she will, when head on, balance with the rudder amidships, taking a certain amount of port and starboard helm as the conditions demand. A ship's steering gear gets its hardest usage when the wind and sea are ahead, as diving becomes a frequent occurrence, and the sea pressures close in on the rudder surface from along the ship's side, causing a kick.

How To "Meet" a Vessel

In writing about a ship's movements it is of great importance to discuss the art of "meeting" your ship when using the rudder. Say, for instance, you are ordered to "port two points," i.e. 22½°, what frequently happens? The ship swings to starboard something like four points or double the amount requested. How is that? Simply bad judgment in not knowing when to counterbalance the effect of the helm. The very instant you notice the ship's head answering the impulse of the rudder, return your wheel to its neutral line and be ready to put it in the opposite direction when the desired point or degree nearly coincides with the lubber's point or ship's head. Only in this way will you make manoeuvres handsomely and stop the ship's head from getting wild. A good quartermaster will describe a graceful curve as neatly on water as one could do on paper with compasses, whilst a bad helmsman will nearly wrench the steering engine off its sole-plate and make the vessel appear as if she was dodging a Hun submarine by the zig-zag method.

In some ways the steam steerer has made a bad helmsman, as there is no manual labor attached to it, whereas in hand gears the fondness for moving rudder too much was against the man's own brawn and muscle. Thus men learned to balance their craft with little work as possible and got better results.

To young men who are taking charge of a watch for the first time it is good policy to form the habit of noticing what the helmsman actually does with his wheel after an order is given. Then, if a mistake is in the making, it is not too late to save the situation, whereas if a vessel once begins to answer her helm, and the order has been misconstrued, it may be very hard to avert damage.

What makes a ship roll? Too much dead weight under the water line. How is that apparent to the sailor? Because we find that a ship does her heaviest rolling either when in ballast or loaded with iron ore, pig iron or some other similar cargo. In ballast or "light ship" the cellular bottom filled with water acts like a loaded dice on the craft, whilst her entire hold is one huge air space. Nothing will prevent a cargo boat in this trim from baling up the ocean or lake with both rails. Is there any attempt by builders to minimize this dangerous movement? Yes, first by giving the ship a good bilge keel, and also by wing ballast tanks.

With a deep-loaded ship carrying iron ore the trouble is much the same with the added danger of shipping heavy seas and breaking in the hatches.

(To be continued)

Quality

Service

FOR COOL BEARINGS

We Manufacture:

Phosphor Bronze Tail Shaft Liners, Pump Liners, Stuffing Boxes, Stern Tube Bushings and Brass Castings of every description.

MECHANICAL AND ELECTRICAL

SHIPS TELEGRAPHS

Rudder Indicators

Shaft Speed Indicators

Electric Whistle Operators

Electric Lighting Equipments, Fixtures, Etc.

Electric and Mechanical Bells

Annunciators, Alarms, Etc.

Loud Speaking Marine Telephones

Installations

Chas. Cory & Son, Inc.

290 Hudson Street - New York City

SHEET METAL WORK

Our factory is equipped to handle special Sheet Metal Work of all kinds up to quarter plate.

Tanks, Buckets, Chutes, Ventilators and Piping are some of the lines we make.

Repairs also promptly and efficiently attended to.

Send us your enquiries.

Geo. W. Reed & Co., Ltd.

37 St. Antoine Street

Montreal

Hoyt
Babbitts
Will Cut Your Costs
Forty Years of Success
HOYT METAL Co TORONTO
Bearings babbitted with Hoyt's "FROST KING" or Hoyt's "NICKEL GEUNINE" give extra long service before they need rebabbitting. Don't use common alloys. Hoyt's Babbitts will give you better satisfaction and cut your babbitting costs.
HOYT'S FROST KING—This general purpose babbitt has won its way into many of the foremost power plants of the world. It represents 40 years of success and has won the favor of Canadian engineers from coast to coast.
HOYT'S NICKEL GENUINE—This famous babbitt is recommended for use in gas engines, gasoline engines and all classes of marine engines. Especially adapted for automobiles. There is no better babbitt made for high-speed work.
Specify HOYT in Your Next Order for Babbitts
Hoyt Metal Company Eastern Avenue and Lewis St. Toronto
London, Eng., New York, U.S.A., St. Louis, Mo., U.S.A.

WE MANUFACTURE A FULL LINE OF

VALVES AND FITTINGS

For SHIPBUILDERS

and solicit inquiries for the following:

SAFETY VALVES
WATER GAUGES AND OTHER BOILER MOUNTINGS
ENGINE FITTINGS
SHIP'S SIDE DISCHARGE AND SUCTION VALVES
BULKHEAD FITTINGS AND

MANIFOLDS

Blue Prints on Application

M—2000
Twin Marine Safety Valve Flanged

JENKINS BROS.
LIMITED

HEAD OFFICE AND WORKS
103 St. Remi Street
MONTREAL, CANADA

EUROPEAN BRANCH
6 Great Queen St., Kingsway
LONDON, W.C. 2, ENG.

M—2002
Marine Water Gauge with M—2005—Protector

THE WAGER FURNACE BRIDGE WALL
A preferred and valuable feature in marine and stationary boilers—endorsed by governments; steamship, freight and passenger steamer companies; railroads; stationary plants; others.
WAGER FURNACE BRIDGE WALL CO., Inc.
OF NEW YORK SINGER BUILDING
Philadelphia, Detroit, Seattle, Portland
San Francisco -:- Vancouver, B. C.

MARINE
Castings

Mason Regulator and Engineering Co.

Limited

Successors to H. L. Peiler & Company

Reilly Marine Evaporator, Submerged Type

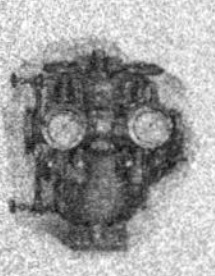

Reilly Multi-screen Feed Water Filter

Reilly Multi-coil Marine Feed Water Heater

Made in Canada By a Canadian Company

Mason No. 126 Style Marine Reducing Valve

We are prepared to supply the well-known auxiliary material shown here. Special attention is directed to our Marine Reducing Valves and Pump Pressure Regulators. Reliable, simple and of "Mason" workmanship. "Reilly" material needs no introduction.

We furnish bulletins and full information on request

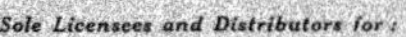

Sole Licensees and Distributors for:

The Mason Regulator Co.
Griscom-Russell Co.
Nashua Machine Co.
Coppus Engineering and Equipment Co.
The Sims Co.

Mason No. 55 Style Pump Pressure Regulator

The Mason Regulator & Engineering Co., Ltd.

Successors to H. L. PEILER & COMPANY

MONTREAL, Office and Factory, 135 DAGENAIS ST.

TORONTO REPRESENTATIVE: Arthur S. Leitch Co., 506 Kent Building, TORONTO

Books for the Engineer's Library

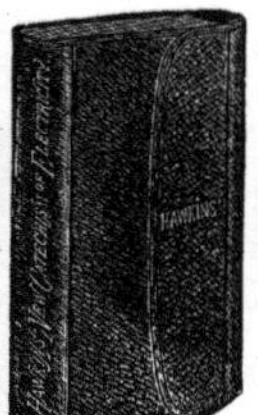

Hawkins' New Catechism of Electricity

It contains 550 pages with 800 illustrations of electrical appliances; it is bound in heavy red leather, with full gold edges and is a most attractive handbook for electricians and engineers. One-third of the book is devoted to the explanation and illustrations of the dynamo, with particular directions relating to its care and management—$2.10 post paid.

Engineers' Examinations

By N. Hawkins, M.E.

It presents in a condensed form the most approved practice in the care and managerment of steam boilers, engines, pumps, electrical and refrigerating machines with examples of how to work the problems relating to the safety valve, strength of boilers and horse power of the steam engine and steam boiler.—$2.10 post paid.

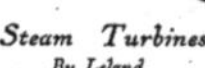

Steam Turbines

By Leland.

A reference work on the development, advantages and disadvantages of the steam turbine; the design, selection, operation and maintenance of steam turbine plants and turbo-generators. 135 pages. Illustrated.—$1.00.

Audel's Gas Engine Manual

A practical treatise relating to the theory and management of gas, gasoline and oil engines, including chapters on producer gas plants, marine motors and automobile engines.—$2.00.

Hydraulic Engineering

By G. D. Hiscox.

This comprehensive book of Hydraulics written by an experienced engineer, is a practical treatise on the properties, power and resources of water for all purposes, including the measurement of streams, the flow of water in pipes or conduits; the horse-power of falling water; turbine and impact water wheels; wave motors, etc. All who are interested in Water Works Development should have a copy. 320 pages, 305 illustrations. Price $4.15 post paid.

Compressed Air

By L. I. Wightman, E.E.

A reference work on the production, transmission and application of compressed air; the selection, operation and maintenance of compressed air machinery; and the design of air power plants. Illustrated.—$1.10.

Modern Steam Engineering in Theory and Practice

By Hiscox.

This book has been specially prepared for the use of the modern steam engineer, the technical students, and all who desire the latest and most reliable information on steam and steam boilers, the machinery of power, the steam turbine, electric power and lighting plants, etc. 450 pages. 400 detailed engravings.—$3.15 post paid.

Boiler Construction

By Kleinhans.

The only book showing how locomotive boilers are built in modern shops. Shows all types of boilers used; gives details of construction; practical facts, such as line of riveting punches and dies, work done per day, allowance for bending and flanging sheets and other data that means dollars to any railroad man. 421 pages, 324 illustrations, six folding plates.—$3.00.

Compressed Air, Its Production, Uses and Application.

By G. D. Hiscox, M.E.

Comprising the physical properties of air from a vacuum to its liquid state, its thermodynamics, compression, transmission and uses as a motive power in the operation of stationary and portable machinery, in mining, air tools, air lifts, pumping of water, acids and oils and the numerous applicances in which compressed air is a most convenient and economical transmitter of power.—Price $5.15 post paid.

Boiler Accessories

By Walter S. Leland, S.B.

Assistant Professor of Naval Architecture, Mass. Institute of Technology, American Society Naval Architects and Marine Engineers. 144 pp., 80 illus. Cloth binding. A treatise giving complete descriptions of the various accessories of the boiler room and engine room essential to economical operation, such as evaporators, pumps, feed-water heaters, injectors, mechanical stokers, etc., with practical instruction in their use.—Price $1.10 post paid.

Technical Book Department

MacLean Publishing Company **143 University Ave., Toronto**

CANADA
FOUNDRIES & FORGINGS
LIMITED
DROP FORGED STEEL
LARGE
SIZES
MARINE BOILER STAY NUTS
FORGED FROM THE SOLID BAR TO LLOYD'S SPECIFICATIONS.
EVERY NUT NEEDS A WRENCH.
C.B.& S.
WELLAND ONT.
FORGED STEEL WRENCHES
EVERY NUT NEEDS A WRENCH.
PRICE ON APPLICATION.
SMITHERIES AT WELLAND, ONT.

BUYERS' DIRECTORY

ACCUMULATORS, HYDRAULIC
Smart-Turner Mach. Co., Hamilton, Ont.

AERATING RESERVOIRS
Spray Engineering Co., Boston, Mass.

AIR PORTS
Mitchell Co., The Robert, Montreal, Que.
Turnbull Elevator Mfg. Co., Toronto, Ont.

ALLOYS, BRASS AND COPPER
Dom. Copper Products Co., Ltd., Montreal, Que.
Mitchell Co., The Robert, Montreal, Que.
Mueller Mfg. Co., H., Sarnia, Ont.
Tallman Brass & Metal Co., Hamilton, Ont.
United Brass & Lead, Ltd., Toronto, Ont.

ANCHORS
Beauchemin & Fils, Sorel, P.Q.
Hopkins & Co., F. H., Montreal, Que.
McNab Co., Bridgeport, Conn.
Henry Rogers Sons & Co., Wolverhampton, Eng.
Wm. C. Wilson & Co., Toronto, Ont.

ASBESTOS GOODS
Wm. C. Wilson & Co., Toronto, Ont.

BABBITT METAL
Aikenhead Hardware, Ltd., Toronto, Ont.
Hoyt Metal Company, Toronto, Ontario.
Tallman Brass & Metal Co., Hamilton, Ont.
Wilkinson & Kompass, Hamilton, Ont.
Wm. C. Wilson & Co., Toronto, Ont.

BAROMETERS
Wilson & Co., Wm. C., Toronto, Canada.

BAR IRON
Mitchells Ltd., Glasgow, Scotland.

BARS, GRATE
Babcock & Wilcox, Ltd., Montreal, Que.

BEARINGS, BRASS
Empire Mfg. Co., London, Ont.
Mueller Mfg. Co., H., Sarnia, Ont.
Mitchell Co., The Robert, Montreal, Que.
Tallman Brass & Metal Co., Hamilton, Ont.
United Brass & Lead Co., Toronto, Ont.

BELLS, SHIPS, ENGINE ROOM, ETC.
Aikenhead Hardware, Ltd., Toronto, Ont.
Clarke & Bros. Co., Montreal, Que.
Cory & Son, Inc., Chas., New York, N.Y.
Empire Mfg. Co., London, Ont.
Mitchell Co., The Robert, Montreal, Que.
Morrison Brass Mfg. Co., James, Toronto, Ont.
Mueller Mfg. Co., H., Sarnia, Ont.
Tallman Brass & Metal Co., Hamilton, Ont.
United Brass & Lead, Ltd., Toronto, Ont.

BELTING, LEATHER
Aikenhead Hardware, Ltd., Toronto, Ont.
Wm. C. Wilson & Co., Toronto, Ont.

BIBBS, COMPRESSION
Empire Mfg. Co., London, Ont.
Mitchell Co., The Robert, Montreal, Que.
Mueller Mfg. Co., H., Sarnia, Ont.
United Brass & Lead, Ltd., Toronto, Ont.

BINNACLES
Hopkins & Co., F. H., Montreal, Que.
Morrison Brass Mfg. Co., James, Toronto, Ont.

BLOCKS, CARGO, HEEL AND TACKLE
Aikenhead Hardware, Ltd., Toronto, Ont.
Hopkins & Co., F. H., Montreal, Que.
Loveridge, Ltd., Docks, Cardiff, Wales.

BLOWERS, TURBO.
Mason Regulator & Engin. Co., Montreal, Que.

BOAT CHOCKS
Corbet Fdry. & Machine Co., Owen Sound, Ont.

BOILER COMPOUND
Wm. C. Wilson & Co., Toronto, Ont.

BOILER CIRCULATORS
Eckliff Circulator Co., Detroit, Mich.

BOILER FEED PUMPS
Can. Ingersoll-Rand Co., Ltd., Sherbrooke, Que.
Darling Bros., Ltd., Montreal, Que.
Goldie & McCulloch Co., Galt, Ont.
Morris Mach. Works, Baldwinsville, N.Y.
Smart-Turner Mach. Co., Hamilton, Ont.
Williams Machinery Co., A. R., Toronto, Ont.

BOILER FITTINGS
Empire Mfg. Co., London, Ont.
Goldie & McCulloch Co., Galt, Ont.
McAvity & Sons Ltd., T., St. John, N.B.
Wager Furnace Bridge Wall Co., Inc., New York, N.Y.

BOILER MOUNTINGS
Jenkins Bros., Montreal.

BOILERS, MARINE
Babcock & Wilcox, Ltd., Montreal, Que.
Can. Fairbanks-Morse Co., Montreal, Que.
Can. Vickers, Ltd., Montreal, Que.
Collingwood Shipbuilding Co., Collingwood, Ont.
Doxford & Sons, William, Sunderland, England.
Engr. & Mach. Wks. of Can., St. Catharines, Ont.
Goldie & McCulloch, Ltd., Galt, Ont.
Hall Engineering Works, Montreal, Que.
Mason Regulator & Engin. Co., Montreal, Que.
Montreal Dry Docks & Shipbuilding Co., Montreal, Que.
National Shipbuilding Co., Goderich, Ont.
Marsh Engineering Works, Belleville, Ont.
Polson Iron Works, Toronto, Ontario.
Port Arthur Shipbuilding Co., Port Arthur, Ont.
Waterous Engine Works Co., Brantford, Ont.
Williams Machinery Co., A. R., Toronto, Ont.

BOOKS, TECHNICAL, MARINE
MacLean Publishing Co., Toronto, Ont.

BOLTS
London Bolt & Hinge Works, London, Ont.
Mitchell Co., The Robert, Montreal, Que.
United Brass & Lead, Ltd., Toronto, Ont.
Wilkinson & Kompass, Hamilton, Ont.

BROKER
Page & Jones, Mobile, Ala., U.S.A.

BRASS SHEETS AND TUBES
Dom. Copper Products Co., Montreal.

BRASS GOODS
Corbet Fdry. & Machine Co., Owen Sound, Ont.
Goldie & McCulloch Co., Galt, Ont.
McAvity & Sons, T., St. John, N.B.
Mitchell Co., The Robert, Montreal, Que.
Mueller Mfg. Co., H., Sarnia, Ont.
Tallman Brass & Metal Co., Hamilton, Ont.

BUCKETS, CLAMSHELL
Beatty & Sons, Welland, Ont.
Morris Crane & Hoist Co., Herbert, Niagara Falls, Ont.

BUCKETS, DUMP
Morris Crane & Hoist Co., Herbert, Niagara Falls, Ont.

BUCKETS, COALING
Beatty & Sons, Welland, Ont.
Engr. & Mach. Wks. of Can., St. Catharines, Ont.
Hopkins & Co., F. H., Montreal, Que.
Marsh Engineering Works, Belleville, Ont.
Reed & Co., Geo., Montreal, Que.

BUSHINGS, BRONZE
Oberdorfer Brass Co., M. L., Syracuse, N.Y.

CABLE
Hopkins & Co., F. H., Montreal, Que.
McNab Co., Bridgeport, Conn.
Wm. C. Wilson & Co., Toronto, Ont.

CABLE, LEAD COVERED AND ARMORED
Standard Underground Cable Co., Hamilton, Ont.

CABLE, ACCESSORIES
Darling Bros., Ltd., Montreal, Que.
Standard Underground Cable Co., Hamilton, Ont.

CAPSTANS
Advance Mach. & Welding Co., Montreal, Que.
Dake Engine Co., Grand Haven, Mich.
Hopkins & Co., F. H., Montreal, Que.
Kennedy & Sons, Wm., Owen Sound, Ont.
Wm. C. Wilson & Co., Toronto, Ont.

CALKING TOOLS, ELECTRIC
Aikenhead Hardware, Ltd., Toronto, Ont.

CALKING TOOLS, PNEUMATIC
Aikenhead Hardware, Ltd., Toronto, Ont.
Can. Ingersoll-Rand Co. Sherbrooke, Que.

CALORIFIERS
Low & Sons, Ltd., Archibald, Glasgow, Scotland

CASTINGS
Beauchemin & Fils, Sorel, P.Q.
Can. Steel Foundries, Ltd., Montreal, Que.
Collingwood Shipbuilding Co., Collingwood, Ont.
Wm. Hamilton Co., Peterboro, Ont.
Kennedy & Sons, Wm., Owen Sound, Ont.
Goldie & McCulloch Co., Galt, Ont.
Marsh Engineering Works, Belleville, Ont.
Mitchell Co., Ltd., Robt., Montreal, Que.
Mueller Mfg. Co., H., Sarnia, Ont.
Tallman Brass & Metal Co., Hamilton, Ont.
United Brass & Lead, Ltd., Toronto, Ont.
Waterous Engine Works Co., Brantford, Ont.

CASTINGS, ALLOY
Magnet Metal & Foundry Co., Ltd., Winnipeg.
Mitchell Co., The Robert, Montreal, Que.
Mueller Mfg. Co., H., Sarnia, Ont.
Oberdorfer Brass Co., M. L., Syracuse, N.Y.
Tolland Mfg. Co., Montreal, Que.
Tallman Brass & Metal Co., Hamilton, Ont.
United Brass & Lead, Ltd., Toronto, Ont.

CASTINGS, ALUMINUM
Empire Mfg. Co., London, Ont.
Magnet Metal & Foundry Co., Ltd., Winnipeg.
Mitchell Co., The Robert, Montreal, Que.
Tallman Brass & Metal Co., Hamilton, Ont.
United Brass & Lead Co., Toronto, Ont.

CASTINGS, BRASS
Crouse-Hinds Co. of Canada, Ltd., Toronto, Ont.
Empire Mfg. Co., London, Ont.
Goldie & McCulloch Co., Galt, Ont.
McAvity & Sons Ltd., T., St. John, N.B.
McNab Co., Bridgeport, Conn.
Mitchell Co., Ltd., Robt., Montreal, Que.
Mueller Mfg. Co., H., Sarnia, Ont.
Oberdorfer Brass Co., M. L., Syracuse, N.Y.
Tolland Mfg. Co., Montreal, Que.
Tallman Brass & Metal Co., Hamilton, Ont.
United Brass & Lead Co., Toronto, Ont.
Waterous Engine Works Co., Brantford, Ont.

CASTINGS, GREY IRON, MALLEABLE, ALUMINUM
Crouse-Hinds Co. of Canada, Ltd., Toronto, Ont.
Darling Bros., Ltd., Montreal, Que.
Engr. & Mach. Wks. of Can., St. Catharines, Ont.
Magnet Metal & Foundry Co., Ltd., Winnipeg.
McAvity & Sons, T., St. John, N.B.
McNab Co., Bridgeport, Conn.
Mitchell Co., The Robert, Montreal, Que.
Mueller Mfg. Co., H., Sarnia, Ont.
Waterous Engine Works Co., Brantford, Ont.

CASTINGS, MANGANESE STEEL

CASTINGS, MANGANESE BRONZE
Oberdorfer Brass Co., M. L., Syracuse, N.Y.
Tallman Brass & Metal Co., Hamilton, Ont.
United Brass & Lead Co., Toronto, Ont.

CELLAR DRAINERS
Mueller Mfg. Co., H., Sarnia, Ont.

CHAINS
Aikenhead Hardware, Ltd., Toronto, Ont.
Hopkins & Co., F. H., Montreal, Que.
Morris Crane & Hoist Co., Herbert, Niagara Falls, Ont.
Henry Rogers, Sons & Co., Wolverhampton, Eng.
Wm. C. Wilson & Co., Toronto, Ont.

CHAIN BLOCKS AND SLINGS
Morris Crane & Hoist Co., Herbert, Niagara Falls, Ont.

CHANDLERY, SHIP
Beauchemin & Fils, Sorel, P.Q.
Hopkins & Co., F. H., Montreal, Que.
Leckie, Ltd., John, Toronto, Ont.

CLAMPS, STEAM AND WATER
Mueller Mfg. Co., H., Sarnia, Ont.

CLOCKS
American Steam Gauge & Valve Mfg. Co., Boston, Mass.
Morrison Brass Mfg. Co., James, Toronto, Ont.
Williams Machinery Co., A. R., Toronto, Ont.

CLOSETS
Mueller Mfg. Co., H., Sarnia, Ont.
United Brass & Lead, Ltd., Toronto, Ont.

COAL
Nova Scotia Steel & Coal Co., New Glasgow, N.S.

COAL HANDLING MACHINERY
Morris Crane & Hoist Co., Herbert, Niagara Falls, Ont.
Waterous Engine Works Co., Brantford, Ont.

COCKS, BILGE, DISCHARGE, INDICATOR,
McAvity & Sons Ltd., T., St. John, N.B.
Mitchell Co., The Robert, Montreal, Que.
Morrison Brass Mfg. Co., James, Toronto, Ont.

COCKS, BASIN
Empire Mfg. Co., London, Ont.
Mitchell Co., The Robert, Montreal, Que.
United Brass & Lead, Ltd., Toronto, Ont.

COMPASSES
Wm. C. Wilson & Co., Toronto, Ont.

COMPRESSORS, AIR
Can. Fairbanks-Morse Co., Montreal, Que.
Canadian Ingersoll-Rand Co., Sherbrooke, Que.
Darling Bros., Ltd., Montreal, Que.
Hopkins & Co., F. H., Montreal, Que.
Smart-Turner Mach. Co., Hamilton, Ont.
Williams Machinery Co., A. R., Toronto, Ont.

CONDENSERS
Darling Bros., Ltd., Montreal, Que.
Goldie & McCulloch Co., Galt, Ont.
Morris Machine Works, Baldwinsville, N.Y.
Smart-Turner Mach. Co., Hamilton, Ont.
Weir Ltd., G. & J., Cathcart, Glasgow, Scotland.
Williams Machinery Co., A. R., Toronto, Ont.

CONDULETS, MARINE
Crouse-Hinds Co. of Canada, Ltd., Toronto, Ont.

CONTRACTORS' SUPPLIES
McAvity & Sons, T., St. John, N.B.

CONSULTING ENGINEER
Watt, J. Murray, Philadelphia, Pa.

CONVEYORS, ASH, COAL
Babcock & Wilcox, Ltd., Montreal, Que.
Hopkins & Co., F. H., Montreal, Que.

COPING MACHINES
Bertram & Sons, Ltd., John, Dundas, Ont.

COPPER TUBES, SHEETS AND RODS
Tallman Brass & Metal Co., Hamilton, Ont.

COPPER TUBES AND SHEETS
Dom. Copper Products Co., Montreal.

COTTON, CALKING
Wilson & Co., Wm. C., Toronto, Canada.

COVERS, CANVAS, FOR HATCHES, LIFE-BOATS, ETC.
Leckie, Ltd., John, Toronto, Ont.
Waterous Engine Works Co., Brantford, Ont.

COUNTERS, REVOLUTION
American Steam Gauge & Valve Mfg. Co., Boston, Mass.

COWLS, SHIPS' VENTILATORS
Pedlar People, Ltd., Oshawa, Ont.

COUPLINGS, AIR HOSE
Cleveland Pneumatic Tool Co. of Canada, Toronto.

CRANES
Aikenhead Hardware, Ltd., Toronto, Ont.
Can. Fairbanks-Morse Co., Montreal, Que.
Hopkins & Co., F. H., Montreal, Que.
Williams Machinery Co., A. R., Toronto, Ont.

CRANES, ELECTRIC
Babcock & Wilcox, Ltd., Montreal, Que.
Morris Crane & Hoist Co., Herbert, Niagara Falls, Ont.
Smart-Turner Mach. Co., Hamilton, Ont.

CRANES, GOLIATH AND PNEUMATIC
Morris Crane & Hoist Co., Herbert, Niagara Falls, Ont.

CRANES, GRANTRY, PORTABLE, JIB
Morris Crane & Hoist Co., Herbert, Niagara Falls, Ont.
Smart-Turner Mach. Co., Hamilton, Ont.

CRANES, OVERHEAD TRAVELLING
Morris Crane & Hoist Co., Herbert, Niagara Falls, Ont.

CRANK SHAFTS
Canada Foundries and Forgings, Welland, Ont.

DAVITS, BOAT
Corbet Fdry. & Machine Co., Owen Sound, Ont.
Hopkins & Co., F. H., Montreal, Que.
Marten, Freeman Co., Toronto, Ont.
Waterous Engine Works Co., Brantford, Ont.

DEAD LIGHTS, BRASS
Goldie & McCulloch Co., Galt, Ont.
Morris Crane & Hoist Co., Herbert, Niagara Falls, Ont.
Mitchell Co., Ltd., Robt., Montreal, Que.

DECK LIGHTS
Mitchell Co., The Robert, Montreal, Que.
Turnbull Elevator Mfg. Co., Toronto, Ont.

DECK PLUGS, ELECTRIC
Crouse-Hinds Co. of Canada, Ltd., Toronto, Ont.
Mitchell Co., The Robert, Montreal, Que.

DECKING FOR SHIPS
Marine Decking & Supply Co., Philadelphia, Pa.

DERRICKS
Aikenhead Hardware, Ltd., Toronto, Ont.
Dake Engine Co., Grand Haven, Mich.
Hopkins & Co., F. H., Montreal, Que.
Marsh Engineering Works, Belleville, Ont.
Morris Crane & Hoist Co., Herbert, Niagara Falls, Ont.

DREDGES
Collingwood Shipbuilding Co., Collingwood, Ont.
Morris Mach. Works, Baldwinsville, N.Y.
Norbom Engineering Co., Philadelphia, Pa.
Polson Iron Works, Toronto.

DRILLS, AIR
Aikenhead Hardware, Ltd., Toronto, Ont.
Can. Ingersoll-Rand Co. Sherbrooke, Que.
Hopkins & Co., F. H., Montreal, Que.

DRILLS, CENTRE
Wilt Twist Drill Co., Walkerville, Ont.

DRILLS, BLACKSMITH AND BIT STOCK
Wilt Twist Drill Co., Walkerville, Ont.

DRILLS, ELECTRIC
Aikenhead Hardware, Ltd., Toronto, Ont.
Wilkinson & Kompass, Hamilton, Ont.

DRILLS, HIGH SPEED
Wilt Twist Drill Co., Walkerville, Ont.

DRILLS, TRACK
Wilt Twist Drill Co., Walkerville, Ont.

DRILLS, RACHET AND HAND
Wilt Twist Drill Co., Walkerville, Ont.

DRILLS, TWIST
Aikenhead Hardware, Ltd., Toronto, Ont.
Williams Machinery Co., A. R., Toronto, Ont.
Wilt Twist Drill Co., Walkerville, Ont.

DRY DOCKS
Can. Vickers, Ltd., Montreal, Que.
Collingwood Shipbuilding Co., Collingwood, Ont.
Doxford & Sons, William, Sunderland, England.
Georgian Bay Shipbuilding & Wrecking Co., Midland, Ont.
National Shipbuilding Co., Goderich, Ont.
Polson Iron Works, Toronto, Ont.
Port Arthur Shipbuilding Co., Port Arthur, Ont.
Yarrows, Limited, Victoria, B.C.

ECONOMIZERS, FUEL
Babcock & Wilcox, Ltd., Montreal, Que.

EJECTORS
Darling Bros., Ltd., Montreal, Que.
Empire Mfg. Co., London, Ont.
Mitchell Co., The Robert, Montreal, Que.
Morrison Brass Mfg. Co., James, Toronto, Ont.
Smart-Turner Mach. Co., Hamilton, Ont.

ELECTRIC LAMPS
Mitchell Co., The Robert, Montreal, Que.
Wm. C. Wilson & Co., Toronto, Ont.

ELECTRO-PLATING
Mitchell Co., The Robert, Montreal, Que.
Tallman Brass & Metal Co., Hamilton, Ont.
United Brass & Lead Co., Toronto, Ont.

ELECTRIC WELDING
Beauchemin & Fils, Sorel, P.Q.

ELEVATING MACHINERY
Darling Bros., Ltd., Montreal, Que.
Goldie & McCulloch, Ltd., Galt, Ont.
Wm. Hamilton Co., Peterboro, Ont.
Morris Crane & Hoist Co., Herbert, Niagara Falls, Ont.
Waterous Engine Works Co., Brantford, Ont.

ELEVATORS
Darling Bros., Ltd., Montreal, Que.
Turnbull Elevator Mfg. Co., Toronto, Ont.

ENGINES, HOISTING
Corbet Fdry. & Machine Co., Owen Sound, Ont.
Hopkins & Co., F. H., Montreal, Que.
Kennedy & Sons, Wm., Owen Sound, Ont.
Marsh Engineering Works, Belleville, Ont.
Port Arthur Shipbuilding Co., Port Arthur, Ont.
Williams Machinery Co., A. R., Toronto, Ont.

ENGINE, INTERNAL COMBUSTION
Doxford & Sons, William, Sunderland, England.

ENGINES, MARINE
Bolinders Co., New York, N.Y.
Can. Fairbanks-Morse Co., Montreal, Que.
Can. Vickers, Ltd., Montreal, Que.
Doxford & Sons, William, Sunderland, England.
Goldie & McCulloch, Ltd., Galt, Ont.
Hopkins & Co., F. H., Montreal, Que.
Iron Works, Ltd., Owen Sound, Ont.
Mason Regulator & Engin. Co., Montreal, Que.
Morris Mach. Works, Baldwinsville, N.Y.
National Shipbuilding Co., Goderich, Ont.
Norbom Engineering Co., Philadelphia, Pa.
Polson Iron Works, Toronto, Ont.
Port Arthur Shipbuilding Co., Port Arthur, Ont.
Trout Co., H. G., Buffalo, N.Y.
Waterous Engine Works Co., Brantford, Ont.
Williams Machinery Co., A. R., Toronto, Ont.

ENGINE STARTERS (AIR)
Scaife & Sons Co., Wm. B., Oakmont, Pa.

ENGINES, STEERING
Corbet Fdry. & Machine Co., Owen Sound, Ont.
Dake Engine Co., Grand Haven, Mich.
Kennedy & Sons, Wm., Owen Sound, Ont.
Wm. C. Wilson & Co., Toronto, Ont.

ENAMELWARE
Mueller Mfg. Co., H., Sarnia, Ont.
United Brass & Lead Co., Toronto, Ont.

EVAPORATORS
Mason Regulator & Engin. Co., Montreal, Que.
McNab Co., Bridgeport, Conn.
Weir Ltd., G. & J., Cathcart, Glasgow, Scotland.

EXTRACTORS, GREASE
American Steam Gauge & Valve Mfg. Co., Boston, Mass.
Darling Bros., Ltd., Montreal, Que.

EYE BOLTS AND NUTS
Canada Foundries and Forgings, Welland, Ont.
Mitchell Co., The Robert, Montreal, Que.
United Brass & Lead, Ltd., Toronto, Ont.

FANS
Aikenhead Hardware Ltd., Toronto, Ont.
Empire Mfg. Co., London, Ont.
Reed & Co., Geo. W., Montreal, Que.
Smart-Turner Mach. Co., Hamilton, Ont.
Williams Machinery Co., A. R., Toronto, Ont.

FENDERS, ROPE
Hopkins & Co., F. H., Montreal, Que.
Leckie, Ltd., John, Toronto, Ont.
Wilson & Co., Wm. C., Toronto, Canada.

FERRO-MANGANESE
Mitchells, Ltd., Glasgow, Scotland.

FILES
Aikenhead Hardware, Ltd., Toronto, Ont.
Williams Machinery Co., A. R., Toronto, Ont.

FIRE BRICKS
Beveridge Paper Co., Montreal, Que.
Mitchells, Ltd., Glasgow, Scotland.
Williams Machinery Co., A. R., Toronto, Ont.

FILTERS, FEED WATER
Darling Bros., Ltd., Montreal, Que.
MacKinnon Steel Co., Sherbrooke, Que.
Mason Regulator & Engin. Co., Montreal, Que.

FITTINGS, MARINE
Hopkins & Co., F. H., Montreal, Que.
McAvity & Sons, T., St. John, N.B.
Mitchell Co., The Robert, Montreal, Que.
United Brass & Lead, Ltd., Toronto, Ont.

FITTINGS, MOTOR BOAT
Empire Mfg. Co., London, Ont.
Mitchell Co., The Robert, Montreal, Que.
Mueller Mfg. Co., H., Sarnia, Ont.
United Brass & Lead, Ltd., Toronto, Ont.

FIXTURES, ELECTRIC
Cory & Son, Inc., Chas., New York, N.Y.
Crouse-Hinds Co. of Canada, Ltd., Toronto, Ont.
Mitchell Co., The Robert, Montreal, Que.
Harvey Hubbell Co. of Canada, Toronto, Can.
Tallman Brass & Metal Co., Hamilton, Ont.

FLAG POLES, STEEL
Dennis Wire & Iron Works Co., London, Ont.

FLOW METERS
Spray Engineering Co., Boston, Mass.

FLUE CLEANERS
Wm. C. Wilson & Co., Toronto, Ont.

FORGES
Aikenhead Hardware, Ltd., Toronto, Ont.
Hopkins & Co., F. H., Montreal, Que.

FLANGING AND EXPANDING MACHINES, PIPE
Lovekin Pipe Expanding & Flanging Mach. Co., Philadelphia, Pa.

FLOODLIGHTS, ELECTRIC
Crouse-Hinds Co. of Canada, Ltd., Toronto, Ont.

FLUORSPAR
Mitchells, Ltd., Glasgow, Scotland.
Scythes & Co., Toronto, Ont.

FORGINGS, ALL KINDS
Aikenhead Hardware, Ltd., Toronto, Ont.
Collingwood Shipbuilding Co., Collingwood, Ont.
Nova Scotia Steel & Coal Co., New Glasgow, N.S.

FORGINGS, STEEL AND IRON
Canada Foundries and Forgings, Welland, Ont.

FURNACE BRIDGE WALLS
Wager Furnace Bridge Wall Co., Inc., 149 Broadway, New York, N.Y.

GAUGES, RECORDING
American Steam Gauge & Valve Mfg. Co., Boston, Mass.
Empire Mfg. Co., London, Ont.

GASKETS
Wm. C. Wilson & Co., Toronto, Ont.

GAUGES COCKS
McAvity & Sons, T., St. John, N.B.
Mitchell Co., The Robert, Montreal, Que.

GAUGE GLASSES
Wm. C. Wilson & Co., Toronto, Ont.

GAUGES, WATER, PRESSURE, COMPOUND AND VACUUM
Aikenhead Hardware, Ltd., Toronto, Ont.
Babcock & Wilcox, Ltd., Montreal, Que.
Empire Mfg. Co., London, Ont.
Jenkins Bros., Montreal.
McAvity & Sons, T., St. John, N.B.
McNab Co., Bridgeport, Conn.
Morrison Brass Mfg. Co., James, Toronto, Ont.

GUARDS, LAMPS
Cleveland Pneumatic Tool Co. of Can., Toronto.

GENERATORS AND CONVERTORS
Can. Fairbanks-Morse Co., Montreal, Que.

GLAZING COMPOSITION
Kuhls, H. B. Fred., 6411 3rd Ave., Brooklyn, N.Y.

GOGGLES
Standard Optical Co., Geneva, N.Y.

GONGS
Clark & Bro., C. O., Montreal, Que.
Mitchell Co., Ltd., Robt., Montreal, Que.

GRAPHITE
Wm. C. Wilson & Co., Toronto, Ont.

GRATINGS
Can. Welding Works, Montreal, Que.
Corbet Fdry. & Machine Co., Owen Sound, Ont.
Canada Wire & Iron Goods Co., Hamilton, Ont.
MacKinnon Steel Co., Sherbrooke, Que.

GRINDERS, ELECTRIC
Wilkinson & Kompass, Hamilton, Ont.

GRINDERS, PNEUMATIC
Can. Ingersoll-Rand Co. Sherbrooke, Que.

GUARDS, MACHINERY
Dennis Wire & Iron Works Co., London, Ont.

GUY RODS AND ANCHORS, ELECTRIC
Crouse-Hinds Co. of Canada, Ltd., Toronto, Ont.

HAMMERS
Canada Foundries and Forgings, Ltd., Welland, Ont.

HARDWARE, MARINE
Hopkins & Co., F. H., Montreal, Que.
Mitchell Co., The Robert, Montreal, Que.

HEADLIGHTS, ELECTRIC
Crouse-Hinds Co. of Canada, Ltd., Toronto, Ont.
Hopkins & Co., F. H., Montreal, Que.
Tallman Brass & Metal Co., Hamilton, Ont.

HEATING EQUIPMENT
Darling Bros., Ltd., Montreal, Que.
Empire Mfg. Co., London, Ont.
Low & Sons, Ltd., Archibald, Glasgow, Scotland

HEATERS, FEED, WATER
Darling Bros., Ltd., Montreal, Que.
Babcock & Wilcox, Ltd., Montreal, Que.
Goldie & McCulloch Co., Galt, Ont.
Mason Regulator & Engin. Co., Montreal, Que.
McNab Co., Bridgeport, Conn.
Weir Ltd., G. & J., Cathcart, Glasgow, Scotland.

HINGES
London Bolt & Hinge Works, London, Ont.
Mitchell Co., Ltd., Robt., Montreal, Que.

HOOKS
Hopkins & Co., F. H., Montreal, Que.
Morris Crane & Hoist Co., Herbert, Niagara Falls, Ont.

HOISTS, ASH
Beatty & Sons, Welland, Ont.
Marsh Engineering Works, Belleville, Ont.
St. Clair Bros., Galt, Ont.
Waterous Engine Works Co., Brantford, Ont.

HOIST BLOCKS
Morris Crane & Hoist Co., Herbert, Niagara Falls, Ont.

HOISTS, CHAIN
Aikenhead Hardware, Ltd., Toronto, Ont.
Can. Fairbanks-Morse Co., Montreal, Que.
Dake Engine Co., Grand Haven, Mich.
Hopkins & Co., F. H., Montreal, Que.
Morris Crane & Hoist Co., Herbert, Niagara Falls, Ont.
Williams Machinery Co., A. R., Toronto, Ont.

HOISTS, CARGO, MOVING, ETC.
Dake Engine Co., Grand Haven, Mich.
Hopkins & Co., F. H., Montreal, Que.
Marsh Engineering Works, Belleville, Ont.
Waterous Engine Works Co., Brantford, Ont.

HOISTING MACHINERY
Beatty & Sons, Welland, Ont.
Can. Ingersoll-Rand Co. Sherbrooke, Que.
Corbet Fdry. & Machine Co., Owen Sound, Ont.
Wm. Hamilton Co., Peterboro, Ont.
Marsh Engineering Works, Belleville, Ont.
Morris Crane & Hoist Co., Herbert, Niagara Falls, Ont.
Waterous Engine Works Co., Brantford, Ont.
Williams Machinery Co., A. R., Toronto, Ont.

HOSE
Wm. C. Wilson & Co., Toronto, Ont.

INDICATORS, ENGINE
American Steam Gauge & Valve Mfg. Co., Boston, Mass.
Cory & Son, Inc., Chas., New York, N.Y.
McNab Co., Bridgeport, Conn.

INDICATORS, SPEED
Aikenhead Hardware, Ltd., Toronto, Ont.
Cory & Son, Inc., Chas., New York, N.Y.

INJECTORS
Aikenhead Hardware, Ltd., Toronto, Ont.
Empire Mfg. Co., London, Ont.
Mitchell Co., The Robert, Montreal, Que.
Morrison Brass Mfg. Co., James, Toronto, Ont.
Williams Machinery Co., A. R., Toronto, Ont.

INGOTS
Broughton Copper Co., Ltd., Manchester, Eng.

INSULATORS, ELECTRIC
Crouse-Hinds Co. of Canada, Ltd., Toronto, Ont.

INSTRUMENTS, NAUTICAL
Leckie, Ltd., John, Toronto, Ont.

IRON AND STEEL
Mitchells Ltd., Glasgow, Scotland.

JACKS
Hopkins & Co., F. H., Montreal, Que.
Morris Crane & Hoist Co., Herbert, Niagara Falls, Ont.

JOINTS, BALL SLIP
Norbom Engineering Co., Philadelphia, Pa.

INSURANCE, MARINE
Toronto Insurance & Vessel Agency, Toronto, Ont.

JOURNAL WEDGES
Canada Foundries and Forgings, Ltd., Welland, Ont.

JUTE PACKING
Stratford Oakum Co., Geo., Jersey City, N.J.

LATHES
Bertram & Sons, Ltd., John, Dundas, Ont.

LEATHER, LACE
Wm. C. Wilson & Co., Toronto, Ont.

LENSES FOR GOGGLES
Standard Optical Co., Geneva, N.Y.

LIFEBUOYS, BELTS AND PRESERVERS
Fosbery Co., Barking, Essex, Eng.
Hopkins & Co., F. H., Montreal, Que.
Wm. C. Wilson & Co., Toronto, Ont.

LIFEBOATS
Hopkins & Co., F. H., Montreal, Que.
Wm. C. Wilson & Co., Toronto, Ont.

LIFE BOAT EQUIPMENT
Eckliff Circulator Co., Detroit, Mich.
Hopkins & Co., F. H., Montreal, Que.
Leckie, Ltd., John, Toronto, Ont.
Wm. C. Wilson & Co., Toronto, Ont.

LIFE JACKETS
Fosbery Co., Barking, Essex, Eng.
Hopkins & Co., F. H., Montreal, Que.
Leckie, Ltd., John, Toronto, Ont.

LIGHTS, ALL KINDS
Can. Fairbanks-Morse Co., Montreal, Que.
Crouse-Hinds Co. of Canada, Ltd., Toronto, Ont.
Hopkins & Co., F. H., Montreal, Que.
Mitchell Co., The Robert, Montreal, Que.
Morrison Brass Mfg. Co., James, Toronto, Ont.
Tallman Brass & Metal Co., Hamilton, Ont.

LIGHTS, SIDE, PORT
Goldie & McCulloch Co., Galt, Ont.
Crouse-Hinds Co. of Canada, Ltd., Toronto, Ont.
Hopkins & Co., F. N., Montreal, Que.
McAvity & Sons Ltd., T., St. John, N.B.
Mitchell Co., The Robert, Montreal, Que.
Tallman Brass & Metal Co., Hamilton, Ont.
Turnbull Elevator Mfg. Co., Toronto, Ont.
Wilson & Co., Wm. C., Toronto, Ont.

LIGHTING SETS
Can. Fairbanks-Morse Co., Montreal.
Cory & Son, Inc., Chas., New York, N.Y.
Crouse-Hinds Co. of Canada, Ltd., Toronto, Ont.
Goldie & McCulloch Co., Ltd., Galt, Ont.
Williams Machinery Co., A. R., Toronto, Ont.
Wilson & Co., Wm. C., Toronto, Canada.

LOADERS, WAGON AND TRUCK
Morris Crane & Hoist Co., Herbert, Niagara Falls, Ont.
Waterous Engine Works Co., Brantford, Ont.

LOCKERS, METAL
Dennis Wire & Iron Works Co., London, Ont.
Reed & Co., Geo. W., Montreal, Que.

LUBRICATORS
Empire Mfg. Co., London, Ont.
Mitchell Co., The Robert, Montreal, Que.

MACHINERY-METAL AND WOOD BORING
Cleveland Pneumatic Tool Co. of Can., Toronto.

MACHINISTS
Corbet Fdry. & Machine Co., Owen Sound, Ont.
Wm. Hamilton Co., Peterboro, Ont.
Hyde Engineering Works, Montreal, Que.
Mitchell Co., The Robert, Montreal, Que.
United Brass & Lead Co., Toronto, Ont.

MANGANESITE, PASTE
Wm. C. Wilson & Co., Toronto, Ont.

MARINE CASTINGS
Wm. Hamilton Co., Peterboro, Ont.
Goldie & McCulloch Co., Galt, Ont.
Mitchell Co., Ltd., Robt., Montreal, Que.
McAvity & Sons, T., St. John, N.B.
Mueller Mfg. Co., H., Sarnia, Ont.
Tallman Brass & Metal Co., Hamilton, Ont.
United Brass & Lead, Ltd., Toronto, Ont.
Waterous Engine Works Co., Brantford, Ont.

MARINE SPECIALTIES
Goldie & McCulloch Co., Galt, Ont.
Hopkins & Co., F. H., Montreal, Que.
McAvity & Sons, T., St. John, N.B.
Mitchell Co., Ltd., Robt., Montreal, Que.
United Brass & Lead, Ltd., Toronto, Ont.

MOTORS
Aikenhead Hardware, Ltd., Toronto, Ont.
Can. Fairbanks-Morse Co., Montreal, Que.
Williams Machinery Co., A. R., Toronto, Ont.

NAUTICAL INSTRUMENTS
Wm. C. Wilson & Co., Toronto, Ont.

NET LIFTERS
Dake Engine Co., Grand Haven, Mich.
Leckie, Ltd., John, Toronto, Ont.

NOZZLES, ALL KINDS
Empire Mfg. Co., London, Ont.
Mitchell Co., The Robert, Montreal, Que.
Tolland Mfg. Co., Boston, Mass.

NUTS
Mitchell Co., The Robert, Montreal, Que.
Wilkinson & Kompass, Hamilton, Ont.
United Brass & Lead, Ltd., Toronto, Ont.

OIL CUPS
Empire Mfg. Co., London, Ont.
Mitchell Co., Ltd., Robt., Montreal, Que.

OILCLOTHING
Wm. C. Wilson & Co. Toronto, Ont.

OIL, LINSEED
Canada Linseed Oil Mills, Montreal, Que.

OILS AND GREASE
Wm. C. Wilson & Co., Toronto, Ont.

OAKUM
Davey & Sons W. O., Jersey City, N.J.
Leckie, Ltd., John, Toronto, Ont.
Stratford Oakum Co., Geo., Jersey City, N.J.
Wm. C. Wilson & Co., Toronto, Ont.

ORNAMENTAL IRONWORK
Canada Wire & Iron Goods Co., Hamilton, Ont.
Can. Welding Works, Montreal, Que.

OPTICAL GOODS
Standard Optical Co., Geneva, N.Y.

OXY-ACETLYENE OUTFITS
Can. Fairbanks-Morse Co., Montreal, Que.
Can. Welding Works, Montreal, Que.
Hopkins & Co., F. H., Montreal, Que.
McAvity & Sons Ltd., T., St. John, N.B.
Williams Machinery Co., A. R., Toronto, Ont.

PAINT
Ault & Wiborg Co. of Can., Ltd., Toronto, Ont.
Wm. C. Wilson & Co., Toronto, Ont.

PAINTS, ANTI-FOULING
Kuhls, H. B. Fred., 6411 3rd Ave., Brooklyn, N.Y.

PAINTS, ANTI-CORROSIVE
Kuhls, H. B. Fred. 6411 3rd Ave., Brooklyn, N.Y.

PAINTING EQUIPMENT, AUTOMATIC
Spray Engineering Co., Boston, Mass.

PACKING, AMMONIA
France Packing Co., Philadelphia, Pa.
Wm. C. Wilson & Co., Toronto, Ont.

PACKING, MARINE
Aikenhead Hardware, Ltd., Toronto, Ont.
France Packing Co., Philadelphia, Pa.
Wm. C. Wilson & Co., Toronto, Ont.

PACKING, METALLIC
Aikenhead Hardware, Ltd., Toronto, Ont.
France Packing Co., Philadelphia, Pa.

PACKING, STEAM
Aikenhead Hardware, Ltd., Toronto, Ont.
France Packing Co., Philadelphia, Pa.
Wm. C. Wilson & Co., Toronto, Ont.

PANELBOARD AND CABINETS, ELECTRIC
Crouse-Hinds Co. of Canada, Ltd., Toronto, Ont.
Preston Woodworking Machy. Co., Preston, Ont.

PATTERNS, WOOD AND METAL
Beauchemin & Fils, Sorel, P.Q.

PIPE, LAPWELD, CAST IRON, RIVETED
Empire Mfg. Co., London, Ont.
Norbom Engineering Co., Philadelphia, Pa.
United Brass & Lead, Ltd., Toronto, Ont.

PIPE RAILING, IRON AND BRASS
Dennis Wire & Iron Works Co., London, Ont.
Mitchell Co., The Robert, Montreal, Que.

PITCH
Reed & Co., Geo., Montreal, Que.

PITCH, PINE
Wilson & Co., Wm. C., Toronto, Canada.

PLANERS
Bertram & Sons, Ltd., John, Dundas, Ont.

PLANERS, STANDARD AND ROTARY
Yates Mach. Co., P. B., Hamilton, Ont.

PLATE CLAMPS
Morris Crane & Hoist Co., Herbert, Niagara Falls, Ont.

PLATE PUNCH TABLES
Norbom Engineering Co., Philadelphia, Pa.

PLUGS, ATTACHMENT
Harvey Hubbel Co. of Canada, Toronto, Can.

PLUMBING EQUIPMENT
Low & Sons, Ltd., Archibald, Glasgow, Scotland.
Empire Mfg. Co., London, Ont.
Mitchell Co., The Robert, Montreal, Que.
Mueller Mfg. Co., H., Sarnia, Ont.
Tallman Brass & Metal Co., Hamilton, Ont.
Wm. C. Wilson & Co., Toronto, Ont.
United Brass & Lead, Ltd., Toronto, Ont.

PROPELLOR BLADES, BRONZE
Corbet Fdry. & Machine Co., Owen Sound, Ont.
Empire Mfg. Co., London, Ont.
Mitchell Co., The Robert, Montreal, Que.
Tallman Brass & Metal Co., Hamilton, Ont.
United Brass & Lead, Ltd., Toronto, Ont.
Yarrows, Limited, Victoria, B.C.

PROPELLER WHEELS
Goldie & McCulloch Co., Galt, Ont.
Kennedy & Sons, Wm., Owen Sound, Ont.
Trout Co., H. G., Buffalo, N.Y.

PIPING, STEAM
Babcock & Wilcox, Ltd., Montreal, Que.

PORT LIGHTS
Goldie & McCulloch Co., Galt, Ont.
Mitchell Co., Ltd., Robt., Montreal, Que.
Mueller Mfg. Co., H., Sarnia, Ont.
Turnbull Elevator Mfg. Co., Toronto, Ont.
Tallman Brass & Metal Co., Hamilton, Ont.
Wm. C. Wilson & Co., Toronto, Ont.
United Brass & Lead, Ltd., Toronto, Ont.

PROPELLER WHEELS
Beauchemin & Fils, Sorel, P.Q.
McNab Co., Bridgeport, Conn.

PULLEYS
Wm. Hamilton Co., Peterboro, Ont.
Smart-Turner Mach. Co., Hamilton, Ont.
Williams Machinery Co., A. R., Toronto, Ont.

PUMP, AIR
Can. Ingersoll-Rand Co. Sherbrooke, Que.
Smart-Turner Mach. Co., Hamilton, Ont.
Weir Ltd., G. & J., Cathcart, Glasgow, Scotland.
Williams Machinery Co., A. R., Toronto, Ont.

PUMPS
Bawden Machine Co., Toronto.
Can. Fairbanks-Morse Co., Montreal, Que.
Canada Foundries & Forgings, Welland, Ont.
Darling Bros., Ltd., Montreal, Que.
Goldie & McCulloch, Ltd., Galt, Ont.
Hopkins & Co., F. H., Montreal, Que.
McAvity & Sons, T., St. John, N.B.
Oberdorfer Brass Co., M. L., Syracuse, N.Y.
Williams Machinery Co., A. R., Toronto, Ont.

PUMPS, CENTRIFUGAL
Can. Ingersoll-Rand Co. Sherbrooke, Que.
Darling Bros., Ltd., Montreal, Que.
Wm. Hamilton Co., Peterboro, Ont.
Goldie & McCulloch Co., Galt, Ont.
Morris Mach. Works, Baldwinsville, N.Y.
Norbom Engineering Co., Philadelphia, Pa.
Smart-Turner Mach. Co., Hamilton, Ont.
Waterous Engine Works Co., Brantford, Ont.
Williams Machinery Co., A. R., Toronto, Ont.

PUMPS, BILGE
Darling Bros., Ltd., Montreal, Que.
Goldie & McCulloch Co., Galt, Ont.
Mitchell Co., Ltd., Robt., Montreal, Que.
Morris Mach. Works, Baldwinsville, N.Y.
Mueller Mfg. Co., H., Sarnia, Ont.

PUMPS, CIRCULATING
Darling Bros., Ltd., Montreal, Que.
Goldie & McCulloch Co., Galt, Ont.
McNab Co., Bridgeport, Conn.
Morris Machine Works, Baldwinsville, N.Y.

PUMPS, FEED WATER
Darling Bros., Ltd., Montreal, Que.
Goldie & McCulloch Co., Galt, Ont.
McNab Co., Bridgeport, Conn.
Morris Mach. Works, Baldwinsville, N.Y.
Weir Ltd., G. & J., Cathcart, Glasgow, Scotland.
Williams Machinery Co., A. R., Toronto, Ont.

PUMPS, LIFT AND FORCE
Darling Bros., Ltd., Montreal, Que.
Low & Sons, A., Glasgow, Scotland.

PUMPS, REES ROTURBO
Goldie & McCulloch Co., Galt, Ont.

PUMPS, HAND AND POWER
Aikenhead Hardware, Ltd., Toronto, Ont.
Darling Bros., Ltd., Montreal, Que.
Mitchell Co., The Robert, Montreal, Que.
Smart-Turner Mach. Co., Hamilton, Ont.
Williams Machinery Co., A. R., Toronto, Ont.

PUMPS, HIGH PRESSURE
Canadian Ingersoll-Rand Co., Sherbrooke, Que.
Darling Bros., Ltd., Montreal, Que.
Morris Mach. Works, Baldwinsville, N.Y.
Smart-Turner Mach. Co., Hamilton, Ont.

PUMPS, STEAM TURBINE
Darling Bros., Ltd., Montreal, Que.
Morris Mach. Works, Baldwinsville, N.Y.
Wm. C. Wilson & Co., Toronto, Ont.

PUMPING MACHINES
Norbom Engineering Co., Philadelphia, Pa.

PUNCHES
Bertram & Sons, Ltd., John, Dundas, Ont.

PUNCHES, SINGLE, DOUBLE AND MULTIPLE
Norbom Engineering Co., Philadelphia, Pa.

PURIFIERS, WATER
Babcock & Wilcox, Ltd., Montreal, Que.

RADIATORS, STEAM, ELECTRIC
Empire Mfg. Co., London, Ont.
Low & Sons, Ltd., Archibald, Glasgow, Scotland.
Wm. C. Wilson & Co., Toronto, Ont.

RAILS, OVERHEAD
Morris Crane & Hoist Co., Herbert, Niagara Falls, Ont.

RANGES
Hopkins & Co., F. H., Montreal, Que.
Wm. C. Wilson & Co., Toronto, Ont.

RECEIVERS, AIR
Can. Ingersoll-Rand Co. Sherbrooke, Que.
Can. Welding Works, Montreal, Que.
Darling Bros., Ltd., Montreal, Que.
MacKinnon Steel Co., Sherbrooke, Que.

REGULATORS, FEED WATER
American Steam Gauge & Valve Mfg. Co., Boston, Mass.

REGULATORS, PRESSURE
Mason Regulator & Engin. Co., Montreal, Que.

RELEASING GEARS
Eckliff Circulator Co., Detroit, Mich.

REPAIRS, MARINE
Corbet Foundry & Mach. Co., Owen Sound, Ont.
Can. Vickers, Ltd., Montreal, Que.
Collingwood Shipbuilding Co., Collingwood, Ont.
Engr. & Mach. Wks. of Can., St. Catharines, Ont.
Georgian Bay Shipbuilding & Wrecking Co., Midland, Ont.
Hyde Engineering Works, Montreal, Que.
Iron Works, Ltd., Owen Sound, Ont.
Kennedy & Sons, Wm., Owen Sound, Ont.
Muir Bros. Dry Dock Co., Port Dalhousie, Ont.
National Shipbuilding Co., Goderich, Ont.
Port Arthur Shipbuilding Co., Port Arthur, Ont.
Yarrow, Limited, Victoria, B.C.

RIGGING SCREWS
Hopkins & Co., F. H., Montreal, Que.

RIGGING, WIRE ROPE
Leckie, Ltd., John, Toronto, Ont.

RIVETS
London Bolt & Hinge Works, London, Ont.
Severance Mfg. Co., S., Glassport, Pa.
Wilkinson & Kompass, Hamilton, Ont.

RIVETERS, PNEUMATIC
Can. Ingersoll-Rand Co., Ltd., Sherbrooke, Que.
Cleveland Pneumatic Tool Co. of Can., Toronto.

RODS, COPPER, BRASS, BRONZE
Standard Underground Cable Co., Hamilton, Ont.
Tallman Brass & Metal Co., Hamilton, Ont.

ROLLS, STRAIGHTENING, BENDING
Bertram & Sons, Ltd., John, Dundas, Ont.

ROOFING
Reed & Co., Geo., Montreal, Que.
Wm. C. Wilson & Co., Toronto, Ont.

ROPE BLOCKS
Aikenhead Hardware, Ltd., Toronto, Ont.
Can. Fairbanks-Morse Co., Montreal, Que.
Morris Crane & Hoist Co., Herbert, Niagara Falls, Ont.

ROPE
Hopkins & Co., F. H., Montreal, Que.
Leckie, Ltd., John, Toronto, Ont.
McNab Co., Bridgeport, Conn.
Stratford Oakum Co., Geo., Jersey City, N.J.
Wm. C. Wilson & Co., Toronto, Ont.

RUBBER COATS
Wm. C. Wilson & Co., Toronto, Ont.

SAW MILL MACHINERY
Wm. Hamilton Co., Peterboro, Ont.
Preston Woodworking Machy. Co., Preston, Ont.
Waterous Engine Works Co., Brantford, Ont.
Yates Machine Co., P. B., Hamilton, Ont.

SAWS, BAND
Preston Woodworking Machy. Co., Preston, Ont.

SCALES, BOILERS, ENGINES
Can. Fairbanks-Morse Co., Montreal, Que.

SCOWS
Collingwood Shipbuilding Co., Collingwood, Ont.
Polson Iron Works, Toronto, Ont.

SCREENS, WIRE
Dennis Wire & Iron Works Co., London, Ont.

SCREWS, COACH
London Bolt & Hinge Works, London, Ont.

SEAM PAINT
Kuhls, H. B. Fred., 6411 3rd Ave., Brooklyn, N.Y.

SEPARATORS, OIL, STEAM
Darling Bros., Ltd., Montreal, Que.
Mason Regulator & Engin. Co., Montreal, Que.
Smart-Turner Mach. Co., Hamilton, Ont.

SHAFTING
Wm. Hamilton Co., Peterboro, Ont.
Mitchell Co., The Robert, Montreal, Que.
Wilkinson & Kompass, Hamilton, Ont.

SHAFTING, BRONZE
Empire Mfg. Co., London, Ont.
Tallman Brass & Metal Co., Hamilton, Ont.

SHEARS
Bertram & Sons, Ltd., John, Dundas, Ont.
Norbom Engineering Co., Philadelphia, Pa.

SHIPBUILDING TOOLS
Aikenhead Hardware, Ltd., Toronto, Ont.
Can. Ingersoll-Rand Co. Sherbrooke, Que.

SHIPS, BUILDERS OF
Can. Vickers, Ltd., Montreal, Que.
Collingwood Shipbuilding Co., Collingwood, Ont.
Doxford & Sons, William, Sunderland, England.
Georgian Bay Shipbuilding & Wrecking Co., Midland, Ont.
National Shipbuilding Co., Goderich, Ont.
Polson Iron Works, Toronto, Ont.
Port Arthur Shipbuilding Co., Port Arthur, Ont.
Yarrows, Limited, Victoria, B.C.

SHIP BROKERS
Page & Jones, Mobile, Ala.
SHIP PLATES
Nova Scotia Steel & Coal Co., New Glasgow, N.S.
SLEDGES
Wilkinson & Kompass, Hamilton, Ont.
SLINGS
Hopkins & Co., F. H., Montreal, Que.
Morris Crane & Hoist Co., Herbert, Niagara Falls, Ont.
SPECIAL MACHINERY
Corbet Fdry. & Machine Co., Owen Sound, Ont.
Wm. Hamilton Co., Peterboro, Ont.
Miller Bros. & Sons, Ltd., Montreal, Que.
Smart-Turner Mach. Co., Hamilton, Ont.
SOCKETS, PULL KEY AND KEYLESS
Harvey Hubbell Co. of Can., Toronto, Can.
SMOOTH-ON
Wm. C. Wilson & Co., Toronto, Ont.
SPIKES
Wm. C. Wilson & Co., Toronto, Ont.
SPRAY COOLING SYSTEMS
Spray Engineering Co., Boston, Mass.
STAMPINGS, SHEET METAL
Pedlar People, Ltd., Oshawa, Ont.
STEAMSHIP AGENTS
Darling Bros., Ltd., Montreal, Que.
Page & Jones, Mobile, Ala.
STEAM SPECIALTIES
Can. Fairbanks-Morse Co., Montreal, Que.
Corbet Fdry. & Machine Co., Owen Sound, Ont.
Dom. Copper Products Co., Montreal.
Empire Mfg. Co., London, Ont.
Mitchell Co., The Robert, Montreal, Que.
STEAM TRAPS
Aikenhead Hardware, Ltd., Toronto, Ont.
American Steam Gauge & Valve Mfg. Co., Boston, Mass.
Darling Bros., Ltd., Montreal, Que.
Empire Mfg. Co., London, Ont.
Mason Regulator & Engin. Co., Montreal, Que.
Mitchell Co., The Robert, Montreal, Que.
Smart-Turner Mach. Co., Hamilton, Ont.
STEEL, HIGH SPEED
Hopkins & Co., F. H., Montreal, Que.
Nova Scotia Steel & Coal Co., New Glasgow, N.S.
Wilkinson & Kompass, Hamilton, Ont.
STEEL SHELVING
Dennis Wire & Iron Works, London, Ont.
STEEL WORK, STRUCTURAL
Babcock & Wilcox, Ltd., Montreal, Que.
Can. Welding Works, Montreal, Que.
Corbet Fdry. & Machine Co., Owen Sound, Ont.
Wm. Hamilton Co., Peterboro, Ont.
MacKinnon Steel Co., Sherbrooke, Que.
STEERING GEARS
Corbet Fdry. & Machine Co., Owen Sound, Ont.
Hopkins & Co., F. H., Montreal, Que.
Engr. & Mach. Wks. of Can., St. Catharines, Ont.
Wm. C. Wilson & Co., Toronto, Ont.
STOCK RACKS FOR BARS, PIPING, ETC.
Mitchell Co., The Robert, Montreal, Que.
Morris Crane & Hoist Co., Herbert, Niagara Falls, Ont.
STOKERS, MECHANICAL
Babcock & Wilcox, Ltd., Montreal, Que.
SUPERHEATERS, STEAM
Babcock & Wilcox, Ltd., Montreal, Que.
Goldie & McCulloch Co., Galt, Ont.
SWITCHBOARDS, ELECTRIC
Crouse-Hinds Co. of Can., Ltd., Toronto, Ont.
TALLOW
Can. Economic Lubricant Co., Montreal, Que.
Wm. C. Wilson & Co., Toronto, Ont.
TANKS, STEEL
Can. Welding Works, Montreal, Que.
Corbet Foundry & Mach. Co., Owen Sound, Ont.
Goldie & McCulloch, Ltd., Galt, Ont.
Hopkins & Co., F. H., Montreal, Que.
MacKinnon Steel Co., Sherbrooke, Que.
Marsh Engineering Works, Belleville, Ont.
Port Arthur Shipbuilding Co., Port Arthur, Ont.
Reed & Co., Geo., Montreal, Que.
Scaife & Sons Co., Wm. B., Oakmont, Pa.
TANKS (AIR, GAS AND LIQUID)
Can. Welding Works, Montreal, Que.
MacKinnon Steel Co., Sherbrooke, Que.
Marsh Engineering Works, Belleville, Ont.
Scaife & Sons Co., Wm. B., Oakmont, Pa.
TAPPING MACHINES
Mueller Mfg. Co., H., Sarnia, Ont.
TELEGRAPHS, SHIPS
McNab Co., Bridgeport, Conn.
Cory & Sons, Inc., Chas., New York, N.Y.
Morrison Brass Mfg. Co., James, Toronto, Ont.
Wilson & Co., Wm., Toronto, Ont.
TELEPHONES, MARINE
Cory & Son, Inc., Chas., New York, N.Y.
McNab Co., Bridgeport, Conn.
TESTERS, METER
Mueller Mfg. Co., H., Sarnia, Ont.
THUMB SCREWS AND NUTS
Canada Foundries & Forgings, Welland, Ont.
Mitchell Co., The Robert, Montreal, Que.
United Brass & Lead, Ltd., Toronto, Ont.
TRACK SYSTEMS
Morris Crane & Hoist Co., Herbert, Niagara Falls, Ont.
TRAVELLING BLOCKS
Morris Crane & Hoist Co., Herbert, Niagara Falls, Ont.
TROLLEYS
Can. Fairbanks-Morse Co., Montreal, Que.
Morris Crane & Hoist Co., Herbert, Niagara Falls, Ont.
TROLLEY HOISTS
Morris Crane & Hoist Co., Herbert, Niagara Falls, Ont.
TRUCKS, HAND, ELECTRIC
Aikenhead Hardware, Ltd., Toronto, Ont.
Can. Fairbanks-Morse Co., Montreal, Que.
TUBES, BOILER
Babcock & Wilcox, Ltd., Montreal, Que.
Broughton Copper Co., Ltd., Manchester, Eng.
Goldie & McCulloch Co., Galt, Ont.
TUBES, COPPER AND BRASS
Mueller Mfg. Co., H., Sarnia, Ont.
Tallman Brass & Metal Co., Hamilton, Ont.
Standard Underground Cable Co., Hamilton, Ont.
TUGS
Polson Iron Works, Toronto, Ont.
Collingwood Shipbuilding Co., Collingwood, Ont.
TURBINES, STEAM
Goldie & McCulloch Co., Galt, Ont.
TURBINES, DIRECT-DRIVING AND GEARED
Doxford & Sons, William, Sunderland, England.
TURNBUCKLES
Canada Foundries & Forgings, Welland, Ont.
Hopkins & Co., F. H., Montreal, Que.
TURNTABLES
Morris Crane & Hoist Co., Herbert, Niagara Falls, Ont.
SPIKES, SMALL RAILROAD
Severance Mfg. Co., S., Glassport, Pa.
UNIONS, ALL KINDS
Dart Union Company, Toronto, Ont.
VALVES, AIR
Mitchell Co., The Robert, Montreal, Que.
Mueller Mfg. Co., H., Sarnia, Ont.
VALVE, DISCS
Wm. C. Wilson & Co., Toronto, Ont.
VALVES
American Steam Gauge & Valve Mfg. Co., Boston, Mass.
Babcock & Wilcox, Ltd., Montreal, Que.
Can. Fairbanks-Morse Co., Montreal, Que.
Darling Bros., Ltd., Montreal, Que.
Empire Mfg. Co., London, Ont.
McAvity & Sons, Ltd., T., St. John, N.B.
Mason Regulator & Engin. Co., Montreal, Que
McNab Co., Bridgeport, Conn.
Norbom Engineering Co., Philadelphia, Pa.
Williams Machinery Co., A. R., Toronto, Ont.
Wm. C. Wilson & Co., Toronto, Ont.
United Brass & Lead, Ltd., Toronto, Ont.
VALVES, FOOT
Aikenhead Hardware, Ltd., Toronto, Ont.
Mitchell Co., The Robert, Montreal, Que.
Smart-Turner Mach. Co., Hamilton, Ont.
VALVES, STOP, REDUCING, SAFETY CHECK, DISCHARGE, SUCTION
Aikenhead Hardware, Ltd., Toronto, Ont.
American Steam Gauge & Valve Mfg. Co., Boston, Mass.
Darling Bros., Ltd., Montreal, Que.
McAvity & Sons Ltd., T., St. John, N.B.
Mitchell Co., The Robert, Montreal, Que.
Morrison Brass Mfg. Co., James, Toronto, Ont.
VALVES, MIXING
Darling Bros., Ltd., Montreal, Que.
Mitchell Co., The Robert, Montreal, Que.
Mueller Mfg. Co., H., Sarnia, Ont.
VALVES, REDUCING, PRESSURE
Mitchell Co., The Robert, Montreal, Que.
Mueller Mfg. Co., H., Sarnia, Ont.
VALVES, STORM
Low & Sons, A., Glasgow, Scotland.
VARNISHES
Aikenhead Hardware, Ltd., Toronto, Ont.
Ault & Wiborg Co. of Can., Ltd., Toronto, Ont.
Leckie, Ltd., John, Toronto, Ont.
Reed & Co., Geo., Montreal, Que.
VENTILATORS, COWL
Can. Welding Works, Montreal, Que.
McNab Co., Bridgeport, Conn.
Mitchell Co., The Robert, Montreal, Que.
VENTILATION EQUIPMENT
Empire Mfg. Co., London, Ont.
Hopkins & Co., F. H., Montreal, Que.
Low & Sons, Ltd., Archibald, Glasgow, Scotland.
WASHERS
London Bolt & Hinge Works, London, Ont.
Wm. C. Wilson & Co., Toronto, Ont.
WATER COLUMNS
Darling Bros., Ltd., Montreal, Que.
Mitchell Co., The Robert, Montreal, Que.
Morrison Brass Mfg. Co., James, Toronto, Ont.
WELDING, ELECTRIC
Can. Welding Works, Montreal, Que.
Hall Engineering Works, Montreal, Que.
Marine Welding Co., Buffalo, N.Y.
Beatty & Sons, M., Welland, Ont.
WATER SOFTENERS
Babcock & Wilcox, Ltd., Montreal, Que.
WATER SUPPLY SYSTEMS
Mueller Mfg. Co., H., Sarnia, Ont.
WATER HEATERS
Darling Bros., Ltd., Montreal, Que.
Empire Mfg. Co., London, Ont.
Morrison Brass Mfg. Co., James, Toronto, Ont.
WHISTLES AND SYRENS
Empire Mfg. Co., London, Ont.
McAvity & Sons, T., St. John, N.B.
McNab Co., Bridgeport, Conn.
Mitchell Co., Ltd., Robt., Montreal, Que.
Morrison Brass Mfg. Co., Jas., Toronto, Ont.
WINCHES, CARGO
Aikenhead Hardware, Ltd., Toronto, Ont.
Corbet Fdry. & Machine Co., Owen Sound, Ont.
Hopkins & Co., F. H., Montreal, Que.
Marsh Engineering Works, Belleville, Ont.
WINCHES, DOCK, SHIP
Beatty & Sons, M., Welland, Ont.
Marsh Engineering Works, Belleville, Ont.
Miller Bros. & Sons, Ltd., Montreal, Que.
Morris Crane & Hoist Co., Herbert, Niagara Falls, Ont.
Wilson & Co., Wm. C., Toronto, Canada.
WINCHES, TOWING
Corbet Foundry & Mach. Co., Owen Sound, Ont.
WINCHES, TRAWL
Beatty & Sons, M., Welland, Ont.
Wm. C. Wilson & Co., Toronto, Ont.
WINDLASSES
Corbet Fdry. & Machine Co., Owen Sound, Ont.
Dake Engine Co., Grand Haven, Mich.
Hopkins & Co., F. H., Montreal, Que.
Wilson & Co., Wm. C., Toronto, Canada.
WIPER CAPS, OILER BOXES, ETC.
Mitchell Co., The Robert, Montreal, Que.
Morrison Brass Mfg. Co., James, Toronto, Ont.
WIRE, COPPER CLAD STEEL
Standard Underground Cable Co., Hamilton, Ont.
WIRE, COPPER, BRASS, BRONZE
Standard Underground Cable Co., Hamilton, Ont.
Tallman Brass & Metal Co., Hamilton, Ont.
WIRE, INSULATED
Standard Underground Cable Co., Hamilton, Ont.
WIRELESS OUTFITS
Marconi Wireless Telegraph Co., Montreal, Que.
WIRE ROPE
Zenith Coal & Steel Products, Ltd., Montreal,
WOOD WORKING MACHINERY
Aikenhead Hardware, Ltd., Toronto, Ont.
Can. Fairbanks-Morse Co., Montreal, Que.
Preston Woodworking Machy. Co., Preston, Ont.
Yates Mach. Co., P. B., Hamilton, Ont.
WOOD BORING TOOLS
Aikenhead Hardware, Ltd., Toronto, Ont.
Can. Ingersoll-Rand Co. Sherbrooke, Que.
WOODITE GAUGE GLASS WASHERS
Wm. C. Wilson & Co., Toronto, Ont.
WRENCHES
Canada Foundries & Forgings, Welland, Ont.
YACHT BROKER
Watt, J. Murray, Philadelphia, Pa.

INDEX TO ADVERTISERS

MORRISON's
MARINE SPECIALTIES
Cover All Demands of the Ship
When a ship is fitted with Morrison's Marine Specialties it is equipped with a line of goods that can be thoroughly relied upon to give long and efficient service.
Every article is guaranteed reliable by thorough factory test. Approved by the Dominion Marine Inspection Department, Lloyds' Survey and Munition Board.
SOME OF OUR LINES:
Stop Valves
Adjustable Check Valves
Non-Return and Equalizing Valves
Check Valves
Gate Valves
Safety Valves
Relief Valves
Pressure Reducing Valves
Injectors
Ejectors
Pressure Gauges
Vacuum Gauges
Ammonia Gauges
Recording Gauges
Engineers' Clocks
Lubricators
Steam Whistles
Sirens
Asbestos Packed Cocks
Gland Cocks
Suction and Discharge Valves
Ships' Telegraphs
Ships' Side Lights
Gong Bells
Gong Pulls
Chain Guide Brackets
Telegraph Chain
Binnacles
Water Service Cocks
Hose Valves and Fittings
Pipe and Fittings, etc., etc.
OUR MARINE CATALOG consists of 60 pages with 150 cuts. It would be of great assistance to you. Pleased to send it on request.
The James Morrison Brass Mfg. Co.
Limited
Toronto Canada
J.M.T.
A GUARANTEE OF QUALITY
Ships' Side Cock with Locking Shield, Bronze
J.M.T. Globe Valve
J.M.T. Reducing Valve
Overboard Discharge Valve, Spring Loaded
Asbestos Packed Boiler Drain Cock, Bronze
Hose-end Gate Valve
Asbestos Packed Salinometer Cock, Bronze
Cylinder Oil Pump
Oil Box, Cast Bronze

CIRCULATES IN EVERY PROVINCE OF CANADA AND ABROAD

MARINE ENGINEERING of Canada

A monthly journal dealing with the progress and development of Merchant and Naval Marine Engineering, Shipbuilding, the building of Harbors and Docks, and containing a record of the latest and best practice throughout the Sea-going World. Published by
The MacLean Publishing Co., Limited
MONTREAL, Southam Building TORONTO, 143-153 University Ave. WINNIPEG, 1207 Union Trust Bldg. LONDON, ENG., 88 Fleet St.

Vol. VIII. Publication Office, Toronto—December, 1918 No. 12

Planning for War —and Peace

In the field of manufacture, as on the field of battle, American strategy has been a big factor in the winning of the war.

Our fighting men at the front, and our fighting men back of the front, down to the humblest of the workers in shop and shipyard, have given heroic demonstrations of their courage, initiative, resourcefulness and spirit of self-sacrifice.

Our war plans have made good. But what about our Peace plans?

Now that the Unconditional Surrender of the Hun looms as an imminent possibility, with its accompanying cessation of war orders and the resumption of intense competition, what does it mean to us industrially? Is it a threat or a promise?

We believe it is a promise—but only to those who continue to seek diligently *now* for better methods of management and production.

Looking over our own record of the war plans in the making and carrying out of which the Keller organization has been privileged to co-operate, it is gratifying to realize that while we have been working for the success of the Allied cause, we have also been working for the post-war success of ourselves and our customers.

By striving steadily to increase our production of Keller-Made Master-Built Pneumatic Tools, and to improve the tools themselves so that they can be depended upon to increase the production of the Shipyards, Shops and Foundries using them, we have not only "done our bit" to render certain the outcome of the great Battle for Democracy, but have at the same time and to the same extent helped to pave the way for the winning of the great Battle for World Trade that is to follow.

Preventable inefficiency is Treason during the war. After the war it will be Business Suicide.

William H. Keller, *President*

KELLER PNEUMATIC TOOL COMPANY, Grand Haven, Mich., U. S. A.

ME-1001-8445

MD-100-H

MN-100-P

No. 8431

HUBBELL SPECIALTIES

No. 8421

Cap. No. 5421
Rec. No. 5547

No. 5765

No. 5702

No. 5735

No. 6915

No. 5406

Cap. No. 5421
Rec. No. 6257

No. 9630

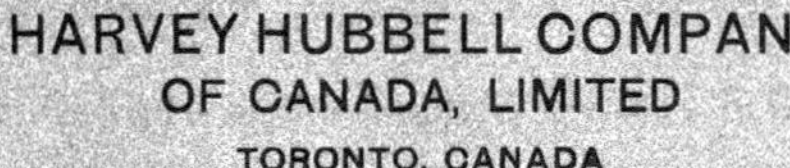

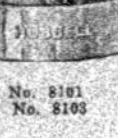

No. 8101
No. 8103

No. 5691

No. 5573

WEIR
Auxiliary Machinery

Characterized by

Originality - - in Design
Quality - - in Material
Accuracy in Workmanship
Efficiency in Performance

FEED PUMPS EVAPORATORS
AIR PUMPS FEED HEATERS
'UNIFLUX' CONDENSING PLANT

Applied To Over Eighty Millions Horse Power

Agents:
PEACOCK BROS., 285 Beaver Hall Hill, Montreal

G. & J. WEIR, LIMITED
CATHCART, GLASGOW - - SCOTLAND

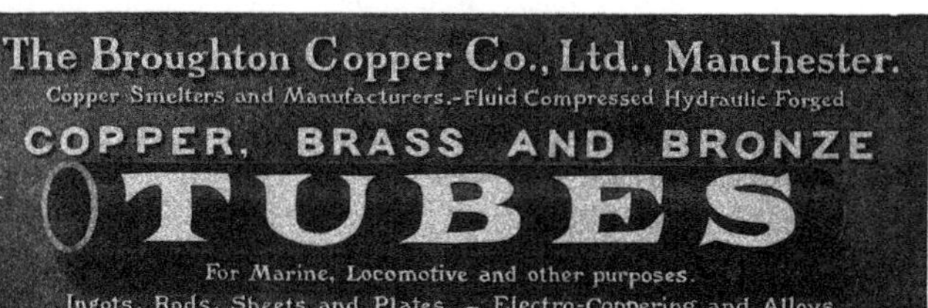

The Broughton Copper Co., Ltd., Manchester.
Copper Smelters and Manufacturers.-Fluid Compressed Hydraulic Forged
COPPER, BRASS AND BRONZE
TUBES
For Marine, Locomotive and other purposes.
Ingots, Rods, Sheets and Plates. - Electro-Coppering and Alloys.

Port Arthur Shipbuilding Company
Limited

PORT ARTHUR, CANADA

Designers and Builders of

Steel Ships, Boilers, Engines, Etc.

Every Modern Facility Available for Repair Work

Dry Dock, 700' x 98' x 16'

PLANT AT PORT ARTHUR

General Offices at Port Arthur, Ontario, Canada

MARINE ENGINEERING of Canada

Volume VIII TORONTO, DECEMBER, 1918 Number 12

The Leading Features of Semi-Diesel Oil Engine

Definition of the Semi-Diesel Engine — Nomenclature — Classification — Compression Pressure — The Disadvantage of High Compression — The Effect of Compression — Fuel Economy— Cycle of Operation—Flexibility

By JAMES RICHARDSON, B.Sc., A.M.I.C.E.*

INTRODUCTION.—On several recent occasions, authorities, when forecasting the lines of development of the oil engine, have expressed the opinion that the so-called semi-Diesel engine would play no inconsiderable part. It might be matter for surprise that publications of technical matter dealing with the semi-Diesel engines are extremely rare in comparison with the vast amount of available data relating to the Diesel engine.

Definition of the Semi-Diesel Engine

The variously-named semi-Diesel, hot-bulb or surface-ignition engine may be defined as an internal-combustion engine, using oil fuel, which has an uncooled portion of the combustion chamber, normally at high temperature, serving to augment the heat generated by the compression pressure and to assist in the vaporization and ignition of the fuel injected at the ignition point of the cycle. From this class should rightly be excluded those oil engines which are not called by their makers Diesel engines, but which rely for ignition, as completely as the Diesel engine, upon the heat generated by the compression of the air charge, and therefore should be so named. The means of injection of the fuel with such engines may vary from the standard air injection system.

Nomenclature

In describing and classifying engines of the semi-Diesel type, it will be necessary generally to adopt the nomenclature and technical expressions familiarized by the literature dealing with the Diesel engine, and convenient often to describe by means of comparison with the better-known Diesel engine.

Classification

The many types of engines of the type under review vary considerably (see Fig. 1) and can be classified according to the cycle of operation upon which they work, whether the two-stroke or the four-stroke cycle, and according to extent to which heat of compression is relied upon for the vaporization and ignition of the injected fuel. In the present stage of development, the chief claim of the semi-Diesel engine to be considered in the forefront of internal-combustion prime movers, is its marked simplicity. Development along probable lines may reasonably and in the near future reveal qualities to gain which a certain degree of simplicity may well be sacrificed. Primarily for reasons of simplicity, the great majority—more than 90 per cent.—of these engines at the present time are designed on the two-stroke cycle principle, limited to its simplest application, and are generally confined to relatively low powers per working cylinder, 125 brake horse-power per cylinder being the maximum attained up to the present time. Fig. 2 shows cross-sections of a Beardmore two-cycle semi-Diesel compression engine and Fig. 3 of a low-compression engine, with references. Fig. 10 shows an external view of a four-cylinder engine of the same make. On the same page are external views of single-cylinder engines by Messrs. Robey and Co., Limited, of London, and by Messrs. Petters, Limited, of Yeovil. The former are made in sizes from 10 to 50 horse-power with single cylinders, and from 10 to 100 horse-power with two cylinders. Fig. 12 shows Messrs. Petters' latest type, which is made in 35 and 50 brake horse-power sizes.

Some of the earlier semi-Diesel engines used air injection of the fuel, but this type was not developed, due primarily to the disadvantage of the extra complication of compressors and their attendant gear. All modern engines of this type work with "solid" or "mechanical" injection of the fuel, because of the simplicity of this system, considered especially in conjunction with the hot bulb for assisting the vaporization and ignition of the injected fuel. Air compressors are again making their appearance on semi-Diesel engines, although not for the purpose of air injection in the ordinary accepted meaning of the term. An air jet is used to cool the combustion chamber and piston, and so to take the place of the water-drip (mentioned later), to increase the efficiency of the scavenging of the main cylinder, and so to make possible engines of relatively high powers, i.e., over 100 brake horse-power per cylinder,

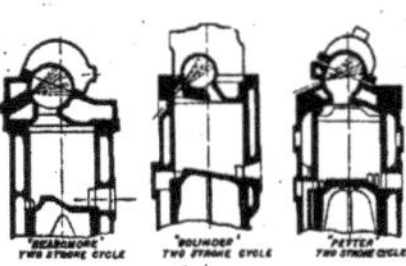
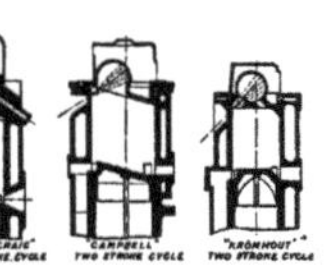

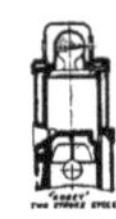
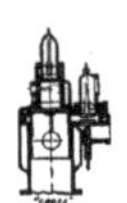

FIG. 1.—NINE REPRESENTATIVE TYPES OF SEMI-DIESEL ENGINES

*Paper read before the Diesel Engine Users' Association, on Thursday, October 24, 1918.

without having recourse to such expedients as separate scavenging pumps, elaborate cooling systems for the main pistons, and so forth.

Compression Pressure

With all internal combustion engines, theory teaches that the higher the compression pressure the less the fuel consumption, the less the heat required from an outside source to attain to the temperature necessary for the first working cycle when starting from cold, the higher the average mean effective pressure reached in the working cylinder, and consequently the smaller the main piston-swept volume for a given indicated horse-power. On the other hand, the lower the compression the more even the turning moment, generally the higher the mechanical efficiency, and the less the effort required to start the engine from rest by way of overcoming the negative work of the first compression stroke.

The semi-Diesel engine in type is a variable compression oil engine, and can be designed to work between, but excluding, the two extreme limits—the higher, that at which the heat of compression alone suffices to ignite and vaporize the injected charge (the Diesel cycle), and the lower, that at which the size of the hot bulb becomes inconveniently large for reasons of strength, when the loss from the hot bulb, by radiation, would be a serious factor, and together with the small power output obtainable with low compression, would

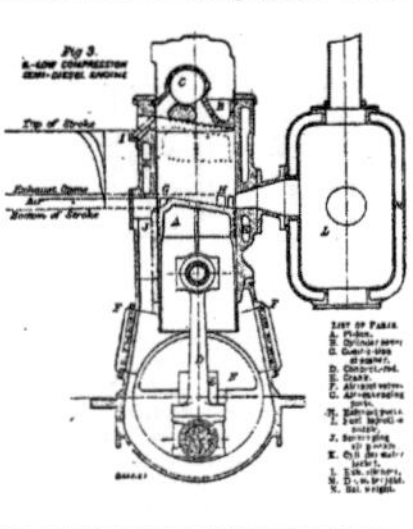

FIG. 3.—LOW COMPRESSION SEMI-DIESEL ENGINE

tend towards an excessively high fuel consumption and a large engine.

The Disadvantage of High Compression

There are disadvantages attendant upon high compressions, and a compromise between the theory that the higher the compression the greater the economy and practical considerations, must be struck. In comparison with steam prime movers, the mechanical efficiency of internal-combustion engines is low, due primarily to the friction of the piston rings. (See Appendix I.) The higher the compression and consequently the maximum pressures the greater this loss on account of the larger number of rings required to ensure satisfactory gas tightness, and the greater the pressure exerted by these rings when forced against the piston walls by the cylinder pressure operating behind the rings.

A further outcome of high pressures is increased piston-ring leakage, and the effects of piston-ring leakage upon economy are very considerable. The higher the compression pressure the greater the heat transfer from the charge in the cylinder to the jacket cooling water.

The Effect of Compression

From these considerations it is clear that there is a compression pressure beyond which practical considerations will cause a diminution rather than an increase in overall efficiency.

With present-day designs of semi-Diesel engines little further economy of fuel consumption is to be sought in this direction. Designers adopt various compressions according to the means foreseen for attaining flexibility and depending on whether the water drip is retained or not. The effect of increase of compression on economy between the limit of 180 lb. per square inch and 450 lb. or 500 lb. per square inch is not, per se, considerable, due to the practical considerations outlined. The principal effect is the possibility of sustaining

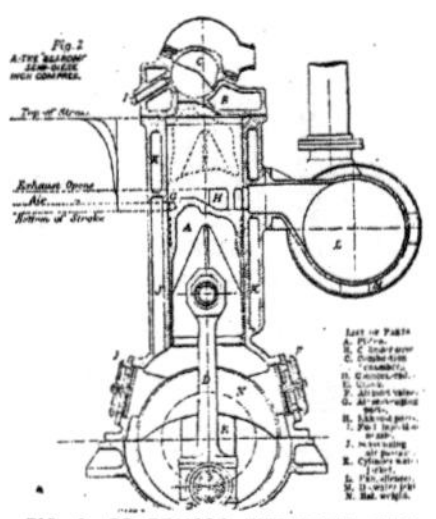

FIG. 2.—BEARDMORE TWO-CYCLE SEMI-DIESEL ENGINE.

higher mean effective pressures with smaller cylinders for a given output, and so attaining somewhat better fuel economy.

Fuel Economy

The fuel economy of semi-Diesel engines is surprisingly good (see Fig. 4), and is accounted for by the cycle of operation being nearer to the more economical explosion cycle than to the constant-pressure burning or Diesel cycle (see Fig. 5).

Cycle of Operation

Actual indicator diagrams do not quite so rigidly follow the theoretical cycle as the gas engine, on account of the difficulty with semi-Diesel engines

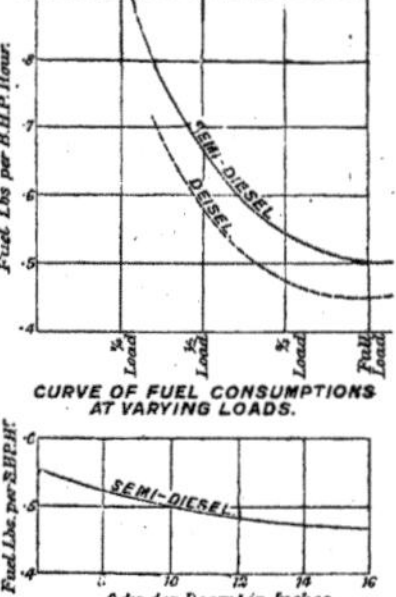

FIG. 4.—FUEL CONSUMPTION OF SEMI-DIESEL ENGINES. FUEL CALORIFIC VALUE OF 18,500 B.T.V. PER POUND.

of regulating the injection and ignition for all loads.

Flexibility

Experience of the operation of internal-combustion engines teaches that this prime mover is primarily a constant speed, and to a somewhat lesser extent a constant load engine. Innumerable designs of details and countless patents have been concerned, with the problem of flexibility and compromise is generally the outcome. Flexibility can be considered under three headings:

(a) Constant mean effective pressure, with varying revolutions and consequently power.

(b) Constant speed of revolution and varying mean effective pressures and power.

(c) Varying speeds and mean effective pressures.

Condition (a) is not required in practice, and cannot normally be met, with maximum or even full load m.e.p., since with a reduction in speed of revolution conditions affecting scavenging efficiency and compression, heat loss, &c., also change, and a small drop in speed of revolution is accompanied by a reduction in m.e.p., and so by a cumulative falling-off in power developed. A low m.e.p. can, of course, be maintained as a constant over a certain range of speed revolution.

Condition (b) constant speed of revolution and varying power as affecting generator engines, &c., require most frequently to be met, and may be considered in detail. At reduced power and m.e.p.'s—

(1) The charge drawn into the crank chamber remains relatively constant in volume or may even be slightly augmented, due to the engine running cooler as the mean effective pressure falls, unless means are provided to throttle the water cooling supply.

(2) The volume of the scavenging charge is approximately the same as at

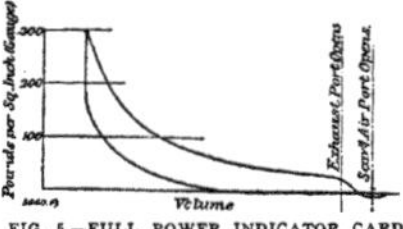

FIG. 5.—FULL POWER INDICATOR CARD OF SEMI-DIESEL ENGINE
Maximum Pressure—300 pounds per Sq. In.
Comp. Pressure—185 pounds per Sq. In.
M.E.P.—49 pounds per Sq. In.
M.E.P. on Brake Horse Power Basis—38 pounds per Sq In.

full power, but may be at a lower temperature and pressure.

(3) The compression pressure will be reduced on account of: (a) Lower scavenging pressure (see 2); (b) less heat abstracted from the cylinder walls, which in turn is due to the less fuel burnt per stroke and so the lower temperature of these walls. Condition (c) requires to be met with various types of machinery, and no difficulty is experienced provided the power of the engine is suitable for its work and a higher m.e.p. is not demanded that can be sustained for the speed of revolution under consideration.

Even where means are provided to throttle the cooling water and the scavenging air at low power, the point is quickly reached where the heat of the bulb is insufficient to vaporize and ignite the charge of injected oil, and the engine will "miss" and stop unless heat be externally applied to the bulb as, for instance, by the blow lamp.

Range of Working

The range of working must be extended to cover from full load or overload to a small load without having recourse to the blow lamp, and for this purpose the water drip has been retained on some designs. At full load water is allowed to enter the working cylinder with the scavenging air and serves by evaporation to take heat from the bulb, so that with a relatively large bulb and a low compression engine, from three-quarters to full power can be satisfactorily developed without overheating of the bulb, and with the water drip cut off the engine will run satisfactorily down to low loads. An overheated bulb will give bad combustion and "coking" of the fuel, and is, besides, a source of danger due to weakening of the metal of the bulb (see annexed Table).

Consumption of Water

The consumption of water through the water drip is very considerable, and varies according to the quality of attention given to the running of the engine, but may reach a value at high powers much in excess of the quantity of fuel burnt. The water should be as pure as possible to cause the minimum harm from deposits on the working surfaces. Water has a deleterious influence on the lubrication of the internal parts, although it is credited with preventing carbonizing of the main piston rings. The water drip, however, is a crude solution of the problem of flexibility, requiring a large supply of fresh water, and with varying loads, regular attendance to the engine, since it is somewhat difficult and calls for complicated gear to connect the water supply with the governor in the same way as is necessary with the fuel supply.

The better solution is to take advantage of another law which is not yet completely explained, viz., that the temperature generated within the cylinder of an internal-combustion engine depends on the load, the compression temperature, and upon the nature of the ignition, whether early, normal or late. Normal ignition may be said to be that ignition which is correct for maximum economy and will give the highest power without trouble, the cleanest exhaust, the sweetest running, &c. Late ignition makes for excessive heat losses to the exhaust and high fuel consumption. Early igniton gives rise to abnormally high temperatures. This last fact is utilized with semi-Diesel engines to counteract the cooling of the bulb with reduced power. By advancing the point of ignition of the fuel charge, as the quantity of fuel is reduced to correspond with the load, the semi-Diesel engine can be made to give satisfactorily running at all loads from full load to no load with the minimum of attention, and without requiring heating of the bulb (see Fig. 6). The governor controls the quantity of fuel to correspond with the load by varying the stroke of the fuel pump, and gears have been designed whereby with reduced quantity of fuel the injection point is advanced according either to B or C in Fig. 6. Scheme C is most necessary for engines requiring to run for long periods at light loads, whilst B suffices generally; C is less easy of attainment by a simple gear.

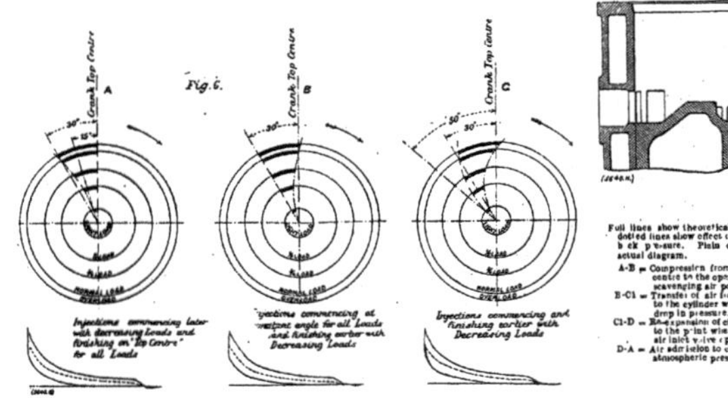

Fig. 6.—Diagram showing control of point of commencement and period of injection for varying quantites of fuel and load, gving corresponding indcator diagrams.

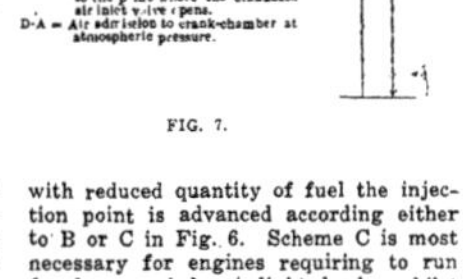

FIG. 7.

Scavenging

The next point of importance is the question of scavenging, which, so far as published data or the results of experimental work are concerned, is almost an unexploited field, in connection with either the two-cycle Diesel or semi-Diesel engine. With two-cycle engines the efficiency of scavenging is lower than with four-cycle engines, which has proved one of the most important deterrents in all spheres of application to that success so often predicted in the past for the two-cycle principle. With two-cycle semi-Diesel engines the amount of air available per working cycle or per revolution for scavenging is limited to the volume swept by the working piston. More air than this cannot be drawn into the crank chamber (unless an induction system to the crank chamber were so designed and fitted, as to give a momentum effect with a slight gain, which subject has not yet been studied for other than high-speed four-cycle engines where the maximum output per units volume is essential). The air, after being drawn into the crank chamber, is impregnated with a certain amount of lubricating oil, as will be discussed later under the heading of "Lubrication," and withdraws a certain heat from the working parts of the engine, especially from the piston. An indicator card and the theory of scavenging is given herewith (see Fig. 7).

Efficiency of Scavenging

Scavenging efficiency can be subdivided under two headings, the efficiency of the pump and the efficiency of scavenging of the working cylinder. As seen from the comparison of the ideal with the working indicator diagram, Fig. 7, there are several losses in the pump:—

(1) Suction loss due to attenuation of the charge.

(2) Compression loss due to leakage through the main bearings.

(3) Loss of volumetric efficiency due to the heating and re-expansion of the clearance air.

To take the three points in order; suction loss requires no explanation. Loss due to leakages through the bearings is now reduced to a minimum with careful design of the air rings and good workmanship. Fig. 8 shows an arrangement which has proved satisfactory. The clearance volume should be kept a minimum although with this type of engine, this volume is always large even when the crankshaft is fitted with balance weights to give better balance and to minimize this clearance volume, and where the clearance for the crank and bottom end is cut fine, since the air must have access to the piston crown for cooling purposes. It is probable that the greatest loss of efficiency is not in the scavenging pump, but concerns the scavenging within the working cylinder.

The effect of the shape, and the size of the scavenging and exhaust passages and ports has not been fully studied, although it can be stated that wide variations in both are possible without any appreciable effect on the performance in practice of the normal semi-Diesel engine, the main considerations being, as would clearly be inferred from the indicator card, Fig. 7, to get rid of the exhaust at a suitable point as rapidly as possible, and that back pressure of exhaust must be reduced to an absolute minimum. Back pressure has an effect in preventing the entrance of scavenging air to the working cylinder, previously compressed to a pressure dependent in some measure upon the speed of revolution. The scavenging air pressure cannot increase to overcome exhaust back pressure as with separate valve-controlled scavenging pumps. The chief effect of back pressure, however, is to decrease the quantity of air drawn into the crank chamber to be compressed. The pressure from which the clearance air in the crank chamber must expand is the back pressure of the exhaust, and until this clearance air has expanded down to the suction pressure no fresh air will be drawn into the crank-case. The volumetric efficiency is entirely dependent upon this factor, as is clearly shown in Fig. 7. Large exhaust ports, ample passages, the close proximity of a large silencer to the cylinder and the minimum of restriction in the exhaust pipes, are necessities.

The two-cycle semi-Diesel engine has almost settled down to a standard design of scavenging and exhaust passages and piston crown without any proof other than that of satisfactory performance, although still at relatively low efficiency. Much experimental work still remains to be done on this subject. This

4-CYLINDER SEMI-DIESEL ENGINE BY WILLIAM BEARDMORE & CO., LTD., GLASGOW.

course involves considerable labor and expense due to the large numbers of variables that have a direct influence, of which but a few will be mentioned. Piston speed must have an effect on the efficiency of both the crank chamber scavenging pump and on the cylinder scavenging. Experience has shown that 800 ft. per minute approximately is the maximum piston speed, above which with present designs the scavenging efficiency falls off somewhat rapidly, the power output from the engine fails to increase with higher revolutions and increased fuel, and the limiting condition of maximum powers are reached.

With this question is intimately associated the subject of the stroke-bore ratio; it can be said that the higher the stroke-bore ratio, the better the conditions of cooling of the piston, because the smaller the diameter of the piston for a given power, and so the shorter the path for the heat to travel from the centre to the cooled walls; also the lower number of revolutions for the desired output of power gives certain advantages for driving types of machinery which are inherently slow speed machines. Furthermore, the author's experience suggests the larger the stroke-bore ratio within the limits of ratio of 1 to 1 and 1.5 to 1, the less, probably, the escape of scavenging air through the exhaust ports, due to a greater quantity of fresh air being entrapped in the combustion chamber, i.e., with a square engine — an engine of approximately equal stroke and bore—a greater percentage of the scavenging air finds its way out through the exhaust ports.

The shape and angle of entrance to the cylinder of the air inlet passages, the type of baffle on the piston crown and the location of the bulb in relation to the path of the scavenging air all have an influence.

Injection

To turn to the question of injection, which depends primarily with a solid injection semi-Diesel engine, on the following factors:—

(1) Turbulence within the cylinder.

(2) Pressure of fuel and rate of injection.

(3) Point of the cycle at which injection occurs.

(4) Fineness of the spray.

(5) Distance of injector from the hot igniting surface.

These points are not given necessarily in order of importance. Turbulence, apart from piston feed which is governed as already stated by consideration of scavenging, is determined by the shape of the piston crown and the combustion chamber, and in the immediate vicinity of the spray by the shape, speed and volume of the spray. The question of turbulence is an exact parallel to that of scavenging and as yet has received little attention, excepting for the experiments arising out of the necessity to burn tar oils in Diesel engines.

The pressure of injection is governed by the piston speed of the fuel pump or by the angle of crank revolution allowed for the injection of the fuel and by the size of the orifice or orifices in the injector. Pressure is necessary more to give momentum to the stationary column of oil than to secure fineness of the injection spray. It is an absolute essential that injection shall be as rapid as possible, and the injection devices so designed that no "after drip" takes place.

ARRANGEMENT OF RINGS FOR SECURING AIR TIGHTNESS OF CRANK CHAMBER SHEWING ALSO SYSTEM OF LUBRICATION OF MAIN AND CRANK PIN BEARINGS. Fig. 8.

FIG. 8.—LIST OF PARTS.

A—Cylinder and Top Chamber.
B—Soleplate.
C.—Crank Shaft.
D.—Main Gearing Cover.
E.—Main Gearing Oil Tubes.
F.—Main Gearing Bushes.
G.—Main Gearing Bushes—White Metal.
H.—Main Gearing Bushes—Oil Tubes.
I.—Centrifugal Oiler.
J.—Airtight Rings.
K.—Airtight Rings Springs.
L.—Airtight Rings Driving Pins.
M.—Balance Weights.

The angle of revolution allowed for the injection period is governed by the ratio of diameter to stroke of the fuel injection pump which injects the fuel through a non-return valve or valves direct into the hot bulb. Obviously with a large diameter and a small stroke fuel pump the period is short and conversely. Practical considerations of design of the pump and governing mechanism determine the period for full power running to be about 30 deg. (see Fig. 6). As regards fineness of spray, no standard has been fixed, although the spray can certainly be "too fine" for rapid ignition.

The next essential is probably that the whole of the oil should be in the combustion chamber in the form of a spray before the first particle touches the hot bulb, which again is a function of rapidity or injection and of the distance through which the oil is thrown, arguing in favor of a long-distance of throw to give rapid ignition and the maximum degree of turbulence in the vicinity of the spray, although a long throw will militate against flexibility. The efficiency of the scavenging will determine to what extent the hot bulb is charged with burnt gases or with fresh air, and will thus have a direct influence upon the speed of ignition and combustion.

Fuel

In view of the simplicity of this type of engine in comparison with the usual four-cycle internal-combustion gas or Diesel engine, it might be matter for surprise that the semi-Diesel engine has not made greater headway in the past than has been the case. The outstanding difference between the semi-Diesel and the Diesel engine has been the small range of working fuels with which it could satisfactorily cope. It is but a few years since the great majority of these engines almost required paraffin or the very lightest of petroleums for their success operation in practice with reasonable costs for upkeep and maintenance. Recently, however, the advantages of simplicity of this engine have been more generally recognized and have led its producers to experiment on the question of utilizing fuel oils of a heavier nature, which have been more readily procurable within the last few years, thus extending greatly the field of application. This movement has been largely responsible for many attendant improvements such as increasing the compression pression.

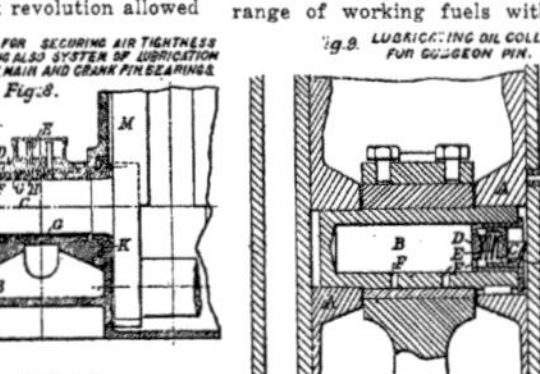

FIG. 9.—LIST OF PARTS.

A.—Piston.
B.—Gudgeon Pin.
C.—Oil Collector.
D.—Oil Collector Casing.
E.—Oil Collector Spring.
F.—Oi Holes.
G.—Oil Tube from Lubricator.

The present stage of development of the semi-Diesel engine permits it to use most of the heavy fuel oils ranging from 0.8 to 0.9 in specific gravity and with flash points from 130 deg. to 250 deg. F. Perhaps the most frequently used oils are "Solar" and "Shale," but paraffin at the one end of the scale and Texas at the other may be said to be quite suitable without special adjustments or contrivances. Very thick oils such as "Mexican," for example, may be used, but require preheating to facilitate pumping, and periodical runs on lighter oils are desirable in such case in order to keep the pipes clear, the pistons clean, and the piston rings free in their grooves. With regard to sulphur, the semi-Diesel is no more sensitive than the Diesel engine. Experiments are being carried out at the present time in order to permit of the use of tar oil.

A note should be made in connection with the subject of burning heavy fuel oils with semi-Diesel engines, that this engine being a "solid" injection engine does not run with such a clean exhaust as is customary with air-injection engines, and the amount of overhauling required for cleaning of piston rings, etc., is on that account greater, and, of course, is increased with the heavier oils as compared with shale oil and such lighter oils. The user must, therefore, balance the gain of cheap and readily-obtained fuels with the extra overhauling which may on that account be required, taking into account the size of the engine as to whether the parts to be handled are of convenient size and weight.

Lubrication

On the subject of lubrication, Fig. 8 shows the means provided for the main and the crank-pin bearings, and Fig. 9 illustrates a satisfactory device for the lubrication of the connecting-rod top end bearing, whereby the oil is collected from the cylinder walls and conveyed to this bearing. The lubrication of the cylinder walls is carried out in exactly the same manner as is customary with Diesel engines, a special lead being provided for the top-end bearing. Forced lubrication to the main, crank-pin and top end bearings cannot be used with semi-Diesel engines, so long as crank-case scavenging is utilized in order to avoid excessive impregnation of the crank chamber air with lubricating oil.

The qualities of lubricating oil desirable for semi-Diesel engines differ in no way from those required by the Diesel engine, and, with careful design of the airtight rings shown in Fig. 8, and good fitting piston rings, the consumption of lubricating oil compares favorably with the figure for two-cycle Diesel engines, and is in the neighborhood of 0.2 lb. per brake horsepower per hour.

Starting

Semi-Diesel engines are started by means of compressed air. Due to the low compression pressure, a comparatively low pressure of starting air is sufficient to ensure reliable starting. The minimum pressure at which the engine will start is from 80 lbs. to 100 lbs. per square inch, and the starting air is generally stored at 200 lbs. per square inch, which, with a suitable volume of storage, gives the requisite number of starts, or a margin for contingencies. Prior to starting the hot bulb is brought to a sufficient temperature to ignite the fuel, which is accomplished by means of a blow lamp in from 10 minutes to 15 minutes. If the engine is provided with special starting plugs of nickel steel screwed in the hot bulb, very much less time is required. These special plugs quickly attain the necessary temperature for ignition. One impulse is generally sufficient to start the engine, so that operating gear for the starting air valves, other than a hand lever, is not customarily fitted with land engines, excepting occasionally in the case of four-cylinder engines. For the compression of starting air a separate hand or power-driven compressor can be installed. The practice with semi-Diesel engines is, during each working stroke to tap off a portion of the working gases through a combined non-return and screw-down valve on the main cylinder, and to pass them to the starting reservoir or reservoirs until such time as any reduction of contents of these reservoirs has been made good.

For multi-cylinder engines which require to start against a load, as for instance those driving pumps or propellers, the same mechanism can well be fitted as with the Diesel engine, the starting air valves being operated either from a main camshaft or through a distributing box, with a secondary camshaft driven from the crankshaft. The quantity of air storage requisite for starting is generally considerably less than is required by Diesel engines, since the hot bulbs have, of necessity, been heated prior to turning the engine, and one revolution on compressed air is sufficient to start the engine.

The quality of reversibility, which is almost exclusively required for marine engines, presents few difficulties where 2-cycle semi-Diesel engines are concerned, and some notes in regard to this subject are given in Appendix II:

Reliability and Regularity in Operation

As would be expected from the extreme simplicity of this prime mover, its reliability and regularity in operation are of a high order. Due to the necessity of restricting the quantity and minimizing the pressure of lubricating oil as already dealt with under the heading of "Lubrication," earlier designs were subject to bearing troubles attributable to the failure of the lubricating system. With modern designs ample bearing surfaces and carefully-designed means for the provision of the necessarily restricted quantity of lubricating oil, these defects have been entirely eliminated.

The system for fuel injection due to the adoption of "solid" or "mechanical" injection is considerably simplified in comparison with air injection engines.

Faulty circulation of cooling water and unsatisfactory designs of castings —more particularly those for the cylinder head and hot bulb—have been the cause of a certain amount of trouble with cracked heads and so forth; but modern designs have practically overcome this earlier source of unreliability.

Conclusion

The rapid extension within the last few years not only of the field of application, but also of the size of engine and power developed per cylinder with semi-Diesel engines, foreshadows considerable developments in the near future. In these developments the influence of the design and practice of the pure Diesel engine will probably play a considerable part, and it may be expected that the lines of design of the Diesel and semi-Diesel engines will become more closely merged.

Practical difficulties would seem almost to confine the semi-Diesel engine to the two-stroke cycle. Developments towards improving the efficiency of scavenging may well be expected. In the United States of America semi-Diesel engines with separate scavenging pumps, crossheads, and so with forced lubrication, are already making their appearance.

The mean effective pressure developed by the semi-Diesel engine under conditions of continuous running are considerably less than those associated with the Diesel engine, due primarily to the question of scavenging and compression already fully discussed. It is not expected that other than a slight increase in mean effective pressures can be looked for in the near future, since the simplicity of the semi-Diesel engine and its relatively low compression pressure will probably be substantially retained. The low mean effective pressure in reducing heat stresses and temperature conditions within the cylinder makes for reliability in operation.

The vexed question of "solid" injection of the fuel becomes a simpler issue when associated with surface ignition. All considerations in the design of the fuel pump operating and controlling gear, and the injection means for semi-Diesel engines, have in the past been subservient to that of simplicity. With a demand for the same degree of flexibility, and a capacity to burn as wide a range of fuels, without recourse to the water drip, as obtains with the Diesel engine, considerable improvements after the war can confidently be anticipated.

APPENDIX I.

Mechanical Efficiency of Internal-Combustion Engines

The following notes have special reference to internal-combustion engines of the trunk piston type; but apply equally, with slight modifications, for crosshead engines.

Mechanical efficiency (the ratio between brake horse-power and indicated horse-power) is affected by the number of auxiliaries which are driven by the main engine.

1. Except in so far as auxiliaries are concerned, the difference between the indicated horse-power and the brake horse-power can be apportioned as follows:—

(a) 50 per cent. is due to piston and piston-ring friction.

(b) 28 per cent. can be attributed to main cylinder pumping losses, suction, exhaust and scavenging.

(c) 22 per cent. is allocated to valve gear and bearing friction, &c., in which are included windage losses and other factors of little importance.

2. Piston friction depends primarily on the following factors:

(a) The quality of the metal of the liner, the piston and the piston rings.

(b) The quality of the lubrication. (Certain tests which have been carried out to prove that a diminution in viscosity of oil increases the mechanical efficiency. In one case the mechanical efficiency was increased by water injection into the combustion chamber.)

(c) The clearance between the piston and the cylinder walls has an influence on efficiency.

(d) The m.e.p. the compression pressure, and the pressure, and the pressure between the liner and the piston, and the liner and the piston rings, can probably have a most suitable value for the reduction of friction loss to a minimum.

(e) The fit and the condition of the piston rings.

(f) The temperature at which the engine runs will have an effect on the lubrication and on the clearance; and it has been substantially proved that there is a temperature of maximum mechanical efficiency.

3. The suction loss, 28 per cent. of the total, is primarily a function of design of ports, valve setting, piston speed and gas speeds.

4. The valve gear and the bearings, 22 per cent. of the total loss, will depend on the design of the engine, the alignment, the efficiency of the lubrication.

In addition to the foregoing, there are records of mechanical efficiency being reduced by increased weight of flywheel.

Generally, mechanical efficiency is adversely affected by increased speed and reduced m.e.p.; and decreased by mal-alignment, &c.

The mechanical efficiency may be affected by the form of the combustion chamber which may produce undue distortion of the piston under working conditions, although this is probably extremely slight; distortion of the piston being more due rather to the condition of the gudgeon pin bearing than to any other cause.

The mechanical efficiency, assuming a constant m.e.p., is practically unaffected by the size of the engine.

In connection with the above, a large number of records of tests of engines have been investigated from Guldner, Supino, D. Clerk, &c.

Halifax, N.S. — The Federal Line intends to put two more steamships on the freight route, New York, Halifax and St. John's, Nfld.

The "Pioneer," the keel of which was laid on July 18 last, is of 8,100 tons deadweight, has twin decks and a speed of eleven and a half knots. The keel of the "Voyageur" was laid in March. She is of 4,350 tons burden.

Buffalo.—In the presence of several Shipping Board officials the bulk freighter "Charles R. Van Hise" was rolled on her side here preparatory to being towed to Montreal. The job was done by two tugs and was highly successful.

The "Van Hise," which was 446 feet long and 50 feet wide, was purchased from the Pittsburgh Steam Ship Company by the Emergency Fleet Corporation. As she was too wide to take through the locks of the Welland Canal the scheme of rolling her on her side was hit upon.

Description of the New C.N.R. Car Ferry "Canora"

Built by Davie Shipbuilding Co.—To Run Between Port Mann and Patricia Bay—Now on Her Voyage to Pacific Coast

THE new car ferry, built by the Davie Shipbuilding and Engineering Co., Levis, Quebec, which was launched in June last, left Quebec on October 1st last for the Pacific Coast. She is going via the Panama Canal, and should now be close to her destination.

This vessel has been built to meet the requirements of the above mentioned railway company for the transportation of passenger and freight railway cars between the railway terminals on Vancouver Island, B.C., and the mainland to points in the vicinity of the City of Vancouver, B.C.

General Design

The type adopted is somewhat similar to that of the car ferries operating on the Great Lakes with the exception of a rolling gate, which is fitted at the stern for the purpose of closing in the space between decks, where the railway cars are carried.

The following are her leading particulars:

Length overall, 308 ft. 0 ins.
Length B.P., 294 ft. 0 in.
Breadth moulded, 52 ft. 0 in.
Depth moulded to car deck, 20 ft. 6 in.
Depth moulded to shelter deck, 38 ft. 8 in.
Draught loaded, 14 ft. 6 in.
Displacement at above draught, 3,400 tons.
Speed on service, 14 miles.
Number of cars carried, 20.

The "Canora" has been constructed under the supervision of Lloyd's Register of Shipping, and is classed 100A, as a train ferry for coast and river service. The general design of the vessel being the work of the railway company's naval architect, Mr. A. Angstrom.

The vessel is constructed on the transverse framing principle, open bottom, type and is sub-divided into six main transverse watertight compartments by five watertight bulkheads. Watertight doors are fitted for communication between the engine and boiler spaces and shaft tunnel.

Water ballast is provided for in peak tanks forward and aft and in trimming tanks on each side of the engine room.

The cars are carried on the main or car deck on three lines of tracks, one line of tracks being on the centre line of vessel and one line each side of centre. The spaces below this car deck are devoted to machinery spaces, crew spaces, stores, holds, coal bunkers and steering compartments.

Above the car deck at a height of eighteen feet there is a complete shelter deck extending the full length and width of the vessel, and on this deck accommodation for passengers and officers is provided.

The Accommodation

The accommodation includes rooms for all officers, large dining saloon, parlor, state rooms for passengers, smoking room, kitchen and pantry, bathrooms and lavatories, and a large observation cabin at the forward end.

Above this accommodation is located the pilot house, and at the stern a pilot house is provided for use in docking the vessel.

The rooms are all tastefully finished and have berths, clothes closets, wash basins, etc., to each room.

The dining saloon is finished in oak panelling and has a large dome over the centre with borrowed lights extending all round dome.

The crew spaces are provided with all necessary accommodation for seamen and firemen, including berths, lockers, etc.

The sanitary arrangements are of the most modern description and provide for a complete service of fresh, salt and hot water throughout vessel.

The heating and ventilating system is most efficient, steam heating being installed in all rooms in accommodation. The ventilation to all spaces is provided by natural means, through patent ventilators carried well above roof of shelter deck accommodation.

For use in case of fire a complete installation of fire extinguishing pipes is provided.

The electric equipment is of the latest type, the generators are placed in the engine room. The main switchboard is located conveniently to the generators, the whole of the wiring and fixtures are of modern design.

Two searchlights are fitted for use when vessel is landing at the slips at night.

The vessel having to go astern for a distance on her run, she has been designed with propellers at both ends; also steering gears and rudders, and in connection with the arrangement the navigating lights, engine room, telegraphs and steering standards are arranged to automatically change over to suit this condition.

The life-saving appliances are sufficient to meet the requirements of all on board and are in accordance with the requirements of the Canadian Government inspection.

Six lifeboats are carried on the shelter and boat decks, with two davits and gear to each boat.

The auxiliary deck machinery includes a large steam windlass situated on the shelter deck for handling the anchor cables, windlass also being provided with drums for use in handling the wire ropes for mooring the ship.

The stern gate is operated by a steam winch at the after end of the shelter deck, gate being carried on girders on this deck.

When gate is closed the stern is completely closed in between the car deck and shelter deck, where the cars are carried.

Machinery

The main propelling machinery has been constructed by Messrs. John Inglis of Toronto, and consists if a four cylinder triple expansion surface condensing engine balanced on the Yarrow, Schlick & Tweedy system, having cylinder 24 in., 38 in., 43 in. and 43 in., with a stroke of 30 in. and indicating about 2,200 horse power.

The engine is arranged to drive a

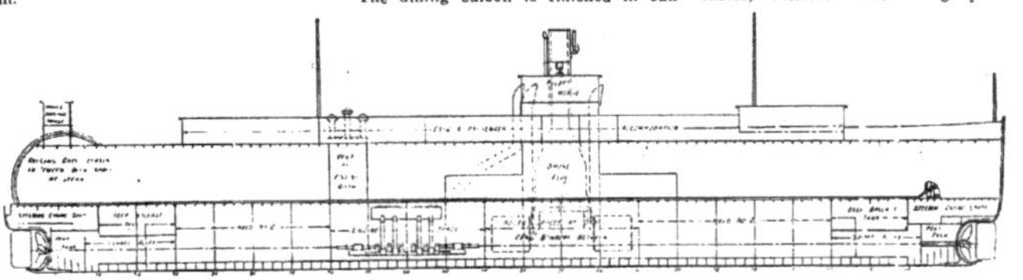

LONGITUDINAL SECTION OF S. S. "CANORA"

ON THE STOCKS

screw propeller at each end of the vessel, the shafting running the full length of the ship.

Steam is supplied by four boilers of Scotch Return Tubular type, 11 ft. 6 in. diameter by 11 ft. 6 in. long, working at a pressure of 175 lbs. per square inch, located in two boiler rooms, one on each side of ship.

Each boiler has two corrugated furnaces, 41 inches diameter, and a complete installation of Howden forced draught is fitted.

The total heating surface for the four boilers is 5,500 square feet.

The surface condenser is of the Weir triangular type, and has a cooling surface of 2,200 square feet.

The circulating pump for main condenser is driven by its own engine and is of centrifugal type.

The auxiliary machinery is of the most up-to-date design and includes two vertical Blake boiler feed pumps, each having capacity for working the four boilers, sanitary pump, fresh water pump, bilge pump and ballast pump.

There is also an evaporating and distilling plant installed of capacity to make up loss in feed water and for drinking and galley supply.

Ash ejectors are fitted in each other room.

Two steam steering gears are provided in separate compartments at each end of the vessel, the valves on gears being operated from pedestals in pilot house by means of control shafting.

THE "LLOYD ROYAL BELGE"

War-scarred Belgium has taken the first step towards her commercial rehabilitation. As plucky in peace as she was in war, she is not waiting for others to help her out, but is helping herself. And in token of her appreciation of Yankee friendship and Yankee worth, she has chosen an American to hew the way for her.

This American is Joseph A. Nash, for four years general manager of the Shipping and Purchasing Department of the Committee for Relief in Belgium. He arrived recently from London in New York in the capacity of manger for the United States and Canada of the Lloyd Royal Belge. This is a purely Belgian enterprise, the first attempt on the part of that country at direct trans-Atlantic connection between her own shores and those of North America, and the first step in the direction of weaving a strong commercial and industrial bond between

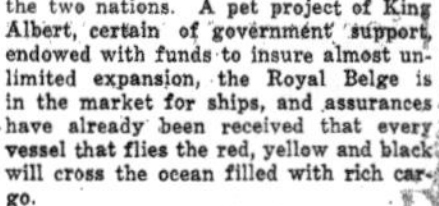

the two nations. A pet project of King Albert, certain of government support, endowed with funds to insure almost unlimited expansion, the Royal Belge is in the market for ships, and assurances have already been received that every vessel that flies the red, yellow and black will cross the ocean filled with rich cargo.

The main offices of the company just now are in Paris, where the line's president, M. Arthur Brys, makes his headquarters while his old home office, at Antwerp, is being prepared for his return. The American branch is located at 11 Broadway, New York City, and from there Mr. Nash is conducting his negotiations.

The Royal Belge was organized as a corporation after the war had broken out. At that time it owned forty ships, and its head, M. Brys, turned them over to war purposes. In his own country, and in France, he is considered one of the most remarkable figures of the day—immensely wealthy, yet always at work; indomitable in his determination to win for Belgium a commanding position as a commercial power and ready to sacrifice his own interests and his own wealth in the service of the nation. The leading men of Belgium look up to him as minor officers in an army look up to their general.

Of the original fleet of forty, only sixteen vessels are left. The other twenty-four were sunk by submarines or mines. Twelve of the remaining ships are temporarily flying the British flag, carrying supplies to France and being engaged in other similar work. The other four are still in the service of the Belgian Relief.

The Royal Belge has a large liquid capital of its own and a great additional sum has been guaranteed for its enlargement. The first trip, as soon as the service has been organized, is to be made between Antwerp and New York; and just as soon as this route has been fully established, the great Belgian port is to send vessels to Philadelphia, Norfolk, Baltimore, Charlestown and even Galveston. Later, Canadian ports may be added.

What Belgium needs first of all, Mr. Nash said, is American machinery. Cotton and textile fibres are next required. as well as immense quantities of tobacco. In return, Belgium will ship us her linens and needlework. Much of this is being manufactured even now, more than 60,000 women and men being engaged in these industries. For, despite the havoc wrought by the Germans, enough has been saved to enable these people to keep at work and produce marketable goods. Glassware, too, is one of Belgium's specialties and its manufacture will be resumed on a large scale just as soon as a proper survey can be taken of what has remained intact of the country's industrial facilities and what is in need of reconstruction.

LAUNCHING

Displacement Deadweight Gross and Tonnage

A Definition of the Various Terms Used in Ship's Measurements —These Terms Are Not Always Clearly Understood—Method of Measuring Register Tonnage Leads to Good Accommodation For Officers and Crews

By T. H. FENNER, Associate Editor

THERE is a considerable amount of misconception among people interested in maritime matters as to the definition of the various kinds of tonnage so frequently mentioned in connection with ships. This is not confined entirely to landsmen, although one would think that men constantly in touch with ships would understand these things. However, it is not so, and the writer has not infrequently heard more or less acrimonious arguments across the messroom table, in the second dog-watch, that couple of hours devoted to spinning yarns and discussing the various pursuits to be embraced in the next port, or if homeward bound, the merits of the different amusements open to the wanderer in his home port. In these days, when so many Canadians are taking an interest in ships, and accounts of buildings and launchings are appearing in the daily press, with the size of the vessels sometimes quoted in terms of deadweight, and again in terms of displacement, gross tonnage, or register tonnage, an explanation of these different terms should be opportune, and fix clearly in the minds of those concerned the real meaning to be attached to the kind of measurement quoted.

Displacement

The term displacement is the most simple of them all, as it expresses exactly what it stands for. It means the actual amount of water displaced by the vessel, when fully loaded, and floating at her normal load line. If we take a box one foot wide by one foot long, and floating with a draught of one foot, the space occupied by it in the water will be just one cubic foot, and one cubic foot of water has been displaced to allow the box to be immersed to that extent. One cubic foot of fresh water weighs 62.5 lbs., and one cubic foot of salt water 64 lbs., so that if floating in salt water the box will have a weight of 64 lbs. The same box floating in fresh water would sink .024 feet deeper before being held up by the water. However, in dealing with ships, we usually take the salt water figure. Provision is made to take into account the difference in draught caused by a vessel going into fresh water by painting both the fresh water load line and the salt water one, so that a ship loading in a fresh water river will, on reaching the ocean, automatically rise to the salt water load line. A cubic foot of salt water weighs 64 lbs., 35 cubic feet weigh one ton, of 2,240 lbs. If we know the length, breadth, and draught of a vessel, together with her block co-efficient, we can at once determine her displacement, or what is exactly the same, her weight. The block co-efficient is the fraction that represents the proportion of the volume of the hull, with its curved sides and tapered ends, to the volume of a rectangular block of the same mean width and length. This co-efficient will vary from .4 in a very fine lined yacht to .8 in a full modelled cargo vessel. Let us take a vessel of ordinary dimensions, say 300 feet long by 45 feet beam and floating at a mean draught of 22 feet. Assume her block co-efficient to be .75, we have then

$$\frac{300 \times 45 \times 22 \times .75}{35} = 6,337 \text{ tons}$$

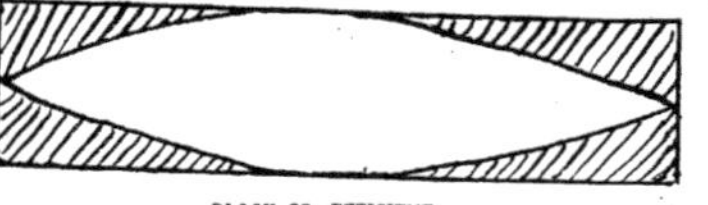

BLOCK CO—EFFICIENT

Therefore the total weight of the hull, stores, coal, and cargo is 6,337 tons, and we say she has a displacement of 6,337 tons.

Deadweight

Deadweight tonnage represents the difference in weight of the vessel, when floating with only her coal, stores, and reserve fresh water aboard, and when loaded with a cargo that brings her down to her load waterline. In other words, it represents her cargo-carrying capacity, upon which in most cases depends her earning ability. Obviously some cargoes will fill a ship's holds without bringing her down to the loadline by a considerable amount. A ship loaded with cotton, for instance, may be full, very full, and yet not have anything like her deadweight on board. In order to get the last available bale of cotton on board it is stowed into all manner of places, even, so it is said, to putting a couple of bales into each officer's and engineer's cabin, and half a dozen in the messroom. That, probably, is an exaggeration, but the writer has been in a vessel loading cotton in New Orleans, where it was packed under the poop, in the alleyways under the bridge deck, and in the small available space under the fo'c's'le head between the sailors' and firemen's fo'c's'les. On the other hand, taking a cargo of steel rails, barbed wire, corrugated iron, etc., from the Bristol Channel to Buenos Ayres, the vessel was down to her load line with the holds far from full. A good example of homogeneous cargo is what is known as general cargo, such as a vessel carries from the United Kingdom or the States out to Australia and New Zealand, comprising case oil, dry goods, pianos, organs, benzine in cases, and various other merchandise.

To take the case of the vessel already quoted, supposing she is loaded with her coal, stores, etc., and is floating at a mean draught of 8 feet, then her displacement in that condition would be 2,314 tons. As her displacement at 22 feet mean draught is 6,337 tons, her deadweight cargo capacity is 6,337 — 2,314 = 4,023 tons, say 4,000 tons.

Gross Tonnage

We now come to the consideration of gross tonnage and here we come to a distinctly different meaning of the word tonnage. Displacement and deadweight tonnage convey their meaning in terms of actual weight, but gross tonnage is a measurement of space, and does not directly have any connection with weight. It is a measurement of capacity. In gross tonnage is included the term underdeck tonnage, or rather the underdeck tonnage is included in the measurement of gross tonnage. One ton of underdeck capacity is equal to 100 cubic feet of space. This is the first part of the vessel measured for space,

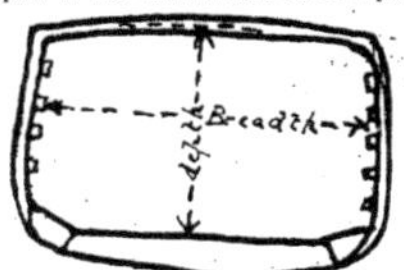

MEASUREMENTS FOR UNDER DECK TONNAGE

and is the total tonnage up to the tonnage deck. The tonnage deck is the top deck in all vessels having less than three decks, and in the case of vessels with more than three decks, up to the second deck from below. The measurement is obtained as follows: The depth of the vessel is taken at any part of the length, from the top of the floors to one-third the camber of the beam of the tonnage deck; the depth of the ceiling is then deducted, this will be from two to three inches. In the case of a vessel having double bottoms the depth is taken from the beam to the top of the tank, and allowance made for the ceiling as before. If the ceiling is not laid directly on the tank top no allowance is made for the depth between ceiling and tank top. In the case of a vessel having double bottom in one part and not in another, which sometimes happens, the two parts are measured separately. The breadths for underdeck tonnage measurements are taken from the inside of the sparring in the holds, or in the case of a vessel having no sparring, from the inside of the framing. The lengths are taken from the points at each end where the framing meets, or where the lines of the sparring come together. The cubic contents are determined from these measurements, the mean breadth being found by Simpson's rules. To find the total gross tonnage, the space occupied by all enclosed erections above the tonnage deck is measured. The usual spaces measured will be the poop, if closed with doors or bulkheads, forecastles, bridges if closed at the ends, engine and boiler room casings, officers' and engineers' quarters. Shelter or awning decks, if not closed at the ends, are not included, and generally the crew's galley, lavatories, and companion ways are excepted. In the case of vessels with a raised quarter deck, the line of the tonnage deck is taken from where the main deck stops and the raised poop begins, and the raised quarter deck is then calculated separately.

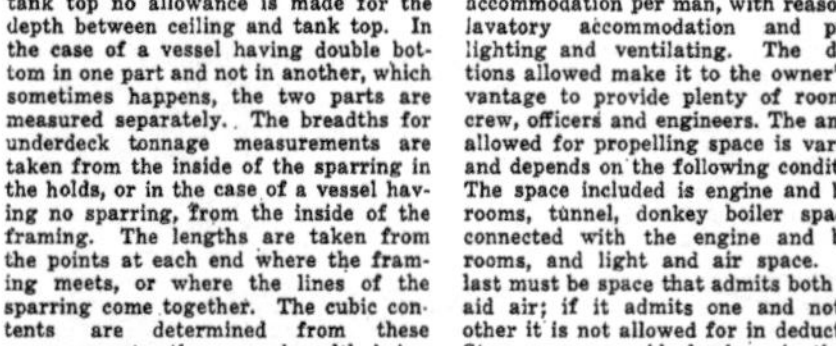

TONNAGE FOR RAISED QUARTER DECK TYPE

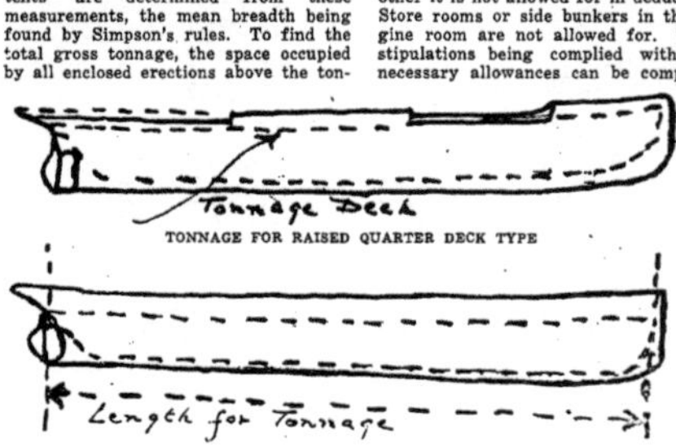

LENGTH MEASUREMENT FOR UNDER DECK TONNAGE

Register Tonnage

Register tonnage is most important to the shipowner as it is on this that the dues are charged at the various ports entered, and in many ports the pilotage fees are based on the register tonnage. The register tonnage is found by making certain deductions from the gross tonnage, comprising closed-in spaces for various uses. Any space that has not been included in the gross tonnage will not be allowed for in the deduction for register tonnage. The spaces included are crew accommodation, master's accommodation, officers' and engineers' rooms and messrooms, bo's'un's stores, sail room, donkey boiler space, if not in engine and boiler room space, chart rooms; passenger accommodation is also deducted. The Board of Trade regulations call for a space of 72 cubic feet and 12 square feet floor space of crew accommodation per man, with reasonable lavatory accommodation and proper lighting and ventilating. The deductions allowed make it to the owner's advantage to provide plenty of room for crew, officers and engineers. The amount allowed for propelling space is variable and depends on the following conditions: The space included is engine and boiler rooms, tunnel, donkey boiler space if connected with the engine and boiler rooms, and light and air space. This last must be space that admits both light aid air; if it admits one and not the other it is not allowed for in deductions. Store rooms or side bunkers in the engine room are not allowed for. These stipulations being complied with, the necessary allowances can be computed.

In screw steamers having a propelling space up to 20 per cent. of the gross tonnage the allowance made is 32 per cent., and in paddle steamers with a propelling space up to 30 per cent. of the gross tonnage, or the Board of Trade may allow 1½ times the actual space in the case of paddle vessels and 1¾ the space in screw steamers. In the case, however, of paddle vessels with 30 per cent. or over of the gross tonnage space, and screw vessels with 20 per cent. or over, the owner is allowed a choice of the two alternative methods. Either method may be of most advantage to him, according to the type of vessel. In large passenger vessels, going short voyages, and not requiring great bunker space, omitting the side bunkers would give him a larger percentage of gross tonnage in the propelling space, and the 1¾ times method would give him a greater reduction. In a cargo steamer, this space could only be obtained by sacrificing cargo space, and would not be such a good proposition. It is a matter for careful consideration in the design and will be governed by the conditions of the trade the vessel will be in. There are various modifications of the deductions, as in the case of the Suez Canal measurements, which differ slightly from those given here. However, without entering on too many details, the explanation here given covers the ground well enough to enable a clear understanding of the various terms used in defining tonnage.

FORD-SMITH CO.'S NEW PLANT

THE Ford-Smith Machine Co., Ltd., of Hamilton, have moved into their new plant on Cavell Ave. This plant is one of the most modern for its size in Canada. While the machine shop is as yet not completely fitted up, the one feature that impresses one all through the plant is the splendid lighting. Every nook and corner is well supplied with daylight. The whole plant seems to be laid out not only for convenience and room for handling large work, but the management has had an eye open for the comfort and welfare of its employees.

The spacious office which has a southern exposure is 60 ft. x 30 ft., is well laid out and well supplied with the most up-to-date equipment. Directly back of this on the east side is the pattern storage room, 60 ft. x 40 ft. This room is thoroughly fireproof and so arranged as to facilitate the handling and storing of patterns.

On the west side is a men's wash room equipped with basins with running hot and cold water. Steel lockers are to be installed here for the men.

Over the wash room is the drafting office. It is well lighted and will be a quiet, cheerful place to work in.

Back of this is the machine shop which is 325 ft. long x 60 ft. wide and ceiling 17 ft. high. The shop is well lighted on both sides by large steel sash windows, with mottled glass to diffuse the light, and is complete with superintendent's office, shipper's office and shipping platform, store room, and tool room. The plant is heated by steam which is produced in a large Spencer heater in the rear of the shop. The shipping department is a raised concrete platform which has a door opening to a railway siding, a covered team-way and side door for small goods. It is complete with platform-scales, office and all necessary equipment.

The floor of the ship is of concrete covered with a heavy maple flooring.

Outside is a spacious yard to be used for storage or extension of buildings. It now has a garage and large bicycle rack.

When the shop is completely fitted up it will be equipped to handle the regular line of Ford-Smith grinders and millers, as well as marine engine work, and other special machinery.

Captain Charles, R.N.R., master of the Cunard liner "Aquitania," is a C.B., this honor having been conferred upon him in recognition of his meritorious services to the Empire during the war.

Novel Method of Building and Launching

Building the French Mine Sweepers at the Canada Car and Foundry Company—Launching Dock 1,400 Feet From Building Ways—Arrangement of Transfer Tables and Launching Trucks

By J. H. ROGERS, Associate Editor

SHIPBUILDING activity throughout the Dominion has been second only to the manufacture of munitions, and at the present time may well be considered on a par to that most important industry. In the immediate future and for some time after the cessation of hostilities the magnitude of shipbuilding will remain paramount among Canadian enterprises. When the building of vessels is mentioned, one generally thinks of what is being accomplished on the seaboard, or those rivers or waterways adjoining or flowing into the Atlantic or Pacific. Naturally, the bulk of the tonnage now under construction, particularly those vessels of large dimensions, are being built at the yards located close to the terminals where ocean vessels are accustomed to travel. This practice, however, is not followed, nor is it necessary, where the general dimensions of the boats will permit of their being taken through the canals that connect the Upper Lakes with the ocean waterways. Many of the vessels of 3,000 tons or less, that are now being built for the Imperial Munitions Board, the Canadian or the French Governments, have been constructed at different yards on the Great Lakes, and afterwards taken down to the St. Lawrence through the inland waterways.

Building berths are invariably erected in close proximity to the point of launching, to facilitate the operation when the boat is completed. This is a necessary essential when vessels are of large size, and it is even advisable for boats of any size, but for those of comparatively small dimensions it is not impossible to have the construction work performed at some distance from the launching point, providing adequate facilities are adopted for removing the hull from the berth to the launching basin.

Early in the present year the Canadian Car and Foundry Co., Limited, felt that their Fort William plant could be utilized for shipbuilding, and immediatey started negotiations with the French Government, which negotiations consummated in an order for twelve mine sweepers, with the understanding that they would all be delivered at Fort William before the close of navigation this year. Considerable work was necessary in making preparations, but after careful consideration the Car Company decided to undertake the construction and delivery of these vessels by the time specified. The decision to do this work at Fort William was determined by the excellent facilities that were available at the Fort William car plant. The work on these twelve vessels has virtually been accomplished by the co-organization of the existing car plant, so that the achievement is all the more remarkable when it is understood that the fulfillment of the contract will be completed well within the schedule time. The last one to leave the stocks has recently passed her dock trials and is expected to reach an ocean terminal before the inland waters freeze over.

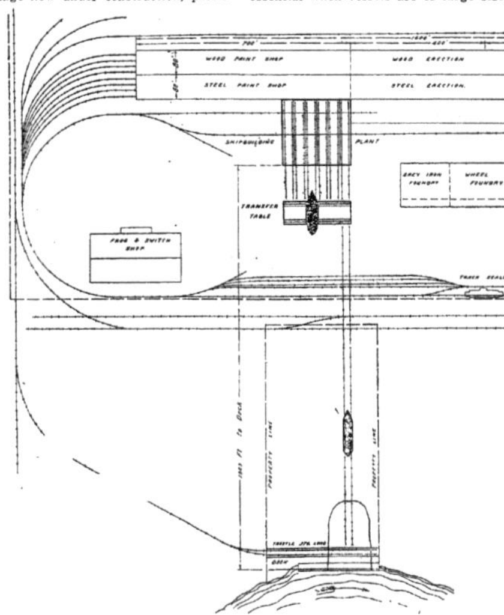

LAYOUT OF FORT WILLIAM PLANT

Peculiar local conditions, however, necessitated much thought and engineering before the problem was satisfactorily solved. As shown by the line drawing, the berths were located fully 1,400 feet from the nearest point of the river, where the launching would take place, and the difference in the two levels was approximately 50 feet. Owing to the current of the river it was deemed advisable to construct a launching basin at the foot of the incline. Part of the old dock was removed and a coffer dam constructed, a basin excavated on the inner side of sufficient depth and area for launching purposes. A channel with an 8 per cent. grade was made from the upper shipbuilding level to the launching basin. The double line of standard railway tracks, that formed a runway for the launching cradles, were extended to the bottom of the basin.

FIG. 1—FIRST OF THE 12 MINE SWEEPERS BEING TAKEN FROM HER BERTH

It will be seen in the property plan that the upper end of the line of tracks, leading to the basin, coincide with the right hand berth of the shed, so that a vessel constructed in this end berth may be moved directly from its building position to the graded track. In the case of vessels in the other berths, however, it is necessary to transfer them from their respective berths to the line of launching tracks. On either side of each berth is located a line of standard gauge tracks that carry the launching cradles. These cradles are constructed of suitable timbers, built up to conform to the outline of the hull at the supporting point, and are supported by special trucks built for the purpose. Each couple is connected with two heavy cross timbers, and in addition the frameworks of the trucks are prevented from spreading by several heavy chains attached to eye bars on either side. Bow and stern cradles are kept at a maximum distance by an arrangement of cables from the framework of each truck.

When a vessel is ready for launching, the cradles are run into position on either side and the wedges inserted to transfer the vessel from its building supports to the cradle trucks. When this has been done the boat is hauled from the berth to its position on the transfer table, by means of locomotives operating on the company's tracks, at right angle to the centre line of the vessel. The transfer table, upon which the vessel is located for side movement to the launching track, is in reality two separate tables, one supporting the bow cradle and the other the stern. This method was adopted owing to the distance between the two cradles.

Intermediate tracks were laid to support the first or stern cradle in its passage to the transfer table. Short lengths of rails were used to connect the permanent tracks with those on the transfer table. The channels for the transfer tables were quite shallow, being about the depth of an ordinary car wheel; the trucks and tables were built up of structural steel. The tracks leading from the transfer table to the head of the incline, and likewise those in the cutting, are laid on long timbers resting on cross ties, which in turn are supported on short piles sunk in the earth. In one of the views is shown the Mantone, the first boat completed, ready to be slid down to the basin. When lowering the vessels down the grade the retarding cables are secured to locomotives and lines of loaded cars, to steady the movement down the incline.

The Mine Sweeper

The general characteristics of this type of mine sweeper are as follows: The overall length is 143 feet and the length between perpendiculars 135 feet, with a moulded breadth of 22½ feet. The displacement, when loaded, is 630 tons, and the approximate weight of the hull when launched is 225 tons. After launching, the engines, boilers and necessary accessories (61 tons), water in the boilers, etc. (21½ tons) and general hull equipment (54 tons), are placed by the builders, and the balance of about 270 tons arranged for by the owners. These vessels are of the standard type of French trawlers, single deck with raised quarter and forecastle decks and steel deck house. The navigating bridge is located on the top of the boiler house and is provided with a steel house for the captain's quarters and accommodation for the wheelsman. The wireless room is located on top of the deck house aft, with lifeboat platforms on either side. Two pole masts are provided, the foremast stepped in a cast housing on the main deck and the mainmast supported by the deck house aft. All the vessels are fitted with six watertight compartments. The stern frame is made of one piece annealed cast steel, the rudder being of the double plate type built up frame. The steam steering gear engine is installed in the upper engine room. The main engines are of the triple expansion type and capable of developing 560 H.P., operating under a boiler pressure of 185 lbs., and a speed of 135 r.p.m., cylinders being 13x22x36x24 inches.

Boilers are of the "Scotch" marine type, 12 feet in diameter and 10 feet 6 inches long. These boilers are fitted with interchangeable Morison corrugated furnaces 3 feet 6 inches in diameter, and operated on the Howden forced draft system. The vessels are electric lighted throughout, current being supplied by means of 7½ K.W., 110 volt, direct connected marine generator sets.

An efficient two-pipe heating system is installed, condensation draining to a trap and thence discharged to a hot well. Pipe coil radiators are fitted in all quarters. Boats are equipped with eleven ventilating cowls, those leading to the boiler room being adjustable from

FIG. 2—ONE OF THE VESSELS ON THE TRANSFER TABLE

FIG. 3—THE PALESTRO AFTER REMOVING FROM TRANSFER TABLE

the fire hold, so as to meet the varying conditions. These vessels are built to the full requirements of Lloyd's Register of shipping and in excess where specified, and will class as +100 A-1 Steam Trawler.

SPEED IN SHIPBUILDING—WHAT IT INVOLVES

The question is often asked, "Why cannot ships be turned out at a much quicker rate?" People who are impatient of shipbuilding progress are apt to overlook the gigantic nature of the work involved in the building of our mercantile marine. While standardization has certainly accelerated production, in so far as numbers of ships are concerned, it cannot possibly accelerate speed of production above a certain degree. The building of a large number of vessels from the same specifications, moulds, and plans can be more rapidly accomplished by standardization, or fabrication, but no matter how many men are employed, no mater how intensified the standardization is, it is a physical impossibility to complete and place into commission a steamer of, say, 5,000 tons gross capacity, or 8,000 tons deadweight, in much less than eight months, from the date of fixing the first plate under present conditions of one working shift a day.

Few people in asking why we cannot turn out ships at a more rapid pace realize the amount of work there is involved in manufacturing the thousand and one parts which go towards the making of a steamer. Engines, boilers, solid steel shafts, these latter weighing many tons, have to be turned on huge lathes and adjusted to a fraction of an inch, steam winches, the steering apparatus, wood work and other internal fittings all have to be made. The steel from which the steamers and their fittings are constructed has to be turned out from the raw material, smelted, rolled and fashioned, and all having been prepared, has to be collected, shaped, and assembled. It is in the making of the component parts that standardization and fabrication have accelerated production. But it is in the actual assembling of the parts that most of the time is occupied.

FIG. 4—VIEW OF THE CRADLE TRUCKS AND THE CUTTING TO THE LAUNCHING SLIP

Let us take, for example, the case of a standard vessel of 5,000 tons gross. In the hull of such a ship no less than 2,300 tons of steel plates and angles have to be used, bolted in position, and fastened together by over half a million iron rivets. Each rivet has to be driven in separately, and it is here that time is taken up. Plate by plate, rivet by rivet, the hull is put into place, adjusted and fastened. Working steadily for nine hours each day, the labor of 400 skilled men and boys is needed constantly for six and a half months if the launch is to take place within the specified time.

Then, when the hull has left the stocks and is afloat in dock, another gang of men and boys take the craft in hand. To these falls the task of fitting the standard engines, which have been made and tested in the shops, the installing of the electric motors and dynamos and the wiring to carry the current for lighting, the fixing of wood-work general furnishing and painting of the interior and exterior of the steamer. Six weeks is the period in which they can complete their work. Thus in eight months one vessel of 5,000 tons gross is ready for commissioning.

If we are to succeed in our building programme, it is absolutely essential that all departments of the shipyard shall be working with clock-like regularity, and that in addition the necessary supplies from outside sources shall be promptly received at the appointed hour. Unless all the departments are adjusted to each other, unless the output of each balances with the other, there cannot possibly be increased production. For example, it is useless to speed up the production of plates if production of frames to which those plates have to be fastened is not equally rapid, nor is it of the slightest advantage to produce plates and frames beyond the capacity of the riveters. Even though these departments were adjusted, and the engineering shops were left out, of what practical use is it building hulls to have them lying idle on the water waiting for engines and other fittings?

Each department of the shipbuilding industry is entirely dependent upon every other department, and unless all are equally adjusted no amount of standardization will produce the desired result. Three main principles are essential to bring about rapid and increased production of mercantile shipping; regular supplies of materials, sufficiency of skilled labor, and what perhaps is most important of all, a close co-operation between employer and employee to get the utmost out of their establishments.

The Electric Arc Used in Steamship Overhauling

Westinghouse Arc Welder Does Good Work in This Line— Repairing a Furnace While Under Steam—Variety of Work Carried On

THE all-round usefulness and adaptability of the electric arc method of welding is well exemplified by the work carried out at the repair shops of the Canada Steamship Lines, at the foot of Yonge Street, Toronto. The machine, which is a Westinghouse arc welder, using direct current of 150 amperes at 75 volts, was installed in 1915, and was one of the first machines to be imported into Canada. The machine shop was started at the same time and has grown into a well fitted up repair shop, where practically all the repairs necessary for the company's steamers can be carried out with ease. The shop and welding outfit are under the supervision of Mr. Noonan, the superintendent engineer of the company, and some very interesting repairs have been performed under his direction. The welding metal used is a soft Swedish iron wire known as Premier welding wire, and during the years of the war, this, like most other steel material, has been hard to get, but since the armistice has been signed Mr. Noonan has been notified by the Steel Company of Canada, who supply this wire, that they can now fill his orders. Last winter no fewer than twenty steamers were laid up at Toronto, and the repairs made at the company's shop, so that there is no lack of work for the machine shop or the welding apparatus; in fact, a welder and helper are kept steadily at work all the year round, and in the summer season, when the boats are all working under conditions of rush, the welding outfit makes possible quick repairs, and eliminates delays. As an instance of this, an especially interesting case might be mentioned. One of the steamers came into port with a furnace leaking, and it was feared that she would be subject to considerable delay. In the ordinary course of events she would have been delayed quite a bit. The cause of the leak was a crack in the furnace tube, and the usual way of repairing this would be to drop the pressure off the boiler and repair the crack by either plug stitching or a patch, according to the condition of the furnace. However, it was decided to repair the furnace by the electric welder, and furthermore, to do it without waiting to lower the steam pressure. This was a bold policy, but such was the confidence felt in the process that no hesitation was expressed by anyone connected with the job. The vessel was brought round to wharf where the machine is installed, and the cable led aboard. The boiler was carrying 125 pounds of steam per square inch, and the job was done with this pressure on. The first difficulty was to stop the water leaking through the crack. and by deft manipulation of the electrode, a thin skin was worked over till the leak was stopped and then the weld was built up in the usual manner. That job held tight all the rest of the season, and was found to be perfectly good when the vessel was laid up at the end of the summer.

Another case was that of a boiler which had undergone extensive repairs, having new combustion chamber back sheets fitted and necessarily new combustion chamber stay bolts. The repairs being completed, the boiler was put under hydraulic test, and at 125 lbs. pressure per square inch, a leak developed in the back head near the ring seam. It was decided to call in the services of the welding machine once more, and the repair was made in practically the same manner as the repair on the furnace. a new piece fitted in. New nuts have been formed on the end of combustion chamber stays by the building-up process, and stay tubes stiffened at the end where corrosion has taken place. Paddle wheel arms are welded in place, badly scored piston rods have been filled in and then turned up as good as new, and many other varieties of repairs are performed daily.

The subjects of our illustrations are repairs on circumferential seams of boilers in one of the company's steamers. The first illustration shows the welder closing up a leaky seam at the forward end of the boiler, close to the floor plates,

WELDING REPAIR TO MARINE BOILER

The first thing to do was to stop the water leaking through the crack, and once this was accomplished, the weld was built up to the required strength. This job was also perfectly successful, no further trouble being experienced. For this kind of work there is no doubt the arc welding is superior to the oxyacetylene flame in the opinion of the men who have these jobs to do, and there is no doubt that the feature the electric arc possesses of applying the heat right at the job itself, and localizing it, has a great advantage over the much larger surface affected by the oxy-acetylene flame. The usual method of repairing a crack or leak is to V the bad spot out so that it can be rebuilt with new metal. If the part is very bad or covers a large area. the piece can be cut out completely and a job which is one of the most awkward to be done by hand caulking. The other illustration shows a completed repair on the circumferential seam at the after end of the boiler, at a part where the marine boiler is peculiarly liable to corrosion. This is on account of wet ashes lying against the shell during the periods of cleaning fires, and at other times, and also to the effect of leaking gaskets on the manhole door, the water running down and along the seam. This seam also is liable to leaks due to the unequal expansion of the boiler shell when raising steam, and even when working. It is not unusual in a boiler which has not been properly circulated, to find this part of the shell and head comparatively cool long after the pressure has been raised to the working point.

We are indebted to the courtesy of Mr. Noonan for the opportunity to obtain these illustrations, and the information furnished by him.

SWITCHBOARD

SALVING A TORPEDOED STEAMER*

THIS vessel was a hurricane decked steamer having a 'tween deck extending fore and aft, the midship portion of which was used as bunker space. The watertight bulkheads extended as far as this 'tween deck, and not to the upper deck, the bulkheads being of the corrugated type. The number of bulkheads, that is, watertight, abaft the engine room were two, one on the after end of No. 6 hold, and the other between Nos. 4 and 5 holds, holds 4 and 5 being divided by a wooden bulkhead.

The torpedo struck the vessel on her port side in No. 5 hold, causing extensive damage, holes being blown in her 'tween deck, and in her starboard side. The tunnel being damaged, considerable quantities of water leaked through the door, making it imperative to shore-up the door to prevent it from bursting. The watertight bulkhead at after end of engine room also leaked badly, and eventually flooded the engine room.

The vessel, having a dangerous list, it was feared that she might fall right over, and the steps taken to prevent it were as follows: The salvage vessel was brought alongside, and a 6 in. wire was passed under her and fastened to the bollards and lower part of the mast of the torpedoed vessel, from the end of the wire on the salvage vessel a 25-ton tackle was made fast, and let to the steamer's mast cross trees, and the fall of the tackle hove taut, by winches, as in Figure 1.

To prevent water flowing along the 'tween deck into No. 6 hold, a bulkhead was erected across the ship, and the seams of same caulked-up at position "A." The stern of the vessel and up to the position "a" 'midships was covered with water at high tide, (see Figure 2.)

An electric submersible pump, capable of delivering 350 tons per hour, was lowered into the engine room and started, the current being supplied from the alternator on the salvage vessel; the engine room was pumped, but this had no apparent effect on the list of the ship. The pump was used intermittently to keep the engine room dry, but fears were again entertained that the vessel was going over, and wires were led out from starboard bow and starboard quarter to the shore and hove taut; these means not proving sufficient, a lighter was brought alongside and made fast, forward of the bridge, the lighter having tanks that could be flooded to sink her about 4 feet. A 9 in. wire was passed round her, the two ends being clamped together with a Bullivant's "Bulldog Grip" clamp, a tackle was made fast to the wire, and the other end to the ship's foremast cross-trees, the fall being taken to the ship's winch, and steam supplied to the deck steam pipe line, with flexible steam pipes from the salvage vessel.

THE TORPEDOED STEAMER

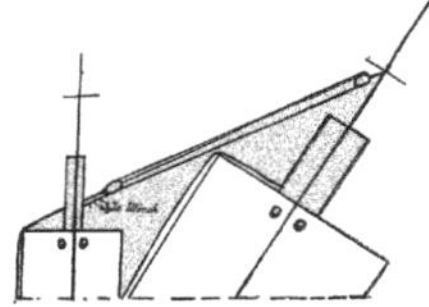

FIG. 1—SHOWING WIRE AND TACKLE WIRE FASTENED TO BOLLARDS ALONGSIDE MAIN MAST

Three hundred tons of coal were discharged from port 'tween deck bunker. The hatches were put on Nos. 4 and 5 holds, and the seams caulked and shored down, from the hatch beams above. The water in the after-peak tank was blown out with compressed air, by fitting a connection to the tank air pipe, water emerging through the sounding pipe.

An 8 in. pump was put into No. 6 hold, and a 4 in. submersible pump, the pumps being started as the tide commenced flowing, large quantities of water leaking through the tunnel into No. 6 hold, the vessel gradually righted herself and floated with a list to port of five degrees; the vessel was then towed in and beached in 10 feet of water low tide. It may be noted that a 4 in. submersible pump delivers 100 tons per hour.

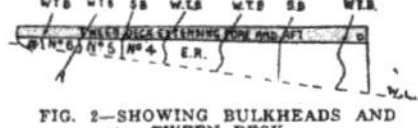

FIG. 2—SHOWING BULKHEADS AND TWEEN DECK

* Transactions of the Institute of Marine Engineers.

Port Colborne. — The Welland Canal closed officially for the season at midnight of the 19th December. The "Grant Morden" will lay here for the winter, and the "War Weasel" will lay at Welland.

LAUNCHING CONCRETE BARGES BOTTOM UP

Having received a large number of enquiries from readers relative to the method of launching some concrete vessels bottom up, we publish a description of this practice, which was given recently in the "Journal" of the Portland Cement Association. The vessel here illustrated is a 200-ton barge, built by the cement shipbuilding yard at Porsgrund, Norway. The lighters are built bottom up, it being easier to cast in this way, the density of the mixture being more easily controlled. The internal mold is made in units that can be easily taken apart, and reassembled for another launch. When the vessel is launched, the centre compartment and the compartments in the bottom of the hull, (which, when launched, are uppermost), are open to the water, but owing to but small vents being left for the air, the water rises slowly, until it gradually runs into the two compartments on each side. Advantage is taken as described in the text under the cut, of the principles of stability of floating bodies to make the barge right itself, which it does in about twenty minutes. It must not be imagined that any form of vessel could be launched this way. It is submerged to its maximum amount. The lower side compartments are never flooded, and thus the vessel is in a state of unstable equilibrium, the centre of gravity being considerably higher than the centre of buoyancy as shown. If now from any cause whatever the vessel heels to one side or the other, the weight of the hull and its contents acting through the centre of gravity and the pressure on the submerged area acting through the centre of buoyancy, form a couple, the moment of which tends to turn the vessel on a longitudinal axis as shown at C. The moment of this couple has its maximum value when the vessel is in the position shown at D, after which it gets gradually less and less until the vessel floats in the correct position as shown at E, when the moment of the couple is again zero, but the vessel is now in a state of stable equilibrium. The slightest list to one side or the other after the vessel has been launched and the compartments filled is thus sufficient to cause her gradually to heel completely over until she floats in a normal position. The flooded compartments are then pumped out, and the mold is removed, this being used again in the construction of another similar vessel.

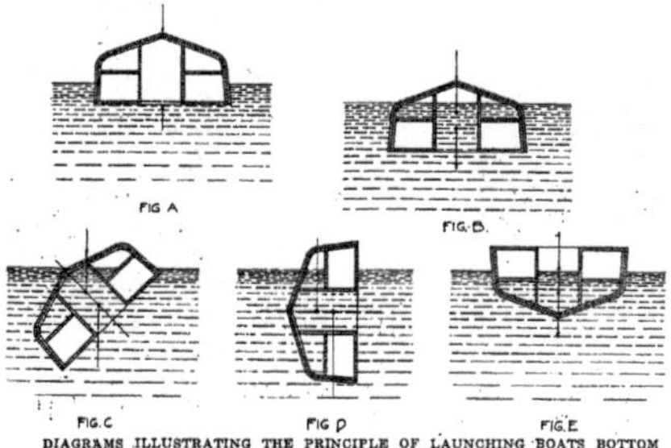

DIAGRAMS ILLUSTRATING THE PRINCIPLE OF LAUNCHING BOATS BOTTOM UP, AS PRACTICED BY THE CONCRETE SHIPBUILDING PLANT AT PORSGUND, NORWAY

is only the special arrangement of compartments allows it in this case:

The diagram A represents a section of the vessel immediately on taking the water. The inner mold divided into compartments is seen, together with the outer reinforced concrete hull of the vessel, and it will be observed that as yet there is no water in the middle compartment of the mold owing to the air not escaping rapidly enough through the vent pipes. As the air escapes, however, the water rises in this compartment, the vessel gradually sinking as it loses its buoyancy, until the water reaches the level of the two upper side compartments, which are also gradually flooded, so that a position is gradually reached as shown at B when the vessel

Toronto.—Promptly at 12 noon on Saturday last "La Quesnoy" left the ways at the Dominion Shipbuilding Co.'s yard. "La Quesnoy" is a steel vessel, of full canal size, and a sister ship of the "St. Mihiel," which is being completed in the same yard. The honor flag won by the company in the Victory Loan Campaign hung from the port bow of "La Quesnoy," and a silk Union Jack presented by her sponsor, flew from the taffrail. The dimensions are as follows: Length o. a., 261 feet; breadth, moulded, 43 feet, and depth, moulded, 28 feet 2 inches. Her engines are triple expansion, 20 x 33½ x 55 x 40 inches stroke. Two Scotch boilers, 14 feet 6 ins. by 12 feet long, built by John Inglis Co.

ELECTRIC WELDING FOR SHIPS

* "Times Engineering Supplement"

Addressing the North-East Coast Institution of Engineers and Shipbuilders, at Newcastle-upon-Tyne, on "Electric Welding for Shipbuilding Purposes," W. S. Abell, Chief Ship Surveyor of Lloyd's Register, stated that the practicability of uniting iron and steel plates by means of electric welding appeared to have been demonstrated about 25 years ago, but it was some considerable time after that before it was suggested that that process might be suitable for use in repairing hulls and boilers.

The first case of any repairs to boilers classed with Lloyd's Register being effected by that process was one in which some small repairs to the combustion chamber seams were carried out in March, 1906, which proved to be satisfactory. The first hull repair was to a cast-steel stern-frame which was broken at the lower part of the stern-post and was welded in place in November, 1907. On examination in May, 1908, the repair appeared to be quite satisfactory, but a year later the stern-post was found to be cracked about 4 in. above the part previously welded, and a new stern-frame was then fitted. During the five years following the introduction of electric welding, the number of repairs so carried out continued to increase, and, by the middle of 1911, about 160 such repairs had been made to boilers and 20 to hulls. Nearly all the boiler cases were confined to welding up cracks and repairing wasted and leaky patches of seams in furnaces, combustion chambers, tube plates, etc., filling up pitted places in the plating in various parts of the boilers, and in some instances repairing the front end plates of the main boilers where badly grooved or cracked in the way of the furnaces. The hull cases related principally to filling the flaws and cracks in rudders, and stern-frames. Since 1911 the use of welding for effecting minor repairs has greatly extended, and in 1917 about 154 cases occurred, of which 140 were to the boilers and the remainder to the hull. Of the latter, six repairs were made to the stern-post, three to the rudder, and five were miscellaneous, including one case in which a previously executed weld had proved to be defective.

Behaviour of Welded Joints

In view of the favorable nature of the experience gained, it was suggested about 18 months ago that the time had arrived when a further step in advance might be made and the process of electric welding adopted throughout the entire structure of a vessel. Before, however, such a course could be approved, it was thought desirable that all the evidence bearing upon the question should be examined with the object of ascertaining the capability of electrically-welded joints to withstand structural stresses. Numerous experiments had been made to find the strength of electric welds in various combinations of plating, but these had nearly always been confined to the determination of the ultimate strength of the welded joints, and, in a few instances, to their ability to resist bending. In these circumstances, it was considered essential that, before any substantial progress could be made, reliable data should be obtained embracing an examination of as many as possible of the properties of electrically-welded joints. The information available regarding the ultimate strength of such joints did not touch at all upon several of the principal problems to be solved before the adoption of electric welding could be sanctioned in ship construction. In particular, information was lacking concerning the elastic properties of welding and welded material and its ability to resist the alternating stresses to which vessels are subject. As is well known, the stress at which a material breaks down under conditions of repeated stress, particularly if of an alternating nature from a maximum tension to a maximum compression, is very much lower than the ultimate strength as measured in the ordinary way, and the behaviour of welded joints under such conditions is a question of considerable importance in its relation to shipbuilding.

In devising tests on material of that kind it was not possible to reproduce exactly the conditions experienced in actual practice, and the object aimed at was, therefore, to arrange a series of experiments which would demonstrate the average reliability of the welds under various conditions of loading. Another source of possible difficulties is that there is not, at least at present, any known method of testing the soundness of a weld after completion. In engineering structures, it is nearly always possible, by examination of the finished work and by suitable mechanical tests, to determine whether the workmanship and material are satisfactory and reliable; and the fact that tests are not possible with welding rendered it all the more necessary that the preliminary experiments should be as thorough as possible. It was found necessary to design several methods of investigating the problems involved. It was decided that the experiments should include a determination by different methods of the modulus of elasticity and the elastic limit of the welding and welded material, together with the incidental information regarding the ultimate strengt hand elongation of the welded joints. A series of alternating stress tests with different forms of specimens tested under various conditions was also devised, together with a number of minor tests, including bending and impact. Complete chemical and microscopic analyses of the welding and welded material were made to ascertain the changes in chemical composition and crystalline structure produced during the welding process.

Results of Tests

The tests showed that the tensile strength of welded joints is greater than that of riveted joints. Such joints may, therefore, be accepted as fully equivalent to the latter. The results of the elasticity tests showed that, in general, the combination of welded and unwelded material behaves practically homogeneously up to at least the elastic limit, a feature of considerable importance in relation to structural stresses. It was only when the results of the alternating stress experiments were considered that the question of the limits of application became a serious problem. These tests showed generally that welded material would not withstand a very large number, say, several millions, of alternations, if the applied stress was greater than about 6½ tons per square inch. The capability to resist alternating stresses is, of course, of the very greatest importance in shipbuilding materials, and would appear, for the present at least, to limit the application of welding to vessels in which that stress is not exceeded. The calculated stresses in ship structures are, in the case of large vessels, rather in excess of that figure, although, from such information as is available, it would appear that the stress actually experienced by the material in a ship is considerably less than that calculated under the usually assumed conditions. It is well, however, to proceed cautiously in the application of a novel method of construction like electric welding, and it will probably be wise to limit its application meantime to vessels of a length of not more than about 300 ft.

The lecturer suggested that, in using electric welding in ship construction, it is desirable that the amount of overhead welding should be limited, and also that as much welding as possible should be done on the ground, before the various members are erected in place, as the operators have then a much better opportunity of securing satisfactory results. The maximum spacing of the service bolts required during the process of erection is a matter which can be decided only after some experience has been gained with vessels of different sizes. It is essential that the spacing should be sufficiently close to enable the surface to be properly drawn together, since the welding does not possess any drawing power, and any neglect to have the surfaces well closed may be very prejudicial to the structure. Probably the most economical arrangement will be found to be a combination of riveting and welding, but the whole problem is one in which there will be gradual development as experience is gained and new methods of erection devised.

Since means of testing the soundness of a weld when completed do not exist at present, it is necessary that the operators employed should be reliable and well trained in the manipulation of the system of welding used. In welding, however, as in many other engineering processes, much can be done to avoid variations in workmanship by the introduction of suitable mechanical devices to assist the workman by improving the means of control over the work. Electrodes are nowadays manufactured with a high degree of uniformity, and, in most sets of welding plants, provision is made to enable the density of the electric current to be varied with the size of the electrode. It is important that the instructions issued regarding the size of the electrode and the corresponding current should be carefully followed and that care should be exercised to see that the metal is properly deposited.

Limits of Use

The improvements in electric welding have opened up considerable possibilities of its extended application in shipbuilding. While it is improbable that, until some practical experience on a fairly large scale has been obtained, welding will be applied to the structural members of vessels exceeding 300 ft. in length, even for smaller vessels there is ample scope for its introduction. Work has already been started by Messrs. Cammell Laird, on a coasting vessel about 150 ft. long, and proposals have been discussed for the building of a number of other welded vessels, ranging from 86 ft. drifters to standard ships 300 ft. long. In addition, numerous proposals, many of which have been adopted, have been considered for using welding in the construction of bulkheads, casings, deck-houses and similar items not subject to the main structural stresses. The extent to which welding will be adopted in the future will in the long run be governed by its relative cost as compared with the present methods of construction, account also being taken of the saving in weight and consequent increase in deadweight capacity due to the elimination of many of the structural elements which will no longer be necessary. On the whole, it appears safe to say that, for vessels of moderate dimensions the use of electric welding will become increasingly common, with possibilities of extension to vessels of larger size as experience is gained and further research leads to a nearer approach to perfection in the process.

Port McNicol.—The largest amounts of grain ever held in storage vessels at this port is here this winter, amounting to 7,927,186 bushels of wheat. The grain is stored in the following steamers: "Ford," 480,000; "Ker," 478,000; "Dalton," 470,000; "Ziesing," 465,000; "Williams," 463,000; "McGonagle," 461,000; "Crawford," 459,332; "Wilkinson," 446,230; "Farrell," 437,800; "Roberts, Jr.," 430,000; "D. R. Hanna," 404,236; "Morse," 396,000; "Wilvin," 391,000; "Stadaconna," 390,000; "Andrew Matthew," 383,000; "Nettleton," 375,000; "Robinson," 367,000; "Sellwood," 363,315; "Chas. Hubbard," 266,773.

The steamer and the motorship have deprived the seafarer of much of the old-time romance attaching to a sailor's life. The advent of wireless has also increased the safety of it. Deprived of these features, a sailor is immediately dependent on his own resources, and Captain Laing gives us a true and seaman-like reflection of sea life under canvas. The present feverish activity in constructing self-propelled craft affords a suitable background for incidents in a career in which gaffs, booms, tar and canvas held sway before being displaced by mechanical unloaders, derrick masts, fuel oil, coal and wire rope.

Part IX.

Time went on, days ran into weeks, and the "Maggie Dorrit" held her head up for the island of Robinson Crusoe fame, namely, Juan Fernandez and Mas-a-fuera, lying some 400 miles W.S.W. of Valparaiso.

Each watch brought its own work aloft and below. The coal baskets were all nicely stropped and siezed, and a few beef tierces sawn in two were also made ready with ropes and spun-yarn for the unloading of our coal cargo into Chilian barges.

It looked as if our trip across the southern ocean would last about five weeks. Fid was again working up Mercator sailings and Hank had got the length of picking out the logarithms that helped to bring about the more or less uncertain results.

Our birds were fed each day at noon, and in lulls when the wind fell for a rest we never got tired of playing with them, although we had no desire to kill and eat any of the feathered sailors on the Chilian voyage. The potato hash, over and above the ordinary allowance, was dealt out, and the colonial chef was a high-stepper.

Jolly-Boat Anticipations

We were looking forward to having a good time in Coquimbo, as the old man would require our services as boatmen. This business of pulling the skipper ashore and aboard, and visiting other vessels for social functions kept the apprentices away from the harder and dirtier work of unloading the coal. At least a first voyager looked at it that way. Fid was also elated as he expected to get some of the boating till Hank and myself proved able to handle the little craft alone.

As we sailed more to the northward the weather became finer, and clear skies, lighter winds and smoother seas made life a little more pleasant. We had no objections to hanging up our oilskin clothes and putting away our heavy leather sea-boots. Very likely they would get their next innings off the Horn, that famous marine milestone that makes all windjammers pay homage while changing from the Pacific into the Atlantic or vice versa.

Making The Chilian Coast

On approaching any foreign port, especially for the first time, sea-apprentices were naturally filled with an anticipation that could hardly be defined. Queries entered one's mind and there was a longing to see the strange land, the natives' faces, and prod into their language, habits and traits. It was with such a line of thought that we sailed towards the coast of one of the most interesting republics in the world—Chile. Being in the vicinity of the island associated with "his man Friday," added not a little to the plastic curiosity inherent to boyhood, either ashore or afloat.

As Fid had been in Chile before, he aired his mongrel Spanish to us, much to our amusement, as we noticed in his interpretations many words which were suspiciously uttered.

Landfall Near Valparaiso

It was announced from aft that we should pick up the land next morning, and sure enough, as Old Sol climbed over the Andes in his eastern glory, the outline of those mountain peaks threw their variegated shadows on the cloudless horizon. This land in the vicinity of Valparaiso was still sixty miles off, as the Chilian coast has, with certain atmospheric conditions, a very long-range effect. The "Maggie Dorrit" was almost sailing parallel with the coast so its approach was very gradual.

On our way north we fell into light winds and had quite a heavy drill on the braces. Our anchors were made ready and all the minor preparations, such as making the vessel look spic and span, were gone through. Towropes and tugboats were never a part of Chilian seamanship, and here it may be said that in all countries colonized by the Spaniards, the commercial progression stands about 200 years behind other places.

It was indeed a pretty experience to sail as we did, right into the anchorage off Coquimbo on a bright afternoon, having made the trip from Australia in forty days, an average daily run of about 150 miles. This was not a smart passage, but it was fair for an old vessel. Our eyes had to be pleased with a landscape of rocks and sand at this part of the coast, and the absence of trees and shrubbery such as we witnessed in Australia was very much noticed.

In sailing into a harbor unattended such as Coquimbo, we started to reduce canvas when about six miles off the anchorage. First of all the ugliest sail to handle on a square rigger—the main sail—was clewed up close to the yard, and if the wind was aft the other lower course or foresail was taken in, too. The mizzen mast was bared. With the rest of the square canvas and a few fore and aft sails the vessel was under perfect and easy control. Everyone was attention, the mate's orders were carried out in a jiffy and a yard at a time was lowered into the lifts, whilst the sail was snugged with clewlines, bunt lines and leech lines. A leadsman is seldom put in the weather chains entering a Chilian harbor; it is sufficient to take a cast just before "rounding to" as the harbors are almost without exception far deeper than an ideal roadstead should be. In this connection, take Valparaiso, one of the greatest and yet one of the most exposed anchorages in the world, and within a stone's throw of the beach you can find thirty and forty fathoms of water, or, say, 200 feet.

As soon as our anchor was down in the outer roadstead the captain of the port came alongside in his trim boat manned with six uniformed natives. After a few preliminary questions this official scaled the Jacob's ladder and went below with our skipper. Such papers as the ship's manifest of stores and cargo, bill of health endorsed in the last port, and ship's register looked over. Then a box of cigars were sampled, like-

Continued on page 56

SEAMANSHIP AND NAVIGATION

Conducted by "The Skipper"*

Articles of direct interest to mariners, discussions of seamanship, navigation rules, and allied topics. Inquiries from readers are invited and will be replied to promptly through these columns, unless otherwise requested.

WOULD a ship deep-loaded with coal act in the same way as the iron ore craft? No, sir, She would roll with much more ease and less danger of bursting things up. Why is this? Because the coal cargo practically fills the ship, therefor a large amount of it is above the water line. This neutralizes the roll or swing and also gives buoyancy to the ship.

Illustrate the other extreme: When a ship can't roll, but is in danger of capsizing. When carrying a deck load of timber the centre of gravity gets too high and the ship "heels" or "lists," frequently to an angle of 25°, and sometimes more. Such a ship will hang down with the use of much helm, and without warning may sit up on an even keel for a short time, only to fall over again on her side. What are the dangers connected with heavy listing? The engineer has his worries, re wing fires, condenser inlet and oiling duties, whilst the master and mates are driven mad trying to figure out the correct deviations for the courses owing to the magnetic disturbance altering the ordinary deviations and errors of the compass.

Diving Movements

What causes diving? Steaming into a head sea. What effect has it on a steamer? It causes propeller "racing," raising the slip of that agent and reducing the vessel's speed very much, especially in ballast. Is diving dangerous? No, generally speaking. What must be attended to in diving? See that the kicking tackles are put on the quadrant. If the steam steering gear is very modern, and no such tackles are carried, see that the spiral buffers are acting properly, and if the gypsie drum of the steering engine has too much play, get after the turn-buckles, for the rudder plays its wildest tricks in a heavy head sea. Of course, diving as well as rolling may at any time cause cargo to shift. In this connection what are the most dangerous cargoes? Grain in bulk and steel rails. Many vessels have come to grief when tumbling around with these cargoes.

With transatlantic ore cargoes is it not possible to prevent some of the dangerous rolling? Yes. A few years ago the writer was bound from Turkey in Asia to Baltimore with chrome ore, and we built two false sides in the two midship hatches to cause the ore to rise. On another voyage from Rio de Janeiro to Philadelphia with ore, we built a raised platform on the skin of the hold, which was another way of lifting the deadweight. The study of a ship's movements brings out many interesting points worthy of the sailor student's deep thought.

Handling a Steamer In

Now that the student has absorbed a little about the tides, winds, currents, and ordinary movements of a ship, let us get down to seamanlike practice in handling a vessel in a river or tideway. Ambitious young men who aspire to become mates and masters must ever bear in mind that there are only two ways of learning the rudiments of their profession. One way is in reading up magazine articles and the other way is in getting tuition from ship's officers on board. Even with both avenues open to the student he alone must do the tall thinking till "old man experience" enters his life.

You are steaming down a river with the ebb tide and wish to make a dock on your port hand.

In this case you must turn round below the dock, ultimately straightening your craft up, stemming the ebb water and heading for the berth. If she is sluggish in turning round, drop the hook for a minute and the nose will soon look upstream.

Why had she to be turned round?

To get her under complete control. No ship is under complete control when going with the tide.

When crawling up close to the dock get the tide on your port bow and attend to the starboard mooring.

Don't be in a hurry with the stern ropes till she falls alongside.

Where the tide was—say over five miles an hour, what extra precaution would you take to ensure a safe berth?

I would drop the offshore anchor ahead of the berth before breasting in. Then, if the bow moorings carried away the ship would sheer off and ride to her hook.

Assuming that you had taken this extra precaution and the head lines parted, what would you immediately do?

Run aft and let rip the stern ropes for fear she got athwart the tide and dragged the anchor.

Say that two steamers are anchored in a tideway and a strong breeze is blowing against the tide; one craft is in ballast whilst the other is deep loaded. How will they act?

The loaded ship will ride with her head to the tide and stern to the wind.

The ship in ballast will be inclined to act in the opposite manner and obey the wind. The reason for this is very apparent when you think out the relationship of the forces at work.

Is there not a large submerged surface for the tide to play on in the loaded ship, and a large part of exposed surface for the wind to act on in the light ship?

In a roadstead devoid of tides and currents, but with a gale blowing, how is it that an anchored ship in ballast behaves so badly?

Because she is too light forward. The ideal draught for a ship at anchor is a little "by the head." Making a vessel draw 15 ft. aft and 5 ft. forward is all right for speed purposes in smooth water, but the fat is in the fire if she has to "heave to" or anchor in a storm.

This is one reason why some of our Great Lake boats and quite a number of ocean craft become the easy prey of foul weather. No man living can handle them in ballast trim when the elements are raging their utmost.

Manoeuvres in a River

You are laying at a river dock, head down, and want to go to a berth three or four miles upstream. The river is narrow and the flood tide running about five miles an hour. How would you act?

Get steam up, drop into the middle of the river and let go the anchor. Don't pay out much chain, only a few fathoms more than the depth of water. Come astern on the engines and she will move upstream under complete command.

When she attempts to break her sheer and get athwart the river give her a shot of cable and bring her up on the brake handsomely. When straightened up heave in the chain till the anchor trips and she will go on her way rejoicing.

If you think she needs a kick astern give it to her. If a kick ahead is required now and again in negotiating a bend, let her have it. Her movements will be as graceful as your seamanship, but with the anchor on the ground and the ship facing the tide she can hardly get out of hand.

Going with the tide in the ordinary way is all right even in a narrow river provided you have no short distance landings to make, but the stern first method just mentioned for short shifts spells safety first. It has been practised on every river in the world that carries the merchant ship on its bosom.

As an incentive to study and to further illustrate the effect of moving waters in relation to seamanship, say that you were coxswain of a four-oared boat and had orders to pull off to a ship anchored in midstream; the river being in spate, how would you proceed?

Pull up the bank, close in, where the water would be slack until the anchored ship was over the small boat's quarter. Then sheer off about three points (34°), no more, and in this sagging condition you would fall right under the large vessel's bows and have a line thrown to you before you drifted astern.

You will find out in practice that in pulling across the tide your little boat will require sufficient helm to prevent her from getting broadside on. In a sailor's understanding, ebb water in a river is water flowing seaward, whilst flood water or tide is water from the sea flowing inland. Be clear on such points in studying.

Broadly speaking, the writer generally

Continued on page 50

The MacLean Publishing Company
LIMITED
(ESTABLISHED 1888)
JOHN BAYNE MACLEAN - - - - - - President
H. T. HUNTER - - - - - - - Vice-President
H. V. TYRRELL - - - - - - - General Manager
PUBLISHERS OF

MARINE ENGINEERING of Canada

A monthly journal dealing with the progress and development of Merchant and Naval Marine Engineering, Shipbuilding, the building of Harbors and Docks, and containing a record of the latest and best practice throughout the Sea-going World.

B. G. NEWTON, Manager. A. R. KENNEDY, Editor.

Associate Editors:

T. H. FENNER J. H. RODGERS (Montreal) W. F. SUTHERLAND

OFFICES

CANADA—
Montreal—Southam Building, 128 Bleury St., Telephone Main 1004.
Toronto—143-153 University Ave., Telephone Main 7324.
Winnipeg—1207 Union Trust Building, Telephone Main 3449.
Eastern Representative—H. V. Tresidder.
Ontario Representative—S. S. Moore.
Toronto and Hamilton Representative—J. N. Robinson.

UNITED STATES—
New York—A. R. Lowe, 111 Broadway, New York, Telephone 8971 Rector
Chicago—A. H. Byrne, 900 Lytton Bldg., 14 E. Jackson Street, 'Phone Harrison 1147
Boston—C. L. Morton, Room 733, Old South Bldg., Telephone Main 1024

GREAT BRITAIN—
London—The MacLean Company of Great Britain, Limited, 88 Fleet Street, E.C., E. J. Dodd, Director, Telephone Central 1960. Address: Atabek, London, England.

SUBSCRIPTION RATE

Canada, $2.00; United States, $2.50; Great Britain, Australia and other colonies, 8s. 4d. per year; other countries, $3.00. Advertising rates on request.

Subscribers who are not receiving their paper regularly will confer a favor by telling us. We should be notified at once of any change in address, giving both old and new.

Office of Publication, 143-153 University Avenue, Toronto, Ontario.

Vol. VIII. DECEMBER, 1918 No. 12

PRINCIPAL CONTENTS

The Wreck of the "Princess Sophia"

THE wreck of the steamer "Princess Sophia," with the loss of 343 lives, on an unlighted reef on the coast of Alaska, last month, has called attention to several undesirable features of the navigation conditions in that part of the world. The most important is the lack of proper lighting, and while the Canadian portion of the route is admitted to be better in this respect than the American, there seems to be room for considerable improvement. The route the steamer was travelling, that from Seattle to Skagway, has a total distance of about 1,150 miles, of which the central portion of 500 is through British Columbia waters and the remaining portions at each end are in United States territorial waters. It is stated that for years Congress has been petitioned to provide more and better lighthouses on this coast, but presumably both in the American and Canadian cases, this district has not the requisite voting power to interest the prevailing Government to spend the money. Now that the necessity of proper aids to navigation has been shown by this appalling loss of life, the powers that be in both countries will, perhaps, move to avoid a repetition of the sad occurrence, or it may be that they have become too much accustomed during the last four years to great losses at sea to be moved by a mere 300 or so.

Another peculiar feature in connection with this disaster, is the state of the laws of both countries with respect to salvage services that may be rendered by vessels of each country to vessels of the other. These laws make it impossible for a Canadian vessel to perform salvage work in American waters, or an American vessel to do so in Canadian waters. The Canadian authorities claim that the law is not meant to prevent the saving of human life, but only to apply to ordinary commercial salvaging operations. It is a very difficult matter for anyone to determine exactly when ordinary commercial operations end, and emergency salvage work entailing the saving of human life begins. A case is cited where a United States vessel in trouble in British Columbia waters, had to wait till permission was received from the B.C. officials before she could transfer her passengers to another vessel of the same line. Now comes the allegation from an eye witness of the wreck of the "Princess Sophia" that the passengers and crew could have been saved. This man, Paul Graham, of Juneau, stated that when the ship struck the Vanderbilt Reef, and before the gale arose that was the cause of her breaking up, the U.S. lighthouse tender "Cedar" and a fishing schooner, the "King and Wings," went to the rescue of the wrecked steamer. They offered to take the passengers and crew off, but the captain of the "Princess Sophia," Capt. F. E. Locke, replied: "We must wait. I cannot use the rescue vessels without orders. The 'Princess Alice' is coming to our assistance. I have wirelessed Vancouver for instructions. We must wait." The two vessels stood by for several hours, till the northerly gale drove them away, and when they returned the "Princess Sophia" was gone, and with her all on board.

It seems incredible that any man having the responsibility of a ship, with her passengers and crew on his hands, could, out of regard for any law, risk all their lives by adhering strictly to the letter, under the circumstances in which he found himself. Ashore on a reef in a dangerous locality, at a time of the year when bad weather was to be normally expected, his first thought would be to secure the lives entrusted to his care, whether he broke the law or not. Surely the old shellback rule of "obey orders if you break owners" would not hold sway in a situation like this. Unfortunately, as in many of these sea tragedies, no one is left to inform us of the captain's reasons for not accepting help, but if there is a law on the statute books that can have the effect of clouding a man's mind as to his imperative duty, the sooner it is removed altogether or made perfectly plain to the most un-legal mind, the better for all concerned in the navigation of these waters. We await the result of the enquiry, which the Commissioner of Wrecks will doubtless hold, with great interest.

Canadian newspaper publishers pay about $3,500,000 a year for their white paper.

Some of the largest Canadian and American newspapers consume from 50 to 100 tons of paper daily.

The daily consumption of newsprint paper in Canada is, approximately, 250 tons; in the United States, 5,750 tons.

Canada's daily output of paper, made into a continuous strip three feet wide, would be long enough to girdle the globe at the equator.

British Cost Is Away Below This Country's

Figures Secured Show That the Shipbuilding Industry in Canada is Up Against a Very Hard Competition by Reason of the Difference in Wages and Cost of Material

Special to MARINE ENGINEER

OTTAWA, November 20.—Under the programme of Canadian shipbuilding authorized by order-in-council last March fifty-five million dollars was to be expended. Up to the date of this article the contracts authorized may be summarized as follows:

Lake type	9	vessels
4,800 ton type	6	"
5,100 " "	8	"
8,100 " "	14	"
10,500 "	2	"
	39	

As mentioned by Sir Joseph Flavelle in an interview given to this paper this week, the policy of the Imperial Munitions Board will be to push its shipbuilding activities to the limit, and as soon as the berths occupied by the Board's vesseds are cleared they will be occupied by ships to be constructed under the Canadian Government programme.

For some time to come the government work will keep all the yards fully occupied. From the very first the Marine Department has been opposed to the increasing of the establishment of new yards. The steel available has not been sufficient to keep existing yards fully occupied, and the Department felt that the best policy was to keep existing yards occupied to their full capacity rather than to increase the number of yards when, owing to the scarcity of labor and material, they would not be running full time. It was also felt that if at any time in future construction should decline, the same thing would apply

The Matter of Price

The prospects for the future largely depend on how far we in Canada can go in the direction of building ships within measurable distance of what they can be built for in the United Kingdom. This applies to the future of shipbuilding in the United States also, because at the present time prices in the Canadian yards compare most favorably with the prices for ships of the same type in American yards.

At present, ships built in Canada are costing about double what they cost in the United Kingdom. This is due to the high cost of material which, in turn, is due to the high cost of labor, mainly.

Steel plates have been available to builders in Great Britain at a price of, approximately, fifty-five dollars a ton. Compare that with the following schedule, showing the delivered cost of steel per long ton from Pittsburg and Chicago to different points in Canada:

From Pittsburg to	Cost per long ton
Vancouver, Victoria and Prince Rupert	$102.25
Port Arthur	89.37
Toronto	78.84
Collingwood	80.19
Montreal	96.89
Three Rivers	81.87
Quebec	81.87
Halifax	84.67

From Chicago to	Cost per long ton
Vancouver, Victoria and Prince Rupert	$100.80
Port Arthur	82.32
Toronto	79.29
Collingwood	81.08
Montreal	99.68
Three Rivers	84.78
Quebec	84.78
Halifax	87.36

So much for the relative cost of materials with ourselves and our great competitors. The next factor is the cost of labor in the shipyards. The following are the wages paid per week to the different classes of shipyard workers in Great Britain and in different points in Canada:

Wages Per Week to Shipyard Workers in Great Britain and Canada, With Per Cent. Increase in Canada Over Great Britain

Trades—	England 50 hrs.	Montreal 50 hrs.	% Inc.	Pt. Arthur 50 hrs.	% Inc.	Pac. Coast 44 hrs.	% Inc.	Pac. Coast % inc. over Montreal
Shipwrights	$16.50	$27.80	68.5	$34.45	108	$36.30	120	30
Joiners	16.50	26.50	60.0	35.77	116	36.30	120	37
Patternmakers	16.50	31.80	92.0	37.00	124	39.30	138	24
Plumbers	16.50	27.80	68.5	34.45	108	33.00	100	19
Blacksmiths	16.05	27.80	73.0	34.45	114	33.00	105	19
Hammermen	13.00	17.50	34.5	35.15	93	24.64	89	41
Painters	16.05	23.85	48.5	29.15	82	20.22	88	26
Machinists	14.40	29.25	100.0	35.75	148	33.00	129	12
Rivetters	15.55	29.15	87.0	35.75	130	33.00	100	12
Holders on	14.15	26.25	85.0	26.50	87	25.52	80	*2¾
Platers	16.05	29.15	81.0	26.50	65	33.00	105	12
Caulkers	15.55	29.15	87.0	35.75	129	33.00	112	13
Electricians	16.50	27.80	68.0	35.77	116	33.00	100	18
Riggers	13.20	23.85	80.0	31.80	141	33.00	150	38
Laborers	12.70	20.00	57.5	21.20	67	21.12	66	5
Engineers	16.09	33.75	110.0	33.75	110	33.00	105	*2
Boiler makers	....	39.25	....	33.75	...	33.00	...	*19
Boiler makers, rivetters	16.50	39.25	137.0	33.75	100	33.00	100	*19
Boiler makers, helpers	12.40	20.00	61.0	22.50	80	23.65	91	18

*Dec.

Approximate Increase—

Montreal over England	80%
Port Arthur over England	108%
Pacific Coast over England	*110%
Pacific Coast over Montreal	*15%

*Not allowing for lower number of working hours.

PROGRESS IN NEW EQUIPMENT

There is Here Provided in Compact Form a Monthly Compendium of Shipbuilding and Marine Engineering Auxiliary Product Achievement

TURNING MARINE THRUST SHAFT

One of the most important parts of a marine engine is the thrust shaft, which transmits the whole thrust of the propeller to the hull, through the medium of the thrust block. It has also to transmit the power from the engine to the propeller, therefore being also subject to torsion. The collars on the thrust shaft have to be turned and faced accurately to ensure an even bearing on all the surface of the thrust shoes. The illustration shows a 13¾ inch diameter, thrust shaft for a large steamer, on the lathe. The engines for which this shaft was built are triple expansion, 25 inch by 42 inch by 67 inch by 42 inch stroke. The flanges of the thrust shaft are 25 inches diameter, collars 22 inches diameter, and the total weight 7,000 lbs. The tail end shaft and all the intermediate shaft lengths were turned up on this lathe, which is the 42 inch triple geared engine lathe of the Canada Machinery Corporation. This lathe has been designed expressly for machining heavy pieces such as described above, and is made especially strong to carry the great weights involved. The bed is made very deep and braced with cross ribs of box design. The headstock is made proportionally strong with the base, and is provided with a four step cone, and back geared drive, in addition to a triple geared drive direct to face plate. The cone has extra wide faces to enable a wide belt to be used, this giving greater power than usually obtainable in this class of lathe. The changes of feed are obtainable through the feed box on the bed below the headstock, and by changing the gears on the head and quadrant plate, any desired feed is obtained. The lathe in the illustration is fitted for motor drive.

TURNING THRUST SHAFT FOR MARINE ENGINE

MILWAUKEE SHAPER

THE MILWAUKEE SHAPER

The tool described here is the latest to be turned out by the Milwaukee Shaper Co., Milwaukee, Wis. The makers claim for this machine extreme simplicity of design, embodying a minimum of working parts, with a consequent absence of jarring or vibrating. The stroke control is actuated by a rack and pinion, and a roller working in a cam on the barrel. It can be adjusted instantly. An indicator is provided on the hand wheel to show when the required length of stroke has been attained. A nut is provided on the hand wheel spindle to lock it when the stroke is set. The table can be swung to any angle by loosening three nuts, and a stop on the apron is provided to bring it back to true position. For some kinds of work the table can be removed, and the work strapped to the apron. The ram is designed with a view to strength, the bearings being made full in width, V shaped and scraped down to a perfectly accurate fit, insuring smooth working, and allowing for the wear to be taken up. The tool head is graduated, and can be swivelled to any angle, while the down feed is provided with a micrometer adjustment. The vise swivel is provided with a graduated base and heavy steel jaws, while for tapered work an extra set of jaws are furnished. The cross rail is designed with extra long, wide bearing surfaces, which are accurately scraped to a true bearing. For taking up wear, a taper key is provided in the apron. Besides the usual swivel table, a tilting table is furnished at a slight extra charge, which can be adjusted from 0 to 7 degrees. The feed arrangement is designed without connecting rods. The gears work on a spline and are always in readiness for action when the table has been set in position. The back gearing is strong, and the change from compound to single can be made very quickly by simply moving a lever at the back of the machine. The column is strong and rigid, the projecting slides giving increased bearing surface to the ram, and the base is cast with strong reinforcing ribs. The base is spread out from the column to make a perfectly stiff, immovable foundation when properly anchored. The table support is provided with an adjustable leg, which can be set to overcome any tendency to spring.

This is a compact and well designed machine, embodying some special features, which should make it a welcome addition to the available tools of this class.

LITTLE DAVID PNEUMATIC CAULKING MACHINE

With the demand for ships and more ships to make good the losses caused by U-boats, and with the shortage of skilled labor, due to the war, a tremendous need arose for mechanical tools to increase and speed up the building of ships, which was a pretty slow process in the days when all the work was done by hand.

The tacking of oakum by mechanical means on the hulls and decks of wooden ships and on wooden decks of steel ships, was never a success until an inventor on the Pacific Coast designed the Little David Pneumatic Caulking Machine, which meets the peculiar requirements necessary, such as coiling or tacking cotton or oakum (either machine or hand spun), to any required depth for final horsing.

The machine is easily handled, weighing only 13¼ lbs. The length over all is 22 inches, and the air consumption at 90 lbs. is 20 cu. ft. of free air per minute. In appearance the Little David pneumatic caulker is similar to a pneumatic riveter, but its action is totally different, being on the same principle as the sand rammer or sewing machine. Guide wheels are used to keep the machine in line with the seam. The stroke being always the same, the depth to which oakum is forced in the seam is regulated by the operator's pressure on the tool.

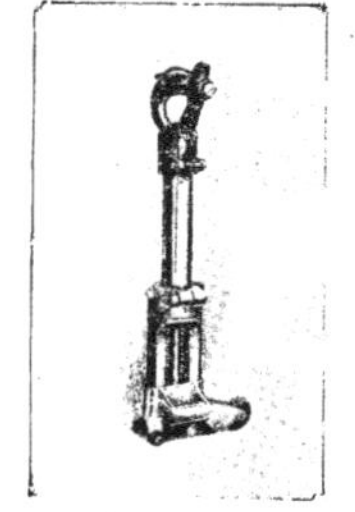

CAULKING MACHINE

The caulking machine operates at the rate of 1,500 tacks per minute, either coiling the oakum or running it straight.

A metal finger at one side over which the oakum passes before it is forced into the seam by the iron, regulates the coiling or tacking. For a long tack the iron is allowed to pass further; for a short tack the normal stroke or position is used.

In standard tack work, such as 5 in. or 6 in. planking, the method employed in certain U. S. yards is to run in the first three strands almost straight to allow cotton or oakum to penetrate farther. The first horsing is then given. The seam is now filled up with two more strands, having these well pinched or coiled, and again horsed. Care must be taken to see that there are no lean places or lumps in the material, otherwise skipping or uneven caulking will result.

All seams on 5 in. or 6 in. planks should be reamed to a depth of 2¼ in. and about 3-16 in. to ¼ in. open at the top.

For the deck with 3 in. planking, ream about 2¼ in. deep, and from ⅛ in. to 3-16 in. open at the top.

Caulking decks, until recently, has always been a slow and tiresome job, but the Little David Caulking Hammer seems to have changed this, as the following test at the Vancouver yard of the G. N. Standifer Corp. proves: 3,100 ft. of single thread oakum was driven home, and time test showed 35 ft. of one seam was completed in 3½ minutes. At the Fraser, Brace and Company's yard, Montreal, where they were using a number of these caulking machines, the following results were obtained in the course of their ordinary work, and without any special test: 41 deck seams, 21 ft. long, were completely finished—that is, both tacked and horsed—in 2½ hours.

In this case two men and two machines were employed, one machine fitted with a 1-16 in. iron for tacking, and the other with a 1-8 in. grooved iron for horsing.

In tacking, the first operation an iron 1-16 in. thick is used. For the horsing of the decks an iron 1-6 inch thick is used, or in some cases we have noted a 3-16 in. iron with a groove, which seemed a decided improvement. A few seconds is all that is necessary to change from one iron to the other.

As stated previously, this No. 23 Little David caulking machine will tack oakum in any part of the hull or deck, but will only do horsing on the deck. The inventor is now endeavoring to design a suitable horsing machine.

It has been estimated that this Little David caulking machine will take the place of ten mechanics. A man by hand will take an average of three days to caulk in a bale of oakum; with a machine the bale will be caulked in an average time of 2½ hours.

This new tool seems to have filled a big vacancy, and from all appearances is here to stay. Wooden ships may not be built very much longer, but the building of steel boats with wooden decks, which require caulking, will be one of Canada's large industries in the future.

Montreal.—The first of the Dominion Government's mercantile marine fleet, the "Canadian Pioneer" and "Canadian Voyageur" were launched from the Vickers' shipbuilding plant by Lady Borden, wife of the Prime Minister, on December 3rd.

Victoria, B.C.—With the recent launching of the 3,000-ton auxiliary schooner "Givenchy" but one more vessel remains

CAULKING DECK PLANKING

to be put into the water at the Foundation Company's Tacoma yard. The "Givenchy" is the nineteenth ship to be completed at Tacoma for the French Government.

CANADA'S MERCANTILE MARINE

Canada's mercantile marine program was inaugurated early this month by the launching from the Canadian Vickers Co.'s yard of the largest steamer yet launched in the Dominion. The vessel, the "Canadian Pioneer," is 8,100 tons d.w. capacity, and her engines, built at the same yard, are triple expansion, 27 inches by 44 inches x 73 inches x 48 inches, of 3,000 I.H.P. The boilers, which are fitted with Howden's forced draft, are 15 ft. 6 in. dia. by 11 ft. 6 in. long, and were also built by the shipbuilders. Most of the auxiliary machinery was built in the company's shops, so that the ship is almost completely a Canadian Vickers product. The secrecy necessary during the war operations having been suspended, the company made the launch an occasion of some ceremony, in accordance with the usage of shipbuilders in the Mother Country. The "Canadian Voyageur," sister ship to the "Pioneer," had been launched some time earlier, without any official ceremony, and at the time of the latter vessel's launching was almost completed and ready for sea. The Minister of Marine, Mr. C. C. Ballantyne, personally congratulated the builders on the splendid performance of their organization in building these ships. The "Canada Voyageur's" keel was laid in March, and the "Canadian Pioneer's" in July last.

Hon. Mr. Ballantyne recalled the negotiations which he, as one of the Harbor Commissioners in 1907-1912, together with Mr. G. W. Stephens and Mr. L. E. Geoffrion, had conducted with the Vickers Company of England, resulting in the establishment of a branch of their industry in the port of Montreal, and the erection of a splendid modern shipbuilding plant, together with a large floating dock that had a lifting capacity of 25,000 tons. It was of interest to know that the thirty acres of land that the Vickers works were situated on had been reclaimed from the river by dredging.

"It is a special pleasure for me this morning to witness the development of steel shipbuilding in Canada," continued the minister, "not only in the Vickers Works, but in the other sixteen yards that are building steel ships for the Government, from Prince Rupert to Halifax. I am pleased to announce that the Government have under construction at the present time thirty-nine steel vessels: Lake type, 9; 4,300-ton type, 6; 5,100-ton type, 8; 8,100-ton type, 14; and 10,-500-ton type, 2.

"Canada requires tonnage, and very badly. It is estimated that owing to losses by enemy submarines the world's tonnage is at least ten million tons short of what it was when war broke out. Canada requires ships as speedily as they can be built, in order that they can be placed on the Atlantic and Pacific oceans, as well as on the Great Lakes, to complete the Government's transportation system and work in conjunction with our Transcontinental system."

Alluding to the announcement already made that a separate steamship company will be created to operate these vessels, Hon. Mr. Ballantyne said that Mr. D. B. Hanna and his board of directors had been given a free hand in the management of the Government's railways, and they would be given an equally free hand in the management of the Canadian Government's mercantile marine.

CANADIAN PIONEER JUST PREVIOUS TO LAYOUT OF FORT WILLIAM PLANT.

Hon. Mr. Ballantyne passed on to emphasize the need for Canada going vigorously after export trade, for which the Government had laid the way by providing ships. The responsibility now rested upon the manufacturers of Canada to rise to the occasion and use every energy to secure export business. He hoped manufacturers and business men would realize that this was the most favorable opportunity they ever had to go after export business. The Government was fully seized with this possibility and at the present time had a trade commission in London to see that the way was made easy for Canada to get her full share of the vast amount of materials that would be required for the restoration of devastated France and Belgium, also to get her share of all those products that would be wanted by those countries which had been engaged in war, to replenish their stocks.

Paying a tribute to the part played by the mercantile marine in the war, the Minister of Marine mentioned the brave men of England's huge mercantile marine who had manned her ships during the war and helped defeat the enemy by the successful transportation of troops, munitions and provisions. "I consider this day is an epoch in the history of Canada," concluded the minister, "and for the first time a mercantile marine flag of Canada has fluttered to the breeze on the "Canadian Pioneer" just launched. With the lead that the Government has given and its determination to assist Canada in every way to do a large export trade, I hope in the very near future that the flag of Canada's merchant marine may be seen in every important port throughout the world, carrying to these distant countries Canada's production of the mine, field, forest and our industries, and bringing back the importations that Canada will find it necessary to make. All this is possible by co-operation of all Canada's diversified interests. I wish every success to the 'Canadian Pioneer,' the 'Canadian 'Voyageur,' and the other Government mercantile marine ships that are to be launched in the very near future."

The "Canadian Pioneer" is of 8,100 tons D.W., twin deck, 11½ knots sea speed, length 400 ft., breadth moulded 52 ft., depth moulded 31 ft., I.H.P. 3,000. A sign that it is the first of a peace fleet was the fact that it was not camouflaged, as are all existing ocean-going ships. It was painted red on the keel up to the 20 ft. draught mark, and black above that, the name being picked out in white lettering on either side of the stern.

CANADIAN PIONEER AFLOAT

Work in Progress in Canadian Shipyards

A record of Work in Process of Completion—Principal Features of Specification—Approximate Launching Dates—Vessels on Order.

BRITISH COLUMBIA

YARROWS, LIMITED, ESQUIMALT, B.C.

One stern-wheel steamer, 165 ft. x 35 ft., for Indian Government service.

At present engaged chiefly on repair work.

H. VOLLMERS, NANAIMO, B.C.

One gasoline tug, 9 tons.

NEW WESTMINSTER CONSTRUCTION & ENGINEERING CO., NEW WESTMINSTER, B.C.

Four wooden steamers, Nos. 10-11-12-13, for undisclosed interests. 250 ft. x 43 ft. 6 in. x 25 ft., moulded depth, 2,800 tons d. w., 1,000 h.p., single screw engines, 12 knots, naval architect, I. Alexander. Launchings March 14, April 14, May 15 and June 15, 1918.

B.C. CONSTRUCTION & ENG. CO., POPLAR ISLAND, NEW WESTMINSTER, B.C.

Four wooden vessels, 1,800 gr. tons, 2,800 tons d.w.

NEW WESTMINSTER MARINE CO., FT. OF FURNESS ST., NEW WESTMINSTER, B.C.

STAR SHIPYARD CO., NEW WESTMINSTER, B.C.

J. A. CROLL, PORT ALBERNI, B.C.

PACIFIC CONSTRUCTION CO., PORT COQUITLAM, B.C.

Taken over the plant of the Coquitlam Shipbuildiug Co.

Gasoline yacht, 50 ft., June, 1918.

Three wooden freight steamers, Nos. 20-21-22, 250 ft. b. p. x 42 ft. 6 in. x 25 ft., 1,800 gr. tons, 2,800 tons d.w., trip. exp. engines, water tube boilers ,12 knots, 1,000 h.p. Two for undisclosed interests, one for builder's account. Launchings May 20, June 15, Sept., 1918.

Four new berths under construction for steel ships up to 10,000 tons.

C. E. BAINTER, PRINCE RUPERT, B.C.

Several wooden fishing steamers, for undisclosed interests.

GRAND TRUNK DRYDOCK SHIP REPAIR CO., LTD., PRINCE RUPERT, B.C.

S. A. MOULTON, PRINCE RUPERT, B.C.

Ten composite vessels for undisclosed interests.

BRITISH-AMERICAN SHIPBUILDING & ENG. CO., VANCOUVER, B.C.

Establishing plant on Kitsilano Reserve.

Twenty wooden vessels being built for undisclosed interests, 3,500 tons each.

JOHN COUGHLAN & SONS, N. VANCOUVER, B.C.

"War Camp," sister ship to "Alaska," launched in March, 1918.

One steel freight steamer, 427 ft. 9 in. x 54 ft. x 29 ft. 9 in., draft 24 ft. 2 in., speed 11½ knots, 5,730 gross tons. Launched Jan. 10, 1918, for B. Stolt Nielsen, Christiania, Norway. Both sold to the Cunard Line.

Three steel steamers, "War Chariot," "War Chief," and "War Noble," Nos. 4-5-6, 425 ft. x 54 ft., 8,800 tons cap., for the Cunard Line.

Six steel freight steamers, 425 ft. x 54 ft. x 24 ft. 2 in., 8,800 tons cap., 1 knots, for undisclosed interests. Delivery Jan., Feb., March, May and July, 1918.

One steel freight steamer, "Alaska," 427 ft. 9 in. x 54 ft. x 29 ft. 9 in., draft 24 ft. 2 in., speed 11½ knots, 5,730 gr. tons, 8,800 d.w. Launched Jan. 10, 1918, for B. Stolt Nielsen, Christiania, Norway. Sold to the Cunard Line.

THE FOUNDATION CO., VANCOUVER, B.C.

FRASER VALLEY SHIPBUILDING CO., VANCOUVER, B.C.

GRANT, SMITH & CO., VANCOUVER, B.C.

Six wooden cargo vessels, 250 ft. x 43 ft. 6 in. x 25 ft., 3,000 tons cap., 9½ knots, for the Canadian Government.

HARRISON & LAMOND, SHIPBUILDERS, LTD., VANCOUVER, B.C.

One wooden auxiliary schooner, 225 ft. x 44 ft. x 21 ft. 4 in., 1,600 gr. tons, 2,550 tons cap., for undisclosed interests.

LYALL SHIPBUILDING CO., NORTH VANCOUVER, B.C.

"War Puget," Feb. 16, 1918, 250 ft., wood, 2,080 tons.

Ten wooden cargo ships, for undisclosed interests. Launchings Jan. 20, Mar. 2, Mar. 23, and April 11, 1918.

Building six wooden vessels for own account.

"War Caribou," April 10, 1918, 3,080tons, wood.

W. R. MANCHION, VANCOUVER, B.C.

STANDARD SHIPBUILDING CO., VANCOUVER, B.C.

Two 3,500-ton vessels for Brazilian Govt., and six for French Govt. of 4,500 tons.

Donohoe reinforced type, wood composite construction.

TAYLOR ENGINEERING CO., VANCOUVER, B.C.

Number of small vessels being constructed at total value of $300,000.

4,500-ton floating drydock, 352 ft. long.

VANCOUVER SHIPYARD, LTD., VANCOUVER, B.C.

One motor freighter, 125 ft. x 24 ft. x 12 ft., wood, 160 h.p., Bolinder's crude oil engine. To be completed Aug., for Taylor Engr. Co.

Also repairing several steamers and schooners.

THE WALLACE SHIPYARDS, LTD., N. VANCOUVER, B.C.

"War Dog," 4,500 tons, steel, May 18, 1918.

"War Power," 4,600 tons, steel, March 23, 1918.

Two steel steamers, 315 ft. x 45 ft. x 27 ft., 4,600 tons d. w., trip. exp. vertical engines ,two S. E. Scotch boilers, 10 knots, single screw, for the Canadian Government. Deliveries Dec., 1917, and Aug., 1918.

Two wooden aux. schooners, 255 ft. o. a., 225 ft. keel x 44 ft. x 21 ft. 4 in., five masts, 1,500 gr. tons, cap. 2,500 tons or 1,500,000 ft. lumber, two hatches, twin 160 h.p. Bolinders Hot-bulb engines, twin screws, for the Canadian Government.

Four wooden cargo vessels, aggregate tonnage of 17,500 tons for undisclosed interests.

CHOLBERG SHIPYARD, VICTORIA, B.C.

4 1,500-ton vessels for Norwegian purchasers. Keels laid Oct. 4th, 10th and 25th respectively.

TAYLOR ENGINEERING CO., VANCOUVER, B.C.

"Teco," 300 tons. Motor coasting freighter fitted with Bolinder 200 h.p. Built for William Mason Rooke.

NEW WESTMINSTER CONSTRUCTION & ENGINEERING CO., NEW WESTMINSTER, B.C.

5 1,500-ton d.w. wooden vessels for the French Government, four of which are in frame. Planking and ceiling progressing in three.

NORTHERN CONSTRUCTION CO., VANCOUVER, B.C.

5 1,500-ton d.w. compound engines, 12 x 24 x 16 stroke, twin screw water tube boilers, 2,000 sq. feet heating surface. Building for French Government.

THE FOUNDATION CO. OF B.C., VICTORIA, B.C.

20 wooden ships for French Government, 3,000 tons d.w., fitted with triple expansion twin screw engines, 1,100 I.H.P. Contract to be completed in December.

WESTERN CANADA SHIPYARDS ,LTD., VANCOUVER, B.C.

"War Nootka," Jan. 4, 1918, 3,080 tons (machinery installed at Ogden Point Assembly Sheds).

Six wooden freight steamers, 250 ft. h.p., 259 ft. o. a., 43½ ft. extreme breadth, molded 42½ ft., 25 ft. molded depth, draft 22 ft., 2,800 tons d. w., for undisclosed interests. "War Nootka," launched January 4, 1918. "War Selkirk" launched March 6, 1918.

"War Tatla," "War Casco, "War Chilkat," "War Tanoo," 4 standard vessels now under construction.

CAMERON-GENOA MILLS CO., VANCOUVER, B.C.

"War Yukon," Jan. 24, 1918, 3,080 tons, wood (machinery installed at Ogden Point Assembling Sheds.

THE FOUNDATION CO. OF B. C., LTD., VICTORIA, B.C.

Building ships for undisclosed interests.

Five wooden steamers, 250 b. p. x 42 ft. 6 in. x 25 ft., 1,800 gr. tons, 2,800 tons d. w., one trip. exp. engine, 1,000 h.p., "Howden" water tube boilers, furnished and installed by owners, 10 knots, for undisclosed interests. Esplen & Son & McNaught, N.Y.C., designers.

No. 1—"War Sonchee," launched Dec. 27, 1917.

No. 2—"War Masset," launched April 11, 1918.

No. 3—"War Babine," ready for launching, awaiting propeller and rudder.

No. 4—"War Camchin," ready for launching, awaiting propeller and rudder.

No. 5—"War Nanoose," ready for launching in about five weeks.

CLARENCE HOARD, VICTORIA, B.C.

Wooden car barge for C.P.R., Jan., 1918.

VICTORIA SHIPBUILDING CO., VICTORIA, B.C.

Constructing wooden ships for British Govt., 3,500 tons, each.

NEW BRUNSWICK

C. T. WHITE & SON, LTD., ALMA, N.B. (Office at Sussex, N.B., also.)

Two schooners aux. power, 143 ft., 900 tons d. w., first to be launched in April; second in June.

JAMES X, LENTEIGNE, LOWER CARAQUET, N.B.

One schooner, 28 tons, wood.

INTERNATIONAL SHIPBUILDING CO., NEWCASTLE, N.B.

Two four-masted wooden auxiliary schooners, 155 ft. keel, 37 ft. depth, 535 net tons. One to be completed in September and the other December, 1918. For builders' account.

EUREKA SHIPBUILDING CO., NORTH HEAD, N.B.

One wooden schooner, "Mollie & Melba," 350 tons reg.

PORT COLBORNE BUILDING & REALTY CO., LTD., REXTON, N.B. (Head Office, Welland, Ont.)

Four-masted wooden schooner. Has cap. of 3 schooners per year.

GRANT & HORNE, ST. JOHN, N.B.

Two wooden cargo steamers, "War Fundy" (launched Aug., 1918) and "War Digby," 250 ft. x 42 ft. 5 in. x 25 ft. 5 in., 1,000 h.p., speed 9¼ knots, 2,800 tons d.w., for undisclosed interests.

MARINE CONSTRUCTION CO., CANADA, LTD., ST. JOHN, N.B.

Four wooden auxiliary, four-masted schooners, 185 ft. o. a., 165 ft. keel x 40 ft., 1,100 d. w. tons, for builder's account. J. Murray Watts, Philadelphia, naval architect. Launching April, 1918.

One wooden schooner, "Dorfontein," 740 reg. tons. Launching June, 1918.

ST. JOHN SHIPBUILDING CO, ST. JOHN, N.B. (Plant at Courtenay Bay.)

Ten five-masted auxl. schooners, oil burning engines, for undisclosed interests.

Establishing plant for steel and wood construction.

ST. MARTIN'S SHIPBUILDING CO., ST. MARTIN'S, N.B.

One wooden schooner, 450 tons, three masts. To be launched in July for builder's account.

PETER & A. A. McINTYRE, ST. JOHN N.B.

One 900-ton schooner under construction.

NEWFOUNDLAND

ANGLO-NEWFOUNDLAND DEVELOPMENT CO., BOTWOOD, NFLD.

Two three-mast aux. schrs., wood, 450 tons, 150 H.P. for the builders' account.

NEWFOUNDLAND SHIPBUILDING CO., HARBOR GRACE, NFLD.

To build ships of 1,200 tons d.w. Steel ships later.

M. E. MARTIN, MORRIS ARM, NR. ST. JOHN'S, NFLD.

Two wooden schooners, 300 tons each, for builder's account.

One wooden schooner, 500 tons, for builder's account.

UNION SHIPBUILDING CO., ST. JOHN, NFLD.

ANNAPOLIS SHIPPING CO., ANNAPOLIS ROYAL, N.S.

"Hilda M. Clark," March 4, 1918, 187 ft., 640 tons reg.

Two schrs. three-masted, wood 500 tons, 170 ft. long. To be completed August and December, 1918.

NOVA SCOTIA

B. L. TUCKER, BASS RIVER, N.S.
One schooner, 350 tons, wood.

JAS. S. CREELMAN, BASS RIVER, N.S.
One three-masted schooner, wood, 170 ft. long, 500 tons. To be completed October, 1918, for builders' account.

HANKINSON SHIPBUILDING CO., BELLIVEAU'S COVE, N.S.
One schooner, 360 tons, wood.

ELI PUBLICOVER, BLANDFORD, N.S.

BRIDGEWATER SHIPPING Co., BRIDGEWATER, N.S.
One tern schooner, 450 tons gr., wood. Adapted for aux. power. To be completed Sept., 1918.
One tern schooner, 275 tons gr., wood. Adapted for aux. power. To be completed Nov., 1918.
Both built on owner's account and for sale.

W. A. NAUGLER, BRIDGEWATER, N.S.
One wooden three-masted schooner, "Wm. A. Naugler," 146 ft. o. a., 116 ft. keel x 32 ft. x 11 ft. 6 in., 360 gr. tons, 600 tons cap., for builder's account.
One schooner, 300 tons, wood.

H. MAC ALONEY, CANNING, N.S.
Two wooden schooners, 400 tons each. For builder's account. Launchings July and October, 1918.

S. M. FIELDS, CAPE D'OR, N.S.
One three-masted schooner, 350 tons.
Keel laid for another three-master.

CHESTER BASIN SHIPBUILDERS, LTD., CHESTER BASIN, N.S.
Two schooners, 100 tons, wood. Launchings Aug., 1917, Dec., 1917.
One four-masted schooner, 60 tons. Launching Sept., 1918.
One two-masted schooner, 100 tons. Launching June, 1918.

MORTIMER PARSONS, CHEVERIE, N.S.
One wooden three-masted schooner, 170 ft. x 38 ft. x 14 ft., 575 gr. tons, 900 tons cap., for builder's account. Launching Sept., 1918.
Two wooden three-masted schooners for builder's account.
One tern schooner, "Donald Parsons," 140 ft. keel x 34 ft. 4 in. x 12 ft. 4 in., 500 gross tons, 825 tons d.w. for Mortimer Parsons. Launched May 11, 1918.
Another vessel, same size, to be launched Nov., 1918, for builder's account.

BAY SHORE SHIPYARD, CHURCH POINT, N.S.
Building two wooden vessels for R. C. Elkin, Ltd., St. John, N.B.

MOISE BELLIVEAU, CHURCH POINT, N.S.
One schooner, 450 tons, wood.

FIDELE BOUDREAU, CHURCH POINT, N.S.
One schooner, 350 tons, wood.

J. E. GASKILL, CHURCH POINT, N.S.
Other yards at Grosses Coques, Little Brook, Meteghan and Port Wade, N.S.
Five wooden schooners from 344 to 387 reg. tons. All sister ships.

COMEAU SHIPBUILDING CO., COMEAUVILLE, N.S.
One three-masted wooden schooner, for builder's account, 450 tons.

J. W. COMEAU, COMEAUVILLE, N.S.
One schooner, 329 tons, Jan., 1918.

J. N. RAFUSE & SONS, CONQUERALL BANKS, DIGBY CO., N.S. (Other plants at Salmon River and Shelburne, N.S.)
Two wooden three-masted schooners, 120 ft. x 32 ft. x 12 ft., 700 tons cap., for builder's account. For the market. Launchings Oct. 25, 1917, and Jan. 10., 1918.
One three-masted wooden schooner, "Integra," 112 ft. x 30 ft. x 11 ft. 6 in., 600 tons cap., for J. O. Williams & Co., St. John's, Newfoundland. Launching Nov. 25, 1917.

E. F. WILLIAMS, DARMOUTH, N.S.
One schooner, 360 tons, wood.

ROBAR BROTHERS, DAYSPRING, N.S.
Schooner for Capt. Ivan Creaser, Dayspring.

MAURICE E. LEARY, DAYSPRING, N.S.
One wooden schooner, 225 tons, 135 ft. o. a. To be launched Aug. 15, 1918, for La Have Outfitting Co., La Have, N.S.

J. NEWTON PUGSLEY & CHAS. ROBERTSON, DILIGENT RIVER, N.S.
One schooner, three-masted, 475 tons.

McLEAN & McKAY, ECONOMY, N.S.
One tern schooner, wood, 400 tons net, 135 ft. keel. To be launched Sept. 1, 1918, for builders' account.

S. J. SOLEY, BOX RIVER, N.S.
One schooner, three-masted, 121 ft. keel, 300 tons. Built of native wood. To be launched Sept., 1918. For builder's account. For sale.

ALLAN & FRASER, FRASERVILLE, N.S.
One schooner, 350 tons, wood.

BERNARD W. MELANSON, GILBERT'S COVE, N.S.
One three-masted schooner, wood, 250 tons net. To be completed November, 1018, for builder's account and for sale.

AMOS BLINN, GROSSES COQUES, N.S.
One schooner, 375 tons. Jan., 1918.
One schooner, 350 tons, wood.

BINN BROS., GROSSES COQUES, N.S.

F. K. WARREN & CO., GROSSES COQUES, N.S. (Offices in Union Bank Chambers, Halifax, N.S., also.)
350-ton tern schooner, under construction Jan., 1918.

HALIFAX SHIPBUILDING CO., HALIFAX, N.S.

J. WILLARD SMITH, HILLSBURN, N.S. (Head Office, St. John, N.B.)
One tern schooner, wood, 140 ft. keel, 35 ft. beam, 450 tons net reg. tons.

J. W. RAYMOND, LITTLE BROOK, N.S.
Building 500-ton wooden schooners for local owners.

R. H. HOWES CONSTRUCTION CO., METEGHAN, N.S.
One 800-ton vessel for undisclosed interests. Launch about Dec. 24.

CONRAD & REINHARDT, LUNENBURG, N.S.
"Excellent." 600-ton, launch in January next. Also 175-ton schooner.

J. WILLARD SMITH, HILLSBURN, N.S.
Wooden three-masted schooner. Launch in April, 1919.

ERNST SHIPBUILDING CO., MAHONE BAY, N.S.
750-ton auxiliary schooner. Completed in January, 1919.

OSMOND O'BRIEN & CO., NOEL, N.S.
"J. Miller." 700 tons. Keel laid April 18. Launched Nov. 20. Completed Nov. 27.

BERNARD & MELANSON, GILBERT'S COVE, N.S.
550-ton vessel for owner's account. Completed Nov. 25.

ANNAPOLIS SHIPPING CO., ANNAPOLIS ROYAL
"Mapleland," 1000-ton schooner, fitted with twin Fairbanks-Morse oil engines, 200 horsepower. Complete about Dec. 10. For builder's account.

L. S. CANNING, WARD'S BROOK, N.S.
"W. C. Smith," 700 tons. To be completed Dec. 18.

McLEAN & McKAY, ECONOMY, N.S.
"Truro Queen," 750 tons, 15 h.p. engine for hoisting sails and anchor. Completed.

HENRY COVEY, INDIAN HARBOR, N.S.

CHARLES GRIFFIN, ISAACS HARBOR, N.S.
One schooner, 40 tons, wood.

LEWIS SHIPBUILDING CO., LEWISTON, N.S.
One schooner, 670 tons. Jan., 1918.

INNOCENT COMEAU, LITTLE BROOK, N.S.
One three-masted schoner, 140 ft. x 35 ft. 6 in. x 17 ft. 6 in., 660 gr. tons, for the Weymouth Shipping Co., Weymouth, N.S., Can. Delivered Jan. 18, 1918.

J. W. RAYMOND, LITTLE BROOK, N.S.
One wooden schooner, No. 4, 140 ft. x 36 ft. x 17 ft., 575 gr. tons, for Jones Bros., Weymouth, Digby Co., N.S.

H. A. FRANK, LIVERPOOL, N.S.
Two wooden schooners, 500 tons, for builder's account.

McKEAN SHIPBUILDING CO., LIVERPOOL, N.S.
One wooden three-masted schooner, 126 ft. keel x 3 ft. x 12 ft. 9 in., 700 tons d.w., for builder's account. Launching May, 1918.

W. F. McKEAN & CO., LIVERPOOL, N.S.
One schooner 400 tons, Jan., 1918.

D. C. MULHALL, LIVERPOOL, N.S.
Two wooden schooners, one 300 gross tons, other 400 gross tons, for undisclosed interests.

ROBIN, JONES & WHITMAN, LIVERPOOL, N.S.
One schooner, 340 tons, wood.

SOUTHERN SALVAGE CO., LIVERPOOL, N.S.
One wooden schooner, three masts, 300 gross tons, 600 tons capacity, for a West Indian firm. Launching Jan., 1918.
One wooden steamer, No. 5, 250 ft. x 43 ft. 6 in. x 25 ft., 2,900 gr. tons, 10 knots, 1,000 h.p., for undisclosed interests. Launching Fall, 1918.
One two-masted schooner, "Win-the-War," 137 ft. o. a. x 25 ft. 2 in. x 11 ft. 6 in. ,187 gr. tons, for the builder's account. Launched Nov., 1917.

CONRAD & REINHARDT, LUNENBURG, N.S.

LUNENBURG MARINE RAILWAY, LUNENBURG, N.S.

FRED A. ROBAR, LUNENBURG, N.S.

J. B. YOUNG, LUNENBURG, N.S.

ERNST SHIPBUILDING CO., MAHONE BAY, N.S.
One two-masted schooner, 120 ft. o. a. x 25 ft. 6 in. x 10 ft. 6 in., 200 tons d.w., for Lunenburg Outfitting Co. April, 1918. Christened "Madeline Adams."
One three-masted schooner 125 ft. keel x 32 ft. x 12 ft., 600 tons, d. w. To be delivered July, 1918.
Two masted schooner, "Agnes D. McGlaston," 130 ft. x 26 ft. 8 in. x 10 ft. 8 in. Launched Nov. 13, 1917.
Keel laid for three-masted wooden schooner, 200 tons.

J. ERNEST & SON, MAHONE BAY, N.S.
One schooner, 520 tons. Jan., 1918.

O. A. HAM, MAHONE BAY, N.S.
One schooner, "Dozle," 128 tons. Launching May 30. One motor boat about 20 tons.

McLEAN CONSTRUCTION CO., MAHONE BAY, N.S. (Leased to Montague Mahaffy, Toronto.)
One three-masted schooner, 325 tons gr. reg., 500 tons cap., West Indies freight type. Will be launched in July.
Keel laid for another three-masted schooner with cargo capacity of 425 tons.

JOHN McLEAN & SONS, MAHONE BAY, N.S.
One schooner, 95 tons, wood.

J. A. BALSOM CO., LTD., MARGARETSVILLE, N.S.
One schooner, 409 tons. Jan., 1918.

CLARE SHIPBUILDING CO., METEGHAN RIVER, N.S.
One wooden schooner for builder's account. 400 tons.

A. H. COMEAU & CO., METEGHAN, N.S.
One schooner, 400 tons, wood.

AGAPIT COMEAU, METEGHAN, N.S.

JOHN F. DEVEAU, METEGHAN, N.S.
A 300-ton schooner was launched recently for Ritcey & Co., Lunenburg, N.S., and named Charles A. Ritcey.
1 schooner, 425 tons. Jan., 1918.
1 schooner, 400 tons, wood.

MILTON SHIPBUILDING CO., YARMOUTH, N.S.
One wooden sailing schooner, 800 tons d.w. Keel laid April 15th, 1918, to be launched December 17th and completed December 31st. Built for sale.

J. E. GASKILL, CHURCH POINT, N.S.
One wooden schooner, 750 tons d.w. Keel laid March, 1918, to be launched December 15th. Built for J. E. Corkill.

THERIAULT SHIPBUILDING CO.
One wooden vessel, 500 tons d.w. Keel laid September 18th, to be launched in May, 1919. Built for owners account.

SIDNEY ST. C. JONES, WEYMOUTH, N.S.
One tern schooner, "Westway," 450 tons d.w. Keel laid April 18, to be launcheed December 30, 1918, and completed in January, 1919. Built for sale. 130 feet by 28 feet 8 inches beam by 10 feet 6 inches deep.

THE HANKINSON SHIPPING CO., BELLEVEAU'S COVE, N.S.
One vessel, 750 tons d.w. "Charles Doucet," fitted for auxiliary engine. Keel laid April 1, to be launched December 17. One vessel 750 tons d.w. Keel laid Aug., 1918, to be launched April, 1919, and completed. Built for owners account.

FAUQUIER & PORTER, HANTSPORT, N.S.
One vessel 1,800 tons d.w. to be fitted with two 100 h.p. crude oil engines (Canadian Fairbanks-Morse), to be launched Dec. 17, 1918, and completed January, 1919. Built for sale. Propose to lay keels for two 500-ton, 3-mast schooners as soon as this schooner is launched.

N.S. SHIPBUILDING & TRANSPORTATION CO., LIVERPOOL, N.S.

"Protea," 400 tons d.w., three-masted schooner with provision for Morse semi-Diesel engines. Keel laid Nov. 18. To be completed May account and sold to Pearson & Walker, Cape Town, S.A.

"Gordon T. Tibbs," 400 tons d.w., three-masted schooner. Keel laid Aug. 30. To be launched Dec. 15 and completed Jan. 19. Built for buildeer's account and sold to G. Tibbs & Sons, St. John's, Nfld.

One beam trawler, 900 tons d.w., 400 i.h.p., with two 200 Fairbanks-Morse semi-Diesel engines. Keel laid Nov. 19. To be completed May next. Built for Rafuse Grey et al Le Havre. Repairing schooner Karmoe for N.Y. owners.

H. ELDERKIN & CO, PORT GREVILLE, N.S.

Two vessels, 825 and 500 tons respectively. Keel laid April and August. To be launched December and January. Built for builder's account.

J. E. GASKILL, METEGHAN, N.S.

THOMAS GERMAN, METEGHAN, N.S.

One schooner, 350 tons, wood.

R. H. HOWES CONSTRUCTION CO., METEGHAN, N.S. (Leased James Cosman's shipyards.)

Building several wooden schooners for own account.

DR. F. H. MACDONALD, METEGHAN, N.S.

One wooden four-masted schooner, "Rebecca L. Macdonald," 201 ft. o. a. x 6 ft. x 16 ft., 800 gr. tons, 1,500 tons d.w., for builder's account. Launching January 1, 1918.

One wooden fishing schooner, No. 3, 84 ft. x 17 ft. x 7 ft., 50 gr. tons, 5 h.p. Launching August, 1918.

One schooner, 544 tons, wood.

METEGHAN RAILWAY & SHIPBUILDING CO., METEGHAN, N.S.

One wooden three-masted schooner, 182 ft. x 35 ft., 450 gr. tons.

One schooner, 470 tons, wood.

CHAS. McNEIL, NEW GLASGOW, N.S.

One wooden 3-masted schooner, 103 ft. x 30 ft. x 10 ft. 9 in., 200 tons, for builder's account. Launching July, 1918.

One wooden 3-masted schooner, 142 ft. x 25 ft. 4 in. x 13 ft., 500 gr. tons, for builder's account. Launching Oct., 1918.

THE NOVA SCOTIA STEEL & COAL CO., NEW GLASGOW, N.S.

Two steel freight steamers, 257 ft. 9 in. x 35 ft. x 20 ft., 1,700 gr. tons, 2,350 tons cap., trip. exp. engines, boilers 10 ft. 6 in. x 11 ft. 6 in., 800 h.p., 8.5 knots, for undisclosed interests. Launching of one Jan, 1918.

Two steel cargo strs. Raised quarter deck type, 248 ft. 9 in. long, 1,649.68 gross tons. Triple expansion engines, 17, 28, 46 x 33 in. stroke. One, the "War Bee," for undisclosed interests to be completed June, 1918. Other for Steel Co., to be completed Oct., 1918.

Two masted schooner, launched May 16, 130 ft. o. a. x 26 ft. x 10 ft. 6 in. Designed by J. S. Gardner.

O'BRIEN BROS., NOEL, N.S.

One schooner, 325 tons, wood.

W. R. HUNTLEY & SON, PARRSBORO, N.S.

One wooden schooner, 175 ft. x 39 ft. x 17 ft., 900 gross tons, for

C. T. WHITE & SON, SUSSEX, N.S.

Two schooners, 325 tons each. Jan., 1918.

One schooner, 490 tons, wood.

One schooner, 850 tons, four masted.

WAGSTAFF & HATFIELD, PARRSBORO, N.S.

One schooner, 400 tons. Jan., 1918.

S. SALTER, PARRSBORO, N.S.

One 200-ton schooner, wood.

DOWLING & STODDART, PORT CLYDE, N.S.

One gas boat, 27 tons, wood.

One schooner, 300 tons, wood.

SWIME BROS., PORT CLYDE, N.S.

Several small motor trawlers, for builder's account.

G. M. COCHRANE, PORT GREVILLE, N.S.

One tern schooner, "Alfredock Hedley," 152 ft. 6 in. x 36 ft. 12 ft. 6 in., 461 tons reg., for Adam B. Mackay, of Hamilton, Ont. Launched Jan., 1918.

One 4-masted schooner, 850 tons, 155 ft. x 37 ft. x 18 ft., 2 decks, wood. To be completed Oct., 1918, for builder's account.

ELLIOTT GRAHAM, PORT GREVILLE, N.S.

Is building a schooner, "Khaki Lad," of about 325 tons, at Port Greville, N.S., for J. W. Kirkpatrick and others. She will be launched early in October, and is reported to have been sold to Newfoundland parties.

One schooner, 360 tons. Jan., 1918.

SMITH CANNING CO., PORT GREVILLE, N.S.

One schooner, 350 tons, wood.

WAGSTAFF & HATFIELD, PORT GREVILLE, N.S.

One schooner, 400 tons, wood.

WILLIAM CROWELL, PORT LATOUR, N.S.

J. W. RAYMOND, PORT MAITLAND, N.S.

One schooner, 375 tons. Jan., 1918.

PORT WADE SHIPBUILDING CO., PORT WADE, N.S. (Head Office, Digby.)

One 350-ton schooner, wood.

JOHN BROWN, PUBLIC LANDING, N.S.

One tow barge, 50 tons, wood.

THE CUMBERLAND SHIPBUILDING CO., PUGWASH, N.S.

Establishing plant for the construction of wooden ships.

MACKENZIE SHIPPING CO., RIVER JOHN, N.S.

One four-masted schooner, 600 tons, wood. About half built. Fitted for aux. engines.

CHARLES McLELLAN, RIVER JOHN, N.S.

One schooner, 100 tons, wood.

W. J. FOLEY, SALMON RIVER, N.S.

Building for J. N. Rafuse & Sons one wooden, 300 tons, under construction. Cap. of plant, 1,000 tons yearly.

J. N. RAFUSE & SONS, SALMON RIVER, N.S. (Other plants at Shelburne and Conquerall Banks.)

Launched recently a three-masted schooner named "Industrial," at W. J. Foley's ship yard at Salmon River, N.S., of the following dimensions: length, 113 ft. breadth, 30 ft., depth, 11½ ft., 325 tons.

ACADIAN SHIPPING CO., SAULNIERVILLE, N.S.

One schooner, 00 tons, wood.

SAULNIERVILLE SHIPBUILDING CO., LTD., SAULNIERVILLE, N.S.

J. LEWIS & SONS, SHEET HARBOR, N.S.

One schooner, wood, four-masted, 725 gr. tons. To be completed Sept., 1918. For private account.

GEO. A. COX, SHELBURNE, N.S.

One schooner, 200 tons, wood.

One schooner, 322 tons, wood.

JOSEPH McGILL SHIPBUILDING & TRANSPORTATION CO., SHELBURNE, N.S.

"Sparkling Glance," 246 tons register, for Harvey & Co., St. John's, Nfld.

One schooner, 160 tons, wood.

W. C. McKAY & SON, SHELBURNE, N.S.

One wooden schooner, 130 ft. o. a. x 26½ ft. x 11 ft. depth of hold, 140 tons register. Launched Dec. 1, 1917.

One three-masted schooner, for Messrs. Hallet, of Newfoundland.

One schooner, 620 tons, wood.

J. N. RAFUSE & SONS, SHELBURNE, N.S. (Other plants at Salmon River and Conquerall Banks.)

One wooden three-masted schooner, 120 ft. x 32 ft. x 12 ft., 700 tons cap., for builder's account. For the market. Launching Jan. 10, 1918.

One wooden three-masted schooner, 118 ft. x 33 ft. x 12 ft., 700 tons cap., for builder's account. For the market. Launching Jan. 15, 1918.

Two three-masted wooden schooners, 122 ft. 6 in. keel x 32 ft. x 12 ft., 275 net tons, for J. O. Williams & Co., St. John's, Newfoundland. One to be built at Salmon River yard and one at Ship Harbor.

THE SHELBURNE SHIPBUILDERS, LTD., SHELBURNE, N.S.

One tern schooner, "Misty Star," 145 ft. x 31 ft. x 11 ft. 6 in., 600 tons d.w., 330 tons gross, for Harvey & Co., St. John's, Newfoundland. Launching Mar. 15, 1918.

One tern schooner, 154 ft. x 32 ft. x 12 ft. 4 in., 400 gr. tons, 700 tons d.w. cap. Launching June, 1918. Not sold yet.

Two schooners, one 349 tons, other 400 tons.

EASTERN SHIPBUILDING CO., SHIP HARBOR, N.S.

One wooden schooner, 600 tons d.w., for J. N. Rafuse & Sons., Halifax, N.S., Can.

One wooden schooner, 300 tons.

STEPHEN MORASH & CO., SHIP HARBOR, N.S.

One wooden schooner, 150 ft. x 37 ft. x 13 ft., four masts, for Canadian interests. Keel laid Feb. 5, 1917.

JAMES E. PETTIS, SPENCER'S ISLAND, N.S.

One schooner, 425 tons, wood, three-masted.

CAPE BRETON SHIPBUILDING CO., SYDNEY, N.S.

THE TUSKET SHIPBUILDING CO., TUSKET, N.S.

AMOS H. STEVENS, TANCOOK, N.S.

ALVIN STEVENS, TANCOOK, N.S.

STANLEY MASON, TANCOOK, N.S.

ALBERT PARSONS, WALTON, N.S.

One schooner, 400 tons. Jan., 1918.

H. T. LeBLANC, WEDGEPORT, N.S.

One steam trawler, 165 ft., 600 tons, for J. N. Rafuse & Sons.

L. S. CANNING, WARDS BROOK, N.S.

One tern schooner, 350 tons, for W. C. Smith & Co., Lunenburg, N.S

WEDGEPORT NAVIGATION & TRANSPORTATION CO., WEDGEPORT, N.S.

T. K. BENTLEY, WEST ADVOCATE, N.S.

One 200-ton steamer.

One 4-masted schooner, wood, 511 tons. To be launched September or October. For owner's account.

J. W. KIRKPATRICK, WEST ADVOCATE, N.S.

One 350-ton schooner, wood.

BOEHNER BROS., WEST LA HAVE, N.S.

Two large fishing schooners, 150 tons each.

One beam trawler, 300 tons, crude oil engines. For local interests.

BEAZLEY BROS., WEYMOUTH, N.S. (Head Office Roy Bldg., Halifax.)

One schooner, 394 tons, three masts.

E. R. GAUDET, WEYMOUTH, N.S.

One 350-ton schooner, three-masted.

E. P. RICE, WEYMOUTH, N.S.

One three-masted schooner, 350 tons.

RICE, WARREN & CO., WEYMOUTH, N.S.

One three-masted schooner, 334 tons.

FALMOUTH SHIPBUILDING & TRANSPORTATION CO., WINDSOR, N.S.

One wooden schooner, 350 gr. tons, 700 tons cap., for undisclosed interests.

NOEL SHIPBUILDING & TRANSPORTATION CO., WINDSOR, N.S.

One wooden schooner, three masts, 138 ft. x 35 ft. x 13 ft., 450 net reg. tons. Oil auxiliary engines. Launching August, 1918, for builder's account.

C. A. NICKERSON, WOODS HARBOR, N.S.

MILTON SHIPBUILDING CO., YARMOUTH, N.S.

Taken over and enlarging old plant of James Jenkins.

350-ton schooner building.

W. O. SWEENY, YARMOUTH, N.S.

100-ton fishing schooner.

YARMOUTH SHIPBUILDING CO., YARMOUTH, N.S.

One 450-ton wooden schooner, 128 ft. keel. To be completed Oct., 1918.

ONTARIO

CAN. ALLIS-CHALMERS (Head Office, Toronto), BRIDGEBURG, ONT.

Six standard freight steamers, 3,500 tons, 261 feet long, for undisclosed interests.

COLLINGWOOD SHIPBUILDING CO., COLLINGWOOD, ONT.

Two steel cargo steamers, Nos. 51-52, 50 ft. x 43 ft. x 25 ft., 2,500 gr. tons, 2,900 tons d. w., triple expansion engines 18-30-50 x 36, 2 boilers 14 ft. x 11 ft., 10 knots, for undisclosed interests. Delivery May and Aug., 1918.

Four deep-sea trawlers, Nos. 53-58, inclusive, 125 ft. x 23 ft. 6 in. x 13 ft. 6 in., 288 gross tons, 12¾-21½-35/24 engines, 1 cylinder 13 ft. 6 in. x 10 ft. 6 in., 10 knots, 500 h.p. for undisclosed interests.

One steel freight steamer of 3,800 tons d. w. for undisclosed interests. Keel laid May 8, 1918.

"War Wizard," May 8, 1918, 3,000 tons d. w., 261 ft. x 43 ft. 6 in. x 20 ft. depth. For undisclosed interests.

"War Witch" same size, now building. To be launched ——

R. MORRILL, COLLINGWOOD, ONT.

"Windsor," Aug. 10, steam tug, 105 ft., for Ontario Gravel & Freighting Co.

CAN. CAR & FOUNDRY CO., FORT WILLIAM, ONT. (Head Office Transportation Bldg., Montreal.)

Twelve steel mine sweepers for French Government.
145 ft. o.a. steel construction. Value $2,500,000.
Plant under construction.

GREAT LAKES DREDGING CO., FORT WILLIAM, ONT.

Two wooden vessels, 2700 tons d.w. cap., 260 ft. o.a., 43 ft. beam. Triple expansion engines of 1,000 h.p. To be launched November, 1918.

"War Sioux" launched May 12, 1918. Keel laid for another twenty minutes later. Launching scheduled for November.

THUNDER BAY CONTRACTING CO., FORT WILLIAM, ONT.

One wooden freight steamer, 261 ft. long, for undisclosed interests.

NATIONAL SHIPBUILDING CO., GODERICH, ONT.

KINGSTON SHIPBUILDING CO., KINGSTON, ONT.

Several steel trawlers for undisclosed interests. First one launched Dec. 22, 1917.

SELBY & YOULDSON, KINGSTON, ONT.

GEORGIAN BAY SHIPBUILDING & WRECKING CO., MIDLAND, ONT.

Tug, 50 tons, wood.

One tug 40 tons, wood.

MIDLAND DRY DOCK CO., MIDLAND, ONT.

Three steel freight steamers, 261 ft. x 43 ft. 6 in. x 23 ft., 3,400 tons d. w., 10 knots, for undisclosed interests. Deliveries, one in July, 1918, and two others before close of navigation, 1918.

Alterations on steamers "Mariska" and "Glenlyon."

"Western Star" being reconstructed.

PORT ARTHUR SHIPBUILDING CO., PORT ARTHUR, ONT.

Six steel trawlers on order. 135 ft. o. a. x 23 ft. 4 in. x 15 ft. 1 in., 294.5 tons, trip. exp. engines, 500 h.p., single end Scotch boiler. Two-masted.

MUIR BROS. DRYDOCK CO., LTD., PORT DALHOUSIE, ONT.

J. W. GEROW, ROSSPORT, ONT.

REID WRECKING CO., SARNIA, ONT.

One steel tug, 157 ft. x 32 ft. x 19 ft., trip. exp. engines, for lake or ocean service. Keel blocks laid March 5. Launching July, 1918. For Reid Wrecking Co.

"Tento," Oct. 22, 1917, 261 ft., 3,500 tons d.w. steel. For Norwegian interests. Transferred to British registry.

POLSON IRON WORKS, TORONTO

Six steel cargo steamers, Nos. 145 to 150 inclusive, for the Imperial Munitions Board. Length 261 feet overall, 43 feet 6 inches beam and 23 feet deep, 2,350 approximate gross tons, 3,500 d.w. capacity; single screw triple expansion engine, 20½ by 33 by 54 by 36-inch stroke. 1,250 h.p., two 14-feet by 12 feet Scotch boilers 180 lbs. per sq. inch w.p.

DAVIS DRY DOCK CO., KINGSTON, ONT.

Building wooden and metallic lifeboats, general repairs.

DOMINION SHIPBUILDING CO., TORONTO, ONT.

"Le Quesnoy," 4,300 tons gross, 1,400 h.p. Triple expansion engines, 2 Scotch marine boilers, to be launched Nov. 23. Built to builder's account, Canadian registry. Also three 3,500-ton steamers on stocks in course of construction and three others same type going through shops. All ships classed.

PORT ARTHUR SHIPBUILDING CO., PORT ARTHUR, ONTARIO

"War Fiend," 3,500 tons d.w., triple expansion engines, 1,300 h.p. Scotch boilers. Launched Oct. 24, 1918. To be completed Nov. Built for the Imperial Munitions Board.

"War Karma," 3,400 tons deadweight. Built for the Imperial Munitions Board. Launched October 26. Will be completed last week of Nov. Triple expansion engines, 20½ by 34 by 56 by 40-inch stroke. 2 Scotch boilers, 15 feet by 11 feet, working pressure 190 lbs. per sq. inch.

TORONTO SHIPBUILDING CO., LTD., TORONTO, ONT. (Toronto Dry Dock Co., Ltd., under same management.)

Two wooden cargo steamers, 250 ft. B.P. 42 ft. 6 in., moulded breadth 25 ft., moulded depth 22 ft., draft 2,500 tons d. w. Triple exp. engines $\frac{20 \times 33 \times 54}{40}$. Howden boiler. Launchings July and Sept., 1918. For undisclosed interests.

THE BRITISH AMERICAN SHIPBUILDING CO., WELLAND, ONT.

Two steel frs., 3,500 tons d. w., 261 ft. O.A., 43 ft. beam, 23 ft. moulded depth, Westinghouse steam turbine engines. Delivery 1918.

WELLAND SHIPBUILDING CO., WELLAND, ONT.

One steel freight steamer, "War Weasel," 261 ft. x 43 ft. 6 in. x 23 ft., 3,300 tons d. w., trip. exp. engines, 14 ft. x 12 in. Scotch boilers, 10 knots, 1,250 h. p. Launching April, 1918.

Two deep frame cargo vessels, "War Badger" and "No. 3," 261 ft. x 43 ft. 6 in. x 23 ft., 3,300 d. w. tons, geared turbines, Howden's forced draught boilers, 10 knots, 1,250 h.p. Launchings June and August, 1918.

PRINCE EDWARD ISLAND

THE CARDIGAN SHIPBUILDING PLANT, CARDIGAN, P.E.I. (Purchased Annandale Lumber Co. plant and removed to Cardigan.)

One three-masted schooner, 325 tons, to be completed November, 1918.

QUEBEC

TIDEWATER SHIPBUILDERS, LTD., THREE RIVERS, QUE.

R. N. LE BLANC, BONAVENTURE, QUE.

J. Z. DEGAGNE, EBOULEMENTS, QUE.

DAVIE SHIPBUILDING & REPAIRING CO., LEVIS, QUE.

Six military barges, 130 feet long.
Eight steel trawlers, 2,035 tons.
One floating crane, 350 tons.
*One steel cargo vessel of 5,000 tons cap.
Building several steel lighters and several wooden drifters.

Steel car ferry "Canora" building for C.N.R., 308 ft. x 52 ft., cap. 20 loaded cars. Speed, 14 knots.

ATLAS CEMENT CONSTRUCTION CO., LTD., MONTREAL, QUE.

One concrete cargo steamer, "Concretia," 125 ft. x 22 ft. x 13 ft., steel ribbed, hull to be from 3 in. to 5 in., thick, for the builder's account.

CANADIAN VICKERS, LTD., MONTREAL, QUE.

Two cargo steamers, 9,400 tons, steel; 1 dredge, 2,364 tons, steel; 12 trawlers, 3,050 tons, steel; 23 drifters, 3,350 tons, wood.

Six freight steamers, 24 ft. draught, 7,000 tons cap., 11 knots, for undisclosed interests. Delivery, 1917, of two for Norwegian interests "Porsanger," 394 ft. 6 in. o. a. x 40 ft. 4 in. x 30 ft., triple exp. engines, launched Nov. 29, 1917.

"War Earl" launched June 8, 1918, 7,000 tons d.w., 380 ft. x 49 ft. Keel of sister ship laid five minutes after this launching.

One steel freight steamer, 2,360 tons, cap., for undisclosed interests.

Three steel cargo vessels, 8,200 tons, to be laid down in May, Aug. and Sept., 1918, for undisclosed interests.

FRASER BRACE & CO., MONTREAL, QUE.

Four wooden cargo steamers, 3,000 tons, for undisclosed interests. Two keels laid during Oct., 1917.

HALL ENGINEERING CO., MONTREAL, QUE.

MONTREAL DRY DOCK & SHIP REPAIRING CO., MONTREAL, QUE.

Operate dock 428 ft. long 30 ft. deep.

MONTREAL SHIPBUILDERS LTD., 37 Belmont St., MONTREAL, QUE. (Associated with Atlas Construction Co.

THE QUEBEC SHIPBUILDING AND REPAIR CO., Board of Trade Bldg., MONTREAL, QUE.

2 Vessels for undisclosed interests 3,000 tons each.

QUINLAN & ROBERTSON, MONTREAL, QUE. (Yards at Quebec.)

Four wooden steamers, for undisclosed interests.

LOUIS GAUNDRY, 12 St. Peter St., QUEBEC, QUE.

QUEBEC SHIPBUILDING & REPAIR CO., ST. LAURENT, QUE.

2 schrs. 1,400 tons and 1,200 tons.

One wooden four-masted auxl. schooner "Martin Connolly," 223 ft. x 42 ft. x 20 ft., 2,100 tons d. w., for undisclosed interests. Launched Oct. 28, 1917.

QUINLAN & ROBERTSON, QUEBEC, QUE.

Four wooden steamers, totalling 6,400 tons, for undisclosed interests.

CANADIAN GOVERNMENT SHIPYARDS, SOREL, QUE.

One steel vessel for undisclosed interests.

H. H. SHEPHERD, SOREL, QUE.

Rebuilding seven drifters. Particulars withheld owing to govt. restrictions.

SINCENNES-McNAUGHTON LINES, SOREL, QUE.

One tug 410 tons, wood.

CAN. GOVERNMENT SHIPYARD, SOREL, QUE.

One steel vessel for undisclosed interests.
Building steam trawlers and wooden drifters for undisclosed interests.

LECLAIRE SHIPBUILDING CO., SOREL, QUE.

Six steel trawlers 125 feet long with single screw triple expansion engines, 500 i.h.p. Two of these launched in November. One more to be launched before ice comes in. Six wooden auxiliary schooners fitted with twin screw "Skandia" engines of 120 h.p. each. These will be all in frame at the end of November and four ready for launching as soon as ice is clear next spring.

NATIONAL SHIPBUILDING CORPORATION, THREE RIVERS, QUE.

"War Mingan" and "War Radnor," launched Oct. 15 and No. 2. Built for the Imperial Munitions Board and will be operated by Anning Bros., Cardiff, Wales. Now building 10 self-propelling barges for France.

THE "DUCT" KEEL IN SEA-GOING STEAMERS*

UNDER the somewhat indefinite title of "Improvements in the Construction of Ships," Mr. E. F. Spanner, R.C.N.C., submitted a paper at the opening meeting of the present session of the Institution of Enginers and Shipbuilders in Scotland, on October 22nd, which has so intimate a bearing on questions of great present-day importance—eg., the maintenance of a ship's floatability after severe damage—that a short summary will be acceptable. The paper was primarily, if not wholly, concerned with propounding the idea, not for the first time proposed, of providing a central, or mid-line duct, along the lower portion of a vessel, for the purpose of dealing with fluid distributed in different compartments along its length. One proposed form of this idea provided for a central pipe running through the vessel, above the inner bottom, but the author briefly dismissed this, as the objections attaching to the existence of a pipe of very large cross section in the holds, piercing the water tight bulkheads, and having numerous branches, were well recognized.

The development of the central duct idea is considered to offer such advantages as regards safety, economy, etc., as should insure it a welcome from those interested in the construction and running of sea-going vessels of the passenger and large cargo types. The author's proposal is to provide the mid-line fore and aft duct by dispensing with the ordinary centre vertical or keel plate in the double bottom, and substituting two such verticals to form a relatively narrow box-keel running the length of the vessel, and having an uninterrupted passage through it from one end to the other, but with vertical doors or valves at specified intervals dividing duct into sections, and with side valves into each water ballast section of the vessel's double bottom, all of which can be operated by pressure, led in pipes within the duct, and operated, if need be, from the engine room.

The author first dealt exhaustively with the form of this duct-keel, and its structural relationship with, and strength effects on, the hull of a vessel, both in service and when docking, etc He then treated fully of its adaptability and utility to the purposes of water ballast and oil fuel carrying in the double bottom compartments, and especially emphasized the important service it is calculated to render in emergency for trimming purposes, as between compartment and compartment, to secure and maintain float-ability in damaged vessels, after the manner made familiar to naval architects and shipowners by Messrs. Brunton Brothers. He briefly described the Brunton system, and remarked that it was thoroughly sound in conception and undoubtedly destined to be widely adopted when some of the practical difficulties were overcome.

Having touched upon most of the principal features of the duct form of keel construction—which the author said had recently been patented by him, jointly with Mr. J. H. Silley, of Messrs. R. and H. Green and Silley Weir, Ltd., of Blackwall, he finished up a paper of great detail by summarizing the advantages claimed for the duct keel vessel. The form of keel proposed was such as to increase the strength of the vessel against longitudinal strains and against strains due to docking or grounding. The duct provided a permanent connection between the pumping plant and the double bottom, and such hold compartments as were desired, enabling water or oil to be pumped into or out of these compartments without the necessity of fitting long leads of suction and filling pipes; also in the case of oil avoiding the necessity of fitting more than a small amount of heating pipe. With the duct fitted there was no need for the piercing of the water-tight bulkheads with numerous holes for the suction and filling pipes. Owing to the manner in which communication was arranged between duct and adjacent compartments, there was no danger of any portion of the vessel being flooded by the entry of water through the duct, unless such flooding was desired by the ship's officers. As the duct keel involved practically no addition to the structural weight of vessel, an amount of deadweight represented by the weight of suction and delivery pipes, joints, bulkhead pieces, valves, strainers, etc., dispensed with, was available for cargo carrying, as was also the space saved by the absence from holds of any ballast or fuel pipes, and of the dunnage necessary for protecting them. The duct would be available, if desired, for arranging of leads of suction and delivery pipes from the pumps to the fresh water tanks, or reserve feed water tanks, so avoiding the necessity of piercing bulkheads to accommodate these leads. The cost of working the form of duct keel described as an integral part of the double bottom of vessels will be inappreciable, the saving in fitting and maintenance of piping considered, and it renders the carrying out of the Brunton system more eminently practicable. The author desired the fullest attention to be drawn to any practical difficulties in the way of the application of the idea. It was desired to establish the soundness or otherwise of the idea as early as possible. In the present circumstances, when heavy demands were being made on the time and energies of all interested in the maintenance of our maritime efficiency, it was essential that impracticable schemes should receive an early quietus. On the other hand, it was reasonable to hope that a welcome would be extended to any idea which offered a sound prospect of success.

* *Marine Engineer* and *Naval Architect.*

HARBOR DEVELOPMENTS IN CANADA

Special to MARINE ENGINEERING

The Marine Department is not losing sight of the fact that hand in hand with the development of Canadian trade must go facilities to handle it. This department is one of those at Ottawa—and they are not too many, which has been forehanded in its preparations for the cessation of hostilities. In addition to the enormous amount of work involved in the building of the terminals at Halifax, and the improvements at Victoria, it has several other large projects in view.

Mr. A. D. Swan, of Montreal, who is considered the best authority on terminals in North America, is now making an extended investigation of the requirements at Vancouver, preliminary to a report to the Government on what is necessary there. It is considered that the very extensive terminals at Victoria should be sufficient to answer the requirements of that port for some time to come.

On the Atlantic coast there is an interesting situation at St. John. The status of that port is not generally known. It is the only port in Canada, owned by the city where it is situated. This privilege was granted to St. John many years ago, I believe, by George III., as a tribute of respect to the United Empire Loyalists from the United States, who had settled at the mouth of the St. John river.

The proposal now is that the Marine Department should take the harbor over from the city for administrative purposes. There may be some sentimental regret over relinquishing or transferring the rights secured to them under royal charter, but the harbor has a bonded indebtedness of a million and a quarter, and the proposal is that the Government should take this over also. The city has been making expenditures on it and collecting the revenues, but it is probable that, apart from the sentimental feeling, there will be relief if it should pass into the hands of the Government. It is understood that legislation will be introduced at the next session of Parliament to the effect above outlined.

Tonnage for Seal Trade.—The British authorities have placed tonnage at the disposal of Newfoundland exporters for the removal of the balance of sealskins and oil of this season's catch.

In Shipbuilding Plants.—Conditions in the shipbuilding industry were reflected in the meeting of the Labor Temple, Toronto, of the shipbuilders and boilermakers. Sixty-five new members were taken into the union and it was decided after hearing the report of the business agent, Herbert Wright, to appoint an assistant.

Sault Ste. Marie, Ont.—Both the American and Canadian locks closed on Monday, December 16, ending navigation for the season of 1918. The last freighter to lock down was the W. Grant Morden, the large Canadian boat.

Sault Ste. Marie.—It is practically certain that the two French mine sweepers, "Cerisoles" and "Inkerman," were lost in Lake Superior during the gale of Nov. 24th. Their crews consisted of altogether 76 officers and men. Search for wreckage along the shores of Lake Superior has so far been without result.

St. John, N.B.—The steamer "Minnedosa," of the C.P.R. fleet, is expected shortly in this port on her maiden voyage. She will be the first large steamer on her maiden voyage since the end of hostilities. She is a 14,000-ton steamer, with a speed of 17 knots, and can accommodate 500 cabin and 1,500 third-class passengers.

London, Eng.—London has been busy during the war equipping and improving her docks and cargo-handling facilities. The Port of London authorities think they have brought their plant to a state of fitness that will be ahead of the existing requirements by a generation, and that the port will become the greatest port in the world is their firm opinion.

New York.—The latest development in the negotiations for the purchase of the British vessels of the International Mercantile Marine, is the return to England of Harold A. Sanderson, chairman of the Board of Directors. The sale was halted by the intervention of an offer from the American Government. Mr. Sanderson will consult with the British Government before any further action is taken.

Parrsboro, N.S.—Letters received here from the crew of the new tern schooner "Celina K. Goldman," from Barbados, state that she had arrived there leaking and would have to discharge and go on the dock for repairs. She was bound from this port for Capetown, South Africa, with a cargo of lumber, and sailed from Spencer's Island on October 16th. She was built at St. Martin's, N.B., this year, was 477 tons register and is on her maiden trip.

Cleveland.— The Canadian freighter, "Grant Morden," has the distinction of making the largest amount of freight earnings for a round trip ever made by a lake steamer. Her last round trip was hard coal, and back with a cargo of grain. The round trip will take about two weeks at the outside, and probably a little less. She took 13,000 tons of coal on her upward trip at $1.15 per ton, and on the run down she has 450,000 bushels of wheat for which she will receive 7 cents per bushel. She will thus earn about $45,000 on the round trip, and even allowing for the extra insurance she would have to pay owing to the lateness of the season, she would make a tidy profit.

Montreal.—The International Mercantile Marine which had such large interests in Europe before the war, is looking around now with a view to reorganize their activities in this field. In this connection Mr. P. V. G. Mitchell is leaving this week for Europe as personal representative of the president of the International Mercantile Marine to make a survey of the conditions, and take charge of the company's business during the reconstruction period. Mr. Mitchell is assistant general manager of the White Star-Dominion Line in Montreal. Antwerp was the chief port of entry and sailing for the International Mercantile Marine in Europe, and Mr. Mitchell's first care will be to see just what the Germans have left of the docks, offices, stores, etc., in that port.

LICENSED PILOTS

ST. LAWRENCE RIVER

Captain Walter Collins, 43 Main Street, Kingston, Ont.; Captain M. McDonald, River Hotel, Kingston, Ont.; Captain Charles J. Martin, 13 Balaclava Street, Kingston, Ont.; Captain T. J. Murphy, 11 William Street, Kingston, Ont.

ST. LAWRENCE RIVER, BAY OF QUINTE, AND MURRAY CANAL

Captain James Murray, 106 Clergy Street, Kingston, Ont.; Capt. James H. Martin, 259 Johnston Street, Kingston, Ont.; John Corkery, 17 Rideau Street, Kingston, Ont.; Captain Daniel H. Mills, 272 University Avenue, Kingston, Ont.

MONTREAL PILOTS' ASSOCIATION

President—Alberic Angers, Montreal.
Secretary—C. B. Hamelin, Champlain, Que.

ASSOCIATIONS

DOMINION MARINE ASSOCIATION

President—A. A. Wright, Toronto. Secretary—Francis King, Kingston, Ont.

GREAT LAKES AND ST. LAWRENCE RIVER RATE COMMITTEE

Chairman—W. F. Herman, Cleveland, Ohio. Secretary—Jas. Morrison, Montreal.

INTERNATIONAL WATER LINES PASSENGER ASSOCIATION

President—O. H. Taylor, New York.
Secretary—M. R. Nelson, 1194 Broadway, New York.

SHIPPING FEDERATION OF CANADA

President—Andrew A. Allan, Montreal; Manager and Secretary—T. Robb, 218 Board of Trade, Montreal; Treasurer—J. R. Binning, Montreal.

SHIPMASTERS' ASSOCIATION OF CANADA

Secretary—Captain E. Wells, 46 St. John Street, Halifax, N.S.

GRAND COUNCIL N.A.M.E. OFFICERS.

A. R. Milne, Kingston, Ont., Grand President.

J. E. Belanger, Bienville, Levis, Grand Vice-President.

Neil J. Morrison, P.O. Box 228, St. John, N.B., Grand Secretary-Treasurer.

J. W. McLeod, Owen Sound, Ont., Grand Conductor.

Lemuel Winchester, Charlottetown, P.E.I., Grand Doorkeeper.

Alf. Charbonneau, Sorel, Que., and J. Scott, Halifax, N.S., Grand Auditors.

1918 Directory of Subordinate Councils, National Association of Marine Engineers.

Name	No.	President	Address	Secretary	Address
Toronto	1	Arch. McLaren	324 Shaw Street	E. A. Prince	49 Eaton Ave.
St. John	2	W. L. Hurder	209 Douglas Avenue	G. T. G. Blewett	86 Murray St.
Collingwood	3	John Osburn	Collingwood, Ont.	Robert McQuade	Collingwood, Ont.
Kingston	4	Joseph W. Kennedy	395 Johnston Street	James Gillie	101 Clergy St.
Montreal	5	Eugene Hamelin	Jeanne Mance Street	M. Lasure	120 Ribard St.
Victoria	6	John E. Jeffcott	Esquimalt, B.C.	Peter Gordon	808 Blanchard St.
Vancouyer	7	Isaac N. Kendall	319 11th St. E., Vanc.	E. Read	232 13th St. W.
Levis	8	Michael Latulippe	Lauzon, Levis, Que.	Arthur Jean	Lauzon, W.
Sorel	9	Nap. Beaudoin	Sorel, Que.	Alf. Charbonneau	Box 204, Sorel, Que.
Owen Sound	10	John W. McLeod	570 4th Ave.	R. J. McLeod	662 8th St.
Windsor	11	Alex. McDonald	28 Crawford Ave.	Neil Maitland	London St. W.
Midland	12	Geo. McDonald	Midland, Ont.	A. E. House	Box 333
Halifax	13	Robert Blair	29 Parrsboro Street	Chas. E. Pearce	Portland St., Dartmouth, N.S.
Sault Ste. Marie	14	Charles H. Innes	27 Euclid Road	Wm. Hindmarch	34 Euclid Rd.
Charlottetown	15	J. A. Rowe	176 King Street	Chas. Cumming	27 Euston St.
Twin City	16	H. W. Cross	436 Ambrose Street	J. W. Farquharson	169 College St.
				A. H. Archand	Champlain, Que.

EMPIRE
Largest Brass Manufacturers in British Empire
Quality
First
MARINE BRASS GOODS
A-810 Square Head Steam Cock
A-2512 Steam Whistle with Valve
A-860 Standard Gate Valve; Solid Wedge; Non-rising Stem
EMPIRE
All goods are factory tested and can be depended upon to give reliable service. They are heavy in weight and made to formula in our own plant.
Back of the Empire "E" (trade mark) stands the guarantee for quality and service of the largest brass manufacturers in the British Empire.
A Few of Our Products:
Steam Traps
Brass Castings
Steam Whistles
Pressure Gauges
Brass Valves
Compression Bibbs
Ship Lavatories and Water Closets
Steam Stops
Lubricators
Brass Fittings
Basin Cocks
E
Quality Mark
E
Quality Mark
NAVAL CASTINGS
Send us your Specifications for Brass, Bronze and Gun Metal Naval Castings
Secure a copy of our finely illustrated catalog--contains the full Empire line
Empire Manufacturing Company
LIMITED
LONDON and TORONTO

FO'C'STLE DAYS

Continued from page 312

wise some favorite tipple to ease the throat in such a parched country. Nearly every transaction in a harbor revolved round a demijohn. The captain of the port in Coquimbo in 1890 was named Hart, and his duties seemed to range from customs affairs to water policeman and magistrate.

At Anchor

Our first night at anchor was spent in looking around the bay from the top of the half-deck house. About a mile to the south was the village with its low buildings, for Chili is a country subject to earthquakes. The mole or pier, where ships' boats landed was noticeable and as the flickering oil lamps were lighted along the water front our view for detail soon vanished. A new song from nature fell weirdly on our ears—the thud and hiss of the breakers and the snort of the sea lions as they frisked around the bay. Now and again the native stevedores could be heard shouting on a coasting steamer which was working cargo after dark, but the most thrilling sound to our young ears was wafted from a pinnace of blue jackets going off to their store ship, H.M.S. "Liffey." This quaint old Nelson ship of wooden wall fame was anchored at double moorings, and had sailed out round Cape Horn to take up her last position in the British navy. We had no mail to read, but looked forward to that bright spot before leaving the coast.

Jolly-Boat Manoeuvres

Next day bright and early we tripped our anchor and kedged inshore a bit where we moored fore and aft and made ready for discharging. Lighters or barges were promised for next day. Fid was given the authority of a coxswain and Hank and myself were his crew. After two hours of fussing around we had the small boat at the accommodation ladder awaiting the captain. The climate was ideal and we were dressed in light garb with badge caps on. The boat was spotlessly clean; varnished back board, fancy cotton covered yoke lines and a cushioned seat for the old man; all reflected credit on the parent craft.

Fid of course took the stroke oar and Hank the next one, whilst I acted as bow-oar and boat hook attender. There were three other British sailing ships in port so we anticipated hearing some giant yarns at the mole. The first day's boating could have been worse, but the fact that no one lost their life, saved the situation. The old man rolled down the gangway and flopped into the sternsheets, nearly capsizing the boat, and skipper-like, blaming it on the boys. When the "shove off" was given I got muddled and shipped the boat-hook for an oar. At this Ally Sloper performance the old man gave a deep sea grunt and mumbled a few nursery rhymes which should only be used in a fog outside the three-mile limit. Soon, however, the "Maggie Dorrit's" jolly-boat was speeding inshore with dignified movement. Fid's strokes were horribly fast, and had the distance been five or six miles instead of about one we should have burst a bloodvessel in trying to imitate a fast reciprocating engine. About half way to the mole, sea lions, a species of seal, began to bob up and down close to the boat. This lucky event caused our stróke oar to take an easier gait.

When "unship your bow oar" was bawled out I threw the blade "apeak" like a veteran salt, but managed to send a little spray into the old man's lap. This was also received with the conventional expressions so dear to all nautical autocrats, and apparently a prerogative of mates and masters of foreign-going ships. "Way enough" was the next order, and then "unrow." None of the work was done splendidly, as boat handling needs practice, but the skipper really got ashore and yelled to us: "Lay off to that buoy till I hail you," to which we responded with hearts full of joy, "aye, aye, sir."

At the buoy we heard the news of the port from the apprentices in other jolly-boats. Although the cool of the morning had long passed, a few elderly natives still clung to their mantillas—a garment not unlike a blanket, with a hole in the centre, through which the head is thrust. Different shades of color were apparent amongst the lower and middle classes, but the upper ten were very similar in complexion and dress to the Spaniards of Bilbao, of iron ore fame, or Seville, which gives its name to luscious oranges that grow on the banks of the Guadalquiver in Európe.

To be continued.

SEAMANSHIP AND NAVIGATION

Continued from page 313

uses the tramp or cargo vessel as a model when laying down his explanations and procedures, for the simple reason that the tramp steamboat represents 85 per cent. of the merchant fleet, and her predicaments are legion, owing mainly to low power, deep loading, and absurd draught when light ship. Any one knows what to do with a ship on a beautiful moonlight night with the sea like a mirror, but it is how to nurse a vessel in heavy weather that is the test of our profession.

The reasons for "heaving to" are numerous, but one of the main points is the fact that a ship "hove to" is in the safest position to fight the elements of wind and sea. On the other hand, a vessel may heave to because, when running with the wind and sea quarterly or dead aft she ships too much water and endangers her vital parts—hatches, engine room, skylight, bridge, superstructure.

Again, sea room is a great factor in the decision of bringing a ship's head to sea and wind. Heavy gales are frequently attended by snow storms, rain or mist, and under such conditions a wary seaman would never dream of keeping before the blast when making for a strange harbor entrance, river bar or narrow sound. especially when uncertain of his position.

Against this cautionary procedure, however, we are always ready to "run" till all is blue when leaving a coast for the open waters. It may be well to remember that the word "run" as used by seamen means that their vessel is before the wind and sea—the opposite to "hove to."

To bring a vessel into a hove to position from a running position in a gale is no easy matter if damage to life and property are to be avoided, for in taking the wind and sea on the beam during the operation many things can go wrong.

She may roll heavy enough to shift her cargo. She may allow one or two seas on board that have poundage enough in them to level everything with the main deck. To those who stand watch on bridge or poop it soon becomes apparent that in all gales there are little lulls in wind and sea which give the observant sailor a chance to "haul her up" in comparative safety.

No one would ever dream of making the change, however, before notifying the engineer, who in turn would tell his men of the coming event and allow the steam to go back a little. The stewards' department should also be notified, that none of them may venture on deck, and that they may further secure dishes and movable stores. Instinct of course tells the deck watch what to do and where to hide, allows Jack to crawl under the fo'c'sle head with a quid in his mouth. Meantime, oil should be allowed to reach the storm sea from the port bow; a closet pipe or 'tween deck scupper is often used for this ordeal and a quiet spell is then watched for.

When the practical eye and sailor judgment think the time is opportune, ring half speed and order the helm a starboard.

Whether you must keep her head on or let the wind and sea a little on either bow are matters which can only be decided by studying the ship's characteristics. For instance, a vessel having a bluff bow like the bilge of a plum-duff is generally best head on, whilst a ship with a sharp nose like a meat chopper is best with a little lee side as she is inclined to dive right under forward and race the screws too heavily aft when her stern flies up in the air with the seesaw motion.

This operation in heavy weather is just the reverse proposition—that of letting a ship "away" after being hove to. The dangers are of course similar but the chances of damage are less. In letting a ship 'off" or "away" from the hove to position, give her full speed the minute you touch the helm. The cautionary procedure as before mentioned is also advisable. Even with the best laid plans and forethought a ship under these manoeuvres may do considerable damage. Fidley doors, engine room skylights, bunker and cargo hatches and funnel guys should be carefully watched from the bridge and conditions in the after wheel house or whale-back be reported by the deck watch.

To "Wear Ship"

This is a sailing ship manoeuvre used on two very different occasions, to change from one tack to another. In ordinary weather a sailing ship changes her tacks by throwing her head into the wind and falling off on the opposite angle, but in very light winds and with a foul bottom ships refuse to do this, therefore they have to go the long way round, some 22 points instead of 10. Again, when canvas is reduced in the close-hauled positions and the wind and sea are stormy, with the speed practically nil, she must also be put on the other tack by "wearing her round" the long, and in this case very dangerous sweep.

It is not uncommon for a man to be swept overboard at such a time, and records are to be found where three or four men have at one time lost their lives through pulling on the braces in the waist of a ship while "waring."

The sails have to be kept in trim for the wind as she changes her head, and thus it is that seamen must be in the well decks when the manoeuvre is being carried out. Whilst the marine world of to-day is almost entirely mechanical, young steamboat officers must at least understand such things as "waring" and "tacking," for the same governing and manoeuvring laws apply to the management of small boats—and who knows the minute that a steamboat sailor will find himself in charge of a lifeboat under oars or canvas. Waring ship then entails both the "keeping away" and the "bringing up" manoeuvres, and I have outlined the dangers of them both.

The Use of Oil on Stormy Waters

Some there are who still believe that the use of oil on rough seas is negligible, but experience and authentic records are enthusiastic over the good results derived from the proper use of oil on board ship, especially when running dead before it or hove to.

When running, allow the oil bags to tow from a cat head position or let the oil percolate through a washroom scupper pipe or other means. This can be done through the medium of oakum or a little waste to allow the slow release of the oil. Canvas bags pricked with a roping needle are good receptacles.

When hove to use any means whereby the oil will get to windward, according to how your ship is fitted and behaving.

The heaviest and thickest oil are most effectual. Coal oil for instance would be of little use, while its parent, crude petroleum, would certainly be valuable, but animal and vegetable oils are best. Where no special oil is carried, the waste oil from the engines will be found useful, and in this respect the writer would ask any sailor who needs visual conversion to the value of oil on rough water to simply look over the ship's side when the bilge pumps are discharging to windward on a stormy day, and if he can't see the calm spots on the sea made by the oil in the bilge water, he must be more than blind to reason.

LIFTING GEAR AT HOG ISLAND

By M. M.

The lifting arrangements adopted at the great Hog Island Shipyard, which has 50 building slips, are on a large and generous scale. The ships to be built are of the "fabricated" type, and the plant used for erecting the steel is much the same as is employed in the States for erecting structural steel work. Whilst at other yards overhead traveling cranes have proved very satisfactory, derricks were adopted at Hog Island. One great advantage was that, whilst special shipbuilding cranes of the pattern hitherto used would have had to be built as special orders involving months of delay, the derricks could be obtained quickly and in large numbers. These derricks are mounted on fixed towers from 33 ft. to 66 ft. high. The derricking boom is 80 ft. long and the vertical strut 43 ft. 6 in. high. There are eight booms per ship over the ways plus an additional derrick near the head of each way. The ship under construction is offset from the centre line between the two rows of towers, thus providing room for a railway track between the hull and the towers on the other side. No part of any two ships under construction is served by less than two books, whilst in some positions three or four hooks are available. Generally the parts to be lifted as received from the bridge yards weigh less than two tons, but the derricks actually provided have a lifting capacity of five tons each. Electric power is supplied for operating the derricks, alternate current motors being used. The current is transmitted at a high voltage from Philadelphia and stepped down to 440 volts for the yard services. All the material arrives by rail and it is intended as far as practicable to run it direct to the ways without entering it first to a storage yard.

Midland, Ont.—The "War Fiend," built at the local shipyard during the past year, left on Tuesday last upon her initial trip to Europe with a cargo of 70,000 bushels of wheat. The vessel was loaded at the Midland elevator, and as far as is known is the first vessel to load here with a cargo for Europe direct without transhipment. The vessel was in charge of Capt. William Taylor.

Quality **Service**

FOR COOL BEARINGS

We Manufacture:

Phosphor Bronze Tail Shaft Liners, Pump Liners,
Stuffing Boxes, Stern Tube Bushings
and Brass Castings of every description.

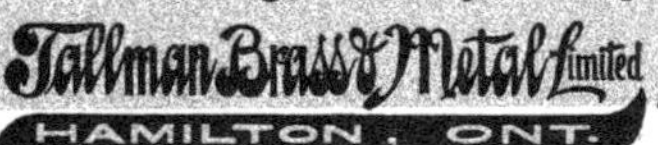

LONGMANS' BOOKS ON SHIPBUILDING

PRACTICAL SHIPBUILDING. A Treatise on the Structural Design and Building of Modern Steam Vessels.

By A. CAMPBELL HOLMS, *Surveyor to Lloyd's Register of Shipping. 2 Vols. Volume I, Text. Medium 8vo; Volume II, Diagrams and Illustrations. Oblong 4to. Sixth Printing.* $20.00 *net.*

The revision recently made was extensive and thorough. Large portions were entirely re-written and the whole brought up-to-date by numerous additions, alterations and amendments. Three new chapters were added on longitudinal framing, damage repairs, and lifeboats and davits. A large amount of new matter was also added in connection with oil vessels, oil fuel, fire extinguishing, freeboard regulations and bulkhead subdivision.

"The book, as a whole, represents the most complete work in existence on the subject of practical shipbuilding."—*Marine Engineering.*

A TEXT-BOOK OF LAYING OFF, OR THE GEOMETRY OF SHIPBUILDING.

By EDWARD L. ATTWOOD, *M.Inst.N.A., R.C.N.C., author of "Warships," "Theoretical Naval Architecture," etc., and* I. C. G. COOPER, *Lecturer in Naval Architecture at Chatham. With Frontispiece and Diagrams. Second Edition. 8vo.* $2.00 *net.*

TEXT-BOOK OF THEORETICAL NAVAL ARCHITECTURE.

By EDWARD L. ATWOOD, *M.Inst.N.A., Member of Royal Corps of Naval Constructors, formerly Lecturer of Naval Architecture at the Royal Naval College, Greenwich. Author of "Warships," etc. Ninth edition. With 150 diagrams and 5 folding tables. Crown 8vo.* $3.50 *net.*

Clearness and conciseness of expression characterize this book. A notable feature in the direct practical application of the methods taught. Two new chapters have been added to this edition, one on launching calculations, and one on the turning of ships.

WARSHIPS. A text-book on the Construction, Protection, Stability, Turning, etc., of War Vessels.

By the same author. With numerous Diagrams. Sixth Edition. Medium 8vo. $4.25 *net.*

Though intended primarily to provide naval officers with authoritative data on the subject, the work will also prove a useful introduction to naval architecture for apprentices and students at dockyards and elsewhere.

The various changes of practice made in recent years will be found embodied in this edition.

STEEL SHIPBUILDER'S HANDBOOK. An Encyclopedia of the Names of Parts, Tools, Operations, Trades Abbreviations, etc., Used in the Building of Steel Ships.

By C. W. COOK, *M.A., B.S. in C.E., Associate Professor of Naval Architecture, University of Southern California. Pocket Size. Leather,* $1.50 *net. Just ready.*

For men working in shipyards, as well as for students in the numerous classes now being conducted in shipbuilding, this handbook will be invaluable. It is the result of a study of the latest and best methods in the large shipbuilding plants all over the country, and contains about 1,600 accurate American definitions, arranged alphabetically and completely cross-referenced. In the back of the book are four plates containing drawings of about 300 parts of ships. The size and shape make it possible to carry the book around for ready reference at all times.

A COMPLETE CLASS-BOOK OF NAVAL ARCHITECTURE. Practical, Laying-Off, Theoretical.

By W. J. LOVETT, *Lecturer on Naval Architecture at the Belfast Municipal Technical Institute. With 173 Illustrations and almost 200 fully worked-out Answers and Questions. 8vo.* $3.50 *net.*

Intended to supply shipwrights, platers, draughtsmen and others in the shipbuilding world with a sufficiency of naval architecture for the ordinary and everyday needs and to enable them afterwards to study higher works on the subject with intelligence and profit.

SHIPYARD PRACTICE. As Applied to Warship Construction.

By NEIL J. McDERMAID, *Member of the Royal Corps of Naval Constructors, late Instructor on Practical Shipbuilding at the Royal Naval College, Devonport. With Diagrams. Second Edition. Medium 8vo.* $5.00 *net.*

"The information given bears internal evidence of accuracy and brings together probably more information than has ever before been made public as to the details and fittings in man-of-war practice."—*Engineering News.*

STRENGTH OF SHIPS.

By ATHOLE J. MURRAY, *Grad. R.N.C., Assist. M.I.N.A. With Diagrams. 3 Folding Plates, and 1 Folding Table. 8vo.* $5.00 *net.*

This book is devoted exclusively to the systematic statement of the strength of the structures and detail fittings of ships. The treatment is not academical, but essentially the outcome of practical experience in ship design.

"* * * a work which fills a distinct gap in the literature of the subject, and is sure to take a leading place amongst kindred text-books."—*Belfast Northern Whig.*

A classified Catalogue of Books in Pure and Applied Science will be sent to any address upon application.

PUBLISHED BY **Longmans, Green & Co.** Fourth Ave. and 30th St. NEW YORK

New Cleveland Pocket-in-Head Riveter
Cleveland Riveting Hammers
For Ship Yards and Boiler Shops
Cleveland Pocket-in-Head Riveting HAMMERS are shorter, weigh less, hit harder, run faster, use less air and have less recoil than any Riveter on the Market. The "Pocket"-in-Head is a "Reservoir" surrounding Main Valve and is filled with Compressed Air, which is discharged in Volume on Piston at each Stroke, greatly increasing speed and power of Blow. Ideal for "Piece-Workers" in Ship Building. Made in 5 sizes: Nos. 40, 50, 60, 80 and 90, with Outside and Inside Latch. Driving capacities from ⅝-in. to 1½-in. Rivets.
Cleveland Metal and Wood Boring Machines
For Ship Building and Boiler Construction
The No. 20-W, illustrated, is speedy and easily handled by one man. Made in 3 sizes—Nos. 10-W, 20-W, 30-W. Capacities, 1" to 4". The No. 20-TC, illustrated, is a lightweight compound geared machine; speeds 150 or 250 R.P.M. Capacity, 1¼ drilling in steel, ⅞" reaming. An ideal one-man machine.
CLECO "Y" HOSE FITTINGS
For Shipyards
BOWES AIR HOSE COUPLINGS
Over 1,500,000 in general use
AIR TIGHT
Cleco clamps attached to Bowes Couplings
CLECO AIR SEATED VALVES
Always Tight
No Packing
Connect with Bowes Coupling.
Write for Bulletins, 34, 38, 39
CLEVELAND PNEUMATIC TOOL CO. of Canada, Ltd., 84 Chestnut St., TORONTO, Ont.
A. R. Williams Machinery Co., Toronto
AGENTS
Williams & Wilson, Montreal

Steel Castings

from ¼ of a pound to 30,000 pounds

SHIP CASTINGS

Steel Propeller Wheels — A Specialty — Steel Stockless Anchors

BEAUCHEMIN & FILS, LIMITED

SOREL, CANADA

We Do Contract Work for Ship Repair and Fitting-Out.

ADVERTISING to be successful does not necessarily have to produce a basketful of inquiries every day.

The best advertising is the kind that leaves an indelible, ineffaceable impression of the goods advertised on the minds of the greatest possible number of probable buyers, present and future.

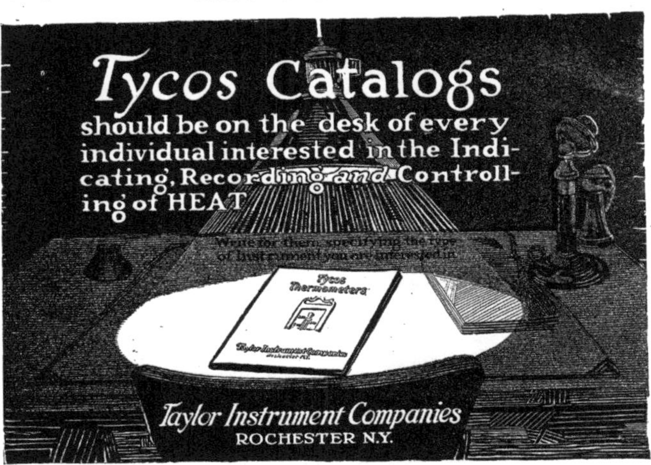
Tycos Catalogs
should be on the desk of every individual interested in the Indicating, Recording and Controlling of HEAT
Taylor Instrument Companies
ROCHESTER N.Y.

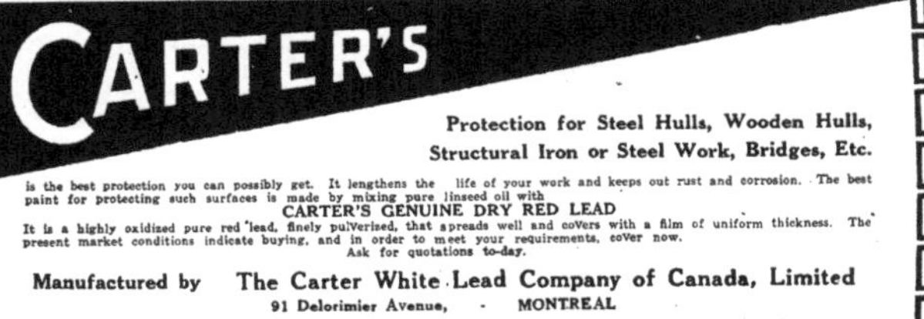
CARTER'S
Protection for Steel Hulls, Wooden Hulls,
Structural Iron or Steel Work, Bridges, Etc.
is the best protection you can possibly get. It lengthens the life of your work and keeps out rust and corrosion. The best paint for protecting such surfaces is made by mixing pure linseed oil with
CARTER'S GENUINE DRY RED LEAD
It is a highly oxidized pure red lead, finely pulverized, that spreads well and covers with a film of uniform thickness. The present market conditions indicate buying, and in order to meet your requirements, cover now.
Ask for quotations to-day.
Manufactured by The Carter White Lead Company of Canada, Limited
91 Delorimier Avenue, - MONTREAL

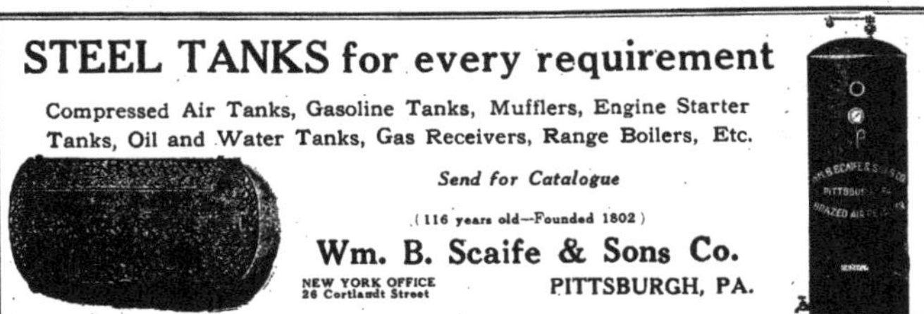
STEEL TANKS for every requirement
Compressed Air Tanks, Gasoline Tanks, Mufflers, Engine Starter Tanks, Oil and Water Tanks, Gas Receivers, Range Boilers, Etc.
Send for Catalogue
(116 years old—Founded 1802)
Wm. B. Scaife & Sons Co.
NEW YORK OFFICE
26 Cortlandt Street
PITTSBURGH, PA.

The Lessons of War

Now that an armistice has been signed and the world-war is seen only in retrospect, we all realize that, despite its hellish hideousness, the war has shown us all more clearly what paths to follow.

The greatest lesson taught by the war is "The Value of Service" as exemplified by the heroic achievements of our soldiers and sailors — by the activities of the Red Cross, the Salvation Army, and other organizations at the front — and by the workers in shipyards, munition plants, and auxiliary industries here in Canada and the United States.

We will make this word "Service" our watchword in peace as it was in war.

MARINE DECKING & SUPPLY CO.

Manufacturers **Contractors** **Engineers**

"LITOSILO" DECKING
"MADESCO" WOOD TACKLE BLOCKS
"MADESCO" BITUMINOUS SOLUTIONS AND ENAMELS

PHILADELPHIA, PA.

Represented in Canada by W. J. Bellingham & Co., Montreal

THE WAGER FURNACE BRIDGE WALL
¶ A preferred and valuable feature in marine and stationary boilers—endorsed by governments; steamship, freight and passenger steamer companies; railroads; stationary plants; others.
WAGER FURNACE BRIDGE WALL CO., Inc.
OF NEW YORK SINGER BUILDING
Philadelphia. Detroit. Seattle. Portland
San Francisco Vancouver, B. C.

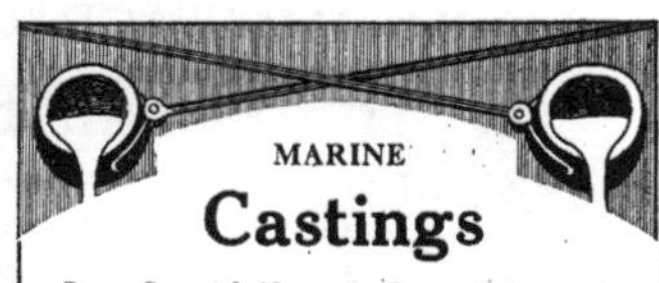

MARINE

Castings

Brass, Gunmetal, Manganese Bronze, Delta Metal, Nickel Alloys, Aluminum, etc.

MARINE AND LOCOMOTIVE ENGINE BEARINGS. MACHINE WORK AND ELECTRO PLATING. METAL PATTERN MAKING

United Brass & Lead, Ltd., Toronto, Ont.

WILKINSON & KOMPASS
TORONTO HAMILTON WINNIPEG
IRON AND STEEL
HEAVY HARDWARE
MILL SUPPLIES
AUTOMOBILE ACCESSORIES
WE SHIP PROMPTLY

Fifty Subscriptions from One Firm

THE International Business Machines Company, of which Mr. Frank E. Mutton is vice-president and general manager, subscribed to 10 copies of THE FINANCIAL POST some months ago—these copies to go to their travelling salesmen. Now this company has increased the number of these subscriptions to 50 because the results of the experimental subscriptions have proved so satisfactory.

FRANK E. MUTTON

Mr. Mutton explained that the object of putting THE POST in the hands of the men of his company was to keep them intelligently acquainted with general business conditions in Canada. He said he knew no better paper than THE POST for the purpose. It would seem that his men have responded fully to effort made to keep them well informed about Canadian business affairs—so much so that the management have added 40 other men to the original 10 to receive THE POST.

When Mr. Mutton was with the National Cash Register Company as its Canadian manager, he was the king of all managers in the matter of sales records. In this position he achieved a big reputation built on solid achievements. He learned salesmanship in a school where competition was of the hottest kind, and where the competitors were brilliant men. Giving Mr. Mutton full credit for superior personal qualities and energy of the most ardent kind, it is taking nothing away from him when it is said that not a little of his success was due to his intimate and sympathetic knowledge of the other man's business. And he taught the men associated with him as salesmen to know the point of view and requirements of the men they called on to sell machines to.

As vice-president and general manager of the International Business Machines Company, Mr. Mutton is putting into operation an idea used by him in past days with brilliant results—he is causing his salesmen to know the business and requirements of their prospective customers. To establish points of contact swiftly and surely is one of the open secrets of successful selling.

In the case of **your** solicitations of customers and desired customers, it is excellent strategy to have your salesmen so well informed about business conditions generally, and about the interests of the men they canvass, that they will be able almost instantly to relate their proposals to the interests of the buyer. When a salesman shows himself intimate with the interests or business or objectives of the man whose order he wants, he is immensely strengthened as a salesman, and his percentage of successful canvasses goes steadily up.

Our definite suggestion to you is: Subscribe to THE POST yourself, and learn from its pages how your salesmen or executives can draw power from this newspaper. Then, having acquired the sought-for knowledge, subscribe to THE POST for each man in your service who can profit you by knowing what is in THE POST each week. If Frank Mutton and other prominent executives are making a success of THE POST as a salesman's aid, it is reasonable to suppose that other managers of salesmen and executives can likewise employ THE POST as a producing agent. And so we ask you to sign and forward the coupon below.

- -

The MacLean Publishing Company, Limited.
Dept. M.E., 143-153 University Ave., Toronto.

Send :: THE FINANCIAL POST OF CANADA (weekly). Subscription price of $3 will be remitted on receipt of invoice in the usual way. Have this copy sent to

When Writing to Advertisers Kindly Mention this Paper.

BUYERS' DIRECTORY

ACCUMULATORS, HYDRAULIC
Smart-Turner Mach. Co., Hamilton, Ont.

AERATING RESERVOIRS
Spray Engineering Co., Boston, Mass.

AIR PORTS
Mitchell Co., The Robert, Montreal, Que.
Turnbull Elevator Mfg. Co., Toronto, Ont.

ALLOYS, BRASS AND COPPER
Dom. Copper Products Co., Ltd., Montreal, Que.
Mitchell Co., The Robert, Montreal, Que.
Mueller Mfg. Co., H., Sarnia, Ont.
Tallman Brass & Metal Co., Hamilton, Ont.
United Brass & Lead, Ltd., Toronto, Ont.

ANCHORS
Beauchemin & Fils, Sorel, P.Q.
Hopkins & Co., F. H., Montreal, Que.
McNab Co., Bridgeport, Conn.
Henry Rogers Sons & Co., Wolverhampton, Eng.
Wm. C. Wilson & Co., Toronto, Ont.

ASBESTOS GOODS
Wm. C. Wilson & Co., Toronto, Ont.

BABBITT METAL
Aikenhead Hardware, Ltd., Toronto, Ont.
Hoyt Metal Company, Toronto, Ontario.
Tallman Brass & Metal Co., Hamilton, Ont.
Wilkinson & Kompass, Hamilton, Ont.
Wm. C. Wilson & Co., Toronto, Ont.

BAROMETERS
Wilson & Co., Wm. C., Toronto, Canada.

BAR IRON
Mitchells Ltd., Glasgow, Scotland.

BARS, GRATE
Babcock & Wilcox, Ltd., Montreal, Que.

BEARINGS, BRASS
Empire Mfg. Co., London, Ont.
Mueller Mfg. Co., H., Sarnia, Ont.
Mitchell Co., The Robert, Montreal, Que.
Tallman Brass & Metal Co., Hamilton, Ont.
United Brass & Lead Co., Toronto, Ont.

BELLS, SHIPS, ENGINE ROOM, ETC.
Aikenhead Hardware, Ltd., Toronto, Ont.
Clarke & Bros. Co., Montreal, Que.
Cory & Son, Inc., Chas., New York, N.Y.
Empire Mfg. Co., London, Ont.
Mitchell Co., The Robert, Montreal, Que.
Morrison Brass Mfg. Co., James, Toronto, Ont.
Mueller Mfg. Co., H., Sarnia, Ont.
Tallman Brass & Metal Co., Hamilton, Ont.
United Brass & Lead, Ltd., Toronto, Ont.

BELTING, LEATHER
Aikenhead Hardware, Ltd., Toronto, Ont.
Wm. C. Wilson & Co., Toronto, Ont.

BIBBS, COMPRESSION
Empire Mfg. Co., London, Ont.
Mitchell Co., The Robert, Montreal, Que.
Mueller Mfg. Co., H., Sarnia, Ont.
United Brass & Lead, Ltd., Toronto, Ont.

BINNACLES
Hopkins & Co., F. H., Montreal, Que.
Morrison Brass Mfg. Co., James, Toronto, Ont.

BLOCKS, CARGO, HEEL AND TACKLE
Aikenhead Hardware, Ltd., Toronto, Ont.
Hopkins & Co., F. H., Montreal, Que.
Loveridge, Ltd., Docks, Cardiff, Wales.

BLOWERS, TURBO.
Mason Regulator & Engin. Co., Montreal, Que.

BOAT CHOCKS
Corbet Fdry. & Machine Co., Owen Sound, Ont.

BOILER COMPOUND
Wm. C. Wilson & Co., Toronto, Ont.

BOILER CIRCULATORS
Eckliff Circulator Co., Detroit, Mich.

BOILER FEED PUMPS
Can. Ingersoll-Rand Co., Ltd., Sherbrooke, Que.
Darling Bros., Ltd., Montreal, Que.
Goldie & McCulloch Co., Galt, Ont.
Morris Mach. Works, Baldwinsville, N.Y.
Smart-Turner Mach. Co., Hamilton, Ont.
Williams Machinery Co., A. R., Toronto, Ont.

BOILER FITTINGS
Empire Mfg. Co., London, Ont.
Goldie & McCulloch Co., Galt, Ont.
McAvity & Sons Ltd., T., St. John, N.B.
Wager Furnace Bridge Wall Co., Inc., New York, N.Y.

BOILER MOUNTINGS
Jenkins Bros., Montreal.

BOILERS, MARINE
Babcock & Wilcox, Ltd., Montreal, Que.
Can. Fairbanks-Morse Co., Montreal, Que.
Can. Vickers, Ltd., Montreal, Que.
Collingwood Shipbuilding Co., Collingwood, Ont.
Doxford & Sons, William, Sunderland, England.
Engr. & Mach. Wks. of Can., St. Catharines, Ont.
Goldie & McCulloch, Ltd., Galt, Ont.
Hall Engineering Works, Montreal, Que.
Mason Regulator & Engin. Co., Montreal, Que.
Montreal Dry Docks & Shipbuilding Co., Montreal, Que.
National Shipbuilding Co., Goderich, Ont.
Marsh Engineering Works, Belleville, Ont.
Polson Iron Works, Toronto, Ontario.
Port Arthur Shipbuilding Co., Port Arthur, Ont.
Waterous Engine Works Co., Brantford, Ont.
Williams Machinery Co., A. R., Toronto, Ont.

BOOKS, TECHNICAL, MARINE
MacLean Publishing Co., Toronto, Ont.

BOLTS
London Bolt & Hinge Works, London, Ont.
Mitchell Co., The Robert, Montreal, Que.
United Brass & Lead, Ltd., Toronto, Ont.
Wilkinson & Kompass, Hamilton, Ont.

BROKER
Page & Jones, Mobile, Ala., U.S.A.

BRASS SHEETS AND TUBES
Dom. Copper Products Co., Montreal.

BRASS GOODS
Corbet Fdry. & Machine Co., Owen Sound, Ont.
Goldie & McCulloch Co., Galt, Ont.
McAvity & Sons, T., St. John, N.B.
Mitchell Co., The Robert, Montreal, Que.
Mueller Mfg. Co., H., Sarnia, Ont.
Tallman Brass & Metal Co., Hamilton, Ont.

BUCKETS, CLAMSHELL
Beatty & Sons, Welland, Ont.
Morris Crane & Hoist Co., Herbert, Niagara Falls, Ont.

BUCKETS, DUMP
Morris Crane & Hoist Co., Herbert, Niagara Falls, Ont.

BUCKETS, COALING
Beatty & Sons, Welland, Ont.
Engr. & Mach. Wks. of Can., St. Catharines, Ont.
Hopkins & Co., F. H., Montreal, Que.
Marsh Engineering Works, Belleville, Ont.
Reed & Co., Geo., Montreal, Que.

BUFFERS, SHIPS' STEERING GEAR SPRING
Loveridge, Ltd., Cardiff, Wales.

BUSHINGS, BRONZE
Oberdorfer Brass Co., M. L., Syracuse, N.Y.

CABLE
Hopkins & Co., F. H., Montreal, Que.
McNab Co., Bridgeport, Conn.
Wm. C. Wilson & Co., Toronto, Ont.

CABLE, LEAD COVERED AND ARMORED
Standard Underground Cable Co., Hamilton, Ont.

CABLE, ACCESSORIES
Darling Bros., Ltd., Montreal, Que.
Standard Underground Cable Co., Hamilton, Ont.

CAPSTANS
Advance Mach. & Welding Co., Montreal, Que.
Dake Engine Co., Grand Haven, Mich.
Hopkins & Co., F. H., Montreal, Que.
Kennedy & Sons, Wm., Owen Sound, Ont.
Wm. C. Wilson & Co., Toronto, Ont.

CALKING TOOLS, ELECTRIC
Aikenhead Hardware, Ltd., Toronto, Ont.

CALKING TOOLS, PNEUMATIC
Aikenhead Hardware, Ltd., Toronto, Ont.
Can. Ingersoll-Rand Co., Sherbrooke, Que.

CALORIFIERS
Low & Sons, Ltd., Archibald, Glasgow, Scotland

CASTINGS
Beauchemin & Fils, Sorel, P.Q.
Can. Steel Foundries, Ltd., Montreal, Que.
Collingwood Shipbuilding Co., Collingwood, Ont.
Wm. Hamilton Co., Peterboro, Ont.
Kennedy & Sons, Wm., Owen Sound, Ont.
Goldie & McCulloch Co., Galt, Ont.
Marsh Engineering Works, Belleville, Ont.
Mitchell Co., Ltd., Robt., Montreal, Que.
Mueller Mfg. Co., H., Sarnia, Ont.
Tallman Brass & Metal Co., Hamilton, Ont.
United Brass & Lead, Ltd., Toronto, Ont.
Waterous Engine Works Co., Brantford, Ont.

CASTINGS, ALLOY
Mitchell Co., The Robert, Montreal, Que.
Mueller Mfg. Co., H., Sarnia, Ont.
Oberdorfer Brass Co., M. L., Syracuse, N.Y.
Tolland Mfg. Co., Montreal, Que.
Tallman Brass & Metal Co., Hamilton, Ont.
United Brass & Lead, Ltd., Toronto, Ont.

CASTINGS, ALUMINUM
Empire Mfg. Co., London, Ont.
Mitchell Co., The Robert, Montreal, Que.
Tallman Brass & Metal Co., Hamilton, Ont.
United Brass & Lead Co., Toronto, Ont.

CASTINGS, BRASS
Crouse-Hinds Co. of Canada, Ltd., Toronto, Ont.
Empire Mfg. Co., London, Ont.
Goldie & McCulloch Co., Galt, Ont.
McAvity & Sons Ltd., T., St. John, N.B.
McNab Co., Bridgeport, Conn.
Mitchell Co., Ltd., Robt., Montreal, Que.
Mueller Mfg. Co., H., Sarnia, Ont.
Oberdorfer Brass Co., M. L., Syracuse, N.Y.
Tolland Mfg. Co., Montreal, Que.
Tallman Brass & Metal Co., Hamilton, Ont.
United Brass & Lead Co., Toronto, Ont.
Waterous Engine Works Co., Brantford, Ont.

CASTINGS, GREY IRON, MALLEABLE, ALUMINUM
Crouse-Hinds Co. of Canada, Ltd., Toronto, Ont.
Darling Bros., Ltd., Montreal, Que.
Engr. & Mach. Wks. of Can., St. Catharines, Ont.
McAvity & Sons, T., St. John, N.B.
McNab Co., Bridgeport, Conn.
Mitchell Co., The Robert, Montreal, Que.
Mueller Mfg. Co., H., Sarnia, Ont.
Waterous Engine Works Co., Brantford, Ont.

CASTINGS, MANGANESE STEEL

CASTINGS, MANGANESE BRONZE
Oberdorfer Brass Co., M. L., Syracuse, N.Y.
Tallman Brass & Metal Co., Hamilton, Ont.
United Brass & Lead Co., Toronto, Ont.

CELLAR DRAINERS
Mueller Mfg. Co., H., Sarnia, Ont.

CHAINS
Aikenhead Hardware, Ltd., Toronto, Ont.
Hopkins & Co., F. H., Montreal, Que.
Morris Crane & Hoist Co., Herbert, Niagara Falls, Ont.
Henry Rogers, Sons & Co., Wolverhampton, Eng.
Wm. C. Wilson & Co., Toronto, Ont.

CHAIN BLOCKS AND SLINGS
Morris Crane & Hoist Co., Herbert, Niagara Falls, Ont.

CHANDLERY, SHIP
Beauchemin & Fils, Sorel, P.Q.
Hopkins & Co., F. H., Montreal, Que.
Leckie, Ltd., John, Toronto, Ont.

CLAMPS, STEAM AND WATER
Mueller Mfg. Co., H., Sarnia, Ont.

CLOCKS
American Steam Gauge & Valve Mfg. Co., Boston, Mass.
Morrison Brass Mfg. Co., James, Toronto, Ont.
Williams Machinery Co., A. R., Toronto, Ont.

CLOSETS
Mueller Mfg. Co., H., Sarnia, Ont.
United Brass & Lead, Ltd., Toronto, Ont.

COAL
Nova Scotia Steel & Coal Co., New Glasgow, N.S.

COAL HANDLING MACHINERY
Morris Crane & Hoist Co., Herbert, Niagara Falls, Ont.
Waterous Engine Works Co., Brantford, Ont.

COCKS, BILGE, DISCHARGE, INDICATOR,
McAvity & Sons Ltd., T., St. John, N.B.
Mitchell Co., The Robert, Montreal, Que.
Morrison Brass Mfg. Co., James, Toronto, Ont.

COCKS, BASIN
Empire Mfg. Co., London, Ont.
Mitchell Co., The Robert, Montreal, Que.
United Brass & Lead, Ltd., Toronto, Ont.

COMPASSES
Wm. C. Wilson & Co., Toronto, Ont.

COMPRESSORS, AIR
Can. Fairbanks-Morse Co., Montreal, Que.
Canadian Ingersoll-Rand Co., Sherbrooke, Que.
Darling Bros., Ltd., Montreal, Que.
Hopkins & Co., F. H., Montreal, Que.
Smart-Turner Mach. Co., Hamilton, Ont.
Williams Machinery Co., A. R., Toronto, Ont.

CONDENSERS
Darling Bros., Ltd., Montreal, Que.
Goldie & McCulloch Co., Galt, Ont.
Morris Machine Works, Baldwinsville, N.Y.
Smart-Turner Mach. Co., Hamilton, Ont.
Weir Ltd., G. & J., Cathcart, Glasgow, Scotland.
Williams Machinery Co., A. R., Toronto, Ont.

CONDULETS, MARINE
Crouse-Hinds Co. of Canada, Ltd., Toronto, Ont.

CONTRACTORS' SUPPLIES
McAvity & Sons, T., St. John, N.B.

CONSULTING ENGINEER
Wall, J. Murray, Philadelphia, Pa.

CONVEYORS, ASH, COAL
Babcock & Wilcox, Ltd., Montreal, Que.
Hopkins & Co., F. H., Montreal, Que.

COPING MACHINES
Bertram & Sons, Ltd., John, Dundas, Ont.

COPPER TUBES, SHEETS AND RODS
Tallman Brass & Metal Co., Hamilton, Ont.

COPPER TUBES AND SHEETS
Dom. Copper Products Co., Montreal.

COTTON, CALKING
Wilson & Co., Wm. C., Toronto, Canada.

COVERS, CANVAS, FOR HATCHES, LIFE-BOATS, ETC.
Leckie, Ltd., John, Toronto, Ont.
Waterous Engine Works Co., Brantford, Ont.

COUNTERS, REVOLUTION
American Steam Gauge & Valve Mfg. Co., Boston, Mass.

COWLS, SHIPS' VENTILATORS
Pedlar People, Ltd., Oshawa, Ont.

COUPLINGS, AIR HOSE
Cleveland Pneumatic Tool Co. of Canada, Toronto.

CRANES
Aikenhead Hardware, Ltd., Toronto, Ont.
Can. Fairbanks-Morse Co., Montreal, Que.
Hopkins & Co., F. H., Montreal, Que.
Williams Machinery Co., A. R., Toronto, Ont.

CRANES, ELECTRIC
Babcock & Wilcox, Ltd., Montreal, Que.
Morris Crane & Hoist Co., Herbert, Niagara Falls, Ont.
Smart-Turner Mach. Co., Hamilton

CRANES, GOLIATH AND PNEUMATIC
Morris Crane & Hoist Co., Herbert, Niagara Falls, Ont.

CRANES, GRANTRY, PORTABLE, JIB
Morris Crane & Hoist Co., Herbert, Niagara Falls, Ont.
Smart-Turner Mach. Co., Hamilton, Ont.

CRANES, OVERHEAD TRAVELLING
Morris Crane & Hoist Co., Herbert, Niagara Falls, Ont.

CRANK SHAFTS
Canada Foundries and Forgings, Welland, Ont.

DAVITS, BOAT
Corbet Fdry. & Machine Co., Owen Sound, Ont.
Hopkins & Co., F. H., Montreal, Que.
Marten, Freeman Co., Toronto, Ont.
Waterous Engine Works Co., Brantford, Ont.

DEAD LIGHTS, BRASS
Goldie & McCulloch Co., Galt, Ont.
Morris Crane & Hoist Co., Herbert, Niagara Falls, Ont.
Mitchell Co., Ltd., Robt., Montreal, Que.

DECK LIGHTS
Mitchell Co., The Robert, Montreal, Que.
Turnbull Elevator Mfg. Co., Toronto, Ont.

DECK PLUGS, ELECTRIC
Crouse-Hinds Co. of Canada, Ltd., Toronto, Ont.
Mitchell Co., The Robert, Montreal, Que.

DECKING FOR SHIPS
Marine Decking & Supply Co., Philadelphia, Pa.

DERRICKS
Aikenhead Hardware, Ltd., Toronto, Ont.
Dake Engine Co., Grand Haven, Mich.
Hopkins & Co., F. H., Montreal, Que.
Marsh Engineering Works, Belleville, Ont.
Morris Crane & Hoist Co., Herbert, Niagara Falls, Ont.

DREDGES
Collingwood Shipbuilding Co., Collingwood, Ont.
Morris Mach. Works, Baldwinsville, N.Y.
Norbom Engineering Co., Philadelphia, Pa.
Polson Iron Works, Toronto.

DRILLS, AIR
Aikenhead Hardware, Ltd., Toronto, Ont.
Can. Ingersoll-Rand Co. Sherbrooke, Que.
Hopkins & Co., F. H., Montreal, Que.

DRILLS, CENTRE
Wilt Twist Drill Co., Walkerville, Ont.

DRILLS, BLACKSMITH AND BIT STOCK
Wilt Twist Drill Co., Walkerville, Ont.

DRILLS, ELECTRIC
Aikenhead Hardware, Ltd., Toronto, Ont.
Wilkinson & Kompass, Hamilton, Ont.

DRILLS, HIGH SPEED
Wilt Twist Drill Co., Walkerville, Ont.

DRILLS, TRACK
Wilt Twist Drill Co., Walkerville, Ont.

DRILLS, RACHET AND HAND
Wilt Twist Drill Co., Walkerville, Ont.

DRILLS, TWIST
Aikenhead Hardware, Ltd., Toronto, Ont.
Williams Machinery Co., A. R., Toronto, Ont.
Wilt Twist Drill Co., Walkerville, Ont.

DRY DOCKS
Can. Vickers, Ltd., Montreal, Que.
Collingwood Shipbuilding Co., Collingwood, Ont.
Doxford & Sons, William, Sunderland, England.
Georgian Bay Shipbuilding & Wrecking Co., Midland, Ont.
National Shipbuilding Co., Goderich, Ont.
Polson Iron Works, Toronto, Ont.
Port Arthur Shipbuilding Co., Port Arthur, Ont.
Yarrows, Limited, Victoria, B.C.

ECONOMIZERS, FUEL
Babcock & Wilcox, Ltd., Montreal, Que.

EJECTORS
Darling Bros., Ltd., Montreal, Que.
Empire Mfg. Co., London, Ont.
Mitchell Co., The Robert, Montreal, Que.
Morrison Brass Mfg. Co., James, Toronto, Ont.
Smart-Turner Mach. Co., Hamilton, Ont.

ELECTRIC LAMPS
Mitchell Co., The Robert, Montreal, Que.
Wm. C. Wilson & Co., Toronto, Ont.

ELECTRO-PLATING
Mitchell Co., The Robert, Montreal, Que.
Tallman Brass & Metal Co., Hamilton, Ont.
United Brass & Lead Co., Toronto, Ont.

ELECTRIC WELDING
Beauchemin & Fils, Sorel, P.Q.

ELEVATING MACHINERY
Darling Bros., Ltd., Montreal, Que.
Goldie & McCulloch, Ltd., Galt, Ont.
Wm. Hamilton Co., Peterboro, Ont.
Morris Crane & Hoist Co., Herbert, Niagara Falls, Ont.
Waterous Engine Works Co., Brantford, Ont.

ELEVATORS
Darling Bros., Ltd., Montreal, Que.
Turnbull Elevator Mfg. Co., Toronto, Ont.

ENGINES, HOISTING
Corbet Fdry. & Machine Co., Owen Sound, Ont.
Hopkins & Co., F. H., Montreal, Que.
Kennedy & Sons, Wm., Owen Sound, Ont.
Marsh Engineering Works, Belleville, Ont.
Port Arthur Shipbuilding Co., Port Arthur, Ont.
Williams Machinery Co., A. R., Toronto, Ont.

ENGINE, INTERNAL COMBUSTION
Doxford & Sons, William, Sunderland, England.

ENGINES, MARINE
Bolinders Co., New York, N.Y.
Can. Fairbanks-Morse Co., Montreal, Que.
Can. Vickers, Ltd., Montreal, Que.
Doxford & Sons, William, Sunderland, England.
Goldie & McCulloch, Ltd., Galt, Ont.
Hopkins & Co., F. H., Montreal, Que.
Iron Works, Ltd., Owen Sound, Ont.
Mason Regulator & Engin. Co., Montreal, Que.
Morris Mach. Works, Baldwinsville, N.Y.
National Shipbuilding Co., Goderich, Ont.
Norbom Engineering Co., Philadelphia, Pa.
Polson Iron Works, Toronto, Ont.
Port Arthur Shipbuilding Co., Port Arthur, Ont.
Trout Co., H. G., Buffalo, N.Y.
Waterous Engine Works Co., Brantford, Ont.
Williams Machinery Co., A. R., Toronto, Ont.

ENGINE STARTERS (AIR)
Scaife & Sons Co., Wm. B., Oakmont, Pa.

ENGINES, STEERING
Corbet Fdry. & Machine Co., Owen Sound, Ont.
Dake Engine Co., Grand Haven, Mich.
Kennedy & Sons, Wm., Owen Sound, Ont.
Wm. C. Wilson & Co., Toronto, Ont.

ENAMELWARE
Mueller Mfg. Co., H., Sarnia, Ont.
United Brass & Lead Co., Toronto, Ont.

EVAPORATORS
Mason Regulator & Engin. Co., Montreal, Que.
McNab Co., Bridgeport, Conn.
Weir Ltd., G. & J., Cathcart, Glasgow, Scotland.

EXTRACTORS, GREASE
American Steam Gauge & Valve Mfg. Co., Boston, Mass.
Darling Bros., Ltd., Montreal, Que.

EYE BOLTS AND NUTS
Canada Foundries and Forgings, Welland, Ont.
Mitchell Co., The Robert, Montreal, Que.
United Brass & Lead, Ltd., Toronto, Ont.

FANS
Aikenhead Hardware Ltd., Toronto, Ont.
Empire Mfg. Co., London, Ont.
Reed & Co., Geo. W., Montreal, Que.
Smart-Turner Mach. Co., Hamilton, Ont.
Williams Machinery Co., A. R., Toronto, Ont.

FENDERS, ROPE
Hopkins & Co., F. H., Montreal, Que.
Leckie, Ltd., John, Toronto, Ont.
Wilson & Co., Wm. C., Toronto, Canada.

FERRO-MANGANESE
Mitchells, Ltd., Glasgow, Scotland.

FILES
Aikenhead Hardware, Ltd., Toronto, Ont.
Williams Machinery Co., A. R., Toronto, Ont.

FIRE BRICKS
Beveridge Paper Co., Montreal, Que.
Mitchells, Ltd., Glasgow, Scotland.
Williams Machinery Co., A. R., Toronto, Ont.

FILTERS, FEED WATER
Darling Bros., Ltd., Montreal, Que.
MacKinnon Steel Co., Sherbrooke, Que.
Mason Regulator & Engin. Co., Montreal, Que.

FITTINGS, MARINE
Hopkins & Co., F. H., Montreal, Que.
McAvity & Sons, T., St. John, N.B.
Mitchell Co., The Robert, Montreal, Que.
United Brass & Lead, Ltd., Toronto, Ont.

FITTINGS, MOTOR BOAT
Empire Mfg. Co., London, Ont.
Mitchell Co., The Robert, Montreal, Que.
Mueller Mfg. Co., H., Sarnia, Ont.
United Brass & Lead, Ltd., Toronto, Ont.

FIXTURES, ELECTRIC
Cory & Son, Inc., Chas., New York, N.Y.
Crouse-Hinds Co. of Canada, Ltd., Toronto, Ont.
Mitchell Co., The Robert, Montreal, Que.
Harvey Hubbell Co. of Canada, Toronto, Can.
Tallman Brass & Metal Co., Hamilton, Ont.

FLAG POLES, STEEL
Dennis Wire & Iron Works Co., London, Ont.

FLOW METERS
Spray Engineering Co., Boston, Mass.

FLUE CLEANERS
Wm. C. Wilson & Co., Toronto, Ont.

FORGES
Aikenhead Hardware, Ltd., Toronto, Ont.
Hopkins & Co., F. H., Montreal, Que.

FLANGING AND EXPANDING MACHINES, PIPE
Lovekin Pipe Expanding & Flanging Mach. Co., Philadelphia, Pa.

FLOODLIGHTS, ELECTRIC
Crouse-Hinds Co. of Canada, Ltd., Toronto, Ont.

FLUORSPAR
Mitchells, Ltd., Glasgow, Scotland.
Scythes & Co., Toronto, Ont.

FORGINGS, ALL KINDS
Aikenhead Hardware, Ltd., Toronto, Ont.
Collingwood Shipbuilding Co., Collingwood, Ont.
Nova Scotia Steel & Coal Co., New Glasgow, N.S.

FORGINGS, STEEL AND IRON
Canada Foundries and Forgings, Welland, Ont.

FURNACE BRIDGE WALLS
Wager Furnace Bridge Wall Co., Inc., 149 Broadway, New York, N.Y.

GAUGES, RECORDING
American Steam Gauge & Valve Mfg. Co., Boston, Mass.
Empire Mfg. Co., London, Ont.

GASKETS
Wm. C. Wilson & Co., Toronto, Ont.

GAUGES COCKS
McAvity & Sons, T., St. John, N.B.
Mitchell Co., The Robert, Montreal, Que.

GAUGE GLASSES
Wm. C. Wilson & Co., Toronto, Ont.

GAUGES, WATER, PRESSURE, COMPOUND AND VACUUM
Aikenhead Hardware, Ltd., Toronto, Ont.
Babcock & Wilcox, Ltd., Montreal, Que.
Empire Mfg. Co., London, Ont.
Jenkins Bros., Montreal.
McAvity & Sons, T., St. John, N.B.
McNab Co., Bridgeport, Conn.
Morrison Brass Mfg. Co., James, Toronto, Ont.

GUARDS, LAMPS
Cleveland Pneumatic Tool Co. of Can., Toronto.

GENERATORS AND CONVERTORS
Can. Fairbanks-Morse Co., Montreal, Que.

GLAZING COMPOSITION
Kuhls, H. B., Fred., 6411 3rd Ave., Brooklyn, N.Y.

GOGGLES
Standard Optical Co., Geneva, N.Y.

GONGS
Clark & Bro., C. O., Montreal, Que.
Mitchell Co., Ltd., Robt., Montreal, Que.

GRAPHITE
Wm. C. Wilson & Co., Toronto, Ont.

GRATINGS
Can. Welding Works, Montreal, Que.
Corbet Fdry. & Machine Co., Owen Sound, Ont.
Canada Wire & Iron Goods Co., Hamilton, Ont.
MacKinnon Steel Co., Sherbrooke, Que.

GRINDERS, ELECTRIC
Wilkinson & Kompass, Hamilton, Ont.

GRINDERS, PNEUMATIC
Can. Ingersoll-Rand Co. Sherbrooke, Que.

GUARDS, MACHINERY
Dennis Wire & Iron Works Co., London, Ont.

GUY RODS AND ANCHORS, ELECTRIC
Crouse-Hinds Co. of Canada, Ltd., Toronto, Ont.

HAMMERS
Canada Foundries and Forgings, Ltd., Welland, Ont.

HARDWARE, MARINE
Hopkins & Co., F. H., Montreal, Que.
Mitchell Co., The Robert, Montreal, Que.

HEADLIGHTS, ELECTRIC
Crouse-Hinds Co. of Canada, Ltd., Toronto, Ont.
Hopkins & Co., F. H., Montreal, Que.
Tallman Brass & Metal Co., Hamilton, Ont.

HEATING EQUIPMENT
Darling Bros., Ltd., Montreal, Que.
Empire Mfg. Co., London, Ont.
Low & Sons, Ltd., Archibald, Glasgow, Scotland

HEATERS, FEED, WATER
Darling Bros., Ltd., Montreal, Que.
Babcock & Wilcox, Ltd., Montreal, Que.
Goldie & McCulloch Co., Galt, Ont.
Mason Regulator & Engin. Co., Montreal, Que.
McNab Co., Bridgeport, Conn.
Weir Ltd., G. & J., Cathcart, Glasgow, Scotland.

HINGES
London Bolt & Hinge Works, London, Ont.
Mitchell Co., Ltd., Robt., Montreal, Que.

HOOKS
Hopkins & Co., F. H., Montreal, Que.
Morris Crane & Hoist Co., Herbert, Niagara Falls, Ont.

HOISTS, ASH
Beatty & Sons, Welland, Ont.
Marsh Engineering Works, Belleville, Ont.
St. Clair Bros., Galt, Ont.
Waterous Engine Works Co., Brantford, Ont.

HOIST BLOCKS
Morris Crane & Hoist Co., Herbert, Niagara Falls, Ont.

HOISTS, CHAIN
Aikenhead Hardware, Ltd., Toronto, Ont.
Can. Fairbanks-Morse Co., Montreal, Que.
Dake Engine Co., Grand Haven, Mich.
Hopkins & Co., F. H., Montreal, Que.
Morris Crane & Hoist Co., Herbert, Niagara Falls, Ont.
Williams Machinery Co., A. R., Toronto, Ont.

HOISTS, CARGO, MOVING, ETC.
Dake Engine Co., Grand Haven, Mich.
Hopkins & Co., F. H., Montreal, Que.
Marsh Engineering Works, Belleville, Ont.
Waterous Engine Works Co., Brantford, Ont.

HOISTING MACHINERY
Beatty & Sons, Welland, Ont.
Can. Ingersoll-Rand Co. Sherbrooke, Que.
Corbet Fdry. & Machine Co., Owen Sound, Ont.
Wm. Hamilton Co., Peterboro, Ont.
Marsh Engineering Works, Belleville, Ont.
Morris Crane & Hoist Co., Herbert, Niagara Falls, Ont.
Waterous Engine Works Co., Brantford, Ont.
Williams Machinery Co., A. R., Toronto, Ont.

HOSE
Wm. C. Wilson & Co., Toronto, Ont.

INDICATORS, ENGINE
American Steam Gauge & Valve Mfg. Co., Boston, Mass.
Cory & Son, Inc., Chas., New York, N.Y.
McNab Co., Bridgeport, Conn.

INDICATORS, SPEED
Aikenhead Hardware, Ltd., Toronto, Ont.
Cory & Son, Inc., Chas., New York, N.Y.

INJECTORS
Aikenhead Hardware, Ltd., Toronto, Ont.
Empire Mfg. Co., London, Ont.
Mitchell Co., The Robert, Montreal, Que.
Morrison Brass Mfg. Co., James, Toronto, Ont.
Williams Machinery Co., A. R., Toronto, Ont.

INGOTS
Broughton Copper Co., Ltd., Manchester, Eng.

INSULATORS, ELECTRIC
Crouse-Hinds Co. of Canada, Ltd., Toronto, Ont.

INSTRUMENTS, NAUTICAL
Leckie, Ltd., John, Toronto, Ont.

IRON AND STEEL
Mitchells Ltd., Glasgow, Scotland.

JACKS
Hopkins & Co., F. H., Montreal, Que.
Morris Crane & Hoist Co., Herbert, Niagara Falls, Ont.

JOINTS, BALL SLIP
Norbom Engineering Co., Philadelphia, Pa.

INSURANCE, MARINE
Toronto Insurance & Vessel Agency, Toronto, Ont.

JOURNAL WEDGES
Canada Foundries and Forgings, Ltd., Welland, Ont.

JUTE PACKING
Stratford Oakum Co., Geo., Jersey City, N.J.

LATHES
Bertram & Sons, Ltd., John, Dundas, Ont.

LEATHER, LACE
Wm. C. Wilson & Co., Toronto, Ont.

LENSES FOR GOGGLES
Standard Optical Co., Geneva, N.Y.

LIFEBUOYS, BELTS AND PRESERVERS
Fosbery Co., Barking, Essex, Eng.
Hopkins & Co., F. H., Montreal, Que.
Wm. C. Wilson & Co., Toronto, Ont.

LIFEBOATS
Hopkins & Co., F. H., Montreal, Que.
Wm. C. Wilson & Co., Toronto, Ont.

LIFE BOAT EQUIPMENT
Eckliff Circulator Co., Detroit, Mich.
Hopkins & Co., F. H., Montreal, Que.
Leckie, Ltd., John, Toronto, Ont.
Wm. C. Wilson & Co., Toronto, Ont.

LIFE JACKETS
Fosbery Co., Barking, Essex, Eng.
Hopkins & Co., F. H., Montreal, Que.
Leckie, Ltd., John, Toronto, Ont.

LIGHTS, ALL KINDS
Can. Fairbanks-Morse Co., Montreal, Que.
Crouse-Hinds Co. of Canada, Ltd., Toronto, Ont.
Hopkins & Co., F. H., Montreal, Que.
Mitchell Co., The Robert, Montreal, Que.
Morrison Brass Mfg. Co., James, Toronto, Ont.
Tallman Brass & Metal Co., Hamilton, Ont.

LIGHTS, SIDE, PORT
Goldie & McCulloch Co., Galt, Ont.
Crouse-Hinds Co. of Canada, Ltd., Toronto, Ont.
Hopkins & Co., F. H., Montreal, Que.
McAvity & Sons Ltd., T., St. John, N.B.
Mitchell Co., The Robert, Montreal, Que.
Tallman Brass & Metal Co., Hamilton, Ont.
Turnbull Elevator Mfg. Co., Toronto, Ont.
Wilson & Co., Wm. C., Toronto, Ont.

LIGHTING SETS
Can. Fairbanks-Morse Co., Montreal.
Cory & Son, Inc., Chas., New York, N.Y.
Crouse-Hinds Co. of Canada, Ltd., Toronto, Ont.
Goldie & McCulloch Co., Ltd., Galt, Ont.
Williams Machinery Co., A. R., Toronto, Ont.
Wilson & Co., Wm. C., Toronto, Canada.

LOADERS, WAGON AND TRUCK
Morris Crane & Hoist Co., Herbert, Niagara Falls, Ont.
Waterous Engine Works Co., Brantford, Ont.

LOCKERS, METAL
Dennis Wire & Iron Works Co., London, Ont.
Reed & Co., Geo. W., Montreal, Que.

LUBRICATORS
Empire Mfg. Co., London, Ont.
Mitchell Co., The Robert, Montreal, Que.

MACHINERY-METAL AND WOOD BORING
Cleveland Pneumatic Tool Co. of Can., Toronto.

MACHINISTS
Corbet Fdry. & Machine Co., Owen Sound, Ont.
Wm. Hamilton Co., Peterboro, Ont.
Hyde Engineering Works, Montreal, Que.
Mitchell Co., The Robert, Montreal, Que.
United Brass & Lead Co., Toronto, Ont.

MANGANESITE, PASTE
Wm. C. Wilson & Co., Toronto, Ont.

MARINE CASTINGS
Wm. Hamilton Co., Peterboro, Ont.
Goldie & McCulloch Co., Galt, Ont.
Mitchell Co., Ltd., Robt., Montreal, Que.
McAvity & Sons, T., St. John, N.B.
Mueller Mfg. Co., H., Sarnia, Ont.
Tallman Brass & Metal Co., Hamilton, Ont.
United Brass & Lead, Ltd., Toronto, Ont.
Waterous Engine Works Co., Brantford, Ont.

MARINE SPECIALTIES
Goldie & McCulloch Co., Galt, Ont.
Hopkins & Co., F. H., Montreal, Que.
McAvity & Sons, T., St. John, N.B.
Mitchell Co., Ltd., Robt., Montreal, Que.
United Brass & Lead, Ltd., Toronto, Ont.

MOTORS
Aikenhead Hardware, Ltd., Toronto, Ont.
Can. Fairbanks-Morse Co., Montreal, Que.
Williams Machinery Co., A. R., Toronto, Ont.

NAUTICAL INSTRUMENTS
Wm. C. Wilson & Co., Toronto, Ont.

NET LIFTERS
Dake Engine Co., Grand Haven, Mich.
Leckie, Ltd., John, Toronto, Ont.

NOZZLES, ALL KINDS
Empire Mfg. Co., London, Ont.
Mitchell Co., The Robert, Montreal, Que.
Tolland Mfg. Co., Boston, Mass.

NUTS
Mitchell Co., The Robert, Montreal, Que.
Wilkinson & Kompass, Hamilton, Ont.
United Brass & Lead, Ltd., Toronto, Ont.

OIL CUPS
Empire Mfg. Co., London, Ont.
Mitchell Co., Ltd., Robt., Montreal, Que.

OILCLOTHING
Wm. C. Wilson & Co., Toronto, Ont.

OIL, LINSEED
Canada Linseed Oil Mills, Montreal, Que.

OILS AND GREASE
Wm. C. Wilson & Co., Toronto, Ont.

OAKUM
Davey & Sons W. O., Jersey City, N.J.
Leckie, Ltd., John, Toronto, Ont.
Stratford Oakum Co., Geo., Jersey City, N.J.
Wm. C. Wilson & Co., Toronto, Ont.

ORNAMENTAL IRONWORK
Canada Wire & Iron Goods Co., Hamilton, Ont.
Can. Welding Works, Montreal, Que.

OPTICAL GOODS
Standard Optical Co., Geneva, N.Y.

OXY-ACETLYENE OUTFITS
Can. Fairbanks-Morse Co., Montreal, Que.
Can. Welding Works, Montreal, Que.
Hopkins & Co., F. H., Montreal, Que.
McAvity & Sons Ltd., T., St. John, N.B.
Williams Machinery Co., A. R., Toronto, Ont.

PAINT
Ault & Wiborg Co. of Can., Ltd., Toronto, Ont.
Wm. C. Wilson & Co., Toronto, Ont.

PAINTS, ANTI-FOULING
Kuhls, H. B. Fred., 6411 3rd Ave., Brooklyn, N.Y.

PAINTS, ANTI-CORROSIVE
Kuhls, H. B. Fred, 6411 3rd Ave., Brooklyn, N.Y.

PAINTING EQUIPMENT, AUTOMATIC
Spray Engineering Co., Boston, Mass.

PACKING, AMMONIA
France Packing Co., Philadelphia, Pa.
Wm. C. Wilson & Co., Toronto, Ont.

PACKING, MARINE
Aikenhead Hardware, Ltd., Toronto, Ont.
France Packing Co., Philadelphia, Pa.
Wm. C. Wilson & Co., Toronto, Ont.

PACKING, METALLIC
Aikenhead Hardware, Ltd., Toronto, Ont.
France Packing Co., Philadelphia, Pa.

PACKING, STEAM
Aikenhead Hardware, Ltd., Toronto, Ont.
France Packing Co., Philadelphia, Pa.
Wm. C. Wilson & Co., Toronto, Ont.

PANELBOARD AND CABINETS, ELECTRIC
Crouse-Hinds Co. of Canada, Ltd., Toronto, Ont.
Preston Woodworking Machy. Co., Preston, Ont.

PATTERNS, WOOD AND METAL
Beauchemin & Fils, Sorel, P.Q.

PIPE, LAPWELD, CAST IRON, RIVETED
Empire Mfg. Co., London, Ont.
Norbom Engineering Co., Philadelphia, Pa.
United Brass & Lead, Ltd., Toronto, Ont.

PIPE RAILING, IRON AND BRASS
Dennis Wire & Iron Works Co., London, Ont.
Mitchell Co., The Robert, Montreal, Que.

PITCH
Reed & Co., Geo., Montreal, Que.

PITCH, PINE
Wilson & Co., Wm. C., Toronto, Canada.

PLANERS
Bertram & Sons, Ltd., John, Dundas, Ont.

PLANERS, STANDARD AND ROTARY
Yates Mach. Co., P. B., Hamilton, Ont.

PLATE CLAMPS
Morris Crane & Hoist Co., Herbert, Niagara Falls, Ont.

PLATE PUNCH TABLES
Norbom Engineering Co., Philadelphia, Pa.

PLUGS, ATTACHMENT
Harvey Hubbel Co. of Canada, Toronto, Can.

PLUMBING EQUIPMENT
Low & Sons, Ltd., Archibald, Glasgow, Scotland.
Empire Mfg. Co., London, Ont.
Mitchell Co., The Robert, Montreal, Que.
Mueller Mfg. Co., H., Sarnia, Ont.
Tallman Brass & Metal Co., Hamilton, Ont.
Wm. C. Wilson & Co., Toronto, Ont.
United Brass & Lead, Ltd., Toronto, Ont.

PROPELLOR BLADES, BRONZE
Corbet Fdry. & Machine Co., Owen Sound, Ont.
Empire Mfg. Co., London, Ont.
Mitchell Co., The Robert, Montreal, Que.
Tallman Brass & Metal Co., Hamilton, Ont.
United Brass & Lead, Ltd., Toronto, Ont.
Yarrows, Limited, Victoria, B.C.

PROPELLER WHEELS
Goldie & McCulloch Co., Galt, Ont.
Kennedy & Sons, Wm., Owen Sound, Ont.
Trout Co., H. G., Buffalo, N.Y.

PIPING, STEAM
Babcock & Wilcox, Ltd., Montreal, Que.

PORT LIGHTS
Goldie & McCulloch Co., Galt, Ont.
Mitchell Co., Ltd., Robt., Montreal, Que.
Mueller Mfg. Co., H., Sarnia, Ont.
Turnbull Elevator Mfg. Co., Toronto, Ont.
Tallman Brass & Metal Co., Hamilton, Ont.
Wm. C. Wilson & Co., Toronto, Ont.
United Brass & Lead, Ltd., Toronto, Ont.

PROPELLER WHEELS
Beauchemin & Fils, Sorel, P.Q.
McNab Co., Bridgeport, Conn.

PULLEYS
Wm. Hamilton Co., Peterboro, Ont.
Smart-Turner Mach. Co., Hamilton, Ont.
Williams Machinery Co., A. R., Toronto, Ont.

PUMP, AIR
Can. Ingersoll-Rand Co. Sherbrooke, Que.
Smart-Turner Mach. Co., Hamilton, Ont.
Weir Ltd., G. & J., Cathcart, Glasgow, Scotland.
Williams Machinery Co., A. R., Toronto, Ont.

PUMPS
Bawden Machine Co., Toronto.
Can. Fairbanks-Morse Co., Montreal, Que.
Canada Foundries & Forgings, Welland, Ont.
Darling Bros., Ltd., Montreal, Que.
Goldie & McCulloch, Ltd., Galt, Ont.
Hopkins & Co., F. H., Montreal, Que.
McAvity & Sons, T., St. John, N.B.
Oberdorfer Brass Co., M. L., Syracuse, N.Y.
Williams Machinery Co., A. R., Toronto, Ont.

PUMPS, CENTRIFUGAL
Can. Ingersoll-Rand Co. Sherbrooke, Que.
Darling Bros., Ltd., Montreal, Que.
Wm. Hamilton Co., Peterboro, Ont.
Goldie & McCulloch Co., Galt, Ont.
Morris Mach. Works, Baldwinsville, N.Y.
Norbom Engineering Co., Philadelphia, Pa.
Smart-Turner Mach. Co., Hamilton, Ont.
Waterous Engine Works Co., Brantford, Ont.
Williams Machinery Co., A. R., Toronto, Ont.

PUMPS, BILGE
Darling Bros., Ltd., Montreal, Que.
Goldie & McCulloch Co., Galt, Ont.
Mitchell Co., Ltd., Robt., Montreal, Que.
Morris Mach. Works, Baldwinsville, N.Y.
Mueller Mfg. Co., H., Sarnia, Ont.

PUMPS, CIRCULATING
Darling Bros., Ltd., Montreal, Que.
Goldie & McCulloch Co., Galt, Ont.
McNab Co., Bridgeport, Conn.
Morris Machine Works, Baldwinsville, N.Y.

PUMPS, FEED WATER
Darling Bros., Ltd., Montreal, Que.
Goldie & McCulloch Co., Galt, Ont.
McNab Co., Bridgeport, Conn.
Morris Mach. Works, Baldwinsville, N.Y.
Weir Ltd., G. & J., Cathcart, Glasgow, Scotland.
Williams Machinery Co., A. R., Toronto, Ont.

PUMPS, LIFT AND FORCE
Darling Bros., Ltd., Montreal, Que.
Low & Sons, A., Glasgow, Scotland.

PUMPS, REES ROTURBO
Goldie & McCulloch Co., Galt, Ont.

PUMPS, HAND AND POWER
Aikenhead Hardware, Ltd., Toronto, Ont.
Darling Bros., Ltd., Montreal, Que.
Mitchell Co., The Robert, Montreal, Que.
Smart-Turner Mach. Co., Hamilton, Ont.
Williams Machinery Co., A. R., Toronto, Ont.

PUMPS, HIGH PRESSURE
Canadian Ingersoll-Rand Co., Sherbrooke, Que.
Darling Bros., Ltd., Montreal, Que.
Morris Mach. Works, Baldwinsville, N.Y.
Smart-Turner Mach. Co., Hamilton, Ont.

PUMPS, STEAM TURBINE
Darling Bros., Ltd., Montreal, Que.
Morris Mach. Works, Baldwinsville, N.Y.
Wm. C. Wilson & Co., Toronto, Ont.

PURIFIERS, WATER
Babcock & Wilcox, Ltd., Montreal, Que.

PUMPING MACHINES
Norbom Engineering Co., Philadelphia, Pa.

PUNCHES
Bertram & Sons, Ltd., John, Dundas, Ont.

PUNCHES, SINGLE, DOUBLE AND MULTIPLE
Norbom Engineering Co., Philadelphia, Pa.

RADIATORS, STEAM, ELECTRIC
Empire Mfg. Co., London, Ont.
Low & Sons, Ltd., Archibald, Glasgow, Scotland.
Wm. C. Wilson & Co., Toronto, Ont.

RAILS, OVERHEAD
Morris Crane & Hoist Co., Herbert, Niagara Falls, Ont.

RANGES
Hopkins & Co., F. H., Montreal, Que.
Wm. C. Wilson & Co., Toronto, Ont.

RECEIVERS, AIR
Can. Ingersoll-Rand Co. Sherbrooke, Que.
Can. Welding Works, Montreal, Que.
Darling Bros., Ltd., Montreal, Que.
MacKinnon Steel Co., Sherbrooke, Que.

REGULATORS, FEED WATER
American Steam Gauge & Valve Mfg. Co., Boston, Mass.

REGULATORS, PRESSURE
Mason Regulator & Engin. Co., Montreal, Que.

RELEASING GEARS
Eckliff Circulator Co., Detroit, Mich.

REPAIRS, MARINE
Corbet Foundry & Mach. Co., Owen Sound, Ont.
Can. Vickers, Ltd., Montreal, Que.
Collingwood Shipbuilding Co., Collingwood, Ont.
Engr. & Mach. Wks. of Can., St. Catharines, Ont.
Georgian Bay Shipbuilding & Wrecking Co., Midland, Ont.
Hyde Engineering Works, Montreal, Que.
Iron Works, Ltd., Owen Sound, Ont.
Kennedy & Sons, Wm., Owen Sound, Ont.
Muir Bros. Dry Dock Co., Port Dalhousie, Ont.
National Shipbuilding Co., Goderich, Ont.
Port Arthur Shipbuilding Co., Port Arthur, Ont.
Yarrow, Limited, Victoria, B.C.

RIGGING SCREWS
Hopkins & Co., F. H., Montreal, Que.

RIGGING, WIRE ROPE
Leckie, Ltd., John, Toronto, Ont.

RIVETS
London Bolt & Hinge Works, London, Ont.
Severance Mfg. Co., S., Glassport, Pa.
Wilkinson & Kompass, Hamilton, Ont.

RIVETERS, PNEUMATIC
Can. Ingersoll-Rand Co., Ltd., Sherbrooke, Que.
Cleveland Pneumatic Tool Co. of Can., Toronto.

RODS, COPPER, BRASS, BRONZE
Standard Underground Cable Co., Hamilton, Ont.
Tallman Brass & Metal Co., Hamilton, Ont.

ROLLS, STRAIGHTENING, BENDING
Bertram & Sons, Ltd., John, Dundas, Ont.

ROOFING
Reed & Co., Geo., Montreal, Que.
Wm. C. Wilson & Co., Toronto, Ont.

ROPE BLOCKS
Aikenhead Hardware, Ltd., Toronto, Ont.
Can. Fairbanks-Morse Co., Montreal, Que.
Morris Crane & Hoist Co., Herbert, Niagara Falls, Ont.

ROPE
Hopkins & Co., F. H., Montreal, Que.
Leckie, Ltd., John, Toronto, Ont.
McNab Co., Bridgeport, Conn.
Stratford Oakum Co., Geo., Jersey City, N.J.
Wm. C. Wilson & Co., Toronto, Ont.

RUBBER COATS
Wm. C. Wilson & Co., Toronto, Ont.

SAW MILL MACHINERY
Wm. Hamilton Co., Peterboro, Ont.
Preston Woodworking Machy. Co., Preston, Ont.
Waterous Engine Works Co., Brantford, Ont.
Yates Machine Co., P. B., Hamilton, Ont.

SAWS, BAND
Preston Woodworking Machy. Co., Preston, Ont.

SCALES, BOILERS, ENGINES
Can. Fairbanks-Morse Co., Montreal, Que.

SCOWS
Collingwood Shipbuilding Co., Collingwood, Ont.
Polson Iron Works, Toronto, Ont.

SCREENS, WIRE
Dennis Wire & Iron Works Co., London, Ont.
Canada Wire & Iron Goods Co.

SHIPS' BUFFERS
Loveridge, Ltd., Cardiff, Wales.

SHIPS' SCREWS
Loveridge, Ltd., Cardiff, Wales.

SCREWS, RIGGING AND STEERING
Loveridge, Ltd., Cardiff, Wales.

SHIPS' VENTILATORS
Loveridge, Ltd., Cardiff, Wales.

SCREWS, COACH
London Bolt & Hinge Works, London, Ont.

SEAM PAINT
Kuhls, H. B. Fred., 6411 3rd Ave., Brooklyn, N.Y.

SEPARATORS, OIL, STEAM
Darling Bros., Ltd., Montreal, Que.
Mason Regulator & Engin. Co., Montreal, Que.
Smart-Turner Mach. Co., Hamilton, Ont.

SHAFTING
Wm. Hamilton Co., Peterboro, Ont.
Mitchell Co., The Robert, Montreal, Que.
Wilkinson & Kompass, Hamilton, Ont.

SHAFTING, BRONZE
Empire Mfg. Co., London, Ont.
Tallman Brass & Metal Co., Hamilton, Ont.

SHEARS
Bertram & Sons, Ltd., John, Dundas, Ont.
Norbom Engineering Co., Philadelphia, Pa.

SHIPBUILDING TOOLS
Aikenhead Hardware, Ltd., Toronto, Ont.
Can. Ingersoll-Rand Co. Sherbrooke, Que.

SHIPS, BUILDERS OF
Can. Vickers, Ltd., Montreal, Que.
Collingwood Shipbuilding Co., Collingwood, Ont.
Doxford & Sons, William, Sunderland, England.
Georgian Bay Shipbuilding & Wrecking Co., Midland, Ont.
National Shipbuilding Co., Goderich, Ont.
Polson Iron Works, Toronto, Ont.
Port Arthur Shipbuilding Co., Port Arthur, Ont.
Yarrows, Limited, Victoria, B.C.

SHIP BROKERS
Page & Jones, Mobile, Ala.

SHIP PLATES
Nova Scotia Steel & Coal Co., New Glasgow, N.S.

SLEDGES
Wilkinson & Kompass, Hamilton, Ont.

SLINGS
Hopkins & Co., F. H., Montreal, Que.
Morris Crane & Hoist Co., Herbert, Niagara Falls, Ont.

SPECIAL MACHINERY
Corbet Fdry. & Machine Co., Owen Sound, Ont.
Wm. Hamilton Co., Peterboro, Ont.
Miller Bros. & Sons, Ltd., Montreal, Que.
Smart-Turner Mach. Co., Hamilton, Ont.

SOCKETS, PULL KEY AND KEYLESS
Harvey Hubbell Co. of Can., Toronto, Can.

SMOOTH-ON.
Wm. C. Wilson & Co., Toronto, Ont.

SPIKES
Wm. C. Wilson & Co., Toronto, Ont.

SPRAY COOLING SYSTEMS
Spray Engineering Co., Boston, Mass.

STAMPINGS, SHEET METAL
Pedlar People, Ltd., Oshawa, Ont.

STEAMSHIP AGENTS
Darling Bros., Ltd., Montreal, Que.
Page & Jones, Mobile, Ala.

STEAM SPECIALTIES
Can. Fairbanks-Morse Co., Montreal, Que.
Corbet Fdry. & Machine Co., Owen Sound, Ont.
Dom. Copper Products Co., Montreal.
Empire Mfg. Co., London, Ont.
Mitchell Co., The Robert, Montreal, Que.

STEAM TRAPS
Aikenhead Hardware, Ltd., Toronto, Ont.
American Steam Gauge & Valve Mfg. Co., Boston, Mass.
Darling Bros., Ltd., Montreal, Que.
Empire Mfg. Co., London, Ont.
Mason Regulator & Engin. Co., Montreal, Que.
Mitchell Co., The Robert, Montreal, Que.
Smart-Turner Mach. Co., Hamilton, Ont.

STEEL, HIGH SPEED
Hopkins & Co., F. H., Montreal, Que.
Nova Scotia Steel & Coal Co., New Glasgow, N.S.
Wilkinson & Kompass, Hamilton, Ont.

STEEL SHELVING
Dennis Wire & Iron Works, London, Ont.

STEEL WORK, STRUCTURAL
Babcock & Wilcox, Ltd., Montreal, Que.
Can. Welding Works, Montreal, Que.
Corbet Fdry. & Machine Co., Owen Sound, Ont.
Wm. Hamilton Co., Peterboro, Ont.
MacKinnon Steel Co., Sherbrooke, Que.

STEERING GEARS
Corbet Fdry. & Machine Co., Owen Sound, Ont.
Hopkins & Co., F. H., Montreal, Que.
Engr. & Mach. Wks. of Can., St. Catharines, Ont.
Wm. C. Wilson & Co., Toronto, Ont.

STOCK RACKS FOR BARS, PIPING, ETC.
Mitchell Co., The Robert, Montreal, Que.
Morris Crane & Hoist Co., Herbert, Niagara Falls, Ont.

STOKERS, MECHANICAL
Babcock & Wilcox, Ltd., Montreal, Que.

SUPERHEATERS, STEAM
Babcock & Wilcox, Ltd., Montreal, Que.
Goldie & McCulloch Co., Galt, Ont.

SWITCHBOARDS, ELECTRIC
Crouse-Hinds Co. of Can., Ltd., Toronto, Ont.

TALLOW
Can. Economic Lubricant Co., Montreal, Que.
Wm. C. Wilson & Co., Toronto, Ont.

TANKS, STEEL
Can. Welding Works, Montreal, Que.
Corbet Foundry & Mach. Co., Owen Sound, Ont.
Goldie & McCulloch, Ltd., Galt, Ont.
Hopkins & Co., F. H., Montreal, Que.
MacKinnon Steel Co., Sherbrooke, Que.
Marsh Engineering Works, Belleville, Ont.
Port Arthur Shipbuilding Co., Port Arthur, Ont.
Reed & Co., Geo., Montreal, Que.
Scaife & Sons Co., Wm. B., Oakmont, Pa.

TANKS (AIR, GAS AND LIQUID)
Can. Welding Works, Montreal, Que.
MacKinnon Steel Co., Sherbrooke, Que.
Marsh Engineering Works, Belleville, Ont.
Scaife & Sons Co., Wm. B., Oakmont, Pa.

TAPPING MACHINES
Mueller Mfg. Co., H., Sarnia, Ont.

TELEGRAPHS, SHIPS
McNab Co., Bridgeport, Conn.
Cory & Sons, Inc., Chas., New York, N.Y.
Morrison Brass Mfg. Co., James, Toronto, Ont.
Wilson & Co., Wm., Toronto, Ont.

TELEPHONES, MARINE
Cory & Son, Inc., Chas., New York, N.Y.
McNab Co., Bridgeport, Conn.

TESTERS, METER
Mueller Mfg. Co., H., Sarnia, Ont.

THUMB SCREWS AND NUTS
Canada Foundries & Forgings, Welland, Ont.
Mitchell Co., The Robert, Montreal, Que.
United Brass & Lead, Ltd., Toronto, Ont.

TRACK SYSTEMS
Morris Crane & Hoist Co., Herbert, Niagara Falls, Ont.

TRAVELLING BLOCKS
Morris Crane & Hoist Co., Herbert, Niagara Falls, Ont.

TROLLEYS
Can. Fairbanks-Morse Co., Montreal, Que.
Morris Crane & Hoist Co., Herbert, Niagara Falls, Ont.

TROLLEY HOISTS
Morris Crane & Hoist Co., Herbert, Niagara Falls, Ont.

TRUCKS, HAND, ELECTRIC
Aikenhead Hardware, Ltd., Toronto, Ont.
Can. Fairbanks-Morse Co., Montreal, Que.

TUBES, BOILER
Babcock & Wilcox, Ltd., Montreal, Que.
Broughton Copper Co., Ltd., Manchester, Eng.
Goldie & McCulloch Co., Galt, Ont.

TUBES, COPPER AND BRASS
Mueller Mfg. Co., H., Sarnia, Ont.
Tallman Brass & Metal Co., Hamilton, Ont.
Standard Underground Cable Co., Hamilton, Ont.

TUGS
Polson Iron Works, Toronto, Ont.
Collingwood Shipbuilding Co., Collingwood, Ont.

TURBINES, STEAM
Goldie & McCulloch Co., Galt, Ont.

TURBINES, DIRECT-DRIVING AND GEARED
Doxford & Sons, William, Sunderland, England.

TURNBUCKLES
Canada Foundries & Forgings, Welland, Ont.
Hopkins & Co., F. H., Montreal, Que.

TURNTABLES
Morris Crane & Hoist Co., Herbert, Niagara Falls, Ont.

SPIKES, SMALL RAILROAD
Severance Mfg. Co., S., Glassport, Pa.

UNIONS, ALL KINDS
Dart Union Company, Toronto, Ont.

VALVES, AIR
Mitchell Co., The Robert, Montreal, Que.
Mueller Mfg. Co., H., Sarnia, Ont.

VALVE, DISCS
Wm. C. Wilson & Co., Toronto, Ont.

VALVES
American Steam Gauge & Valve Mfg. Co. Boston, Mass.
Babcock & Wilcox, Ltd., Montreal, Que.
Can. Fairbanks-Morse Co., Montreal, Que.
Darling Bros., Ltd., Montreal, Que.
Empire Mfg. Co., London, Ont.
McAvity & Sons, Ltd., T., St. John, N.B.
Mason Regulator & Engin. Co., Montreal, Que.
McNab Co., Bridgeport, Conn.
Norbom Engineering Co., Philadelphia, Pa.
Williams Machinery Co., A. R., Toronto, Ont.
Wm. C. Wilson & Co., Toronto, Ont.
United Brass & Lead, Ltd., Toronto, Ont.

VALVES, FOOT
Aikenhead Hardware, Ltd., Toronto, Ont.
Mitchell Co., The Robert, Montreal, Que.
Smart-Turner Mach. Co., Hamilton, Ont.

VALVES, STOP, REDUCING, SAFETY CHECK, DISCHARGE, SUCTION
Aikenhead Hardware, Ltd., Toronto, Ont.
American Steam Gauge & Valve Mfg. Co., Boston, Mass.
Darling Bros., Ltd., Montreal, Que.
McAvity & Sons Ltd., T., St. John, N.B.
Mitchell Co., The Robert, Montreal, Que.
Morrison Brass Mfg. Co., James, Toronto, Ont.

VALVES, MIXING
Darling Bros., Ltd., Montreal, Que.
Mitchell Co., The Robert, Montreal, Que.
Mueller Mfg. Co., H., Sarnia, Ont.

VALVES, REDUCING, PRESSURE
Mitchell Co., The Robert, Montreal, Que.
Mueller Mfg. Co., H., Sarnia, Ont.

VALVES, STORM
Low & Sons, A., Glasgow, Scotland.

VARNISHES
Aikenhead Hardware, Ltd., Toronto, Ont.
Ault & Wiborg Co. of Can., Ltd., Toronto, Ont.
Leckie, Ltd., John, Toronto, Ont.
Reed & Co., Geo., Montreal, Que.

VENTILATORS, COWL
Can. Welding Works, Montreal, Que.
McNab Co., Bridgeport, Conn.
Mitchell Co., The Robert, Montreal, Que.

VENTILATORS, SHIPS'
Loveridge, Ltd., Cardiff, Wales.

VENTILATION EQUIPMENT
Empire Mfg. Co., London, Ont.
Hopkins & Co., F. H., Montreal, Que.
Low & Sons, Ltd., Archibald, Glasgow, Scotland

WASHERS
London Bolt & Hinge Works, London, Ont.
Wm. C. Wilson & Co., Toronto, Ont.

WATER COLUMNS
Darling Bros., Ltd., Montreal, Que.
Mitchell Co., The Robert, Montreal, Que.
Morrison Brass Mfg. Co., James, Toronto, Ont.

WELDING, ELECTRIC
Can. Welding Works, Montreal, Que.
Hall Engineering Works, Montreal, Que.
Marine Welding Co., Buffalo, N.Y.
Beatty & Sons, M., Welland, Ont.

WATER SOFTENERS
Babcock & Wilcox, Ltd., Montreal, Que.

WATER SUPPLY SYSTEMS
Mueller Mfg. Co., H., Sarnia, Ont.

WATER HEATERS
Darling Bros., Ltd., Montreal, Que.
Empire Mfg. Co., London, Ont.
Morrison Brass Mfg. Co., James, Toronto, Ont.

WHISTLES AND SYRENS
Empire Mfg. Co., London, Ont.
McAvity & Sons, T., St. John, N.B.
McNab Co., Bridgeport, Conn.
Mitchell Co., Ltd., Robt., Montreal, Que.
Morrison Brass Mfg. Co., Jas., Toronto, Ont.

WINCHES, CARGO
Aikenhead Hardware, Ltd., Toronto, Ont.
Corbet Fdry. & Machine Co., Owen Sound, Ont.
Hopkins & Co., F. H., Montreal, Que.
Marsh Engineering Works, Belleville, Ont.

WINCHES, DOCK, SHIP
Beatty & Sons, M., Welland, Ont.
Marsh Engineering Works, Belleville, Ont.
Miller Bros. & Sons, Ltd., Montreal, Que.
Morris Crane & Hoist Co., Herbert, Niagara Falls, Ont.
Wilson & Co., Wm. C., Toronto, Canada.

WINCHES, TOWING
Corbet Foundry & Mach. Co., Owen Sound, Ont.

WINCHES, TRAWL
Beatty & Sons, M., Welland, Ont.
Wm. C. Wilson & Co., Toronto, Ont.

WINDLASSES
Corbet Fdry. & Machine Co., Owen Sound, Ont.
Dake Engine Co., Grand Haven, Mich.
Hopkins & Co., F. H., Montreal, Que.
Wilson & Co., Wm. C., Toronto, Canada.

WIPER CAPS, OILER BOXES, ETC.
Mitchell Co., The Robert, Montreal, Que.
Morrison Brass Mfg. Co. James, Toronto, Ont.

WIRE, COPPER CLAD STEEL
Standard Underground Cable Co., Hamilton, Ont.

WIRE, COPPER, BRASS, BRONZE
Standard Underground Cable Co., Hamilton, Ont.
Tallman Brass & Metal Co., Hamilton, Ont.

WIRE, INSULATED
Standard Underground Cable Co., Hamilton, Ont.

WIRELESS OUTFITS
Marconi Wireless Telegraph Co., Montreal, Que.

WIRE ROPE
Zenith Coal & Steel Products, Ltd., Montreal.

WOOD WORKING MACHINERY
Aikenhead Hardware, Ltd., Toronto, Ont.
Can. Fairbanks-Morse Co., Montreal, Que.
Preston Woodworking Machy. Co., Preston, Ont.
Yates Mach. Co., P. B., Hamilton, Ont.

WOOD BORING TOOLS
Aikenhead Hardware, Ltd., Toronto, Ont.
Can. Ingersoll-Rand Co. Sherbrooke, Que.

WOODITE GAUGE GLASS WASHERS
Wm. C. Wilson & Co., Toronto, Ont.

WRENCHES
Canada Foundries & Forgings, Welland, Ont.

YACHT BROKER
Watt, J. Murray, Philadelphia, Pa.

Your men will smile
Behind
STOCO
SAFETY GOGGLES
The Stoco Safety Goggle solves the eye protection problem. It not only affords the greatest possible degree of safety but it is also comfortable—Your men will wear it.
The exclusive design of the Stoco Safety goggle insures strength and adequate protection between and around the eyes. The frame will stand a lot of hard usage and the lenses are made from optical glass specially treated for strength. Made with headbands or easy cable ear-bows at $9.00 per dozen.
Single sample will be sent on request without charge to Safety Engineers or department heads.
STANDARD OPTICAL CO.
GENEVA, N.Y.

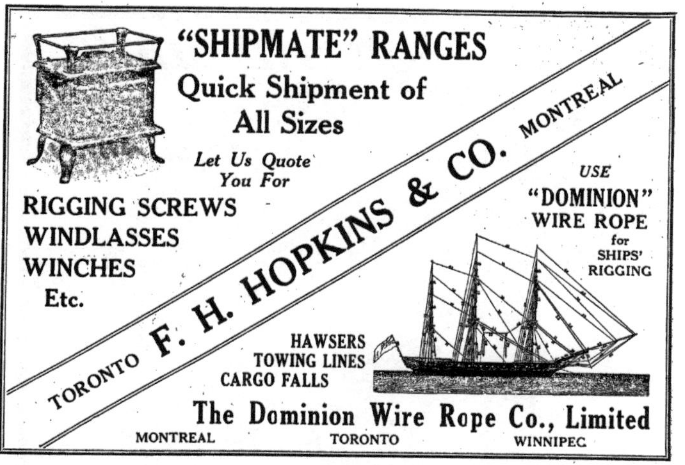

INDEX TO ADVERTISERS

Lightning Source UK Ltd.
Milton Keynes UK
UKHW010954211118
332724UK00008B/129/P